ENVIRONMENTAL SCIENCE, ENGINEERING AND TECHNOLOGY

HANDBOOK OF PHYTOREMEDIATION

Environmental Science, Engineering and Technology

Additional books in this series can be found on Nova's website under the Series tab.

Additional E-books in this series can be found on Nova's website under the E-books tab.

ENVIRONMENTAL SCIENCE, ENGINEERING AND TECHNOLOGY

HANDBOOK OF PHYTOREMEDIATION

IVAN A. GOLUBEV
EDITOR

Nova Science Publishers, Inc.
New York

For permission to use material from this book please contact us:
Telephone 631-231-7269; Fax 631-231-8175
Web Site: http://www.novapublishers.com

Additional color graphics may be available in the e-book version of this book.

LIBRARY OF CONGRESS CATALOGING-IN-PUBLICATION DATA

Handbook of phytoremediation / editor: Ivan A. Golubev.
p. cm.
ISBN 978-1-61728-753-4 (hardcover)
1. Phytoremediation. I. Golubev, Ivan A.
TD192.75.H36 2010
628.5--dc22

2010026939

Published by Nova Science Publishers, Inc. † New York

CONTENTS

PREFACE

Phytoremediation is the use of green plants and their associated microbiota, soil amendments, and agronomic techniques to remove, contain, or render harmless environmental contaminants. It is an emerging technology which offers a potentially cost-effective and environmentally sound alternative to the environmentally destructive physical methods which are currently practiced for the cleanup of contaminated groundwater, terrestrial soil, sediments, and sludge. This handbook presents current research from around the globe in the study of phytoremediation including such topics as the application of phytoremediation technologies for water decontamination from persistent organic pollutants; phytoremediation of uranium contaminated soils; phytoremediation using constructed mangrove wetlands; the phytoextraction capability of maize and sunflowers; and the phytoremediative processes occurring in salt marshes.

Chapter 1 - Phenolic compounds present in the drainage from several industries are harmful pollutants and represent a potential danger to human health. Conventional treatments for phenol removal from industrial wastewaters have several limitations so, there is a need to look for alternative and environmental friendly technologies to complement or substitute the conventional ones. In recent years, phytoremediation has been recognized as a cheap and eco-friendly alternative technology which could be used for the remediation of organic contaminants, such as phenolics. Despite most phytoremediation studies were performed with soil-grown or hydroponically grown plants; more recently some results were obtained with the help of in vitro cell and tissue cultures, such as hairy roots. They have been used as tools for screening the potencialities of different plant species to tolerate, accumulate and remove high concentrations of phenols with high efficiency. In addition, using different plant model systems it could be established that plants metabolize a number of phenolic compounds by common metabolic pathways. Uptake of phenolics depends on the plant species as well as on their physico-chemical properties. While the main metabolites detected from phenolic´s transformation are polar conjugates, some plant species could incorporate large amounts of these chemicals and associated metabolites, as bound residues, through reactions catalized by oxido-reductases. Hence, cell wall is considered one of the important detoxification sites of phenolic compounds in plants. In addition, plant roots produce and exude high amounts of oxido-reductive enzymes, such as peroxidases, which are associated with the non specific oxidative polymerisation of phenolic free radicals in the cell wall. So, these enzymes may play an important role in polymerising reactions and, also, they are likely to be the key enzymes in the removal of phenol and chlorophenols. Moreover, different peroxidase isoenzymes might play different roles in the removal process. So, in this chapter, the use of plants as enzyme sources, as well as partially purified oxidases is discussed, as good alternatives for remediation purposes.

On the other hand, with the application of genetic engineering, it is feasible to manipulate plant capabilities to tolerate, accumulate, and/or metabolize pollutants, and thus to create an appropriate plant for environmental cleanup. Therefore, this chapter also examines and discusses the recent advances in enhancing phytoremediation of phenolic compounds through transgenic plant research.

Current knowledge, the areas which need to be explored and perspectives are presented to improve the efficiency and to asses the feasibility of phenolics′ phytoremediation.

Chapter 2 - Pollution by persistent organic pollutants (pesticides, pharmaceuticals, petroleum hydrocarbons, PAHs, PCBs, etc.) is an environmental problem that is recognized worldwide. In order to address this problem, cost effective technologies have been developed and evaluated for the decontamination of soil and water resources. Phytoremediation is a promising technology that uses plants and the associated rhizosphere microorganisms to remove, transform/detoxify, or accumulate organic and inorganic pollutants present in soils, sediments, surface or ground water, wastewater, and even the atmosphere. In fact, as a result of their sedentary nature, plants have evolved diverse abilities for dealing with toxic compounds in their environment. They, therefore, possess a variety of pollutant attenuation mechanisms that makes their use in remediating contaminated land and water more feasible than physical and chemical remediation. Currently, phytoremediation is used for treating many classes of organic xenobiotics including petroleum hydrocarbons, chlorinated solvents, polycyclic aromatic hydrocarbons, pesticides, explosives, pharmaceutical compounds and their metabolites, and it involves several decontamination mechanisms. There are several different types of phytotechnologies such as, for instance, treatment constructed wetlands. The aim of this work is to present a review on the application of phytoremediation technologies for water decontamination from persistent organic pollutants, with special emphasis focused on the removal of a class of emergent pollutants that has recently been receiving a lot of attention, the pharmaceutically active compounds. Within the realm of phytotechnologies, constructed wetlands for wastewater treatment deserve a special focus as these systems have been used with success for the removal of several different types of organic xenobiotics.

Chapter 3 - Environmental uranium contamination based on human activity is a serious problem worldwide. Soil contaminated with uranium poses a long-term radiation hazard to human health through exposure via the food-chain and other pathways. This chapter is an overview of processes and modern techniques for remediation of soils contaminated with uranium, with special attention on phytoremediation. Phytoremediation takes advantage of plant to extract, sequester pollutants in soil, water, air with an aim of pollutants removal and transformation into harmless forms.

The objective of this chapter is to develop better understanding of plants behavior and the degree of affinity towards the adoption of uranium for hyperaccumulators plants based on review of international research. To understand the mechanism of uranium uptake in plants and accumulation, a necessary prerequisite is application of radiophytoremediation on the "real" scale. For this purpose, the authors investigated these processes using three different aspects with selected cultivated plants:

Vegetative tests in pots, of fully controlled conditions, with corn plants that were grown on two types of soil, pseudogley and chernozem, together with its phytotoxic effect on the plant development, height, yield and seed germination.

Greenhouse experiments with tailings from the closed uranium mine Kalna on southeast of Serbia. Three series of experiments were conducted in plastic-house. First, with three plant species (corn, sunflower and green peas), were grown in pots on the four substrate variants, tailings in mixture with sand. Substrate was irrigated with drinking water and „uranium water", which issue out from the mine. Another experiment was conducted in order to investigate the uptake of U in several kinds of root - crops, bulbous and tuberous plants: carrot, onion, potatoes, radish, red beet and sugar beet. Content of uranium was determined in leaves and root (surface root layer and edible part were peeled). Also the authors investigated uranium adoption in four genotypes of: corn, sunflower and soy bean.

Vegetation test on real, native, conditions on tailings, from the closed uranium mine Kalna. The experiment was carried out on the elementary plots one square meter in size, with: bean, cabbage, lettuce, corn, onion, potatoes, spinach, and sunflower.

Well-organized use of phytotechnology means integrated management strategy for contaminated site which includes: proper selection of plants (uranium hyperaccumulators), improving mobility of uranium with amendments (organic agents) and application sequestering agents for immobilization and transformation of excess uranium, which plant didn't accept.

Chapter 4 - The interest in phytoremediation has been rapidly increasing in the last twenty years. A relevant number of scientific papers have investigated several aspects of the matter, first exploring the physiological processes and then the molecular characteristics of the plants to find the genes responsible for the metal (hyper)tolerance.

Since 1998, our research group has had a number of projects concerning phytoremediation financed with public funds. In 2005, the authors designed the first Italian in situ experiment of phytoremediation. This trial took place within an area included into the polluted area Laguna di Grado e Marano (Grado and Marano lagoon) which belongs to the national priority list (Ministry Decree 468/2001). The experimental site was located on the property of an Italian chemical company in Torviscosa (Udine). Several aspects of phytoremediation were investigated, such as: (i) phytoextraction potential of Sorghum bicolor and Helianthus annuus; (ii) the growth of Populus spp. and Salix spp. and trace element uptake; (iii) strategies for the enhancement of metal absorption from the soil and for increasing the translocation rate in plants; (iv) metals' mobility and their availability to plants and pedofauna. All the aspects were investigated both under pot and field trial conditions.

More recently, the authors worked on metallophytes and hyperaccumulators. Such species, being able to tolerate and accumulate high amounts of several elements, were proposed for phytostabilization of heavily polluted soils and mine tailings. The fertility of heavily polluted soils and mine tailings is always very low. Properly designed agronomic practices are expected to support plant growth and biomass yield. Pot experiments testing the effects of different levels of fertilization on the growth of Thlaspi caerulescens on polluted soils and mine tailings were done. In the summer of 2007, a field survey was conducted at the former lead/zinc mining site in Cave del Predil (Julian Alps) to investigate the presence of metallophytes.

The learned lessons are consistent with the views prevailing in the scientific debate. After a decade of research, phytoextraction seems not feasible at the present state of knowledge. To the contrary, phytostabilization to decrease metal mobility is a realistic alternative.

Further research at field scale and efforts in discovering new hyperaccumulators and/or metal tolerant populations of native species must be done to promote phytoremediation to become a practical option for the remediation of polluted soils.

Chapter 5 - Phytoremediation advantages are widely known nowadays. It is a method applicable for large areas with low concentration of pollutants treatment or areas where only the finishing step of cleaning is required. Very often these kinds of places represent great problems because there is no possibility to take all the soil to the landfills, and often they are part of agricultural fields. There are many studies dealing with application of a variety of plants for the treatment of soils contaminated by heavy metals or organics. Plants growing on these contaminated soils developed several ways of coping with the toxicity of pollutants including avoiding their accumulation, different detoxification mechanisms or even metal excretion from their body. This work is focused on heavy metal contamination cleanup by phytoremediation with the aim to describe some of the possible ways to assess the stress of plants. There are several factors which can be used in the plant stress assessment such as reduction of biomass production, plant growth inhibition, changes in photosynthesis, germination inhibition, and production of antioxidant enzymes. Knowledge of these factors brings us closer to understanding the molecular mechanisms of heavy metal accumulation by plants and it indirectly helps further application of phytoremediation as well as has numerous additional biotechnological implications. For instance, health-threatening human deficiencies in trace metals appear to be widespread in developing countries and possibly worldwide but engineering of plants accumulating essential metals such as Zn or Se in their edible parts might help in enriching human diets for these important elements.

Chapter 6 - Soil is the vital medium in the natural environment. Its pollution has grown into a global issue. Metals contamination is one of the heaviest environmental problems in soil.

This paper will review the status of soil contamination, its risks and sources in the beginning. Human activities broke the soil balance with low background toxic metal level, and shrank the area of agricultural soil globally. Both essential and unessential metals ruin the balance of the ecosystem, with increasing economic loss and human health damage. Soil pollution was caused not only by naturally generated mechanisms including earthquakes, volcanic eruption, but also by anthropogenic inputs such as industrial development, fossil fuel burning, mining, metallurgy, electroplating, waste disposal, long-term application of sewage sludge, fertilizer application, etc.

Then the review will summarize the phytoremediation technique for soil contamination. Phytoremediation of soil pollution is a popular method to remove toxic pollutants from soil with low cost and environmental sustainability; it is composed of phytoextraction and phytostability. On one hand, phytoextraction is defined as the use of hyperaccumulating plants to transport metals from the soil and concentrate them in plants that can be harvested. Up till now, about 400 hyperaccumulators have been documented in the world. Phytoextraction situations are separately reviewed for several toxic heavy metals including cadmium, arsenic, lead, zinc, etc. On the other hand, phytostabilization technology takes advange of plants to reduce leaching of the pollutants by eliminating or minimizing the mobile and bioavailable fractions of metals in the soil. Various plant species could be good candidates to improve scenery remediation in abandoned sites, by shaping efficient vegetation cover.

Finally, the mechanism of metal remediation in soil was also evaluated in this paper. Chelants, bio-microbes and genetic procedures have been applied to assist increasing the accumulation of metals by plants, after that some comments were given on limits of phytoremediation application. The truth is that phytoremediation has not become a commercially available technology in the field yet even if it could be promising.

Chapter 7 - Organic and inorganic pollutants in the soil are one of the major environmental problems of present days. The traditional removal techniques do not provide any acceptable remedies for the removal of metal as well as organic pollutants from the system. Soil amendments are usually a cost effective and environment friendly technology. But disposal of solid wastes on land leads to contamination of both soil and groundwater. Bioremediation is an up coming environmental friendly technology that uses microbes as well as plants to clean up the toxic metals and other pollutants from the soil of the contaminated environment. The use of metal tolerant microbes and plants for the removal of toxic metal from the polluted system is a low cost technology. The specific microbial and plant species use to remove specific contaminants which have discussed in this paper. Metal-accumulating species can concentrate different metals up to 100 to 1000 times in their body which is very much species and site specific. The phytoremediation of heavy metals is divided into four sub-sections: (1) Phytoextraction: the use of plants to remove the toxic metals from the soil into the harvestable parts of plants, (2) phytofiltration: the use of plants root to accumulate the toxic metals from the water system (3) Phytostabilization: the use of metal tolerant plants to remove the bio-available toxic metal from the soil and (4) Phytovolatilazation: the use of plants to take up contaminants from the soil, transforming them into volatile form and transpiring them into the atmosphere. The harvestable parts like root and shoot, which are rich in metals, can easily be reclaimed and recycled after harvesting the plants from the contaminated site. Bioremediation technologies can be generally classified as in situ and ex situ. This paper reviews the mobility, bioavailability and responses of microbes and plants in presence of metals and other pollutants in the system. Bioremediation may be employed to bother specific contaminants such as degradation of chlorinated hydrocarbons, heavy metals, oil spills, crude oil, nitrate and sulfate by indigenous or exogenous bacteria as well as plants. In general, bioremediation is a very promising and emerging technology for the removal of different kind of pollutants from the soil and water, which can be, approaching commercialization for near future.

Chapter 8 - Human activities and industrial development lead to a deterioration of the environment that affects, to a greater or lesser extent, all countries. In Asturias (Spain), mining, steel mills and the chemical industry have produced wastes with high concentrations of heavy metals, with the consequent risks for the environment and human health. This problem requires an efficient and technologically feasible solution.

Phytoremediation is considered an effective, low-cost and environmental friendly technology for cleaning up heavy metal-polluted sites. It is based on the capacity of some plants, called hyperaccumulators, for taking these metals from the soil and accumulating them above a threshold value in their harvestable tissues.

One of the strategies that can be followed when working in phytoremediation is the use of native hyperaccumulator plants of high biomass, mainly those adapted to the climatic and soil conditions of the polluted site. According to this, the aim of our work was the identification of plants that spontaneously grow in different heavy metal-polluted soils of our region. After measuring the metal content of these plants, the authors selected the species according not only to their metal accumulation capacity, but also to the amount of biomass, percentage cover/aggregation, frequency of appearance in polluted areas or having special characteristics that make plants prone to hyperaccumulate metals, such as being nitrophilous or resistant to other types of stress. The authors tested in the greenhouse the effect of the heavy metals on plant growth and development and their maximum accumulation capacity. Thus, the plants selected were Dittrichia viscosa and Betula celtiberica for Cd, Melilotus alba for Pb, Anthyllis vulneraria for Zn, and Carex pendula for Hg. Later, they selected through in vitro culture the most accumulator plantlets of some of these species for further cloning and use in phytoremediation programs, so they obtained clone DV-A of D. viscosa, clone BC-K of B. celtiberica, and clone MA-X of M. alba.

Chapter 9 - Phytoremediation is an emerging technology that employs the use of higher plants to clean-up metal contaminated environments; when applied, there is a need for constant monitoring of metal concentrations in soil, water and biological materials in order to evaluate the success of the applied technology and to control metal uptake in plant tissues in order to prevent accumulation of unwanted toxic metals in food chains.

In view of the growing needs of global environmental protection and also to minimize the relevant research costs, it is important that in phytoremediation studies and their application the analytical procedures for determination of elemental concentrations in soil, water and biological materials are accurate, reliable and reproducible, but on the other hand rapid and cheap, with simple sample preparation. Therefore in this chapter the main characteristics, sample preparation protocols, and applications of X-ray fluorescence-based analytical techniques for "bulk" sample analyses, namely energy dispersive X-ray fluorescence spectrometry (EDXRF) and total reflection X-ray fluorescence spectrometry (TXRF), are presented. Although EDXRF and TXRF are far less popular methods for analyses of element concentrations in soil, water, air and biological materials than, for example, atomic absorption spectroscopy (AAS) and/or inductively coupled plasma atomic-emission spectroscopy (ICP-AES), they are much cheaper, simpler and environmentally friendlier, which is particularly advantageous from the economic and environmental protection points of view.

Chapter 10 - Secondary saline and/or sodic soils in irrigation regions constrains crop yield, and their reclamation is important to meet the need of an increasing population in many countries where arable acreage is limited. Because of their relatively low salinity/sodicity levels compared with primary saline/sodic soils and the high expenditure of traditional reclamation activity, secondary saline/sodic soils should adopt different reclamation strategies and techniques.

To improve soil structure and maintain a reduction or at least balance of salts in secondary saline/sodic soils is a base for sustainable production activity of agriculture. The strategies and techniques on the sustainable nonchemical remediation of secondary saline/sodic soils, combining with farmland management measures such as irrigation and drainage, field engineering, agronomy, and rainfall utilization were discussed in the article. The status quo of remediation of secondary saline/sodic soils in China and the related field management measures were also presented. Lots of experimental results and reclamation practical activities indicated that by the aid of suitable management measures, the phytoremediation or bioremediation of secondary saline/sodic soils without lots of chemical amendment application is feasible for technology, acceptable to farmers for economic benefit, and sustainable to environments.

Chapter 11 - Heavy metals (HMs) as such as cadmium (Cd), copper (Cu) and zinc (Zn) have been widespread in soils by human activities (for example, mining, smelting and agriculture). These metals can affect the environmental quality and the health of people. The risk associated to their occurrence and the possibility to cleanup them using phytoremediation systems is increasing the interest for understanding the biological basis of metal tolerance and accumulation process in plants.

Species belonging to the Populus genus (poplars) are suitable candidates for phytoremediation. These trees have a high biomass production, extensive roots, high rates of transpiration and easy propagation. Also, the wide genetic diversity comprised within this genus and the development of multiple biotechnologies and information resources allow a genetic improvement based on traditional and biotechnological approaches. Studies carried out in different experimental conditions show that poplars exposed to Cu, Cd and Zn exhibit distinct tolerance levels and metal accumulation patterns. This response depends on specific genotypes. Some of them have been proposed as candidate for phytostabilization and phytoextraction.

Exposition of poplars to toxic concentrations of Cd, Cu and Zn triggers different effects on growth, biomass partitioning, metal allocation, photosynthesis, carbohydrate and nitrogen metabolism, reactive oxygen species (ROS) production, among others. Plants dispose different homeostatic mechanisms for coping with metal excess. These operate at different levels and their regulation determinates the ability of plant to restrict the metal uptake and (or) root to shoot transport, and compartmentalization. Biological mechanisms underlying metal homeostasis and tolerance in poplars and other tree species are only partially understood. Metal uptake in roots can be regulated by the exudation of organic acid anions, the binding effect of the cell wall and the flux of ions through plasmalem metal transporters. In cytoplasm, metals are chelated and/or transported toward organelles by peptidic chelators. Simultaneously, excesses of metallic ions can be directed to vacuole or apoplast by membrane transporters. Metals are mobilized through the xylem from roots to aerial structures in a process driven by transpiration. Inside leaf cells, a regulated network of membrane transporters and chelators directs metals to their final destination. A further defensive line against metal induced ROS involves enzymes and reducing metabolites. Response to metal stress also includes expression of general defense proteins and signaling elements as such as calcium and ethylene.

Chapter 12 - Phytoremediation is the "use of green plants and their associated microbiota, soil amendments, and agronomic techniques to remove, contain, or render harmless environmental contaminants" (Cunningham et al., 1996). It is an emerging technology which offers a potentially cost-effective and environmentally sound alternative to the environmentally destructive physical methods which are currently practiced for the cleanup of contaminated groundwater, terrestrial soils, sediments, and sludge (Shimp et al., 1993; Schnoor et al., 1995; Salt et al., 1998; Frick et al., 1999; Banks et al., 2000; Ke et al., 2003a, b; Bert et al., 2009).

Chapter 13 - There is evidence that many legume species of the flowering plant family Fabaceae may be efficiently used in phytoremediation of heavy metal polluted soils, particularly for revegetation and phytostabilization of mine soils. For such purposes, a number of legume species were used and this chapter gives an updated glimpse on scientific experiences dealing with microbial effects on several legume species growing in heavy metal polluted soils. Legume species are able to form symbiosis with various beneficial microorganisms, such as nitrogen-fixing nodule bacteria, arbuscular mycorrhizal fungi and plant growth-promoting bacteria. Such plant microbe associations have implications in plant growth, nutrition and disease control. The symbioses between legumes and microorganisms provide nutrients for the plant, stimulate plant growth, exert antistress effects on plants, improve soil fertility, and restore ecosystem biodiversity and functions. This makes legumes very tempting subjects for phytoremediation purposes, particularly for the development of ecologically friendly phytostabilization technologies, since many of HM polluted soils are characterized by low nutrients and degenerated biocenosis. Moreover, symbiotrophic microorganisms possess a number of mechanisms which may be involved in improving tolerance of plants to environmental stresses, including those caused by heavy metals. As a consequence, the use of legume species for phytoremediation purposes should be considered in the context of their interactions with symbiothrophic microorganisms. Several plant species from the family Fabaceae and their performances in combination with microorganisms on heavy metal polluted soils or hydroponics are reported in this chapter. Particular attention is drawn on the effects of symbiotrophic microorganisms on legumes in the presence of heavy metals in conditions of monoinoculation and in combined inoculations. Intraspecific variability of plant species in their interactions with microorganisms is also discussed as well as the perspectives for phytoremediation with genetically modified legumes and symbiotrophic microorganisms. Successful attempts to increase tolerance to and accumulation of HMs by legume plants via genetic modifications and selection are mentioned. Although the presence of literature reports on the use of legume plants for phytoremediation purposes, it is undoubtedly wise to state that their potential for phytoremediation has not yet been adequately explored. Aim of this chapter is the discussion of advantages and problems in the application of legume-microbe systems for restoration and phytoremediation of polluted soils.

Chapter 14 - Phytoremediation which involves the use of plants and rhizospheric organisms for the removal of pollutants is an emerging technology for the clean up of contaminated sites. The removal of textile dyes mediated by plants has been one of the most neglected areas of phytoremediation research. Dyes, which are primary constituents of the wastes from textile industry effluents, constitute a group of recalcitrant compounds, many of which are known to have toxic and carcinogenic effects. Hence, the review focuses on the studies of the mechanisms adopted by plants in the removal of textile dyes and the future scope for research in this area which will help in broadening the horizons of phytoremediation technologies. Plant species many a times referred to as 'green livers', are known to possess a wide range of detoxifying and biotransforming enzymes some of which may also be secreted extracellularly in the rhizosphere and can bring about the transformation of organic pollutants such as textile dyes.

The use of in vitro plants for phytoremediation studies can help to explore the enzymatic status and the products of metabolism of the dye, thus providing a new dimension to phytoremediation studies. The use of transgenic plants with microbial genes can combine the advantages of both plant and microbial systems for enhanced dye degradation. Biotechnological approaches involving the development of hairy roots and suspension cultures may find good utility in phytoremediation studies. The ultimate aim of phytoremediation involves applying these well studied plant systems at the contaminated sites which may constitute the development of constructed wetlands for on-site treatment of industrial effluents.

Chapter 15 - Metallophytes have the ability to tolerate extreme metal concentrations. This unique property commends them to be exploited in technologies such as biogeochemical and biogeobotanical prospecting as well as phytoremediation. Although there are many publications on metallophytes and their potential use in phytoremediation, in Botswana such studies are in their infancy, albeit the country having numerous mining activities. This paper discusses the chemical studies of metallophytes from mineralized zones and other vulnerable areas in Botswana as well as their potential use in phytoremediation. The metallophytes dealt with include Helichrysum candolleanum, Blespharis aspera, Tephrosia longipes and Indigofera melanadenia some of whose capacity for multiple metal accumulation is investigated. A number of analytical methods have been applied in these studies. These include attractive sample preparation techniques such as microdialysis and solid phase extraction as well as chromatographic methods such as size exclusion chromatography and online liquid chromatography-solid phase extraction-nuclear magnetic resonance which are particularly employed for speciation studies. These techniques have demonstrated a lot of potential for metallophytes research.

Chapter 16 - High concentrations of metallic elements as Cd, Pb and Cr can cause harmful effects to the environment. These highly toxic pollutants constitute a risk for the aquatic and terrestrial life, especially plants, animals and humans. They are associated to diverse bioavailable geochemical fractions, such as the water-soluble fraction and the exchangeable fraction, and to non-available fractions such as those associated with the crystalline net of clays and silica minerals.

Depending upon its chemical and physical properties different mechanisms of metals toxicity in plants can be distinguish, such as production of reactive oxygen species from the auto-oxidation, blocking and/or displacement of essential functional groups or metallic ions of biomolecules, changes in the permeability of cellular membranes, reactions of sulphydryl groups with cations, affinity for reactions with phosphate groups and active groups of ADP or ATP, substitution of essential ions, induction of chromosomal anomalies and decrease of the cellular division rate.

To deal with heavy metal pollution, remediation using plants, including woody species, is becoming a widespread practice. Phytoremediation is an environmentally friendly technology and the use of woody species presents advantageous characteristics as an economic and ecologically viable system that becomes an appropriate, practical and successful technology.

Phytoremediator woody species, with (i) high biomass production, (ii) deep root system, (iii) high growth rate, (iv) high capacity to grow in soils with low nutrient availability and (v) high capacity to allocate metals in the trunk, can be an alternative for the recovery of degraded soils due to excess of metallic elements.

Chapter 17 - Plant growth promoting rhizobacteria (PGPR) Pseudomonas P. aureofaciens, P. chlororaphis, P. fluorescens and their plasmid-bearing variants: destructors of polycyclic aromatic hydrocarbons (PAH) (naphthalene, phenanthrene), strains resistant to heavy metals (cobalt, nickel) and metalloids (arsenic), and multifunctional ones combined both characteristics, were used to estimate their impact in the phytoremediation process. All used bacterial strains that possessed ability to produce phytohormone indole acetic acid, various antifungal compounds, and suppressed phytopathogens. The PGPR strain's ability to degrade naphthalene and phenanthrene was shown to be stable in the rhizosphere at different conditions. The introducing of PGPR destructors in the rape rhizosphere increased the naphthalene biodegradation efficiency up to 90% in comparison with control without bacteria at gnotobiotic system in 7 days cultivation. The arsenite resistant PAH-destructors P. aureofaciens BS1393(pBS216,pKS1) and P. chlororaphis PCL1391(pBS216,pKS1) also promoted mostly complete naphthalene degradation at the same experiments supplemented arsenite (15 mg/kg). It was shown, that the most active strains P. fluorescens 38a(pBS216) and P. aureofaciens OV17(pOV17) in the barley rhizosphere decreased the phenanthrene concentration 2 and 3 times respectively in 28 days in pot experiments. The impact of rhizosphere strains in plant accumulation of heavy metals/metalloids was tested in pot experiments. The cobalt-nickel resistant strain P. aureofaciens BS1393(pBS501) promoted growth of barley plants and protected from chlorosis contrary to the sensitive strain P. aureofaciens BS1393 in soil containing 235–940 mg Ni/kg. In one month growing the nickel accumulation in plant biomass increased by 5.6 and 2.5 times in the case of sensitive and resistant strain, respectively, compared to non-treated plants. The sorghum plants, inoculated by the resistant P. aureofaciens BS1393(pKS1) and phosphate-dissolving P. aureofaciens BS1393(pUCP22:gltA) strains accumulated arsenic in plant biomass on an average of 25% more than non-treated plants in one month growing on arsenic contaminated soil (100 mg/kg). Nevertheless, the amount of bacteria in the plant rhizosphere varied, depending on bacterial species, plasmids occurrence and experiment conditions, but PGPR inoculation of plants protected them against PAH and metal/metalloid phytotoxicity, promoted seed germination and plant biomass.

Chapter 18 - A large number of sites worldwide are contaminated by arsenic (As) as a result of human activities as well as from natural sources. Arsenic is a vital environmental and health concern due to its known chronic and epidemic toxicity. The main arsenic exposures to humans are through water pathway and food contamination originates from natural processes. Many of the available remediation technologies lost economic favor and public acceptance because of some unavoidable limitations of those technologies. Therefore, phytoremediation, a plant-based green technology, becomes an emerging and alternative technology that aims to extract or inactivate As in the environment. However, two approaches have been proposed in literature for the phytoremediation of arsenic: continuous or natural phytoremediation, and chemically enhanced phytoremediation. The first one is based on the use of natural hyperaccumulator plants having the ability to accumulate very high concentration of As in their shoots with exceptionally higher tolerance to As toxicity. On the other hand, As uptake in high biomass crop plants is increased using some chelating ligands in chemically enhanced phytoremediation technology.

Freshwater and seawater around the world have been contaminated by As from various anthropogenic activities and natural sources over time. Therefore, remediation of As-contaminated aquatic systems is important as it is for terrestrial system. Aquatic macrophytes could be used to remediate the aquatic system. The use of aquatic macrophytes or other floating plants in phytoremediation technology is commonly known as phytoextraction. This cleanup process involves biosorption and accumulation of As. Recently, aquatic macrophytes and some other small floating plants such as Spirodela polyrhiza L., Lemna spp., Azolla pinnata, Salvinia natans, Eichhornia crassipes have been investigated for the remediation of As-contaminated aquatic systems. Compared to the As-phytoremediation in terrestrial system, less work has been done in aquatic systems. In this chapter, process and prospect of As phytoremediation by aquatic macrophytes is discussed.

Chapter 19 - Because plant roots are in direct contact with pollutants in contaminated soil or water, their responses to toxic substances are of particular importance in phytoremediation and phytomining research. Genetically transformed hairy roots offer many practical advantages in experimental studies, such as ease of initiation, culture, and maintenance, indefinite propagation of material derived from the same parent plant, and genotypic and phenotypic stability. Hairy roots have been applied mainly in metabolic studies of xenobiotic biotransformations and degradation in plants, and for determining the responses of plant tissues to toxic heavy metals. The aim of this chapter is to review the applications of hairy roots in phytoremediation and phytomining research. Experimental results are also presented to demonstrate the capacity of hairy root cultures to hyperaccumulate heavy metals such as cadmium and nickel, allowing practical examination of the biological mechanisms responsible for elevated heavy metal tolerance in hyperaccumulator plant species.

Chapter 20 - Phytoremediation is the use of plants to remove contaminants from the environment or render them harmless. Current engineering-based technologies to clean up soils are costly, and most considerations usually state that soil phytoremediation will be cheaper than alternatives such as soil washing. However, phytoremediation is a comparatively new field and not all of its applications are well understood. Most metal-contaminated soils contain more than one metal. For example, combinations of Pb and Zn are common in urban soils, while Pb, Zn, Cd and Cu are all often present in the vicinity of a metallurgic smelter. There will be minimal economic value in a technology that can efficiently remove one metal from a soil but leave most of another behind. However, most of the experiments on phytoremediation only address a single metal contaminant. Two field surveys were carried out in order to understand the multiple-metal effect on phytoremediation.

A 2-ha survey was performed over 2 years to study how plants such as eucalypts would remove lead and zinc from the abandoned mine at Sanguinheiro (40°30'N and 8°18'W in Portugal). The average comparison of metal content in leaves is summarized as follows (cf. remediation zone vs. background zone): Pb – 2.9 vs. 3.6 mg kg-1 and Zn – 29.7 vs. 14.1 mg kg-1. Another 8-ha survey using willow was also performed over 2 years under conditions of continuous metal deposition near the Monchegorsk smelter (68°02'N and 34°48'E in the most northern part of the European fringe of the Russian Federation). The average comparison of metal content in leaves is summarized as follows (cf. remediation zone with fertilization vs. background zone): K – 6781 vs. 7635 mg kg-1, Mn – 43 vs. 845 mg kg-1, P – 2303 vs. 2856 mg kg-1, and Zn – 109 vs. 161 mg kg-1. The results obtained from the metal analysis (Cu, Ni, Fe, Pb, etc.) indicate high efficiencies of phytoremediation (i.e. preferable effect of phytoextraction), but a clear relation of leaf chemistry with soil chemistry could not be obtained.

Both field tests in Portugal and Russia suggest that the root system is more important than the leaf system in the evaluation of remediation efficiency. The data presented in this chapter may help the planning of a commercial application of phytoremediation in cases of multiple-metal stress. However, long-term observation is also necessary to confirm reliable feasibility for underpinning the design of a large-scale phytoremediation project.

Chapter 21 - Phytoextraction of heavy metals (HMs) is a promising technology that uses plants to remove pollutants from soil. Two high biomass yield crops, maize and sunflower, with their ability to accumulate HMs, have been widely used to remediate contaminated soils. Nine commercial cultivars of maize and three of sunflower were characterized for their

Genetic Bio-Diversity (GBD) using two different molecular approaches: Random Amplified Polymorphic DNA (RAPD) and Amplified Fragment Length Polymorphism (AFLP).

A pot experiment was subsequently carried out to estimate the phytoextraction capacity of three cultivars for each plant species grown on multi-metal (copper and zinc) artificially contaminated soil. The HM accumulation was estimated in all three plant organs: root, stem and leaf.

The results of the molecular analysis showed a considerable GBD among all tested cultivars. Moreover, a highly significant difference was observed among cultivars for their HM extraction capability. In both species, the highest metal concentration was detected in roots, followed by stems and leaves; sunflower cultivars exhibited the highest potential for the removal of HMs from a multi-metal polluted soil.

Chapter 22 - Selenium is essential for many organisms but is toxic at elevated concentrations. The window between nutritious and toxic levels of Se is narrow, and both Se deficiency and toxicity are problems worldwide. For plants Se serves no known essential function, and uptake of Se by plants can lead to toxicity due to the similarity of Se to sulfur (S) and the incorporation of Se into S proteins. However, many plants readily take up Se and can benefit from increased Se due to increased growth and/or as an elemental defense.

In relation to Se, plants can be classified into three categories: 1) non-Se accumulators 2) Se accumulators, and 3) Se hyperaccumulators. Non-Se accumulators do not accumulate Se, or only accumulate trace concentrations of Se, even when growing on seleniferous soils, Se accumulators can accumulate up to 1,000 mg Se kg-1 and Se hyperaccumulators accumulate upwards of 1,000 mg Se kg-1 and as much as 15,000 mg Se kg-1.

Elevated tissue Se levels can protect plants from a variety of herbivores and pathogens, including fungi, arthropods and mammals. This elemental plant defense may act as a convenient pesticide when using plants for Se phytoremediation, and may also help prevent toxic Se concentrations from entering the ecosystem. Selenium as a defense has been disarmed in at least one instance, by a population of diamondback moth (Plutella xylostella), and probably has been disarmed on other occasions. Understanding the mechanisms that have led to the disarmament of Se as a defense is important to better understand how plant Se may enter higher trophic levels. In addition, many decomposers in seleniferous environments appear to have evolved Se tolerance, resulting in increased decomposition rates of Se-rich plant material and possibly faster release of Se into soil. Selenium may also influence pollination. There is evidence that Se accumulation changes flower phenotype characteristics and that important reproductive tissues, such as pistils, stamens, nectar and pollen, accumulate Se. Another interesting ecological aspect of plant Se accumulation is the role of rhizosphere and endophytic-microbes in Se (hyper) accumulation; there is evidence that rhizosphere microbes can increase plant Se accumulation and volatilization.

Investigating the ecological implications of Se accumulation in plants is crucial to managing phytoremediation of Se-polluted sites. Moreover, studies on the effects of Se on plant ecology may serve as a model for ecological implications of plant accumulation of other elements during phytoremediation or production of fortified foods.

Chapter 23 - Contamination of soil and water has grown as an environmental problem along with the increase of human activities. Among the main pollutants involved, agrochemicals are of major concern, since millions of tons are applied every year for crops and forestry, expecting that nature would take care of them. Although there are many effective physical-chemical methods for soil and water decontamination, the application cost of such techniques and the wide expanse of moderately polluted areas make them inappropriate. In this context, phytoremediation has arisen as an environmentally friendly, low cost and effective alternative for this kind of pollution. Nevertheless, the effectiveness of the process depends on the particular characteristics of the soil, the contaminant and the environmental conditions and their interactions, which makes phytoremediation a site-specific technology. The field-scale applicability of the results obtained at lab research mainly depends on the accuracy of the selected experimental system. In this way, there are two divergent positions: on one hand, a simplified system (cell cultures, organ cultures, hydroponics) where the variables are reduced at minimum and fully controlled, gives precise information about the mechanisms involved in the remediation process. On the other hand, a complex experimental system (microcosms) gives information closely related to real scale, but having less control over the experimental variables involved. They have designed and optimized experimental systems of different complexity for studying phytoremediation of soils contaminated with agrochemicals. Azinphos–methyl, 2,4-dichlorophenoxyacetic acid, 2,4-dichlorophenoxybutyric acid and atrazine were selected since they are among the most controversial agrochemicals, because of their toxicity and potential as environmental pollutant. In the designed experimental systems, the biodegradation potential of model and novel tolerant plant species and their influence on soil microflora was observed. At the same time, the systems were used to investigate the mechanisms involved in plant tolerance to herbicides. The soil, contaminant, microflora and plant interactions observed in lab scale experiments and the degradation profiles of the different agrochemicals will be discussed. Conclusions about the influence of experimental system complexity on mechanisms elucidation and reliability of the scaling-up will be presented.

Chapter 24 - The practice of phytoremediation to remove unwanted elements from soils can be turned to a different application – to generate nanoparticles of a wanted element. The same processes are at work but the goal is different. In phytoremediation the task is to remove a contaminant from soil, whereas in phytomining it is to concentrate a valuable element and for phytosynthesis it is to synthesise a particular form, for example nanoparticles.

The understanding of the formation of nanoparticles (generally noble metals) by plants also contributes to the understanding of the uptake and accumulation of specific elements by plants, which may then be applied to phytoremediation and to phytomining. This chapter describes the use of plants to produce silver nanoparticles and the understanding that has been developed around the mechanism underlying the nanoparticle formation.

Chapter 25 - Phytoremediation utilizes different plant species as a media of containment, destruction, or extraction of contaminants from different matrices including soil and water. Plants require essential metals i.e. Cu, Mn, Fe, Zn, Mo, etc. for growth and as such they are capable of accumulating these metals. Plants can also accumulate Cd, Cr, Pb, Co, Ag, Se, Hg, etc., which are apparently non-essential for their growth and survival. This metal-accumulating property of plants has made them very popular in recent days in the remediation of metal-contaminated soil. This approach of remediation has the benefit of cost savings compared to the conventional treatment options. Plants capable of concentrating metal pollutants at enhanced rate - the hyperaccumulators - are commonly used for metal-polluted soil remediation. But, the bioavailability of the metals limits the performance of hyperaccumulators since a large proportion of metals in contaminated soils exist in 'non-labile' state. There came the application of synthetic chelants to enhance the mobility and phytoavailability of metals to remediating plants. Various chelants are available which forms bioavailable and water-soluble stable metal complexes facilitating phytoextraction of these metals at enhanced rates by plants. While chelants are used because of their powerful metal solubilizing properties, it is the same characteristic which gives them the potential of becoming an eco-environmental threat. Environmental concerns are evoked due to the high persistency and poor photo-, chemo- and biodegradability of metal-chelant complexes. Different approaches have been proposed to combat the eco-environmental concerns raised by the use chelants in phytoremediation. Within the scope of this chapter, the authors will focus on the chelant assisted phytoremediation approaches for the removal of heavy metal contaminants from soil and eco-environmental consequences associated with it.

Chapter 26 Salt marshes located in estuaries frequently receive large inputs of nutrients (Caçador, et al., 1993; Tobias et al., 2001), and also of particulate and dissolved organic matter.

Salt marsh plants retain suspended particles and associated anthropogenic metals transported by the tides.

This high nutrient input makes salt marsh one of the most productive ecosystems of the planet. In highly industrialized estuaries, along with this nutrient input there is also a large input of heavy metals (Figure 1) which will be accumulated in salt marsh sediments (Caçador at al., 1996; Doyle and Otte, 1997).

Chapter 27 - In Taiwan, many heavy metals (HMs)-contaminated arable soils have been founded since 1980. Agricultural irrigation system was mixed with river waste water contaminated with HM is the primary reason for the contamination of cropping lands. Soil turnover/attenuation technique, which mixes the surface 30 cm layer of contaminated soils with deeper clean soil layer, was the most popular technique to be used to dilute the HM-contaminated soils to meet the soil regulation of HM in the soil contamination site. Phytoextraction technique was also regarded as another candidate technique to remove the HMs from the HMs-contaminated sites. Seedlings of various native garden flowers of Taiwan were planted either in-situ in HM-contaminated sites or in pot experiments artificially spiked soils to investigate their tolerance and removal capacity from the sites. These sites were primary contaminated with cadmium (Cd), lead (Pb), zinc (Zn), or mixed-combined with them. The total removal of HMs plays an important role prior to conduct a successful phytoextraction and decontamination. Although some of the selected plant species can accumulate higher concentration of HM in their shoots, they are small biomass and thus just can remove little amounts of HM from the contaminated soils.

The accumulation and growth of a specific plant in-situ grown in contaminated site is quite different compared with that of pot experiments. This paper summarizes the total removal capacity of high potential super accumulator garden flowers and estimated the period needed to cleanup the Cd from the contamination sites.

Chapter 28 - Of the various physico-chemical and biological technologies that have been used for remediation of heavy metals (HMs) contaminated soils, all methods are expensive and totally destroy physical, chemical and biological properties of treated soils, reduce yield of plant growth and disrupt ecosystems. Therefore, it is best to develop suitable, natural, cheaper and in situ technologies to recover degraded land. Phytoremediation is an alternative to physico-chemical methods and is emerging as a promising environmentally friendly method for detoxification and /or deactivation and removal of elements from polluted soils. It is possible to improve the capabilities of plants in different types of phytoremediation processes by inoculating with appropriate soil microorganisms especially arbuscular mycorrhizal fungi (AMF). Some AMF species occur naturally and form a symbiosis with plant roots in the HMs polluted soils. In some cases, AMF have generally such a strong influence on plant biomass and can increase HMs uptake and root-to shoot transport (phytoextraction), while in other cases AMF contribute to HMs stabilization within the soil/root and reduce their uptake (phytostabilization). In this chapter, some knowledge concerning the role of AMF in phytoextraction and phytostabilization of HMs contaminated soil was summarized and discussed.

Chapter 29 - Plant growth-promoting bacteria are soil bacteria that are involved in a beneficial association with plants; these bacteria use a variety of mechanisms to facilitate plant growth. The major mechanisms used by plant growth-promoting bacteria include the functioning of the enzyme 1-aminocyclopropane-1-carboxylate (ACC) deaminase which cleaves the compound ACC, the immediate precursor of the phytohormone ethylene in all higher plants, and synthesis of the plant hormone indoleacetic acid (IAA). Plant growth-promoting bacterial strains that contain ACC deaminase and produce IAA provide a wide range of different plant species with a significant level of protection from the damage caused by various environmental stresses including heavy metals and the presence of organic environmental contaminants. Here the authors discuss how bacterial ACC deaminase and IAA work synergistically to facilitate plant growth during the phytoremediation of metals and/or organics.

Chapter 30 - Large areas of agricultural land have been contaminated with potentially toxic metals like Pb by smelting activities in the last centuries (Loska et al., 2004). The possible negative impacts on the environment and human health demand the need for remediation of contaminated sites. Conventional remediation techniques for heavily contaminated soils like excavation or soil-flushing are very cost intensive and not appropriate for large areas of low or medium contaminated agricultural land. Furthermore, these technologies result in a removal of topsoil and in many cases also subsoil needed for agricultural production or in the decrease of its fertility. Discussion and research has therefore focused on in-situ remediation technologies which seem to be cost-effective and environmentally acceptable. The use of plants for the remediation of potentially toxic metals, so-called phytoextraction, could be an alternative.

Chapter 31 - Three pre-selected poplar clones and two soil HCH degrader micro-organisms have been experimentally applied in a contaminated agricultural soil in the basin of Fiume Sacco near to Rome for its reclamation. The aim was to successfully associate soil cleaning by rhizoremediation with an economically sustainable biomass for energy production of large poplar plantations. Plants and micro-organisms were selected for the best association with bacteria to obtain 1) the maximum HCH concentration reduction in soil, 2) the minimum plant contamination with HCH, and 3) the maximum biomass production. Results showed that an association between all these traits is possible in a specific poplar clone inoculated with a selected HCH degrader bacterium. The need for a pre-remediation phase in situ to select best candidate plants and bacteria with lowest HCH accumulation in its organs is emphasized. Rhyzoremediation associated with the safe thermo-convertible biomass production is confirmed as a sustainable recovery of soils interdicted to food-agricultural activities.

In: Handbook of Phytoremediation
Editor: Ivan A. Golubev

ISBN: 978-1-61728-753-4

Chapter 1

PHYTOREMEDIATION OF PHENOLIC COMPOUNDS: RECENT ADVANCES AND PERSPECTIVES

Elizabeth Agostini**, Melina A. Talano, Paola S. González, Ana L. Wevar Oller and María I. Medina
Departamento de Biología Molecular, FCEFQyN,
Universidad Nacional de Río Cuarto, (UNRC), Ruta Nacional 36 Km 601,
CP 5800, Río Cuarto, Córdoba, Argentina

ABSTRACT

Phenolic compounds present in the drainage from several industries are harmful pollutants and represent a potential danger to human health. Conventional treatments for phenol removal from industrial wastewaters have several limitations so, there is a need to look for alternative and environmental friendly technologies to complement or substitute the conventional ones. In recent years, phytoremediation has been recognized as a cheap and eco-friendly alternative technology which could be used for the remediation of organic contaminants, such as phenolics. Despite most phytoremediation studies were performed with soil-grown or hydroponically grown plants; more recently some results were obtained with the help of in vitro cell and tissue cultures, such as hairy roots. They have been used as tools for screening the potencialities of different plant species to tolerate, accumulate and remove high concentrations of phenols with high efficiency. In addition, using different plant model systems it could be established that plants metabolize a number of phenolic compounds by common metabolic pathways. Uptake of phenolics depends on the plant species as well as on their physico-chemical properties. While the main metabolites detected from phenolic´s transformation are polar conjugates, some plant species could incorporate large amounts of these chemicals and associated metabolites, as bound residues, through reactions catalized by oxido-reductases. Hence, cell wall is considered one of the important detoxification sites of phenolic compounds in plants. In addition, plant roots produce and exude high amounts of oxido-reductive enzymes, such as peroxidases, which are associated with the non specific oxidative polymerisation of phenolic free radicals in the cell wall. So, these enzymes may play an important role in polymerising reactions and, also, they are likely to be the key enzymes in the removal of phenol and chlorophenols. Moreover, different peroxidase isoenzymes

* Tel.: +54 358 4676537; fax: +54 358 4676232; *E-mail address:* eagostini@exa.unrc.edu.ar (E. Agostini).

might play different roles in the removal process. So, in this chapter, the use of plants as enzyme sources, as well as partially purified oxidases is discussed, as good alternatives for remediation purposes.

On the other hand, with the application of genetic engineering, it is feasible to manipulate plant capabilities to tolerate, accumulate, and/or metabolize pollutants, and thus to create an appropriate plant for environmental cleanup. Therefore, this chapter also examines and discusses the recent advances in enhancing phytoremediation of phenolic compounds through transgenic plant research.

Current knowledge, the areas which need to be explored and perspectives are presented to improve the efficiency and to asses the feasibility of phenolics′ phytoremediation.

Keywords: phytoremediation; phenolic compounds; peroxidases; laccases; removal efficiency; phenolic metabolism; transgenic plants.

1. Phenolic Compounds: Characteristics, Environmental Impact and Toxicity

Phenol and its halogenated derivatives are considered as high priority pollutants because of their toxicity and possible accumulation in the environment. The generic terms "phenols" and "phenolics" are frequently used to describe those alcohol derivatives of benzene. They are mainly of anthropogenic origin, due to their wide utilization in several industries. The basic information concerning some physical and chemical properties of phenol, which was selected as a representative phenolic contaminant, is included in Table 1. Besides its toxicity, phenol is very soluble in water and can confer bad odour and taste to food and drinking water, making it unfit for use. In relation to halophenols, there are four classes (fluoro, chloro, bromo and iodo) and each class comprises 19 congeners from mono- through penta-halogenated, with physico-chemical properties such as solubility, volatility, and the octanol–water partition coefficient (Kow) varying systematically as a function of halogen content and substitution pattern (Garg et al., 2001). Among them, chlorophenols have the highest industrial value and are the most studied until now. The molecular structures of phenol, alkylphenols and some selected chlorophenols [4-chlorophenol (4-CP); 2,4- dichlorophenol (2,4-DCP); 2,4,6-trichlorophenol (2,4,6-TCP) and pentachlorophenol (PCP)] are shown in Figure 1. By comparison to phenol and chlorophenols, relatively little seems to be known regarding natural and anthropogenic sources, environmental concentrations, distribution and fate of the simple bromo-, fluoro-, and iodophenols (Rayne et al., 2009). It is likely that these three contaminant classes will gain widespread interest by the environmental research community in the near future because of advances in sample collection, processing, and analytical methods. They will also gain importance because of their potential to form either more toxic compounds-such as halogenated dibenzo-p dioxins and furans under combustion conditions and photochemical processes-or through the need to develop remediation technologies. There is also increasing concern over phenolic environmental pollutants with endocrine activity, such as alkylphenols (Figure 1), especially octylphenol and nonylphenol that are metabolites of non-ionic surfactants and they are found in considerable amounts in sewage sludges (Sweetman, 1994). Moreover, nonylphenol and its derivatives are frequently found in crop plants which may produce a strong impact on food quality (Harvey et al., 2002).

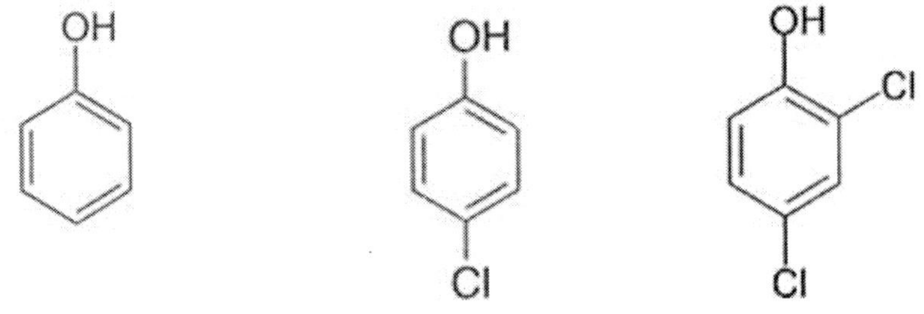

Phenol 4-chlorophenol 2,4-dichlorophenol

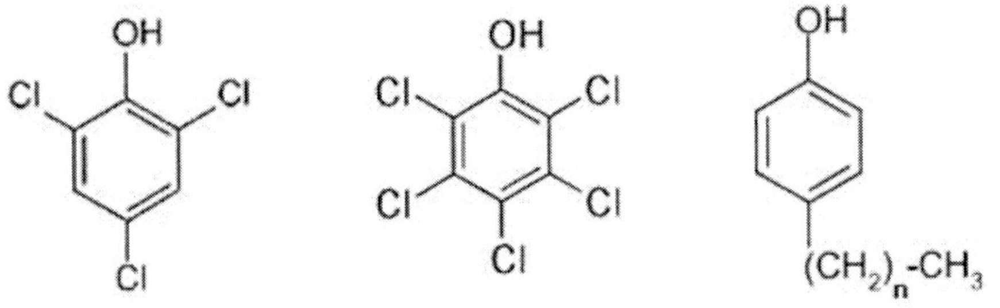

2,4,6-trichlorophenol pentachlorophenol alkylphenol

Figure 1. Structure of selected phenolic compounds.

The worldwide production of phenol has been fearly constant since the 1980s (WHO, 1994). Nowadays, industrial phenol production is over three million tons per year, being used mostly in petrochemical industry, synthesis of resins, dyes, pharmaceuticals, perfumes, pesticides, tanning agents, solvents or lubricating oils (Iurascu et al., 2009). Because of their high rates of production and usage, phenol and various halophenols are widely found in environmental samples, particularly in aquatic systems (surface water, rivers, lakes, etc) and in the surrounding soils, where they are introduced from industrial effluents (WHO, 1994). Water flowing on the surface or penetrating into the depths of soil, could lead to significant contamination of groundwaters and sediments, where high phenol levels have been reported. In addition, phenol has been detected in rain, but data are very scarce. Atmospheric phenols and nitrophenols, directly emmitted through combustion processes of vehicles or by different industries, have also received an special interest over the last years (Schummer et al., 2009). Furthermore, phenolic contaminants can be introduced to the environment via agricultural run-off and as a result of partial degradation of other aromatic organic contaminants, such as polycyclic aromatic hydrocarbons (PAHs), polychlorinated biphenyls (PCBs) and certain surfactants (Huang et al., 2005).

Quantitative data available for phenols in industrial wastewaters are generally expresed in terms of total concentration and show a wide range of variability depending on the procedence. For instance, phenol concentrations for refinery effluents are around 50 mg L^{-1} for distillation units, in the range of 50-500 mg L^{-1} for catalytic cracking and visbreaking processes and up to 500 mg L^{-1} in the spent caustic solutions. Likewise, waste solutions generated from coal conversion processes usually contain 200-600 mg L^{-1} of phenols (Nayak and Singh, 2007).

Phenols are toxic, carcinogenic, mutagenic and teratogenic (Autenrieth et al., 1991). So, they are considered priority pollutants in the US Environmental Protection Agency (EPA) list and their discharge in the aquatic environment is becoming increasingly restrictive. In addition, phenol is included in the class 2 water hazardous pollutants list in several countries (Iurascu et al., 2009). A phenol concentration of 1 mg L^{-1} or greater, affects aquatic life and may represent a risk to human health. Therefore, in most cases a stringent effluent discharge limit of less than 0.5 mg L^{-1} is imposed. In this sense, the US EPA and the WHO have established a limit concentration of 1 μg L^{-1} for phenolic compounds in drinking water (Srivastava et al., 2006) while the European Community defined a limit of 5 μg L^{-1} (Schummer et al., 2009 and references there in). However, phenols are frequently found in higher concentrations than those established by Regulatory Organizations.

For example, in Southern Finland, concentrations ranging from 25 to 55 mg L^{-1} of chlorophenols were found in the groundwater from a contaminated aquifer (Quan et al., 2003 and references there in) while in some rivers of Argentina levels from 0.4 to 2.28 mg L^{-1} were detected (Paisio et al., 2009). Moreover, Schummer et al., (2009) collected extensive data about the concentration of phenols and nitrophenols in rainwater samples from urban and rural areas of Eastern France.

They concluded that the concentrations of these pollutants are about 10 times higher than those of pesticides and 1000 times higher than those of PAHs on the same sites and at the same period.

Phenolic toxicity has been studied on selected microbes (e.g. protozoa, yeast and bacteria), algae, duckweed and numerous invertebrates and vertebrates. Human consumption of phenol contaminated water can cause severe pain, blood changes, liver injury and muscular effects, and even death (Flocco et al., 2002; Aksu, 2005). In addition, chronic toxic effects on human include vomiting, difficulty in swallowing, anorexia, liver and kidney damage, headache and other mental disturbances (Srivastava et al., 2006). A probable oral lethal dose to humans is 50–500 mg kg^{-1}. Similarly, chronic effects on animals include shortened lifespan, reproductive problems, lower fertility and changes in behaviour (Flocco et al., 2002).

In areas of petroleum industry it was frequently observed that phenols induced genotoxic effects in animals and human (Paisio et al., 2009 and references therein) and depending on the organism tested, the acute toxicity of phenol, estimated by the LC_{50} value, varied from 6.5 to 1840 mg L^{-1}. For instance, the aquatic toxicity of phenol (LC_{50}) is 12 mg L^{-1} for *Daphnia magna*, 178 mg L^{-1} for *Xenopus* and 183.70 mg L^{-1} for *Rhinella arenarum* embryos (Iurascu et al., 2009; Bernardini et al., 1996; Paisio et al., 2009). Despite the fact that phenol can produce lethal and teratogenic effects on some amphibian species (Paisio et al., 2009), the most important effects reported in short- term animal studies were neurotoxicity, liver and kidney damage, respiratory effects and growth retardation (WHO, 1994). As it could be seen, the high chronic toxicity of phenolic compounds negatively affects aquatic and terrestrial ecosystems, interrupting comunity stability. In addition, this hazardous pollutants can enter in food chains through agricultural products or drinking water. Thus, the removal of such compounds from water and soils is of relevant significance.

Table 1. Some physical and chemical properties of phenol

Common name	Phenol
Chemical formula	C_6H_6O
Relative molecular mass	94.11 g mol^{-1}
Relative density	1.071 g mL^{-1}
Melting point	43 °C
Boiling point	181.75 °C
Relative vapour density (air =1)	3.24
Solubility in water (16°C)	67 g L^{-1}
Log Kow	1.46
Henry's Law Constant	3.97×10^{-7} atm-$m^3 mol^{-1}$ (25 °C)
pKa	9.994

2. Phenolics Treatment Technologies

Environmental problems associated with the presence of phenolics in natural waters and soils have resulted in the development of several methods for the removal of such compounds. These include physico-chemical treatment processes and biological methods.

2.1. Physico-Chemical Methods

There is a large number of physico-chemical technologies available for the treatment of phenol and its derivatives, but none of them is applicable to all situations. These processes are based on the principles of adsorption, precipitation and coagulation, chemical oxidation, sedimentation, filtration, osmosis, ion exchange, etc.

Adsorption technology is currently being used extensively for the removal of phenolic compounds and there are many adsorbents in use (Srivastava et al., 2006; Nayak and Singh et al., 2007). The costs of these adsorbents strongly depends on local availability, processing requirements, treatment conditions, and both recycle and lifetime issues. Activated carbon is one of the most widely used adsorbents (Dabrowski et al., 2005). However, this method only removes few milligrams of phenolics per gram of activated carbon, it is quite expensive and the higher the quality the greater the cost. In addition, there are some disadvantages associated with both chemical and thermal regeneration of spent carbon, which is expensive, impractical on a large scale, produces additional effluent and results in considerable loss of the adsorbent (Aksu, 2005). Thus, this situation has stimulated research into specialty absorbents that may facilitate a cheap and effective chemical regeneration process (Nayak and Singh, 2007). In this way, Lin and Juang (2009) have recently compared different resins used for phenol removal and they found that Amberlite IRA-420 and HiSiv1000 resins were the best. On the other hand, natural materials such as bagasse, red mud, zeolites, clay, or certain waste products such as eucalyptus barks, chitin, rice husk, coal, carbonized sewage sludge from industrial operations, have been explored for their technical feasibility to remove phenol and its derivatives from contaminated water due to their low-cost and local availability (Aksu,

2005; Srivastava et al., 2006; Kuleyin et al., 2007; Lin and Juang, 2009). Other low-cost adsorbents such as agricultural wastes have been studied, but less extensively. Further work is necessary to find low-cost adsorbents with a high adsorption capacity for phenols. In addition, little information exists containing full cost and application comparisons of various adsorbents. Moreover, the obtained by products remain to be explored in order to asses if they are less toxic than the parent ones.

Thus, chemical oxidation, when economically and technologically viable, is the preferred option for phenolic removal, because it is not limited to a simple transference of contaminants from one phase to another. Chemical oxidation involves the total or partial destruction of pollutants to carbon dioxide and water or eventually to harmless end products. In this way, ozone is gaining acceptance as oxidising agent for phenolic compounds since it does not introduce strange substances to the aqueous matrix. As a consequence, ozone can be used to reduce the phenolic content of wastewaters as a pre-treatment step if a post-biological polishing stage is to be applied. Another alternative for the removal of phenols from wastewater is the so called Advanced Oxidation Processes (AOPs), which operates at near ambient temperature and atmospheric pressure. (Gimeno et al., 2005; Iurascu et al., 2009). Some of these processes combine the use of ozone and other agents (hydrogen peroxide, UV radiation, high pH, etc.) to generate highly active hydroxyl radicals. The addition of appropriate catalysts, such as the photocatalytic oxidation and perovskite type catalysts could be used to optimize phenolic removal (Gimeno et al., 2005; 2007; Carbajo et al., 2007). More recently, a new heterogeneous photo- assisted Fenton conversion of phenol has been proposed. The results have shown that almost complete conversion of phenol was possible after only 5 min (Iurascu et al., 2009).

Other chemical treatments of phenols include: chlorination, deep-well injection, incineration and solvent extraction. However, they have several disadvantages which limited their use. For example, chlorination, is not recommended because it can result in the formation of chlorinated phenols and other by products, which have been reported as toxic and non biodegradable (Iurascu et al., 2009). Moreover, solvent extraction methods are expensive and deep-well injection may lead to contamination of groundwater.

The main drawback of all the above mentioned technologies relies on the economy of the process and, in some cases, on the low mineralization level achieved, involving the need of a final polishing stage.

2.2. Biological Methods

2.2.1. Bioremediation and Biosorption

Bioremediation, i.e. the use of living organisms to manage or remediate polluted soils and water, is a well known biotechnological tool to degrade contaminants into non-or less-toxic compounds. Compared with traditional physico-chemical methods, bioremediation is generally the safest and least disruptive treatment. Regarding phenolic microbial remediation, there is a wide variety of pure and mixed cultures of microorganisms capable of degrading these compounds under both aerobic and anaerobic conditions. It is well described that phenolics are degraded and mineralized by microorganisms because they are used as a source of energy and carbon skeletons for cell protein synthesis, both in terrestrial and aquatic environments. The reactions of ring-fission are frequently catalized by intracellular mono or dioxygenases and the final products are molecules able to enter the Tricarboxilic Acids Cycle (Harvey et al., 2002). Although microbial degradation of phenols is seen as a cost effective method, it is limited by the intrinsic properties of these compounds owing to their toxicity and, frequently, by their slow rate of biodegradation. Other factors affecting biodegradation could be low pollutant bioavailability (mass transfer), aeration, scarce nutrient level at contaminated sites (might require bio-stimulation) as well as problems with thermal conditions (Alcalde et al., 2006). All these limiting factors should be addressed on a case-by-case basis to obtain the maximum microbial growth for decontamination purposes. However, several reports describing microbial bioremediation technology, as an appropriate process for decontamination of phenolics and a detailed description of these processes could be found in the literature (Cai et al., 2007; Dong et al., 2008; Field and Sierra-Alvarez, 2008; Indu Nair et al., 2008; Dos Santos et al., 2009; Cordova-Rosa et al., 2009; Liu et al., 2009).

On the other hand, microorganisms such as bacteria, fungi, yeast and also algae, and plants, can remove some pollutants from aqueous solutions through passive sorption and such process is called biosorption, which takes place essentially at cell wall level. In recent years, a number of studies have focused on the biosorption of phenols, chloro- and nitro-phenols. Depending on the phenolic compound and the species of microorganism used, different binding capacities have been determined (Jianlong et al., 2000; Calace et al., 2002; Rao and Viraraghavan, 2002; Aksu and Gönen, 2004). Despite the fact that biosorption is a promising alternative to replace or supplement present treatments for the removal of low concentrations of phenolics, its use is still in a research stage. Thus, more studies are needed to develop practical applications (Aksu, 2005).

2.2.2. Phytoremediation as a Promising Alternative Technology

As it could be seen, all the methods mentioned above had several limitations and, sometimes, they are disruptive to the environment. So, it has become necessary to look for environmentally friendly treatment technologies to complement or substitute the conventional ones. In recent years, phytoremediation, which was defined as the use of green plants to remove, contain or render harmless organic or inorganic environmental contaminants (Cunningham and Ow, 1996), has been recognized as a cheap and eco-friendly alternative technology that can be tried out for the remediation of organic contaminants. A variety of pollutant attenuation mechanisms possessed by plants makes their use in remediating contaminated land and water more feasible than physico-chemical remediation (Gerhardt et al., 2009). The main objective of scientists, agronomists, and engineers dealing with phytoremediation is to exploit by the most rational way possible the potential of this natural process. Remediation technologies based on plants represent an attractive alternative because they are independent of an external energy supply, they have more public acceptance than the use of chemical methods, they are not invasive and have many advantages, which were described in detail by Pilon-Smits (2005) As regards their direct roles in remediation processes, plants use several different strategies for dealing with environmental chemicals: phytoextraction, phytodegradation, phytovolatilization and rhizodegradation (Gerhardt et al 2009; Abhilash et al., 2009).

Enhancement of phytoremediation process is one of the challenges of current research. Selection, traditional breeding and genetic engineering focus on the optimization of pollutant tolerance, root and shoot biomass, root architecture and morphology, pollutant uptake properties, degradation capabilities for organic pollutants etc. (Wenzel, 2009). In addition, plants increase the amount of organic carbon in the soil, which can stimulate microbial activity and augment the rhizospheric degradation of the pollutants. According to this, other approaches are directed to the management of microbial consortia, which includes not only rhizospheric bacteria but also endophytes, their selection and engineering, their beneficial effects on plants, or the modification of pollutant bioavailability. Additional strategies include proper management of the soil and the optimization of some agricultural factors.

Several plant-based experimental systems have been studied for phytoremediation purposes. In increasing order of complexity, they are plant cell cultures such as callus and cell suspensions (Harvey et al., 2002), differentiated organ cultures such as roots, hairy roots and shoots (Mackova et al., 2001; Suresh et al., 2005; Singh et al., 2006), explants such as leaf disks and excised roots and whole plants in hydroponic culture, in potted soil under greenhouse cultivation and in the field (Flocco et al., 2002; Singh et al., 2008; Schröder et al., 2008; Doran, 2009). It is important to note that plant tissue cultures cannot represent or simulate many aspects of whole plant cultivation, and a proper design of experiments and interpretation of results are required to avoid experimental artefacts and to obtain the maximum benefit from using plant tissue culture models. However, *in vitro* plant model systems have been a very useful tool for studying the uptake of organic compounds without the interference of soil matrix. Regarding hydroponic cultures, they allow the control and reproduction of experimental conditions and, also, plant roots can be exposed homogenously to the test compound, avoiding local variations, which may occur in soils. So, they have been successfully used for studying the removal of several organic compounds, including phenolics (Narayanan et al., 1999; Ucisik and Trapp, 2006; Doty et al., 2007; Singh et al., 2008), as well as heavy metals (Liu et al., 2007) and radionucleides (Ramaswami et al., 2001; Soudek et al., 2006).

In addition, aseptic *in vitro* cultures such as hairy roots, have proved to be a suitable model system to study xenobiotic detoxification and the activity of central detoxification enzymes, without the interference of soil and microbes. Hairy roots offer the important advantages of greater genotypic and phenotypic stability than dedifferentiated cultures, thus providing a more reliable and reproducible experimental system over time (Doran, 2009). It is well known that they are able to metabolize per se hazardous compounds by common metabolic pathways (Pletsch et al., 1999; Nepovim et al., 2004). Furthermore, the organized nature of hairy root cultures provides an added advantage, making them more amenable for cultivation in bioreactors to study the process in a large scale (Suresh et al., 2005). Many investigations have demonstrated that hairy roots derived from different plant species could be used for the treatment of several contaminants such as PCBs (Mackova et al., 1997; Mackova et al., 2001), pesticides like DDT (Suresh et al., 2005) and nitroaromatic compounds like 2,4-dinitrotoluene; 2,4,6- trinitrotoluene (TNT) and aminotoluenes (Nepovim et al., 2004). In addition, hairy roots of different plant species were succesfully used to remove phenol, 2,4-DCP and other chlorophenols (Gonzalez et al., 2006; Santos de Araujo et al., 2006; Coniglio et al., 2008; Sosa Alderete et al., 2009; Talano et al., 2010). Table 2 summarizes different aseptic *in vitro* cultures used for phenolic remediation.

A critical overview of the application of plant tissue cultures in phytoremediation research is provided by Doran (2009). The author concluded that *in vitro* cultures are not a replacement for soil-cultivated plants; instead, they are a powerful auxiliary model system. So, the results derived from tissue cultures can be used to predict the responses of plants to environmental contaminants, and to improve the design and reduce the cost of subsequent conventional whole plant experiments.

Table 2. Summary of different plant tissue cultures used for phenolic compounds removal

Plant tissue culture	Plant species	Phenolic compound	References
Cell suspension	*Triticum aestivum, Glyxine max*	PCP	Harms and Langebartels, 1986
Cell suspension	*Triticum aestivum*	PCP	Schäfer and Sandermann, 1988
Cell suspension	*Daucus carota, Atriplex hortensis*	Nonylphenol	Bokern and Harms, 1997; Bokern et al., 1998
Hairy roots	*Daucus carota*	Phenol and chloroderivatives	Santos de Araujo et al., 2002
Hairy roots	*Brassica napus*	2,4-DCP	Agostini et al., 2003
Transgenic hairy roots	*Solanum lycopersicon*	Phenol	Wevar Oller et al., 2005
Hairy roots	*Solanum lycopersicon*	Phenol	González et al., 2006
Hairy roots	*Brassica juncea, Beta vulgaris, Raphanus sativus, Azadirachta indica*	Phenol	Singh et al., 2006
Hairy roots	*Daucus carota* L., *Ipomoea batatas* L. *Solanum aviculare*	Phenol and chloroderivatives	Santos de Araujo et al., 2006
Hairy roots	*Brassica napus*	Phenol	Coniglio et al., 2008
Transgenic hairy roots	*Nicotiana tabacum*	Phenol	Sosa Alderete et al., 2009
Hairy roots	*Nicotiana tabacum*	2,4-DCP	Talano et al., 2010
Cell suspension	*Nicotiana tabacum*	2,4-DCP	Laurent et al., 2007

3. Uptake, Metabolism and Degradation of Phenolic Compounds by Plants

3.1. Uptake of Phenol and its Derivatives

Higher plants seem to be organisms with the inherent capacity to absorb contaminants with different chemical structures from soil, water and air. The intensity of the absorption depends on bioavailability which is one of the most limiting factors in phytoremediation of organic pollutants. Pollutant bioavailability is understood as the result of many interacting factors associated with contaminant characteristics (molecular mass, concentration, polarity, etc), soil properties (content of humic substances, clay and mineral content, pH, water content, porosity), temperature, some other physical, chemical and agronomical factors, as well as the plant species used and the associated microbial community (Kristich and Schwarz, 1989; Reid et al., 2000). In this context, hydrophobic and nonpolar organic matter is of particular importance for the binding of organic pollutants, such as phenolics, anilines and PAHs to the soil matrix, which is known to progress as the contact time increases, rendering pollutants less bioavailable. This phenomenon is known as "ageing" and determines the entrapment of the pollutant within humic complexes, nano- and micropores (Reid et al., 2000; Semple et al., 2003; Harvey et al., 2002). Tabak et al. (1994) studied the bioavailability and biodegradation kinetics of phenol in surface and subsurface soils and developed a predictive model for biodegradation kinetics applicable to soil systems.

Plants play a direct role in the removal of a contaminant by (1) sorption on plant tissues and/or (2) uptake and subsequent translocation, metabolization, storage or volatilization. Sorption to roots would be considered the first step, because when pollutants present in soil water or groundwater come into contact with roots, they may sorb or bind to the root structure and cell walls. Such sorption should be relatively reversible or not, depending on different variables. The sorption of contaminants to the root surface has been reported in several plant species and it can be estimated in control experiments using dead or inactivated biomass. For instance, phenolic compounds are partially adsorbed onto roots and this process of non-specific binding by physical sorption to plant tissues contributed for phenol removal from the liquid medium (Dec and Bollag, 1994; Santos de Araujo et al., 2006; Coniglio et al., 2008; Sosa Alderete et al., 2009). However, the use of appropriate controls, have determined that the contribution of this process to the overall phenol removal is low and sometimes depreciable (Agostini et al., 2003; Coniglio et al., 2008). In addition, the sorption process is usually estimated by the so-called Root Concentration Factor (RCF). Briggs et al., (1982) defined RCF as the ratio of organic chemical sorbed onto the root (mg Kg^{-1} fresh root tissue) respect to the compound concentration in hydroponic solution (mg L^{-1}). Thus RCF has units of L Kg^{-1}. The RCF describes the potential of a given xenobiotic to accumulate in the plant root, without differentiating between surface accumulation and uptake into the root tissue (Schröder and Collins, 2002). RCF is heavily dependent on the octanol-water partitioning coefficient Kow, and specially with log Kow, because log RCF was correlated with log Kow via a least square regression equation (Briggs et al., 1982; Burken and Schnoor, 1998). The log Kow gives an indication of compound hydrophobicity that predetermines the effectiveness of absorption and translocation of a contaminant in plants. It is known that contaminants with a log Kow >3.5 are well adsorbed on soil granules or plant root surfaces and do not penetrate into the plant. These hydrophobic pollutants are candidates for phytostabilization and/or rhizosphere bioremediation. However, this non-specific binding is not the only mechanism involved since specific sorption at chemical sites and enzymatic transformations by membrane-bound proteins are other mechanisms of potential importance (Dietz and Schnoor, 2001). For instance, in a study performed with hybrid poplars, RCF values for some typical contaminants, including phenol and PCP were determined. It was concluded that PCP (RCF 30 L Kg^{-1}) was highly sorbed to root tissues because of its hydrophobicity (log Kow 5.05) while phenol (RCF 11.6 L Kg^{-1}) bound slightly to roots, because it is a moderately hydrophobic compound. Phenol specific sorption and enzymatic transformation could be other mechanisms involved (Dietz and Schnoor, 2001).

Apart from the sorption processes above mentioned, the uptake or absorption of hazardous phenolics by plants is considered as an important process, which mainly contributes to real phenolic removal. This uptake is primarily carried out through plant roots and leaves. However, roots absorb substances, together with water, less selectively than leaves. Root absorption is performed in two phases: in the first fast phase, substances diffuse (passive uptake) from the surrounding medium into the root, while in the second they slowly accumulate in the tissue (Korte et al., 2000). It is accepted that an optimmum hydrophobicity may exist, which allows the pollutant to bind to the lipid bilayer of the membrane but not too strongly, to facilitate its transport. So, an optimal uptake is reached by compounds with log Kow in the range between 1 and 3.5 (Dietz and Schnoor, 2001; Pascal-Lorber et al., 2008). However, there are indications that the log Kow alone is not an absolute predictor of the compound uptake by plants. Moreover, Schröder et al. (2008) concluded that the uptake of xenobiotics is dependent on their physico-chemistry and specially on the relation between log Kow and the dissociation constant, pKa. They were also able to demonstrate that there is a sound correlation between log Kow and uptake rates.

In the case of phenol, its log Kow is 1.46, so it can be easily absorbed by roots as well as 4-CP (log Kow 2.85); 2,4-DCP (log Kow 3.05), and 2,4,6-TCP (log Kow 3.7) (Pascal-Lorber et al., 2008; Weyens et al., 2009). As it was demonstrated in *Lemna gibba* (Barber et al., 1995), almost 90% of the supplied phenol disappeared over 8-day- growth period. Moreover, Ucisik and Trapp (2006), demonstrated a clear relation between uptake and removal in willow trees (*Salix viminalis*). They concluded that phytoremediation of phenol would be best with concentrations in water or soil solution of less than 250 mg L^{-1}, at which phenol degradation by willows or associated bacteria is rapid and efficient and the toxic effects on trees are negligible. In addition, in hairy root cultures derived from other plant species such as *Brassica napus*, *Daucus carota*, *Solanum lycopersicon*; *Ipomoea batatas*, *Solanum aviculare* and *Nicotiana tabacum* a rapid uptake and metabolization of phenol and various chlorophenols were reported (Agostini et al., 2003; Santos de Araujo et al., 2002; Gonzalez et al., 2006; Santos de Araujo et al., 2006; Talano et al., 2010).

Although log Kow value of PCP (log Kow 5.05), indicates that it could be probably less absorbed by root cells, there are some reports which show that roots can uptake this compound (Harvey et al., 2002). According to some data (Qiu et al., 1994) 21% of PCP from the soil was found in the roots of grasses and 15% in the shoots, after 155 days of cultivation whereas in *Eichhornia crassipes*, an aquatic plant widely used for wastewater treatment, PCP uptake by the plant was rapid and reached a nearly steady state between 24 and 48 h of exposure (Roy and Hänninen, 1994). In another study, several plants showed PCP uptake ability (Bellin and O'Connor, 1990). These and other examples clearly demonstrate that uptake may depend not only on the pollutant's lipophilicity and dissociation constant, but also on specific inherent properties of the root itself and the transport tissues involved (Schröder et al., 2008). Thus, uptake and translocation of various organic pollutants can differ among plant species and thereby, conclusions concerning any contaminant uptake by a particular plant species cannot be applied to others, even to those belonging to the same genus.

Other important contaminant properties controlling their fate in the environment include the vapour pressure and the Henry constant. The vapour pressure indicates whether or not a pollutant is easily volatilized in dry soil conditions, while the Henry constant provides a better measure of the volatilization potential in wet and flooded soil. Based on the high solubility of phenol in water, its low vapour pressure and its Henry's low constant (Table 1), Flocco et al. (2002), concluded that the volatilization of phenol should not be considered an important mechanism for the disappearance of this contaminant from a solution.

It is important to note that a fraction of the contaminant can undergo microbiological transformations by phyllospheric and, mainly, by rhizospheric microorganisms. Sandhu et al. (2007) provided the first direct evidence that leaf-associated microbial communities in phyllosphere can degrade phenol from the air (phylloremediation). This finding indicated that bacteria on leaves could potentially contribute to phenolic natural attenuation. On the other hand, rhizospheric microorganisms can produce metabolites and intermediates derived from contaminant transformation which can penetrate into roots. Plants can also release extracellular degradative enzymes into the rhizosphere (Schnoor et al., 1995). Reports are available on the degradation of phenolic compounds by secreted plant laccases (Wang et al., 2004; Sonoki et al., 2005), which will be discussed below. Other plant-derived enzymes with potential to contribute to the degradation of phenolic compounds in the rhizosphere include dehalogenases involved in dehalogenating chlorinated compounds and peroxidases (Adler et al., 1994; Susarla et al., 2002). In this context, current knowledge of the relative importance and efficiency of plant extracellular enzymes in the presence of degrading microorganisms is still very limited, but taking into account the half-life of these enzymes it could be suggested that they may actively degrade organic pollutants for a few days following their release from plant tissues (Schnoor et al., 1995).

Apart from the direct release of degradative enzymes, plants are able to stimulate the activities of microbial degrader communities. For example, allelopathic chemicals secreted by roots can induce the synthesis of specific enzymes in degrader organisms and thus enhance rhizodegradation of pollutants. In addition, plant roots exude compounds that can serve as cometabolites which is very important especially when microorganisms cannot use the pollutant as a sole carbon source. Furthermore, root-exuded compounds may also selectively support specific microbial strain growth. So, as it is well described in several reviews recently published (Wood, 2008; Kamaludeen and Ramasamy, 2008; Gerhardt et al., 2009; Wenzel, 2009), rhizoremediation is considered as a very promising alternative. Moreover, endophytic bacteria are likely to interact more closely with their host and, hence, they have considerable biotechnological potential to improve the applicability and efficiency for phytoremediation (Doty, 2008; Weyens et al., 2009). However, to our knowledge, at present, there are not many studies including rhizoremediation or plant-endophyte interactions for phenolic compound remediation.

3.2. Translocation and Distribution

Once absorbed by roots and/or leaves, contaminants are translocated to different plant cells by the transpiration stream and assimilate flow, by the same physiological process used to transport nutrients. The uptake into the hydraulic system of the plant and thus the passage into stem and leaves may be quantified by calculating the transpiration stream concentration factor, TSCF (Burken, 2003 and references therein). This parameter is also considered as a measure of uptake efficiency in rooted vascular plants. It was defined as the ratio of the pollutant concentration in the transpiration stream of the plant respect to the concentration in soil water and depends on physico-chemical properties, chemical speciation and the plant itself (Dietz and Schnoor, 2001). TSCF can vary from zero (no uptake) to 1.0 (uptake at the same concentration as the soil water concentration). Data obtained from studies carried out with hybrid poplars indicate TSCF values of 0.48 and 0.04 for phenol and PCP, respectively (Dietz and Schnoor, 2001).

In addition, transpiration rate (T, liters per day) is another key variable which determines the rate of pollutant uptake, translocation and distribution for a given phytoremediation application and depends on the plant type, leaf area, nutrients, soil moisture, temperature, etc. In this sense, plants with high transpiration rates, like hybrid poplars and willows, show rapid uptake of pollutants, so they are usually employed in phytoremediation of several contaminants. So, the determination of such parameters (TSCF and T) is very important to study plant uptake efficiency for various pollutants, allowing a more accurate prediction of treatment times required for remediation.

The processes of xylem-loading of hazardous chemicals, like phenolics and distribution in leaf tissue have not been well investigated, but are thought to be analogous to herbicide movement in plants. It might be assumed that metabolism of phenolics, like other xenobiotics in plants, is confined to root and leaf tissues, and it hardly takes place during transport in the plant vascular system (Schröder et al., 2008). Both, plant roots and leaves have been described to possess elaborate detoxification mechanisms for organic xenobiotics.

3.3. Metabolism of Hazardous Phenolic Compounds in Plants

3.3.1. Phenolic Compounds Metabolism and the Green Liver Model

It has generally been accepted that several enzyme systems, not necessarily physiologically connected, form a metabolic cascade for the detoxification, breakdown and final storage of organic xenobiotics (Schröder et al., 2008). Detoxification mechanisms described for phenolic compounds, resemble more the reactions in the animal liver than the bacterial metabolism, following the "*green liver*" model proposed for the metabolism of other organic pollutants by Sandermann (1994). This network of reactions can be subdivided into three distinct phases: tranformation (phase I), conjugation (phase II) and compartmentation (phase III). Recently, the last phase has been categorized into two independent phases, one confined to transport and storage in the vacuole, and a second one involving final reactions, such as cell wall binding or excretion (Schröder et al., 2007; Abhilash et al., 2009).

3.3.2. Transformation (Phase I)

The first metabolic step is the transformation of the initial substrate and generally includes several enzymatically catalized oxidations. However, this step is not essential for some pollutants. Pollutant transformation increases its solubility and provides an opportunity for conjugation, the next step in the removal process. In many plants, a number of transformations take place simultaneously and different enzymes can be identified. Despite the fact that biochemical processes accompanying phenolic detoxification in plants are not well investigated, there are many examples in the literature indicating that the activities of peroxidase isoenzymes, laccases and other phenol-oxidases are a crucial step in phenolic metabolism (Dec and Bollag, 1994; Agostini et al., 2003; Coniglio et al., 2008). Peroxidases catalyze a reaction known as oxidative coupling, which is involved in the detoxification of phenol in aqueous solutions while in soils this coupling may occur with the humic material (Flocco et al., 2002; Coniglio et al., 2008). For instance, phenol was completely removed from the incubation medium by aseptically grown *Vetiveria zizanoides* plantlets and this was associated with inherent production of peroxidase and H_2O_2 (Singh et al., 2008). In addition, a significant increase in peroxidase activity was detected in alfalfa roots exposed to phenol, at 10 and 30 days of exposure (Flocco et al., 2002). In these experiments, as well as in other studies, the products of plant bioconversion of this pollutant remain unidentified (Singh et al., 2008). It is important to note that not only the level but also the isoenzyme pattern of peroxidases could be modified by environmental stress, such as that produced by phenolics at high concentrations (Agostini et al., 2003). These aspects will be discussed with more detail in following sections.

Regarding phenolic full transformation, there are only few reports indicating the degradation of different phenolic compounds to CO_2 (mineralization) or regular cell metabolites (Ugrekhelidze et al., 1999). Considering that organic compounds are rarely mineralized in plants (Sandermann, 1992; Schnoor et al., 1995; Schröder and Collins, 2002) and, sometimes, only a small amount of toxicant present in the cell is mineralized while the rest undergoes conjugation, various authors have suggested that this conversion is depreciable or even that does not take place in the metabolic pathway of phenolic compounds (Harvey et al., 2002; Pascal-Lorber et al., 2008). This is in accordance with the knowledge that only few enzymes, present in plants, are able to catalyze ring opening reactions of organic compounds, in contrast to the degradative metabolism of microorganisms. However, few works found in the literature showed oxidation of benzene and phenol by crude enzyme extracts of many plants, which formed muconic acid after ring cleavage, with catechol as an intermediate. Then, further oxidation of muconic acid may lead to the formation of fumaric acid (Durmishidze et al., 1969; Chrikishvili et al., 2005). In addition, Ugrekhelidze et al. (1999) found that a small amount of phenol molecules assimilated through wheat and mung bean roots could be transformed via aromatic ring cleavage and bibasic carbonic acid formation. This process, which involves the mineralization of a pollutant, is usually known as deep oxidation and is one of the most desirable ecological features of plants. However, in nature it rarely occurs. Depending on plant species, the contaminant´s nature and its concentration, a relatively small proportion of the environmental contaminant penetrating into the plant cell undergoes deep oxidation. (Kvesitadze et al., 2006).

3.3.3. Conjugation (Phase II)

Once transformation occurs, conjugation with endogenous compounds (mono-, oligo- and polysaccharides, proteins, peptides, amino acids, organic acids, lignin, etc.) is predominantly the next step in the detoxification or metabolism of pollutants (Phase II). However, some pollutants are conjugated without being preceded by transformation. The formation of conjugates leads to an enhancement of the hydrophilicity of organic contaminants, and consequently to an increase in their mobility. Such characteristics simplify further compartmentation of the transformed toxic compounds.

The process of conjugation is usually carried out by enzymes, such as O-glucosyl transferases (OGT; EC 2.4.1.7), N-glucosyltransferases (EC 2.4.1.71), N-malonyltransferases (EC 2.3.1.114), glutathione S-transferase (EC 2.8.1.18), etc. This process leads to the formation of peptide, ether, ester, thioether or other bonds of a covalent nature. Hydroxyl, NH^{2-}, SH^- and COOH functions on a molecule usually trigger glycosyl-transfer mediated by glycosyltransferases, whereas the presence of conjugated double bonds or halogen-functions proceeds to glutathione conjugation (catalyzed by glutathione S-transferase). In this context,

Schröder and Collins (2002) have pointed out that to know the type of primary conjugation is crucial, because this will determine the final fate of the compound in phytoremediation.

Regarding phenolics, they can be conjugated with carbohydrates such as glucose and glucuronic acid, in different proportions depending on the plant species. In fact, *in vitro* glycosilation of simple phenols was demonstrated several times. For example, this typical detoxification mechanism has been reported for the metabolism of phenol, 2,4-DCP and 2,4,5-TCP in duckweed (*Lemna gibba*). In this work, β-glucoside conjugates were detected as final products of phenolic metabolism, which were progressively dehalogenated (Ensley et al., 1997). However, when the metabolic fate of 2,4-DCP was investigated in six macrophytes, the 2,4-DCP-glucoside conjugate was described as an intermediate metabolite (Pascal-Lorber et al., 2004). Once this intermediate metabolite is formed, more complex monoglucoside esters, either malonyl or acetyl, are detected in these macrophytes. These authors also described an unusual glucosyl-pentose conjugate as the 2,4-DCP major metabolite in *Lemna minor* and *Glyceria maxima*. Conjugation with pentose would prevent further saccharide chain elongation. Moreover, soluble β-D glucoside and O-malonyl-β-D-glucoside conjugates were detected after metabolism of PCP by wheat and soybean plants, which translocate and accumulate them in vacuoles (Schmitt et al., 1985). Similarly, Pascal-Lorber et al. (2003) reported that both plant species shared a common metabolism for $[^{14}C]$-2,4-DCP since the malonylated glucoside conjugates were found as the final major metabolites. Conjugation with malonic acid is a common process, specific to plants, and may represent a signal for sequestration of glycoside conjugates in vacuoles. In addition, Day and Saunders (2004) have characterized more complex 2,4-DCP and 2,4,5-TCP glycosides such as a glucose-apiose, a hydroxymethyl-3-tetrose conjugate, in *Lemna minor*. Lately, the presence of complex glycosides has also been described in edible plants, such as spinach, radish and lettuce (Table 3) and glucuronide conjugates have only been characterized in spinach treated with 2,4-DCP and 2,4,5-TCP (Pascal-Lorber et al., 2008). It is important to point out that these glucuronide-conjugates have rarely been described in plants, and there are only few reports in the literature (Bokern, et al., 1996; Laurent et al., 2007). More examples of phenolic-conjugates described in different plant species are presented in Table 3.

There are findings which suggested that different plant species have distinct OGTs to detoxify different xenobiotics, rather than utilize the same enzyme in each case (Brazier et al., 2003). For instance, an OGT has been partially purified from wheat shoots (Brazier et al., 2003). This enzyme was characterised as a monomeric 53 kDa protein and was distinct from other OGTs previously identified in wheat (Schmitt et al., 1985). Among the xenobiotic phenols tested, the purified enzyme preparation showed at least a 10-fold preference for 2,4,5-TCP instead of 4-nitrophenol, PCP and 2,4,6-TCP. Interestingly, the OGT from soybean had a similar substrate preference for synthetic phenols as the OGT described by Brazier et al. (2003). Contrarily, the 43 kDa OGT described by Schmitt et al. (1985) was more active in conjugating PCP. However, low substrate specificity and cross reactivity of OGT isoenzymes (i.e. use of xenobiotic as well as natural substrates) were also reported. The metabolism of 2,4,5-TCP in *Arabidopsis* is a remarkable example. This compound is used by at least six members of the glucosyltransferase superfamily (Sandermann, 2004 and references therein).

Table 3. Examples of several phenolics conjugates from different plant and cell cultures species

Phenolic compound	Plant/Tissue Culture	Plant species	Conjugate type	References
PCP	Cell suspensions	Wheat (*Triticum aestivum*) Soybean (*Glycine max*)	β-D-glucosides o-malonyl- β-D glucosides	Schmitt et al., 1985
Phenol Nitrophenol Hydroxiphenols	Cell suspensions	Gardenia jasminoides Ellis	β-D-monoglucosides	Misukami et al., 1987
4 n-nonylphenol	Cell suspensions	Wheat (*Triticum aestivum*)	Glucuronide-conjugates	Bokern et al., 1996
Phenol 2,4-DCP 2,4,5-TCP	Aquatic plant	Duckweed (*Lemna gibba*)	β-D-glucosides	Ensley et al., 1997
Phenol	Sterile seedlings	Mung bean (*Phaseolus aureus*) Wheat (*Triticum vulgare*)	Peptide conjugates	Ugrekhelidze et al., 1999
4 n-nonylphenol	Plant	*Lemna minor*	Deoxipentose conjugates	Thibaut et al., 2000
2,4-DCP	Cell suspension	Wheat (*Triticum aestivum*) Soybean (*Glycine max*)	DCP-(malonyl)-glucosides	Pascal-Lorber, et al., 2003
2,4-DCP	Plants	*Lemna minor*	2,4-dichorophenyl-β-D-glucopyranoside 2,4-dichorophenyl-β-D-(6-O-malonyl)-glucopyranoside 2,4-dichorophenyl-β-D-glucopyranosyl-(6,1)-β-D-apiofuranoside	Day and Saunders, 2004
2,4-DCP	Aquatic plant	*Myriophyllum spicatum* *Hippuris vulgaris* *Mentha aquatica* *Glyceria maxima*	DCP-(malonyl)-glucoside	Pascal-Lorber et al., 2004
2,4-DCP	Aquatic plant	*Salvinia natans*	DCP-(acetyl)-glucoside	Pascal-Lorber et al., 2004
2,4-DCP	Aquatic plant	*Lemna minor*	DCP-(pentosyl)-glucoside	Pascal-Lorber et al., 2004
Phenol	Sterile seedlings	Ryegrass (*Lolium perenne* L.)	Phenol-peptide conjugates	Chrikishvili et al., 2005
Phenol Chlorophenols	Hairy roots	*Daucus carota* *Ipomoea batatas* *Solanum aviculare*	Polar conjugates (possible with sugars or proteins)	Santos de Araujo et al., 2006

Phenolic compound	Plant Tissue Culture	Plant species	Conjugate type	References
2,4-DCP	Cell suspensions	Tobacco (*Nicotiana tabacum*)	DCP-glucoside conjugates DCP-(6-O-malonyl)-glucoside DCP-(6-O-acetyl)-glucoside DCP-α1,6-glucosyl-pentose DCP-triglycoside-glucuronic acid	Laurent et al., 2007
4-CP	Plant	Radish (*Raphanus sativus*)	4-CP-(acetyl)-hexose (major) 4-CP-(malonyl)-hexose	Pascal-Lorber et al., 2008
2,4-DCP	Plant	Radish (*Raphanus sativus*)	2,4-DCP-(acetyl)-hexose (major)	Pascal-Lorber et al., 2008
2,4,5-TCP	Plant	Radish (*Raphanus sativus*)	2,4,5-TCP-hexose-sulfate 2,4,5-TCP-(malonyl)-hexose-sulfate	Pascal-Lorber et al., 2008
2,4,5-TCP	Plant	Lettuce (Lactuca sativa L.)	2,4,5-TCP-(acetyl)-hexose (major) 2,4,5-TCP-deoxypentose-(malonyl) hexose	Pascal-Lorber et al., 2008
2,4-DCP	Plant	Spinach (*Spinacea oleracea*)	2,4-DCP-(malonyl) hexose-hexuronic acid (major)	Pascal-Lorber et al., 2008

Moreover, a glucosyltransferase isoenzyme mixture from tobacco leaves can be efficiently used to glucosylate 2,4,5-TCP, PCP and 4-nitrophenol (Harvey et al., 2002).

Thus, different plant species can differ in the degree of cross-reactivity of their detoxifying enzymes (Sandermann, 2004). In some cases when phenols are glycosylated, the existence of di- and triglycosides has been demonstrated. For instance, diglycoside (gentiobioside) and triglycosides were formed from exogenous hydroquinone in wheat embryos (Harborne, 1977). In addition, disaccharide conjugates, formed by glycosil extension, were also described in the metabolic pathway of 2,4,5-TCP in radish (Pascal-Lorber et al., 2008).

Contrarily, phenol was not glycosylated in intact plants of maize (*Zea mays*), pea (*Pisum sativum* L.) and pumpkin (*Cucurbita pepo*) (Arziani et al., 2002). In some annual plant seedlings, phenol was not glycosylated either, but was conjugated with low-molecular-weight peptides, forming phenol-peptide conjugates. A study of [1–6-^{14}C] phenol metabolism in sterile seedlings of mung bean (*Phaseolus aureus*) and wheat (*Triticum vulgare*) demonstrated that phenol formed conjugates with low-molecular-mass peptides (Ugrekhelidze et al., 1999). However, among the peptides participating in conjugation, glutatione and homoglutatione were not found. Other monophenols also formed conjugates with peptides in plants, namely α-naphthol in maize, pea, and pumpkin seedlings (Ugrekhelidze et al., 1980; Ugrekhelidze et al., 1983); *o*-nitrophenol in pea seedlings (Ugrekhelidze et al., 1980; Ugrekhelidze et al., 1983); and a hydroxyl derivative of 2,4-D in maize, pumpkin, and pea seedlings (Arziani et al., 1983; 2002). It was observed that in some plants treated with phenol, the low molecular-mass peptides concentration increases (Ugrekhelidze et al., 1983). Besides, phenols are covalently bound to peptides via hydroxyl groups. It is important to note that peptides participating in the conjugation of phenols considerably differ in their aminoacid composition. According to the existing information, in some plants, conjugation with low-molecular-mass peptides seems to be an important detoxification pathway for monophenols (Arziani et al., 2002).

Furthermore, direct conjugation to lignin can occur. Lignin is a phenolic, structurally nonrepeating macromolecule, which is active in conjugation reactions, and often plays the role of a carrier of xenobiotics and their primary transformants (Sandermann, 1994). Such compounds are incorporated into the lignin structure by being covalently coupled with the biopolymer. It has been shown that tautomeric forms of the lignin monomer coniferyl alcohol (quinone-methyl) couple xenobiotics with amino and hydroxyl groups. In this sense, it is interesting to mention that if PCP is hydroxylated, upon which it acquires a second hydroxyl group, this intermediate can be conjugated with lignin, forming an insoluble compound, which is removed from the cell and stored in the cell wall (Sandermann, 1994). Also, coniferyl alcohol is easily conjugated with 1,2-dihydroxy-3,4,5,6-tetrachlorobenzene, which is an intermediate of PCP hydroxylation (Sandermann, 1987).

In addition, Bokern and Harms (1997), investigating the metabolism of [^{14}C] 4-nonylphenol in suspension cultures of 12 plant species, found that in 7 of the cultures, the xenobiotic is conjugated with lignin. However, lignin is not the only biopolymer involved in binding with xenobiotics. In the leaves, xylan and lignin are the preferred compounds, whereas in the stems pectin and lignin are the main components which bind xenobiotics.

Regarding PCP metabolism, in the aquatic plant *Eichhornia crassipes* the major by products were identified as *ortho-* and *para-* substituted chlorohydroxyphenols (chlorocatechols and hydroquinones), -anisoles, and -veratroles. Partially dechlorinated products of PCP were also detected. A major portion of the absorbed PCP and metabolites was found in bound/conjugated form. A significant increase was also observed in the activity of glutathione S-transferase, a major conjugating enzyme, and in the activities of superoxide dismutase and ascorbate peroxidase in PCP exposed plants (Roy and Haenninen, 1994). On the other hand, Schäfer and Sandermann (1988) identified tetrachlorocatechol as a primary metabolite of PCP in cell suspension cultures of wheat.

Unlike deep oxidation, conjugation does not lead to complete detoxification of the xenobiotic, which preserves its basic molecular structure and hence reduces only partially its toxicity. So, conjugation is not the most successful pathway of xenobiotic detoxification from an ecological point of view. Conjugates of toxic compounds are especially hazardous upon entering the food chain, because enzymes of the digestive tract of warm-blooded animals can hydrolyze conjugates and release the xenobiotics or products of their partial transformation, which in some cases, are more toxic than the parent xenobiotic. So, it is important to perform an adequate characterization of the structure of different conjugates in order to evaluate their bioavailability and the risk that they represent for human and animal health.

3.3.4. Compartmentation (Phase III)

Once conjugates are formed, they can be sequestered or compartmentalized, which is known as phase III of pollutant metabolism. Soluble conjugates (coupled with peptides, sugars, amino acids, etc.) are accumulated in vacuoles. This process takes place with the participation of ATP-binding cassette (ABC) transporters (Schröder et al., 2007). Metabolites stored in the vacuoles could be further processed before exportation to cell wall. However, very little is known about these processes (Pascal-Lorber et al., 2008). Insoluble conjugates (coupled with protein, lignin, starch, pectin, cellulose, xylan and other polysaccharides), are moved out of the cell via exocytosis and are accumulated in the apoplast or cell wall. This may lead to the formation of so-called 'bound residues' because of their inability to be extracted by chemical methods. These conjugates may be covalently bound to stable tissues in the plant (Trapp and Karlson, 2001). Hence, the main objective of compartmentation is essentially to remove toxic compounds from metabolic tissues. In this sense, plants have a greater ability to compartmentalize the products of metabolism and detoxification within internal structures and between organs, producing different localization between *in vitro* and *in vivo* systems.

When xenobiotic chemicals are applied to plant cells, the original compound, the primary products of its metabolism and product conjugates, are distributed between extractable and nonextractable fractions of the biomass. Generally, these nonextractable or 'bound residues' of plant cells cannot be released from the plant matrix by extraction with solvents, probably because of covalent association with lignin, hemicellulose or pectin in the plant cell (Harvey et al., 2002). In fact, incorporation of metabolites in covalent and noncovalent linkage with proteins, lignin, pectin, polysaccharides, cellulose, hemicellulose, starch, and cutin has been reported (Sapp et al., 2003). Bound residues seem to be found in those species that are most tolerant to organic pollutants and the pattern of binding depends on the plant species and the physico-chemical properties of the compound (Harvey et al., 2002). These bound residues have attracted considerable interest and concern in recent years because persistence of chemical residues in edible plants may allow toxic components to enter the food chain (Sandermann, 2004; Trapp et al., 2001). However, there is increasing evidence to suggest that the formation of the bound residue fraction is one of the most important detoxification pathways in plant cells (Harvey et al., 2002).

Plant cell suspensions have been used to investigate the properties of bound residues (Sandermann et al., 1983; Sapp et al., 2003). The advantages of *in vitro* cultures for these studies include elimination of microbial effects and measurement artefacts due to photosynthetic refixation of ^{14}C into natural nonextractable cell components (Sandermann, 2004). However, the question is whether plant cells in culture incorporate metabolites into bound residues in the same way or to the same extent as in plants. Several experiments suggest that measurements of bound components in plant tissue cultures may underestimate the characteristic levels of whole plants, while in other studies cultured plant cells have been found to generate similar levels of bound residues to those found in whole plants (Langebartels and Harms, 1986; Schmidt et al., 1993). In any case, plant tissue cultures have been recommended for initial experiments on bound residues to minimize the expense of greenhouse or field trials (Sandermann, 2004). In this way, Harms et al. (2003), studied the formation of bound residues using 14 different cell cultures derived from different plant species and exposed to 4-nonylphenol. They concluded that the formation of bound residue would be species-specific and the capacity to form such residues may be associated with higher tolerance to the pollutant. Talano et al. (2010) using tobacco hairy roots capable to remove 2,4-DCP with high efficiency, showed the possible fate of a lignin-like polymer in the xylem of roots as a result of this pollutant transformation. Although the results obtained were an indirect evidence of 2,4-DCP final product, it is one of the few reports which showed an *in vivo* localization of these bound residues and, moreover the possible chemical nature of them.

In plants, due to their lack of an efficient excretory system, xenobiotic conjugates finally are sequestered in plant storage compartments, mainly vacuoles, or are integrated as bound residues in cell walls. However, there are few reports which described the existence of a possible excretory system in many plants. In this sense, environmental pollutants, such as phenolics, absorbed by the roots can also be excreted via leaves, although this excretion is uncommon as compared to root excretion. (Korte et al., 2000). Seidel and Kickuth (1967) described that plants kept on phenol solution excreted this pollutant by the leaves of bulrush (*Scirpus lacustris* L.). In this case, the excretion occured so rapidly that after 90 min phenol could be measured in the air near the leaves and after several hours it could be detected even by smell. The inference from this and other analogous studies (Schröder et al., 2007) is that some plants can excrete derivatives of the pollutants absorbed from the soil or groundwater and gradually dilute them into the air or into the soil. This could mean that some plants possess an excretion system for unwanted compounds. Moreover, in plant cell cultures, substantial proportions of specific metabolites are often found in the culture medium, suggesting a possible excretion of xenobiotics (Canivenc et al.,1989; Groeger and Fletcher, 1988; Laurent and Scalla, 1999).

In summary, there are several potential mechanisms for uptake, metabolism and degradation of phenolics in plants. They are represented in Figure 2. Possible mechanisms include sorption and uptake of the pollutant and/or its metabolites into the roots, microbial transformation performed by rhizospheric and endophytic microorganisms, several transformations catalized by extra and/or intracellular enzymes, xylem transfer of the compounds to the leaves, foliar uptake from the air, probably phloem transfer and bound residue formation, among other processes. All these mechanisms contribute to the phytoremediation of phenolics from the environment.

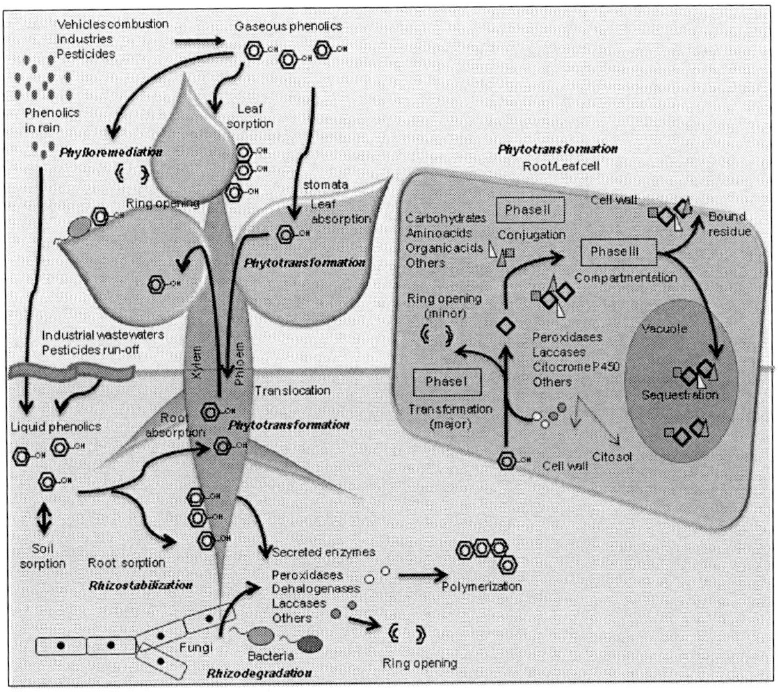

Figure 2. Schematic representation of proposed mechanisms involved in phenolic transformations in plant-soil-air-environments. Phenols from different sources can be stabilized or degraded in the rhizosphere and phyllosphere, sorbed and/or absorbed by roots and/or leaves, translocated and metabolized inside the plant cells.

The degradation pathways of hazardous phenolic compounds seem to be significantly complicated and, at present, they are still unknown in several plant species. In part, the complicated appearance is undoubtedly due to the present lack of essential information. However, it is absolutely clear that in plants as well as in microbes, intracellular enzymatic degradation of contaminants is mainly carried out by oxidative enzymes. Thus, in the remediation process, the knowledge of the variety of enzymes and levels of their activities are the main basis of any kind of phyto- or bioremediation technology. In this sense, an overview of the use of oxidative plant enzymes, as an alternative, in phenol remediation process will be described in the next section.

4. Use of Enzymes for Phenolic Compounds Removal

In the past years, enzymes have become an attractive remediation alternative technique to conventional ones, since they provide a system simpler than a whole organism (Sutherland et al., 2004). In this context, recent biotechnological advances have allowed the production of cheaper and more readily available enzymes through better isolation and purification procedures (Durán, 1997). The potential advantage of the enzymatic treatment as compared with conventional ones include: (1) applicability to a wide variety of recalcitrant compounds including those that are toxic for microorganisms; (2) ability to accomplish treatment over wide ranges of contaminant concentrations, pH and temperature; (3) insensitivity to variations in contaminant concentrations; (4) its low volume sludge/residual production; and (5) high reaction rate (Aitken et al., 1994; Karam and Nicell, 1997). Despite these advantages, enzymatic processes have some problems which may limit their application. The major drawback to the extensive use of many enzymes compared to chemical catalysts is their relatively low stability and their often high cost of purification (Villeneuve et al., 2000). However, new developments in the design of enzymatic "cocktails" application for the biotreatment of wastewaters are being performed (Alcalde et al., 2006).

Many xenobiotics can be biodegraded through enzymatic transformation, for example, polycyclic aromatic hydrocarbons (PHAs), polynitrate aromatic compounds, pesticides such as organochlorine insecticides, aromatic amines and phenols (Alcalde et al., 2006 and references therein). Historically, the most studied enzymes in remediation are reductases, phosphatases, dehalogenases and oxidases (Wolfe and Hoehamer, 2003).

Regarding phenolic compounds removal, the most implicated enzymes include several oxidative enzymes such as peroxidases, laccases and tyrosinases. Reports in this context will be presented below, to show the advances in the use of these enzymes from different sources, for phenol compound removal.

4.1. Peroxidases

Peroxidases (EC 1.11.1.7) are heme-containing oxidoreductases widely distributed in living organisms including microorganism, animals and plants, which catalyze the oxidation of a great number of aromatic compounds such as phenol and derivatives, in the presence of H_2O_2 as oxidizing reagent (Klibanov et al., 1980).

Plant peroxidases are secreted via the endoplasmic reticulum to the apoplast or the vacuole (Hiraga et al., 2001). They are mainly found ionically or covalently bound to cell wall polymers, or they occur as soluble proteins in the intercellular spaces of plant tissues. There are also many reports about secreted plant peroxidases (Agostini et al., 1997; Talano et al., 2003; Talano et al., 2006) as well as from fungi (Aitken et al., 1994), which would be of great importance in removal processes of xenobiotics present in water and soils. Peroxidases have been involved in a broad range of physiological processes such as lignin and suberin biosynthesis, auxin catabolism, biotic and abiotic stress response and senescence, as well as scavenging of hydrogen peroxide (H_2O_2) (Hamid and Rehman, 2009 and references therein). Multiple molecular forms of these enzymes are found in higher plants (isoperoxidases) with different molecular mass and isoelectric point, which determine different groups such as acidic, neutral and basic peroxidase isoenzymes (Lagrimini et al., 1987; González et al., 2008).

Klibanov and colleagues (Klibanov et al., 1980; Klibanov et al., 1983) were the first in proposing the use of plant peroxidases to remove phenolic compounds from aqueous solutions. In this reaction H_2O_2 oxidizes the enzyme into a catalytically active form which is capable of reacting with the phenolic substrate. As a result of this, a phenoxy radical is formed. Two equivalents of phenol are converted by each equivalent of enzyme into highly reactive radical species. These reactive radicals, generated through the catalytic reaction of peroxidases, spontaneously polymerize to form insoluble polymers (Klibanov et al., 1980). Further enzymatic approaches based on peroxidase activity for both the remediation of phenolic compounds (Wagner and Nicell, 2002; Coniglio et al., 2008; González et al., 2008;

Alemzadeh and Nejati, 2009) and for the remediation and/or decolorization of several dyes with aromatic structure, present in wastewater industrial effluent have been reported (Husain, 2009).

As it is well known, in phenol removal process, peroxidases can be inactivated by three possible mechanisms: 1- Adsorption of polymerized phenol on peroxidases resulting in hindering the access of a substrate to the enzyme active site. 2- Irreversible reactions between the enzyme and phenyl or phenoxy radicals that occur by one-electron oxidation of phenolic substrates during the catalytic cycle. 3- Suicide-peroxide inactivation which is a significant and dominant type of inactivation in diluted phenol solution (up to about 0.2 mM) (Nazari et al., 2007). In spite of these limitations, there is a great interest in developing competitive biocatalysts for industrial applications by improving enzyme activity, stability and recycling capacity. Such improvements have been approached by chemical and physical modifications like immobilization and protection strategies, as well as genetic modification of native enzymes (Mateo et al., 2007). Some researchers have suggested the addition of protective compounds such as polyethylene glycol (PEG) to decrease the adsorption of polymers onto the enzyme's active site and increase the lifetime of the active enzyme (Kinsley and Nicell, 2000). Other additives proposed to avoid enzyme inactivation are surfactants (Tween 20, Triton X-100), chitosan gel or activated carbon (Tonegawa et al., 2003). Immobilization of enzymes has improved phenol and chlorophenols removal, with promising results and it seems to be more suitable when large amount of wastewater need to be processed. Many materials and different methods have been used for such immobilization like glass beads, polymers, ion exchange resins, magnetite and aluminum-pillared clay (Caromori and Fernandes, 2004; Shukla and Devi et al., 2005; Cheng et al., 2006).

Here we present several approaches about the use of peroxidases, in particular horseradish peroxidase (HRP), soybean peroxidase (SBP) and those from other sources, involved in phenolic removal.

4.1.1. Horseradish Peroxidases (HRP)

Since the demonstration in 1983 by Klibanov et al., that HRP was effective to remove phenol and other toxic pollutants, HRP is one of the most extensively studied enzymes. This is not only due to historical reasons, but also because of its availability, relatively easy extraction and purification as well as the growing number of applications. A great deal of research has been carried out to demonstrate the effectiveness of HRP and recent investigations are summarized in Table 4.

There are many studies that focused in the optimization of HRP-catalyzed phenol removal, through the evaluation of variables such as pH, temperature, H_2O_2 and enzyme concentration. HRP has proved to achieve maximal removal efficiency between 25 and 40 °C at neutral pH, although it was still quite active in a pH range between 6 and 8 (Bódalo et al., 2006). Ghasempur et al. (2007) found the optimal conditions for an efficient phenol removal process with HRP using a collection of mathematical and statistical techniques useful for modeling the effects of several independent variables, denominated Response Surface Methodology (Mayer and Montgomery, 2002).

HRP-catalyzed polymerization process has been effective for reducing not only phenols but also phenol halogenated derivatives, such as chlorophenols (2-chlorophenol (2-CP); 4-chlorophenol (4-CP); 2,4-dichlorophenol (2,4-DCP); 2,6-dichlorophenol (2,6-DCP); 2,4,6-trichlorophenol (2,4,6-TCP); pentachlorophenol (PCP); etc.

Chlorophenols removal, mediated by HRP was studied by Wagner and Nicell (2002). They found a rapid (3 h) and efficient (95% and >96%) removal of 1 mM phenol and 0.6 mM 2,4-DCP, respectively, with addition of H_2O_2. This research group also studied the toxicity of post-removal solutions through Microtox^{TM} assay. They found that, during phenol, 2,4-DCP, 2-methylphenol, 2-CP and 4-CP removal, toxic products were formed in the HRP-catalyzed reaction, however these products had a tendency to react over time to produce non-toxic products (Wagner and Nicell, 2002). Besides, the toxicity was correlated with high values of absorbance at 400 nm corresponding to the wavelength peak for quinones, which were associated with solution toxicity. This work showed that strategies such as, supplementation of additional H_2O_2 after the completion of enzymatic phenol oxidation or lowering the rate of H_2O_2 addition, overcome problems related to the formation of toxic compounds over the course of the reaction (Wagner and Nicell, 2002). Moreover, Ghioureliotis and Nicell (1999) observed an accumulation of toxic soluble products during the removal of phenol using HRP and H_2O_2. As it could be seen, efficient HRP phenol polymerization is not always accompanied by a considerable reduction of the solution toxicity. So, the formation of significant quantities of products which exhibit toxicity represents an important challenge to the eventual application of enzymatic treatments using HRP for wastewater decontamination.

These results point out the importance of the characterization of the reaction products, formed during the oxidation of phenol and/or chlorophenols. In this context, Laurenti et al. (2003) studied the final products of chlorophenols removal catalyzed by HRP by UV-visible and mass spectrophotometry. They found different and specific products depending on the molecular structure of chlorophenolic substrate. However, the formation of dimmers (biphenyl groups) and *p*-quinones with Cl^- ions liberation, was a repetitive characteristic of the transformation of 2,6-DCP (Laurenti et al., 2002) and 2,4-DCP (Laurenti et al., 2003). The specific reaction mechanism and deshalogenation event, thoroughly explained in the above mentioned papers, is supported by experimental data about chloride release and HCl formation, as was also reported by other authors (Dec et al., 2003; Talano et al., 2010). In addition, Laurenti et al. (2002) found that the oxidation of 2,6-DCP in the presence of HRP and H_2O_2 generated several products and the mixture became more complex as the reaction took longer time. As a result, trimmers, tetramers and high molecular weight products were generated. These species, being less soluble than primary products, tended to precipitate and could be easily separated from the liquid reaction phase, which could make this process an interesting tool for detoxifying wastewaters with chlorophenols (Laurenti et al., 2002; Laurenti et al., 2003).

Regarding PCP removal, Zhang and Nicell (2000) reported that HRP was capable of catalyzing an efficient oxidation and detoxification of 0.05 mM PCP in the presence of H_2O_2 with a 2:1 (H_2O_2/PCP) stoichiometry, reaching the maximum efficiency in a pH range 4-5. The products identified in treated solutions were mostly dimmers, which showed higher toxicity than residual PCP. These authors concluded that unidentified soluble products contribute to this excessive toxicity.

In order to improve the efficiency of phenol removal by HRP, different strategies have been used. Entezari and Pétrier (2004) found that a gradually enzyme addition to a reactor together with a combined removal method mediated by HRP/H_2O_2 and ultrasound produced a more efficient phenol removal and an acceleration of phenol rate degradation, with no precipitation. Since polymers produced during phenol removal could be responsible of the inactivation of enzyme, the absence of precipitated polymers during ultrasound treatment leads to a reduction of HRP dose requirements, which is important for economic feasibility of this method. In this context, Nazari et al. (2007) showed high removal efficiencies for phenol, guaiacol, 2-CP, 4-CP and 2,4-DCP (more than 98%) in concentrations from 50 to 100 mg L^{-1} with HRP activated and stabilized by Ni^{+2}. This and other metal ions can coordinate to active site residues leading to activation of enzymes and produce an increment in long-term stability of HRP (Mahmoudi et al., 2002).

Besides, to optimize removal efficiency with HRP, chemical modification of the enzyme, was carried out. The chemical modification of HRP with phthalic anhydride and glucosamine hydrochloride increased their thermostability (10 and 9-fold, respectively), which is an important aspect since the temperature of wastewaters is often high. Moreover, these chemical modifications increased the removal efficiency of phenolics, since both modified enzymes showed greater affinity and specificity by phenol than native HRP (Liu et al., 2002).

Concerning HRP protection strategies, PEG has been used as an additive with positive results (Entezari and Pétrier, 2004; Bódalo et al., 2006; González et al., 2008). Dalal and Gupta (2007) proposed a combined method of enzyme protection and immobilization. These authors immobilized HRP by bioaffinity layers using lectin Concanavalin A bound to Sephadex beads which were used for the treatment of wastewaters containing *p*-chlorophenol (1 mM). In the presence of PEG, immobilized HRP completely removed *p*-chlorophenol in 60 min while with free HRP the removal was only 45% in the same time. Moreover, with immobilized HRP 1200-fold less enzyme was required in the presence of PEG for an efficient *p*-chlorophenol removal.

Furthermore, Gómez et al. (2006) used HRP immobilized on glutaraldehyde-activated aminopropyl glass beads. They found a beneficial effect of the immobilization on the stability of HRP and obtained higher phenol transformation than that achieved with free enzyme. In addition, Alemzadeh and Nejati (2009) studied the immobilization of HRP in another support, such as porous calcium alginate, for phenol removal. These authors emphasized the possibility of recycling the capsules with enzyme up to four cycles without serious changes in their catalytic performance.

4.1.2. Soybean Peroxidases (SBP)

SBP emerged in 1990s, since alternative and cheaper peroxidase sources became necessary. The seed coat of soybean is a rich source of SBP (Gillikan and Graham, 1991) and since soybean shells are a by-product of the food industry, their use could significantly reduce the overall cost of the enzymatic treatment.

SBP has proved to be effective in removing phenolic compounds from wastewaters, as shown in Table 4 (Caza et al., 1999; Flock et al., 1999). Kinsley and Nicell (2000) found phenol removal efficiencies greater than 95%, for concentrations from 1 to 10 mM of the pollutant, using SBP and PEG. This enzyme has also been used for phenol degradation in soil (Geng et al., 2004). In that study, soybean seed hulls constituted the source of SBP, which demonstrates the great potential of this available and inexpensive material for remediation of soils contaminated with phenolic compounds.

Comparative studies of the phenol removal process using both HRP and SBP are frequently described. In this context, Bódalo et al. (2006) found that HRP acted faster than SBP for phenol removal but it was more susceptible to inactivation, so the addition of a sufficient amount of PEG was necessary and also critical for enzyme protection. Later, Bódalo et al. (2008) found that immobilized SBP removed 5% more of 4-CP than HRP with the same enzyme concentration and in a shorter time. Post-removal solution toxicity that results after using SBP or HRP could be different. Ghioureliotis and Nicell (1999) using SBP found that, the quantity of toxic residual products formed after phenol removal was lower than using HRP. Based on these results both HRP and SBP seem to be suitable for eliminating phenolics from wastewaters. However, the final choice between these enzymes to carry out this process, must be based on several aspects such as: enzyme stability, toxicity of post-removal solutions, catalytic reaction rate, economic factors and others.

4.1.2. Other Sources of Peroxidases

Although there is a great number of studies focused on the use of HRP or SBP to obtain an efficient removal of phenolic compounds, new sources of peroxidases from different plant species are being used (Table 4). Moreover, due to the high cost of purification as well as the complexity of obtaining process, new forms of enzyme applications have currently emerged, like crude extracts, minced tissue, roots and hairy roots (López-Molina et al., 2003; Govere et al., 2007; Duarte-Vázquez et al., 2003; González et al., 2008; Coniglio et al., 2008).

In this context, López-Molina et al. (2003) have studied the effectiveness in phenolic compounds removal using extracts from artichoke (*Cynara scolymus* L.), which contained several peroxidase isoenzymes and polyphenol oxidases (PPO). The addition of adequate and not exceeding H_2O_2 concentration as well as proper agitation were studied and the results showed that using a mixture of enzymes (peroxidases and PPO) such as those found in artichoke extracts, made wastewater treatment more effective than using either peroxidases or PPO alone. Furthermore, the authors concluded that artichoke extracts were simple and cheap to produce from a low value source (López-Molina et al., 2003). Other plant materials, such as minced horseradish roots, could be a viable alternative for phenol transformation present in manures with deodorization effect, since peroxidases can polymerize phenolic odorants and hence reduce malodor (Govere et al., 2007).

Peroxidases from other sources such as turnip (*Brassica napus*) roots (Duarte-Vázquez et al., 2003), bitter gourd (*Momordica charantia*) (Akhtar and Husain, 2006), tomato (*Solanum lycopersicon*) hairy roots (González et al., 2008) and turnip hairy roots (Coniglio et al., 2008) have been proposed for phenol phytoremediation assays. Interesting results have been obtained with hairy roots as a source of peroxidase isoenzymes (Agostini et al., 2003; Santos de Araujo et al., 2004; González et al., 2008; Coniglio et al., 2008; Talano et al., 2010). Santos de Araujo et al. (2004) studied the kinetic behavior of peroxidase pools from hairy root extracts of carrot, sweet potato and kangaroo apple for phenol, catechol, 2-CP and 2,6-DCP, in order to compare their ability to detoxify phenols.

In our studies, efficient phenol and 2,4-DCP removal has been obtained through the use of peroxidases from tomato and turnip hairy roots. Moreover, the involvement of particular isoenzyme groups from hairy roots could be established through the analysis of qualitative and quantitative changes in peroxidase isoenzymes profiles, after consecutive removal cycles. Peroxidase isoenzymes involved in phenol removal may show variation in substrate preference and catalytic efficiency towards phenol (Coniglio et al. 2008). So, the major participation of acidic peroxidases from *B. napus* hairy roots for 2,4-DCP removal (Agostini et al., 2003) as well as for phenol removal (Coniglio et al., 2008) could be determined.

Table 4. Summary of different approaches about the use of peroxidases and laccases for phenolic compounds removal, discussed in the text

Enzymes	Plant sources	Pollutant	Findings	References
Horseradish peroxidases				
HRP	Horseradish	phenol	Formation of toxic soluble products, PEG addition did not reduce toxicity	Ghioureliotis and Nicell, 1999
	Horseradish	PCP	Efficient detoxification with 2:1 H_2O_2/PCP ratio, in a pH range 4-5	Zhang and Nicell, 2000
	Horseradish	phenol, 2,4-DCP, 2-methylphenol, 2-CP, 4-CP	High removal efficiencies but toxic products formation. Supplementation of additional H_2O_2 or lowering rate of H_2O_2 addition reduced toxicity	Wagner and Nicell 2002 Wagner and Nicell, 2002
	Horseradish	phenol	Efficient HRP/ H_2O_2/ultrasound removal method. PEG addition accelerated degradation	Entezari and Pétrier, 2004
	Horseradish	phenol	Removal efficiency (85%), optimal pH 6-8 and temperature 25-40° C, effective protection by PEG	Bódalo et al., 2006
	Horseradish	phenol, guaiacol, 2-CP, 4-CP, 2,4-DCP	High removal efficiencies with HRP activated and stabilized with Ni^{2+}	Nazari et al., 2007
	Horseradish	phenol	Optimization of process through a Response Surface Methodology (RSM)	Ghasempur et al., 2007
	Horseradish	2,6-DCP	Characterization of reaction products. Dimmers, trimmers and higher molecular weight products, which precipitate, were formed	Laurenti et al., 2002
HRP modified (phthalic anhydride and glucosamine hydrochloride)	Horseradish	phenol	Chemical modification of the enzyme increased HRP thermostability and phenol removal efficiency	Liu et al., 2002
HRP immobilized by bioaffinity layers (lectin Concanavalin A)	Horseradish	*p*-chlorophenol	Immobilized HRP completely removed *p*-CP in comparison with only 45% with free enzyme. With immobilized HRP less enzyme was required	Dalal and Gupta, 2006
HRP immobilized (glutaraldehyde-activated aminopropyl glass beads)	Horseradish	phenol	Higher stability of immobilized HRP and higher phenol transformation than with free enzyme	Gómez et al., 2006

Table 4 (Continued)

Enzymes	Plant sources	Pollutant	Findings	References
HRP immobilized (calcium alginate)	Horseradish	phenol	Reuse of capsules with enzyme up to four cycles without serious changes in their catalytic performance	Alemzadeh and Nejati, 2009
Soybean peroxidases				
SBP	Soybean	phenol	95% removal with PEG addition.	Kinsley and Nicell, 2000
SBP soybean seed hulls	Soybean	phenol in soils	High removal efficiency	Geng et al., 2004
SBP immobilized	Soybean	4-CP	Immobilized SBP removed more efficiently 4-CP and in a shorter time than HRP	Bódalo et al., 2008
Other peroxidase sources				
Extracts with peroxidases and polyphenol oxidases	Artichoke (*Cynara scolymus* L.)	phenol, 4-CP	Efficient removal using extracts containing peroxidases and polyphenol oxidases	López-Molina et al., 2003
Peroxidases	*Brassica napus* hairy roots	2,4-DCP	Efficient removal (98-99 %) in optimal pH range (4-8)	Agostini et al., 2003
Peroxidase extracts	*Brassica napus* hairy roots	phenol	Acidic peroxidases were more resistant to consecutive cycles of removal, higher affinity and catalityc efficiency than basic peroxidases	Coniglio et al., 2008
Peroxidase extracts and partially purified	Tomato hairy roots	phenol, 2,4-DCP	Total and basic peroxidases removed more efficiently both substrates than acidic peroxidases. PEG addition increased phenol removal	González et al., 2008
Immobilized partially purified peroxidases	Bitter gourd (*Momordica charantia*)	phenols/related compounds	Higher removal by immobilized peroxidase than with free peroxidases. Immobilized peroxidases were active for longer time.	Akhtar and Husain, 2006
Immobilized peroxidase extracts in spheres of calcium alginate	Turnip (*Brassica napus*) roots	phenol and real wastewaters effluents	Effective phenol removal and increased immobilized enzyme stability with PEG addition.	Quintanilla-Guerrero et al., 2008
Laccases				
Laccase immobilized (microporous polypropylene hollow fibber membranes)	*Rhus vernicifera*	3,4-dimethylphenols, 4-ethylphenol	Efficiently degraded (50-100%) within 48 h	Moeder et al., 2004
Laccase immobilized (polypropylene membrane)	*Rhus vernicifera*	phenol	Effective phenol degradation. Immobilized laccase retained higher activity than free enzyme.	Georgieva et al., 2008

In contrast, González et al. (2008) found that the group of basic peroxidase isoenzymes from tomato hairy roots would be the most likely involved in 2,4-DCP and phenol removal. We established that the addition of PEG (100 mg L^{-1}) increased phenol removal efficiency as well as retained post-removal peroxidase activity using tomato peroxidase isoenzymes (González et al., 2008).

It is noteworthy that, the studies in establishing and understanding the enzymatic mechanism of contaminant degradation, are important for the selection of candidate enzymes that might be produced in large amounts and used as catalysts for contaminant break down (González et al., 2006).

Immobilization and protection strategies have also been applied with peroxidases from other sources. Duarte-Vázquez et al. (2003) found a 99 % of bisphenol, 3-CP and *m*-cresol removal by turnip peroxidases using PEG as additive. On the other hand, Quintanilla-Guerrero et al. (2008) immobilized turnip peroxidases by entrapment in spheres of calcium alginate and they were assayed for the detoxification of synthetic phenolic solutions and real wastewaters effluents from paint factories. Furthermore, with the addition of PEG, turnip peroxidase stability increased, reaction time was reduced from 3 h to 10 min and more effective phenol removal was achieved. Immobilized peroxidases purified from bitter gourd were also effectively used for the treatment of wastewaters contaminated with phenols or other related compounds and they were active for longer time of incubation than free peroxidases (Akhtar and Husain, 2006).

4.2. Laccases

Laccases (E.C. 1.10.3.2) are copper-containing, secretory and cell wall localized glycoproteins (Gianfreda et al., 1999). They catalyze the oxidation of phenolic substrates such as *o*- and *p*-diphenols, aminophenols, polyphenols, polyamines, lignins and aryl diamines as well as some inorganic ions coupled to the reduction of molecular O_2 to water (Solomon et al., 1996). In a typical laccase reaction, a phenolic substrate is subjected to a one-electron oxidation giving rise to an aryloxy-radical. This active specie can be converted to a quinone in the second stage of the oxidation. The quinone as well as the free radical product undergoes non-enzymatic coupling reactions leading to polymerization (Horak et al., 1999). Laccases are characterized by low substrate specificity. Simple diphenols such as hydroxiquinone and catechols are good substrates for the majority of laccases, but guaiacol and 2,6-dimethoxiphenol are generally better substrates (Gagne and Blase, 1997).

Laccases have received much attention from researchers in the last decades due to their ability to oxidize highly recalcitrant environmental pollutants, which make them very useful for their application to several biotechnological processes (Rodriguez Couto et al., 2006). However, the occurrence of laccases in higher plants appears to be far more limited than in fungi, hence most reports related with biotechnological application including phenolic removal processes involve fungal laccases (Itoh et al., 2000; Zhang et al., 2009).

In plants, the presence and characterization of laccases have been documented in the tree *Rhus vernicifera* (Huttermann et al., 2001). This enzyme has been extensively studied in relation with its capacity and kinetic properties for phenolic compounds transformation (Ji et al., 1988; Okasaki et al., 2000). Moreover, Moeder et al. (2004) reported that laccases of the same tree, immobilized on microporous polypropylene hollow fibber membranes, efficiently degraded compounds like 3,4-dimethylphenols and 4-ethylphenol within 48 h. More recently, Georgieva et al. (2008) studied catalytic activity of an immobilized laccase from *Rhus vernicifera* and its capacity for phenol degradation in a bioreactor. The immobilization was performed on a polypropylene membrane chemically modified with chromic acid and the immobilized enzyme retained about 52% of its maximum activity, while the free enzyme retained only 37%. These results demonstrated the possible application of laccases for the treatment of industrial effluents polluted by phenols (Table 4).

4.3. Tyrosinases

Tyrosinases (monophenol monoxygenase) (EC 1.14.18.1) are widely distributed from bacteria to mammals and even present different characteristics in different organs of the same organisms, such as in roots and leaves of higher plants (Burton, 1994). It is well known that tyrosinases catalyze two different oxygen-dependent reactions that occur consequently: the *o*-hydroxylation of monophenols to yield *o*-diphenols (cresolate activity) and the subsequent oxidation of *o*-diphenols to *o*-quinones (catecholase activity) (Fenoll et al., 2000). Typical substrates for tyrosinase besides phenols are *p*-hydroxy- and 3,4-dihydroxyphenylpropionic acids and caffeic acid (Kahn et al., 1999). Tyrosinases have been found in several fruits and vegetables, such as tomato, potato, apple, pear, spinach and strawberry (Selinheimo et al., 2007 and references therein). At present, there is an increasing interest in using tyrosinases in industrial applications. They have many interesting applications in food and non-food processes, especially due to their crosslinking abilities (Selinheimo et al., 2007). To our knowledge, the use of plant tyrosinases for phenol removal processes has not been reported. However, several reports about fungal tyrosinases implicated in this process, have been published (Setharam and Saville, 2003; Girelli et al., 2006; Amaral et al., 2009).

The studies presented here clearly demonstrate that enzymes represent an alternative strategy for phenolic compounds removal. However, enzymatic reactions may not always be applied to high scale due to the high cost of enzyme purification. So, recently the use of transgenic plants provides a promising tool in the field of phytoremediation to decontaminate polluted environments. In the following section, the recent advances in the use of transgenic plants for their potential in removing phenolic compounds are presented.

5. Transgenic Plants for Remediation of Phenolic Compounds

There are several approaches that may lead to enhance phytoremediation of phenol and similar small organic contaminants such as screening studies to identify the most suitable plant species or varieties and optimization of agronomic practices to maximize biomass production and, consequently, phenol degradation. Agronomic practices like fertilization may also affect this process by influencing microbial density and composition in the rhizosphere (Pilon-Smits, 2005). However, it is clear that the most promising approach is the use of genetic engineering methods to develop transgenic plants for phytoremediation. This powerful technology allows the manipulation of a plant's capacity to tolerate, accumulate, and/or metabolize pollutants, and thus to confer superior degradation abilities in plants. The most important advantage of genetic engineering is that it allows a fast introduction of genes from other species and consequently properties into plants that could not be introduced via conventional breeding. This advantage is very important because, generally, plants lack the catabolic pathway for the mineralization of pollutants compared to microorganisms (Eapen et al., 2007). In fact, phenol mineralization has mainly been observed in presence of microbes and it was attributed entirely to microbial metabolic activity (Bokern et al., 1998), therefore this difficulty can be successfully overcome by transgenic technology. In addition, the genes of interest can be expressed in plants with important properties for remediation and/or with a better development in certain ecosystems (Sonoki et al., 2005; Karavangeli et al., 2005; Eapen et al., 2007; Macek et al., 2007).

In this section, we present what has been achieved so far in this field and then we focus on the design and creation of transgenic plants for phenol phytoremediation.

During the last two decades, numerous publications have reported the development of several transgenic plants with potential application in environmental remediation. Since the development of the first transgenic plants for remediation of heavy metal contaminated soil in 1989 (Misra and Gedama, 1989), genetically modified plants have been obtained for phytoremediation of organic pollutants such as explosives (French et al., 1999; Hooker and Skeen, 1999; Hannink et al., 2001 and 2007; Travis et al., 2007; Rylott et al., 2006; Van Aken, 2009), chlorinated solvents (Doty et al., 2000 and 2007), and herbicides (Gullner et al., 2001; Karavangeli et al., 2005; Kawahigashi, 2009). In addition, plants specifically engineered for phenol remediation have appeared since 2002. Most of these investigations consisted in single step transformations with the introduction of one gene of interest, which was responsible for the enhanced metabolization of xenobiotics or resulted in the increased resistance of pollutants. However, one of the most recent developments in transgenic technology, called multigene co-transformation or gene stacking (Li et al., 2003; Halpin, 2005), allows the insertion, in one single step, of multiple genes for the complete degradation of the xenobiotics within the plant system, and probably it will change this actual scenario.

Basically, transgenic technology has focused in two principal strategies to improve pollutants removal (1) the manipulation of phase I of metabolic activity to enlarge *in planta* degradation rates, or to impart novel metabolic activity, and (2) the enhanced secretion of reactive enzymes from roots leading to accelerate *ex planta* degradation of organic contaminants (James and Strand, 2009). Although many investigations that have pursued the increase of *in planta* degradation rates could be assayed in phenol phytoremediation, the principal strategy applied with this class of contaminant has been the *ex planta* degradation. Probably, to avoid the potential accumulation of toxic metabolites due to the incomplete phenol metabolizing that occurs in plants. In the first case, since cytochrome P450-mediated oxidation reactions are the most important in phase I of *in planta* transformations, overexpression of many P450 proteins have been largely applied to enhance phytoremediation of organic compounds (James and Strand, 2009). For example, the mammalian isoform P450 2E1 (CYP2E1), implicated in the metabolism of several xenobiotic contaminants, has been expressed in tobacco plants (*Nicotiana tabacum* cv. Xanthii) under the *Mac* promoter resulting in a marked increase in metabolism of Trichloroethylene (TCE) and ethylene dibromide compared to control vectors in hydroponic reactors (Doty et al., 2000). The transformed tobaccos also metabolized vinyl chloride, benzene, toluene, chloroform, and bromodichloromethane. Considering the structural similarities of these chemicals with phenols, it is possible that CYP2E1 transgenic tobacco plants could also degrade phenols, but this possibility has not been assayed. On the other side, *ex planta* phytoremediation techniques include overexpression of genes for extracellular enzymes such as laccases and peroxidases from plants, fungal or microbial species. An advantage of this strategy is that it may overcome mass transfer limitations since contaminants do not have to be taken up by the roots and, consequently, minimize the problem of introduction of contaminants into the food chain. Van Aken (2009) considers that this is the major limitation of phytoremediation, that is, the threat that accumulated toxic compounds would contaminate the food chain. Furthermore, in this strategy, phenol and other classes of xenobiotics could be degraded simultaneously since the activity of most secreted enzymes, like laccases or peroxidases, is generally nonspecific. Relative to peroxidases, Iimura et al. (2002) have investigated the ectopic expression of fungal peroxidases in plant systems. They expressed a manganese peroxidase gene (MnP) from *Coriolus versicolor* in tobacco, which could remediate PCP. They found that MnP activity in liquid medium was 50 times greater compared to controls, resulting in approximately 2-fold reduction of PCP.

More recently, they transformed aspen (*Populus* sp.) with a MnP from the fungus *Trametes versicolor* and obtained a more-rapid removal of bisphenol A from hydroponic media compared to controls (Iimura et al., 2007). Regarding laccases, in a pioneering work, Wang et al. (2004) developed a transgenic *Arabidopsis* which expressed a secretory laccase, LAC1, from cotton (*Gossypium arboretum*) under the activity of the CaMV 35S promoter. The LAC1 plants showed enhanced resistance to phenolic compounds such as 2,4,6-TCP. Similarly, an extracellular fungal laccase from *Coriolus versicolor* was expressed in tobacco resulting in an enhanced degradation of bisphenol A and PCP in hydroponics (Sonoki et al., 2005).

However, degradation in soils was not examined. Then, in 2008, Hirai et al. reported a more-efficient expression of a fungal laccase (*scL*) from *Schizophyllum commune* in tobacco. They used a mutagenized *scL* (*scL12*) sequence with a decrease in the CpG-dinucleotide motif to avoid the problem with such sequences that are particulary unfavourable to efficient expression in plants. Transgenic *scL 12* plants were able to remove TCP more effectively than control plants.

Transgenic technology also exploits other *in vitro* plant-based experimental systems that are available to the phytoremediation researcher, mainly cell suspensions and hairy roots that have widely been applied in numerous studies focus on the identification of plant capabilties to tolerate, assimilate, detoxify, metabolize, and store a wide variety of organic and heavy metal pollutants (Doran, 2009).

Regarding phenol phytoremediation, transgenic hairy roots and cell suspensions have been obtained. In our laboratory, we developed transgenic tomato (*Solanum lycopersicon* Mill. cv. Pera) hairy root lines, using successive transformation with *A. tumefaciens* and *A. rhizogenes*. These hairy roots overexpressed *tpx1*, a native peroxidase driven by the CaMV 35S promoter (Wevar Oller et al., 2005).

The overexpression of *tpx1* resulted in higher, *in vivo* and *in vitro* peroxidase activity and one of the transgenic hairy root lines tested removed phenol with an efficiency higher than wild type cultures. Even when both this gene and *tpx2* were ectopically expressed in transgenic tobacco hairy root cultures, phenol removal increased (Sosa Alderete et al., 2009), showing the versatility of the system. Recently, Sakamoto et. al achieved the heterologous expression of a laccase gene, *lcc1*, from the fungus *Lentinula edodes* in tobacco BY-2 cells to produce large amounts of enzyme for the degradation and detoxification of environmental pollutants (Sakamoto et al., 2008).

Collectively, all these studies demonstrated that extracellular enzymatic activity is increased through genetic manipulation what leads to the degradation of important pollutants and to the improved ability for plants to grow in other phytotoxic environments.

Thus, *ex situ* secretion of laccase and peroxidase enzymes may be a valuable tool in phytoremediation of phenol and other small organic pollutants. A summary of recent advances in development of transgenic plants and hairy roots used for phenolic remediation is provided in Table 5.

Another important facet suitable for improvement through transgenic technology is plant-microorganism interactions for rhizoremediation, including endophytic and rhizospheric bacteria, and mycorrhizal fungi. It has been proposed that transgenic plants could initiate xenobiotic degradation and release the metabolites for further degradation by rhizobacteria, and this could be applied to phenol compounds. Although this is still a relatively new approach for remediation, the inverse condition, genetically modified rhizobacteria in association with wild type plants has been successfully applied. For example, the association of *Chinese chieve* and recombinant *Pseudomonas gladioli* M-2196 harbouring the genes encoding PCP-degrading enzymes of *Sphingobium chlorophenolicum* resulted in a decrease of 40% in the amount of PCP in soil (Nakamura et al., 2004). Plant exudates contain small molecules that act as signals to initiate the root colonization. He et al. observed that PCP degradation was higher in the rhizosphere of ryegrass that in far-root soil (He et al., 2005). Concerning endophytes, the incursion of transgenic technology in this field is even most recent, with the first transgenic endophytes for phytoremediation developed in 2004 for toluene degradation (Barac et al., 2004).

Table 5. Selected examples of transgenic plants and plant hairy roots which remediate phenolic compounds

Gene	Source	Target plant	Effect	Reference
Mn-peroxidase	*C. versicolor*	*N. tabacum*	Remediation of PCP	Iimura et al., 2002
Laccase *LAC1*	*G. arboreum*	*A. thaliana*	Remediation of 2,4,6-TCP and phenolic allelochemicals	Wang et al., 2004
Laccase	*C. versicolor*	*N. tabacum*	Remediation of PCP	Sonoki et al., 2005
Peroxidase *tpx1*	*S. lycopersicum*	*S. lycopersicum*	Remediation of phenol	Wevar Oller et al., 2005
Peroxidase *tpx1* and *tpx2*	*S. lycopersicum*	*N. tabacum*	Remediation of phenol	Sosa Alderete et al., 2009

Recent studies on endophytes, reveal new possibilities to future application, e.g. Wang et al., (2006) reported the production of laccase by *Monotospora sp.*, an endophyte fungus of *Cynodon dactylon*. This knowledge opens new perspectives in transgenic technology for phenol phytoremediation.

6. Opportunities in the Design and Creation of Transgenic Plants for Phenolics Phytoremediation

Nowadays, the progress in the development of plant molecular biology tools virtually allows the design and creation of plants *a la carte*. All the steps in this process are very well known, from the isolation and purification of a segment of DNA to transformation and selection protocols. The selected segment of DNA can come from any organism, from bacteria to mammals, as it was already mentioned; the gene product can be targeted to certain cellular compartments (e.g. chloroplast, vacuole, mitochondrion, or apoplast) by adding specific targeting information in the gene construct, and furthermore, inducible promoters allow the induced expression of the transgene only under certain environmental conditions (stress-induced, light-induced). Typically, one single transgene is used in the presented investigations and now, the optimization of phytoremediation using genetic modifications may require the introduction of several genes for transport, multistep metabolic pathways, and sequestration (James and Strand, 2009). Multigene co-transformation or transgene stacking, in which multiple traits are conferred to plants by the expression of two or more foreign genes in one single transformation step, has been successfully used to develop engineered plants for agricultural applications. However, some problems, such as silencing, have been encountered (Li et al., 2003; Halpin, 2005). This important advance in transgenic technology will allow the development of transgenic plants with better abilities for phytoremediation, for example, genes concerned in the uptake by roots and genes involved in metabolic phases of the 'green liver' model.

Owing to the crucial role of the root system in most phytoremediation processes, recent studies remark its importance as a fundamental parameter to consider when attempting improving phytoremediation ability of either *in planta* metabolic activity or *ex planta* enzymatic secretions (Mohammadi et al., 2007). At least three aspects should be considered: the root-specific expression of transgenes, the optimization of root growth and health and, finally, the root-size according to the plant species. In this sense, the root-specific expression of transgenes is clearly a strategy that could maximize the efficacy of phytoremediation. This could be achieved by means of different promoters that direct the expression of the gene only in roots. In general, phytoremediation related plant transformations have largely utilized the CaMV 35S promoter to drive constitutive expression in most plant tissues. However, there is evidence that transgene expression under CaMV 35S promoter in root tissue may be less than in leaves (Wilde et al., 1992; Kajita et al., 1994). Other promoters such as ubiquitin 3 (UBQ3) from *Arabidopsis* (Studart-Guimaraes et al., 2006) or the *rolD* of *Agrobacterium rhizogenes* (Elmayan and Tepfer, 1995) are active in roots and may be valuable for obtaining high root-specific activities (James and Strand, 2009).

Research has also focused on ethylene plant production in response to stress induced by pollutants. Ethylene inhibits root growth and is considered a major obstacle to improving phytoremediation efficiency in plants. Bacterial 1-aminocyclopropane-1-carboxylate (ACC) deaminase regulates ethylene levels in plants by metabolizing its precursor ACC into a-ketobutyric acid and ammonia (Bernard, 2005). Thus, transgenic plants that express bacterial ACC deaminase genes could reduce ethylene levels, resulting in a more extensive root system (Arshad et al., 2007). Furthermore, a combination of the genes related to different phases of phenol degradation with the gene for ACC deaminase may improve the phytoremediation activity of transgenic plants. Although *A. thaliana* is a well characterized laboratory model plant, they are not suitable for phytoremediation applications, given its small stature and shallow root system. Poplar and aspen (*Populus sp.*), as well as willows (*Salix sp.*), on the contrary, are widely distributed, fast-growing, high biomass plants ideal for phytoremediation applications (Schnoor, 2000), thus, the development of transgenic trees is the main objective of many phytoremediation projects. As mentioned above, aspen (*Populus seiboldii* x *Populus gradientata*) transformed with a MnP from the fungus *Trametes versicolor* resulted in a more-rapid removal of bisphenol A from hydroponic media (Iimura et al., 2007). There are more examples related to other organic pollutants, as transformed poplar (*Populus tremula* x *Populus alba*) with rabbit CYP2E1 under the CaMV 35S promoter hairy root cultures exposed to TCE which resulted in the production of chloral and trichloroethanol higher than controls (Banerjee et al., 2002). Then, cuttings from the same plants showed more-rapid uptake of TCE, VC, CT, chloroform, and benzene from hydroponics (Doty et al., 2007).

The main drawback in transgenic technology is that the physiological mechanisms by which pollutants enter plant roots are still partially understood. If we know which molecular mechanisms are involved in uptake, tolerance and accumulation processes, and which genes control these mechanisms, we can manipulate them to our advantage. Thus, it is unclear whether increased *in planta* metabolism due to genetic modification will result in increased uptake of organic contaminants in field applications, or whether transport may limit the overall rates.

Considering the applicability of these transgenic plants for environmental cleanup, even when results from laboratory and greenhouse studies look promising for several transgenic plants, field studies will be the ultimate test to establish their phytoremediation potential, their competitiveness, and risks associated with their use. This is because in most studies the effects on excreted proteins or cofactors of inhibitory substances, or contaminant sorption and availability, in complex soil environments, has yet to be determined. In fact, numerous efforts to translate laboratory and greenhouse results to the field have proven challenging (Gerhardt et al., 2009). Another important barrier to field application of transgenic plants for bioremediation arises from the possible risk of horizontal gene transfer to related wild or cultivated plants (Davison, 2005). Therefore, it is likely that next generations of transgenic plants will involve systems that prevent the spread of genes. Indeed, a range of molecular strategies have been designed to potentially impede transgene movement, collectively called Genetic Use Restriction Technologies (GURTs) since 2000 approximately. GURTs, like introduction of transgenes into chloroplastic DNA or the use of conditional lethality genes, have been reviewed (Hills et al., 2007), and at present many of these technologies are still largely at a theoretical stage of development (Hills et al., 2007). In future research, taking advantage of the advances in biotechnology and 'omic' technologies, the development of novel transgenic plants for efficient phytoremediation of xenobiotic pollutants, field testing and commercialization will soon become a reality (Eapen et al., 2007).

CONCLUSION

Phytoremediation is a multicomponent process, which combines the use of plants and, in many cases, the associated microorganisms to remediate polluted environments. The complexity and heterogeneity of sites often polluted with several metals, metalloids and organic compounds requires the design of integrated phytoremediation systems that combine different processes and approaches. Based on the examples presented in this chapter, it is evident that an appropriate selection of plants and microorganisms together with the investigation of both enzymology and gene technology can offer many advantages to improve phenolic phytoremediation. Therefore, there are several strategies currently followed by modern phytoremediation technologies.

The probably induction mechanisms of the enzymes involved in phenolic metabolism, their overall intracellular distribution and the regulation of their activities seem to be especially promising avenues of further investigation and wide potential application. Moreover, some purified or partially purified enzymes may behave as powerful catalysts in the remediation of harmful phenolics. In order to implement an enzyme-based treatment for phenol removal, isolation, purification and production costs should be considered. As it was mentioned, immobilization and protection approaches will be very important in cost reduction. In this context, the use of plant materials (roots, tissues, etc.) as enzyme sources, constitute a good alternative. Besides, enzymes from plants and rhizospheric microorganisms acting together could potentially increase the advantages of phenolic phytoremediation from soils and water. In this sense, the roles of root exudates and arbuscular mycorrhizal fungi on plant capabilities to uptake and metabolize phenolics from contaminated soils and wastewaters, is poorly understand.

Engineering rhizobacteria and endophytes could also enhance degradation of organic pollutants and offer novel opportunities to address multiply contaminated sites. So, it is necessary to develop and study new phenolic phyto/rhizoremediation technologies. However, it is obvious that the complexity of interactions in the plant–microbe–soil pollutant system requires substantial research efforts to improve our understanding of the rhizosphere processes involved and to exploit this technology.

Despite the fact that the role of different plants and *in vitro* cultures derived from them in phenolic remediation has been investigated, little is known about the overall processes taking place in the plant cell after penetration of the contaminants, especially concerning the mechanisms determining cell adaptation and survival. Considering that physiological effects of phenolic compounds on plant cells are complex and even unknown, it is necessary to characterize this aspects, through more experimental data. By understanding the processes that occur in plants exposed to phenolics, a better knowledge of the potential ecological impacts can be gained. In addition, the toxicity of transformation products derived from many phenolic compounds is relatively unknown. Thus, studies identifying these products and determining their toxicities are necessary, mainly due to their probable food chain effects.

Although an extensive knowledge is now available on genes and enzymes involved in phenolic removal, one of the most important challenges is how to use this basic scientific information to improve the efficiency of phytoremediation in the field. These could imply the application of genetically modified plants. Genetic engineering of plants is necessary for at least two reasons. Firstly, plant metabolism rarely mineralizes hazardous phenolics, thus, some transgenic plants and, also microorganisms will be necessary to achieve this goal. Secondly, genetic engineering could be useful to produce hybrid or the novo enzymes to transform or even mineralize complex compounds, which are difficult to metabolize. In this chapter, we have presented some examples in which successful results were obtained using the above mentioned technology. However, it is expected that considering the recent advances in genetics, proteomics and metabolomics, novel genes expressing detoxifying enzymes could be cloned and expressed into plants allowing the host plant to have a wider range of phytoremediation capabilities. These plant improvements may have great potential for field applications assuming public acceptance of the use of more genetically modified organisms. Emphasis should be put on evaluating results obtained in simplified experiments, such as those performed with *in vitro* cultures, hydroponics or pot plants, and on applying these findings to heterogeneous and polluted field sites, and also on the functioning of phyto/rhizoremediation systems under various ecological conditions.

ACKNOWLEDGMENTS

E.A.; M.A.T. and P.S.G. are members of the research career from Consejo Nacional de Investigaciones Científicas y Técnicas (CONICET) (Argentina). A.L.W.O. has a post-doctoral fellowship from CONICET. Financial support for author´s research was provided by PPI (SECyT- UNRC), CONICET, PICTO (FONCyT-SECyT-UNRC) and Ministerio de Ciencia y Tecnología de la Provincia de Córdoba.

REFERENCES

Abhilash, PC; Jamil, S; Singh, N. Transgenic plants for enhanced biodegradation and phytoremediation of organic xenobiotics. *Biotech. Adv.*, (2009) 27, 474-488.

Adler, PR; Arora, R; El Ghaouth, A; Glenn, DM; Solar, JM. Bioremediation of phenolic compounds from water with plant root surface peroxidases. *J. Environ. Qual.*, (1994) 23, 1113-1117.

Agostini, E; Coniglio, MS; Milrad, S; Tigier, H; Giulietti, A. Phytoremediation of 2,4-dichlorophenol by *Brassica napus* hairy roots cultures. *Biotechnol. Appl. Biochem.*, (2003) 37,139-144.

Agostini, E; de Forchetti, SM; Tigier, HA. Production of peroxidases by hairy roots of *Brassica napus*. *Plant Cell Tiss. Organ Cult.*, (1997) 47, 177-182.

Aitken, MD; Massey, IJ; Chen, T; Heck, PE. Characterisation of reaction products from the catalysed oxidation of phenolic pollutants. *Water Res.*, (1994) 28, 1879-1889.

Akhtar, S; Husain Q. Potential applications of immobilized bitter gourd (*Momordica charantia*) peroxidase in the removal of phenols from polluted water. *Chemosphere*, (2006) 65, 1228-1235.

Aksu, Z. Application of biosorption for the removal of organic pollutants: a review. *Process Biochem.*, (2005) 40, 997-1026.

Aksu, Z; Gönen, F. Biosorption of phenol by immobilized activated sludge in a continuous packed bed: prediction of breakthrough curves. *Process. Biochem.*, (2004) 39, 599-613.

Alcalde, M; Ferrer, M; Plou, FJ; Ballesteros, A. Environmental biocatalysis: from remediation with enzymes to novel green processes. *Trends Biotechnol.*, (2006) 24, 281-287.

Alemzadeh, I; Nejati, S. Phenols removal by immobilized horseradish peroxidase. J. Hazard. *Mater.*, (2009) 166, 1082-1086.

Amaral, P; García, D; Cardoso, M; Mendes, M; Coelho, MA; Pessoa, F. Enzymatic reactions in near critical CO_2: the effect of pressure on phenol removal by tyrosinase. *Int. J. Mol. Sci.,* (2009) 10, 5217-5223.

Arshad, M; Saleem, M; Hussain, S. Perspectives of bacterial ACC deaminase in phytoremediation. *Trends Biotechnol.*, (2007) 25, 356-362.

Arziani, B; Ugrekhelidze, D; Kvesitadze, G. Detoxication mechanism of exogenous monatomic phenols in pea seedlings. Ecotoxicol. *Environ. Saf.*, (2002) 51, 85-89.

Arziani, B; Ugrekhelidze, D; Mithaishvili, T. Detoxification of 2,4-dinitrophenol in plants. *Physiol. Rastenni*, (1983) 30, 1040-1042.

Autenrieth, RL; Bonner, JS; Akgerman, A; Okaygun, M; McCreary, EM. Biodegradation of phenolic wastes. *J. Hazard. Mater.*, (1991) 28, 29-53.

Barac, T; Taghavi, S; Borremans, B; Provoost, A; Oeyen, L; Colpaert, JV; Vangronsveld, J; van der Lelie D. Engineered endophytic bacteria improve phytoremediation of water-soluble, volatile, organic pollutants. *Nat. Biotechnol.,* (2004) 22, 583-588.

Barber, JT; Sharma, HA; Ensley, HE; Polito MA; Thomas, DA. Detoxification of phenol by the aquatic angiosperm *Lemna gibba*. *Chemosphere*, (1995) 31, 3567-3574.

Bellin, CA; O'Connor, GA. Plant uptake of pentachlorophenol from sludge-amended soils. *J. Environ. Qual.*, (1990) 19, 598-602.

Bernadini, G; Spinelli, O; Presutti, C; Vismara, C; Bolzacchini, E; Orlando, M; Settimi, R. Evaluation of the developmental toxicity of pesticide MCPA and its contaminants phenol and chlorocresol. *Environ. Toxicol. Chem.*, (1996) 15, 754-760.

Bernard, RG. Modulation of plant ethylene levels by the bacterial enzyme ACC deaminasa. *FEMS Microbiol. Lett.*, (2005) 251, 1-7.

Bódalo, A; Gómez, JL; Gómez, E; Bastida, J; Máximo MF. Comparison of commercial peroxidases for removing phenol from water solutions. *Chemosphere*, (2006) 63, 626-632.

Bódalo, J; Bastida, MF; Maximo, MC; Montiel, M; Gomez, M; Murcia, D. A comparative study of free and immobilized soybean and horseradish peroxidases for 4-chlorophenol removal: protective effects of immobilization. *Bioproc. Biosyst. Eng.*, (2008), 31, 587-593.

Bokern, M.; Nimtz, M.; Harms, H. Metabolites of 4-n-nonylphenol in wheat cell suspension cultures. *J. Agric. Food Chem.*, (1996) 44, 1123-1127.

Bokern, M; Harms, HH. Toxicity and metabolism of 4-n-nonylphenol in cell suspension cultures of different plant species. *Environ. Sci. Technol.*, (1997) 31, 1849-1854.

Bokern, M; Raid, P; Harms, H. Toxicity, uptake and metabolism of 4-n-nonylphenol in root cultures and intact plants under septic and aseptic conditions. *Environ. Sci. Pollut. Res.,* (1998) 5, 21-27.

Brazier, M; Cole, DJ; Edwards, R. Partial purification and characterisation of a 2,4,5-trichlorophenol detoxifying O-glucosyltransferase from wheat. *Phytochemistry*, (2003) 64, 419-424.

Briggs, GG; Bromilow, RH; Evans, AA. Relationships between lipophilicity and root uptake and translocation of non-ionized chemicals by barley. *Pestic. Sci.*, (1982) 13, 495-504.

Burken, JG. Uptake and metabolism of organic compounds: green-liver model. In: McCutcheon SC, Schnoor JL (eds) *Phytoremediation: Transformation and control of contaminants*. Wiley Interscience, Jhon Wiley and Sons, Hoboken, New Jersey; 2003; 59-84.

Burton, SG. Biocatalysis with polyphenol oxidase-a review. *Catal. Today*, (1994) 22, 459-487.

Cai W; Li J; Zhang Z. The characteristics and mechanisms of phenol biodegradation by *Fusarium sp. J. Hazard. Mater.*, (2007) 148, 38-42.

Calace, N; Nardi, E; Petronio, BM; Pietroletti, M. Adsorption of phenols by papermill sludges. *Environ. Pollut.*, (2002) 118, 315-319.

Canivenc, MC; Cagnac, B, Cabanne, F; Scalla, R. Induced changes of chlorotoluron metabolism in wheat cell suspension cultures. Plant Physiol. *Biochem.*, (1989) 27, 193-201.

Carbajo, M; Beltrán, FJ; Gimeno, O; Acedo, B; Rivas, FJ. Ozonation of phenolic wastewaters in the presence of a perovskite type catalyst. *Appl. Catal. B, Environ.*, (2007) 74, 203-210.

Caromori, SS; Fernandes, KF. Covalent immobilization of horseradish peroxidase on to poly(ethylene terephetalate)-poly (aniline) composite. *Process Biochem.*, (2004) 39, 883-888.

Cheng, J; Ming, S; Zuo, P. Horseradish peroxidase immobilized on aluminium-pillared interlayered clay for the calalytic oxidation of phenolic wastewaters. *Water Res.*, (2006) 40, 283-290.

Chrikishvili, DI; Lomidze, EP; Mithaishvili, TI. Phenol conjugation with peptides and final transformations of conjugates in English ryegrass seedlings. *Appl. Biochem. Microbiol.* (2005) 41, 596-599.

Coniglio, MS; Busto, VD; González, PS; Medina, MI; Milrad, S; Agostini, E. Application of *Brassica napus* hairy root cultures for phenol removal from aqueous solutions. *Chemosphere*, (2008) 72, 1035-1042.

Cordova-Rosa, SM;. Dams, RI; Cordova-Rosa, EV; Radetski, MR; Corrêa, AXR; Radetski CM. Remediation of phenol-contaminated soil by a bacterial consortium and *Acinetobacter calcoaceticus* isolated from an industrial wastewater treatment plant. *J. Hazard. Mater.*, (2009) 164, 61-66.

Cunningham, S; Ow, OW. Promises and prospects of phytoremediation. *Plant Physiol.*, (1996) 110, 715-719.

Dabrowski, A; Podkoscielny, P; Hubicki, Z; Barczak, M. Adsorption of phenolic compounds by activated carbon-a critical review. *Chemosphere*, (2005) 58, 1049-1070.

Dalal, S; Gupta, MN. Treatment of phenolic wastewater by horseradish peroxidase immobilized by bioaffinity layering. *Chemosphere*, (2007) 67, 741-747.

Davison, J. Risk mitigation of genetically modified bacteria and plants designed for bioremediation. *J. Ind. Microb. Biotechnol.*, (2005) 32, 639-650.

Day, JA; Saunders, FM. Glycosidation of chlorophenols by *Lemna minor*. Environ. *Toxicol. Chem.*, (2004) 23, 613-620.

Dec, J; Bollag, JM. Dehalogenation of chlorinated phenols during oxidative coupling, *Environ. Sci. Technol.*, (1994) 28, 484-490.

Dec, J; Haider, K; Bollag, JM. Release of substituens from phenolic compounds during oxidative coupling reactions. *Chemosphere*, (2003) 52, 549-556.

Dietz, A; Schnoor, JL. Advances in phytoremediation. *Environ. Health Perspect.*, (2001) 109, 163-168.

Dong, X; Hong, Q; He, L; Jiang, X; Li, S. Characterization of phenol-degrading bacterial strains isolated from natural soil. *Int. Biodeterior Biodegrad.*, (2008) 62, 257-262.

Doran, PM. Application of plant tissue cultures in phytoremediation research: incentives and limitations. *Biotechnol. Bioeng.*, (2009) 103, 60-76.

Dos Santos, VL; De Souza Monteiro, A; Telles Braga, D; Matos Santoro, M. Phenol degradation by *Aureobasidium pullulans* FE13 isolated from industrial effluents. *J. Hazard. Mater.*, (2009) 161, 1413-1420.

Doty, SL. Enhancing phytoremediation through the use of transgenics and endophytes. *New Phytol.* (2008) 179, 318-333.

Doty, SL; James, CA; Moore, AL; Vajzovic, A; Singleton, GL; Ma, C; Khan, Z; Xin, G; Kang, JW; Park, JY; Meilan, R; Strauss, SH.; Wilkerson, J; Farin, F; Strand, SE. Enhanced phytoremediation of volatile environmental pollutants with transgenic trees. *Proc. Natl. Acad. Sci. USA*, (2007) 104, 16816-16821.

Doty, SL; Shang, TQ; Wilson, AM; Tangen, J; Westergreen, AD; Newman, LA; Strand, SE; Gordon, MP. Enhanced metabolism of halogenated hydrocarbons in transgenic plants containing mammalian cytochrome P450 2E1. *Proc. Natl. Acad. Sci. USA*, (2000) 97, 6287-6291.

Duarte-Vázquez, MA; Ortega-Tovar, M; García-Almendárez, B; Regalado, C; Whitaker, JR. Removal of aqueous phenolic compounds from a model system by oxidative polymerization with turnip (*Brassica napus* L. var. purple top white globe) peroxidase. *J. Chem. Technol. Biotechnol.*, (2003) 78, 42-47.

Dunford, HB. *Peroxidase in Chemistry and Biology*, In: Everse, J; Everse, KE (eds) vol. II, CRC Press, Boca Raton, FL; 1991; 1-17.

Durán, N. *Oxidative enzymes applied to bioremediation*. In: Esposito E. (ed). Proceedings of the first National Meeting of Environmental Applied Microbiology, Campinas, S.P., Brazil; 1997; 183.

Durán, N; Esposito, E. Potential applications of oxidative enzymes and phenoloxidase-like compounds in wastewater and soil treatment: a review. *Appl. Catal. B: Environ.*, (2000) 28, 83-99.

Durmishidze, S; Ugrekhelidze, D; Djikiya, A; Tsevelidze, D. The intermediate products of enzymatic oxidation of benzene and phenol. Dokladi Akademii Nauk SSSR; 1969; 466-469.

Eapen, S; Singh, S; D'Souza, SF. Advances in development of transgenic plants for remediation of xenobiotic pollutants. *Biotechnol. Adv.*, (2007) 25, 442-451.

Elmayan, T; Tepfer, M. Evaluation in tobacco of the organ specificity and strength of the *rolD* promoter, domain-a of the 35S promoter and the $35S^2$ promoter. *Transgenic. Res.*, (1995) 4, 388-396.

Ensley, HE; Sharma, HA; Barber, JT; Polito, MA. *Metabolism of chlorinated phenols by Lemna gibba, duckweed*. In: Kruger, EL; Anderson, TA; Coats, JR (eds). Phytoremediation of Soil and Water Contaminants. American Chemical Society, Washington DC, ACS Symposium Series 664; 1997; 238-253.

Entezari, MH; Pétrier, C. A combination of ultrasound and oxidative enzyme: sono-biodegradation of phenol. *Appl.. Catal. B: Environ.*, (2004) 53, 257-263.

Fenoll, LG; Rodriguez-López, JN; Varon, R; García-Ruíz, PA; García-Canovas, F; Tudela. J. Action mechanism of tyrosinase on *meta*- and *para*-hydroxylated mono phenols. *Biol. Chem.*, (2000) 381, 313-320.

Field, JA; Reyes Sierra-Alvarez, E. Microbial degradation of chlorinated phenols. Rev. *Environ. Sci. Biotechnol.*, (2008) 7, 211-241.

Flocco, CG; Lo Balbo, A; Carranza, MP; Giulietti, AM. Removal of phenol by alfalfa plants (*Medicago sativa* L.) grown in hydroponics and its effect on some physiological parameters. *Acta Biotechnol.*, (2002) 22, 43-54.

Flock, C; Bassi, A; Gijzen, M. Removal of aqueous phenol and 2-chlorophenol with purified soybean peroxidase and raw soybean hulls. *J. Chem. Technol. Biotech.*, (1999) 74, 303-309.

French, CE; Rosser, SJ; Davies, GJ; Nicklin, S; Bruce, NC. Biodegradation of explosives by transgenic plants expressing pentaerythritol tetranitrate reductase, *Nat. Biotechnol.*, (1999) 17, 491-494.

Gagne, F; Blase, C. Evaluation of industrial wastewater quality with a chemiluminescent peroxidase activity assay. *Environ. Toxicol. Water Qual.*, (1997) 12, 315-320.

Garg, R; Kurup, A; Hansch, C. Comparative QSAR: on the toxicology of the phenolic OH moiety. *Crit. Rev. Toxicol.*, (2001) 31, 223-245.

Geng, Z; Bassi, AS; Gijzen M. Enzymatic treatment of soils contaminated with phenol and chlorophenols using soybean seed hulls. *Water Air Soil Pollut.*, (2004) 154, 151-166.

Georgievaa, S; Godjevargovaa, T; Portacciob, M; Leporeb, M; Mitab, DG. Advantages in using non-isothermal bioreactors in bioremediation of water polluted by phenol by means of immobilized laccase from *Rhus vernicifera*. *J. Mol. Catal. B: Enzymatic.*, (2008) 55, 177-184.

Gerhardt, KE; Huang, XD; Glick, BR; Greenberg, BM. Phytoremediation and rhizoremediation of organic soil contaminants: potential and challenges. *Plant Sci.*, (2009) 176, 20-30.

Ghasempur, S; Torabi, S-F; Ranaei-Siadat, S-O; Jalali-Heravi, M; Ghaemi, N; Khajeh, K. Optimization of peroxidase-catalyzed oxidative coupling process for phenol removal from wastewater using response surface methodology. *Environ. Sci. Technol.*, (2007) 41, 7073-7079.

Ghioureliotis, M; Nicell, JA. Assesment of soluble products of peroxidase-catalyzed polymerization of aqueous phenols. *Enzyme Microb. Technol.*, (1999) 25, 185-193.

Gianfreda, L; Xu, F; Bollag, JM. Laccases: A useful group of oxidoreductive enzymes. *Bioremediation J.*, (1999) 3, 1-26.

Gillikan, JW; Graham, JS. Purification and developmental analysis of the major anionic peroxidase from the seed coat of *Glycine max*. Plant Physiol., (1991) 96, 214-220.

Gimeno, O; Carbajo, M; Beltrán, FJ; Rivas, FJ. Phenol and substituted phenols AOPs remediation. *J. Hazard. Mater.*, (2005) 119, 99-108.

Girelli, AM; Mattei, E; Messina, A. Phenols removal by immobilized tyrosinase reactor in on-line high performance liquid chromatography. *Anal. Chim. Acta*, (2006) 580, 271-277.

Gómez, JL; Bódalo A; Gómez, E; Bastida, J; Hidalgo, AM; Gómez, M. Immobilization of peroxidases on glass beads: an improved alternative for phenol removal, *Enzyme Microb. Technol.*, (2006) 39, 1016-1022.

González, PS; Agostini, E; Milrad, SR. Comparison of the removal of 2,4-dichlorophenol and phenol from polluted water, by peroxidases from tomato hairy roots, and protective effect of polyethylene glycol. *Chemosphere*, (2008) 70, 982-989.

González, PS; Capozucca, C; Tigier, H; Milrad, S; Agostini, E. Phytoremediation of phenol from wastewater, by peroxidases of tomato hairy root cultures. Enzyme. *Microb. Technol.*, (2006) 39, 647-653.

Govere, EM; Tonegawa, M; Bruns, MA; Wheeler, EF; Kephart, KB; Voigt, JW; Dec, J. Using minced horseradish roots and peroxides for the deodorization of swine manure: A pilot scale study. *Bioresource Technol.*, (2007) 98, 1191-1198.

Groeger, AG; Fletcher, JS. The influence of increasing chlorine content on the accumulation and metabolism of polychlorinated biphenyls (PCBs) by Paul's Scarlet rose cells. *Plant Cell Rep.*, (1988) 329-332.

Gullner, G; Komives, T; Rennenberg, H. Enhanced tolerance of transgenic poplar plants overexpressing gamma-glutamylcysteine synthetase towards chloroacetanilide herbicides. *J. Exp. Bot.*, (2001) 52, 971-979.

Halpin, C. Gene stacking in transgenic plants-the Challenger for 21st Century plant biotecnology. *Plant Biotechnol. J.*, (2005) 3, 141-155.

Hamid, M; Rehman, KU. Potential applications of peroxidases. *Food Chem.*, (2009) 115, 1177-1186.

Hannink, N; Rosser, SJ; French, CE; Basran, A; Murray, JAH; Nicklin, S; Bruce, NC. Phytodetoxification of TNT by transgenic plants expressing a bacterial nitroreductase. *Nat. Biotechnol.*, (2001) 19, 1168-1172.

Hannink, N; Subramanian, M; Rosser, SJ; Basran, A; Murray, JAH; Shanks, JV; Bruce, NC. Enhanced transformation of TNT by tobacco plants expressing a bacterial nitroreductase. *Int. J. Phytorem.*, (2007) 9, 385-401.

Harborne, JB. Flavonoid sulphates: a new class of natural product of ecological significance in plants. *Prog. Phytochem.*, (1977) 4, 189-208.

Harms, H; Bokern, M; Kolb M, Bock C. *Transformation of organic contaminants by different plant systems.* In: Phytoremediation: Transformation and control of contaminants, McCutcheon, SC; Schnoor, JL (eds). Wiley Interscience, John Wiley and Sons, Inc, Hoboken, New Jersey; 2003; 285-316.

Harvey, PJ; Campanella, BF; Castro, PM; Harms, H; Lichtfouse, E; Schäffner, AR; Smrcek, S; Werck-Reichhart, D. Phytoremediation of polyaromatic hydrocarbons, anilines and phenols. *Environ. Sci. Pollut.*, (2002) 9, 29-47.

He, Y; Xu, JM; Tang, CX; Wu, YP. Facilitation of pentachlorophenol degradation in the rhizosphere of ryegrass (*Lolium perenne* L.). *Soil Biol. Biochem.*, (2005) 37, 2017-2202.

Hills, MJ; Hall, L; Arnison, PG; Good, AG. Genetic use restriction technologies (GURTs): strategies to impede transgene movement. *Trends Plant Sci.*, (2007) 12, 177-183.

Hiraga, S; Sasaki, K; Ito, H; Ohashi, Y; Matsui, H. A large family of class III peroxidases. *Plant Cell Physiol.*, (2001) 42, 462-468.

Hirai, H; Kashima, Y; Hayashi, K; Sugiura, T; Yamagishi, K; Kawagishi, H; Nishida, T. Efficient expression of laccase gene from white-rot fungus *Schizophyllum commune* in a transgenic tobacco plant. *FEMS Microbiol. Lett.*, (2008) 286, 130-135.

Hooker, BS; Skeen, RS. Transgenic phytoremediation blast onto the scene. *Nat. Biotechnol.*, (1999) 17, 428.

Horak, D; Karpisek, M; Turkova, J; Benes, M. Hydrazidefunctionalized poly (2-hydroxyethyl metacrylate) micro spheres for immobilization of horseradish peroxidase. *Biotechnol. Prog.*, (1999) 15, 208-215.

Huang, Q; Tang, J; Webet, WJ. Precipitation of enzyme-catalyzed phenol oxidative coupling products: Background ion and pH effects. *Water Res.,* (2005) 39, 3021-3027.

Husain, Q. Peroxidase mediated decolorization and remediation of wastewater containing industrial dyes: a review. *Environ. Sci. Biotech.* (2009) 115, 1569-1705.

Huttermann, A; Mai, C; Kharazipour, A. Modification of lignin for the production of new compounded materials. *Appl. Microbiol. Biotechnol.*, (2001) 55, 387-394.

Iimura, Y; Ikeda, S; Sonoki, T; Hayakawa, T; Kajita, S; Kimbara, K; Tatsumi, K; Katayama, Y. Expression of a gene for Mn peroxidase from *Coriolus versicolor* in transgenic tobacco generates potential tools for phytoremediation. *Appl. Microbiol. Biotechnol.*, (2002) 59, 246-251.

Iimura, Y; Yoshizumi, M; Sonoki, T; Uesugi, M; Tatsumi, K; Horiuchi, KI; Kajita, S; Katayama, Y. Hybrid aspen with a transgene for fungal manganese peroxidase is a potential contributor to phytoremediation of the environment contaminated with bisphenol A. *J. Wood Sci.*, (2007) 53, 541-544.

Indu Nair, C; Jayachandran, K; Shashidhar ,S. Biodegradation of phenol (review). *African J. Biotech.*, (2008) 7 (25), 4951-4958.

Itoh, K; Fujita, M; Kumano, K; Suyama, K; Yamamoto, H. Phenolic acids affect transformations of chlorophenols by a *Coriolus versicolor* laccase. *Soil. Biol. Biochem.*, (2000) 32, 85-91.

Iurascu, B; Siminiceanu, I; Vione, D; Vicente, MA; Gil, A. Phenol degradation in water through a heterogeneous photo- fenton procesas catalyzed by Fe- treated laponite. *Water Res.,* (2009) 43, 1313-1322.

James, CA; Strand, SE. Phytoremediation of small organic contaminants using transgenic plants. *Curr. Opin. Biotechnol* , (2009) 20, 37–241.

Ji, L; Wang, G; Peng, Q. (1988) *Immobilized laccase from Rhus vernicifera by adsorption of concentration gel and its properties.* Wuhan Daxue. Xuebao. Ziran. Kexueban. 4, 63-66 [Chem Abst 1997;111:92839].

Jianlong, YQ, Horan, N; Stentiford, E. Bioadsorption of pentachlorophenol (PCP) from aqueous solution by activated sludge biomass. *Bioresource Technol.*, (2000) 75,157-61.

Kahn, V; Ben-Shalom, N; Zakin, V. p-hydroxyphenylpropionic acid (phppa) and 3,4-dihydroxyphenylpropionic acid (3,4-dppa) as substrates for mushroom tyrosinase. *J. Food Biochem.*, (1999) 23, 75-94.

Kajita, S; Osakabe, K; Katayama, Y; Kawai, S; Matsumoto, Y; Hata, K; Morohoshi, N. *Agrobacterium*-mediated transformation of poplar using a disarmed binary vector and the overexpression of a specific member of a family of poplar peroxidase genes in transgenic poplar cell. *Plant Sci.*, (1994) 103, 231-239.

Kamaludeen, SPB; Ramasamy, K. Rhizoremediation of metals: harnessing microbial communities. *Indian J. Microbiol.*, (2008) 48, 80-88.

Karam, J, Nicell, JA. Potential application of enzymes in waste treatment. *J. Chem. Tech. Biotechnol.*, (1997) 69, 141-153.

Karavangeli, M; Labrou, NE; Clonis, YD; Tsaftaris, A. Development of transgenic tobacco plants overexpressing maize glutathione-*S*-transferase I for chloroacetanilide herbicides phytoremediation. *Biomolec. Eng.*, (2005) 22, 121-128.

Kawahigashi, H. Transgenic plants for phytoremediation of herbicides. *Curr. Opin. Biotechnol.*, (2009) 20, 225-230.

Kinsley, C; Nicell, JA. Treatment of aqueous phenol with soybean peroxidase in the presence of polyethylene glycol. *Bioresource Technol.*, (2000) 73, 139-146.

Klibanov, AM; Alberti, BN; Morris, ED; Felshin, LM. Enzymatic removal of toxic phenols and anilines from wastewaters. *J. Appl. Biochem.*, (1980) 2, 414-421.

Klibanov, AM; Tu, TM; Scott, KP. Peroxidase catalyzed removal of phenols from coal conversion wastewater. *Science*, (1983) 221, 259-261.

Korte, F; Kvesitadze, G; Ugrekhelidze, D; Gordeziani, M; Khatisashvili, G; Buadze, O; Zaalishvili, G; Coulston, F. Review: organic toxicants and plants. *Ecotoxicol. Environ. Saf.*, (2000) 47, 1-26.

Kristich, MA; Schwarz, OJ. Characterization of ^{14}C-naphthol uptake in excised root segments of clover (*Trifolium pratense* L.) and fescue (*Festuca arundinacea* Screb.). *Environ. Monit. Assess*, (1989) 13, 35-44.

Kuleyin; A. Removal of phenol and 4-chlorophenol by surfactant-modified natural zeolite. *J. Hazard. Mater.*, (2007) 144, 307-315.

Kvesitadze, G; Khatisashvili; G; Sadunishvili, T;·Ramsden, JJ. T*he fate of organic contaminats in the plant cell.* In: Kvesitadze, G; Ramsden, JJ. (eds) Biochemical mechanisms of detoxification in higher plants. Basis of phytoremediation.. Springer Berlin, Heidelberg, New York; 2006; 103-165.

Lagrimini, LM; Burkhart, W; Moyer, M; Rothstein, S. Molecular cloning of complementary DNA encoding the lignin-forming peroxidase from tobacco: Molecular analysis and tissue-specific expression. *Proc. Natl. Acad. Sci. USA*, (1987) 84, 7542-7546.

Langebartels, C; Harms, H. Plant cell suspension cultures as test system for an ecological evaluation of chemicals. Growth inhibition effects and comparison with the metabolic fate in intact plants. *Angew. Botanik*, (1986) 60, 113-123.

Laurent, F; Canlet, C; Laurent, L; Pascal-Lorber, S. Metabolic fate of [^{14}C]-2,4-dichlorophenol in tobacco cell suspension cultures. *Environ. Toxicol. Chem.*, (2007) 26, 2299-2307.

Laurent, FMG; Scalla, R. Metabolism and cell wall incorporation of phenoxyacetic acid in soybean cell suspension culture. *Pestic. Sci.*, (1999) 55, 3-10.

Laurenti, E; Ghibaudi, E; Todaro, G; Ferrari, RP. Enzymatic degradation of 2,6-dichlorophenol by horseradish peroxidase: UV-visible and mass spectrophotometric characterization of the reaction products. *J. Inorg. Biochem.*, (2002) 92, 75-81.

Laurenti,, E; Ghibaudi, E; Ardissone, S; Ferrari, RP. Oxidation of 2,4-dichlorophenol catalyzed by horseradish peroxidase: characterization of the reaction mechanism by UV-visible spectroscopy and mass spectrometry. *J. Inorg. Biochem.*, (2003) 95, 171-176.

Li, L; Zhou, Y; Cheng, X; Sun, J; Marita, JM; Ralph, J; Chiang, VL. Combinatorial modification of multiple lignin traits in trees through multigene cotransformation. *Proc. Natl. Acad. Sci. USA*, (2003) 100, 4939-4944.

Lin, SH; Juang, RS. Adsorption of phenol and its derivatives from water using synthetic resins and low-cost natural adsorbents: A review. *J. Environ. Manage.*, (2009) 90, 1336-1349.

Liu, CP; Shen, ZG; Li, XD. Accumulation and detoxification of cadmium in *Brassica pekinensis* and *B. chinensis*. *Biol. Plantarum*, (2007) 51, 116-120.

Liu, JZ; Song, HY; Weng, LP; Ji, LN. Increased thermostability and phenol removal efficiency by chemical modified horseradish peroxidase. *J. Mol. Cat. B: Enzym.*, (2002) 18, 225-232.

Liu, YJ; Zhang, AN; Wang, XC. Biodegradation of phenol by using free and immobilized cells of *Acinetobacter sp.* XA05 and *Sphingomonas sp.* FG03. *Biochem. Eng. J.*, (2009) 44, 187-192.

López-Molina, D; Hiner, ANP; Tudela, J; García-Cánovas, F; Rodríguez-López, JN. Enzymatic removal of phenols from aqueous solution by artichoke (*Cynara scolymus* L.) extracts. *Enzym. Microb. Technol.*, (2003) 33, 738-742.

Macek, T; Kotrba, P; Svatos, A; Novakova, M; Demnerova, K; Mackova, M. Novel roles for genetically modified plants in environmental protection. *Trends Biotechnol.*, (2007) 26, 146-152.

Mackova, M; Chroma, L; Kucerova, P; Burkhard, J; Demnerova, K; Macek, T. Some aspects of PCB metabolism by horseradish cells. *Int. J. Phytoremediat.*, (2001) 3, 401-414.

Mackova, M; Macek, T; Ocenaskova, J; Burkhard, J; Pazlarova, J. Biodegradation of polychlorinated biphenyls by plant cells. *Int. Biodet. Biodegr.*, (1997) 39, 317-325.

Mahmoudi, K; Nazari, A; Moosavi-Movahedi, A; Saboury, AA. Enthalpy analysis of horseradish peroxidase in the presence of Ni^{2+}: a stabilization study. *Thermochim. Acta*, (2002) 385, 33-39.

Mateo, C; Palomo, JM; Fernandez-Lorente, G; Guisan, JM; Fernandez-Lafuente, R. Improvement of enzyme activity, stability and selectivity via immobilization techniques. *Enzyme Microbial Technol.*, (2007) 40, 1451-1463.

Mayer, RH; Montgomery, DC. *Response Surface Methodology: Process and product optimization using designed experiments, Experimental designs for fitting response surfaces*. I. Wiley, New York. 2002; 325-348.

Misra, S; Gedamu, L. *Heavy-metal tolerant transgenic Brassica napus L. and Nicotiana tabacum L. plants. Theor. Appl. Genet.*, (1989) 78, 161-168.

Moeder, M; Martin, C; Koeller, G. Degradation of hydroxylated compounds using laccase and horseradish peroxidase immobilized on microporous polypropylene hollow fiber membranes. *J. Membr. Sci.*, (2004) 245, 183-190.

Mohammadi, M; Chalavi, V; Novakova-Sura, M; Laiberte, JF; Sylvestre, M. Expression of bacterial biphenyl-chlorobiphenyl dioxygenase genes in tobacco plants. *Biotechnol. Bioeng.*, (2007) 97, 496-505.

Nakamura,T; Motoyama, T; Suzuki, Y; Yamaguchi, I. Biotransformation of pentachlorophenol by *Chinese chive* and a recombinant derivative of its rhizosphere-competent microorganism, *Pseudomonas gladioli* M-2196. *Soil Biol. Biochem.*, (2004) 36, 787-795.

Narayanan, M; Erickson, LE; Davis, LC. Simple plant based design strategies for volatile organic pollutants. *Env. Progr.*, (1999) 18, 231-242.

Nayak, PS; Singh, BK. Removal of phenol from aqueous solutions by sorption on low cost clay. *Desalination*, (2007) 207, 71-79.

Nazari, K; Esmaeili, N; Mahmoudi, A; Rahimi, H; Moosavi-Movahedi, AA. Peroxidative phenol removal from aqueous solutions using activated peroxidase biocatalyst. *Enzyme Microb. Technol.*, (2007) 41, 226-233.

Nepovim, A; Podlipná, R; Soudek, P; Schröder, P; Vanek, T. Effects of heavy metals and nitroaromatic compounds on horseradish glutathione S-transferase and peroxidase. *Chemosphere*, (2004) 57, 1007-1015.

Okazaki, S; Goto, M; Wariishi, H; Tanaka, H;.Furusaki, S. Characterization and catalytic properties of surfactant-laccase complex in organic media. *Biotechnol. Prog.*, (2000) 16, 583-588.

Paisio, CE, Agostini, E; González, PS; Bertuzzi, ML. Lethal and teratogenic effects of phenol on *Buffo arenarum* embryos. *J. Hazard. Mater.*, (2009) 167, 64-68.

Pascal-Lorber, S; Despoux, S; Rathahao, E; Canlet, C; Debrauwer, L; Laurent, F. Metabolic fate of [^{14}C] chlorophenols in radish (*Raphanus sativus*), lettuce (*Lactuca sativa*), and spinach (*Spinacia oleracea*). *J. Agric. Food Chem.*, (2008) 56, 8461-8469.

Pascal-Lorber, S; Rathahao, E; Cravedi, JP; Laurent, F. Metabolic fate of [^{14}C]-2,4-dichlorophenol in macrophytes. *Chemosphere*, (2004) 56, 275-284.

Pascal-Lorber,S; Rathahao, E; Cravedi, JP, Laurent F. Uptake and metabolic fate of [^{14}C]-2,4-dichlorophenol and [^{14}C]-2,4-dichloroaniline in wheat (*Triticum aestivum*) and soybean (*Glycine max). J. Agric. Food Chem.*, (2003) 51, 4712-4718.

Pilon-Smits E. Phytoremediation. *Annu. Rev. Plant Biol.*, (2005) 56, 15-39.

Pletsch, M; Santos de Araujo, B; Charlwood, BV. Novel biotechnological approaches in environmental remediation research. *Biotechnol. Adv.*, (1999) 17, 679-687.

Qiu, X; Shah, SI; Kendall, EW; Sorensen, DL; Sims, RC; Engelke, MC. Grass-Enhanced bioremediation for clay soils contaminated with polynuclear aromatic hydrocarbons. In: Anderson, TA; Coats, JR (eds). Bioremediation through rhizosphere technology. Washington; ACS Symp Ser. *Am Chem Soc*; 1994; 563, 142-157.

Quan, X; Shi, H; Wang, J; Qian, Y. Biodegradation of 2,4-dichlorophenol in sequencing batch reactors augmented with immobilized mixed culture. *Chemosphere*, (2003) 50, 1069-1074.

Quintanilla-Guerrero, F; Duarte-Vázquez, MA; García-Almendarez, BE; Tinoco, R; Vázquez-Duhalt, R; Regalado, C. Polyethylene glycol improves phenol removal by immobilized turnip peroxidase. *Bioresour. Technol.*, (2008) 99, 8605-8611.

Ramaswami, A; Carr, P; Burkhardt, M. Plant uptake of uranium: hydroponic and soil system studies. *Int. J. Phytorem.*, (2001) 3, 189-201.

Rao, JR; Viraraghavan, T. Biosorption of phenol from an aqueous solution by *Aspergillus niger* biomass. *Bioresour. Technol.*, (2002) 85, 165-71.

Rayne, S; Forest, K; Friesen KJ. Mechanistic aspects regarding the direct aqueous environmental photochemistry of phenol and its simple halogenated derivatives. *A review. Environ. Int.*, (2009) 35, 425-437.

Reid, BJ; Jones, KC; Semple, KT. Bioavailability of persistent organic pollutants in soils and sediments-a perspective on mechanisms, consequences and sediments-a perspective on mechanisms, consequences and assessment. *Environ. Pollut.*, (2000) 108, 103-112.

Rodríguez Couto, S; Toca Herrera, JL. Industrial and biotechnological applications of laccases: *A review. Biotech. Adv.*, (2006) 24, 500-513.

Roy, S; Hänninen, O. Pentachlorophenol: uptake/elimination kinetics and metabolism in an aquatic plant *Eichhornia crassipes. Environ. Toxicol. Chem.*, (1994) 13, 76-773.

Rylott, EL; Jackson, RG; Edwards, J; Womack, GL; Seth-Smith, HMB; Rathbone, DA; Strand, SE; Bruce, NC. An explosive-degrading cytochrome P450 activity and its targeted application for the phytoremediation of RDX. *Nat. Biotechnol.*, (2006) 24, 216-219.

Sakamoto, Y; Nakade, K; Yano, A; Nakagawa, Y; Hirano, T; Irie, T; Watanabe, H; Nagai, M; Sato, T. Heterologous expression of *lcc1* from *Lentinula edodes* in tobacco BY-2 cells results in the production an active, secreted form of fungal laccase. *Appl. Microbiol. Biotechnol.*, (2008) 79, 971-980.

Sandermann H. Higher plant metabolism of xenobiotics: the "green liver" concept. *Pharmacogenetics* (1994) 4, 225-241.

Sandermann H., Plant metabolism of xenobiotics, *Trends Biochem. Sci.* (1992) 17, 82-84.

Sandermann H. Pestizid-rückstände in nahrungspflanzen. Die rolle des pflanzlichen metabolismus. *Naturwissenschaften*, (1987) 74, 573-578.

Sandermann, H. Molecular ecotoxicology of plants. *Trends Plant Sci.*, (2004) 9, 406-413.

Sandermann, H; Scheel, D; Trenck, T. Metabolism of environmental chemicals by plants-copolymerization into lignin. *J. Appl. Polym. Sci., Appl. Polym. Symp.* (1983) 37, 407-420.

Sandhu, A; Halverson, LJ; Beattie, GA. Bacterial degradation of airborne phenol in the phyllosphere. *Environ. Microbiol.*, (2007) 9, 383-392.

Santos de Araujo, B; Charlwood, BV; Pletsch, M. Tolerance and metabolism of phenol and chloroderivatives by hairy root cultures of *Daucus carota* L. *Environ. Pollut.*, (2002) 117, 329-335.

Santos de Araujo, B; Dec, J; Bollag, JM; Pletsch, M. Uptake and transformation of phenol and chlorophenols by hairy root cultures of *Daucus carota*, *Ipomoea batatas* and *Solanum aviculare*. *Chemosphere*, (2006) 63, 642-651.

Santos de Araujo, B; Omena de Oliveira, J; Salgueiro Machado, S; Plestch, M. Comparative studies of the peroxidases from hairy roots of *Daucus carota*, *Ipomea batatas* and *Solanum aviculare*. *Plant Sci.*, (2004) 167, 1151-1157.

Sapp, M; Ertunc, T; Bringmann, I; Schaffer, A; Schmidt, B. Characterization of the bound residues of the fungicide cyprodinil formed in plant cell suspension cultures of wheat. Pest. *Manag. Sci.* (2003) 60, 65-74.

Schäfer, W; Sandermann, H. Metabolism of pentachlorophenol in cell suspensions cultures of wheat (*Triticum aestivum* L.). Tetrachlorocatechol as a primary metabolism. *J. Agric. Food Chem.*, (1988) 36, 370-377.

Schmidt, B; Rivero, C; Thiede, B; Schenk, T. Metabolism of 4-nitrophenol in soybean excised leaves and cell suspension cultures of soybean and wheat. *J. Plant Physiol.*, (1993), 141, 641-646.

Schmitt, R; Kaul, J; Trenck, T; Schaller, E; Sandermann, H. β-DGlucosyl and O-malonyl-β-D-glucosyl conjugates of pentachlorophenol in soybean and wheat: Identification and enzymatic synthesis. *Pestic. Biochem. Physiol.*, (1985) 24, 77-85.

Schnoor, J. Degradation by plants-Phytoremediation. In: Rehm, HJ; Reed, G (eds). *Biotechnology*. Wiley-VCH, Germany; 2000; 371-384.

Schnoor, JL; Licht, LA; McCutcheon, SC; Wolfe, NL; Carreira, LH. Phytoremediation of organic and nutrient contaminant. *Environ. Sci. Technol.* (1995) 29, 318-323.

Schröder, P; Collins, CJ. Conjugating enzymes involved in xenobiotic metabolism of organic xenobiotics in plants. *Int. J. Phytorem.*, (2002) 4, 1-15.

Schröder, P; Daubner, D; Maier, H; Neustifter, J; Debus, R. Phytoremediation of organic xenobiotics-Glutathione dependent detoxification in *Phragmites* plants from European treatment sites. *Bioresour. Technol.*, (2008) 99, 7183-7191.

Schröder, P; Scheer, CE; Diekmann, F; Stampfl, A. How plants cope with foreign compounds: Translocation of xenobiotic glutathione conjugates in roots of barley (*Hordeum vulgare*). *Environ. Sci. Pollut. Res.*, (2007) 14, 114-122.

Schummer, C; Groff, C; Al Chami, J; Jaber, F; Millet, M. Analysis of phenols and nitrophenols in rainwater collected simultaneously on an urban and rural site in east of France. *Sci. Total Environ.*, (2009) 407, 5637-5643.

Seetharam, GB; Saville BA. Degradation of phenol using tyrosinase immobilized on siliceous supports. *Water Res.*, (2003) 37, 436-440.

Seidel, K; Kickuth, R. *Exkretion von Phenol in der Phylosphäre von Scirpus lacustris L. Naturwissenschaften*, (1967) 52, 517-525.

Selinheimo, E; NiEidhin D, Steffensen, C; Nielsen, J; Lomascolo, A; Halaouli, S; Record, E; O'Beirne, D; Buchert, J; Kruus, K. Comparison of the characteristics of fungal and plant tyrosinases. *J. Biotechnol.* (2007) 130, 471-480.

Semple, KT; Morriss, WWJ; Paton, GI. Bioavailability of hydrophobic organic contaminants in soils: fundamental concepts and techniques for analysis. *Eur. J. Soil Sci.*, (2003) 54, 809-818.

Shukla, SP; Devi, S. Covalent coupling of peroxidase to a copolymer of acrylamide (AAm)-2-hydroxyethyl methaacrylate (HEMA) and its use in phenol oxidation. *Process Biochem.*, (2005) 40, 147-154.

Singh, S; Melo, JS; Eapen, S; D'Souza, SF. Phenol removal using *Brassica juncea* hairy roots: role of inherent peroxidase and H2O2. *J. Biotechnol.*, (2006) 123, 43-49.

Singh, S; Melo, JS; Eapen, S; D'Souza, SF. Potential of vetiver (*Vetiveria zizanoides* L. Nash) for phytoremediation of phenol. *Ecotoxicol. Environ. Saf.* (2008) 71, 671-676.

Solomon, EI; Sundaram, UM; Machonkin, TE. *Multicopper oxidases and oxygenases*. Chem. Ver., (1996) 96, 2563-2605.

Sonoki, T; Kajita, S; Ikeda, S; Uesugi, M; Tatsumi, K; Katayama, Y; Iimura, Y.Transgenic tobacco expressing fungal laccase promotes the detoxification of environmental pollutants. *Appl. Microbiol. Biotechnol.*, (2005) 67, 138-142.

Sosa Alderete, LG; Talano, MA; Ibañez, SG; Purro, S; Agostini, E; Medina, MI. Establishment of transgenic tobacco hairy roots expressing basic peroxidases and its application for phenol removal. *J. Biotechnol.*, (2009) 139, 273–279.

Soudek, P; Valenová, S; Vavrikova, Z; Vanek, T. ^{137}Cs and ^{90}Sr uptake by sunflower cultivated under hydroponic conditions. *J. Environ. Radioactivity*, (2006) 88, 236-250.

Srivastava, VC; Swamy, MM; Mall, ID; Prasad, B; Mishra, IM. Adsorptive removal of phenol by bagasse fly ash and activated carbon: equilibrium, kinetics and thermodynamics. *Colloids Surf. A: Physicochem. Eng. Asp.*, (2006) 272, 89-104.

Studart-Guimaraes, C; Lacorte, C; Brasileiro, ACM. Evaluation of heterologous promoters in transgenic *Populus tremula* x *P. alba* plants. *Biol. Plantarum* (2006) 50, 15-20

Suresh, B; Sherkhane, PD; Kale, S; Eapen, S; Ravishankar, GA. Uptake and degradation of DDT by hairy root cultures of *Cichorium intybus* and *Brassica juncea*. *Chemosphere*, (2005) 61, 1288-1292.

Susarla, S; Medina, VF; McCutcheon, SC. Phytoremediation: an ecological solution to organic chemical contamination. *Ecol. Eng.*, (2002) 18, 647-658.

Sutherland, TD; Horne, I; Russell, RJ; Oakeshott, JG. Enzymatic bioremediation: from enzyme discovery to applications. Clinical Exp. Pharmacol. *Physiol.*, (2004) 31, 817-821.

Sweetman, AJ. Development and application of a multi-residue analytical method for the determination of n-alkanes, linear alkylbenzenes, polynuclear aromatic hydrocarbons and 4-nonylphenol in digested sewage sludges. *Water Res.*, (1994) 28, 343-353.

Tabak, HH; Gao, C; Lai, L; Yan, X; Pfanstiel, S; Kim, IS; Govind, R. Determination of bioavailability and biodegradation kinetics of phenol and alkylphenols in Soil. In: Tedder, WD; Pohland, FG (eds). *Emerging Technologies in Hazardous Waste Management* IV. ACS Symposium Series; 1994; 51-77.

Talano, MA; Agostini, E; Medina, MI; Reinoso, H; Tordable, MC; Tigier, HA; Milrad de Forchetti, S. Changes in ligno-suberization of cell walls of tomato hairy roots produced by salt treatment: the relationship with the release of a basic peroxidase. *J. Plant Physiol.*, (2006) 163, 740-749.

Talano, MA; Frontera, S; González, P; Medina, MI; Agostini, E. Removal of 2,4-diclorophenol from aqueous solutions using tobacco hairy root cultures. *J. Hazard. Mater.*, (2010) 176, 784-791.

Tonegawa, M; Dec, J; Bollag, JM. Use of additives to enhance the removal of phenols from water treated with horseradish and hydrogen peroxide. *J. Environ. Qual.*, (2003) 32, 1222-1227.

Trapp, S; Karlson, U. Phytoremediation of organic pollutants. *J. Soils Sediments* (2001) 1, 37-43.

Travis, ER; Hannink, NK; Van der Gast, CJ; Thompson, IP; Rossen, SJ; Bruce, NC. Impact of transgenic tobacco on trinitrotoluene (TNT) contaminated soil community. *Environ. Sci. Technol.*, (2007) 41, 5854-5861.

Ucisik AS, Trapp S. Uptake, removal, accumulation, and phytotoxicity of phenol in willow trees (*Salix viminalis*). *Environ. Toxicol, Chem.*, (2006) 25, 2455-2460.

Ugrekhelidze, D; Arziani, B. *Peptide conjugates of α-naphthol and onitrophenol in plants.* Bull. Georg. Acad. Sci., (1980) 100, 686-689.

Ugrekhelidze, D; Arziani, B; Mithaishvili, T. Peptide conjugates of exogenous monoatomic phenols in plants. *Physiol. Rastenni*, (1983) 30, 102-107.

Ugrekhelidze, D; Kvesitadze, G; Arziani, B; Mithaishvili, T; Phiriashvili, V. Detoxication of phenol in annual plant seedlings. *Ecotoxicol. Environ. Saf.*, (1999) 42, 119-124.

Van Aken, B. Transgenic plants for enhanced phytoremediation of toxic explosives. *Curr. Opin. Biotechnol.,* (2009) 20, 231-236.

Villeneuve, P; Muderhwa, JM; Graille, J; Haas, MJ. Customizing lipases for biocatalysis: a survey of chemical, physical and molecular biological approaches. *J. Mol. Catal. B: Enzymatic*, (2000) 9, 113-148.

Wagner, M; Nicell, JA. Detoxification of phenolic solutions with horseradish peroxidase and hydrogen peroxide. *Water Res.*, (2002) 36, 4041-4052.

Wang, GD; Li, QJ; Luo, B; Chen, XY. *Ex planta* phytoremediation of trichlorophenol and phenolic allelochemicals via an engineered secretory laccase. *Nat. Biotechnol.*, (2004) 22, 893-897.

Wang, JW; Wu, JH; Huang, WY; Tan, RX. Laccase production by *Monotospora sp.*, an endophytic fungus in *Cynodon dactylon*. *Biores. Technol.*, (2006) 97, 786-789.

Wenzel, WW. Rhizosphere processes and management in plant-assisted bioremediation (phytoremediation) of soils. *Plant Soil*, (2009) 321, 385-408.

Wevar Oller, AL; Agostini, E; Talano, MA; Capozucca, C; Milrad, SR; Tigier, HA; Medina, MI. Overexpressionof a basic peroxidase in transgenic tomato (*Lycopersicon esculentum* Mill. cv. Pera) hairy roots increases phytoremediation of phenol. *Plant Sci.*, (2005) 169, 1102-1111.

Weyens, N; Van der Lelie, D; Taghavi, S; Vangronsveld, J. Phytoremediation: plant-endophyte partnerships take the challenge. *Curr. Opin. Biotechnol.*, (2009) 20, 248-254.

Wilde, HD; Meagher, RB; Merkle, SA. Expression of foreign genes in transgenic yellow-poplar plants. *Plant Physiol.*, (1992) 98, 114-120.

Wolfe, NL; Hoehamer, CF. Enzymes used by plants and microorganisms to detoxify organic compounds. In: McCutcheon, SC; Schnoor, JL (eds). *Phytoremediation: Transformation and Control of Contaminants*. John Wiley and Sons, Hoboken, NJ; 2003; 159-187.

Wood, TK. Molecular approaches in bioremediation. *Curr. Opin. Biotechnol.*, (2008) 19, 572-578.

WHO (World Health Organization), (1994) *Environmental Health Criteria* 161: Phenol.

Zhang, G; Nicell, JA. Treatment of aqueous pentachlorophenol by horseradish peroxidase and hydrogen peroxide, *Water Res.*, (2000) 34, 1629-1637.

Zhang, J; Xu, Z; Chen, H; Zong, Y. Removal of 2,4-dichlorophenol by chitosan-immobilized laccase from *Coriolus versicolor*. *Biochem. Eng. J.*, (2009), 45, 54-59.

In: Handbook of Phytoremediation
Editor: Ivan A. Golubev

ISBN: 978-1-61728-753-4

Chapter 2

PHYTOREMEDIATION: AN OPTION FOR REMOVAL OF ORGANIC XENOBIOTICS FROM WATER

***Ana Dordio*[1,2*] *and A. J. Palace Carvalho*[1,3]**
[1]Department of Chemistry, University of Évora, Évora, Portugal
[2]Institute of Marine Research, University of Évora, Évora, Portugal
[3]Centro de Química de Évora, University of Évora, Évora, Portugal

ABSTRACT

Pollution by persistent organic pollutants (pesticides, pharmaceuticals, petroleum hydrocarbons, PAHs, PCBs, etc.) is an environmental problem that is recognized worldwide. In order to address this problem, cost effective technologies have been developed and evaluated for the decontamination of soil and water resources. Phytoremediation is a promising technology that uses plants and the associated rhizosphere microorganisms to remove, transform/detoxify, or accumulate organic and inorganic pollutants present in soils, sediments, surface or ground water, wastewater, and even the atmosphere. In fact, as a result of their sedentary nature, plants have evolved diverse abilities for dealing with toxic compounds in their environment. They, therefore, possess a variety of pollutant attenuation mechanisms that makes their use in remediating contaminated land and water more feasible than physical and chemical remediation. Currently, phytoremediation is used for treating many classes of organic xenobiotics including petroleum hydrocarbons, chlorinated solvents, polycyclic aromatic hydrocarbons, pesticides, explosives, pharmaceutical compounds and their metabolites, and it involves several decontamination mechanisms. There are several different types of phytotechnologies such as, for instance, treatment constructed wetlands. The aim of this work is to present a review on the application of phytoremediation technologies for water decontamination from persistent organic pollutants, with special emphasis focused on the removal of a class of emergent pollutants that has recently been receiving a lot of attention, the pharmaceutically active compounds. Within the realm of phytotechnologies, constructed wetlands for wastewater treatment deserve a special focus as these systems have been used with success for the removal of several different types of organic xenobiotics.

* Tel: +351 - 266 745343; E-mail address: *avbd@uevora.pt*

1. Introduction

The rising life standards and demographic pressures exert stress on the water supply and on the quality of drinking water. Demand management and water and wastewater reuse may help mitigate some of this pressure. In wastewater management, new challenges are caused by new chemicals of concern, including many notably toxic substances and endocrine disrupters, which often pass through wastewater treatment plants (WWTPs) without being efficiently removed. Such substances may potentially cause serious impacts on aquatic ecosystems and human health.

Several thousands of different organic compounds are currently used in the modern society. These compounds include pesticides, organic solvents, explosives, dyes, phenols, petroleum hydrocarbons and a new class of emergent pollutants, pharmaceuticals. Many of these organic pollutants are persistent in the environment and potentially show adverse ecotoxicological effects. While some of these classes of pollutants are known to be present in the environment for a long time already and their toxic effects are well studied, awareness of the contamination of the aquatic environment by pharmaceuticals has only arisen more recently. This is due to the fact that these pollutants are found in that medium generally at very low concentration levels (μg L-1 – ng L-1), and it was not until the last decade that analytical methodologies and instrumentation have become available with sufficiently low detection and quantification limits to allow for the environmental monitoring of these substances (Fent et al., 2006; Petrovic and Barceló, 2007; Aga, 2008; Barceló and Petrovic, 2008; Miège et al., 2009; Bolong et al., 2009; Kümmerer, 2009). The ecotoxicity associated with this class of pollutants is also largely unassessed but potential for chronic effects caused by long term exposure and for cumulative and even synergistic action exists (Fent et al., 2006; Kim et al., 2007; Zounkova et al., 2007; Farré et al., 2008; Cooper et al., 2008; Bolong et al., 2009). Nonetheless, increasing amounts of pharmacologically active substances are consumed yearly in human and veterinary medicine which are essential for the diagnosis, treatment, or prevention of diseases.

Through excretion or disposal of unused or expired products, pharmaceuticals and their metabolites are continuously introduced into the sewage system. As many of these compounds receive inefficient treatment in WWTPs (which were not designed to deal with this type of pollutant), they eventually are released into the environment. This is considered to be the main route for contamination of the aquatic environment by pharmaceuticals (Nikolaou et al., 2007; Aga, 2008; Kasprzyk-Hordern et al., 2009; Kümmerer, 2009). Over the last years, in numerous monitoring studies, residues of lipid regulating drugs, analgesics and anti-inflammatory drugs, antibiotics, hormones, antidiabetics, neuroactive compounds and beta-blocker drugs have all been detected worldwide in wastewaters, surface waters, ground waters and even drinking waters (Fent et al., 2006; Aga, 2008; Barceló and Petrovic, 2008; Miège et al., 2009; Kümmerer, 2009).

The foreseeable environmental consequences of high environmental loads of pharmaceuticals points to the urgent need for finding ways to retain and remove these pollutants from wastewaters before they reach the receiving water bodies. Optimization of the conventional WWTP processes has been attempted by increasing hydraulic and solid retention times (Tauxe-Wuersch et al., 2005; Maurer et al., 2007; Vieno et al., 2007; Aga, 2008; Zhang et al., 2008; Miège et al., 2009). In addition, some advanced technologies have been evaluated for their capability to decontaminate waters polluted with pharmaceuticals. However, despite the sometimes high removal efficiencies attained, these processes are generally not cost-effective on a large scale (Fent et al., 2006). In fact, there is still a great need for finding applicable technologies with higher efficiencies at reasonable cost of operation and maintenance.

A phytoremediation technology for wastewater treatment that has been gaining increasing popularity over the last decades, known as constructed wetlands systems (CWS), exploits the well-known ability of natural wetlands to depurate water and attempts to optimize the processes responsible for this depuration. These systems are becoming an alternative to conventional wastewater treatment processes or are being integrated in WWTPs as a secondary or tertiary treatment step. Low cost and low maintenance are some of their most attractive characteristics (Kadlec and Wallace, 2009).

CWS may be considered, nowadays, a more mature technology for the removal of bulk pollutants such as suspended solids, organic matter, pathogens and nutrients (Vymazal et al., 1998; Kadlec and Wallace, 2009; Vymazal, 2009). Meanwhile, focus is now moving towards the removal of more specific and recalcitrant compounds for which the conventional wastewater treatment systems are not effective. In fact, CWS are currently being used more frequently for the cleanup of specific pollutant types such as organic xenobiotics and new challenges have been emerging such as the removal of pharmaceuticals and other micropollutants which present new environmental problems to be solved.

2. Persistent Organic Pollutants

The fast growth in chemical and agrochemical industries during the last century have resulted in the release of a large number of new chemical compounds into the environment. In fact, a lot of different organic compounds are now used in the day-to-day life of human beings and many of these are frequently being detected in numerous environmental monitoring studies (Ballschmiter, 1992; Daughton and Ternes, 1999; Jones and de Voogt, 1999; Gavrilescu, 2005; Doble and Kumar, 2005; Fent et al., 2006; Petrovic and Barceló, 2007; Aga, 2008; Barceló and Petrovic, 2008; Corcoran et al., 2010; Lofrano et al., 2010; El Shahawi et al., 2010; Perelo, 2010).

Over the last decades, there has been an increasing focus on a subset of harmful organic chemicals, mostly xenobiotics (synthetic compounds of anthropogenic origin that do not exist naturally in biological systems), commonly classified as Persistent Organic Pollutants (POPs) (Jones and de Voogt, 1999; Gavrilescu, 2005; El Shahawi et al., 2010). POPs are chemical substances that persist in the environment, bioaccumulate throughout the food chain, and pose a risk of causing adverse effects to human health and the environment because, when found at levels higher than background, can be toxic to biotic communities (Jones and de Voogt, 1999; El Shahawi et al., 2010). These POPs include polycyclic aromatic hydrocarbons, petroleum hydrocarbons, chlorinated solvents, explosives, dyes, phenols and pesticides (Jones and de Voogt, 1999; Gavrilescu, 2005; El Shahawi et al., 2010).

Recent advances in analytical techniques and the use of advanced instrumentation (which significantly lowered the detection and quantification limits for analyses of organic compounds in complex environmental matrices) have aided in detecting low levels of other toxic organics in the environment which came to be known as the class of emerging pollutants. An ever growing list of compounds has been detected over the last decade in raw and treated wastewater, biosolids and sediments, receiving waters, ground water and drinking water. Among these substances, pharmaceuticals have become one of the most important subset of compounds in the class of emerging pollutants (Aga, 2008; Miège et al., 2009; Kümmerer, 2009; El Shahawi et al., 2010).

Distributed into different parts of the environment, organic xenobiotics can be transported over long or short distances and can also undergo a variety of reactions and transformations (El Shahawi et al., 2010). Because of these many competing interactions the fate of such pollutants is not easy to predict and, in many cases, their ecotoxicological effects are difficult to assess (El Shahawi et al., 2010). Nevertheless, the established possibility of long-range transport of these substances to regions where they have never been used or produced and the consequent threats they may pose to the environment of the whole globe, has motivated the international community to call, at several occasions, for urgent global actions to be taken with the aim of reducing the release of these chemicals.

Pollution of soils and surface or ground waters occurs as a result of wastewater point sources, improper use and disposal or accidental release of organic chemicals into the environment. In recent years, numerous strategies and technologies have been developed for wastewater treatment or remediation of contaminated areas.

Some advanced wastewater treatment technologies that have been evaluated for the removal of POPs include advanced oxidative processes, activated carbon adsorption, membrane filtration and membrane bioreactors (Fent et al., 2006; Radjenovic et al., 2007; Kim et al., 2007; Esplugas et al., 2007; Snyder et al., 2007; Aga, 2008; Benner et al., 2008). In overall, several techniques have been developed and used for the decontamination of soils and natural waters. Some in situ processes include washing with detergent; extraction of topsoil using vacuum, steam, or hot air stripping; flooding (raising low density hydrophobic liquids to the surface above the water table); etc. Ex situ techniques include excavating the contaminated soil or pumping liquid and subjecting it to chemical oxidation, solvent extraction, adsorption, thermal desorption, etc., and later returning the treated soil or liquid back to its original place (Susarla et al., 2002; Doble and Kumar, 2005; Gan et al., 2009).

Despite the sometimes high removal efficiencies attained, these processes are not, however, widely used mainly for reasons of cost effectiveness (Zodrow, 1999; Susarla et al., 2002; Doble and Kumar, 2005; Fent et al., 2006; Gan et al., 2009). Consequently, there is a growing need for alternative treatment processes for removing POPs from soils, natural waters and wastewaters that have higher efficiencies at reasonable costs of operation/maintenance.

Phytotechnologies have been successfully used for removal of many POPs from contaminated soils, waters and wastewaters (Salt et al., 1998; Macek et al., 2000; Dietz and Schnoor, 2001; Susarla et al., 2002; Haberl et al., 2003; Pilon-Smits, 2005; Eapen et al., 2007; Olette et al., 2008; Imfeld et al., 2009; Gan et al., 2009; Aken et al., 2009). This type of approach attempts to exploit the ability of plants to adsorb, uptake and concentrate or metabolize organic xenobiotics, as well as to release root exudates that enhance compound biotransformation and microbial degradation. In particular for the removal of organic xenobiotics from raw wastewaters and WWTPs effluents, the implementation of some phytotechnologies such as constructed wetlands systems (CWS) has become a popular option (Haberl et al., 2003; Olette et al., 2008; Imfeld et al., 2009). These systems are increasingly being used to provide a form of secondary or tertiary treatment for wastewaters, and have already been used with success to remove from contaminated waters several POPs of various classes of pollutants (Williams, 2002; Haberl et al., 2003; Grove and Stein, 2005; Imfeld et al., 2009), among which some of the most important ones are briefly described below:

a) Polycyclic Aromatic Hydrocarbons

Polycyclic aromatic hydrocarbons (PAHs) consist of a large group of several hundred organic compounds characterized by containing two or more fused aromatic rings. These compounds are important pollutants because of their ubiquitous presence in the environment and the fact that some of them are considered as dangerous substances due to their toxic and mutagenic or carcinogenic potential (Kadlec and Wallace, 2009; Haritash and Kaushik, 2009; Perelo, 2010). For this reason, several of these which are considered especially harmful are included in the European Union and United States Environmental Protection Agency (EPA) priority pollutant list (Parrish et al., 2004; Gan et al., 2009; Perelo, 2010).

Both natural and anthropogenic sources contribute PAHs to the environment. Those compounds are formed during the incomplete combustion of almost any organic material (combustion of fossil fuels, forest fires, volcanic activities, automobile exhausts, etc.) but other sources are their synthesis by microorganisms, fungi, plants, and animals. Crude oil and other petroleum based products also contribute significant amounts of PAHs to the environment (Haritash and Kaushik, 2009; Gan et al., 2009).

The hazards associated with the PAHs can be overcome by the use of conventional methods which involve removal, alteration, or isolation of the pollutant. Such techniques involve excavation of contaminated soil and its incineration or containment, thermal desorption, solvent extraction or chemical oxidation. These technologies are expensive, and in many cases transfer the pollutant from one phase to another (Haritash and Kaushik, 2009; Gan et al., 2009). Phytotechnologies have been successfully used for removal of PAHs from contaminated soils, waters and wastewaters (Salt et al., 1998; Susarla et al., 2002; Parrish et al., 2004; Pilon-Smits, 2005; Eapen et al., 2007; Kadlec and Wallace, 2009; Gan et al., 2009; Sun et al., 2010; Perelo, 2010). Main mechanisms of removal should involve rhizodegradation as some PAHs are hardly taken up by plants as a result of their physicochemical characteristics (e.g. water solubility) and, thus, appear less suitable for phytodegradation (Macek et al., 2000; USEPA, 2000; Susarla et al., 2002; Newman and Reynolds, 2004; Imfeld et al., 2009; Vymazal, 2009; Fountoulakis et al., 2009).

b) Petroleum Hydrocarbons

Crude oil is a lipophilic mixture that consists of more than 17000 organic compounds and is regarded as the most complex, naturally occurring mixture of organic substances. It mainly consists of alkanes, cycloalkanes, and PAHs. Contamination of the environment by total petroleum hydrocarbons (TPHs) arises from natural as well as anthropogenic sources and is potentiated by the widespread use of so many petroleum-based products in the modern society. Human-mediated sources of TPHs include offshore oil production, marine transportation, atmospheric or aerial depositions from combustion of coal and gas flaring, direct ocean dumping, coastal, municipal and industrial wastes, and runoff (Knight et al., 1999; Doble and Kumar, 2005).

Only a relatively small number of TPHs is well characterized for toxicity. The health effects of some fractions can be well characterized based on their components or representative compounds, for example the fraction of light aromatics and, in particular, the BTEX (benzene, toluene, ethylbenzene, and xylenes) fraction. These monoaromatic hydrocarbons, which are typical constituents of TPHs wastewaters, are commonly found in some petroleum-based fuels. Of all of the BTEX compounds, benzene is of most concern, because it is the most toxic and a well-known human carcinogen. The benzene ring is a chemical structure that is very common in nature, which together with its high thermodynamic stability, provides for a significant persistence in the environment; therefore, many aromatic compounds are major environmental pollutants. Benzene contamination is, thus, a significant problem. This hydrocarbon is very soluble and mobile, especially in ground and surface waters and it is poorly biodegraded in the absence of oxygen.

Decontamination of soils and waters polluted with TPHs has been attempted by techniques involving soil solidification (binding hydrocarbon to soil), flooding, excavation of contaminated soil and subjecting it to incineration or chemical oxidation, washing with detergent, solvent extraction, adsorption, etc.. However, once again, these solutions for decontamination of polluted sites are generally expensive. Phytotechnologies for remediation of soils and natural waters has become a low-cost alternative to the conventional approaches that is been increasingly used in the latest years (USEPA, 2000; Susarla et al., 2002; Newman and Reynolds, 2004; Pilon-Smits, 2005; Eapen et al., 2007; Weishaar et al., 2009; Perelo, 2010). In particular, the application of CWS for the treatment of wastewater contaminated with TPHs has been seen as an alternative solution that has been revealed frequently effective (Vymazal et al., 1998; Knight et al., 1999; Omari et al., 2003; Gessner et al., 2005; Kadlec and Wallace, 2009; Imfeld et al., 2009; Vymazal, 2009).

c) Chlorinated Solvents

The term chlorinated solvents refers to a family of organics containing one or more chlorine atoms, which include derivatives of methane, ethane, and ethene but also polychlorinated biphenyls (PCBs). Common uses of chlorinated solvents include dry cleaning, degreasing operations, polymer manufacturing and as a chemical intermediate (Susarla et al., 2002; USEPA, 2004; Amon et al., 2007; Aken et al., 2009; Perelo, 2010).

Trichloroethylene (TCE), perchloroethylene (PCE) and polychlorinated biphenyls (PCBs) are among the most predominant chlorinated solvents which are present in the environment as contaminants (Bourg et al., 1992; Susarla et al., 2002; USEPA, 2004; Amon et al., 2007; Perelo, 2010). TCE, mainly used as a metal cleaning agent and in specialty adhesives, is a probable carcinogen and can affect kidneys, liver, lungs, and heart rate. PCE, which is also used as a metal cleaning agent and in dry cleaning, on the other hand is not classified as a carcinogen but has been known to affect the central nervous system and to cause irritation of the skin, eyes, and upper respiratory system (USEPA, 2004). PCBs are synthetic oils that do not readily react at room temperature and are primarily used as coolants and/or insulators. They are classified as probable carcinogens by the EPA and the International Agency for Research on Cancer. PCB contamination is an ecological concern, because some by-products from burning them at low temperatures (e.g. dioxins) are highly toxic and carcinogenic (USEPA, 2004; Perelo, 2010).

Most chlorinated solvents are only slightly soluble in water and, with the exception of vinyl chloride, have densities greater than that of water. This combination leads to the formation of dense non-aqueous phase liquid deposits which act as a slow releasing, continuous source of chlorinated solvents (Amon et al., 2007). For this reason they tend to remain in the environment for long periods of time and take a long time to remediate. Traditional methods for remediating chlorinated solvent contamination include natural attenuation, soil vapor extraction, air sparging and pump and treat approaches (USEPA, 2004; Amon et al., 2007). Phytoremediation mechanisms that have been successful in containing and/or remediating chlorinated solvents include rhizodegradation, phytodegradation, phytovolatilization and hydraulic control using hybrid poplar and willow trees (Salt et al., 1998; Macek et al., 2000; USEPA, 2000; Susarla et al., 2002; Newman and Reynolds, 2004; Pilon-Smits, 2005; Eapen et al., 2007; Aken et al., 2009). In the treatment of this class of POPs, the use of CWS has also revealed to be an efficient low-cost solution in several studies (Haberl et al., 2003; Amon et al., 2007; Kadlec and Wallace, 2009; Imfeld et al., 2009; Vymazal, 2009).

d) Explosives

There are three main classes of explosives: nitroaromatics, nitramines and nitrate esters. The contamination of the environment by explosives, especially by nitroesters and nitroaromatics, is a worldwide environmental problem. Contamination of soil and waters with explosives is largely due to manufacturing, storage, testing and inappropriate waste disposal of explosive chemicals and, therefore, most contaminated sites are located at ammunition factories and other places where these compounds were handled. This involved open detonation and burning of explosives at army depots, evaluation facilities, artillery ranges, and ordnance disposal sites (Hannink et al., 2002).

The primary explosives at hazardous waste sites are 2,4,6-trinitrotoluene (TNT), hexahydro-1,3,5-trinitro-1,3,5-triazine (Royal Demolition eXplosive-RDX) and octahydro-1,3,5,7-tetranitro-1,3,5,7-tetrazine (High Melting eXplosive-HMX). TNT and RDX are listed as "priority pollutants" and "possible human carcinogens" by the U.S. Environmental Protection Agency (EPA) (Etnier, 1989; Hannink et al., 2002; Van Aken, 2009). TNT is a nitroaromatic constituent of many explosives. In a refined form, TNT is stable and can be stored over long periods of time. It is relatively insensitive to blows or friction. It is readily acted upon by alkalis to form unstable compounds that are very sensitive to heat and impact. Health effects due to exposure to TNT include anemia, abnormal liver function, skin irritation, and cataracts (Hannink et al., 2002; USEPA, 2004; Van Aken, 2009). RDX is a nitramine widely used as an explosive and as a constituent in plastic explosives. RDX can cause seizures when large amounts are inhaled or eaten (Etnier, 1989; Hannink et al., 2002; USEPA, 2004). Long-term health effects on the nervous system due to low-level exposure to RDX are not known. HMX is a nitramine that explodes violently at high temperatures. It is used in nuclear devices, plastic explosives and rocket fuels. Insufficient studies on the effects of HMX to the health of humans and animals have been performed (USEPA, 2004).

Incineration, landfilling and pump and treat systems are traditional methods applied to remove explosives contamination from soil and groundwater (Hannink et al., 2002; USEPA, 2004; Van Aken, 2009). These approaches are expensive and can cause air pollution with ash generation (Susarla et al., 2002; Van Aken, 2009). Phytoremediation mechanisms that have been successful in containing and/or remediating explosives contamination include phytoextraction, phytodegradation and phytostabilization using different type of plants in constructed wetlands (Salt et al., 1998; Macek et al., 2000; USEPA, 2000; Hannink et al., 2002; Susarla et al., 2002; Newman and Reynolds, 2004; Pilon-Smits, 2005; Eapen et al., 2007; Kadlec and Wallace, 2009; Imfeld et al., 2009; Vymazal, 2009; Van Aken, 2009).

Axenic studies have shown that plants are capable of transforming TNT without microbial contribution, but very little accumulation of TNT has been found in plant material (Hughes et al., 1996; Vanderford et al., 1997; Susarla et al., 2002). Therefore, plant-enhanced degradation, or phytoremediation, of TNT by aquatic macrophytes has been proposed as a promising groundwater treatment process (USEPA, 2000).

e) Pesticides

Heavy usage of pesticides over the years (mostly via direct land application) has resulted in their ubiquitous dispersal, most typically in aquatic environments (Chaudhry et al., 2002). The intensive use of pesticides has been a public concern for decades owing to their potential risk to human health and the environment.

Pesticides can enter the water bodies via diffuse sources or via point sources. Diffuse source pesticide inputs result from field applications of pesticides and can arise not only from agricultural practices but also from some non-agricultural practices as well, affecting both surface and ground waters. These include tile drain overflow, baseflow seepage, surface and subsurface runoff, erosion, spray drift, deposition after volatilization, leaching through the soils and infiltration through river banks and beds (Reichenberger et al., 2007). Major point sources of pesticides and their treatment-resistant metabolites are WWTP effluents, farmyard runoff (either directly into streams or into the sewer system) and accidental spills (Gerecke et al., 2002; Neumann et al., 2002; Bailey et al., 2005; Gomez et al., 2007; Reichenberger et al., 2007).

Mitigation of pesticide contamination inputs into water bodies includes both the reduction of diffuse source and of point source inputs, and remediation of contaminated areas. Traditional methods of pesticide soil remediation include excavation and/or chemical oxidation processes (i.e. photocatalysis, ozonation, iron-catalyzed Fenton's reaction) or thermal processes (i.e. low temperature thermal desorption, incineration).

Phytotechnologies have been increasingly used over the latest years to remediate the more persistent pesticides (Salt et al., 1998; Macek et al., 2000; USEPA, 2000; Chaudhry et al., 2002; Susarla et al., 2002; Newman and Reynolds, 2004; Pilon-Smits, 2005; Xia and Ma, 2006; Bouldin et al., 2006; Eapen et al., 2007; Henderson et al., 2007; Olette et al., 2008; Dosnon-Olette et al., 2009). Difficulties remain, including the potential phytotoxicity of some compounds (i.e. herbicides) that were originally developed to destroy plant material. Typically the mechanisms involved in pesticide phytoremediation are phytodegradation, rhizodegradation, and phytovolatilization.

Constructed wetlands are gaining recognition as potential best management practices for the reduction of pesticide concentrations in agricultural runoffs (Schulz, 2004). Generally, their success can be attributed to their diversity of function, as they improve the potential for the range of pesticide transport and degradation processes. CWS have been applied mostly to treat wastewater from point sources such as WWTP effluents (Moore et al., 2002; Haberl et al., 2003; Braskerud and Haarstad, 2003; Ralf et al., 2003; Rose et al., 2006; Blankenberg et al., 2006; Matamoros et al., 2008b; Vymazal, 2009; Moore et al., 2009). However, field studies have shown that vegetated drainage ditches (Moore et al., 2001; Dabrowski et al., 2006) and constructed wetlands (Schulz and Peall, 2001) are also highly effective in reducing pesticide concentrations and toxicity in non-point-derived wastewaters and runoff events. As a result, vegetated agricultural water bodies have been proposed as efficient buffer zones for the protection of more sensitive receiving waters from agricultural runoffs by Moore et al. (2000).

Among the several classes of organic xenobiotics, pharmaceuticals have been attracting much attention of the international scientific community recently, although they have been contaminating many water bodies for already quite a long time (Garrison et al., 1976; Hignite and Azarnoff, 1977). The low concentrations of these compounds in the environment (typically present at trace levels, from low μg L-1 to ng L-1) associated to the unavailability, until recently, of suitably sensitive methods of analysis for these low concentration ranges, has been the main reason for the late interest on the environmental problems posed by these compounds.

Chemicals used in human and veterinary medicine are being continuously introduced in the environment, mainly due to improper disposal of unused or expired drugs, through metabolic excretion and manufacturing processes. Some of these pharmaceutical residues are discharged directly in the environment without going through appropriate treatment, but even those receiving appropriate disposal in WWTPs in many cases are not effectively removed by the conventional wastewater treatment processes (Halling-Sørensen et al., 1998; Daughton and Ternes, 1999; Heberer, 2002; Fent et al., 2006).

2.1. Pharmaceutical Compounds

Human pharmaceuticals comprise a wide ranging class of bioactive compounds with substantial variability in chemical structures, functions, behavior and activity. Their development and use aims specific biological effects and most of them are polar compounds because they are supposed to be transported within organisms through an aqueous medium. The molecular weights of the chemical molecules range typically from 200 to 1000 Da (Kümmerer, 2009).

Despite being consumed worldwide in increasing quantities, there are still not enough data available on the total use of pharmaceuticals. The consumption and application of human pharmaceuticals may vary considerably from country to country due to differences in the prevalence of diseases, treatment habits and options, or simply for market reasons. For instance, some pharmaceuticals are, in some countries, sold over the counter without prescription, while in others they are only available by prescription.

In the European Union over 3000 different pharmaceutically active substances are used in human medicine (Fent et al., 2006; Hummel et al., 2006; Ternes et al., 2007) which can be divided in several different pharmaceutical classes such as analgesics and anti-inflammatory drugs, beta-blockers, lipid regulators, neuroactive compounds and antibiotics. Many of these pharmaceuticals, from all pharmaceutical classes, are frequently detected in monitoring studies worldwide (Heberer, 2002; Fent et al., 2006; Nikolaou et al., 2007; Petrovic and Barceló, 2007; Aga, 2008; Kasprzyk-Hordern et al., 2009; Miège et al., 2009; Kümmerer, 2009).

2.1.1. Human Pharmaceuticals Sources, Fate and Effects in the Environment

In recent years, the occurrence and fate of pharmaceutical residues in the environment has gained much scientific and public attention. Over the last decade, scientists have established a large, diverse, and sometimes unexpected variety of routes through which human pharmaceuticals cross (and are distributed to) various environmental compartments. Figure 1 presents a representation of possible pathways for human pharmaceuticals in the environment.

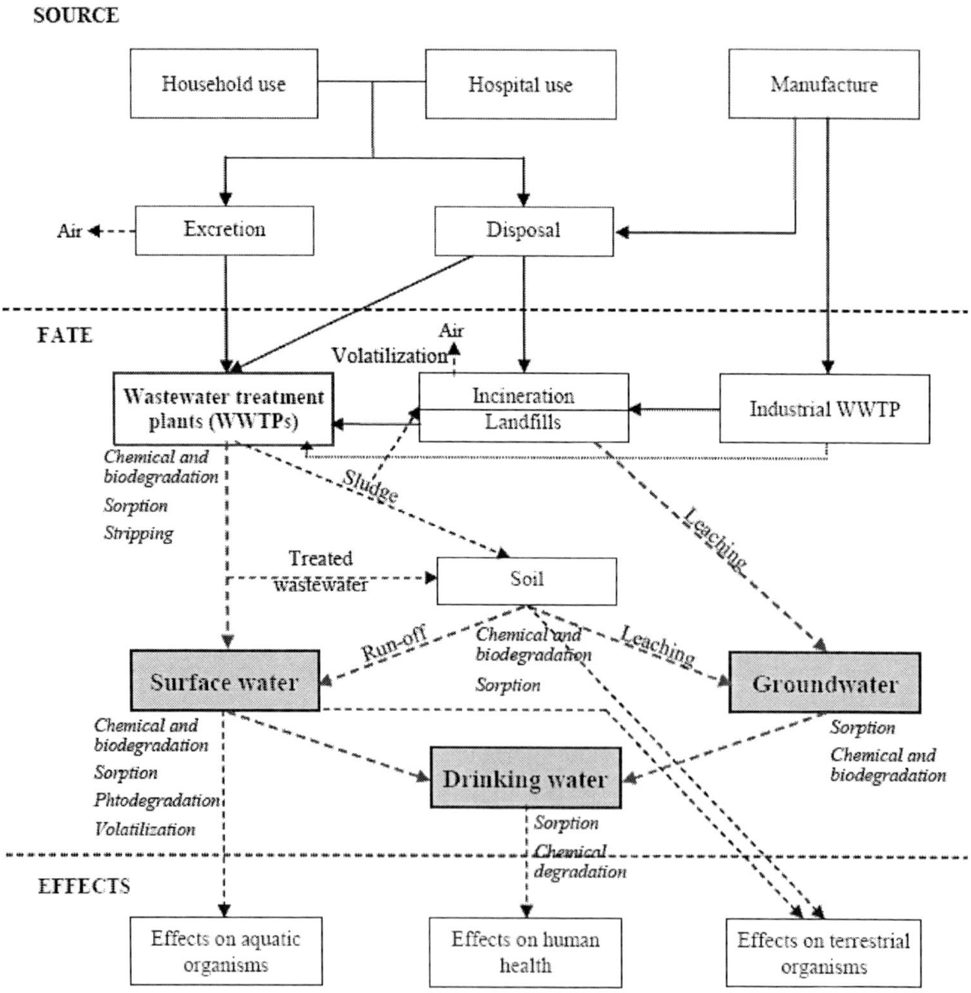

Figure 1. Pathways of human pharmaceuticals in the environment (adapted from Heberer (2002)).

The primary route of entry for the human pharmaceuticals, their metabolites and transformation products into the environment is through wastewater point sources (Nikolaou et al., 2007; Aga, 2008). Pharmaceuticals enter the sewage system either through excretion of non-metabolized products and their metabolites or through the use of the sewage system to dispose of excess medications. In the WWTPs these compounds generally evade efficient removal by the conventional wastewater treatment processes.

Once released into the environment via the discharge of treated wastewater, pharmaceuticals are subjected to the same potential type of transport, transfer and transformation/degradation processes as other organic contaminants. Thus, the interaction of pharmaceuticals with soil, surface and ground water is similarly complex. Transport and transformation processes of pharmaceuticals in the aquatic environment may involve sorption, hydrolysis, biological transformation/degradation, redox reactions, photodegradation, volatilization and precipitation/dissolution (Petrovic and Barceló, 2007; Aga, 2008; Kümmerer, 2008; Farré et al., 2008; Kümmerer, 2009). These processes occur continuously in the environment and influence the presence and mobility of pharmaceuticals in aquatic ecosystems. Behavior of drugs under any of these pathways for partitioning, degradation or transformation may contribute to reduce their concentrations in the environment or remove them entirely and thereby reduce their potential to impact human health and aquatic life.

Pharmaceutical compounds that are marketed in large quantities and are water soluble or slightly soluble, yet resistant to degradation through biological or chemical processes, have the greatest potential to reach steady-state levels in the environment and to be detected in surface and ground waters and in drinking water supplies (Jjemba, 2006; Petrovic and Barceló, 2007; Aga, 2008).

A major difference between pharmaceuticals and other "traditional" environmental POPs (e.g. chlorinated solvents, pesticides, PCBs, PAHs, explosives) is that pharmaceutical compounds, in general, have passed through the human digestive tract and possibly through a conventional WWTP. Two consequences of this pre-exposure to biochemical metabolism are that many drugs will enter the aquatic environment in a modified form and, on the other hand, those that are unaltered, consequently share a resistance to biotic transformation. This allows certain inferences to be made regarding the importance of various abiotic transformation processes for pharmaceutical compounds in the aquatic environment (Arnold and McNeill, 2007).

Given the water solubility of many pharmaceuticals, the abiotic processes most likely to transform them and to more permanently remove them from the aquatic environment include hydrolysis and photodegradation (Petrovic and Barceló, 2007; Aga, 2008; Kümmerer, 2008). However, considering the passage of pharmaceuticals through the digestive tract and their relatively long-residence time in aqueous environments within the WWTPs, hydrolysis reactions likely play a less important role in the aquatic fate of many pharmaceuticals that reach the environment (Arnold and McNeill, 2007). On the other hand, direct photodegradation by sunlight may be an important elimination process for pharmaceuticals with absorbances in the 290-800 nm region (Velagaleti, 1997; Andreozzi et al., 2003).

In any case, the extent of the diverse abiotic and biotic processes that may potentially have an influence on the short-term behavior and long-term fate of a pharmaceutical in the environment are controlled by many factors related both with the pharmaceutical properties and with the environmental conditions. Some of the most important physical and chemical properties of the pharmaceuticals that affect their fate in the environment are the molecular structure, polarity, ionization constant, water solubility, octanol-water partition coefficient, sorption distribution coefficient and the compound's half-life.

In addition to the compound's properties, the fate of pharmaceuticals is also determined by the environmental conditions. Some of those factors include the temperature, sunlight, pH, content of organic matter in soils and sediments and redox conditions.

Evidence being accumulated over the latest years supports the case that, under ordinary conditions, pharmaceuticals have such physicochemical properties which makes them in many cases refractory to degradation and transformation and, consequently, do indeed have the potential to reach the environment (Halling-Sørensen et al., 1998; Fent et al., 2006; Nikolaou et al., 2007; Petrovic and Barceló, 2007; Aga, 2008; Kümmerer, 2009). However, little is known about the impending human or ecological hazards that can arise from the cumulative exposure to the "cocktail" of pharmaceuticals and metabolites present in the different environmental compartments (notwithstanding the low concentrations at which they are observed to occur).

Human pharmaceuticals are designed to target specific metabolic and molecular pathways and, as side-effect, when introduced in the environment they may affect analogous pathways in animals having identical or similar target organs, tissues, cells or biomolecules. Even in animals lacking or having different receptors for drugs, dissimilar modes of action may occur.

It is important to recognize that, for many drugs, their specific modes of action are not well known and often not only one but many different modes of actions occur. Therefore, the ecotoxicity of most pharmaceuticals is difficult to assess (Fent et al., 2006). In addition, the metabolites and degradation by-products of pharmaceuticals are also of concern, because many of them have a toxicity which in many cases is similar to or even higher than the parent compounds(Fent et al., 2006).

The current literature about ecotoxicological effects of human pharmaceuticals deals mainly with the short-term exposure acute toxicity evaluated in standardized tests and generally focused on aquatic organisms. Acute toxicity values are in the mg L-1 dose range for most of the pharmaceuticals detected in the environment (Halling-Sørensen et al., 1998), but reported levels in surface water are at least three orders of magnitude below (Fent et al., 2006; Nikolaou et al., 2007; Aga, 2008).

It is, on the other hand, more difficult to assess (but more relevant) whether long-term chronic effects have any environmental significance as these toxicity data is generally lacking (Fent et al., 2006). Nonetheless, some primary effects can be identified which derive from the presence of pharmaceuticals and related substances in the environment including cumulative impacts, endocrine disruption, development of antibiotic-resistant bacteria and genotoxic effects (Daughton and Ternes, 1999; Bendz et al., 2005; Fent et al., 2006; Zounkova et al., 2007; Farré et al., 2008; Fent, 2008; Cooper et al., 2008; Bolong et al., 2009).

Besides toxicity, the element of persistence is of particular importance when considering the environmental significance of pharmaceuticals. Unlike POPs like pesticides, many pharmaceuticals are not lipophilic, so they do not bioaccumulate in the environment. However, some of those are "pseudo persistent pollutants" due to their continuous introduction in the environment. While not persistent in terms of a long half-life, these chemicals are constantly entering the environment, resulting in long-term exposure for the aquatic ecosystem.

Overall, the ecotoxicity of pharmaceuticals can be characterized as a game of risk. The possibility of negative impacts is present, and a number of researchers are trying to quantify the risk posed by various pharmaceuticals (Crane et al., 2006; Emblidge and DeLorenzo, 2006; Hernando et al., 2006b; Enick and Moore, 2007; Cooper et al., 2008; Cunningham et al., 2009).

Risk assessments rely on models that predict the physical, chemical, and biological properties and the corresponding ecotoxicity potential of non-assessed compounds by comparing them to assessed compounds. Sanderson et al. (2004) have prioritized drug classes in terms of their predicted toxicity, ranking sedatives and anti-psychotics as high priority, while anti-epileptics were ranked lower on the priority list. For specific pharmaceutical compounds Hernando et al. (2006b) and Cooper et al. (2008) calculated risk quotients from known toxicology data, and identified a set of high risk pharmaceuticals among which are ibuprofen, carbamazepine, naproxen, diclofenac and ketoprofen.

2.1.2. Human Pharmaceutical Occurrence in the Environment

The environmental occurrence of pharmaceuticals was first reported in 1976 by Garrison et al., who detected clofibric acid in treated wastewater in the USA at concentrations from 0.8 to 2 μg L−1. In Europe, the first comprehensive studies of the occurrence of pharmaceuticals in rivers and streams were reported in the mid 1980s by Watts et al. (1983), Waggott (1981), and Richardson and Bowron (1985).

In Canada, ibuprofen and naproxen were also detected in wastewaters in 1986 (Rogers et al., 1986; Nikolaou et al., 2007; Hao et al., 2007). After these findings, the occurrence of pharmaceuticals in environmental samples has been investigated in several countries: Brazil (Stumpf et al., 1999), Canada (Lishman et al., 2006; Hao et al., 2006; Comeau et al., 2008), UK (Ashton et al., 2004; Zhang and Zhou, 2007; Kasprzyk-Hordern et al., 2008), France (Andreozzi et al., 2003; Rabiet et al., 2006; Leclercq et al., 2009), Germany (Ternes, 1998; Heberer, 2002; Weigel et al., 2004; Hernando et al., 2006a; Osenbrük et al., 2007), Greece (Andreozzi et al., 2003; Koutsouba et al., 2003), Italy (Andreozzi et al., 2003; Zuccato et al., 2005), Spain (Hernando et al., 2006a; Carballa et al., 2008; Kuster et al., 2008), Sweden (Andreozzi et al., 2003; Bendz et al., 2005; Zorita et al., 2009), USA (Stackelberg et al., 2004; Benotti and Brownawell, 2007; Palmer et al., 2008; Benotti et al., 2009), among others.

The occurrence of pharmaceuticals in different environmental compartments, especially waters, has been already reviewed by several authors (Halling-Sørensen et al., 1998; Daughton and Ternes, 1999; Kümmerer, 2001; Jones et al., 2001; Heberer, 2002; Fent et al., 2006; Nikolaou et al., 2007; Petrovic and Barceló, 2007; Khetan and Collins, 2007; Aga, 2008; Kümmerer, 2008; Kümmerer, 2009). Many of the compounds that have become ubiquitous in surface waters and treated wastewater are mostly from the classes of the anti-inflammatories, antibiotics, blood lipid regulators, beta-blockers or neuroactive drugs (Nikolaou et al., 2007; Aga, 2008; Miège et al., 2009).

In all countries with developed medical care systems, some other compounds such as X-ray contrast media can also be expected to be present at appreciable concentrations in wastewaters (Heberer, 2002; Putschew and Jekel, 2007).

Among the most consumed drugs, those harder to biodegrade typically tend to be more frequently present in treated wastewaters and environmental samples. Nevertheless, the presence of some of the easily biodegradable pharmaceuticals may also still occur in effluents of WWTPs, even when removal efficiencies obtained with conventional wastewater treatment processes are high. In fact, due to heavy influent loads, some of the most consumed pharmaceuticals may, even after substantial removal during the treatment, subsist in significant amounts in the effluent that is discharged to the receiving water bodies.

The occurrence of metabolites or transformation products of pharmaceuticals has not yet been studied in much detail apart from some specific compounds (e. g. clofibric acid, 10,11-dihydro-10,11-epoxycarbamazepine, 3 – hydroxycarbamazepine, salicylic acid, hydroxyl-ibuprofen, carboxy-ibuprofen) (Farré et al., 2008; Miège et al., 2009; Leclercq et al., 2009).

2.1.3. Pharmaceutical Removal in Wastewater Treatment Plants

Recent studies have clearly shown that the removal of pharmaceutical compounds in municipal WWTPs is often incomplete. In fact, in many cases up to 90% of the initial amounts of pharmaceuticals entering the WWTPs may remain in effluent after treatment (Fent et al., 2006; Aga, 2008; Cooper et al., 2008). As a consequence, a significant fraction of the pharmaceuticals and their metabolites entering the WWTPs are discharged with the final effluent into the aquatic environment.

Conventional WWTPs were designed to remove bulk pollutants and not to deal with pharmaceuticals or other trace pollutants. Due to the highly variable physical and chemical properties of these compounds, the efficiencies by which they are removed may vary substantially. It is also mostly unknown whether WWTPs could be cost-effectively modified to reduce pharmaceutical discharges. Typically, there is very little elimination of organic micropollutants from the preliminary treatment of wastewater, and it is also unlikely that many pharmaceuticals will be removed during screening or primary sedimentation (Jones et al., 2005). As there is little biological activity, any pollutant removal at this stage will rely on both the tendency of the individual drugs to adsorb to solids and the degree of suspended solids removal in the primary sedimentation tanks (Zhang et al., 2008). Usually, there is little change in dissolved polar organics (such as pharmaceuticals) at this point, so little to no loss of polar drugs may be expected here. In general, elimination of any compound by sorption to sludge is considered relevant only when the log Kd for that compound is higher than 2.48 (i.e. Kd > 300 L kg-1) (Joss et al., 2005).

Activated sludge and trickling filters are the more common types of secondary biological treatment used in conventional WWTPs. Losses of drugs in both treatments may occur by the same mechanisms as other organic micropollutants, which include sorption to and removal in sludge and/or chemical degradation/transformation (such as hydrolysis) and biotransformation/biodegradation (aerobic, anoxic and anaerobic). In activated sludge processes, little loss by volatilization during aeration (stripping) is expected due to the low volatility of most pharmaceuticals (Larsen et al., 2004; Miège et al., 2009). In fact, it is found that Henry coefficients of over ~ 10−3 are required for significant stripping in a bioreactor with fine bubble aeration (Larsen et al., 2004).

Drugs remaining in the wastewater after primary and secondary treatment may be eliminated by tertiary or advanced treatments. Advanced treatment techniques such as chemical oxidation (e.g. ozonation) and membrane treatment (e.g. ultrafiltration) have been shown to be able to remove pharmaceuticals. How effectively they do so varies with the treatment conditions employed but, in some cases, it was possible to achieve levels below detection limits in drinking water treatment works (Ikehata et al., 2006; Esplugas et al., 2007; Guil et al., 2007). However, in most countries only a small number of WWTPs include these adaptations.

The issue of emergent pollutants such as pharmaceuticals and the need for regulating water quality parameters for this type of contamination have been raised several times by specialists (Robinson et al., 2007; Bolong et al., 2009). In fact, environmental agencies worldwide are evolving towards a greater awareness to this problem, electing these new substances as priority pollutants and requiring new environmental risk assessments to be carried out as part of the process of approving new substances for public use (Kot-Wasik et al., 2007). In this context, it is foreseeable that wastewater treatment requirements become more stringent in the coming years in terms of the limiting concentrations of many of these substances in the WWTPs effluents. To meet these new requirements, many of the existing conventional WWTPs will have to be adapted or reformed in the coming years. Consequently, there is a growing need for alternative wastewater treatment processes for removing pharmaceuticals from waters that have higher efficiencies at reasonable costs of operation/maintenance.

An alternative low-cost wastewater treatment option for removal of pharmaceuticals from wastewater may be the use of phytotechnologies such as constructed wetlands systems which have already shown high efficiencies in removing some pharmaceuticals (Matamoros et al., 2005; Matamoros et al., 2007a; Conkle et al., 2008; Matamoros et al., 2008a; Park et al., 2009; Matamoros et al., 2009a; Dordio et al., 2009a; Dordio et al., 2010) and other organic recalcitrant compounds (e.g. pesticides, poliaromatic hydrocarbons, explosives, chlorinated solvents, petroleum hydrocarbons) from contaminated waters (Haberl et al., 2003; Matamoros et al., 2007b; Matamoros et al., 2008b; Imfeld et al., 2009).

3. Phytoremediation Technologies

Phytoremediation is a broad term that has been in use since the early 1990s to refer to a group of technologies that use plants and its associated microorganisms, enzymes and water consumption to remove, retain, immobilize or transform/degrade pollutants, primarily of anthropogenic origin, from soil, sludges, sediments, water and wastewater and even the atmosphere (Salt et al., 1998; Macek et al., 2000; USEPA, 2000; Dietz and Schnoor, 2001; Susarla et al., 2002; Pilon-Smits, 2005; Vangronsveld et al., 2009). Phytoremediation is appealing because it is relatively inexpensive and aesthetically pleasing to the public, compared to alternate remediation strategies (e.g. involving excavation, or chemical in situ stabilization/conversion) (Zodrow, 1999; Macek et al., 2000; Susarla et al., 2002; Eapen et al., 2007).

Phytoremediation can be performed both in situ and ex situ. In the latest years efforts have focused on accelerating degradation of organic pollutants, usually in concert with root rhizosphere microorganisms, or remove hazardous heavy metals from soils or water (USEPA, 2000).

Several mechanisms have been identified by which plants can reduce pollutants availability in various environmental compartments. These include phytoextraction, phytovolatilization, phytodegradation, rhizodegradation, rhizofiltration, phytostabilization, and hydraulic control (Salt et al., 1998; ITRC, 1999; Macek et al., 2000; USEPA, 2000; Dietz and Schnoor, 2001; Susarla et al., 2002; Singh and Jain, 2003; Pilon-Smits, 2005):

- phytoextraction/phytoaccumulation – this process involves the removal of pollutants by the roots of plants and subsequent transport to aerial plant parts; pollutants accumulated in stems and leaves are harvested with accumulating plants and removed from the site;
- phytodegradation – this consists on the conversion of organic pollutants into compounds with reduced toxicity through the action of internal or secreted enzymes;
- phytovolatilization – through this process soluble pollutants are taken up with water by the roots, transported to the leaves and volatilized into the atmosphere through the stomata; amounts of pollutants transpired are proportional to the water flow and usually relatively low;
- rhizodegradation – breakdown of organic pollutants through microbial enzymatic activity is called rhizodegradation; the types of plants growing in the contaminated area influence the amount, diversity and activity of microbial populations;
- rhizofiltration – this mechanism of pollutants retention involves either adsorption or absorption by plants roots; consequently, large root surface areas are usually required for these processes;
- phytostabilization – through this process, accumulation by plant roots or precipitation in the soil by root exudates immobilizes and reduces the availability of soil pollutants; plants growing on polluted sites also stabilize the soil and can serve as a groundcover thereby reducing wind and water erosion and direct contact of the pollutants with animals;
- hydraulic control – containment of pollutants within a site can also be achieved by limiting the spread of a contaminant plume through plant evapotranspiration.

In the soil/water-plant-atmosphere continuum, a specific contaminant can be remediated at specific points along this continuum by different phytoremediation mechanisms. This is shown in Figure 2.

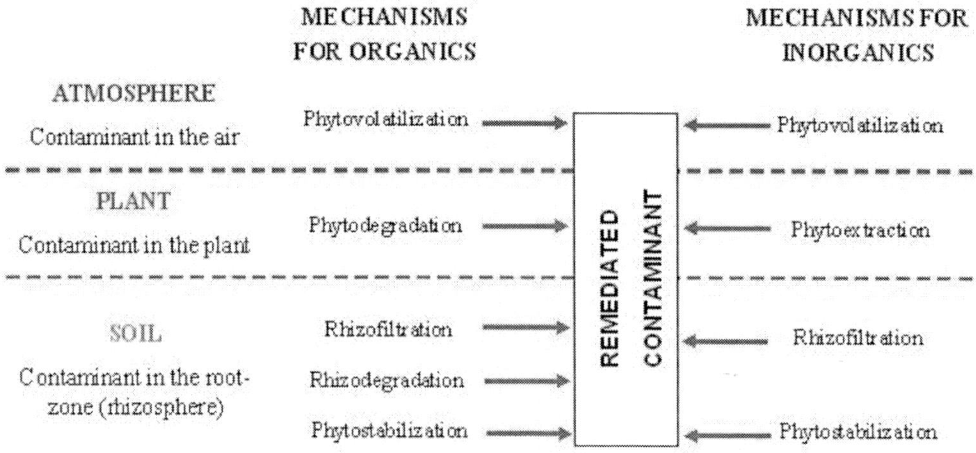

Figure 2. Contaminant fate in the Soil/Water-Plant-Atmosphere Continuum (adapted from ITRC (1999)).

As any other technology, phytoremediation has both advantages and limitations. Several benefits can be obtained from the use of phytoremediation (ITRC, 1999; Macek et al., 2000; USEPA, 2000; Pilon-Smits, 2005):

- Phytoremediation installations provide improved aesthetics and receive a better acceptance from the public as they are less invasive and destructive than other technologies. They additionally may provide habitat to animals and promote biodiversity.
- Phytoremediation is a low-cost technology compared to other treatment methods. Studies have indicated that implementing phytoremediation may result in cost savings of 50 to 80 % over traditional technologies.
- It is easy to implement and maintain, and it is effective for a variety of organic and inorganic compounds.
- It reduces the amount of dust emission and may promote better air quality in the vicinity of the site. Vegetation also helps reduce erosion by wind or water.

Phytoremediation technology, however, has its limitations and is not applicable or successful for all sites and situations. Some disadvantages must be noted (ITRC, 1999; Macek et al., 2000; USEPA, 2000; Pilon-Smits, 2005):

- Extremely high concentrations of pollutants may not allow plants to grow or cause them to die. So, phytoremediation is more effective or limited to lower concentrations of contaminants.
- For phytoremediation to be successful, contaminants must be within reach of plants roots, therefore it is restricted to sites which shallow contamination, with effective depth limited by the size of the rooting zone.
- Phytoextraction can cause contaminants to accumulate in plant tissues. The potential for ecological exposure through consumption of contaminated plants by animals and possible effect on the food chain is an environmental concern. Harvesting of contaminated biomass may be required.
- Phytovolatilization may remove contaminants from the subsurface but might then cause increased airborne exposure. In general, the fate of contaminants is often unclear, which may raise important issues with the potential for the dissemination of some pollutants in the environment.
- If non-native species are selected for phytoremediation, the consequences of introducing them in the ecosystem may be unknown or unexpected.
- Phytoremediation may take a longer time to achieve remediation goals (sometimes several year) than is required by other treatment technologies. For instance, a tree stand may take several growing seasons to be established and for contaminant concentrations to be reduced.
- The success of phytoremediation may be seasonal, depending on location. Its effectiveness may be variable, affected by climate conditions.

Several different types of phytotechnology applications can be considered in relation to the different ways the plants and their rhizosphere organisms are used. They can be used as filters in constructed wetlands or in a hydroponic setup, the latter consisting of a rhizofiltration treatment. Trees can be used as a hydraulic control barrier to create an upward water flow in the root zone, preventing contamination to leach down or to prevent a contaminated ground water plume from spreading horizontally (USEPA, 2000; Pilon-Smits, 2005). Plants can also be used to stabilize pollutants in soil, either simply by preventing erosion, leaching or runoff, or by converting pollutants to less bioavailable forms, for example through precipitation. Other uses may include phytoextraction of pollutants followed by harvesting of the aerial plant parts, and the plant material can subsequently be used for nonfood purposes (e.g. wood, cardboard) or ashed and diposed in a landfill. In the case of valuable metals, recycling of the accumulated element can be carried out, thus comprising a technology which is termed phytomining (Pilon-Smits, 2005).

In several phytotechnologies a given mechanism for the pollutant's removal may be predominant but the various possible mechanisms are not mutually exclusive. On the contrary, several mechanisms can occur simultaneously, either concurrently or cooperatively, in the case of some phytotechnologies. For instance, in a more complex phytotechnology application such as a constructed wetlands system, rhizodegradation and phytostabilization, phytoextraction, phytodegradation amd phytovolatilization can all contribute in varying degrees to the overall effect of the system on the pollutants removal.

4. Constructed Wetlands Systems

Constructed wetlands systems (CWS) are engineered systems designed and constructed to make use of the natural processes involving wetland vegetation, soil and their associated microbial assemblages to assist in wastewater treatment. The concerted action of all these components (support matrix, vegetation and microbial populations), through a variety of chemical, physical and biological processes, is responsible for the depuration of wastewaters achieved in a CWS. These systems take advantage of many of the same processes that occur in natural wetlands, but do so within a more controlled environment. CWS can be classified in several different types, as depicted in Figure 3, according to the type of water flow regime: free water surface (FWS) constructed wetlands, vertical subsurface flow (VSSF) constructed wetlands, and horizontal subsurface flow (HSSF) constructed wetlands; and according to the type of vegetation, especially in respect to the way it is anchored to the bottom of the system: CWS with free-floating plants, CWS with submerged plants, and CWS with emergent plants.

CWS have been used to treat a variety of wastewaters (municipal, industrial and agricultural) and including urban runoff (USEPA and USDA-NRCS, 1995; Cooper et al., 1996; Vymazal et al., 1998; Sundaravadivel and Vigneswaran, 2001; Haberl et al., 2003; Stottmeister et al., 2003; Scholz and Lee, 2005; Kadlec and Wallace, 2009; Vymazal, 2009).

In the past, CWS have been used mainly as wastewater treatment alternatives or complementary to the conventional treatment for domestic wastewaters of small communities. Thus, CWS have been mostly applied in the removal of bulk wastewater pollutants such as suspended solids, organic matter, excess of nutrients and pathogens.

More recently, CWS applications for dealing with more specific pollutants, such as organic xenobiotics, have been meeting a larger interest and have been the subject of an increasing number of studies.

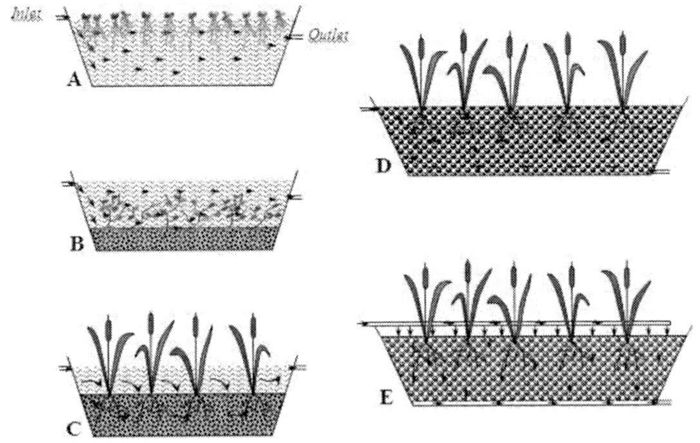

Figure 3. Different types of CWS (A, FWS with free-floating plants; B, FWS with submerged plants; C, FWS with emergent plants; D, Horizontal SSF; E, Vertical SSF) (adapted from Dordio et al. (2008)).

In many of such studies, CWS have been proving to be efficient and cost-effective solutions for the removal of some organic xenobiotics such as pesticides, azo dyes, explosives and petroleum hydrocarbons (Williams, 2002; Haberl et al., 2003; Braskerud and Haarstad, 2003; Low et al., 2008; Davies et al., 2008; Imfeld et al., 2009; Vymazal, 2009; Moore et al., 2009; Tang et al., 2009). However, their evaluation for the removal of pharmaceuticals, their metabolites and transformation products is currently a very active topic of research and can be considered as work still in progress (Gross et al., 2004; Dordio et al., 2007; Matamoros et al., 2007b; Conkle et al., 2008; Matamoros et al., 2008a; Matamoros et al., 2008b; Park et al., 2009; Matamoros et al., 2009a; Dordio et al., 2009a; Matamoros et al., 2009b; Dordio et al., 2009b; Dordio et al., 2010). Ultimately, the optimization of CWS for the removal of more specific target compounds requires a basic knowledge of the processes involved in the removal of the pollutants and the interactions between those and the CWS components. New trends in CWS research are moving towards the study of such processes and interactions and focus on the selection and optimization of the CWS components for more specific applications.

4.1. Organic Xenobiotics Removal in CWS

A specialized use of CWS for the removal of specific organic compounds or classes of compounds has been developing as a growing type of CWS applications in comparison to the treatment of bulk pollutants. Significant work exists with wastewater contaminated with organic xenobiotics from the petroleum industry, food processing industry, pesticide contaminated agricultural runoff, landfill leachates, and waters containing surfactants, solvents or mineral oils (Vymazal et al., 1998; Williams, 2002; Haberl et al., 2003; Imfeld et al., 2009; Vymazal, 2009). Very little is commonly known, about the exact pathways of the organic xenobiotics removal in CWS. Given the diversity of chemical characteristics of these compounds, which despite being classified under a common designation of xenobiotics are in fact formed by widely unrelated families of chemical substances, it is of no surprise that several very diverse mechanisms are responsible for their removal. The same observation can be made specifically for pharmaceuticals as this subset of xenobiotics is equally varied in terms of their chemical properties. A comprehensive description of organic xenobiotics removal in CWS is, thus, not an easy task to accomplish and these systems have been and still are largely operated as a "black box". Although much of the design of CWS in the past has been done with little knowledge of the roles played by each component and how its function could be optimized, nowadays the knowledge that has been accumulating is beginning to be applied. A much greater variety of plant species, matrix materials and constructed wetlands designs is now, being introduced. The goals of the target contaminants to remove in CWS are also becoming progressively more ambitious.

4.1.1. Main Removal Processes in CWS

Organic xenobiotics removal by CWS involves several interdependent processes which may be classified as abiotic (physical or chemical) or biotic (carried out by living organisms such as plants and microorganisms). These processes are basically the same occurring in natural wetlands and also identical to those responsible for the fate of xenobiotics in the environment. The way in which CWS differ from natural wetlands is that CWS are engineered systems where these processes occur in a controlled environment and conditions are optimized in order to maximize pollutants removal. The primary abiotic and biotic processes that participate in removing organic xenobiotics from contaminated water in a CWS are described in Table 1.

Table 1. Abiotic and biotic processes involved in xenobiotics removal in CWS (Pilon-Smits, 2005; Reddy and DeLaune, 2008)

Processes	Description
Abiotic	
Sorption	Including adsorption and absorption, the chemical processes occurring at the surface of plants roots and solid matrix that result in a short-term retention or long-term immobilization of xenobiotics.
Hydrolysis	The chemical breakdown of organics by the action of water, a process which is pH-dependent.
Photodegradation	Degradation of organic xenobiotics by the action of sunlight
Redox reactions	Modification, which sometimes may be quite substantial, of xenobiotics due to the action of oxidizing or reducing agents. Redox reactions are also frequently brought about by biotic agents (e.g. bacteria), or enzymatically catalyzed.
Precipitation	For those compounds which can exist in several forms with different water solubilities, the conversion into the most insoluble forms.
Filtration	Removal of particulate matter and suspended solids.
Volatilization	Release of some organic xenobiotics as vapors, which occurs when these compounds have significant vapor pressures.
Biotic	
Aerobic/anaerobic biodegradation	Metabolic processes of microorganisms, which play a significant role in organic xenobiotics removal in CWS.
Phytodegradation	Breakdown of organic xenobiotics having first been taken up by plants.
Rhizodegradation	Enhancement of microbial degradation of some organic xenobiotics by the stimulus provided by substances released in roots exudates.
Phytovolatilization	Uptake and transpiration of volatile organics through the aerial plant parts.

Removal of pollutants from water may be accomplished through storage in the wetland solid matrix and in the vegetation, or through losses to the atmosphere. A basic understanding of how these processes operate in wetlands is extremely helpful for assessing the potential applications, benefits and limitations of CWS.

a) Abiotic Processes

A wide range of physical and chemical processes are involved in the abiotic removal of contaminants in CWS. The most important abiotic removal process occurring in CWS, at the surface of plants roots and solid media, is sorption, resulting in a short-term retention or long-term immobilization of the contaminants (Vymazal et al., 1998; Reddy and DeLaune, 2008; Kadlec and Wallace, 2009). The type of materials that compose the support matrix will have a strong influence over the occurrence of sorption processes. The chemical characteristics of the solid matrix determines its capacity to sorb pollutants (Muller et al., 2007; Reddy and DeLaune, 2008b) but retention in abiotic components is also a function of several characteristics of the wastewater, such as its dissolved organic matter content, pH and electrolyte composition, as well as of the pollutant itself. In addition, a good hydraulic conductivity of the support matrix, which avoids the occurrence of overland flows and preferential channeling, is crucial for a good and uniform contact of the wastewater with the CWS media and, thus, for the efficiency of the system (Vymazal et al., 1998; Kadlec and Wallace, 2009).

Other common abiotic processes such as hydrolysis, photodegradation, redox reactions and volatilization (Table 1) can also contribute, in varied extents, to the removal of some particular classes of compounds, depending on their specific properties. However, with exception of sorption, abiotic processes are not, in general, major removal processes for most organic compounds such as pharmaceuticals because either the conditions in CWS or the properties of the compounds are not suitable. For example, photodegradation can only be significant in FWS systems, if plant density is not too high (such that it does not cause too much shade) and if the compounds are photosensitive. In SSF systems photodegradation does not occur in appreciable extent as the water level is below the solid matrix surface and, therefore, exposure of the pollutants to sunlight is very limited in this type of systems. The process of volatilization is also of modest importance for substances with low volatility, as is the case of most pharmaceuticals. Where CWS are used as a tertiary treatment stage after conventional secondary treatment in a WWTP, organic compounds do not suffer, in most cases, appreciable hydrolysis either, as they have been already subjected to such type of processes in secondary treatment stage (and, in case of pharmaceuticals, previously in the human digestive system as well). Therefore, the organic xenobiotics present in CWS influents are those that have resisted such hydrolysis processes or they are the transformation products of those substances that did not resist it.

b) Biotic Processes

CWS are biological systems in which biological processes play a major role in the removal of pollutants. The two biotic components which are responsible in CWS by the biological contribution to the removal of organics are the wetland vegetation and the microbial populations.

The plants growing in natural wetlands (often called wetland plants or macrophytes), are also typically the plant species used in constructed wetlands as these are well adapted to the water saturated conditions found in these systems (Brix, 1994; Brix, 1997; Kadlec and Wallace, 2009). Major roles of macrophytes in CWS include the filtration of large debris; provision of surface area for microorganisms development and release of exudates by roots (normally including organic acids, sugars, amino acids, vitamins and enzymes (Macek et al., 2000; Alkorta and Garbisu, 2001)) that stimulate microorganisms' growth; enhancement of the hydraulic conductivity of the support matrix (roots and rhizomes growth help to prevent clogging in the matrix); transport and release of oxygen through the roots (which increases aerobic degradation and nitrification); and attenuation of the wastewater pollutants load (nutrient and xenobiotics uptake) (Brix, 1994; Sundaravadivel and Vigneswaran, 2001; Haberl et al., 2003; Kadlec and Wallace, 2009).

In CWS, an increasingly important role is being attributed to plants in removing poorly or non-biodegradable organic xenobiotics through their capacity to sorb them in their roots and even uptake them and sequester/transform them in their tissues (Macek et al., 2000; Korte et al., 2000; Dietz and Schnoor, 2001; Pilon-Smits, 2005; Collins et al., 2006). In fact , direct uptake by plants is a widely recognized process for inorganic pollutants removal. In the case of some organic substances, plant uptake has been observed to also play a significant role among biotic processes, especially for those compounds with a moderate hydrophobicity ($0.5 < \log Kow < 3.5$) (Dietz and Schnoor, 2001; Schroder and Collins, 2002; Pilon-Smits, 2005). Nevertheless, action of microorganisms is generally accepted to be the major route for organic xenobiotics elimination in wetlands. However, even in these microbial processes, plants do play a relevant role, through the influence of exudates released in the rhizosphere which have the effect of stimulating the development and the activity of microorganisms (apart from also contributing, in some cases, to the catalytic degradation of pollutants) (Macek et al., 2000; Singer et al., 2003; Pilon-Smits, 2005).

Although microorganisms may also provide a measurable amount of contaminant uptake and storage, it is their metabolic processes that play the most significant role in the decomposition of organic compounds through the transformation of complex molecules into simpler ones (Reddy and DeLaune, 2008). This provides an important biological mechanism for removal of a wide variety of organic compounds. However, the efficiency and rate of organic compounds degradation by microorganisms is highly variable for different compounds types.

The characteristics of the biotic components (vegetation and microorganisms) obviously also have a tremendous influence on the CWS behavior. Microorganisms populations develop naturally in CWS and are exposed to similar factors as those affecting their development in WWTPs. However, the characteristics of these microbial populations can be modified by inoculation of the CWS with strands that are more adequate for the purpose of the system. Important characteristics of both microorganisms and plants are their tolerance to the more toxic pollutants (at typical wastewater concentrations) and their capacity to, respectively, biodegrade or uptake them. In the case of the vegetation, other factors related with the CWS design such as plant density and layout of the specimens (e.g. the way specimens of different species may be intermixed when planted in the beds) all have to be considered and carefully planned as their influence may range from subtle differences in the system's behavior to more pronounced impacts in the overall efficiency (Kadlec and Wallace, 2009). In particular, the cycles of vegetative activity of some species in addition to variations of climate conditions may lead to significant seasonal changes in the system's efficiency, which in some cases may be mitigated by using polycultures of vegetation (Kadlec and Wallace, 2009).

4.1.2. Pharmaceutical Removal in CWS

Studies conducted so far on the removal of pharmaceuticals in CWS have shown the potential of these systems to remove a wide variety of compounds (Table 2). However, there is still ample work of optimization to be carried out on these systems, which must be based on a better understanding of how the several processes involved perform their functions and interoperate. A more profound characterization of the roles played by each CWS component in the overall pharmaceuticals removal efficiency, related with the properties of each substance involved, is also necessary to guide an optimal selection of each component. The available studies on this subject, although providing valuable information, are still scarce and further work is still necessary.

4.2. The Role of Plants in Organic Xenobiotics Removal

Plants play an important role in the biotic processes of organics removal in CWS, as described previously involving several processes of which many details still remain to be known or fully understood. Subjected to the direct or indirect action of plants, xenobiotics can be stabilized or degraded in the rhizosphere, adsorbed or accumulated in the roots and transported to the aerial parts, volatilized or degraded inside the plant tissues (Figure 4).

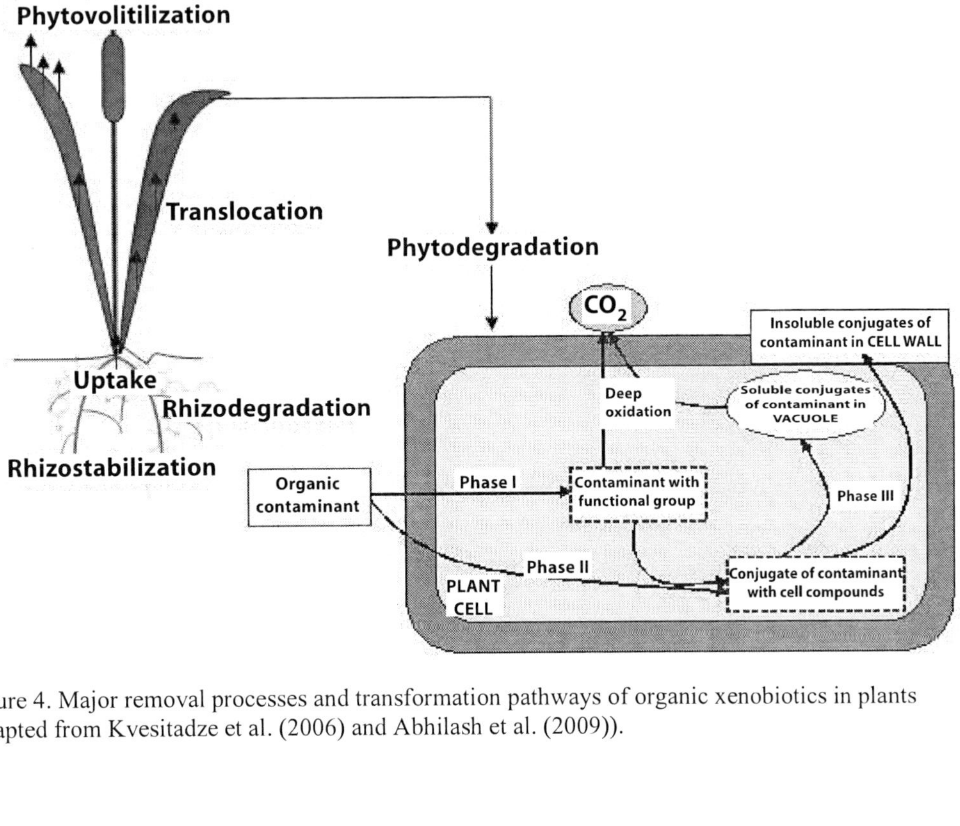

Figure 4. Major removal processes and transformation pathways of organic xenobiotics in plants (adapted from Kvesitadze et al. (2006) and Abhilash et al. (2009)).

Table 2. Pharmaceutical removal in different types of CWS

Organic compound	Physico-chemical properties			Type of CWS [c]	Type of substrate /plant [d]	% Removed [e]	Removal processes suggested by authors	References
	S_W [*, a] (25°C) (mg L^{-1})	logK_{ow} [*,b] (25° C)	pKa [*]					
Ibuprofen	21	3.97	4.9	HSSF	Gravel/ *Phragmites australis*	48 (deep) 81 (shallow)	Microbial degradation, sorption	(Matamoros et al., 2005)
				HSSF	Gravel/ *Phragmites australis*	71	ibid.	(Matamoros and Bayona, 2006)
				VSSF	Gravel/ *Phragmites australis*	99	ibid.	(Matamoros et al., 2007a)
				HSSF	Gravel/ *Phragmites australis*	52	ibid.	(Matamoros et al., 2008a)
				HSSF	n.d.	65	ibid.	(Matamoros et al., 2009a)
				VSSF	n.d.	89	ibid.	
				FWS	*Typha* spp. + *Phragmites australis*	95 (Winter) 96 (Summer)	ibid.	(Matamoros et al., 2008b)
				Lagoon + VSSF	Gravel/ *Hydrocottle* spp. + *Phragmites australis*	> 99	n. d.	(Conkle et al., 2008)
				CWS (microcosms)	Expanded clay/ *Typha* spp	82 (Winter) 96 (Summer)	Microbial degradation, plant uptake, sorption	(Dordio et al., 2010)
Carbamazepine	17.7	2.45	14	HSSF	Gravel/ *Phragmites australis*	26 (deep) 16 (shallow)	Sorption	(Matamoros et al., 2005)
				HSSF	Gravel/ *Phragmites australis*	16	ibid.	(Matamoros and Bayona, 2006)
				VSSF	Gravel/ *Phragmites australis*	26	ibid.	(Matamoros et al., 2007a)
				HSSF	Gravel/ *Phragmites australis*		ibid.	(Matamoros et al., 2008a)
				HSSF	n.d.	38	ibid.	(Matamoros et al., 2009a)
				FWS	*Typha* spp. + *Phragmites australis*		ibid.	(Matamoros et al., 2008b)
				FWS	*Acorus* + *Typha* spp.	65	Plant uptake	(Park et al., 2009)
				CWS (microcosms)	Expanded clay/ *Typha* spp	88 (Winter) 97 (Summer)	Sorption, plant uptake	(Dordio et al., 2010)
Clofibric acid	583	2.57	3.18	*HSSF*	Gravel/ Phragmites australis	n.r.	-	(Matamoros et al., 2005)
				FWS	Typha spp. + Phragmites australis	32 (Winter) 36 (Summer)	n.d	(Matamoros et al., 2008b)
				CWS (microcosms)	Expanded clay/ Typha spp	48 (Winter) 75 (Summer)	Sorption, plant uptake	(Dordio et al., 2010)

Organic compound	Physico-chemical properties[c]			% Removed [e]	Type of CWS [c]	Type of substrate /plant [d]	% Removed [e]	Removal processes suggested by authors
	S_W [*, a] (25°C) (mg L^{-1})	Type of CWS	Type of substrate /plant [d]					
Atenolol	13300	0.16	9.6	*Lagoon + VSSF*	Gravel/ Hydrocottle spp. + Phragmites australis	> 99	n.d.	(Conkle et al., 2008)
				FWS	Acorus + Typha spp.	97	n.d	(Park et al., 2009)
				CWS (microcosms)	Expanded clay/ Typha spp or Phragmites australis	> 92	Sorption, plant uptake	(Dordio et al., 2009a)
Naproxen	15.9	3.18	4.15	*HSSF*	Gravel/ Phragmites australis	85	Microbial degradation, sorption	(Matamoros and Bayona, 2006)
				VSSF	Gravel/ Phragmites australis	89	ibid.	(Matamoros et al., 2007a)
				HSSF	n.d.	45	ibid.	(Matamoros et al., 2009a)
				VSSF	n.d.	92	ibid.	
				FWS	Typha spp. + Phragmites australis	52 (Winter) 92 (Summer)	ibid.	(Matamoros et al., 2008b)
Ketoprofen	51	3.12	4.45	*HSSF*	Gravel/ Phragmites australis	38	Sorption	(Matamoros and Bayona, 2006)
				HSSF	n.d.	90	ibid.	(Matamoros et al., 2009a)
				VSSF	n.d.	n.r.	ibid.	
				FWS	Typha spp. + Phragmites australis	97 (Winter) 99 (Summer)	Photodegradation	(Matamoros et al., 2008b)
Diclofenac	2.4	4.51	4.15	*HSSF*	Gravel/ Phragmites australis	15	Sorption	(Matamoros and Bayona, 2006)
				VSSF	Gravel/ Phragmites australis	73	ibid.	(Matamoros et al., 2007a)
				HSSF	n.d.	21	ibid.	(Matamoros et al., 2009a)
				FWS	Typha spp. + Phragmites australis	73 (Winter) 96 (Summer)	ibid.	(Matamoros et al., 2008b)
Caffeine	21600	-0.07	10.4	*HSSF*	Gravel/ Phragmites australis	97	Microbial degradation, sorption	(Matamoros and Bayona, 2006)
				VSSF	Gravel/ Phragmites australis	99	ibid.	(Matamoros et al., 2007a)
				HSSF	n.d.	97	ibid.	(Matamoros et al., 2009a)
				VSSF	n.d.	99	ibid.	
				Lagoon + VSSF	Gravel/ Hydrocottle spp. + Phragmites australis	> 99	n.d.	(Conkle et al., 2008)
Salicylic acid	2240	2.26	2.97	*HSSF*	Gravel/ Phragmites australis	96	Microbial degradation, sorption	(Matamoros and Bayona, 2006)

Table 2. (Continued)

Organic compound	Physico-chemical properties[c] S_W[*, a] (25°C) (mg L^{-1})	Type of CWS	Type of substrate /plant [d]	% Removed [e]	Type of CWS [c]	Type of substrate /plant [d]	% Removed [e]	Removal processes suggested by authors
				VSSF	Gravel/ Phragmites australis	98	ibid.	(Matamoros et al., 2007a)
				HSSF	n.d.	95	ibid.	(Matamoros et al., 2009a)
Sulfamethoxazole	610	0.89		*Lagoon + VSSF*	Gravel/ Hydrocottle spp. + Phragmites australis	91	n.d.	(Conkle et al., 2008)
				FWS	Acorus + Typha spp.	30		(Park et al., 2009)
Metoprolol	16900	1.88	9.6	*Lagoon + VSSF*	Gravel/ Hydrocottle spp. + Phragmites australis	92	n.d.	(Conkle et al., 2008)
Sotalol	5510	0.24		*Lagoon + VSSF*	Gravel/ Hydrocottle spp. + Phragmites australis	30	n.d.	(Conkle et al., 2008)
Acetaminophen	14000	0.46	9.38	*Lagoon + VSSF*	Gravel/ Hydrocottle spp. + Phragmites australis	100	n.d.	(Conkle et al., 2008)
Gemfibrozil	10.9	4.77		*Lagoon + VSSF*	Gravel/ Hydrocottle spp. + Phragmites australis	64	n.d.	(Conkle et al., 2008)

[*] PHYSPROP, 2009;

[a] S_w=Water solubility;

[b] K_{ow}=Octanol-water partition coefficient;

[c] CWS: Constructed wetlands system; HSSF: Horizontal subsurface flow; VSSF: Vertical subsurface flow; FWS: Free Water Surface;

[d] n.d.: not detailed;.

[e] n.r.: not removed.

Many organic pollutants can be readily taken up by plants but, as consequence of being xenobiotic, there are no specific transporters for these compounds in the plant membranes. Therefore, they move into and within plant tissues via diffusion (or passive uptake) (Dietz and Schnoor, 2001; Pilon-Smits, 2005; Collins et al., 2006). The flux is driven by the water potential gradient created throughout the plant during transpiration, which depends on the plants characteristics and the CWS environmental conditions. Translocation of the compounds is highly dependent on their concentrations and physicochemical properties such as water solubility, log Kow and pKa (Korte et al., 2000; USEPA, 2000; Alkorta and Garbisu, 2001).

An optimal hydrophobicity may exist such that the organic compound has a tendency to bind to the lipid bilayer of the membrane but not too strongly so that transport can still occur. Direct uptake by roots is usually an efficient removal mechanism for moderately hydrophobic organic chemicals (log Kow = 0.5 – 3.5) (USEPA, 2000; Dietz and Schnoor, 2001; Pilon-Smits, 2005). In general, hydrophobic chemicals (log Kow > 3.5) are bound so strongly to the lipophilic root solids and cell walls that they cannot easily enter and be translocated in the plant. On the other hand, chemicals that are quite water soluble (log Kow < 0.5) are not sufficiently sorbed to roots nor effectively transported through the lipid bilayer of plant membranes. Some studies, however, indicate that the log Kow value of a chemical may not be the sole factor determining its tendency to be taken up and some compounds have been shown to be able to penetrate plant membranes despite a low log Kow (Renner, 2002). The capacity of a compound to be removed from water by a given plant, may also depend on other factors such as initial pollutant concentration, the anatomy and the root system of the plant (Chaudhry et al., 2002). In addition, very hydrophobic chemicals (log Kow > 3.5) are also candidates for phytostabilization and/or rhizosphere bioremediation by virtue of their long residence times in the root zone (USEPA, 2000; Dietz and Schnoor, 2001; Pilon-Smits, 2005).

Uptake has primary control over translocation, metabolism and phytotoxic action because the total amount of xenobiotic available for these processes is determined by the amount of compound absorbed by the plant. Metabolism influences both xenobiotic uptake and their phytotoxic action by either rendering the compound less or more active (Dietz and Schnoor, 2001; Kvesitadze et al., 2006).

4.2.1. What Happens to Organic Xenobiotics Once Taken Up by the Plant?

Organic xenobiotics taken up by roots are translocated into different organs of the plants as a result of the physiological processes involved in the transport of nutrients. The main forces involved in this transport are related to the transpiration stream, i.e. transport of water and dissolved substances from roots to shoots, passing through vessels and tracheids located in the xylem (Kvesitadze et al., 2006).

The importance of the transpiration stream for the uptake and translocation of organics by plants is expressed in the following equation (Briggs et al., 1983; Dietz and Schnoor, 2001):

$$U = (TSCF)\ (T)\ (C) \qquad (1)$$

where U is the rate of organic compound assimilation (mg day-1); T, the rate of plant transpiration, (L day-1); C, the organic compound concentration in the water phase (mg L-1); TSCF, the transpiration stream concentration factor, is a dimensionless ratio between the concentration of the organic compound in the liquid of the transpiration stream (xylem sap) and the bulk concentration in the root zone solution (Dietz and Schnoor, 2001; Doucette et al., 2005; Kvesitadze et al., 2006).

The TSCF has been extensively used in modeling of uptake and translocation of organic compounds in plants. With the possible exception of some hormone-like chemicals such as the phenoxy acid herbicides, there is no evidence of active uptake (TSCF > 1) of xenobiotic organic chemicals (Doucette et al., 2005). A chemical is said to be excluded (TSCF < 1) when its uptake is not directly proportional to water uptake (TSCF = 1), although the mechanism of uptake is still thought to be a passive process. However, factors such as membrane permeability and xylem sap solubility of the contaminant may limit the extent or kinetics of passive uptake (Doucette et al., 2005). Sorption and rapid metabolism of contaminants within the plant may also reduce xylem concentrations and keep measured TSCF values from reaching one (Doucette et al., 2005).

For organic chemicals, several empirical relationships between TSCF and the log Kow of the xenobiotic have been reported in which these values are related by characteristic bell-shaped gaussian curves (Briggs et al., 1983; Hsu et al., 1990; Sicbaldi et al., 1997; Burken and Schnoor, 1998; de Carvalho et al., 2007; Paraiba, 2007). These, again, suggest an optimal lipophilicity (corresponding to the maxima of the Gaussian curves) for uptake and translocation and infer that compounds which are either highly polar (log Kow < 0.5) or are highly lipophilic (log Kow > 3.5) will not be significantly taken up by plants. However, laboratory and field experiments with some xenobiotics such as 1,4-dioxane, MTBE, sulfolane and diisopropanolamine also suggest that these predictive schemes may not be applicable for some non-ionizable, highly water soluble organics (Aitchison et al., 2000; Rubin and Ramaswami, 2001; Groom et al., 2002; Chard et al., 2006).

4.2.2. Plant Detoxification Processes

Organic xenobiotics which penetrate the plant cells are exposed to plant's metabolic transformations that may lead to their partial or complete degradation or through which they may be transformed in less toxic compounds and bound in plant tissues (Korte et al., 2000; Kvesitadze et al., 2006).

Metabolism of foreign compounds in plant systems is generally considered to be a "detoxification" process that is similar to the metabolism of xenobiotic compounds in humans, hence the name "green liver" that is used to refer to these systems (Sandermann, 1994). Once an organic xenobiotic is taken up and translocated, it undergoes one or several phases of metabolic transformation, as is illustrated by the diagram in Figure 4.

Three possible phases of metabolic transformation of organic compounds in higher plants can be identified (Sandermann, 1994):

Phase I – Functionalization: involves a conversion/activation (oxidation, reduction or hydrolysis) of lipophilic xenobiotic compounds (Komives and Gullner, 2005; Eapen et al., 2007); in this phase the molecules of the hydrophobic compound acquire a hydrophilic functional group (e.g. hydroxyl, amine, carboxyl, sulphydryl) through enzymatic transformations. The polarity and water solubility of the compound increase as a result of these processes, which also causes an increased affinity to enzymes catalyzing further transformation (conjugation or deep oxidation (Korte et al., 2000; Kvesitadze et al., 2006)) by the addition or exposure of the appropriate functional groups. In the case of a low concentration, oxidative degradation of some xenobiotics to common metabolites of the cell and CO2 may take place. Following this pathway, a plant cell not only detoxifies the compound but also assimilates the resulting carbon atoms for cell needs. In case of a high concentration, full detoxification is not achieved and the contaminant is exposed to conjugation (Korte et al., 2000).

During this phase several different groups of enzymes are known to play an important role (Sandermann, 1992; Sandermann, 1994; Macek et al., 2000; Eapen et al., 2007). In plants, oxidative metabolism of the xenobiotics is mediated mainly by cytochrome P450 monooxygenase which is of crucial importance in the oxidative processes to bioactivate the xenobiotics into chemically reactive electrophilic compounds which subsequently form conjugates during Phase II. Peroxidases are another important group of enzymes, which help in the conversion of some of the xenobiotics. Peroxygenases may also be involved in the oxidation of some compounds. Nitroreductase is needed for the degradation of nitroaromatics and laccase for breaking up aromatic ring structures.

Phase I reactions are the first step needed to ultimately make a xenobiotic less toxic; those reactions modify the molecule to be ready for Phase II and Phase III reactions which further detoxify the chemical. However, if it already has a functional group suitable for Phase II metabolism, the compound can directly be used for Phase II without entering Phase I.

Phase II – Conjugation: involves conjugation of xenobiotic metabolites of Phase I (or the xenobiotics themselves when they already contain appropriate functional groups) to endogenous molecules (proteins, peptides, amino acids, organic acids, mono-, oligo- and polysaccharides, pectins, lignin, etc.) (Coleman et al., 1997; Korte et al., 2000; Dietz and Schnoor, 2001; Eapen et al., 2007); as result of conjugation, compounds of higher molecular weight are formed with greatly reduced biological activity and usually reduced mobility. The end products of Phase II are usually less toxic than the original molecules or Phase I derivatives.

Conjugation is catalyzed by transferases. Enzymes such as glutathione-S-transferases, glucosyl transferase and N-malonyl transferases are associated with Phase II (Eapen et al., 2007).

Conjugation of Phase I metabolites takes place in the cytosol, but it is harmful to accumulate these compounds in cytosol (Eapen et al., 2007).

Phase III – Compartmentalization: involves modified xenobiotics getting compartmentalized in vacuoles or getting bound to cell wall components such as lignin or hemicellulose (Coleman et al., 1997; Dietz and Schnoor, 2001; Eapen et al., 2007). In this phase (a potential final step in the non-oxidative utilization of xenobiotics) the conjugates are removed from vulnerable sites in cytosol and transported to sites where they may not interfere with cellular metabolism: soluble conjugates (with peptides, sugars, amino acids, etc.) are accumulated in vacuoles, whereas insoluble conjugates (coupled with pectin, lignin, xylan and other polysaccharide) are taken out of the cell and accumulated in plant cell walls (Sandermann, 1992; Sandermann, 1994; Eapen et al., 2007).

Phase III reactions are unique to plants because they do not excrete xenobiotics as animals do. Plants therefore, need to somehow remove the xenobiotic within their own system. ATP driven vacuolar transporters are the main enzymes involved in this phase and further processing of conjugates may take place in the vacuolar matrix (Eapen et al., 2007).

It is assumed that Phase III products are no longer toxic; however, this area of xenobiotic fate in plants is poorly understood, especially with reference to the identity of the sequestered products and any subsequent fate in herbivores who might consume those plants.

Metabolism of pesticides has already been extensively studied (Chaudhry et al., 2002; Coleman et al., 2002; Eapen et al., 2007). More recently, the metabolism of non-agricultural xenobiotics such as trichloroethylene (TCE), TNT, glyceroltrinitrate (GTN), PAHs, PCBs and other organic compounds has also been studied (Görge et al., 1994; Salt et al., 1998; Alkorta and Garbisu, 2001; Hannink et al., 2002; Eapen et al., 2007). It has been shown that most of these compounds are metabolized, but only a few chemicals appear to be fully mineralized. Studies applied to pharmaceutical substances, however, are very scarce until now (Huber et al., 2009) in spite of the great interest that such data represents for phytoremediation applications.

4.2.3. Phytotoxicity of Organic Xenobiotics

All plants have defense mechanisms to protect them from the negative effects of small quantities of foreign compounds. The relative rates of organic xenobiotics uptake, translocation, and metabolism usually determines whether or not these compounds will induce a plant response, which can be inferred from the plant's physiological, biochemical and molecular responses.

Physiological responses such as growth reduction, chlorosis and necrosis of tissues can usually be observed easily. Quantitative parameters can also be evaluated, which can provide an assessment of physiological toxic responses to xenobiotic stress, namely the plant's relative growth rate (RGR) or the photosynthetic pigments concentration in plant tissues. Alteration of these two parameters, have been observed for plants exposed to xenobiotics, sometimes compromising plant viability (Mishra et al., 2006).

The photosynthetic apparatus is one of the most important targets of stress in plants. Indeed, most of the metabolic responses induced by stress conditions have consequences on the aptitude of the plant to maintain an efficient light energy conversion (Rmiki et al., 1999). Alteration in the chlorophylls (total, a and b) and carotenoids contents have been reported in plants subjected to stress conditions (Ferrat et al., 2003). In general stressed plants tend to increase their carotenoid content to provide protection against the formation of free oxygen radicals. A decrease in total chlorophyll and in the ratio chlorophyll/carotenoids are often observed (Ferrat et al., 2003). These variations in photosynthetic pigments under exposure to trace metals and organic xenobiotics such as herbicides have been observed for various species (Ferrat et al., 2003).

Biochemical alterations are also induced by the presence of organic xenobiotics which lead to a production of reactive oxygen species (ROS). These chemical species are partially reduced forms of atmospheric oxygen (O2). They typically result from the excitation of the triplet O2 to form singlet oxygen (O21) or from the transfer of one, two or three electrons to O2 to form, respectively, superoxide radicals (•O2-), hydrogen peroxide (H2O2) or O23- (which dismutates into water and hydroxyl radicals, •OH) (Wojtaszek, 1997; Mittler, 2002; Apel and Hirt, 2004; Smirnoff, 2005). These are usually produced by plants as by-products of various metabolic pathways (such as photosynthesis and respiration) localized in different cellular compartments (predominantly in chloroplasts, mitochondria and peroxisomes) (Apel and Hirt, 2004).

Under physiological steady state conditions these molecules are scavenged by different antioxidative defense components that are often confined to particular compartments (Apel and Hirt, 2004). However, their overproduction can be triggered by external stress factors such as xenobiotics exposure. When exposed to xenobiotics, plants activate pathways to metabolize these foreign compounds which produce large amounts of ROS and may perturb the normal equilibrium between production and scavenging of ROS. As a result of these disturbances, intracellular levels of ROS may rapidly rise, thus posing a threat to the cell viability (Mittler, 2002; Masella et al., 2005).

The rapid and transient production of high quantities of ROS consequently results in what is called "oxidative burst" (Wojtaszek, 1997; Apel and Hirt, 2004). ROS, unlike the atmospheric oxygen, are capable of unrestricted oxidation of various cellular components and, if not controlled, have the ability to damage biomolecules (e.g. membrane lipid peroxidation, protein oxidation, enzyme inhibition, and DNA and RNA damage) and that can ultimately lead to the oxidative destruction of the cell (Mittler, 2002; Masella et al., 2005).

For these reasons, the ROS levels within the cells must be strictly controlled and kept within a narrow range. Plants cells have developed mechanisms to monitor and scavenge excessive amounts of ROS. Plants tolerance to pollutants is related to their capacity to cope with ROS over production.

Despite these problems, a steady-state of ROS is required within the cells because they also act as a signal for the activation of stress response and defense pathways. Thus, ROS can be viewed as cellular indicators of stress, and as secondary messengers involved in the stress-response signal transduction pathway. As such, the measurement of antioxidant enzymes activities has been used frequently for assessing environmental stress induced in plants by various pollutants (Mittler, 2002; Apel and Hirt, 2004).

Two different mechanisms are involved in ROS control: one that will enable the fine modulation of low levels of ROS for signaling purposes, and one that will enable the detoxification of excess ROS, especially during stress. Mechanisms of ROS detoxification exist in all plants and can be categorized as non-enzymatic (e.g. by flavanones, anthocyanins, carotenoids and ascorbic acid (AsA)) or as enzymatic. Major enzymatic ROS scavengers in plants include superoxide dismutase (SOD), ascorbate peroxidase (APX), glutathione peroxidase (GPX), and catalase (CAT) whose pathways are represented in Figure 5 (Mittler, 2002; Apel and Hirt, 2004; Geoffroy et al., 2004; Ashraf, 2009). The enzymes involved are present in different cell compartments and their expression is genetically controlled and regulated both by developmental and environmental stimuli, according to the needs for removing the ROS produced in the cells (De Gara et al., 2003).

SOD acts as the first line of defense against ROS, dismutating superoxide (O2-) to H2O2 (Figure 5a). APX, GPX, and CAT subsequently detoxify H2O2.

CAT, present in the peroxisomes of nearly all aerobic cells, can protect the cell from H2O2 by catalysing its decomposition into O2 and H2O (Mittler, 2002; Apel and Hirt, 2004) (Figure 5b). In contrast to CAT, APX requires an ascorbate and glutathione (GSH) regeneration system, the ascorbate-glutathione cycle (Figure 5d).

Detoxifiying H2O2 to H2O by APX occurs by oxidation of ascorbate to monodehydroascorbate (MDA) (Equation 1 in Figure 5d), which can be regenerated by MDA reductase (MDAR) using NAD(P)H as reducing agent (Equation 2 in Figure 5d). MDA can spontaneously dismutate into ascorbate and dehydroascorbate (DHA). Ascorbate regeneration is mediated by dehydroascorbate reductase (DHAR) driven by the oxidation of GSH to oxidized glutathione (GSSG) (Equation 3 in Figure 5d).

Finally, glutathione reductase (GR) can regenerate GSH from GSSG using NAD(P)H as a reducing agent (Equation 4 in Figure 5d). Like APX, GPX also detoxifies H2O2 to H2O, but uses GSH directly as a reducing agent (Equation 1 in Figure 5c). The GPX cycle is closed by regeneration of GSH from GSSG by GR using NAD(P)H (Equation 2 in Figure 5c) (Wojtaszek, 1997; Mittler, 2002).

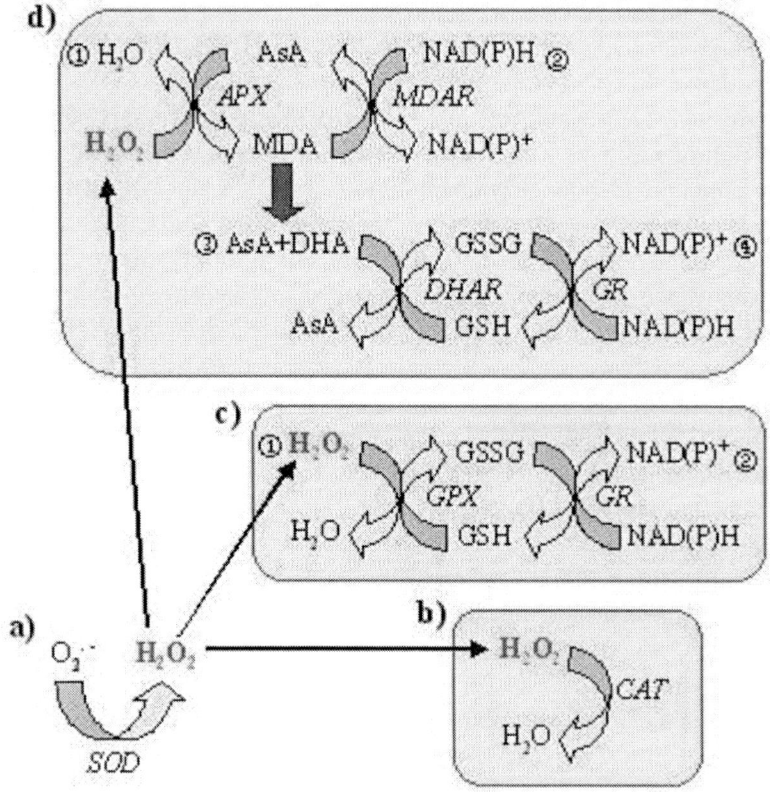

Figure 5. The main cellular pathways for ROS scavenging in plants. (a) Superoxide dismutase (SOD). (b) Catalase (CAT). (c) The glutathione peroxidase cycle. (d) The ascorbate-glutathione cycle. Abbreviations: AsA, ascorbate; APX, ascorbate peroxidase; MDA, monodehydroascorbate; MDAR, monodehydroascorbate reductase; DHA, dehydroascorbate; DHAR, dehydroascorbate reductase; GSH, glutathione; GSSG, oxidized glutathione; GPX, glutathione peroxidase; GR, glutathione reductase. (Adapted from Mittler (2002)).

Unlike most organisms, plants have multiple genes encoding SOD and APX. Different isoforms are specifically targeted to chloroplasts, mitochondria, peroxisomes, as well as to the cytosol and apoplast (Table 3). Whereas GPX is located in cytosol, CAT is located mainly in peroxisomes (Table 3).

The extent of oxidative stress in a cell is determined by the concentrations of superoxide, hydrogen peroxide, and hydroxyl radicals. Therefore, the balance of SOD, APX, and CAT activities will be crucial for suppressing toxic ROS levels in a cell. Changing the balance of scavenging enzymes will induce compensatory mechanisms (Mittler, 2002; Apel and Hirt, 2004; Smirnoff, 2005).

In Table 3 a summary is presented of the several enzymes and reactions involved in the enzymatic ROS control and detoxification processes.

Table 3. Detoxifying enzymes and respective ROS scavenging reactions (Wojtaszek, 1997; Mittler, 2002; Blokhina et al., 2003; Ashraf, 2009)

Enzyme	EC number	Location	Reaction catalyzed
Superoxide dismutase	1.15.1.1	Chol, Cyt, Mit, Per, Apo	O2 .- + O2 .- + 2H+ → 2 H2O2 + O2
Catalase	1.11.1.6	Per	2 H2O2 → O2 + 2H2O
Ascorbate peroxidase	1.11.1.11	Chol, Cyt, Mit, Per, Apo	2Asa + H2O2 → 2MDA + 2H2O (2MDA → AsA + DHA)
Guaiacol peroxidase	1.11.1.7	Cyt	Donor + H2O2 ↔ oxidized donor + 2H2O
Glutathione peroxidase	1.11.1.12	Cyt	2GSH + H2O2 ↔GSSG + 2H2O
Glutathione S-transferase	2.5.1.18	Cyt, Mit	RX+GSH ↔ HX + R-S-GSH*
MDA reductase	1.6.5.4	Chl, Mit, Per, Cyt	NADPH + MDA ↔ NAD(P)+ + AsA
DHA reductase	1.8.5.1	Chl, Mit, Per	2GSH + DHA ↔ GSSG + AsA
Glutathione reductase	1.6.4.2	Chl, Cyt, Mit	2NADPH + GSSG ↔ 2NADP+ + 2GSH

*Abbreviations: Apo, apoplast; Chl, chloroplast; Cyt, cytosol; Mit, mitochondria; Per, peroxisome; * R may be an aliphatic, aromatic or herocyclic group; X may be a sulfate, nitrite or halite group.

In the processes of detoxification and response against oxidative stress, phenolic compounds, which are ubiquitous plant secondary metabolites, may have an important role both in enzymatic as well as non-enzymatic mechanisms. Many phenolic compounds are used as substrates for antioxidant enzymes (and for the guaiacol peroxidase in particular) (Smirnoff, 2005). In addition, phenolic compounds have antioxidant properties on their own because their chemical structure enables them to quench radicals, which makes them also important players in the non-enzymatic control of the oxidative burst (Smirnoff, 2005).

The alteration of the levels of phenolic compounds is, therefore, a quantitative parameter which is complimentary to the evaluation of antioxidant enzymes activities to assess the oxidative stress induced by the plants exposure to xenobiotics.

The toxic effects caused in plants (including Typha spp.) by specific types of xenobiotic organic compounds (especially pesticides, but including other classes of compounds as well) have been extensively studied (Langan and Hoagland, 1996; Wilson et al., 2000; Amaya-Chavez et al., 2006; Olette et al., 2008). However, once again, oxidative stress induced by pharmaceuticals is still poorly studied and characterized, thus contributing to the general scarcity of data available for adequate design of phytoremediation solutions for pharmaceuticals contamination (Pomati et al., 2004; Boxall et al., 2006; Kong et al., 2007; Dordio et al., 2009b).

CONCLUSION

Pollution caused by POPs is a matter of considerable concern as many of these substances have significant ecotoxicity and thus may negatively impact the environment and even present risks to human health. However, as awareness of the problem has arisen during the last decades, the realization that the conventional wastewater treatment processes were inefficient to deal with these types of pollutants and the need to also remediate sites that were already significantly contaminated has led to a search for new processes and technologies that could adequately reduce the presence or release of most of these substances to the environment. Some of the more efficient or traditionally used technologies that have been available have high costs which limit their application in most situations. As a low cost alternative, plants and associated rhizosphere microorganisms have over time been employed to treat wastewaters as well as remediate contaminated sites, in a variety of phytotechnologies where the targets to be cleaned up and the mechanisms in effect during the treatment may differ substantially.

Phytotechnologies have several advantages among which are the aesthetically pleasant installations and the usually good public acceptance, but the typically long periods required to achieve the remediation goals and the limitation to low concentrations of the pollutants due to the possible toxicity to the plants may in some cases deter its applicability. Despite some of its limitations, phytotechnologies have nevertheless been used with considerable success in the removal of varied types of POPs, both from contaminated soils and water. Several different types of mechanisms may be involved to different extents in the removal processes, some of which may be of the abiotic type but the most important ones being possibly of a biotic nature, due either from direct action of plants or from degradation by microorganisms stimulated by the plants.

Among the several types of phytotechnologies, constructed wetlands systems have been gaining an increasing popularity. The concerted action of plant species adapted to water saturated environments, microorganisms characteristic of these systems, and minerals of the support matrix, has shown good capabilities to deal with POPs through a variety of physical, chemical and biological processes. These systems are engineered in order to optimize these pollutant removal processes within a controlled environment. In order to achieve higher efficiencies, a good understanding of the mechanisms involved in the removal processes and the role played by each component are necessary so that components can be conveniently selected and design and operating conditions can be optimized. Some studies have already been conducted with this focus on understanding how CWS work and can be optimized. However, there is still a substantial amount of work that can and needs to be carried out in this area.

A subset of POPs which is part of a group of so-called emergent pollutants and has been raising special concern in the latest years is that of pharmaceutically active compounds. The need to treat this special class of compounds which were designed to provoke a biochemical effect and thus may cause particularly negative (and unpredicted) impacts has led to the study of CWS applications for the removal of this type of POP. This recent hot topic in wastewater treatment research shows the possibilities that are still open to the development of phytotechnologies, and applications of CWS in particular, for low cost solutions of emergent environmental problems. Studies conducted so far have shown the good perspectives that are presented for the use of these technologies. However, investigation of the mechanisms involved in the removal of these substances from the contaminated media, once again, may prove essential to attain the level of maturity of a wider application of these systems on a larger scale with reasonable results.

REFERENCES

Abhilash, P. C.; Jamil, S.; Singh, N. *Biotechnol. Adv.* 2009, 27, 474-488.

Aga, D. S. *Fate of Pharmaceuticals in the Environment and in Water Treatment Systems*; CRC Press: Boca Raton, FL, 2008.

Aitchison, E. W.; Kelley, S. L.; Alvarez, P. J. J.; Schnoor, *J. L. Water Environ. Res.* 2000, 72, 313-321.

Aken, B. V.; Correa, P. A.; Schnoor, *J. L. Environ. Sci. Technol.* 2009.

Alkorta, I.; Garbisu, C. *Bioresour. Technol.* 2001, 79, 273-276.

Amaya-Chavez, A.; Martinez-Tabche, L.; Lopez-Lopez, E.; Galar-Martinez, M. Chemosphere 2006, 63, 1124-1129.

Amon, J. P.; Agrawal, A.; Shelley, M. L.; Opperman, B. C.; Enright, M. P.; Clemmer, N. D.; Slusser, T.; Lach, J.; Sobolewski, T.; Gruner, W.; Entingh, *A. C. Ecol. Eng.* 2007, 30, 51-66.

Andreozzi, R.; Raffaele, M.; Nicklas, P. *Chemosphere* 2003, 50, 1319-1330.

Apel, K.; Hirt, H. *Annu. Rev. Plant Biol.* 2004, 55, 373-399.

Arnold, W. A.; McNeill, K. In *Analysis, fate and removal of pharmaceuticals in the water cycle*; Petrovic, M.; Barceló, D.; Eds.; Elsevier: Amsterdam, Netherlands, 2007; pp 361-385.

Ashraf, M. *Biotechnol. Adv.* 2009, 27, 84-93.

Ashton, D.; Hilton, M.; Thomas, K. V. *Sci. Total Environ.* 2004, 333, 167-184.

Bailey, H. C.; Elphick, J. R.; Krassoi, R.; Mulhall, A. M.; Lovell, A. J.; Slee, D. *J. Environ. Toxicol. Chem.* 2005, 24, 1773-1778.

Ballschmiter, K. Angew. *Chem. Int. Edit.* 1992, 31, 487-515.

Barceló, D.; Petrovic, M. *Emerging contaminants from industrial and municipal waste.* Ocurrence, Analysis and Effects; Springer-Verlag: Berlin, Germany, 2008.

Bendz, D.; Paxéus, N. A.; Ginn, T. R.; Loge, F. J. *J. Hazard. Mater.* 2005, 122, 195-204.

Benner, J.; Salhi, E.; Ternes, T.; von Gunten, U. *Water Res.* 2008, 42, 3003-3012.

Benotti, M. J.; Brownawell, B. J. *Environ. Sci. Technol.* 2007, 41, 5795-5802.

Benotti, M. J.; Trenholm, R. A.; Vanderford, B. J.; Holady, J. C.; Stanford, B. D.; Snyder, S. A. *Environ. Sci. Technol.* 2009, 43, 597-603.

Blankenberg, A. G.; Braskerud, B.; Haarstad, K. Int. J. *Environ. Anal. Chem.* 2006, 86, 225-231.

Blokhina, O.; Virolainen, E.; Fagerstedt, K. V. *Ann. Bot.* 2003, 91, 179-194.

Bolong, N.; Ismail, A. F.; Salim, M. R.; Matsuura, T. *Desalination* 2009, 239, 229-246.

Bouldin, J. L.; Farris, J. L.; Moore, M. T.; Smith, J.; Cooper, C. M. *Chemosphere* 2006, 65, 1049-1057.

Bourg, A. C.; Mouvet, C.; Lerner, D. N. Q. J. *Eng. Geol. Hydrogeol.* 1992, 25, 359-370.

Boxall, A. B. A.; Johnson, P.; Smith, E. J.; Sinclair, C. J.; Stutt, E.; Levy, L. S. *J. Agric. Food Chem.* 2006, 54, 2288-2297.

Braskerud, B. C.; Haarstad, K. *Water Sci. Technol.* 2003, 48, 267-274.

Briggs, G. G.; Bromilow, R. H.; Evans, A. A.; Williams, M. *Pestic. Sci.* 1983, 14, 492-500.

Brix, H. *Water Sci. Technol.* 1994, 29, 71-78.

Brix, H. *Water Sci. Technol.* 1997, 35, 11-17.

Burken, J. G.; Schnoor, J. L. *Environ. Sci. Technol.* 1998, 32, 3379-3385.

Carballa, M.; Omil, F.; Lema, J. M. *Chemosphere* 2008, 72, 1118-1123.
Chard, B. K.; Doucette, W. J.; Chard, J. K.; Bugbee, B.; Gorder, K. *Environ. Sci. Technol.* 2006, 40, 4788-4793.
Chaudhry, Q.; Schröder, P.; Werck-Reichhart, D.; Grajek, W.; Marecik, R. *Environ. Sci. Pollut. Res.* 2002, 9, 4-17.
Coleman, J. O. D.; BlakeKalff, M. M. A.; Davies, T. G. E. *Trends Plant Sci.* 1997, 2, 144-151.
Coleman, J.; Frova, C.; Schröder, P.; Tissut, M. *Environ. Sci. Pollut. Res.* 2002, 9, 18-28.
Collins, C.; Fryer, M.; Grosso, A. Environ. *Sci. Technol.* 2006, 40, 45-52.
Comeau, F.; Surette, C.; Brun, G. L.; Losier, R. *Sci. Total Environ.* 2008, 396, 132-146.
Conkle, J. L.; White, J. R.; Metcalfe, C. D. *Chemosphere* 2008, 73, 1741-1748.
Cooper, E. R.; Siewicki, T. C.; Phillips, K. *Sci. Total Environ.* 2008, 398, 26-33.
Cooper, P. F.; Job, G. D.; Green, M. B.; Shutes, R. B. E. Reed Beds and Constructed Wetlands for Wastewater Treatment; WRc Publications: Medmenham, Marlow, UK, 1996.
Corcoran, J.; Winter, M. J.; Tyler, *C. R. Crit. Rev. Toxicol.* 2010, 40, 287-304.
Crane, M.; Watts, C.; Boucard, *T. Sci. Total Environ.* 2006, 367, 23-41.
Cunningham, V. L.; Binks, S. P.; Olson, M. *J. Regul. Toxicol. Pharmacol.* 2009, 53, 39-45.
Dabrowski, J. M.; Bennett, E. R.; Bollen, A.; Schulz, R. *Chemosphere* 2006, 62, 204-212.
Daughton, C. G.; Ternes, T. A. *Environ. Health Perspect.* 1999, 107, 907-938.
Davies, L. C.; Pedro, I. S.; Ferreira, R. A.; Freire, F. G.; Novais, J. M.; Martins-Dias, S. *Water Sci. Technol.* 2008, 58, 2017-2023.
de Carvalho, R. F.; Bromilow, R. H.; Greenwood, R. *Pest Manag. Sci.* 2007, 63, 798-802.
De Gara, L.; de Pinto, M. C.; Tommasi, F. Plant Physiol. *Biochem.* 2003, 41, 863-870.
Dietz, A. C.; Schnoor, *J. L. Environ. Health Perspect.* 2001, 109, 163-168.
Doble, M.; Kumar, A. In *Biotreatment of Industrial Effluents*; Butterworth-Heinemann: Burlington, 2005; pp 241-253.
Dordio, A.; Pinto, J.; Dias, C. B.; Pinto, A. P.; Carvalho, A. J. P.; Teixeira, D. M. Int. *J. Environ. Anal. Chem.* 2009a, 89, 835-848.
Dordio, A.; Carvalho, A. J. P.; Teixeira, D. M.; Dias, C. B.; Pinto, A. P. *Bioresour. Technol.* 2010, 101, 886-892.
Dordio, A.; Carvalho, A. J. P.; Pinto, A. P.; Costa, C. T. In *Pesticide Research Trends;* Tennefy, A. B.; Ed.; Nova Science Publishers: Hauppauge, NY, USA, 2008.
Dordio, A. V.; Duarte, C.; Barreiros, M.; Carvalho, A. J. P.; Pinto, A. P.; da Costa, C. T. *Bioresour. Technol.* 2009b, 100, 1156-1161.
Dordio, A. V.; Teimão, J.; Ramalho, I.; Carvalho, A. J. P.; Candeias, A. *J. E. Sci. Total Environ.* 2007, 380, 237-246.
Dosnon-Olette, R.; Couderchet, M.; Eullaffroy, P. *Ecotox. Environ. Safe.* 2009, 72, 2096-2101.
Doucette, W. J.; Chard, J. K.; Moore, B. J.; Staudt, W. J.; Headley, J. V. *Microchem. J.* 2005, 81, 41-49.
Eapen, S.; Singh, S.; D'Souza, S. F. *Biotechnol. Adv.* 2007, 25, 442-451.
El Shahawi, M. S.; Hamza, A.; Bashammakh, A. S.; Al Saggaf, W. T. *Talanta* 2010, 80, 1587-1597.
Emblidge, J. P.; DeLorenzo, *M. E. Environ. Res.* 2006, 100, 216-226.
Enick, O. V.; Moore, M. M. *Environ. Impact Assess. Rev.* 2007, 27, 707-729.

Esplugas, S.; Bila, D. M.; Krause, L. G.; Dezotti, M. J. *Hazard. Mater*. 2007, 149, 631-642.

Etnier, E. L. *Regul. Toxicol. Pharmacol*. 1989, 9, 147-157.

Farré, M. l.; Pérez, S.; Kantiani, L.; Barceló, D. *Trac-Trends Anal. Chem*. 2008, 27, 991-1007.

Fent, K. In Pharmaceuticals *in the environment: sources, fate, effects and risks*; Kümmerer, K.; Ed.; Springer-Verlag: Berlin, Germany, 2008; pp 175-203.

Fent, K.; Weston, A. A.; Caminada, D. *Aquat. Toxicol*. 2006, 76, 122-159.

Ferrat, L.; Pergent-Martini, C.; Romeo, M. *Aquat. Toxicol*. 2003, 65, 187-204.

Fountoulakis, M. S.; Terzakis, S.; Kalogerakis, N.; Manios, T. Ecol. Eng. 2009, 35, 1702-1709.

Gan, S.; Lau, E. V.; Ng, H. K. *J. Hazard. Mater*. 2009, 172, 532-549.

Garrison, A. W.; Pope, J. D.; Allen, F. R. In *Identification and Analysis of Organic Pollutants in Water*; Keith, L. H.; Ed.; Ann Arbor Science Publishers: Ann Arbor, MI, USA, 1976; pp 517-556.

Gavrilescu, M. *Eng. Life Sci*. 2005, 5, 497-526.

Geoffroy, L.; Frankart, U.; Eullaffroy, P. *Environ. Pollut*. 2004, 131, 233-241.

Gerecke, A. C.; Scharer, M.; Singer, H. P.; Muller, S. R.; Schwarzenbach, R. P.; Sagesser, M.; Ochsenbein, U.; Popow, G. *Chemosphere* 2002, 48, 307-315.

Gessner, T. P.; Kadlec, R. H.; Reaves, R. P. *Ecol. Eng*. 2005, 25, 457-469.

Gomez, M. J.; Martinez Bueno, M. J.; Lacorte, S.; Fernandez-Alba, A. R.; Aguera, A. *Chemosphere* 2007, 66, 993-1002.

Görge, E.; Brandt, S.; Werner, D. *Environ. Sci. Pollut. Res*. 1994, 1, 229-233.

Groom, C. A.; Halasz, A.; Paquet, L.; Olivier, L.; Dubois, C.; Hawari, *J. Environ. Sci. Technol*. 2002, 36, 112-118.

Gross, B.; Montgomery-Brown, J.; Naumann, A.; Reinhard, M. *Environ. Toxicol. Chem*. 2004, 23, 2074-2083.

Grove, J. K.; Stein, O. R. *Water Res*. 2005, 39, 4040-4050.

Guil, M. D. H.; Petrovic, M.; Radjenovic, J.; Fernández-Alba, A. R.; Fernández-Alba, A. R.; Barceló, D. In *Analysis, fate and removal of pharmaceuticals in the water cycle*; Petrovic, M.; Barceló, D.; Eds.; Elsevier: Amsterdam, Netherlands, 2007; pp 451-474.

Haberl, R.; Grego, S.; Langergraber, G.; Kadlec, R. H.; Cicalini, A. R.; Martins-Dias, S.; Novais, J. M.; Aubert, S.; Gerth, A.; Thomas, H.; Hebner, A. *J. Soil Sediment*. 2003, 3, 109-124.

Halling-Sørensen, B.; Nors Nielsen, S.; Lanzky, P. F.; Ingerslev, F.; Holten Lützhøft, H. C.; Jørgensen, S. E. *Chemosphere* 1998, 36, 357-393.

Hannink, N. K.; Rosser, S. J.; Bruce, N. C. *Crit. Rev. Plant Sci*. 2002, 21, 511-538.

Hao, C. Y.; Lissemore, L.; Nguyen, B.; Kleywegt, S.; Yang, P.; Solomon, K. *Anal. Bioanal. Chem*. 2006, 384, 505-513.

Hao, C.; Clement, R.; Yang, P. *Anal. Bioanal. Chem*. 2007, 387, 1247-1257.

Haritash, A. K.; Kaushik, C. P. *J. Hazard. Mater*. 2009, 169, 1-15.

Heberer, T. Toxicol. *Lett*. 2002, 131, 5-17.

Henderson, K. L.; Belden, J. B.; Coats, J. R. *Environ. Sci. Technol*. 2007, 41, 4084-4089.

Hernando, M. D.; Heath, E.; Petrovic, M.; Barceló, D. *Anal. Bioanal. Chem*. 2006a, 385, 985-991.

Hernando, M. D.; Mezcua, M.; Fernández-Alba, A. R.; Barceló, D. *Talanta* 2006b, 69, 334-342.

Hignite, C.; Azarnoff, D. L. *Life Sci.* 1977, 20, 337-341.
Hsu, F. C.; Marxmiller, R. L.; Yang, A. Y. S. *Plant Physiol.* 1990, 93, 1573-1578.
Huber, C.; Bartha, B.; Harpaintner, R.; Schroder, P. Environ. *Sci. Pollut. Res.* 2009, 16, 206-213.
Hughes, J. B.; Shanks, J.; Vanderford, M.; Lauritzen, J.; Bhadra, R. *Environ. Sci. Technol.* 1996, 31, 266-271.
Hummel, D.; Loffler, D.; Fink, G.; Ternes, T. A. *Environ. Sci. Technol.* 2006, 40, 7321-7328.
Ikehata, K.; Naghashkar, N. J.; El Din, M. G. *Ozone-Sci. Eng.* 2006, 28, 353-414.
Imfeld, G.; Braeckevelt, M.; Kuschk, P.; Richnow, H. H. *Chemosphere* 2009, 74, 349-362.
ITRC. *Phytoremediation Decision Tree;* Interstate Technology and Regulatory Cooperation Work Group - Phytoremediation Work Team: Washington, DC, USA, 1999.
Jjemba, P. K. *Ecotox. Environ. Safe.* 2006, 63, 113-130.
Jones, K. C.; de Voogt, P. *Environ. Pollut.* 1999, 100, 209-221.
Jones, O. A. H.; Voulvoulis, N.; Lester, J. N. *Environ. Technol.* 2001, 22, 1383-1394.
Jones, O. A. H.; Voulvoulis, N.; Lester, J. N. *Crit. Rev. Environ. Sci. Technol.* 2005, 35, 401-427.
Joss, A.; Keller, E.; Alder, A. C.; Göbel, A.; McArdell, C. S.; Ternes, T.; Siegrist, H. *Water Res.* 2005, 39, 3139-3152.
Kadlec, R. H.; Wallace, S. D. *Treatment Wetlands*; CRC Press: Boca Raton, FL, USA, 2009.
Kasprzyk-Hordern, B.; Dinsdale, R. M.; Guwy, A. *J. Water Res.* 2008, 42, 3498-3518.
Kasprzyk-Hordern, B.; Dinsdale, R. M.; Guwy, A. *J. Water Res.* 2009, 43, 363-380.
Khetan, S. K.; Collins, T. *J. Chem. Rev.* 2007, 107, 2319-2364.
Kim, S. D.; Cho, J.; Kim, I. S.; Vanderford, B. J.; Snyder, S. A. *Water Res.* 2007, 41, 1013-1021.
Knight, R. L.; Kadlec, R. H.; Ohlendorf, H. M. *Environ. Sci. Technol.* 1999, 33, 973-980.
Komives, T.; Gullner, G. Z. *Naturforsch.* (C) 2005, 60, 179-185.
Kong, W. D.; Zhu, Y. G.; Liang, Y. C.; Zhang, J.; Smith, F. A.; Yang, A. *Environ. Pollut.* 2007, 147, 187-193.
Korte, F.; Kvesitadze, G.; Ugrekhelidze, D.; Gordeziani, M.; Khatisashvili, G.; Buadze, O.; Zaalishvili, G.; Coulston, F. *Ecotox. Environ. Safe.* 2000, 47, 1-26.
Kot-Wasik, A.; Debska, J.; Namiesnik, *J. Trac-Trends Anal. Chem.* 2007, 26, 557-568.
Koutsouba, V.; Heberer, T.; Fuhrmann, B.; Schmidt-Baumler, K.; Tsipi, D.; Hiskia, A. *Chemosphere* 2003, 51, 69-75.
Kümmerer, K. *Chemosphere* 2001, 45, 957-969.
Kümmerer, K. *Pharmaceuticals in the environment: sources, fate, effects and risks*; Springer-Verlag: Berlin, Germany, 2008.
Kümmerer, K. *J. Environ. Manage.* 2009, 90, 2354-2366.
Kuster, M.; López de Alda, M. J.; Hernando, M. D.; Petrovic, M.; Martín-Alonso, J.; Barceló, D. *J. Hydrol.* 2008, 358, 112-123.
Kvesitadze, G.; Khatisashvili, G.; Sadunishvili, T.; Ramsden, J. J. In *Biochemical Mechanisms of Detoxification in Higher Plants;* Kvesitadze, G.; Khatisashvili, G.; Sadunishvili, T.; Ramsden, J. J.; Eds.; Springer: Berlin, Germany, 2006; pp 103-165.
Langan, M. M.; Hoagland, K. D. *Bull. Environ. Contam. Toxicol.* 1996, 57, 307-314.
Larsen, T. A.; Lienert, J.; Joss, A.; Siegrist, H. J. *Biotechnol.* 2004, 113, 295-304.
Leclercq, M.; Mathieu, O.; Gomez, E.; Casellas, C.; Fenet, H.; Hillaire-Buys, D. *Arch. Environ. Contam. Toxicol.* 2009, 56, 408-415.

Lishman, L.; Smyth, S. A.; Sarafin, K.; Kleywegt, S.; Toito, J.; Peart, T.; Lee, B.; Servos, M.; Beland, M.; Seto, P. *Sci. Total Environ.* 2006, 367, 544-558.

Lofrano, G.; Nikolaou, A.; Kostopoulou, M.; Pagano, G.; Belgiorno, V.; Napoli, R. M. A. In *Xenobiotics in the Urban Water Cycle: Mass Flows, Environmental Processes, Mitigation and Treatment Strategies;* Fatta-Kassinos, D.; Bester, K.; Kümmerer, K.; Eds.; Springer: Dordrecht, Netherlands, 2010; pp 129-145.

Low, D.; Tan, K.; Anderson, T.; Cobb, G. P.; Liu, J.; Jackson, W. A. Ecol. Eng. 2008, 32, 72-80.

Macek, T.; Mackova, M.; Kas, *J. Biotechnol. Adv.* 2000, 18, 23-34.

Masella, R.; Di Benedetto, R.; Vari, R.; Filesi, C.; Giovannini, *C. J. Nutr. Biochem.* 2005, 16, 577-586.

Matamoros, V.; Arias, C.; Brix, H.; Bayona, *J. M. Environ. Sci. Technol.* 2007a, 41, 8171-8177.

Matamoros, V.; Garcia, J.; Bayona, *J. M. Environ. Sci. Technol.* 2005, 39, 5449-5454.

Matamoros, V.; Arias, C.; Brix, H.; Bayona, *J. M. Water Res.* 2009a, 43, 55-62.

Matamoros, V.; Bayona, *J. M. Environ. Sci. Technol.* 2006, 40, 5811-5816.

Matamoros, V.; Caselles-Osorio, A.; García, J.; Bayona, J. M. Sci. *Total Environ.* 2008a, 394, 171-176.

Matamoros, V.; García, J.; Bayona, J. M. *Water Res.* 2008b, 42, 653-660.

Matamoros, V.; Hijosa, M.; Bayona, J. M. *Chemosphere* 2009b, 75, 200-205.

Matamoros, V.; Puigagut, J.; García, J.; Bayona, J. M. *Chemosphere* 2007b, 69, 1374-1380.

Maurer, M.; Escher, B. I.; Richle, P.; Schaffner, C.; Alder, A. C. *Water Res.* 2007, 41, 1614-1622.

Miège, C.; Choubert, J. M.; Ribeiro, L.; Eusèbe, M.; Coquery, M. *Environ. Pollut.* 2009, 157, 1721-1726.

Mishra, S.; Srivastava, S.; Tripathi, R. D.; Kumar, R.; Seth, C. S.; Gupta, D. K. *Chemosphere* 2006, 65, 1027-1039.

Mittler, R. *Trends Plant Sci.* 2002, 7, 405-410.

Moore, M. T.; Bennett, E. R.; Cooper, C. M.; Smith, S.; Shields, F. D.; Milam, C. D.; Farris, J. L. *Agric. Ecosyst. Environ.* 2001, 87, 309-314.

Moore, M. T.; Cooper, C. M.; Smith, S.; Cullum, R. F.; Knight, S. S.; Locke, M. A.; Bennett, E. R. *Environ. Pollut.* 2009, 157, 250-256.

Moore, M. T.; Rodgers, J. H.; Cooper, C. M.; Smith, S. Environ. Pollut. 2000, 110, 393-399.

Moore, M. T.; Schulz, R.; Cooper, C. M.; Smith, S.; Rodgers, J. H. *Chemosphere* 2002, 46, 827-835.

Neumann, M.; Schulz, R.; Schafer, K.; Muller, W.; Mannheller, W.; Liess, M. *Water Res.* 2002, 36, 835-842.

Newman, L. A.; Reynolds, C. M. *Curr. Opin. Biotechnol.* 2004, 15, 225-230.

Nikolaou, A.; Meric, S.; Fatta, D. *Anal. Bioanal. Chem.* 2007, 387, 1225-1234.

Olette, R.; Couderchet, M.; Biagianti, S.; Eullaffroy, P. *Chemosphere* 2008, 70, 1414-1421.

Omari, K.; Revitt, M.; Shutes, B.; Garelick, H. *Water Sci. Technol.* 2003, 48, 275-281.

Osenbrük, K.; Gläser, H. R.; Knöller, K.; Weise, S. M.; Möder, M.; Wennrich, R.; Schirmer, M.; Reinstorf, F.; Busch, W.; Strauch, G. *Water Res.* 2007, 41, 3259-3270.

Palmer, P. M.; Wilson, L. R.; O'Keefe, P.; Sheridan, R.; King, T.; Chen, C. Y. *Sci. Total Environ.* 2008, 394, 90-102.

Paraiba, L. C. *Chemosphere* 2007, 66, 1468-1475.
Park, N.; Vanderford, B. J.; Snyder, S. A.; Sarp, S.; Kim, S. D.; Cho, *J. Ecol. Eng.* 2009, 35, 418-423.
Parrish, Z. D.; Banks, M. K.; Schwab, A. P. *Int. J. Phytoremediat.* 2004, 6, 119-137.
Perelo, L. W. *J. Hazard. Mater.* 2010, 177, 81-89.
Petrovic, M.; Barceló, D. *Analysis, fate and removal of pharmaceuticals in the water cycle*; Elsevier: Amsterdam, Netherlands, 2007.
Pilon-Smits, E. *Annu. Rev. Plant Biol.* 2005, 56, 15-39.
Pomati, F.; Netting, A. G.; Calamari, D.; Neilan, B. A. *Aquat. Toxicol.* 2004, 67, 387-396.
Putschew, A.; Jekel, M. In *Analysis, fate and removal of pharmaceuticals in the water cycle;* Petrovic, M.; Barceló, D.; Eds.; Elsevier: Amsterdam, Netherlands, 2007; pp 265-278.
Rabiet, M.; Togola, A.; Brissaud, F.; Seidel, J. L.; Budzinski, H.; Elbaz-Poulichet, *F. Environ. Sci. Technol.* 2006, 40, 5282-5288.
Radjenovic, J.; Petrovic, M.; Barceló, D. *Trac-Trends Anal. Chem.* 2007, 26, 1132-1144.
Ralf, S.; Matt, T. M.; Erin, R. B.; Jerry, L. F.; Sammie, S., Jr.; Charles, M. C. *Environ. Toxicol. Chem.* 2003, 22, 1262-1268.
Reddy, K. R.; DeLaune, R. D. *Biogeochemistry of wetlands: science and applications*; CRC Press: Boca Raton, FL, USA, 2008.
Reichenberger, S.; Bach, M.; Skitschak, A.; Frede, H. G. *Sci. Total Environ.* 2007.
Renner, *R. Environ. Sci. Technol.* 2002, 36, 410A-413A.
Richardson, M. L.; Bowron, J. M. J. Pharm. *Pharmacol.* 1985, 37, 1-12.
Rmiki, N.-E.; Lemoine, Y.; Schoefs, B. In *Handbook of plant and crop stress*; Pessarakli, M.; Ed.; CRC Press, 1999; pp 465-482.
Robinson, I.; Junqua, G.; Van Coillie, R.; Thomas, O. Anal. Bioanal. *Chem.* 2007, 387, 1143-1151.
Rogers, I. H.; Birtwell, I. K.; Kruzynski, G. M. *Water Poll. Res. J. Can.* 1986, 21, 187-204.
Rose, M. T.; Sanchez-Bayo, F.; Crossan, A. N.; Kennedy, I. R. *Chemosphere* 2006, 63, 1849-1858.
Rubin, E.; Ramaswami, A. *Water Res.* 2001, 35, 1348-1353.
Salt, D. E.; Smith, R. D.; Raskin, I. Annu. Rev. Plant Physiol. *Plant Mol. Biol.* 1998, 49, 643-668.
Sandermann, H. *Trends Biochem. Sci.* 1992, 17, 82-84.
Sandermann, H. *Pharmacogenet.* 1994, 4, 225-241.
Sanderson, H.; Johnson, D. J.; Reitsma, T.; Brain, R. A.; Wilson, C. J.; Solomon, K. R. Regul. *Toxicol. Pharmacol.* 2004, 39, 158-183.
Scholz, M.; Lee, B. h. *Int. J. Environ. Stud.* 2005, 62, 421-447.
Schroder, P.; Collins, C. *Int. J. Phytoremediat.* 2002, 4, 247-265.
Schulz, R. *J. Environ. Qual.* 2004, 33, 419-448.
Schulz, R.; Peall, S. *K. C. Environ. Sci. Technol.* 2001, 35, 422-426.
Sicbaldi, F.; Sacchi, G. A.; Trevisan, M.; Delre, A. A. M. *Pestic. Sci.* 1997, 50, 111-119.
Singer, A. C.; Crowley, D. E.; Thompson, I. P. *Trends Biotechnol.* 2003, 21, 123-130.
Singh, O. V.; Jain, R. K. *Appl. Microbiol. Biotechnol.* 2003, 63, 128-135.
Smirnoff, N. *Antioxidants and reactive oxygen species in plants*; Blackwell Publishing Ltd: Oxford, UK, 2005.

Snyder, S. A.; Adham, S.; Redding, A. M.; Cannon, F. S.; DeCarolis, J.; Oppenheimer, J.; Wert, E. C.; Yoon, Y. Desalination 2007, 202, 156-181.
Stackelberg, P. E.; Furlong, E. T.; Meyer, M. T.; Zaugg, S. D.; Henderson, A. K.; Reissman, D. B. *Sci. Total Environ*. 2004, 329, 99-113.
Stottmeister, U.; Wiessner, A.; Kuschk, P.; Kappelmeyer, U.; Kastner, M.; Bederski, O.; Muller, R. A.; Moormann, *H. Biotechnol. Adv*. 2003, 22, 93-117.
Stumpf, M.; Ternes, T. A.; Wilken, R. D.; Silvana, V. R.; Baumann, *W. Sci. Total Environ*. 1999, 225, 135-141.
Sun, T. R.; Cang, L.; Wang, Q. Y.; Zhou, D. M.; Cheng, J. M.; Xu, *H. J. Hazard. Mater*. 2010, 176, 919-925.
Sundaravadivel, M.; Vigneswaran, S. *Crit. Rev. Environ. Sci. Technol*. 2001, 31, 351-409.
Susarla, S.; Medina, V. F.; McCutcheon, *S. C. Ecol. Eng*. 2002, 18, 647-658.
Tang, X.; Eke, P. E.; Scholz, M.; Huang, S. *Bioresour. Technol*. 2009, 100, 227-234.
Tauxe-Wuersch, A.; De Alencastro, L. F.; Grandjean, D.; Tarradellas, *J. Water Res*. 2005, 39, 1761-1772.
Ternes, T. A. *Water Res*. 1998, 32, 3245-3260.
Ternes, T. A.; Bonerz, M.; Herrmann, N.; Teiser, B.; Andersen, H. R. *Chemosphere* 2007, 66, 894-904.
USEPA. *Introduction to phytoremediation*; EPA/600/R-99/107; Office of Research and Development: Cincinnati, OH, USA, 2000.
USEPA. *Phytoremediation field studies database for chlorinated solvents, pesticides, explosives, and metals;* Office of Superfund Remediation and Technology Innovation: Washington, DC, USA, 2004.
USEPA; USDA-NRCS. *A Handbook of Constructed Wetlands*. Volume 1: General Considerations; USEPA Region III with USDA-NRCS: Washington, DC, USA, 1995.
Van Aken, B. *Curr. Opin. Biotechnol*. 2009, 20, 231-236.
Vanderford, M.; Shanks, J. V.; Hughes, *J. B. Biotechnol. Lett*. 1997, 19, 277-280.
Vangronsveld, J.; Herzig, R.; Weyens, N.; Boulet, J.; Adriaensen, K.; Ruttens, A.; Thewys, T.; Vassilev, A.; Meers, E.; Nehnevajova, E.; van der, L. D.; Mench, M. Environ Sci. *Pollut. Res. Int*. 2009, 16, 765-794.
Velagaleti, *R. Drug Inf. J*. 1997, 31, 715-722.
Vieno, N.; Tuhkanen, T.; Kronberg, *L. Water Res*. 2007, 41, 1001-1012.
Vymazal, *J. Ecol. Eng*. 2009, 35, 1-17.
Vymazal, J.; Brix, H.; Cooper, P. F.; Green, M. B.; Haberl, R. *Constructed wetlands for wastewater treatment in Europe*; Backhuys Publishers: Leiden, The Netherlands, 1998.
Waggott, A. In *Chemistry in water reuse*; Cooper, W. J.; Ed.; Ann Arbor Science Publishers: Ann Arbor, MI, USA, 1981; pp 55-99.
Watts, C. D.; Crathorne, M.; Fielding, M.; Steel, C. P. In *Analysis of organic micropollutants in water*; Angeletti, G.; Bjørseth, A.; Eds.; D. Reidel Pub. Co.: Dordrecht, Netherlands, 1983; pp 120-131.
Weigel, S.; Kallenborn, R.; Hühnerfuss, *H. J. Chromatogr*. A 2004, 1023, 183-195.
Weishaar, J. A.; Tsao, D.; Burken, J. G. *Int. J. Phytoremediat*. 2009, 11, 509-523.
Williams, *J. B. Crit. Rev. Plant Sci*. 2002, 21, 607-635.
Wilson, P. C.; Whitwell, T.; Klaine, *S. J. Arch. Environ. Contam. Toxicol*. 2000, 39, 282-288.
Wojtaszek, P. *Biochem*. J. 1997, 322, 681-692.

Xia, H.; Ma, X. *Bioresour. Technol.* 2006, 97, 1050-1054.
Zhang, Y.; Geißen, S. U.; Gal, C. *Chemosphere* 2008, 73, 1151-1161.
Zhang, Z. L.; Zhou, *J. L. J. Chromatogr.* A 2007, 1154, 205-213.
Zodrow, *J. J. Remediat.* J. 1999, 9, 29-36.
Zorita, S.; Mårtensson, L.; Mathiasson, *L. Sci. Total Environ.* 2009, 407, 2760-2770.
Zounkova, R.; Odraska, P.; Dolezalova, L.; Hilscherova, K.; Marsalek, B.; Blaha, *L. Environ. Toxicol. Chem.* 2007, 26, 2208-2214.
Zuccato, E.; Castiglioni, S.; Fanelli, *R. J. Hazard. Mater.* 2005, 122, 205-209.

In: Handbook of Phytoremediation
Editor: Ivan A. Golubev

ISBN: 978-1-61728-753-4

Chapter 3

PHYTOREMEDIATION OF URANIUM CONTAMINATED SOILS

Mirjana D. Stojanović ***and Jelena V. Milojković***
Institute for Technology of Nuclear and Other Mineral Raw Materials,
Belgrade, Serbia

ABSTRACT

Environmental uranium contamination based on human activity is a serious problem worldwide. Soil contaminated with uranium poses a long-term radiation hazard to human health through exposure via the food-chain and other pathways. This chapter is an overview of processes and modern techniques for remediation of soils contaminated with uranium, with special attention on phytoremediation. Phytoremediation takes advantage of plant to extract, sequester pollutants in soil, water, and air with an aim of pollutant removal and transformation into harmless forms.

The objective of this chapter is to develop a better understanding of plants behavior and the degree of affinity towards the adoption of uranium for hyperaccumulators plants based on review of international research. To understand the mechanism of uranium uptake in plants and accumulation, a necessary prerequisite is the application of radiophytoremediation on the "real" scale. For this purpose, we investigated these processes using three different aspects with selected cultivated plants:

1. Vegetative tests in pots of fully controlled conditions, with corn plants that were grown on two types of soil, pseudogley and chernozem, together with its phytotoxic effect on the plant development, height, yield, and seed germination.
2. Greenhouse experiments with tailings from the closed uranium mine Kalna on the southeast of Serbia. Three series of experiments were conducted in plastic-house. First, three plant species (corn, sunflower, and green peas) were grown in pots on the four substrate variants, tailings in mixture with sand. The substrate was irrigated with drinking water and "uranium water", which issues out from the mine. Another experiment was conducted in order to investigate the uptake of U in several kinds of

[*] Institute for Technology of Nuclear and Other Mineral Raw Materials, 86 Franchet d'Esperey St, Belgrade 11 000, Serbia, Tel. +381 11 3691 722, Fax. +381 11 3691 583, E-mail address: m.stojanovic@itnms.ac.rs

roots - crops, bulbous, and tuberous plants: carrot, onion, potatoes, radish, red beet, and sugar beet. Content of uranium was found in leaves and roots (surface root layer and edible parts were peeled). Also we investigated uranium adoption in four genotypes of corn, sunflower, and soy bean.

3. Vegetation test on real, native conditions on tailings, from the closed uranium mine Kalna. The experiment was carried out on the elementary plots one square meter in size, with bean, cabbage, lettuce, corn, onion, potatoes, spinach, and sunflower.

Well-organized use of phytotechnology means an integrated management strategy for contaminated sites which include proper selection of plants (uranium hyperaccumulators), improving mobility of uranium with amendments (organic agents), and application sequestering agents for immobilization and transformation of excess uranium, which the plants didn't accept.

Keywords: phytoremediation, uranium, contaminated soils, hyperaccumulators plants, soil amendments.

INTRODUCTION:

1. URANIUM IN THE ENVIRONMENT - CHARACTERISTICS, SOURCES, CONSEQUENCES

Sources of uranium in the environment originate from natural geological –geochemical processes and human (anthropogenic) activities. Natural sources include excessive weathering of mineral and metal ions from rocks, displacement of certain contaminants from groundwater or subsurface layers of soil, atmospheric deposition from volcanic activity, and transport of continental dusts (McIntyre, 2003).

Widespread use of nuclear energy, application of weapons with depleted uranium, nuclear testing, coal combustion, oil and gas production, production and application of phosphoric fertilizer, mineral processing and formation radioactive waste landfill, improper waste storage practices, and uranium tailings are the main anthropogenic sources of uranium entering the environment. All these human activities resulted in soil contamination with uranium, ie. "Technologically-Enhanced Naturally Occurring Radioactive Material." – TENORM (NRC, 1999).

Use of phosphoric fertilizers are the main anthropogenic source of the uranium input in the environment (about 73% of the total input uranium). On the basis of the U concentration in phosphate fertilizers, McBride and Spiers (2001) estimated that 50 years of the application of a specific phosphate fertilizer (e.g., 100 kg ha−1 year−1 as P2O5) would lead to the addition of 2.4 kg of U per hectare to the topsoil, corresponding to an increase of about 1 mg kg−1 in the soil . Around 1,500 t of mineral fertilizers based on phosphorus are applied per annum in Serbia. It is estimated that around 210 kg of uranium (30 g/ha) are in this way introduced into the environment. (Tunney, et. al., 2009.; Stojanović et. al., 2006). Centuries of mining and milling of uranium and other elements have resulted in the generation of significant quantities of radioactive waste materials, considered a menace to public health and environmental quality. Generally, a mine capable of producing 100,000 tonnes of uranium ore annually will simultaneously produce 100,000–600,000 t of waste tailings (Gavrilescu et al., 2009). As a consequence, there may be a risk for ecosystems, agro-systems, and health.

As a consequence of the past war activities, a large area (in Kosovo and some locations in Serbia and two wars in Iraq) was contaminated by depleted uranium (DU) and toxic heavy metals. Depleted uranium (DU) ammunition was used on a relatively limited scale during NATO strikes on Serbia and Kosovo in 1999. According to available data, the bombing of 112 sites in Kosovo and Metohija, and 12 locations in southern Serbia introduced about 10 tonnes of DU into the environment (Zunić et al., 2008; Rajković and Đorđević, 2006). Anthropogenic sources of uranium entering in the environment are presented on Figure 1.

To compare this with natural levels, it can be recalled that 1 kg of soil typically contains a few micrograms of uranium (UNSCEAR, 2000). Considering that, the task for all of us is the minimizing of dangerous effects of depleted uranium and keeping it from penetrating the nutrition chain. Otherwise, this invisible threat will take effect endlessly with all of its dangerous consequences. There is an urgent need for remediation of this contamination in order to prevent its possible long-term effects, not only on the population in the contaminated regions, but also on the neighboring countries. Therefore, it is necessary that together with permanent monitoring of environmental contamination, selection of cost effective remediation technology appropriate for large areas such as contaminated water and soil is used. Conventional remediation techniques such as excavation, treatment (soil washing, chelating), conditioning, and disposal of low-level radioactive waste are necessary for heavily contaminated sites. However, for a large area of contaminated soil and aquifer sediments, in situ remediation is appealing since it is much less disruptive to the ecosystem and hydrology, reduces the risk of worker exposure during remediation, and is typically less expensive than conventional technologies. In situ remediation involves minimizing the mobility of contaminants by transferring them to stable, non-labile phases via chemically induced transformation (Igwe et al., 2005; AbdEl-Sabour, 2007). Phytoremediation, as a form of remediation technology, is used in respect to plants to partly or substantially remediate selected contaminants in contaminated soil, sludge, sediment, ground water, surface water, and waste water.

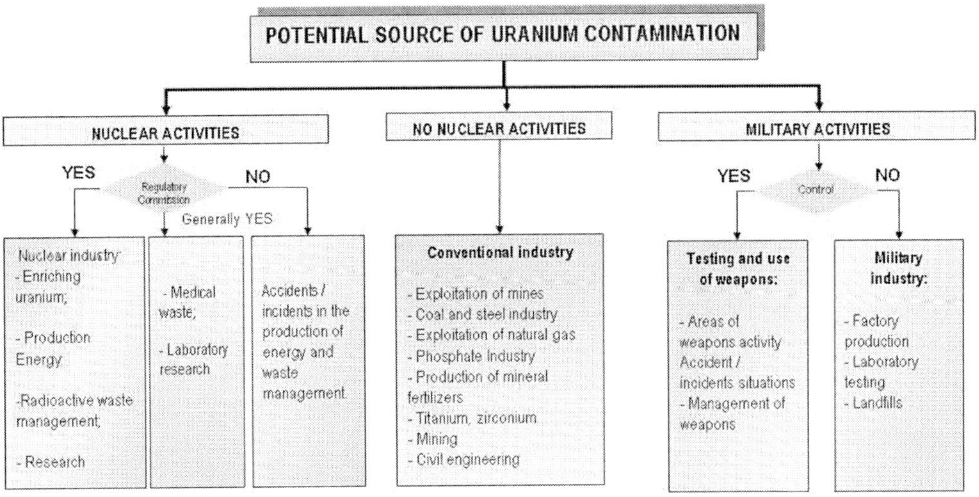

Figure 1. Potential anthropogenic source of uranium entering in the environment.

It utilizes a variety of plant biological processes and the physical characteristics of plants to aid in the site remediation. Phytoremediation is a continuum of processes, with the different processes occurring in differing degrees for the different conditions, media, contaminants, and plants. There are different forms of phytoremediation: phytoextraction, phytostabilization, phytotransformation, phytodegradation, phytostimulation rhizodegradation, phytovolatilization, and rhizofiltration (Grubišić, et al. 2006).

1.1. Characteristics and Occurrence of Uranium in the Environment

Natural uranium is a mixture of three types (or isotopes) of uranium, written as U-234, U-235, and U-238, 99.27% is U-238. The element undergoes radioactive decay, leading to a long series of 13 different radionuclides before finally reaching a stable state as Pb-206. These radionuclides emit alpha or beta radiation and some also emit gamma radiation of widely varying energies.

Uranium is most abundant among the naturally occurring actinides. Its concentration in the earth's crust may range from 1 to 4 mg kg-1 in sedimentary rocks, to ten or even hundreds of milligrams per kilogram in phosphate-rich deposits and uranium-ore deposits. Uranium is a natural chemotoxic and radiotoxic heavy metal (Stojanovic, 2006).

Compared with other cations, uranium is classified as fairly mobile in oxidation conditions over the entire range of pH, and immobile in reduction conditions. The potential risk of uranium soil contamination is a global problem. Depleted, enriched, and natural uranium contamination in soil and water has been identified at many sites worldwide, so that measures for preventing assimilation by plants should be considered a preliminary step towards the remediation of contaminated areas (Gavrilescu, 2009).

Generally, the majority of radionuclides released into the environment finally accumulate in either the upper layer of soils or in the interstitial system of sediments in aquatic systems. As a consequence, there may be a risk for ecosystems, agro-systems, and health.

Uranium can be found in soil as sorbed (both on soil particles and pore water), complexed, precipitated, and reduced forms, all of which have various impacts on mobility and fate in the soil environment (Gavrilescu et al., 2009). Uranium speciation is closely related to soil properties (especially pH). It is most mobile as uranium (VI), which predominantly exists as UO2+ and as soluble carbonate complexes in solution.

Between pH 4.0 and 7.5, the pH range of most soils, uranium (VI) primarily exists in hydrolysed forms and is readily taken up by plants from the exchangeable and soluble fractions of the soil, while negligible amounts of uranium(VI) can remain in soluble and exchangeable forms for a significant amount of time, thereby limiting the amount available for plant uptake.

Uranium can be retained bound or immobilised in soil, or can be mobilised by different mechanisms. This provides the basis for certain remediation technologies, their combination determining the mobility and fate of uranium. Uranium is retained by soil in three ways: by adsorption onto the surface of mineral particles, by complex formation with humus in organic particles, and by precipitation reactions.

The mobility of uranium in soil is mainly controlled by complex formation and redox reactions; complex formation leads to mobile species or precipitation of U-bearing minerals. Redox reactions change the solubility between the two major oxidation states, U(IV)–U(VI).

The reduction of U(VI) to U(IV) immobilizes uranium, whereas the oxidation of U(IV) to U(VI) mobilizes uranium and there is dissolution of U(IV)compounds.

Uranium speciation is closely related to soil properties, being dependent on several factors such as pH, redox potential, temperature, soil texture, amount of organic and inorganic compounds, moisture, and microbial activity. The dependency of the speciation distribution on pH and carbon dioxide concentration in a closed system is shown in soluble forms can migrate with soil water, be taken up by plants or aquatic organisms, or can be volatilised.

1.2. Global Cycle of Uranium in Nature

Knowledge of the global cycle of uranium is not intended to only determine the level of contamination and to recognize consequences, but to achieve the acquisition of knowledge with which we can safely predict all factors that affect its fasten transport and thus to develop models of environmental protection (Figure 2).

Distribution of uranium in the lithosphere and hydrosphere is performed in conditions of complex chemical and physical-chemical natural processes, including mechanisms of degradation of minerals that contain uranium.

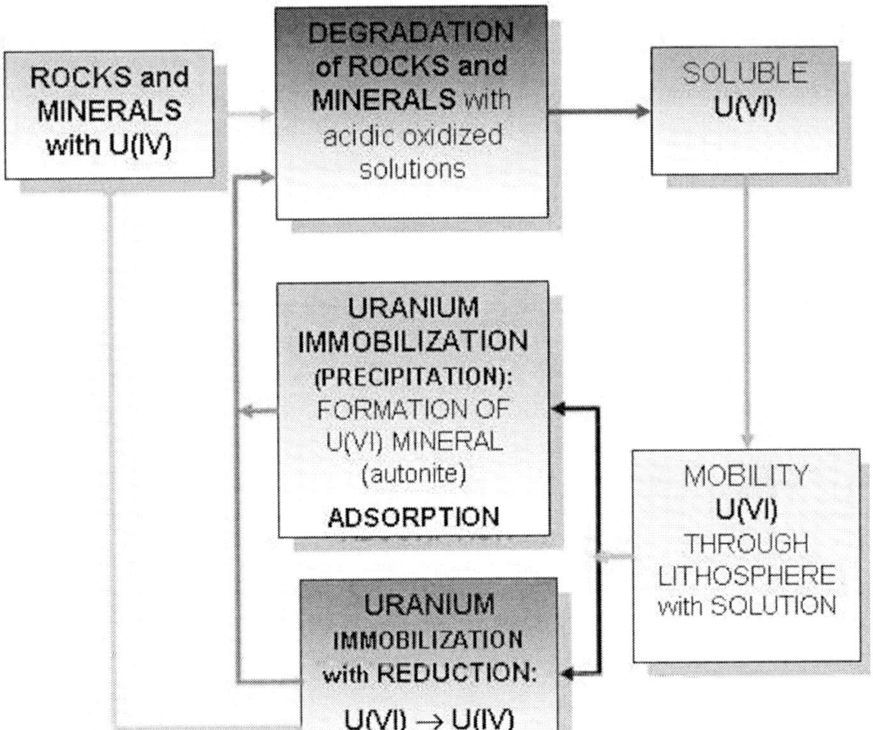

Figure 2.Global cycle of uranium in nature.

Solubility of uranium in the soil primarily depends on the environmental pH, redox potential, and the material and mineral composition of the solid phase, concentration of inorganic compounds, the quantity and type of organic compounds in soil and soil solution, soil temperature, pressure, moisture content, and microbial activities.

In distribution of uranium in the system of the soil - water is mainly in soluble or suspended form, diffusions or mass transfer.

Processes that remove uranium from the soil solution were precipitation, coprecipitation, adsorption, microbial reduction, and the embedding of biological systems.

The dominant factor that affects the release of uranium from petrogenic and accessory minerals granitoids are acidic hydrothermal solutions. They are a medium that, in addition to leaching uranium, affect the degradation (alteration) granitite rocks in which uranium is located. Hydrothermal solutions are a mixture of magmatic, warm ground and meteor, cold surface water, rich in oxygen, which promotes the oxidation of U(IV) to U(VI). Uranium(VI) is soluble in the form uranyl ion, UO22 +, which complexes can be easily transported hydrothermal solutions.

The cycle of mobilization of uranium in nature begins with U(IV)oxidation as long as the complexes of uranium in water-stable phase flow process of expansion of uranium through nature. The process of contamination of natural uranium stops when uranium is reduced or fixed. However, with changing conditions in nature, uranium can be fixed to restart uranium, and so the cycle will start again. Processes of uranium precipitation with reduction is of great importance for nature and man; it excludes uranium from water flows and thus suspends its process of spreading, and contamination of the environment.

Fixation of uranium as precipitation from solution is the only way for nature to protect from the spread of uranium and its radioactive products. Fixation of uranium can be described by two main mechanisms: precipitation (including oxidoreduction) and adsorption (Stojanović, 2006).

1.2.1. Distribution of Uranium in Function of Chemical and Physical-Chemical Processes

Uranium may be present in soil as precipitated, sorbed, complexed, and reduced forms, hich impact its mobility, and fate in the subsurface soil environment. (Zhou and Gu 2005).

The major U species that exists under an oxidative environment is divalent uranyl ion (UO22+). This positively charged UO22+ is adsorbed on the negatively charged sites of soil components, and these sites increase with soil pH. The U sorption capacity of soil, therefore, increases with soil pH. However, when carbonate concentration increased with increases in pH, U became mobile in soil because of the formation of a soluble and negatively charged carbonate-U complex. UO22+ is also sorbed on Fe and Al sesquioxides. Under a reductive environment, the major U species is insoluble UO2. In addition to soil composition, the types and concentrations of coexisting ions, soil pH, and redox conditions control the mobility of U in soil. This implies that agricultural practices have significant effects on the mobility of U in soil. In the case of the use of acidic soil for agricultural purposes in Japan, pH reclamation with liming is necessary. Rice is usually cultivated under submerged soil conditions, and the paddy water is drained prior to harvest; as a result, the paddy field soil undergoes alternating changes between reduced and oxidized conditions. These unique agricultural practices in Japan affect the fate of soil U added due to the intensive use of phosphate fertilizers (Yamaguchi et al., 2009).

In nature, uranium is oxidized due to the entry of oxygen, with an increase of its fugacity, which comes from surface water through cracks. These solutions are also slightly acidic, because they contain CO_2, which with water forms carbonic acid (H_2CO_3). Affinity for oxygen is such, that the first sulfide will be oxidized to sulfate, quadricvalet (quadrovalent) uranium to hexavalent, and only at a higher redox potential always present in the water, Fe^{2+} ion to Fe^{3+} ions. Those changes are usually presented with Eh–pH diagram for system U–C–O–H (Langmuir, 1978).

The reduction of U(VI) to U(IV) by abiotic and biotic processes, as well as its re-oxidation has received considerable attention because the oxidation state of uranium has a significant effect on its mobility in the natural environment. Uranium exists in solution predominantly as UO_2^{2+} and as soluble carbonate complexes $(UO_2)_2CO_3(OH)_3^-$, $UO_2CO_3^{\circ}$, $UO_2(CO_3)_2^{2-}$, $UO_2(CO_3)_3^{4-}$, and possibly $(UO_2)_3(CO_3)_6^{6-}$ (Duff and Amrhein, 1996).

Between pH 4.0 and 7.5, the pH range of most soils, U(VI) exists primarily in hydrolyzed forms. Uranium (VI), i.e., uranyl, uranium will exist in the +6 oxidation state under oxidizing to mildly reducing environments. Uranium (IV) is stable under reducing conditions and is considered relatively immobile because U(IV) forms sparingly soluble minerals, such as uraninite (UO_2) (Gavrilescu et al., 2009).

The assessments of the uranium solubility and speciation (nature and concentration species) are predicted from thermodynamic data, taking into account the presence of inorganic ligands in the groundwaters studied, mainly $[OH]^-$, $[HCO_3]^-$, $[CO_3]^{2-}$, $[H_2PO_4]^-$ $[HPO_4]^{2-}$, $[PO_4]^{3-}$, $[SO_4]^{2-}$ (in case of disposal in rock-salt formation), and the properties of these waters (redox potential) (Dozol and Hagemann, 1993).

Numerous investigations of the adsorption of uranium on soils and minerals have shown that carbonate complexing appreciably reduces adsorption of uranium, leading to its release from soils (Pabalan et al., 1998)

In addition to dissolved carbonate, uranium can also form stable complexes with other naturally occurring inorganic and organic ligands such as phosphate complexes [$UO_2HPO_4^0$ (aq) and $UO_2PO_4^-$]. Complexes with sulfate, fluoride, and possibly chloride are potentially important uranyl species where concentrations of these anions are high. However, their stability is considerably less than the carbonate and phosphate complexes (Grenthe et al., 1992.).

2. Uranium Removal and Soil Remediation

The objective of any remedial action is to reduce the risks to human health, the environment, and property to acceptable levels by removing or reducing the source of contamination or by preventing exposure to it. Once the decision has been made that remedial action is necessary, there are various options possible for achieving the objective. The ambient activity of radionuclides is one of the most important criteria that determines the need for remedial action.

Various strategies have been proposed for the remediation of contaminated environments in order to reduce the detrimental effects of uranium on ecosystems and local communities. These strategies include physical, chemical, and biological technologies. Chemical-physical technologies can enable efficient decontamination of the uranium-polluted soil and groundwater. Bioremediation utilizes characteristics of plants or microorganisms to remove or immobilize uranium in soils or waters. The ability of some microorganisms to reduce uranium(VI) to uranium(IV) could be used to remediate the contaminated environment, but microbes that can do this more efficiently should be isolated so that this technique can be more widely applied. The metabolic activity of bacteria, algae, fungi, and plants, which can modify pH, promote extra-cellular binding, transformation, and formation of complexes or precipitates, can affect uranium speciation, and thus uranium mobility. Phytoremediation has proved to be one of the best alternatives for the remediation of uranium-contaminated soils, or the ecological restoration of areas contaminated by uranium mine tailings. Further research is still necessary to find uranium hyperaccumulators that allow more efficient phytoextraction, and to find tolerant plant species for phytostabilisation.

An assessment of the potential efficacy of available technologies prior to their application is needed. This requires a knowledge of the contaminant distribution, soil characteristics, and adhesion / absorption characteristics of contaminants on soil particles; an evaluation of the physical, chemical, and biological processes with the potential to remediate radioactive contaminated soils; ranking of the available technologies based on experience and the ease of implementation; an evaluation of technologies from an engineering prospective to determine the potential for scale-up as well as the cost effectiveness; the identification of secondary waste treatment requirements for full-scale implementation; and the identification of difficulties and the additional research needed before limitations in technology can be overcome.

Clearly, there are both advantages and disadvantages with any remediation technology, and each technology may be applicable in certain circumstances only, determined from data gathered during the phase of site characterization. Such data is used to determine the initial need for site remediation, plans for further remediation, and implementation of remedial actions as well as to ensure that there is compliance regarding the residual concentrations of radionuclides in the environment post-remediation.

2.1. Methods and Techniques for Uranium Removal

Radionuclides and heavy metals are retained by soil in three ways:

- Adsorption onto the surface of mineral particles
- Complexation by humic substances in organic particles
- Precipitation reaction

The mobility of uranium in soil is mainly controlled by complexation and redox reactions:

- Complexation leads to mobile species or precipitation of U bearing minerals
- Redox reactions change the solubility between the two major oxidation states: U(IV)–U(VI):

- Reduction of U(VI) to U(IV) immobilizes uranium
- Oxidation of U(IV) to U(VI) mobilizes uranium because of the dissolution of U(IV)bearing minerals

Remediation technologies available for treating uranium contaminated soils and groundwater could be applied as either ex situ or in situ techniques (Suthersan and Payne, 2005; AbdEl-Sabour 2007; Charbonneau, 2009).

According to Gaverilescu et al, (2009) we could classify methods and techniques for uranium removal as: natural attenuation, physical processes, chemical methods, biological methods, and electrokinetic methods. These processes and techniques for uranium removal are presented on Figure 3.

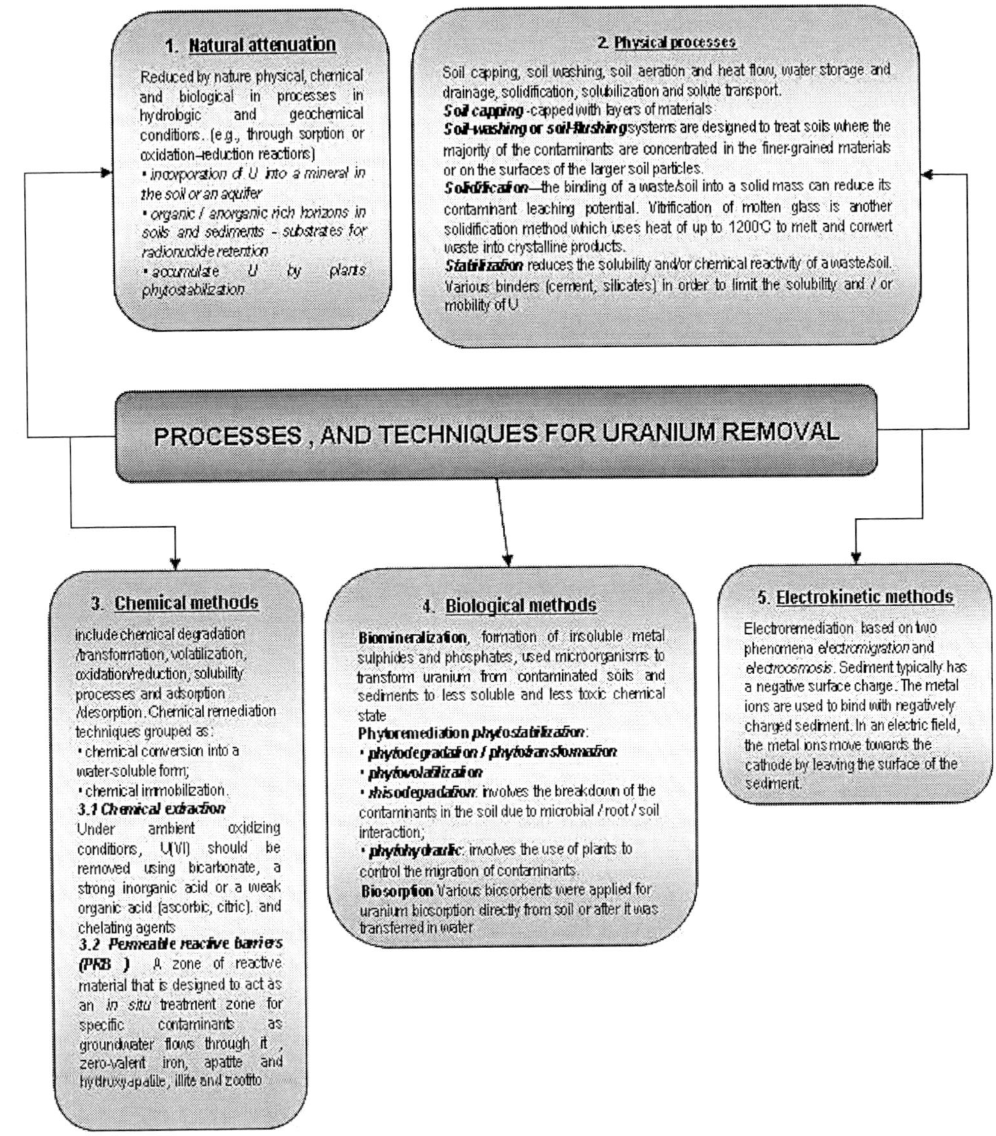

Figure 3. Processes and techniques for uranium removal.

Each one of the above fundamental technical choices will direct decision makers to substantially different paths with regard to their subsequent choices, actions, and potential results, making significantly different technological options for application available within a remediation program, which involves multidisciplinary environmental research on characterization, monitoring, modelling, and technologies for remediation (Gaverilescu et al, 2009). Any measurable remediation objective have to consider several factors, which could induce an impact on the decision making process , like basic evaluation criteria that include engineering and non-engineering reasons for ensuring the achievability of the "cheaper, smarter, and cleaner" soil remediation philosophy, such as (Gaverilescu et al, 2009):

- Cleanup goals
- Form and concentration of pollutants
- Volume and physical/chemical properties of the polluted soils
- Remediation effectiveness
- Designated use of the cleaned site
- Cost associated with the remediation program
- Occupational safety and health risks associated with the technology
- Potential secondary environmental impacts (collateral damage)
- Prior experience with the application of the technology
- Sustainability of any necessary institutional control
- Socio - economic considerations

3. Phytoremediation

USEPA has defined phytoremediation *as the use of plants for containment, degradation. or extraction of xenobiotics from water or soil substrates* (USEPA, 2000), or as "The use of vegetation to contain, sequester, remove, or degrade inorganic and organic contaminants in soils, sediments, surface waters, and groundwater."(Tsao, 2003).

Phytoremediation involves the use of plants to extract, sequester, and/or detoxify the pollutants present in soil, water, and air. For long-time projects and adequate pollutants, phytoremediation is considered the cheaper and simpler option available for soil cleanup (Fellet et al., 2007; Susarla et al., 2002). This technique takes advantage of the natural abilities of plants to take up (absorb) and accumulate metals and radionuclides (McIntyre, 2003). These plants could be used in an efficient way if they are adapted to a wide range of environmental conditions. Plants for phytoremediation are tolerant plants, having heavy metal hyper accumulation potential, which could be beneficial in phytoremediation for cleanup of soil and water. On the other hand, tolerant food crops, if exposed to heavy metals in their growth medium, may be dangerous as carriers of toxic metals in the food chain leading to food toxicity (Gavrilescu et al., 2009).

Plants for phytoremediation of U-contaminated soils could be selected by using a mathematical model related to plant characteristics (e.g. biomass and planting density) to predict a long-term U-removal rate from the soil (Hashimoto et al., 2005).

Plant-assisted remediation of soil can generally occur through one or more of the following mechanisms (Dushenkov et al., 1999; Gavrilescu et al., 2009):

- *Phytostabilization*: involves the use of plants to contain or immobilize contaminants in the soil by:
  - Absorption and accumulation by roots
  - Adsorption onto root surface

- Precipitation within the root zone
- *Phytodegradation/phytotransformation*: involves the breakdown of contaminants through:
  - Metabolic processes(internally)
  - Release of enzymes into the soil
- *Phytovolatilization*: the uptake and transpiration into the atmosphere of a contaminant by a plant
- *Rhisodegradation*: involves the breakdown of the contaminants in the soil due to microbial/root/soil interaction
- *Phytohydraulics*: involves the use of plants to control the migration of contaminants

Radionuclide bioavailability mostly depends on (Dushenkov et al., 1999):

- Type of radionuclide deposition
- Time of deposition
- Soil characteristics

Phytoextraction aims at removing the toxic trace elements from the soil through uptake by plants or by volatilisation. Phytoextraction seems to be the most applied phytoremediation procedure. Benefit of this technique is that pollutants are actually removed from soil. (Duquène et al., 2009; Gavrilescu et al., 2009; Diaz and Kirkham, 2007) Phytoextraction efficiency could be enhanced by methods such as genetic engineering (Dhankher et al., 2002), microbial activities (de Souza et al., 1999), fertilizers (Bennett et al. 1998), and an addition of amendments. Taking into consideration the effects of various soil amendments on uranium desorption from soil to soil solutions, there are a number of reports in literature about physiological characteristics of uranium uptake and accumulation in plants, and techniques on how to trigger uranium hyperaccumulation in plants (Mkandawire et al., 2005; Vandenhove et al., 2001). Those investigations specify that soil organic matter sequestered uranium, rendering it largely unavailable for plant uptake (Finneran et al., 2002). Moreover, the uranyl (UO22+) cation is the chemical species of U most readily accumulated in plant shoots (Ebbs et al., 1998).

3.1. Uranium Transfer Factors and Effect of Uranium Content on Plants

The transfer of radionuclides from soils to plants is dependent on three classes of factors (Gavrilescu et al., 2009):

- Quantity factor (that is the total amount of potentially available elements)
- Intensity factor (the activity, the ionic ratios of elements in the soil solution, presence of other species (nitrogen, phosphorous)
- Reaction kinetics (the rate of element transfer from solid to liquid phases and to plant roots).

The rate of phytoextraction of inorganic contaminants depends on the net soil-to-plant transfer rate. Radioecologists have long measured concentration ratios and transfer factors

(TFs). TFs were developed primarily as part of "empirical" environmental models at the dawn of the nuclear age more than 50 years ago. Soil-to-plant TFs for radionuclides taken either for a single species on many soils, many species on a single soil, or many species on many soils, have been shown empirically by radioecologists to be very variable, lognormally distributed, time dependent, and concentration dependent. Uranium is radionuclide with low Soil –to Plant Transfer (Willey, 2007).

Linear relationships between total radionuclide concentration in the hydroponic solution and total amount of the radionuclide in the plant roots is distinguished (Shtangeeva and Ayrault, 2004; Rodríguez et al., 2006). Results obtained from growing plants in a hydroponics medium do not reflect real situations existing in a field. Soil and liquid media are absolutely different systems and mechanisms of metal uptake by plants growing in nutrient solutions and in soils may be rather different. Each plant and soil combination may have a unique curvilinear relationship. Detailed descriptions of site-specific soils must be created to screen plants for radionuclide extraction capability (Shtangeeva, 2008).

Shahandeh and Hosssner, (2002) evaluated the influence of specific soil fractions on U bioavailability from contaminated soils. Two of the plant species, Sunflower and Indian mustard, were selected to compare the degree of U removal from different soil types utilizing different sources, forms, and rates of U. Effects of soil type were examined with one U mine tailing soil and eight cultivated soils: four acid soils and four calcareous soils contaminated with different rates (100 to 600 mg U(VI) kg^{-1}) as uranyl nitrate. There was a direct relationship between the rate of soil contamination and U accumulation in shoots or roots. Uranium concentration in shoots or roots of sunflower varied with soil type, regardless of soil U (VI) contamination rate. According to the uranium concentration in shoots and roots, they concluded that sunflower plants grown on calcareous soils can accumulate more uranium than one grown on acid soils. This may be because of forming uranium carbonate complexes. Calcareous soils containing free carbonate and uranyl ions are complexed with the carbonate radical, forming highly mobile, anionic complexes. Uranium solubility and mobility were probably limited in some acid (clay) soils due to the presence of highly adsorptive Fe and Mn oxides.

Efficiency of uranium extraction decreased sharply from hydroponics to sandy loam and organic - rich soil, indicating that soil organic matter sequestered uranium, rendering it largely unavailable for plant uptake (Ramaswami et al., 2001).

Effects of different concentrations of uranium tailings conditioned with garden soil on growth and biochemical parameters in sunflowers showed the necessity of additions of garden soil before re-vegetation. The conditioning can improve the quality of uranium tailings and provide a better environment by alteration in nutritional status (Jagetiya and Purohit, 2006).

The influence of U on plant growth could be measured in terms of a Tolerance Index (TI) and Grade of Growth Inhibition (GGI).

Tolerance Index = [(Mean biomass of plant species with uranium treatment)/ (Mean biomass of control plant species)] × 100.

Tolerance Index was used (Baker et al., 1994) for evaluation of heavy metal uptake, accumulation, and tolerance for a number of metallophyte species.

The Grade of Growth Inhibition (GGI) represents the effects of uranium tailing concentration on dry mass (Leita et al., 1993).

Grade of Growth Inhibition = [(Dry mass of control plants − Dry mass of uranium treated plants)/ (Dry mass of control plants)] × 100.

In numeros publications there is contradictory information on the toxicity of soil uranium in plants. Canon (1952), Morishima (1976), Sheppard et al. (1992), Jagetiya and Purohit (2005), and Stojanović et al. (2009) reported that low levels of uranium concentration stimulated plant growth while Aery and Jain (1997), and Hafez and Ramadan (2002) showed detrimental effects of uranium. Investigation by Jagetiya and Purohit (2006) on survival of sunflower plants (variety Sungold double orange) over 100 days on higher tailing concentrations (up to 75%) showed that sunflowers may be helpful in the revitalization of uranium mining waste. In that study, the influence on plant growth was measured in terms of the Tolerance Index (TI) and Grade of Growth Inhibition (GGI), and it demonstrates that sunflowers can tolerate uranium to a certain level and hence can be used to filter contaminated runoff in hazardous radioactive waste sites.

3.1.1. Plant - Uranium Hyperaccumulators

There are several attributes ascribed to the ideal candidate plant species for phytoremediation of metals. First, the plants should have either a low biomass with a high metal capacity or a high biomass plant with an enhanced metal uptake potential. Specifically, the plant should have a sufficient capacity to accumulate the metal of concern within the harvestable biomass at a level greater than 1% (for some metals, greater than 1000 mg kg^{-1}). Furthermore, the plant should have a sufficient capacity to tolerate the site conditions and accumulate multiple metal contaminants. Finally, the species should be fast growing and have a suitable plant phenotype for easy harvest, treatment, and disposal (McIntyre, 2003).

According to the PHYTOREM data base, sunflowers are recognized as hyperaccumulators of uranium. PHYTOREM was developed by the Environment of Canada and this database consist of 775 plants with capabilities to accumulate or hyperaccumulate one or several of 19 key metallic elements. Species were considered as hyperaccumulators if they took up greater than 1,000 mg/kg dry weight of most metals. Sunflowers had a content of uranium of more than 15,000 mg kg^{-1} dry weight. Plant hyperaccumulators like sunflowers (*Helianthus annuus)* have the highest phytoremediation potential since there are also crop plants with well established cultivation methods (McIntyre, 2003). The index of tolerance and the bioaccumulation coefficient were two indices used for screening plants and evaluating metal uptake and phytotoxicity effects (Dushenkov et al., 1995; Nanda-Kumar et al., 1995). *Uncinia leptostachya* and *Coprosma arborea* were considered unusual U accumulators, whose U contents were around 3 mg kg^{-1} a.w. (Peterson, 1971). Furthermore, the leaves of black spruce (*Picea mariana*) were reported to contain U in excess of 1,000 mg kg^{-1} dry weight (Chang et al., 2005). In the investigations of Shahandeh and Hosssner (2002), thirty four plant species were screened for uranium (U) accumulation from U contaminated soil. Plant species used for extraction of U(VI) from contaminated soils were dicotyledonous and monocotyledonous plants: (field crops, cool and warm season grasses, and the *Brassica* family). Plant species selection was based on the agronomic importance of the crop, dry matter production, and apparent tolerance to heavy metals. They found a significant difference in accumulation between plant species, and the sunflower and Indian mustard plants showed the highest uranium accumulation .

In the investigation of Hashimoto et al. (2005), 32 plant species from 5 families were screened in order to find plants capable of U accumulation in the shoot tissue. Sand culture methods were used and plants accumulated from 4 to 416 mg of U per kg dry tissue weight. Plant species in Chenopodiaceae and Fabaceae had the highest mean U concentrations while plants in Poaceae accumulated less U than the dicotyledonous plants tested. Based on the results of the sand culture screening, Hashimoto et al. (2005) reported that plants for phytoremediation of U-contaminated soils could be selected by using a mathematical model related to plant characteristics (e.g. biomass and planting density).

3.1.2. Distribution of U in different plant parts

There are numerous reports in literature that concentrations of U in roots are significantly higher than in above-ground parts (shoots) (Chang et al., 2005; Stojanović et al. 2009).

In general, roots serve as a natural barrier, preventing the transport of many trace metals, including radionuclides to upper plant parts. Moreover, the rate of uranium translocations from roots to shoots is probably species-dependant. It may be different for different species and even cultivars (Shtangeeva, 2008). Shahandeh and Hossner (2002) reported that U concentration in roots of different plants collected from the same site were 30–50 times higher than U concentration in shoots, and Stojanović et al., (2009) reported content of uranium in roots of corn plants ten times higher compared to the shoots. Between other plant species tested by the authors, sunflowers and Indian mustard had the highest root U concentrations, and wheat and ryegrass had the lowest U concentrations in roots (Shtangeeva, 2008).

3.2. Time of Uranium Deposition

Element concentrations in the plant tissues can vary with time, for instance, during vegetation season (Myung and Thornton, 1997; Otero and Macias, 2002). Certain variations in the plant uranium concentrations over shorter time (days or even hours) could be expected (Shtangeeva, 2008).

Investigations by Shtangeeva (2008) on temporal variations of U in two native plants, couch-grass Elytrigia repens L. and plantain Plantago major L, showed that diurnal variations of U in roots and leaves of couch-grass sampled from soil rich with U were significant, with highest U concentration at times when soil temperature was the highest. However, maximum values of U concentration in roots and leaves of plantains sampled simultaneously from the same site were registered 4 hours later. Short-term variations in U concentrations in plantains could not be explained by the changes in soil temperature. Certain differences between these plants in U uptake would be expected because couch-grass and plantains belong to two different classes: Monocotyledoneae (Monocots) and Dicotyledoneae (Dicots). Short-term dynamics of U radionuclide plant concentrations are rather significant, regular and species – specific (Shtangeeva, 2008).

3.3. Role of Amendments in Uranium Phytoremediation

3.3.1. Improving Phytoremediation with Organic Agents

There are two general approaches to phytoextraction: continuous and chemically enhanced phytoextraction. The first approach uses naturally hyperaccumulating plants with the ability to accumulate an exceptionally high metal content in the shoots (Leštan, 2006). Another method is the application of synthetic and natural organic agents as a means of improving the mobilization of uranium and increases the efficiency of phytoextraction. A key to the success of U phytoextraction is to increase soil U availability to plants.

There could be several problems identified with applications of U hypraccumulator plants: (1) plants take up the more available metal fraction, but less available fractions cannot be extracted, and (2) hyperaccumulators often have low biomass, which results in a low amount of metal extracted from the site. Therefore, increasing availability of metals is an alternative that has to be studied (Diaz and Kirkham, 2007). Availability of uranium from soil to plants is improved by applying some methods such as chelation, complexation aiming to solubilize, detoxify, and enhance U accumulation by plants (Duquène et al., 2006).

In literature, there have been numerous reports about amendments in phytoremediation compounds that increase the uptake of uranium by varius plants.

Amendments could be organic compounds such as synthetic chelating agents (ethylenediaminetetraacetic acid (EDTA), N-hydroxyethyl-ethylenediamine-N,N',N'-triacetic acid (HEDTA), diethylenetrinitrilopentacetic acid (DTPA)), natural fulvic acid, humic acid, and more natural low molecular weight organic acids (citric, malic, oxalic, and acetic acid).

The most frequently used is EDTA, which has been reported as more effective than other synthetic chelators for several heavy metals. The use of chelating agents to increase metal availability to plants, with the aim of extracting them from soil is called "assisted," "induced," or "enhanced" phytoextraction. However, increasing metal mobility in soil also increases the risk of pollutants leaching into groundwater (Diaz and Kirkham 2007).

Theoretically, the metal-chelating efficiency of chelating agents depends on the stability constant (logK) of the metal-complex formation. Martell and Smith (2003) compiled an extensive database of stability constants for different metals and chelating agents. They tested the most important chelating agents for enhanced phytoextraction of metals from the soil and their stability constant of complex formation (logK at T 20-25oC and ionic strength 0.1-1.0). For low bioavailable UO22+ in the soil, the value for LogK is more than 25 for the following: ethylenediamine tetraacetic acid (EDTA), trans-1,2-diaminocyclohexane-N,N,N′,N′ - tetraacetic acid (CDTA), dietylenetriamine pentaacetic acid (DTPA), and less than 10 for nitrilotriacetic acid (NTA), ethylenebis(oxyethylenetrinitrilo)-N,N,N',N' tetraacetic acid (EGTA), and citric acid.

The ability of a chelating agent to facilitate phytoextraction does not necessarily always relate to this theoretical affinity for metals (U) (Leštan, 2006).

Solubilization of U from contaminated soil by synthetic chelates (HEDTA- N-hydroxylethylene diamine triacetic acid) and organic acids (citric, oxalic) with hydroponic screening experiments, indicated that citric acid solubilized over100 times more U than the other amendments. Using citric acid as the principle soil amendment yields the possibility to develop an effective phytoremediation strategy for U-contaminated soils (Ebbs et al., 1998; Ebbs at al., 2001). Huang et al., (1998) found that some organic acids can be added to soils to increase U desorption from soil to soil solution and to trigger a rapid U accumulation in plants. Betwen tested organic acids (acetic acid, citric acid, and malic acid) , citric acid was most effective in enhancing U accumulation in plants. Shoot U concentrations of Brassica juncea and Brassica chinensis grown in a U-contaminated soil (total soil U, 750 mg kg-1) increased from less than 5 mg kg-1 to more than 5,000 mg kg-1 in citric acid-treated soils and the authors of the study claim that this is the highest shoot U concentration reported for plants grown on U-contaminated soils. The authors suggest that the strong mobilization of U by citric acid is due to the formation of citrate–uranyl complexes rather than the decreased pH, and found a close correlation between the U and the Fe and Al concentrations in the soil solution after the addition of citric acid, which they explained by the dissolution of Fe and Al sesquioxides and hence release of U from soil material to the soil solution. In a carbonate solution, U can form carbonate or hydroxide complexes, which are highly soluble. Elless and Lee (1998) state that for U solubility in soils, U-bearing minerals are more important than sorption–desorption processes.

In a pot study (with a soil U concentration of 280 mg kg-1), the addition of 0.95 g kg-1 of citric acid enhanced the soluble U concentration in the soil 35-fold, whereas the addition of several artificial chelating agents (EDTA, HEDTA, and DTPA) at the same molar concentrations (5 mmol kg-1) had negligible effects (Huang et al., 1998).

3.3.2. Uranium Immobilization Agents

Addition of chelating agents in order to enhance phytoextraction may promote leaching of the pollutants (uranium) into groundwater. Therefore, there is a need for application of sequestering agents, such as apatite, zeolite, or clay that will enable hydrological control, immobilization, and transformation of excess uranium, which plants didn't accept. Furthermore, sequestering agents can be used as ground cover in perennial phytoremediation, for adsorption of uranium, which can leach from fallen leaves in autumn.

A proposal for successful phytoremediation is presented on Figure 4.

Mechanisms of immobilization with sequestering agents (apatite, zeolite, clay, zerovalent iron, etc.) can be described rather as a reductive precipitation process than a simple sorption process in which uranyl is distributed between the solution and solid phases according to adsorption affinity and capacity of the adsorbent surfaces.

The understanding of U(VI) removal mechanisms through either reductive precipitation or sorption/co-precipitation has important environmental implications, because the reduced U(IV) could be potentially re-oxidized when it's exposed to the air or dissolved O_2 in a few hours or days. Similarly, the sorbed U(VI) species could be desorbed and therefore remobilized as groundwater in geochemistry changes.

Knox et al. (2008) evaluated the influence of three types of phosphate (rock phosphate, biological phosphate, and calcium phytate) and two microbial amendments (*Alcaligenes piechaudii* and *Pseudomonas putida*) on U mobility in two sediments. All tested phosphate amendments reduced aqueous U concentrations more than 90%, probably due to formations of insoluble phosphate precipitates. The addition of *A. piechaudii* and *P. putida* were found to reduce U concentrations 63% and 31%, respectively. Uranium removal in phosphate treatments were significantly reduced in the presence of those two microbes.

Uranium may react with apatite to form mineral phases of the autunite group, a diverse group of over 40 minerals, having the general formula: $M(UO_2PO_4)_2{\cdot}nH_2O$ (Bostick et al., 2000). It is well-known that apatite minerals react with many transition and heavy metals, metalloids, radionuclides, and to rapidly form secondary phosphate precipitates that are stable over a wide range of geochemical conditions (Arey et al., 1999). Bostick et al. (2000) demonstrated that ground fish bone (biological apatite) was highly effective for the removal of soluble uranium from synthetic groundwater, and that crystalline autunite (calcium uranyl phosphate) is formed at high loadings of uranium.

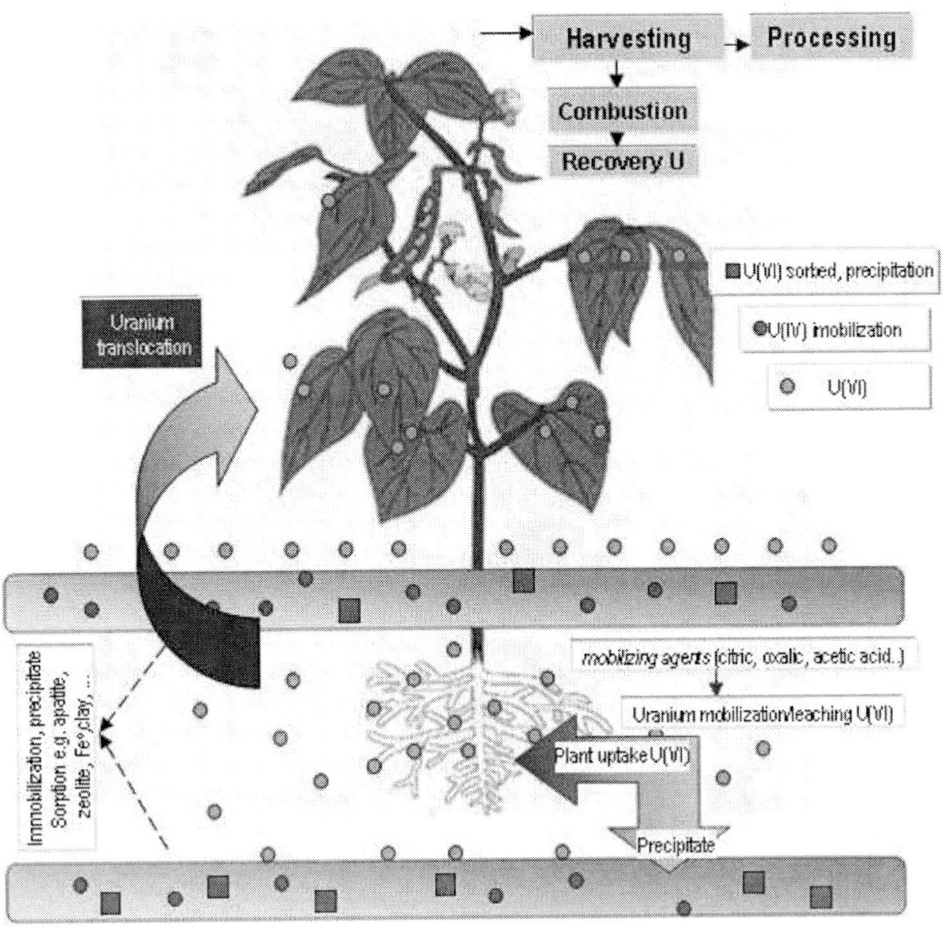

Figure 4. Schematic phytoremediation technique: plant-immobilization agents -mobilization agents.

In the 300 Area of the Hanford Site, an innovative polyphosphate material was tested for its ability to reduce uranium concentrations in the groundwater, with promising initial results. Uranium in groundwater is of concern in the area immediately adjacent to the Columbia River. The project is testing and demonstrating whether this innovative material can be injected into the subsurface to sequester the uranium in place as an insoluble uranium phosphate mineral, thus reducing the uranium concentrations in groundwater. (Wellman et al., 2008).

Phosphate-induced metal stabilization (PIMS) using apatite stabilizes uranium in situ by chemically binding it into the new low-solubility phase(Ksp=10-49). Uranium-phosphate-autunite is stable across a wide range of geological conditions for millions of years. Laboratory studies were conducted to quantify different forms of apatite sequestered by uranium contaminants, natural phosphates from Lisina deposit (14.43 % P2O5), phosphate concentrate samples with 34.95 % P2O5, and mechanochemically activated natural apatite. The results show that the mineral apatite 'Lisina' is very effective for the treatment of contaminated soils in situ immobilization of U. Largest efficiency showed the phosphate concentrate (Stojanović et al., 2008). Matijašević et al., (2006) investigated the adsorption of uranium(VI) on heulandite/ clinoptilolite rich zeolitic tuff modified with diferent hexadecyltrimethylammonium (HDTMA) ion. The results reported that organozeolites are effective for the removal of uranium(VI) from aqueous solutions, contaminated soils, and ground water systems.

Stojanović et al. (2009) recommended synergistic mixtures of zeolite and apatite as reactive remediation agents. Their application consists of reactive permeable barriers directly mixed with contaminated land in combination with suitable agrotechnical measures for the correction of pH, added as a liner in the contaminated sites. Reductive precipitation of U(VI) to U(IV) species by zerovalent iron reactive barriers is the dominant mechanism for uranium removal (Phillips et al., 2000).

Sequential extraction techniques, whereby a sequential series of increasingly more harsh extracts are used to operationally define how strongly a contaminant is sorbed to a soil (Tessier et al., 1979), are combined with these approaches. Together, these studies provide information about: bioavailability potential mobility, chemical liability, and sorption process of contaminants (Arey et al., 1999). The batch extraction methods provide a way for rapidly screening numerous alternative treatment scenarios, especially for evaluating contaminant mobility. However, limitation of such methods demands use of experiment's plant growth and bioassays to assess biological availability (Hinton et al., 1998).

4. The Potential of Some Cultivated Plants in Phytoremediation of Uranium

Research with the purpose of environmental protection and reduction of ionizing radiation on the regional level (Serbia), was conducted by researchers from the Institute for Technology of Nuclear and Other Mineral Raw Materials , Belgrade. Investigations have lasted for several decades, where it was determinated the degree of accumulation of cultivated plants mostly proposed human nutrition for their application for phytoremediation. Screening plant species had the aim of planting exposed sites in Serbia.

To understand the mechanism of uranium uptake in plants and accumulation, a necessary prerequisite is the application of radiophytoremediation on the "real" scale. For this purpose, we investigated these processes using three different aspects with selected cultivated plants.

Obtained results from many years of research are useful for further investigations of the significance of U in the life of plants and their application in phytoremediation.

4.1. Corn Plants as Uranium Accumulator (Vegetative Tests under Fully Controlled Conditions on Two Types of Soil)

Vegetative test of fully controlled conditions were applied to find out the coefficient of uranium accumulation in tissues of corn plants that were grown on two types of soil, together with uranium phytotoxic effects on the plant development, height, yield, and seed germination. Uranium was added to soils in the amounts of 10–1000 mg kg-1 (Stojanović et al., 2009).

Vegetation experiments were carried out on two types of soil: pseudogley site Varna-Šabac and experimental chernozem fields of the Institute of corn from Zemun fields. As for the test cultures, the corn varieties ZPSC 633rd were used.

Chernozem is between the neutral and alkali type of soil. Soil is well provided with affordable and accessible phosphorous, potassium, and a lot of humus medium provided with the total nitrogen. Pseudogley soil has a lot of acidic qualities without lime, medium qualities provided with affordable and accessible phosphorous potassium, and poor qualities with humus and medium provided with the total nitrogen. Plastic pots were filled with 3 kg of the soil, which was homogenized before seeding compounding with NPK fertilizer, 13:13:15, in the amount of 800 kg / ha, to ensure equable supply of the plants with most important nutrients. Two series were made in different time intervals in the following variations:

I Series	II Series
NPK Ø	NPK Ø
NPK + U (10 mg Ukg^{-1})	NPK + U (100 mg Ukg^{-1})
NPK + U (25 mg Ukg^{-1})	NPK + U (250 mg Ukg^{-1})
NPK + U (50 mg Ukg^{-1})	NPK + U (500 mg Ukg^{-1})
NPK + U (100 mg Ukg^{-1})	NPK + U (1000 mg Ukg^{-1})

First experiment last for 40 and second for 45 days.

Uranium (VI) was added in solution in the form of UO2(NO3)2 6H2O. The experiment was set with 10 grains of corn by the tailings in five repetitions. Ten days after from the germinate date that the plants were grouped, every tailing had six properly formed plants. Plants were picked in the development phase (7-9 leaves). Uranium content was determined in plant organs, roots, and overground parts (shoot) of corn plants. Phytotoxic effects of uranium were monitored through the level of seed germination, the percentage of survival, plant height, and contribution dry masses.

Uranium content was determined by the fluorometric method by employing 26-000 Jarrell Ash Division instrument. (detection limit 0.005mg kg-1, rang 0.05mg kg-1 – 5mg kg-1, correlation coefficient R>0.997) (Stojanović et al., 1993).

4.1.1. Contents of Uranium Accumulated in Corn

Average values of uranium content in roots and shoots of cultivated corn on pseudogley and chernozem under different doses of uranium are presented in table 1.

Experiments were placed in five repetitions. The analysis of variance and LSD test for the level of risk of 5% and 1% was performed in relation to the chernozem. Concentration of uranium in the root of the corn grown on pseudogley was significantly higher in relation to the chernozem. Other differences were not statistically significant.

From the database it can be concluded that the uranium content in the roots and shoots of corn grow proportionally with increasing doses of added uranium to both types of soil. The uranium content in all treatments were higher in roots than in the shoots. In the first experiment on chernozem uranium content in the root was 9.4, and on pseudogley it was 9.16 times higher compared to the shoot, while in the second experiment the ratio was higher and amounted to 12.8 in chernozem and 13, 57 to pseudogley. In the first experiment on chernozem with concentration of 100 mg kg-1 U, its content in the root was 63.58 mg kg-1, and when the experiment was done on pseudogley its content was 73.98 mg kg-1. In the second experiment, chernozem concentration of U was 100.19 mg kg-1, and 57.59% more in relation to the first experiment and 129.38 mg kg-1 on pseudogley, which is 74.88% more than in the first experiment. As the experiment didn't last the same amount of time, it is clear that when the root system is reflected in both types of soil it hasn't achieved "physiological threshold."

Table 1 Content of uranium U (mg kg^{-1}) in the root and shoot of corn which was cultivated on pseudogley and chernozem.

experiment	TREATMENT	pseudogley			chernozem		
		Root X_R	Shoot X_S	X_R/X_S	Root X_R	Shoot X_S	X_R/X_S
I	NPK Ø	0.11	0.01	11.00	0.08	<0.01	8.00
	NPK+10mg U kg^{-1}	7.60	0.84	9.03	4.48	0.64	7.56
	NPK+25mg U kg^{-1}	12.21	2.07	5.90	11.83	1.61	7.34
	NPK+50mg U kg^{-1}	32.01	3.50	9.14	27.66	2.31	11.0
	NPK+100mg U kg^{-1}	73.98	6.09	10.7	63.38	4.85	13.1
		average content X_R/X_S: 9.16			average content X_R/X_S:9.40		
II	NPK Ø	0.12	0.01	12.00	0.09	<0.01	9.00
	NPK+ 100 mg U kg^{-1}	129.38	6.96	18.58	100.19	5.99	16.70
	NPK+ 250 mg U kg^{-1}	389.80	33.49	11.63	384.54	24.42	15.57
	NPK+ 500 mg U kg^{-1}	718.04	59.39	12.10	557.40	48.09	11.59
	NPK+1000 mgU kg^{-1}	1744.4	128.59	13.56	1206.10	110.22	10.93
		average content X_R/X_S: 13.57			average content X_R/X_S: 12.80		

The accumulation of uranium in the first and second experiment was higher in the root system on pseudogley, which explains the higher mobility of the uranium in soils with acidic reactions, which increases its availability for plants.

Content of uranium in the shoots of corn in the first experiment of 100 mg kg-1 was 4.85 mg kg-1 and 6.09 on the pseudogley . In the second series of experiments on the pseudogley, the uranium content was 6.96 mg/kg, ie 14% more than in the first experiment, and 5.99 mg kg-1 in chernozem, which is 23% more in relation to the content from the first experiment.

The results indicate that the uranium content in the corn plant is proportionate to the length of the experiment. Far less increase in uranium in the ground part in relation to the root; it points towards the "physiological threshold," acceptation of uranium in shoots.

On figures 5 and 6, a linear relationship between uranium content in the root and shoot of corn in the function of its content in both types of soil examined can be seen. Linear dependence included maximum amounts of uranium from 1,000 mg kg-1 added in soil, which suggests that maize neither in root or in the over ground part has not reached "the threshold of physiological" adoption. In the graphic we can see that the adoption of uranium in much larger root systems in relation to the shoots of both types of soil. The uranium content in the floral organs in all treatments was higher in pseudogley and the team, so its toxic effect in relation to the chernozem is as well.

On the other hand, the characteristics of corn roots is such that it cannot achieve "physiological threshold" of uranium adoption with high doses of uranium in the soil of 1,000 mg kg^{-1}, and that it has a high coefficient of accumulation, leads us to the assumption that the corn can be used for land remediation or **rhizofiltration** waste effluents contaminated with uranium.

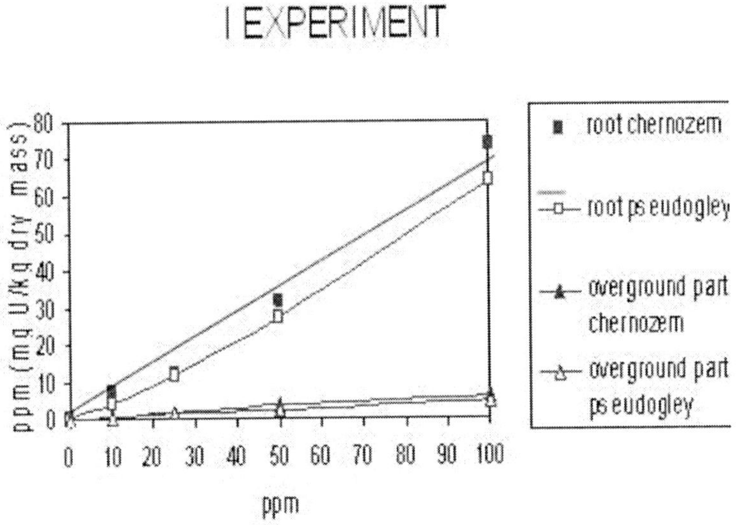

Figure 5. Content U (mg kg^{-1}) in the root and shoot parts of corn in the function of its content in the soil (experiment I).

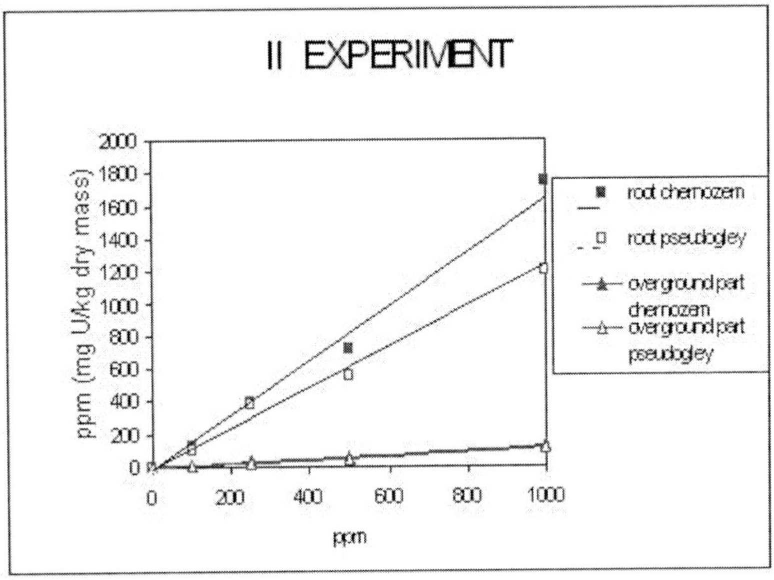

Figure 6 Content U (mg kg^{-1}) in the root and shoot of corn in the function of its content in the soil (experiment).

4.1.2. Impact of Uranium on Contribution of Dry Mass of Corn

In the first experiment adding uranium in portions of 10, 25, 50, and 100 mg kg^{-1} had the consequence of increasing the root mass on both types of soil in relation to the control (Table 2). Stimulating effects of uranium on the contribution of maize roots is confirmed on both types of soil, but the effect of that was stronger on pseudogley than on chernozem. These results are interpreted by the fact that in the acidic environment U is mobile, which increases its availability to root system.

In all variants, contribution of dry mass of shoots grown on chernozem was smaller in comparison to the control, unlike pseudogley, where the yield of overground parts was higher. The concentration of uranium 250, 500 and 1,000 mg kg^{-1} influenced on the reduction of root contribution and shoots of corn on both types of soil.

Phytotoxic effects of uranium on the production of dry mass of plants were being increased with increasing doses of uranium.

Table 2. U influence on contribution of dry mass of corn (g/pot) grown on pseudogley and chernozem (comparative results)

Treatments	chernozem				pseudogley			
	root		shoot		root		shoot	
	X	%	X	%	X	%	X	%
NPK Ø	3.74	100.00	3.81	100.00	3.01	100.00	3.35	100.00
NPK+10mg U kg^{-1}	4.07	108.82	3.70	97.11	3.36	111.63	3.94	117.62
NPK+25mg U kg^{-1}	4.31	115.24	3.09	81.11	3.91	129.90	3.60	107.63
NPK+50mg U kg^{-1}	4.12	110.16	3.68	96.58	3.76	124.1	3.98	118.80
NPK+100mg U kg^{-1}	4.06	108.56	3.72	97.63	3.77	125.25	4.06	121.19
NPK Ø	3.55	100.00	3.58	100.00	2.86	100.00	3.28	100.00
NPK+ 100 mg U kg^{-1}	3.68	103.66	3.31	92.46	2.99	104.54	3.43	104.57
NPK+ 250 mg U kg^{-1}	2.53	71.27	2.61	72.91	2.29	80.70	2.78	84.76
NPK+ 500 mg U kg^{-1}	2.41	67.88	2.47	68.99	1.86	65.03	2.50	76.21
NPK+1000 mgU kg^{-1}	1.63	45.91	1.87	49.72	1.29	45.10	1.37	41.76

In all variations of the experiment, production of dry mass overground parts on chernozem was smaller in comparison to the control, unlike pseudogley, where the yield was higher in overground parts in all the variants.

4.1.3. Impact of Uranium on Seed Germinability

Phytotoxic effects of uranium from uranil-nitrate was determined on the basis of the criteria of reducing the seed germinability, survival percentage, and height of plants at the end of experiments. Time of germination of the seeds on chernozem was on average between 4-5 days and on pseudogley between 6-7 days, regardless of the treatment and series of experiments. In the first series of experiments, significant differences were not found in terms of germination between soil types and treatments in relation to the control. Uranium doses greater than 100 mg kg-1 influenced the reduction of the percentage of germination in relation to the control in both soil types.

Thus, when U concentration was 250 mg kg-1, the percentage of seed germination on chernozem was 91.6%, 89.3% on the pseudogley, at 500 ppm 79.2% on chernozem, or pseudogley 76.6% and in 1,000 mg kg-1 U, 64.6% on chernozem, and 63.8% on pseudogley. A higher percentage of seed germination in doses of 250, 500, and 1,000 mg kg-1 U were on chernozem. In the same experiment, the phytotoxic effect of U through the criteria of plants survival was examined.

In the first series of experiments, the percentage of survival was 100% in both types of soil. In the second series of experiments, the percentage of surviving plants was 100% at concentrations of 100 and 250 mg kg-1.When the uranium concentration was 500 ppm, the percentage of plant survival on chernozem was 95% and 5% more in relation to the pseudogley, and with 1,000 ppm it was 88.3% on chernozem or 11.6% more in relation to pseudogley. Toxic effects of uranium were higher in acidic soils at concentrations of 500 and 1,000 mg kg-1.

4.1.4. Impact of Uranium on Height of Corn Plants

The third indicator of phytotoxic effects of uranium is the reduction of plant height, while noxiousness increases with increasing content of uranium in the soil. Thus, in the first series of experiments, height of plants already decreases at 25 mg kg-1 U and with 100 mg kg-1 U, the average height of plants was 46 cm on chernozem, or 43 cm on pseudogley, which is 15% and 16% lower compared to the control.

Table 3 Impact of uranium on height of corn plants

Experiment	Treatment	HEIGHT plants			
		chernozem		pseudogley	
		cm	%	cm	%
I	NPK Ø	54	100.0	50	100.0
	NPK+10mg U kg^{-1}	54	100.0	50	100.0
	NPK+25mg U kg^{-1}	48	88.8	45	90.0
	NPK+50mg U kg^{-1}	50	92.6	47	94.0
	NPK+100mg U kg^{-1}	46	85.0	43	84.0
II	NPK Ø	52	100.0	49	100.0
	NPK+ 100 mg U kg^{-1}	45	86.5	39	79.0
	NPK+ 250 mg U kg^{-1}	46	88.5	38	77.6
	NPK+ 500 mg U kg^{-1}	45	86.5	38	77.6
	NPK+1000 mgU kg^{-1}	35	67.3	26	53.1

In the second series of experiments, increasing content of uranium significantly influenced the reduction of height of plants, and in relation to the chernozem the acidity of pseudogley is contributed to the reduction of the height of plants. So the uranium concentration of 1,000 mg kg-1 on chernozem yielded a plant height of 35 cm compared to the 52 cm that was the height of the control, and the pseudogley was 26 cm compared to the 49 cm of the control (Table 3). Phytotoxic effects of uranium confirmed Aleksahina (1985) on rice.Vegetation experiments were conducted on meadowy chernozem with uranium inputs of 100, 500, and 1,000 mg kg^{-1}. In mature milk wax with uranium concentrations of 100 ppm, the percentage of survival was 92% for 500 mg kg^{-1} and 52% for 1000 mg kg^{-1} 16%.

Comparing results of phytotoxic effects of the percentage of survival of rice that is received from Aleksahina , with the results obtained from corn, it can be concluded that corn is more resistant in relation to the rice. In the same paper, Aleksahina points to phytotoxic effects of uranium through the reduction of the height of plants. So in the stage of formatting flowering heads of rice plants, its height in the control variant was 100 cm; when the concentration U in the soil was 100 mg kg^{-1} then the height was 85 cm, 500 mg/kg with 71 cm, and 62 cm with 1,000 mg kg^{-1}. Comparing the results of percentile reduction in the height of rice and maize with increased content of uranium on chernozem, it can be concluded that they are the same trend and share similar results. Hemo and radiotoxic effects of uranium in concentrations of up to 100 mg kg^{-1} in the soil are manifested only through the reduction of plant growth, which points to the great resistance of corn plants. All other criteria for phytotoxic effects of uranium are expressed in concentrations above 100 mg U kg^{-1}, and most of the 1,000 mg U kg^{-1} in both types of soil. The general conclusion is that increasing the concentration of uranium in the soil inhibits the growth of plants and reduces seed germination and the percentage of survival due to its chemical and **radiotoxicity**. With increasing content of uranium in the soil, its phytotoxic effect increases according to the following series:

plant height >seed germination> survival of plants

The phytotoxic effect is increased with increasing contents of uranium and it is expressed strongest in the concentration of 1,000 mg kg-1. In all variants, phytotoxic effects of uranium on the maize plants were stronger on pseudogley in relation to the chernozem. This explains the higher uranium mobility in acidic soils, which increased its availability for plants. The content of uranium in roots was on average ten times higher compared to the overground part. Uranium content in the roots and overground parts of corn is shown as a linear dependence of added uranium in both types of soil examined. Linear dependence is included in max. dose of 1,000 mg kg-1 uranium, which means that the plant parts haven't reached "the threshold of physiological" accumulation of uranium. For this reason, maize can be used for remediation in uranium contaminated soil. Uranium content in the roots and overground parts of corn plants grown in pseudogley was significantly higher than those on chernozem.

4.2. The Effect of the Uranium Content in the Tailings and "Uranium Water" on Some Cultivated Plants

Large amounts of solid wastes (tailings) were gained from the exploitation and treatment of uranium ore from the closed uranium mine Gabrovnica-Kalna, in southeast f Serbia. From 1953 to 1962, there was an exploitation of uranium. Then great amounts of tailings were made that today cover an area of about 0.1 km^2 and contain about 15,33 mg U/ kg. The objective of this study was to better the understanding of U uptake and accumulation in cultivated plants and whether different contents of uranium (U) in the substrate affect its concentration in plants and their dry matter mass. In the experiment, three plant species (corn NSSC 231, sunflower N.S. Dukat, and green peas Smederevska Palanka) were grown in pots during 40 days on two substrate variants and tailings mixed with sand. Each variant was suffused by drinking water and "uranium water" (UW), which issue out from the mine, containing 0,053 mg U/ dm^3. Drinking water (DW) and sand uncontained the uranium. So it was four vegetational variants for growing plants, which differed in the uranium content:

1. tailings with UW.
2. tailings and sand (w/w, 1:1) with UW.
3. tailings with DW.
4. tailings and sand (w/w, 1:1) with DW.

At the end of the 40 day period the plants were analyzed. The above-ground parts were separated from the roots, dry matter mass was measured, and the U concentration in the plant part was determined by the fluorimetric method. Concentration of U in above-ground parts in all three investigated plant species significantly increased when the content of U is the highest in the substrate (Figure 7). The highest concentration of U in plants occurred in the plants grown on substrate variant 1. Plants grown on substrate variant 2 had significantly higher concentrations of U than those grown on substrate variant 4. Sunflowers had higher concentration of U compared to maize and peas.

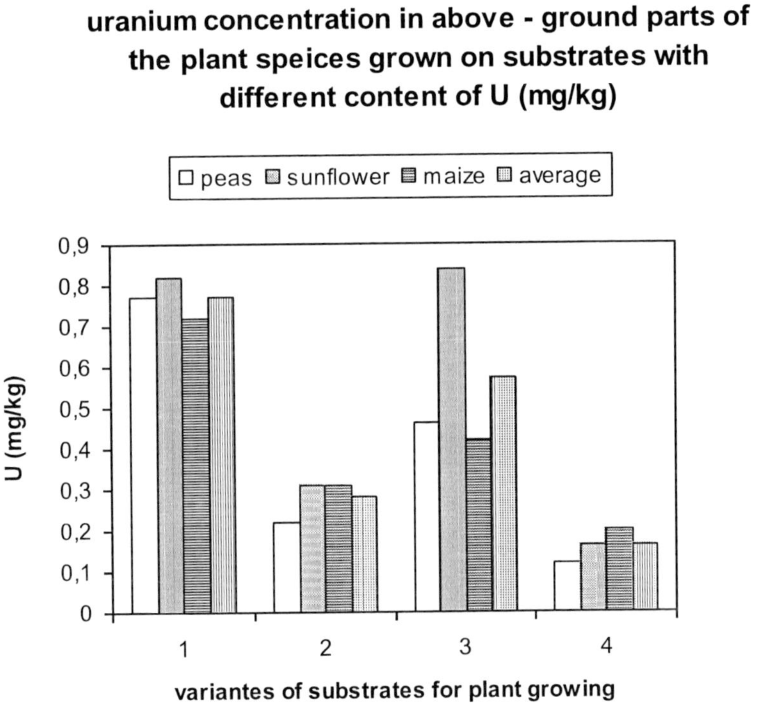

Figure 7. Uranium concentration in above-ground parts of the plant species grown on substrates with different content of U (mg/kg).

Dry matter mass was higher in variants where UW was used, both in plants grown on tailings only and in those grown on the mixture of tailings and sand (Figure 8). Average dry matter mass was higher in plants grown on the substrate variant 2 and somewhat lower in plants grown on substrate variant 1, in comparasion with plants grown on substrate variant 1 with plants grown on substrate variant 3 and 4 to which only DW was added.

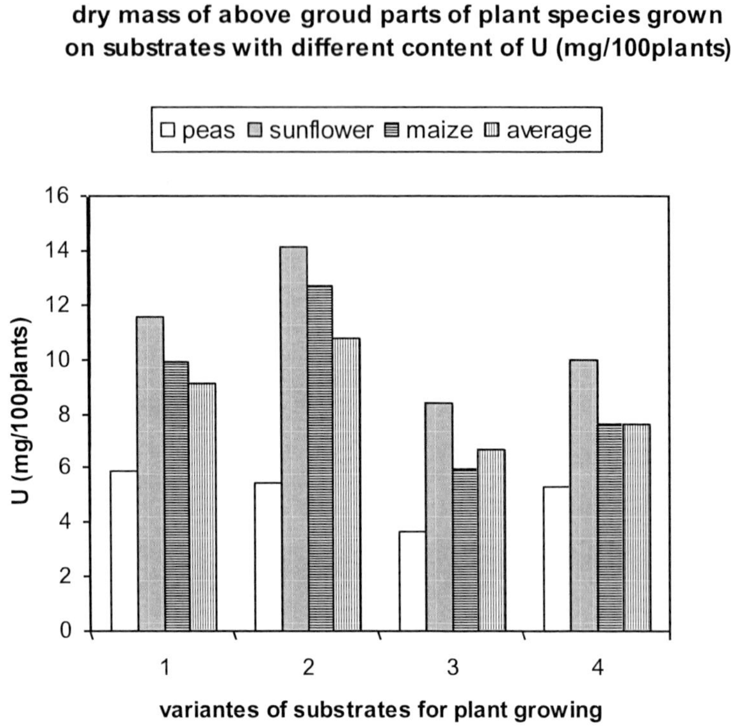

Figure 8. Dry mass of above-ground parts of the plant species grown on substrates with different content of U (mg/100 plants).

In all three investigated plant species, dry matter mass was higher in plants grown on substrate variants to which sand was added, probably due to better aeration in the plant's root zone. The dry matter mass was the highest in sunflowers, then maize, and it was the lowest in green peas. These differences were highly significant. Concentration of U was significantly higher in roots than in above-ground parts. The obtained results point to the conclusion that U concentration depends on its content in the substrate as well as on the plant species (Figure 9). Uranium concentration in roots follow the order of its concentration in the above ground parts, depending on the substrate variant. The highest concentration of U was found in the roots of plants grown on substrate variant 1, then 2, 3, and 4. In comparison to other plant species, U concentration was the highest in sunflowers, both in roots and above-ground parts. Peas had no significant difference in U compered with two other species. Average root dry matter mass coincides with dry matter mass in the above-ground parts, though there were no significant diferences between the plants grown on substrate variants 3 and 4 (Figure 10). The highest root dry matter mass is held by maize, then peas, and lastly sunflowers. The differences between plant species were very significant. However, it is the fact that there are no available results about several plant species grown under the same conditions which could be comparatively analyzed from this aspect. These investigations can be very useful because they indicate that intense adoption of uranium and other undesired elements exibited by certain plant species could be utilized for decontamination of water and soil. According to Adler (1996), the Chernobyl sunflower project was initiated in 1994. Prior to this project, Phytotech studies showed that sunflowers reduced the content of uranium in water from 350 ppb to 5 ppb, in 24 h.

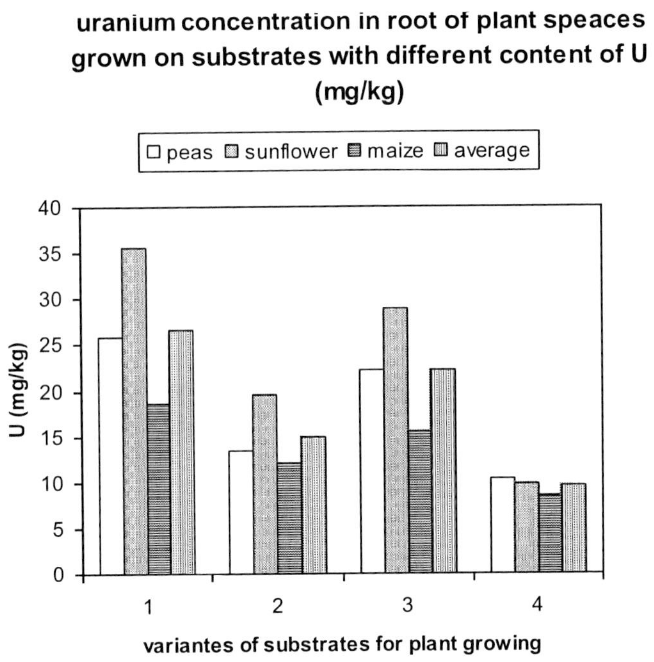

Figure 9. Uranium concentration in root of the plant species grown on substrates with different content of U (mg/kg).

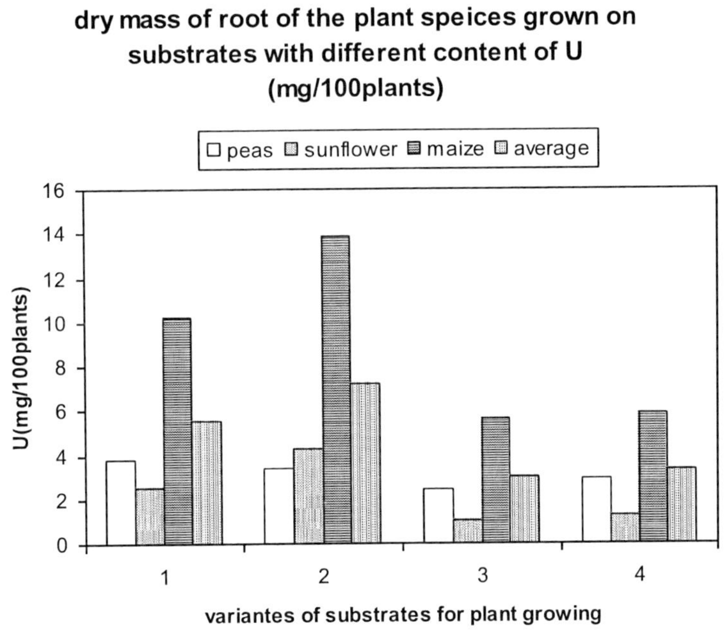

Figure 10. Dry mass of root of the plant species grown on substrates with different content of U (mg/100 plants).

In the summer of 1996, Phytotech and DOE researchers began a project using sunflowers to remove uranium from contaminated springs at the Oak Ridge (Tenn.) National Laboratory. Effects of different concentrations (25, 50, 75, and 100%) of uranium tailings conditioned

(Umra mine -India) with garden soil on growth and biochemical parameters in sunflowers were studied. The shoot and root length, fresh and dry mass, as well as leaf area and chlorophyll contents showed a significant negative correlation with the applied uranium tailing concentrations. Yellowing of leaves was recorded in all the tailing concentrations. Soluble proteins (leaf) showed significant enhancement as the concentration of uranium tailing increased indicating a breakdown of structural insoluble proteins. Survival of sunflower plants over 100 days on higher tailing concentrations (up to 75%) showed that sunflowers may be helpful in revitalization of uranium mining waste (Jagetiya and Purohit, 2006). Obtained results point to a very interesting phenomenon. The comparison of the results obtained from investigated plants grown on variants 1 and 3, tailings with and withought UW, show that addition of UW increased not only the concentration of U but also dry matter mass, both in above-ground parts and roots. The same phenomenon was observed for variants 2 and 4, where sand was added to barren soil, when either UW or DW was added. Differences in concentration of U and dry matter mass were larger when sand was added to the tailings. Shown results of dry matter mass undoubtedly point to the phenomenon of radiostimulation in particular cases, which becomes obvious when the values for dry matter mass and U concentration are compared (variants 1 and 3, and 2 and 4). This phenomenom was noticed a long time ago as concerning the influence of certian doses of radiation in seed and plants.

Discussion: Obtained results shows that the U concentration in the investigated plant species depended on the U concentration in the substrate on which the plants were grown. The U concentration was different in sunflowers, maize, and peas, both in above-ground parts and roots. Concentration of U was significantly higher in roots than in above-ground parts. Sunflowers had higher concentration of U compared to maize and peas.The dry matter mass was highest in sunflowers, then maize, and it was the lowest in green peas. These differences were highly significant. Sunflowers could be used for phytoremediation of soil contaminated with uranium. Substrate properties determined the tolerance and accumulation of U in plants. Results could be useful for further investigations of the significance of U in the life of plants and in plant choice on areas contaminated with uranium from anthropogenic resources and used for phytoremediation.

4.3. The Reaction of the Plant Species and their Genotypes on Uranium Uptake

Reports have shown that various plant species as well as genotypes, cultivars, and lines within species differ in uptake, accumulation, and distribution of mineral elements (Clark, 1983; Clark and Brown, 1974; 1980; Clark et al., 1982; Sarić, 1983). Information about reactions of the plant species and their genotypes on uranium uptake is limited. Only Chen et al., (2005), demonstrated mycorhizal effects on uranium uptake from uranium contaminated soil by two genotypes of plant hairless barley and fortify differences.

For investigations of the uranium adoption in various genotypes, we used three plant species, each represented by four genotypes:

1. Corn-inbred lines: 316207, 326037, MO17HZ and B73H7
2. Sunflower-4 sorts coded from 1 to 4
3. Soy bean-4 sorts: Kolubara, Vera, Ravanica and Balkan

The seeds of plant species mentioned above were obtained from the Institute for Field and Vegetable Crops-Novi Sad. This investigation was performed in a plastic-house where the plants were cultivated in the size 59x32x23cm, that were filled with soil (first series) and tailings from the uranium mine in Gabrovnica-Kalna (second series). In order to distinguish the differences in uranium adoption in various genotypes, the most reliable way was to perform the experiment under determined conditions (water culture, glass-house, and plastic-house) and this was why we have used a plastic-house for our tests. The pots were filled with the soil taken from the field, some 20 km far from the uranium mine and without any possibility of the pollution from the water streams coming from the mine. Other series of pots were filled with the waste from the uranium mine Gabrovnica-Kalna. The pots were filled up with 15 kg of soil (0.66 mg U/kg) or waste (15.3 mg/kg). All pots with plants were placed in the plastic-house and grown for 60 days. Afterwards the plants were taken, washed, dried, and prepared for the chemical analyses. Uptake and concentration of U is specific for particular plant species.

Table 4. Concentration of uranium (mg kg^{-1}) in the shoot (overground parts) of different plant species

SOY BEAN	I sort (Kolubara)		II sort (Vera)		III sort (Ravanica)		IV sort (Balkan)	
	soil	waste	soil	waste	soil	waste	soil	waste
	0,19	1,80	0,16	1,76	0,14	1,56	0,21	1,28
	Total average (soil)		0,175		Total average (waste)		1,600	
CORN	I sort (316207)		II sort (326037)		III sort (MO17HZ)		IV sort (B73H7)	
	soil	waste	soil	waste	soil	waste	soil	waste
	0,13	1,58	0,07	1,39	0,08	0,93	0,06	0,92
	Total average (soil)		0,086		Total average (waste)		1,205	
SUNFLOWER	I sort		II sort		III sort		IV sort	
	soil	waste	soil	waste	soil	waste	soil	waste
	0,40	0,95	0,41	0,96	0,37	1,03	0,17	0,91
	Total average (soil)		0,338		Total average (waste)		0,963	

4.3.1. Uranium Concentration in Shoot (Overground Part)

The average values of uranium concentration in different plant species (Table 4) and their investigated genotypes vary to a great extent. On average, all four inbred lines of corn had the lowest uranium concentration when grown on soil; higher concentrations were found in sunflowers and the highest in soy beans. Considering that the plants are grown on waste, the lowest uranium concentration was found in sunflowers, then in corn, and the highest in soy beans. These differences are very significant and originate from the biological properties of the species and genotypes.

4.3.2. Uranium Concentration in Root

Obtained results (Table 5) show that the difference in uranium concentration between investigated plant species and genotypes existed not only in the overground part, but also in the root.

Table 5. Concentration of uranium (mg kg^{-1}) in the root of different plant specie

SOY BEAN	I sort(Kolubara)		II sort(Vera)		III sort(Ravanica)		IV sort(Balkan)	
	soil	waste	soil	waste	soil	waste	soil	waste
	0,48	9,97	0,51	8,30	0,56	8,74	0,60	7,80
	Total average (soil)		0,538		Total average (waste)		8,703	
CORN	I sort(316207)		II sort(326037)		III sort(MO17HZ)		IV sort(B73H7)	
	soil	waste	soil	waste	soil	waste	soil	waste
	0,40	8,52	0,37	9,05	0,38	7,70	0,20	8,66
	Total average (soil)		0,338		Total average (waste)		8,483	
SUNFLOWE R	I sort		II sort		III sort		IV sort	
	soil	waste	soil	waste	soil	waste	soil	waste
	0,51	2,79	0,48	2,77	0,45	2,65	0,53	2,45
	Total average (soil)		0,493		Total average (waste)		2,665	

Obtained results definitely point out that uranium adoption, or more precisely uranium concentration, is influenced by its content in specific environments where the plant is grown, specific plant species, and especially their genotypes.

These differences are connected to genetic properties of investigated genotypes that have to be treated in further investigations. Interesting phenomena is found in sunflower behavior, where the larger differences between the genotypes were noticed in plants grown on soil rather than in the plants grown on waste.

The reason for this can be found in the fact that sunflowers uptake uranium more intensively in comparison to the other two investigated species. This can be justified with the results given by Adler (1996), which showed that sunflowers are the best decontaminator of uranium from polluted environments among investigated plant species.

This very interesting plant species is convenient for decontamination of specific elements and isotopes, or heavy metals due to the specific biological properties of the plants, particularly some of the plant segments and the properties of the investigated element.

4.3.3. Relationship Between the Content of Uranium in Plants and Substrate

The greatest differences of uranium concentration in the overground parts of plants grown on soil are registered in investigating the genotypes of soy beans, depending on the uranium content in substratum (1.425 mg/kg), then in corn (1.119 mg/kg), and at the end in sunflowers (0.625 mg/kg).

In the plants grown on waste, the differences between the uranium concentrations in roots of investigated plant species were as follows: 8.155 mg/kg in soy beans, 8.145 mg/kg in corn and 2.172 mg/kg in sunflowers.

It can be concluded that the sequence of plant species regarding the uranium concentration, depending on its content in the substratum, was the same in the overground part and in roots.

The reason for the low differences in uranium concentration in sunflowers grown on soil and waste is in the fact that sunflowers uptake uranium intensively, even at the lowest possible contents in the substratum.

Conclusion : This investigation was performed in plastic-houses where the plants were cultivated and filled with soil (first series) and waste from the uranium mine in Gabrovnica-

Kalna (second series) under determined conditions. Obtained results show that the difference in uranium concentration between investigated plant species and genotypes existed not only in the overground part, but also in the roots.

Obtained results definitely point out that uranium adoption, or more precisely uranium concentration, is influenced by its content in the specific environment where the plant is grown, plant species, and especially their genotypes.

4.4. Concentration of Uranium in Root-Crops, Bulbous and Tuberous Plants

Continuing our investigations related to the uptake and concentration of uranium (U) in cultivated plants depending on their biological properties, we wanted to investigate uptake, distribution, and concentration of U in several kinds of root-crops,bulbous and tuberous plants grown on tailings in natural conditions.

The folowing plant species were used in the experiment: carrot (Daucus carota L,cv Nantes), onion (Allium cepa L., cv Kupusinski jabučar), potato (Solanum tuberosum L., cv Desire), radish (Ranphanus sativus var. Maior cv. Smederevska Palanka), and sugarbeet (Beta vulgaris L.,cv Dana). They were grown on a tailing of the Gabrrovnica- Kalna uranium mine located in Serbia.

Samples for analyses were collected at the end of August. Plants were separated into leaves for all plant species as well as bulbs (onion), tubers (potato), and thickened roots (carrot, radish, redbeet, and sugarbeet). All plant organs were rinsed thoroughly with destilled water, blotted dry and dried. In order to be able to determine the surface contamination of bulbs, tubers, and thickened roots, a layer was peeled (4-5 mm) and analysed separately from the uncontaminated remainder of the underground plant organ.

The obtained results show that there are differences in the concentration of U between the investigated plant species both in the above and underground organs (root). Potatoes had the highest concentration in the shoot. Differences between potato and other investigated plant species were significant. There were no significant differences in the concentration of U between radish and carrots. The differences that were found between all other species were highly significant.

The concentration of U in thickened roots, tubers, and bulbs were different for each investigated species. Red beets had the highest concentration of U in thickened roots and the differences in comparison with other plant species were highly significant. There were no significant differences in U concentrations between sugarbeets, potatoes, and onions.

In order to obtain the differences in the concentrations of U in the possibly contaminated outer cover of thickened roots and bulbs, they were peeled (4-5 mm layer). The obtained results show the U concentrations to be significantly different for the peel and remainder. A significantly higher concentration of U was found in the peel (Table 6). The main differences in U conentrations between the investigated parts of the thickened roots, tubers, and bulbs was 10 times higher in the peripheral peel than in the remainder. These differences were specific for each plant species and they ranged from 4 times (red beet) to 28 times (potato) higher in the peel than in the remainder.

Table 6. U concentrations in the thickened roots, bulbs and tubers of plant species

plant	I	II	I:II	III	III:II
species	peel	remainder		leaves	
redbeet	1.58	0.39	4.08	1.14	2.98
radish	1.98	0.35	5.72	0.49	1.40
carot	1.68	0.31	5.43	0.70	2.26
sugarbeet	1.01	0.11	8.96	0.89	7.84
potato	2.74	0.09	27.95	1.20	12.30
onion	/	0.08	-	0.28	-
average	1.78	0.22	10.43	0.78	3.54
X					
LSD 0.5%	0.449	0.085		0.235	
0.1%	0.615	0.117		0.325	

Legend:

I - peel-peeled outer covering, the thickness of about 4-5 mm.

II - remainder the remainder left after peeling the outer covering of thickened roots, bulbs and tubers.

Comparing the obtained values for ratios of U concentrations in leaves and thickened roots, tubers, and bulbs (peel and remainder), it becomes clear that there are big differences depending on plant species. These results point not only to specificity of U uptake but also to the specificity of its distribution in different plant species.

Results show that there are differences in concentration of U in thickened roots, tubers, and bulbs of different plant species. We can be sure of that because the differences were not only found in the surface layer and peel, but also in the remainder. Ratio values are not only different and specific for particular plant species but also for different radionuclides. The obtained results are of great theoretical importance and are also significant from the aspect of environmental protection, that is in solving the problems arising in the domain of the food chain.

4.5. Testing the Accumulation of Uranium by Cultivated Plants on Uranium Tailing- "*In Situ*"

Much attention was paid to the determining of contents of uranium on the terrain surrounding certain mines in the first place in water, soil, and in the autochthonic flora. Sheard (1986) found the largest content of Pb210, Po210, and U in lichen and moss in uraniferous regions. In non-uraniferous regions, the highest accumulation of R226 was found in shrub species and vascular plants. The highest level of these radionuclides in non-ranilevel s in non-uraniferous regions was found in the soil. Ibrahim and Whicler (1987, 1992) determined concentration ratios of plant/soil values in native vegetation in the followin order of nuclides: U238>Th230>Po210> Ra>226>PB210, in various sites around a uranium mining and milling operations in the Western United States.

However, a great number of research was done on the content of uranium in cultivated plants. Most noted that data on the content of uranium in cultivated plants are those obtained from the experiments in which the parameters of soil characteristics and biological characteristics of plant and plant organs were taken into consideration. Lai (1983) found large differences in uranium concentrations depending not only on plant species but also on the cultivars. These results were obtained from plants grown in different regions of India. Frindik (1998) investigated the content of uranium in the samples of soils, vegetables, cereals, and fruits. He proved that the content of uranium depended on the plant species as well as on the locality.

Thirty four plant species were screened for uranium (U) accumulation from U contaminated soil. There was a significant difference in U accumulation among plant species. Sunflower (Helianthus annuus) and Indian mustard (Brassica juncea) accumulated more U than other plant species. Sunflower and Indian mustard were selected as potential U accumulators for further study in one U mine tailing soil and eight cultivated soils (pH range4.7 to 8.1) contaminated with different rates (100 to 600 mg U(VI) kg-1). Uranium fractions of contaminated soils (exchangeable, carbonate, manganese (Mn), iron (Fe), organic, and residual) were determined periodically over an 8-week incubation period. Uranium accumulated mainly in the roots of plant species. The highest concentration of U was 102 mg U kg-1 in plant shoots and 6200 mg U kg-1 in plant roots. Plant performance was affected by U contamination rates, especially in calcareous soils. Plants grown in soils with high carbonate-U fractions accumulated the most U in shoots and roots. The lowest plant U occurred in clayey acidic soils with high Fe, Mn, and organic U-fractions. The effectiveness of U remediation of soils by plants was strongly influenced by soil type. Soil properties determined the tolerance and accumulation of U in plants (Shahandehand and Hossner, 2002).

The aim of these investigations was to determine whether there were any differences in the uptake; that is the content of uranium in the cultivated biologically wide apart plant species, and also how the content of uranium varies in the plants grown on the deposit of uranium mines. Investigation of this problem is important from several aspects. Besides determining the content of uranium in plants, it is also necessary to find out all the changes in plants themeselves caused by uranium, that is to determine its phytotoxicity. If a high level of uranium content is found in certain plant species, that particular species could be used in the experiments as a model for determining the phytotoxicity of this element. Finally, plants with higher levels of uranium would not be grown in the vicinity of uranium mines, nor would they be used as food . All the above mentioned points are clocely connected to the problems of the protection of the human environment.

Uptake and accumulation of U has been studied in plants native to uranium mine sites, but not in cultivated plants which are commonly consumed by humans. Our study was conducted to better understand uptake and accumulation of U in beans (Phaseolus vulgaris), cabbage (Brassica oleracea), lettuce (Lactuca sativa), maize (Zea mays), onions (Allium cepa), potatoes (Solanum tuberosum), spinach (Spinacia oleracea), and sunflowers (Helianthus annuus) grown on a deposit of the Kalnna-Gabrovnica uranium mine located in Serbia during two years, with an average uranium content of 17 ppm (Sarić et.al.,1995).

Eight plant species were investigated with the aim of determining whether and what differences in uranium content can be found among the cultivated plant species i.e.tehir cultivars. All plant species, i.e. their cultivars, were planted on deposit during the last decade of April during two years. The experiment was carried out in four iterations on the elementary plots one square meter in size. During the vegetative period of plants, nitrogen, phosphorous, and potassium mineral fertilizers were applied. Plant specimens from salad, spinach, and onions were picked at the end of June. In the beginning of August, vegetative organs were taken from other plant species, while the fruit of granins were picked in September, during the phase of full maturation (Figure 11, 12 and 13).

Figure 11. Experimental field in closed uranium mine in Kalna (potato and green beans).

Figure 12. Experimental field in closed uranium mine in Kalna (lettuce).

Figure 13. Experimental field in closed uranium mine in Kalna (maize).

Obtained results of the contents of uranium for all plant species are given separately for roots and above-ground parts of plants. Separate values were obtained for corncob and corn grains, beans and the seedcase, potato tubers, and onion heads. Also, the values for the content of uranium in leaves for corn, sunflower, beans, potatoes, and cabbage were obtained form the leaves of different age groups.

Content of uranium in the roots (Table 7) showed that differences between plant species were rather prominent. The highest content of uranium was found in the roots of onions, significantly higher during the whole period of investigation, then salad sunflowers. The lowest content of uranium was found in the roots of cabbage and potatoes.

Table 7. Content of uranium in the root of different plant species (ppm)

Rang of plant		First year		Second year
1.	Onion	8,38	Onion	15,78
2.	Salad	8,07	Spinach	10,18
3.	Sunflower	5,22	Salad	9,12
4.	Beans	5,15	Sunflower	7,85
5.	Spinach	4,46	Maize	6,13
6.	Maize	4,36	Potato	5,39
7.	Cabbage	4,28	Beans	2,95
8.	Potato	1,89	Cabbage	2,02
5%		2,24		2,04
LSD				
1%		3,04		4,01

Content of uranium in the above-ground plant parts is shown in Table 8. The above-ground parts of the investigated plants also showed a large scale difference in the content of uranium. Salad and spinach had significantly higher concentrations of uranium in comparisson to other investigated plant species. The biggest differences between the plant species depending on the year of investigation were found in potatoes and sunflowers. It is characteristic that salad and spinach have high contents of uranium in both plant organs, while cabbage was characterized by low uranium content in both roots and the above-ground parts. The grading list made according to the content of uranium in the roots and above-ground parts of the investigated plants show that onions have the highest content of uranium in the roots, while the uranium content in the above-ground parts falls in the group of plants with the lowest level of uranium content.

In Contrast to the root and above-ground parts, content of uranium was much lower (0,03 ppm) in corn grain and in single beans (0,02 ppm). In potato tubers, it was 0,08 and in the onion heads it was 0,07 ppm. Concentration of uranium in corncob was 0,04 and in bean seedcases it was 0,07 ppm.

Content of uranium in plant leaves of different age groups is presented in Table 9. Obtained results show that the uranium content in the older leaves of the five differences were strongly expressed in potatoes, sunflowers, beans, cabbage, and corn when the investigations were made during the period of full maturity of corn and sunflowers.Not only was the content of uranium higher in the older leaves, but the differences in the content between the young and old leaves were bigger than those in the first stage of investigation.

Table 8 Content of uranium in the aboveground parts of (ppm)

Rang of plant		First year		Second year
1.	Salad	1,12	Salad	1,19
2.	Spinach	0,87	Potato	1,06
3.	Beans	0,44	Spinach	0,74
4.	Potato	0,39	Beans	0,54
5.	Cabbage	0,24	Sunflower	0,49
6.	Maize	0,21	Maize	0,48
7.	Sunflower	0,14	Cabbage	0,15
8.	Maize	0,33	Onion	0,13
5%		0,18		0,40
LSD				
1%		0,24		0,54

Table 9. Content of uranium in young and old leaves of some plant species

Leaves	Maize	Sunflower	Beans	Cabbage	Potato
		I stage of plant growth			
Young	0.07	0.24	0.53	0.23	0.39
Old	0.15	0.35	0.76	0.41	0.66
5%	0.04	0.07	0.15	0.06	0.38
LSD					
1%	0.08	0.14	0.27	0.12	0.65
		II stage of plant growth			
Young	0.11	0.40			
Old	0.19	1.04			
5%	0.04	0.19			
LSD					
1%	0.08	0.36			

The fact that the uranium content is higher in the roots than in the above-ground parts of plants speaks about its mobility. The value for the uranium content in the salad root was high while the value for the above-ground parts was the highest among the investigated plants. Typical examples of immobilization of uranium in roots is onion, for which root values were extremely high for both years of investigation, while the values for the above-ground parts were extremely low. Characteristic results of the distribution of uranium in certain root tissues were obtained by Mordechai et al. (1988) for Azolla. In the comparison of our results to those obtained by Mordechai, the most interesting data is that which shows that uranium present in the root of Azolla was not detected in the above-ground parts. Keeping in mind the results obtained for Azolla, which is an aquatic plant, our results show that uptaken uranium is transferred to and distributed in all organs of investigated cultivated plants and that it accumulated in different amounts.

This can be seen from the values obtained for generative organs and leaves of plants of different ages. According to the results obtained by Weisshaar (1993), content of natural uranium radionuclide, as is Pb210, was the highest in the most widely spread plants cultivated for food (salad and spinach), also shown by our results. According to Sheard (1986), high content of U in lichen and moss shows that the primary source of U for these plants is not the soil. The low content of U in vascular plants form non-uraniferous regions that show that it is the result of the content of U in ground water and soil solution. However, the question of the amount of U, from the soil but uptaken by above ground plant organs due to the atmospheric precipitation remains unsolved.

It is known that translocation of certain elements of mineral nutrition is different not only between the root and the above-ground parts but also between different leaves. So, the transport of Ca is much weaker from older to younger leaves in comparison with the transport of N or P, i.e. the uptake Ca accumulates more in older than in younger leaves. Our results on the uranium content in younger and older leaves of plants during certain phases of ontogenetic development show that its content is higher in older than in younger leaves, which makes uranium, concerning this characteristic, similar to Ca. The differences in the intensity of translocation and retranslocation are in the first place dependant on the specific role of particular elements in the physiological and biochemical processes.

Generally, content of uranium in roots for all investigated plants was higher than in above-ground parts. Older leaves accumulated more U than younger leaves. This indicates that uptake and translocation of U is plant species dependent.

Plants showed resistance survival in real conditions "In situ" and indicated a possibility of their use in phytoremediation of uranium contaminated soils.

CONCLUSION

Widespread use of nuclear energy, application of weapons with depleted uranium, nuclear testing, production and application of phosphoric fertilizer, coal combustion, oil and gas production, mineral processing and formation radioactive waste landfill, improper waste storage practices, and uranium tailings are the main anthropogenic sources of environmental uranium contamination. As a consequence , there may be a risk for ecosystems, agro-systems, and health because uranium is a natural radiotoxic and chemotoxic heavy metal. Solving this global problem requires appropriate management and strategy for uranium contaminated soils, and that includes the application of currently innovative available remediation technologies based on chemico physical and biological methods.

The presence of contaminants such as metals in the soil and environment has prompted governments worldwide to initiate environmental laws, policies, and programs to address concerns especially when the environment exceeds natural ambient levels. Selection of a remediation option for a site contaminated with uranium is complex, time consuming, and site specific (McInture, 2003). So, there is a need for the development of new in situ applicable remediation technologies, with simple applications without obstructing the ecosystem.

Applications of this technique must be economically suitable especially in underdeveloped countries where these problems are present.

Phytoremediation is an emerging technology with considerable promise for remediating and restoring contaminated sites. The continued urgency for contaminated site clean up in developed and developing countries alike, demands that phytoremediation be given careful, serious, and immediate consideration as an effective, promising, and innovative environmental technological solution. Phytoremediation is expected, in certain situations, to demonstrate superior economic, technical, and environmental advantages over traditional physical, chemical, and thermal remediation techniques (Mclnture, 2003). This technique makes more environmental sense than soil washing or removing the contaminated soils (Willey, 2007).

This chapter shows the behavior and uranium uptake by plants used in human nutrition: corn, soy beans, sunflowers, carrots, onions, potatoes, radishes, sugarbeets, beans, cabbage, lettuce, maize, and spinach. This is very important because these plants are cultivated with well established methods. Studies are also helpful in terms of screening and applying the plants for phytoremediation (biological techniques) like hyperaccumulator plants. Knowing the properties of plants is not enough for successful phytoremediation of contaminated areas.

Successfully applied phytoremediation technology is correlated with the degree of contamination in the area, physical-chemical properties of soil, hydrogeological and morphological characteristics, and properties of native flora. For this reason there is not a universal technology, because each site is specific and requires a special approach. In many cases soil cleanup goals depend on the concentration of pollutants.

Like every technique, phytoremediation has its advantages and disadvantages. The first advantage is the fact that phytoremediation is the least harmful method because it uses natural organisms, with minimal disturbance to the environment. Also, there is a greater public acceptance of this method. Furthermore, this is the most cost-effective technique in comparison to some traditional (conventional) ways of cleaning the soil. There is of course the possibility of re-exploitation of metals – used plants may produce recyclable metal-rich plant residues. Disadvantages of this method lie in that it is limited to the surface layer, which in turn depends on the depth of the roots of the plants (usually 15 – 30 cm deep). Also elemental (uranium) concentrations in the plant tissues can vary with time during the vegetation season. The major disadvatege of the phytoremediation tehnique is the time requirement. Phytoremediation is frequently slower than traditional techniques (physical, chemical, or thermal), requiring several growing seasons for site clean up. For this reason, the technology is not an appropriate solution when the target contaminant presents an imminent danger to human health or the environment. More over, there are problems with the leaching the pollutants in groundwater, which is entirely possible to prevent, as well as the possibility of pollutants entering the food chain (Saier and Trevors, 2010; Vaněk et al., 2010; Mclnture, 2003; Chang et al. 2005). A great deal of research focus and investment can therefore be expected into the management of the soil–plant system in the next 50 years (Willey, 2007).

Remediation of areas contaminated with uranium requires a multidisciplinary approach and combination of various technical measures for complete control of pollutants and prevention of uranium entering in the food chain, with the aim of protecting the population from ionizing radiation.

REFERENCES

AbdEl-Sabour, M.F. (2007). Remediation and bioremediation of uranium contaminated soils, *Electronic Journal of Environmental, Agricultural and Food Chemistry, 6(5),* 2009-2023.

Adler, T. (1996). Using plants to tackle polluted water and soil. *Science news*, *150,* 42-43.

Aery, N.C. & Jain, G.S. (1997). Effect of uranyl nitrate on seed germination and early seedling growth of *Triticum aestivum* L. *Biologia. Bratislava, 52*, 115–119.

Aleksahina R.M. (1985) *Radioekologia oročeanogo zemledenija*, Moskva, p186.

Alkorta, I. Hernandez-Allica, J. Becerril, J.M. Amezaga, I. Albizu, I. Onaindia, M. & Garbisu, C. (2004). Chelate-enhanced phytoremediation of soils polluted with heavy metals. *Reviews in Environmental Science and Biotechnology***,** *3*, 55-70.

Arey, J.S. Seaman, J.C. & Bertsch, P.M. (1999). Immobilization of U (VI) in contaminated sediments by apatite addition. *Environmental Science and Technology, 33,* 337–342.

Baker, A.J.M. McGrath, S.P. Sidholi, C.M.D & Reeves, R.D. (1994). The possibility of in situ heavy metal decontamination of polluted soils using crops of metal-accumulating plants. *Resources***,** *Conservation and Recycling***,** *11*, 41–49.

Bennett, FA. Tyler, EK. Brooks, RR. Gregg, PEH. & Stewart, RB. Fertilization of hyperaccumulators to enhance their potential for phytoremediation and phytomining. In: Brooks RR, editor. *Plants that Hyperaccumulate Heavy Metals*, Cambridge, CABI Publishing; 1998; 249–259.

Bostick, C.B. Fendorf, S. Barnett, O.M. Jardine, P. M. & Brooks, C.S. (2002). Uranyl Surface Complexes Formed on Subsurface Media from DOE Facilities. *Soil Science Society of America Journal, 66,* 99-108.

Canon, H. L. (1952). The effect of uranium-vanadium deposits on the vegetation of Colorado Plateau. *American Journal of Science, 250*, 735–770.

Chang, P. Kim, K.W. Yoshida, S. & Kim, S.Y. (2005). Uranium accumulation of crop plants enhanced by citric acid, *Environmental Geochemistry and Health, 27*, 529–538.

Charbonneau, M. (2009) Current Problems and Recent Advances of in situ Bioremediation of Uranium contaminated sites MMG 445 Basic Biotechnology eJournal [http://ejournal.vudat.msu.edu/index.php/mmg445/article/viewArticle/375], 5(1).

Chen, B. Roos, P. Borggard, O.K. Zhu Y.G. & Jakobsen, I. (2005). Mycorrhiza and root hairs in barley enhance acquisition of phosphorus and uranium from phosphate rock but mycorrhiza decreases root to shoot uranium transfer, *New Phytologist*, *165,* 591-598.

de Souza, M.P. Huang, C.P. Chee, N. & Terry, N. (1999). Rhizosphere bacteria enhance the accumulation of selenium and mercury in wetland plants. *Planta, 209(2)*, 259–263.

Dhankher, O.P. L., Rosen, Y.B.P. Shi, J. Salt, D. Senecoff, J.F. Sashti, N.A. & Meagher, R.B. (2002). Engineering tolerance and hyperaccumulation of arsenic in plants by combining arsenate reductase and gamma-glutamylcysteine synthetase expression. *Nature Biotechnology, 20*, 1140–1145.

Diaz MF. and Kirkham BM. Testing the Manipulation of Soil Availability of Metals In: Willey N., Walker JM. editors, *Methods in Biotechnology, 23*: *Phytoremediation: Methods and Reviews.* Totowa, NJ: Humana Press Inc., 2007; 121-129.

Dozol, M. & Hagemann, R. (1993). Radionuclide migration in groundwaters, review of the behaviour of actinides, *Pure and Applied Chemistry*, *65*, 1081–1102.

Duff, M.C. & Amrhein, C. (1996). Uranium(VI) adsorption on geothite and soil in soil carbonate solutions, *Soil Science Society of America, 60,* 1393–1400.

Duquène, L. Vandenhove, H. Tack, F. van der Avoort, E. Wannijn, J. van Hees, M. Phytoavailability of uranium: influence of plant species and soil characteristics, In: Merkel BJ, Hasche-Berger A, editors. *Uranium in the Environment. Mining Impact and Consequences*. Netherlands: Springer Verlag -Berlin Heidelberg; 2006; 469–476.

Duquène, L. Vandenhove, H. Tack, F. Meers, E. Baeten, J. & Wannijn, J. (2009). Enhanced phytoextraction of uranium and selected heavy metals by Indian mustard and ryegrass using biodegradable soil amendments, *Science of the total environment, 407,* 1496 – 1505.

Dushenkov, S. Mikheev, A. Prokhnevsky, A. Ruchko, M. & Sorochinsky, B. (1999). Phytoremediation of radiocesium-contaminated soil in the vicinity of Chernobyl, Ukraine, *Environmental Science and Technology, 33,* 469–475.

Dushenkov, V. Nanda Kumar, P. B. A. Motto, H. & Raskin, I. (1995). Rhizofiltration: the Use of Plants to Remove Heavy Metals from Aqueous Streams. *Environmental Science and Technology, 29*, 1239 - 1245.

Ebbs, S.D. Brady, D.J. & Kochian L.V. (1998). Role of U speciation in the uptake and the translocation of uranium by plants, *Journal of Experimental Botany, **49***, 1183–1190.

Ebbs, S.D. Norvell, W.A. & Kochian, L.V. (1998). The effect of acidification and chelating agents on the solubilization of uranium from contaminated soil. *Journal of Environmental Quality*, *27*, 1486–1494.

Ebbs, S.D. Brady, D. Norvell, W.A., & Kochian, L. (2001). Uranium speciation, plant uptake, and phytoremediation. *Practice Periodical of Hazardous, Toxic and Radioactive Waste Management, **5,*** 130–135.

Elless, M.P. & Lee, S.Y. (1998). Uranium solubility of carbonate-rich uranium-contaminated soils. *Water, Air, and Soil Pollution, 107,* 147–162.

Fellet, G. Marchiol, L. Perosab, D. & Zerbia, G. (2007). The application of phytoremediation technology in a soil contaminated by pyrite cinders. *Ecological Engineering*, *31,* 207–214.

Finneran, K.T. Anderson, R.T. Nevin, K.P. & Lovley, D.R. (2002). Potential for bioremediation of uranium-contaminated aquifers with microbial U(VI) reduction. *Soil and Sediment Contamination*, *11,* 339–357.

Gavrilescu, M. (2009). Overview on remediation of uranium contaminated environments. Lab International [http://www.labint-online.com/featured-articles/overview-on-remediation-of-uranium-contaminated-environments/index.html], (25/09/2009).

Gavrilescu, M., Pavel, L. V. & Cretescu, I. (2009). Characterization and remediation of soils contaminated with uranium, *Journal of Hazardous Materials, 163,* 475–510.

Grenthe, I. Fuger, J. Konings, RJM. Lemire, RJ. Muller, AB. Nguyen-Trung, C. Wanner, J. *The Chemical Thermodynamics of Uranium*, Amsterdam: Elsevier. 1992.

Grubišić, M. Stojanović, M. & Adamović V. Station and important plants of soils and waste water – phytoremediation. In: Stojanović M, editor. *Radionuclide contamination of Serbian soil and remediation possibility*. Belgrade, Serbia: ITNMS; 2006; 291-317.

Hafez, M.B. & Ramadan, Y.S. (2002). Treatment of radioactive and industrial wastes by *Eichornia crassipes. Journal of Radioanalytical and Nuclear Chemistry, 252(3),* 537–540.

Hashimoto, Y. Lester, B.G. Ulery, A.L. & Tajima, K. (2005) Screening of potential phytoremediation plants for uranium-contaminated soils: a sand culture method and metal-removal model, *Journal of Environmental Chemistry, 15,* 771–781.

Hinton, G.T. Knox, A.S. Kaplan, D.I. & Sharitz, R. (2005). Phytoextraction of uranium and thorium by native trees in a contaminated wetland. *Journal of Radioanalytical and Nuclear Chemistry, **264**,* 417–422.

Huang,W.J. Blaylock, J.M. Kapulnik, Y. & Ensley B.D. (1998). Phytoremediation of Uranium-Contaminated Soils: Role of Organic Acids in Triggering Uranium Hyperaccumulation in Plants, *Environmental Science and Technology, 32(13),* 2004–2008.

Ibrahim, S.A. & Whicker, F.W. (1987). Plant accumulation and plant/soil concentration ratios of Pb210 and Po210 at various sites within a uranium mining and milling operation. *Environmental and Experimental Botany, 27(2),* 202-213.

Ibrahim, S.A. & Whicker, F.W. (1992). Comparative Plant Uptake and Environmental Behavior of U.Series Radionuclides at a Uranium Mine-mill. *Journal of Radioanalytical and Nuclear chemistry-articles, 156(2),* 253-267.

Igwe, J.C. Nnorom, I.C. & Gbaruko, B.C. (2005). Kinetics of radionuclides and heavy metals behavior in soils: Implications for plant growth. *African Journal of Biotechnology, 4,* 1541-1547.

Jagetiya, B. L. & Purohit, P. (2006). Effects of various concentrations of uranium tailings on certain growth and biochemical parameters in sunflower. *Biologia, Bratislava, 61(1),* 103—107.

Knox, A.S. Brigmon, R.L., Kaplan, D.I. & Paller, M.H. (2008). Interactions among phosphate amendments, microbes and uranium mobility in contaminated sediments. *Science of The Total Environment, 395(2-3),* 63-71.

Langmuir D. (1978). Uranium solution-mineral equilibria at low temperatures with applications to sedimentary ore deposits. *Geochimica et Cosmochimica Acta, 42,* 547–569.

Leita, L. Nobili, M.D.E. Mondini, C. & Garcia, M.T. (1993). Response of leguminosae to cadmium exposure. *Journal of Plant Nutrition, 16,* 2001–2012.

Leštan, D. Enhanced heavy metal phytoextraction, In: Mackova M et al. editors, *Phytoremediation Rhizoremediation,* Springer the Netherlands; 2006; 115 - 132.

Luo, W. & Baohua,G.B.(2009). Dissolution and Mobilization of Uranium in a Reduced Sediment by Natural Humic Substances under Anaerobic Conditions. *Environmental Science and Technology, 43(1),* 152–156.

Martell, A.E. & Smith, R.M. (2003). NIST critically selected stability constants of metal complexes; Version 7.0. NIST: Gaithersburg.

Matijašević, S. Daković, A. Tomašević, Č.M., Stojanović, M. & Ileš D. (2006) Uranium(VI) adsorption on surfactant modified heulandite/clinoptilolite rich tuff. *Journal of the Serbian Chemical Society, 71(12),* 1323–1331.

McBride, M.B. & Spiers, G. A. (2001). Trace element content of selected fertilizers and dairy manures as determined by ICP-MS. *Communications in Soil Science and Plant Analysis, 32(1-2),* 139-156.

McIntyre, T. Phytoremediation of Heavy Metals from Soils In: Scheper T., Tsao DT. editors, *Advances in Biochemical Engineering/Biotechnology, 78.* New York: Springer-Verlag Berlin Heidelberg; 2003; 97-123.

Mkandawire, M. Taubert, B. & Dudel G.E. (2005). Resource manipulation in uranium and arsenic attenuation by *Lemna gibba* L. (duckweed) in tailing water of a former uranium mine. *Water, Air, and Soil Pollution, 166*, 83–101.

Mordechai, S. Elisha, T.O. Eberhardt, F. & Huttermann, A. (1988). Localization and Toxic Effects of Cadmium, Copper, and Uranium in Azolla. *Plant Physiology, 88*, 30-36.

Morishima, H. Koga, H. Kawai, H. Honda, Y. & Katsurayama, K. (1976). Experimental studies on the movement of uranium in the soil to vegetables. *Radioisotopes, 25,* 773-780.

Myung, C.J. & Thornton, I. (1997). Environmental contamination and seasonal variation of metals in soils, plants and waters in the paddy fields around a Pb–Zn mine in Korea. *Science of the Total Environment, 198,* 105–121.

Nanda-Kumar, P. B. A. Dushenkov, V. Motto, H. & Raskin, I. (1995). Phytoremediation: the use of plants to remove heavy metals from soils. *Environmental Science and Technology, 29,* 1232-1238.

NRC 1999. Evaluation of Guidelines for Exposures to Technologically Enhanced Naturally Occurring Radioactive Materials. National Research Council, National Academy Press.Washington D.C. 1999.

Otero, X.L. & Macias, F. (2002). Variation with depth and season in metal sulfides in salt marsh soils. *Biogeochemistry, 61*, 247–268.

Pabalan, R.T. Turner D.R. Nertetti, F.P. & Prikryl, J.D. Uranium(VI) sorption onto selected mineral surfaces, key geochemical parameters, in: Jenne EA., editor. *Adsorption of Metals, Geomedia.* San Diego CA: Academic Press; 1998;

Pande, S.K. Deshmukh, A.N. & Shrivastava, P.K. (1993). The significance of the dormant stage in the growth-cycle of deciduous plants for biogeochemical uranium prospecting, India. *Journal of Geochemical Exploration,* ***46***, 365–374.

Peterson, PJ. (1971) Unusual accumulations of elements by plants and animals. *Sci Prog ,59,* 505–526.

Phillips, H.D. Gu, B. Watson, B.D. Roh, Y.L. Liang, L. & Lee, S.Y. (2000). Performance Evaluation of a Zerovalent Iron Reactive Barrier: Mineralogical Characteristics *Environmental Science and Technology, 34 (19),* 4169–4176.

Rajković, BM; Đorđević, A. Possibility of Serbian soil contamination by depleted uranium after NATO bombing 1999, In: Stojanović M, editor. *Radionuclide contamination of Serbian soil and remediation possibility.* Belgrade, Serbia: ITNMS; 2006; 167-219.

Ramaswami, A. Carr, P. & Burkhardt, M. (2001). Plant-uptake of uranium: hydroponic and soil system studies. *International Journal of Phytoremediation, 3*, 189–201.

Rodríguez, B.P. Tomé, V. F. Fernández, P.M. & Lozano, J.C. (2006). Linearity assumption in soil-to-plant transfer factors of natural uranium and radium in *Helianthus annuus* L. *Science of the Total Environment, 316*, 1–7.

Rothbaum, P. McGaveston, D.A. Wall, T. Johnston, A.E. & Mattingly, G.E.G. (1979). Uranium accumulation in soils from long-continued applications of superphosphate. *European Journal of Soil Science*, *30*, 147–153.

Saier, H.J. & Trevors T.J. (2010). Phytoremediation. *Water, Air, and Soil Pollution*, *205(1),* 61-63.

Sarić, M.R. Stojanović, M. & Babić, M. (1995) Uranium in plant species grown on Natural barren soil. *Journal of Plant Nutrition, 18(7),* 1509–1518.

Shahandeh, H. & Hossner, L. R. (2002). Role of soil properties in phytoaccumulation of Uranium. *Water, Air, and Soil Pollution, 141*, 165–180.

Sheard, J.W. (1986). Distribution of uranium series radionuclides in upland vegetation of northern Saskatchewan. II. Patterns of accumulation among species and localities. *Canadian Journal of Botany, 64,* 2453–2463.

Sheppard, S.C. Evenden, W.G. & Anderson, A.J. (1992). Multiple assays of uranium in soil. *Environmental Toxicology and Water Quality, 7*, 275–294.

Shtangeeva, I. & Ayrault, S. (2004). Phytoextraction of thorium from soil and water media. *Water, Air, and Soil Pollution, 154*, 19–35.

Shtangeeva, I. (2008). Uptake of uranium and thorium by native and cultivated plants. *Journal of Environmental Radioactivity, 30*, 1-6.

Stojanović, M. Milojković, J. Grubišić, M. & Ileš, D. (2009). Application alumosilicate minerals for remediation of uranium contaminated land. The first international congress "Engineering, materials and management in the process industry", Jahorina, Republika Srpska, 523-526.

Stojanović, M. Grubišić, M. Stevanović, D. Milojković, J. & Ileš, D. (2008). Remediation of the Serbian soils contaminated by radionuclides in the function of the sustainable development. *Chemical Industry and Chemical Engineering Quarterly, 14(4),* 265-267.

Stojanović, M. Mrdaković- Popić, J. Stevanović, D. & Martinović, Lj. (2006). Phosphorus Fertilizers As Source Of Uranium In Serbian Soils. *Agronomy for Sustainable Development, 26,* 179-183.

Stojanović, M. Babić, M. Stevanović, D. & Martinović Lj. (2006). Effect of long-term application of phosporus fertializers on uranium contaminated Serbian soils. In: Stojanović M, editor. *Radionuclide contamination of Serbian soil and remediation possibility*. Belgrade, Serbia: ITNMS; 2006; 67-117.

Stojanović, M. & Martinović, Z. Overview of analytical methods for the determination of uranium. In: Sarić M. editor. *The effect of use of phosphorus fertilizers on uranium contamination 5*. Belgrade: Serbian Academic of Sciences and Art; 1993; 1-18.

Stojanović, M. Stevanović, D. Iles, D. Grubišić, M. & Milojković, J. (2009). The Effect of the Uranium Content in the Tailings on Some Cultivated Plants. *Water, Air, and Soil Pollution, 200,* 101-108.

Stojanović, M., Stevanović, D. Iles, D. Grubišić, M. & Milojković, J. (2009). Phytotoxic effect of uranium on the growing up and development plant of corn. *Water, Air, and Soil Pollution*, DOI 10.1007/s11270-009-0208-4 (Online) 1573-2932.

Susarla, S. Medina, V.F. & McCutcheon, S.C. (2002). Phytoremediation: an ecological solution to organic chemical contamination. *Ecological Engineering, 18,* 647–658.

Suthersan S.S. and F.C. Payne, (2005) *In situ Remediation Engineering*. CRC Press / ITPS.

Tessier, A. Campbell, P.G.C. & Bisson, M. (1979). Sequential extraction procedure for the speciation of particulate trace metals. *Analytical Chemistry, 51*, 844–850.

Tsao, T. David. Overview of Phytotechnologies, In: David T. Tsao *Phytoremediation*, Springer-Verlag Berlin Heidelberg; 2003: 1-50.

Tunney, H. Stojanović, M. Mrdaković - Popić, J. McGrath, D. & Zhang, C. (2009). Relationship of soil phosphorus with uranium in grassland mineral soils in Ireland using soils from a long-term phosphorus experiment and a National soil database. *Journal of Plant Nutrition and Soil Science, 172(3)*, 346–352.

UNSCEAR (1993) (United Nations Scientific Committee on the Effects of Atomic Radiation). 1993. Report to the General Assembly, with scientific annexes, New York.

UNSCEAR, 2000. *Report to General Assembly, with Scientific Annexes, Sources and Effects of Ionizing Radiation*. United Nations, New York.

USEPA, Introduction to Phytoremediation, EPA 600/R-99/107, 2000, U.S. Environmental Protection Agency, Office of Research and Development, Cincinnati, OH, 2000.

Vandenhove, H. Van Hees, M. & Van Winckel, S. (2001). Feasibility of phytoextraction to clean up low-level uranium-contaminated soil. *International Journal of Phytoremediation, 3,* 301–320.

Vaněk, T; Podlipna, R; Soudek P. General Factors Influencing Application of Phytotechnology Techniques. In: Kulakow PA, Pidlisnyuk, VV editors. *Application of Phytotechnologies for Cleanup of Industrial, Agricultural, and Wastewater Contamination.* Netherlands: Springer; 2010; 1-13.

Wall, J.D. & Krumholz, L.R. (2006).Uranium reduction. *Annual Review of Microbiology, 60,* 149–166.

Weisshaar, R. (1993). Pb210 in Foodstuffs. *Deutsche lebensmittel rundschau, 89,* 205-208.

Wellman DM JS Fruchter, VR Vermeul, and MD Williams.2008.Challenges Associated with Apatite Remediation of Uranium in the 300 Area Aquifer . PNNL-17480, Pacific NorthwestNational Laboratory, Richland, WA. http://www.pnl.gov/main/publications/external/technical_reports/PNNL-17480.pdf

Willey, N. Soils Contaminated With Radionuclides: Some Insights for Phytoextraction of Inorganic Contaminants. In: Walker JM, editor. *Methods in Biotechnology, 23: Phytoremediation: Methods and Reviews.* Totowa, NJ: Humana Press Inc.; 2007; 305-317.

Yamaguchi, N. Kawasaki, A. & Iiyama, I. (2009). Distribution of uranium in soil components of agricultural fields after long-term application of phosphate fertilizers. *Science of the Total Environment, 407(4),* 1383-1390.

Zhou, P. & Gu, B. (2005). Extraction of oxidized and reduced forms of uranium from contaminated soils: effects of carbonate concentration and pH. *Environmental Science and Technology, 39,* 4435–4440.

Zunić, S.Z. Mietelski, J.W. Blazej, S. Gaća, P. Tomankiewicz, E. Ujić, P.Celiković, I. Cuknić, O. & Demajo, M. (2008). Traces of DU in samples of environmental bio-monitors (non-flowering plants, fungi) and soil from target sites of the Western Balkan region, *Journal of Environmental Radioactivity, 99,* 1324–1328.

In: Handbook of Phytoremediation
Editor: Ivan A. Golubev
ISBN: 978-1-61728-753-4

Chapter 4

A Decade of Research on Phytoremediation in North-East Italy: Lessons Learned and Future Directions

Luca Marchiol*[*], *Guido Fellet, Filip Poščić and Giuseppe Zerbi
Dipartimento di Scienze Agrarie e Ambientali, Università di Udine,
Via delle Scienze 208, I-33100 Udine, Italy

Abstract

The interest in phytoremediation has been rapidly increasing in the last twenty years. A relevant number of scientific papers have investigated several aspects of the matter, first exploring the physiological processes and then the molecular characteristics of the plants to find the genes responsible for the metal (hyper)tolerance.

Since 1998, our research group has had a number of projects concerning phytoremediation financed with public funds. In 2005, we designed the first Italian in situ experiment of phytoremediation. This trial took place within an area included into the polluted area Laguna di Grado e Marano (Grado and Marano lagoon) which belongs to the national priority list (Ministry Decree 468/2001). The experimental site was located on the property of an Italian chemical company in Torviscosa (Udine). Several aspects of phytoremediation were investigated, such as: (i) phytoextraction potential of *Sorghum bicolor* and *Helianthus annuus*; (ii) the growth of *Populus* spp. and *Salix* spp. and trace element uptake; (iii) strategies for the enhancement of metal absorption from the soil and for increasing the translocation rate in plants; (iv) metals' mobility and their availability to plants and pedofauna. All the aspects were investigated both under pot and field trial conditions.

More recently, we worked on metallophytes and hyperaccumulators. Such species, being able to tolerate and accumulate high amounts of several elements, were proposed for phytostabilization of heavily polluted soils and mine tailings. The fertility of heavily polluted soils and mine tailings is always very low. Properly designed agronomic practices are expected to support plant growth and biomass yield. Pot experiments testing the effects of different levels of fertilization on the growth of *Thlaspi caerulescens* on polluted soils and mine tailings were done. In the summer of 2007, a field survey was

[*] Phone +39 432 558611; Fax +39.432.558603; E-mail: marchiol@uniud.it

conducted at the former lead/zinc mining site in Cave del Predil (Julian Alps) to investigate the presence of metallophytes.

Our learned lessons are consistent with the views prevailing in the scientific debate. After a decade of research, phytoextraction seems not feasible at the present state of knowledge. To the contrary, phytostabilization to decrease metal mobility is a realistic alternative.

Further research at field scale and efforts in discovering new hyperaccumulators and/or metal tolerant populations of native species must be done to promote phytoremediation to become a practical option for the remediation of polluted soils.

1. INTRODUCTION

In Europe, the number of sites requiring remediation in 2007 was estimated at 250,000 by the European Environment Agency (EEA, 2007). In the same year, in Italy, more than 1 million hectares, divided into 54 different sites, were included in the national list of polluted sites which represents 3% of the national territory. Thirteen thousand sites are likely to be included in the list and 4,400 of these have already proven to be contaminated (ISPRA, 2008). More than 170,000 ha are marine sites. Within the National Polluted Sites List, lay the main industrial areas. According to some recent estimates, 30 billion € are necessary to remediate the national polluted sites. Almost 20,000 ha of these polluted sites are located in two regions in the North East of the country: Veneto and Friuli Venezia Giulia.

The awareness of the negative impacts that the anthropogenic activities have been causing on the environment and the consequent huge costs that bare upon the countries to deal with the resulting environmental and health issues motivated the scientific community to develop alternative low cost technologies. Such alternatives include the phytotechnologies. Phytotechnologies are low impact approaches for either in situ or ex situ soil and water treatment and have become increasingly attractive to the environmental agencies and commercial practitioners (Raskin et al. 1997). In general, these techniques may be referred to as "gentle" remediation options that do not have a significant negative impact on the soil function and structure (Bardos et al. 2008).

The settlement and the maintenance of greeneries in polluted areas, other than the landscape aesthetical aspects, offer functional advantages such as (i) erosion prevention that might be responsible for the spreading of the pollutants, (ii) a better definition of a hydro-balance and, hence, a (iii) limitation of a possible leaching of the contaminants into the groundwater.

A great deal of progress has been achieved at the experimental level. Several comprehensive reviews by Salt et al. 1995, Chaney et al. 1997, McGrath and Zhao 2003, Pilon-Smits 2005, Chaney et al. 2007, Vangronsveld et al. 2009 and Wu et al, 2010 summarized many important aspects of this novel plant-based technology and the achievements of the scientific community.

The most fascinating application amongst the phytotechnologies is phytoextraction that has been defined as "the use of pollutant-accumulating plants to remove metals or organics from soil by concentrating them in the harvestable parts" (Salt et al. 1999). Significant efforts were devoted to find the group of higher plants to be used in phytoextraction. The efficiency of phytoextraction follows a simply rule: the rate of metal removal depends upon the biomass harvested and the metal concentration in the harvested biomass. Therefore, the most debated issue regards the choice of the most effective species: hyperaccumulators vs. biomass species. However, neither the hyperaccumulators nor the biomass species possess the features required; it is thought that the "Holy Grail" of phytotechnologies is currently hidden in the potential of molecular biology (Baker and Whiting, 2002).

Despite the intensive research in the last decade, another widening gap between science and practicality lies in the fact that very few field trials to demonstrate the feasibility of the phytotechnologies have been realized. So far, unrealistic field scale extrapolations from experimental data from lab and greenhouse trials have raised doubts about the feasibility of metal phytoextraction (Dickinson et al. 2009).

An inventory of the field trials performed in Europe in the years 2000–2008 indicated that 25 field trials took place in 9 European countries (SUMATECS, 2009). The phytoextraction potentials were evaluated, studying biomass species and, to a lesser extent, hyperaccumulators (Table 1).

The field trial established in Italy was managed by our group. Since 1998, we have been managing projects concerning phytoremediation. In the framework of a research project financed by the Italian Ministry of Research and in cooperation with groups from other Italian universities, in 2005, we designed an *in situ* experiment of phytoextraction using biomass crops. The trial took place in an industrial site included within the polluted area *Laguna di Grado e Marano* (Grado and Marano lagoon) belonging to the clean up National Priority List. Up to now and to our best knowledge, neither pilot scale trials of phytoremediation has been established in Italy, nor has any paper been published in scientific journals reporting information of similar experimental activities.

Unlike phytoextraction, phytostabilization is not intended to remove metal contaminants from a site, but rather to stabilize them by the accumulation in roots or the precipitation within the rhizosphere, reducing the risk to human health and the environment. It is applicable in scenarios where there is a potential risk of human health impacts, and the exposure to hazardous substances may be reduced to acceptable levels by containments. That is the case of highly polluted areas, where the removal of metals by phytoextraction using hyperaccumulators or crops is not efficient due to the slowness of the process (Dickinson et al. 2009). Phytostabilization is also advantageous when decontamination strategies are impractical because of the extent of the contaminated area or the lack of adequate funding (Santibáñez et al. 2008). It may also serve as an interim strategy to reduce the risk at sites where complications delay the selection of the most appropriate technique. A typical scenario in which phytostabilization could be considered is represented by the anthropogenic metalliferous sites (e.g., abandoned mining sites, smelter sites) where the presence of wastes and mine tailings can result in severe pollution and have anaesthetic impacts on the local environment.

With regard to the plants for phytostabilization, there is a general agreement about the potential of native metallophytes. Such plants must not accumulate metals into their aboveground biomass or else, they must localize the metal accumulation to the root tissues. Moreover, being able to grow in unfertile soils, such plants always have a high metal tolerance and are highly adapted to the local environmental conditions (Frérot et al. 2006).

In addition to the research on phytoextraction, a couple of years ago we began working on phytostabilization.

Table 1. Overview of metal phytoextraction field trials in Europe (modified from SUMATECS 2009)

Element	Plant species	State/Location	Reference
Cd, Cr, Cu, Ni, Pb, Zn	*S. viminalis*	B/Menen	Vervaeke et al. 2003; Meers et al. 2005
Cd, Zn	*B. napus*	B/Balen, Buden	Grispen et al. 2006
Cd, Cu, Zn	*Q. robur*, *P. alba*, *A. pseudoplatanus*	B/Deinze	Vandecasteele et al. 2008
Cd, Cu, Zn	*S. viminalis*, *N. tabacum*, *H. annuus*, *B. juncea*, *Z. mays*, *T. caerulescens*	CH/Dornach, Caslano	Keller et al. 2003; Hammer and Keller 2003
Cd, Cu, Zn	*B. pendula*, *S. viminalis*, *A. incana*, *F. excelsior*, *S. mougeotii*	CH/Le Locle	Rosselli et al. 2003
Cd, Pb	*Z. mays*	CZ/Pribram	Neugschwandtner et al. 2008
Pb	*Pelargonium* cvs.	F/Bazoches, Toulouse	Arshad et al. 2008
Cd, Zn	*T. caerulescens*	F/La Bouzule	Schwartz et al. 2003
As, Cu, Cd, Co, Pb, Zn	*H. annuus*, *S. bicolor*	I/Torviscosa	Marchiol et al. 2005; Vamerali et al. 2009
Cd, Zn	*B. napus*	NL/Budel	Grispen et al. 2006
Cd, Cr, Cu, Ni, Pb, Zn	*S. viminalis*	S/Uppsala, Enkóping, lake Malaren	Klang-Westin and Eriksson 2003; Dimitriou et al. 2006
Pb, Zn, Cu, Cd	*B. carinata*, *B. juncea*	SP/Aznalcollar	del Rio et al. 2000
Pb, Sb, Tl, Zn	*O. europea*, *P. alba*, Mediterranean shrubs	SP/Aznalcollar	Dominguez et al. 2008
Cd, Zn	*T. caerulescens*, *A. halleri*	UK/Bedfordshire	McGrath et al. 2006
As, Cd, Cu, Ni	*Betula* spp.	UK/Liverpool	French et al. 2006
Cu	*A. cordata*, *A. incana*, *A. glutinosa*, *C. monogyna*, *S. caprea*	UK/Merseyside, Manchester	Dickinson 2000
Cd, Cu, Ni, Zn	*Salix* spp. *T. caerulescens*	UK/Nottingham	Pulford et al. 2002; Maxted et al. 2007a,b;
Cd, Cu, Zn	*Salix* spp.	UK/Warrington	King et al. 2006

As in the previous case, we found a case study consisting of an abandoned mining site in the Alps. The first aim of this research was to characterize the native vegetation of the lead and zinc mine area in order to identify new plant species of potential use in phytostabilization. At first, we performed a survey to assess the level of contamination of the site, to determine the level of accumulation of elements in tissues and to identify metal tolerant species. This research is currently ongoing and aims also to develop and manage a future revegetation of the mine tailings and other mine wastes dumping sites.

The following chapters illustrate the main activities and achievements reached in the last decade of research by our group.

2. Phytoextraction of Metals and Metalloids in an Industrial Site

An intense and thorough investigation to find a polluted site suitable for on site experiments of phytoextraction in Italy, brought the research team of Udine in contact with the chemical enterprise Caffaro srl. The company owns an industrial plant located in Torviscosa, a little town in the province of Udine (Italy). The industrial site is know by the Italian environmental agencies and the Italian Environmental Ministry for its severe pollution due to the processes that took place within its perimeter. The experimental site of Torviscosa appeared to be particularly suitable for studies on phytoextraction, for the presence of several metals in the soil layer explored by the roots of the plants.

The chemical plant of Torviscosa was established by SNIA in 1938. Initially, the main activity was the production of cellulose as primary component for synthetic fibers. During the following decades the plant had been revamped in order to produce primary base and fine chemicals. The Italian decree 468/2001 included the site and the surrounding areas in the National Priority List of polluted sites under the name of Laguna di Grado e Marano.

The soil of the experimental site is polluted by several heavy metals and As. The main source of the pollution was an industrial facility for the sulphur recovery, that roasted pyrite ore, which contains primarily pyrite (FeS_2), smaller amounts of chalcopyrite ($CuFeS_2$), sphalerite (ZnS), magnetite (Fe_3O_4), As and several trace metals (mainly Cd, Cu and Zn). The industry, which generated pyrite cinders as a by-product of sulphuric acid manufacture, ceased the activity in the late 1970s. The pyrite cinders had been dumped on a 5 ha area next to the very facility. This resulted in a thick layer of wastes, which was covered with a thinner layer of coarse material named topsoil. Over the years, the site had been colonized by a vegetation cover of ruderal species.

Following is the research activity described in its several steps, starting from the preliminary lab scale tests to the field ones.

2.1. Pyrite Cinders Pot Trials: Testing the Crops

Before the field scale experiments, a pot trial was planned in order to observe the impacts in terms of toxicity of heavy metals and metalloids to the plants growth and survival. Being the substrate contaminated by several heavy metals, it was essential to previously test the plants ex situ to observe their behavior when in contact with the pollutants. Besides, it is known that different species have different uptake vocation for different heavy metals. For these reasons, it is of great importance and interest in this scenario, to preliminarily investigate the species before proceeding with the field trials. And so was done for Torviscosa.

The pot experiment was performed in controlled conditions and took place in two different sessions. The first one involved four crops – *Glycine max*, *Helianthus annuus*, *Sorghum bicolor* and *Zea mays* – which were chosen for their economic importance and diffusion in the local areas and for their high biomass production. For the second one, two species were selected: *S. bicolor* and *H. annuus* (Fellet et al. 2007).

The trial results gave insight on the crops responses to the fertilization as an agronomical practice to be applied on the assisted in situ phytoextraction experiment.

In the spring of 2004, deals of topsoil and pyrite cinders were collected from the experimental site of Torviscosa and characterized.

The topsoil and the pyrite cinders collected at the experimental site were also used to prepare different substrates for the experiments. Two topsoil-pyrite cinders mixtures were prepared and used to fill the pots. Topsoil and cinders were mixed respectively in 1:1 (v/v) ratio (P50%) and in 1:1,5 ratio (P66%).

The first session of the experiment aimed to examine the response of the four crops to the experimental substrates. Seeds of plants of *G. max* cv. Sapporo, *H. annuus* cv. 289x978, *S. bicolor* cv. Isadei and *Z. mays* cv. PR34F02 were sown in 2 L pots containing topsoil as control and P50% and P66% mixtures as treatments. Plants were grown during 40 days on a laboratory bench lit by lamps which gave 500 μmols m^{-2} s^{-1} of photosynthetically active radiation (PAR) to the plant top with a 12:12 h photoperiod. The pots were rotated randomly and daily to equalize their light exposure. Ambient temperature was maintained at 25±2 °C. Each pot was irrigated every two days with distilled water.

In the same growth conditions, the second session of the experiment took place. It was designed to identify the possible relationships between the nutritional state of the plants and the uptake of the heavy metals and As. The growth period was 50 days.

Seeds of *H. annuus* cv. 289x978 and *S. bicolor* cv. Isadei were sown in 2 L pots containing the P66% mixture. The same species were used by Madejon et al. (2003) and Murillo et al. (1999), in a soil polluted by pyrite cinders in Spain. Two levels of fertilization were defined.

The controls received no fertilization (No Fert), while the other plants did (+Fert). In particular, *S. bicolor* received an amount of ammonium nitrate, calcium phosphate and potassium chloride equivalent to respectively 125 kg ha^{-1} N, 40 kg ha^{-1} P_2O_5 and 100 kg ha^{-1} K_2O); *H. annuus* received 150 kg ha^{-1} N, 60 kg ha^{-1} P_2O_5 and 290 kg ha^{-1} K_2O.

At the end of both the experiments, the plants were carefully harvested, washed with deionised water, divided into roots, shoots and leaves and oven dried at 105 °C for 24 h.

2.1.1. Characterization of Polluted Soil

The samples collected from the site and the ones from the substrates prepared for the pot experiment were air-dried and screened by means of a 2 mm sieve for the characterization (Table 2).

Table 2. Physical and chemical characteristics of the substrates

Parameter	Topsoil	P50%	P66%
Sand 2-0.2 mm (% w/w)	56	36	27
Silt 0.2-0.02 mm (% w/w)	21	37	41
Silt (% w/w)	0	5	12
Clay (% w/w)	23	22	20
pH (H_2O)	8.6	7.14	7.22
Organic C (%)	1.09±0.03[(†)]	0.68±0.04	0.58±0.02
CEC (Cmol kg^{-1})	5.82±0.24	3.32±0.04	2.33±0.14
EC (mS/cm)	0.31±0.01	1.93±0.01	1.88±0.01
Active $CaCO_3$ (%)	4.22±0.40	1.74±0.09	0.99±0.001
N (%)	0.73±0.05	0.58±0.16	0.42±0.06
P Olsen (mg kg^{-1})	11.8±1.55	53.5±0.82	62.7±0.77
Exchangeable K (mg kg^{-1})	17.3±0.87	17.2±0.83	10.6±1.04

[(†)] Mean standard error.

Amongst the data reported, the pH values measured in all the substrates are surprisingly above the neutrality. Considering the chemical properties of the pyrite cinders, this evidence indicates that these wastes lost most of their acidic potential.

From an environmental point of view, this means a lower mobility of the pollutants along the soil profile. On the other hand, this suggested that the bioavailable fraction of the metals could not be very high.

In order to determine the heavy metals contents, the soil samples were oven dried (105 °C for 24 h). Subsequently, the dry samples were acid digested with a microwave oven (CEM, MARSXpress) according to the EPA method 3051 (USEPA, 1995a). After the mineralization, the samples were filtered (0.45μm PTFE) and diluted.

Total content of As, Cd, Cr, Cu, Ni, Pb and Zn content in the substrates were determined by an ICP-OES (Varian Inc., Vista MPX). The analysis for As were done separately with a continuous flow vapour generation system that provided improved detection limits for the element (VGA-77, Varian Inc.).

A similar protocol was used to analyze the plant samples. In fact, the EPA method 3052 (USEPA, 1995b) was adopted. Method 3052 is considered to be a rapid multi-element, microwave assisted acid digestion suitable for ICP-OES analyses (Varian Inc., Vista MPX).

Table 3 reports the concentration of heavy metals in the pyrite cinder and in the experimental substrates; the thresholds for the heavy metals in soils, according to the decree DM 471/1999 (currently 152/2006), are reported as term of comparison.

According to this regulation, when the concentrations of pollutants measured in a soil are above the thresholds, the soil is classified as potentially polluted and further investigations are mandatory in order to check the degree of hazardousness of the very site in comparison with the surrounding areas and to define the type of intervention required for the case.

Table 3. Concentration of heavy metals (mg kg^{-1}) in the pyrite cinders and in the experimental substrates and, as reference, the thresholds established by the law DM 471/1999 for residential$^{(a)}$ and industrial areas$^{(b)}$, respectively

Substrate	As	Cd	Cr	Cu	Ni	Pb	V	Zn
Pyrite Cinders	964	9.84	3.41	3,290	4.03	278	7.96	1,448
Topsoil	19.9	0.83	14.1	72.8	16.4	22.4	20.9	96.5
P50%	586	6.43	6.15	1,589	8.18	203	11.2	989
P66%	718	7.53	5.95	1,943	6.94	244	9.90	1,322
DM 471/99$^{(a)}$	20	2	150	120	120	100	90	150
DM 471/99$^{(b)}$	50	15	800	600	500	1,000	250	1,500

2.1.2. Experimental Design and Data Analysis

Both the sessions were set up in a randomized block design. To evaluate the phytoextraction potential of the species, the following parameters were calculated: (i) the plant average concentration of the plants fractions (mg kg^{-1}), (ii) the bioconcentration factor ($BCF=[Me]_{roots}/[Me]_{soil}$), and (iii) the translocation factor ($TF=[Me]_{shoots}/[Me]_{roots}$) (Zhao et al. 2003). The experimental data was subjected to a two-way analysis of variance (ANOVA). The comparisons between treatments used the Student-Newmann-Keuls's test ($p<0.05$). Each treatment consisted of 8 replicates. As far as the statistical analysis of the ICP data is concerned, in every sample for which the element was not detectable by the instrument, the element concentration was assumed to be one-half of the respective detection limit (Nadal et al. 2004). The detection limits were: 3 μg L^{-1} for As, 1.5 μg L^{-1} for Pb, 0.9 μg L^{-1} for Cu, 0.4 μg L^{-1} for Co and 0.2 μg L^{-1} for Cd and Zn.

2.1.3. Soil Fertility and Plant Growth

The soil analysis showed clearly that all the substrates were very poor in organic C (OC) and the nutrient content of the topsoil was rather low (Table 2). However, the addition of pyrite cinders reduced to almost half the content of the OC in the topsoil. The substrates were also low in CEC, total N and exchangeable K. The contents of exchangeable Ca, Fe and Mg increased in the mixtures due to the fraction of pyrite cinders.

The heavy metals and As concentrations measured in the sole cinders were very high (Table 3). The mixtures resulted multi-contaminated by heavy metals and As, as expected. In particular, taking as reference the thresholds of the decree 471/1999, the concentration of As and Cu are higher than the limits for the industrial areas, while the values of Cd, Pb and Zn exceed the limits for residential areas but not the thresholds set for industrial soils.

At the end of the growing period, for both the sessions of the experiment, no macroscopic evidences of metal toxicity on the plants tissues were observed: all the species tolerated the highest dose of pyrite cinders. The total biomass of *G. max*, *S. bicolor*, *Z. mays*, and *H. annuus* measured at the end of the first session is showed in figure 1. In agreement with the expectations, the ANOVA indicated a significant effect of the factor "species" ($p<0.001$) but this was probably due to the different ecological *habitus* of the plants. On the contrary, the factor "treatment" was statistically significant at $p=0.038$. The plants of *Z. mays* grown in the substrate containing the highest rate of pyrite cinders surprisingly produced more biomass than others. No differences were recorded for the other species.

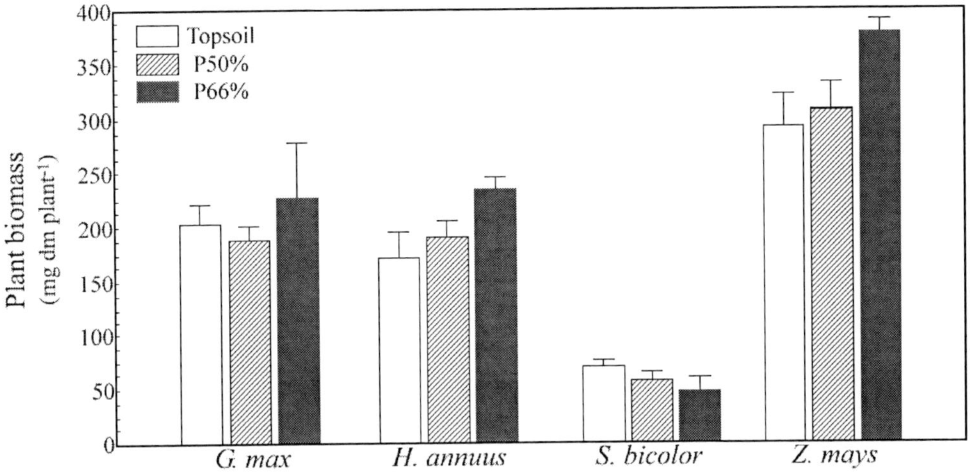

Figure 1. Total biomass of *G. max*, *H. annuus*, *S. bicolor* and *Z. mays* grown on topsoil, P50%, and P66% substrates in session 1. Vertical bars represent the standard error for eight replicates.

In the second session *H. annuus* and *S. bicolor* were studied. Figure 2 reports the dry weight of the plant biomass of both control and fertilized plants. A part from the differences between the species, the fertilization did not result in an increase of the total plant biomass. In fact, the dry matter accumulation in the plant tissues was lower in fertilized plants than in non fertilized ones although the difference was statistically not significant.

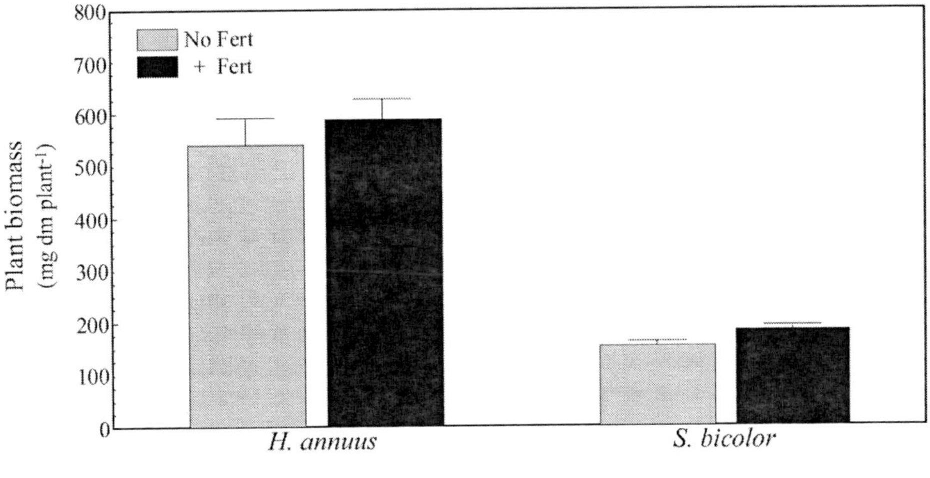

Figure 2. Total biomass of *H. annuus* and *S. bicolor* grown on unfertilized and fertilized P66% substrate in session 2. Vertical bars represent the standard error for eight replicates.

2.1.4. Heavy Metals and Arsenic in Plant Fractions

The average concentration of As and heavy metals found in the biomass of the four species are the result of the process of uptake and translocation (Figure 3). Both experimental factors – "species" and "substrate" – affected significantly the plant growth for all the elements. It was observed that *G. max* and *S. bicolor* accumulated much more As and heavy metals than *H. annuus* and *Z. mays* (Figure 3).

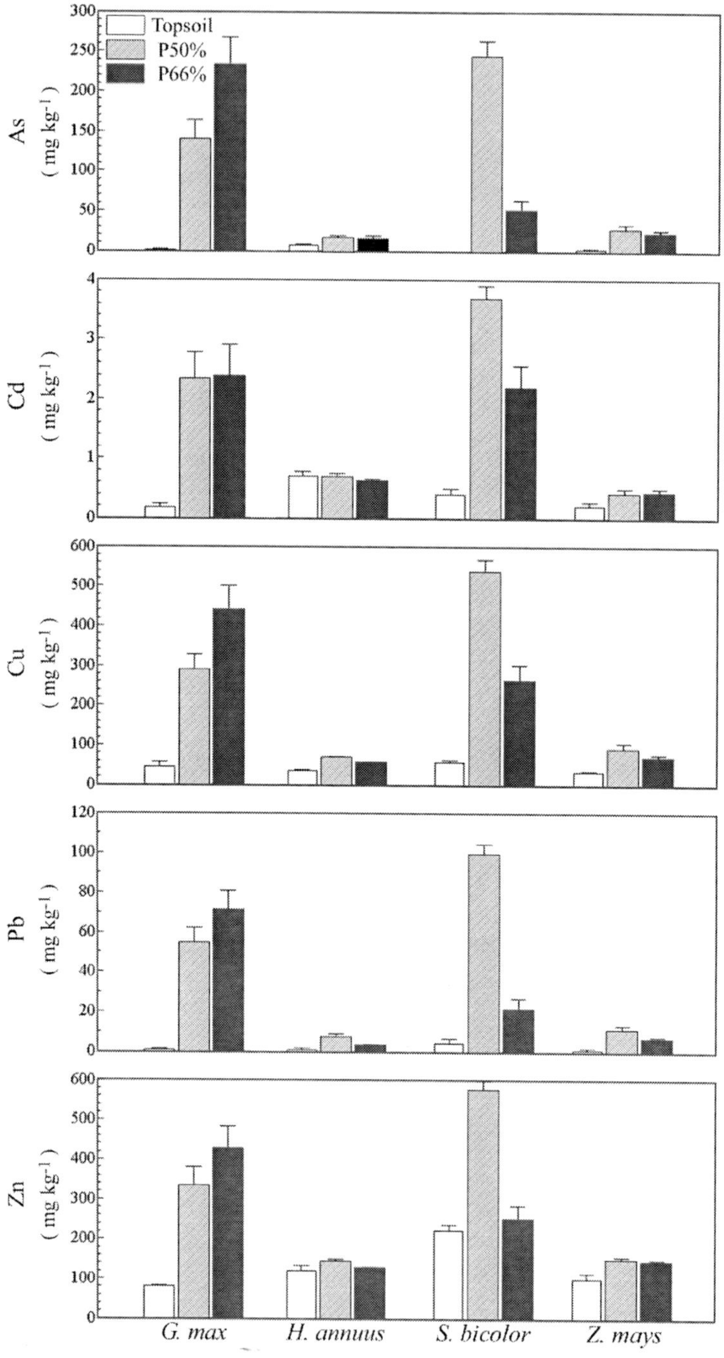

Figure 3. Average concentration (mg kg^{-1}) of As, Cd, Cu, Pb and Zn in plants of *G. max*, *H. annuus*, *S. bicolor* and *Z. mays* grown on topsoil, P50%, and P66% substrates in session 1. Vertical bars represent the standard error for eight replicates.

The concentration of As recorded in *G. max* increased of about 1.7 folds passing from P50% to P66%. On the opposite, *S. bicolor* seemed to exclude As when the concentration rose towards the highest level of soil contamination. In fact, the plants grown in P66% had an

As concentration more than 4.8 times lower than the P50% ones (Figure 3). The plants of *H. annuus* and *Z. mays* showed no response to the treatment having about the same As content. A similar behavior can be extended to Cu and Pb, even if some differences in the absolute values of concentration of the elements were recorded, in particular for *S. bicolor* P66%. The plants of *G. max* did not respond positively to the increase of Cd and Zn concentration in the substrate. Perhaps, the plants reached their tolerance thresholds, ceasing the metal uptake process (Figure 3).

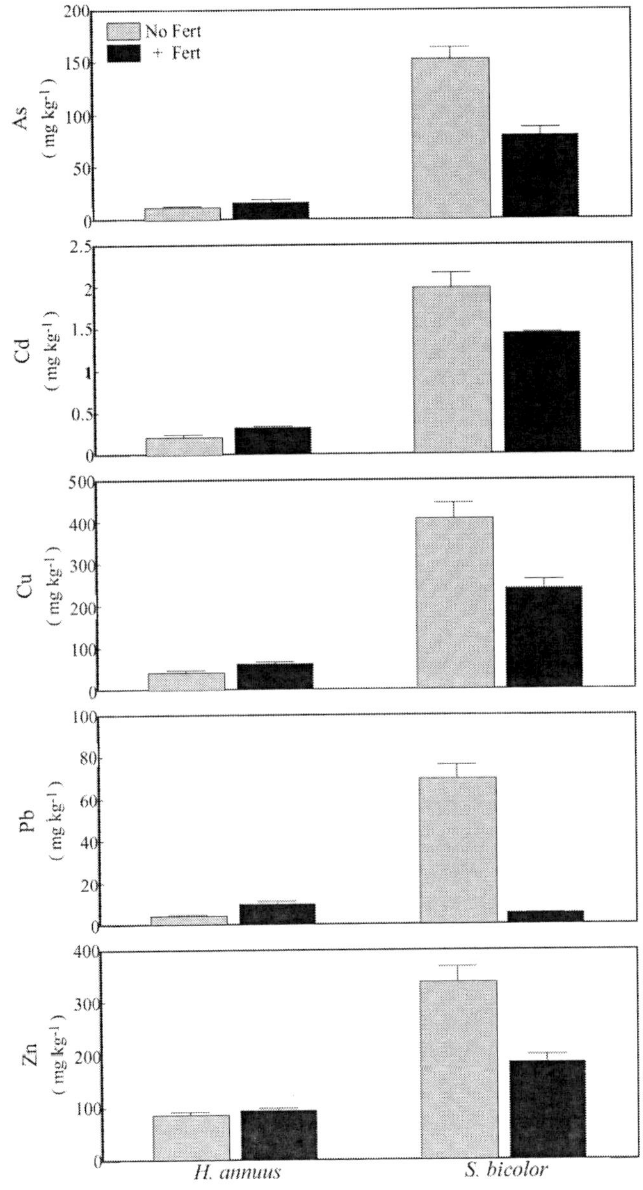

Figure 4. Average concentration (mg kg^{-1}) of As, Cd, Cu, Pb and Zn in plants of *H. annuus* and *S. bicolor* grown on unfertilized and fertilized P66% substrate in session 2. Vertical bars represent the standard error for eight replicates.

In figure 4, the concentration of As and heavy metals found in plants of *H. annuus* and *S. bicolor* at the end of the second session of the experiment are showed. Both the experimental factors – species and fertilization – affected significantly the metal content, while the interaction "species x fertilization" indicated that these species behave differently in response to the treatment.

The amelioration of the substrate fertility did not result in an increase of the concentration of As and heavy metals in the plants of *S. bicolor* (Figure 4); in particular the concentration of As, Cd, Cu and Zn in fertilized plants was respectively 2, 1.5, 1.7 and 1.8 times lower than the control plants.

On the contrary, the concentration of heavy metals in the plants of fertilized *H. annuus* was higher than unfertilized ones. The highest increase was recorded for Pb which was 3 times higher than control; the concentration of the other elements was on average 1.3 times higher than controls.

2.1.5. Bioconcentration and Translocation Factor

The mobility of the heavy metals from the polluted substrate into the roots of the plants, and the ability to translocate the metals from roots to the harvestable aerial part, were evaluated respectively by means of the bioconcentration factor (BCF) and the translocation factor (TF). BCF is defined as the ratio of metal concentration in the roots to that in the soil while TF is the ratio of metal concentration in the shoots to the roots (Yoon et al. 2006).

Table 4 reports the BCF and TF calculated for the plants observed in session 1.

The P50% substrate allowed the plants to bioconcentrate much more As and heavy metals than P66%. In fact, for each element the BCF of P50% plants was significantly higher ($p<0.001$) than others (Table 4). The species behaved differently; *G. max* and *S. bicolor* had the highest BCF of As and Cd respectively, while the BCF of all the elements were < 0.5 for *H. annuus* and *Z. mays*. These species, however, having the highest TF for all the metals, resulted to be the most efficient in the translocation of such elements. A part from the case of Cd, this process seemed not to be influenced by the level of pollution of the substrate (Table 4). Table 5 reports the BCF and TF calculated for the plants of session 2.

The BCF values are all <1, indicating that the elements concentrations never exceed the respective soil level. However, some differences have been observed among the species and comparing the treated and untreated plants. The average value of BCF of *S. bicolor* is always and significantly higher than *H. annuus*. BFC calculated on the experimental data are coherent with those presented by Mattina et al. (2003).

Compared to the control plants, the BCF in treated plants of *S. bicolor* decreased of a percentage comprised between 19 and 34%. Conversely, in *H. annuus* the BCF always responded positively to the fertilization, even at a lower absolute value (Table 5).

The TF values indicate the rate of translocation of the metal absorbed by the root system to the aboveground biomass. In general, TF did not follow the same trend observed for BCF. On average the TF is greater in *H. annuus* than *S. bicolor*. Higher translocation was observed in *H. annuus* for Cd, Cu and Zn overall, while lower values resulted for *S. bicolor*. In the case of TF, the interaction between the experimental factors resulted significant, too. With regard to As and Cd, the fertilization slightly increased the translocation factor of *S. bicolor*. The same treatment caused the decrease of the translocation root-shoot of As (-76%), Cu (-56%) and Zn (-47%) in *H. annuus* (Table. 5).

Table 4. Results of *Student-Newman-Keuls* test (p=0.05) and ANOVA for the bioconcentration factor (BCF) and the translocation factor (TF) calculated for plants of *G. max*, *S. bicolor*, *Z. mays* and *H. annuus* grown during the first session of the pot experiment on the substrates P50%, and P66%. BCF = =$[Me]_{roots}/[Me]_{soil}$; TF=$[Me]_{shoot}/[Me]_{roots}$

		BCF					TF				
Factors		As	Cd	Cu	Pb	Zn	As	Cd	Cu	Pb	Zn
Treatment	P50%	0.55 a	0.78 a	0.47 a	0.62 a	0.70 a	0.04 a	0.11 b	0.06 a	0.98 a	0.12 a
	P66%	0.37 b	0.47 b	0.34 b	0.35 b	0.44 b	0.02 a	0.21 a	0.06 a	0.78 a	0.12 a
Species	*G. max*	0.81 a	0.97 a	0.57 a	0.81 a	0.81 a	0.02 a	0.05 b	0.04 bc	0.38 b	0.04 c
	S. bicolor	0.58 b	0.98 a	0.63 a	0.69 a	0.83 a	0.04 a	0.15 b	0.03 c	0.7 b	0.06 c
	Z. mays	0.22 c	0.31 b	0.23 b	0.23 b	0.42 b	0.01 a	0.04 b	0.05 b	0.75 b	0.11 b
	H. annuus	0.22 c	0.24 b	0.19 b	0.20 b	0.23 c	0.04 a	0.41 a	0.12 a	1.70 a	0.27 a
ANOVA (P values)	Sp.	***	***	***	***	***	ns	***	***	***	***
	Trea.	***	***	***	***	***	ns	***	ns	ns	ns
	Sp. X Trea.	***	*	***	***	***	ns	ns	***	***	***

Table 5. Results of *Student-Newman-Keuls* test (p=0.05) and ANOVA for the bioconcentration factor (BCF) and the translocation factor (TF) calculated for unfertilized (No Fert) and fertilized (+ Fert) plants of *S. bicolor* and *H. annuus* observed in session 2. BCF = =$[Me]_{roots}/[Me]_{soil}$; TF=$[Me]_{shoot}/[Me]_{roots}$

		BCF					TF				
Factors		As	Cd	Cu	Pb	Zn	As	Cd	Cu	Pb	Zn
Treatment	No Fert	0.34 a	0.39 a	0.32 a	0.41 a	0.37 a	0.02 a	0.10 a	0.07 a	0.03 b	0.31 a
	+ Fert	0.27 b	0.36 a	0.30 a	0.17 b	0.31 a	0.02 a	0.09 a	0.03 b	0.21 a	0.16 b
Species	*S. bicolor*	0.44 a	0.58 a	0.47 a	0.37 a	0.55 a	0.02 a	0.06 b	0.01 b	0.21 a	0.05 b
	H. annuus	0.17 b	0.18 b	0.15 b	0.21 a	0.16 b	0.02 a	0.13 a	0.1 a	0.04 b	0.43 a
ANOVA (P values)	Sp.	***	***	***	***	***	ns	ns	***	***	***
	Trea.	*	ns	ns	***	ns	ns	ns	*	***	*
	Sp. X Trea.	***	***	***	***	***	ns	ns	*	***	*

2.1.6.Foremost Outcomes for the Field Trial

Relevant and interesting differences in the average concentration of the metals found in the tissues of the plants were detected. In the perspective of a full scale application of phytoremediation, this is not "per se" very conclusive since it only gives an indication about the capability of the crops to uptake the pollutants occurring in the soil. For the implications of the following field experiment, the most important evidences were those regarding the BCF and TF. On the other hand these parameters are considered the basic test that can be used to evaluate the potential of plants for phytoremediation. In general, being the BCF's values < 1 the crops are not as efficient as hyperaccumulators in the uptake of As and heavy metals. The BCF values ranged from 0.192 to 0.979. These values are coherent with those reported by Yoon et al (2006) on native plants collected in pollutes areas. The behavior by *G. max* resulted very interesting in terms of both BCF and biomass growth. Considering the BCF, it comes out that the species has a similar behavior to *S. bicolor* with the exception for As. Taking into account the capacity of this plant to translocate the pollutants, together with its BCFs, it appears that soybean can preferentially uptake and accumulate Pb to the aboveground biomass rather than the other pollutants. Andrade et al. (2004) provided data about the increase of Pb uptake by *G. max* that was recorded after the establishment of a double symbiosis with an arbuscular mycorrhizal fungus and *Bradyrhizobium*. No inoculation was used in the pot experiment, so it was possible to suppose that *G. max* could perform better. Data from the second session of the pot experiment showed that *S. bicolor* is more efficient than *H. annuus* in the translocation of As, Cd, Cu and Pb; Zn appeared to be more translocated towards the shoots in *H. annuus*. Although phytoremediation is still a new technology, in the last few years a lot of basic research has been carried out in an attempt to understand how plants take up large quantities of metals, together with the mechanisms of metal translocation from roots to shoots, storage and detoxification. However more applied projects in the field are needed to clarify the real potential of this technology, since the knowledge needed for proper cultivation of plants is still lacking. One of the general principles of phytoremediation is to match the proper plant species to the contaminated site. This means that consideration must be given to soil condition, micro-climate, pests and diseases as well as the contaminant to be cleaned up (Licht and Isebrands, 2005). Agronomists will have to further provide solutions to applied aspects of crop management practices which have the potential to maximize the cleanup efficiency of plants. Planting practices, seedbed conditions, irrigation, crop rotations, crop cycle duration, and, finally, methods and scheduling of the harvest should be investigated. To date, for all these aspects there is a lack of experience in field conditions particularly for multi-contaminated soils. Several questions need appropriate clarification and research.

3. Field Trial in Torviscosa (Italy)

Subsequently the pot experiment, the field trial took place. *Sorghum bicolor* and *Helianthus annuus* were chosen as first species to be studied on site. The in situ phytoremediation trial in Torviscosa was established in the early Spring 2005. One amongst the specific aims of the work was to observe the concentration of the metals in plants during the crop cycle in order to establish the amount of metal removable by the crops. This information is of particular interest in evaluating the clean up duration, the number of annual croppings required to reach the target of soil remediation and the costs of the whole process. In fact, to evaluate the extensive application of phytoremediation, we need to estimate the mass of elements that is expected to be transferred from the contaminated soil to the plant biomass. This data rise considering (i) the initial level of contamination and (ii) the target value to be reached after remediation (such as the one defined by the local legislation). Finally, the output of metals obtained by a single crop cycle, expressed in grams per hectare per year, must be multiplied by the number of clean up cycles. Some examples of such calculations were reported by Robinson et al. (1999) and Keller et al. (2003) for remediation of contaminated soils and by Ernst (2005) for mine wastes.

3.1. Soil Preparation and Experimental Design

The preparation of the site begun with the removal of the vegetation that grew on the topsoil layered on the pyrite cinders. Once the area was rid of the plants, the 30 cm topsoil layer was sieved to separate the coarse material bigger then 5cm. This resulted on a reduction of the topsoil thickness to 25 cm. By ploughing the experimental area to a depth of 30 cm, the seedbed resulted to be composed by a mixture of the two waste materials. Other agronomic practices were performed to give more uniformity to the substrate before the seeding. The overall site area measured 2000 m^2 and it was the set of different experiments, conducted autonomously or in cooperation with other Italian research units whose role in the project is discussed on paragraph 3.4. Half of the area was used by the research team of Udine. The element characterization, carried out following the same protocols described for the pot trials experiment, revealed the following average element concentrations: As 310 mg kg^{-1}, Cd 4.30 mg kg^{-1}, Co 51 mg kg^{-1}, Cu 1,530 mg kg^{-1}, Pb 233 mg kg^{-1} e Zn 980 mg kg^{-1}. The concentration of As and Cu exceeded the limits for industrial sites; the values of Cd, Co, Pb and Zn are higher than the limits fixed for green and residential areas. For the purposes of the field research, both limit values were considered, even if the experimental site was located inside an industrial area. Table 6 reports the results of the soil physical and chemical characterization.

At the end of the soil preparation, the plots were traced. According to the experimental design, the site was divided in 27 plots of 15.7 m^2 (4.5 x 3.5 m) each. The field trial was arranged in a randomized block design with two factors — species and treatment — and three replications.

The experiment dealt with the nutritional state of the polluted soil without any other practice to enhance the phytoextraction process. The plants growing in the field trial were subjected to three treatments: the Ctrl-treatment consisted of the pure polluted soil; in the Fert-treatment the native soil was subjected to mineral fertilization, while the Org treatment was established by amending the native soil with cow manure. Fert-treated *S. bicolor* and *H. annuus* received 150 kg N ha^{-1} ($(NH_2)_2CO$ in granules) but this amount was split in two times: 100 kg N ha^{-1} at sowing and 50 kg N ha^{-1} after 7 weeks. Phosphorus was provided to both crops as calcium phosphate ($CaHPO_4$) at the dose of 60 kg CaHPO ha^{-1}; potassium fertilization was done with granular KCl at the dose of 60 kg K_2O ha^{-1} and 290 kg K_2O ha^{-1} for *S. bicolor* and *H. annuus* respectively.

Table 6. Basic parameters measured in the soil of the field trial of Torviscosa (Italy)

Parameter		
Sand (2 - 0.05 mm)	(% w/w)	69.4
Silt (0.02 - 0.002 mm)	(% w/w)	25.4
Clay (< 0.002 mm)	(% w/w)	5.17
pH (H_2O)		7.75
Organic C	(%)	9.10
CEC	(Cmol kg^{-1})	5.5
EC	(dS m^{-1})	2.65
P Olsen	(mg kg^{-1})	7.78
Exchangeable K	(mg kg^{-1})	132
Exchangeable Ca	(mg kg^{-1})	3,041
Exchangeable Mg	(mg kg^{-1})	198
Exchangeable Na	(mg kg^{-1})	13.3
Total Fe	(%)	9.02

Organic fertilization was provided by adding the native soil with 90 t ha^{-1} of mature cow manure which resulted containing 316 mg C_{org} g_{ss}^{-1}; 19.5 mg N g_{ss}^{-1}; 8.26 mg P_{tot} g_{ss}^{-1}; 171 mg Cu_{tot} g_{ss}^{-1}; 1,178 mg Zn_{tot} g_{ss}^{-1}. The fertilization was done immediately before the seeding. On may the 5th, 2005 seeds of *H. annuus* cv. Carnia were sowed at the density of 8 plants m^{-2}. Seeds of *S. bicolor* cv. Isadei were sowed broadcast at the density of 300 seeds m^{-2}. The plots were watered during the crop cycle in two drought periods using a sprinkler irrigation system which supplied the plots with about 7 mm h^{-1}. Weeds were removed manually during the crop cycle.

3.2. Plants Development: Biomass Production and Trace Elements Removal

Six samplings were done during the crop cycle. Plants of *S. bicolor* were collected by sampling 0.4 m^2 of plots. In the case of *H. annuus*, two sub-samples per plot were collected by sampling three adjacent plants in 0.75 m in two different areas of the plot for a total of 6 plants. The dry biomass of the plants fractions were measured after drying for 24 h at 105 °C in a forced-air oven the fresh samples. The plants biomass in terms of tons per hectare was estimated. Being the native soil very poor, the biomass production during the growth cycle of the *S. bicolor* and *H. annuus* was 10-20 fold higher in those plots that received a nutrient supply (both in mineral and organic form) than in the ones growing in the control soil.

In the case of *S. bicolor*, the ANOVA revealed a significant effect of the treatments. The biomass production was higher in Fert soil than in Org ($p < 0.001$). At the 6th sampling date, which corresponded to 112 days after sowing, the highest productions of aboveground biomass were recorded: 1.54, 22.1 and 16.9 tons of dry matter per hectare for Ctrl, Fert and Org respectively. *Helianthus annuus* behaved in an opposite way if compared with *S. bicolor*.

In fact, the highest biomass was recorded for the plants grown in the manure amended plots. At the 6th sampling date, the mean aboveground biomass were 0.37, 3.38 and 6.30 tons of dry matter per hectare for Ctrl, Fert and Org respectively.

Table 7. Concentrations of As, Cd, Co, Cu, Pb and Zn in roots and shoots of *S. bicolor* and *H. annuus* recorded in plant samples collected 112 DAS. Data are means of six replicates ± standard error. Significance levels of Species (Sp), treatments (Trea) and the interaction Species x Treatments (Sp x Trea) are shown

Species	Treatments	Roots						Shoots					
		As	Cd	Co	Cu	Pb	Zn	As	Cd	Co	Cu	Pb	Zn
		(mg kg^{-1})	(mg kg^{-1})	(mg kg^{-1})	(mg kg^{-1})	(mg kg^{-1})	(mg kg^{-1})	(mg kg^{-1})	(mg kg^{-1})	(mg kg^{-1})	(mg kg^{-1})	(mg kg^{-1})	(mg kg^{-1})
Sorghum bicolor	Ctrl	67.5±18	1.75±0.31	9.42±3	594±121	60.1±26	265±70	5.23±0.93	0.20±0.06	0.47±0.05	28.6±4.5	2.73±0.38	86.4±19
	Fert	47.8±13	1.71±0.11	7.14±0.4	535±36	60.1±4.1	328±105	7.28±1.30	0.26±0.05	0.91±0.03	29.6±6.3	4.10±0.48	55.5±3.6
	Org	32.5±9.8	1.35±0.4	5.39±0.5	468±67	22.7±2	466±110	13.1±4.3	0.25±0.05	1.77±0.63	48.9±14	6.08±2.9	115±2.3
Helianthus annuus	Ctrl	48.6±7.17	2.31±0.68	7.48±1.54	837±141	42.9±4.63	242±43.4	0.62±0.18	0.64±0.08	0.55±0.18	36.2±12	2.52±0.82	118±9.09
	Fert	142±88.6	1.57±0.1	8.73±1.27	706±195	54.0±7.74	222±23.6	1.93±0.07	0.34±0.06	0.71±0.13	42.0±5.22	3.53±0.88	112±6.27
	Org	12.2±2.2	0.73±0.12	1.99±0.25	336±26.1	15.7±4.65	121±22.4	0.98±0.39	0.22±0.09	0.19±0.02	23.2±0.81	0.53±0.13	140±33.4
ANOVA	*Sp.*	*n.s.*	*n.s.*	*n.s.*	*n.s.*	*n.s.*	*	***	*	*	*n.s.*	*n.s.*	*
	Trea.	*n.s.*	*	*	*	*	*n.s.*	*n.s.*	*n.s.*	*n.s.*	*n.s.*	*n.s.*	*n.s.*
	Sp. X Trea.	*n.s.*	*n.s.*	*n.s.*	*n.s.*	*n.s.*	*n.s.*	*n.s.*	*	*	*n.s.*	*n.s.*	*n.s.*

* = P<0.05 ; ** = P<0.01; *** = P<0.001.

As far as the metals accumulation is concerned, table 7 reports the concentrations which were measured in the roots and in the shoots of *S. bicolor* and *H. annuus* collected at the moment of the higher biomass yield. The root concentration values resulted higher than those observed in the shoots. As a consequence, the TF – which is typically >1 in the case of hyperaccumulators – was < 1 for all the elements for the two crops.

The amount of metals that the crops could have hypothetically extracted from the polluted site of Torviscosa was calculated by multiplying the metal concentration measured in the harvestable fraction (Table 7) and the biomass weight.

Figure 5 reports the values of metal removal by harvesting the sole aboveground biomass and the whole plant. The data from the field trial are compared to those discussed by Madejón et al. (2003) and Keller et al. (2003). The papers provided information about phytoextraction of heavy metals by crops cultivated in field trials established in the area affected by the pyrite tailings spill in Aznalcóllar (Spain), and at the experimental site of Dornach (Switzerland) respectively. In the lands contaminated by the mine sludges of Aznalcóllar, *H. annuus* and *S. bicolor* were studied (Murillo J.M. et al.,1999) while in the polluted soil of Dornach *Brassica juncea*, *Zea mays*, *Nicotiana tabacum*, stands of *Salix* sp. and also the hyperaccumulator *Thlaspi caerulescens* were cultivated over two years.

3.2.1. Arsenic

The highest value of metal output was reached by *S. bicolor* which removed from the soil 219 g ha^{-1} of As. This value is about 10 folds higher than the metal output obtained by *H. annuus*. The absence of nutrients in Ctrl soil resulted in a relevant difference among the soil treatments; in Ctrl soil the mean removal of As by *S. bicolor* was 7.48 g ha^{-1}, Fert and Org plants allowed the removal of respectively 158 and 219 g ha^{-1} of As. Our data has the same magnitude than those provided by Madejón et al. (2003).

A noticeable increase of the metal removal could be obtained by harvesting not only the plant shoots but also the roots. In this way it would be possible to remove from the contaminated soil about approximately 370 g ha^{-1} of As, 2,4 folds more than the result obtained by harvesting the plant shoots (figure 5). In the field trial, by harvesting the aboveground biomass of *H. annuus* we obtained an As removal of respectively 0.18, 6.74 and 3.96 g ha^{-1} for Ctrl, Fert and Org plants were obtained (figure 5). In the experiments performed in the area of Aznalcóllar, after the mine spill, in growing plants of *H. annuus* a removal of As of approximately 3 g ha^{-1} was observed.

3.2.2. Cadmium

On average, *S. bicolor* permitted a removal of Cd higher than *H. annuus*. Harvesting sorghum at the moment of the highest biomass production, we removed 5.62 g ha^{-1} of Cd with Fert plants, while in the case of Org plants, the removal was of 4.31 g ha^{-1}. The maximum Cd output obtained by plants of H. annuus was about of 1.59 in both Fert and Org plants (Figure 5). The control plants, affected by a severe nutritional stress, produced a very low biomass and therefore their removal of Cd was negligible.

Keller et al. (2003) reported the following Cd removal from the Dornach experimental site: *B. juncea* 6.95 g ha^{-1}, *N. tabacum* 41.7 g ha^{-1}, *Z. mays* 9 g ha^{-1} and *T. caerulescens* 179 g ha^{-1}. A part the hyperaccumulator *T. caerulescens*, the values of maize and Indian mustard are coherent with our data on sorghum. Figure 5 also points out that by harvesting the whole plants of *S. bicolor* the output of Cd could be increased respectively by about 2.5 and 1.7 folds.

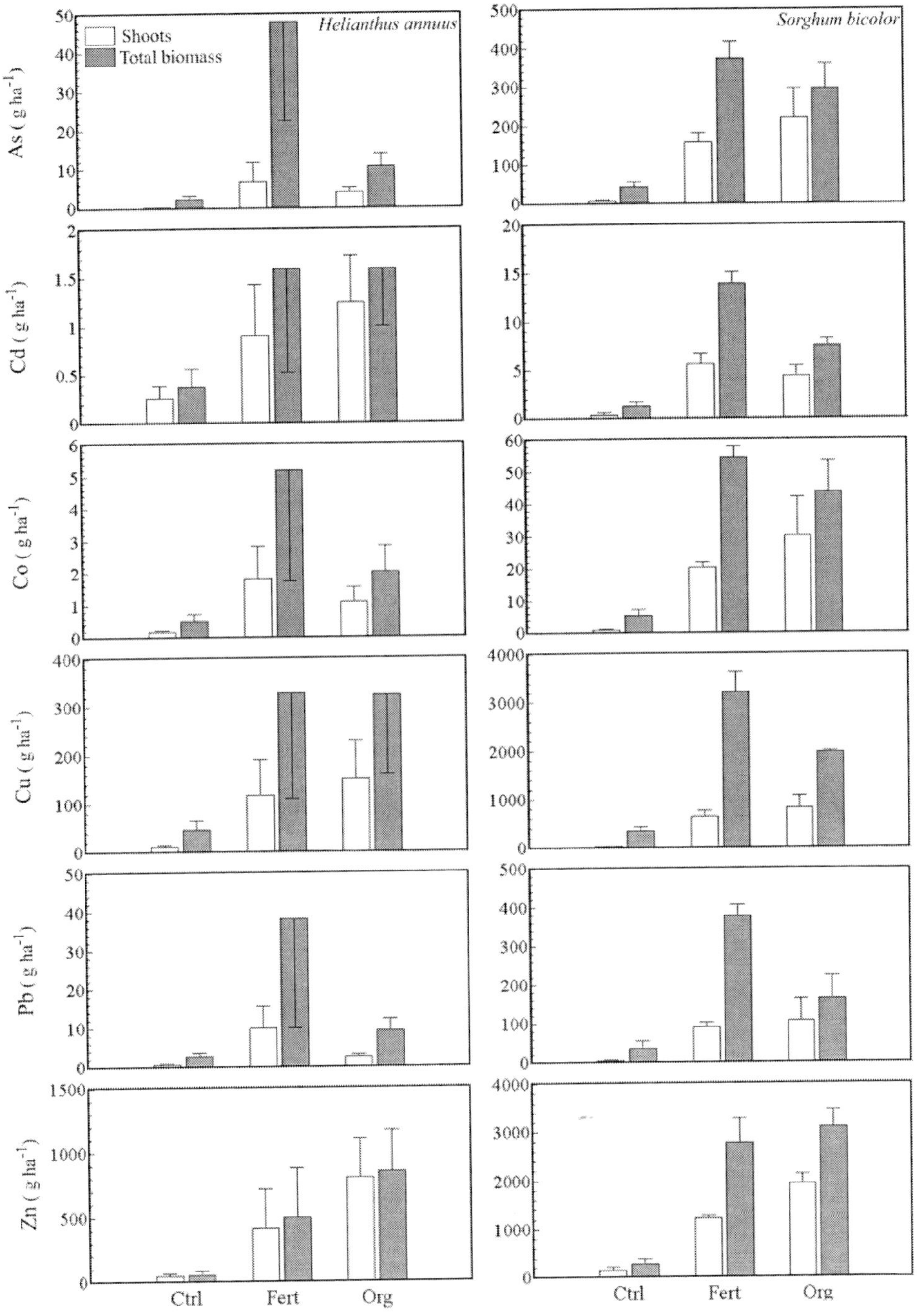

Figure 5. Removal of trace metals by the shoots and the whole plant of *S. bicolor* and *H. annuus* grown in respectively control, mineral fertilized and organic amended soil. Vertical bars for each bar represent the standard error for three replicates.

3.2.3. Cobalt

No data from field trials on Co phytoremediation have been found in literature. In our experiment, a single growth cycle of *S. bicolor* allowed a Co removal of respectively 20.2 g ha^{-1} and 30.2 g ha^{-1} for Fert and Org plants. The metal output calculated for *H. annuus* resulted one order of magnitude lower than S. bicolor (Figure 5). Recalculating the output of Co, considering the contribution of the roots, it can be appreciated an increase in the removal of the metal of about 2.7 and 1.4 folds by respectively Fert and Org plants of S. bicolor.

3.2.4. Copper

The maximum removal of Cu by plants of *S. bicolor* reached respectively 820 and 644 g ha^{-1} in Org and Fert plants. Lower values were observed for *H. annuus*; Org plants removed 151 g ha^{-1} of Cu and the Fert ones removed 116 g ha^{-1} of Cu (Figure 5). The hypothetical harvest of the entire plants of Fert *S. bicolor* could allow a removal of Cu of 3,215 g ha^{-1}, which is about 5 times higher than the result obtained by harvesting the sole aboveground biomass. In the case of Org plants of sorghum the removal of Cu was more than doubled (Figure 5). The same result was reached by the plants of *H. annuus*.

The removal of copper calculated on the base of data collected by Keller et al. (2003) at the experiment of Dornach showed that *T. caerulescens* had an efficiency in the removal of Cu lower than the crops. In more detail, the amounts of Cu extracted from the soil were: 146 g ha^{-1} for *B. juncea*, 474 g ha^{-1} for *N. tabacum*, 163 g ha^{-1} for *Z. mays*, finally 50 g ha^{-1} for *T. caerulescens* (Keller et al, 2003). This low value of Cu removal is not surprising since this species is highly specialized for Zn. On the other hand, this feature is one of the limitations of hyperaccumulators.

3.2.5. Lead

Combining the biomass yield with the concentration of Pb in the shoots it turned out that the highest removal of the element, 107 g ha^{-1}, would have been obtained by harvesting the Org plants of *S. bicolor*, while Fert plant removed 91 g ha^{-1} of Pb (Figure 5). Hypothesizing the harvest of the total biomass of *S. bicolor* the Pb removal would rise up to respectively 378 g ha^{-1} and 164 g ha^{-1} for Fert and Org plants (Figure 5). The values of Pb removal calculated for the plants of *H. annuus* were significantly lower than *S. bicolor*. Currently no realistic perspective of large scale application of phytoremediation in Pb polluted soils can be predicted.

3.2.6. Zinc

The total amount of Zn removed from the soil of Torviscosa by *S. bicolor* calculated for the plants in correspondence of the maximum biomass growth gave the values of respectively 1,223 and 1,944 g ha^{-1} for Fert and Org plants. These values could be, on average, about doubled in the case of the harvest of the entire plant biomass. Due to the low biomass of Ctrl plants, the Zn removal was 148 g kg^{-1} (Figure. 5). The metal removal calculated for *H. annuus* suffered also from a certain variability. The values were 410 and 804 g kg^{-1} respectively for Fert and Org plants. Observing plants of sunflower, Madejón et al. (2003) estimated a Zn removal of 2,050 g ha^{-1} when considering only the harvesting of the shoots of sunflower. Considering the also the roots, the value increased up to 2,140 g ha^{-1}. The output of metal computed from the data collected on crops by Keller et al. (2003) showed the following values: *B. juncea* 894 g ha^{-1}, *N. tabacum* 1,834 g ha^{-1}, *Z. mays* 1,998 g ha^{-1}. In the same experiment the hyperaccumulator *T. caerulescens* resulted more efficient as expected, showing a metal removal of 5,052 g ha^{-1}.

3.3. Use of Sorghum and Sunflower for Phytoremediation

Phytoextraction is a long-term remediation effort, requiring many cropping cycles to reduce metal concentrations to acceptable levels. The time required for remediation is dependent on the type and extent of metal contamination and the length of the growing season. However, the most important feature that the plants should have is the efficiency of metal removal.

In the case of Torviscosa, the highest amounts of trace elements removed by *S. bicolor* equates to only 0.03%, 0.056%, 0.024%, 0.225%, 0.018% and 0.082% of the total soil As, Cd, Co, Cu, Pb and Zn content, respectively. The metal removal by *H. annuus* had the same order of magnitude. The removal of trace elements was negligible in comparison to the amount in the soil. Therefore, under the experimental conditions of Torviscosa, *S. bicolor* and *H. annuus* showed a very poor potential. On the other hand, the experimental design did not consider any specific practice to enhance the bioavailability of the trace metals. Only the nutrients supply for sorghum and sunflower were increased recording a positive feedback by the crops in terms of element removal, but other agronomic practices were not taken into account. All of the factors that must be considered in successful agriculture also should be considered during phytoremediation. From this point of view a lot of work is still needed.

Among the few experiments dealing with the in situ performances of hyperaccumulators it was observed that hyperaccumulator plants in certain conditions were not able to express their potential.

For example, by testing the field performances of the hyperaccumulators *T. caerulescens* and *Arabidopsis halleri*, McGrath et al. (2006) found that *A. halleri* did not perform as expected, removing only 0.02% and 0.066% of the total soil content of Cd and Zn. These values are not so different than those observed in the field experiment in Torviscosa which involved non hyperaccumulator plants.

As regard *T. caerulescens*, perhaps the most studied metal hyperaccumulator, it was demonstrated that the biomass yields are highly variable, ranging between 0.5 and 13.4 t ha^{-1} (Felix 1997; Schwartz 1997; McGrath et al. 2000; Hammer and Keller 2003; McGrath et al. 2006), depending on different ecological site characteristics.

However, the fundamental traits of plants useful for phytoremediation, summarized by Chaney et al. (2000), should be expressed in a wide range of environmental conditions to transform phytoremediation from a natural phenomenon to a sustainable technology for soil clean up.

3.4. Other Research Activities on the Site of Torviscosa

The experimental site of Torviscosa offered the opportunity to develop different research activities planned and carried out by four research teams. Besides the research team from

Udine, the others were from the universities of Milan, Padua and Pisa. In this paragraph, a short description of their studies is reported.

The research unit of Milan focused around the assessment of the ecological health of soils by evaluating different biological activities as monitoring instruments. Short term plant bioassays (germination, root elongation and growth tests) were tested on soil sampled in the field trials at the beginning of the phytoremediation process and after one year. Barley root elongation and lettuce biomass production bioassays were better predictors of the restoration following organic amendments and cropping than chemical data alone (Zaccheo et al. 2007).

Nematodes were chosen as bioindicators of the effectiveness of land reclamation by following the natural colonization of the polluted ecosystem. The analysis of the nematofauna (general composition, trophic structure and biodiversity) highlighted that the pyrite cinders were not inhabited by soil fauna and in the covering soil the presence of the nematodes was low. In the early stages of the remediation process agronomic practices (i.e. amending and cropping) increased nematode abundance and biodiversity comparing to the initial situation, as suggested by the Ecological and Maturity Indices (Corsini et al, 2007).

Moreover, the potential of the polluted soil to release redox sensitive contaminants like As was studied in microcosms kept in oxic and anoxic conditions. The addition of low molecular weight organic compounds (glucose, citric acid) and more complex compounds (plant material) enhanced the bacterial activity and in submerged conditions induced marked changes in redox potential.

Free oxygen was initially consumed by respiring microorganisms, that drove the microcosms towards anaerobic conditions. Facultative anaerobic and anaerobic bacteria reduced the alternative electron acceptors (NO_3^-, Mn^{4+}, Fe^{3+}) causing a decrease of the soil redox potential resulting in leaching of As together with Fe and Mn (Corsini et al. 2010a). These results suggest that flooding of soils containing a readily utilizable carbon source might induce soil redox potentials that allow mobilization of As, which can be up taken by plants and eventually translocated into the food chain.

Further studies in a greenhouse aimed to evaluate the possibility of As phytoremediation with willows (*Salix purpurea* L.) under flooded and not flooded conditions. Plants accumulated As in leaves up to 10-fold the initial concentration during the first 15 days after transplanting. However, no significant increase in As content after 56 days was recorded for plants in both flooded and not flooded conditions (Corsini et al. 2010b). These findings could be explained by the fact that the plant available As in the soil became phytotoxic. Although *S. purpurea* demonstrated aptitude for growing in flooded condition, its use for cleaning of As contaminated soil is not suitable and phytostabilization could be considered a more suitable strategy.

In conclusion, a soil polluted by pyrite cinders under flooded conditions can represent a temporary sink for the As that can be easily released to the soil solution and possibly enter the food chain from the water-soil system. The presence of plants in the site may help to reduce the environmental risks throughout phytostabilization processes. Therefore, the management of contaminated soils must provide a constant monitoring of the environmental conditions in order to limit As release. The research studies of the group from Padua, aimed to evaluate whether the woody biomass phytoremediation is a realistic management option (Vamerali et al. 2009). Comparing ploughing and sub-soiling (0.35 m depth) as different agronomic practices, the growth of *Populus* spp. and *Salix* spp. and their trace element uptakes were investigated in both pot and field trials. The species differences were marginal and the species selection was not critical. Impaired above-ground productivity and low translocation of trace elements showed that bioavailable contaminant stripping was not feasible.

The most significant finding was of coarse and fine roots proliferation in surface layers that provided a significant sink for trace elements. The team came to the conclusion that phytostabilization and effective immobilization of metals and As may be achieved at the site by soil amelioration combined with woody species establishment.

Confidence to achieve a long-term and sustainable remediation requires a more complete quantification of root dynamics and a better understanding of rhizosphere processes.

Moreover, fodder radish (*Raphanus sativus* L. var. *oleiformis* Pers.) was cultivated in pyrite cinders in a series of pot trials (Bandiera et al. 2009) by the same research team. The effects of the application of humic acids (HA) to the cultures were evaluated at various concentrations (0.1 and 1 g kg^{-1} of pyrite cinders) and for different methods of application (mixed with cinders before sowing; applied on the top of substrate after sowing; foliar spraying), in comparison with untreated controls. HA increased the bioavailability of various heavy metals, especially when applied at high concentrations. Low concentrations of HA, either mixed with cinders or applied on foliage, slightly enhanced shoot growth. High levels of HA were generally phytotoxic, probably because of the ensuing higher heavy metal concentrations in the shoots. Foliar spraying was partially able to attenuate the phytotoxicity of high dosages of HA, as revealed by the increased rate of leaf expansion. A negative correlation was found between shoot biomass and concentration of heavy metals, and higher removals were generally associated with smaller doses of HA and foliar application. Within the range of concentrations tested, humic substances led to a reduction in root diameter and an increase in specific root length (SRL, length per unit weight of roots), significant only at the highest dosage, together with a tendency towards enhancement of translocation of heavy metals to the shoot – both positive traits for phytoremediation purposes.

Last but not least, the research unit from the University of Pisa, studied the potential of nine different plant species to grow in the presence of metals (As, Cd, Cu, Pb and Zn) and to accumulate them in the shoots by germination and root length tests, and successively by hydroponic experiments (Quartacci et al. 2007). Among the species, *Brassica carinata* was the one that accumulated the highest amounts of metals in shoots without suffering a significant biomass reduction. To further evaluate the potential of *B. carinata* for chelante-enhanced phytoextraction of a natural, multiple metal polluted soil (As, Cd, Cu, Pb and Zn), both hydroponic and pot experiments were carried out with nitrilotriacetic acid (NTA) or (S,S)-ethylenediamine disuccinic acid (EDDS) as complexing agents. The hydroponic study with solutions containing the five metals together showed that accumulation of Cd, Cu, Pb and Zn in shoots was higher following EDDS addition compared to NTA. Ethylenediamine disuccinic acid was more effective than NTA in desorbing Cu, Pb and Zn from the soil, whereas As and Cd were poorly extracted. *B. carinata* plants were grown for 4 weeks in the metal-contaminated soil and then the soil was amended with 5 mmol kg^{-1} NTA or EDDS.

All plants were harvested 1 week after amendment. In comparison to NTA, EDDS was more effective in enhancing the concentrations of Cu, Pb and Zn in *B. carinata* shoots (2- to 4-fold increase compared to the control). One week after chelate addition, the DTPA-extractable metal concentrations in the polluted soil were lower in the EDDS treatment in comparison with the NTA amendment. Even though *B. carinata* showed a reduced growth and a relatively low metal uptake, it demonstrated the ability to survive and tolerate the presence of more metals simultaneously. Further studies focused on the removal of metals

(As, Cd, Cu, Pb and Zn) from a multiple metal-contaminated soil by growing *B. carinata* in succession to spontaneous metallicolous populations of *Pinus pinaster*, *Plantago lanceolata* and *Silene paradoxa* (Quartacci et al. 2009). The results showed that the growth of the metallicolous populations increased the extractable metal levels in the soil, which resulted in a higher accumulation of metals in the above-ground parts of *B. carinata*. Root exudates of the three metallicolous species were analyzed to elucidate their possible role in the enhanced metal availability. The presence of metals stimulated the exudation of organic and phenolic acids as well as flavonoids. It was suggested that root exudates played an important role in solubilising metals in soil and in favoring their uptake by roots.

4. Phytostabilization Perspectives at a Mining Site in Julian Alps (Italy)

Phytostabilization consists in using metallicolous plants to establish a persistent plant cover in order to prevent pollution from spreading by erosion, water percolation, leaching and from toxic dust dispersal by wind (Ernst, 1996). A typical scenario in which phytostabilization could be considered is represented by the anthropogenic metalliferous sites (e.g. abandoned mining sites, smelter sites) where the presence of mine tailings and wastes results in severe pollution and poses aesthetic impacts on the local environment (Wong 2003).

Metalliferous soils provide very restrictive habitats for plants due to the phytotoxicity which results in severe selection pressures. The unique plants with an ability to tolerate metal toxicity and thrive on metalliferous soils are the metallophytes (Baker et al. 2010). Such environments represent the typical ecological niche for metallophytes. Hence they should be considered as a relevant reservoir of biodiversity and therefore biodiversity conservation strategies should focus also on environments altered by anthropogenic impacts (Bizoux et al. 2008). Since 2007, our group is involved in a project that aims at the environmental restoration of the mine soils and tailings at Cave del Predil (Tarvisio), which is a former mining site in Julian Alps. Preliminary studies have been planned before testing at a field scale some strategies of land restoration. As first step, a field survey was carried out to find out the metallophytes occurring in this alpine valley and to evaluate the potential for phytoremediation of such species. Afterwards, a series of pot experiments started with the ultimate aim of revegetating and phytostabilizating the mine tailings of Cave del Predil. This work is in part still in progress.

4.1. Metallophytes at the Former Lead/Zinc Mining Site of Cave Del Predil, Julian Alps

Since the Roman age, the Raibl mining site, which is located in the North East of Italy, was exploited. This lead/zinc mine for many years was considered as the most productive Italian one (Francescutti, 1991). The village of Cave del Predil (901 m above sea level, 46° 26' 26" N, 13°34' 16" E) is situated in the Rio del Lago valley, on the northern side of the Julian Alps (NE Italy), very close to the Austrian and Slovenian borders (Figure 6).

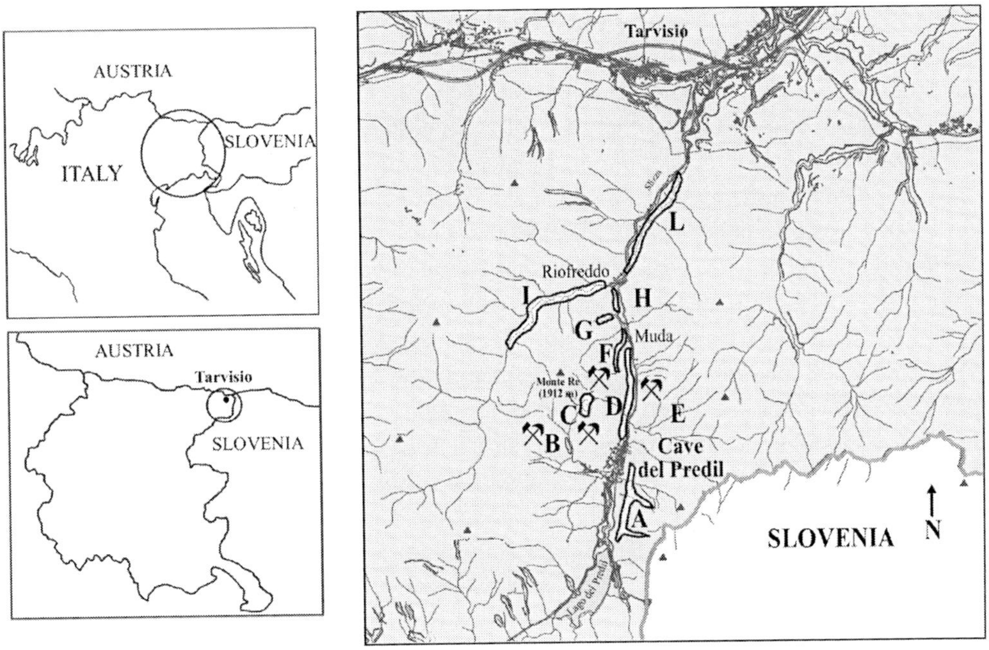

Figure 6. Soil and plants collecting sites.

The mean annual rainfall is about 2,100 mm and the monthly mean temperature ranges between 2.4 °C (January) and 16.4 °C (August). The valley belongs to the basin of the river Danube. The Rio del Lago stream flows from the Predil lake and encounters the Rio Freddo stream forming the Slizza stream that descends through the Austrian border, towards the river Gail.

The mining activity followed the ore fields of lead and zinc occurring mainly as sphalerite (PbS) and galena (ZnS). To a lesser extent, other sulphides such as marcasite (FeS_2), pyrite (FeS_2), pirrothine (FeS), arsenopyrite (FeAsS) and cinnabar (HgS) were extracted. The minerals were accompanied by a variety of gangue minerals among which the most important were dolomite and barytine. Moreover, series of faults within these geological formation results in occurrences of lead-zinc sulphide outcrops (Di Colbertaldo, 1948; Reeves and Brooks, 1983).

The mining activity ceased in 1991. Currently, the most relevant environmental issue of the mining area is the management of the mine tailings that cover an area of about 20 ha. The mining wastes, neither stabilized, nor protected, are completely barren and subjected to significant wind and water erosion. It is believed that pollutants may reach the floodplain and therefore the food chain. A very rare plant community – *Thlaspietum cepaeifolii* – was described as strictly related with the mine sites of Cave del Predil by Ernst (1974) and Hartl et al. (1992). The hyperaccumulator *Thlaspi rotundifolium* (L.) Gaudin subsp. *cepaeifolium* (Wulf.) Rouy et Fouc is one of the dominant members of the plant community which also includes a metal resistant ecotype of *Minuartia verna*, *Silene vulgaris*, *Biscutella laevigata*, *Poa alpina* subsp. *alpina*, *Dianthus sylvestris*, and a series of companion species.

The first evidence of Pb/Zn hyperaccumulation by *T. rotundifolium* subsp. *cepaeifolium* was reported by Reeves and Brooks (1983). In herbarium specimens collected at the Rio del Lago valley, a Pb concentration of 8,200 mg kg^{-1} (0.82%) was measured. At the time, that was the highest level of lead ever observed in any flowering plant. As well as Pb, *T. rotundifolium* subsp. *cepaeifolium* was classified also as Zn hyperaccumulator having reached in its shoots up to 17,300 mg kg^{-1} (1.73%) (Reeves and Brooks, 1983). More recently, Wenzel and Jockwer (1999) found up to 1,930 mg kg^{-1} of Pb and up to 9,060 mg kg^{-1} of Zn in the shoots of *T. rotundifolium* subsp. *cepaeifolium* collected at Arnoldstein (Austrian Alps) on the banks of the river Gail, confirming Pb but not Zn hyperaccumulation. Conversely, Reeves and Baker (2000) noticed a concentration of 21,000 mg kg^{-1} of Zn in the aboveground biomass.

The soils supporting *T. rotundifolium* subsp. *cepaeifolium* is characterized by Pb contents ranging from 600 mg kg^{-1} (Bleiberg ob Villach, Austria) to 4,000 mg kg^{-1} (Jauken, Austria), Zn contents ranging from 7,000 mg kg^{-1} (Črna, Slovenia) to 140,000 mg kg^{-1} (Jauken, Austria) and small amounts of Cu which results always less than 100 mg kg^{-1} (Ernst, 1974). These sites are comprised in an area extending along the Southern Limestone Alps from Carnic Alps and Gailtal Alps eastwards, to Julian Alps and Karawanken which corresponds with the worldwide distribution of *T. rotundifolium* subsp. *cepaeifolium* (Jalas et al. 1999).

The occurrence of *Alyssum wulfenianum* in the Rio del Lago valley was noticed by Vergnano Gambi et al. (1979). Reeves and Baker (1983) in their work did not verify *sensu stricto* hyperaccumulation of Pb or Zn. Nevertheless, some of the concentrations reported were very high (860 and 2,500 mg kg^{-1} respectively). Ernst (1974) did not include *A. wulfenianum* Bernh in the *Thlaspietum cepaeifolii*. On the other hand, Grabherr and Mucina (1993) proposed this plant as characteristic of the community and the Atlas Florae Europaeae (Jalas et al. 1999) reported that *A. wulfenianum* and *T. rotundifolium* subsp. *cepaeifolium* have a very similar distribution. No other data on the metallophytes of the Rio del Lago valley have been published after Reeves and Brooks (1983).

In the early '80, a worldwide search for new hyperaccumulators was carried out analyzing plant tissues collected from herbariums. This simple approach revealed the metal concentration in the plants but, obviously, did not provide any information about the soil. For this reason, in order to investigate other species belonging to the *Thlaspietum cepaeifolii* and potentially for phytoremediation, a field survey in the Rio del Lago valley was carried out. The results of the plant survey demonstrated that among the species belonging to the *Thlaspietum cepaeifolii* community, *M. verna* and *B. laevigata* were the better distributed. Moreover, they were the sole species also growing on the mine tailings. For these reasons, they were considered in the study, beyond the hyperaccumulators *A. wulfenianum* and *T. rotundifolium* subsp. *cepaeifolium*. *Minuartia verna* is a very common metallophyte, although not a hyperaccumulator, whose widespread presence has been recorded in a number of mine sites in Europe where studies on metallophytes have been done: Aachen-Liége orefield, Belgium (Smith, 1979), Central and Northern Pennines, U.K. (Smith, 1979), St. Laurent le Minier, France (LaCoste et al. 1999), Harz region, Germany (Ernst et al. 2004) and Gail valley, Austria (Wenzel and Jockwer, 1999). *Biscutella laevigata* is a mountain brassica species also occurring in calcareous dry grasslands, in plains and distributed in Europe from Eastern Pyrenean Mountains to the Transylvanian Alps (Wierzbicka et al. 2004).

The soil and plant samples were collected in summer 2007 from 10 sites at the altitude of 752–1450 m above sea level. Three main environmental conditions were identified: (i) native soil not impacted by mining operations (areas A, E; G); (ii) mining sites and mine tailings (areas B, C, D, F); (iii) banks of three alpine streams (areas H, I, L) (Figure 6). The river banks were explored for the presence of accumulation of water-borne metalliferous gravels and sands noticed by Reeves and Brooks (1983). The metal content in soil and plant samples was determined following the procedures described in paragraph 2.1.1. The bioavailable fractions of the metals was obtained by DTPA-extraction (Martens and Lindsay, 1990).

4.1.1. Soil Conditions and Plant Ecology

The substrates collected were coarse and poor in structure whereas a loam soil was found only in site E, as expected. All soil samples, as well as the mine tailings, are slightly alkaline. Organic carbon content varied widely; the lowest values were recorded in sites H, I and L and in the sites where the mining activities took place (sites B, C, D, and F). To the contrary, native soils (A, G), which were not disturbed by the mining operations, had an average OC percentage of 2.87 and the forest soil (E) showed the highest OC content (13.4%). Total nitrogen was rather low, ranging from 0.04 to 0.22%; the highest value (1.06%) was detected in the forest soil (Table 8).

Table 8. Characterization of the collected substrates

	Area	pH	OC	Total N	Sand	Silt	Clay
			(%)	(%)	(%)	(%)	(%)
Native soils	A	7.89	2.46	0.18	71.7	20.3	8
	E	7.52	13.4	1.06	46.6	20.9	34.3
	G	8.48	3.28	0.22	82	12.7	5.25
Mining soils	B	7.67	2.10	0.17	69.7	20.5	9.81
	C	8.20	0.32	0.04	72.8	19	8.19
	D	8.08	0.56	0.06	58.7	28.3	13
	F	7.96	0.63	0.07	80	12	8
Stream banks	H	8.49	0.38	0.04	83.6	10.5	5.88
	I	7.66	0.88	0.07	76.9	13.4	9.75
	L	8.25	0.43	0.05	82.1	11.3	6.63

Tables 9 and 10 report the geometric mean and the range values of respectively the total and the DTPA extractable metal concentration in the substrates. As expected, the soil samples collected at the mining sites had very high metal concentrations. In spite of a broad data variability, it is noticeable that also in native soils, anomalous values of metal concentration were recorded, particularly for Pb, Tl and Zn.

The highest percentages of bioavailable forms of elements were observed in native soils (on average 25%, 23.5% and 21.6% respectively for Cd, Pb and Zn), while in all the samples collected from the mining areas and the stream banks such percentage fell below 5%. This was probably due to the different mineral composition of the substrates. An independent behaviour was found in the case of Tl. In several samples the bioavailable forms were not detectable. By averaging the data, it was found that the bioavailable fraction constituted less than 0.1% of the total Tl. This could be considered realistic and confirmed evidences by Zbíral et al (2002).

Table 9. Total Cd, Pb, Tl and Zn in substrates collected in the sampling sites. The values are the geometric mean; in parenthesis are reported the range data

	Area	Cd (mg kg^{-1})	Pb (mg kg^{-1})	Tl (mg kg^{-1})	Zn (mg kg^{-1})
Native soils	A	0.78 (0.46–1.28)	34.2 (4.50–176)	158 (131–178)	44 (10.2–156)
	E	6.13 (3.42–11)	669 (331–1,353)	94.2 (85.8–103)	885 (185–4,230)
	G	0.80 (0.37–15.1	49.1 (8.15–3,657)	160 (121–264)	93.2 (19.3–3,923)
Mining soils	B	107 (0.88–235)	11,758 (1561–19,484)	169 (150–181)	24,569 (837–41,857)
	C	9.58 (3.64–19.7)	3,818 (1,778–12,431)	756 (221–4549)	17,900 (7,57–42,513)
	D	32.1 (10.4–216)	7,450 (2,255–59,950)	390 (165–1,411)	18,318 (5,351–66,563)
	F	11.2 (6.03–7.8)	1,707 (448–2,530)	236 (162–310)	10,061 (2,302–16,295)
Stream banks	H	19.8 (1.74–81.6)	4,629 (164–36,124)	343 (173–862)	8899 (557–33,381)
	I	1.77 (0.23–26)	4,016 (n.d.[†]–69,387)	270 (n.d. –447)	376 (7.49–14,688)
	L	22.8 (4.73–271)	3,874 (657–26,994)	204 (115–488)	8487 (1,979–31,341)

[†] Not detectable.

Table 10. DTPA extractable Cd, Pb, Tl and Zn in substrates collected in the sampling areas. The values are the geometric mean; in parenthesis are reported the range data

	Area	DTPA Cd (mg kg^{-1})	DTPA Pb (mg kg^{-1})	DTPA Tl (mg kg^{-1})	DTPA Zn (mg kg^{-1})
Native soils	A	0.15 (0.02–0.63)	11.4 (1.51–87.7)	0.08 (n.d. – 0.08)	11.6 (1.59–71)
	E	0.73 (0.5–1.06)	101 (43–238)	n.d.	65.7 (13.4–323)
	G	0.20 (0.04–3.31)	22.9 (1.49–584)	0.07 (n.d. –0.07)	23 (2.68–56)
Mining soils	B	2.35 (0.09–4.37)	128 (28.4–194)	n.d.	349 (65.4–61)
	C	0.35 (0.18–0.62)	34.9 (8.33–78.5)	0.59 (n.d. –2.08)	382 (155–550)
	D	1.57 (0.52–15.9)	62.5 (4.75–1153)	0.55 (0.09–9.78)	463 (309–486)
	F	1.36 (0.46–2.84)	53.9 (20.4–116)	1.54 (0.24–7.57)	547 (361–61)
Stream banks	H	1.08 (0.16–4.45)	296 (39.3–1484)	0.23 (n.d. –2.28)	249 (49.6–022)
	I	0.06 (0.02–1.73)	605 (n.d.[†] –875)	0.07 (n.d. –0.17)	10.1 (n.d. –59)
	L	1.01 (0.22–8.65)	160 (13.2–1299)	0.19 (n.d. –3.82)	274 (66.3–1050)

[†] Not detectable.

The relation between total metal content in substrates and the amount of the bioavailable fraction is described in figure 7. The interpolation carefully describes the relationship in the case of Cd (R^2=0.67) and Zn (R^2=0.79). Such relationship was weaker for Pb (R^2=0.32), while in the case of Tl the regression (R^2=0.05) did not describe any link between the soil total content of the element and the bioavailable fraction.

Plant metal hyperaccumulation was evaluated on the basis of the criteria proposed by Baker and Brooks (1989) for Cd, Pb and Zn (100, 1,000 and 10,000 mg kg^{-1}, respectively) and by LaCoste et al. (1999) for Tl (500 mg kg^{-1}) and the bioconcentration factor (BF=Me_{shoots}/Me_{soil}) following Zhao et al. (2003).

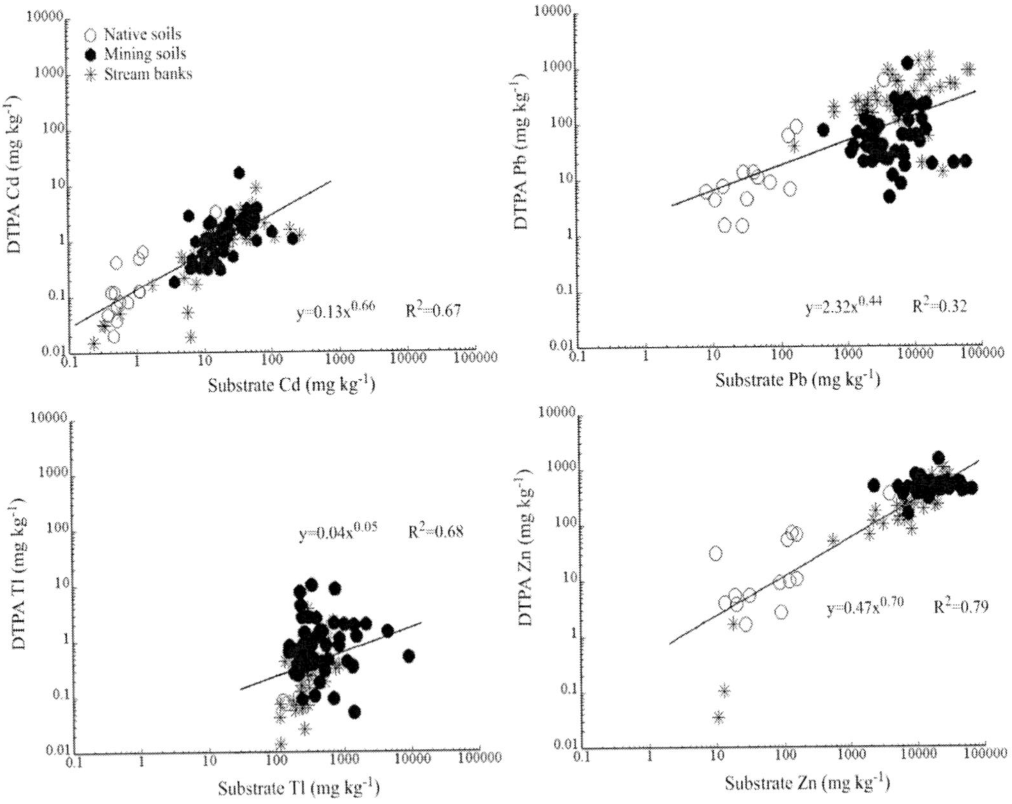

Figure 7. Relationships between total and DTPA extractable fraction of Cd, Pb, Tl e Zn measured in the substrates sampled in native soils, mining sites and in the stream banks.

In table 11 the range of concentrations of Cd, Pb, Tl and Zn in the roots and shoots of the species are reported. For reference, a series of literature data on the same species collected in mining sites in Carinthia, Austria (Wenzel and Jockwer, 1999), France (LaCoste et al. 1999; Reeves et al. 2001), Germany (Ernst et al. 2004) and Poland (Wierbicka et al. 2004) and serpentine soils in Italy (LaCoste et a., 1999; Lombini et al. 1998) are reported as well.

None of the species showed a Cd concentration in their fractions higher than 100 mg kg^{-1}, on the other hand, the Cd concentrations in the substrates were not extreme, although anomalous if compared to the ones in normal soils. However, in *M. verna* concentrations up to 62.7 and 24.7 mg kg^{-1} of Cd were detected respectively in roots and shoots, whereas in the same fractions the maximum values of Cd concentration in *T. rotundifolium* subsp. *caepaeifolium* were respectively 47.9 and 78 mg kg^{-1} (Table 11). The Pb concentrations recorded in plant shoots were in the following ranges: *A. wulfenianum* 28–826 mg kg^{-1}, *B. laevigata* 0.75–8,774 mg kg^{-1}, *M. verna* 40–12,574 mg kg^{-1} and *T. rotundifolium* subsp. *caepaeifolium* 29.2–2,817 mg kg^{-1} (Table 11). *Minuartia verna* showed the highest Pb accumulation in the shoots, with a higher concentration of this element than the hyperaccumulation threshold for 14 specimens out of 33. In the shoots of 17 out of 33 specimens of the Pb-hyperaccumulator *T. rotundifolium* subsp. *caepaeifolium* the concentration of Pb was higher than 1,000 mg kg^{-1}. An unexpected high concentration of Pb (8,774 mg kg^{-1}) was measured in the shoots of a single specimen of *B. laevigata* (Table 11).

Table 11. Range of concentrations of Cd, Pb, Tl and Zn observed in plants of *A. wulfenianum*, *B. laevigata* subsp. *laevigata*, *M. verna* subsp. *verna* and *T. rotundifolium* subsp. *cepaeifolium* collected in the mining site of Cave del Predil (Italy), and found, for reference, in literature

Species	n[†]	Location	Cd (mg kg^{-1})	Pb (mg kg^{-1})	Tl (mg kg^{-1})	Zn (mg kg^{-1})	Reference
Alyssum wulfenianum	13	Cave del Predil, Italy (L)[‡]		15-860		31–2,500	Reeves and Brooks 1983
	23	Cave del Predil, Italy (R)	2.20–10.9	253–2,401	n.d.[§] –9,688	354–5,670	
	23	Cave del Predil, Italy (S)	0.59–5.95	28–826	31.5–2,441	194–3,720	
Biscutella laevigata subsp. *laevigata*	3	Les Aviniéres, France (L)			244 – 308		Anderson 1999
	34	Les Aviniéres, France (W)			20–15,199		LaCoste et al. 1999
	15	Tuscany, Italy (W)			<1–2		LaCoste et al. 1999
	2	Gailitz, Austria (L)			295–495		Leblanc et al. 1999
	1	Arnoldstein, Austria (S)	78.3	1,090		4,870	Wenzel and Jockwer 1999
	5	Mount Prinzera, Italy [¶] (W)				53	Lombini et al. 1999
	5	Boleslaw, Poland (S)			0.21		Wierbicka et al. 2004
	54	Cave del Predil, Italy (R)	0.01–8.45	0.53–4,916	13.6–9,984	16.8–5,692	
	54	Cave del Predil, Italy (S)	0.01–4.75	0.75–8774	33.4–32,661	12.7–3,669	

Species	n(†)	Location	Cd	Pb	Tl	Zn	Reference
			(mg kg^{-1})	(mg kg^{-1})	(mg kg^{-1})	(mg kg^{-1})	
Minuartia verna subsp. *verna*	n a.(††)	Bredelem, Germany (L)	49.5	99.5		5,427	Ernst et al. 2004
	n.a.	Oker, Germany (L)	42.5	240		1,988	Ernst et al. 2004
	n.a.	Pierrefitte, France (W)	< 4	62		1,932	Reeves et al. 2001
	n.a.	St Dalmas-de-Tende, France (W)	11	14		1,251	Reeves et al. 2001
	3	Arnoldstein, Austria (S)	32.6–59	1,290–2,180		5,550–7,700	Wenzel and Jockwer 1999
	33	Cave del Predil, Italy (R)	0.60–62.7	151–9,329	24–4,055	16.8–27,521	
	33	Cave del Predil, Italy (S)	0.55–24.7	40–12,574	18.3–4,007	24.1–18,372	
Thlaspi rotundifolium subsp. *cepaeifolium*	14	Cave del Predil, Italy (L)	–	130–8,200	–	2,350–17,300	Reeves and Brooks, 1983
	25	Cave del Predil, Italy (R)	1.30–47.9	282–14,435	53.1–4,268	71.7–19,568	
	25	Cave del Predil, Italy (S)	1.35–78	29.2–2,817	16.2–14,376	28.1–11,573	

(†) Specimens; (‡) L= leaves, R=roots, S=shoots, W=whole plant; (§) Not detectable; (¶) *Biscutella laevigata* subsp. prinzerae; (††) Not available data.

The most extensive investigation on Tl hyperaccumulation was reported by LaCoste et al. (1999). The aboveground biomass of specimens of *B. laevigata* was collected in the mining districts of St.Laurent le Minier (France) and Rocca San Silvestro (Italy). The Tl hyperaccumulation was verified only in the French specimens of *B. laevigata* and the highest value of Tl concentration was 15,199 mg kg^{-1} that is well above the limit of hyperaccumulation. Leblanc et al. (1999) reported a maximum Tl concentration of 495 mg kg^{-1} in leaves of *B. laevigata* sampled in Gailitz (Austria). Unfortunately, that paper did not provide information about the environmental conditions of the sampling sites. However, Gailitz is the German name of the Slizza creek. This belongs to the Gailitz's basin and flows northwards from Italy towards Austria. Therefore, it is likely that the specimens of *B. laevigata* referenced by Leblanc et al were collected not farther than 10 km NE from our sampling sites in the Rio del Lago valley.

A detailed study on the vegetation of metal rich sites in the Austrian Alps by Wenzel and Jockwer (1999) provided others evidences of metal accumulation by *B. laevigata*. In specimens collected from a soil containing 43.7 mg kg^{-1} of Cd, 4,010 mg kg^{-1} of Pb and 15,000 mg kg^{-1} of Zn, were found up to 78.3 mg kg^{-1} of Cd, up to 1090 mg kg^{-1} of Pb and 4,870 mg kg^{-1} of Zn in their shoots. This work is of particular interest, covering different Austrian regions; among them, the most southern site lies in the Gailitz valley, very close to the Italian border.

As expected, being a Tl-hyperaccumulator, a number of specimens of *B. laevigata* (25 out of 54, 46%) showed a shoot Tl-concentration well above the threshold of 500 mg kg^{-1}. To our knowledge, the highest Tl concentration ever recorded in tissues of higher plants is equal to 19,600 mg kg^{-1} LaCoste et al. (1999).

As regard our data, the average values of Tl concentrations in roots and shoots were 540 and 3,332 mg kg^{-1}, whereas the highest Tl concentration recorded in roots and shoots of *B. laevigata* were respectively 9984 and 32,661 mg kg^{-1} (Table 11).

Also in the shoots of other species known neither as Tl-tolerant nor Tl-accumulator, several cases of Tl concentrations well above the threshold of 500 mg kg^{-1} were recorded (Table 11).

In figure 8 are showed the average values of the BFs of Cd, Pb, Tl and Zn calculated for all the specimens of *A. wulfenianum*, *B. laevigata*, *M. verna* and *T. rotundifolium* subsp. *caepaeifolium*. Most plants have a bioconcentration factor for heavy metals of less than 0.2 (McGrath and Zhao, 2003). For the hyperaccumulators this parameter is usually greater than 1, and in some cases it falls within the range 50–100.

Three significant evidences rise from figure 8. Firstly, from an ecological point of view it should be noted that the species considered where always present in mining soils – which were also the most metal polluted ones – and in the sands and gravels along the water streams, whereas in the native soils was found only *B. laevigata*. Secondly, even though expected to be hyperaccumulators or highly metal tolerant *A. wulfenianum*, *M. verna* and *T. rotundifolium* subsp. c*epaeifolium* did not confirm this trait in terms of BF for Pb and Zn that resulted < 1. Third, *B. laevigata* meets the second condition of hyperaccumulators; the average values BF for Tl calculated for plants collected in the polluted substrates (mining soils and stream banks) reached values of 28 and 3.9 respectively. Surprisingly even the BFs for Tl of the other species were > 1 (Figure 8).

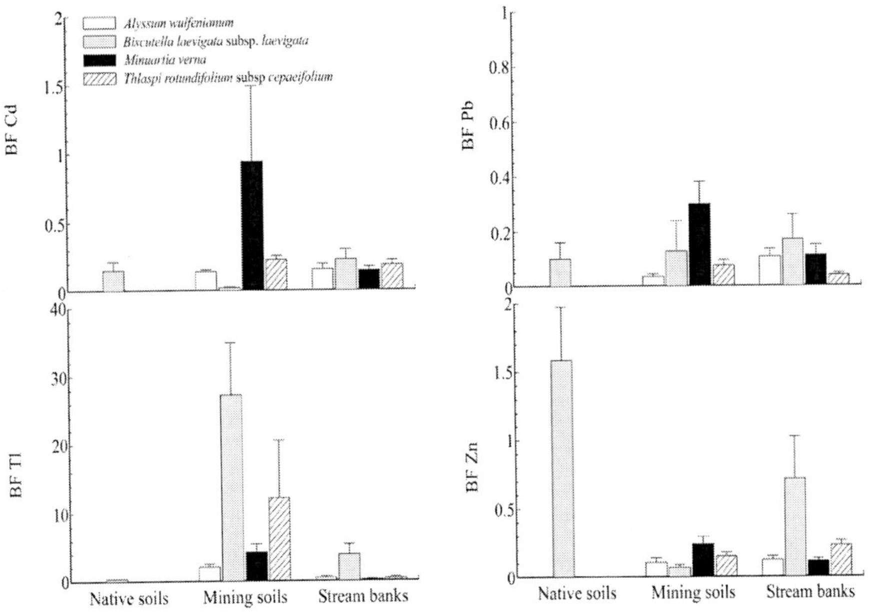

Figure 8. Bioconcentration factor (BF) of Cd, Pb, Tl e Zn as function of total metal content in substrate calculated for *A. wulfenianum*, *B. laevigata* subsp. *laevigata*, *M. verna* and *T. rotundifolium* subsp. *caepaeifolium*.

4.1.2. Really Hyperaccumulators?

The presence in the mining site of Cave del Predil of *A. wulfenianum* and *T. rotundifolium* subsp. *caepaeifolium*, described by Reeves and Brooks (1983) was confirmed. Such species are distributed in both the most polluted sites of the valley and in the contaminated sediments of the streams. Lead hyperaccumulation was strongly confirmed in particular for *T. rotundifolium* subsp. *cepaeifolium*; the highest roots and shoots Pb concentrations were respectively 14,435 and 2,817 mg kg^{-1}.

Experimental data were carefully analyzed in order to identify other metal hyperaccumulation other than *A. wulfenianum* and *T. rotundifolium* subsp. *cepaeifolium*. Such species were *B. laevigata* and *M. verna*.

The metal hyperaccumulation – formerly defined by simply considering the metal concentration in the plant biomass – was evaluated taking into account the bioconcentration factor. Yanqun et al. (2005) proposed that in searching for hyperaccumulators a further condition should be considered: the concentration of a metal in the shoots of hyperaccumulators must be 10-500 times higher than in the same species collected in uncontaminated soils. In our case, the last condition could not be monitored since the data references from unpolluted sites were missing.

The most relevant result provided by our work is the evidence of Tl hyperaccumulation by the studied species. Until now, the hyperaccumulation of Tl was demonstrated only for *B. laevigata* and *Iberis intermedia* (Leblanc et al. 1999). We confirmed this distinctive feature of *B. laevigata*, but we observed Tl hyperaccumulation in *A. wulfenianum*, *M. verna* and *T. rotundifolium* subsp. *cepaeifolium* as well.

Presumably these species developed into populations which are highly adapted to the local environmental conditions – climate, altitude, metal outcrops and soil pollution – that *per se* are likely to be favourable to this effect, even over short periods of time (Ernst, 1974; Bradshaw and McNeill, 1981). That is why the ecological value of metalliferous habitats in evolutionary terms is considered highly valuable (Smith, 1979; Whiting et al. 2002).

Further work is going to be planned at Cave del Predil attempting to pinpoint the genetic basis and morphological traits of the local population of metallophytes. This is potentially useful in general terms of phytoremediation and in the selection of more efficient plants, particularly those able to hyperaccumulate multiple metals.

4.2. Growth of *Thlaspi caerulescens* on Compost Amended Mine Tailings

The environmental impact of the mine tailings disposal sites is enormous. Generally, these sites if neither reclaimed nor protected can spread dangerous materials via wind dispersion and water erosion. These environments are barren or have minimal vegetation since the physicochemical characteristics of these sites are not favourable to most plant species because of the occurrence of many growth-limiting factors. Common chemical and physical limitations to the plant growth include: extreme pH values – generally low although some tailings may be alkaline – high salt content, lack of required nutrients, metal toxicities, high bulk density, lack of soil structure, slow water infiltration, low water retention, and low air permeability (Wong, 2003; Mendez and Maier, 2008). However, the toxic effects of residual quantities of metals and the lack of essential plant nutrients of such wastes are the most inhibitory factors to plants (Barrutia et al. 2009). To overcome these limitations, some strategies have been proposed to promote the establishing of a sustainable plant community (Tordoff, 2000).

Phytostabilization is also advantageous when decontamination strategies are impractical because of the extent of the contaminated area or the lack of adequate funding (Santibanez et al. 2008; Alvarenga et al. 2009). Unlike other options of phytoremediation, phytostabilization is not intended to remove metal contaminants from a site, but rather to stabilize them by accumulation in roots or precipitation within the root zone, reducing the risk to the human health and the environment.

Research has since shown that the use of biosolids and fertilizers is a key factor in phytostabilization of mine tailings to facilitate the plant establishment and growth (Tordoff, 2000). The expected effect of soil amendments with biosolids is an increase of the organic matter and water content of the soil and a rise of the pH (Chaney et al. 2007; Nwachukwu and Pulford 2008). As a consequence, organic matter, by binding the metals, decreases their leaching and bioavailability. Farrel et al. (2009) demonstrated that the addition of composted wastes to an acidic, heavy-metal contaminated soil greatly increases the soil pH and plant growth, whilst greatly reducing the accumulation of pollutants in the soil.

As regards the plants suitable for phytostabilization, only the species that have evolved through natural selection and that got adapted to contaminated environments are able to colonize such sites and should be considered for successful reclamation of mine tailings (Whiting et al. 2004; Frérot et al. 2006).

The previous work described the potential for phytoremediation of some metallicolous species collected at the mining site of Cave del Predil (NE Italy). Before testing the suitability of local species for the revegetation of the mine tailings, different schemes of amendment and nitrogen fertilization were studied in order to ameliorate the fertility of the polluted substrate.

A greenhouse pot trial was designed to evaluate the growth of *Thlaspi caerulescens* and the patterns of metal distribution in plant tissues in Pb/Zn mine tailings amended with municipal waste compost and inorganic N fertilizer. The objectives of this study were to observe: (i) the differences in plant growth and metal accumulation in response to different conditions of fertility and (ii) the response of *T. caerulescens*, which is known as Cd-Zn hyperaccumulator, on a multi-contaminated substrate. *T. caerulescens* which is a Cd/Zn hyperaccumulator and a model species for research on phytoremediation was used in this experiment in order to compare our experimental data with the literature ones.

4.2.1. Experimental Design

During the survey previously described, a certain amount of Pb/Zn mine tailings was collected from the tailings dump at he former mining site of Cave del Predil, Tarvisio, Italy. Tailing samples were taken from the upper 30 cm. Subsequently, the material was air dried, homogenized mechanically and sieved at 2 mm. Municipal solid waste compost (MSWC) was collected from Iris compost plant (Moraro, Gorizia, Italy).

In spring 2008, the pot experiment was established following a factorial design with 9 treatments (1 control and 8 combinations of organic amendment and mineral fertilization), 5 replicates (pots). Each pot had 3 plants.

Table 12. Chemical characterization of mine tailings and the substrate mixtures

Substrate		OC (%)	N tot (%)	pH	EC ($mS\ cm^{-1}$)
C	Compost	25.6	2.76	7.85	7.81
T	100 % Mine tailings	0.12	0.047	8.28 c[(†)]	0.42 d
TC_1	T + 4% MSWC			8.88 b	2.07 c
TC_2	T + 12% MSWC (w/w)			9.19 a	5.8 a
TF_1	T + 70 kg N ha^{-1}			8.35 c	0.35 d
TF_2	T + 140 kg N ha^{-1}			8.34 c	0.33 d
TC_1F_1	T + 4% MSWC (w/w) + 70 kg N ha^{-1}			8.74 b	2.25 c
TC_1F_2	T + 4% MSWC (w/w) + 140 kg N ha^{-1}			8.72 b	2.22 c
TC_2F_1	T + 12% MSWC (w/w) + 70 kg N ha^{-1}			9.1 a	5.36 a
TC_2F_2	T + 12% MSWC (w/w) + 140 kg N ha^{-1}			8.88 b	4.27 b

[(†)] Different letters indicate significant differences after ANOVA and Student-Newmann-Keuls's test ($P<0.05$).

The MSWC were combined with the mine tailings at rates of 4 and 12% (w/w), respectively equivalent to 54 and 162 t ha^{-1}. Two levels of nitrogen fertilization equivalent to 70, and 140 kg N per hectare were respectively provided by irrigation with a solution of $Ca(NO_3)_2$ (calcium nitrate).

To ensure an optimal plant growth, phosphorus (10 mg kg^{-1}) and potassium (26 mg kg^{-1}) were provided with a solution containing KCl (potassium chloride) and KH_2PO (potassium dihydrogen phosphate) following Sirguey et al. (2006).

The compost/N fertilizer dose combination were also considered. As expected, the soil pH rose significantly in response to the MSWC-dose, whereas the lower values were recorded for T (100% mine tailings), TF_1 and TF_2 substrates (tailings + N fertilization) (Table. 12).

The mine tailings and MSWC were analyzed for the following parameters: pH (solid/ water slurry at the ratio of 1/2.5); electrical conductivity (EC; saturated paste extract); total C and N were determined by an elemental analyzer (Carlo Erba CHN 1 500). The metal content in soil and plant samples was determined following the procedures described in paragraph 2.1.1.

The bioavailable fractions of the metals was obtained by DTPA-extraction for Cd, Pb (Martens and Lindsay, 1990) and NH_4NO_3-extraction for Tl (Gryschko et al. 2005). The accuracy of the ICP analysis was checked by running up the standard solutions every 15 samples. Data analysis was carried out by one way analysis of variance and Student-Newmann-Keuls's test at $P<0.05$. Total and bioavailable Cd, Pb, Tl and Zn found in mine tailings and the substrate mixtures are reported in table 13.

Except for a certain data variability, in the tailings/compost mixtures a dilution effect of the metals was observed due to the increase of the compost dose. This effect, of course, was not observed in the case of fertilization but it appeared again in the compost/N fertilizer dose combination treatments.

With regard to the metal bioavailability, it was expected that the higher the compost dose in the mixtures the lower the bioavailable fraction. This was observed only in the case of Cd. Data variability hidden the effect of the compost amendment in the case of other metals.

Table 13. Total Cd, Pb, Tl and Zn found in compost, mine tailings and the substrate mixtures. Metal bioavailable fraction found in mine tailings and tailings/compost mixtures

Substrate	Total				Bioavailable			
	Cd (mg kg^{-1})	Pb (mg kg^{-1})	Tl (mg kg^{-1})	Zn (mg kg^{-1})	Cd (mg kg^{-1})	Pb (mg kg^{-1})	Tl (mg kg^{-1})	Zn (mg kg^{-1})
C	< d.l.(†)	84	< d.l.	220	–	–	–	–
T	27.7 abc(‡)	5,851 ab	303 a	17,937 abc	0.036 a	1.2 a	72.8 c	27.2 a
TC_1	25.6 bc	4,637 bc	262 ab	15,047 bcd	0.038 a	1.51 a	85 abc	29.1 a
TC_2	23.1 c	3,583 c	251 ab	13,569 cd	0.031 a	0.91 a	86.9 abc	25.3 a
TF_1	30.9 ab	7,292 a	280 a	20,176 a	0.014 bc	1.71 a	95 ab	29.7 a
TF_2	27 abc	6,954 a	299 a	17,594 abc	0.025 ab	0.75 a	95.7 ab	21.6 a
TC_1F_1	33.4 a	6,219 ab	264 ab	18,324 ab	0.024 ab	1.53 a	77.3 bc	28 a
TC_1F_2	27.4 abc	4,729 bc	263 ab	17,655 abc	0.01 c	1.11 a	98.6 a	19.3 a
TC_2F_1	27.9 c	3,763 c	224 bc	16,356 abc	0.027 ab	1.32 a	79.6 abc	25.3 a
TC_2F_2	21.3 c	3,560 c	194 c	12,366 d	0.01 c	1.27 a	97.5 a	21.1 a

(†)Not detectable; (‡) Different letters indicate significant differences after ANOVA and Student-Newmann-Keuls's test ($P<0.05$).

4.2.2. Plant Response

Seeds of *T. caerulescens* (Ganges population, Montpellier, France) were germinated in a controlled environment (16 h/8 h day/night cycle, PAR 500 µmol m^2 s^{-1}, 20 °C). Sixty day old seedlings were transplanted in 2 l pots containing the Pb/Zn tailings in a greenhouse. Each pot was watered three times per week reaching the water holding capacity. At the end of the growth cycle which lasted 90 days, the plants were carefully removed from each pot and the roots were accurately washed. Plants fractions (roots and shoots) were weighed and the leaf area was measured by a Leaf Area Meter (LiCor 3000).

Figure 9 shows the leaf area and the biomass of the plant fractions. The effects of the treatments are clearly demonstrated: a significant increase in plant canopy surface was observed as well as the increase of the below- and above-ground biomass in compost amended mine tailings.

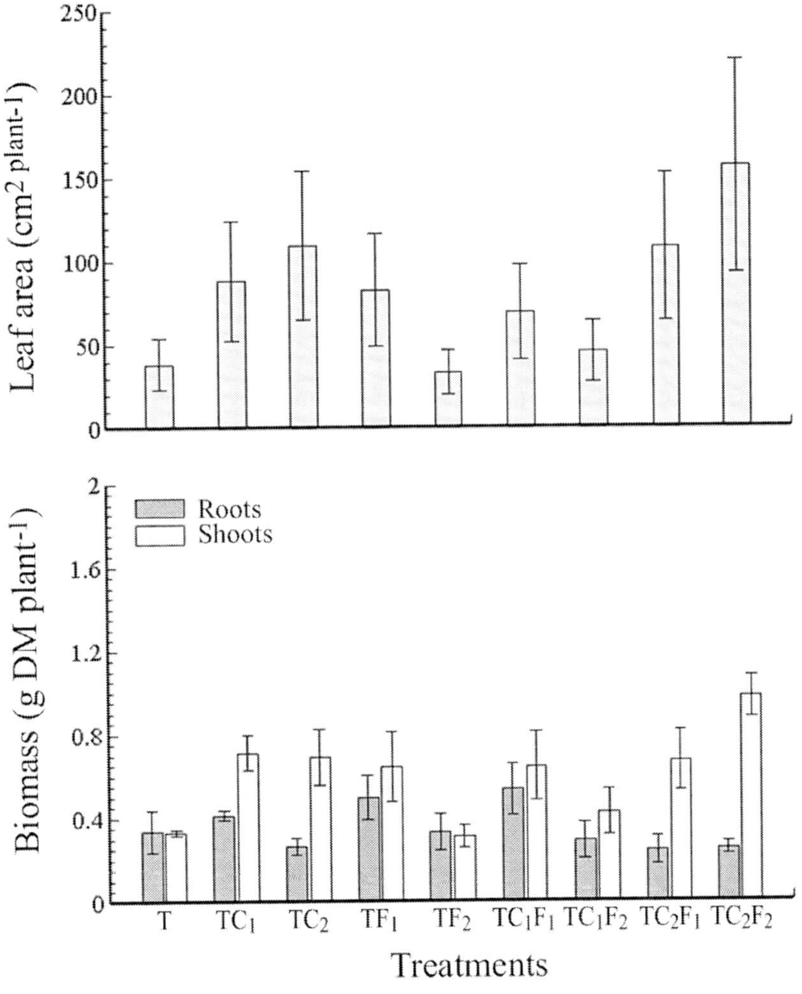

Figure 9. Leaf area and biomass weight of roots and shoots recorded in plants of *T. caerulescens* grown in Pb/Zn mine tailings amended with municipal waste compost and N fertilized.

The N fertilization alone gave a smaller support to the plant growth: the plants that received only the mineral N supply showed a lower biomass growth, thus confirming the fundamental role of organic matter. Indeed, better results in terms of aboveground biomass were recorded in the TC_2F_2 mixture, where the positive effects of organic amendment was combined with the N supply. The metal accumulation in the plant fractions is showed in figure 10. It should be emphasized that the peaks of metal accumulation in the root tissues were always in correspondence to the N supply without compost (TF_1 and TF_2 treatments) (Figure 10).

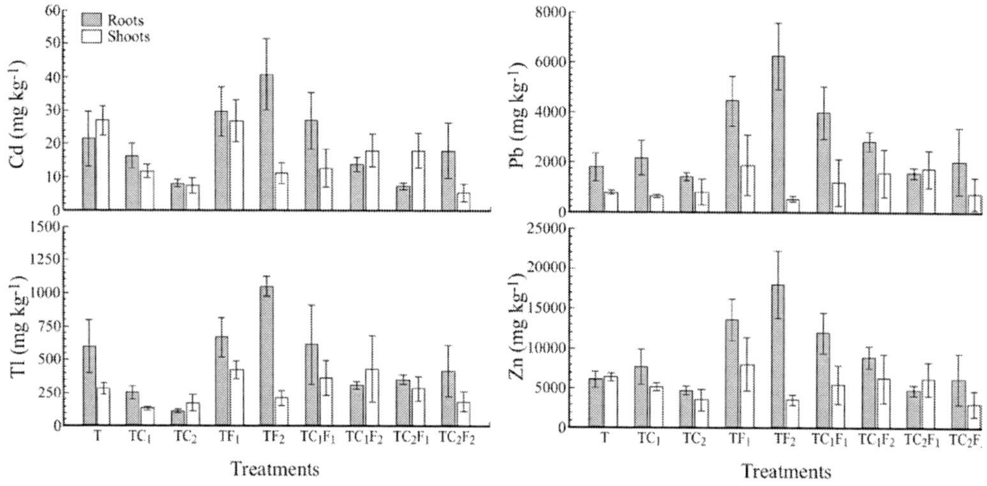

Figure 10. Concentration of Cd, Pb, Tl and Zn in roots and shoots of *T. caerulescens* grown in Pb/Zn mine tailings amended with municipal waste compost and N fertilized.

Even if the Cd concentration in the mine tailings was not very high, *T. caerulescens* responded to the treatments lowering the metal accumulation in their fractions. The ANOVA confirmed that this reduction was significantly affected by the treatments, occurring in TC_1 and TC_2.

Regarding the other metals, evidences of hyperaccumulation were observed. In roots of N-fertilized *T. caerulescens* the concentrations were up to about 6,000 mg kg^{-1} of Pb, that is six times greater than the hyperaccumulation threshold (Baker and Brooks, 1989). Also, in this case the metal concentration in roots was significantly affected by the treatments at ANOVA. The mineral fertilization promoted a significant increase in the metal accumulation by plant fractions whereas lower values were recorded in plants grown on compost amended mine tailings. For the shoots, the effects of the treatments were less clear.

Thallium concentration recorded in plants grown in T, TF_1, TF_2 substrates exceeded the threshold of hyperaccumulation of 500 mg kg^{-1} (Lacoste et al. 1999). A significant reduction of Tl concentration in the roots and shoots relative to the control occurred in TC_2 treatment.

The ANOVA revealed significant effects of the experimental treatments on the accumulation of Zn in roots. The higher levels of statistical significance were found for TF_1, TF_2, and TC_1F_1, whereas the other treatments produced no significant effects. The Zn hyperaccumulation threshold, which is equal to 10,000 mg kg^{-1} (Baker and Brooks, 1989) was exceeded in the roots of TF_1, TF_2 and TC_1F_1 plants (Figure 10). The N-supply presumably stimulated the plant growth but at the same time, this was severely limited by the presence of the pollutants. *Thlaspi caerulescens* accumulated high amounts of metals but limited itself in growth due to the adverse environmental conditions. The compost amendment supplied organic matter to a sterile substrate and, indirectly, lowered the metal concentrations by a dilution effect, making the soil environment explored by the plant roots less hostile. Our results agreed to those in literature and confirmed that compost addition in polluted soils decreases the plant accumulation of heavy metals (Pérez-de-Mora et al. 2006; Tandy et al. 2009). This research will continue for the next two years, aiming to study the potential of native pseudometallophytes for phytostabilization. A series of pot experiments based on a factorial scheme, will be set up to evaluate the potential for phytostabilization of *Biscutella laevigata*, *Silene vulgaris*, *Anthyllis vulneraria* and *Poa alpina*. The improvement of the substratum will be achieved through combinations of different biosolids (waste compost and green compost), other amendants (biochar) and fertilization.

5. Lessons Learned

After a decade of research in this field, considering what has been gained in terms of experience, considering the results from our activities and the literature's state of the art, we can better assess and accept some theories published in literature whereas reject other ones.

The most significant work that we managed in this period was the field experiment of metal phytoextraction. That work was an exciting challenge that make us very proud for testing our expertise at a field scale to counteract a problem for the good of the community. This activity had also a relevance from a scientific point of view, being the first Italian *in situ* trial of phytoremediation.

Unfortunately, we experienced some discouraging administrative problems with the local administrations. The field experiment was considered by the public administration as a technological plant of waste treatment. Hence, a lot of permissions were requested and a lot of time was spent to manage the administrative tricks. Finally, we were not allowed to maintain the field research for more than three years.

Considering that the public administration is usually the main beneficiary for innovative and less expensive clean up technologies, we wish that in the next future there will be a less hostile relationships between science and administration in Italy. Perhaps, this problem should be addressed at EU level making it simpler to perform such kind of research activity at large-scale. On the other hand if this will not happen, the phytotechnologies will fail to become a viable option for the soil clean up. The efforts and investments devoted to the lab research should be oriented supporting field scale investigations considering that sometimes the complexity is so high as to weaken the validity of the results obtained by lab experiments.

In agreement with Chaney et al. (2007), we support the idea that phytoremediation of moderately polluted soils is possible not only through the use of hyperaccumulator species but also with biomass species (short rotation forestry or biomass crops).

However, the excessive length of the process - calculated considering the efficiency of the plants and the amount of metal/s to be removed from the soil (in relation to the soil clean up target) is the main limiting factor. To one hand this duration, that was referred by Van Nevel et al. (2007) as the Achilles'heel of phytoremediation, is particularly unacceptable if we consider a land where there is an urgent need for other land use options. On the other hand the negative public "hype" of phytoremediation due to the duration of the process is perhaps perceived by the public even worse than what it is because of incorrect early expectations. However, it can not be that an industrial site, exposed to industrial pollution for 100 years, is reclaimed in a few months spending little money. Currently, phytoremediation technologies are neither fast nor efficient as the market of soil clean up technologies expects. Also, the economic case to support it is often marginal and the time required for phytoremediation may be unrealistic. The commercial application of phytoremediation as a practical site solution is yet considered not feasible (Onwubuya et al; 2009). The initial enthusiasms on phytoremediation recorded in the middle '90 was disappointed and the green business lacks. It is likely that the actual complexity of the problem was underestimated. However, in our opinion, the significant amount of experience gained worldwide should not be lost. Further investments of intellectual and financial resources will overcome the current problems restoring a real applicative potential to phytoremediation.

Conclusion

A lot of work is expected to be done to improve the efficiency of the process, particularly focusing on the most appropriate agronomic management of it. This work must be done in field conditions, otherwise misleading results and indications may arise, bringing negative consequences, and the prospects of phytoremediation would likely to be further damaged.

Further progress on phytoextraction is expected in the following research lines:

- definition of agronomic practices to improve the phytoextraction process efficiency of hyperaccumulators and biomass agronomic species (McGrath, 2006; Chaney et al. 2007), also by changing the metal bioavailability (Evangelou et al. 2007; Wang et al. 2006, Zaccheo et al. 2007);
- research on the microbial symbiosis or fungal symbiosis that may improve the metal uptake (Abou-Shanab et al. 2006, Wang et al. 2007);
- use of hormones or other substances that modify the plant metabolism improving the metal translocation (Israr and Sahi 2008);
- breeding between hyperaccumulators and tolerant species and genetic engineering of plants to create high efficiency and high biomass plants (Zhang et al. 2006);
- increase the resolution and accuracy of the genetic mapping of loci for metal hyperaccumulation and hypertolerance (Kramer, 2010).
- further use of the biomass to improve the sustainability of the technique (Dickinson et al. 2009, Meers et al. 2010).

Regarding the phytostabilization, Whiting et al. (2004) pointed out some key problems concerning the role of metallophytes in the ecological restoration of polluted sites that need research:

- identifying and understanding the metal-tolerance in local metallophytes native to the specific mining area;

- overcoming the typical slow growth rate of stress-tolerant species and improving the sward ground cover in metal tolerant grasses;
- reducing fertilizer inputs and identifying the nitrogen fixing metallophytes to promote "low-maintenance" vegetative cover;
- developing metallophytes with multiple metal-tolerance systems to use on heterogeneous wastes;
- developing metal-excluding plants to minimize the transfer of the metals into the food chain from the restored sites.

ACKNOWLEDGMENTS

This decade of research on phytoremediation was supported by the Italian Ministry of Research and University (PRIN 1998, PRIN 2003, PRIN 2005, PRIN 2008), Region Friuli Venezia Giulia (L.R. 26/05, 2006-2007). We gratefully acknowledge Mike Frank Quartacci (University of Pisa, I) Teofilo Vamerali (University of Parma, I), Valentino Casolo (University of Udine, I) and Patrizia Zaccheo (University of Milano, I). All the people involved in the PRIN projects are acknowledged as well.

The field experiment of Torviscosa would not have been possible without the material support of Caffaro srl which allowed the establishment of the experimental trials within their property, and the technical support provided by Patrizia Simeoni (University of Udine, I), Marta Plazzotta (ARPA-FVG, I), Domenico Perosa (Caffaro srl).

The helpful contribution of Diego Chiabà, Carla Danelutti, Ilaria Cecon, Federico Zamboni, Chiara Odorico, Paola Ranieri, Moreno Vidoni, Antonino Danelutto and Sabina Moroldo is acknowledged.

A special thank you is due to Rufus Chaney (USDA-ARS, Beltsville, MD, USA) for his valuable comments and suggestions regarding the work on metallophytes.

REFERENCES

Abou-Shanab, R., Ghanem, N., Ghanem, K., Al-Kolaibe, A. (2007). Phytoremediation potential of crop and wild plants for multi-metal contaminated soils. *Research Journal of Agriculture and Biological Sciences*, 3, 370-376.

Alvarenga, P., Gonçalves, A.P., Fernandes, R.M., de Varennes, A., Vallini, G., Duarte, E., Cunha-Queda, A.C. (2009). Organic residues as immobilizing agents in aided phytostabilization: (I) Effects on soil chemical characteristics. *Chemosphere*, 74, 1292-1300.

Anderson, C. W. N., Brooks, R. R., Chiarucci, A., LaCoste, C. J., Leblanc, M., Robinson, B. H., Simcock, R., Stewart, R. B. 1999. Phytomining for nickel, thallium and gold. *Journal of Geochemical Exploration*, 67, 407-415.

Andrade, S.A.L., Abreu, C.A., de Abreu M.F., Silveira, A.P.D. (2004). Influence of lead additions on arbuscular mycorrhiza and *Rhizobium* symbioses under soybean plants. *Applied Soil Ecology*, 26, 123-131.

Arshad, M., Silvestre, J., Pinelli, E., Kallerhoff, J., Kaemmerer, M., Tarigo, A., Shahid, M., Guiresse, M., Pradere, P., Dumat, C. (2008). A field study of lead phytoextraction by various scented *Pelargonium* cultivars. *Chemosphere*, 71, 2187-2192.

Baker, A.J.M., Whiting, S.N. (2002). In search of the Holy Grail – a further step in understanding metal hyperaccumulation? *New Phytologist*, 155, 1-4.

Baker, A.J.M., Brooks, R.R. (1989). Terrestrial higher plants which hyperaccumulate metallic elements – a review of their distribution, ecology and phytochemistry. *Biorecovery*, 1, 81-126.

Baker, A.J.M., Ernst, W.H.O., Van der Ent, A., Malaisse, F., Ginocchio. R. Metallophytes: *the unique biological resource, its ecology and conservational status in Europe, Central Africa and Latin America.* In: Batty L:C., Hallberg, K. and Jarvis, A. P. editors. Ecology of Industrial Pollution: Remediation, Restoration and Preservation. Cambridge University Press, U.K., 2010; 7-39.

Bandiera, M., Mosca, G., Vamerali, T. (2009). Humic acids affect root characteristics of fodder radish (*Raphanus sativus* L. var. *oleiformis* Pers.) in metal-polluted wastes. *Desalination*, 246, 78-91.

Bardos, P., Andersson-Skold, Y., Blom, S., Keuning, S., Pachon, C., Tack T. (2008). *Brownfields, bioenergy and biofeedstocks, and green remediation.* Proceedings of the 10th International UFZ-Deltares / TNO conference on Soil:Water systems (CONSOIL), Special Sessions, Milan, Italy, 3-10.

Barrutia, O., Epelde, L., García-Plazaola, J.I., Garbisu, C., Becerril, J.M. (2009). Phytoextraction potential of two *Rumex acetosa* L. accessions collected from metalliferous and non-metalliferous sites: Effect of fertilization. *Chemosphere*, 74, 259-264.

Bizoux, J.-P. Daïnou, K., Raspé, O., Lutts, S., Mahy, G. (2008). Fitness and genetic variation of *Viola calaminaria*, an endemic metallophyte: implications of population structure and history. *Plant Biology*, 10, 684-693.

Chaney, R.L., Malik, M., Li, Y.M., Brown, S.L., Brewer, E.P., Angle, J.S., Baker, A.J.M. (1997). Phytoremediation of soil metals. *Current Opinions in Biotechnology*, 8, 279-284.

Chaney, R.L., Li, Y.M., Brown, S.L., Homer, F.A., Malik, M., Angle, J.S., Baker, A.J.M., Reeves, R.D., Chin, M. *Improving metal hyperaccumulator wild plants to develop commercial phytoextraction systems: approaches and progress.* In: Terry, N., Bañuelos, G.S. editors. Phytoremediation of Contaminated Soil and Water. Proceedings of the Symposium on Phytoremediation, Fourth International Conference on the Biogeochemistry of Trace Elements. Boca Raton (USA), CRC Press , 2000, 129-158.

Chaney, R.L., Angle, J.S., Broadhurst, C.L., Peters, C.A., Tappero, R.V., Sparks, D.L. (2007). Improved understanding of hyperaccumulation yields commercial phytoextraction and phytomining technologies. *Journal of Environmental Quality*, 36, 1429-1443.

Corsini, A., Zaccheo, P., Crippa, L., Bruini, V., Manachini, B. *Nematodes colonization of pyrite cinders in a phytoremediation study.* Proceedings of the 1st International Symposium on Nematodes as Environmental Bioindicators. Heriot-Watt University, Edinburgh, UK. 12-13 June 2007.

Corsini, A., Cavalca, L., Zaccheo, P., Crippa, L., Andreoni, V. (2010a). Impact of glucose on microbial community of a soil containing pyrite cinders: role of bacteria in arsenic mobilization under submerged condition. *Soil Biology & Biochemistry*, doi: 10.1016/j.soilbio.2009.12.010.

Corsini, A., Crippa, L., Zaccheo, P. (2010b). *Biological mobilization of arsenic in a pyrite cinders polluted soil*. Proceedings XXXIX Annual Meeting of the European Society for New Methods in Agricultural Research (ESNA), Brno, 25-29 August 2009. In press.

del Río, M., Font, R., Fernández-Martínez, J.M., Domínguez, J., de Haro, A. (2000). Field trials of *Brassica carinata* and *Brassica juncea* in polluted soils of the Guadiamar river area. *Fresenius Environmental Bulletin*, 9, 328-332.

Di Colbertaldo, D. Il giacimento piombo zincifero di Raibl in Friuli (Italia). Roma. Pubbl. S.A. Miniere Cave di Predil (Raibl). 1948.

Dickinson, N.M. (2000). Strategies for sustainable woodlands on contaminated soils. *Chemosphere*, 41, 259-263.

Dickinson, N.M., Baker, A.J.M., Doronila, A., Laidlaw, S., Reeves, R.D. (2009). Phytoremediation of inorganics: realism and synergies. *International Journal of Phytoremediation*, 11, 97-114.

Dimitriou, I., Aronsson, P., Weih, M. (2006). Stress tolerance of five willow clones after irrigation with different amounts of landfill leachate. *Bioresource Technology*, 97, 150-157.

Domínguez, M.T., Marañón, T., Murillo, J.M., Schulin, R., Robinson R.D. (2008). Trace element accumulation in woody plants of the Guadiamar Valley, SW Spain: a large-scale phytomanagement case study. *Environmental Pollution*, 152, 150-159.

EEA. 2007. *Progress in management of contaminated sites*. Report CSI 015, European Environment Agency, Copenhagen, Denmark. Online report.

Ernst, W. H.O. Schwermetallvegetation der Erde. Stuttgart, Fischer; 1974.

Ernst W.H.O. 1996. Bioavailability of heavy metals and decontamination of soils by plants. *Applied Geochemistry*, 11, 163-167.

Ernst, W.H.O., Knolle, F., Kratz, S., Schnung, E. (2004). Aspects of toxicology of heavy metals in the Harz region – a guided excursion. Landbauforschunf Völkenrode, 54, 53-71.

Ernst, W.H.O. (2005). Phytoextraction of mine wastes - Options and impossibilities. *Chemie der Erde-Geochemistry*, 65, 29-42.

Evangelou, M.W.H., Ebel, M., Schaeffer, A. (2007). Chelate assisted phytoextraction of heavy metals from soil. *Chemosphere*, 68, 989-1003.

Farrell, M., Perkins, W.T., Hobbs, P.J., Griffith, G.W., Jones, D. L. (2009). Migration of heavy metals in soil as influenced by compost amendments. *Environmental Pollution*, 158, 55-64.

Felix H. (1997) Field trials for *in situ* decontamination of heavy metal polluted soils using crops of metal-accumulating plants. *Zeitschrift für Pflanzenernährng und Bodenkunde*, 160, 525-529.

Fellet, G., Marchiol, L., Perosa, D., Zerbi, G. (2007). The application of phytoremediation technology in a soil contaminated by pyrite cinders – a preliminar pot experiment. *Ecological Engineering*, 31, 207-214.

Francescutti, M. Mille anni di storia della Val Canale e della sua foresta. Udine: Chiandetti; 1991.

French, C.J., Dickinson, N.M., Putwain, P.D. (2006). Woody biomass phytoremediation of contaminated brownfield land. *Environmental Pollution*, 141, 387-395.

Frérot, H., Lefébvre, C., Gruber, W., Collin, C., Dos Santos, A., Escarré, J. (2006). Specific interactions between local metallicolous plants improve the phytostabilization of mine soils. *Plant and Soil*, 282:53-65.

Grabherr, G., Mucina, L. *Die Pflanzengesellschaften Osterreichs*. Teil. II-Naturlichee waldfreie Vegetation. Jena; Gustav Fischer Verlag; 1993.

Grispen, V.M.J., Nelissen, H.J.M., Verkleij, J.A.C. (2006). Phytoextraction with *Brassica napus* L.: a tool for sustainable management of heavy metal contaminated soils. *Environmental Pollution*, 144, 77-83.

Gryschko, R., Kuhnle, R., Terytze, K., Breuer, J., Stahr, K. 2005. Soil Extraction of Readily Soluble Heavy Metals and As with 1 M NH_4NO_3-Solution – Evaluation of DIN 19730. *Journal of Soils and Sediments*, 5, 101-106.

ISPRA. Annuario dei dati ambientali 2008. Roma: Istituto Superiore per la Protezione e la Ricerca Ambientale; 2008.

Israr, M., Sahi, S.V. (2008). Promising role of plant hormones in trans location of lead in *Sesbania drummondii* shoots. *Environmental Pollution*, 153, 29-36.

Hammer, D., Keller, C. (2003). Phytoextraction of Cd and Zn with Thlaspi caerulescens in field trials. *Soil Use and Management*, 19, 144-149.

Hartl, H., Kniely, G., Leute, G.H., Niklfeld, H., Perko M. Verbreitungsatlas der Farn- und Blütenpflanzen Kärntens., Klagenfurt: Naturwissenschaftlicher Verein für Kärnten; 1992.

Jalas, J., Suominen, J., Lampinen, R., Kurtto, A. (1999). *Atlas Florae Europaeae. Distribution of Vascular Plants in Europe*. Vol. 12. - The Committee for Mapping the Flora of Europe & Societas Biologica Fennica Vanamo, Helsinki.

Keller, C., Hammer, D., Kayser, A., Richner, W., Brodbeck, M., Sennhauser, M. (2003). Root development and heavy metal phytoextraction efficiency: comparison of different plant species in the field. *Plant and Soil*, 249, 67-81.

King, R.F., Royle, A., Putwain, P.D., Dickinson, N.M. (2006). Changing contaminant mobility in a dredged canal sediment during a three-year phytoremediation trial. *Environmental Pollution*, 143, 319-326.

Klang-Westin, E, Eriksson, J. (2003). Potential of *Salix* as phytoextractor for Cd on moderately contaminated soils. *Plant and Soil*, 249, 127-137.

Krämer, U. 2010. Metal hyperaccumulation in plants. *Annual Review Plant Biology*, 61, 28.1-28.18.

Licht, L.A., Isebrands, J.G. (2005). Linking phytoremediated pollutant removal to biomass economic opportunities. *Biomass and Bioenergy*, 28, 203-218.

LaCoste, C., Robinson B., Brooks R.R., Anderson, C., Chiarucci, A., Leblanc, M. (1999). The phytoremediation potential of thallium-contaminated soils using *Iberis* and *Biscutella* species. *International Journal of Phytoremediation* 4, 327-338.

Leblanc, M., Petit, D., Deram, A., Robinson, B.H. Brooks, R.R. (1999). The phytomining and environmental significance of hyperaccumulation of thallium by two plant species. *Economic Geology*, 94, 109-114.

Lombini, A., Dinelli, E., Ferrari, C., Simoni, A. (1998). Plant–soil relationships in the serpentinite screes of Mt. Prinzera (Northern Apennines, Italy). *Journal of Geochemical Exploration* 64, 19-33.

Ma, L.Q., Komar, K.M., Tu, C., Zhang, W.H., Cai, Y., Kennelley, E.D. (2001). A fern that hyperaccumulates arsenic — a hardy, versatile, fast growing plant helps to remove arsenic from contaminated soils. *Nature*, 409, 579.

Madejón, P., Murillo, J.M., Maranón, T., Cabrera, F., Soriano, M.A. (2003). Trace element and nutrient in sunflower plants two years after the Aznalcóllar mine spill. *Science of the Total Environment*, 307, 239-257.

Marchiol, L., Fellet, G., Perosa, D., Zerbi, G. (2007). Removal of trace metals by *Sorghum bicolor* and *Helianthus annuus* in a site polluted by industrial wastes: a field experience. *Plant Physiology and Biochemistry*, 45, 379-387.

Martens, D.C., Lindsay, W.L. Testing soils for copper, iron, manganese, and zinc. In: Westerman, R.L. editors. Soil Testing and Plant Analysis. *Soil Science Society of America*, Madison, 1990; 229-264.

Mattina, M.J.I., Lannucci-Berger, W., Musante, C., White, J.C. (2003). Concurrent plant uptake of heavy metals and persistent organic pollutants from soil. *Environmental Pollution,* 124, 375-378.

Maxted, A.P., Black, C.R., West, H.M., Crout, N.M.J., McGrath, S.P., Young, S.D. (2007a). Phytoextraction of cadmium and zinc from arable soils amended with sewage sludge using *Thlaspi caerulescens*: development of a predictive model. *Environmental Pollution*, 150, 363-372.

Maxted, A.P., Black, C.R., West, H.M., Crout, N.M.J., McGrath, S.P., Young, S.D. (2007b). Phytoextraction of cadmium and zinc by *Salix* from soil historically amended with sewage sludge. *Plant and Soil*, 290, 157-172.

McGrath, S.P., Zhao, F.J. (2003). Phytoextraction of metals and metalloids from contaminated soils. *Current Opinion in Biotechnology*, 14, 277-282.

McGrath, S.P., Lombi, E., Gray, C.W., Caille, N., Dunham, S.J., Zhao, F.J. (2006). Field evaluation of Cd and Zn phytoextraction potential by the hyperaccumulators *Thlaspi caerulescens* and *Arabidopsis halleri*. *Environmental Pollution*, 141,115-125.

McGrath, S.P., Dunham, S.J., Correl, R.L. *Potential for phytoextraction of zinc and cadmium from soils using hyperaccumulator plants*. In: Terry N., Banuelos G. editors. Phytoremediation of contaminated soil and water. Boca Raton FL, CRC Press 2000; 109-128.

Meers, E., Lamsal, S., Vervaeke, P., Hopgood, M., Lust, N., Tack, F.M.G. (2005). Availability of heavy metals for uptake by *Salix viminalis* on a moderately contaminated dredged sediment disposal site. *Environmental Pollution*, 137, 354-364.

Meers, E., Van Slycken, S., Adriaensen, K., Ruttens, A., Vangronsveld, J., Du Laing, G., Witters, N., Thewys, T., Tack, F.M.G. (2010). The use of bio-energy crops (*Zea mays*) for 'phytoattenuation' of heavy metals on moderately contaminated soils: A field experiment. *Chemosphere*, 78, 35-41.

Mendez, M.O., Maier, R. M. (2008). Phytostabilization of mine tailings in arid and semiarid environments – An emerging remediation technology. *Environmental Health Perspectives*, 116, 278-283.

Murillo, J.M., Maranón, T., Cabrera, F., López, R. (1999). Accumulation of heavy metals in sunflower and sorghum plants affected by the Guadiamar spill. *Science of the Total Environment*, 242, 281-292.

Nadal, M., Schuhmacher, M., Domingo, J.L. (2004). Metal pollution of soils and vegetation in an area with petrochemical industry. *Science of the . Total Environment*, 321, 59-69.

Neugschwandtner, R.W., Tlustoš, P., Komárek, M., Száková, J. (2008). Nutrient mobilization and nutrient contents of Zea mays in response to EDTA additions to heavy-metal-contaminated agricultural soil. *Journal of Plant Nutrition and Soil Science*, 172, 520-527.

Nwachukwu, O.I., Pulford, I.D. (2008). Comparative effectiveness of selected adsorbant materials as potential amendments for the remediation of lead-, copper and zinc-contaminated soil. *Soil Use and Management*, 24, 199-207.

Onwubuya, K., Cundy, A., Puschenreiter, M., Kumpiene, J., Bone, B., Greaves, J., Teasdale, P., Mench, M., Tlustos, P., Mikhalovsky, S., Waite, S., Friesl-Hanl, W., Marschner, B., Müller, I. (2009). Developing decision support tools for the selection of "gentle" remediation approaches. *Science of the Total Environment*, 407, 6132-6142.

Pérez-de-Mora, A., Madrid, F., Cabrera, F., Madejón, E. (2007). Amendments and plant cover influence on trace element pools in a contaminated soil. *Geoderma*, 139, 1-10.

Pilon Smits, E. (2005). Phytoremediation. *Annual Review Plant Biology*, 56,15-39.

Pulford. D., Riddell-Black, D., Stewart, C. (2002). Heavy metal uptake by willow clones from sewage sludge-treated soil: the potential for phytoremediation. *International Journal of Phytoremediation*, 4, 59-72.

Quartacci, M.F., Irtelli, B., Baker A.J.M., Navari-Izzo, F. (2007). The use of NTA and EDDS for enhanced phytoextraction of metals from a multiply contaminated soil by *Brassica carinata*. *Chemosphere* 68: 1920-1928.

Quartacci, M.F., Irtelli, B., Gonnelli, C., Gabbrielli, R., Navari-Izzo, F. (2009). Naturally-assisted metal phytoextraction by Brassica carinata: Role of root exudates. *Environmental Pollution* 157, 2697-2703.

Raskin, I., Smith, R.D., Salt, D.E. (1997). Phytoremediation of metals: using plants to remove pollutants from the environment. *Current Opinion in Biotechnology*, 8, 221-226.

Reeves, R.D., Baker, A.J.M. (2000). *Metal-accumulating plants*. In: Raskin, I., Ensley, B.D. editors. Phytoremediation of Toxic Metals: Using Plants to Clean Up the Environment. John Wiley and Sons Inc., New York, pp. 193-229.

Reeves, R.D., Brooks, R.R. (1983). Hyperaccumulation of lead and zinc by two metallophytes from mining areas of central Europe. *Environmental Pollution* (Series A) 31, 277-285.

Reeves, R.D., Schwartz, C., Morel, J.L., Edmondson, J. (2001). Distribution and metal-accumulating behaviour of *Thlaspi caerulescens* and associated metallophytes in France. *International Journal of Phytoremediation*, 3, 145-172.

Robinson, B.H., Brooks, R.R., Clothier, B.E. (1999). Soil amendments affecting nickel and cobalt uptake by *Berkheya codii*: potential use for phytomining and phytoremediation. *Annals of Botany,* 84, 689-694.

Rosselli, W., Keller, C., Boschi K. (2003). Phytoextraction capacity of trees growing on a metal-contaminated soil. *Plant and Soil*, 256, 265-272.

Salt, D.E., Blaylock, M., Kumar, N.P.B.A., Dushenkov, V., Ensley, D., Chet, I., Raskin, I. (1995). Phytoremediation: a novel strategy for the removal of toxic metals from the environment using plants. *Biotechnology*, 13, 468-474.

Salt, D.E., Smith, R. D., Raskin, I. 1999. Phytoremediation. *Annual Review* Plant Physiology *Plant Molecular Biology*, 49: 643-68.

Santibáñez, C., Verdugo C., Ginocchio, R. (2008). Phytostabilization of copper mine tailings with biosolids: Implications for metal uptake and productivity of *Lolium perenne*. *Science of the Total Environment*, 395, 1-10.

Schwartz C., Phytoextraction des métaux pollués par la plante hyperaccumulatrice *Thlaspi caerulescens*. ENSAIA thesis Nancy France (1997).

Schwartz, C., Echevarria, G., Morel, J.L.(2003). Phytoextraction of cadmium with *Thlaspi caerulescens*. *Plant and Soil*, 249, 27-35.

Sirguey, C., Schwartz, C., Morel, J.M. (2006). Response of Thlaspi caerulescens to nitrogen, phosphorus, and sulfur fertilization. *International Journal of Phytoremediation*, 8, 149-161.

Smith, R.F. (1979). The occurrence and need for conservation of metallophytes on mine wastes in Europe. *Environmental Geochemistry Health* 4, 131-147.

SUMATECS 2009. Sustainable management of trace element contaminated soils – Development of a decision tool system and its evaluation for practical application. Final Research Report. *www.snowman-era.net*

Quartacci M. F., B. Irtelli, C. Gonnelli, R. Gabbrielli, F. Navari-Izzo. (2009). Naturally-assisted metal phytoextraction by Brassica carinata: *Role of root exudates Environmental Pollution* 157 : 2697–2703.

Tandy, S., Healey, J. R., Nason, M. A., Williamson, J. C., Jones, D. L. (2009). Remediation of metal polluted mine soil with compost: Co-composting versus incorporation. *Environmental Pollution*, 157, 690-697.

Tordoff, G.M., Baker, A.J.M., Willis, A.J. (2000). Current approaches to the revegetation and reclamation of metalliferous mine wastes. *Chemosphere*, 41, 219-228.

USEPA. 1995a. EPA Method 3051: *Microwave assisted acid digestion of sediments, sludges, soils, and oils,* in: Test Methods for Evaluating Solid Waste, 3rd edition. Washington, DC.

USEPA. 1995b. EPA Method 3052: *Microwave assisted acid digestion of siliceous and organically based matrices*, in: Test Methods for Evaluating Solid Waste, 3rd edition. Washington, DC.

Vamerali, T., Bandiera, M., Coletto, L., Zanetti, F., Dickinson, N.M., Mosca, G. (2009). Phytoremediation trials on metal- and arsenic-contaminated pyrite wastes (Torviscosa, Italy). *Environmental Pollution*, 175, 887-894.

Van Nevel, L., Mertens. J., Oorts. K., Verheyen, K. (2007). Phytoextraction of metals from soils: How far from practice? *Environmental Pollution*, 150, 34-40.

Vandecasteele, B., Samyn, J., De Vos, B., Muys, B. (2008). Effect of tree species choice and mineral capping in a woodland phytostabilisation system: A case-study for calcareous dredged sediment landfills with an oxidised topsoil. *Ecological Engineering*, 32, 263-273.

Vangronsveld, J., Herzig, R., Weyens, N., Boulet, J., Adriaensen, K., Ruttens, A., Thewys, T., Vassilev, A., Meers, E., Nehnevajova, E., van der Lelie, D., Mench, M. (2009). Phytoremediation of contaminated soils and groundwater: lessons from the field. *Earth and Environmental Science*, 16: 765-794.

Vergnano Gambi, O., Brooks, R. R., Radford, C. C. (1979). L'accumulo di nichel nelle specie italiane del genere *Alyssum*. Webbia, 33, 269-77.

Vervaeke, P., Luyssaert, S., Mertens, J., Meers, E., Tack, F.M.G., Lust, N. (2003). Phytoremediation prospects of willow stands on contaminated sediment: a field trial. *Environmental Pollution*, 126, 275-282.

Wenzel, W.W., Jockwer, F. (1999). Accumulation of heavy metals in plants grown on mineralised soils of the Austrian Alps. *Environmental Pollution*, 104, 145-155.

Whiting, S. N., Reeves, R. D., Richards, D., Johnson, M. S., Cooke, J. A., Malaisse, F., Paton, A., Smith, J. A. C., Angle, J. S., Chaney, R. L., Ginocchio, R., Jaffré, T., Johns, R., McIntyre, T., Purvis, O. W., Salt, D. E., Schat, H., Zhao, F. J., Baker, A. J. M. (2004).

Research priorities for conservation of metallophyte biodiversity and their potential for restoration and site remediation. *Restoration Ecology*, 12, 106- 116.

Whiting, S.N., Reeves, R.D., Baker, A.J.M., 2002. Conserving biodiversity: mining metallophytes and land reclamation. *Mine Environmental Management*, 10, 11-16.

Wierzbicka, M., Szarek-Łukaszewska, G., Grodzińska, K. (2004). Highly toxic thallium in plants from the vicinity of Olkusz (Poland). *Ecotoxicological Environmental Safety*, 59, 84-88.

Wang, C., Li. X. C., Wang, P. F., Zou, L. M., Ma, H.T. (2006). Total contents and extractable fraction of metals in sludges from urban wastewater treatment plants of China. *Pedosphere*, 16, 756-761.

Wang, F.Y., Lin, X.G., Yin, R. (2007). Inoculation with arbuscular mycorrhizal fungus *Acaulospora mellea* decreases Cu phytoextraction by maize from Cu-contaminated soil. *Pedobiologia*, 51, 99-109.

Wong, M.H. (2003). Ecological restoration of mine degraded soils, with emphasis on metal contaminated. *Chemosphere*, 50, 775-780.

Wu, G., Kang, H., Zhang, X., Shao, H.; Chu, L., Ruan, C. (2010). A critical review on the bio-removal of hazardous heavy metals from contaminated soils: Issues, progress, eco-environmental concerns and opportunities. *Journal of Hazardous Materials*, 174, 1-8.

Yanqun, Z., Yuan, L., Jianjun, C., Haiyan, C., Li, Q., Schwartz, C. (2005). Hyperaccumulation of Pb, Zn and Cd in herbaceous grown on lead-zinc mining area in Yunnan, China. *Environmental International*, 31, 755-762.

Yoon, J., Cao, X. D., Zhou, Q. X., Ma, L. Q. (2006). Accumulations of Pb, Cu and Zn in native plants growing on a contaminated Floride site. *Science of the Total Environment*, 368: 456-464.

Zaccheo, P., Crippa, L., Marchiol, L., Dell'Amico, E., Colombo, M., Andreoni, V. Valutazione della compromissione biologica di un suolo contenente ceneri di pirite in un progetto di fitorisanamento. In: SISS, Società Italiana Scienza del Suolo, editor. Atti Convegno Nazionale Suolo Ambiente Paesaggio. Modena (Italia). 2007. ISBN/ISSN: 978-88-902831-0-9.

Zbíral, J., Pavlíčkova, J., Havlíková, S., Čižmárová, E., Němec, P., Sáňka, M., Kubáň, V., Medek P. (2002). Comparison of several soil extractants for determination of thallium. *Communications Soil Science Plant Analysis*, 33, 3303-3312.

Zhang, R.-Q., Tang, C.-F., Wen, S.-Z., Liu, Y.-G., Li, K.-L. 2006. Advances in research on genetically engineered plants for metal resistance. *Journal of Integrative Plant Biology*, 48, 1257-1265.

Zhao, F.J., Lombi, E., McGrath, S.P. (2003). Assessing the potential for zinc and cadmium phytoremediation with the hyperaccumulator Thlaspi caerulescens. *Plant Soil*, 249, 37-43.

In: Handbook of Phytoremediation
Editor: Ivan A. Golubev

ISBN: 978-1-61728-753-4

Chapter 5

PHYTOREMEDIATION OF HEAVY METAL CONTAMINATED SOILS – PLANT STRESS ASSESSMENT

***Jana Kadukova*[1*] *and Jana Kavuličova*[2]**
[1]Technical University of Kosice, Faculty of Metallurgy,
Department of Non-Ferrous Metals and Waste Treatment,
Letna 9, 04200 Kosice, Slovakia.
[2]Technical University of Kosice, Faculty of Metallurgy,
Department of Chemistry, Letna 9, 04200 Kosice, Slovakia.

ABSTRACT

Phytoremediation advantages are widely known nowadays. It is a method applicable for large areas with low concentration of pollutants treatment or areas where only the finishing step of cleaning is required. Very often these kinds of places represent great problems because there is no possibility to take all the soil to the landfills, and often they are part of agricultural fields. There are many studies dealing with application of a variety of plants for the treatment of soils contaminated by heavy metals or organics. Plants growing on these contaminated soils developed several ways of coping with the toxicity of pollutants including avoiding their accumulation, different detoxification mechanisms or even metal excretion from their body. Our work is focused on heavy metal contamination cleanup by phytoremediation with the aim to describe some of the possible ways to assess the stress of plants. There are several factors which can be used in the plant stress assessment such as reduction of biomass production, plant growth inhibition, changes in photosynthesis, germination inhibition, and production of antioxidant enzymes. Knowledge of these factors brings us closer to understanding the molecular mechanisms of heavy metal accumulation by plants and it indirectly helps further application of phytoremediation as well as has numerous additional biotechnological implications. For instance, health-threatening human deficiencies in trace metals appear to be widespread in developing countries and possibly worldwide but engineering of plants accumulating essential metals such as Zn or Se in their edible parts might help in enriching human diets for these important elements.

[*] Tel. 00421-55-6022426, fax. 00421-55-6022428, E-mail: jana.kadukova@tuke.sk, kadukova@post.sk

1. Introduction

Accumulation of different compounds from soil, water or even from the atmosphere is a typical feature of plants as the organisms dependent on mineral nutrition. However, in the case where high concentration of metals are present in the soil, their accumulation is not always considered good or necessary, especially in medicinal plants or agricultural crops. But even this "bad feature" can be used for a people benefit. Nowadays, phytoremediation—biotechnology using plants for cleaning up the environment—has become widely applied.

First mention of plant accumulation of high amounts of metals was described in 1885 by Baumann for *Thlaspi caerulescens* and *Viola calaminaria* (Lasat, 2000) and slowly increased knowledge of this phenomenon. But it took almost one hundred years until it was used as an environmental counterbalance to industrial processes. The idea of using plants that hyperaccumulate metals to selectively remove and recycle excessive soil metals was introduced in 1983, gained public exposure in 1990, and has increasingly been examined as a potential practical and more cost-effective technology than the soil replacement, solidification and washing strategies presently used (Chaney et al., 1997). The term phytoremediation was used for the first time in 1991 and means plant-based action (phyto – plant, remediation – to recover) (Raskin et al., 1994).

2. Phytoremediation

Phytoremediation has been defined as the use of green plants and their associated rhizospheric microorganisms, soil amendments, and agronomic techniques to remove, degrade, or detoxify harmful environmental pollutants (Ouyang, 2002; Schwitzguébel, 2002). Plants ideal for phytoremediation should fulfil four main requirements (Schnoor, 1997):

1. they must be fast growing and have high biomass,
2. have deep roots,
3. have easily harvestable aboveground portion,
4. accumulate large amounts of metals (~ 1000 mg/kg) in aboveground biomass.

Phytoremediation presents many advantages compared to other remediation techniques (Raskin et al., 1994; Schwitzguébel, 2002; Dercová et al., 2005):

- it can be performed with minimal environmental disturbance;
- it is applicable to a broad range of contaminants, including many metals with limited alternative options and radionuclides;
- possibly less secondary air and water wastes are generated than with traditional methods;
- organic pollutants may be degraded to CO_2 and H_2O, removing environmental toxicity;
- it is cost-effective for large volumes of water having low concentrations of contaminants;
- topsoil is left in usable condition and may be reclaimed for agricultural use;

- soil can be left at the site after contaminants are removed, rather than having to be disposed or isolated;
- it is cost-effective for large areas having low to moderately contaminated surface soils;
- plant uptake of contaminated groundwater can prevent off-site migration;
- does not need to use heavy vehicles and devices which damage soil.

However, several drawbacks and limitations of phytoremediation also exist:

- a long time is often required for remediation;
- the treatment is generally limited to soils at a meter from the surface and groundwater within a few meters of the surface;
- climatic or hydrologic conditions may restrict the rate of growth of plants that can be utilized;
- the ground surface at the site may have to be modified to prevent flooding or erosion;
- contaminants may still enter the food chain through animals/insects that eat plant material containing contaminants;
- soil amendments may be required.

Phytoremediation is best applied at sites with shallow contamination of organic, nutrient, or metal pollutants. It is well-suited for use at very large field sites where other methods of remediation are not cost-effective or practicable (Schnoor, 1997). Plants can be used to treat most classes of contaminants—toxic metals, radionuclides and recalcitrant organic pollutants, like chlorinated pesticides, organophosphate insecticides, petroleum hydrocarbons (BTEX), polynuclear aromatic hydrocarbons (PAHs), sulfonated aromatics, phenolics, nitroaromatics and explosives, polychlorinated biphenyls (PCBs), and chlorinated solvents (TCE, PCE). This method is often complementary to traditional bioremediation techniques, based on the use of microorganisms only (Alkorta, Garbisu, 2001; Ouyang, 2002; Schwitzguébel, 2002; Abhilash et al., 2009).

Remediation of inorganic contaminants differs from the case with organic compounds. Because organic compounds can be mineralized, but the remediation of inorganic contamination must either physically remove the contaminant from the system or convert it into a biologically inert form (Cunningham, Ow, 1996). But even in the case that there is no other possibility as to landfill harvested plants (e.g., metals accumulated in plant body have no use or there is no way how to economically extract metal from the plant, etc.) phytoremediation brings advantages because of significant reduction of contaminated material for landfilling. Besides soil treatment, plants have been successfully used for treatment of wastewater (municipal and industrial wastewater) (Majer Newman et al., 2000; Dunne et al., 2005; Chavan et al., 2007; Zurita et al., 2009; Khan et al., 2009) and even some information is known for treatment of the atmosphere (Liua et al., 2007).

2.1. Phytoremediation Techniques

With the increasing amount of information about phytoremediation and development of new applications, several phytoremediation techniques can be distinguished (Schwitzguébel,

2002; Pulford, Watson, 2003; Yang et al., 2005; Mackova et al., 2006; Gerhardt et al., 2009). The most common are listed in Table 1.

Phytoremediation techniques suitable for heavy metal contaminated soils or water clean up basically include phytostabilization, phytoextraction, rhizofiltration and phytovolatilization. Besides plants rhizospheric microorganisms play an important role in metal removal from soil (Ike et al., 2007).

Phytostabilization is usually applied on soils which are so heavily contaminated that removal of metals using plants would take an unrealistic amount of time. In such a case, it is best to choose fast growing plants which can grow in metal contaminated and nutrient deficient soil and which can immobilize heavy metals through absorption and accumulation by roots or precipitation within the rhizosphere not translocating them into shoots (Wong, 2003; Pietramellara et al., 2009). In the present time, it is often used to revegetate mine tailings to minimize wind and water erosion of tailings in a cost-effective way. Phytostabilization also improves the chemical and biological characteristics of the contaminated soils (Alvarenga et al., 2008; Chen et al., 2008; Grandlic et al., 2009).

Table 1. Phytoremediation techniques

TECHNIQUE	CHARACTERISTICS
phytoextraction (phytoaccumulation)	the use of pollutant-accumulating plants to remove metals or organics from soil by concentrating them in harvestable plant parts
phytotransformation	the partial or total degradation of complex organic molecules or their incorporation into plant tissues
rhizoremediation (plant-assisted bioremediation, phytostimulation)	the release of plant exudates/enzymes into the root zone stimulates the microbial and fungal degradation of organic pollutants
rhizofiltration	the use of plant roots to absorb or adsorb pollutants, mainly metals, but also organic pollutants, from water and aqueous waste streams, concentrate and precipitate them
phytostabilization	the use of plants to reduce the mobility and bioavailability of pollutants in the environment, preventing their migration to groundwater or their entry into the food chain
phytovolatilization	the use of plants to volatilize pollutants or metabolites
removal of aerial contaminants	uptake of various volatile organics by leaves
dendroremediation	the use of trees to evaporate water and to extract pollutants from the soil
hydraulic control	the control of the water table and the soil field capacity by plant canopies.

Phytoextraction is a method used for extracting metals from the soils by concentrating them in harvestable plant parts. It can be divided into two methods—induced phytoextraction and continuous phytoextraction (Salt et al., 1998). Induced phytoextraction uses plants producing a big amount of biomass in which metal accumulation is enhanced by addition of a chemical, such as EDTA, NTA, EDDS, etc. (Kos, Leštan, 2003; Quartacci et al., 2007; Saifullah et al., 2009). Continuous phytoextraction uses plants with natural abilities to accumulate high levels of metals—hyperaccumulators (McGrath et al., 2002). Hyperaccumulating plants have an extraordinary ability to accumulate heavy metals, translocate and concentrate them in roots and aboveground shoots or leaves (Schnoor, 1997;

Clemens et al., 2002). Metal translocation from the roots to the shoots for the purpose of harvesting is one of the key goals of phytoextraction research (Jarvis, Leung, 2001). An important issue in phytoextraction is whether the metals can be economically recovered from the plant tissue or whether disposal of the waste is required. Several methods have been studied for contaminated plants disposal such as incineration, liquid extraction, ashing or direct disposal at a hazardous waste site. They application depends on the cost of the process and availability of appropriate technology (Sas-Nowosielska et al., 2004)

Interesting possibilities brings the use of plant with special characteristics to make the process more efficient. For example, Mediterranean halophytic shrub or tree *Tamarix smyrnensis* uses salt glands to excrete the excess of salt from soil. In the case that metals are present in soil are these metals excreted through salt glands in the form of non-toxic crystals. This method is called *phytoexcretion* (Kadukova, Kalogerakis, 2007; Manousaki et al., 2008).

Rhizofiltration is primarily for treatment of extracted groundwater, surface water, and wastewater with low contaminant concentrations. It is defined as the use of plants, both terrestrial and aquatic, to absorb, concentrate, and precipitate contaminants from polluted aqueous sources in their roots (Schwitzguébel, 2002; Eapen et al., 2003; Peng et al., 2008). Advantage of rhizofiltration is that contaminants do not have to be translocated to the shoots, thus, species other than hyperaccumulators may be used. This method is suitable for treatment of water with low metal contamination (Schwitzguébel, 2002; Rousseau et al., 2004; Choo et al., 2006), for treatment of water contaminated with radionuclides (Soudek et al., 2007; Vera Tomé et al., 2008).

Phytovolatilization involves the use of plants to take up contaminants from the soil, transforming them into volatile forms and transpiring them into the atmosphere (Schnoor, 1997). Genetically modified *Arabidopsis thaliana* and *Lyriodendron tulipifera* can grow in soil with higher mercury concentration and transfer it from Hg^{2+} to Hg^0 (Špirochová et al., 2001). Using this method is necessary to control release of formed volatile compounds to the atmosphere.

Selection of plants for phytoremediation of metals depends on the particular application - for phytostabilization, rhizofiltration, phytovolatilization or phytoextraction. Plants that accumulate toxic metals can be grown and harvested economically, leaving the soil or water with a greatly reduced level of toxic metal contamination. Dried, ashed or composted plant residues, highly enriched in heavy metals may be isolated as hazardous waste or recycled as bio-metal ore (Raskin et al., 1994; Koppolu et al., 2003).

3. Metal Stress

High heavy metal concentrations in soils could be very toxic for plants and affect the plant metabolism. Metal phytotoxicity occurs when metals move from soil to plant roots and are further transported to various sites in the shoots. The phytotoxicity symptoms of different metals were discussed in several articles (Kabata-Pendias, 2001; Peralta-Videa et al., 2009; Yadav 2009). Toxic effects depend on the kind of metal as well. For example, the phytotoxicity of some relatively common heavy metals such as Cd, Cu, Hg and Ni is substantially greater than that of Pb and Zn (Raskin et al., 1994).

3.1. Plant Adaptation to Metal Stress

In the case that toxic metals are present in the environment plants do not have the possibility to leave so they were forced to develop relevant mechanisms for adaptation to metal stress. In general, two responses can be distinguished in the presence of metals - metal sensitivity and metal resistance. Sensitivity to metals results in injury or death of plants. Resistance means that in spite of metal toxicity plant reacts in a way that allows it to survive high concentration of metals, and to produce the next generation of plants. Resistance includes avoidance, which describes mechanisms for external protection of the plant from metal stress and tolerance, in which the plant is able to survive internal stress imposed by high metal concentrations (Orcutt, Nilsen, 2000).

Some plants are even able to accumulate high metal concentrations in their body – they are called hyperaccumulators. Content of specific metals in their tissues exceeds levels that are actually required for normal growth and development. First time the term "hyperaccumulator" was used by Brooks and his colleagues in 1977 (Brooks et al., 1977) to describe plants able to accumulate high amounts of nickel. Hyperaccumulators can concentrate metals in their aboveground tissues to levels far exceeding those present in the soil or in the non-accumulating species growing nearby.

One of the definitions suggests that a plant containing more than 0.1% of Ni, Co, Cu, Cr and Pb or 1% of Zn in its leaves and stems on the dry weight basis, can be considered a hyperaccumulator, irrespective of the metal concentration in the soil (Raskin et al., 1994). But there are plants in nature containing high metal concentrations in their tissues which do not reach concentration of these metals in surrounding soil. For this reason one more requirement was added into the definition of hyperaccumulator – plant has to accumulate metal in the amount exceeding its amount in soil (Yang et al., 2005). Several plants belong among hyperaccumulators. *Agrostis stolonifera* accumulates 300-times more of arsenic from soil than other plants growing in the same area; *Minuartia verna* contained 1000-times more of cadmium than was in soil (Domažlická, et al., 1994). Very good nickel hyperaccumulators are plants belonging to the species *Alyssum* and *Thlaspi* (Raskin et al., 1994). *Pteris vittata* is good hyperacumulation of arsenic (Zhang et al., 2004). *Sonchus asper* has hyperaccumulation capacity to Pb and Zn, *Corydalis pterygopetala* has hyperaccumulation capacity to Zn and Cd (Yanqun et al., 2005). *Sedum alfredii* was found to be a hyperaccumulator of Cd (Sun et al., 2007). The list of hyperaccumulators contains many species today.

Plants that can grow on soils that are contaminated with high levels of metals are known as metallophytes (Banásová, 1996; Hronec, 1996). They have specific biological mechanisms that enable them to tolerate high metal concentrations. Some of them can grow only on metal containing soil – they are absolute metallophytes. Other species may occur on contaminated as well as uncontaminated soils (Nessner Kavamura, Esposito, 2010).

Plants that are tolerant to metalliferous soils can be divided into three groups according to the metal concentration found in their tissues (Baker, Brooks, 1989):

1. accumulator
2. indicator
3. excluder

The accumulator is a plant able to uptake of very high levels of ions to the extent of exceeding the levels in the soil. The indicator can take up metals at a linear rate relative to the concentration of metal in the soil. The excluder takes up metals but restricts increased concentrations in the shoots until a critical level is reached, above which metal concentrations start to increase in the shoots. For example, copper exclusion was found as the mechanism responsible for Cu tolerance in *Malva sylvestris* (Boojar, Goodarzi, 2007). Fig. 1 illustrates these three types of tolerant plants.

Plants differ among species with respect to how much of a specific metal they can accumulate in their shoots. A single plant species may also differ with respect to uptake, transport and accumulation of different metals. The place where the metals are stored is different in different plants species, as well. These differences are usually based on leaf/root ratios for specific elements which are species specific and metal specific (Orcutt, Nilsen, 2000).

Hyperaccumulation is one of the fundamental characteristics for plant generally used for phytoremediation. Almost all metal-hyperaccumulating species known today were discovered on metalliferous soils, either natural or man-made, often growing together with metal excluders. Actually, almost all metal hyperaccumulating plants are endemic to these soils, suggesting that hyperaccumulation is an important ecophysiological adaptation to heavy metal stress and one of the manifestations of heavy metal resistance (Raskin et al., 1994; Sun et al., 2007; Maestri et al., 2010).

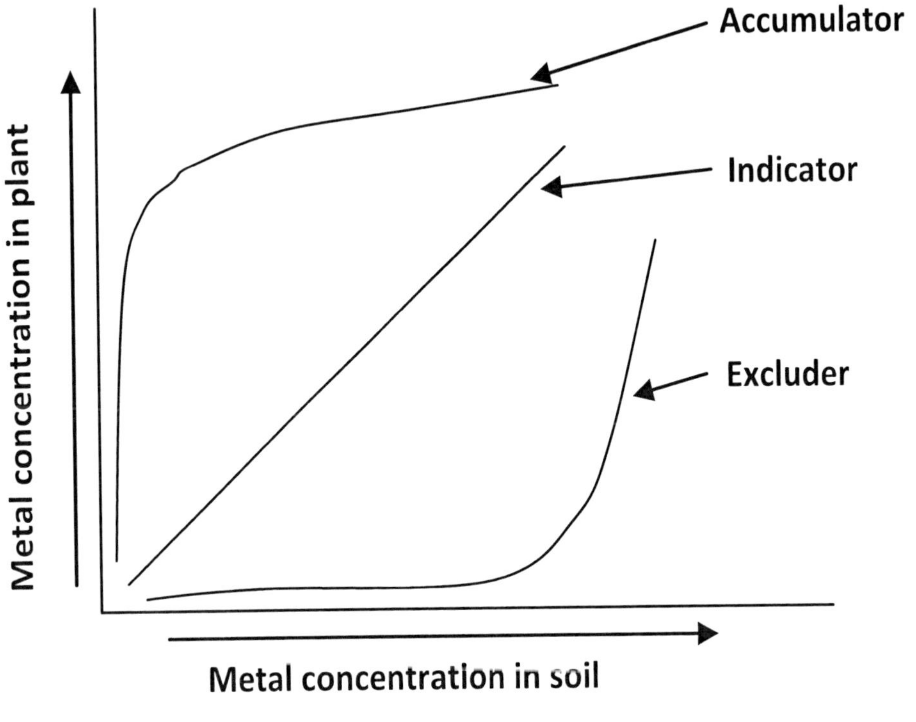

Figure 1. Three types of tolerant plants (adapted from Orcutt, Nilsen, 2000).

The list of hyperaccumulating plants is expanding rapidly. The basic characteristics of hyperaccumulating plants are (Chaney et al., 1997):

- plant must be able to tolerate high levels of the element in root and shoot cells: hypertolerance is the first property that makes hyperaccumulation possible,
- plant must have ability to translocated an element from the roots to the shoots at high rates,
- there must be a rapid uptake rate for the element at level that occur in soil solution.

Uptake and sequestration of toxic materials represents an interesting biological puzzle. The metals that are hyperaccumulated by plants generally are viewed as toxic in relatively low doses; the logical extension of this principle is that the metal levels in these plants might render them relatively toxic to other organisms with which they interact (Boyd, Martens, 1998).

3.2. Mechanisms of Metal Accumulation by Plants

The metal accumulation in various parts of the plant depends upon availability and the species of metals in soils, solubility, their translocation potential, and the type of plant species (Lasat, 2002; Sinha et al., 2009). Molecular understanding of plant metal accumulation leads to better understanding of phytoremediation and widening of its applications. The phenomenon of metal accumulation in plants can be except of cleaning up environment used also in other fields for example in human nutrition. Engineering Zn accumulation in edible plant parts might help in enriching diets for Zn, foliar application of Se increase the antioxidant activity of tea and supply human body with Se which is necessary for the protection against several diseases and cancer (Xu et al, 2003). Conversely, most of the toxic non-essential elements such as Cd enter the human body via plant-derived material (i.e. food or tobacco smoke). Identification of mechanisms governing the metal accumulation process could result in development of plants with lower metal content (Clemens et al., 2002).

Ability of organisms, including microorganisms, to survive, depended, besides other factors also, on their ability to cope with toxic properties of various materials present in their environment.

Organisms have adapted to higher toxicity in environment by evolving mechanisms to maintain low intracellular concentrations of toxic pollutants. On the other hand, many of the elements, including metals, are as trace elements essential for normal development of organisms. They play an important role in organisms. Metals which may not be as yet identifiable as serving a beneficial biologic function are referred to as nonessential. Certain concentration of essential metals is necessary for optimal growth of organisms, but an oversupply results in toxic effects and lethality in the end. Nonetheless, there is no doubt that all metals are potentially hazardous to living organisms, and not necessarily at large exposure levels (Förstner, Wittmann, 1979).

The concentration of essential elements in the environment is usually lower than requirement of organisms, so they had to develop mechanisms able to sequestrate and concentrate these elements from the environment (Wood, Wang, 1985). But because control of accumulation is imperfect, organisms have to cope with exposure to unwanted elements. A lack of specificity of uptake and distribution systems also leads to the accumulation of metals such as Cd, As or Sb, which is generally considered nonessential (Wood, Wang, 1985, Clemens et al., 2002).

Metal accumulation by plants can be divided into three steps:

- Mobilization, root uptake and sequestration
- Translocation
- Tissue distribution and storage

1. Mobilization, Root Uptake and Sequestration

Very important step for metal accumulation is its sequestration by roots as growth media of plants – nutrient solutions, soil etc. are the main sources of metals. Kabata-Pendias (2001) distinguished three basic processes responsible for metal uptake by plant roots:

- Cation exchange by roots
- Transport inside cells by chelating agents or other carriers
- Rhizosphere effects

The initial contact a metal has with a plant is usually in roots and at the membrane level, although heavy atmospheric deposition of elements can cause the effect to be initiated in the shoots (Clemens et al., 2002). The actual metal bioavailability depends on the soil and metal physical - chemical properties as well as metal content, water content and other elements in the rhizosphere (Yang et al., 2005) and for some metals can be limited because of low solubility and strong binding to soil particles (Clemens, 2006). For example, iron is mainly present in the form of insoluble hydroxides, but availability of Zn is less restricted, whereas the bioavailability of some of the target metals in phytoremediation, particularly Pb, is limited. Low soil availability of metals can be a major factor limiting the application of phytoremediation. But plants can actively contribute to metal availability. They enhance the metal accumulation by two basic mechanisms – acidification of the rhizosphere and the exudation of carboxylates (Zhao et al., 2001; Clemens et al., 2002). Acidification of rhizosphere soil was found in several plant species accumulating Cu, Ni, Zn and Cd (Yang et al., 2005). Root exudates of plants are composed mainly of amino acids (e.g., aspartic, glutamic, prolinic) and vary with plant species (and varieties), microorganism association, and conditions of plant growth (Kabata-Pendias, 2001). For example, species from the family *Poaceae* secrete phytosiderophores (chelating agents) to solubilize soil Fe and accumulate the intact chelate into root cells (Chaney et al., 1997). Roots can reduce soil-metal ions by specify plasma membrane bound metal reductases, e.g. pea plants deficient in Fe and Cu have an increased ability to reduce Fe^{III} an Cu^{II} that is coupled with an increased uptake of Cu, Mn, Fe and Mg (Raskin et al., 1994). Iron accumulation by roots of dicotyledonous plants is based upon reduction of exogenous ferric iron to ferrous iron by reductases and the subsequent transport across the root plasma membrane by Fe^{II} transporters. Similar carrier systems transport a broad range of divalent cations (Cu^{2+}, Ni^{2+}, Zn^{2+} and Cd^{2+}) into the plant root cells (Briat, Lebrun, 1999). Except of it, root-colonizing bacteria, as well as mycorrhiza, have a large impact on the availability of heavy metals for plant uptake (Fitz, Wenzel, 2002; Glick, 2003). For instance, soil bacteria significantly enhance Se and Hg accumulation by the saltmarsh bulrush (*Scirpus robustus*) and rabbit-foot grass (*Polypogon monspeliensis*) (de Souza et al., 1999).

Application of chelators such as ethylenedimanetetraacetic acid (EDTA), ethyleneglycoltetraacetic acid (EGTA), ethylenediaminedisuccinic acid (EDDS) to the soil to enhance metal bioavailability, especially Pb, was studied (Wu at al., 1999; Barona et al., 2001; Kos, Leštan, 2003; Hornik et al., 2009), but the risk of this technique is that metal-chelate complexes are very soluble and leach into the groundwater easily (Römkens et al., 2002; Madrid et al., 2003).

Substances pass into roots only through the cuticle-free unsuberized cell walls. Therefore, roots absorb substances far less selectively than leaves. Environmental contaminants enter the roots together with water, like nutrients (Kvesitadze et al., 2006). Solubilized metal ions may enter the root either via the extracellular (apoplastic) or via intracellular (symplastic) pathways. Apoplastic binding of metallic cations such as copper and zinc can contribute significantly to the total cation content of the roots and may also serve as a transient metal storage pool (Briat, Lebrun, 1999). Apoplastic transport is limited by the high cation exchange capacity of the cell walls, unless the metal ion is transported as a non-cationic metal chelate (Raskin et al., 1994). Cation uptake selectivity from the soil solution depends upon specific sites located in the plasma membrane of individual cells. For example, surface chemical properties of root cell wall, depending on its $RCOO^-$/RCOOH composition can account for the different manganese and copper toxicity tolerance exhibited by two tobacco genotypes (Briat, Lebrun, 1999).

Symplastic transport requires that metal ions move across the plasma membrane, which usually has a large negative resting potential. This negative potential provides a strong electrochemical gradient for the inward movement of metal ions (Raskin et al., 1994). Transport systems and intracellular high-affinity binding sites then mediate and drive uptake across the plasma membrane. Several organic acids have been identified as positive reagents to accelerate the absorption of heavy metals by roots (Wu et al., 2010). In roots of the grass *Deschampsia caespitose*, citric acid concentration increased when Zn-tolerant ecotypes were exposed to Zn excess (Rauser, 1999).

From yeast studies, it is apparent that most of the cation transporters show a rather broad substrate range, enabling even non-essential metals such as Cd to enter cells. However, for phytoremediation, specificity and affinity of transport systems have to be considered in relation to relative abundance of different substrates. Non-target elements such as Ca might out-compete target elements (Clemens et al., 2002). Even externally added chelates enhanced Pb desorption from soil to soil solution, facilitated Pb transport into xylem and increased Pb translocation from roots to shoots (Barona et al., 2001).

2. Translocation

Metal absorbed by roots are translocated to different plant organs by the same physiological processes as are used to transport nutrients (Kvesitadze et al., 2006). Generally, long-distance transport of metals in higher plants is carried out by the vascular tissues (xylem and phloem) and is partly related to the transpiration (Kabata-Pendias, 2001).

The apoplast continuum of the root epidermis and cortex is readily permeable for solutes. The cell walls of the endodermal cell layer act as a barrier for apoplastic diffusion into the vascular system. Solutes have to be taken up into the root symplasm before entering xylem. Xylem-loading system controls metal accumulation from external solutions to xylem stream. This process is mediated by membrane transport proteins (Cabañero, Carvajal, 2007; Mori et al., 2009).

Chelatation with some ligands probably determines the transport of metal. For example nicotianamine is known as necessary for the redistribution of Fe, Zn, and Mn via the phloem and is required for Cu transport in the xylem. Takahashi et al. (2003) found that it may also be required for the intracellular regulation of metal binding proteins, such as Zn-finger proteins. Chelation with other ligands, for example histidine and citrate, appears to route metals primarily to the xylem (Clemens et al., 2002). Organic acids, especially citrate, and amino acids are the main metal chelators for some metals such as Fe, Pu, Ni and Cd in xylem. In the case of nickel, increasing free histidine in the xylem sap enhanced translocation of this metal to the shoots, and could explain nickel hyperaccumulation in some plants (Briat, Lebrun, 1999). By contrast, chelation with other ligands, such as phytochelatins or metallothioneins, might route metals predominantly to root sequestration (Clemens et al., 2002).

Xylem-unloading processes are the first step in controlled distribution and detoxification of metals in the shoot (Raskin et al., 1994; Clemens et al., 2002). Except of xylem transport which is predominantly transporting metal ions and water, part of the metals can be in the form of assimilates re-distributed from leaves to the parts of a plant below (shoot axes, roots) and above (shoot tops, fruits) the leaves, passing through sieve tubes in the phloem (Kvesitadze et al., 2006). Metal transport among plant organs depends on the kind of metal. In general, Ag, B, Li, Mo, and Se are easily transported from roots to abovegrounds parts ; Mn, Ni, Cd, and Zn are moderately mobile ; and Co, Cu, Cr, Pb, Hg, and Fe are strongly bound in root cells (Kabata-Pendias, 2001).

3. Tissue Distribution and Storage

Metals reach the apoplast of leaves in the xylem sap, from where they have to be scavenged by leaf cells. Transporters mediate uptake into the symplast, and distribution within the leaf occurs via the apoplast or the symplast. Metal sequestering occurs inside every plant cell, maintaining the concentrations within the specific physiological ranges in each organelle and ensuring delivery of metals to metal-requiring proteins. Excess of heavy metals is sequestered in leaf cell vacuoles (Boojar, Goodarzi, 2007; Mleczek et al., 2009). The distribution pattern varies with plant species and element. Furthermore, trichomes apparently play a major role in storage and detoxification of metals (Clemens et al., 2002).

Intracellular binding and sequestration drive the passage of transition metal ions across the plasma membrane. Several processes known to contribute to metal tolerance are associated with metal accumulation at the same time. Exposure to toxic heavy metals or to high concentration of micronutriens induces the synthesis of phytochelatins as one of the principal responses of plants, animals and many fungi (Winge et al., 1998; Malmström, Leckner, 1998; Bang, Pazirandeh, 1999; Gardea-Torresdey et al., 2004). Phytochelatins are low-molecular-weight thiols which bind metal ions by using the thiol group as the ligand. They consist of only three amino acids – Glu, Cys and Gly and have various sizes with the general structure (γ-Glu-Cys)nGly (n = 2 to 11), synthesized de novo from glutathione in very short time after the exposure to toxic metals (Cobbett, 2000). Phytochelatins loaded with heavy metals are pumped at the expense of ATP into the vacuoles. Because of the acidic environment in the vacuole, the heavy metals are probably liberated from the phytochelatins and finally deposited there (Heldt, 1997; Cobbett, 2000). Process of the phytochelatin synthesis is shown in Figure 2.

Phytochelatins serve as a mechanism for sequestration of heavy metals in the plant vacuole and present an example of biomineralization (Buchanan et al., 2000). Phytochelatin deficiency is correlated with Cd, Cu and As hypersensitivity. Zhang et al. (2005) have found that treatment with Cd and As resulted in a strong increase of the phytochelatin contents in the roots of garlic seedlings.

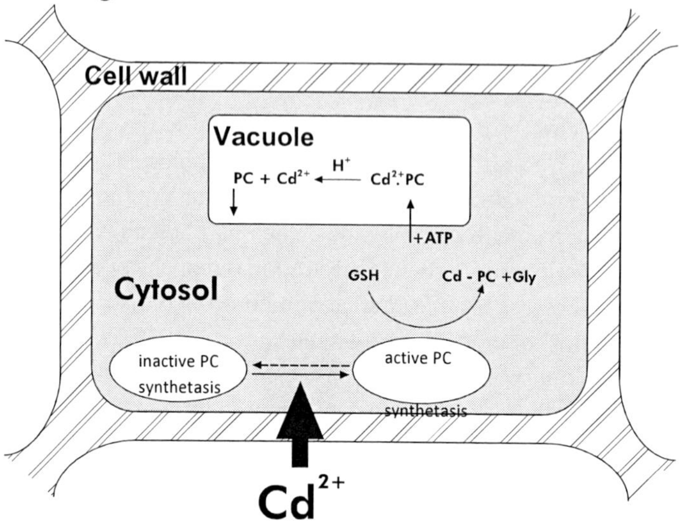

Figure 2. Process of phytochelatin synthesis in the presence of Cd (adapted from Zenk, 1996).

Other potential mediators of metal sequestration and accumulation include members of the Cation Diffusion Facilitator family (CDF). In eukaryotic systems, they have been implicated in moving Zn, Cd and Co from the cytosol into cellular compartments. Studies on various hyperaccumulators and their respective close relatives show a possible correlation between differences in either expression levels or substrate specificities of CDFs and the hyperaccumulation phenotype (Williams et al., 2000; Clemens et al., 2002; Blaudez et al., 2003). Also other classes of proteins are probably connected with metal transport in cells – the heavy metal (or CPx-type) ATPases, the natural resistance associated macrophage protein family of proteins, zinc-iron permease (ZIP) family proteins etc. (Yang et al., 2005) and ferritins (Briat, Lebrun, 1999).

3.3. Metal Phytotoxicity

The phytotoxicity symptoms seen in the presence of excessive amounts of heavy metals may be due to a range of interactions at the cellular/molecular level (Hall, 2002). Three different molecular mechanisms (Figure 3) of heavy metal toxicity in plants can be outlined according to their distinct chemical and physical properties:

a) stimulated generation of reactive oxygen species (ROS) by autoxidation and Fenton reaction that modify the antioxidant defence and elicit oxidative stress,
b) blocking of essential functional groups in biomolecules, direct interaction with proteins due to their affinities for thiol-, histidyl- and carboxyl-groups, causing binding of metals to target structural, catalytic and transport sites of the cell. For example Hg^{2+} ions inhibit the activities of antioxidant enzymes especially of glutathione reductase, and also raise a transient depletion of GSH,

c) displacement of essential metal cations from specific binding sites, causing functions to collapse. For example, Cd^{2+} replaces Ca^{2+} in the photosystem II (PS II) reaction centre, causing the inhibition of PS II photoactivation (Schützendübel, Polle, 2002; Hall, 2002; Rai et al., 2004; Sharma, Dietz, 2009).

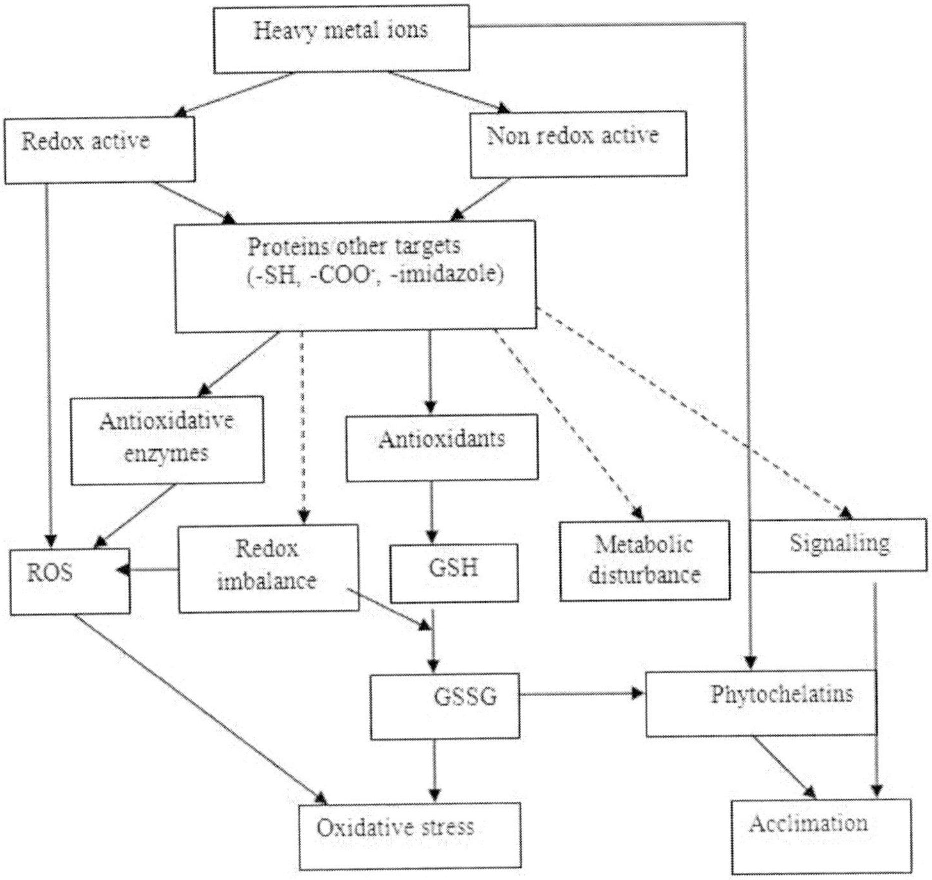

Figure 3. Mechanisms of heavy metal induced oxidative stress and related cellular processes (adapted from Rama Devi, Prasad, 2004; Sharma, Dietz, 2009).

Heavy metals present in the environment can be divided into two groups: non-redox-active heavy metals, such as Zn and Cd and the redox-active heavy metals Fe, Cu, Cr, V and Co (Mithöfer et al., 2004). Redox-active heavy metals directly contribute to the formation of ROS. A general feature of heavy metals is to bind to proteins and other targets, or to compete for binding site in several functional groups (-SH means thiols, -COO− means carboxylic acids and imidazole means histidyl residues). This leads to the changes in functions of targeted proteins, which imply changes in cell metabolism, or trigger signaling events, which can lead to acclimation (Sharma, Dietz, 2009). The biosynthesis of phytochelatins (PCs) is induced by heavy metals such as Cd, Hg, Pb, Zn, Cu and Ag. Among these, Cd generally has the highest induction ability and is strongly conjugated by PCs (Tsuji et al., 2002; Hall, 2002). Activated responses to acclimation evoke cycle of feedback with different sites of heavy metals actions. This results in the repairing of damaged macromolecules, strengthening of the antioxidant defence system and decreasing of heavy metals contents in plasmatic compartments. In view of the differences in the chemical properties of metals and their concomitant distinct behaviour in biological systems these mechanisms might not exclusively account for their toxicity (Sharma, Dietz, 2009).

3.3.1. Heavy Metal-Induced Oxidative Stress

One of the typical heavy metals toxicity symptoms in plants is the generation of reactive oxygen species (ROS) causing oxidative damage to plants. These ROS include superoxide radical ($O_2^{\bullet-}$), hydroxyl radical ($OH^{\bullet-}$) and hydrogen peroxide (H_2O_2) that are produced as by products during membrane linked electron transport activities as well as by a number of metabolic pathways. ROS are partially reduced forms of molecular oxygen - O_2 (Briat, Lebrun, 1999; Mittler, 2002; Shao, et al., 2008). Ground state triplet molecular oxygen (Figure 4) may be converted to the much more reactive ROS forms either by energy transfer or by electron transfer reactions (Apel, Hirt, 2004). Transition metals, such as Fe and Cu, have catalytic function in these reactions (Florence 1984; Briat, Lebrun, 1999).

Under abiotic and biotic stresses, the increased generation of ROS initiates signaling responses that include enzyme activation, programmed cell death and cellular damage (Mittler, 2002; Neil et al., 2002; Pitzschke, Hirt, 2006). Production of ROS is connected not only with stress but also with common metabolic activities such as respiration and photosynthesis localized in mitochondria, chloroplasts, and peroxisomes (Figure 5). Among other sources of ROS identified in plants belong NADPH oxidases, amine oxidases and cell-wall-bound peroxidases in the apoplast (Mittler, 2002; Laloi et al., 2004). ROS serve also as signaling molecules, for example in the recognition of attack by fungal pathogens and herbivores (Bohnert, Sheveleva 1998; Mittler, 2002).

Production and removal of ROS must be strictly controlled because they can cause damage to the biomolecules (e.g. membrane lipids, proteins, chloroplast pigments, enzymes, nucleic acids) and disturbances in signalling processes. In the case of stress induced by a variety of environmental stressors such as soil salinity, drought, extremes of temperature and heavy metals the equilibrium between ROS production and scavenging may be perturbed (Møller, 2001; Rodrígez-López et al., 2000; Mittler 2002; Rios-Gonzales et al., 2002; Zhu et al., 2004; Rhoads et al., 2006).

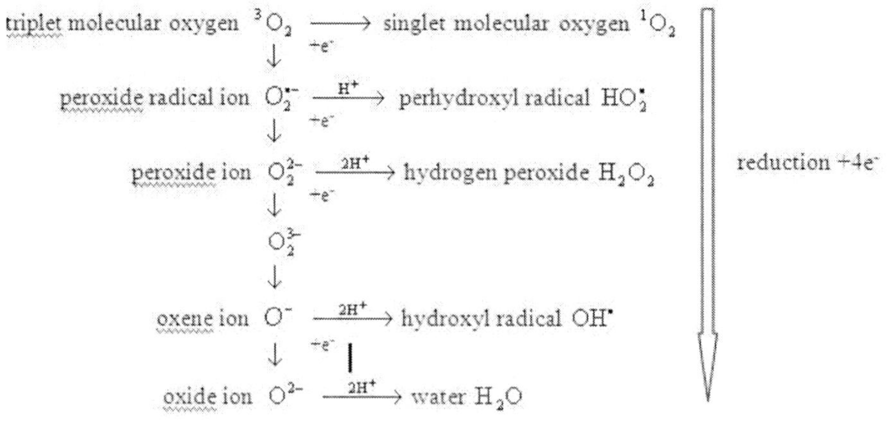

Figure 4. Generation of ROS by energy transfer or sequential univalent reduction of ground state triplet oxygen (adapted from Apel, Hirt, 2004).

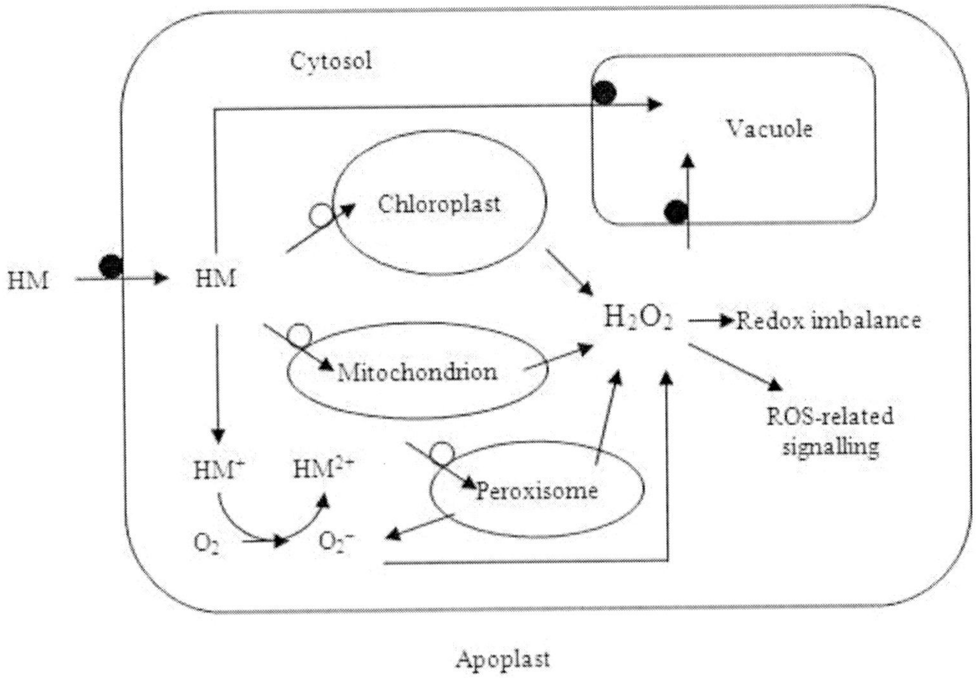

Figure 5. Pathways of heavy metal-dependent ROS generation (adapted from Sharma, Dietz, 2009).

The cellular redox perturbation seems to be an essential prerequisite for development of heavy metals - dependent phytotoxicity symptoms (Sharma, Dietz, 2009). The rapid increase in ROS concentration is called "oxidative burst" (Apel, Hirt, 2004).

To combat the oxidative damage plants have the antioxidant defense system comprising of enzymes catalase, peroxidases, superoxide dismutases and the nonenzymic constituents such as α-tocopherol, ascorbate and reduced glutathione which remove, neutralize and scavenge the ROS.

Redox-active heavy metals cause the conversion of H_2O_2 to the highly reactive $OH^{\bullet}$ molecule via the Fenton reaction (1, 2). During this process metal ion is oxidized and consequently reduced with superoxide radicals $O_2^{\bullet-}$. Direct formation of $OH^{\bullet}$ from H_2O_2 and $O_2^{\bullet-}$ was suggested by Haber–Weiss (3) (Mithöfer et al., 2004; Hernández, et al., 2008; Sharma, Dietz 2009). Chemical reactions involved in hydroxyl radical $OH^{\bullet}$ generation are described as follows:

Fenton reaction:

$$Fe^{2+}/Cu^{+} + H_2O_2 \rightarrow Fe^{3+}/Cu^{2+} + OH^{\bullet} + OH^{-} \quad (1)$$

$$Fe^{3+}/Cu^{2+} + O_2^{\bullet-} \rightarrow Fe^{2+}/Cu^{+} + O_2 \quad (2)$$

Haber-Weiss reaction:

$$H_2O_2 + O_2^{\bullet-} \xrightarrow{Fe^{2+}, Fe^{3+}} OH^{\bullet} + OH^{-} + O_2 \quad (3)$$

ROS react with different cellular constituents. Their reactivity in cell depends on chemical reactivity, redox potential, half-life and mobility within the cellular compartments (Šlesak et al., 2007; Qiu et al., 2008).

The most reactive and short-lived are the $OH^{\bullet}$ radicals. This kind of ROS initiates radical chain reactions and probably causes irreversible chemical modifications of various cellular components, oxidation of biomolecules within diffusion distance, reversible as well as irreversible oxidative modifications of proteins (Taulavuori et al., 2005).

Another ROS is the hydroperoxyl radical ($HO_2^{\bullet}$), protonated form of superoxide radical $O_2^{\bullet -}$, that is probably responsible for the lipid peroxidation - an autocatalytic process that changes membrane structure and function resulting in an increase of the plasma membrane permeability which lead to leakage of potassium ions and other solutes and may finally cause cell death (Chaoui et al., 1997; Verma, Dubey, 2003; Rai et al., 2004).

Likewise, DNA is oxidized mainly by $OH^{\bullet}$ and $^1O_2$, which have been reported to affect guanine, but less by H_2O_2 and $O_2^{\bullet -}$ and it also reacts with reactive nitrogen species. Oxidized polyunsaturated fatty acids are precursors for signaling molecules like jasmonic acid, oxilipins and volatile derivatives (Briat, Lebrun, 1999; Apel, Hirt, 2004; Mithöfer et al., 2004; Sharma, Dietz 2009).

3.3.2. ROS Detoxification – Antioxidant Systems

Various stressors destroy the equilibrium between ROS production and detoxification. To resist oxidative damage, the antioxidant enzymes and certain metabolites present in plants play an important role leading to adaptation and ultimate survival of plants during periods of stress (Pitzschke, Hirt, 2006; Shao et al., 2008).

The role of antioxidant defence mechanisms is to interact with active forms of oxygen and keep ROS at a low level and prevent them from exceeding toxic thresholds (enzymatic antioxidant systems) and regenerate oxidized antioxidants (non-enzymatic systems) (Larson, 1988; Apel, Hirt, 2004). Various antioxidant responses and degrees of tolerance to metal-induced oxidative stress are exhibited by different metal accumulating species (Qiu et al., 2008).

Non-Enzymatic ROS Scavenging Mechanisms

Non-enzymatic antioxidants include the major cellular redox buffers ascorbate, glutathione reductase (GR) and glutathione (GSH), as well as tocopherol, flavonoids, alkaloids and carotenoids (Foyer, Noctor 2005; Shao et al., 2008; Hernández et al., 2008; Sharma, Dietz 2009). GR and GSH are important components of the ascorbate-glutathione pathway responsible for the removal of H_2O_2 in the different cellular compartments. GSH is oxidized by ROS forming oxidized glutathione (GSSG), ascorbate is oxidized to monodehydroascorbate (MDA) and dehydroascorbate (DHA). Through the ascorbate-glutathione cycle GSSG, MDA and DHA can be reduced reforming GSH and ascorbate. In responses to abiotic stress, plants increase the activity of GSH biosynthetic enzymes and GSH levels (Zhang, Kirkham, 1996; Yadav, 2009).

Enzymatic ROS Scavenging Mechanisms

Enzymatic ROS scavenging mechanisms in plants include ascorbate peroxidase (APX), glutathione peroxidase (GPX), superoxide dismutase (SOD) and catalase (CAT). These enzymes are involved in the detoxification of $O_2^{\bullet-}$ radicals and H_2O_2, thereby preventing the formation of $OH^{\bullet}$ radicals (Zhang, Kirkham, 1996).

SODs act as the first line of defence against ROS, dismutating superoxide to H_2O_2. CAT, APX and GPX subsequently detoxify H_2O_2. CAT converts hydrogen peroxide into water. In contrast to CAT, APX converts H_2O_2 into water by the ascorbate-glutathione cycle (an ascorbate is the reducing agent and glutathione (GSH) is the part of regeneration system). Similarly to APX, GPX also converts hydrogen peroxide into water by the glutathione-peroxidase cycle using equivalents from GSH as reducing agents. Oxidized GSSG is again converted into GSH by glutathione reductase (GR) (Table 2) (Apel, Hirt, 2004; Santos de Araujo et al., 2004). The ASC–GSH cycle is another important antioxidant mechanism involved in H_2O_2 detoxification that it is composed by the enzymes APX, GR, DHAR and MDHAR, and by GSH and ASC (Noctor, Foyer, 1998).

Among the number of enzymes regulating H_2O_2 intracellular levels CAT, APX and GPX are considered to be the most important (Noctor, Foyer, 1998; Milone et al., 2003).

CAT is less efficient than peroxidases (POD) in H_2O_2 scavenging because of its low substrate affinity. Therefore as long as the stress is not too strong for the plant's defence capacity, the main response to heavy metals is an increase in SOD and POD activities (Siedlecka, Krupa, 2002). Peroxidases are widely accepted as "stress enzymes". The measurement of their activity is one of the most common parameters in metal stress evaluation. Increasing of their activity was well documented under a variety of stressful condition, such as water stress (Yordanov, 2000), chilling, salinity (Rios-Gonzalez, 2002), γ-radiation (Verma, Dubey, 2003) and under toxic levels of metals (Baccouch et al., 1998; Shah et al., 2001).

Furthermore, GPX participates in the lignin biosynthesis and might build up a physical barrier against poisoning of the heavy metals (Rai et al., 2004). Peroxidases are ubiquitous enzymes found in virtually all green plants, the majority of fungi and aerobic bacteria (Kvesitadze et al., 2006). The extent of oxidative stress in a cell is determined by the amounts of superoxide, H_2O_2 and hydroxyl radicals. So instead of only POD activity measurement would the measurement of SOD, APX and CAT activities together provide more specific picture about the extent of oxidative stress in cell level (Apel, Hirt, 2004).

Table 2 The principal modes of enzymatic ROS scavenging by superoxide dismutase (SOD), catalase (CAT), ascorbat peroxidase, and the glutathione peroxidase (GPX) (adapted from Apel, Hirt 2004)

Superoxide Dismutase: $O_2^{\bullet-} \xrightarrow{SOD} H_2O_2$	4)
Catalase. $H_2O_2 \xrightarrow{CAT} H_2O + 1/2O_2$	5)
Ascorbate peroxidase: $H_2O_2 + \text{ascorbate} \xrightarrow{APX} H_2O + MDA$	6)
Glutathione peroxidase: $H_2O_2 + GSH \xrightarrow{GPX} H_2O + GSSG$	7)

4. Stress Evaluation

4.1. Germination

Germination assay is a basic procedure to determine heavy metals toxic effects on plants (An, 2006; Labra et al., 2006; Di Salvatore et al., 2008). It is well documented that germination process is highly disturbed by metal stress; however, there are not much explanation on the molecular mechanism of the inhibition of seed germination in the presence of metals (Ahsan et al., 2007). Seed germination and the early seedling growth are more sensitive to metal pollution because some of the defense mechanisms have not developed (Liu et al., 2005; Xiong, Wang, 2005). Delay in germination has been often observed after heavy metals exposure (Espen et al., 1997; Bansal et al., 2002; El-Ghamery et al., 2003; Rodriguez, Alatossava, 2010).

In the study of Rahoui et al. (2010) was examined the effect of cadmium on the germination of seeds and the growth of early seedlings of pea (*Pisum sativum*). Nutrient loss in germination medium resulting in the drastic decrease in their availability for an adequate metabolism acts, at least in part, to lower the germination rate and to delay the subsequent embryo growth in Cd-treated seeds. From results of Ahsan et al. (2007) it is obvious that copper has a detrimental effect on rice seed germination and that 1.5 mM of copper is the end point for rice seed germination. The germination rate of the rice seeds significantly decreased with the increase of copper concentrations, ranging from 1.0 mM to 2.0 mM. The shoot growth of the germinating seeds was greatly inhibited at 0.2 mM copper, and further decreased at higher copper concentrations. In addition, no root formation was observed at copper concentrations higher than 0.5 mM. Relationship between the concentration of As in wheat seedlings and germination percentage, relative root length, relative shoot height exhibited significant negative correlation (Liu et al., 2005). Li et al. (2007) observed negative effects of arsenic on seed germination and the growth of roots and shoots only at higher arsenic concentrations (5 – 20 mg.kg-1) but at low arsenic concentrations (0 – 1 mg.kg-1) stimulation of seed germination was reported. Peralta et al. (2001) found that 40 mg/kg of Cr^{VI} reduced the ability of lucerne seeds (*Medicago sativa*) to germinate by 23% and decreased their growth in the contaminated medium. In according to Zeid (2001) the reduced germination of seeds under Cr stress could be caused by a depressive effect of Cr on the activity of amylases and on the subsequent transport of sugars to the embryo axes. Sensitivity of the seed germination and root elongation test can be improved by replacing filter paper with agar (Di Salvatore et al., 2008).

Wierzbicka, Obidzińska (1998) found that the negative effect of metal on seed germination is strongly related to the seed coat permeability to metal ions. The most negative effect on seed germination have Ni, Cd and Al, they inhibit germination even in low concentrations. They are followed by Pb and Cu and the less toxic is Zn which in low concentrations supports seed germination of *Helianthus annuus* (Chakravarty, Srivastava, 1992).

Adaptation of plants can be observed also in the seed germination level. For example, seeds of *Echinochloa colona* collected from contaminated area of minewaste dumps showed higher germination rate in metal containing solutions than seeds of the same plant from uncontaminated areas. Higher germination was observed for these seeds in solutions with metal in comparison with metal free solution (it was opposite to the seeds collected in uncontaminated area (Rout et al., 2000).

4.2. Reduction in Growth

Toxic metals are usually responsible for the reduction of plant growth (Chaoui et al., 1997). The reduction in growth can be expressed as reduced growth rate or decreased biomass production. This reduction can be due to specific toxicity of the metal to the plant, antagonism with other nutrients in the plant, or inhibition of the root penetration in the soil (Begonia et al, 1998).

To test the phytotoxicity of metals Leita et al. (1993) suggested calculation of the Grade of Growth Inhibition (GGI).

$$GGI = [(C-T) / C],$$

where C and T represent the dry weight of tissues of control (C) and metal-treated plants (T). For plants without stress, where growth is not inhibited the GGI = 0, i.e., 100% growth.

The decrease of dry biomass weight was reported for cadmium treated pepper (León et al., 2002), bean (Chaoui et al., 1997), wheat (Milone et al., 2003). But there were reported also cases when low level of cadmium had positive effect on plants growth although they are very poorly discussed in literature (Arduini et al., 2004). Two possible mechanisms were suggested. Low cadmium levels hyperpolarize the plasma membranes at the root surface, thus increasing the trans-membrane potential, which is an energy source for cation uptake. Moreover, cadmium has been found to induce genes related to mammalian cell proliferation, which could increase growth, though they are also considered responsible for Cd-induced carcinogenesis (Arduini et al., 2004). Marchiol et al. (2004) did not observed any obvious symptoms of metal toxicity during the treatment with the mixture of Cd, Pb, Cr, Ni, Zn and Cu but they have recorded growth reduction and decrease in biomass production in both plant, *Brassica napus* and *Raphanus sativus*, treated with the metal mixture.

No growth and biomass inhibition but even higher amount of biomass in plants treated with lead observed also Begonia et al. (1998) for Indian mustard (*Brassica juncea*). The Indian mustard plants treated with lead in concentration 100 and 250 mg/l produced 20% more biomass than control plants. And at very high lead concentration in solution (500 mg/l) the biomass production decreased very little, only by 7% in comparison with control. It is possible that the small amount of Pb translocated to the shoot was not enough to elicit a reduction in shoot biomass. Sekhar et al. (2004) recorded only slight decrease in growth of *Hemidesmus indicus* root and shoots treated with Pb concentration from 100 to 10 000 mg/kg.

For *Pteris vittata* treated with the As and Cd and As and Pb was observed stimulated growth and 12 times higher biomass production in comparison with controls treated only with arsenic (Fayiga et al., 2004). The authors suggested that growth stimulation might result from added N nutrition since all metals were added as nitrate salts. Kadukova et al. (2008) observed stimulation of biomass production in Pb, Cd and Pb + Cd treated *Tamarix smyrnensis*. The authors suggested that growth stimulation by Cd and Pb might also result from added N nutrition. On the basis of published experimental results it is obvious that the plant growth does not need to be always reduced in the presence of heavy metals so the determination of only growth rate or biomass production can not be considered as the only one parameter to evaluate plant stress.

4.3. Photosynthetic Pigments

Chlorophylls are the basic photosynthetic pigments. They are essential for radiance fixation and consequent transformation of energy in photosynthesis. Carotenoids serve as accessorial pigments. They include carotenes and xanthophylls. They function in pigmentation and as antioxidants (Havaux, 1998; Škerget et al., 2005). The content and composition of photosynthetic pigments are important indicators of the status of photosynthetic apparatus and depends on the plant species, mineral nutrition and growth conditions (Masarovičová, Repčák, et al., 2008).

A common response of plants to metal stress is a decrease of the chlorophyll content in leaves of plants and subsequently the reduction of photosynthesis that finally leads to a lower biomass production (Monteiro et al., 2009). The reduction of chlorophyll content in the presence of heavy metals may be due to an inhibition of chlorophyll biosynthesis (Xiong, 1997). Reduction in the levels of photosynthetic pigments, including chlorophylls *a* and *b* and carotenoids, on exposure to heavy metals have been observed in many species for Cu (MacFarlane, Burchett, 2001; Prasad et al., 2001), Zn (Ghnaya et al., 2009; Radić et al., 2010), Cd (Wu et al., 2003; Ekmekçi et al. 2008), Pb (Mishra et al., 2006; Cenkci et al., 2010). Ghnaya et al. (2009) found a decreased of chlorophyll *a*, chlorophyll *b* content and the total amount of chlorophyll in *Brassica napus* treated with different Cd and Zn concentrations. Likewise, Monferrán et al. (2009) observed symptoms of changes in the photosynthetic apparatus in *Potamogeton pusillus* after exposure to copper, except of other changes they have observed a decrease in chlorophyll *a* and chlorophyll *b*. Zaman and Zereen (1998) observed significant decrease in chlorophyll *a*, chlorophyll *b* content and the total amount of chlorophyll in radish plants treated with different Pb and Cd concentrations. The reduction in chlorophyll content in turn, at least partly, would lead to a decrease of biomass (Sinha et al., 1993). In shoot tissues of *Bacopa monnieri* was noticed a negative significant correlation of Fe uptake with chlorophyll content. Adverse effect of Fe on the chlorophyll contents might be due to the strong oxidation of the photochemical apparatus, reduction in chloroplast density and size, phosphorus deficiency or reduced Mn transport (Sinha et al., 2009). The exposure of pea *Pisum sativum* seedlings to Cd resulted in a reduction of chlorophyll and carotene content in leaves. The deleterious effect of Cd became more pronounced with increasing concentrations. In plants exposed to 7 mg Cd/kg growth media, chlorophyll *a*, chlorophyll *b* and carotene decreased by 50.6%, 51.9% and 45.3%, respectively, compared to control plants (Hattab et al., 2009). Similarly in green algae *Scenedesmus quadricauda* slight decrease of chlorophyll content was observed after exposition to Cu (Kováčik et al., 2010).

In higher plants two kinds of chlorophyll exist – chlorophyll *a* and chlorophyll *b*. Their content in plants is usually in the ratio 3:1 (chlorophyll *a* dominates). The changes in the ratio of chlorophyll *a/b* can also reflect the negative influence of heavy metals to the photosynthetic apparatus of plants (Porra, 2002). It can be connected with the fact that chlorophyll *a* is more sensitive to some heavy metals than chlorophyll *b* (Pandey, Sharma, 2002). A significant increase was seen in the chlorophyll *a/b* ratio for Zn alone, suggesting that the chlorophyll *b* pool is more sensitive to Zn exposure (Macfarlane, Burchett 2001). Dazy et al. (2008) found that Cr exposure disturbed the cellular redox status, chlorophyll content and chlorophyll *a/b* ratio in *Fontinalis antipyretica* apices. Effects on chlorophyll contents and chlorophyll *a/b* ratios were also shown even at low Cr concentrations.

Similarly as in the case of growth reduction heavy metals not always cause the reduction in chlorophyll content. For example, Xiong (1997) did not observe significant decrease of chlorophyll content in plants *Sonchus oleraceus* treated with 800 and 1600 mg/l of lead. Only at very high Pb concentration in soil (3200 mg/l) 18% of the decrease of chlorophyll content was observed. Also Gupta et al. (2009) did not report changes in chlorophyll content in the leaves of *Sesamus indicum* treated with the mixture of cadmium, copper and zinc. The chlorophyll *a*, chlorophyll *b* and carotene contents in seedling of *Pisum sativum* exposed to copper decreased only after addition of the highest Cu concentration -700 mg/kg (Hattab et al., 2009). Similarly only higher concentration of arsenic in soil (10 and 50 mg/kg) led to decrease of chlorophyll *a*, chlorophyll *b* and carotenoids contents in clover shoots. After addition of zinc and cadmium together with arsenate in the heavy metal mixture variant chlorophyll content did not decreased so significantly as if only arsenate was added (Mascher et al., 2002). Exposition of *Sedum alfredii* leaves to 1000 μM of Cd in the nutrient solution resulted in the increase of total chlorophyll, chlorophyll *a* and *b* content on the basis of fresh weight, but the ratio of chlorophyll *a* to chlorophyll *b* decreased (Zhou, Qiu, 2007). Dazy at al. (2008) found that two different forms of chromium had opposite effect on chlorophyll content of *Fontinalis antipyretica* Cr^{III} as nitrate salt seemed to decrease total chlorophyll content whereas Cr^{III} as chloride salt and Cr^{VI} lead to chlorophyll accumulation. Cr^{III} modified chlorophyll *a/b* ratio but Cr^{VI} had no effect on that parameter.

Jiang et al. (2007) observed increasing of chlorophyll content after applying of external P in the complex pollution of Cd and Zn.

The decrease in carotenoid contents was found in rapeseed cultivar (*Brassica napus*) treated with Zn^{2+} and Cd^{2+}. According to carotenoid content measurement these cultivars were more sensitive to the stress of Cd^{2+} then Zn^{2+}. This drastic reduction of carotenoids may be the result of a strong production of ROS (Ghnaya et al., 2009). In the study of Singh et al. (2008) carotenoid contents in *Beta vulgaris* plants decreased significantly with increasing concentrations of fly ash as compared to the control at 35 days after sowing. Increasing concentration of Cu significantly inhibited carotenoid concentration in *Withania somnifera* (Khatun et al., 2008).

In the case that metal were added into soil carotenoids decrease was observed. But in plants growing in polluted areas higher carotenoid content was found. It can be connected with the fact that carotenoid serve as antioxidants what could protect these plants against harmful effects and be the part of adaptation mechanisms (Sillanpää et al., 2008).

4.4. Antioxidant Enzymes

Induction in the activities of antioxidant enzymes is a general strategy adopted by plants to overcome oxidative stress due to the imposition of environmental stresses (Shah et al., 2001). Mechanisms of the decreasing of the oxidative stress in plants by antioxidant enzymes are described above in the section 3.3.2. Measurement of the activity of antioxidant enzymes can provide useful information about the stress level in plants. The level of anionic isoenzymes of guaiacol dependent peroxidase represents quantitative and qualitative changes caused by the metals and analyses of its profile can be considered as a metabolic indicator of heavy metal stress (Baccouch, 1998; Verma, Dubey, 2003; Qiu, et al., 2008). Significant increases in peroxidase activity were found with Cu and Zn at concentrations lower than those inducing visible toxicity (Macfarlane, Burchett, 2001). Cd has been found to trigger oxidative stress in tobacco cells within minutes of exposure, possibly signaling the first defenses against heavy metals at the plasma membrane (Boominathan, Doran, 2003). Ni induced high POD activity in leaves of *Zea mays* (Baccouch et al., 1998). Verma and Dubey (2003) found 1.2 – 5.6 times increase in peroxidase activity in roots of rice treated with lead. In study of Kavuličova et al. (2009) guaiacol peroxidase activity strongly responded to metal exposure. Specific activity of peroxidase was found to be significantly higher (1.5 times) at plants treated even with low metal content in soil in comparison with control plants. At high metal content (5 times higher concentration above standard in soil) was guaiacol POD activity 3.5 times higher than in control. Results indicate a considerable enhancement in the activity of guaiacol peroxidase, suggesting that this antioxidant enzyme acts to reduce the impact of metal toxicity.

In some plants activity of peroxidase in shoots was found to be lower than in roots. For example, in tomato after Cu treatment high peroxidase activity was recorded only in roots and stems but not in shoots (Mazhoudi et al., 1997). Shah et al. (2001) observed the increase of guaiacol peroxidase activity in roots and shoots of rice treated with Cd. They recorded higher POD activity in shoots than in roots. On the contrary Chaoui et al. (1997) did not observed guaiacol- dependent peroxidase activity increase in leaves of beans after Cd treatment.

Israr, Sahi (2006) investigated the effect of mercury on the levels of antioxidant enzymes SOD, APX and GR in the *Sesbania drummondi* cell cultures. The *Sesbania* cell cultures tolerated Hg up to a content 40 μM and higher contents caused toxicity to the cells. The activities of APX, SOD were markedly increased in response to Hg treatments. The activity of SOD in plants *Ceratophyllum demersum* increased 3.6-times in plants supplemented with Zn (200 μM) over the control but just 1.5-times in plants treated with Cd (10 μM) (Aravind, Prasad, 2003). In *Bruguiera gymnorrhiza* plants under heavy metal stress (different concentration of Pb^{2+}, Cd^{2+}, and Hg^{2+}), CAT activity was not affected in the leaves, but increased in the roots. In *Kandelia candel*, CAT activity increased in both leaves and roots (Zhang et al., 2007).

At low metal concentration when no toxicity symptoms were observed no changes in antioxidant enzyme activity were observed for Mn treated barley plants and slight but significant increase of SOD and membrane-bound APX activities and decrease of soluble APX and CAT activities for plants treated with Cu. But when barley plants were exposed to Mn and Cu concentration which caused visible toxicity symptoms in both cases decrease of SOD and increase of CAT and GPX activities were reported. Activity of APX after treatment with Cu did not change and after Mn treatment decreased (Demirevska-Kepova, et al., 2004). In cyanobacteria *Spirulina platensis* the increase of SOD activity was observed when different heavy metals (Cu, Zn, Pb) were added into solution. The highest increase was observed when Pb was added (Choudhary et al., 2007). In the presence of Pb CAT activity in *Sesbania* plants was elevated by 265% and SOD activity by 180%, but no significant changes in their activities were observed in the presence of Pb + EDTA, Pb + DTPA or Pb + HEDTA (Ruley et al., 2004). Cadmium induced an oxidative stress in pea plants characterized by a reduction in reduced GSH content, CAT (close to 50%) and CuZn-SOD activities. The effect of cadmium on APX was different from that of CAT, showing a slight increase in activity (Romero-Puertas et al., 2007). Cd treated pepper expressed decrease of the CAT and SOD and increase of GPX (León et al., 2002). Zhang et al. (2005) observed the decrease of CAT and SOD activities in garlic treated with 5 and 10 mM $CdCl_2$ at the beginning of the experiment but activity of CAT slowly recovered to the extent of control or even increased over the control. When exposed to 1mM $CdCl_2$ CAT and SOD activities increased. POD activities at 5 and 10 mM Cd firstly dramatically increased and then slowly decreased but still were higher than in controls, at 1 mM POD activities gradually increased. High Cu concentration (2.5 mM) in media increased SOD activity in red cabbage seedlings but decreased CAT activity. On the contrary, lower Cu concentration slightly but statistically significantly increased CAT activity and did not change SOD activity (Posmyk, et al., 2009). Suppression of the activities of SOD, POD and CAT led to H_2O_2 burst, lipid peroxidation, cell death and growth inhibition in Cd treated rice (Guo et al., 2009). Significant and dose-related increases of SOD and CAT activities were observed in leaves of plants growing in Elizabeths' river basin in Wyoming, USA contaminated mainly by Cd, Cu, Hg and Zn (Dazy et al., 2009). Similarly Mobin and Khan (2007) observed CAT and SOD activity increase in Cd treated *Brassica juncea*. An increase of SOD, APX and CAT activities in *Phragmites australis* treated with Cd was observed in all parts of the plant (roots, stolons and leaves) in spite of the fact that most of the cadmium was accumulated in roots (Iannelli et al., 2002).

However, it is important to point out that under extreme conditions of stress, a plant may be too weak to produce enough antioxidant enzymes to protect itself. As such, low activity of antioxidant enzyme may not always indicate low stress and it should be always connected with the evaluation of plant biomass production (Fayiga et al., 2004).

4.5. Antioxidants

Except of antioxidant enzymes also other compounds may protect plants against oxidative stress. Glutathione, phytochelatins, flavonoids, tocoferol and ascorbate belong among them. The measurement of their content in plants can also be used as the indicator of heavy metal induced stress (Bharagava et al. 2008).

Glutathione (GSH) represents one of the major sources of non-protein thiols in most plant cells. Thiol group is important in the formation of mercaptide bond with metals. This reactivity along with the relative stability and high water solubility of GSH makes it an ideal compound to protect plants against stresses including oxidative stress caused by heavy metal (Foyer, Noctor 2005; Mendoza-Cózatl, Moreno-Sánchez, 2006). Non-growth inhibiting Cd concentration in growth medium caused increase of Cd concentration in leaves of tobacco seedlings resulting in slight decrease of glutathione which was observed only during first hours of Cd^{2+}-exposure and, by day 2, was completely recovered to control levels (Vögeli-Lange, Wagner, 1996). The total glutathione pool remained relatively unaffected by Zn exposure to *Avicennia marina* plants (Caregnato et al., 2008). The significant decrease of non-protein thiol content in barley leaves treated with 15 μM Cu was probably due to enhanced thiol transport to the roots where the excess Cu is initially immobilized. Visible toxicity symptomps at 150 μM – 1500 μM Cu treatment corresponded with drastic increase of these thiols (Demirevska-Kepova, et al., 2004). The amount of total glutathione and GSH in Zn treated *Phaseolus vulgaris* was significantly higher at the beginning of the experiment but then decreased but the GSSG content increased slowly during the experiment. Ratio GSSG/GSH was higher but its enhancement was not significant (Cuypers et al., 2001). GSH concentration increase was observed in *Sedum alfredii* plants treated with Pb. Authors of this study suggested that GSH played an important role in Pb chelatation and detoxification (Gupta et al., 2010). GSH in *Sedum alfredii* increased also in the presence of Cd (Sun et al., 2007). Similarly high content of GSH was found in plants *Thlaspi caerulescens* and *Thlaspi praecox* growing on Cd and Zn contaminated soil (Pongrac et al., 2009) and seeds of the bean treated with the Cd (Szőllősi et al., 2009). But it was reported that the increase in GSH content above a control level does not always influence metal tolerance in plants but the decrease below this level may weaken the tolerance (Ohlsson et al., 2008).

Phytochelatins (PCs) are a set of heavy metal binding peptides. They play an important role in heavy metal detoxification as well as in the maintenance of ionic homeostasis. PCs are synthesized inductively by exposure to heavy metals such as Hg, Cu, Zn, Pb and Ni. During the exposure of plants to such metals PCs are synthesized from GSH (Yadav 2009; Yadav, 2010). The increase of phytochelatin content in maize seedling growing in media with different heavy metals depended on kind of the metal. Its formation was stimulated most effectively by Cd, less by Zn and Cu and negligibly by Ni. Total glutathione declined with Cd and Zn exposure, however, with excess Cu the roots contained 45% more total glutathione than did the controls. The reactions to excess Cd differed along the length of roots. In the 1 cm apical region a high production of PCs occurred with a moderate loss of total glutathione. In the mature region, PC content was 2.5-fold less than in apices, several unidentified thiols accumulated, and total glutathione levels declined drastically (Tukendorf, Rauser, 1990). Concentration of PCs in *Brassica juncea* plants treated with Cd and Cd-EDTA increased significantly during first days of treatment but then sharply decreased. Enhanced level of PCs indicates the plant capacity to detoxify the metal via chelatation and sequestration in vacuoles. At higher Cd concentrations after almost one month exposure, plants experienced toxicity besides accumulation of PCs, probably due to GSH depletion (Seth et al., 2008). In the presence of Cd and Pb synthesis of phytochelatins was observed in *Phaeodactylum tricornutum*. The increase of PCs concentration was followed by the consumption of GSH in cells. Over 50% of the cellular glutathione was converted into phytochelatins after 2 h exposure (Morelli, Scarano, 2001). Similarly the increase of PCs concentration was observed in wheat seedlings cultivated in medium with Cd (Lindberg et al., 2007). Application of heavy metals to cell suspension cultures and whole plants of *Silene vulgaris* and tomato induced the formation of heavy metal–phytochelatin-complexes with Cu and Cd and the binding of Zn and Pb to lower molecular weight substances (Leopold et al., 1999). Srivastava et al. (2006) found that PC concentration positively correlates with no visible toxicity symptomps in *Hydrilla verticillata* exposed to copper. Visible toxicity symptoms were found in plants which did not synthetise phytochelatins.

However, phytochelatin biosynthesis is apparently not necessary for metal tolerance in *Salix*, as PC was not found in this system either in tolerant or sensitive clones, before or after metal treatment (Ohlsson et al., 2008). No PCs were detected also in plant *Sedum alfredii* treated with Pb and Cd although significant amount of Pb was accumulated in roots and shoots of these plants (Sun et al., 2007, Gupta et al, 2010).

Flavonoids play an important role in plants as flower pigments (Gidda, Varin, 2006). In plants, flavonoids are involved in many processes, including plant-pathogen interactions, pollination, and seed development. Flavonoids have been suggested to act as antioxidants, protecting plants from oxidative stress (Hernández et al., 2008). In spinach was evidenced H_2O_2 oxidation by flavonol glycosides by intact chloroplasts (Takahama, 1984).

Other compounds with antioxidant activity might still be found such as leucasin from *Leucas aspera* (Meghashri et al., 2010).

CONCLUSION

In summary, phytoremediation processes hold great promise as a means to clean-up polluted soils and water. Currently, the most advanced and effective phytoremediation technology is phytoextraction of heavy metals from soils using hyperaccumulating plants. Studying heavy metal-plant interactions should improve our understanding of the mechanisms of ion uptake, accumulation and resistance. Further development of phytoremediation requires an integrated multidisciplinary research effort that combines plant biology, soil biochemistry, soil microbiology as well as agricultural and environmental engineering. A very important part of the research with the direct implication for phytoremediation but also plant physiology is the evaluation of plant stress in the presence of heavy metals. There are several parameters which can be used as stress indicators such as germination, plant growth, photosyntetic pigments, antioxidant enzymes, etc. Symptoms of stress in plants depend on the particular metal (but a metal combination can act differently), plant species but also on preliminary adaptation and other factors. So it is important for stress evaluation to assess several "stress indicators" together.

ACKNOWLEDGMENT

This work was supported by Slovak grant agency – project VEGA1/0134/09.

REFERENCES

Abhilash, P.C., Jamil, S., Singh, N. (2009). Transgenic plants for enhanced biodegradation and phytoremediation of organic xenobiotics. *Biotechnology Advances*, 27, 474-488.

Ahsan, N., Lee, D.-G., Lee, S.-H., Kang, K. Y., Lee, J. J., Kim, P. J., Yoon, H.-S., Kim, J.-S., Lee, B.-H. (2007). Excess copper induced physiological and proteomic changes in germinating rice seeds. *Chemosphere*, 67, 1182-1193.

Alkorta, I., Garbisu, C. (2001). Phytoremediation of organic contaminants in soil. *Bioresource Technology*, 79, 273-276.

Alvarenga, P., Gonçalves, A.P., Fernandes, R.M., de Varennes, A., Vallini, G., Duarte, E., Cunha-Queda, A.C. (2008). Evaluation of composts and liming materials in the phytostabilization of a mine soil using perennial ryegrass. *Science of the total environment*, 406, 43–56.

An, Y.J. (2006). Assessment of comparative toxicities of lead and copper using plant assay. *Chemosphere*, 62,1359–1365.

Apel, K., Hirt, H. (2004). Reactive Oxygen Species: Metabolism, Oxidative Stress, and Signal Transduction. *Annual Review of Plant Biology*, 55, 373-399.

Aravind, P., Prasad, M. N. V. (2003.) Zinc alleviates cadmium-induced oxidative stress in Ceratophyllum demersum L.: a free floating freshwater macrophyte. *Plant Physiology and Biochemistry*, 41, 391-397.

Arduini, I., Masoni, A., Mariotti, M., Ercoli, L. (2004). Low cadmium application increase miscanthus growth and cadmium translocation, *Environmental and Experimental Botany*. 52, 89-100.

Baccouch, S., Chaoui, A., El Ferjani, E. (1998). Nickel-induced oxidative damage and antioxidant responses in Zea mays shoots, *Plant Physiology and Biochemistry*, 36, 689-694.

Baker, A.J.M., Brooks, R.R. (1989). Terrestrial higher plants which hyperaccumulate metallic elements. A review of their distribution, ecology and phytochemistry. *Biorecovery*, 1, 2, 81-126.

Banásová, V. (1996). Rastliny na substrátoch s vysokým obsahom ťažkých kovov. Zborník zo seminára: Ťažké kovy v ekosystéme, E'96, BIJO Slovensko, s r.o., 81-94.

Bang, S.S, Pazirandeh, M. (1999). Biochemical characteristics of heavy metal uptake by Escherichia coli NCP immobilized in kappa-carrageenan beads. *Biohydrometallurgy and the Environment toward the Mining of the 21st Century*, Proceedings of the International Biohydrometallurgy Symposium IBS´99, Ed.: R.Amils, A.Balester, Elsevier, 193-199.

Bansal, P., Sharma P., Goyal, V. (2002). Impact of lead and cadmium on enzyme of citric acid cycle in germinating pea seeds. *Biologia Plantarum*, 45, 125-127.

Barona, A., Aranguiz, I., Elías, A. (2001). Metal assotiations in soil before and after EDTA extractive decontamination: implications for the effetiones of further clean-up procedures. *Environmental Pollution*, 113, 79-85.

Begonia, G.B., Davis, C.D., Begonia, M.F.T., Gray, C.N. (1998). Begonia, Growth Responses of Indian Mustard [Brassica juncea (L.) Czern.] and its Phytoextraction of Lead from a Contaminated Soil, *Bulletin of Environmental Contamination Toxicology*, 61, 38-43.

Bharagava, R.N., Chandra, R., Rai, V. (2008). Phytoextraction of trace elements and physiological changes in Indian mustard plants (Brassica nigra L.) grown in post methanated distillery effluent (PMDE) irrigated soil. *Bioresource Technology*, 99, 8316-8324.

Blaudez, D., Kohlerb, A., Martin, F., Sanders, D., Chalot, M. (2003). Poplar Metal Tolerance Protein 1 Confers Zinc Tolerance and Is an Oligomeric Vacuolar Zinc Transporter with an Essential Leucine Zipper Motif. *The Plant Cell*, 15, 2911-2928.

Bohnert, H. J., Sheveleva, E. (1998). Plant stress adaptations - making metabolism move. *Current Opinion in Plant Biology*, 1, 267-274.

Boojar, M.M.A., Goodarzi, F. (2007). The copper tolerance strategies and the role of antioxidative enzymes in three plant species grown on copper mine. *Chemosphere*, 67, 2138-2147.

Boominathan, R., Doran, P.M. (2003). Cadmium Tolerance and Antioxidative Defenses in Hairy Roots of the Cadmium Hyperaccumulator, Thlaspi caerulescens. *Biotechnology and Bioengineering*, 83, 158-167.

Boyd, R.S., Martens, S.N. (1998). The significance of metal hyperaccumulation for biotic interaction. *Chemoecology*, 8, 1-7.

Briat, J.-F., Lebrun, M. (1999). Plant responses to metal toxicity. *Plant biology and pathology*, 322, 43-54.

Brooks, R.R., Lee, J., Reeves, R.D., Jaffre, T. (1977). Detection of nickeliferous rocks by analysis of herbarium speciments of indicator plants. *Journal of Geochemical Exploration*, 7, 49-57.

Buchanan, B.B., Gruissem, W., Jones, R.L. (2000). *Biochemistry and Molecular Biology of Plants*. Courier Companies, Inc., 843-845.

Cabañero, F.J., Carvajal, M. (2007). Different cation stresses affect specifically osmotic root hydraulic conductance, involving aquaporins, ATPase and xylem loading of ions in Capsicum annuum, L. plants. *Journal of Plant Physiology*, 164, 1300-1310.

Caregnato, F. F., Koller, C. E., MacFarlane G. R., Moreira J. C. F. (2008). The glutathione antioxidant system as a biomarker suite for the assessment of heavy metal exposure and effect in the grey mangrove, Avicennia marina (Forsk.) Vierh. *Marine Pollution Bulletin*, 56, 1119-1127.

Cenkci, S., Ciğerci, I. H., Yıldız, M., Özay, C., Bozdağ, A., Terzi, H. (2010). Lead contamination reduces chlorophyll biosynthesis and genomic template stability in Brassica rapa L. *Environmental and Experimental Botany*, 67, 467-473.

Chakravarty B., Srivastava, S. (1992). Toxicity of some heavy metals in vivo and in vitro in Helianthus annuus. *Mutation Research Letters*, 283, 287-294.

Chaney, R.L., Malik, M., Li, Y.M., Brown, S.L., Brewer, E.P., Angle, J.S., Baker, A.J.M. (1997). Phytoremediation of soil metals. *Current Opinion in Biotechnology*, 8, 279-284.

Chaoui, A., Mazhoudi, S., Ghorbal, M.H., El Ferjani, E. (1997). Cadmium and zinc induction of lipid peroxidation and effects of antioxidant enzyme activities on beans (Phaseolus vulgaris L.). *Plant Science*, 127, 139-147.

Chavan, P.V., Dennett, K.E., Marchand, E.A., Gustin, M.S. (2007). Evaluation of small-scale constructed wetland for water quality and Hg transformation. *Journal of Hazardous Materials*, 149, 543-547.

Chen, B., Roos, P., Zhu, Y-G., Jakobsen, I. (2008). Arbuscular mycorrhizas contribute to phytostabilization of uranium in uranium mining tailings. *Journal of Environmental Radioactivity*, 99, 801-810.

Choo, T.P., Lee, C.K., Low, K.S., Hishamuddin, O. (2006). Accumulation of chromium (VI) from aqueous solutions using water lilies (Nymphaea spontanea), *Chemosphere*, 62, 961-967.

Choudhary, M., Jetley, U. K., Khan, M. A., Zutshi, S., Fatma, T. (2007). Effect of heavy metal stress on proline, malondialdehyde, and superoxide dismutase activity in the cyanobacterium Spirulina platensis-S5. *Ecotoxicology and Environmental Safety*, 66, 204-209.

Clemens, S., Palmgren, M.G., Krämer, U. (2002). A long way ahead: understanding and engineering plant metal accumulation. *Trends in plant science*, 7, 7, 309-315.

Clemens, S. (2006). Toxic metal accumulation, responses to exposure and mechanisms of tolerance in plants. *Biochimie*, 88, 1707-1719.

Cobbett, C.S. (2000). Phytochelatins and Their Roles in Heavy Metal Detoxification. *Plant Physiology*, 123, 825-832.

Cunningham, S.D., Ow, D.W. (1996). Promises and Prospects of Phytoremediation. *Plant Physiology*, 110, 715-719.

Cuypers, A., Vangronsveld, J., Clijsters, H. (2001). The redox status of plant cells (AsA and GSH) is sensitive to zinc imposed oxidative stress in roots and primary leaves of Phaseolus vulgarit. *Plant Physiology and Biochemistry*, 39, 657-664.

Dazy, M., Béraud, E., Cotelle, S., Meux, E., Masfaraud, J.-F., Férard, J.-F. (2008). Antioxidant enzyme activities as affected by trivalent and hexavalent chromium species in Fontinalis antipyretica Hedw. *Chemosphere*, 73, 281-290.

Dazy, M., Béraud, E., Cotelle, S., Grévilliot, F., Férard, J.-F., Masfaraud, J.-F. (2009). Changes in plant communities along soil pollution gradients: Responses of leaf antioxidant enzyme activities and phytochelatin contents. *Chemosphere*, 77, 376-383.

Demirevska-Kepova, K., Simova-Stoilova, L., Stoyanova, Z., Hölzer, R., Feller U. (2004). Biochemical changes in barley plants after excessive supply of copper and manganese. *Environmental and Experimental Botany*, 52, 253-266.

de Souza, M.P., Chee, H.N., Terry, N. (1999). Rhizosphere bacteria enhance the accumulation of selenium and mercury in wetland plants. *Planta*, 209, 259-263.

Dercová, K., Makovníková, J., Barančíková, G., Žuffa, J. (2005). Bioremediácia toxických kovov kontaminujúcich vody a pôdy. *Chemické listy*, 99, 682-693.

Di Salvatore, M., Carafa, A. M., Carratù, G. (2008). Assessment of heavy metals phytotoxicity using seed germination and root elongation tests: A comparison of two growth substrates, *Chemosphere*, 73, 1461-1464.

Domažlická, E., Vodičková, H., Mader, P. (1994). Fytochelatiny. Biologické listy, 59, 2, 81-92.

Dunne, E.J., Culleton, N., O'Donovan, G., Harrington, R., Olsen, A.E. (2005). An integrated constructed wetland to treat contaminants and nutrients from dairy farmyard dirty water. *Ecological Engineering*, 24, 221-234.

Eapen, S, Suseelan, K.N., Tivarekar, S., Kotwal, S.A., Mitra, R. (2003). Potential for rhizofiltration of uranium using hairy root cultures of Brassica juncea and Chenopodium amaranticolor. *Environmental Research*, 91, 127-33.

Ekmekçi, Y., Tanyolaç, D., Ayhan, B. (2008). Effects of cadmium on antioxidant enzyme and photosynthetic activities in leaves of two maize cultivars. *Journal of Plant Physiology*, 165, 600-611.

El-Ghamery, A. A., El-Kholy, M. A., El-Yousser, M. A. A. (2003). Evaluation of cytological effects of Zn2+ in relation to germination and root growth of Nigella sativa L. and Triticum aestivum L. *Mutation Research/Genetic Toxicology and Environmental Mutagenesis*, 537, 29-41.

Espen, L., Pirovano, L., Cocucci, S. M. (1997).Effects of Ni2+ during the early phases of radish (Raphanus sativus) seed germination. *Environmental and Experimental Botany*, 38, 187-197.

Fayiga, A. O., Ma, L. Q., Cao, X., Rathinasabapathi, B. (2004). Effects of heavy metals on growth and arsenic accumulation in the arsenic hyperaccumulator Pteris vittata L. *Environmental Pollution*, 132, 289-296.

Fitz, W.J., Wenzel, W.W. (2002). Arsenic transformations in the soil-rhizosphere-plant system: fundamentals and potential application to phytoremediation. *Journal of biotechnology*, 99, 259-278.

Florence, T. M. (1984). The production of Hydroxyl Radical from Hydrogen peroxide. Journal *of Inorganic Biochemistry*, 22, 221-230.

Foyer, C. H., Noctor, G. (2005). Redox homeostasis and antioxidant signalling: a metabolic link between stress perception and physiological response. *Plant Cell*, 17, 1866-1875.

Förstner, U., Wittmann, G.T.W. (1979). *Metal Pollution in the Aquatic Environment*. 1st ed., Springer-Verlag, Berlin, New York.

Gardea-Torresdey, J.L., Peralta-Videa, J.R., Montes, M., de la Rosa, G., Corral-Diaz, B. (2004). Bioaccumulation of cadmium, chromium and copper by Convolvulus arvensis L.: impact on plant growth and uptake of nutritional elements. *Bioresource Technology*, 92, 229-235.

Gerhardt, K. E., Huang, X.-D., Glick, B. R., Greenberg, B. M. (2009). Phytoremediation and rhizoremediation of organic soil contaminants: Potential and challenges. *Plant Science*, 176, 20-30.

Ghnaya, A. B., Charles, G., Hourmant, A., Hamita, J. B., Branchard, M. (2009). Physiological behaviour of four rapeseed cultivar (Brassica napus L.) submitted to metal stress. *Comptes Rendus Biologies*, 332, 363-370.

Gidda, S. K., Varin, L.. (2006). Biochemical and molecular characterization of flavonoid 7-sulfotransferase from Arabidopsis thaliana. *Plant Physiology and Biochemistry*. ,44, 628-636.

Glick, B.R. (2003). Phytoremediation: synergistic use of plants and bacteria to clean up the environment. *Biotechnology Advances*, 21, 383-393.

Grandlic, C.J., Palmer, M.W., Maier, R.M. (2009). Optimization of plant growth-promoting bacteria-assisted phytostabilization of mine tailings. *Soil Biology and Biochemistry*, 41, 1734-1740.

Guo, B., Liang, Y., Zhu, Y. (2009). Does salicylic acid regulate antioxidant defense system, cell death, cadmium uptake and partitioning to acquire cadmium tolerance in rice? *Journal of Plant Physiology*, 166, 20-31.

Gupta, A. K., Sinha, S. (2009). Antioxidant response in sesame plants grown on industrially contaminated soil: Effect on oil yield and tolerance to lipid peroxidation. *Bioresource Technology*, 100, 179-185.

Gupta, D. K., Huang, H. G., Yang, X. E., Razafindrabe, B. H. N., Inouhe, M. (2010). The detoxification of lead in Sedum alfredii H. is not related to phytochelatins but the glutathione. *Journal of Hazardous Materials*, 177, 437-444.

Hall, J. L. (2002). Cellular mechanisms for heavy metal detoxification and tolerance. In *Journal of Experimental botany*, 53, 1-11.

Hattab, S., Dridi, B., Chouba, L., Kheder, M. B., Bousetta, H. (2009). Photosynthesis and growth responses of pea Pisum sativum L. under heavy metals stress. *Journal of Environmental Sciences*, 21, 1552-1556.

Havaux, M. (1998). Carotenoids as membrane stabilizers in chloroplasts. *Trends Plant Science*, 3, 147-151.

Heldt, H-W. (1997). *Plant Biochemistry and Molecular Biology*. Oxford University Press, 296-297.

Hernández, I., Alegre, L., Van Breusegem, F., Munné-Bosch, S. (2009). How relevant are flavonoids as antioxidants in plants? *Trends in Plants Science*, 14, 125-132.

Hornik, M., Pipiska, M., Maresova, J., Augustin, J. (2009). *Uptake and Translocation of Metal Complexes in Vascular Plants*. Proceedings of 1st Conference on Biotechnology and Metals, Kosice, Slovakia, 29-32.

Hronec, O. (1996). Ťažké kovy a ich pohyb v pôdach a rastlinách. Zborník zo seminára: Ťažké kovy v ekosystéme, E'96, BIJO Slovensko, s r.o., 41-49.

Iannelli, M. A., Pietrini, F., Fiore, L., Petrilli, L., Massacci, A. (2002). Antioxidant response to cadmium in Phragmites australis plants. *Plant Physiology and Biochemistry*, 40, 977-982.

Ike, A., Sriprang, R., Ono, H., Murooka, H.,Yamashita,M.: (2007). Bioremediation of cadmium contaminated soil using symbiosis between leguminous plant and recombinant rhizobia with the MTL4 and the PCS genes. *Chemosphere*, 66, 1670-1676.

Israr, M., Sahi, S. V. (2006). Antioxidative responses to mercury in the cell cultures of Sesbania drummondii. *Plant Physiology and Biochemistry*, 44, 590-595.

Jarvis, M.D., Leung, D.W.M. (2001). Chelated lead transport in Chamaecytisus proliferus (L.f.) link ssp. proliferis var. palmensis (H.Christ): an ultrastructural study. *Plant Science*, 161, 433-441.

Jiang, H.M., Yang, J.C., Zhang, J.F. (2007). Effects of external phosphorus on the cell ultrastructure and the chlorophyll content of maize under cadmium and zinc stress. *Environmental Pollution*, 147, 750-756.

Kabata-Pendias, A. (2001). Trace Elements in Soils and Plants. Third Edition, CRC Press, 2001.

Kadukova, J., Kalogerakis, N. (2007). Lead accumulation from non-saline and saline environment by Tamarix smyrnesis Bunge. *European Journal of Soil Biology*, 43, 216-223.

Kadukova, J., Manousaki, E., Kalogerakis, N. (2008). Pb and Cd Accumulation and Phyto - Excretion by Salt Cedar (Tamarix smyrnensis Bunge). *International Journal of Phytoremediation,* 1, 10, 2008, 31-46.

Kavuličova, J., Kadukova, J., Podracky, J., Ivánova, D. (2009). *Effect of heavy metals on oxidative stress in Linum usitatissimum.* 1st Internal Conference on Biotechnology and Metals, 57-60.

Khan, S., Ahmad, I., Shah, T., Rehman, S., Khaliq, A. (2009). Use of constructed wetland for the removal of heavy metals from industrial wastewater. *Journal of Environmental Management*, 90, 3451–3457.

Khatun, S., Ali, M. B., Hahn, E.-J., Paek, K.-Y. (2008). Copper toxicity in Withania somnifera: Growth and antioxidant enzymes responses of in vitro grown plants. *Environmental and Experimental Botany*, 64, 279-285.

Koppolu, L., Agblevor, F.A., Clements, L.D. (2003). Pyrolysis as a technique for separating heavy metals from hyperaccumulators.Part II: Lab-scale pyrolysis of synthetic hyperaccumulator biomass. *Biomass and Bioenergy*, 25, 651-663.

Kos, B., Leštan, D. (2003). Induced Phytoextraction/Soil Washing of Lead Using Biodegradable Chelate and Permeable Barriers. *Environmental Science and Technology*, 37, 624-629.

Kováčik, J., Klejdus, B., Hedvabny, J., Bačkor, M. (2010). Effect of copper and salicylic acid on phenolic metabolites and free amino acids in Scenedesmus quadricauda (Chlorophyceae). *Plant Science*, 178, 3, 703-711.

Kvesitadze, G., Khatisashvili, G., Sadunishvili, T., Ramsden, J.J. (2006). *Biochemical Mechanisms of Detoxification in Higher Plants*. Springer-Verlag Berlin Heidelberg.

Labra, M., Gianazza, E., Waitt, R., Eberini, I., Sozzi, A., Regondi, S., Grassi, F., Agradi, E. (2006). Zea mays L. protein changes in response to potassium dichromate treatments. *Chemosphere*, 60, 1234-1244.

Laloi, C., Apel, K., Danon, A. (2004). Reactive oxygen signalling: the latest news. *Current Opinion in Plant Biology,* 7, 323-328.

Larson, R. A. (1988). The antioxidants of higher plants. *Phytochemistry*, 27, 969-978.

Lasat, M.M. (2002). Phytoextraction of toxic metals, a review of biological mechanisms. *Journal of Environmental Qualily*, 31, 109-120.

Lasat, M.M. (2000). Phytoextraction of Metals from Contaminated Soil: A Review of Plant/Soil/Metal Interaction and Assessment of Pertinent Agronomic Issues. *Journal of Hazardous Substance Research*, 2, 1-25.

Leita, L., Nobili, M. D., Mondini, C., Garcia, M. T. B. (1993). Response of Leguminosae to cadmium exposure. *Journal of Plant Nutrition*, 16, 2001-2012.

Leopold, I., Günther, D., Schmidt, J., Neumann, D. (1999). Phytochelatins and heavy metal tolerance. *Phytochemistry*, 50, 1323-1328.

León, A. M., Palma, J. M., Corpas, F. J., Gómez, M., Romero-Puertas, M. C., Chatterjee, D., Mateos, R., M., del Río, L. A., Sandalio, L. M. (2002). Antioxidative enzymes in cultivars of pepper plants with different sensitivity to kadmium. *Plant Physiology and Biochemistry*, 40, 813-820.

Li, C-X, Feng, S-L., Shao, Y., Jiang, L-N., Lu, X-Y., Hou, X-L. (2007). Effects of arsenic on seed germination and physiological activities of wheat seedlings. *Journal of Environmental Sciences*, 19, 725-732.

Lindberg, S., Landberg, T., Greger, M. (2007). Cadmium uptake and interaction with phytochelatins in wheat protoplasts. *Plant Physiology and Biochemistry*, 45, 47-53.

Liu, X., Zhang, S., Shan, X., Zhu, Y.G. (2005). Toxicity of arsenate and arsenite on germination seedling growth and amylolytic activity of wheat. *Chemosphere*, 61, 293-301.

Liua, Y-J., Mub, Y-J., Zhub, Y-G., Dinga, H., Arens, N.C. (2007). Which ornamental plant species effectively remove benzene from indoor air? *Atmospheric Environment*, 41, 650-654.

MacFarlane, G. R., Burchett, M. D. (2001). Photosynthetic Pigments and Peroxidase Activity as Indicators of Heavy Metal Stress in the Grey Mangrove, Avicennia marina (Forsk.) Vierh. *Marine Pollution Bulletin*, 42, 233-240.

Mackova, M., Dowling, D., Macek, T. (2006). Phytoremediation, Rhizoremediation. *Focus on Biotechnology*, 9A, Spinger, Berlin, New York.

Madrid, F., Liphadzi, M.S., Kirkham, M.B. (2003). Heavy metal displacement in chelate-irrigated soil during phytoremediation. *Journal of Hydrology*, 272, 107-119.

Maestri, E., Marmiroli, M., Visioli, G., Marmiroli, N. (2010). Metal tolerance and hyperaccumulation: Costs and trade-offs between traits and environment. *Environmental and Experimental Botany*, 68, 1-13.

Majer Newman, J., Clausen, J.C., Neafsey, J.A. (2000). Seasonal performance of a wetland constructed to process dairy milkhouse wastewater in Connecticut. *Ecological Engineering*, 14, 181-198.

Malmström, B.G., Leckner, J. (1998). The chemical biology of copper. *Current Opinion in Chemical Biology*, 2, 286-292.

Manousaki, E., Kadukova, J., Papadantonakis, N., Kalogerakis, N. (2008). Phytoextraction and phytoexcretion of Cd by the leaves of Tamarix smyrnensis growing on contaminated non saline and saline soils. *Environmental Research*, 106, 3, 326-332.

Marchiol, L., Assolari, S., Sacco, P., Zerbi, G. (2004). Phytoextraction of heavy metals by canola (Brassica napus) and radish (Raphanus sativus) grown on multicontaminated soil. *Environmental Pollution* 132, 21-27.

Masarovičová, E., Repčák, M., Erdelský, K., Gašparíková, O., Ješko, T., Mistrík, I. (2008). Fyziológia rastlín, Univerzita Komenského Bratislava.

Mascher, R., Lippmann, B.,Holzinger, S., Bergmann, H. (2002.) Arsenate toxicity: effects on oxidative stress response molecules and enzymes in red clover plants. *Plant Science*, 163, 961-969.

Mazhoudi, S., Chaoui, A., Ghorbal, M. H., El Ferjani, E. (1997). Response of antioxidant enzymes to excess copper in tomato (Lycopersicon esculentum, Mill.). *Plant Science*, 127, 129-137.

McGrath, S.P., Zhao, F.J., Lombi, E. (2002). Phytoremediation of metals, metalloids, and radionuclides. *Advances in Agronomy* 75, 1-56.

McKersie, B.D., Leshem, Y.Y. (1994). Stress and stress coping in cultivated plants. *Biologia plantarum,* 256.

Meghashri, S., Vijay Kumar, H., Gopal, S. (2010). Antioxidant properties of a novel flavonoid from Leaves of Leucas aspera. *Food Chemistry*, Article in Press.

Mendoza-Cózatl, D. G., Moreno-Sánchez, R. (2006). Control of glutathione and phytochelatin synthesis under cadmium stress. Pathway modeling for plants. *Journal of Theoretical Biology*, 238, 919-936.

Milone, M.T., Sgherri, C., Clijsters, H., Navari-Izzo, F. (2003). Antioxidative responses of wheat treated with realistic concentration of cadmium. *Evironmental and experimental botany*, 50, 265-276.

Mishra, S., Srivastava, S., Tripathi, R. D., Kumar, R., Seth, C. S., Gupta, D. K. (2006). Lead detoxification by coontail (Ceratophyllum demersum L.) involves induction of phytochelatins and antioxidant system in response to its accumulation. *Chemosphere*, 65, 1027-1039.

Mithöfer, A., Schulze, B., Boland, W. (2004). Biotic and heavy metal stress response in plants:evidence for common signals. *Federation of European Biochemical Societies Letters*, 566, 1-5.

Mittler, R. (2002). Oxidative stress, antioxidants and stress tolerance. *Trends in Plant Science*, 7, 405-410.

Mleczek, M., Rissmanna, I., Rutkowski, P., Kaczmarek, Z., Golinski, P. (2009). Accumulation of selected heavy metals by different genotypes of Salix. *Environmental and Experimental Botany,* 66, 289-296.

Mobin, M., Khan, N. A. (2007). Photosynthetic activity, pigment composition and antioxidative response of two mustard (Brassica juncea) cultivars differing in photosynthetic capacity subjected to cadmium stress. *Journal of Plant Physiology*, 164, 601-610.

Møller, I. M. (2001). Plant mitochondria and oxidative stress: electron transport, NADPH turnover, and metabolism of reactive oxygen species. *Annual Rewiew of Plant Physiology and Plant Molecular Biology,* 52, 561-591.

Monferrán, M. V., Agudo, J. A. S., Pignata, M. L., Wunderlin, D. A. (2009). Copper-induced response of physiological parameters and antioxidant enzymes in the aquatic macrophyte Potamogeton pusillus. *Enviromental pollution*, 157, 2570-2576.

Monteiro, M. S., Santos, C., Soares, A. M. V. M., Mann, R. M. (2009). Assessment of biomarkers of cadmium stress in lettuce. *Ecotoxicology and Enviromental safety*, 72, 811-818.

Morelli, E., Scarano, G. (2001). Synthesis and stability of phytochelatins induced by cadmium and lead in the marine diatom Phaeodactylum tricornutum, *Marine Environmental Research*, 52, 383-395.

Mori, S., Uraguchi, S., Ishikawa, S., Araoa, T. (2009). Xylem loading process is a critical factor in determining Cd accumulation in the shoots of Solanum melongena and Solanum torvum. *Environmental and Experimental Botany*, 67, 127-132.

Neil, S., Desikan, R., Hancock, J. (2002). Hydrogen peroxide signalling. *Current Opinion in Plant Biology,* 5, 388-395.

Nessner Kavamura, V., Esposito, E. (2010). Biotechnological strategies applied to the decontamination of soils polluted with heavy metals. *Biotechnology Advances*, 28, 61-69.

Noctor, G., Foyer, C. H. (1998). Simultaneous Measurement of Foliar Glutathione, γ-Glutamylcysteine, and Amino Acids by High-Performance Liquid Chromatography: Comparison with Two Other Assay Methods for Glutathione. *Analytical Biochemistry*, 264, 98-110.

Ohlsson, A. B., Landberg, T., Berglund, T., Greger, M. (2008). Increased metal tolerance in Salix by nicotinamide and nicotinic acid. *Plant Physiology and Biochemistry*, 46, 655-664.

Orcutt, D.M., Nilsen, E.T. (2000). *The Physiology of Plants under Stress, Soil and Biotic Factors*. JohnWileyandSons.

Ouyang, Y. (2002). Phytoremediation: modelling plant uptake and contaminant transport in the soil-plant-atmosphere continuum. *Journal of Hydrology*, 266, 66-82.

Pandey, N., Sharma, Ch. P. (2002). Effect of heavy metals Co2+, Ni2+ and Cd2+ on growth and metabolism of cabbage. *Plant Science*, 163, 753-758.

Peng, K., Luo, Ch., Lou, L., Li, X., Shen, Z. (2008). Bioaccumulation of heavy metals by the aquatic plants Potamogeton pectinatus L. and Potamogeton malaianus Miq. and their potential use for contamination indicators and in wastewater treatment. *Science of the total environment*, 392, 22-29.

Peralta, J. R., Gardea-Torresdey, J. L., Tiemann, K. J., Gomez, E., Arteaga, S., Rascon, E., Parsons, J. G. (2001). Uptake and effects of five heavy metals on seed germination and plant growth in alfalfa (Medicago sativa L.). *Bulletin of Environmental Contamination and Toxicology*, 66, 727-734.

Peralta-Videa, J.R., Lopez, M.L., Narayan, M., Saupe, G., Gardea-Torresdey, J. (2009) The biochemistry of environmental heavy metal uptake by plants: Implications for the food chain. *The International Journal of Biochemistry and Cell Biology*, 41, 1665-1677.

Pietramellara, G., Nannipieri, P., Renella, G. (2009). Microbial biomass, respiration and enzyme activities after in situ aided phytostabilization of a Pb- and Cu-contaminated soil. *Ecotoxicology and Environmental Safety*, 72, 115-119.

Pitzschke, A., Hirt, H. (2006). Mitogen-Activated protein Kinases and Reactive Oxygen Species Signaling in Plants. *Plant Physiology*, 141, 351-356.

Pongrac, P., Zhao, F.-J., Razinger, J., Trámec, A., Regvar, M. (2009). Physiological responses to Cd and Zn in two Cd/Zn hyperaccumulating Thlaspi species. *Environmental and Experimental Botany*, 66, 479-486.

Porra, R. J. (2002). The chequered history of the development and use of simultaneous equations for the accurate determination of chlorophylls a and b. *Photosynthesis Research*, 73, 149-156.

Posmyk, M. M., Kontek, R., Janas, K.M. (2009). Antioxidant enzymes activity and phenolic compounds content in red cabbage seedlings exposed to copper stress. *Ecotoxicology and Environmental Safety*, 72, 596-602.

Prasad, M. N. V., Malec, P., Waloszek, A., Bojko, M., Strzalka, K. (2001). Physiological responses of Lemna trisulca L. (duckweed) to cadmium and copper bioaccumulation. *Plant Science*, 161, 881-889.

Pulford, I.D., Watson, C. (2003). Phytoremediation of heavy metal-contaminated land by trees – a review. *Environment International*, 29, 529-540.

Qiu, R.-L., Zhao, X., Tang, Y.-T., Yu, F.-M., Hu, P.-J. (2008). Antioxidative response to Cd in a newly discovered cadmium hyperaccumulator, Arabis paniculata F. *Chemosphere*, 74, 6-12.

Quartacci, M.F., Irtelli, B., Baker, A.J.M. Navari-Izzo, F. (2007). The use of NTA and EDDS for enhanced phytoextraction of metals from a multiply contaminated soil by Brassica carinata. *Chemosphere*, 68, 1920-1928.

Radić, S., Babić, M., Škobić, D., Roje, V., Pevalek-Kozlina, B. (2010). Ecotoxicological effects of aluminum and zinc on growth and antioxidants in Lemna minor L. *Ecotoxicology and Environmental Safety*, 73, 336-342.

Rahoui, S., Chaoui, A., Ferjani E. E. (2010). Membrane damage and solute leakage from germinating pea seed under cadmium stress. *Journal of Hazardous Materials,* Article in Press.

Rai, V., Vajpayee, P., Singh, S.N., Mehrotra, S. (2004). Effect of chromium accumulation on photosynthetic pigments, oxidative stress defence systems, nitrate reduction, proline level and eugenol content of Ocimum tenuiflorum L. *Plant Science*, 167, 1159-1169.

Rama Devi, S., Prasat, M. N. V. (2004). *Membrane Lipid Alterations in Heavy Metal Exposed Plants*. In Prasat, M. N. V. Heavy Metal Stress: from biomolecules to Ecosystems in Plants. Berlin, Heidelberg: Springer Verlag, 462.

Raskin, I., Nanda Kumar, PBA, Dushenkov, S., Salt, D.E. (1994). Bioconcentration of Heavy Metals by Plants. *Current Opinion in Biotechnology*, 5, 3, 285-290.

Rauser, W. E. (1999). *Structure and Function of Metal Chelators Produced by Plants: The case for organic acids, amino acids, phytin, and metallothioneins Cell Biochemistry and Biophysics*, 31, 19-48.

Rhoads, D. M., Umbach, A. L., C. Subbaiah, C. C., James N. Siedow, J. N. (2006). Mitochondrial reactive oxygen species. *Contribution to oxidative stress and interorganellar signaling. Plant Physiology*, 141, 357-366.

Rios-Gonzalez, K., Erdei, L., Lips, S.H. (2002). The activity of antioxidant enzymes in maize and sunflower seedlings as affected by salinity and different nitrogen sources. *Plant Science*, 162, 923-930.

Rodrígez-López, J. N., Espín, J. C., del Amor, F., Tudela, J., Martínez, V., Cerdá, A., García-Cánovas, F. (2000). Purification and Kinetic Characterization of an Anionic Peroxidase from Melon (Cucumis melo L.) Cultivated under Different Salinity Conditions. Journal of *Agricultural and Food Chemistry*, 48, 537-1541.

Rodriguez, L. M., Alatossava, T. (2010). Effects of copper on germination, growth and sporulation of Clostridium tyrobutyricum. *Food Microbiology*, Article in Press.

Romero-Puertas, M. C., Corpas, F. J., Rodríguez-Serrano, M., Gómez, M., del Río, L. A., Sandalio, L. M. (2007). Differential expression and regulation of antioxidative enzymes by cadmium in pea plants. *Journal of Plant Physiology*, 164, 1346-1357.

Römkens, P., Bouwman, L., Japenga, J., Draaisma, C. (2002). Potentials and drawbacks of chelate-enhanced phytoremediation of soils. *Environmental Pollution*, 116, 109-121.

Rousseau, D.P.L., Vanrolleghem, P.A., De Pauwa, N. (2004). Model-based design of horizontal subsurface flow constructed treatment wetlands: a review. *Water Research*, 38, 1484-1493.

Rout, G. R., Samantaray, S., Das, P. (2000). Effects of chromium and nickel on germination and growth in tolerant and non-tolerant populations of Echinochloa colona (L.) Link. *Chemosphere*, 40, 855-859.

Ruley, A. T., Sharma, N. C., Sahi, S. V. (2004). Antioxidant defense in a lead accumulating plant, Sesbania drummondii. *Plant Physiology and Biochemistry*, 42, 899-906.

Saifullah, Meers, E., Qadir, M., de Caritat, P., Tack, F.M.G., Du Laing, G., Zia, H. (2009). EDTA-assisted Pb phytoextraction. *Chemosphere*, 74, 1279-1291.

Salt, D.E., Smith, R.D., Raskin, I. (1998). *Phytoremediation.* Annual Review of Plant Physiology and Plant Molecular Biology, 49, 643-668.

Santos de Araujo, B., Omena de Oliveira, J., Machado, A.S., Pletsch, M. (2004). Comparative studies of the peroxidases from hairy roots of Daucus carota, Ipomoea batatas and Solanum aviculare. *Plant Science*, 167, 1151-1157.

Sas-Nowosielska, A., Kucharski, R., Małkowski, E., Pogrzeba, M., Kuperberg, J.M., Kryński, K. (2004). Phytoextraction crop disposal—an unsolved problem. *Environmental Pollution*, 128, 373-379.

Schnoor, J.L. (1997). *Phytoremediation.* Technology evaluation report, Ground-Water Remediation Technologies Analysis Center, Iowa.

Schwitzguébel, J-P. (2002). Hype or Hope: The Potential of Phytoremediation as an Emerging Green Technology. *Federal Facilities Environmental Journal*, 109-125.

Schützendübel, A., Polle, A. (2002). Plant responses to abiotic stresses: heavy metal induced oxidative stress and protection by mycorrhization. *Journal of Experimental Botany*, 53, 1351-1365.

Sekhar, K.Ch., Kamala, C.T., Chary, N.S., Balaram, V., Garcia, G. (2005). Potential of Hemidesmus indicus for phytoextraction of lead from industrially contaminated soils. *Chemosphere*, 58, 507-514.

Seth, C. S., Chaturvedi, P. K., Misra, V. (2008). The role of phytochelatins and antioxidants in tolerance to Cd accumulation in Brassica juncea L. *Ecotoxicology and Environmental Safety*, 71, 76-85.

Shah, K., Kumar, R.G., Verma, S., Dubey, R.S. (2001). Effect of cadmium on lipid peroxidation, superoxide anion generation and activities of antioxidant enzymes in growing rice seedlings. *Plant Science*, 161, 1135-1144.

Shao, H.-B., Chu, L.-Y., Lu, Z.-H., Kang, C.-M. (2008). Primary antioxidant free radical scavenging and redox signaling pathways in higher plant cells. *International Journal of Biological Sciences*, 4, 8-14.

Sharma, S. S, Dietz K.-J. (2009). The relationship between metal toxicity and cellular redox imbalance. *Trends in Plant Science*, 14, 43-50.

Siedlecka, A., Krupa, Z. (2002). Simple method of Arabidopsis thaliana cultivation in liquid nutrient medium. *Acta Physiologiae Plantarum*, 24, 163-166.

Sillanpää, S., Salminen, J.-P., Lehikoinen, E., Toivonen, E., Eeva, T. (2008). Carotenoids in a food chain along a pollution gradient. *Science of the Total Environment*, 406, 247-255.

Singh, A., Sharma, R. K., Agrawal, S. B. (2008). Effects of fly ash incorporation on heavy metal accumulation, growth and yield responses of Beta vulgaris plants. *Bioresource Technology*, 99, 7200-7207.

Sinha, S.K., Srivastava, H.S., Tripathi, R.D. (1993). Influence of some growth regulators and cations on the inhibition of chlorophyll biosynthesis by lead in maize. *Bulletin of Environmental Contamination Toxicology*, 51, 241-246.

Sinha, S., Basant, A., Malik, A., P. Singh, K. P. (2009). Iron-induced oxidative stress in a macrophyte: A chemometric approach. *Ecotoxicology and Environmental Safety*, 72, 585-595.

Srivastava, S., Mishra, S., Tripathi, R. D., Dwivedi, S., Gusta, D. K. (2006). Copper-induced oxidative stress and responses of antioxidants and phytochelatins in Hydrilla verticillata (L.f.) Royle. *Aquatic Toxicology*, 80, 405-415.

Soudek, P., Petřík, P., Vágner, M., Tykva, R., Plojhar, V., Petrová, Š., Vaněk, T. (2007). Botanical survey and screening of plant species which accumulate 226Ra from contaminated soil of uranium waste depot. *European Journal of Soil Biology*, 43, 251-261.

Sun, Q., Ye, Z.H., Wang, X.R., Wong, M.H. (2007). Cadmium hyperaccumulation leads to an increase of glutathione rather than phytochelatins in the cadmium hyperaccumulator Sedum alfredii. *Journal of Plant Physiology*, 164, 1489-1498.

Szőllősi, R., Varga, I. S., Erdei, L., Mihalik, E. (2009). Cadmium-induced oxidative stress and antioxidative mechanisms in germinating Indian mustard (Brassica juncea L.) seeds. *Ecotoxicology and Environmental Safety*, 72, 1337-1342.

Škerget, M., Kotnik, P., Handolin, M.,. Hraš, A.R., Simonič, M., Knez, Ž. (2005). Phenols, proanthocyanidins, flavones in some plant materials and their antioxidant activities. *Food Chemistry*, 89, 191-198.

Šlesak, I., Libik, M., Karpinska, B., Karpinski, S., Miszalski, Z. (2007). The role of hydrogen peroxide in regulation of plant metabolism and cellular signalling in reponse to environmental stresses. *Acta Biochimica Polonica*, 54, 39-50.

Špirochová, I.: Punčochářová, J., Kafka, Z., Kubal, M., Soudek, P., Vaněk, T. (2001). Studium kumulace ťežkých kovu v rostlinách. *Chemické listy,* 95, 335-336.

Takahama, U. (1984). Hydrogen peroxide-dependent oxidation of flavonols by intact spinach chloroplasts, *Plant Physiology*, 74, 852-855.

Takahashi, M., Terada, Y., Nakai, I., Nakanishi, H., Yoshimura, E., Mori, S., Nishizawa, N.K. (2003). Role of Nicotianamine in the Intracellular Delivery of Metals and Plant Reproductive Development. *The Plant Cell*, 15, 1263-1280.

Taulavuori, K., Prasad, M. N. V., Taulavuori, E., Laine, K. (2005). Metal stress consequences on frost hardiness of plants at northern high latitudes: a review and hypothesis. *Environmental Pollution,* 135, 209-220.

Tsuji, N., Hirayanagi, N., Okada, M., Miyasaka, H., Hirata, K., Zenk, M. H., Miyamoto, K. (2002). Enhancement of tolerance to heavy metals and oxidative stress in Dunaliella tertiolecta by Zn-induced phytochelatin synthesis. *Biochemical and Biophysical Research Communications*, 293, 653-659.

Tukendorf, A., Rauser, W. E. (1990). Changes in glutathione and phytochelatins in roots of maize seedlings exposed to kadmium. *Plant Science*, 70, 155-166.

Vera Tomé, F., Blanco Rodríguez, P., Lozano, J.C. (2008). Elimination of natural uranium and 226Ra from contaminated waters by rhizofiltration using Helianthus annuus L. *Science of the total environment*, 393, 351-358.

Verma, S., Dubey, R. S. (2003). Lead toxicity induces lipid peroxidation and alters the activities of antioxidant enzymes in growing rice plants. *Plant Science*, 164, 645-655.

Vögeli-Lange, R., Wagner, G. J. (1996). Relationship between cadmium, glutathione and cadmium-binding peptides (phytochelatins) in leaves of intact tobacco seedlings. *Plant Science,* 114, 11-18.

Wierzbicka, M., Obidzińska, J. (1998). The effect of lead on seed imbibition and germination in different plant species, *Plant Science*, 137, 155-171.

Williams, L. E., Pittman, J.K., Hall, J.L. (2000). Emerging mechanisms for heavy metal transport in plants. *Biochimica et Biophysica Acta*, 1465, 104-126.

Winge, D.R., Jensen, L.T., Srinivassan, Ch. (1998). Metal-ion regulation of gene expression in yeast. *Current Opinion in Chemical Biology*, 2, 216-221.

Wong, M.H. (2003). Ecological restoration of mine degraded soils, with emphasis on metal contaminated soils. *Chemosphere*, 50, 775-780.

Wood, J. M., Wang, H-K. (1985). *Microbial Resistance to Heavy Metals. In Environmental Inorganic Chemistry* (Proceedings), VCH Publishers, Inc. Deerfield Beach, Florida, 487-512.

Wu, G., Kang, H., Zhang, X., Shao, H., Chu, L., Ruan, C. (2010). A critical review on the bio-removal of hazardous heavy metals from contaminated soils: Issues, progress, eco-environmental concerns and opportunities. *Journal of Hazardous Materials*, 174, 1-8.

Wu, F., Zhang, G., Dominy, P. (2003). Four barley genotypes respond differently to cadmium: lipid peroxidation and activities of antioxidant capacity, *Environmental and Experimental Botany* , 50, 67-78.

Wu, J., Hsu, F.C., Cunningham, S.D. (1999). Chelate-Assisted Pb Phytoextraction: Pb Availability, Uptake, and Translocation Constraints. *Environmental Science and Technology*, 33, 1898-1904.

Xiong, Z.-T. (1997). Bioaccumulation and physiological effects of excess lead in a roadside pioneer species Sonchus oleraceus L. *Environmental Pollution*, 97, 275-279.

Xiong, Z. T.,Wang, H. (2005). Copper toxicity and bioaccumulation in Chinese cabbage (Brassica pekinensis Rupr.), *Environmental Toxicology*, 20, 188-194.

Xu, J., Yang, F., Chen, L., Hu, Y., Hu, Q. (2003). Effect of Selenium on Increasing the Antioxidant Activity of Tea Leaves Harvested during the Early Spring Tea Producing Season. *Journal of Agricultural and Food Chemistry*, 51, 1081-1084.

Yadav, S. K. (2009). Heavy metals toxicity in plants: An overview on the role of glutathione and phytochelatins in heavy metal stress tolerance of plants. *South African Journal of Botany*, Article in Press.

Yadav, S. K. (2010). Heavy metals toxicity in plants: An overview on the role of glutathione and phytochelatins in heavy metal stress tolerance of plants. *South African Journal of Botany*, Article in Press.

Yang, X., Feng, Y., He, Z., Stoffella, P.J. (2005). Molecular mechanisms of heavy metal hyperaccumulation and phytoremediation. *Journal of Trace Elements in Medicine and Biology*, 18, 339-353.

Yanqun, Z., Yuan, L., Jianjun, C., Haiyan, C., Li, Q., Schvartz, C. (2005). Hyperaccumulation of Pb, Zn and Cd in herbaceous grown on lead–zinc mining area in Yunnan, China. *Environment International*, 31, 755-762.

Yordanov, I., Velikova, V., Tsonev, T. (2000). Plant responses to drought, acclimation, and stress tolerance.*Photosynthetica* 38, 171-186.

Zaman, M. S., Zereen, F. (1998) Growth Responses of Radish Plants to Soil Cadmium and Lead Contamination. *Bulletin of Environmental Contamination Toxicology*, 61, 44-50.

Zeid, I.M. (2001). Responses of Phaseolus vulgaris to chromium and cobalt treatments, *Biologia Plantarum*, 44, 111-115.

Zenk, M.H. (1996). Heavy metal detoxification in higher plants – a review. *Gene* 179, 21-30.

Zhang, F.-Q., Wang, Y.-S., Lou, Z.-P., Dong, J.-D. (2007). Effect of heavy metal stress on antioxidative enzymes and lipid peroxidation in leaves and roots of two mangrove plant seedlings (Kandelia candel and Bruguiera gymnorrhiza). *Chemosphere,* 67, 44-50.

Zhang, H., Jiang, Y., He, Z., Ma, M. (2005). Cadmium accumulation and oxidative burst in garlic (Allium sativum). *Journal of Plant Physiology*, 162, 977-984.

Zhang, J., Kirkham, M. B. (1996). Enzymatic responses of the ascorbate-glutathione cycle to drought in Sorghum and sunflower plants. *Plant Science*. 113, 139-147.

Zhang, W., Cai, Y., Downum, K.R., Ma, L.Q. (2004). Thiol synthesis and arsenic hyperaccumulation in Pteris vittata (Chinese brake fern). *Environmental Pollution*, 131, 337-345.

Zhao, F.J., Hamon, R.E., McLaughlin, M.J. (2001). Root exudates of the hyperaccumulator Thlaspi caerulescens do not enhance metal mobilization. *New Physiologist*, 151, 613-620.

Zhou, W., Qiu, B. (2007). Effects of cadmium hyperaccumulation on physiological characteristics of Sedum alfredii Hance (Crassulaceae). *Plant Science*, 169, 737-745.

Zhu, Z., Wei, G., Li, J., Qian, Q., Yu, J. (2004). Silicon alleviates salt stress and increases antioxidant enzymes activity in leaves of salt-stressed cucumber (Cucumis sativus L.). *Plant Science*, 167, 527-533.

Zurita, F., De Anda, J., Belmont, M.A. (2009). Treatment of domestic wastewater and production of commercial flowers in vertical and horizontal subsurface-flow constructed wetlands. *Ecological Engineering*, 35, 861-869.

Reviewed by
Ing. Alena Luptákova, PhD.,
Institute of Geotechnics, Slovak Academy of Sciences,
Watsonova 45, 043 53, Košice,
Slovakia, E-mail: luptakal@saske.sk

In: Handbook of Phytoremediation
Editor: Ivan A. Golubev

ISBN: 978-1-61728-753-4

Chapter 6

REVIEWS ON SOIL POLLUTION, RISKS, SOURCES AND PHYTOREMEDIATION INVOLVING METAL CONTAMINANTS

Yan-Ju Liu [1,2,*], ***Qing-Jun Liu*** [1] ***and Hui Ding*** [1]

[1]Department of Environmental Pollution Analysis and Control, Beijing Center for Physical and Chemical Analysis, 100089, P.R. China

[2]Division of Environmental Health and Risk Management, School of Geography, Earth and Environmental Sciences, The University of Birmingham, Edgbaston, Birmingham B15 2TT, United Kingdom

ABSTRACT

Soil is the vital medium in the natural environment. Its pollution has grown into a global issue. Metals contamination is one of the heaviest environmental problems in soil.

This paper will review the status of soil contamination, its risks and sources in the beginning. Human activities broke the soil balance with low background toxic metal level, and shrank the area of agricultural soil globally. Both essential and unessential metals ruin the balance of the ecosystem, with increasing economic loss and human health damage. Soil pollution was caused not only by naturally generated mechanisms including earthquakes, volcanic eruption, but also by anthropogenic inputs such as industrial development, fossil fuel burning, mining, metallurgy, electroplating, waste disposal, long-term application of sewage sludge, fertilizer application, etc.

Then the review will summarize the phytoremediation technique for soil contamination. Phytoremediation of soil pollution is a popular method to remove toxic pollutants from soil with low cost and environmental sustainability; it is composed of phytoextraction and phytostability. On one hand, phytoextraction is defined as the use of hyperaccumulating plants to transport metals from the soil and concentrate them in plants that can be harvested. Up till now, about 400 hyperaccumulators have been documented in the world. Phytoextraction situations are separately reviewed for several toxic heavy metals including cadmium, arsenic, lead, zinc, etc. On the other hand, phytostabilization technology takes advange of plants to reduce leaching of the pollutants by eliminating or minimizing the mobile and bioavailable fractions of metals in the soil. Various plant

* Email: liuyanju@hotmail.com

species could be good candidates to improve scenery remediation in abandoned sites, by shaping efficient vegetation cover.

Finally, the mechanism of metal remediation in soil was also evaluated in this paper. Chelants, bio-microbes and genetic procedures have been applied to assist increasing the accumulation of metals by plants, after that some comments were given on limits of phytoremediation application. The truth is that phytoremediation has not become a commercially available technology in the field yet even if it could be promising.

Keywords: soil pollution; metals; phytoremedation; chelant; genetic procedure.

Soil plays an important role of medium for material and energy exchanges among the atmosphere, hydrosphere, biosphere and lithosphere (Qishlaqi, 2009). After being polluted, it does bring adverse effects to plants, animals and humans (DalCorso et al., 2008). At Ardai, for example, mature mountain pine forest with significant scenic value was severely damaged and killed due to the toxic emission far away from a smelter 10–20 km (Braanaas et al., 1970). In Asia, large amounts of hazardous materials were considered closely related to humans' reproductive toxicity, neurotoxicity, teratogenesis, immunotoxicity and carcinogenesis (Pringle and Rondinelli, 1998). Up till now, soil pollution has been spreading over many countries of the world (Franzius, 1994). Soil contaminants can input into subsurface arising from agriculture, municipal landfills, informal dumps, and leaking storage tanks (Pringle and Rondinelli, 1998) and thus raise potential risks to the ecosystem. As one of the most heavy soil pollution issues, metals contamination in soil and its risk, sources, together with phytoremediation will be reviewed in this paper. The mechanism of phytoremediation and genetic technology are also summarized.

1. Soil Metals Contamination, Risks, Source and Control

1.1. Metals Contamination in Soil and Its Risks

The realistic situation of soil with low background toxic metal level has been sharply broken under types of human activities. As a typical example, around 22,000 t (metric ton) cadmium has been released worldwide over the past five decades (Singh et al., 2003). Metal-contaminated soil has been reported at more than 50,000 sites in the USA (Ensley, 2000), 80,000 in Germany (Franzius 1994), and similar areas in other countries. There were ca. 100,000 ha of land contaminated by heavy metals in Europe and in the USA, with the shrinkage of about 10,000 ha for agricultural land, even applying the less strict thresholds of the present soil protection regulation (Lewandowski et al., 2006). Soil arsenic reached up to 186 mg/kg in El Paso, TX (EPA, 2003) and 2330 mg/kg at the Tresavean copper mine in Cornwall, UK. In China, arsenic content in arsenic-contaminated soils reached 800–45000 mg/kg (Wang et al., 2006a), far higher than the limit for agriculture soil (only 40 mg/kg requested by GB15618-1995 of China). The situation does not seem better for other elements. In mining-affected agricultural soil of Bangladesh, potassium and calcium show the highest concentrations in the range of 10,400–24,400 and 12,600–77,000 mg/kg, respectively, with the mean values as 2–5 orders high as the world normal averages (Bhuiyana et al., 2010). Iron and manganese were tested up to 140,221 mg/kg and 12,114 mg/kg respectively in soil near Lipu manganese mine (Wang et al., 2009). Even for surface soil from some parks of Beijing, lead has an obvious concentration of 207.5 mg/kg (Zhang et al., 2009).

It is true that some metals and their metalloids, including zinc, calcium, chromium, copper, iron, potassium, magnesium, manganese and sodium, etc., are essential micronutrients for plant growth, animal and human development (Underwood, 1977). Low dose consumption of these elements, e.g., selenium in amounts <0.05–0.10 mg/kg, even brings animals deficiency disease (Underwood, 1977), but an excessive dose can adversely become toxic (Ouzounidou et al., 1994; Marschner, 1995; Marmiroli and Maestri, 2008; Yang et al., 2008; Comino et al., 2009), e.g. when selenium exceeds the feed level at 4–5 mg/kg (Underwood, 1977). The potential toxicity of selenium exists in many regions of the western United States, where plant tissues remove selenium from contaminated soil by accumulation and harvest (Banuelos et al., 1997) , then transfer it to livestock feeding on it (Burk, 1994). In China, selenium in paddy rice (0.116 mg/kg) and cabbage (0.05 mg/kg) collected from Xuzhou exceeded dietary allowance designed for humans (55 μg) (Huang et al. 2009a), which brought potential toxicity to humans.

Adversely, most heavy metals or their metalloids have no biological role (Bruins et al., 2000), and are a pervasive concern due to their toxicity, abundance, persistence and subsequent accumulation in the environment (Dong et al., 2010). Heavy metals have polluted many sites in industrial countries and pose risks to ecological systems (Dermont et al., 2008; Wei et al., 2007a) and human health (Berndes et al., 2004) especially when exceeding a safe level (Dudka and Miller, 1999). When above a limited concentration, heavy metals could cause heavy damage to life. For instance, cadmium, a toxic metal, will lead to leaf chlorosis, root putrescence and growth inhibition (Solis-Dominguez et al., 2007). Chromium can cause the reduction in photosynthetic pigments, protein, cysteine, ascorbic acid, nitrate reductase activity and non-protein thiol contents of plants based on the experiments of plant species *Ocimum tenuiflorum* (Rai et al., 2004). Lead, copper and nickel in some rhizospheric soils have shown toxicity to livestock when their concentrations (e.g. 10,171, 4125 and 917 mg/kg) exceeded normal level of 30, 25 and 50 mg/kg each (Robinson et al., 2008). Soil heavy metals easily entered the human body through the food chain (Cheng, 2003; Zhu et al., 2008), e.g., rice grain accumulated heavy metal from contaminated soil (Wang et al., 2003) and posed a serious risk to human health. A cross-country investigation of China indicated that some heavy metals, e.g., cadmium, mercury and lead, in soils and crops exceeded Chinese government levels in some regions (Wang et al., 2001). As an example, near Lechang Lead/Zinc Mine, Guangdong, China, lead daily intakes by local residents were 2.6 mg for adults and 1.2 mg for children, higher than the allowable level (Yang et al., 2004a). In addition, direct ingestion of soil, e.g., by children, may also contribute largely to the accumulation of heavy metal in humans (Ruby et al., 1996; Möller, et al., 2005), and even cause human cancers (Abernathy et al., 1999; ATSDR, 2005).

In summary, both essential and nonessential metals can bring damage to cell membranes, alter enzyme specificity, disrupt cellular functions and damage the structure of DNA at high toxic level (Bruins et al., 2000), which have attracted and will continuously create more public concerns (Welch et al., 2000; Polizzotto et al., 2006).

1.2. Soil Pollution Sources

Except natural disasters such as volcanic activities and earthquakes, anthropogenic inputs associated with industrial and agricultural activities involving mining, metallurgy, electroplating, fossil fuel combustion, waste disposal, fertilizer application and wastewater irrigation, are the principle sources of heavy metals in soil (Huang et al., 2007; Fornes et al., 2009).

Traffic is a main contributor to heavy metal pollution in urban and other traffic-related soil. Lead and zinc in traffic affected soil can be 100 times as high as those in soil unaffected by traffic (Kadi, 2009). When considering metal contamination by the roadside, the highest heavy metal concentrations were usually observed at 10–30 m from the road edge (Hafen and Brinkman, 1996; Zupancic, 1999; Al-Chalabi et al., 2000; Lin et al., 2000; Sutherland and Tolosa, 2001; Ramakrishnaiah and Somashekar, 2002; Fakayode and Olu-Owolabi, 2003; Kocher et al., 2005; Li et al., 2005; Li et al., 2007). For railroads, the highest level of lead, zinc, and cadmium, could be on the road edge, and then decreased with increasing distance from the road (Ma et al., 2009a). As to highways, the major pollutant lead reaches its highest level at 20 to 40 m away from the road (Han et al., 2009). In addition, some studies showed that the richest abundance of soil nematodes was found at 5 m while the lowest at 20 m away from the highway (Han, et al., 2009), implying that the worst living condition for life is at the point of 20 meters away from the highway. This could be a side support for the heaviest lead pollution at the point. Even worse, metal contamination, once raised by traffic, will accumulate in soil gradually, for instance, Rome urban soils had significantly higher Pt concentration (up to six times greater) in 2001 than those in 1992, as the result of automobile catalytic converters use during the last decade (Petrucci et al., 2000).

Mining is a historical topic as a contributor of heavy metal pollution to the air, soil and water over the world. After metals were released, they exchanged actively in these medium. Mining-related works such as tailings and acid mine drainage probably raise lots of heavy metal contamination problems about copper, zinc, cadmium and lead, etc. (Zhou et al., 2007). High selenium level in agricultural soils could be primarily caused by neighboured coalmines and power plants (Huang et al., 2009). As a worldwide soil pollution source, mining take an important role for soil heavy metal pollution in Spain (Fornes et al., 2009). The situation is probably worse in China, for example, the Xiangtan Manganese Mine in the middle of Hunan Province, has been mined since 1913 (Liu et al., 2006) and keeping its disturbance to the related ecosystem since then. The previous part in this paper has provided enough data for metals pollution raised by mining.

Many other human activities can also cause soil pollution. For example, burning brown coal with high contents of pyrite and heavy metals contaminated the neighboured soil by the deposition of the fly ash (Kapička et al., 1999). In addition, military also devoted to soil metals pollution, in Swiss, a military country, there were over 400 tons of lead entering soils annually at some 2000 military shooting ranges (MSRs) (Robinson et al., 2008). The "hot spots" of soil metal contamination in United States were primarily raised by industrial and military activities (USEPA 1995; Williford and Bricka, 2000). Agricultural activities including improperly fertilizing sewage sludge can pollute the soil (Wang et al., 2001); moreover, metals in agricultural waste can also enter soils and sediments (Yalcin, 2007). In Irish, agricultural soil was contaminated both by organ chlorine insecticide residues, PCBs and trace elements, cadmium, chromium, copper, mercury, nickel, lead and zinc (McGrath, 1995).

Practically, soil was polluted by mixed metal sources in most cases. For instance, arsenic could be introduced into the environment through a by-product of mining and smelting processes, coal combustion etc. (Doušova et al., 2006). The soils contaminated with radionuclide could be the result of proximity to radio-mineral outcrops, or anthropogenic development of nuclear technology (Tang and Willey, 2003). In Tianjin city of China, analysis of 188 surface soil samples indicated that soil contaminants of the city were mixingly contributed by coal, wood and gas combustion, petroleum spill, vehicle emission, biogenic conversion, etc. (Ye et al., 2006).

After soil was polluted by metals on its surface, the contaminants can spread over by penetration usually down to at least 50-80 cm in soil (Gabrera et al., 1999), either in solution phase of the spill (most of copper, zinc and cadmium), or as part of the solid phase (other metals) (Simón, 1999).

1.3. Soil Pollution Control

The condition of soils is getting more deteriorate with the progress of civilization (Wilden et al., 2001). It has brought unstoppable heavy economic loss and health problems. In China, about 12 million tons grain crops contaminated with heavy metals every year and it causes direct economic losses of more than 20 billion CNY (about 3 billion dollars) (MSEPA, 2006), not to mention the health damage. Therefore it is necessary to perform various interventions for cleanup of toxic metals from the contaminated soils (Liu et al., 2009), so as to achieve soil reclamation and revitalization (Schaaf et al., 2004; Luster et al., 2008).

Remediation methods for metal-contaminated soils involve ex-situ physical and chemical methods such as solidification, stabilization of metals, electro kinetics, soil rinsing, pyrometallurgical separation, excavation, thermal treatment chemical oxidation, photo catalytic degradation, and integrated remediation technologies etc (Mulligan et al., 2001; Gan et al., 2009). In general, they are costly and unaffordable for developing countries. Considering abandoned industrial sites and their remediation market in the United States and European Union alone, the remediation costs expected to exceed $20 billion annually (Boyajian and Carreira, 1997). If using soil washing, excavation, or pump and treat systems, the estimated soil remediation fees will be $7 to $8 billion per year in U.S. (Bennett et al. 2003a). The remediation cost for historic mine sites releasing acid mine drainage has been estimated to be more than AUD$100 000 per hectare (Harries, 1997). The remediation expense could be more expensive if the soil was contaminated also by organic, e. g. remediation cost of incineration, landfill, thermal desorption, and chemical dehalogenation can reach US $50–$1000 per ton (Dàvila et al., 1993).

Soil remediation methods mentioned above are often incompatible with maintaining soil structure and fertility (Pulford and Watson, 2003). They sometimes bring adverse effect if their byproducts are more unsafe than the parent compounds to the soil. When remediating pesticide pollution, the use of chemical reagent such as a strong acid, alkali, oxidants, or catalytic reduction usually causes deactivation (Lu et al., 1999; Chiron et al., 2000). In addition, more problems might be raised when disposing the organics emissions by thermal processes, e. g. some initial trials might need spaces big enough for storing large quantities unwanted pollutants (Johnston and Stringer, 1992).

Soil remediation techniques previously designed were costly and more risky. This pushed scientists searching for non-invasive solutions, which could keep the original soil structure when removing the contaminants from soil (Bungart and Huttl, 2004; Eapen et al., 2007). As a result, ex-situ and in-situ bioremediation or phytoremediation were provided on the stage of soil remedial technologies (Pringle and Rondinelli, 1998; Peer et al., 2005; Luo, 2009). Phytoremediation plays more beneficial role than other soil cleaning methods (Chaney et al., 1997; Meagher, 2000) mentioned above. With the performance in-situ and solar-driven (LeDuc and Terry, 2005), the phytoremediation can be up to 1,000-fold cheaper than those conventional remediation methods such as excavation and reburial (Memon and Schröder, 2009).

It is also environmentally-friendly (Shi and Cai, 2009), sustainable (McGrath et al., 2002; Salt et al., 1998), and with a low maintenance request (Cunningham et al., 1995). The phytoremediation can bring farmers amazing economic earnings, up to about 14,600 and 14,850 € ha^{-1} respectively over a period of 20 years, based on the substitution cost and hedonic price analysis (Lewandowski et al., 2006).

3 Phytoremediation

3.1. Definition and Classifications

Phytoremediation refers to using plants (Yang, 2008) and their associated microorganisms (Schwitzguébel, 2002) to degrade, transform, assimilate, metabolize, or detoxify hazardous pollutants from soil. It is widely viewed as the ecologically responsible alternative to the environmentally destructive physical remediation methods (Meagher, 2000) and can be applied in treating many kinds of soil contaminants including heavy metals, radionuclide, petroleum hydrocarbons, chlorinated organic compounds, pesticides, explosives, etc (Yang, 2008).

A practicable procedure of phytoremediation should start from cultivating a highly metal-tolerant plant species which is a suitable metal accumulator, be followed by irrigating with a diluted metal-chelating agent to increase metal bioavailability and thus enhance plant uptake, and be ended with harvesting and off-site removing the cultivated plants (Hayes et al., 2003).

The main pathway of phytoremediation composes of phytoextraction and phytostabilization. With the absorption of a wide range of soil metals, many wetland plants can colonize heavily metal-polluted areas and be ideal for both phytostabilization and phytoextraction (Deng et al., 2004a).

3.2. Phytoextraction

Initially, phytoextraction was defined as using hyperaccumulating plants to transport metals from the soil and concentrate them in plants that can be harvested (Kidd et al., 2009).

Hyperaccumulator should contain metal concentration larger than 1,000 ppm, or hold metal enrichment factor (EF) higher than 1 (Notes: EF=plant concentration/soil concentration).

Hyperaccumulators was first applied by Jaffre and his co-workers when they observed the accumulation of nickel in *Sebertia accuminata* (Jaffre et al., 1976). They are featured of loosely habitat requirements and high efficiency of soil phytoremediation (Cunningham et al., 1995; Lebrun, 2001; Garcia et al., 2005). Usually they are endemic to metalliferous soils and able to tolerate and accumulate metals to very high value in the above-ground tissues (approx. 100 times as high as that of nonaccumulator plant species) (Milner and Kochian, 2008). Being a hyperaccumulator, the metal content in plant was differently requested for kinds of metals, e. g. the threshold for arsenic hyperaccumulation is 1000 μg/g (Robinson et al., 2006) in leaf or aboveground tissue. The limit could be >10,000 μg/g for manganese and zinc. For metals of copper, cobalt, chromium, nickel, lead, the number have to be higher than 1000 μg/g. Cadmium value for Cd-hyperaccumulator was demanded higher than 100 μg/g (Greger, 1999; Baker and Brooks, 1989; Baker et al., 2000).

Searching for hyperaccumulators started with the research of selecting seldom naturally occurring studied plants (Krämer, 2005). Up to date, only about 400 hyperaccumulators have been documented in the world, representing less than 0.2% of all angiosperms (Baker and Brooks, 1989; McGrath and Zhao, 2003). Based on previous studies, many hyperaccumulating plants reserved for some soil pollutants such as nickel (320 families), cobalt (30 families), copper (34 families), while only one family was tested to accumulate metals of silver, gold, thallium, uranium (Baker et al., 2000; Reeves and Baker, 2000). For instance, *Leersia hexandra* (Zhang et al., 2007), *Thlaspi praecox* (Tolra et al., 2006), *Thlaspi caerulescens*, *Solanum nigrum* (Wang et al., 2008) and *Arabidopsis halleri* (Alvarez-Ayuso, 2008) are hyperaccumulators of cadmium. *Chengiopanax sciadophylloides* (Mizuno et al. 2006), *Phytolacca acinosa* (Xu et al., 2006) and *Phytolacca acinosa* (Xue et al., 2005) are hyperaccumulators for manganese. Meanwhile, *Potentilla griffithii* are Zn-hyperaccumulaor (Qiu et al., 2006).

The remediation potential of hyperaccumulators relies upon their growth rates (i.e., biomass production) and metal accumulation rate (g metal per kg of plant tissue). Based this point, hyperaccumulators such as *Thlaspi caerulescens* or *Alyssum bertolonii* taking up one or two metals (Robinson et al., 1998) could be ideal choice to remediate soil contamination only if they have also a highest biomass. Practically, most hyperaccumulators have low products. This could be compensated by high biomass plants (Baker and Brooks, 1989; Reeves and Baker, 2000; Alvarez-Ayuso, 2008), even if they are usually not metal-specific and contain low to average heavy metal concentrations (Keller et al., 2003). Thus phytoremediation could be a promising technique of removing soil pollutants where hyperaccumulators or accumulators are used to take up large quantities of pollutant metals (Salt et al., 1998; Roosens et al., 2003). Some hyperaccumulators cannot be well considered as good choice for removing pollutants due to their poor growth, but really a good one after increasing their biomass. For instance, though willow well accumulates cadmium and zinc in leaves when growing in a heavily mixed polluted soil (Jensen et al., 2009), it is not a perfect plant due to its limited growth. However, this plant has been proved growing well on moderately polluted soils, where it extracted 0.13% of total cadmium and 0.29% of the total zinc per year (Jensen et al., 2009).

The process of extracting metals from soil by hyperaccumulating plants can be called phytomining (Harris et al., 2009). It could produce 'bio-ore' if the metals have economic value (Sheoran et al., 2009). Hyperaccumulator *Berkheya coddii* of nickel and *Brassica juncea* of Gold (Au) are considered to be profitable for phytomining in Australia (Harris et al., 2009).

Being a technique to remediate heavy metals contaminated soil, phytoextraction has attracted wide attentions (Takenaka et al., 2009). When growing in polluted soil, some plant species e.g. cultivated species (wheat, barley or sunflower) were able to extract heavy metals, at a less or greater extent (Bejerre and Schierup, 1985; Chaney, 1989; Baker et al. 1994, Simón, 1998). Those plants, e.g. *Elsholtzia splendens* (Peng et al., 2005) and *Brassica nigra* (Indian mustard plants) (Bharagava et al., 2009), both metal-tolerant and well accumulating in contaminated soil, could be a better remedial candidate than those species only with strong accumulating ability. Many heavy metals have been investigated for their phytoremediation in soil; some typical studies will be summarized as follows and in Table 1.

Cadmium: Cadmium is one typical toxic heavy metal in soil. Its accumulation investigation has been widely processed both in labs and fields. Based on the threshold of cadmium hyperaccumulator of above 0.01% cadmium dry tissue (or 100 μg/g) (Baker and Brooks, 1989), *Thlaspi caerulescens* (Zhao et al., 2002) and *Arabidopsis halleri* (Zhao et al., 2006), Liyu No. 6 (Wang et al., 2007), and genus *Thalspi* (Kirkham, 2006) are all reported in the list of Cd-hyperaccumulator. These species could play an important role in the treatment of soils stressed by cadmium. In lab experiment, plant species indicated different abilities to accumulate cadmium. For example, garlic can well remove cadmium from solutions and accumulate it in its root, with only a small amount of cadmium to their bulbs and shoots (Jiang et al., 2001). In a hydroponic and pot experiment, oilseed rape Chuanyou II-10 is the most effective cadmium-phytoextractor among 21 varieties of oilseed rape and Indian mustard (Wang and Su, 2005). Furthermore, cadmium accumulators are prospective for field application in many samples. Phytoextraction of cadmium by *Averrhoa carambola* (carambola) might be a feasible option to clean up agricultural soils slightly contaminated by this toxic metal. High cadmium accumulating rice varieties could be practical choice for moderate cadmium polluted paddy fields (IBARAKI et al., 2009). India rice Chokoukoku reduced the total soil cadmium content by 38%, removing 883 g cadmium ha^{-1} after growing for 2 years without irrigation after drainage, adaptable to remediate the low to moderate levels of soil cadmium contamination (Murakami et al., 2009). *Thlaspi caerulescens* can be used to clean up moderately cadmium contaminated soils and remove 1.3% of the metal from the locality where cadmium exceeded allowable limits (McGrath et al., 2006). One-year-old tree carambola attained high-biomass of 18.6 t/ha shoot yield and extracted 213 g cadmium /ha after growing 170 days (Li et al., 2009). When Salix cultivation expansion is available, Salix-based cadmium management will also likely take place (Berndes et al., 2004). In addition, different plant ecotype might play variable roles on cadmium extraction. For instance, hyperaccumulating ecotype of *Sedum alfredii* can accumulate more than 6000 mg/g cadmium in shoots (Yang et al., 2004b), whereas its contrasting non-hyperaccumulating ecotype showed neither tolerance nor hyperaccumulation ability to cadmium (Xiong et al., 2004).

Table 1. Reviews on metal accumulations by plants (Notes: NA: non available; T: tolerant; MT: moderately tolerant; ST: strong tolerant; S: sensitive)

Experiment types	Metal names	Treat (or soil) concentration	Plant names	Metal conten in plants	References	Reaction to pollution
field site	As	184mg/kg	*Poa annua*	aboveground: 5000mg/kg	Ma et al. , 2001	T
mine tailings	As	49.6mg/kg	*Baccharis sarothroides*	shoots:36.9mg/kg ; roots:44.6mg/kg	Haque et al., 2008	
Pot experiment	As	pot soil: 220mg	*Pteris vittata*	green fronds:c.2100μg/g	Koller et al., 2008	
arsenic mine	As	total:51-261 mg/kg; bioavailable:4-48mg/kg	*Pteris vittata*	fronds: 3–704 mg/kg	Wei and Chen, 2006	
arsenic mine	As	total:39-299 mg/kg; bioavailable:7-48mg/kg	*Pteris cretica*	fronds: 149–694 mg/kg,	Wei and Chen, 2006	
arsenic-contaminated sites	As	1262 mg/kg	*Pteris multifida*	fronds: 301-2142 mg/kg	Wang et al., 2006a	
arsenic-contaminated sites	As	47,235 mg/kg	*Pteris oshimensis*	fronds: 1977-6244056 mg/kg	Wang et al., 2006a	
field investigation	As	NA	*Pteris multifida*	fronds: up to 2061 mg/kg; roots:692.7 mg/kg	Du et al., 2005	
mining areas	As	total : 814 mg/kg	*Athyrium yokoscense*	old fronds:242 mg/kg	Van et al., 2006	T
modified Hoagland solution	Cd	25mg /L	*Lonicera japonica*	stems:344.49±0.71μg/g DW; shoots:286.12±9.38μg/g DW	Liu et al., 2009	T
old silver mine	Cd	up to 37.86μg/g	*Evodiopanax innovans*	leaves: up to118μg/g	Takenaka et al., 2009	T
substrates	Cd	200mg/kg	*Cannabis sativa*	roots: 4052.8±225.6μg/g	Shi and Cai, 2009	MT
substrates	Cd	200mg/kg	*Ricinus communis*	2517.9±178.7μg/g	Shi and Cai, 2009	MT
substrates	Cd	200mg/kg	*Arachis hypogaea*	2124.5±52.1μg/g	Shi and Cai, 2009	MT

Table 1. (Continued)

Experiment types	Metal names	Treat (or soil) concentration	Plant names	Metal conten in plants	References	Reaction to pollution
substrates	Cd	200mg/kg	*Carthamus tinctorius*	1076.6±29.9μg/g	Shi and Cai, 2009	MT
field investigation	Cd	NA	*Thlaspi caerulescens*	shoots: 10 000 ppm	Milner and Kochian, 2008	T
metal-contaminated sites	Cd	total: 0.80-678mg/kg; labile:0.002-2.4mg/kg	*Salix caprea*	leaves: up to 116 mg/kg	Unterbrunner et al., 2007	
contaminated site	Cd	up to 300mg /kg	*Populus trichocarpa* x *P. deltoides*	leaves: up to 209 mg/kg	Robinson et al., 2000	
mine tailings	Cd	52.4 mg Cd/kg	*Salix phylicifolia*	up to 12.5 mg /kg	Robinson et al. , 2000	
contaminated soil	Cd	25-50 mg/kg	*Rorippa globosa*	stems and leaves, >100 mg/kg	Wei et al., 2008	
spill-affected sites	Cd	1.44 mg/kg	*Populus alba*	leaves: up to 3mg/kg	Domínguez et al., 2008	
land fill site	Cd	up to 15.8 mg /kg DW	*Poplar Bachte*	7.9–26.7mg /kg DW	Vandecasteele et al., 2008	
abandoned mine site	Cd	total:8 mg/kg; available:3.5mg/kg	*Cynodon dactylon*	14 ± 10mg/kg	Archer and Caldwell, 2004	T
abandoned mine site	Cd	total:9 mg/kg; available:3mg/kg	*Juncus usitatus*	26 ± 10 mg/kg	Archer and Caldwell, 2004	T
hydroponic method	Cd	10 mg/kg DW substrate	*cardoon cv. Peralta*	shoots:242 mg/kg	Hernández-Allica et al., 2008	grow affected
field trial	Cd	Total:7.2±0.92 mg /kg; Extractable:0.52±0.03mg /kg	*Viola baoshanensis*	shoots:28 mg /kg DW; total:0.17 kg /ha	Zhuang et al., 2007	

Experiment types	Metal names	Treat (or soil) concentration	Plant names	Metal conten in plants	References	Reaction to pollution
pot experiments	Cd	180 mg /kg	*Borago officinalis*	109 mg /kg	Evangelou et al., 2007	
pot experiments	Cd	180 mg /kg	*Sinapis alba*	123 mg /kg	Evangelou et al., 2007	
pot experiments	Cd	100 mg /kg	*Phacelia boratus*	42 mg/kg	Evangelou et al., 2007	
old smeltery	Cd	85.9mg/kg	*Ambrosia trifida*	shoots: 12.8mg/kg ; roots:14.0 mg/kg	Cui et al., 2007	
old smeltery	Cd	54.4mg/kg	*Solanum nigrum*	shoots: 19.1mg/kg; roots:4.0 mg/kg	Cui et al., 2007	
old smeltery	Cd	20.9mg/kg	*Polygonum lapathifolium*	shoots: 5.5mg/kg; roots:3.4 mg/kg	Cui et al., 2007	
old smeltery	Cd	130.3mg/kg	*Physalis angulata*	shoots: 12.2mg/kg ; roots:28.4 mg/kg	Cui et al., 2007	
old smeltery	Cd	197.3mg/kg	*Helianthus tuberosus*	shoots: 9.7mg/kg; roots:13.0 mg/kg	Cui et al., 2007	
old smeltery	Cd	50.8mg/kg	*Conyza canadensis*	shoots: 18.7mg/kg; roots:8.2 mg/kg	Cui et al., 2007	
old smeltery	Cd	14.6mg/kg	*Abutilon theophrasti*	shoots: 5.4mg/kg; roots:2.1mg/kg	Cui et al., 2007	
old smeltery	Cd	11.2mg/kg	*Chenopodium acuminatum*	shoots: 8.5mg/kg ; roots:2.6 mg/kg	Cui et al., 2007	
lead–zinc mining area	Cd	1896mg/kg	*Corydalis petrophila*	shoots: 329.8 mg/kg ; roots: 301.0 mg/kg	Zu et al., 2005	
lead/zinc mine	Cd	1.32 and 2 mg/kg	*Viola baoshanensis*	shoots: 456 to 2310 mg/kg; roots: 233 to 184mg/kg	Wei et al., 2004	
solution	Cd	50 mg/L	*Viola baoshanensis*	shoots:4825 mg/kg; roots:233 to 1846 mg/kg	Wei et al., 2004	S

Table 1. (Continued)

Experiment types	Metal names	Treat (or soil) concentration	Plant names	Metal conten in plants	References	Reaction to pollution
lead/Zn mine area	Cd	NA	*Sedum jinianum*	shoots:103-478 mg/kg DM	Xu et al., 2009	
nutrient	Cd	100 mu mol L-1	*Sedum jinianum*	shoots:5083 mg/kg DM	Xu et al., 2009	
pot soil	Cd	2.4mg/kg	*Sedum jinianum*	shoots: 16.4 mg/kg DM	Xu et al., 2009	
pot soil	Cd	9.2 mg/kg	*Sedum jinianum*	shoots: 79.8 mg/kg DM	Xu et al., 2009	
abandoned mining site	Cd	total: 28.1mg /kg; water-soluble: 1.37 mg /kg; exchangeable:16.8 mg /kg	*Athyrium yokoscense*	shoots : 0.9 g/kg	Chen et al., 2009	
abandoned mining site	Cd	total: 28.1mg /kg; water-soluble: 1.37 mg /kg; exchangeable:16.8 mg /kg	*Arabis flagellosa*	shoots : 0.3 g/kg	Chen et al., 2009	
mining areas	Cd	total:10.5 mg/kg	*Athyrium yokoscense*	fronds:up to 1095 mg /kg,	Van et al., 2006	T
heavy-metal polluted area	Cd	5.7–17.5 mg/kg	*Crassocephalum crepidioides*	aboveground: 14.6–78.6 mg/kg	Yamato et al., 2008	
pot culture	Cd	5 mg/kg	*Crassocephalum crepidioides*	aboveground: 121.2 mg kg	Yamato et al., 2008	
pot culture	Cd	2 mg/kg	*Crassocephalum crepidioides*	aboveground tissue: 106.1 µg	Yamato et al., 2008	
pot experiment	Cd	6 mg/kg	*Thlaspi caerulescens*	shoots: up to 582 pg/pot	Nishiyama et al., 2005	
mine tailings	Cr	84.5mg/kg	*Baccharis sarothroides*	shoots:105.5mg/kg ; roots:57.1mg/kg	Haque et al., 2008	
waste amended soil	Cr	400.37±8.35µg/g	*Typha angustata*	aboveground:6.31±0.3 µg/g; belowground:15.4±1.9µg/g	Bose et al., 2008	

Experiment types	Metal names	Treat (or soil) concentration	Plant names	Metal conten in plants	References	Reaction to pollution
field survey	Cr	34-186 mg/kg	*Leersia hexandra*	leaves:2978 mg/kg DW	Zhang et al., 2007	ST
mine tailings	Cu	526.4mg/kg	*Baccharis sarothroides*	shoots:1214.1mg/kg; roots:818.3mg/kg	Haque et al., 2008	
waste amended soil	Cu	245.87±4.33μg/g	*Typha angustata*	aboveground:1.73±0.04 μg/g; belowground:4.18±0.5μg/g	Bose et al., 2008	
old smeltery	Cu	3944.4mg/kg	*Ambrosia trifida*	shoots: 26.4mg/kg; roots:71.2 mg/kg	Cui et al., 2007	
old smeltery	Cu	1831.6mg/kg	*Solanum nigrum*	shoots: 16.8mg/kg ; roots:34.4 mg/kg	Cui et al., 2007	
old smeltery	Cu	788.4mg/kg	*Polygonum lapathifolium*	shoots: 14.3mg/kg; roots:19.0 mg/kg	Cui et al., 2007	
old smeltery	Cu	2719.6mg/kg	*Physalis angulata*	shoots: 47.9mg/kg ; roots:49.5 mg/kg	Cui et al., 2007	
old smeltery	Cu	12531mg/kg	*Helianthus tuberosus*	shoots: 21.1mg/kg; roots:191.3 mg/kg	Cui et al., 2007	
old smeltery	Cu	895.8mg/kg	*Conyza canadensis*	shoots: 10.2mg/kg ; roots:18.0 mg/kg	Cui et al., 2007	
old smeltery	Cu	711.5mg/kg	*Abutilon theophrasti*	shoots: 37.2mg/kg ; roots:32.5 mg/kg	Cui et al., 2007	
old smeltery	Cu	376.4mg/kg	*Chenopodium acuminatum*	shoots: 28.1mg/kg; roots:27.7 mg/kg	Cui et al., 2007	
mining areas	Cu	total: 3464 mg/kg	*Athyrium yokoscense*	roots: 375, 2040 and 1165 mg /kg	Van et al., 2006	T
mercury-contaminated soil	Hg	500ppm	*Pteris vittata*	shoots: 540 mg/kg	Su et al., 2009	no visual stress symptoms
mercury-contaminated soil	Hg	1000 ppm	*Pteris vittata*	shoots: 1469 mg/kg	Su et al., 2009	no visual stress symptoms

Table 1. (Continued)

Experiment types	Metal names	Treat (or soil) concentration	Plant names	Metal conten in plants	References	Reaction to pollution
waste amended soil	Mn	1062±18.2μg/g	*Typha angustata*	belowground:119.21±5.8μg/g; aboveground:43.86±3.2 μg/g	Bose et al., 2008	
mine tailings	Ni	39.7mg/kg	*Baccharis sarothroides*	shoots:30.9mg/kg ; roots:96.8mg/kg	Haque et al., 2008	
waste amended soil	Ni	201.22±3.77μg/g	*Typha angustata*	aboveground:1.8±0.04 μg/g; belowground:5.3±0.8μg/g	Bose et al., 2008	
a field survey	Ni	total:58.9 mu g//g; available:0.143 mu g/g	*Teucrium polium*	shoots:13.21 mu g/g	Sinegani and Dastjerdi, 2009	
a field survey	Ni	total:58.9 mu g/g; available:0.143 mu g/g	*Alyssum bracteatum*	shoots:10.98 mu g/g	Sinegani and Dastjerdi, 2009	
a field survey	Ni	total:58.9 mu g/g; available:0.143 mu g/g	*Ebenus stellata*	shoots:8.84 mu g/g	Sinegani and Dastjerdi, 2009	
mine tailings	Pb	207.4mg/kg	*Baccharis sarothroides*	shoots:107.3mg/kg; roots:151.9mg/kg	Haque et al., 2008	
waste amended soil	Pb	123.63±3.25μg/g	*Typha angustata*	aboveground:2.75±0.03 μg/g; belowground:7.21±0.9μg/g	Bose et al., 2008	
Pb-polluted soil	Pb	Extractable for: HCl: 6,643 ± 169 mg/kg; CH_3COONH_4: 832 ± 18 mg/kg; Water:0.679 ± 0.012mg/kg	*Fagopyrum esculentum*	leaves:8,000 mg/kg DW; stems:2,000 mg/kg DW; roots:3,300 mg/kg DW	Tamura et al., 2005	without significant damage
hydroponic method	Pb	500 mg/kg DW substrate	*maize* cv. *Ranchero*	shoots:18 753 mg/kg	Hernández-Allica et al., 2008	grow affected

Experiment types	Metal names	Treat (or soil) concentration	Plant names	Metal conten in plants	References	Reaction to pollution
Soil	Pb	Total :10 600±800 mg/kg	cabbage	shoots: up to 5010 and4620 mg/kg DW	Shen et al., 2002	
old smeltery	Pb	9385.1mg/kg	*Ambrosia trifida*	shoots: 111.9mg/kg ; root:516.5 mg/kg	Cui et al., 2007	
old smeltery	Pb	7121.3mg/kg	*Solanum nigrum*	shoots: 87.4mg/kg ; roots:183.8 mg/kg	Cui et al., 2007	
old smeltery	Pb	4828.9mg/kg	*Polygonum lapathifolium*	shoots: 56.3mg/kg; roots:64.8 mg/kg	Cui et al., 2007	
old smeltery	Pb	3705.8mg/kg	*Physalis angulata*	shoots:331.3mg/kg; roots:527.3 mg/kg	Cui et al., 2007	
old smeltery	Pb	3044mg/kg	*Helianthus tuberosus*	shoots: 54.2mg/kg; roots:490.3 mg/kg	Cui et al., 2007	
old smeltery	Pb	2069.8mg/kg	*Conyza canadensis*	shoots: 18.7mg/kg; roots:43.8 mg/kg	Cui et al., 2007	
old smeltery	Pb	1004.3mg/kg	*Abutilon theophrasti*	shoots: 61.4mg/kg; roots:38.7 mg/kg	Cui et al., 2007	
old smeltery	Pb	1004.3mg/kg	*Chenopodium acuminatum*	shoots: 97.9mg/kg; roots:101.9 mg/kg	Cui et al., 2007	
lead–zinc mining area	Pb	4455.6mg/kg	*Crisium chlorolepis*	shoots: 1198.8 mg/kg; roots: 629.3mg/kg	Zu et al., 2005	
lead–zinc mining area	Pb	6204.6mg/kg	*Taraxacum mongolicum*	shoots: 1065.0 mg/kg; roots: 1016.4mg/kg	Zu et al., 2005	
mining area	Pb	total: 343 mg/kg	*Athyrium yokoscense*	roots: 375, 2040 and 1165 mg/kg	Van et al., 2006	T
field experiment	Se	2843–4345 μg/kg	*Brassica napus*	leaves:157-209 mg/kg; grains:64-201 mg/kg; stems:42-93 mg/kg	Dhillon and Dhillon, 2009	
semi-arid Keban mining area	Strontium	398mg/kg	*Euphorbia macroclada*	shoots: 453 mg/kg; roots :243 mg/kg	Sasmaz et al., 2009	

Table 1. (Continued)

Experiment types	Metal names	Treat (or soil) concentration	Plant names	Metal conten in plants	References	Reaction to pollution
semi-arid Keban mining area	Strontium	398mg/kg	*Verbascum cheiranthifolium*	149and106mg/kg	Sasmaz et al., 2009	
semi-arid Keban mining area	Strontium	469mg/kg	*Astragalus gummifer*	278and 223 mg/kg	Sasmaz et al., 2009	
old silver mine	Zn	up to 124.9μg/g	*Evodiopanax innovans*	leaves: up to 1040μg/g	Takenaka et al., 2009	T
field investigation	Zn	NA	*Thlaspi caerulescens*	shoots:30 000 ppm	Milner and Kochian, 2008	T
mine tailings	Zn	51.7mg/kg	*Baccharis sarothroides*	shoots:55.2mg/kg; roots:40.1mg/kg	Haque et al., 2008	
metal-contaminated sites	Zn	total:15.0-61 600 mg/kg labile:0.06-119 mg/kg	*Salix caprea*	leaves: 4680 mg / kg	Unterbrunner et al., 2007	
mine tailings	Zn	14,500 mg/kg	*Salix borealis*	1130 mg /kg	Robinson et al. , 2000	
waste amended soil	Zn	320.42±7.43μg/g	*Typha angustata*	aboveground:2.11±0.07 μg/g; belowground:5.72±1.1μg/g	Bose et al., 2008	
spill-affected sites	Zn	457 mg/kg	*Populus alba*	leaves: up to 410 mg/ kg	Domínguez et al., 2008	
land fill site	Zn	25-1785 mg /kg DW	*hybrid poplar*	574–1520mg /kg DW	Vandecasteele et al., 2008	
hydroponic method	Zn	100 mg Zn/kg DW substrate	*rapeseed* cv. *Karat*	shoots: 10 916 mg/kg	Hernández-Allica et al., 2008	T
field trial	Zn	Total:1,050±89; Extractable:93±7.8mg /kg	*Sedum alfredii*	shoot:6,279 mg /kg DW; total:32.7 kg/ha	Zhuang et al., 2007	
old smeltery	Zn	4775.7mg/kg	*Ambrosia trifida*	shoots: 264.6mg/kg; roots:103.6 mg/kg	Cui et al., 2007	

Experiment types	Metal names	Treat (or soil) concentration	Plant names	Metal conten in plants	References	Reaction to pollution
old smeltery	Zn	1785.9mg/kg	*Solanum nigrum*	shoots: 94.7mg/kg ; roots:50.4 mg/kg	Cui et al., 2007	
old smeltery	Zn	1434.2mg/kg	*Polygonum lapathifolium*	shoots: 168.5mg/kg; roots:69.9 mg/kg	Cui et al., 2007	
old smeltery	Zn	4116mg/kg	*Physalis angulata*	shoots: 180.8mg/kg; roots:310.5 mg/kg	Cui et al., 2007	
old smeltery	Zn	9161.7mg/kg	*Helianthus tuberosus*	shoots: 205.9mg/kg; roots: 200.4 mg/kg	Cui et al., 2007	
old smeltery	Zn	736.8mg/kg	*Conyza canadensis*	shoots: 74.4mg/kg; roots:20.9 mg/kg	Cui et al., 2007	
old smeltery	Zn	1234.2mg/kg	*Abutilon theophrasti*	shoots: 158.2mg/kg; roots:56.1 mg/kg	Cui et al., 2007	
old smeltery	Zn	900mg/kg	*Chenopodium acuminatum*	shoots: 154.9mg/kg; roots:104.7 mg/kg	Cui et al., 2007	
lead–zinc mining area	Zn	8755.3mg/kg	*Incarvillea sp.*	shoots: 7004.3 mg/kg; roots: 6050.1mg/kg	Zu et al., 2005	
lead–zinc mining area	Zn	9166.7mg/kg	*Corydalis pterygopetala*	shoots: 5959.9 mg/kg; roots: 5402.3mg/kg	Zu et al., 2005	
lead–zinc mining area	Zn	11630.9 mg/kg	*Arabis alpinal var. parviflora*	shoots:5256.5 mg/kg; roots: 4075.1 mg/kg	Zu et al., 2005	
lead–zinc mining area	Zn	13031.5mg/kg	*Arabis alpinal var. parviflora*	shoots: 5632.8 mg/kg roots: 4508.7mg/kg	Zu et al., 2005	
a field survey	Zn	Mean total:259.7 mu g/g; available: 5.067 mu g/g	*Stachys inflate*	shoots:556.88 mu g/g	Sinegani and Dastjerdi, 2009	
a field survey	Zn	total:259.7 mu g/g; available: 5.067 mu g/g	*Ebenus stellata*	shoots:508.8 mu g/g	Sinegani and Dastjerdi, 2009	
a field survey	Zn	total:259.7 mu g/g; available: 5.067 mu g/g	*Astragalus glaucanthus*	shoots:449.53 mu g/g	Sinegani and Dastjerdi, 2009	

Table 1. (Continued)

Experiment types	Metal names	Treat (or soil) concentration	Plant names	Metal conten in plants	References	Reaction to pollution
lead/Zn mine area	Zn	NA	*Sedum jinianum*	shoots:4165-8349 mg/kg DM	Xu et al., 2009	
pot soil	Zn	619 mg/kg	*Sedum jinianum*	shoots:1560 mg/kg DM	Xu et al., 2009	
pot soil	Zn	4082mg/kg	*Sedum jinianum*	shoots: 15,558 mg/kg DM	Xu et al., 2009	
field trial	Zn	980 mg/kg	*Sorghum bicolor*	2000 g/ha	Marchiol et al., 2007	
field trial	Zn	980 mg/kg	*Helianthus annuus*	1000 g/ha	Marchiol et al., 2007	
abandoned mining sit	Zn	total: 2771 mg /kg; Water-soluble: 48.1 mg /kg; exchangeable:357 mg /kg	*Athyrium yokoscense*	shoots : 4.0 g/kg	Chen et al., 2009	
abandoned mining site	Zn	total: 2771 mg /kg; water-soluble: Zn 48.1 mg /kg;	*Arabis flagellosa*	shoots : 24 g/kg	Chen et al., 2009	
mining areas	Zn	total: 2422 mg/kg	*Athyrium yokoscense*	roots: 375, 2040 and1165 mg/kg	Van et al., 2006	T

Arsenic: Fern is a popular group to investigate arsenic accumulation. For instance, Australian native fern *Pteris umbrosa* have shown to be arsenic hyperaccumulators (Koller et al. 2007; Zhao et al. 2002), perennial Chinese Brake Fern *Pteris vittata* can accumulate up to 2.3% arsenic when soil arsenic content was up to 1500 mg/kg, possibly remediating arsenic-contaminated soil (Ma et al., 2001). *Calamagrostis arundinacea* also accumulated high amount of arsenic in roots and shoots when growing in contaminated arsenic soil (Zabłudowska et al., 2009). *Pteris multifida* might accumulate the maximum arsenic in fronds when the soil-available phosphorus was 55 mg/kg and the soil pH was 8.34 (Du et al., 2005). *Pteris vittata* was more efficient in accumulating arsenic, and *Pteris umbrosa* can translocate arsenic to the fronds efficiently (Koller et al., 2008). In addition, although some aquatic species contain arsenic less than 1000 mg/kg, the threshold for arsenic hyperaccumulation, e.g. *Rorippa nasturtium-aquaticum* and *Mentha* sp. holding respective arsenic concentrations of up to 138 and 88 mg/kg FW, they still present a human health risk when inputting into food chain (Robinson et al., 2006).

Lead: Lots of plants could remove lead from the contaminated soil. Generally, sunflower was suited to phytoremediation of moderately Pb-contaminated soil by phytoextraction (Lin et al., 2009). Buckwheat (Tamura et al., 2005), *Ricinus communis*, *Tephrosia candida* and *Debregeasia orientalis* (Liu et al., 2008a) were reported to have a great potential to remediate different level of Pb-contaminated soils. In field trial, radish (*Raphanus sativus*) is ideal for remediating lead-polluted topsoils through accumulating most extracted lead in the roots (208.1 mg/kg) and shoots (27.25 mg/kg), with the yields up to 20 t/ha if seeding and havest reach up to five times a year (Kapourchal et al., 2009).

Zinc: Zinc can be accumulated in different parts of the plants. *Festuca arundinacea* predominantly absorbs zinc in its root tissues, whereas Indian mustard *Brassica juncea* contained higher concentrations in shoots (Batty and Anslow, 2008). *Lupinus albus* cv. *Multolupa* (Lupin) can be considered as potential phytoremediator and promising representative for the revegetation of degraded landfill areas with slightly acid or neutral soils polluted with zinc (Pastor et al., 2003). Sometime the potential to extract zinc and other metals from contaminated soils might be limited by nutrient disorders and toxic effects of zinc that suppress plant growth (Hamlin and Barker, 2006).

Other metals: Phytoremediation studies have been also processed for many other kinds of metals contaminants in soil. As an example, sunflower (*Helianthus annuus*) is potential to accumulate copper without being overly sensitive to its toxicity (Lin et al., 2003). *Ocimum tenuiflorum* accumulated high amount of chromium in a concentration and duration dependent manner (Rai et al., 2004). The barley plants absorbed more nickel from the medium polluted soil than heavily one no matter what the chemical form of nickel was supplied (Molas and Baran, 2004). *Beta vulgaris* and plant species in Asteraceae could efficiently accumulate radiocesium (Tang and Willey, 2003), and Indian mallow directly uptake and phytotransform TNT (Chang et al., 2004). Moreover, *Phytolacca acinosa* and *Castanea henryi* can be used as potential species for phytoremediation of Mn-contaminated soil (Wang et al., 2009).

Co-existed metals: Generally, metal contaminants are co-existed in soil and some plants can remove one metal together with another one from soil, as shown in quantities of studies. Some plants can accumulate two metals, e. g. *Sedum jinianum*, could take a role of cadmium hyperaccumulator and also hold a high capacity to accumulate zinc in shoots (Xu et al., 2009). Hybrid poplar (Vandecasteele et al., 2008), *Pteris umbrosa* (Takenaka et al., 2009) and

Picris divaricata (Tang et al., 2009a) could also accumulate very high cadmium and zinc at the same time. Instead, Irankoh indigenous plant species could accumulate large quantity of cadmium and lead (Sinegani and Dastjerdi, 2008). *Cynodon dactylon*, *Hirsfeldia incana*, *Malva nicaeensis* and *Silybum marianum* were efficient in removing lead or zinc (Río-Celestino et al., 2006), furthermore, *Diploschistes muscorum* (Sarret et al., 1998) and some ecotype of *Sedumal fredii* (Yang et al., 2004a) can hyperaccumulate the both. What is more, some plants can accumulate three or more kinds of metals at the same time, e. g. wild *Arabis paniculata* is a hyper-tolerant and well accumulated plant for lead (2300mg/kg DW), zinc (20,800 mg/kg) and cadmium (434mg/kg) (Tang et al., 2009). Spinach (*Spinacea oleracea*), cabbage (*Brassica oleracea*), a grass–legume mix (red fescue, *Festuca rubra*; ryegrass, *Lolium perenne*), and bean (*Vicia faba*) accumulated lead, cadmium and zinc to various degrees (Archer and Caldwell, 2004). Willow can even accumulate seven heavy metals of cadmium, cobalt, chromium, copper, nickel, lead and zinc at different capacity (Mleczek et al., 2009), and *Sorghum bicolor* absorb arsenic, cadmium, copper, lead and zinc (Fellet et al., 2007), *Athyrium yokoscense* extract arsenic, cadmium, lead and copper (Van et al., 2006, Table 1). *Pteridium aquilinum* As-2 (EU476184) could also accumulate several kinds of metals in contaminated areas (Chang et al., 2009). A natural perennial fern *Dicropteris dichotoma* could hyperaccumulate several light rare earth elements such as La, Ce, Pr and Nd (LREEs) in acidic soil (Shan et al., 2003). Vetiver can even phytoextract both heavy metals and organic wastes from soil. It accumulates up to 0.4% lead in shoot and 1% lead in root, up to 1% zinc in shoot and root, and simultaneously biodegrades organic wastes (2,4,6-trinitroluene, phenol, ethidium bromide, benzo[a]pyrene, atrazine) (Danh et al., 2009). After accumulation, some plants can efficiently transfer metals to above ground parts, e.g. salix and betula transfer zinc and cadmium to leaves and twigs (Rosselli et al., 2003). When treated with different concentrations of metals, plant might accumulate special metals, for example, when growing on copper, lead, manganese, and zinc contaminated soil, *Festuca rubra* (creeping red fescue) had the highest accumulation ability for copper, while *Helianthus annuus* (sunflower) did for lead and zinc, and *Poa pratensis (Kentucky bluegrass)* for manganese (Padmavathiamma and Li, 2009). In soil co-contaminated by cadmium, lead and 2,4,6-trinitrotoluene (TNT), barnyard grass (*Echinochloa crusgalli*), sunflower (*Helianthus annuus*), Indian mallow (*Abutilon avicennae*) and *Indian joinivetch* (*Aeschynomene indica*) likely and rapidly remove TNT (Lee et al., 2007).

In addition, only those metal-tolerant phytoaccumulators could be efficient for soil phytoremediation. As an example, *Poa pratensis*, *Gnaphalium afine* and *Pteris vittata* were lead-tolerant, *Conyza canadensis*, *Poe pmtensi* and *Gnaphalium afine* were cadmium-tolerant in mining site (Liu et al., 2006). Being heavy metal tolerant plant, *Brassica nigra* are well addressed in regions at low and moderate level of lead, zinc and cadmium (Angelova and Ivanov, 2009).

3.3. Phytostabilization

Phytostabilization technology takes advantage of plants to prevent metal migration and immobilize them in soil (Pivetz, 2001). It aims to reduce contaminant and leaching of the pollutants by eliminating or minimizing the mobile and bioavailable fractions of trace elements in the soil (Ruttens et al., 2006).

Before appearance of phytostabilization, chemical stabilization technique was used to reduce contaminants in polluted soil by adopting inexpensive amendments (Lee et al., 2009). Soil amendments material such as liming agents, phosphates and apatite, iron, aluminum and manganese oxyhydroxides, organic amendments, industrial waste products have been widely used in phytostabilisation experiments (Bernal et al., 2007).

As an eco-substitute of chemical amendment technique, phytostabilization usually adopts the high tolerant and low accumulating plants for toxic metals (Vangronsveld et al., 1995a). For instance, acid soil tolerant species such as *Cynodon dactylon* (couch), *Juncus usitatus* (common rush) and *Lomandra longifolia* (spinyheaded mat rush) were identified as the potential used in phytostabilisation programs (Archer and Caldwell, 2004). Metal tolerant species e.g. *Carduus pycnocephalus* (Asteraceae), *Dasypyrum villosum* (Poaceae), *Ferula communis* (Apiaceae), *Silybum marianum* (Asteraceae), *Sinapis arvensis* (Brassicaceae) and *Stipa austroitalica* (Poaceae) can most be considered as excluder to promote metals stabilization and soil conservation (Brunetti et al., 2009). Moreover, pioneer plants will be ideal species for the phytostabilization of mine tailings (Lei and Duan, 2008), e. g. *Atriplex halimus subsp. schweinfurthii* are potential for use in the phytostabilization of Cd-contaminated salt soils (Nedjimi and Daoud, 2009).

In practical application, a well-developed plant cover can protect the contaminated soil from leaching of toxic materials. For instance, Rhodes grass is well suited to the revegetation of low levels tailings–polluted soils (Keeling and Werren, 2005). *Leucaena leucocephala* can grow on poor soils with chromites overburdens so as to provide an efficient cover preventing toxic leaching (Rout et al., 1999). *Jatropha curcas* was another candidate to recover and reclamation metalloid and metal contaminated soil system (Yadav et al., 2009), and *Alnus rugosa* inoculated with Frankia was for tailings recovery and forest ecosystem reestablishment (Mehta, 2005). Ordinarily, reclamation strategy using tree species is more realistic, integrated, low-cost, ecologically sound and sustainable than using other plant types (Dickinson, 2000). Even taking both rules of accumulation and phytostabilization, the panicked golden rain tree *Koelreuteria paniculata* and the common elaeocarpus tree *Elaeocarpus decipens* could act as both accumulators in remediation and delegates improving scenery recovery in abandoned mining areas (Tian et al., 2009). Moreover, establishment of vegetation cover (Kidd et al., 2009) may be more appropriate for phytostabilization when combining with organic amendments (Pichtel and Bradway, 2008). For instance, red-mud addition can reduce lettuce's uptake of cadmium, lead, and zinc at 86%, 58%, and 73%, respectively (Lee et al., 2009).

4. Mechanism of Phytoremediation

Either enhancing (for phytoextraction) or reducing (for phytostabilisation) the bioavailability of metal contaminants in the rhizosphere could significantly improve the efficiency of remediation techniques (Kidd et al., 2009). Phytoremediation efficiency is affected by both soil and plants factors (Hooda, 2007). Soil acidification helps to raise metal accumulation in plants (Clemente et al., 2005) by increasing metal bioavailability and mobilization (Darling and Thomas, 2003). For instance, lower pH increases solubility of Cu^{2+} (Baker and Senft, 1995) and also favours cadmium accumulation (Kirkham, 2006). *Pteris multifida* might accumulate the maximum arsenic in fronds when the soil pH was 8.34 (Du et al., 2005). In addition, synthesis of phytochelatins, metallothioneins and enzymes involving in stress response (DalCorso et al., 2008) could be vital to prevent metal absorption of plants or to detoxify metal ions.

4.1. Chelants

Chelants technology usually assisted phytoextraction by involving the process of transferring metals from the bulk soil to the plant root surfaces, uptake into the roots and translocation to the shoots (Tandy et al., 2006). For instance, soil-solution zinc can be increased 97.69% by chelating/desorption (Collins et al., 2002).

As a common chelant, ethylenediaminetetraacetic acid (EDTA) was added to the soil and promoted metal bioavailability and then plants' accumulation. After EDTA addition, lead concentration was logically increased in jack beans (*Canavalia ensiformis*) (Gabos, 2009), in cabbage shoots (Shen et al., 2002), and in root and shoot tissues of vetiver grass (*Vetiveria zizanioides*) (Andra et al., 2009), with an increase up to 15-24 folds higher than control. Appropriate level of EDTA application can increase cadmium accumulation in rainbow pink (*Dianthus chinensis*) (Lai and Chen, 2005), and in stems and leaves of *Solanum nigrum* (Sun et al., 2009). Some heavy metal uptake was enhanced by 117% in root, 62% in stem, 86% in leaves of *Jatropha curcas* when EDTA was applied at 0.3 g/kg (Jamil et al., 2009).

Other chelants can also be assistants for metals' bioavailability and thus uptake by plants. This could be certified in quantities of examples. Organic manure increased copper bioavailability and uptake in spring wheat (*Triticum aestivum*) (Saifullah et al., 2009). Trisodium nitrilotriacetic acid (NTA) increased the lead uptake in roots and shoots of *Brassica Juncea* (Singh et al., 2009). S, S-ethylenediamine disuccinic acid (EDDS) was most effective in increasing uranium, lead and copper concentrations in shoots of Indian mustard (Duquènea, et al., 2009). The addition of KCl with adjustment of pH to 5.5 can enhance cadmium uptake by the waterlily *Nymphaea aurora* (Nymphaeaceae) (Schor-Fumbarov et al., 2003). EDTA and EDTA-IAA-KN (indole-3-acetic acid-kinetin) both significantly increased zinc and manganese in leaves of alfalfa (Lopez, 2009). Ethylenegluatarotriacetic acid (EGTA) and sodium dodecyl sulfate (SDS) could promote cadmium accumulation in shoots and roots of *Calendula officinalis* and *Althaea rosea* (Liu et al., 2008b).

Moreover, different species of chelants play various functions in phytoremediation of different metal contamination in soil. EDTA was more efficient than chlorides in solubilizing metals (especially lead) from the soil matrix (Komárek et al., 2008), for instance, being more efficient than propylenediaminetetraacetic acid (PDTA) for lead accumulation in green onion (*Allium fistulosum*) (Cho et al., 2009), ammonium nitrate and ammonium sulphate in promoting lead accumulation in sweet sorghum (Zhuang et al., 2009), tartrate and glutamate for metal lead bioavailability in soil (Doumett et al., 2008). EDDS mobilized more nickel and copper than EDTA, while EDTA mobilized more cadmium and lead than EDDS (Meers et al., 2005). Occasionally, combination of chelants might bring a synergistic effect, both extractable lead and zinc concentration could be increased greatly when EDTA and S were combined (Cui et al., 2004). Moreover, combination of manure, sulfuric acid and Diethylentriamene pentaacetate (DTPA) turned sunflower (*Helianthus annuus*) to be higher extracting potential for lead and zinc removal from polluted soil (Solhi et al., 2005).

Adverse to chelants application objective, strong solubilization of metals might not practically result in the increase of metal uptake by plants (Wu et al., 2006). Application of EDTA was most effective in enhancing lead uptake in cabbage shoots, meanwhile decreased the lead uptake in oat (*Avena sativa*), canola (*Brassica napus*) and barley (Shen et al., 2002). Chelant EDTA could help to increase zinc accumulation by *Brassica juncea* but not for oat or barley (Ebbs and Kochian, 1998). Even together with sulphuric acid and potassium chloride, EDTA could not significantly increase the uptake of cadmium by *Thlaspi caerulescens* (Maxted et al., 2007). Even though chelants increased metals solubility, no big raise of metal uptake was observed in some plants e. g. in the maize (*Zea mays*) shoots (Lombi et al., 2001). This might be resulted by the formation of less bioavailable chelates, as an example, chelates Cu–EDDS (S,S-ethylenediaminedisuccinic acid) might be a less bioavailable form to plants e.g. to garland chrysanthemum (*Chrysanthemum coronarium*) (WEI et al., 2007b). Sometimes, chelants could even result in the decrease of metal uptake and translocation by plants, e. g. presence of phosphates generally reduces the uptake of uranium and its translocation from the roots to the shoots in *Helianthus annuus* (Tomé et al., 2009).

4.2. Bio-Microbes

Plant-soil-animal interactions might influence metal mobility in soil (Ma et al., 2003), and relevant plant-microorganism associations are able to significantly enhance the extraction rate of metals by plants (Lebeau et al., 2008). In the process, the microbiota from the rhizosphere played an important role (Marques, 2009). This could be observed in some bacteria which helped plants to grow and accumulate heavy metals (Jiang et al., 2008). In a glasshouse trial, dip-site rhizosphere microbes could promote arsenic accumulation in grass *Agrostis tenuis* on contaminated dip-site soil (Chopra et al., 2007). Arbuscular mycorrhizal (AM) fungi are beneficial to increase arsenic accumulation in Chinese brake fern (*Pteris vittata*) (Agely et al., 2005), which easily grew in metal polluted soils with the AM help (Redon et al., 2008). In addition, earthworm increased soil bioavailable metal concentrations, as an example, diethylenetriamine-pentaacetic acid-Zn (DTPA-Zn) resulted in a direct increase of zinc uptake by plants (Wang et al., 2006b). Therefore the presence of earthworms and sequential plant harvesting, e. g. *Lolium multiflorum* and earthworms of *Pheretima* sp., could be viable remediation strategy for Pb/Zn tailings (Cheng and Wong, 2008). In addition, the enzymatic processes in plants are involved in phytodegradation of pesticides (Chaudhry et al., 2002), and trichloroethylene (TCE)-degrading bacteria also assisted in degrading TCE (Weyens et al., 2009).

4.3. Genetic Procedure

Except chelate and rhyzospheric microbiota, plant genetic engineering might also increase phytoremediation efficiency (Marques, 2009), as an instrumental (Kotrba et al., 2009) and promising potentiality (Hooda, 2007). It takes advantage of transgenic technology for applying the ability of selected species and plant genotypes to absorb specific types of pollutants, in quantities many times exceeding their natural contents in plants (Raskin et al., 1997; Meagher, 2000; McGrath and Zhao, 2003; Deng et al., 2004b).

Genetic technology is to harbor the genes signaturing tolerance and hyperaccumulation from identified hyperaccumulator plant species into the transgenic plants (Shan and Nongkynrih, 2007). It is reported that hyperaccumulation capacities of plants, e.g. *Thlaspi caerulescens*, can be significantly distinguished based on molecular data (Basic et al., 2006), and most important genes were found to be closely correlated to hyperaccumulation and tolerance of plant species (Maestri et al., 2010). For example, phytoremediation related genes have been identified and characterized in *Thlaspi caerulescens* (Milner and Kochian, 2008), *Arabidopsis thaliana*, rice, poplar and *Chlamydomonas reinhardtii* (Memon and Schröder, 2009), et al. Furthermore, a number of transgenic plants have been generated in an attempt to modify the tolerance, uptake or homeostasis of trace elements (Krämer and Chardonnens, 2001). This technology has been provided with some pleasant results. For instance, transgenic Indian mustard (*Brassica juncea*) could remove 6% zinc and 25% cadmium from the metal polluted soil (Bennett et al., 2003b). With inoculation of nickel mobilizing strains, *Psychrobacter* sp. SRA1 and *Bacillus cereus* SRA10 could increase the efficiency of nickel phytoextraction (Ma et al., 2009b).

Further studies have been focused on the mechanism of the transgenic phytoaccumulation. Previous studies showed that transgenic plants with the substantial improvement in selenium accumulation overexpressed both sulfurylase (APS) and selenocysteine methyltransferase (SMT) (LeDuc et al., 2006). Similar features were also observed in three transgenic Indian mustard (*Brassica juncea*) with high Se-accumulating ability. They overexpressed enzymes adenosine triphosphate sulfurylase (APS), ç-glutamyl-cysteine synthetase (ECS), and glutathione synthetase (GS) (Bañuelos et al., 2005), or selenocysteinelyase (cpSL) and selenocysteine methyltransferase (SMT) (Bañuelos et al., 2007).

5. Phytoremediation Prospects

As an aesthetically pleasing, solar energy driven and cost effective technology, phytoremediation has not become a commercially available technology in the field, except alone for low-level contamination or as a final polishing step following the pretreatment of the high concentrations of contaminants (Yang, 2008). The phyteremediation technology is either hindered by the slow growth rates of metal hyperaccumulators, or the limited solubility of metals in soils (Baker and Brooks, 1989; Prasad, 2003). Up till now, limited knowledge was obtained on mechanism of metal translocation and accumulation in plants involving the interactions between rhizosphere and plants (Hooda, 2007). Phytoremediation technology could also bring adverse effects to the environment in the process of improving the metal removal when combining with other technology.

In the first point, when chelants were used to increase soil soluble metal levels as to help their uptake in plants, the chelant bound metal (e.g., Pb) might be less toxic than free metal, and hence might induce less stress on plants (Andra et al., 2009) and on the environment (Shibata et al., 2007). Whereas, the metal-solubilising effect of chelant e.g., EDTA, was shorter-lived in the less contaminated and more highly calcareous soil (Walker et al., 2003). Even if toxic availability is decreased by adding amendments, it is still in the soil and keeping potential hazards (Kirkham, 2006). Furthermore, chelants addition technique might create water pollution problems (Lombi et al., 2001) due to the serious leaching of the metals and chelants themselves (Romkens et al., 2001). Phytoremediation is usually limited to contaminated topsoil where plants can grow (Qishlaqi, 2009), with a slow process around 5 to 15 years and shutdown in winter due to unavailable growth.

Secondly, when improving the possibility of removing metal contaminants from soil by plants, genetic engineering might also generate ecosystem risk.

In addition, phytoremediation has to be considered in two ways including phytoaccumulation removing contaminants from soil and identifying crop varieties that do not accumulate toxic metal in grains or fruits (Qishlaqi, 2009). For instance, Indian mustard plants (*Brassica nigra*) are well adopted to tolerate and accumulate high quantities of trace elements; plants containing high trace element content are not fit for consumption by human beings and animals (Bharagava et al., 2008).

ACKNOWLEDGMENTS

Thanks for the support of China Scholarship Council (2008106667), Knowledge Innovation Program of Beijing Academy of Science and Technology (IG200807C1) and National Natural Science Funds (Grant No. 40875082 and No. 40745027).

REFERENCES

Abernathy, CO; Liu, Y; Longfellow, D; et al. Arsenic: health effects, mechanisms of actions, and research issues. *Environ Health Perspect*, 1999. 107, 593-595.

Agely, AA; Sylvia, DM; Ma, LQ. Mycorrhizae Increase Arsenic uptake by the hyperaccumulator Chinese Brake Fern (Pteris vittata L.). *J Environ Qual*, 2005. 34, 2181-2186.

Agency for Toxic Substance and Disease Registry (ATSDR). *Toxicological profile: arsenic.* Department of Health and Human Services, Atlanta, GA. 2005.

Al-Chalabi, AS; Hawker, D. Distribution of vehicular lead in roadside soils of major road of Brisbane, Australia. *Water Air Soil Poll,* 2000. 118(3-4), 299-310.

Alvarez-Ayuso, E. Cadmium in soil-plant systems: an overview. *Int J Environ Poll*, 2008. 33 (2-3): 275-291.

Andra, SS; Datta, R; Sarkar, D; et al. Analysis of phytochelatin complexes in the lead tolerant vetiver grass [*Vetiveria zizanioides* (L.)] using liquid chromatography and mass spectrometry. *Environ Poll,* 2009. 157, 2173-2183.

Angelova, V; Ivanov, K. Bio accumulation and distribution of heavy metals in black mustard (*Brassica nigra* Koch). *Environ Monit Assess*, 2009. 153, 449-459.

Archer MJG; Caldwell, RA. Response of six Australian plant species to heavy metal contamination at an abandoned mine site. *Water Air Soil Poll*, 2004. 157, 257-267.

Baker, AJM; Brooks, RR. Terrestrial higher plants which hyperaccumulate metallic elements—a review of their distribution, ecology and phytochemistry. *Biorecovery*, 1989. 1, 81-126.

Baker, AJM; McGrath, SP; Sidoli, CMD; et al. The possibility of in situ heavy metal decontamination of polluted soils using crops of metal accumulating plants. *Resources, Conservation and Recycling*, 1994. 11, 41-49.

Baker, AJM; McGrath, SP; Reeves, R; et al. Metal hyperaccumulator plants: a review of the ecology and physiology of a biological resource for phytoremediation of metal-polluted soils. In: Terry, N., Banuelos, G. (Eds.), *Phytoremediation of Contaminated Soil and Water*. Lewis Publishers, Boca Raton, 2000. pp. 85-107.

Baker, DE; Senft, JP. Copper. In: Alloway, BJ (Ed.), *Heavy Metals in Soils*, second ed. Blackie Academic and Professional, London, 1995. pp. 179-205.

Bañuelos G, LeDuc DL, Pilon-Smits EA, Terry N. Transgenic Indian mustard overexpressing selenocysteine lyase or selenocysteine methyltransferase exhibit enhanced potential for selenium phytoremediation under field conditions. *Environ Sci Technol*. 2007. 41(2), 599-605.

Bañuelos, G; Terry, N; LeDuc, DL; et al. Field trial of transgenic Indian mustard plants shows enhanced phytoremediation of selenium-contaminated sediment. *Environ Sci Technol*, 2005. 39, 1771-1777.

Banuelos, GS; Ajwa, HA; Terry, N; et al. Phytoremediation of selenium laden soils: A new technology. *J Soil Water Conserv*, 1997. 52(6), 426-430.

Basic, N; Salamin, N; Keller, C; et al. Cadmium hyperaccumulation and genetic differentiation of *Thlaspi caerulescens* populations. *Biochem System Ecol*, 2006. 34, 667-677.

Batty, LC; Anslow, M. Effect of a polycyclic aromatic hydrocarbon on the phytoremediation of zinc by two plant species (*Brassica juncea* and *Festuca arundinacea*). *Int J Phytorem*, 2008. 10 (3), 236-251.

Bejerre, GK; Schierup, HH. Uptake of heavy metal by oat as influenced by soil type and addition of cadmium, lead, zinc and copper. *Plant Soil*, 1985. 88,: 57-69.

Bennett, LE; Burkhead, JL; Hale, KL; et al. Analysis of transgenic Indian mustard plants for phytoremediation of metal-contaminated mine tailings. *J Environ Qual*, 2003b. 32, 432-440.

Bennett, LE; Burkhead, JL; Hale, KL; et al. Analysis of transgenic Indian mustard plants for phytoremediation of metal contaminated, mine tailings. *J Environ Qual*, 2003a. 32, 432-440.

Bernal, MP; Clemente, R; Walker, DJ. The role of organic amendments in the bioremediation of heavy metal-polluted soils. In: Gore, R.W. (Ed.), *Environmental Research at the Leading Edge*. Nova Science Publishers, Inc., New York, 2007. pp. 1-57.

Berndes, G; Fredrikson, F; Börjesson, P. Cadmium accumulation and salix-based phytoextraction on arable land in Sweden. *Agri Ecosys Environt*, 2004. 103, 207-223.

Bharagava, RN; Chandra, R; Rai, V. Phytoextraction of trace elements and physiological changes in Indian mustard plants (*Brassica nigra* L.) grown in post methanated distillery effluent (PMDE) irrigated soil. *Biores Tech*, 2008. 99, 8316-8324.

Bhuiyana, MAH; Parvez, L; Islam, MA; et al. Heavy metal pollution of coal mine-affected agricultural soils in the northern part of Bangladesh. *J Hazard Mater*, 2010. 173, 384-392.

Bose, S; Vedamati, J; Rai, V; et al. Metal uptake and transport by *Tyaha angustata* L. grown on metal contaminated waste amended soil: An implication of phytoremediation. *Geoderma*, 2008. 145, 136-142.

Boyajian, GE; Carreira, LH. Phytoremediation: A clean transition from laboratory to marketplace? *Nat Biotechnol*, 1997. 15(2), 127-128.

Braanaas, T; Vettismorki, F. Fluoride damage of pine forest in Vettismorki. *Tidsskr Skogbruk*, 1970. 78, 379-401.

Bruins, MR; Kapil, S; Oehme, FW. Microbial resistance to metals in the environment. *Ecotoxicol Environ Safety*, 2000. 45,198-207.

Brunetti, G; Soler-Rovira, P; Farrag, K; et al. Tolerance and accumulation of heavy metals by wild plant species grown in contaminated soils in Apulia region, Southern Italy. *Plant Soil*, 2009. 318, 285–298.

Bungart, R; Hüttl, RF. Growth dynamics and biomass accumulation of 8-yearold hybrid poplar clones in a short-rotation plantation on a clayey-sandy mining substrate with respect to plant nutrition andwater budget. *Eur J Forest Res*, 2004. 123, 105–115.

Burk, RF. *Selenium in Biology and Human Health*. New York: Springer. 1994. 58-64.

Chaney, RL; Malik, M; Li, YM; et al. Phytoremediation of soil metals. *Environ Biotechnol*, 1997. 8, 279-284.

Chaney, RL; Toxic element accumulation in soils and crops: protecting soil fertility and agricultural food-chains. In: Bar-Yosef G, Barrow NH, Goldshmid J (eds) *Inorganic contaminants in wadose zone*. Springer, Berlin. 1989. pp 140-158.

Chang, J-S; Yoon, I-H; Kim, K-W. Heavy metal and arsenic accumulating fern species as potential ecological indicators in As-contaminated abandoned mines. *Ecol Indic*, 2009. 9, 1275-1279.

Chang, Y-Y; Kwon, Y-S; Kim, S-Y; et al. Enhanced degradation of 2,4,6-Trinitrotoluene (TNT) in a soil column planted with Indian mallow (*Abutilon avicennae*). *J Biosci Bioeng*, 2004. 97(2), 99-103.

Chaudhry, Q; Schröder, P; Werck-Reichhart, D; et al. Prospects and limitations of phytoremediation for the removal of persistent pesticides in the environment. *Environ Sci and Pollut Res*, 2002. 9 (1), 4- 17.

Chen, Z; Setagawa, M; Kang, Y; et al. Zinc and cadmium uptake from a metalliferous soil by a mixed culture of *Athyrium yokoscense* and *Arabis flagellosa*. *Soil Sci Plant Nutr*, 2009. 55, 315-324.

Cheng, S. Heavy metal pollution in China: origin, pattern and control. *Environ Sci Pollut Res*, 2003. 10 (3), 192 - 198.

Cheng, JM; Wong, MH. Effects of earthworm (*Pheretima* sp.) on three sequential ryegrass harvests for remediating Lead/Zinc mine tailings. *Int J Phytorem*, 2008. 10 (3), 173-184.

Chiron, S; Fernandez-Alba, A; Rodriguez, A; et al. Pesticide chemical oxidation: State-of-the-art. *Water Res* (Oxford) , 2000. 34, 366-377.

Cho, Y; Bolick, JA; Butcher, DJ. Phytoremediation of lead with green onions (*Allium fistulosum*) and uptake of arsenic compounds by moonlight ferns (*Pteris cretica* cv. *Mayii*). *Microchem J*, 2009. 91, 6-8.

Chopra, BK; Bhat, S; Mikheenko, IP; et al. The characteristics of rhizosphere microbes associated with plants in arsenic-contaminated soils from cattle dip sites. *Sci Total Environ*, 2007. 378, 331-342.

Clemente, R; Walker, DJ; Bernal, MP. Uptake of heavy metals and As by *Brassica juncea* grown in a contaminated soil in Aznalcóllar (Spain): The effect of soil amendments. *Environ Poll*, 2005. 138, 46-58.

Collins, RN; Merrington, G; Mclaughlin, MJ; et al. Uptake of intact zinc-ethylenediaminetetraacetic acid from soil is dependent on plant species and complex concentration. *Environ Toxicol Chem*, 2002. 21(9), 1940-1945.

Comino, E; Fiorucci, A; Menegatti, S; et al. Preliminary test of arsenic and mercury uptake by *Poa annua*. *Ecol Eng*, 2009. 35, 343-350.

Cui, S; Zhou, Q; Chao, L. Potential hyperaccumulation of Pb, Zn, Cu and Cd in endurant plants distributed in an old smeltery, northeast China. *Environ Geol*, 2007. 51, 1043-1048.

Cui, Y; Wang, Q; Dong, Y; et al. Enhanced uptake of soil Pb and Zn by Indian mustard and winter wheat following combined soil application of elemental sulphur and EDTA. *Plant and Soil*, 2004. 261, 181-188.

Cunningham, SD; Berti, WR; Huang, JW. Phytoremediation of contaminated soils. *Trends Biotechnol*, 1995. 13, 393-397.

DalCorso, G; Farinati, S; Maistri, S; et al. How Plants Cope with Cadmium: Staking All on Metabolism and Gene Expression. *J Integr Plant Biol*, 2008. 50 (10), 1268-1280.

Danh, LT; Truon, P; Mammucari, R; et al. A choice plant for phytoremediation of heavy metals and organic wastes. *Int J Phytorem*, 2009. 11(8), 664-691.

Darling, CTR; Thomas, VG. The distribution of outdoor shooting ranges in Ontario and the potential for lead pollution of soil and water. *Sci Total Environ*, 2003. 313, 235-243.

Dàvila, B; Whitford, KW; Saylor, ES. *Technology alternatives for the remediation of PCB-contaminated soil and sediment.* US Environmental Protection Agency, EPA/540/S-93/506, 1993.

Deng, H; Ye, ZH; Wong, MH. Accumulation of lead, zinc, copper and cadmium by 12 wetland plant species thriving in metal-contaminated sites in China. *Environ Poll*, 2004a. 132, 29-40.

Deng, H; Ye, ZH; Wong, MH. Accumulation of lead, zinc, copper and cadmium by 12 wetland plant species thriving in metal-contaminated sites in China. *Environ Pollut*, 2004b. 132, 29-40.

Dermont, G; Bergeron, M; Mercier, G; et al. Metal-contaminated soils: remediation practices and treatment technologies. *Practice Peiodical of Hazardous, Toxic, and Radioactive Waste Management*, 2008.188-209.

Dhillon, SK; Dhillon, KS. Phytoremediation of selenium-contaminated soils: the efficiency of different cropping systems. *Soil Use and Management*, 2009. 25, 441-453

Dickinson, NM. Strategies for sustainable woodland on contaminated soils. *Chemosphere*, 2000. 41,259-263.

Domínguez, MT; Marañón, T; Murillo, JM; et al. Trace element accumulation in woody plants of the Guadiamar Valley, SW Spain: A large-scale phytomanagement case study. *Environ Poll*, 2008. 152, 50-59.

Dong, X; Li, C; Li, J; et al. A novel approach for soil contamination assessment from heavy metal pollution: A linkage between discharge and adsorption. *J Hazard Mater*, 2010. 175(1-3):1022-1030.

Doumett S; Lamperi, L; Checchini, L; et al. Heavy metal distribution between contaminated soil and *Paulownia tomentosa*, in a pilot-scale assisted phytoremediation study: Influence of different complexing agents. *Chemosphere*, 2008. 72, 1481-1490.

Doušova, B; Grygar, T; Martaus, A; et al. Sorption of As(V) on aluminosilicates treated with FeII nanoparticles. *J Colloid Interf Sci,* 2006. 302, 424-431.

Du, WB; Li, ZA; Zou, B; et al. Pteris multifida Poir., A new arsenic hyperaccumulator: characteristics and potential. *Int J Environ Poll*, 2005. 23 (4), 388-396.

Dudka, S; Miller, WP; 1999. Permissible concentrations of arsenic and lead in soils based on risk assessment. *Water Air Soil Pollut*. 113, 127-132.

Duquènea, L; Vandenhovea, H; Tackb, F; et al. Enhanced phytoextraction of uranium and selected heavy metals by Indian mustard and ryegrass using biodegradable soil amendments. *Sci Total Environ*, 2009. 407, 1496-1505.

Eapen, S; Singh, S; D'Souza, SF. Advances in development of transgenic plants for remediation of xenobiotic pollutants. *Biotechnol Adv*, 2007. 25, 442-451.

Ebbs, S; Kochian, L. Phytoextraction of Zinc by Oat (*Avena sativa*), Barley (*Hordeum vulgare*), and Indian Mustard (*Brassica juncea*). *Environ Sci Technol*, 1998. 32, 802-806.

Ensley, BD. Rationale for use of phytoremediation pp. 3-11. In: Raskin I, Ensley BD (eds) Phytoremediation of toxic metals: using plants to clean up the environment. *J. Wiley and Sons,* New York, USA, 2000. 304 pp.

EPA, 2003. The source of anomalous lead and arsenic concentrations in soils from the El paso community-El paso, Texas, *<http://www.epa.gov/earth1r6/6sf/pdffiles /finalreport.pdf>*.

Evangelou, MWH; Kutschinski-Klöss, S; Ebel, M; et al. Potential of *Borago officinalis*, *Sinapis alba* L. and *Phacelia boratus* for Phytoextraction of Cd and Pb from Soil. *Water Air Soil Pollut*, 2007. 182, 407-416.

Fakayode, SO; Olu-Owolabi, BI. Heavy metal contamination of roadside topsoil in Osogbo, Nigeria: Its relationship to traffic density and proximity to highways. *Environ Geol*, 2003. 44(2), 150-157.

Fellet, L; Marchiol, D; Perosa, GZ. The application of phytoremediation technology in a soil contaminated by pyrite cinders. *Ecol Eng*, 2007. 31, 207-214.

Fornes, F; García-de-la-Fuente, R; Belda, RM. Manuel Abad 'Alperujo' compost amendment of contaminated calcareous and acidic soils: Effects on growth and trace element uptake by five Brassica species. *Biores Tech*, 2009. 100, 3982-3990.

Franzius, V. Aktuelle Entwicklungen zur Altlastenproblematik in der Bundesrepublik Deutschland. *Umwelt Technologie Aktuell*, 1994. 6, 443-449.

Gabos, MB; de Abreu CA; Coscione AR. EDTA assisted phytorremediation of a Pb contaminated soil: metal leaching and uptake by Jack beans. *Scientia Agricola*, 2009. 66 (4), 506-514.

Gabrera, F; Clemente, L; Barrientos, ED; et al. Heavy metal pollution of soils affected by the Guadiamar toxic flood. *Sci Total Environ*, 1999. 242, 117-129.

Gan, S; Lau, EV; Ng, HK. Remediation of soils contaminated with polycyclic aromatic hydrocarbons (PAHs). *J Hazard Mater*, 2009. 172, 532-549.

Garcia, G; Zanuzzi, AL; Faz, A. Evaluation of heavy metal availability prior to an in situ soil phytoremediation program. *Biodegradation*, 2005. 16, 187-194.

Greger, M. *Metal availability, uptake, transport and accumulation in plants*. In M. N. V. Prasad (Ed.), Heavy metal stress in plants (pp. 1-27). Berlin: Springer. 1999.

Hafen, MR; Brinkman, R. Analysis of lead in soils adjacent to an interstate highway in Tampa, Florida. *Environ Geochem Health*, 1996. 18, 171-179.

Hamlin, RL; Barker, AV. Phytoextraction potential of Indian mustard at various levels of zinc exposure. *J Plant Nutri*, 2006. 29 (7), 1257-1272.

Han D; Zhang X; Tomar VVS; et al. Effects of heavy metal pollution of highway origin on soil nematode guilds in North Shenyang, China. *J Environ Sci,* 2009. 21, 193-198.

Haque, Z; Peralta-Videa, JR; Jones, GL; et al. Screening the phytoremediation potential of desert broom (*Baccharis sarothroides* Gray) growing on mine tailings in Arizona, USA. *Environ Poll,* 2008. 153, 362-368.

Harries, J. Acid Mine Drainage in Australia: Its Extent and Potential Future Liability, Supervising Scientist Report 125, Supervising Scientist, Canberra. 1997.

Harris, AT; Naidoo, K; Nokes, J; et al. Indicative assessment of the feasibility of Ni and Au phytomining in Australia. *J Clean Prod*, 2009. 17, 194-200.

Hayes, WJ; Chaudhry, TM; Buckney, RT; et al. Phytoaccumulation of trace metals at the Sunny Corner Mine, New South Wales, with suggestions for a possible remediation strategy. *Australas J Ecotoxicol.*, 2003. 9(1), 69-82.

Hernández-Allica, J; Becerril, JM; Garbisu, C. Assessment of the phytoextraction potential of high biomass crop plants. *Environ Poll*, 2008. 152, 32-40.

Hooda, V. Phytoremediation of toxic metals from soil and waste water. *J Environ Bio*, 2007. 28 (2), 367-376 Suppl.

Huang, L; Cai, LS; Jia, Y. The source and distribution of and precaution countermeasures for the harmful heavy metal in Chinese environment. *Sci-tech Inform Dev Econ*, 2007. 17, 189-191 (in Chinese)

Huang, S; Hua M; Feng, J; et al. Assessment of selenium pollution in agricultural soils in the Xuzhou District, Northwest Jiangsu, China. *J Environ Sci*, 2009. 21, 481-487.

Ibaraki T; Kuroyanagi, N; Murakami, M. Practical phytoextraction in cadmium-polluted paddy fields using a high cadmium accumulating rice plant cultured by early drainage of irrigation water. *Soil Sci Plant Nutr*, 2009. 55, 421-427.

Jaffre, T; Brooks, RR; Lee, J; et al. Sebertia acuminata: a nickelaccumulating plant from New Caledonia. *Science*, 1976. 193, 579-580.

Jamil, S; Abhilash, PC; Singh, N; et al. Jatropha curcas: A potential crop for phytoremediation of coal fly ash. *J Hazard Mat*, 2009. 172, 269-275

Jensen, JK; Holm, PE; Larsenc JNMB; et al. The potential of willow for remediation of heavy metal polluted calcareous urban soils. *Environ Poll*, 2009. 157, 931-937.

Jiang, C; Sheng, X; Qian, M; et al. Isolation and characterization of a heavy metal-resistant *Burkholderia* sp. from heavy metal-contaminated paddy field soil and its potential in promoting plant growth and heavy metal accumulation in metal-polluted soil. *Chemosphere*, 2008, 72, 157-164.

Jiang, W; Liu, D; Hou, W. Hyperaccumulation of cadmium by roots, bulbs and shoots of garlic (*Allium sativum* L.), *Biores Tech*, 2001. 76, 9-13.

Johnston, PA; Stringer, RL. Unwanted pesticides-Options for disposal. *Pestic Outlook*, 1992. 31, 33-37.

Kadi, MW. 2009. Soil pollution hazardous to environment: a case study on the chemical composition and correlation to automobile traffic of the roadside soil of Jeddah city, Saudi Arabia. *J Hazard Mat*, 2009. 168, 1280-1283.

Kapička, A; Petrovský, E; Ustjak, S; et al. Proxy mapping of fly-ash pollution of soils around a coal-burning power plant: A case study in the Czech Republic. *J Geochem Expl*, 1999. 66, 291-297.

Kapourchal, SA; Kapourchal, SA; Pazira, E; et al. Assessing radish (*Raphanus sativus* L.) potential for phytoremediation of lead-polluted soils resulting from air pollution. *Plant Soil Environ*, 2009. 55(5), 202-206.

Keeling, SM; Werren, G. Phytoremediation: The Uptake of Metals and Metalloids by Rhodes Grass Grown on Metal-Contaminated Soil. *Remediation J*, 2005. 15(2): 53-61.

Keller, C; Hammer, D; Kayser, A; et al. Root development and heavy metal phytoextraction efficiency: comparison of different plant species in the Weld. *Plant Soil*, 2003. 249, 67-81.

Kidd, P; Barcelo, J; Bernal, MP; et al. Trace element behaviour at the root-soil interface: Implications in phytoremediation. *Environ Exp Bot*, 2009. 67, 243-259.

Kirkham, MB. Cadmium in plants on polluted soils: Effects of soil factors, hyperaccumulation, and amendments. *Geoderma*, 2006. 137, 19–32.

Kocher, B; Wessolek, G; Stoffregen, H. Water and heavy metal transport in roadside soils. *Pedosphere*. 2005.15(6), 746–753.

Koller, CE; Patrick, JW; Rose, RJ; et al. Arsenic and heavy metal accumulation by *Pteris vittata* L. and *P. umbrosa* R. Br.. *Bull Environ Contam Toxicol*, 2008. 80, 128–133.

Koller, CE; Patrick, JW; Rose, RJ; et al. *Pteris umbrosa* R.Br. as an arsenic hyperaccumulator: Accumulation, partitioning and comparison with the established hyperaccumulator *Pteris vittata*. *Chemosphere*, 2007. 66, 1256-1263.

Komárek, M; Tlustoš, P; Száková, J; et al. The use of poplar during a two-year induced phytoextraction of metals from contaminated agricultural soils. *Environ Poll*, 2008. 151, 27-38.

Kotrba, P; Najmanova, J; Macek, T; et al. Genetically modified plants in phytoremediation of heavy metal and metalloid soil and sediment pollution. *Biotech Advances*, 2009. 27, 799-810.

Krämer, U. Phytoremediation: novel approaches to cleaning up polluted soils. *Curr Opin Biotechnol,* 2005. 16, 133-141.

Krämer, U; Chardonnens, AN. The use of transgenic plants in the bioremediation of soils contaminated with trace elements. *Appl Microbiol Biotechnol*, 2001. 55, 661-672.

Lai, H-Y; Chen, Z-S. The EDTA effect on phytoextraction of single and combined metals-contaminated soils using rainbow pink (*Dianthus chinensis*). *Chemosphere*, 2005. 60, 1062-1071.

Lebeau, T; Braud, A; Jézéquel, K. Performance of bioaugmentation-assisted phytoextraction applied to metal contaminated soils: A review. *Environ Poll*, 2008. 153, 497-522.

Lebrun, M. Phytoremediation of toxic metals using plants to clean up the environment. *Plant Sci*, 2001. 160, 1073-1075.

LeDuc, DL. Norman terry phytoremediation of toxic trace elements in soil and water. *J Ind Microbiol Biotechnol*, 2005. 32, 514-520.

LeDuc, DL; AbdelSamie, M; Móntes-Bayon, M; et al. Overexpressing both ATP sulfurylase and selenocysteine methyltransferase enhances selenium phytoremediation traits in Indian mustard. *Environ Poll*, 2006. 144, 70-76.

Lee, I; Baek, K; Kim, H; et al. Phytoremediation of soil co-contaminated with heavy metals and TNT using four plant species. *J Environ Sci Health A Tox Hazard Subst Environ Eng*, 2007. 42 (13), 2039-2045.

Lee, S-H; Lee, J-S; Choi, YJ; et al. In situ stabilization of cadmium-, lead-, and zinc-contaminated soil using various amendments. *Chemosphere*, 2009. 77, 1069-1075.

Lei D; Duan C. Restoration potential of pioneer plants growing on lead-zinc mine tailings in Lanping, southwest China. *J Environ Sci*, 2008. 20, 1202-1209.

Lewandowski, I; Schmidt, U; Londo, M; et al. The economic value of the phytoremediation function—assessed by the example of cadmium remediation by willow (*Salix* ssp). *Agric Syst*, 2006. 89 (1), 68–89.

Li, B; Lin, YS; Zhang, XF; et al. Heavy metal pollution of the soils and wheat grains alongside the Shanghai-Nanjing expressway. *Rural Eco-Environ*, 2005. 21(3), 50–53. (In Chinese)

Li, FR; Kang, LF; Gao, XQ; et al. Traffic-related heavy metal accumulation in soils and plants in northwest China. *Soil Sediment Contam*. 2007. 16(5), 473-484.

Li, JT; Liao, B; Dai, ZY; et al. Phytoextraction of Cd-contaminated soil by carambola (*Averrhoa carambola*) in field trials.*Chemosphere*, 2009. 76, 1233-1239.

Lin, C; Liu, J; Liu, L; et al. Soil amendment application frequency contributes to phytoextraction of lead by sunflower at different nutrient levels. *Environ Exp Bot*, 2009. 65, 410-416.

Lin, J; Jiang, W; Liu, D. Accumulation of copper by roots, hypocotyls, cotyledons and leaves of sunflower (*Helianthus annuus* L.). *Biores Tech*, 2003. 86, 151-155.

Lin, J; Qiu, QR; Chen, JA; et al. Assessment on pollution of heavy metals and metalloid in soil along road. *J Environ Health*, 2000. 17(5), 284-286. (in Chinese)

Liu, J-N; Zhou, Q-X; Sun T; et al. Identification and Chemical Enhancement of Two Ornamental Plants for Phytoremediation. *Bull Environ Contam Toxicol*, 2008b. 80, 260-265.

Liu, X; Gao, Y; Sardar K; et al. Accumulation of Pb, Cu, and Zn in native plants growing on contaminated sites and their potential accumulation capacity in Heqing, Yunnan. *J Environ Sci*, 2008a. 20, 1469-1474.

Liu, Y-G; Zhang, H-Z; Zeng G-M; et al. Heavy metal accumulation in plants on Mn mine tailings. *Pedosphere*, 2006. 16(1), 131-136.

Liu, Z; He, X; Chen, W; et al. Accumulation and tolerance characteristics of cadmium in a potential hyperaccumulator—Lonicera japonica Thunb. *J Hazard Mat*, 2009. 169, 170-175.

Lombi, E; Zhao, FJ; Dunham, SJ; et al. Phytoremediation of Heavy Metal-Contaminated Soils: Natural Hyperaccumulation versus Chemically Enhanced Phytoextraction . *J Environ Qual*, 2001. 30, 1919-1926.

Lopez, ML; Peralta-Videa, J; Parsons, J; et al. Effect of Indole-3-acetic acid, kinetin, and ethylenediaminetetraacetic acid on plant growth and uptake and translocation of lead, micronutrients, and macronutrients in alfalfa plants. *Int J Phytorem* , 2009. 11 (2), 131-149.

Lu, MC; Chen, N; Chang, CP. Oxidation of dichlorvos with hydrogen peroxide using ferrous ion as catalyst. *J Hazard Mater*, 1999. 65, 277-288.

Luo, YM. Current Research and Development in Soil Remediation Technologies. *Prog Chem*, 2009. 21 (2-3), 558-565.

Luster, J; Menon, M; Hermle, S; et al. Initial changes in refilled lysimeters built with metal polluted topsoil and acidic or calcareous subsoils as indicated by changes in drainage water composition. *Water Air Soil Pollut*, 2008. 8, 163-176.

Ma, J-H; Chu, C-J; Li, J; et al. Heavy metal pollution in soils on railroad side of Zhengzhou-Putian Section of Longxi-Haizhou railroad, China. *Pedosphere*, 2009a. 19(1), 121-128.

Ma, LQ; Komar, KM; Zhang, TC; et al. A fern that hyperaccumulates arsenic. *Nature*, 2001. 409, 579.

Ma, Y; Dickinson, NM; Wong, MH. Interactions between earthworms, trees, soil nutrition and metal mobility in amended Pb/Zn mine tailings from Guangdong, China. *Soil Biol Biochem*, 2003. 35, 1369-1379.

Ma, Y; Rajkumar, M; Freitas, H. Improvement of plant growth and nickel uptake by nickel resistant-plant-growth promoting bacteria. *J Hazard Mater*, 2009b. 166, 1154-1161.

Maestri, E; Marmiroli, M; Visioli, G; et al. Metal tolerance and hyperaccumulation: costs and trade-offs between traits and environment. *Environ Exp Bot*, 2010. 68, 1-13.

Marchiol, L; Fellet, G; Perosa, D; et al. Removal of trace metals by *Sorghum bicolor* and *Helianthus annuus* in a site polluted by industrial wastes: A field experience. *Plant Physiol Biochem*, 2007. 45, 379-387.

Marmiroli, N; Maestri, E. Trace elements contamination and availability: human health implications implications of food chain and biofortification. In: Prasad, M.N.V. (Ed.) *Trace Elements as Contaminants and Nutrients: Consequences in Ecosystems and Human Health*. John Wiley and Sons Inc., New York, 2008. pp. 23-53.

Marques, APGC; Rangel, AOSS; Castro, PML. Remediation of heavy metal contaminated soils: phytoremediation as a potentially promising clean-up technology. *Crit Rev Environ Sci Tech*, 2009. 39 (8), 622-654.

Marschner, H. *Mineral nutrition of higher Plants*, Ed. 2. Academic Press, San Diego. 1995.

Maxted, AP; Black, CR; West, HM; et al. Phytoextraction of cadmium and zinc from arable soils amended with sewage sludge using *Thlaspi caerulescens*: Development of a predictive model. *Environ Poll*, 2007. 150, 363-372.

McGrath, D. Organic micropollutant and trace element pollution of Irish soils. *Sci Total Environ*, 1995.164, 125-133.

McGrath, SP; Lombi, E; Gray, CW; Caille, N.; Dunham, SJ; Zhao, FJ. Field evaluation of Cd and Zn phytoextraction potential by the hyperaccumulators *Thlaspi caerulescens* and *Arabidopsis halleri*. *Environ Poll*, 2006. 141, 115-125.

McGrath, SP; Zhao, FJ. Phytoextraction of metals and metalloids from contaminated soils. *Curr. Opin. Biotechnol.* , 2003. 14, 277-282.

McGrath, SP; Zhao, FJ; Lombi, E. Phytoremediation of metals, metalloids and radionuclides. *Adv Agron,* 2002. 75, 1-56.

Meagher, RB. Phytoremediation of toxic elemental and organic pollutants. *Curr Opin Plant Biol*, 2000. 3, 153-162.

Meers, E; Ruttens, A; Hopgood M; et al. Potential of *Brassic rapa*, *Cannabis sativa*, *Helianthus annuus* and *Zea mays* for phytoextraction of heavy metals from calcareous dredged sediment derived soils. *Chemosphere*, 2005. 61, 561-572.

Mehta, P. Evaluating the potential of alder-Frankia symbionts for the remediation and revegetation of oil sands tailings. *Masters Abstracts International*, 2005. 45(04), 99.

Memon, AR; Schröder, P. Implications of metal accumulation mechanisms to phytoremediation. *Environ Sci Pollut Res*, 2009. 16, 162-175.

Milner, MJ; Kochian, LV. Investigating heavy-metal hyperaccumulation using *Thlaspi caerulescens* as a model system. *Ann Bot*, 102: 3-13, 2008.

Minister of the State Environmental Protection Administration(MSEPA), Conference on China's Soil Contamination Problem, 2006, *http://www.sepa.gov.cn/xcjy /zwhb/200607 /t20060718 78512.htm.*

Mizuno, T; Hirano, K; Hosono, A; et al. Continual pH lowering and manganese dioxide solubilization in the rhizosphere of the Mn-hyperaccumulator plant *Chengiopanax sciadophylloides*. *Soil Sci Pl Nutr*, 2006. 52, 726-733.

Mleczek, M; Łukaszewski, M; Kaczmarekb, Z; et al. Efficiency of selected heavy metals accumulation by *Salix viminalis* roots. *Environ Exp Bot*, 2009. 65, 48-53.

Molas, J; Baran, S. Relationship between the chemical form of nickel applied to the soil and its uptake and toxicity to barley plants (*Hordeum vulgare* L.). *Geoderma*, 2004. 122, 247-255.

Möller, A; Müller, HW; Abdullah, A; et al. Urban soil pollution in Damascus, Syria: concentrations and patterns of heavy metals in the soils of the *Damascus Ghouta*. *Geoderma*, 2005. 124, 63-71.

Mulligan, CN; Yong, RN; Gibbs, BF. Remediation technologies for metalcontaminated soils and groundwater: an evaluation. *Eng Geol*, 2001. 60, 193-207.

Murakami, M; Nakagawa, F; Ae, N; et al. Phytoextraction by Rice Capable of Accumulating Cd at High Levels: Reduction of Cd Content of Rice Grain. *Environ Sci Technol*, 2009. 43, 5878-5883.

Nedjimi, B; Daoud, Y. Cadmium accumulationin *Atriplex halimus* subsp. *schweinfurthii* and its influence on growth, proline, root hydraulic conductivity and nutrient uptake. *Flora*, 2009. 204, 316-324.

Nishiyama, Y; Yanai, J; Kosaki, T. Potential of *Thlaspi cuerulescens* for cadmium phytoremediation: comparison of two representative soil types in Japan under different planting frequencies. *Soil Sci. Plant Nutr*, 2005. 51 (6), 827 -834.

Ouzounidou, G; Symeonidis, L; Babalonas, D; et al. Comparative responses of a copper-tolerant and a copper-sensitive population of *Minuartia hirsuta* to copper toxicity. *J Plant Physiol*, 1994. 144, 109-115.

Padmavathiamma, PK; Li, LY. Phytoremediation of metal-contaminated soil in temperate humid regions of British Columbia, Canada. *Int J Phytorem*, 2009. 11 (6), 575-590.

Pastor, J; Hernández, AJ; Prieto, N; et al. Accumulating behaviour of *Lupinus albus* L. growing in a normal and a decalcified calcic luvisol polluted with Zn. *J Plant Physiol*, 2003. 160, 1457-1465.

Peer, WA; Baxter, IR; Richards, EL; et al. In: Tamas, M., Martinoia, E. (Eds.), Phytoremediation and hyperaccumulator plants. Topics in current genetics. *Mol Biol Metal Homeos Detoxification* , 2005.14, 299-340.

Peng, HY; Yang, XE; Jiang, LY; et al. Copper phytoavailability and uptake by *Elsholtzia splendens* from contaminated soil as affected by soil amendments. J Environ Sci Health *A Tox Hazard Subst Environ Eng*, 2005. 40 (4), 839-856.

Petrucci, F; Bocca, B; Alimonti, A; et al.Determination of Pd, Pt and Rh in airborne particulate and road dust by highresolution ICP-MS: a preliminary investigation of the emission from automotive catalysts in the urban area of Rome. *J Anal At Spectrosc*, 2000. 15, 525 -528.

Pichtel, J; Bradway, DJ. Conventional crops and organic amendments for Pb, Cd and Zn treatment at a severely contaminated site. *Biores Tech*, 2008. 99, 1242-1251.

Pivetz, BE. *Phytoremediation of contaminated soil and ground water at hazardous waste sites*. Ground Water Issue, EPA/540/S-01/ 500, USEPA, Office of Solid Waste and Emergency Response, Washington, D. C. 2001

Polizzotto, ML; Harvey, CF; Li, G; et al. Solid-phases and desorption processes of arsenic within Bangladesh sediments. *Chem Geo,* 2006. 228, 97-111.

Prasad, MNV. Phytoremediation of metal-polluted ecosystems: Hype for commercialization. *Russ J Plant Physiol*, 2003. 50(5), 686-700.

Pringle, JL; Rondinelli DA, Managing Asia's environmental crisis: EQM opportunities for remediation and technology transfer. *Environ Qual Manag*., 1998. 7(3), 41.

Pulford, ID; Watson, C. Phytoremediation of heavy metal-contaminated land by trees—a review. *Environ Int*, 2003. 29, 529-40.

Qishlaqi, A; Moore, F; Forghani, G. Characterization of metal pollution in soils under two landuse patterns in the Angouran region, NW Iran; a study based on multivariate data analysis. *J Hazard Mater*, 2009. 172, 374-384.

Qiu, RL; Fang, XH; Tang, YT; et al. Zinc hyperaccumulation and uptake by *Potentilla griffithii* hook. *Int J Phytorem*, 2006. 8, 299-310.

Rai, V; Vajpayee, P; Singh, SN; et al. Effect of chromium accumulation on photosynthetic pigments, oxidative stress defense system, nitrate reduction, proline level and eugenol content of *Ocimum tenuiflorum* L. *Plant Science*, 2004. 167, 1159-1169.

Ramakrishnaiah, H; Somashekar, RK. Heavy metal contamination in roadside soils and their mobility in relations to pH and organic carbon. *Soil Sediment Contam*, 2002. 11(5), 643-654.

Raskin, I; Smith, RD; Salt, DE. Phytoremediation of metals: using plants to remove pollutants from the environment. *Plant Biotechnol*, 1997. 8, 221-226.

Redon, P-O; Guiristain, TB; Leyval, C. Influence of Glomus intraradices on Cd partitioning in a pot experiment with *Medicago truncatula* in four contaminated soils. *Soil Biol Biochem*, 2008. 40, 2710-2712.

Reeves, RD; Baker, AJM. *Metal-accumulating plants*. In: Raskin, I., Ensley, B.D. (Eds.), Phytoremediation of toxic metals using plants to clean up the environment. John Wiley and Sons Inc., NY, 2000. pp. 193-230.

Río-Celestino, MD; Font, R; Moreno-Rojasc, R; et al. Uptake of lead and zinc by wild plants growing on contaminated soils. *Ind Crops Prod*, 2006. 24, 230-237.

Robinson, B; Kim, N; Marchetti, M; et al. Arsenic hyperaccumulation by aquatic macrophytes in the Taupo Volcanic Zone, New Zealand. *Environ Exp Bot*, 2006. 58, 206-215.

Robinson, BH; Bischofberger, S; Stoll, A; et al. Plant uptake of trace elements on a Swiss military shooting range: Uptake pathways and land management implications. *Environ Poll*, 2008. 153, 668-676.

Robinson, BH; Leblanc, M., Petit, D., et al. The potential of *Thlaspi caerulescens* for phytoremediation of contaminated soils. *Plant Soil*, 1998. 203, 47-56.

Robinson, BH; Mills, TM; Petit, D; et al. Natural and induced cadmium-accumulation in poplar and willow: Implications for phytoremediation. *Plant and Soil*, 2000. 227, 301-306.

Romkens, P; Japenga, J; Harmsen, J. Phytoremediation of soils: Evaluation of feasibility and risks related to chelate addition. *Phytoremmediation, Wetland and Sediments*, 2001. 6 (5), 137-144.

Roosens, N; Verbruggen, N; Meerts, P; Ximenez-Embun, P; Smith, J. Natural variation in cadmium tolerance and its relationship to metal hyperaccumulation for seven populations of *Thlaspi caerulescens* from Western Europe. *Plant Cell Environ*., 2003. 10, 1657-1672.

Rosselli, W; Keller, C; Boschi, K. Phytoextraction capacity of trees growing on a metal contaminated soil. *Plant and Soil*, 2003. 256, 265-272.

Rout, G.R., Samantaray, S., Das, P. Chromium, nickel and zinc tolerance in *Leucaena leucocepahla* (K8). *Sivae Genetica*, 1999. 48, 3-4.

Ruby, MV; Davis, A; Schoof, R; et al. Estimation of lead and arsenic bioavailability using a physiologically based extraction test. *Environ Sci Technol*, 1996. 30, 422-430.

Ruttens, A; Colpaert, JV; Mench, M; et al. Phytostabilization of a metal contaminated sandy soil. II. Influence of compost and/or inorganic metal immobilizing soil amendments on metal leaching. *Environ Pollut*, 2006. 144, 533-539.

Saifullah, GA; Qadir, M. Lead phytoextraction by wheat in response to the EDTA application method. *Inter J Phytorem*, 2009. 11 (3), 268-282.

Salt, DE; Smith, RD; Raskin, I. Phytoremediation. *Ann Rev Plant Physiol Plant Mol Biol*, 1998. 49, 643-648.

Sarret, G; Manceau, A; Cuny, D; et al. Mechanism of lichen resistance to metallic pollution. *Environ Sci Technol*, 1998. 32, 3325-3330.

Sasmaz, A; Sasmaz, M. The phytoremediation potential for strontium of indigenous plants growing in a mining area. *Environ Exp Bot*, 2009. 67, 139-144.

Schaaf, W; Wecker, B; Pan, T; et al. Changes in top soil properties of forest soils in north-eastern Germany due to long-term element accumulation. *Plant Soil*, 2004. 264, 85-95.

Schor-Fumbarov, T; Keilinl Z; Tel-Or, E. Characterization of cadmium uptake by the water lily *Nymphaea aurora*. *Int J Phytorem*, 2003. 5 (2), 169-179.

Schwitzguébel, J-P; Hype or Hope: The potential of phytoremediation as an emerging green technology. *Federal Facilities Environmental Journal*/Spring, 2002. 109

Shan, K; Nongkynrih, JM. Metal hyperaccumulation and bioremediation. *Biol Plantarum*, 2007. 51 (4), 618-634.

Shan, X; Wang, H; Zhang, S; et al. Accumulation and uptake of light rare earth elements in a hyperaccumulator *Dicropteris dichotoma*. *Plant Sci*, 2003. 165, 1343-1353.

Shen, Z-G; Li, X-D; Wang, C-C; et al. Lead Phytoextraction from Contaminated Soil with High-Biomass Plant Species. *J Environ Qual*, 2002. 31,1893-1900.

Sheoran V; Sheoran AS; Poonia, P. Phytomining: A review. *Minerals Engineering*, 2009. 22, 1007-1019.

Shi, G; Cai, Q. Cadmium tolerance and accumulation in eight potential energy crops. *Biotech Advances,* 2009. 27, 555-561.

Shibata, M; Konno, T; Akaike, R; et al. Phytoremediation of Pb contaminated soil with polymer-coated EDTA. *Plant Soil*, 2007. 290, 201-208.

Simón, L. Cadmium accumulation and distribution in sunflower plant. *J Plant Nutr*, 1998. 21, 341-345.

Simón, M; Ortiz, I; García, I; et al. Pollution of soils by the toxic spill of a pyrite mine (Aznalcollar, Spain). *Sci Total Environ*, 1999. 242, 105-115.

Sinegani, AAS; Dastjerdi, FS. The accumulation of zinc and nickel in Irankoh indigenous plant species on a contaminated land. *Soil Sedi Contam*, 2009. 18 (4), 525-534.

Sinegani, AAS; Dastjerdi, FS. The potential of Irankoh indigenous plant species for the phytoremediation of cadmium and lead contaminated land. *Soil Sedi Contam*. 2008. 17(2), 181-188.

Singh, OV; Labana, S; Pandey, G; et al. Phytoremediation: an overview of metallic ion decontamination from soil. *Appl Microbiol Biotechnol*, 2003. 61, 405-412.

Singh, SK; Ramprakash, SA; Phytoextraction of lead from Pb contaminated soil by *Brassica juncea* as affected by chelating agents. *Environ Ecol*, 2009. 27(2A), 971-974.

Solhi, M; Shareatmadari, H; Hajabbasi, MA. Lead and zinc extraction potential of two common crop plants, Helianthus annuus and Brassica napus. *Water Air Soil Poll*, 2005. 167, 59-71.

Solis-Dominguez, FA; Gonzalez-Chavez, MC; Carrillo-Gonzalez, R; et al. Accumulation and localization of cadmium in *Echinochloa polystachya* grown within a hydroponic system, *J Hazard Mater*, 2007. 141, 630-636.

Su, Y; Han, FX; Shiyab, S; et al. *Accumulation of mercury in selected plant species grown in soils contaminated with different mercury compounds*. ICEM2007: Proceedings of the 11th. International Conference on Environmental Remediation and Radioactive Waste Management, PTS A and B, 2009, pp. 1001-1007

Sun, Y; Zhou, Q; Wang, L; et al. The Influence of Different Growth Stages and Dosage of EDTA on Cd Uptake and Accumulation in Cd-Hyperaccumulator (*Solanum Nigrum* L.). *Bull Environ Contam Toxicol*, 2009. 82, 348-353.

Sutherland, RA; Tolosa, CA. Variation in total and extractable elements with distance from roads in an urban watershed, Honolulu, Hawaii. *Water Air Soil Poll*, 2001. 127(1-4), 315-338.

Takenaka, C; Kobayashi, M; Kanaya, S. Accumulation of cadmium and zinc in *Evodiopanax innovans*. *Environ Geochem Health*, 2009. 31, 609-615.

Tamura, H; Honda, M; Sato, T; et al. Pb hyperaccumulation and tolerance in common buckwheat (*Fagopyrum esculentum* Moench). *J Plant Res*, 2005. 118, 355–359.

Tandy, S; Schulin, R; Nowack, B. The influence of EDDS on the uptake of heavy metals in hydroponically grown sunflowers. *Chemosphere*, 2006. 62, 1454–1463.

Tang, S; Willey, NJ. Uptake of 134Cs by four species from the Asteraceae and two varieties from the Chenopodiaceae grown in two types of Chinese soil. *Plant and Soil*, 2003. 250, 75–81.

Tang, Y-T; Qiu, R-L; Zeng, X-W; et al. Lead, zinc, cadmium hyperaccumulation and growth stimulation in *Arabis paniculata* Franch. *Environ Exp Bot*, 2009b. 66, 126-134

Tang, YT; Qiu, RL; Zeng, XW; et al. Zn and Cd hyperaccumulating characteristics of *Picris divaricata* Vant. *Int J Environ Pollut*, 2009a. 38, 26-38.

Tian, D; Zhu, F; Yan, W; et al. Heavy metal accumulation by panicled goldenrain tree (*Koelreuteria paniculata*) and common elaeocarpus (*Elaeocarpus decipens*) in abandoned mine soils in southern China. *J Environ Sci*, 2009. 21, 340-345.

Tolra, R; Pongrac, P; Poschenrieder, C; et al. Distinctive effects of cadmium on glucosinolate profiles in Cd hyperaccumulator *Thlaspi praecox* and non-hyperaccumulator *Thlaspi arvense*. *Plant and Soil*, 2006. 288, 333-341.

Tomé, FV; Rodríguez, PB; Lozano, JC. The ability of *Helianthus annuus* L. and *Brassica juncea* to uptake and translocate natural uranium and 226Ra under different milieu conditions. *Chemosphere*, 2009. 74, 293-300.

Underwood, EJ. Selenium. In: *Trace elements in human and animal nutrition*. 4th Edn. Academic Press, New York. 1977. pp. 302-346.

Unterbrunner, R; Puschenreiter, M; Sommer, P; et al. Heavy metal accumulation in trees growing on contaminated sites in Central Europe. *Environ Poll*, 2007. 148, 107-114.

USEPA. *Contaminants and remedial options at selected metalcontaminated sites*. EPA/540/R-95/512, Office of Research and Development, Washington, D.C. 1995.

Van TK; Kang, Y; Fukui, T; et al. Arsenic and heavy metal accumulation by *Athyrium yokoscense* from contaminated soils. *Soil Sci Plant Nutr*, 2006. 52, 701-710.

Vandecasteele, B; Samyn, J; De Vos, B; et al. Effect of tree species choice and mineral capping in a woodland phytostabilisation system: A case-study for calcareous dredged sediment landfills with an oxidised topsoil. *Ecol Eng*, 2008. 32, 263-273.

Vangronsveld, J; Sterckx, J; Van Assche, F; et al. Rehabilitation studies on an old nonferrous waste dumping ground, effects of revegetation and metal immobilization by beringite. *J Geochem Explor,* 1995a. 52, 221-229.

Walker, DJ; Clemente, R; Roig, A; et al. The effects of soil amendments on heavy metal bioavailability in two contaminated Mediterranean soils. *Environ Poll*, 2003, 122, 303-312.

Wang, DQ; Liu, J; Zhang, AL; et al. Bioaccumulation of heavy metals by dominant plants growing in an abandoned manganese mine in Guangxi, South China. *Prog Environ Sci and Tech.* 2009. II(A), 1988-1991.

Wang, HB; Ye, ZH; Shu, WS; et al. Arsenic uptake and accumulation in fern species growing at arsenic-contaminated sites of southern China: Field surveys. *Int J Phytorem*, 2006a. 8 (1), 1-11.

Wang, JQ; Su, DC, Distribution of cadmium in oilseed rape and Indian mustard grown on cadmium contaminated soil. *J Environ Sci-CHINA*, 2005. 17 (4), 572-575.

Wang, L; Zhou, Q; Ding, L; et al. Effect of cadmium toxicity on nitrogen metabolism in leaves of *Solanum nigrum* L. as a newly found cadmium hyperaccumulator. *J Hazard Mater*, 2008. 154, 818-825.

Wang, M; Zou, J; Duan, X; et al. Cadmium accumulation and its effects on metal uptake in maize (*Zea mays* L.) . *Biores Tech*, 2007. 98, 82-88.

Wang, QR; Cui, YS; Liu, XM; et al. Soil contamination and plant uptake of heavy metals at polluted sites in China. *J Environ Sci Health A Tox Hazard Subst Environ Eng*, 2003. 38 (5), 823-838.

Wang, QYD; Cui, Y; Liu, X. Instances of soil and crop heavy metal contamination in China. *Soil Sedi Contam*, 2001. 10 (5): 497-510.

Wang. D; Li, H; Wei, Z; et al. Effect of earthworms on the phytoremediation of zinc-polluted soil by ryegrass and Indian mustard. *Biol Fertil Soils*, 2006b. 43, 120-123.

Wei L; Luo C; Wang C; et al. Birodegradable chelating agent ethylenediaminedisuccinic acid reduces uptake of copper through alleviation of copper toxicity in hydroponically grown Chrysanthemum coronarium L. *Environ Toxicol Chem*, 2007b. 26(4), 749-754.

Wei, C-Y; Chen, T-B. Arsenic accumulation by two brake ferns growing on an arsenic mine and their potential in phytoremediation. *Chemosphere*, 2006. 63, 1048-1053.

Wei, L; Shu, WS; Lan, CY. *Viola baoshanensis*, a plant that hyperaccumulates cadmium. *Chinese Sci Bull*, 2004. 49 (1): 29-32.

Wei, S; Zhou, Q; Saha, UK. Hyperaccumulative Characteristics of Weed Species to Heavy Metals. *Water Air Soil Pollut,* 2008. 192, 173-181.

Wei, Z; Xi, B; Zhao, Y; et al. Effect of inoculating microbes in municipal solid waste composting on characteristics of humic acid. *Chemosphere*, 2007a. 68, 368-374.

Welch, A.H., Westjohn, D.B., Helsel, D.R., Wanty, R.B., 2000. Arsenic in ground water of the United States: occurrence and geochemistry. *Ground Water* 38, 589-604.

Weyens, N; Taghavi S; Barac T; et al. Bacteria associated with oak and ash on a TCE-contaminated site: characterization of isolates with potential to avoid evapotranspiration of TCE. *Environ Sci Pollut Res*, 2009. 16, 830-843.

Wilden, R; Schaaf, W; Huttl, RF. Element budgets of two afforested mine sites after application of fertilizer and organic residues. *Ecol Eng*, 2001. 17, 253-273.

Williford, CW; Bricka, RM. Physical separation of metalcontaminated soils. Chapter 7, *Environmental restoration of metalscontaminated soils*, 1st Ed., I. K. Iskandar, ed., CRC Press LLC, Boca Raton, Fla. 2000. 121-165.

Wu, Q; Deng, J; Long X; et al. Christophe. Selection of appropriate organic additives for enhancing Zn and Cd phytoextraction by hyperaccumulators. *J Environ Sci*; 2006. 18(6), 1113-1118.

Xiong,YH; Yang, XE; Ye, ZQ; et al. Characteristics of cadmium uptake and accumulation by two contrasting ecotypes of *Sedum alfredii* Hance. *J Environ Sci Health*, 2004. 39, 2925-40.

Xu, XH; Shi, JY; Chen YX; et al. An investigation of cellular distribution of manganese in hyperaccumlator plant *Phytolacca acinosa* Roxb. using SRXRF analysis. *J Environ Sci-China*, 2006. 18(4), 746-751.

Xu; LS; Zhou, SB; Wu, LH; et al. Cd and Zn tolerance and accumulation by *Sedum jinianum* in East China. *Int J Phytorem*, 2009. 11(3), 283-295.

Xue SG; Chen, YX; Baker AJM; et al. Manganese uptake and accumulation by two populations of *Phytolacca acinosa* Roxb. (Phytolaccaceae). *Water Air Soil Poll*, 2005. 160, 3-14.

Yadav, SK; Juwarkar, AA; Kumar, GP; et al. Bioaccumulation and phyto-translocation of arsenic, chromium and zinc by *Jatropha curcas* L.: Impact of dairy sludge and biofertilizer. *Biores Tech*, 2009. 100, 4616-4622.

Yalcin, MG; Narin, I; Soylak, M. Heavy metal contents of the Karasu Creek Sediments. Nigde, Turkey. *Environ Monit Assess*, 2007. 128, 351-357.

Yamato, M; Yoshida, S; Iwase, K. Cadmium accumulation in *Crassocephalum crepidioides* (Benth.) *S. Moore* (Compositae) in heavy-metal polluted soils and Cd-added conditions in hydroponic and pot cultures. *Soil Sci Plant Nutr*, 2008. 54, 738-743.

Yang, L. Phytoremediation: An ecotechnology for treating contaminated sites. *Amer Soc Civ Eng*, 2008. 12(4), 290.

Yang, Q; Lan, C; Shu, W. Copper and zinc in a paddy field and their potential ecological impacts affected by wastewater from a lead/zinc mine, P. R. China. *Environ Monit Assess*, 2008. 147, 65-73.

Yang, QW; Shu, WS; Qiu JW; et al. Lead in paddy soils and rice plants and its potential health risk around Lechang Lead/Zinc Mine, Guangdong, China. *Environ Inter*, 2004a. 30, 883- 889

Yang, XE; Long, XX; Ye, HB; et al. Cadmium tolerance and hyperaccumulation in a new Zn-hyperaccumulating plantspecies (*Sedum alfredii* H). *Plant Soil*, 2004b. 259, 181-9.

Ye, B; Zhang Z; Mao T. Pollution sources identification of polycyclic aromatic hydrocarbons of soils in Tianjin area, China. *Chemosphere*, 2006. 64, 525-534.

Zabłudowska, E; Kowalska, J; Jedynak, Ł; et al. Search for a plant for phytoremediation - What can we learn from field and hydroponic studies? *Chemosphere*, 2009. 77, 301-307.

Zhang X-H; Liu, J; Huang, H-T; et al. Chromium accumulation by the hyperaccumulator plant *Leersia hexandra* Swartz. *Chemosphere*, 2007. 67, 1138-1143.

Zhang, S; Peng, P; Huang, W; et al. PCDD/PCDF pollution in soils and sediments from the Pearl River Delta of China. *Chemosphere*, 2009. 75, 1186-1195.

Zhao, FJ; Hamon, RE; Lombi, E; et al. Characteristics of cadmium uptake in two contrasting ecotypes of the hyperaccumulator *Thlaspi caerulescens*. *J Exp Bot*, 2002. 53, 535-43.

Zhao, FJ; Jiang, RF; Dunham, SJ; et al. Cadmium uptake, translocation and tolerance in the hyperaccumulator *Arabidopsis halleri*. *New Phytol*, 2006. 172, 646-654.

Zhou J-M; Dang Z; Cai M; et al. Soil Heavy Metal Pollution Around the Dabaoshan. Mine, Guangdong Province, China. *Pedosphere*, 2007. 17(5), 588-594.

Zhu, Y-G; Williams, PN; Meharg, AA. Exposure to inorganic arsenic from rice: a global issue? *Environ Poll*, 2008. 154, 169-171.

Zhuang, P; Shu W; Li, Z; et al. Removal of metals by sorghum plants from contaminated land. *J Environ Sci*, 2009. 21, 1432-1437.

Zhuang, P; Yang, QW; Wang, HB; et al. Phytoextraction of heavy metals by eight plant species in the field. *Water Air Soil Pollut*, 2007. 184, 235-242.

Zu, Y; Li, Y; Chen, J; et al. Hyperaccumulation of Pb, Zn and Cd in herbaceous grown on lead–zinc mining area in Yunnan, China. *Environ Inter*, 2005. 31, 755-762.

Zupancic, N. Lead contamination in the roadside soil of Siovenia. *Environ Geochem Health*, 1999. 2(1), 37-50.

In: Handbook of Phytoremediation
Editor: Ivan A. Golubev

ISBN: 978-1-61728-753-4

Chapter 7

PHYTOREMEDIATION: A PROMISING TECHNOLOGY OF BIOREMEDIATION FOR THE REMOVAL OF HEAVY METAL AND ORGANIC POLLUTANTS FROM THE SOIL

Sutapa Bose[1], Vivek Rai[2], Subarna Bhattacharya[3], Punarbasu Chaudhuri[4] and A. K. Bhattacharyya[5]

[1]Lamont-Doherty Earth Observatory, Columbia University, New York 10964, USA
[2] Columbia University Medical Center, New York10032, USA
[3]Department of Environmental Studies, Rabindra Bharati University, Kolkata, India
[4]Department of Environmental Science, University of Calcutta, Kolkata, India
[5]School of Environmental Sciences, Jawaharlal Nehru University, New Delhi, India

ABSTRACT

Organic and inorganic pollutants in the soil are one of the major environmental problems of present days. The traditional removal techniques do not provide any acceptable remedies for the removal of metal as well as organic pollutants from the system. Soil amendments are usually a cost effective and environment friendly technology. But disposal of solid wastes on land leads to contamination of both soil and groundwater. Bioremediation is an up coming environmental friendly technology that uses microbes as well as plants to clean up the toxic metals and other pollutants from the soil of the contaminated environment. The use of metal tolerant microbes and plants for the removal of toxic metal from the polluted system is a low cost technology. The specific microbial and plant species use to remove specific contaminants which have discussed in this paper. Metal-accumulating species can concentrate different metals up to 100 to 1000 times in their body which is very much species and site specific. The phytoremediation of heavy metals is divided into four sub-sections: (1) Phytoextraction: the use of plants to remove the toxic metals from the soil into the harvestable parts of plants, (2) phytofiltration: the use of plants root to accumulate the toxic metals from the water system (3) Phytostabilization: the use of metal tolerant plants to remove the bio-available toxic metal from the soil and (4) Phytovolatilazation: the use of plants to take up contaminants from the soil, transforming them into volatile form and transpiring them into the atmosphere. The harvestable parts like root and shoot, which are rich in metals, can easily be reclaimed and recycled after harvesting the plants from the contaminated site. Bioremediation technologies can be generally classified as *in situ* and *ex situ*. This paper reviews the mobility, bioavailability and responses of microbes and plants in presence of metals and other pollutants in the system. Bioremediation may be employed to bother specific contaminants such as degradation of chlorinated hydrocarbons, heavy metals, oil spills, crude oil, nitrate and sulfate by indigenous or exogenous bacteria as well as plants. In general, bioremediation is a very promising and emerging technology for the removal of different kind of pollutants from the soil and water, which can be, approaching commercialization for near future.

Keywords: Bioremediation, phytoremediation, heavy metals, soil pollution.

INTRODUCTION

A major environmental concern due to unsafe dispersal of industrial and urban wastes produced by human activities is the main cause of soil contamination. Controlled and uncontrolled disposal of waste, accidental and process spillage, mining and smelting of metalliferous ores, sewage and industrial sludge application to agricultural soils are responsible for the movement of contaminants into uncontaminated natural sites as dust or leachate and contribute towards contamination of our ecosystem. A wide range of inorganic and organic compounds cause pollution, include heavy metals, combustible and hazardous wastes, explosives and petroleum products. Major component of inorganic contaminates are heavy metals [1] they present a different problem than organic contaminants (Table 1). Microbial bioremediation has been somewhat successful for the degradation of certain organic contaminants, but is ineffective at addressing the challenge of toxic metal contamination, particularly in soil. Although organic molecules can be degraded, toxic metals can only be remediated by removal from soil.

Phytoremediation is defined as the use of green plants to remove pollutants from the environment or to render them harmless. The basic idea that plants can be used for environmental remediation is very old and cannot be traced to any particular source. It provides a critical summary of the present state of the art, with particular emphasis on phytoremediation of soil heavy metal contaminants.

Table 1. Soil concentration ranges and regulatory guidelines for some toxic metals

Metal	Soil concentration range[a] (mg kg^{-1})	Regulatory limits[b] (mg kg^{-1})
Pb	1.00-6,9000	600
Cd	0.10-345	100
Cr	0.05-3,950	100
Hg	<0.01-1,800	270
Zn	150.0-5,000	1,500

[a] [2], [b] [3] Nonresidential direct contact soil cleanup criteria (NJDEP, 1996).

2. Soil Pollution and Heavy Metals

Soil pollution has been continuously increasing as a result of industrial activities, and contamination of soil constitutes a severe environmental problem all over the world. Nowadays, people mostly think about the magnitude of the pollution in the soils calls for immediate action. Unfortunately, the enormous costs associated with the removal of pollutants from soils by means of traditional physicochemical methods have been encouraging companies to ignore the problem. The common approaches used to treat metal-polluted soils are fixation (chemical processing of the soil to immobilize the metals) and leaching (using acid solutions or proprietary leachants to desorb and leach metals from soil followed by the return of clean soil residue to the site) [4]. Another common technique for remediation of heavy metal-contaminated soil is 'soil washing'. The use of washing solutions containing synthetic agents, which may enhance the release of metals from the soil aggregates [5]. However, many of these agents have poor photo-, chemo-, and biodegradability, which may have a negative impact on the soil-living microorganisms and plant growth [6]. Apart from minimizing the impact of future incidents by means of controlling pollution input, it is imperative to organize innovative technologies which could economically remediate polluted soils [7].

The primary sources of metal pollution are the burning of fossil fuels, mining and smelting of metalliferous ores, downwash from power lines, municipal wastes, fertilizers, pesticides and sewage [8]. The danger of heavy metals is aggravated by their almost indefinite persistence in the environment. Although some metals are essential for life (i.e., they provide essential cofactors for metalloproteins and enzymes), at high concentrations they can act in a deleterious manner by blocking essential functional groups, displacing other metal ions, or modifying the active conformation of biological molecules [9]. In addition, they are toxic for both plants and microorganisms. In fact, many metals affect directly various physiological and biochemical processes causing reduction in growth, inhibition of photosynthesis and respiration, and degeneration of main cell organelles[10]. Some metals are accumulated in roots (especially Pb), probably due to some physiological barriers against metal transport to aerial parts, while others are easily transported in plants [11].

3. Microorganisms and Metal Contamination

Microbes can, in principle, solubilize metals, thereby increasing their bioavailability, or immobilize them and reduce their bioavailability. Nonradioactive As, Cd, Cu, Hg, Pb and Zn and radioactive Sr, Cs and U are the most environmentally important metallic pollutants. Although organic molecules can be degraded, toxic metals can only be remediated by removal from soil. Microorganisms can detoxify metals by valence transformation, extracellular chemical precipitation, or volatilization. In fact, some microorganisms can enzymatically reduce a variety of metals in metabolic processes [12]. Some bacteria gain energy for growth by coupling the oxidation of simple organic acids and alcohols, hydrogen, or aromatic compounds, to the reduction of Fe(III) or Mn(IV). The reduction of the toxic selenate and selenite to the insoluble and much less toxic elemental selenium may be exploited to enhance removal of these anions from contaminated sites [13, 14, 15,16, 17]. The more toxic form of chromium, Cr(VI), can also be detoxified by bacteria and its enzymatic mechanism responsible for the reduction of Cr(VI) to Cr(III) is currently being studied and may ultimately lead to a commercial bioremediation process [15, 18, 19, 20]. Microorganisms also contribute in the cycling of carbon as the change in carbon distribution would also affect the distribution of metals in the soil solution. In addition to this, microorganisms may have an indirect influence on metal mobility as they affect pH and the redox potential [21].

This reflects that microorganisms in general would have a potential to modify the chemical state and thereby the distribution and mobility of many metals in a soil environment. Three saprotrophic fungi (A. niger, Penicillium bilaiae, and a Penicillium sp.) were used for bioremediation of heavy metals (lead)-contaminated soils [22]. Khan et al. [23] suggested that they play a protective role, restricting the uptake of metals by plants by immobilizing the metals in the fungal tissue. However, there have also been reports of some plants accumulating higher metal concentrations, in some cases up to toxic levels, due to the presence of arbuscular mycorrhizae[24]. The presence of periphyton associated with P. australis in freshwater wetlands enhanced the ability of the reed to accumulate and retain metals [25].

4. Bioremediation

Bioremediation is a technique to use of living organisms and plant to reduce or transform the contaminants in less toxic form. The bioremediation strategy can be applied in situ or ex situ, depending on the site that they will be applied. In situ is the treatment done in the site of the contamination and ex situ, takes place the removal of soil or water to successive treatment [26, 27, 28]. The use of microbes in soil remediation processes of organic contaminants has been successful [29, 30, 31]. Bioremediation of metal contaminated soil has not been applied in a technical scale. There is an extensive range of techniques of bioremediation that have been developed in the last few years table 2.

Biosorption, the binding of metals to cell surfaces with the use of Trichoderma reesei was analysed to see the characteristics of adsorption and desorption of cadmium and copper ions [32]. Bioleaching of metals, another example of the use of micoorganisms in the decontamination of aqueous solutions and soils that consists of the recovery of metals by some microorganisms capable of dissolving them from the environment [33]. White et al., [34] showed that the metals (Cd, Co, Cr, Mn, Ni and Zn) were significantly leached from soil sulfur-oxidizing bacteria followed by precipitation of the leachate metals by sulfate-reducing bacteria.Kumar and Nagendran [35] reported that this methodology is useful for an efficient removal of heavy metals from soil and the soil can be further disposed off safely.

Thiobacillus is able to perform the microbiological leaching of metals such as copper, silver, uranium and zinc by the components oxidation, causing the release of protons that can replace the adsorbed metals to soil particles, or else, the oxidation followed by electron transfer to oxygen inducing the solubilization of metals [36]. The bioremediation of chromium-contaminated soil by indigenous microorganisms Pseudomonas fluorescens was found to be a promising to remove chromium from industrial waste dumping site [37].

The technique of landfarming can be applied to soils contaminated by crude oil [38, 39, 40, 41]. It involves the use of natural processes occurring in soil to transform the contaminants [42]. Al-awadhi et al., [38] reported the reduction of more than 80% in the content of polycyclic aromatic hydrocarbon (PAH) in soil after 15 months of treatment. Recently Jacques et al., [43] have characterized a microbial consortia isolated from landfarming processes of oil, efficient in the degradation of PAHs. In the same way, composting can also be applied in the decontamination of soils with PAHs. In situ bioremediation of oily sludge contaminated soil by biostimulation of indigenous microbes through adding manure was conducted at the Shengli oilfield in northern China. After bioremediation for 360 days, total petroleum hydrocarbon (TPH) content was reduced by 58.2% in the treated plots compared with only 15.6% in the control plot [44] (Liu et al., 2010). In situ bioremediation of soil from desert mining area by using native soil bacteria consortium were well studied in Atacama Region [45].

Ahtiainen et al., [46] have noticed a significant reduction in the concentration of total PAH as well as in its toxicity after five months of treatment. In an another study, the volatile solutes transportation phenomenon in soils during the application of bioventing and demonstrated that volatilization has an important role in the first days followed by biodegradation after this period [47]. The simulation of decontamination of several organic pollutants by bioventing was performed by Sui et al.,[48]. The decontamination of diesel-oil-contaminated soil by the constant use of motor vehicles in ski runs, compared the process of natural attenuation to biostimulation with fertilizer (N-P-K), showing that the fertilized soil had a loss of approximately 72% of the contamination against 50% of the unfertilized soil. However, the study concluded that biostimulation was more effective during the first summer, with the positive effects less pronounced at the end of the research. Natural attenuation had promoted high levels of oil degradation for the time being, proving to be effective for longer periods [49]. Margesin and Schinner [50] have used biostimulation of microorganisms by the addition of inorganic nutrients to the decontamination of wastewater contaminated with anionic surfactants and fuel oil. There is also the use of alternative technologies [51] (Ethilene Diamine Tetraacetic Acid) was combined with some plant species in the attempt to increase the extraction of heavy metals like cadmium, lead and zinc from soil. The technique was efficient in the phytoextraction of metals by Brassica rapa, being observed a higher concentration of metals in shoots, thus significantly reducing the concentration of metals in roots. However, the use of EDTA was toxic to Trifolium pratense and inhibited the formation of mycorrhiza. In case of contamination by selenium, the Enterobacter cloacae is highly active in the reduction of toxic selenium to their less toxic and insoluble form, being a viable alternative of bioremediation [52]. Another interesting technique is the application of the methanotrophic bacterium Methylosinus trichosporium in biofilters to the decontamination of water contaminated by trichloroethylene [53]. A micro-algal/bacterial biofilter has also been used for the detoxification of copper and cadmium metal wastes [54]. Industry effluents can also be treated using biofilters. A bacterium Chromobacterium violaceum used to treat an effluent contaminated by silver nanoparticles, effectively reducing its concentration [55]. These techniques not only represent an emerging technology but also present a great advantage of being cost effective when compared to the traditional remediation methods due to the use of indigenous microorganisms with a versatile metabolism. A microbial surfactant (biosurfactant), sophorolipid, was investigated for its potential to enhance bioavailability and, hence, the biodegradation of crude oil [56]. However, in several studies there was a need to inoculate bacteria together with nutrients to increase the degradation potential, generating a more effective result.

Table 2. Different mechanisms and applications of bioremediation

Techniques	Principle	Applications
Natural attenuation	Existing indigenous microorganisms capable of degrading the contaminants in its natural environment	Diesel-oil-contaminated soil [49] PAH contaminated soil [73]
Bioaugmentation	Addition of exogenous microorganisms with capability of degrading the contaminants that are persistent to the indigenous microflora	Pentachlorophenol-containing aquifers[74] polycyclic aromatic hydrocarbons [57]
Biostimulation	Addition of nutrients that stimulate the growth and development of indigenous microorganisms, escalating their metabolic activity, thus uplifting the degradation	Uranium contaminated aquifer [75], diesel-oil-contaminated soil [49] effluent contaminated by SDS and diesel-oil [51]
Bioleaching	Specific microorganisms like Thiobacillus ferrooxidans and T. thiooxidans encourage the metals solubilization	Heavy metal contaminated soil [34,35], dissolution of metals [76]
Biofilters	use of bacteria in filters to decontamination of polluted water and wastes	Waste contaminated by Cu^{2+} and Cd^{2+} [54], effluent contaminated by silver nanoparticles [55], water contaminated by trichloroethylene [56]
Biopiling	The material to be treated is piled over an aerated system and nutrients are added to it	Hydrocarbon-contaminated soil [77]
Biosorption	Adsortion of metals and other ions of an aqueous solution by the use of microorganisms	Biosorption of cadmium by Trichoderma reesei [32], biosorption of Pb (II) and Cd (II) ions from aqueous solution [78]
Bioventing	Combination of venting of soil to remove the volatile compounds with bioremediation that uses the oxygen to degrade the organic contaminants	Hydrocarbon-contaminated soil [79], many organic compounds [48], volatile organic compounds [47], petroleum refinery [80]
Composting	Nutrients addintion to soil that is mixed to increase aeration and activation of indigenous microorganisms	Creosote-contaminated soil [46].
Phytoremediation	Use of plants to extract, sequestrate or decontaminate terrestrial or aquatic environments	Soil contaminated by atrazine [82] and arsenic [83], phytoextraction of metals from soil [51, 84, 7, 85]

Techniques	Principle	Applications
Landfarming	Soil is organized in lots and is periodically turned over by agricultural practices to stimulate the degradation by indigenous microorganisms	Oil-contaminated soil [38, 40, 41]
Rhizoremediation	The plant releases exudates that will increase the rhizospheric microorganisms that will help plant growth and the degradation of contaminants	[86, 87]

A microbial consortium composed by six bacteria: Mycobacterium fortuitum, Bacillus cereus, Microbacterium sp., Gordonia polyisoprenivorans, Microbacteriaceae bacterium, naphthalene-utilizing bacterium and a fungus identified as Fusarium oxysporum to degrade and mineralize different concentrations of anthracene, phenanthrene and pyrene in soil [57]. The bioaugmentation resulted in more than 90% of degradation of PAHs. with added nutrients and a microbial consortium, which improved the physico-chemical conditions of oil-contaminated soil [58]. But, it is extremely difficult to any strain, introduced or not, to achieve dominance, unless the spatial isolation is eliminated [59]. Recent researches have studied the use of the bioremediation technique in soils contaminated with herbicides [60, 61, 62]) and with heavy metals [63, 64] that apply plants in the extraction of soil metals [65, 66,67,68]. In the case of soil remediation, there are several factors that should be considered, such as soil characteristics; type and concentration of contaminants [69, 70]. It is interesting to use combined remediation methods as reported by Lei et al., [71]. They have used biopiling and biofilter in the attempt to decontaminate a soil with hydrocarbons. Besides, heavy metal contaminated soils are hard to decontaminate as heavy metals cannot be easily destroyed [72]. In this way, heavy metal should be cleaned using some approaches simultaneously, like phytoremediation in combination with some microorganisms — rhizoremediation.

Not all contaminants, however, are easily treated by using microorganisms. For example, heavy metals such as cadmium and lead are not readily absorbed or captured by organisms. The assimilation of metals such as mercury into the food chain may worsen matters. The possibility of using plants in environmental remediation has increased in the past few years, researches have been conducted and the method became an emerging alternative to the restoration of contaminated sites [88-99]. However, this methodology requires an understanding of the intrinsic factors that contribute to the success of the technique that will be discussed latter.

5. Phytoremediation: A Potential Tool of Bioremediation

Phytoremediation has been used effectively to remediate inorganic and organic contaminants in soil and groundwater. The idea of using plants to extract metals from contaminated soil was reintroduced and developed by Utsunamyia [100] and Chaney [101. The first field trial on Zn and Cd phytoextraction was conducted by Baker et al. [102]. Phytoremediation is currently divided into the following areas: (a) Phytofiltration: the use of plants and plant root associated with microorganisms to absorb and degrade the pollutants, mainly metals and organic pollutants, from water and aqueous waste streams; (b) phytostabilization: the use of plants to reduce the bioavailability of pollutants in the environment; (c) phytoextraction: the use of pollutant-accumulating plants to remove metals or organics from soil by concentrating them in the harvestable parts; (d) phytovolatilization: the use of plants to volatilize pollutants; and the use of plants to remove pollutants from air. The different types of phytoremediation is summarized in Table 3.

Table 3 Different mechanisms of phytoremediation [1]

Process	Mechanisms	Contaminants
Phytofiltration	Rhizosphere accumulation	Organics, inorganic
Phytostabilization	Complexation	Inrganic
Phytoextraction	Hyperaccumulation	Inorganic
Phytovolatilization	Volatilization	Organics, inorganic

Table 4 summarizes the cost of different remediation technologies. Among the listed remediation technologies, phytoextraction is one of the lowest cost techniques for contaminated soil remediation. There is a need to develop suitable cost-effective biological soil remediation techniques to remove contaminants without affecting soil fertility.

Before application of any kind of phytoremediation strategy in a contaminated site, ecological evaluation has to be analyzed. Specially, scientist should consider the local ecological relationship of the studied plant/s.

Table 4. Cost of different remediation technology [103]

Process	Cost (US$/ton)	Other factors
Vitrification	75–425	Long-term monitoring
Land Filling	100-500	Transport/excavation/ monitoring
Chemical treatment	100-500	Recycling of contaminant
Electrokinetics	20-200	Monitoring
Phytoextraction	5-40	Phyomass disposal

5.1. Phytofiltration

Phytofiltration is the use of plant roots (rhizofiltration) or seedlings (blastofiltration) to absorb or adsorb pollutants, mainly metals, from water and aqueous waste streams [104]. Mechanisms involved in the biosorption process include chemisorption, complexation, surface and pore adsorption-complexation, ion exchange, microprecipitation, hydroxide condensation onto the biosurface, and surface adsorption [105].

Plant roots or seedlings grown in aerated water mixed with polluted effluents absorb, precipitate and concentrate toxic metals from it [106, 107]. Rhizofiltration uses terrestrial plants instead of aquatic plants because the former feature much larger fibrous root systems covered with root hairs with extremely large surface areas.

Table 5. Summary of recent works on Phytofiltration

Plant Species	Metals	Treatments	References
B. juncea, H. annuus	Cu, Cd, Cr, Ni, Pb, and Zn	Roots of hydroponically grown terrestrial plants used to remove toxic elements from aqueous solutions	[108]
Sunflower plants	U	Rhizofiltration of U in water by roots of sunflower plants	[117]
Water Hyacinth	As, Cd Cr, Cu, Ni, and Se	The abilities of water hyacinth to take up and translocate six trace elements – As, Cd, Cr, Cu, Ni, and Se were studied under controlled conditions	[110]
Duckweed	Hg	Effects of pH, copper and humic acid	[113]
Duckweed (Lemna minor L.) and water velvet (Azolla pinnata)	Fe and Cu	Solutions enriched with 1·0, 2·0, 4·0, and 8·0 ppm of these 2 metal ions, renewed every 2 days over a 14-day test period.	[120]
Fern (Pteris Sp.)	As	Solar-powered hydroponic Technique, small scale clean-up.	[107, 118]
Eichhornia crassipes, Azolla filiculoides, Lemna minor, Lemna gibba, Ceratophyllum demersum, Nymphaea spontanea, Nymphaea alba L., V. Spiralis, Nelumbo nucifera, Myriophyllum spicatum, Potamogeton lucens, Salvinia herzogoi, Myriophyllum brasillensis, Cabomba sp., Myriophylhum aquaticum, Ludwigina palustris and Mentha aquatic, Scapania undulata and floating macrophytes Pistia stratiotes	As , Metals	Hydroponic technique, industrial wastewater and aqueous solution	[121]
E. splendensand E. argyi	Cu	Hydroponic technique, contaminated water	[122]
Limnocharis flava (L.) .	Cd	Cd Low level Cd contaminated water	[123]

The process involves raising plants hydroponically and transplanting them into metal-polluted waters where plants absorb and concentrate the metals in their roots and shoots [4, 108-110]. Plant roots can solubilize heavy metals by acidifying their soil environment with protons extruded from the roots. A lower pH releases soil bounded metal ions into the soil solution. Solubilized metal ions may enter the root either via the extracellular or intracellular pathways. As they become saturated with the metal contaminants, roots or whole plants are harvested for disposal [109, 110].

Dushenkov et al. [108] elucidated that the translocation of metals to shoots would decrease the efficiency of rhizofiltration by increasing the amount of contaminated plant residue required for disposal. However, Zhu et al. [110] suggest that the efficiency of the process can be increased by using plants with a heightened ability to absorb and translocate metals. Many aquatic plants have the ability to remove heavy metals from water, including water hyacinth (Eichhornia crassipes, [110, 111], pennywort (Hydrocotyle umbellata L., [112], and duckweed (Lemna minor L., [113]. However, these plants have limited potential for rhizofiltration because they are not efficient in removing metals as a result of their small, slow growing roots [108]. The high water content of aquatic plants complicates their drying, composting, or incineration. P. australis has high capability to remove metals from wetlands [114-116]. Sunflower (Helianthus annus L.) and Indian mustard (Brassica juncea Czern.) are the most potential terrestrial plants for removing metals from water solution. The roots of Indian mustard are effective in capturing Cd, Cr, Cu, Ni, Pb, and Zn [108], whereas sunflower removes Pb, U [117] and 137Cs from hydroponic solutions A novel phytofiltration technology has been proposed by Sekhar et al. [119] for removal and recovery of lead (Pb) from wastewaters. Plants used for phytofiltration should be able to accumulate and tolerate significant amounts of the target metals, in conjunction with easy handling, low maintenance costs, and a minimum of secondary waste requiring disposal. It is also desirable for plants to produce significant amounts of root biomass or root surface area. Table 5 is showing the examples of phytofiltration.

5.2. Phytostabilization

Use of certain plant species to immobilize contaminants in soil, through absorption and accumulation by roots, adsorption onto roots or precipitation within the root zone and physical stabilization of soils is called phytostabilization. Phytostabilization reduces the mobility of contaminants and prevents migration to groundwater or atmosphere. The process can re-establish a vegetative cover at sites where natural vegetation is lacking due to high metal concentrations [124-125].

The main approaches to re-vegetation are : a. Metal-tolerant species may be used to restore vegetation to such sites, thereby decreasing the potential migration of contaminants through wind, transport of exposed surface soils, leaching of soil and contamination of groundwater [126] b. Soil amendments can reduce the metal contaminaton to groundwater [127] which optimize the Soil agronomic condition. The soil may require lime addition, fertilization (nitrogen, phosphorous, potassium, and other mineral nutrients), carbon addition, and soil conditioners, such as aged manure, sewage sludge, compost, straw or mulch [128-129]. Unlike other phytoremediative techniques, phytostabilization is not anticipated to remove metal contaminants from a site, but somewhat to stabilize them by accumulation in roots or precipitation within root zones, reducing the risk to human health and the environment The interference to site activities may be less than with more intrusive soil remediation technologies. The heavy metal uptake potential largely varies with plant species, metal availability in the system and other environmental conditions.

The metal accumulation by various wetland plants were studied in many parts of the world, such as, Salvania natanas [130], Lemna polyrrhiza [131, Ceratophylum demersum L., Spirodela polyrrhiza (L.) Schleid, Bacopa monnieri, Hygrorrhiza aristata [132], 1995),

Eichornia crassipes [133-135], Typha latifolia and Phragmites australis [136-139]. Phytostabilization is most effective for treating a wide range of sites where large areas are subject to surface contamination [140-141].

Table 6. Summery of the recent papers of phytostabilization

Plant Species	Metals	Treatments	References
B. juncea	Cd	Soil amendments – liming materials, phosphate compounds and biosolids	146
B. juncea	Zn, Cu, Mn, Fe, Pb and Cd	organic amendments (cow manure and compost) and lime	[149, 150]
Lolium italicum and Festuca arundinaceae	Pb and Zn	Compost at two rates (10%, and 30% v/v)	[151]
H. hirta and Z. fabago	Pb, Zn and Cu	Characterization of soil and plant samples from a mine tailing located in South-East Spain for further phytostabilisation	[152]
Atriplex lentiformis(Torr.) S. Wats	As, Cu, Mn, Pb, Zn	Greenhouse study using compost	[127]
Pistacia terebinthus Bieberstein	Cu	Field study using chicken fertilizer and 1:1 soil and mine waste	[153]
Loblolly (Pinus taeda) and Virginia (Pinus virginiana) pine trees	Fe, Mn, Al	A plot design consisting of three subsurface treatments (1) ripping and compost amended, (2) ripping only, and (3) control	[154]
Lolium perenne L	Zn, Pb, and Cd	Mine waste combined with synthetic (Calcinit + urea + PK14% + calcium carbonate) or organic (cow slurry) amendment,	[128]
Wheat plant Pea Plants	Zn, Cu, Mn, Ni, Cd, Cr, Pb	A greenhouse experiment: Soil amended with industrial sludge (10 %, 20%, 30%), as well as lime treatments (0.5% and 1%)	[98-99]
Canna indica L.	Cr, Fe, Cd, Ni,Cu, Zn, Mn, Pb	A pot experiment: Soil amended with industrial sludge (10 %, 20%, 30%)	[129]

However, some highly contaminated sites are not suitable for phytostabilization, because plant growth and survival is impossible. Phytostabilization has advantages over other soil-remediation practices as it is less expensive, easier to implement, and preferable aesthetically [141-142]. Yoon et al. [143] evaluated the potential of 36 plants (17 species) growing on a contaminated site and found that plants with a high bio-concentration factor (BCF, metal concentration ratio of plant roots to soil) and low translocation factor (TF, metal concentration ratio of plant shoots to roots) have the potential for phytostabilization. The lack of appreciable metals in shoot tissue also eliminates the necessity to treat harvested shoot residue as a hazardous waste [109]. Smith and Bradshaw [144] led to the development of two cultivars of Agrostis tenuis Sibth and one of Festuca rubra L which are now commercially available for phytostabilizing Pb-, Zn-, and Cu-contaminated soils. Stabilization also involves soil amendments to promote the formation of insoluble metal complexes that reduce biological availability and plant uptake, thus preventing metals from entering the food chain [145-146, 141]. One way to facilitate such immobilisation is by altering the physicochemical properties of the metal-soil complex by introducing a multipurpose anion, such as phosphate, that enhances metal adsorption via. anion-induced negative charge and metal precipitation [146]. Addition of sludge, or compost together with lime to raise soil pH [99,129], is a common practice for immobilizing heavy metals and improving soil conditions, to facilitate re-vegetation of contaminated soils [147].

Soil acidification, due to the oxidation of metallic sulphides in the soil, increases heavy metal bioavailability; but liming can control soil acidification; also, organic materials generally promoted fixation of heavy metals in non-available soil fractions. Revegetation of mine tailings usually requires amendments of phosphorus, even though phosphate addition can mobilize arsenic (As) from the tailings. Leachates and uptakes of As were found to be higher with an organic fertilizer amendment than superphosphate, particularly in combination with barley [148].

Active phytoremediation followed by natural attenuation, was effective for remediation of the pyrite-polluted soil [149, 150]. Root-to-shoot translocation factors were smaller in amended versus control plants, indicating a reduction in the risk of metals entering the food chain through phytostabilization [98-99, 128-129]. Recent research results on phytostabilization are summarized in Table 6.

5.3. Phytoextraction

Phytoextraction or phytoaccumulation, involves the uptake and translocation of heavy metal or inorganics from the soil by plant roots into easily harvestable shoot which must be disposed of properly after they are harvested. The term phytoremediation is a concept, whereas phytoextraction is a specific clean-up technology. Plant-based environmental remediation technology, or phytoremediation, has been widely pursued in recent years as an in situ, cost-effective potential strategy for the cleanup of trace-metals from contaminated sites [4]. The development of a commercially feasible technology (phytoextraction) depends on several factors including: identifying or creating an ideal phytoextraction plant, optimizing soil and crop management practices, and developing methods for biomass processing and metal extraction [155]. There are three sets of ratios of concentrations, which should ideally be considered in plant uptake studies. (a) Root/soil: this ratio gives information concerning the uptake of an element by the root from the soil. It suggests the bioavailability of that element and its uptake by the root from the soil gives some information as to whether it is accumulated or excluded by the root system. (b) Leaf/root: this ratio shows if there is free movement between root and aerial parts, or if the element is accumulated in either roots or leaves. (c) Leaf/soil: this ratio depends on ratios (1) and (2). In plants without leaves the root/shoot concentration ratio is measured [156]. Depending on the following three arbitrary ratios of elemental concentrations in plants and soils, plants can be categorized as: (a) Accumulator: plant/soil ratio>1.5 (b) Concentration indicator: plant/soil ratio between 1.5 and 0.5 (c) Excluder- concentration indicator: plant/soil ratio between 0.5 and 0.1. [157-158]. The three categories of plants in response to metal accumulations in their body has shown in figure I [159].

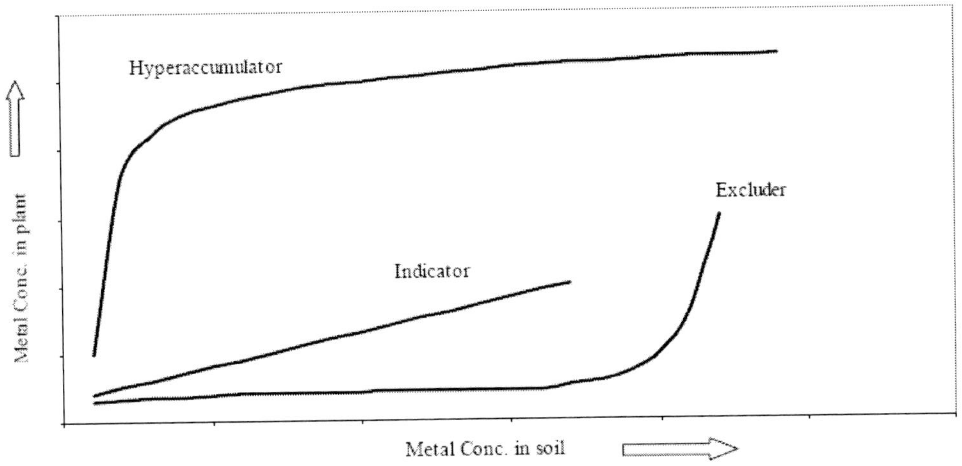

Figure 1. Conceptual response strategies of metal concentrations in plant tops in relation to increasing total metal concentrations in the soil[159].

Some plants which grow on metalliferous soils have developed the ability to accumulate massive amounts of indigenous metals in their tissues without symptoms of toxicity are calles hyperaccumulator [160-162]. Examples of commonly reported hyperaccumulators are given in Tables 7. Plants that grow on soils with metal concentrations that are normally toxic are “metal tolerant”, or “metallophytes”. Some of these plants exclude the toxic metals from their tissue, other assimilate the metals present to such a degree that they are termed “accumulator”. Accumulators are defined as plants with metal concentration in their tissues greater than that of soil [163]. Some plants have a natural ability to absorb and accumulate trace elements in their tissues. They have adaptations that enable them to survive and to reproduce in soils heavily contaminated with Zn, Cu, Pb, Cd, Ni, and As. Such plant species are divided into two main groups: the so-called pseudometallophytes that grow on both contaminated and non-contaminated soils, and absolute metallophytes that grow only on metal contaminated and naturally metal rich soils [164]. Over 400 hyperaccumulator plants have been reported, including members of the Asteraceae, Brassicaceae, Caryophyllaceae, Cyperaceae, Cunouniaceae, Fabaceae, Flacourtiaceae, Lamiaceae, Poaceae, Violaceae, and Euphobiaceae [165] Table 7 is showing the accumulation potentials of heavy metals in few important plants. These plants are selected and planted at a site based on the metals present and site conditions. After they have grown for several weeks or months, the plants are harvested. Planting and harvesting may be repeated to reduce contaminant levels to permissible limits. The time required for remediation depends on the type and extent of metal contamination, the duration of the growing season, and the efficiency of metal removal by plants, but it normally ranges from 1 to 20 years. This technique is suitable for remediating large areas of land contaminated at shallow depths with low to moderate levels of metal-contaminants [166, 167].

Table 7. Examples of Phytoaccumulators and their accumulation potential [173]

Plants	Metal	Content (mg kg^{-1})	Reference
T. caerulescens	Zn	39,600 (shoots)	[174]
T. caerulescens	Cd	1,800	[175]
Ipomea alpine	Cu	12,300	[175]
Sebertia acuminate	Ni	25% by wt. dried sap	[176]
Haumaniastrum robertii	Co	10,200	[177]
A. racemosus	Se	14,900	[178]
P. vittata	As	27,000	[179]
Berkheya coddii	Ni	5,500	[180]
Iberis intermedia	Ti	3,070	[181]

Table 8. Sumary of recent work on Phytoextraction

Plant species	Metal	Results	Reference
Pistia stratiotes	Ag, Cd, Cr, Cu, Hg, Ni, Pb and Zn	All elements accumulated mainly in the root system.	[196]
Spartina plants	Hg	Organic Hg was absorbed and transformed into an inorganic form (Hg+, Hg2+) and accumulated in roots	[197]
H. annuus	Pb	Pb concentrated in the leaf and stem indicating the prerequisites of a hyperaccumulator plant	[198]
H. indicus	Pb	Heavy metal mainly accumulated in roots and shoots	[199]
Sesbania drummondii	Pb	Pb accumulated as lead acetate in roots and leaves, although lead sulfate and sulfide were also detected in leaves, whereas lead sulfide was detected in root samples. Lead nitrate in the nutrient solution biotransformed to lead acetate and sulfate in its tissues. Complexation with acetate and sulfate may be a lead detoxification strategy in this plant species	[200]
Lemna gibba	As	A preliminary bioindicator for As transfer from substrate to plants. Used for As phytoremediation of mine tailing waters because of its high accumulation capacity	[201]
P. vittata, P. cretica, P. longifolia and P. umbrosa	As	Suitable for phytoremediation in the moderately contaminated soils	[202]

Plant species	Metal	Results	Reference
Alyssum	Ni	Majority of Ni is stored either in the leaf epidermal cell vacuoles, or in the basal portions of the numerous stellate trichomes. The metal concentration in the trichome basal compartment was the highest ever reported for healthy vascular plant tissue, approximately 15–20% dry weight	[203]
Solanum nigrum and C. Canadensis	Cd	High concentration of Cd accumulated. Tolerant to combined action of Cd, Pb, Cu and Zn	[204]
T. caerulescens	Cd	High Cd-accumulating capability, acquiring Cd from the same soil pools as non-accumulating species.	[205]
Arabis gemmifera	Cd and Zn	Hyperaccumulator of Cd and Zn, with phytoextraction capacities almost equal to T. caerulescens	[206]
Sedum alfredii	Cd	Mined ecotype had a greater ability to tolerate, transport, and accumulate Cd, compared to non-mined ecotype	[207]
Stanleya pinnata	Se	Adapted to semi-arid western U. S. soils and environments. Uptake, metabolism and volatilization of Se	[208]
Austromyrtus bidwilli. P. acinosa Roxb	Mn	Australian native hyperaccumulator of Mn, grows rapidly, has substantial biomass, wide distribution and a broad ecological amplitude	[209-210]
Brassica juncea	Pb, Cd, Cu, Ni, Zn	simultaneous accumulation of several metals after applying metal chelates (EDTA, EGTA etc.)	[155]
Thlaspi caerulescens Silene vulgaris, respectively	Cd Zn	Continous phytoextraction	[182]
V. baoshanensis S. alfredii, R. crispus V. baoshanensis, S. alfredii, Rumex K-1, and R. crispus	Zn, Cd Pb	Higher bioaccumulation factors were found for Cd in V. baoshanensis and for Zn in S. alfredii and these resulted in greater extractions of Cd and Zn, respectively. The extraction ability of R. crispus to remove Cd and Zn was considerable, due to its higher biomass. Addition of EDTA enhanced the accumulation of Pb in shoots of V. baoshanensis, S. alfredii, Rumex K-1, and R. Crispus	[211]

The basic strategies of phytoextraction are (a) development of chelate-assisted phytoextraction, which can be called induced phytoextraction; and (b) long-term continuous phytoextraction. If metal availability is not adequate for sufficient plant uptake, chelates or acidifying agents may be added to the soil to release them [65, 168-169]. Several chelating agents, such as EDTA (ethylene diamine tetra acetic acid), EGTA (ethylene glycol-O,O′-bis-[2-amino-ethyl]-N,N, N′,N′,-tetra acetic acid), EDDHA (ethylenediamine di o-hyroxyphenylacetic acid), EDDS (ethylene diamine disuccinate) and citric acid, have been used to enhance phytoextraction by mobilizing metals and increasing metal accumulation in different studies [170-171]. However, there is a potential risk of leaching of metals to groundwater, and a lack of reported detailed studies regarding the persistence of metal-chelating agent complexes in contaminated soils [172]. Some of the recent reports on phytoextraction are summarized in Table 8.

The best known metal hyperaccumulator may be Thlaspi caerulescens (alpine pennycress). While most plants show toxicity symptoms at Zn accumulation of about 100 ppm, T. caerulescens was shown to accumulate up to 26,000 ppm without showing any damage [182]. Many hyperaccumulators, including T. caerulescens, have been shown to colonize metal-rich soils such as calamine soil (soil enriched in Pb, Zn, and Cd). Ebbs et al. [183] reported that B. juncea, while having one-third the concentration of Zn in its tissue, is more effective at removing Zn from soil than Thlaspi caerulescens, a known hyperaccumulator of Zn. The advantage is due primarily to the fact that B. juncea produces ten-times more biomass than T. caerulescens.

Metal hyperaccumulator species, capable of taking up metals in the thousands of ppm, possess additional detoxification mechanisms. Understanding the mechanisms of rhizosphere interaction, uptake, transport and sequestration of metals in hyperaccumulator plants will lead to designing novel transgenic plants with improved remediation traits ([184]. For example, research has shown that in T. goesingense, a Ni hyperacccumulator, high tolerance was due to Ni complexation by histidine, which rendered the metal inactive [185-186]. Sequestration in the vacuole has been suggested to be responsible for Zn tolerance in the shoots of the Zn-hyperaccumulator T. caerulescens [169, 187]. Cadmium, a potentially toxic metal, has been shown to accumulate in plants where it is detoxified by binding to phytochelatins [188-190].

Liu et al. [191] performed a survey of Mn mine tailing soils with high concentrations of Mn, Pb, and Cd and eight plants growing on mine. It was found that Poa pratensis, Gnaphalium affine, Pteris vittata, Conyza Canadensis and Phytolacca acinosa possessed specially good metal-enrichment and metal-tolerant traits.

The effectiveness of phytoextraction of heavy metals in soils also depends on the availability of metals for plant uptake [192]. The rates of redistribution of metals and their binding intensity are affected by the metal species, loading levels, aging and soil properties [193].

Generally, the solubility of metal fractions is in the order: exchangeable > carbonate specifically adsorbed > Fe–Mn oxide > organic sulfide > residual.

Kufka and Kuras [194] reported that the process of metal uptake and accumulation by different plants depend on the concentration of available metals in soils, solubility sequences, the plant species growing on these soils and soil pH, EC, CEC, OC, etc.

The use of hyperaccumulators in phytoextraction is limited by some factors (a) Hyperaccumulators often take up a specific metal. (b) Most hyperaccumulators grow slowly and have small biomass. (c) The plants often grow in remote areas and are rare; in certain cases, their habitat is threatened by mining and other development activities. (d) Little is known about their breeding potential, pest management, agronomic characteristics and physiology (Cunningham, 1995). Therefore, using wild plants as a seed source is also unreliable. Moreover, the selection and testing of multiple hyperaccumulator plants could enhance the rate of phytoremediation, giving this process a promise one for bioremediation of environmental contamination [195].

Table 9. Major factors limiting the success and applicability of phytoextraction [214]

Plant-based biological imitation	Regulatory limitations	Other limitations
1) Low plant tolerance	1) Lack of cost and performance data	1) Contaminant beneath root zone
2) Lack of contaminant translocation from root shoot	2) Regulators unfamiliarity with the technology	2) Lengthy process
3) Small size of remediating plants	3) Disposal of contaminated plant waste	3) Contaminant in biologically unavailable form
	4) Risk of food chain contamination	4) Lack of remediating plant Species

The limitations of phytoextraction have shown in table 9. Phytoextraction involves repeated cropping of plants in contaminated soil until the metal concentration drops to an acceptable level. The waste volume can be reduced by thermal, microbial, physical or chemical means. If phytoextraction could be combined with biomass generation and its commercial utilization as an energy source, then it could be turned into a profitable operation, with the residual ash available to be used as an ore [65, 212-213]. Phyto-mining includes the generation of revenue by extracting soluble metals produced by the plant biomass ash, also known as bio-ore.

5.4. Phytovolatilizaton

Phytovolatilization involves the use of plants to take up contaminants from the soil, transforming them into volatile form and transpiring them into the atmosphere. Phytovolatilization occurs as growing trees and other plants take up water and the organic and inorganic contaminants. Some of these contaminants can pass through the plants to the leaves and volatilise into the atmosphere at comparatively low concentrations [215]. In recent researches have focused on naturally-occurring or genetically-modified plants capable of absorbing elemental forms of these metals from the soil, biologically converting them to gaseous species within the plant, and releasing them into the atmosphere. Selenium phytovolatilization has received the most attention in these days [215-218] because this element is a serious problem in many parts of the world where there are Se-rich soil [177]. The release of volatile Se compounds from higher plants was first reported by Lewis et al. [219]. Terry et al. [216] report that members of the Brassicaceae are capable of releasing up to 40 g Se ha−1 day −1 as various gaseous compounds. Some aquatic plants, such as cattail

(Typha latifolia L.), have potential for Se phytoremediation [220]. The volatilization of Se and Hg is also a enduring site solution, because the inorganic forms of these elements are removed, and gaseous species are not likely to redeposit at or near the site [221-222].In addition, sites that utilize this technique may not need much management after the original planting. This remediation method has the additional benefits of negligible site disturbance, less erosion, and no requirement for dispose of contaminated plant matter [222]. The transfer of Hg (O) to the atmosphere would not contribute significantly to the atmospheric pool. This technique appears to be a promising tool for remediating Se- and Hg- contaminated soils. Volatilization of arsenic as dimethylarsenite has also been hypothesized as a resistance mechanism in marine algae. Ma et al. [223] recently discovered the first known and extremely efficient arsenic hyperaccumulating plant, Pteris vittata which is also effective at volatilizing As; it removed about 90% of the total uptake of As from As-contaminated soils in the greenhouse, where the environment was similar to the subtropics [224]. Studies on arsenic uptake and distribution in higher plants indicate that arsenic predominantly accumulates in roots and that only small quantities are transported to shoots. However, plants may enhance the biotransformation of arsenic by rhizospheric bacteria, thus increasing the rates of volatilization [4].

However, phytovolatilization should be avoided for sites near population centres and at places with unique meteorological conditions that promote the rapid deposition of volatile compounds. Hence the consequences of releasing the metals to the atmosphere need to be considered carefully before adopting this method as a remediation tool.

Conclusion

Phytoremediation combined to rhizoremediation are possible alternatives that could be applied as one can be a complement to the other. However, it is important to first apply grasses with fast growth until the levels of the soil contaminants are lower, followed by the removal of such vegetation, incineration and recovery of the metals [225]. This measure makes possible the use of tree species to then phytostabilize the contaminants. Phytoremediation of metals and organics may be approaching commercialization.

Additional, short-term advances in phytoremediation are likely to come from the selection of more efficient plant varieties and soil amendments and from optimizing agronomic practices used for plant cultivation. Major longterm improvements in phytoremediation should come when scientists isolate genes from various plant, bacterial, and animal sources, which can enhance the metal accumulation or degradation of organics. In addition, manipulating rhizospheric bacteria to enhance their role in phytoremediation can increase the efficiency of the future phytoremediation efforts. However, biology only cannot accompany make phytoremediation work. The highly integrated nature of phytoremediation requires synergy with many other disciplines. However, it is clear that the utilization of the remarkable potential of green plants as well as its microbes in rhizosphere to accumulate elements and compounds from the environment and to perform biochemical transformations is becoming a new frontier of science. The processes that affect metal availability, metal uptake, translocation, chelation, degradation, and volatilization need to be investigated in detail. Fast growing plants with high biomass and good metal uptake ability are needed. In most of the contaminated sites hardy, tolerant, weed and microbial species exist and bioremediation through these and other non-edible species can restrict the contaminant from being introduced into the food web. Several methods of plant disposal have been described but data regarding these composting and compaction can be treated as pre-treatment steps for volume reduction, but care should be taken to collect leachate resulting from compaction.

As heavy metal contaminated soils are hard to decontaminate as heavy metals cannot be easily destroyed. In this way, they should be cleaned using some approaches simultaneously, like phytoremediation in combination with some microorganisms — rhizoremediation.

Finally, it is important to keep in mind that a variety of remediation approaches may be required to accomplish all reclamation approaches at a contaminated site [226]. The type of approach or approaches chosen will most likely be site-specific and depend on the desired speed of reclamation as well as the cost dedicated to the reclamation effort.

REFERENCES

[1] Ghosh M. , Singh S.P. 2005. A review on phytoremediation of heavy metals and utilization of its byproducts. *Applied Ecology and Environmental Research* 3(1): 1-18.

[2] Riley, R.G., J.M. Zachara, and F.J. Wobber, 1992. *Chemical contaminants on DOE Lands and Selection of Contaminated Mixtures for Subsurface Science Research*, US-DOE Off, Energy Res Subsur Sci Prog, Washington, DC.

[3] NJ DEP. 1996. Soil Cleanup Criteria. New Jersey Department of Environmental Protection. *Proposed Cleanup Standards for Contaminated Sites*, NJAC 7:26D.

[4] Salt, D.E., M. Blaylock, P.B.A. Nanda Kumar, V. Dushenkov, B.D. Ensley, I. and I. Raskin. 1995. Phytoremediation: A novel strategy for the removal of toxic metals from the environment using plants. – *Biotechnol.* 13: 468-474.

[5] Peters, W. R. 1999. Chelant extraction of heavy metals from contaminated soils. *Journal of Hazardous Materials*, 66, 151–210.

[6] Greman, H., Velikonja-Bolta, S., Vodnik, D., Kos, B., Lestan, D. 2001. EDTA enhanced heavy metal phytoextraction: metal accumulation, leaching and toxicity. P*lant and Soil*, 235, 105–114.

[7] Garbisu, C., Alkorta, I. 2001. Phytoextraction: A cost-effective plant-based technology for the removal of metals from the environment. *Bioresource Technology*, 77, 229–236.

[8] Alloway, B. J. 1995. *Soil processes and the behavior of metals*. In: Alloway B. J. (Ed), Heavy metals in soils (pp. 38–57). London: Blackie.

[9] Collins, Y.E., Stotzky, G., 1989. *Factors a€ecting the toxicity of heavy metals to microbes*. In: Beveridge, T.J., Doyle, R.J. (Eds.), Metal Ions and Bacteria. Wiley, Toronto, Canada, pp 31-90.

[10] Vangronsveld, J., Clijsters, H., 1994. *Toxic e€ect of metals*. In: Farago, M.G., Weinheim, V.C.H. (Eds.), Plant and the Chemical Elements, New York, Basel, Cambridge, Tokyo, pp. 149-177.

[11] McBride, M.B. 1995. Toxic metal accumulation from agricultural use of sludge: Are USEPA regulations protective? *J. Environ. Qual.* 24:5-18.

[12] Lovley, D.R., 1993. Dissimilatory metal reduction. *Annu. Rev. Microbiol.* 47, 263-290.
[13] Combs, G.F., Garbisu, C., Yee, B.C., Yee, A., Carlson, D.E., Smith, N.R., Magyarosi, A.C., Leighton, T., Buchanan, B.B., 1996. Bioavailability of selenium accumulated by selenite-reducing bacteria. Biol. *Trace Element Res.* 52, 209-225.
[14] Garbisu, C., Gonzalez, S., Yang, W.H., Yee, B.C., Carlson, D.E., Yee, A., Smith, N.C., Otero, R., Buchanan, B.B., Leighton, T., 1995a. Physiological mechanisms regulating the conversion of selenite to elemental selenium by Bacillus subtilis. *BioFactors* 5, 29-37.
[15] Garbisu, C., Ishii, T., Smith, N.R., Yee, B.C., Carlson, D.E., Yee, A., Buchanan, B.B., 1995b. *Mechanisms regulating the reduction of selenite by aerobic Gram (+) and ()) bacteria.* In: Hinchee, R.E., Means, J.L., Burris, D.R. (Eds.), Battelle Press, Columbus, OH, pp. 125-131.
[16] Garbisu, C., Ishii, T., Leighton, T., Buchanan, B.B., 1997a. Bacterial reduction of selenite to elemental selenium. *Chem. Geol.* 132, 199-204.
[17] Garbisu, C., Alkorta, I., Carlson, D.E., Leighton, T., Buchanan, B.B., 1997b. Selenite bioremediation potential of indigenous microorganisms from industrial activated sludge. *Microbiol. SEM* 13, 437-444.
[18] Garbisu, C., Llama, M.J., Serra, J.L., 1997c. Effect of heavy metals on chromate reduction by Bacillus subtilis. *J. Gen. Appl. Microbiol.* 43, 369-371.
[19] Garbisu, C., Alkorta, I., Llama, M.J., Serra, J.L., 1998. Aerobic chromate reduction by Bacillus subtilis. *Biodegradation* 9, 133-141.
[20] Ishibashi, Y., Cervantes, C., Silver, S., 1990. Chromium reduction in Pseudomonas putida. *Appl. Environ. Microbiol.* 56, 2268-2270.
[21] Ledin, M. 2000. Accumulation of metals by microorganisms—processes and importance for soil systems. *Earth Science Reviews*, 51, 1–31.
[22] Arwidsson, Z., Johansson, E., von Kronhelm T., Allard B., van Hees P. 2009. Remediation of metal contaminated soil by organic metabolites from fungi I—production of organic acids. *Water, Air, and Soil pollution.* doi:10.10007/s11270- 009-0067-z.
[23] Khan A.G. 2005. Role of soil microbes in the rhizospheres of plants growing on trace metal contaminated soils in phytoremediation, *J Trace Elem Med Biol* 18 355–364.
[24] Weissenhorn, I., Leyval, C. 1995. Root colonization of maize by a Cd-sensitive, a Cd-tolerant Glomus mosseae, cadmium uptake in s, culture. *Plant, Soil*, 175, 233-238.
[25] Lakatos G, Kiss M, Mezzaros I. Heavy metal content of common reed (Phragmites australis/Cav./Trin. ex Steudel) and its periphyton in Hungarian shallow standing waters. *Hydrobiologia* 1999;415:47– 53.
[26] K. Gruiz and E. Kriston, In situ bioremediation of hydrocarbons in soil, *J Soil Contam* 4 (1995), pp. 163–174.
[27] Vidali M., 2001. Bioremediation. An overview, *Pure Appl Chem* 73: 1163– 1172.
[28] Saadoun I.M.K., Al-Ghzawi Z.D., 2005. Bioremediation of petroleum contamination, *Biorem Aquat Terr Ecosyst* , 173–212.
[29] Laine, M. M., Jorgensen, K. S. 1997. Effective and safe composting of chlorophenol-contaminated soil in pilot scale. *Environmental Science and Technology*, 31(2), 371–378.

[30] Jorgensen, K. S., Puustinen, J., and Suortti, A.-M. (2000). Bioremediation of petroleum hydrocarbon-contaminated soil by composting in biopiles. *Environmental Pollution,* 107, 245–254.

[31] Riffaldi, R., Levi-Minzi, R., Cardelli, R., Palumbo, S., Saviozzi, A. 2006. Soil biological activities in monitoring the bioremediation of diesel oil-contaminated soil. *Water, Air, and Soil Pollution*, 170, 3–15.

[32] Kim S.K., Park C.B., Koo Y.-M., Yun H.S. 2003. Biosorption of cadmium and copper ions by *Trichoderma reesei* RUT C30, *J Ind Eng Chem* 9: 403–406.

[33] Garcia Jr. O. 1997. *Microrganismos e metais*. In: Melo IS, Azevedo JL, editors. Microbiologia Ambiental. Jaguariúna: EMBRAPA, 440.

[34] White C., Shaman A.K., Gadd G.M.1998. An integrated microbial process for the bioremediation of soil contaminated with toxic metals, *Nat Biotechnol* 16: 572–575.

[35] Kumar R.N., Nagendran R. 2009. Fractionation behavior of heavy metals in soil during bioleaching with *Acidithiobacillus thiooxidans*, *J Hazard Mater* 169: 1119–1126.

[36] Rulkens WH, Grotenhuis JTC, Tichy R. *Methods for Cleaning Contaminated Soils and Sediments.* 1995. In: Salomons W, Förstner U, Mader P, editors. Heavy Metals - Problems and Solutions. Berlin: Springer-Verlag Berlin Heidelberg, 412.

[37] Jeyalakshmi D, Kanmani S., 2008, Bioremediation of chromium contaminated soil by Pseudomonas fluorescens and indigenous microorganisms, *J Environ Sci Eng.* 50(1):1-6.

[38] Al-awadhi N., Al-Daher R., Elnawawy A., Balba M.T. 1996. Bioremediation of oil-contaminated soil in Kuwait. I. Landfarming to remediate oil-contaminated soil, *J Soil Contam* 5: 243-260.

[39] Jacques R.J.S., Okeke B.C., Bento F.M., Teixeira A.S., Peralba M.C.R., Camargo F.A.O. 2008. Microbial consortium bioaugmentation of a polycyclic aromatic hydrocarbons contaminated soil, *Biores Technol* 99: 2637–2643.

[40] Mancini O., Cuccu W., Molinari M. 2005. *Landfarming and phytoremediation in an urban area: a case study*, In situ and on-site bioremediation, Proceedings of the International In Situ and On-Site Bioremediation Symposium, 8th, Baltimore, *MD,* United States, June 6–9.

[41] Saadoun I.M.K., Al-Ghzawi Z.D., 2005. Bioremediation of petroleum contamination, *Biorem Aquat Terr Ecosyst*, 173–212.

[42] Pope DF, Matthews JE. 1992. *Bioremediation using the land treatment concept.* USEPA/600/R-93/164. Robert S. Kerr, Environmental Research Laboratory. US Environmental Protection Agency, Ada, OK.

[43] Jacques R.J.S., Okeke B.C., Bento F.M., Peralba M.C.R., Camargo F.A.O. 2007. Characterization of a polycyclic aromatic hydrocarbon-degrading microbial consortium from a petrochemical sludge landfarming site, *Biorem J* 11: 1–11.

[44] Liu W, Luo Y, Teng Y, Li Z, Ma L.Q. 2010. *Bioremediation of oily sludge-contaminated soil by stimulating indigenous microbes Environ Geochem Health,* 32:23–29.

[45] Godoy-Faúndez A, Antizar-Ladislao B, Reyes-Bozo L, Camaño A, Sáez-Navarrete C, 2008. Bioremediation of contaminated mixtures of desert mining soil and sawdust with fuel oil by aerated in-vessel composting in the Atacama Region (Chile), *J Hazard Mater*, 151(2-3):649-57.

[46] Ahtiainen J., Valo R., Järvinen M., Joutti A. 2002. Microbial toxicity tests and chemical analysis as monitoring parameters at composting of creosote-contaminated soil, *Ecotoxicol Environ Saf* 53: 323–329.
[47] Suko, T. Fujikawa , T. Miyazaki 2006. Transport phenomena of volatile solute in soil during bioventing technology, *J ASTM Int* 3.
[48] Sui H., Li X., Jiang B. and Huang G. 2007. Simulation of remediation of multiple organic contaminantssystem by bioventing, *Huagong Xuebao (Chinese Edition)* 58: 1025–1031.
[49] Margesin R., Schinner F. 2001. Bioremediation (natural attenuation and biostimulation) of diesel-oil-contaminated soil in an Alpine glacier skiing area, *Appl Environ Microbiol* 67: 3127–3133.
[50] Margesin R., Schinner F. 1998. Low-temperature bioremediation of a waste water contaminated with anionic surfactants and fuel oil, *Appl Microbiol Biotechnol* 49: 482–486.
[51] Grman H., Velikonja-Bolta S., Vodnik D., Kos B, Letan D. 2001. EDTA enhanced heavy metal phytoextraction: metal accumulation, leaching and toxicity, *Plant Soil* 235: 105–114.
[52] Frankenberger Jr WT, Arshad M. 2001. Bioremediation of selenium-contaminated sediments and water. *BioFactors*, 14(1–4):241–54.
[53] R.T. Taylor, A.G. Duba, W.B. Durham, M.L. Hanna, K.J. Jackson and M.C. Jovanovich *et al.*, In situ bioremediation of trichloroethylene-contaminated water by a resting-cell methanotrophic microbial filter, *Conference of the International in situ bioremediation, Ontario (Canada)* (1992), pp. 20–24.
[54] S. Loutseti, D.B. Danielidis, A. Economou-Amilli, C. Katsaros, R. Santas and P. Santas, The application of a micro-algal/bacterial biofilter for the detoxification of copper and cadmium metal wastes, *Biores Technol* 100 (2009), pp. 2099–2105.
[55] N. Duran, P.D. Marcato, G.I.H. De Souza, O.L. Alves and E. Esposito, Antibacterial effect of silver nanoparticles produced by fungal process on textile fabrics and their effluent treatment, *J Biomed Nanotech* 3 (2007), pp. 203–208.
[56] Kang S W, Kim B Y, Shin D J, Kim E K. Enhanced Biodegradation of Hydrocarbons in Soil by Microbial Biosurfactant, Sophorolipid. *Appl Biochem Biotechnol* (2010) 160:780–790.
[57] R.J.S. Jacques, B.C. Okeke, F.M. Bento, A.S. Teixeira, M.C.R. Peralba and F.A.O. Camargo, Microbial consortium bioaugmentation of a polycyclic aromatic hydrocarbons contaminated soil, *Biores Technol* 99 (2008), pp. 2637–2643.
[58] S. Mishra, J. Jyot, R.C. Kuhad and B. Lal, Evaluation of inoculum addition to stimulate *in situ* bioremediation of oily-sludge-contaminated soil, *Appl Environ Microbiol* 67 (2001), pp. 1675–1681.
[59] Zhou J, Xia B, Treves DS, Wu LY, Marsh TL, O'neill RV, Palumbo AV, Tiedje JM. Spatial and resource factors influencing high microbial diversity in soil. *Applied and Environmental Microbiology* 2002;68(1):326–34.
[60] F.R. Pires, C.M. Souza, A.A. Silva, S.O. Procópio and L.R. Ferreira, Fitorremediação de solos contaminados com herbicidas, *Planta Daninha* 21 (2003), pp. 335–341.
[61] M. Kumar and L. Philip, Bioremediation of endosulfan contaminated soil and water — optimization of operating conditions in laboratory scale reactors, *J Hazard Mater* 136 (2006), pp. 354–364.

[62] I. Robles-González, E. Ríos-Leal, R. Ferrera-Cerrato, F. Esparza-García, N. Rinderkenecht-Seijas and H.M. Poggi-Varaldo, Bioremediation of a mineral soil with high contents of clay and organic matter contaminated with herbicide 2, 4-dichlorophenoxyacetic acid using slurry bioreactors: effect of electron acceptor and supplementation with an organic carbon source, *Process Biochem* 41 (2006), pp. 1951–1960.

[63] D.R. Lovley and J.D. Coates, Bioremediation of metal contamination, *Curr Opin Biotechnol* 8 (1997), pp. 285–289.

[64] A. Ikea, R. Sripranga, H. Onoa, Y. Murookaa and M. Yamashita, Bioremediation of cadmium contaminated soil using symbiosis between leguminous plant and recombinant rhizobia with the MTL4 and the PCS genes, *Chemosphere* 66 (2007), pp. 1670–1676.

[65] S.D. Cunningham and D.W. Ow, Promises and prospects of phytoremediation, *Plant Physiol* 110 (1996), pp. 715–719.

[66] A.C. Dietz and J.L. Schonnor, Advances in phytoremediation, *Environ Health Perspect* 109 (2001).

[67] E. Arthur, P. Rice, P. Rice, T. Anderson, S. Baladi and K. Henderson *et al.*, Phytoremediation — an overview, *Crit Rev Plant Sci* 24 (2005), pp. 109–122.

[68] S. Wei, J.A.T. Silva and Q. Zhou, Agro-improving method of phytoextracting heavy metal contaminated soil, *J Hazard Mater* 150 (2008), pp. 662–668.

[69] R. Boopathy, Factors limiting bioremediation technologies, *Biores Technol* 74 (2000), pp. 63–67.

[70] V. Sheoran, A.S. Sheoran and P. Poonam, Remediation techniques for contaminated soils, *Environ Eng Manag. J* 7 (2008), pp. 379–387.

[71] J. Lei, J.L. Sansregret and B. Cyr, Biopiles and biofilters combined for soil cleanup, *Pollut Eng* 26 (1994), pp. 56–58.

[72] S.D. Cunningham, W.R. Berti and J.W. Huang, Phytoremediation of contaminated soils, *Trends Biotechnol* 13 (1995), pp. 393–397.

[73] P. Conte, A. Agretto, R. Spaccini and A. Piccolo, Soil remediation: humic acids as natural surfactants in the washings of highly contaminated soils, *Environ Pollut* 135 (2005), pp. 515–522.

[74] K.E. Stormo and R.L. Crawford, Preparation of encapsulated microbial cells for environmental applications, *Appl Environ Microbiol* 58 (1992), pp. 727–730.

[75] R.T. Anderson, H.A. Vrionis, I. Ortiz-Bernad, C.T. Resch, P.E. Long and R. Dayvault *et al.*, Stimulating the in situ activity of *Geobacter*species to remove uranium from the groundwater of a uranium-contaminated aquifer, *Appl Environ Microbiol* 69 (2003), pp. 5884–5891.

[76] D. Mishra, D.-J. Kim, D.E. Ralph, J.-G. Ahn and Y.-A. Rhee, Bioleaching of metals from spent lithium ion secondary batteries using*Acidithiobacillus ferrooxidans*, *Waste Manage* 28 (2008), pp. 333–338.

[77] W. Mohn, C. Radziminski, M.-C. Fortin and K. Reimer, On site bioremediation of hydrocarbon-contaminated arctic tundra soils in inoculated biopiles, *Appl Microbiol Biotechnol* 57 (2001), pp. 242–247.

[78] A. Sari and M. Tuzen, Kinetic and equilibrium studies of biosorption of Pb(II) and Cd(II) from aqueous solution by macrofungus (*Amanita rubescens*) biomass, *J Hazard Mater* 164 (2009), pp. 1004–1011.

[79] J. Lee, H. Bae, J. Jeong, J.-Y. Le, Y.-Y. Yang and I. Hwang *et al.*, Functional expression of a bacterial heavy metal transporter in *Arabidopsis*enhances resistance to and decreases uptake of heavy metals, *Plant Physiol* 133 (2003), pp. 589–596.
[80] M.J. Zenker, G.R. Brubaker, D.J. Shaw and S.R. Knight, Passive bioventing pilot study at a former petroleum refinery, *In Situ and On-Site Bioremediation, Proceedings of the International In Situ and On-Site Bioremediation Symposium, 8th, Baltimore, MD, United States* (2005) June 6-9.
[81] I. Brunner, J. Lustera, M.S. Günthardt-Goerga and B. Frey, Heavy metal accumulation and phytostabilisation potential of tree fine roots in a contaminated soil, *Environ Pollut* 152 (2008), pp. 559–568.
[82] J.G. Burken and J.L. Schnoor, Phytoremediation: plant uptake of atrazine and role of root exudates, *J Environ Eng* 122 (1996), pp. 958–963.
[83] P. Visoottiviseth, K. Francesconi and W. Sridokchan, The potential of Thai indigenous plant species for the phytoremediation of arsenic contaminated land, *Environ Pollut* 118 (2002), pp. 453–461.
[84] Weis J. S., Weis P., Metal uptake, transport and release by wetland plants: implications for phytoremediation and restoration, *Environment International* 30 (2004) 685– 700.
[85] Raskin I, Smith RD, Salt DE. 1997. Phytoremediation of metals: using plants to remove pollutants from the environment. *Curr. Opin. Biotechnol.* 8:221–26.
[86] C. Leyval, K. Turnau and K. Haselwandter, Effect of heavy metal pollution on mycorrhizal colonization and function: physiological, ecological and applied aspects, *Mycorrhiza* 7 (1997), pp. 139–153.
[87] R. Benisse, M. Labat, A. Elasli, F. Brhada, F. Chandad and P.P. Liegbott *et al.*, Rhizosphere bacterial populations of metallophyte plants in heavy metal-contaminated soils from mining areas in semiarid climate, *World J Microbiol Biotechnol* 20 (2004), pp. 759–766.
[88] A. Azadpour and J.E. Matthews, Remediation of metal-contaminated sites using plants, *Remediation* 6 (1996), pp. 1–18.
[89] C. Garbisu, J. Hernandez-Allica, O. Barrutia, I. Alkorta and J.M. Becerril, Phytoremediation: a technology using green plants to remove contaminants from polluted areas, *Rev Environ Health* 17 (2002), pp. 173–188.
[90] O.V. Singh, S. Labana, G. Pandey, R. Budhiraja and R.K. Jain, Phytoremediation: an overview of metallic ion decontamination from soil,*Appl Microbiol Biotechnol* 61 (2003), pp. 405–412.
[91] A. Vassilev, J.-P. Schwitzguebel, T. Thewys, D. Van Der Lelie and J. Vangronsveld, The use of plants for remediation of metal-contaminated soils, *Sci World J* 4 (2004), pp. 9–34.
[92] D.G. Paquin, S. Campbell and Q.X. Li, Phytoremediation in subtropical Hawaii — a review of over 100 plant species, *Remed J* 14 (2004), pp. 127–139.
[93] K. Shah and J.M. Nongkynrih, Metal hyperaccumulation and bioremediation, *Biol Plant* 51 (2007), pp. 618–634.
[94] M. Rajkumar and H. Freitas, Influence of metal resistant-plant growth-promoting bacteria on the growth of *Ricinus communis* in soil contaminated with heavy metals, *Chemosphere* 71 (2008), pp. 834–842.

[95] M.I. Lone, Z.-H. He, J. Stoffella and X. Yang, Phytoremediation of heavy metal polluted soils and water: progresses and perspectives, *J Zhejiang Univ Sci B* 9 (2008), pp. 210–220.

[96] X. Xiao, T. Chen, Z. An, M. Lei, Z. Huang and X. Liao *et al.*, Potential of *Pteris vittata* L. for phytoremediation of sites co-contaminated with cadmium and arsenic: the tolerance and accumulation, *J Environ Sci* 20 (2008), pp. 62–67.

[97] P.K. Padmavathiamma and L.Y. Li, Phytoremediation technology: hyper-accumulation metals in plants, *Water Air Soil Pollut*184 (2007), pp. 105–126.

[98] Bose S, Bhattacharyya AK. Heavy metal accumulation in wheat plant grown in soil amended with industrial sludge. Chemosphere. 2008 Jan;70(7):1264-72.

[99] Bose S, Chandrayan S, Rai V, Bhattacharyya AK, Ramanathan AL. Translocation of metals in pea plants grown on various amendment of electroplating industrial sludge. *Bioresour Technol.* 2008 Jul;99(10):4467-75.

[100] Utsunamyia, T. (1980). *Japanese patent application no.* 55-72959.

[101] Chaney, R. L. (1983). *Plant uptake of inorganic waste constitutes.* In J. F. Parr, P. B. Marsh, and J. M. Kla (Eds.), Land treatment of hazardous wastes (pp. 50–76). Park Ridge, NJ: Noyes Data Corp.

[102] Baker, A. J. M., Reeves, R. D., and McGrath, S. P. (1991). *In situ decontamination of heavy metal polluted soils using crops of metal accumulating plants – A feasibility study.* In R. E. Hinchee and R. F. Olfenbuttel (Eds.), In-situ bioremediation (pp. 539–544). Stoneham, M. A: Butterworth-Heinemann.

[103] Glass, D. J. (1999). *U.S. and international markets for phytoremediation*, 1999–2000 (p. 266). Needham, MA: D. Glass Associates.

[104] Prasad, M. N. V, and Freitas, H. (2003). Metal hyperaccumulation in plants – Biodiversity prospecting for phytoremediation technology. *Electronic Journal of Biotechnology*, 6, 275–321.

[105] Torresdey Gardea J. L., G., Rosa, de la and J. R. Peralta-Videa. 2004. Use of phytofiltration technologies in the removal of heavy metals: A review, *Pure Appl. Chem.*, Vol. 76, No. 4, pp. 801–813, 2004.

[106] Dushenkov, S., and Kapulnik, Y. (2000). *Phytofilitration of metals.* In I. Raskin and B. D. Ensley (Eds.), Phytoremediation of toxic metals – Using plants to clean-up the environment (pp. 89–106). New York: Wiley.

[107] Elless, P. M., Poynton, Y. C., Williams, A. C., Doyle, P. M., Lopez, C. A., Sokkary, A. D., et al. (2005). Pilot-scale demonstration of phytofiltration for drinking arsenic in New Mexico drinking water. *Water Research*, 39(16), 3863–3872.

[108] Dushenkov, V., Kumar, P. B. A. N., Motto, H., and Raskin, I. (1995). Rhizofiltration: The use of plants to remove heavy metals from aqueous streams. *Environmental Science and Technology*, 29, 1239–1245.

[109] Flathman, P. E., and Lanza, G. R. (1998). Phytoremediation: Current views on an emerging green technology. *Journal of Soil Contamination*, 7(4), 415–432.

[110] Zhu, Y. L., Zayed, A. M., Quian, J. H., De Souza, M., and Terry, N. (1999). Phytoaccumulation of trace elements by wetland plants: II. Water hyacinth. *Journal of Environmental Quality*, 28, 339–344.

[111] Kay, S. H., Haller, W. T., and Garrard, L. A. (1984). Effect of heavy metals on water hyacinths [Eichhornia crassipes (Mart.) Solms]. *Aquatic Toxicology*, 5, 117–128.

[112] Dierberg, F. E., Débuts, T. A., and Goulet, J. R. N. A. (1987). *Removal of copper and lead using a thin-film technique*. In K. R. Reddy andW. H. Smith (Eds.), Aquatic plants for water treatment and resource recovery (pp. 497–504). Magnolia.

[113] Mo, S. C., Choi, D. S., and Robinson, J. W. (1989). Uptake of mercury from aqueous solution by duckweed: The effect of pH, copper, and humic acid. *Journal of Environmental Health*, 24, 135–146.

[114] Kraus ML. *Wetlands: toxicant sinks of reservoirs? Wetland hydrology*. Proceedings of the National Wetland Symposium, Sept. 1987, Chicago, IL, USA. Berne, NY: Association of State Wetland Managers; 1987. p. 192– 6.

[115] Keller B, Lajtha K, Cristofor S. *Trace metal concentrations in the sediments and plants of the Danube delta, Romania*. Wetlands 1998;18:42–50.

[116] Windham L, Weis JS, Weis P. Lead uptake, distribution, and effects in two dominant salt marsh macrophytes, Spartina alterniflora (cordgrass) and Phragmites australis (common reed). *Mar. Pollut. Bull.* 2001b;42:811 – 6.

[117] Dushenkov, S., Vasudev, D., Kapulnik, Y., Gleba, D., Fleisher, D., Ting, K. C., et al. (1997a). Removal of uranium from water using terrestrial plants. *Environmental Science and Technology*, 31(12), 3468–3474.

[118] Huang, J.W., Poynton, C.Y., Elless, M., Kochian, L.V. 2004. Phytofiltration of arsenic from drinking water using an arsenic-hyperaccumulating fern. *Environmental Science and Technology*. 38:3412-3417.

[119] Sekhar, K. C., Kamala, C. T., Chary, N. S., Sastry, A. R. K., Rao, T. N., and Vairamani, M. (2004). Removal of lead from aqueous solutions using an immobilized biomaterial derived from a plant biomass. *Journal of Hazardous Materials*, 108, 111–117.

[120] Jain, S. K., Vasudevan, P., Jha, N. K. (1989). Removal of some heavy metals from polluted water by aquatic plants: Studies on duckweed and water velvet. *Biological Wastes*, 28(2), 115–126.

[121] Anawar, H.M. Garcia-Sanchez, A., Alam, M. Tari Kul, Rahman, M. Majibur Phytofiltration of water polluted with arsenic and heavy metals *International Journal of Environment and Pollution*, Volume 33, Numbers 2-3, 11 July 2008 , pp. 292-312(21).

[122] Tian S , Peng H , Yang X , Lu L, Zhang L 2008. *Bulletin of Environmental Contamination and Toxicology* 81 (1), 85-89.

[123] Abhilash P.C., Pandey V C, Srivastava P, Rakesh P.S., Chandran S, Singh N, Thomas A.P. 2009, Phytofiltration of cadmium from water by Limnocharis flava (L.) Buchenau grown in free-floating culture system *J. Haz. Mat*, 170(2-3), 791-97.

[124] Munshower FF. 1994. *Practical Handbook of Disturbed land Revegetation*. Boca Raton, FL:Lewis Publishing,

[125] Tordoff GM, Baker AJM, Willis AJ. 2000. Current approaches to the revegetation and reclamation of metalliferous mine wastes. *Chemosphere* 41:219-228,

[126] Stoltz, E., and Greger, M. (2002). Accumulation properties of As, Cd, Cu, Pb and Zn by four wetland plant species growing on submerged mine tailings. *Environmental and Experimental Botany*, 47(3), 271–280.

[127] Mendez MD, Glenn EP, Maier RM. 2007. Phytostabilization potential of quailbush for mine tailings: growth, metal accumulation, and microbial community changes. *J Environ Qual* 36:245-253.

[128] Epelde L, Becerril J M., Mijangos I, Garbisu C, Evaluation of the Efficiency of a Phytostabilization Process with Biological Indicators of Soil Health *J Environ Qual* 38:2041-2049 (2009).

[129] Bose S, Jain A, Rai V, Ramanathan AL. Chemical fractionation and translocation of heavy metals in Canna indica L. grown on industrial waste amended soil. *J Hazard Mater*. 2008 Dec 15;160(1):187-93.

[130] Zayed, A., Gowthaman, S., Teryy, N., 1998. Phytoaccumulation of trace elements by wetland plants: I. Duckweed. *Journal of Environmental Quality* 27, 715–721.

[131] Sharma, S.S., Gaur, J.P., 1995. Potential of Lemna Polyrrhiza for removal of heavy metals. *Ecological Engineering* 4, 37–43.

[132] Rai, U.N., Sinha, S., Tripathi, R.D., Chandra, P., 1995.Waste water treatability potential of some aquatic macrophytes: Removal of heavy metals. *Ecological Engineering* 5, 5–12.

[133] Vesk, P.A., Nockold, C.E., Allway, W.G., 1999. Metal localisation in water hyacinth roots from an urban wetland. *Plant, Cell and Environment* 22, 149–159.

[134] Mehra, A., Farago, M.E., Banerjee, D.K., 2000. A study of Eichornia crassipes growing in the overbank and floodplains soils of the river Yamuna in Delhi, India. *Environmental Monitoring and Assessment* 60, 25–45.

[135] Singhal, V., Rai, J.P.N., 2003. Biogas production fromwater hyacinth and channel grass used for phytoremediation of industrial effluents. *Bioresource Technology* 86, 221–225.

[136] Babcock, M.F., Evans, D.W., Alberts, J.J., 1983. Comparative uptake and translocation of trace elements from coal ash by Typha latifolia. *Science of the Total Environment* 28, 203–214.

[137] Ye, Z.H., Whiting, S.N., Lin, Z.Q., Lytle, C.M., Qian, J.H., Terry, N., 2001. Removal and distribution of iron, manganese, cobalt and nickel within a Pennsylvania constructed wetland treating coal combustion by-product lechate. *Journal of Environmental Quality* 30, 1464–1473

[138] Batty, L.C., Paul, L.Y., 2004. Growth of Phragmites australis (Cav.) Trin ex. Steudel in mine water treatment wetlands: effects of metal and nutrient uptake. *Environmental Pollution* 132, 85–93.

[139] Tiwari, S., Kumari, B., Singh, S.N., 2008. Evaluation of metal mobility/immobility in fly ash induced by bacterial strains isolated from the rhizospheric zone of Typha latifolia growing on fly ash dumps. *Bioresource Technology* 99, 1305–1310.

[140] Cunningham, S. D., Shann, J. R., Crowley, D. E., and Anderson, T. A. (1997). *Phytoremediation of contaminated water and soil.* In E. L. Kruger, T. A. Anderson, and J. R. Coats (Eds.), Phytoremediation of soil and water contaminants. ACS Symposium series 664 (pp. 2–19). Washington, DC: American Chemical Society.

[141] Berti, W. R., and Cunningham, S. D. (2000). *Phytostabilization of metals.* In I. Raskin and B. D. Ensley (Eds.), Phytoremediation of toxic metals: Using plants to clean-up theenvironment (pp. 71–88). New York: Wiley.

[142] Schnoor, J. L. (2000). *Phytostabilization of metals using hybrid poplar trees.* In I. Raskin and B. D. Ensley (Eds.), Phytoremediation of toxic metals: Using plants to cleanup the environment (pp. 133–150). New York: Wiley.

[143] Yoon, J., Cao, X., Zhou, Q., and Ma, L. Q. (2006). Accumulation of Pb, Cu, and Zn in native plants growing on a contaminated Florida site. *Science of The Total Environment*, 368(2–3), 456–464.

[144] Smith, R. A. H., and Bradshaw, A. D. (1992). Stabilization of toxic mine wastes by the use of tolerant plant populations. *Transactions of the Institution of Mining and Metallurgy*, 81, A230–A237.

[145] Adriano, D. C., Wenzel, W. W., Vangronsveld, J., and Bolan, N. S. (2004). Role of assisted natural remediation in environmental cleanup. *Geoderma*, 122, 121–142.

[146] Bolan, N. S., Adriano, D. C., and Naidu, R. (2003). Role of phosphorus in (im)mobilization and bioavailability of heavy metals in the soil-plant system. *Reviews of Environmental Contamination and Toxicology*, 177, 1–44.

[147] Williamson, A., and Johnson, M. S. (1981). Reclamation of metalliferous mine wastes. In N. W. Lepp (Ed.), *Effect of heavy metal pollution on plants,* vol. 2 (pp. 185–212). Barking, Essex, UK: Applied Science Publishers.

[148] Mains, D., Craw, D., Rufaut, C. G., and Smith, C. M. S. (2006b). Phytostabilization of gold mine tailings from New Zealand. Part 2: Experimental evaluation of arsenic mobilization during revegetation. *International Journal of Phytoremediation*, 8(2), 163–183.

[149] Clemente, R., Almela, C., and Bernal, P. M. (2006). A remediation strategy based on active phytoremediation followed by natural attenuation in a soil contaminated by pyrite waste. *Environmental Pollution*, 143(3), 397–406.

[150] Clemente, R., Walker, J. D., Roig, A., and Bernal, P. M. (2003). Heavy metal bioavailability in a soil affected by mineral sulphides contamination following the mine spillage at Aznalc′ollar (Spain). *Biodegradation*, 14, 199–205.

[151] Rizzi, L., Petruzzelli, G., Poggio, G., and Vigna, G. (2004). Soil physical changes and plant availability of Zn and Pb in a treatability test of phytostabilization. *Chemosphere*, 57(9), 1039–1046.

[152] Conesa, M. H., Faz, A., and Arnaldos, R. (2006). Initial studies for the phytostabilization of a mine tailing from the Cartagena– La Union Mining District (SE Spain). *Chemosphere*, 66(1), 38–44.

[153] Johansson L, Xydas C, Messios N, Stoltz E, Greger M. 2005. Growth and Cu accumulation by plants grown on Cu containing mine tailings in Cyprus. *Appl Geochem* 20:101-107,

[154] Barton C, Marx D, Adriano D, Koo B J, Newman L, Czapka S, Blake. Phytostabilization of a landfill containing coal combustion waste *J. Environmental Geosciences*, v. 12, no. 4 (December 2005), pp. 251–265.

[155] Blaylock, M.J, Salt, D.E., Dushenkov, S., Zakharova, O., Gussman, C., 1997. Enhanced accumulation of Pb in Indian mustard by soil applied chelating agents. *Environ. Sci. Technol.* 31, 860±865.

[156] Farago M. E and Mehra A. 1991. "Uptake of elements by the copper-tolerant plant *America maritime*". *Metal Compounds in Environment and Life*, 4: 163-169.

[157] Farago M. E, Cole M., Xiao X and Vaz M. C 1992. "Preliminary assessment of metal bioavailability to plants in the Neves Corvo area Portugal". *Chemical Speciation and Bioavailability*. Vol 4 (1), 19-27.

[158] Baker, A. J.M. (1981): "Accumulators and Excluders-strategies in the response of plants to heavy metals". *J. Plant nutrition*. Vol 3 (1-4), 643.

[159] Raskin, I, Kumar, P.B.A.N., Dushenkov, S. and Salt, D. (1994): Bioconcentration of heavy metals by plants. – *Current Opinion Biotechnology* 5; 285-290.

[160] Reeves, R. D., and Brooks, R. R. (1983). Hyperaccumulation of lead and zinc by two metallophytes from a mining area of Central Europe. *Environmental Pollution Series A,* 31, 277–287.

[161] Baker, A. J. M., and Brooks, R. R. (1989). Terrestrial higher plants which hyper accumulate metallic elements – Review of their distribution, ecology, and phytochemistry. *Biorecovery*, 1, 81–126.

[162] Entry, J. A., Watrud, L. S., and Reeves, M. (1999). Accumulation of 137Cs and 90Sr from contaminated soil by three grass species inoculated with mycorrhizal fungi. *Environmental Pollution*, 104, 449–457.

[163] Förstner, U. 1991, in G. H. Bolt *et al.* (eds.), *Interaction at the Soil Colloid – Soil Solution Interface*, NATO ASI series, Kluwer Academic Publishers, Dordrecht, The Netherlands, pp. 543–582.

[164] Baker, A. J.M. (1987): Metal Tolerance. *New Phytologist* 106, 93-111.

[165] Baker, A. J. M., McGrath, S. P., Reeves, R. D., and Smith, J. A. C. (2000). *Metal hyperaccumulator plants: A review of the ecology and physiology of a biological resource for phytoremediation of metal-polluted soils.* In N. Terry and Q. Banuelos (Eds.), Phytoremediation of contaminated soil and water (pp. 85–197). Boca Raton, FL: Lewis.

[166] Kumar, P. B. A. N., Dushenkov, V., Motto, H., and Raskin, I. (1995). Phytoextraction: The use of plants to remove heavy metals from soils. *Environmental Science and Technology*, 29(5), 1232–1238.

[167] Blaylock, M. J., and Huang, J. W. (2000). *Phytoextraction of metals*. In I. Raskin and B. D. Ensley (Eds.), Phytoremediation of toxic metals: Using plants to clean-up the environment (pp. 53–70). New York: Wiley.

[168] Huang, J. W., Chen, J., Berti, W. R., and Cunningham, S. D. (1997). Phytoremediation of lead contaminated soil: Role of synthetic chelates in lead phytoextraction. *Environmental Science and Technology*, 31(3), 800–805.

[169] Lasat, M. M., Fuhrmann, M., Ebbs, S. D., Cornish, J. E., and Kochian, L. V. (1998). Phytoremediation of a radio cesium contaminated soil: evaluation of cesium-137 bioaccumulation in the shoots of three plant species. *Journal of Environmental Quality*, 27(1), 165–168.

[170] Tandy, S., Schulin, R., and Nowack, B. (2006). The influence of EDDS on the uptake of heavy metals in hydroponically grown sunflowers. *Chemosphere*, 62(9), 1454–1463.

[171] Cooper, E. M., Sims, J. T., Cunningham, S. D., Huang, J. W., Berti, W. R. (1999). Chelate-assisted phytoextraction of lead from contaminated soil. *Journal of Environmental Quality*, 28, 1709–1719.

[172] Lombi, E., Zhao, F. J., Dunham, S. J., and MacGrath, S. P. (2001a). Phytoremediation of heavy metal-contaminated soils: Natural hyperaccumulation versus chemically enhanced phytoextraction. *Journal of Environmental Quality*, 30, 1919–1926.

[173] Padmavathiamma P.K., Li L. Y., Phytoremediation Technology: Hyper-accumulation Metals in Plants, *Water Air Soil Pollut* (2007) 184:105–126.

[174] Reeves, R. D., and Brooks, R. R. (1983). Hyperaccumulation of lead and zinc by two metallophytes from a mining area of Central Europe. *Environmental Pollution Series A,* 31, 277–287.

[175] Baker, A. J. M., and Walker, P. L. (1989). *Ecophysiology of metal uptake by tolerant plants*. In A. J. Shaw (Ed.), Heavy metal tolerance in plants: Evolutionary aspects (pp. 155– 177). Boca Raton, FL: CRC.

[176] T., Brooks, R. R., Lee, J., and Reeves, R. D. (1976). Sebertia acumip. A nickel-accumulating plant from New Caledonia. *Science*, 193, 579–580.

[177] Brooks, R. R. (ed) (1998). *Plants that hyperaccumulate heavy metals* (p. 384). Wallingford: CAB International.

[178] Beath, O. A., Eppsom, H. F., and Gilbert, G. S. (1937). Selenium distribution in and seasonal variation of vegetation type occurring on seleniferous soils. J*ournal of the American Pharmaceutical Association*, 26, 394–405.

[179] Wang, J., Zhao, F., Meharg, A. A., Raab, A., Feldmann, J., McGrath, P. S. (2002). Mechanisms of arsenic hyperaccumulation in Pteris vittata. Uptake kinetics, interactions with phosphate, and arsenic speciation. *Plant Physiology*, 130, 1552–1561.

[180] Robinson, B. H., Brooks, R. R., Howes, A. W., Kirkman, J. H., Gregg, P. E. H. (1997). The potential of the highbiomass nickel hyperaccumulator Berkheya coddii for phytoremediation and phytomining. *Journal of Geochemical Exploration*, 60, 115–126.

[181] Leblanc, M., Petit, D., Deram, A., Robinson, B., and Brooks, R. R. (1999). The phytomining and environmental significance of hyperaccumulation of thallium by Iberis intermedia from southern France. *Economic Geology*, 94(1), 109–113.

[182] Brown, S.L., Chaney, R.L., Angle, J.S., Baker, A.J.M., 1995. Zinc and calcium uptake by hyperaccumulator *Thlaspi caerulescens* grown in nutrient solution,. *Journal of the American Soil Science Society*, 59, 12-133.

[183] Ebbs SD, Lasat MM, Brady DJ, Cornish J, Gordon R, Cochian LV. 1997. Phyto extraction of cadmium and zinc from a contaminated site.. *Journal of Environmental Quality*, 26, 1424-1430.

[184] Eapen, S., and D'Souza, S. F. (2005). Prospects of genetic engineering of plants for phytoremediation of toxic metals. *Biotechnology Advances*, 23, 97–114.

[185] Krämer, U., R.D. Smith, W. Wenzel, I. Raskin, and D.E. Salt, 1997. The role of metal transport and tolerance in nickel hyperaccumulation by Thlaspi goesingense Halacsy. *Plant Physiol* 115: 1641-1650.

[186] Krämer, U., J.D. Cotter-Howells, J.M. Charnock, A.J.M. Baker, and A.C. Smith, 1996. Free histidine as a metal chelator in plants that accumulate nickel. *Nature* 379: 635-638.

[187] Lasat, M.M., A.J.M. Baker, and L.V. Kochian, 1996. Physiological characterization of root Zn2+ absorption and translocation to shoots in Zn hyperaccumulator and nonaccumulator species of Thlaspi. *Plant Physiol* 112: 1715-1722.

[188] Wagner, G.J., 1984. Characterization of a cadmium-binding complex of cabbage leaves. *Plant Physiol* 76: 797-805.

[189] Steffens, J.C., 1990. The heavy metal-binding peptides of plants. *Annu Rev Plant Physiol Mol Biol* 41:553-575.

[190] Cobbett, C.S., and P.B. Goldsbrough, 1999. *Mechanisms of metal resistance: Phytochelatins and metallothioneins. In* Phytoremediation of Toxic Metals: Using Plants to Clean Up the Environment, eds. I. Raskin and B.D. Ensley, pp 247-269, John Wiley and Sons Inc, New York, NY.

[191] Liu, Y. G., Zhang, H. Z., Zeng, G. M., Huang, B. R., Li, X. (2006). Heavy metal accumulation in plants on Mn mine tailings. *Pedosphere*, 16(1), 131–136.
[192] Li, Y. M., Chaney, R. L., Angle, J. S., Baker, A. J. M. (2000). *Phytoremediation of heavy metal contaminated soils*. In D. L. Wise et al. (Eds.), Bioremediation of contaminated soils. New York: Marcel Dekker.
[193] Han, F. X., Banin, A., Kingery, W. L., Triplrtt, G. B., Zhou, L. X., Zheng, S. J., et al. (2003). New approach to studies of heavy metal redistribution in soil. *Advances in Environmental Research*, 8, 113–120.
[194] Z. Kufka, M. Kuras, *Heavy metals in soils contaminated from different sources*, in: N. Cheremissinoff Paul (Ed.), Ecological Issues and Environmental Impact Assessment, Gulf Publishing Company, Houston, TX, 1997, pp. 175–180.
[195] Suresh, B., Ravishankar, G. A. (2004). Phytoremediation – A novel and promising approach for environmental clean-up. *Critical Reviews in Biotechnology*, 24, 97–124.
[196] Odjegba, V. J., Fasidi, I. O. (2004). Accumulation of trace elements by Pistia stratiotes: Implications for phytoremediation. *Ecotoxicology*, 13, 637–646.
[197] Tian, J. L., Zhu, H. T., Yang, Y. A., He, Y. K. (2004). Organic mercury tolerance, absorption and transformation in Spartina plants. Zhi Wu Sheng Li Yu Fen Zi Sheng Wu Xue Xue Bao (*Journal of Plant Physiology and Molecular Biology*), 30, 577–582.
[198] Boonyapookana, B., Parkplan, P., Techapinyawat, S., DeLaune, R. D., Jugsujinda, A. (2005). Phytoaccumulation of lead by sunflower (Helianthus annuus), tobacco (Nicotiana tabacum), and vetiver (Vetiveria zizanioides). *Journal of Environmental Science and Health A*, 40, 117–137.
[199] Chandra Sekhar, K., Kamala, C. T., Chary, N. S., Balaram, V., Garcia, G. (2005). Potential of Hemidesmus indicus for phytoextraction of lead from industrially contaminated soils. *Chemosphere*, 58, 507–514.
[200] Sharma, N. C., Gardea-Torresdey, J. L., Parsons, J., Sahi, S. V. (2004). Chemical speciation and cellular deposition of lead in Sesbania drummondii. *Environmental Toxicology and Chemistry*, 23, 2068–2073.
[201] Mkandawire, M., Dudel, E. G. (2005). Accumulation of arsenic in Lemna gibba L. (duckweed) in tailing waters of two abandoned uranium mining sites in Saxony, Germany. *Science of the Total Environment*, 336, 81–89.
[202] Caille, N., Swanwick, S., Zhao, F. J., McGrath, S. P. (2004). Arsenic hyperaccumulation by Pteris vittata from arsenic contaminated soils and the effect of liming and phosphate fertilisation. *Environmental Pollution*, 132, 113–120.
[203] [203]. Broadhurst, C. L., Chaney, R. L., Angle, J. S., Maugel, T. K., Erbe, E. F., Murphy, C. A. (2004). Simultaneous hyperaccumulation of nickel, manganese, and calcium in Alyssum leaf trichomes. *Environmental Science and Technology*, 38, 5797–5802.
[204] Wei, S. H., Zhou, Q. X., Wang, X., Cao, W., Ren, L. P., Song, Y. F. (2004). Potential of weed species applied to remediation of soils contaminated with heavy metals. *Journal of Environmental Science- China*, 16, 868–873.
[205] Schwartz, C., Echevarria, G., Morel, J. L. (2003). Phytoextraction of cadmium with Thlaspi caerulescens. *Plant Soil,* 24, 27–35.

[206] Kubota, H., Takenaka, C. (2003). Arabis gemmifera is a hyperaccumulator of Cd and Zn. *International Journal of Phytoremediation*, 5, 197–120.
[207] Xiong, Y. H., Yang, X. E., Ye, Z. Q., He, Z. L. (2004). Characteristics of cadmium uptake and accumulation by two contrasting ecotypes of Sedum alfredii Hance. J*ournal of Environmental Science and Health*. Part A, Toxic/ Hazardous Substances and Environmental Engineering, 39, 2925–2940.
[208] Parker, D. R., Feist, L. J., Varvel, T. W., Thomason, D. N., Zhang, Y. Q. (2003). Selenium phytoremediation potential of Stanleya pinnata. *Plant Soil*, 249, 157–165.
[209] Bidwell S. D., Woodrow, I. E., Batianoff, G. N., Sommer- Knudsen, J. (2002). Hyperaccumulation of manganese in the rainforest tree Austromyrtus bidwillii (Myrtaceae) from Queensland, Australia. *Functional Plant Biology*, 29, 899–905.
[210] Xue, S. G., Chen, Y. X., Reeves, R. D., Baker, A. J., Lin, Q., Fernando, D. R. (2004). Manganese uptake and accumulation by the hyperaccumulator plant Phytolacca acinosa Roxb. (Phytolaccaceae). *Environmental Pollution*, 131, 393–399.
[211] Zhuang P., Yang Q. W., Wang H. B., Shu W. S., *Phytoextraction of Heavy Metals by Eight Plant Species in the Field Water Air Soil Pollut* (2007) 184:235–242.
[212] Brooks, R. R. (ed) (1998). *Plants that hyperaccumulate heavy metals* (p. 384). Wallingford: CAB International.
[213] Comis, D. (1996). Green remediation: Using plants to clean the soil. *Journal of Soil and Water Conservation*, 51(3), 184–187.
[214] Lasat, M., 2000. Phytoextraction of metals from contaminated soil: a review of plant/soil/ metal interaction and assessment of pertinent agronomic issues. *J. Hazardous Substance Research*, 2, 5–25.
[215] Mueller, B., Rock, S., Gowswami, Dib, Ensley, D. (1999): *Phytoremediation Decision Tree. – Prepared by - Interstate Technology and Regulatory Cooperation Work Group*; pp 1-36.
[216] Terry, N., Carlson, C., Raab, T. K., and Zayed, A. (1992). Rates of selenium volatilization among crop species. *Journal of Environmental Quality*, 21, 341–344.
[217] Banuelos, G. S., Cardon, G., Mackey, B., Ben-Asher, J.,Wu, L. P., Beuselinck, P., (1993). Boron and selenium removal in B-laden soils by four sprinkler irrigated plant species. *Journal of Environmental Quality*, 22(4), 786–797.
[218] McGrath, S. P. (1998). *Phytoextraction for soil remediation*. In R. R. Brooks (Ed.), Plants that hyperaccumulate heavy metals: Their role in phytoremediation, microbiology, archaeology, mineral exploration and phytomining (pp. 261–288). New York: CAB International.
[219] Lewis, B. G., Johnson, C. M., and Delwiche, C. C. (1966). Release of volatile selenium compounds by plants: Collection procedures and preliminary observations. *Journal of Agricultural and Food Chemistry*, 14, 638–640.
[220] Pilon-Smits, E. A. H. (2005). Phytoremediation. *Annual Review of Plant Biology*, 56, 15–39.
[221] Atkinson, R., Aschmann, S. M., Hasegawa, D., Eagle-Thompson, E. T., Frankenberger, J. R. (1990). Kinetics of the atmospherically important reactions of dimethylselenide. *Environmental Science and Technology*, 24, 1326–1332.
[222] Heaton, A. C. P., Rugh, C. L., Wang, N., Meagher, R. B. (1998). Phytoremediation of mercury- and methyl mercurypolluted soils using genetically engineered plants. *Journal of Soil Contamination*, 74, 497–510.

[223] Ma LQ, Komar KM, Tu C, Zhang WH, Cai Y, Kennelley ED (2001) A fern that hyperaccumulates arsenic: a hardy, versatile, fast-growing plant helps to remove arsenic from contaminated soils. *Nature* 409: 579-579.

[224] Sakakibara, Masayuki; Watanabe, Aya; Inoue, Masahiro; Sano, Sakae; and Kaise, Toshikazu (2007) "Phytoextraction And Phytovolatilization Of Arsenic From As-Contaminated Soils By Pteris vittata," *Proceedings of the Annual International Conference on Soils, Sediments, Water and Energy*: Vol. 12, Article 26.

[225] Kavamura V. N., Esposito E., Biotechnological strategies applied to the decontamination of soils polluted with heavy metals, 2010. *Biotechnology Advances, Volume 28, Issue 1, 61-69.*

[226] Hrudey, S. E. and Pollard, S. J. 1993. The challenge of contaminates sites: remediation approaches in North America. *Environmental Review.* 1 : 55-72.

In: Handbook of Phytoremediation
Editor: Ivan A. Golubev
ISBN: 978-1-61728-753-4

Chapter 8

LOOKING FOR NATIVE HYPERACCUMULATOR SPECIES USEFUL IN PHYTOREMEDIATION

***R. Fernández, I. Carballo, H. Nava, R. Sánchez-Tamés, A. Bertrand and A. González*[*]**

Departamento de Biología Organismos y Sistemas, Oviedo University, Catedrático Rodrigo Uría s/n, 33071 Oviedo, Spain

ABSTRACT

Human activities and industrial development lead to a deterioration of the environment that affects, to a greater or lesser extent, all countries. In Asturias (Spain), mining, steel mills and the chemical industry have produced wastes with high concentrations of heavy metals, with the consequent risks for the environment and human health. This problem requires an efficient and technologically feasible solution.

Phytoremediation is considered an effective, low-cost and environmental friendly technology for cleaning up heavy metal-polluted sites. It is based on the capacity of some plants, called hyperaccumulators, for taking these metals from the soil and accumulating them above a threshold value in their harvestable tissues.

One of the strategies that can be followed when working in phytoremediation is the use of native hyperaccumulator plants of high biomass, mainly those adapted to the climatic and soil conditions of the polluted site. According to this, the aim of our work was the identification of plants that spontaneously grow in different heavy metal-polluted soils of our region. After measuring the metal content of these plants, we selected the species according not only to their metal accumulation capacity, but also to the amount of biomass, percentage cover/aggregation, frequency of appearance in polluted areas or having special characteristics that make plants prone to hyperaccumulate metals, such as being nitrophilous or resistant to other types of stress. We tested in the greenhouse the effect of the heavy metals on plant growth and development and their maximum accumulation capacity. Thus, the plants selected were *Dittrichia viscosa* and *Betula celtiberica* for Cd, *Melilotus alba* for Pb, *Anthyllis vulneraria* for Zn, and *Carex pendula* for Hg. Later, we selected through *in vitro* culture the most accumulator plantlets of some of these species for further cloning and use in phytoremediation programs, so we obtained clone DV-A of *D. viscosa*, clone BC-K of *B. celtiberica*, and clone MA-X of *M. alba*.

[*] Area de Fisiología Vegetal, Dpto. B.O.S. Facultad de Biología, Universidad Oviedo, C/ Catedrático Rodrigo Uría s/n, Tel.: + 34 985 10 48 13 Fax: + 34 985 10 48 67, E-mail address: aidag@uniovi.es

1. Introduction

Intensive industrial development has degraded the environment in Asturias (Spain). Soil contamination with heavy metals is one of the most serious environmental problems. The most dangerous damage in the slag heaps of Asturias is produced by mining of lead, zinc, cadmium and mercury. Remediation of these soils is a challenge. Disposal of the toxic substances of the soil in conventional ways raises many technical and economic troubles. In these polluted zones, many attempts to restore the original vegetation or to plant species with commercial interests have failed due to adverse soil characteristics: toxic levels of heavy metals, predominance of rocks with different granulometry that impairs water availability, scarce nutrients, or lack of microorganisms beneficial as mycorrhizal fungus (Jasper et al., 1992). Because of the toxicity of high heavy metal concentration, land reclamation demands careful selection of plants adapted to grow on these soils. In spite of this, a reduced number of species can spontaneously colonize these polluted soils. Accumulators and/or hyperaccumulators are specialized plants that grow naturally on mineralized or metal-contaminated soils and are able to accumulate and tolerate high metal concentrations in shoots (Chaney, 1983; Baker and Brooks, 1989; Brooks, 1998; Reeves and Baker, 2000). The term hyperaccumulator was first used by Brooks et al. (1977) to describe plants that contain >1000 mg kg-1 nickel in their dried leaves and afterward extended to Cu (Malaisse et al., 1978), Co (Brooks et al., 1980) and Pb (Reeves and Brooks, 1983) but a 10000 mg kg-1 threshold was suggested by Baker and Brooks (1989) for Mn and Zn, which are normally present at higher and more widely varying concentration (about 27–150 mg kg-1). Normal Cd levels are so low (0.05–0.2 mg kg-1) that species capable of concentrating this element to >100 mg kg-1 should be regarded as Cd hyperaccumulators (Baker et al., 1994). The defined levels of these elements are one order of magnitude greater than those found in non-accumulator species (Salt and Kramer, 2000). Field and laboratory experiments on plant uptake and phytotoxicity have demonstrated that plants accumulate and tolerate moderate levels of inorganic Hg from both soil and atmospheric sources. However, Hg hyperaccumulating plants have not been identified so far.

Phytoremediation is a term applied to a group of technologies that use plants to reduce, remove, degrade, or immobilize environmental toxins, with the aim of restoring polluted sites (Peer et al., 2006).

Nowadays, phytoremediation efforts have been focused on the use of plants to accelerate degradation of organic contaminants, or remove hazardous heavy metals from soils or water. Any plant used as a phytoremediator must be able to tolerate high concentrations of the toxic substance, and its success depends on the selection of the appropriate plants. Hyperaccumulator plants are found in 45 different families, and 397 metal accumulating taxa have been identified to date with the highest occurrence among the Brassicaceae (Reeves and Baker, 2000). These plants are quite varied, from perennial shrubs and trees to small annual herbs.

Plants native to the target area should be considered since they are adapted to the local climate, insects, and diseases (Ensley, 2000). In order to make this technology feasible, the plants must extract large concentrations of heavy metals into their roots, translocate the heavy metals to aboveground biomass, produce a large quantity of plant biomass and also have an easy cultivation and harvest, preferably several times per year (Tong et al., 2004; Gosh and Singh, 2005).

Since most metal hyperaccumulators are small plants with low biomass, efforts are being made to find new hyperaccumulators, and plant culture approaches are being used to increase crop plant biomass.

The present work was conducted to address the following points:

1) Identify plant species and populations growing in contaminated areas, collecting information about their abundance and biomass.
2) Determine the content of Zn, Cd, Pb and Hg in soil samples from contaminated areas and of some plant species that grow in those areas, selecting the most interesting ones for phytoremediation purposes.
3) Study the metal uptake and accumulation capacity of the selected species in controlled greenhouse conditions.
4) Select and clone the most accumulator plantlets of those species through *in vitro* culture for their further use in phytoremediation.

2. Materials and Methods

2.1. Analysis of Soils from Polluted Areas

Based on information from the "Dirección General de Calidad Ambiental de la Consejería de Medio Ambiente del Principado de Asturias", we selected 7 locations in Asturias (northern Spain) contaminated with Zn, Cd, Pb and Hg. Selected sites were:

1) Lugones (L), a partial marshy land over 17000 m2 with an artificial lake, near a cuppermill, at 150 m of altitude, taking, having high concentration of Zn, Cd, Cu, As, Cr and Ni. Sampling in this area was made in 2 zones with different vegetation: L1, dry area between the factory and a small lake; L2, aquatic and wet zone, beside the factory.
2) La Granda (G) slag and ashes dump from a blast furnace, at 200 m altitude, taking over 80000 m2, having high soil concentration of Zn. Two zones were identified: G1, mug area; and G2, dry zone.
3) Peña –Terronal (P) wastes from a cinnabar mine at La Peña (Mieres), at an altitude of 240 m, over 20000 m2, having high soil concentration of Hg and Cd. Wastes are homogeneous, so no division in zones was made to collect samples.
4) Los Rueldos (LR) cinnabar slag heap, at 425 m of altitude, with a pronounced slope (170%) having high concentration of As, Hg and Pb. Two zones were distinguished for sampling, LR1 low zone; and LR2 high zone.
5) Muñón Cimero (MC) a slag heap of abandoned mine of cinnabar at 200 m from Lena, at 700 m altitude, reaching 1200 m2, having high soil concentration of Hg and As. Two different zones were identified: MC1 low slag heap area; MC2 in high slag heap area.

6) Maramuñiz (M) two slag heaps from abandoned cinnabar mine at an altitude of 600 m, with a 500 m2 surface, having high soil Hg and As concentration. Each heap was made a sampling zone, MR1 and MR2.
7) Control was a soil sampled from Redes natural reserve.

From each site three soil samples were taken from the plant root zone (15-20 cm depth), mixed and oven-dried at 40 °C for 48 h, grounded and passed through a 2 mm sieve and stored in plastic containers. For estimation of total metal content, soil samples (100 mg) were digested with HNO3: H2O2: HF (3:1:1, v/v) in order to liquidize the samples completely (Fargasova, 1994, Montaser, 1998). After cooling, two consecutive dilutions with doubly de-ionised water were made, first a dilution 1:50, followed by dilution 1:20 to which 10 μg Rh kg-1 concentrated HNO3 was added as internal standard. Triplicate soil samples were finally analysed for heavy metals content by inductively coupled plasma-mass spectrometry (ICP-MS) using an Agilent-7500c instrument connected to a PC. Blanks, in which the same procedure was followed but without adding soil material were also run.

To study the metal mobility in soils, a European Community Bureau of References (BCR) procedure (Rauret et al., 2000) was made with the sifted fraction divided in exchangeable and reducible, immediately available for the plant, and oxidizable and residual, available in the long term. For estimation of metal content in fraction samples we followed the same procedure as in soil samples (Fargasova, 1994; Montaser, 1998). The reference values for contaminated soils are those of the Dutch legislation (Visser, 1994).

2.2. Selection of Species Growing in Polluted Areas

2.2.1. Vegetation

In spring and autumn of 2003 we collected and identified the specimens found in all of these areas in an adequate phenologic state (Díaz-González et al., 1994). Vegetation inventories were perfomed following the classical sigmatist method (Braun-Blanquet and Pavillard, 1928). The study area was chosen from the soils sampled in section 2.1 according to their ecological and topographical characteristics. Their relative cover was set with the abundance-dominancy index, and plant sociability was determined following the indexes by Braun-Blanquet and Pavillard (1928). Then, a bibliographic research was made in order to highlight from the species found in these polluted areas the families related to heavy metal accumulator plants and the leguminous or nitrophilous species. We also marked the plants with good growth and high biomass, desirable characters in phytoremediation, and the species present in more than one inventory, as this could indicate tolerance to high levels of heavy metals. Plant percentage cover (Braun-Blanquet scale) is defined as the horizontal surface of the plant compared to the total surface of the inventory, and it is classified as: + = < 1%; 1 = 1% - 10%; 2 = 10% - 25%; 3 = 25% - 50%; 4 = 50% - 75%; 5 = 75% - 100%

Plant aggregation is the increase in the number of plants surrounding the mother plant, and their ranks are:

1 = growing solitary
2 = forming clumps or dense groups

3 = forming small parches or cushions
4 = forming larger carpets
5 = growing in large, almost pure population stands

The resulting inventories were analysed following the sintaxonomic scheme of Rivas Martínez et al. (2001).

2.2.2. Metal Accumulation in Plants from Polluted Sites

From all the plants collected in the polluted areas, we selected some of them according to the frequency of appearance in slag heaps, plant biomass, relationship between the species or its family with heavy metal accumulation and nitrophilous character that could promote heavy metal accumulation.

Plant samples of these species were rinsed once with water, twice with doubly de-ionized water (Milli-Q 185 Plus system, 18MΩ cm-1), oven dried at 70 ºC for 48 hours, powdered with liquid nitrogen and stored at 4 ºC until analysis. Dry samples (100 mg) of plant material were digested with 3 mL of concentrated and high-purified HNO3 (68%) in a microwave oven at 240W for 6 min according to Montaser (1998) and Fernández et al. (2008).

Triplicate plant samples were analysed for heavy metals by inductively coupled plasma-mass spectrometry (ICP-MS) using an Agilent-7500c instrument as before. The method was validated with the measurement of a certified reference sample (BCR-CRM 100 beech leaves).

2.3. Survival, Growth and Metal Accumulation under Greenhouse Conditions

We studied plant survival, growth and development and the maximum metal accumulation capacity of selected species from polluted areas when growing in greenhouse with known amounts of heavy metals, natural light and a temperature ranging 12-27 ºC. These species were Dittrichia viscosa, Betula celtiberica and Lotus corniculatus for Cd, Melilotus alba, D. viscosa and Equium vulgare for Pb, D. viscosa, B celtiberica and Anthyllis vulneraria for Zn, and Carex pendula for Hg.

In each 7 cm diameter-pot we placed 20 g of perlite and watered them with 60 mL of nutrient solution (Hoagland and Arnon, 1950) diluted at 25% with pH 5.7. Different concentrations of nitrates of each metal were added to the medium: 0, 1, 10, 50 mg Cd kg-1; 0, 10 and 100 mg Pb kg-1; 0, 10 and 100mg Zn kg-1; and 0, 1, 10 and 30 mg Hg kg-1. In the case of Pb nitrate, EDTA in equimolar concentration was also added to prevent Pb precipitation by increasing its solubility. Five pots per treatment were used and in each pot six seeds of each species were placed after their imbibition for 24 h in a 10% Hoagland and Arnon solution.

Plants were watered each 3-4 days with milli-Q water and each 10 days with 25% Hoagland and Arnon solution. After 3 weeks the germination percentage was determined and at 4 weeks three plants per pot were left. Plants were collected 3 months later and carefully washed with water, weighed and measured root and shoot length, number of leaves and roots, and oven dried at 70º C for 48 hours for obtaining dry weight. Then they were powdered with liquid nitrogen and analysed for metals by ICP- MS as described in section 2.1.

2.4. *In Vitro* Selection if the Most Accumulator Clone

Seeds of D. viscosa, B. celtiberica, M. alba, A. vulneraria and C. pendula were collected from plants growing on polluted areas. All seeds were imbibed for one hour in 1:10 dilution Hoagland solution (Hoagland and Arnon, 1950), then sterilized in 70% ethanol for 30 seconds followed by 3 one-minute rinses with sterile distilled water, and subsequently rinsed in a 20% commercial bleach dilution (50 g sodium hypochlorite L-1) for 15 minutes, followed by 3 five-minute rinses with sterile distilled water. Each seed was introduced into a culture tube containing half-strength Hoagland solution, but with Fe and vitamins substituted by those of Murashige and Skoog medium (MS) (1962). Sucrose at 30 g L-1 and agar at 0.7 g L-1 were added and the pH of the medium adjusted to 5.7. Culture tubes were kept in darkness at 25º C for 3 days in a growth chamber and then changed to 16 h photoperiod. Each germinated seed turned into a seedling, and apical shoot segments (10-25 mm) of these seedlings were transferred individually to a vessel with semi-solid medium under sterile conditions for subsequent cloning. An exception had to be made for C. pendula and A. vulneraria, since their particular development did not permit us obtaining apical shoot segments and the whole plant was divided and transferred. D. viscosa and B. celtiberica were grown on half-strength Murashige and Skoog medium (Murashige and Skoog, 1962) at pH 5.7. M. alba was grown on MS medium at pH 6.1 but with Fe substituted by 0.2 g L-1 of a commercial iron chelate (sequestrene 138 Fe G-100, Ciba Geigy) for preventing chlorosis (MSS medium). Finally, A. vulneraria was cultured in MS at pH 6.5 and C. pendula in MS at pH 5.7. Explants of each seedling will constitute a clone.

To study the effect of heavy metals on plant growth and their accumulation in plants, apical shoot segments of 2-2.5 cm long of clones of these species were transferred to medium containing each metal nitrate at the highest concentration that plants had tolerated in the greenhouse (chapter 3.3): D. viscosa to MS medium with 10 mg Cd kg-1, B. celtiberica to ½ MS medium with 10 mg Cd kg-1, M. alba to a MSS medium with 200 mg Pb kg-1, A. vulneraria to MS medium with 100 mg Zn kg-1, and C. pendula to MS with 20 mg Hg kg-1. In the case of lead, EDTA was also added to the culture medium. Plants cultured in metal-free medium were used as control. After 40 days in culture dry weight of shoot and root of the clones was measured. In order to select the greatest accumulator clone of each species, analyses for metal were made by ICP-MS as described in section 2.1. Shoot-to-root ratio (S/R) of metal accumulation was calculated by the following equation (Baker, 1981):

$$S/R = \frac{\text{mg metal kg}^{-1}\text{ dry wt. in shoots}}{\text{mg metal kg}^{-1}\text{ dry wt. in roots}}$$

2.5. Statistical Analysis

Data were analysed and tested for homogeneity of variance and for normal distribution using Sigmastat 3.5 from Systat®. ANOVA was used for the greenhouse growth data. For metal uptake experiments in the greenhouse and in vitro 3 analyses with 3 repeats were carried out for each plant and the Mann and Whitney's test was applied. The level of significance was 0.05 in all cases.

3. Results and Discussion

3.1. Analysis of Soils from Polluted Areas

Characteristics of the soils used in this study are shown in Table 1. Lugones reached values above 8000 mg Zn kg-1, 900 mg Pb kg-1 and 13 mg Cd kg-1 and so exceeded the threshold established by Kabata-Pendias and Pendias (2001) and Greger (2004) for a non-polluted soil and the intervention levels stated by Dutch legislation for a polluted soil. High concentrations of Hg were detected in L2 sample. At La Granda there were high levels of Zn only in mud zones, and Cd concentration was above normal values.

Table 1. Heavy metals in soil from contaminated areas

Slag heap Zone	mg kg^{-1}			
	Cd	Pb	Zn	Hg
Lugones				
L1	13.57*	941*	8824*	ND
L2	8.95	475	1233*	41*
La Granda				
G1	1.9	123	1010*	0.8
G2	ND	27	41	0.6
Peña Terronal				
P	ND	167	801*	914*
Los Rueldos				
LR1	ND	1922*	25	183*
LR2	ND	1545*	21	155*
LR3	ND	22667*	251	1302*
Muñón Cimero				
MC2	10.3	48	48	181*
Maramuñiz				
M	ND	14	41	ND
Parque de Redes (control)	0.012	9.89	0.14	ND
Non-polluted soil (Greger, 2004)	1-1	50-50	100-150	0.15-0.15
Intervention levels in Dutch legislation (Visser, 1994)	12	570	720	10

ND not detected. * Levels above critical limits (Visser, 1994).

High levels of Zn appeared at La Peña Terronal, and also Hg, as in Los Rueldos and Muñón Cimero due to historical cinnabar mining activities. At Los Rueldos, Pb contamination was the highest in all the samples assayed. Neither Maramuñiz nor control samples had any heavy metal level above critical limits. The BCR procedure gave us information about metal availability for plants and their concentration in each fraction (Table 2). At Lugones most part of Zn, Cd and Pb appeared in exchangeable (A) and reducible (B) fractions, which are available for plants. Sixty percent of Cd was available at La Granda, whereas only half of Pb and one third of Zn were in A and B fractions. At Muñón Cimero, Zn appeared at a very low concentration, but was the most mobile metal and half of Cd and a quarter of Pb were also available. The recovery percentage of Hg through the BCR method was low, and so was Hg concentration in every fraction, probably because this metal is extremely volatile and the BCR procedure implies some heating.

Table 2. Percentages of Zn, Cd, Pb and Hg found in each BCR fraction of soil from the studied areas (A: exchangeable fraction; B: reducible fraction; C: oxidable fraction; D: residual fraction)

Slag heap	Fraction	% total Cd	% total Pb	% total Zn	% total Hg
Lugones	A	58.75	28.75	49.37	0.97
	B	29.50	59.02	46.12	37.14
	C	3.23	3.05	0.90	1.21
	D	8.52	9.18	3.61	60.68
La Granda	A	29.63	0.39	13.65	9.02
	B	32.34	52.76	17.10	81.21
	C	2.15	1.63	0.07	9.77
	D	35.88	45.22	69.18	0.00
Muñón Cimero	A	15.32	0.28	41.48	0.72
	B	37.84	24.24	43.10	1.71
	C	24.96	3.26	15.42	74.38
	D	21.88	72.22	0.00	23.19

The total amount of metals present in the soils and their availability for plants emphasized the need of phytoremediation in those areas. Lugones would be a good place to make a study with Zn, Cd and Pb whereas La Granda and Muñón Cimero would be good for Zn and Hg respectively. Los Rueldos would be used in Pb and Hg studies.

3.2. Selection of Plant Species Growing in Polluted Areas

3.2.1. Vegetation

The inventories obtained in the slag heaps are constituted by 104 different species (Table 3). Twenty seven percent are grasses, coming from seeds of the meadows nearby. There are also 26% of nitrophilous or subnitrophilous including weeds of crops and roadside plants. These species are opportunists and produce a large number of seeds, and so tend to be secondary succesions.

Table 3. Species present in the slag heaps

1	Acer pseudoplatanus L.	T	53	Lonicera periclymenum L.	S
2	Anagallis arvensis L.	N	54	Lotus corniculatus L.	P
3	Anarrhinum bellidifolium (L.) Willd.	F	54	Malus domestica Borkh.	T, C
4	Anthyllis vulneraria L.	P	56	Matthiola incana (L.) R. Br.	C
5	Aster squamatus (Sprengel) Hieron.	N	57	Medicago lupulina L.	P
6	Betula celtiberica Rothm. and Vasc	T	58	Melilotus alba Medicus	N
7	Bidens aurea (Aiton) Sherff	C	59	Parentacetalia viscosa L. (Caruel)	P
8	Blackstonia perfoliata (L.) Hudson	N	60	Petrorhagia nanteuili (Burnat) P. W. Ball	P
9	Brachypodium pinnatum (L.) Beauv. spp rupestre (Host) Schübler and Martens	P	61	Picris hieraciodes L. s.l.	F
10	Calamintha nepeta ssp. nepeta	F	62	Pimpinella tragium Vill.	F
11	Calluna vulgaris (L) Hull	S	63	Pinus sp.	T, C
12	Carex pendula Hudson	A	64	Piptatherum miliaceum (L.) Cosson	N
13	Castanea sativa Mill	T	65	Plantago lanceolata L.	P
14	Centaurea debeauxi Gren. and Gordon	P	66	Plantago maior L.	N
15	Centaurium erytrhaea Rafn	N	67	Polygala vulgaris L.	P
16	Chaenorhinum origanifolium (L.) Kosteletaky	F	68	Polygonum persicaria L.	N
17	Clinopodium vulgare L.	F	69	Populus nigra L.	T
18	Conyza Canadensis (L.) Cronq.	N	70	Potentilla erecta (L) Raüsch	F
19	Cornus sanguinea L.	S	71	Potentilla reptans L.	N
20	Cortaderia selloana (Schultes and Schultes fil.) Ascherson and Graebner	C	72	Potentilla sterilis (L.) Garcke	P
21	Corylus avellana L.	T	73	Prunella vulgaris L.	P
22	Crataegus monogyna Jacq.	T	74	Pteridium aquilinum (L.) Kuhn	P
23	Crepis capillaries (L.) Wallr.	P	75	Quercus robur L.	T
24	Crepis vesicaria ssp. Taraxacifolia (Thuill. Thell. ex Schinz)	P	76	Ranunculus bulbosus L.	P
25	Daboecia cantabrica (Hudson) C. Koch	S	77	Reichardia picroides (L.) Roth	N
26	Dactylis glomerata L.	P	78	Rosa gr. canina L.	F, S
27	Daucus carota L.	P	79	Rubia peregrina L. s.l.	F
28	Desmazeria rigida (L.) Tutin	N	80	Rubus sampaioanus Sudre ex Samp	F
29	Digitaria sanguinalis (L.) Scop	N	81	Rubus ulmifolius Schott	F
30	Dittrichia viscosa (L.) W. Greuter	N, C	82	Rumex acetosa L.	P
31	Echium vulgare L.	P	83	Rumex obtusifolius L.	N
32	Equisetum ramosissimum Desf.	N	84	Salix atrocinerea Brot.	T
33	Equisetum telmateia Ehrh.	A	85	Salix caprea L	T
34	Erica cinerea L.	S	86	Sambucus nigra L.	N, T
35	Erica vagans L.	S	87	Sanguisorba minor Scop.	P
36	Eucalyptus globules Labill.	T, C	88	Scabiosa columbaria L.	P
37	Eupatorium cannabinum L.	A	89	Scirpus sylvaticus L.	A
38	Festuca rubra L.	P	90	Scrophularia scorodonia L.	N
39	Foeniculum vulgare Miller	F	91	Sedum album L.	N
40	Hedera helix L.	S	92	Silene inaperta L.	N
41	Helianthemum nummularium (L.) Miller	C	93	Solanum dulcamara L.	F
42	Hieracium gr. murorum L.	F	94	Sonchus tenerrimus L.	N

Table 3. (Continued)

43	Hirschfeldia incana (L.) Lagrèze-Fossat	N	95	Sporobolus tenacissimus (L. fill) Beauv.	N
44	Holcus lanatus L.	P	96	Teucrium scorodonia L.	F
45	Hypericum humifusum L.	N	97	Trachelium caeruleum L.	P, C
46	Inula conyza DC.	N	98	Trifolium dubium Sibth.	P
47	Juncus bulbosus L. spp bulbosus	A	99	Typha domingensis (Pers.) Steudel	A
48	Juncus effusus L.	A	100	Ulex cantabricus Alvarezandal.	F
49	Laurus nobilis L.	T	101	Ulmus minor L.	T,F
50	Leontodon hispidus L.	P	102	Urtica dioica L.	N
51	Leontodon taraxacoides (Vill) Merat.	P	103	Verbascum thapsus L.	N
52	Linum catharticum L.	P	104	Viola gr. riviniana Reichenb.	F

In these inventories there are 7% aquatic species or related to moist soils, due to the fact that some of the samples were taken from quagmires or lands that are flooded temporary or permanently. There are also 17% of herbaceous communities, vivacious, whose primary localization are the fringe forest and 21% of trees and shrubs, mostly riparian communities, as birches and willows, abundant in the studied areas. Lastly, we found a reduced group of crop and introduced plants (9%). As it was mentioned in the materials and methods, each slag heap was divided into several plots according to soil characteristics (mud, wet, dry and aquatic).

We identified the species growing in each of these plots, making a total of 13 inventories (Tables 4-9). The presence of many plants of the same species in one slag heap could imply tolerance to soil pollution. Thus, in a first selection, we choose 32 species (Table 10) according to their percentage cover/aggregation, frequency of appearance, biomass and all the characters already cited in section 2.2.2 of materials and methods. Most of the plants in this list are herbaceous, but there are also some woody species, such as *Betula celtiberica, Salix caprea* and *Salix atrocinerea*, and some ferns and aquatic plants as *Typha domingensis, Eupatorium cannabinum* and *Carex pendula*.

Table 4. Lugones inventory. The first number in the percentage cover/aggregation denotes cover level, the second denotes aggregation rank, and + is for a cover of the species less than 1% of the total inventory

L1 Inventory (aquatic plot) Altitude: 150 m Slope: 0% Cover: 50% Sampled area: 100 m^2		L2 Inventory (wet plot) Altitude: 150 m Slope: 0% Cover: 40% Sample area: 25 m^2	
Species	Cover/ aggregation	Species	Cover aggregation
Centaurea nigra	2.1	Typha domingensis	4.4
Conyza canadensis	2.1	Equisetum telmateia	2.3
Brachypodium pinnatum	1.3	Eupatorium cannabinum	2.3
Equisetum telmateia	1.3	Carex pendula	1.2
Echium vulgare	1.2	Conyza canadensis	1.1

Mentha suaveolens	1.2	Centaurea nigra	1.1
Anagallis arvensis	1.1	Dittrichia viscosa	+
Blackstonia perfoliata	1.1	Medicago lupulina	+
Daucus carota	1.1	Picris hieraciodes	+
Equisetum ramosissimum	1.1	Plantago lanceolata	+
Lotus corniculatus	1.1	Rubia peregrina	+
Rumex acetosa	1.1	Rubus ulmifolius	+
Sanguisorba minor	1.1	Acer pseudoplatanus	+
Hirschfeldia incana	+	Cornus sanguinea	+
Hypericum humifusum	+	Equisetum ramosissimum	+
Picris hieraciodes	+	Pteridium aquilinum	+
Potentilla reptans	+	Rosa grupo canina	+
Silene inaperta	+	Scirpus sylvaticus	+
		Ulmus minor	+

Table 5. La Granda inventory. The first number in the percentage cover/aggregation column denotes cover level, the second denotes aggregation rank, and + is for a cover of the species less than 1% of the total inventory

G1 Inventory (mud plot) Altitude: 200 m Slope: 15% Cover: 80% Sampled area: 100 m2		G2 Inventory (dry plot) Altitude: 200 m Slope: 15% Cover: 90% Sample area: 100 m2	
Species	Cover/ aggregation	Species	Cover/ aggregation
Hedera helix	2.2	Anthyllis vulneraria	3.3
Betula celtibérica	1.2	Cortaderia selloana	2.3
Laurus nobilis	1.1	Calamintha nepeta	2.2
Inula conyza	1.1	Dittrichia viscosa	2.2
Daucus carota	1.1	Foeniculum vulgare	2.2
Dactylis glomerata	1.1	Melilotus alba	2.2
Bartsia latifolia	1.1	Salix caprea	1.3
Agrostis stolonifera	1.1	Aster squamatus	1.1
Verbascum thapsus	+	Blackstonia perfoliata	1.1
Scrophularia scorodonia	+	Conyza canadensis	1.1
Salix atrocinerea	+	Dactylis glomerata	1.1
Rubus ulmifolius	+	Medicago lupulina	1.1
Rubia peregrina	+	Petrorhagia prolifera	1.1
Pteridium aquilinum	+	Pipthaterum miliaceum	1.1
Picris hieraciodes	+	Reichardia picroides	1.1
Lotus corniculatus	+	Solanum dulcamara	1.1
Lonicera periclymenum	+	Sonchus tenerrimus	1.1
Holcus lanatus	+	Sporobolus tenacissimus	1.1
Eucaliptus	+	Trachelium caeruleum	1.1

Table 5. (Continued)

G1 Inventory (mud plot) Altitude: 200 m Slope: 15% Cover: 80% Sampled area: 100 m2		G2 Inventory (dry plot) Altitude: 200 m Slope: 15% Cover: 90% Sample area: 100 m2	
Species	Cover/ aggregation	Species	Cover/ aggregation
		Desmazeria rigida	+
		Digitaria sanguinalis	+
		Populus nigra	+
		Trifolium dubium	+

Table 6. Peña-Terronal inventory. The first number in the percentage cover/aggregation denotes cover level, the second denotes aggregation rank, and + is for a cover of the species less than 1% of the total inventory

P Inventory Altitude: 240 m Slope: 0% Cover: 60% Sampled area: 100 m^2	
Species	Cover/aggregation
Salix caprea	1.1
Festuca rubra	1.1
Matthiola incana	1.1
Salix atrocinerea	1.1
Anarrhinum bellidifolium	+
Blackstonia perfoliata	+
Dactylis glomerata	+
Dittrichia viscosa	+
Holcus lanatus	+
Piptatherum miliaceum	+
Plantago lanceolata	+
Quercus robus	+

3.2.2. Metal Accumulation in Plants from Polluted Areas

Another important issue in plant selection is their metal accumulation capacity. From the species selected in table 10 we analysed by ICP-MS the ones with a percentage cover/aggregation higher than 1.1. An exception was made with Picris hieracioides, that in spite of having a percentage cover/aggregation of 2.2, it was not analysed because it was collected at Maramuñiz, which turned to be a metal-free site, and in the other inventories it had a (+) percentage cover/aggregation. Another exception was made with Lotus corniculatus, although it had a 1.1 percentage cover/aggregation, it was found in almost all the areas sampled. Some species were analysed in different plots because they grew simultaneously in several areas of the slag heap (Table 11).

Table 7. Los Rueldos inventory. The first number in the percentage cover/aggregation denotes cover level, the second denotes aggregation rank, and + is for a cover of the species less than 1% of the total inventory

LR1 Inventory (low plot) Altitude: 425 m Slope: 40 - 50% Cover: 80% Sampled area: 1000 m^2		LR2 Inventory (middle plot) Altitude: 425 m Slope: 50% Cover: 95% Sample area: 1000 m^2	
Species	Cover/aggregation	Species	Cover/aggregation
Betula celtiberica	3.3	Betula celtiberica	3.3
Calluna vulgaris	3.3	Calluna vulgaris	3.3
Castanea sativa	2.2	Pteridium aquilinum	3.3
Erica cinerea	2.2	Ulex cantabricus	2.2
Quercus robur	1.2	Daboecia cantabrica	2.2
Ulex cantabricus	1.1	Castanea sativa	1.2
Rubus ulmifolius	1.1	Quercus robur	1.2
Pteridium aquilinum	1.1	Rubus sp.	1.1
Daboecia cantabrica	+		

Table 8. Muñón Cimero inventory. The first number in the percentage cover/aggregation column denotes cover level, the second denotes aggregation rank, and + is for a cover of the species less than 1% of the total inventory

MC1 Inventory (roadside plot) Altitude: 700 m Slope: 30% Cover: 60% Sampled area: 200 m^2		MC2 Inventory (high right plot) Altitude: 700 m Slope: 30% Cover: 40% Sampled area: 200 m^2	
Species	Cover/aggregation	Species	Cover/aggregation
Centaurea nigra	2.3	Holcus lanatus	2.2
Equisetum telmateia	2.3	Acer pseudoplatanus	1.1
Dactylis glomerata	2.2	Anarrhinum bellidifolium	1.1
Plantago lanceolata	2.2	Salix caprea	1.1
Lotus corniculatus	1.1	Centaurea nigra debeauxi	+
Bidens aurea	+	Crepis capillaris	+

Table 8. (Continued)

MC3 Inventory (low plot) Altitude: 700 m Slope: 30% Cover: 50% Sampled area: 200 m^2		MC4 Inventory (high left plot) Altitude: 700 m Slope: 30 % Cover: 60% Sampled area: 200 m^2	
Species	Cover/ aggregation	Species	Cover/ aggregation
Festuca rubra	+	Juncus effusus	+
		Leontodon taraxacoides	+
		Rubus ulmifolius	+
		Thrincia hirta	+
Populus nigra	2.2	Holcus lanatus	3.3
Salix atrocinerea	2.2	Plantago lanceolata	2.2
Betula celtibérica	1.2	Anarrhinum bellidifolium	1.1
Malus domestica	1.2	Blackstonia perfoliata	1.1
Pinus sp.	1.2	Centaurea nigra debeauxi	1.1
Salix caprea	1.2	Centaurium erythraea	1.1
		Daucus carota	1.1
		Festuca rubra	1.1
		Leontodon taraxacoides	1.1
		Dactylis glomerata	1.1

Table 9. Maramuñiz inventory. The first number in the percentage cover/aggregation rate column denotes cover level, the second denotes aggregation rank, and + is for a cover of the species less than 1% of the total inventory

M1 Inventory Altitude: 600 m Slope: 60 % Cover: 60% Sampled area: 300 m^2		M2 Inventory Altitude: 600 m Slope: 30 % Cover: 60% Sample area: 300 m^2	
Species	Cover/ aggregation	Species	Cover/ aggregation
Clematis vitalva	3.3	Salix atrocinerea	2.3
Rubus ulmifolius	3.3	Salix caprea	2.3
Erica vagans	2.2	Clematis vitalba	2.2
Picris hieraciodes	2.2	Anthyllis vulneraria	1.1
Pteridium aquilinum	2.2	Chaenorhinum origanifolium	1.1
Teucrium scorodonia	2.2	Corylus avellana	1.1
Salix caprea	1.3	Helianthemum nummularium	1.1
Centaurium erytrhaea	1.1	Inula conyza	1.1
Clinopodium vulgare	1.1	Lotus corniculatus	1.1
Leontodon taraxacoides	1.1	Medicago lupulina	1.1

M1 Inventory Altitude: 600 m Slope: 60 % Cover: 60% Sampled area: 300 m^2		M2 Inventory Altitude: 600 m Slope: 30 % Cover: 60% Sample area: 300 m^2	
Species	Cover/ aggregation	Species	Cover/ aggregation
Ranunculum bulbosus	1.1	Scabiosa columbaria	1.1
Rosa canina	1.1	Sedum album	1.1
Crataegus monogyna	+	Teucrium scorodonia	1.1
Dactylis glomerata	+	Betula pubescens	+
Festuca rubra	+	Brachypodium pinnatum ssp rupestre	+
Hypericum sp	+	Centaurea nigra debeauxi	+
Leontodon hispidus	+	Clinopodium vulgare	+
Linum catharticum	+	Hieracium grupo murorum	+
Medicago lupulina	+	Holcus lanatus	+
Pimpinella tragium	+	Picris hieraciodes	+
Plantago lanceolata	+	Prunella vulgaris	+
Potentilla sterilis	+	Rubus ulmifolius	+
Prunella vulgaris	+	Viola gr. riviniana	+

Table 10. Species selected from slag heaps with interesting characters for phytoremediation. N denotes nitrophilous character, and L indicates a leguminous species

Species	Cov/ Agre	Appearance frecuency in inventory	More common in zone	Size (cm)	Relation with metals	N and/or L
Typha domingensis (Pers.) Steudel *	4.4	1	L	250		
Betula celtiberica Rothm. and Vasc *	3.3	6	LR	Tree	Zn, Pb	
Pteridium aquilinum (L) Kuhn*	3.3	5	LR	180	Zn, Pb, Cu	
Anthyllis vulneraria L.*	3.3	2	G	<100	Cd, Pb, Ni	L
Centaurea debeauxi Gren and Gordon*	2.3	6	MC	<100		N
Equisetum telmateia Ehrh.*	2.3	4	L	100	Au, Zn, Pb	
Eupatorium cannabinum L.*	2.3	1	L	150	Cd, Co, Cr	
Salix atrocinerea Brot. *	2.2	5	MC	Tree	Zn, Mn, Cd	
Holcus lanatus L.*	2.2	5	MC	<100	Cd, Zn, Pb	
Dittrichia viscosa (L.) W. Greuter*	2.2	3	G	100	As, Cd, Pb,	N
Foeniculum vulgare Miller*	2.2	1	G	120		N

Table 10. (Continued)

Species	Cov/ Agre	Appearance frecuency in inventory	More common in zone	Size (cm)	Relation with metals	N and/or L
Picris hieraciodes L. s.l.	2.2	4	M	100		
Melilotus alba Medicus*	2.2	1	G	150		L
Conyza canadensis (L.) Cronq.*	2.1	3	L	100		N
Salix caprea L.*	1.2	5	MC	Tree	Zn, Pb, Cd	
Carex pendula Hudson *	1.2	1	L	150	Cu, Co	
Echium vulgare L*	1.2	1	L	<100	Pb, Zn, Cu	N
Malus domestica Borkh.*	1.2	1	S	Tree		
Picris hieraciodes L. s.l.	2.2	4	M	100		
Piptatherum miliaceum (L.) Cosson	1.1	2	G	150	Pb, Zn	N
Lotus corniculatus L.*	1.1	5	G, L, M,MC	<100	Se, B, Cu, Zn	L
Medicago lupulina	1.1	4	G, M	<100	Ni, Zn, Pb	N, L
Acer pseudoplatanus L.	1.1	2	MC	Tree	Cd, Zn	
Equisetum ramosissimum Desf.	1.1	2	L	150	Zn, Pb, As	
Solanum dulcamara L.	1.1	1	G	250		N
Aster squamatus (Sprengel)	1.1	1	G	100		N
Corylus avellana L.	1.1	1	M	Tree	Fe, Zn, Cu	
Laurus nobilis L.	1.1	1	G	Tree		N
Rumex acetosa L.	1.1	1	L	<100	Cd, Pb	N
Juncus bulbosus L.ssp *bulbosus*	+	1	MC	<100	Ni	
Juncus effusus L.	+	1	MC	150	Ni,As, Zn	
Scirpus sylvaticus L.	+	1	L	<100	Cu, Co	
Silene inaperta L.	+	1	L	<100	Pb, Zn	

* Denotes selection for further analyses of heavy metal content.

Table 11. Heavy metal accumulation of selected species at contaminated areas

Slag Heap	Zone	Species	mg kg-1			
			Cd	Pb	Zn	Hg
Lugones	L1	Centaurea debeauxi	0.08	5.67	48	ND
		Conyza canadensis	2.29	ND	142	ND
		Echium vulgare	ND	0.82	80	ND
		Lotus corniculatus	ND	ND	52	ND
	L2	Carex pendula	ND	1.89	99	28.3
		Typha dominguensis	ND	1.68	71	ND

		Pteridium aquilinum	0.3	4.49	193	ND
		Eupatorium cannabinum	1.90	2.91	242	0.8
		Equisetum telmateia	ND	2.01	182	ND
	G1	Betula celtiberica	0.28	3.5	89	ND
	G2	Foeniculum vulgare	1.07	9	478	ND
La Granda		Anthyllis vulneraria	0.05	0	75	0
		Ditrichia viscosa	2.71	21	799	ND
		Melilotus alba	0.34	47.4	184	0.2
	LR1	Betula celtiberica	0	1.92	105	ND
Los Rueldos		Pteridium aquilinum	0	2.37	42.8	0.4
	LR2	Betula celtiberica	0.04	18.6	226	ND
		Pteridium aquilinum	0.83	55	49	ND
	S3	Salix caprea	1.1	1.67	284	9
		Salix atrocinerea	2.9	1.1	215	11
Muñón Cimero		Malus domestica	0.2	0.5	24	ND
	S4	Betula celtiberica	8.43	4.79	361	0.1
		Holcus lanatus	0.2	2.52	48	10
Amount of metal in non-contaminated plants (mg kg-1)			0.05-0.2	5-10	27-150	0.1
Amount of metal toxic for plants (mg kg-1)			5-30	30-300	100-400	1-3

ND not detected.

For the correct interpretation of the results, it must be considered that the amount of metal accumulated by each species depends on the level of soil contamination they grew on, and this could give an idea of their accumulating capacities. Metal concentrations considered as normal in plants are: 27-150 mg Zn kg-1, 5-10 mg Pb kg-1, 0.05- 0.2 mg Cd kg-1 and 0.1 mg Hg kg-1, whereas excessive or toxic levels are: 100-400 mg Zn kg-1, 30-300 mg Pb kg-1, 5-30 mg Cd kg-1 and 1-3 mg Hg kg-1 (Kabata-Pendias and Pendias, 2001; Greger, 2004).

According to their cadmium content, Betula celtiberica from Muñón Cimero was the most Cd accumulator species, followed by Salix atrocinerea also from Muñón Cimero and Dittrichia viscosa from La Granda (Table 11). Melilotus alba, from La Granda has the highest Pb content, more than 2-fold of the Pb found in D. viscosa also from La Granda and B. celtiberica from los Rueldos (Table 11). Many species accumulated Zn at toxic levels or even higher (Table 11), such as D. viscosa and Foeniculum vulgare from La Granda and B. celtiberica and Salix caprea from Muñón Cimero. Most of the plants studied did not accumulate much Hg and the highest levels were found in Carex pendula from Lugones (28 mg Hg kg-1 dry wt.). In Muñón Cimero, S. atrocinerea, S. caprea and Holcus lanatus accumulated Hg above toxic levels (Table 11). Based on these studies, we selected B. celtiberica as woody and D. viscosa as herbaceous for Cd; M. alba, D. viscosa and B. celtiberica for Pb; D. viscosa and B. celtiberica for Zn and for Hg we only selected C. pendula, since it was by far the most accumulator.

3.3. Survival, Growth and Metal Accumulation of Selected Plants under Greenhouse Conditions

Once selected from the polluted areas the species that accumulated the highest amount of metal, and also had a great percentage cover/aggregation and frequency of appearance, we study which was the highest metal concentration that these plants could tolerate and how much metal they could accumulate. In all cases, the selected plants were compared with less accumulator plants (Lotus corniculatus, Echium vulgare and Anthyllis vulneraria) but with high percentage cover/aggregation and/or frequency of appearance in the same areas in order to check their accumulation capacity at different metal concentrations.

3.3.1. Studies with Cadmium

Germination of *D. viscosa* when grown with 50 mg Cd kg^{-1} did not differ from the control and the survival rate was over 75% at all the assayed conditions (Table 12).

Shoot length and number of leaves diminished when 10 and 50 mg Cd kg-1 were added to the substrate; however root length was only reduced at 50 mg Cd kg-1. At 10 and 50 mg Cd kg-1 the fresh and dry weight were also reduced (Table 12). Root length can be used as an index of metal tolerance (Wilkins, 1978; Mádico et al., 1992), since root is in contact with the metal in the substrate and it will be affected more rapidly by high metal concentrations than other parts of the plant.

The response of non-tolerant species to the increase of metal concentration in the substrate is growth inhibition due to metal toxicity. In the case of Cd, plant growth and root elongation inhibition are symptoms of metal-induced stress (Wilkins, 1978; Küpper et al., 2004), because Cd changes the permeability of the cellular membrane in roots, and so changes ion fluxes, inhibits respiration in some species (Obata et al., 1997; Llamas et al, 2000), and starts oxidative stress routes (Dat et al., 2000). The metal content in plants increased as long as increased Cd concentration in the substrate. D. viscosa grown at 50 mg Cd kg-1 reached 266 mg Cd kg-1 dry wt. in the whole plant (Table 12).

In the case of B. celtiberica, germination and survival were similar to the control in all treatments (Table 12). Shoot length and number of leaves were reduced when growing at 10 mg Cd kg-1 whereas root length was reduced at 50 mg Cd kg-1.

According to the tolerance index of Wilkins (1978), B. celtiberica and D. viscosa had similar Cd tolerance levels because their root length diminished at the same Cd concentration. In other woody species, such as beech and pine (Schützendübel et al., 2001), and in other herbaceous species, such as Thlaspi caerulescens and Silene vulgaris (Schat et al., 2002) root length was also inhibited as a response to exposure to Cd. Apart from a high metal content, accumulator plants show also great metal tolerance (Wójcik et al., 2005), and this implies the activation of intracellular detoxification mechanisms.

The cost of this defence is high, so these plants often have lower growth and biomass (Barceló and Poschenrieder, 2003). When B. celtiberica was cultured at 50 mg Cd kg-1 metal content in plants increased as rose the Cd concentration added to the substrate, reaching values above 600 mg Cd kg-1 dry wt. in the whole plant (Table 12). Germination and survival of Lotus corniculatus were similar to the ones of D. viscose, except at 50 mg Cd kg-1, where germination increased but survival decreased (Table 12).

Shoot length did not differ much among the different Cd concentrations assayed, except at 1 mg kg-1 in substrate, where shoot length, and fresh and dry weight increased compared to control plants. Root growth only diminished at 50 mg kg-1, as happened in the other two species tested. However, Lotus corniculatus accumulated only 31 mg Cd kg-1 dry wt., whereas Cd content in D. viscosa and B.celtiberica was 8 and 20-fold higher respectively.

Thus, we discarded Lotus corniculatus, that although appeared frequently in the slag heaps it had a low percentage cover/aggregation and it did not accumulate much Cd. Considering the results we can also confirm that this species do not increase its accumulation capacity at higher metal levels when grown under greenhouse conditions.

Finally, we selected B. celtiberica and D. viscosa for further in vitro studies.

3.3.2. Studies with Lead

Germination of Melilotus alba was not affected by the addition of Pb (Table 13) being their survival maximum at 100 mg Pb kg-1. The shoot and root length, number of leaves and fresh and dry weight were not affected by Pb at the assayed concentrations (Table 13). Plants rapidly respond to absorbed Pb, through inhibition of growth, change in branching pattern, stunted growth of the plant, water imbalance, disturbed mineral nutrition and chlorosis (Sharma and Dubey, 2005). Inhibition of root growth, is one of the primary effects of Pb toxicity, propably due to the inhibition of cell division in the root tip by perturbing the aligment of microtubules (Eun, et al., 2000; Yang et al., 2000), but this effect was not observed in M. alba, so this species seems to be tolerant to Pb. Biochemical tolerance to Pb is related to the capacity of plants to restrict Pb to the cell walls, synthesis of osmolytes and activation of the antioxidant defence system. On the contrary, Wierzbicka and Potocka (2002) suggest that constitutional tolerance to lead may be associated with the water requirements of a plant species, so that plants growing in moist soils show the highest tolerance, but these differences are not the result of differences in the amount of lead taken up by the particular species, nor their ability to distribute lead in tissues and cells.

As far as Pb content is concerned accumulation in the plant at 100 mg Pb kg-1 is 15-fold than at 10 mg Pb kg-1 (Tab. 13).

When D. viscosa is cultured with Pb, neither germination nor survival were affected (Table 13). Moreover, no differences in growth were found among treatments (Table 13). Pb accumulation was lower than in M. alba (Table 13), although it reached noxious values of 30-300 mg Pb kg-1 (Kabata-Pendias and Pendias, 2001).

Pb inhibited germination in B. celtiberica at low concentration (10 mg Pb kg-1 in substrate), whereas survival was only reduced at 100 mg Pb kg-1 in substrate. However no differences in growth were found among the Pb concentrations assayed (Table 13). The highest Pb accumulation in this species was found at 100 mg Pb kg-1 in substrate (Table 13), exceeding by far toxic levels.

Echium vulgare shows low Pb accumulation levels in the slag heaps. Pb did not affect the survival and even increased the germination of these plants (Table 13). Shoot length was similar to control at all assayed concentrations, whereas root length only decreased at 100 mg Pb kg-1 in substrate. However, fresh and dry weights and number of leaves were reduced in all Pb-treated plants with no differences among treatments. Plants accumulated Pb gradually up to 585 mg Pb kg-1 dry wt. when were cultured on a substrate with 100 mg Pb kg-1 (Table 13). From these results we can conclude that the accumulation capacity of M. alba is much higher than that of the other species tested, so it was selected for further studies.

Table 12. Germination (G), survival (S), growth and Cd concentration of *Dittrichia viscosa*, *Betula celtiberica* and *Lotus corniculatus* after 3 months in the greenhouse. Different letters indicate significant differences at $p<0.05$

	mg Cd kg-1 substrate	G (%)	S (%)	Length (cm)		N. Leaves	Weight (mg)		mg Cd kg-1 dry wt. plant
				Shoot	Root		Fresh	Dry	
Dittrichia viscosa	0	37 a	86 a	4.43 ± 0.35 a	10.36 ± 0.48 a	8.97 ± 0.48 a	226.58 ± 25.42 a	36.60 ± 0.75 a	1.90 ± 0.60 a
	1	37 a	76 a	4.59 ± 0.54 a	11.29 ± 0.79 a	9.70 ± 0.51 a	273.93 ± 40.67 a	34.40 ± 3.30 a	34.40 ± 6.80 b
	10	43 a	86 a	3.04 ± 0.35 b	10.07 ± 0.88 a	7.72 ± 0.31 b	187.00 ± 24.67 b	25.40 ± 2.27 a	186.40 ± 97.80 c
	50	41 a	91 a	0.79 ± 0.09 c	6.70 ± 0.53 b	5.89 ± 0.28 c	57.70 ± 6.42 b	7.70 ± 0.50 b	266.80 ± 20.70 c
Betula celtiberica	0	23 a	100 a	10.89 ± 0.67 a	11.96 ± 0.85 a	9.83 ± 0.44 a	1004.50 ± 118.62 a	180.19 ± 21.28 a	2.84 ± 0.46 a
	1	24 a	95 a	10.58 ± 0.79 a	12.71 ± 0.92 a	9.25 ± 0.41 a	1331.75 ± 218.65 a	244.83 ± 40.20 a	14.40 ± 5.88 b
	10	32 b	96 a	7.39 ± 0.62 b	11.29 ± 0.74 a	5.79 ± 0.33 b	1112.29 ± 147.04 a	175.07 ± 23.14 a	96.17 ± 15.45 c
	50	23 a	94 a	1.59± 0.30 c	6.58 ± 0.38 b	4.16 ± 0.33 c	150.87 ± 24.49 b	23.62 ± 4.94 b	619.53 ± 17.29 d
Lotus corniculatus	0	37 a	82 a	3.18 ± 0.46 a	4.03 ± 0.37 a	6.97 ± 0.61 a	62.10 ± 9.82 a	11.60 ± 0.96 a	0.49 ± 0.12 a
	1	47 a	87 a	5.23 ± 0.93 b	5.47 ± 0.59 a	7.53 ± 0.51 a	169.50 ± 37.85 b	24.00 ± 2.97 b	7.82 ± 2.16 b
	10	47 a	93 a	3.53 ± 0.36 a	2.73 ± 0.35 b	7.23 ± 0.56 a	74.60 ± 9.68 a	13.20 ± 0.87 a	29.36 ± 1.84 c
	50	53 b	67 b	2.43 ± 0.25 a	1.10 ± 0.11 c	6.60 ± 0.46 a	30.10 ± 2.83 c	6.40 ± 0.60 a	31.17 ± 3.22 c

Table [illegible]. Germination (G), survival (S), growth and Pb concentration of *Melilotus alba*, *Dittrichia viscosa*, *Betula celtiberica* and *Equium vulgare* after 3 months in greenhouse. Different letters indicate significant differences at p<0.05

	mg Pb kg-1 substrate	G (%)	S (%)	Length (cm)		N. Leaves	Weight (mg)	
				Shoot	Root		Fresh	Dry
Melilotus alba	0	23 a	69 a	10.91 ± 1.66 a	13.48 ± 1.45 a	8.88 ± 2.34 a	475.75 ± 115.30 a	84.40 ± 26.74 a
	10	20 a	69 a	11.87 ± 0.83 a	14.06 ± 1.22 a	7.57 ± 1.29 a	563.57 ± 146.40 a	137.50 ± 61.08 a
	100	23 a	95 b	13.88 ± 0.96 a	14.02 ± 0.91 a	10.30 ± 1.89 a	910.03 ± 205.10 a	215.80 ± 35.19 a
Dittrichia viscosa	0	41 a	75 a	9.62 ± 0.90 a	8.85 ± 0.54 a	12.1 ± 0.74 a	501.60 ± 72.78 a	259.25 ± 60.63 a
	10	39 a	100 b	8.35 ± 1.11 a	9.81 ± 0.85 a	11.29 ± 0.87 a	441.57 ± 96.96 a	207.02 ± 64.62 a
	100	38 a	100 b	6.33 ± 0.17 a	9.87 ± 0.43 a	10.67 ± 0.33 a	402.33 ± 65.82 a	194.07 ± 0.00
Betula celtiberica	0	36 a	100 a	11.18 ± 0.95 a	11.93 ± 0.73 ab	9.09 ± 0.65 a	1151.31 ± 159.02 ab	206.60 ± 28.56 ab
	10	9 b	100 a	12.53 ± 0.96 a	14.27 ± 1.26 b	8.73 ± 0.52 ab	964.05 ± 46.24 ab	174.49 ± 8.37 ab
	100	8 b	67 b	12.46 ± 1.94 a	11.50 ± 0.98 ab	7.56 ± 0.63 b	1207.33 ± 198.73 b	220.31 ± 36.16 b
Equium vulgare	0	12 a	100 a	12.31 ± 0.46 a	6.33 ± 0.85 a	14.52 ± 0.35 a	4674.32 ± 142.61 a	622.31 ± 19.04 a
	10	12 a	100 a	13.22 ± 0.96 a	5.29 ± 0.24 a	10.73 ± 0.24 b	3424.03 ± 620.12 b	416.98 ± 65.6 b
	100	28 b	100 a	12.83 ± 0.87 a	4.41 ± 0.27 b	11.17 ± 1.10 b	3319.52 ± 755.66 b	474.52 ± 70.83 b

Table 14. Germination (G), survival (S), growth and Zn concentration of *Dittrichia viscosa*, *Betula celtiberica* and *Anthyllis vulneraria* after 3 months in greenhouse. Different letters indicate significant differences at p<0.05

	mg Zn kg^{-1} substrate	G (%)	S (%)	Length (cm)		N. Leaves	Weight (mg)		mg Zn kg^{-1} dry wt. plant
				Shoot	Root		Fresh	Dry	
Dittrichia viscosa	0	42 a	100 a	4.81 ± 0.33 a	11.69 ± 0.65 a	9.51 ± 0.34 a	178.82 ± 15.62 a	27.82 ± 2.89 a	47.72 ± 15.17 a
	10	39 a	93 a	7.91 ± 0.43 b	8.29 ± 0.66 b	9.14 ± 0.38 a	233.93 ± 31.06 b	35.81 ± 4.74 a	329.04 ± 28.29 b
	100	27 b	80 b	0.49 ± 0.07 c	1.75 ± 0.24 c	5.15 ± 0.37 b	10.95 ± 2.07 c	1.58 ± 0.40 b	1980.97 ± 90.70 c
Betula celtiberica	0	6 a	100 a	5.03 ± 0.37 a	10.76 ± 0.46 a	6.00 ± 0.00 a	541.67 ± 18.62 a	90.32 ± 12.58 a	20.68 ± 10.8 a
	10	16 ab	100 a	5.78 ± 0.47 a	10.91 ± 0.34 a	6.41 ± 0.30 a	103.85 ± 11.56 a	92.13 ± 10.33 a	227.04 ± 18.1 b
	100	32 b	94 a	5.49± 0.48 a	12.08 ± 0.81 a	6.75 ± 0.33 c	221.73 ± 39.22 b	30.52 ± 5.39 a	657.03 ± 20.39 d
Anthyllis vulneraria	0	32 a	100 a	9.02 ± 0.62 a	10.52 ± 0.63 a	9.00 ± 1.00 a	682.50 ± 119.02 a	119.16 ± 20.83 a	53.87 ± 5.66 a
	10	32 a	100 a	10.83 ± 0.68 a	11.41 ± 0.93 a	11.15 ± 1.58 a	1094.41 ± 155.32 b	220.24 ± 31.17 b	229.66 ± 22.74 b
	100	12 b	100 a	9.43 ± 1.07 a	10.30 ± 0.73 c	14.21 ± 0.75 b	1233.03 ± 226.48 c	196.84 ± 33.92 b	2078.28 ± 196.56 d

Table 15. Germination (G), survival (S), growth and Hg concentration of *Carex pendula* after 3 months in greenhouse. Different letters indicate significant differences at p<0.05

mg Hg kg^{-1} substrate	G (%)	S(%)	Length (cm)		N. Leaves	Weight (mg)		mg Pb kg^{-1} dry wt. plant
			Shoot	Root		Fresh	Dry	
0	50 a	100 a	10.56 ± 0.5 a	11.01 ± 0.52 a	7.47 ± 0.27 a	210.93 ± 22.85 a	49.41 ± 4.4 a	1.00 ± 0.29 a
1	87 c	100 a	12.09 ± 0.46 b	12.39 ± 0.59 a	7.00 ± 0.19 a	216.32 ± 19.96 a	47.82 ± 5.4 a	2.31 ± 0.34 b
10	67 b	100 a	12.23 ± 0.43 b	12.81 ± 0.6 a	7.59 ± 0.17 a	188.64 ± 17.92 a	43.39 ± 1.61 a	16.03 ± 2.65 c
30	87 c	87 b	6.55 ± 0.36 c	7.71 ± 0.46 b	6.36 ± 0.15 b	80.75 ± 7.38 b	15.21 ± 0.86 b	111.29 ± 17.68 d

3.3.3. Studies with Zinc

Germination and survival in D. viscosa decreased when the Zn concentration in substrate rose (Table 14). Despite that the culture in 10 mg Zn kg-1 increases shoot length and fresh weight, at 100 mg Zn kg-1 all the parameters tested were drastically inhibited (Table 14). Metal accumulators plants have higher requirements of this essential micronutrient, and some species even show a positive growth response to a moderate increase of Zn in the substrate (Baker et al., 1994).

Barceló et al., (1996) working with Brassicaceae stated that the Zn concentration in substrate that allows to reach an optimum range will promote growth of plants, they will even grow more than in control conditions, mostly in hyperaccumulator plants. These two statements agree with the increase in growth found in D. viscosa at 10 mg Zn kg-1 in substrate whereas at 100 mg Zn kg-1 there is a clear metal toxicity. Zn concentration in D. viscosa exceeded 1900 mg Zn kg-1 dry wt. (Table 14) when plants grew at 100 mg Zn kg-1 in substrate although their growth was decreased. This Zn concentration was much higher than the toxic values for a plant (100-400 mg Zn kg-1, Kabata-Pendias and Pendias, 2001). Zn promoted germination in Betula celtiberica and no differences were found when compared with the control in all the parameters tested (Table 14). According to Wilkins (1978) B. celtiberica tolerate up to 100 mg Zn kg-1 in substrate and no changes were measured in root length. These results are in line with those of Utriainen et al. (1997) in B. celtiberica exposed to Zn and Cu.B. celtiberica accumulated Zn over the threshold values, reaching more than 600 mg Zn kg-1 dry wt.

Anthyllis vulneraria did not accumulate Zn in the slag heaps. Germination was reduced by 70% at 100 mg Zn kg^{-1}, but survival was 100% in all cases (Table 14). Growth was promoted by Zn at all the concentrations assayed (Table 14). Analysis by ICP-MS revealed that in *A. vulneraria* the Zn content increased as it rose in soil (Table 14), exceeding 2000 mg Zn kg^{-1} dry wt. of plant without a reduction in growth, so metal was not toxic for the plant, on the contrary that in *D. viscosa*, whose growth was greatly reduced. From the tested species *A. vulneraria* is the most tolerant accumulator plant for Zn, so we selected it for further analyses.

3.3.4. Studies with Mercury

Germination of C. pendula increased with Hg addition, but survival slightly diminished when plants were cultured with 30 mg Hg kg-1 (Table 15). None of the growth parameters measured was affected by the low Hg concentrations, except for shoot length that increased. At 30 mg Hg kg-1, all parameters decreased (Table 15). So the upper limit of Hg tolerance in these plants was 10 mg Hg kg-1. Hg is considered to be one of the most toxic metals because it specifically inactivates enzymes, particularly the cytochrome oxidases of the respiratory pathway. Besides, Hg can bind to SH-group-containing enzymes, structurally altering proteins, and destroying cell membranes and DNA (Kvesitadze et al., 2006). Therefore, Hg can disturb almost any function in which proteins are involved. As the concentration of Hg in the soil increased, the Hg accumulation in plant rose reaching 111 mg Hg kg-1 when was cultured at 30 mg Hg kg-1 in soil, that is 35-fold higher than the toxic value for a plant (Greger, 2004). In summary, plants selected in the greenhouse as the most accumulator ones and that will be used later for obtaining the most accumulator clone by in vitro culture were: Dittrichia viscosa and Betula celtiberica for Cd, Melilotus alba for Pb, Anthyllis vulneraria for Zn and Carex pendula for Hg.

3.4. *In Vitro* Selection of the Most Accumulator Clone

Once selected the most accumulator species, it was necessary to have enough plants for soil phytoremediation. If we start this process from many different seeds, it could happen that metal accumulation was not the same accumulation in all plants, because of the genetic variability. Greger and Landberg (1999) found a huge difference in metal uptake and distribution of Cd, Zn and Cu in different clones of Salix vitaminalis. Thus, we decided to clone the most accumulator through micropropagation and so increase the number of plants at our disposal for further experiments.

3.4.1. Dittrichia Viscosa and Betula Celtiberica for Cd

Clones of D. viscosa differed in Cd accumulation (Figure 1a), being DV-A clone the most accumulator one, with more than 1400 mg kg-1 dry wt. in the whole plant, 7-fold more Cd that measured in greenhouse conditions (Table 12). So, under in vitro conditions these plants increased the accumulation capacity obtained in the greenhouse. As it concerns the pattern of Cd distribution, in roots accumulated more than in shoots, except in the DV-G clone, whose accumulation was almost equal (Figure. 1a). For defining D. viscosa as a possible hyperaccumulator we focused on Cd shoot accumulation, as a plant is classed as hyperaccumulator when it accumulates more than 100 mg Cd kg-1 dry wt. in shoots (Reeves and Baker, 2000). Shoots of D. viscosa clones accumulated Cd ranging from about 300 mg kg-1 dry wt. in DV-B to over 1200 mg kg-1 dry wt. in DV-A and DV-F clones. A similar natural variation in Cd accumulation was also observed among accessions of Brassica napus L. (Grispen et al., 2005). These clones showed a biomass similar among them but lower than the control (Figure 1a).

In B. celtiberica clones, differences in Cd content were found (Figure 1b) but with lower values than in D. viscosa, being BC-K clone the highest Cd accumulator with 660 mg Cd kg-1 dry wt. in the whole plant. As in D. viscosa, Cd accumulation was much higher than in greenhouse (Table 12) and Cd was located preferably in roots.

Focusing on Cd shoot content, the BC-N and BC-K clones had the highest Cd accumulation (approximately 460 mg Cd kg-1 dry wt.) and also high biomass (Figure 1b) with little reduction of growth. Moreover B. celtiberica can also be mycorrhized with Paxillus filamentosus, and this condition improves both growth and Cd uptake (Fernández et al., 2008). Variation in the accumulation of Cd among different clones of Salix has also been reported, and the uptake ability of each willow clone was stable along time (Landberg and Greger, 1996; Greger and Landberg, 1999; Vysloužilová et al., 2003). Up till now there have been described a few Cd hyperaccumulator species, the best known being Thlaspi caerulescens (Ganges ecotype), which accumulates more than 10000 mg Cd kg-1 dry wt. in shoots (Lombi et al., 2000; Vassilev et al., 2002). However, T. caerulescens generates low biomass, which limits its use in phytoremediation (Brooks, 1998). Arabidopsis halleri can also accumulate more than 1000 mg Cd kg-1 dry wt. in leaves when cultured on a hydroponic system containing Cd (Cosio et al., 2004), but lower values have been reported for most Cd hyperaccumulators. In our case, D. viscosa and B. celtiberica (Figure 1a, 1b) cultured in vitro reached and exceeded the hyperaccumulation values of 100 mg Cd kg-1 dry wt. in shoots.

In hyperaccumulator plants not all the absorbed metal is retained in the roots (Page and Feller, 2005) and so the relation between the Cd content in shoots and in roots (S/R) should be >1 (Baker et al., 1994; Zhao et al., 2000).

For D. viscosa, the maximum S/R was 0.9 for the DV-A and 1 for the DV-G clone (Figure 1a), showing a good transport to the shoot. On the contrary, S/R values for B. celtiberica were lower (0.3-0.4) (Figure 1b), but it confirms that there was Cd transport to the shoots. In other woody species such as Salix (Vysloužilová et al., 2006) and Populus (Fisherová et al., 2006) there is also Cd transport to the shoot, but in Alnus, Fraxinus and Sorbus Cd is usually retained in the roots (Roselli et al., 2003).

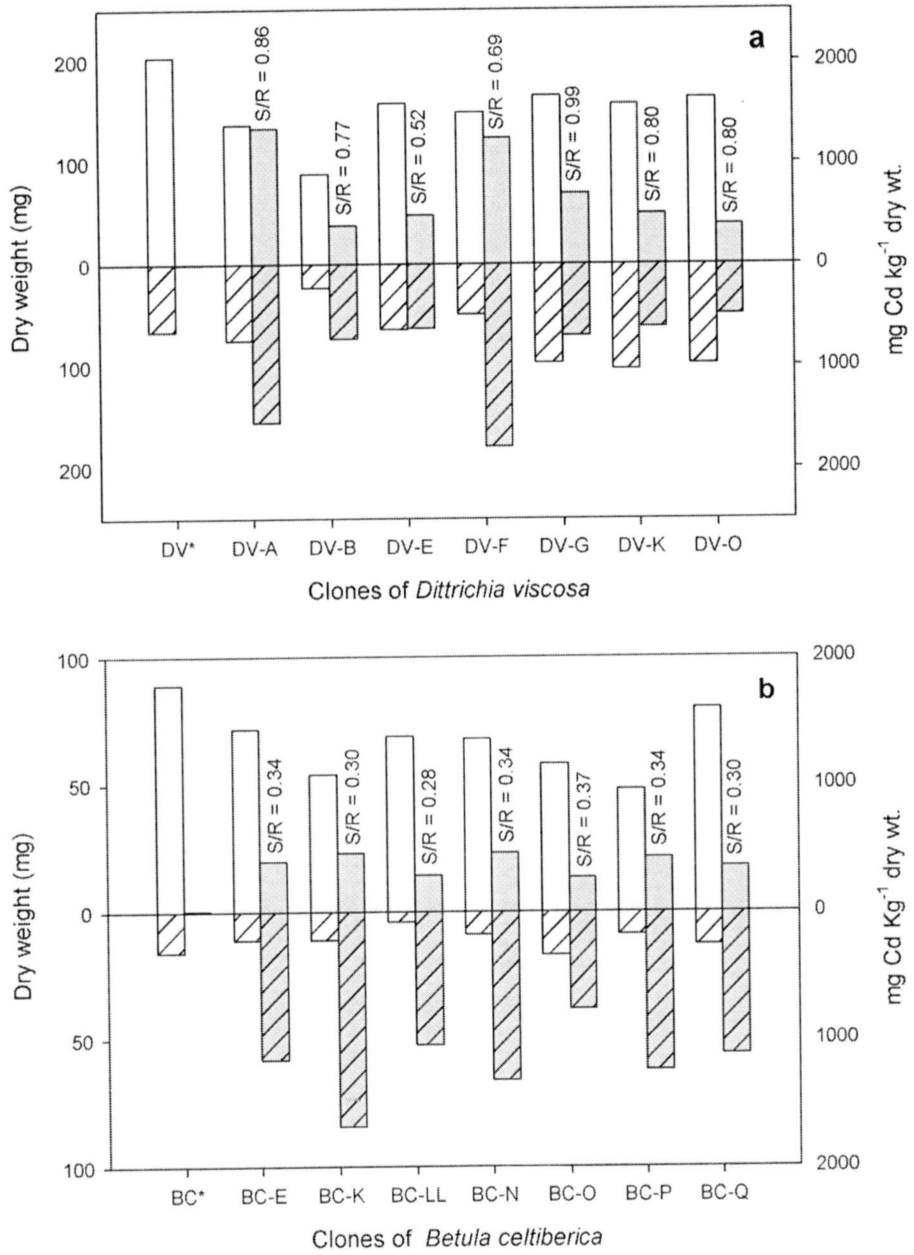

Figure 1. Dry weight (white) and Cd accumulation (grey) in shoots (plain) and roots (striped) of (a) *D. viscosa* cultured for 40 days in MS medium (DV*) or MS plus 10 mg Cd kg^{-1} and in (b) *B.celtiberica* clones cultured for 40 days in ½ MS medium (BC*) or ½ MS plus 10 mg Cd kg^{-1}.

3.4.2. Melilotus Alba for Pb

Analysis of Pb content in M. alba showed different accumulation capacity among clones. MA-X, the most accumulator clone reached more than 1000 mg Pb kg-1 dry wt. in shoots (Figure 2). A plant is classed as Pb hyperaccumulator when it accumulates more than 1000 mg Pb kg-1 dry wt. in shoots (Brooks, 1998).

Up till now, only Thlaspi rotundifolium, which accumulates 8200 mg Pb kg-1 dry wt. (Reeves and Brooks, 1983), and Phragmites australis, which can accumulate more than 4800 mg Pb kg-1 dry wt. (Ye et al., 1997), are said to be Pb hyperaccumulators. However, some species such as Zea mays, Brassica juncea, and Ambrosia artemisiifolia reach concentrations of Pb in amounts that can be toxic for most plants (375, 247 and 96 mg Pb kg-1 dry wt. respectively) although they do not reach the hyperaccumulation level (Huang and Cunningham, 1996).

Pb in plants grown in uncontaminated and unmineralized areas appears to be quite constant, ranging from 0 to 10 mg Pb kg-1 dry wt. and averaging 2 mg Pb kg-1 dry wt. (Kabata-Pendias and Pendias, 2001). Pb in M. alba roots exceeded that of shoots therefore relation S/R was very low (Figure 2).

This is in line with the results of Cunningham et al. (1995) who proved that Pb accumulation is always higher in roots due to the low mobility of this metal through the xylem. Also in Salix rubens cultured in a Pb-polluted soil the largest Pb accumulation in leaves (12.8 mg Pb kg-1 dry wt.) represented only 2% of that in the roots (Vysloužilová et al., 2006). The low transport rate through the xylem was also observed by other authors (Stolz and Greger, 2002; Pugh et al., 2002). Pb retention is based on extracellular precipitation or binding to ion exchangeable sites in the galacturonic and glucuronic carboxyl groups which restricts its transportation via apoplast (Rudakova et al., 1998; Sharma and Dubey, 2005).

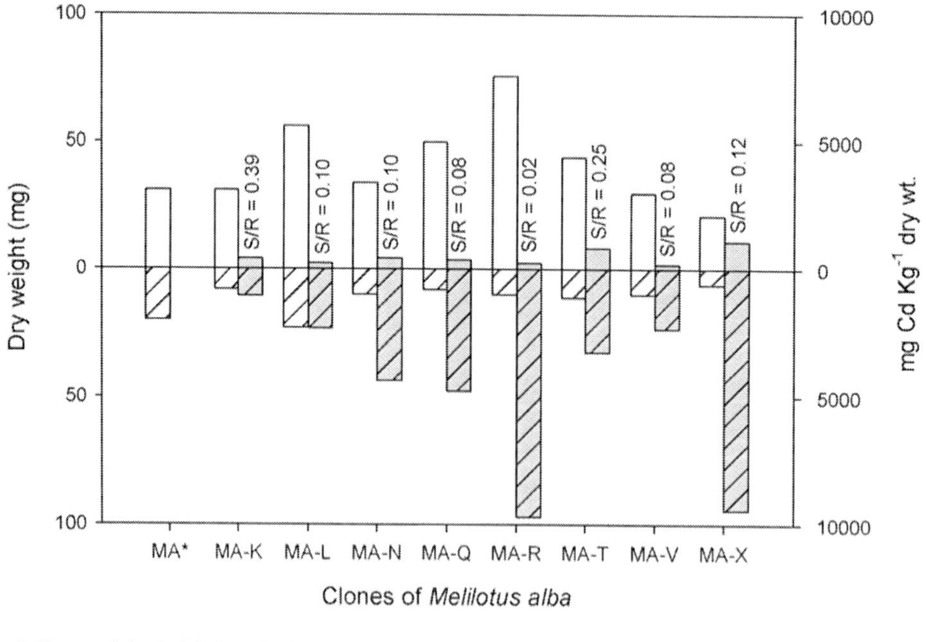

Figure 2. Dry weight (white) and Pb accumulation (grey) in shoots (plain) and roots (striped) of *M. alba* clones cultured for 40 days in MSS medium (MA*) or MSS plus 200 mg Pb kg^{-1}.

The addition of synthetic chelates, such as H-EDTA or EDTA, in combination with low pH, effectively prevents cell wall retention of lead, making it available for translocation to shoots (Jarvis and Leung, 2002). Pb moves predominantly into the root apoplast and subsequently moves in a radial manner across the cortex and accumulates near the endodermis. It appears that casparian strips of the endodermis are the major limiting factor restricting Pb transportation towards the central cylinder. This may in part account for the reports of higher accumulation of Pb in roots than in shoots (Verma and Dubey, 2003).

However, recent studies showed that lead accumulation may be promoted by the addition of some plant hormones. In Sesbania drummondii it was enhanced by 654 and 415% when treated with 100 mM indolacetic acid and 100 mM naphthaleneacetic Acid respectively (Israr and Sahi, 2008). These authors suggest a protective role of these growth hormones against Pb toxicity in this species by overpowering the damage caused by Pb, which leads to a better Pb acumulation. Studies on the mechanism of Pb toxicity suggest that Pb bonds to nucleic acids and causes aggregation and condensation of chromatin, as well as inhibiting replication, transcription and ultimately cell division and plant growth (Johnson, 1998).

3.4.3. Anthyllis Vulneraria for Zinc

The most Zn accumulator clones of A. vulneraria were AV-C, AV-D and AV-M (Fig. 3), ranging from 562 to 399 mg Zn kg-1 dry wt. in the whole plant. Zn concentration has lower values than those found in the greenhouse, probably because the plant grew slower in in vitro culture, so it would take more time for the same accumulation. In shoots AV-C, AV-D and AV-M clones 619, 469 and 378 mg Zn kg-1 dry wt. respectively were reached exceeding the threshold of toxicity. Thlaspi caerulescens can accumulate 25000-30000 mg Zn kg-1 dry wt. in shoots without symptoms of toxicity or growth reduction (Brown et al., 1995), exceeding hyperaccumulation level fixed in 10000 mg Zn kg-1 dry wt. in shoots (Reeves y Baker, 2000). Another Zn hyperaccumulator species is Arabidopsis halleri, which accumulate 32000 mg Zn kg-1 dry wt. in shoots (Zhao et al., 2000).

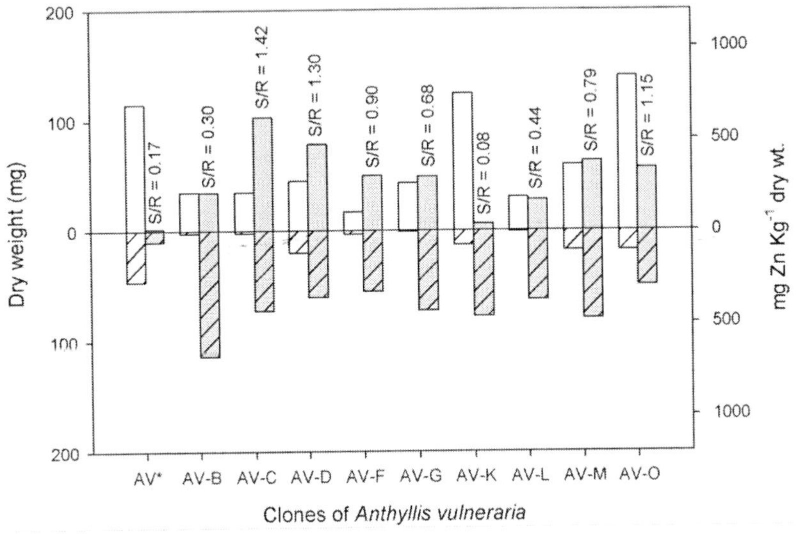

Figure 3. Dry weight (white) and Cd accumulation (grey) in shoots (plain) and roots (striped) of A. vulneraria clones cultured for 40 days in MS medium (MA*) or MS plus 100 mg Zn kg-1.

However, only AV-C, AV-D and AV-O clones had a value of S/R higher than 1, and for AV-M this value was 0.8 (Figure 3). This showed a good transport from root to shoot through the xylem as Zn is an essential element which enters through the root and it is distributed to all plant aerial organs where it promotes the function of some enzymes and participates in the metabolic route from tryptophan to indolacetic acid (Barceló et al., 2001). Zinc can also be redistributed through the floem from old to young leaves (Page y Feller, 2005). Furthermore, high S/R values were also found by Zhao et al. (2000) for the hyperaccumulator A. halleri.

3.4.4. Carex Pendula for Mercury

Hg content of C. pendula was very similar among clones, exceeding 100 mg Hg kg-1 dry wt. in shoots. However, biomass was greatly reduced because the Hg added to the culture medium (20 mg Hg kg-1) was very toxic for the plant as it was the 30 mg Hg kg-1 used in greenhouse. Hg accumulation was always higher in roots than in shoots (Figure 4). Low translocation of Hg to the shoots is probably due to the roots having strong affinities for Hg, whereby most of the Hg is trapped in the roots (Wang and Greger, 2004).

These authors measured homogeneous Hg accumulation in Salix among clones assayed but lower than the detected in C. pendula. On the contrary, Israr et al (2006) described a concentration of 998 mg Hg kg-1 dry wt. in shoots in Sesbania drummondii when exposed to 100 mg Hg L-1, which was the upper limit of Hg tolerance in those seedlings. Up till now the level for hyperaccumulation has not been defined (Peer at al 2006). Organomercurials and ionic Hg are toxic to plants, and Greger (2004) established Hg concentration in non-polluted plants in 0.1 mg Hg kg-1 dry wt. in shoots, whereas toxic levels of 1-3 mg Hg kg-1 dry wt. affects all plant physiology (Patra y Sharma, 2000).

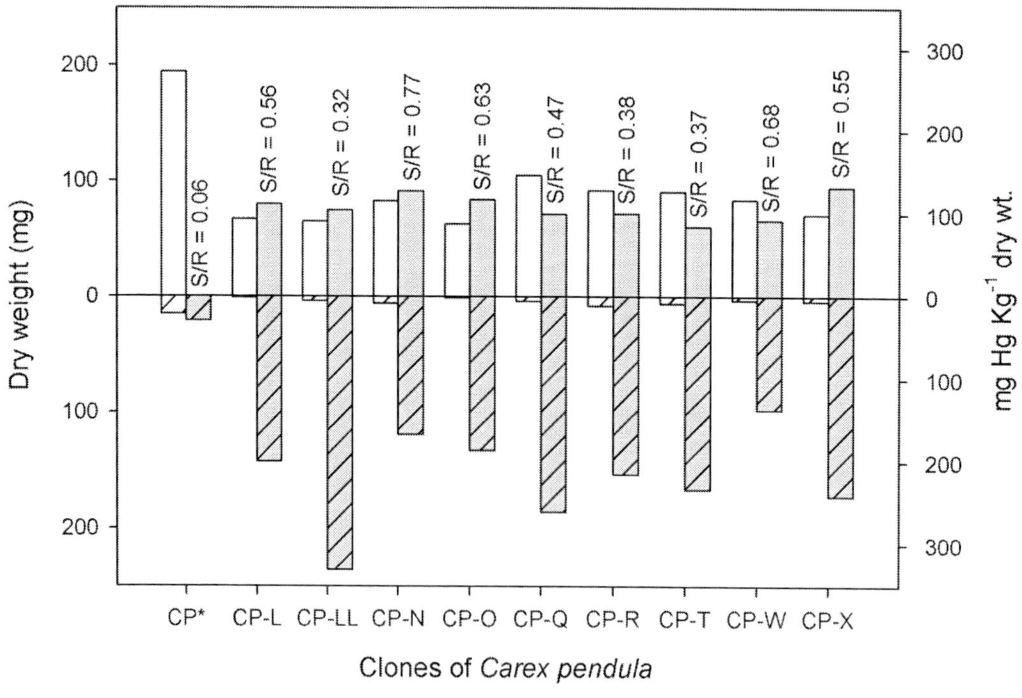

Figure 4. Dry weight (white) and Cd accumulation (grey) in shoots (plain) and roots (striped) of *C. pendula* clones cultured for 40 days in MS medium (CP*) or MS plus 30 mg Hg kg^{-1}.

In all cases Hg is mainly accumulated in roots, where most of it is retained in the cell wall and only a small percentage is translocated to the shoots (Patra and Sharma, 2000; Wang and Greger, 2004; Israr et al., 2006) although for Caulanthus sp. (Leonard et al., 1998) and Rumex induratus (Moreno-Jiménez et al, 2006) values of S/R > 0.9 have been measured. Hg accumulation in plants has been studied in Pisum sativum L., Mentha spicata L., water lettuce, and Norway spruce. They all absorbed Hg from solution most of it is accumulated in the roots (Godbold and Hütterman, 1988, Skinner et al., 2007). Mercury accumulation has also been observed in aquatic plants although at lower values (Coquery and Welbourn, 1994; Kalac and Svoboda, 2000; Skinner et al. 2007). One way of enhancing Hg uptake is the addition of ammonium thiosulphate to the medium.

Moreno et al., (2005) show that this compound mobilises non soluble Hg forms in substrates increasing substantially the content of Hg in the shoots when compared to water-treated plants, which could be advantageous for the remediation of polluted soils using hyperaccumulator plants.

CONCLUSION

In summary, an exhaustive plant selection must be made prior to the application of a phytoremediation program. Selection of plants from the populations of the metal-polluted sites for their recovery is now the focus, as they are adapted to soil and climatic conditions to the zone, which should made phytoremediation a much easier task.

We have established some criteria for this selection, such as metal concentration in the polluted areas and in plants, percentage cover/aggregation and frequency of appearance, biomass and nitrophilous or leguminous characters.

Based on the results of this work, these parameters for choosing the appropriate native species have been successful, since the selected plants not only maintain their accumulation capacity, but they also improved this capacity when grown in the greenhouse.

Moreover these plants tolerate the presence of heavy metals, and although at some metal concentrations growth was reduced, plants did not show other symptoms of toxicity such as chlorosis and necrosis and can survive in polluted soils.

In order to obtain the most accumulator plants in sufficient amounts but minimize the genetic variability, we searched for the most accumulator clone through in vitro culture. Although we were conscious that this technique does not simulate natural environment, it turned out to be a very useful tool in our studies.

At this point, we got some species very promising for soil-recovering purposes: Dittrichia viscosa and Betula celtiberica for Cd and Melilotus alba for Pb. These species accumulated high amounts of metal in vitro, with the advantage that we can produce an unlimited number of plants with the same characteristics of the selected clone.

Obviously the accumulator capacity of DV-A clone of D. viscosa, BC-K of B. celtiberica and MA-X clone of M. alba must be checked under natural conditions, as this improved accumulation can be due to the method of culture used.

Thus, these plants have been cultured in a greenhouse with artificially metal-polluted soil, and, although these assays are not finished yet, some preliminary analyses confirm the metal accumulation capacity of these clones.

Furthermore, these clones have been cultured in a slag heap with heavy metals and organic contaminants for a long-term study, and up to now their growth is not affected by this multi-context pollution.

Many studies about the responses of plants to heavy metals are focused on model species, but unfortunately most of them have handicaps that limit their feasible usefulness in phytoremediation.

On the contrary, working with native populations that can really be used for this purpose is an interesting approach to these techniques. Thus, much research must be done in order to understand the mechanisms involved in the desirable characters shown by these species but it will, without a doubt, improve the success of further phytoremediation programmes.

REFERENCES

Baker, A.J.M. (1981). Accumulators and excluders – strategies in the response of plants to heavy metals. *J. Plant. Nutr.*, 3, 643-654.

Baker, A.J.M., and Brooks, R.R. (1989). Terrestrial higher plants which hyperaccumulate metal elements – A review of their distribution, ecology and phytochemistry. *Biorecovery*, 1, 81-126.

Baker, A.J.M., Reeves, R.D., and Hajar, A.S.M. (1994). Heavy metal accumulation and tolerance in British populations of the metallophyte *Thlaspi caerulescens* J. and C. Presl (Brassicaceae). *New Phytol.*, 127(1), 61-68.

Barceló, J., and Poschenrieder, C. (2003). Phytoremediation: principles and perspectives. *Contributions to Science*, 2 (3), 333-344.

Barceló, J., Nicolás, G., Sabater, B., and Sánchez Tamés, R. (2001). *Nutrición Mineral*. In: Fisiología Vegetal (pp. 121-137). Madrid: Ed. Pirámide.

Barceló, J., Tolrá, R.P., and Poschenrieder, C. (1996). Zinc hyperaccumulation in Thlaspi caerulescens. I. Influence on growth and mineral nutrition. *J. Plant Nutr.*, 19 (12), 1531-1540.

Braun-Blanquet, J., and Pavillard, J. (1928). *Vocabulaire de Sociologie Végétale*. Montpellier.

Brooks, R. R. (Ed.) (1998). *Plants that hyperaccumulate heavy metals*: Their role in phytoremediation, microbiology, archaeology, mineral explotation and phytomining (p. 380). Oxon: CAB international.

Brooks, R.R., Lee, J., Reeves, R.D. and Jeffre, T. (1977). Detection of nickeliferous rocks by analisis of herbarium specimens of indicator plants. *J. Geochem. Explor.*, 7, 49-57.

Brooks, R.R., Reeves, R.D., Morrison, R.S., and Malaisse, F. (1980). Hyperaccumulation of copper and cobalt: A review. *Bull. Soc. R. Bot. Belg.*, 113, 166-172.

Brown, S.L., Chaney, R.L., Angle, J.S., and Baker, A.J.M. (1995). Zinc and cadmium uptake by hyperaccumulator Thlaspi caerulescens grown in nutrient solution. *Soil Sci. Soc. Am. J.,* 59, 125-133.

Chaney, R.L. (1983). Plant uptake of inorganic waste constitutes. In: J.F. Paar, P.B. Marsh, and J.M. Kle (Eds.), *Land treatment of hazardous wastes* (pp. 50-76) Park Kidge: Noyes Pata Corp.

Coquery, M., and Welbourn. P.M. (1994). Mercury uptake from contaminated water and sediment by the rooted and submerged aquatic macrophyte *Eriocaulon septangulare. Arch. Environ. Contam. Toxicol.*, 26 (3), 335-341.

Cosio, C., Martinoia, E., and Keller, C. (2004). Hyperaccumulation of cadmium and zinc in *Thlaspi* caerulescens and Arabidopsis halleri at the leaf cellular level. *Plant Physiol.*, 134, 716-725.

Cunningham, S.D., Berti, W.R., and Huang, J.W. (1995). Phytoremediation of contaminated soils. *Trends Biotechnol.* 13 (9), 393-397.

Dat, J., Vandenabeele, S., Vranová, E., Van Montagu, M, Inzé, D., and Van Breusegem, F. (2000). Dual action of the active oxygen species during plant stress responses. *Cell. Mol. Life Sci.,* 57, 779-795.

Díaz González, T., Fernández-Prieto, J.A., Nava, S., and Fernández Casado, M.A. (1994). Herbario FCO. *Itenera Geobot.*, 8, 529-600.

Ensley, B.D. (2000). *Rationale for use of phytoremediation* In: I. Raskin, and B.D. Ensley (Eds.), Phytoremediation of Toxic Metals: Using Plants to Clean-Up the Environment (pp. 3-11). New York: John Wiley and Sons INC.

Eun, S-O., Youn, H. S., and Lee, Y. (2000). Lead disturbes microtubule organization in the root meristem of Zea mays. *Physiol. Plant.*, 110, 357-365.

Fargasova, A. (1994). Effect of Pb, Cd, Hg, As, and Cr on germination and root growth of *Sinapis alba* seeds. *Bull. Environ. Contam. Tox.*, 52 (3), 452-456.

Fernández, R., Bertrand, A., Casares, A., García, R., González A., Tamés, R.S. (2008). Cadmium accumulation and its effect on the in vitro growth of woody fleabane and mycorrhized white birch. *Environ. Pollut.*, 152, 522-529.

Fisherová, Z., Tlustoš, P., Száková, J., and Šichorová, K. (2006). A comparison of phytoremediation capability of selected plant species for given trace elements. *Environ. Pollut.*, 144 (1), 93-100.

Godbold, D.L., and Hütterman, A. (1988). Inhibition of photosynthesis and transpiration in relation to mercury induced root damage in spruce seedlings. *Physiol. Plant.*, 74, 270-275.

Gosh, M., and Singh J.P. (2005). A review on phytoremediation of heavy metals and utilization of its byproducts. *Appl. Ecol. Environ. Res.*, 3 (1), 1-18.

Greger, M. (2004). Metal availability, uptake, transport and accumulation in plants. In: M.N.V. Prasad, *Heavy metal stress in plants: from biomolecules to ecosystems* (pp. 1-27). Berlin-Heidelberg: Springer-Verlag.

Greger, M., and Landberg, T. (1999). Use of willow in phytoextraction. *Int. J. Phytoremed.*, 1, 115-123.

Grispen V.M.J., Nelissen, H.J.M, and Verkleij J.A.C. (2005). Phytoextraction with Brassica napus L.: A tool for sustainable management of heavy metal contaminated soils. *Environ. Pollut.* 144, 77-83.

Hoagland, D.R., and Arnon, D.I. (1950). The culture method for growing plants without soil. *Calif. Agric. Exp. Circle*, 347, 1-32.

Huang, J.W., and Cunningham, S.D. (1996). Lead phytoextraction: species variation in lead uptake and translocation. *New Phytol.* 134, 75-84.

Israr, M., and Sahi, S. (2008). Promising role of plant hormones in translocation of lead in Sesbania drummondii shoots. *Environ. Pollut.*, 153, 29-36.

Israr, M., Sahi, S., Datta, R., and Sarkar, D. (2006). Bioaccumulation and physiological effects of mercury in *Sesbania drummondii*. *Chemosphere*, 65, 591-598.

Jarvis, M.D., and Leung, D.W.M. (2002). Chelated lead transport in *Pinus radiata*: an ultrastructural study. *Environ. Exp. Bot.*, 48, 21-32.

Jasper, D.A., Abbott, L.K., and Robsona D. (1992). Soil disturbance in native ecosystems: The decline and recovery of infectivity of mycorrhizal fungi. In: *Mycorrhizas in ecosystems* (pp 151-155). Oxon: CAB international.

Johnson, F.M. (1998). *The genetic effects of environmental lead. Mutat. Res.*, 410, 123-140.

Kabata-Pendias, A., and Pendias, H. (2001). *Trace elements in soils and plants*. Florida: CRC. Press.

Kalac, P., and Svoboda, L. (2000). A review of trace element concentrations in edible mushrooms. *Food Chem.*, 69, 273-281.

Küpper, H., Mijovilovich, A., Meyer-Klaucke, W., and Kroneck, P.M.H. (2004). Tissue- and age-dependent differences in the complexation of cadmium and zinc in the cadmium/zinc hyperaccumulator *Thlaspi caerulescens* (Ganges Ecotype) revealed by X-ray absorption spectroscopy. *Plant Physiol.*, 134, 748-757.

Kvesitadze, G., Khatisashvili, G., Sadunishvili, T., and Ramsden, J.J. (2006). Contaminants in the environment. In: *Biochemical mechanisms of detoxification in higher plants* (p. 24). Berlin-Heidelberg: Springer-Verlag

Landberg, T., and Greger, M. (1996). Differences in uptake and tolerance to heavy metals in *Salix* from unpolluted and polluted areas. *App. Geochem.*, 11, 175-180.

Leonard, T.D, Taylor, G.E.Jr., Gustin, M.S., and Fernández, G.C.J. (1998). Mercury and plants in contaminated soils: 1. Uptake, partioning and emission to the atmosphere. *Environ. Toxicol. Chem.*, 17, 2063-2071.

Llamas, A., Ullrich, C.I., and Sanz, A. (2000). Cd^{2+} effects on transmembrane electrical potential difference, respiration and membrane permeability of rices (*Oryza sativa* L.) roots. *Plant Soil*, 219, 21-28.

Lombi, Z., Zhao, F.J., Dunham, S.J., and McGrath, S.P. (2000). Cadmium accumulation in populations of *Thlaspi caerulescens* and Thlaspi goesingense. *New Phytol.*, 145, 11-20.

Mádico, J., Poschenrieder, C., Vázquez, M.D., and Barceló, J. (1992). *Effects of high zinc and cadmium concentrations on the metallophye Thlaspi caerulescens* J. et C. Presl. (Brassicaceae). Suelo y Planta, 2, 495-504

Malaisse, F., Grégoire, J., Brooks, R.R., Morrison, R. S., and Reeves, R.D. (1978). Aeollanthus biformifolius De Wild.: A hyperaccumulator of copper from Zaïre. *Science*, 199 (4331), 887-888.

Montaser, A. (1998). *Inductively coupled plasma mass spectometry*. New York: Wiley-VCH.

Moreno, F.B., Anderson, C.W.N., Stewart, R.B., Robinson, B.H., Nomuro, R., Gohmsei, M., and Meech, J.A. (2005). Effect of thioligands on plant-Hg accumulation and volatilisation from mercury-contaminated mine tailings. *Plant Soil*, 275, 233-246.

Moreno-Jiménez, E., Gamarra, R., Carpena-Ruiz, R.O., Millán, R., Peñalosa, J.M., and Esteban, E. (2006). Mercury bioaccumulation and phytotoxicity in two wild plant species of Almadén area. *Chemosphere*, 63, 1969-1973.

Murashige, T., and Skoog, F. (1962). A revised medium for rapid growth and bioassays with tobacco tissue cultures. *Physiol. Plant*, 15 (3), 473-479.

Obata, H., Inoue, N., Marsuoka, M., and Umebayashi, M. (1997). Effects of cadmium on the fluxes of K^+ and H^+ in excised roots. *Soil Sci. Plant Nutr.*, 43 (2), 439-442.

Page, V., and Feller, U. (2005). Selective transport of zinc, manganese, nickel, cobalt and cadmium in the root system and transfer to the leaves in young wheat plants. *Ann. Bot.*, 96 (3), 425-434.

Patra, M., and Sharma, A. (2000). Mercury toxicity in plants. Bot. Rev., 66 (3), 379-421.

Peer, W.A., Baxter, I.R., Richards, E.L., Freeman, J.L., and Murphy, A.S. (2006). Phytoremediation and hyperaccumulator plants. In: M.J. Tamás, and E. Martinoia (Eds.) *Molecular biology of metal homeostasis and detoxification: From Microbes to man* (pp. 299-340), Berlin-Heidelberg: Springer-Verlag.

Pugh, R.E., Dick, D.G., and Fredeen, A.L. (2002). Heavy metal (Pb, Zn, Cd, Fe and Cu) contents of plant foliage near the anvil range lead/zinc mine, Faro, Yukon Territory. *Ecotox. Environ. Safety*, 52, 273-279.

Rauret, G., López-Sánchez, J.-F., Sahuquillo, A., Barahona, E., Lachica, M., Ure, A.M., Davidson, C.M., Gomez, A., Lu Èck, D., Bacon, J., Yli-Halla, M., Muntaub, H., and Quevauviller, Ph. (2000). Application of a modified BCR sequential extraction (three-step) procedure for the determination of extractable trace metal contents in a sewage sludge amended soil reference material (CRM 483), complemented by a three-year stability study of acetic acid and EDTA extractable metal content. *J. Environ. Monit.*, 2, 228-233.

Reeves, R., and Baker, A. (2000). *Metal accumulating plants.* In: I. Raskin, I., and B.D. Ensley (Eds.), Phytoremediation of toxic metals: Using plants to clean up the environment (pp. 193-229). New York: B. John Wiley and Sons, Inc.

Reeves, R.D., and Brooks, R.R. (1983). European species of *Thlaspi* L. (Cruciferae) as indicators of nickel and zinc. *J Geochem. Explor.*, 18, 275-283.

Rivas-Martínez, S., Fernández-González, F., Loidi, J., Lousa, M., and Penas, A. (2001). Syntaxonomical checklist of vascular plant communities of Spain and Portugal to association level. *Itinera Geobot.*, 14, 5-341.

Roselli, W., Keller, C., and Boschi, K. (2003). Phytoextraction capacity of trees growing in a metal contaminated soil. *Plant Soil,* 256, 265-272.

Rudakova, E.V., Karakis, K.D., and Sidorshina, E.T. (1998). The role of plant cell walls in the uptake and accumulation of metal ions. *Fiziol. Biochim. Kult. Rast.*, 20, 3-12.

Salt, D.E., and Krämer, U. (2000). *Mechanisms of metal hyperaccumulation in plants.* In: I. Raskin, and B.D. Ensley (Eds.), Phytoremediation of Toxic Metals: Using Plants to Clean-Up the Environment (pp. 231-246.), New York: John Wiley and Sons INC.

Schat, H., Llugany, M., Vooijs, R., Hartley-Whitaker, J. and Bleeker, P. (2002). The role of phytochelatins in constitutive and adaptative heavy metal tolerances in hyperaccumulator and non-hyperaccumulator metallophytes. *J. Exp. Bot.*, 53 (379), 2381-2392.

Schützendübel, A., Schwanz, P., Teichmann, T., Gross, K., Langenfeld-Heyser, R., Godbold, D.L., and Polle, A. (2001). Cadmium-induced changes in antioxidative systems, hydrogen peroxide content, and differentiation in Scots pine roots. *Plant Physiol.*, 127, 887-898.

Sharma, P., and Dubey, R.S. (2005). Lead toxicity in plants. *Braz. J. Plant Physiol.*, 17 (1), 35-52.

Skinner, K., Wright, N., and Porter-Groff, E. (2007). Mercury uptake and accumulation by four species of acquatic plants. *Environ. Pollut.*, 145, 234-237.

Stolz, M., and Greger, M. (2002). Accumulation properties of As, Cd, Cu, Pb and Zn by four wetland plant species growing on submerged mine tailings. *Environ. Exp. Bot.,* 4, 271-280.

Tong, Y-P., Kneer, R., and Zhu, Y-G. (2004). Vacuolar compartmentalization: a second-generation approach to engineering plants for phytoremediation. *Trends Plant Sci.*, 9 (1), 7-9.

Utriainen, M.A., Kärenlampi, L.V., Kärenlampi, S.O., and Schat, H. (1997). Differential tolerance to copper and zinc of micropropagated birches tested in hydroponics. *New Phytol.*, 137, 543-549.

Vassilev, A., Vangrosveld, J., and Yordanov, I. (2002). Cadmium phytoextraction: present state, biological backgrounds and research needs. *Bulg. J. Plant Physiol.* 28, 68-95.

Verma, S., and Dubey, R.S. (2003). Lead toxicity induces lipid peroxidation and alters the activities of antioxidant enzymes in growing rice plants. *Plant Sci.*, 164 (4), 645-655.

Visser, W.J.F. (1994). *Contaminated Land Policies in some Industrialised Countries*. The Hague: Technical Soil Protection Committee (TCP).

Vysloužilová, M., Száková, J., Tlustoš, P., and Pavlíková, D. (2003). Cadmium and zinc phytoextraction potential of seven clones of *Salix* spp. planted on heavy metal contaminated soils. *Plant Soil Environ.*, 49 (12), 542-547.

Vysloužilová, M., Puschenreiter, M., Wieshammer, G., and Wenzel, W.W. (2006). Rhizosphere characteristics, heavy metal accumulation and growth performance of two willow (*Salix* × *rubens*) clones. *Plant Soil Environ.,* 52, (8), 353-361

Wang, Y., and Greger, M. (2004). Clonal differences in mercury tolerance, accumulation and distribution in willow. *J. Environ. Qual.*, 33 (5), 1779-1785.

Wierzbicka, M., and Potocka, A. (2002) Tolerance to lead of plants growing on dry or moist soils. *Acta Biol. Crac. Ser. Bot.*, 44, 21-28.

Wilkins, D.A. (1978).The measurement of tolerance to edaphic factors by means of root growth. *New. Phytol*, 80, 623-633.

Wójcik M., Vangronsveld, J., and Tukiendorf, A. (2005). Cadmium tolerance in *Thlaspi caerulescens* I. Growth parameters, metal accumulation and phytochelatin synthesis in response to cadmium. *Environ. Exp. Bot.*, 53, 151-161.

Yang, Y.Y., Jung, J.Y., Song, W.Y., Suh, H.S., and Lee, Y. (2000). Identification of rice varieties with high tolerance or sensitivity to lead and characterization of the mechanism of tolerance. *Plant Physiol.*, 124, 1019-1026.

Ye, Z.H., Baker, A.J.M., Wong, M.H., and Willis, A.J. (1997). Zinc, lead and cadmium tolerance, uptake and accumulation by the common reed, Phragmites australis (Cav.) *Trin. Ex Steudel. Ann. Bot.*, 80, 363-370.

Zhao, F.J., Lombi, E., Breedon, T., and McGrath, S. P. (2000). Zinc hyperaccumulation and cellular distribution in Arabidopsis halleri. *Plant Cell Environ.*, 23, 507-514.

In: Handbook of Phytoremediation
Editor: Ivan A. Golubev

ISBN: 978-1-61728-753-4

Chapter 9

USE OF X-RAY FLUORESCENCE-BASED ANALYTICAL TECHNIQUES IN PHYTOREMEDIATION

Marijan Nečemera, **Peter Kump**≠a and **Katarina Vogel-Mikuš**‡b*

[a]Jožef Stefan Institute, Ljubljana, Slovenia

[b]Department of Biology, Biotechnical Faculty, University of Ljubljana, Ljubljana, Slovenia

ABSTRACT

Phytoremediation is an emerging technology that employs the use of higher plants to clean-up metal contaminated environments; when applied, there is a need for constant monitoring of metal concentrations in soil, water and biological materials in order to evaluate the success of the applied technology and to control metal uptake in plant tissues in order to prevent accumulation of unwanted toxic metals in food chains.

In view of the growing needs of global environmental protection and also to minimize the relevant research costs, it is important that in phytoremediation studies and their application the analytical procedures for determination of elemental concentrations in soil, water and biological materials are accurate, reliable and reproducible, but on the other hand rapid and cheap, with simple sample preparation. Therefore in this chapter the main characteristics, sample preparation protocols, and applications of X-ray fluorescence-based analytical techniques for "bulk" sample analyses, namely energy dispersive X-ray fluorescence spectrometry (EDXRF) and total reflection X-ray fluorescence spectrometry (TXRF), are presented. Although EDXRF and TXRF are far less popular methods for analyses of element concentrations in soil, water, air and biological materials than, for example, atomic absorption spectroscopy (AAS) and/or inductively coupled plasma atomic-emission spectroscopy (ICP-AES), they are much cheaper, simpler and environmentally friendlier, which is particularly advantageous from the economic and environmental protection points of view.

* e-mail: marijan.necemer@ijs.si

≠ e-mail: peter.kump@ijs.si

‡ e-mail: katarina.vogel@bf.uni-lj.si

BASIC PRINCIPLES OF X-RAY FLUORESCENCE

X-Ray Radiation

X-rays are electromagnetic waves with a spectrum spanning wavelengths from about 80 nm (about 15 eV) down to about 0.001 nm (about 1.2 MeV), overlapping to some extent the region of γ-rays (**Table 1**). Electromagnetic radiation (usually above 1 MeV) generated by nuclear processes is usually called *γ-radiation* while the radiation below 80 nm wavelength generated by electrons slowed down in the outer field of an atomic nucleus or by transitions between bound states of electrons in the electronic shells of an atom is called *X-radiation*.

Table 1. Spectrum of electromagnetic radiation from γ-rays to TV and FM radio waves (IR – infrared light; UV – ultraviolet light)

	Wavelengths	Energy (eV)		
	0.00001 nm	100 MeV		
	0.0001 nm	10 MeV	γ-rays	
	0.001 nm	1 MeV		*Hard X-rays*
Medical X-rays	0.01 nm	100 keV	X-rays	
	0.1 nm	10 keV		*Soft X-rays*
	1 nm	1 keV		
	10 nm	100 eV	UV	
	100 nm	10eV		
	1 μm	1 eV	Visible light	
	10 μm	0.1 eV	IR	
	100 μm	0.01 eV		
	1 mm	1 meV	microwaves	
Cellular phones	1 cm	10 meV		
Radio emissions	1 m	100 meV		
	10 m	1 μeV	TV and FM radio waves	

The history of X-ray fluorescence (XRF) dates back to the accidental discovery of X-rays in 1895 by the German physicist Wilhelm Conrad Roentgen. While studying cathode rays in a high-voltage, gaseous-discharge tube, Roentgen observed that even though the experimental tube was encased in a black cardboard box, a barium-platinocyanide screen, which was lying adjacent to the experiment, emitted fluorescent light whenever the tube was in operation. This was possible because the energies of X-ray photons are of the same order of magnitude as the binding levels of inner-shell electrons (K, L, M, N,..), and therefore they can be used to excite and/ or probe these atomic levels (Janssens, 2004).

Interaction of X-Rays with Matter

When an X-ray beam passes through matter, some photons are absorbed inside the material or scattered away from the original path (Figure 1).

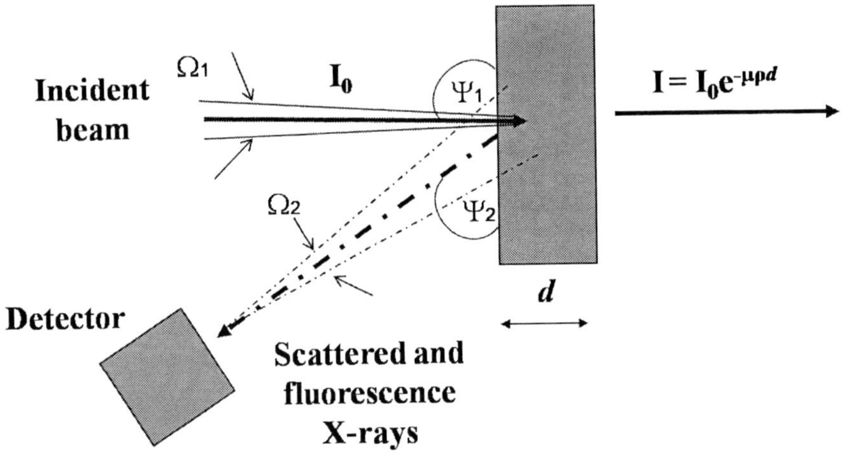

Figure 1. Interactions of X-rays with matter.

The intensity (I_0) of an X-ray beam passing through a layer of thickness (d) and density (ρ) is reduced to an intensity (I) according to the Lambert-Beer law (1)

$$I = I_0\, e^{-\mu\rho d} \quad (1)$$

The number of photons per second (the intensity) is reduced, but their energy remains generally unchanged. The term (μ) is called the *mass attenuation coefficient* and has the dimensions of $cm^2\ g^{-1}$. The product ($\mu_L = \mu\rho$) is called linear absorption coefficient and is expressed in cm^{-1}. $\mu(E)$ is sometimes also called the *total cross-section for X-ray absorption* at energy (E).

X-Ray Fluorescence

X-ray fluorescence is induced by the excitation (ionization) of atoms in the tightly bound inner K (for the elements $20 < Z < 50$) and L ($Z > 50$) atomic shells with energies that must exceed the binding energies of the K and L electrons, through the process called the *photoelectric effect* (Markowicz, 1993). The atoms can be excited with X–ray photons, by irradiation from an *X-ray source* (e.g. X-ray tube, radioisotope source). In the photoelectric absorption process an X-ray photon from the X-ray source is completely absorbed by an atom and an electron (called a *photoelectron* (Figure 2a)) is ejected from one of the inner (K or L) shells (Figure 3). Part of the excitation photon energy is used to overcome the binding energy (Φ) of the electron and the rest is transferred to the electron in the form of kinetic energy. After the interaction with an X-ray photon, the atom (in fact ion) is left in a highly excited state since a vacancy has been created in one of its inner (K or L) shells. This atom almost immediately returns to a more stable electronic configuration in the process called *relaxation or deexcitation,* where electrons from outer shells (L, M or N) fill the vacancy created in the K or L shells (Figure 3). Since during relaxation the electrons pass from a higher to a lower energy state, the difference between the energies can be emitted in the form of *characteristic X-ray photons* (K, L, or M) (Figure 2a, 3), or the energy can be absorbed by an electron in one of the outer shells (L, M, or N) (Figure 2b, 3), which is then ejected from the atom as an *Auger electron* (Figure 2b). Characteristic K, L and M X-ray photons are typical of particular elements, and this provides the basis for distinguishing between elements in X-ray spectrometry (Markowicz, 1993). The ratio between the number of emitted characteristic X-rays and the total number of inner shell vacancies in a particular atomic shell that gave rise to them, is called the *fluorescence yield* of that shell (e.g. ω_K). For light elements ($Z<20$) predominately Auger electrons are produced during relaxation after K-shell ionization ($\omega_K < 0.2$) (Figure 2b), while medium and high Z elements preferentially relax in a radiative manner ($0.2 < \omega_K < 1.0$) (Figure 2a).

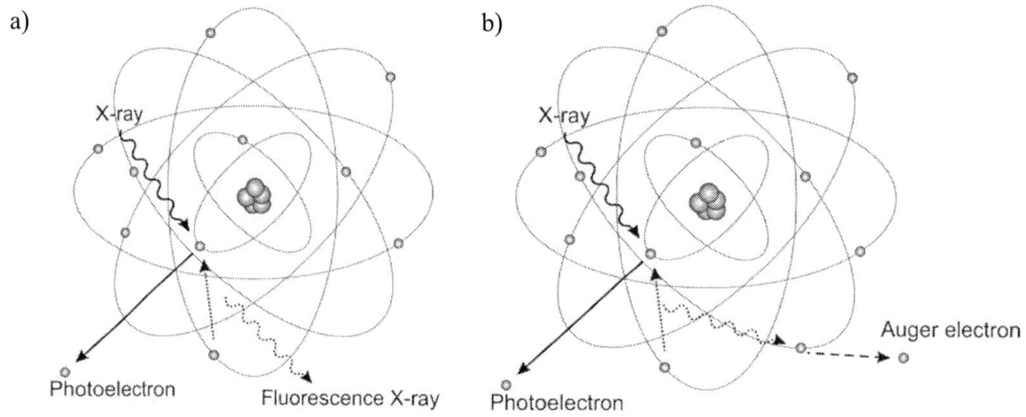

Figure 2. Interaction of an atom with X-rays a) *photoelectric effect*; b) *Auger effect*

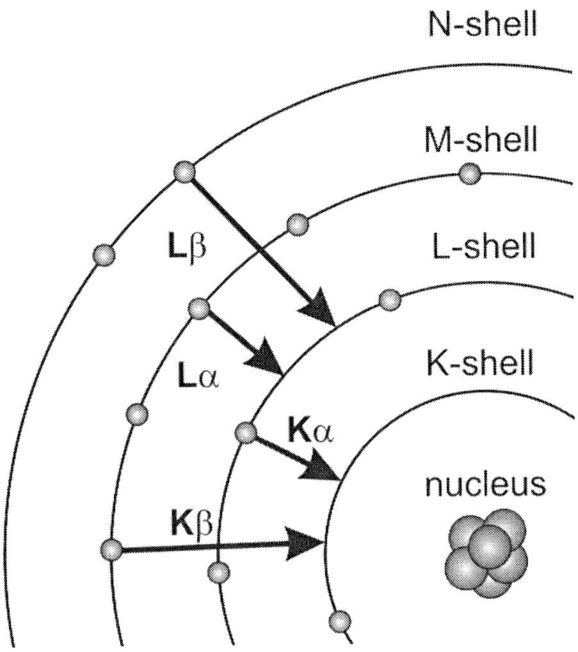

Figure 3. Electronic transitions in an excited calcium atom, where one of the electrons in the lower orbital is missing.

Photoelectric absorption can only occur if the energy of the photon (E) is equal or higher than the binding energy (Φ) of an electron. For example, an X-ray photon with an energy of 17.4 keV (e.g. from an X-ray tube with a Mo anode) can eject a K-electron ($\Phi_K = 7.112$ keV) or an L_3-electron ($\Phi_{L3} = 0.706$ keV) from a Fe atom, but a 5.9 keV photon (e.g. from a radioisotopic source of Fe-55) can only eject an electron from the L-shell of the Fe atom. Since photoelectric absorption can occur at each of the (excitable) energy levels of the atom, the *total photoelectric cross-section* σ_i is the sum of the (sub)shell-specific contributions (2)

$$\sigma_i = \sigma_{i,K} + \sigma_{i,L} + \sigma_{i,M} + \cdots$$

$$= \sigma_{i,K} + (\sigma_{i,L1} + \sigma_{i,L2} + \sigma_{i,L3}) + (\sigma_{i,M1} + \sigma_{i,M2} + \cdots\cdots + \sigma_{i,M5}) + \cdots \quad (2)$$

In the case for example of Mo (Figure 4) at high energy, e.g. > 50 keV, the probability of ejecting a K-electron is rather low and that of ejecting an L_3-electron is even lower. As the energy of the X-ray photon decreases, the cross-section increases, i.e., more vacancies are created. At a binding energy $\Phi_{KMo} = 19.99$ keV, there is an abrupt decrease in the cross-section, because X-rays with lower energy can no longer eject electrons from the K-shell. However, these photons continue to interact with the (more weakly bound) electrons in the L- and M-shells. The discontinuities of the photoelectric cross-section are called *absorption edges*. The ratio of the cross-section just above and just below the absorption edge is called the jump ratio (r). As XRF is the result of selective absorption of radiation, followed by spontaneous emission, an efficient absorption process is required. An element can therefore be determined with high sensitivity by means of XRF when the exciting radiation has its maximum intensity at an energy just above the K or L-edge (for heavier elements) of that element.

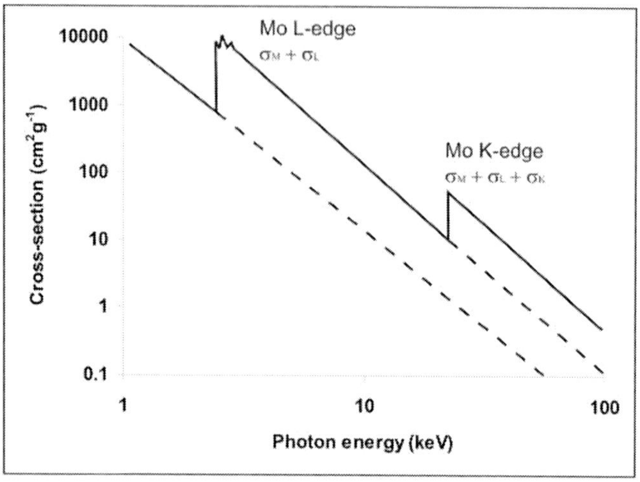

Figure 4. Variation of σMo as a function of X-ray photon energy. The K, L1, L2 and L3 absorption edges are clearly visible (after Janssens, 2004).

Soon after the discovery of X-rays, in 1913, Henry Moseley established the relation between the atomic number (Z) and the specific X-ray wavelength (λ) of an element (3),

$$1/\lambda = K\,(Z - s)^2 \tag{3}$$

where K and s are constants; s is the shielding constant and takes value close to one, and K has a different value for each of the line series considered (e.g. the K_α - lines, the L_α - lines,..(Figure 3)). Each unique atom has a number of available electrons that can take part in the energy transfer and since millions of atoms are typically involved in the excitation of a given specimen, all possible de-excitation routes are taken.

Application of X-Ray Fluorescence Analysis to Environmental Monitoring and Phytoremediation

The processes described above represent the fundamentals of various X-ray based fluorescence analytical techniques, ranging from standard energy dispersive X-ray fluorescence analysis (EDXRF) enabling "bulk" elemental analyses in different materials, to highly sophisticated particle induced and synchrotron-based X-ray fluorescence techniques such as, for example, micro-PIXE and XRF micro-spectroscopy enabling spatial localization of particular elements in different samples (Vogel-Mikuš et al., 2007, 2008a, b, 2010, Kaulich et al., 2009).

Increasing anthropogenic activities such as metalliferous mining and smelting, metallurgical industries, use of fertilizers and soil amendments in high-production agriculture, and land disposal techniques for municipal or solid wastes has led to the widespread introduction of minor and trace elements, both metals and metalloids, into the environment, causing acute and diffuse contamination of soil and waters. Such elements otherwise occur in natural and perturbed environments in small amounts, but when present in sufficient bioavailable concentration, they are toxic to living organisms. In contrast to many organic pollutants, trace elements possess a long residence time in the soil system because they cannot be degraded (see Kidd et al., 2009).

In view of developing strategies for soil protection, there is a need to develop new, viable and *in situ* technologies for soil protection and remediation. Therefore last two decades has seen the emergence of eco-friendly, milder soil remediation techniques using different plant species. These techniques, collectively known as *phytoremediation*, are generally considered to be less invasive, more cost-effective and restorative of soil structure and functions compared to conventional, civil-engineering methods (based on techniques such as leaching of the pollutant, solidification/ stabilization, size selection and pyrometallurgical processes, electrokinetic treatment, chemical oxidation/ reduction of a pollutant, and excavation) (Thompson-Eagle and Frankenberger, 1990; Wenzel et al., 1999; Lombi et al., 2000; Mulligan et al., 2001; Barceló and Poschenrieder, 2003; McGrath and Zhao, 2003; Kidd et al., 2009). According to several authors the cost of remediation per contaminated hectare of soil using conventional techniques ranges between 0.27 and 1.6 million dollars, while phytoremediation costs from about 10–1000 times less (McNeill and Waring, 1992; Glass, 1999; Cunningham and Berti, 2000; Kid et al., 2009). Phytoremediation is based on the use of plants and their associated microorganisms to remove, stabilize, or detoxify pollutants. Over the last decade these plant-based technologies have been gaining public and regulatory support and several comprehensive reviews are available summarizing their most important aspects (Salt et al., 1998; Chaney et al., 1997; Wenzel et al., 1999; Navari-Izzo and Quartacci, 2001; McGrath et al., 2002; McGrath and Zhao, 2003; Prasad and Freitas, 2003; Pilon- Smits, 2005, Kidd et al., 2009).

In phytoremediation two basic techniques can be applied: *Phytoextraction*, which aims to remove trace elements from the soil through their uptake and accumulation by plants, and *phytostabilization,* which aims to establish a vegetation cover and promote *in situ* inactivation of trace elements by combining the use of metal-tolerant plants and soil amendments that help reduce the mobility and toxicity of pollutants and, at the same time, may increase soil fertility and improve plant establishment. In phytoextraction the plant biomass can then be harvested, thereby removing the metals from the site and the harvested plant material may be used to recover valuable metals or can be burnt and the ash disposed of under controlled conditions or in some cases, when it does not contains toxic metals such as Pb, Cd, Cr, Hg or As, recycled as a fertilizer (Keller et al., 2005). This technique is only effective if the plants accumulate large concentrations of metals/ metalloids in their shoots and have a reasonable biomass production (McGrath and Zhao, 2003). Phytoextraction is best suited for the remediation of diffusely polluted areas, where pollutants occur at a relatively low concentration and superficially (Rulkens et al., 1998). On the other hand, plants for phytostabilization should retain the metals at the root level and have restricted transport to aerial parts (excluder behaviour, Baker, 1981) to avoid further transfer into the food chain (Wenzel et al., 1999). This technique primarily targets contaminants that form insoluble compounds and or are strongly adsorbed, and is applicable in situations where soil properties, (e.g. high contents of clay or organic matter), a priori favour immobilization (Cunningham et al., 1995). The stabilization of pollutants can be an important remediation option for large areas with high and multi-elemental contamination (Kidd et al. 2009).

Prior to application of phytoremediation at metal polluted sites, extensive studies and determination of element concentrations in soil and plant materials are needed in order to determine the most suitable remediation technique to be applied and afterwards constant monitoring of soil and plant metal concentrations is needed in order to follow the success of the applied technique. Because in environmental monitoring of metal pollutants and application of phytoremediation techniques "bulk" analyses of soil, water and biological materials are especially important, in this chapter special attention will be given to standard energy dispersive X-ray fluorescence (EDXRF) and total reflection X-ray fluorescence (TXRF) analyses. EDXRF is the cheapest analytical technique of all because of the simple sample preparation and is especially suitable for soil and plant material analyses, while TXRF is especially convenient for elemental analysis in natural waters, because of its high sensitivity.

Standard Energy-Dispersive X-Ray Fluorescence Analysis

XRF Instrumentation

There are many types of XRF spectrometers available on the market today, most of which can be separated into two main categories: wavelength-dispersive XRF (WDXRF) and energy-dispersive XRF (EDXRF). In WDXRF, the characteristic radiation emitted from the sample is separated into wavelengths using a diffraction device. The energy resolution in the WDXRF spectra is governed by the appropriate use of diffraction crystals in each region of the spectrum. Usually in WDXRF spectrometers, the analysis of different elements is carried out in a sequential way by synchronously scanning the orientation of the monochromatic device and the detector. However, in multi-channel spectrometers, the use of several diffraction devices or detector set-ups allows one to measure several elements simultaneously (although the number of channels is limited). Unlike WDXRF systems, conventional EDXRF spectrometers consist of only two basic units – the excitation source and the spectrometer or detection system (Figure 5). In this case, the resolution of the EDXRF system depends directly on the resolution of the detector. Typically, a semiconductor detector of high intrinsic resolution is employed [Si(Li)]. The use of this type of detector allows one to record an electronic signal (voltage pulse) processed by the preamplifier and amplifier, which is proportional to the energy of the detected photon dissipated within the sensitive volume of the detector. An analogue to digital converter (ADC) and a multi-channel analyzer are then used to sort, integrate, store, and display the detected pulses into a X-ray spectrum (Margui et al. 2009). Interestingly, using this configuration, all of the X-rays emitted by the sample are collected at a very high rate irrespective of their size. In addition, this configuration also enables high speed acquisition and display of spectral data. EDXRF systems are therefore classified according to the type of *excitation source*, the *geometry of excitation* and the *type of energy dispersive detector* installed.

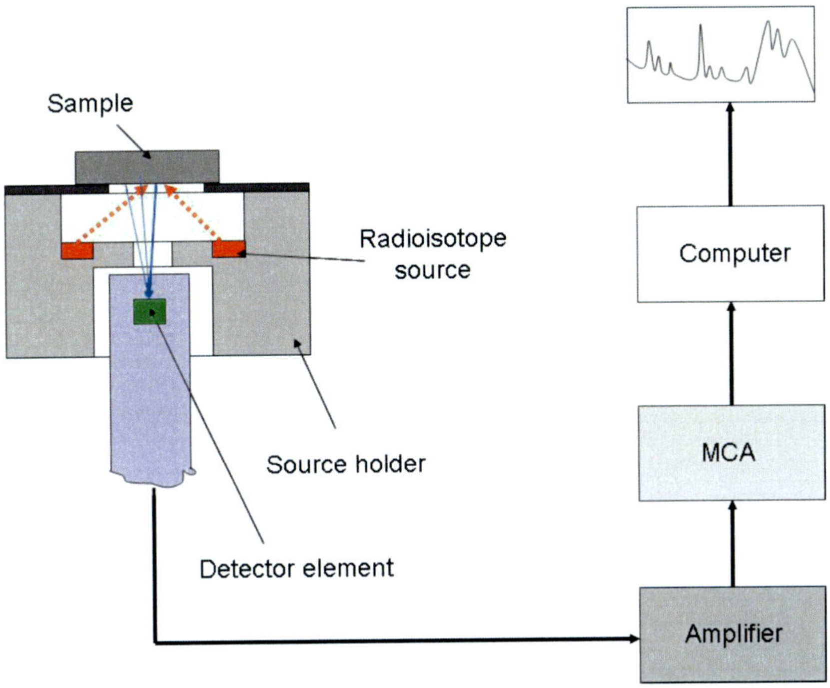

Figure 5. Diagram of an XRF system composed of a radioisotopic source, energy dispersive detector, amplifier and multi-channel analyzer.

Excitation Sources

Excitation of the elements in the sample can be performed using almost monochromatic *radioisotopic sources*, or polychromatic radiation from an *X-ray tube*. Table 2 lists radioisotope sources typically used in EDXRF analysis.

Table 2. Commonly used radioisotopic sources used in EDXRF analysis (adopted after Kalnicki and Singhvi, 2001)

Isotope	Half-life	Useful radiation	Energy (keV)	X-rays excited efficiency
Fe-55	2.7 years	Mn-K X-rays	5.9	Al-Cr
Co-57	270 days	Fe-K X-rays	6.4	<Cr
		γ-rays	14.4;122;136	
Cd-109	1.3 years	Ag-K X-rays	21.1	K-Tc
Am-241	470 years	Np-L X-rays	14, 21	Sn-Tm
		γ-rays	59	

The most commonly used sources include Fe-55, Co-57, Cd-109, and Am-241. Each of these emits radiation at specific energy levels and therefore efficiently excites elements within a specific atomic number range. As a result, no single radioisotopic source is sufficient for exciting the entire range of elements of interest in environmental analysis, and many instruments use two or three sources to maximize element range. The half-life of a source is important, especially for Fe-55, Co-57, and Cd-109 sources. With half-lives as short as 270 days some sources may have to be replaced after a few years when their intensity decreases to a level too low to provide adequate excitation of the elements of interest (Kalnicky and Singhvi, 2001).

Alternatively, an *X-ray tube* can be used to irradiate the sample with characteristic and continuum X-rays. X-ray tubes can be air cooled (low power, 3 W - 50 W) or water cooled (high power, 2 kW), with different anodes, such as Ag (22.1 keV), Rh (20.2 keV), Mo (17.4 keV), and Cr (5.4 keV).

XRF measurements may be performed in different geometries. When thick specimens are excited by the continuous X-ray spectrum from an X-ray tube, the sensitivity of the method is lowered because of the relatively high background from the scattered continuous radiation from the sample, or its substrate, in the spectral region of the fluorescent radiation. A secondary target irradiation geometry can be used to partly monochromatize the X-ray tube radiation, and thus to decrease this background scattering (Jaklevic and Giauque 1993).

Detectors

The X-ray detector converts the energies of the X-ray fluorescence photons into voltage pulses that can be counted to provide a measurement of the total X-ray flux. X-ray detectors are typically "proportional" devices where the absorbed energy of the incident X-ray photon determines the size of the output voltage. A polychromatic fluorescence beam of radiation incident upon the detector produces a spectrum, with a pulse height distribution proportional to the energy distribution of the incident fluorescence beam. A multichannel analyzer collects the spectrum and enables discrimination between characteristic fluorescence lines of different elements, depending on the resolution of the detector (Kalnicky and Singhvi, 2001).

The three most common types of detectors are: *the gas flow proportional detector, the scintillation detector*, and *semiconductor detectors (Si(Li), SiPIN, SiDrift and Hyperpure Ge detectors).* These detectors differ in resolution and intrinsic efficiency. Resolution is the ability of the detector to separate X-rays of different energies, and is important for minimizing spectral interferences and overlap, while the efficiency depends on the absorption of X-rays in the sensitive volume of the detector. Semiconductor detectors have the best resolution and are preferred for EDXRF instruments. These detectors may require liquid nitrogen as a coolant or employ electronic cooling via built-in Peltier elements.

The selection of detector is very important. In cheaper spectrometers, a radioactive source and a proportional detector (gas) can be used. However, one shortcoming of such a device is the poor energy resolution of the detector (800–1.000 eV at 5.9 keV), making the quantification of the insufficiently resolved characteristic X-ray lines in the measured spectrum quite difficult. This problem can be overcome by using semiconductor detectors such as Si(Li), Si PIN or Si drift (SDD) detectors, which have energy resolutions of around 120 eV to 140 eV at 5.9 keV.

Sampling and Sample Preparation for EDXRF

Sampling and sample preparation represent two critical steps in the analysis of environmental samples. To accurately characterize site conditions, the samples collected must be representative of the site or area under investigation (Margui et al., 2009). Representative soil or vegetal sampling ensures that a sample or group of samples accurately reflects the concentration of the contaminant(s) of concern at a given time and location. Analytical results from representative samples reflect the variation in contaminant presence and concentration range throughout a site. Parameters affecting the variability of the results of representative samples include: (1) geologic and plant material variability, (2) contaminant concentration variability, (3) collection and preparation variability, and (4) analytical variability.

Use of analytical techniques such as AAS or ICP-AES usually involves sample-preparation procedures for total destruction of the matrix by chemical treatment. Sample dissolution is usually a demanding, time-consuming step that sometimes limits application of the analytical procedure in environmental studies and quality-control processes. Commonly, dry ashing (involving combustion of the sample) and wet digestion (involving digestion with strong acids) have been used to destroy the organic matter and dissolve the analytes in such matrices (Margui et al., 2009). Compared to dry-ashing methods, wet-mineralization procedures using acid digestion present a wide range of options, depending on the choice of reagents and their mixtures as well as the devices used for the procedure (Margui et al., 2009). Especially demanding is wet digestion of soil and plant samples containing silicon, as in such cases, the use of hydrofluoric acid is essential to achieve total destruction of the matrix and thus determination of the total concentrations of analytes. In the last 15 years classical open systems (digestions at atmospheric pressure) using conventional sources of heating (e.g., sand baths and hot plates) were gradually replaced by digestion procedures using microwave ovens and closed vessels, since in open systems there were problems with losses of volatile elements, such as Hg, Cd and others. In addition, microwave assisted procedures can also shorten the digestion procedure and reduce the amount of reagents employed, as well as avoiding analyte losses and contamination from other samples or from the surroundings (Margui et al., 2009). To sum up, the choice of the best decomposition procedure for soil and vegetal samples should be preceded by verification of the procedure for each specific matrix and analyte under study. This becomes quite difficult in environmental studies where several or many soil and plant samples are used as pollution indicators for different metals, and in such cases, the application of techniques that obviate matrix destruction become even more attractive. For this reason, study of the suitability of other methods for direct and multi-elemental analysis of soil and vegetal samples has increased in recent years, from instrumental neutron activation analysis (INAA) to simple X-ray fluorescence-based techniques. INAA is based on measuring the radioactivity produced by neutron reactions on naturally-occurring nuclides (Nečemer et al. 2008), but the most serious shortcomings of INAA are its high costs, the availability of a nuclear reactor for irradiation and the rather long time of analysis imposed by the waiting (cooling) periods for the decay of interfering activated short-lived radionuclides, although the methods is very sensitive and accurate. In addition, INAA does not allow determination of some environmentally-important elements (e.g., Pb) (Margui et al., 2009), so when summing up the advantages and disadvantages of INAA, the method is not very suitable for routine application in environmental studies. Most X-ray fluorescence (XRF) techniques, on the other hand, comply with the desirable features for analysis of soil and vegetal specimens, including:

i. the possibility of performing analysis directly on solid samples;
ii. multi-element capability;
iii. the possibility of performing qualitative, semi-quantitative and quantitative determinations;
iv. a wide dynamic range;

v. high throughput;
vi. low cost per determination.

The main drawbacks of XRF instrumentation restricting its more frequent use for environmental purposes have been its limited sensitivity for some important pollutant elements (e.g., Cd, Pb and Hg) and a somewhat poorer precision and accuracy compared to atomic spectroscopic techniques. Nevertheless, there have recently been improvements in XRF instrumentation (e.g., development of spectrometers using digital signal processing and enhancement of X-ray production with better designs for excitation-detection), which have added the advantage of increased instrumental sensitivity, thus allowing improvements in both precision and productivity. These improvements have therefore increased the possibility of the use of XRF spectroscopy as a technique in the environmental field (Margui et al., 2009).

Both solid and liquid samples can be analysed by EDXRF. In the case of solid samples, no special destructive chemical treatment of the sample is necessary. Determination of the composition of solid samples, without any sample preparation, is possible for samples that are homogeneous in all three dimensions and with a flat surface. This is the case for direct analysis of metals and alloys and that has been one of the main applications of XRF. However, most environmental solid materials (e.g. soil and vegetal materials) require sample pre-treatment to make them homogeneous and to ensure the quality and the reproducibility of measurements (Margui et al., 2009). Commonly, this procedure is based on crushing or grinding the materials into fine powder followed by pelletization at high pressure, which is the most frequent method of preparing soil and vegetal samples for analysis by XRF techniques (Margui et al., 2009). Prior to grinding, soil or vegetal samples are usually oven dried at 60-100°C, or in the case of vegetal samples freeze dried to remove their water content. In reducing the soil or vegetal material to a fine powder, grinding or milling is usually employed with concomitant problems and contamination arising from the grinding matrix. Particularly for trace elements, precautions should be taken by choosing suitable materials (e.g., agate, silicon carbide, boron carbide, and tungsten carbide). Agate grinding may introduce significant contamination into biological material by Ti, V, Cr, Mn, Fe and Pb (Margui et al., 2009), but this is the case in all analytical procedures, because for AAS and ICP-AES analysis the samples should also be well ground and homogenized prior to acid digestion. For EDXRF analysis approximately 100-200 mg of solid, well ground and homogenized (soil or vegetal) material is sufficient; however any inhomogeneity of the pulverized solid sample can have a large influence on the accuracy of the measurement, especially in the case of lighter elements (Nečemer et al. 2008). In addition care should be taken during pellet preparation to assure a uniform thickness in order to avoid bias due to inhomogeneity.

The accuracy of the measurement of solid soil and vegetal samples by EDXRF can also be influenced by sample moisture. Sample dilution tends to decrease the apparent concentration as the moisture level increases. This effect is most severe for analytes with low energy X-ray lines (less than 5 keV), and may be negligible for elements with higher energy X-ray lines (for example, Pb). To some extent, the dilution effect may be counteracted by the reduced matrix absorption for the analyte X-ray lines when water replaces the higher atomic number (and, therefore, more absorbing) soil/ sediment matrix. The direction and magnitude of the bias introduced by moisture is, therefore, dependent on the analyte X-ray line energy and the composition of the sample. The overall error may be minor when the moisture content is small (5–20%), but it may be a major source of error when the soil is saturated with water. Soil/ sediment samples should be therefore dried when moisture content is greater than 20% (Margui et al., 2009), as described previously.

For liquid samples, 100–1.000 ml of solution is required, and the elements are concentrated from the sample by precipitation. Several precipitation agents are available for this purpose. For instance, the reagent ammonium pyrrolidine dithiocarbamate (APDC) can be used for the precipitation of Cu, Fe, Ni and Pb. Note that any particular precipitating agent can selectively precipitate only certain specific elements. Therefore, only these specific elements can be determined in liquid samples using this method of sample preparation. The precipitated elements are separated from the liquid phase by filtration, and the precipitate that is deposited on the filter is measured directly by the EDXRF system. Due to the preconcentration of the precipitated elements, the limits of detection for these elements decrease to a few 10 $\mu g\ l^{-1}$, which cannot be achieved in the analysis of solid samples. This approach is especially suitable for monitoring contaminating elements in water.

Quantitative Elemental Analysis by ED X-Ray Fluorescence Spectroscopy

Basic Principles

As previously described, X-ray fluorescence spectroscopy is based on X-ray excitation of atoms in the sample material by the photo-effect process, followed by radiative decay of the excited atoms, i.e. by emission of characteristic X-rays of the particular atoms. In the relaxation process the radiative and Auger transitions compete and the fluorescent yield determines the probability of radiative transition. The fluorescent yield favours the radiative decay of heavier atoms. The intensities of radiative transitions of atoms in the sample, measured by the X-ray spectrometer, are then used in qualitative and quantitative analyses of the elemental composition of the sample. The measured intensities of characteristic X-rays depend on the mode of excitation (radioisotope or X-ray tube excitation), on the fundamental physical constants which determine the probabilities of photo-effect excitation of atoms, on the probability of radiative decay (fluorescent yield), on absorption of radiation emitted within the sample and penetrating towards the detector, and finally on detector efficiency and also on the geometry of the excitation-detection experiment. The process of X-ray fluorescence is well understood and presented in many books and papers (Tertian and Claisse, 1982; Rousseau, 1984; He and Van Espen, 1991; Van Dyck et al., 1986). The relation between the measured characteristic intensities and concentrations of respective atoms in the sample is established using the above mentioned fundamental parameters and experimental conditions. In the case of monochromatic excitation by energy E_1 in the K shell of atoms, the respective relation is as follows:

$$I_i = S_i \cdot c_i \cdot T_i(c_1, c_2, \ldots\ldots\ldots c_n) \cdot H_i(c_1, c_2, \ldots\ldots\ldots c_n) \qquad (4)$$

where

I_i: measured fluorescent intensity

S_i: elemental sensitivity (slope of calibration curve in the case of a thin or diluted sample)

T_i: absorption correction factor (depends on sample composition)
H_i: enhancement correction factor (depends on sample composition)
c_i: concentration of element "i"

$$S_i = G \cdot K_i \tag{5}$$

$$G = A_0 \cdot \Omega_1 \cdot \Omega_2 \cdot \overline{\operatorname{cosec}\psi_1} \tag{6}$$

$$K_i = \sigma_i^{ph}(E_1)(1 - \frac{1}{J_k})_i \cdot \omega_i^k \cdot f_i^{K\alpha} \cdot \varepsilon_{rel}(E_i) \tag{7}$$

A_0: activity of the excitation source [photons s^{-1} str^{-1}]

ψ_1, ψ_2: effective incident and take off angles, respectively

Ω_1, Ω_2: differential solid angles for the incident (primary) and emerging (characteristic radiation), respectively

$\sigma_i^{ph}(E_1)$: photo effect cross-section at energy E_1 in element "i"

$(1 - \frac{1}{J_k})_i$: relative probability for excitation of K-shell of "i"

ω_i^k: fluorescent yield for K-shell of element "i"

$f_i^{K\alpha}$: relative transition probability for K_α X-ray of "i"

$\varepsilon_{rel}(E_i)$: relative detector efficiency for X-rays of energy E_i

The combined absorption of primary and fluorescence X-rays in the sample is determined as follows:

$$\overline{a_{i,s}} = \mu_s(E_1).\overline{\operatorname{cosec}\psi_1} + \mu_s(E_i).\overline{\operatorname{cosec}\psi_2} \tag{8}$$

μ_i (E_1): absorption cross section in element "i" at energy E
μ_s (E_i): absorption cross section in sample at energy E

The expressions for the absorption and enhancement correction factors *Ti* and *Hi* are as follows:

$$T_i(c_1, \ldots c_n) = \frac{1 - \exp(-\overline{a_{i,s}}.\rho.d)}{\overline{a_{i,s}}} \tag{9}$$

$$H_i(c_1, c_2, \ldots c_n) = 1 + \Sigma_k\, \rho_{i,k}(c_1, \ldots c_i, \ldots c_k, \ldots c_n) \cdot c_k \tag{10}$$

In the case of a polychromatic excitation defined by the distribution $w(E_j)$ of primary X-rays, which is usually calculated (Pella et al., 1985), the excitation and absorption or enhancement of fluorescent radiation must be treated together. Factorisation of the basic equation is impossible, and the evaluation of particular concentrations becomes more complicated. The basic equation becomes:

$$I_i = G \cdot K_i \cdot \sum_j [\sigma_i^{ph}(E_j) \cdot w(E_j) \cdot T_i(E_i, E_j) \cdot (1 + \sum_k \rho_{i,k}(E_j) \cdot c_k)] \quad (11)$$

where $\rho_{i,k}$ are contributions to the enhanced intensity of element "i" by excitation of the fluorescent radiation of elements "k". The summation is performed over all the elements in the sample which could enhance element "I". This factor depends on the composition of the sample through absorption in all elements of the sample. In this case the constant Ki is no longer expressed as above but is defined as:

$$K_i = (1 - \frac{1}{J_k})_i \cdot \omega_i^k \cdot f_i^{K_\alpha} \cdot \varepsilon_{rel}(E_i) \quad (12)$$

Exact expressions for the above mentioned correction factors, as well as for the expressions of Ki for L-series X-rays can be found in Tertian and Claisse, (1982); Rousseau, (1984); He and Van Espen, (1991); Van Dyck et al., (1986).

Starting Quantitative XRF Analysis (Principal Problems and Necessary Assumptions)

The theoretical background of the X-ray fluorescence process is well known and supported by the above equations. The measured fluorescent intensities represent starting data for the quantification procedure. But there is a basic problem, namely, when using the above equations the concentration of the element can be determined from the measured intensity only, if the composition of the sample is known. Namely the fluorescent intensity depends not only on the concentration of the respective element but also on concentrations of all other elements in the sample, which attenuate the excitation and fluorescent radiation in the sample before it excites the atoms within the sample and when the emitted fluorescent radiation penetrates towards the surface in the direction of the detector. In this case, but only if all the elements in the sample respond by a fluorescent signal in the spectrum, the concentrations can be obtained from the respective set of equations (4) by iteration (the number of unknowns - *Ci* in this case equals the number of equations). But in almost all other cases a problem exists due to the unknown part of the sample, that part which does not give a response in the spectrum with fluorescent lines (light elements like H, C, O, F, and sometimes also Na, Mg, Al, Si, etc., depending on the excitation, low fluorescence yield, and detector efficiency). These elements, which comprise the so-called residual or dark or low-Z matrix, additionally attenuate the excitation and measured fluorescent radiation. Different approaches to quantification are therefore applied to solve this problem:

i. In the case of known composition of the residual or dark or low Z part of the sample matrix (i.e. oxides, cellulose, aluminosilicates, etc.), the concentrations can be calculated by iteration of a set of equations considering additional absorption in the selected matrix;

ii. Use of measured intensities of the scattered excitation radiation in the spectrum enables assessment of the composition of the sample matrix (Van Dyck and Van Grieken, 1980);

iii. Additional measurement of absorption performed on the sample by the transmission-emission method (Markowicz and Van Grieken, 1993).

The first approach leads to semi-quantitative analysis. Namely in most cases the selected composition of the residual matrix is only a more or less good guess or approximation and therefore it is in principle not possible to say that the result is quantitative (determined namely from measured quantities and fundamental constants).

The second approach uses the scattered primary radiation from the sample, which is usually measured together with the fluorescence. Scattering is in principle a rather complicated physical process, dependent on the geometry of the experiment, on the thickness and on the average atomic number of the atoms in the sample. This correction is usually applied to rather thin samples, for which the absorption corrections are rather small and therefore uncertainty of these corrections does not very much affect the uncertainty of the results.

On the other hand the absorption process is a straightforward process and yields good experimental results. In two of the models of our approach to quantitative analysis we utilized absorption measurements in the sample by the transmission-emission method at a single energy, usually at 8.04 keV or 17.44 keV, corresponding to application of Cu or Mo radiators (Markowicz and Van Grieken, 1993). By the iteration of the set of equations (4) for the measured elements and including the absorption in a selected residual matrix, the concentrations of measured elements could be obtained. The absorption in this particular sample at an energy of 8.04 or 17.44 keV is calculated and compared with the measured value. If the values do not coincide, further iteration, selecting gradually larger absorption in the selected residual matrix then leads towards the final values of concentrations of the measured elements and also to a correct residual matrix, so that the measured absorption at a particular energy in the sample coincides with the calculated one.

Calibration of the XRF System for Quantitative Analysis

In any model or approach to quantitative analysis it is first necessary to calibrate the XRF system, in order to evaluate the geometrical constant G as defined in equation (6). For this purpose a set of thin samples (Van Espen and Adams, 1981) or thick pure metals or samples of stable chemical compounds (Yap et al., 1987) are measured and the sensitivities Si are calculated using equations (5) to (9), employing the known compositions of selected samples. It should be mentioned that the uncertainty of such a calibration can greatly influence the quantification of the results. To evaluate the uncertainty of the calibration, we proceeded in a somehow different way from that of other authors Markowicz et al., (1992), and evaluated the geometrical constant G by equations (5-7) from the calculated sensitivities. The evaluated constant G by definition should be the same for all calibrated elements. Therefore the calculated standard deviation of the average value of G in principle determines the uncertainty of the calibration procedure, which includes the uncertainty of the measured intensities, uncertainties of fundamental parameters in constant Ki, as well as uncertainties in the absorption coefficients used in the calculations (Mc Master and Delgrand, 1986). This uncertainty represents a part of the total uncertainty of the complete quantification procedure. In most cases this method of calibration yields an average geometrical constant G with an uncertainty of 2% to 5%. In the case of the polychromatic excitation, the uncertainty can also reach 5% to 10%, due to the uncertainty of the calculated distribution of the excitation X-rays $w(E_j)$ (Pella et al., 1985). In the quantification procedure the geometrical constants for particular calibrated elements are usually used rather than the average geometrical constant.

A Case Study of the Quantification Procedure

Quantification starts with the measurement of the spectrum of an unknown sample, which was prepared as a pressed pellet. The spectrum of the NIST 1573a tomato leaves standard, excited by a Cd-109 radioisotope source, and measured by an X-ray spectrometer with a Si(Li) detector, is shown on Figure 6. The measured complex spectrum in which some elemental lines overlap (i.e. K and Ca or Mn and Fe) was analysed using the AXIL spectrum fitting program (Van Espen and Janssens, 1993).The calculated results were the intensities of all measured elements, which could then be further used in different models of the quantification package developed by P. Kump under the name QAES (Quantitative Analysis of Environmental Samples). This package includes models for the analysis of thin samples:

1. Thin layers on a thick substrate
2. Aerosols collected on a filter
3. Water sample precipitated on a millipore filter

For the analysis of intermediate and thick samples the following models are available:

Absorption in the total matrix

1. Measurement of absorption in the sample
2. Measurement of absorption...(independent of slope parameter)
3. List of experimental absorption parameters

Absorption in residual matrix

4. List of absorption parameters for typical low-Z matrices
5. List of experimental absorption parameters for standards
6. Interelement coefficients method (valid for thick samples)

"Model 6" also contains some other models appropriate for the analysis of thick samples:

6.1 Empty residual matrix (all sample constituents are measurable)
6.2 Element composition of residual matrix is known
6.3 List of absorption parameters for typical low-Z matrices
6.4 List of experimental absorption parameters for standards
6.5 Internal standard based analysis by matching residual matrix

As an example "Model 2" was applied to the sample of Tomato leaves. After the fluorescent sample spectrum was measured, an additional measurement of the absorption in the sample itself, using a Cu radiator and transmission-emission method was performed. This measurement gave the ratio of measured intensities, with and without the sample, of the Cu K-series X-rays at the energy of 8.04 keV. The program then evaluated, by applying the known geometrical constants, the sensitivities *Si* of all measured elements and also selected an appropriate residual matrix (in the case of inorganic samples usually oxygen or alumino-silicate and for organic samples cellulose or carbohydrate). Since the analysed sample was an organic matrix, the cellulose matrix was selected and the iteration started by using the reduced absorption in this matrix set as 0.1 $\mu_{cell}[E_i]$. The program evaluated the absorption correction and also enhancement correction in this matrix and obtained the first approximation to the elemental concentrations of measured elements. The iteration then continued and stopped after the consecutive iterations of elemental concentrations for this particular selected dark matrix converged. Because the measured element concentrations and residual matrix were now known, the program evaluated the absorption in the sample at 8 keV and compared it with the measured absorption. If the values did not coincide, the program gradually increased the absorption as e.g. to 0.2 $\mu_{cell}[E_i]$ in the selected residual matrix and repeated the iteration under these conditions. These iteration procedures continued until the calculated absorption in the sample coincided with the measured absorption at 8 keV. The program then printed the final results and also showed a graph comparing the absorption in the evaluated residual matrix (crosses) with typical matrices of low-Z elements. The output of the numerical results and a residual matrix graph for this particular sample are shown in Table 3 and on Fig. 7.

It is evident from the graph that the residual matrix determined by the program (white line with crosses) coincides with the absorption in/ for the cellulose matrix. The numerical output defines the evaluated matrix with two parameters, which correspond to the slope (A1) and intersect (A0) of the corresponding calculated absorption curve on the graph. The program also saves the evaluated parameters for the residual matrix and in the case that analysis of similar samples is required, the quantification "Model 5" is selected, and the »known« residual matrix obtained is used for these other samples as well. In Table 4 the results obtained by this quantification (Model 5) in which we used the parameters obtained for a residual cellulose matrix are presented. It can be seen that the results are similar.

The results presented in the Tables 3 and 4 also show the absorption correction for the sample analysed. The values of the corrections T in [g cm^{-2}] are in principle equivalent to the effective sample thickness, which in a sense represents that part of the sample which contributes all the fluorescent radiation to the respective spectral line. It is obvious that this effective sample thickness is different for different elements and it is also much smaller than the overall sample thickness shown on the right hand side of the first row in the table.

As mentioned above, the QAES quantitative analysis package also contains other quantification models. "Model 6.1" is applied for the analysis of alloys for which all the elements in the sample usually register in the spectrum by fluorescent lines and the residual matrix is empty.

"Model 6.5" also employs an analysis based on an internal standard. The sample material can be spiked with an internal standard of an element which preferably does not appear in the sample. It is important that after addition of the internal standard, the sample is thoroughly homogenized prior to preparation of the pressed pellet for analysis. The program then starts with some appropriate residual matrix (it could be oxygen, cellulose, aluminosilicate, etc.), considering at first the reduced absorption in this matrix, and then by an iterative procedure obtains the concentrations of measured elements, including the internal standard. Comparing the value obtained for the internal standard with the established value, the iterations continue until these values coincide. In this way as well as the concentrations of the measured elements the residual matrix and its parameters are also obtained similarly to the model, which utilized the measured absorption in the sample by the transmission-emission method.

As already mentioned, the uncertainty of the calibration procedure is included in the overall uncertainty of the final concentration results. Additionally the quantification procedure contributes to the overall uncertainty through the uncertainty of the measured fluorescent intensities, the uncertainty of the AXIL spectrum analysis procedure, and the uncertainty in measurements of sample absorption by the transmission-emission method. The overall uncertainty is quoted in the analytical results. In Table 4 the last column also quotes the certified and uncertified concentration values of NIST 1573a tomato leaves standard.

Usually the accuracy of the results is within the estimated uncertainty. The certified values of the standard reference materials are assumed to be so-called conventional true values and therefore the accuracy of the analysis results can be estimated from the certified values of the standard reference material. Beside the uncertainties of the measured fluorescent intensities and the uncertainties of fundamental physical constants applied in quantification, the most serious uncertainty originates from the inhomogeneity of the sample. As well as sample inhomogeneity, the particle size of the sample material also plays an important role. In the quantification models applied these sources of uncertainty have not been considered. But in general in many applications, the EDXRF technique offers fast and non-destructive sample analysis with, for most purposes, acceptable uncertainty and sufficient accuracy, and can be therefore efficiently used in research studies as well as in industry and in environmental monitoring and control of metal pollutants.

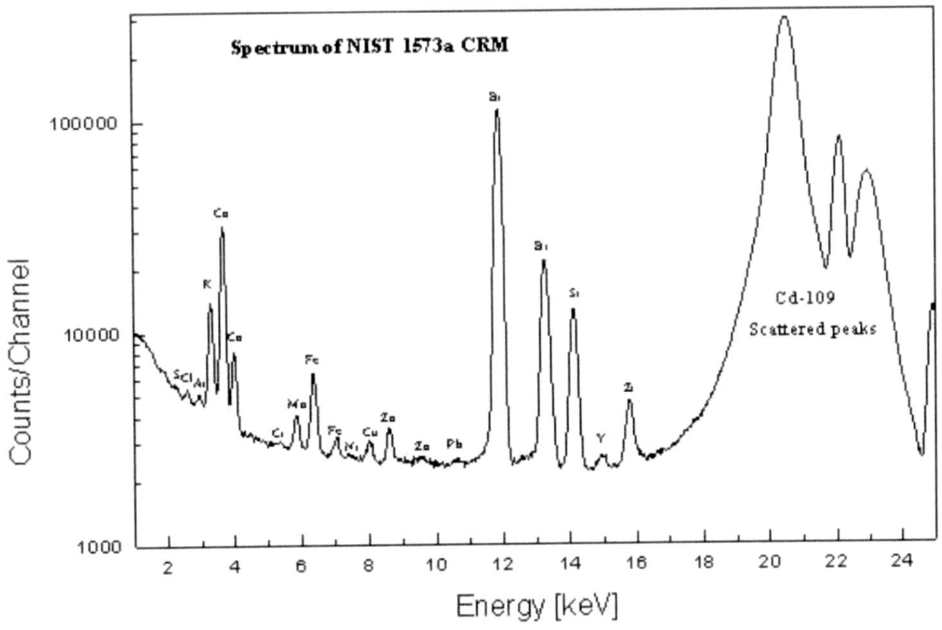

Figure 6. Spectrum of certified reference material NIST 1573a Tomato Leaves excited with Cd-109 radioisotopic source.

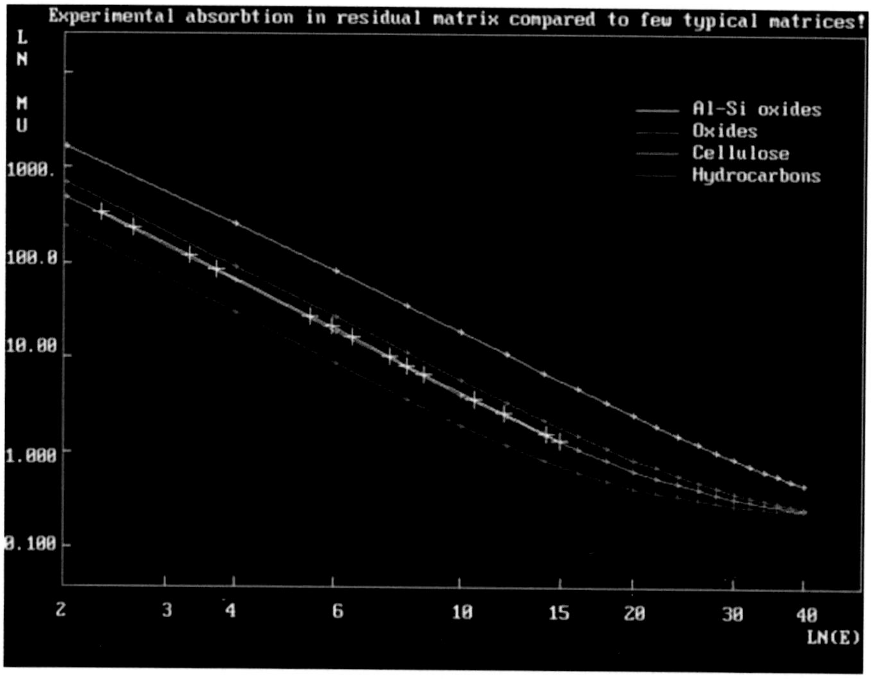

Figure 7. Experimental absorption in residual matrix measured in the certified reference material Tomato leaves NIST 1573a excited with a Cd-109 radioisotopic source (bold white curve with crosses), compared to some typical matrices.

Table 3. Measured concentrations of certified reference material Tomato leaves NIST 1573a excited with Cd-109 radioisotopic source and analysed with "Model 2" - Measurement of absorption...(independent of slope parameter); Sample: ST1573CW, matrix:[ARES = 4102.771]; weight [g cm-^{2}]: 0.160.

El.	E[keV]	Int[cs^{-1}]	S	T [g cm^{-2}]	Conc [g g^{-1}]	Uncert. [g g^{-1}]
S	2.307	0.013	7.89E+02	0.0027	5.66E-03	1.30E-03
Cl	2.622	0.027	1.74E+03	0.0038	3.66E-03	4.45E-04
K	3.312	0.884	4.23E+03	0.0072	2.74E-02	1.23E-03
Ca	3.69	2.615	7.14E+03	0.0076	4.75E-02	2.12E-03
Ti	4.508	0.008	1.24E+04	0.0092	7.26E-05	2.21E-05
Cr	5.411	0.008	2.15E+04	0.015	2.60E-06	5.32E-06
Mn	5.895	0.124	2.72E+04	0.019	2.47E-04	1.21E-05
Fe	6.4	0.355	3.48E+04	0.0237	3.56E-04	1.96E-05
Ni	7.472	0.004	5.27E+04	0.0354	2.17E-06	7.57E-07
Cu	8.041	0.038	6.02E+04	0.0424	1.49E-05	9.83E-07
Zn	8.631	0.117	7.05E+04	0.0499	3.31E-05	1.69E-06
Pb	10.54	0.006	5.76E+04	0.0734	1.34E-06	2.45E-07
Br	11.91	14.61	1.22E+05	0.0873	1.37E-03	6.09E-05
Sr	14.14	1.490	1.67E+05	0.1032	8.62E-05	3.85E-06
Y	14.93	0.043	1.79E+05	0.1078	2.25E-06	1.74E-07

*When analysing the same sample with a Fe-55 source (the following were considered): Absorption intercept: A0 = 9480 [Reduce ARES = 39 at C = 0.15 %]; Absorption slope: A1 = -2.967; RESIDUAL MATRIX characterised by: Absorption intercept: A0 = 4103 (4103) [C = 8.65 %]; Absorption slope: A1 = -2.967 (-2.840) [CHI = 0.00] Residual matrix most probably of: ORGANIC ORIGIN and composed of CELLULOSE OR CARBOHYDRATES! Total matrix [%] = 17.83 (0.31)

Table 3. Measured concentrations in the certified reference material NIST 1573a Tomato Leaves excited with a Cd-109 radioisotopic source and analysed with "Model 5" - List ofexperimental absorption parameters for standards. Sample: NIST 1573 CW matrix:[ARES = 4120]; weight [g cm^{-2}]: 0.160.

El.	E[keV]	Int[cs^{-1}]	S	T [g cm^{-2}]	Conc [g g^{-1}]	Uncert. [g g^{-1}]	C ref [g g^{-1}]
S	2.307	0.013	7.89E+02	0.0027	5.76E-03	1.33E-03	9.60E-03*
Cl	2.622	0.027	1.74E+03	0.0038	3.69E-03	4.48E-04	6.60E-03*
K	3.312	0.884	4.23E+03	0.0072	2.72E-02	1.22E-03	2.70E-02
Ca	3.69	2.615	7.14E+03	0.0077	4.70E-02	2.10E-03	5.05E-02
Ti	5.895	0.132	2.72E+04	0.0192	2.50E-04	1.24E-05	2.46E-04
Cr	6.400	0.384	3.48E+04	0.0239	3.57E-04	2.07E-05	3.68E-04
Mn	7.472	0.007	5.27E+04	0.0356	2.21E-06	1.28E-06	1.59E-06
Fe	8.041	0.069	6.02E+04	0.0426	1.47E-05	1.55E-06	4.70E-06
Ni	8.631	0.125	7.05E+04	0.0503	3.50E-05	1.74E-06	3.09E-05
Cu	10.54	0.02	5.76E+04	0.0739	1.66E-06	6.28E-07	-
Zn	11.91	14.60	1.22E+05	0.0881	1.35E-03	6.03E-05	1.30E-03*
Pb	14.14	1.520	1.67E+05	0.1043	8.70E-05	3.89E-06	8.50E-05*
Br	14.93	0.043	1.79E+05	0.109	2.23E-06	1.72E-07	-

* Uncertified values

RESIDUAL MATRIX characterised by: Absorption intercept: A0 = 4120 (4120) [C = 8.59 %]; Absorption slope: A1 = -2.970 (-2.970) [CHI = 0.00]; Residual matrix most probably of: ORGANIC ORIGIN and composed of CELLULOSE OR CARBOHYDRATES!

3.5. Detection Limits

In most cases in quantitative XRF analysis it is important to determine the limit of detection (LOD). It is usually accepted that the minimal detectable intensity of a spectral line must exceed by a factor 3 the standard deviation of the integrated background under the spectral line. According to this definition the minimal detectable mass or LOD in grams for a definite element is expressed as:

$$LOD\ [g] = 3.\ (\sqrt{B} / t) / S \tag{13}$$

where sensitivity S is expressed as the signal count rate per gram of sample, B is the integrated background in counts under the spectral line, and t is the time of measurement. It is important to stress that the sensitivity and therefore the LOD depend very much on the absorption in the sample matrix. But the explicit dependence of the LOD on the time of measurement is not quite appropriate. Therefore the above relation can be written in somewhat different form, which shows that the accepted LOD as such is quite arbitrary and also requires some additional data to really correctly determine the limit of detection:

$$LOD\ [g] = 3.\ (I_B / S)\ .\ (\sqrt{B}/B) \tag{14}$$

From this expression it is evident that the LOD depends on the relative standard deviation of the background under the spectral line and not only on the measuring time. But since the sensitivity S is also defined as the signal count rate of the spectral line corresponding to the sample mass m_0 (I_0 / m_0), the LOD at 33% relative standard deviation of the background can be expressed as:

$$LOD\ [g] = (I_B / I_0).\ m_0 \tag{15}$$

which is logical and confirms that the LOD depends only on the ratio of background to signal count rates (Kump, 1997).

Total Reflection X-Ray Fluorescence

TXRF Excitation Module

The basic fundamentals of total reflection XRF spectrometry (TXRF) are similar to those of EDXRF, although they have quite different excitation modes. In TXRF systems, the samples are first deposited as dried liquid residues on an optically smooth substrate, which is usually quartz. They are then excited by a well-collimated X-ray beam at an angle smaller than the angle of total reflection for the substrate (<1.8 mrad for quartz) (Klockenkämper 1997; Kump et al. 1997). In this case, the majority of the incident X-ray radiation is totally reflected from the quartz surface, and only a minor part of it is absorbed by the deposited sample to excite fluorescence. The penetration of the incident X-ray beam into the reflecting material is drastically reduced under these conditions, and the scattered and fluorescence radiation contributed by the carrier in this geometry is therefore negligible. Consequently, the background radiation due to scattering on a small amount of sample is very low, significantly increasing the sensitivity of TXRF when compared to standard XRF (Schwenke and Knoth 1993; Kump et al. 1997).

The sensitivity level, however, still strongly depends on the atomic number of the element, although it does extend down to a few ppb (1 $\mu g\ kg^{-1}$) dry weight. To achieve the described excitation conditions, a special total reflection module is required to shape the excitation beam from the X-ray tube into a suitable form that will excite a small amount of the dried sample residue placed on the quartz substrate. Although there are many expensive commercially available TXRF spectrometers, there are also several cheaper laboratory-built systems that exist worldwide. EDXRF systems, which usually have a fine-focus Mo anode X-ray tube, an X-ray generator, a semiconductor X-ray detector and spectroscopy electronics, may be equipped with a total reflection module provided by the Atom Institute (Vienna). In this way, a cheap alternative to commercial TXRF can be built and used for multi-element analyses of different environmental samples.

Sample Preparation and Quantification

For analyses by TXRF, the samples must be in liquid form. A solid ground sample thus requires destructive treatment using wet or dry digestion procedures. This process usually utilizes a decomposition procedure with a small amount of ground material (0.1-0.2 g), and involves the application of a mixture of mineral acids followed by microwave digestion. The resulting solutions can be analysed by TXRF after the addition of an internal standard, which is usually a Ga, V or Y in the form of a standard AAS solution. A small amount of this decomposed sample solution (10 μl) is then pipetted onto a quartz substrate, dried, and measured. As in the case of EDXRF, the TXRF method enables multi-element analysis. Usually, with a Mo anode excitation tube, elements from $Z = 16$ (S) to $Z = 92$ (U) can be determined in the concentration range of a few percent to a few mg kg $^{-1}$. The determination of lighter elements like Na and Mg is possible in a vacuum and with the application of a Si drift detector. Since only a very small amount of dry sample is analysed by TXRF, the relative sensitivity is about one to two orders of magnitude beter than for EDXRF, although the absolute sensitivity of TXRF is very good and reaches a few picograms, which should be contrasted with a few micrograms for EDXRF. The main advantage of TXRF over EDXRF is the possibility of rapidly analysing a larger number of liquid samples (i.e. natural waters, different soil extracts for determining the fractions of metals extracted by $CaCl_2$ or NH_4-acetate) and samples that may be prepared by a simple procedure (i.e. by the dilution of soluble samples, like bee honey in water or juices and vines) (Nečemer et al., 2009). TXRF is also very suitable for the analysis of very small amounts of biological samples, like plant xylem sap, where only a small amount of material is available. In the later case TXRF is actually the only method which can provide multi-element analysis on a very small amount of sample.

Advantages snd Disadvantages of the Use of X-Ray Fluorescence-Based Techniques in Phytoremediation Studies and Their Application

X-ray fluorescence techniques are not only suitable for laboratory analyses but with the development of various portable analysers (Kalnicki and Singhvi, 2001) there is also the possibility to analyse soil and vegetal samples directly at one site, which is especially convenient in phytoremediation studies and their application. Though portable analysers are generally less sensitive than laboratory XRF systems, the results are still sufficient in most of cases to meet the site action level requirements (Kalnicki and Singhvi, 2001). However, it has to be noted that results obtained with portable XRF systems are typically surface measurements only; therefore, sampling location, preparation, and homogenization are important for in situ measurements. In addition, the results of analyses with portable analyzers can also be influenced by physical matrix effects due to variations in the physical character of the sample, such as for example moisture content and inhomogeneity (Kalnicki and Singhvi, 2001).

As summarized in Table 5 laboratory EDXRF and TXRF systems are especially convenient for determination of samples from metal polluted environments and vegetal samples of metal hyperaccumulating plants, which contain high concentrations of metals. It should also be mentioned that TXRF and EDXRF are not the most appropriate techniques for the analysis of low concentrations of light elements (e.g., Na, Mg Al, and Si). In addition, the determination of low concentrations of Cd in plant material by TXRF using Mo anode excitation is often only very approximate, because plant material usually contains high amounts of potassium, and the fluorescent K lines (3.34 keV) strongly interfere with the Cd L lines (3.13 keV). On the other hand, the preparation of solid samples by wet digestion requires the application of time-consuming chemical procedures that require various highly pure and costly mineral acids, equipment and qualified personnel for sample preparation and handling.

Table 5. Sensitivity of different XRF and TXRF systems useful for application in phytoremediation studies and applications as calculated from soil and vegetal sample measurements

EDXRF	Elements	LOD for soil samples Expressed in mg kg^{-1} or % (where indicated)	LOD for vegetal samples Expressed in mg kg^{-1} or % (where indicated)
Radioisotopic source Fe-55 (5.9 keV)	Al – Cr	Al 2.1%, Si 0.64%, S 783, Ti 60	Al 2.3%, Si 0.7%, S 0.18%, Ti 70
Radioisotopic source Cd-109 (22.1 keV)	K – Tc	K 655, Cr 39, Fe 16, Zn 5, Pb 5, Th 1.4	K0.13%, Cr 77, Fe 30,Zn 9, Pb 6, Th 2.5
Radioisotopic source Am-241 (26 and 60 keV)	Pb-Tm	Pb 20, Sr 3, Cd 0.9, La 1.2	Pb 45, Sr 7, Cd 2, La 2
X-ray tube – polychromatic excitation (U=50kV)	Si-Cd	Si 4%, K 550, Ca 325, Cu 23, Y 19, Cd 120	Si 6%, K 190, Ca 140, Cu 13, Y 10, Cd 80
TXRF X-ray tube – monochromatic excitation (Mo K_{α})	S-Y	S 965, Ca 150, Cr 44, Fe 28, Pb 16, Sr 8	S 400, Ca 75, Cr 24, Fe 16, Pb 17, Sr 5

CONCLUSIONS

1. X-ray fluorescence spectroscopy is based on X-ray excitation of atoms in the sample material by the photo-effect process, followed by radiative decay of the excited atoms, i.e. by emission of their characteristic X-rays. The intensities of radiative transitions of atoms in the sample measured by the X-ray spectrometers are then used in qualitative and quantitative analyses of the elemental composition of the sample.
2. Energy dispersive X-ray fluorescence (EDXRF) analysers can be classified according to the type of excitation sources, the geometry of excitation and the type of energy dispersive detectors installed.
3. In standard EDXRF only simple sample preparation including grinding, homogenization and pelletization is needed for solid samples such as soil or vegetal material, while in much more sensitive total reflection X-ray fluorescence (TXRF) spectroscopy the samples must be mineralized using sophisticated dry ashing or wet digestion procedures. On the other hand, TXRF can be very useful for element analyses in water due to its high sensitivity.
4. The measured fluorescent intensities represent the starting data for the quantification procedure. However, fluorescent intensities do not depend solely on the

concentration of the respective element, but also on the concentrations of all other elements in the sample, which attenuate the excitation and fluorescent radiation in the sample before it excites the atoms within the sample and when the emitted fluorescent radiation penetrates towards the surface in the direction of the detector. In environmental samples such as soil and vegetal material part of the sample usually does not respond by fluorescent lines in the spectrum (e.g. light elements like H, C, O, F, and sometimes also Na, Mg, Al, Si, etc., depending on the excitation, the fluorescence yield, and detector efficiency), thus constituting the so-called residual or dark or low-Z matrix, which additionally attenuates the excitation and measured fluorescent radiation. Different approaches in quantification can be applied to solve this problem, with experimental determination of the matrix composition by additional measurement of absorption in the sample by the transmission-emission method being the most reliable.

5. Although XRF-based techniques (especially standard EDXRF) are not as sensitive as for example AAS or ICP-AES, they are much cheaper and environmentally friendlier, and due to the simplicity of the sample preparation procedures they enable analyses of large numbers of samples, which is a common requirement in environmental monitoring and phytoremediation studies.

REFERENCES

Baker, AJM. Accumulators and excluders—strategies in the response of plants to heavy metals. *J. Plant Nutr.,* 1981, vol. 3, p. 643–654.

Barceló, J; and Poschenrider, C. Phytoremediation: principles and perspectives. *Contributions to Science*, 2003, vol. 2, p. 333–444.

Chaney, RL; Malik, M; Li, YM; Brown, SL; Brewer, EP; Angle, JS; Baker, AJM. Phytoremediation of soil metals. *Curr. Opin. Biotechnol.*, 1997, vol. 8, p. 279–284.

Cunningham, SD; Berti,WR; Huang, JW. Phytoremediation of contaminated soils. *Trends Biotechnol.*, 1995, vol. 13, p. 393–397.

Cunningham, SD; Berti, WR. Phytoextraction and phytostabilization: technical, economic, and regulatory considerations of the soil-lead issue. In: Terry N, Banuelos G, Editors. *Phytoremediation of Contaminated Soil and Water*. Boca Raton FL: Lewis Publishers; 2000; p. 359–375.

Glass, DJ. *U.S. and International Market for Phytoremediation.* 1999–2000 D., Needham, M.A., Glass Assoc. Inc., 1999.

He, F; Van Espen, P. General Approach for Quantitative Energy Dispersive X-Ray Fluorescence Analysis Based on Fundamental Parameters. *Anal. Chem.*, 1991, vol. 63, p. 2237-2224.

Jaklevic, JM; Giaque, RD. Energy Dispersive X-ray Fluorescence Analysis Using X-ray Tube Excitation. In: Van Grieken RE, Markowicz AA, editors. *Handbook of X-Ray Spectroscopy: Methods and Techniques.* New York: M. Dekker, Inc.; 1993; p.151-179.

Janssens, K. X-ray based methods of analysis. In: Janssens K, editor. *Non-destructive microanalysis of cultural heritage materials.* Amsterdam: Elsevier; 2004; p. 129-226.

Kalnicky, DJ; Singhvi, R. Field portable XRF analysis of environmental samples. *J. Hazard. Mat.*, 2001, vol. 93, p. 93-122.

Kaulich, B; Gianoncelli, A; Beran, A; Eichert, D; Kreft, I; Pongrac, P; Regvar, M; Vogel-Mikuš; K; Kiskinova, MP. Low-energy X-ray flourescence microscopy opening new opportunities for bio-related research. *Journal of the Royal Society interface*, 2009, vol. 6, p. S641-S647.

Keller, C; Marchetti, M; Rossi, L; Lugon-Moulin, N. Reduction of cadmium availability to tobacco (*Nicotiana tabacum*) plants using soil amendments in low cadmium-contaminated agricultural soils: a pot experiment. *Plant Soil,* 2005, vol. 276, p. 69–84.

Kidd, P; Barceló, J; Bernal, MP; Navarri-Izzo, F; Poschenrieder, C; Shilev, S; Clemente, R; Monterosso, C. Trace element behaviour at the root-soil interface: Implications in phytoremediation. *Environ. Exp. Bot.*, 2009, vol. 67, p. 243-259.

Klockenkämper, R. *Total-reflection X-ray fluorescence analysis*. New York: John Wiley; 1997.

Kump, P; Nečemer, M; Veber, M. Determination of trace elements in mineral water using total reflection X-ray fluorescence spectrometry after preconcentration with ammonium pyrrolidinedithiocarbamate. *X-ray spectrom.*, 1997, vol. 26, p. 232-236.

Kump, P. Some considerations on the definition of the limit of detection in X-ray fluorescence spectrometry, *Spectrochim. Acta Part B*, 1997, vol. 52, p. 405-408.

Lombi, E; Zhao, EJ; Dunham, SJ; McGrath, SP. Cadmium accumulation in populations of Thlaspi caerulescens and Thlaspi goesingense. *New. Phytol.*, 2000, vol. 145, p. 11-20.

Markowicz, A; Haselberger, N; Mulenga, P; Accuracy of calibration Procedure for Energy-Dispersive X-Ray Fluorescence Spectrometry. *X-Ray Spectrom.*, 1992, vol. 21, p. 271-276.

Markowicz, AA. X-ray physiscs. In: Van Grieken RE, Markowicz AA, editors. *Handbook of X-Ray Spectroscopy: Methods and Techniques*. New York: M. Dekker, Inc.; 1993; p. 1-71.

Markowicz, AA; Van Grieken, RE. Quantification in XRF Analysis of Intermediate - Thickness Samples. In: Van Grieken RE, Markowicz AA, editors. *Handbook of X-Ray Spectroscopy: Methods and Techniques.* New York: M. Dekker, Inc.; 1993; p.339-358.

Marguí, E; Queralt, M; Hidalgo, M. Application of X-ray fluorescence spectrometry to determination and quantification of metals in vegetal material, *Trends Anal. Chem.*, 2009, vol. 28, p. 362- 372.

McGrath, SP; Zhao, FJ; Lombi, E. Phytoremediation of metals, metalloids, and radionuclides. *Adv. Agron.*, 2002, vol. 75, p. 1–56.

McGrath, SP; Zhao, FJ. Phytoextraction of metals and metalloids from contaminated soils. *Curr. Opin. Biotechnol.*, 2003, vol. 14, p. 277–282.

McMaster, WH; Delgrand, NK; Mallet, JH; Hubbell, JH. *Compilation of X-Ray cross sections,* UCRL report 50174. Lawrence Livermore Lab., University of California, USA: 1968.

McNeill, K.R; Waring, S. Vitrification of contaminated soils. In: Rees JF, editor. Contaminated Land Treatment Technologies. *Elsevier Applied Science for SCI*; 1992.

Mulligan, CN; Yong, RN; Gibbs, BF. Remediation technologies for metal contaminated soils and groundwater: an evaluation. *Eng. Geol.*, 2001, vol. 60, p. 193-207.

Navari-Izzo, F; Quartacci, MF. Phytoremediation of metals: tolerance mechanisms against oxidative stress. *Minerva Biotecnologica*, 2001, vol. 13, p. 23–83.

Nečemer, M; Kump, P; Ščancar, J; Jačimovič R; Simčič, J; Pelicon, P; Budnar, M; Jeran, Z; Pongrac, P; Regvar, M; Vogel-Mikuš, K. Application of X-ray fluorescence analytical techniques in phytoremediation and plant biology studies. *Spectrochim. Acta part B,* 2008, vol. 63, p. 1240-1247.

Nečemer, M; Košir, IJ; Kump, P; Kropf, U; Korošec, M; Bertoncelj, J; Ogrinc, N; Golob, T. Application of total reflection X-ray spectrometry in combination with chemometric methods for determination of the botanical origin of Slovenian honey. *J. agric. food chem.*, 2009, vol. 57, no. 10, p. 4409-4414.

Pella, PA; Liangyuan Feng; Small, JA. Analytical Algorithm for Calculation of Spectral Distributions of X-Ray Tubes for Quantitative X-ray Fluorescence Analysis. *X-Ray Spectrom.*, 1985, vol. 14 , p. 125-135.

Pilon-Smits, E. Phytoremediation. *Annu. Rev. Plant Biol.*, 2005, vol. 56, p. 15–39.

Prasad, MNV; Freitas, H. Metal hyperaccumulation in plants—biodiversity prospecting for phytoremediation technology. *Electron. J. Biotechnol,* 2003, vol. 6, p. 285–321.

Rousseau, RM. Fundamental Algorithm between Concentration and Intensity in XRF Analysis, 1 – Theory. *X-Ray Spectrom.*, 1984, vol. 13, p. 115-120.

Rulkens, WH; Tichy, R; Grotenhuis, JTC. Remediation of polluted soil and sediment: perspectives and failures.*Water Sci. Technol.*, 1998, vol. 37, p. 27–35.

Salt, DE; Smith, RD; Raskin, I. Phytoremediation. *Annual Review of Plant Physiology and Plant Molecular Biology*, 1998, vol. 49, p. 643–668.

Schwenke, H; Knoth, J. Total Reflection XRF. In: Van Grieken, RE; Markowicz, AA (Eds.) *Handbook of X-Ray Spectroscopy: Methods and Techniques*. New York: M. Dekker, Inc.; 1993; p.453-488.

Tertian, R; Claisse, F. *Principles of Quantitative X-Ray Fluorescence Analysis,* London-Philadelphia- Rheine: Heyden & Son Ltd.; 1982.

Thompson-Eagle, ET; Frankenberger, JWT. Protein-mediated selenium biomethylation in evaporation pond water. *Environ. Toxicol. Chem.*, 1990, vol. 9, p. 1453- 1560.

Van Dyck, PM; Van Grieken, RE. Absorption Correction via Scattered Radiation in Energy-Dispersive X-Ray Fluorescence Analysis for Samples of Variable Composition and Thickness. *Anal. Chem.*, 1980, vol. 52, 1859-1864.

Van Dyck, PM; Torok, SB; and Van Grieken, RE. Enhancement Effect in X-ray Fluorescence Analysis of Environmental Samples of Medium Thickness. *Anal. Chem.,* 1986, vol. 58, p. 1761-1766.

Van Espen, P; Adams, F. Calibration of Tube excited Energy-Dispersive X-Ray Spectrometers with Thin Film Standards and with Fundamental Constants. *X-Ray Spectrom.*, 1981, vol. 10, p 64-68.

Van Espen, PJM; Janssens, KHA. Spectrum evaluation. In: In: Van Grieken RE, Markowicz, AA, editors. *Handbook of X-Ray Spectroscopy: Methods and Techniques*. New York: M. Dekker, Inc.; 1993; p. 181-293.

Vogel-Mikuš, K; Pongrac, P; Kump, P; Nečemer, M; Simčič, J; Pelicon, P; Budnar, M; Povh, B; Regvar, M. Localisation and quantification of elements within seeds of Cd/Zn hyperaccumulator Thlaspi praecox by micro-PIXE. *Environ. Pollut*, 2007, vol. 147, p. 50-59.

Vogel-Mikuš, K; Regvar,M; Mesjasz-Przybyłowicz, J; Przybyłowicz WJ; Simčič, J; Pelicon, P; Budnar, M. Spatial distribution of cadmium in leaves of metal hyperaccumulating Thlaspi praecox using micro-PIXE. *New Phytol.*, 2008a, vol. 179, p. 712-721.

Vogel-Mikuš, K; Simčič, J; Pelicon, P; Budnar, M; Kump, P; Nečemer, M; Mesjasz-Przybyłowicz, J; Przybyłowicz WJ; Regvar, M. Comparison of essential and non-essential element distribution in leaves of the Cd/Zn hyperaccumulator Thlaspi praecox as revealed by micro-PIXE. *Plant Cell Environ.*, 2008b, vol. 31, p. 1484-1496.

Vogel-Mikuš, K; Arčon, I; Kodre, A. Complexation of cadmium in seeds and vegetative tissues of the cadmium hyperaccumulator Thlaspi praecox as studied by X-ray absorption spectroscopy. *Plant soil.,* 2010, vol 13, p. 439-451.

Wenzel, WW; Adriano, DC; Salt, D; Smith, R. Phytoremediation: a plant-microbe-based remediation system. In: Adriano, AB; Bollag, JM; Frankenberger WT, Sims RC, editors. Bioremediation of contaminated soils. Madison WI: *ASA Monograph* 37; 1999; p. 273-303.

Yap, CT; Kump, P; Tang, SM; Wijesinghe, L. Calibration of the Radioisotope-Excited X-Ray Spectrometer with Thick Standards. *Appl. Spectrosc.*, 1987, vol. 41, p. 80-85.

In: Handbook of Phytoremediation
Editor: Ivan A. Golubev

ISBN: 978-1-61728-753-4

Chapter 10

SUSTAINING REMEDIATION OF SECONDARY SALINE AND/OR SODIC SOILS IN CONJUNCTION WITH FIELD MANAGEMENT[1]

Fa Hu Li[2], Pei Ling Yang and Jin Rong Guo
China Agricultural University,
Beijing 100083, China

ABSTRACT

Secondary saline and/or sodic soils in irrigation regions constrains crop yield, and their reclamation is important to meet the need of an increasing population in many countries where arable acreage is limited. Because of their relatively low salinity/sodicity levels compared with primary saline/sodic soils and the high expenditure of traditional reclamation activity, secondary saline/sodic soils should adopt different reclamation strategies and techniques.

To improve soil structure and maintain a reduction or at least balance of salts in secondary saline/sodic soils is a base for sustainable production activity of agriculture. The strategies and techniques on the sustainable nonchemical remediation of secondary saline/sodic soils, combining with farmland management measures such as irrigation and drainage, field engineering, agronomy, and rainfall utilization were discussed in the article. The status quo of remediation of secondary saline/sodic soils in China and the related field management measures were also presented. Lots of experimental results and reclamation practical activities indicated that by the aid of suitable management measures, the phytoremediation or bioremediation of secondary saline/sodic soils without lots of chemical amendment application is feasible for technology, acceptable to farmers for economic benefit, and sustainable to environments.

[1] Supported by the Studies of Safety Evaluation and Pollution Prevention Technology and Demonstration for Groundwater Resources in Beijing (No. D07050601510000) and the Program of State Natural Science Fund of China (No. 40971130).

[2] Corresponding author. College of Water Conservancy and Civil Engineering, China Agricultural University, No. 17, Qinghua East Avenue, Haidian District, Beijing 100083, China. E-mail: lifahu@cau.edu.cn

Keywords: Amendment; Drainage System; Groundwater Table; Irrigation Area; Manure; Phytoremediation; Salinity; Sodicity.

INTRODUCTION

With the increasing population and industrialization (or urbanization) development in the world, cropland area per capita and available fresh water resources in agriculture are decreasing, and food safety is becoming an increasing threat for many regions especially developing countries (Pimentel et al., 1999; Qadir et al., 2003). Because of limited arable land area and eco-environmental protection, cultivated farmland area cannot be enlarged unlimitedly, so to increase the yield output per unit acreage is our only choice.

There is a large amount of cultivated fields that has a middle or low yield output in the world because of various reasons, and soil salinization or/and sodification is one of the major reasons (Kendirli et al., 2005; Eynard et al., 2006; Rengasamy, 2006a; Imada et al., 2009). It is estimated that about 6–23% (800 × 106 – 1500 × 106 ha) of land area in the world and 20–33% (77 × 106 ha) of irrigated farmland area are salt-affected (Keren, 2000; Qadir et al., 2000; Eynard et al., 2006; Qadir et al., 2006a; Imada et al., 2009). Sodium-affected land area amounts for about 30% of total land area (Rengasamy, 2006a) and there is about 210 × 106 – 556 × 106 ha of sodic soil worldwide (Levy, 2000). Although these estimated data have a relative great variable range, its total amount yet is vast. If the agricultural yield per unit acreage in these farmlands with low output can increase 10%, it is enough to feed thousands million of people. Therefore, the improvement of saline/sodic soils possibly is a potential important contribution to meet the requirement of increasing food supply especially in developing countries.

Saline/sodic soil reclamation is feasible no matter from theory or from reclamation activity at present. There is no great obstacle on theory development of saline/sodic soil reclamation to guide practical production activity. Many successful reclamation activities occurring all over the world verify this. However, the area of secondary saline/sodic soils in the world is not decreased, and on the contrary it is enlarging (Eynard et al., 2006; Clyma et al., 2008; Ghafoor et al., 2008; Qureshi et al., 2008; Qadir et al., 2009), which embarrasses the wisdom of human beings in the contemporary era. The reason for the increasing area of saline/sodic soils maybe is associated with land management (Armstrong et al., 1996), fresh water resource scarcity (Minhas, 1996; Fang and L., 1997; Willardson et al., 1997; Sharma and Minhas, 2005), and irrigation and drainage management (Fang and Chen, 2007; Qureshi et al., 2008; Qadir et al., 2009). Other issues related with reclamation activity such as social consciousness and fiscal investment, etc., also contribute to this tendency (Qadir et al., 2001; Qadir et al., 2006b). The improved technologies on saline/sodic soil reclamation are needed so as to satisfy the requirements on public environmental consciousness and affordable ability by farmers (Qadir et al., 2006b). Saline/sodic soil reclamation theories and technologies were reviewed in this article, and the emphasis was focused on the sustainable reclamation technologies of secondary saline/sodic soils. The status in quo about saline/sodic soil reclamation in China also was introduced and discussed. As a great country on population and agricultural production, the current situation on soil reclamation practices may have some implications on land management for other developing countries.

TRADITIONAL RECLAMATION TECHNIQUES OF SALINE AND/OR SODIC SOILS

Excessive salts and Na+ in soils deteriorates soil production conditions and decreases crop yields. Soil salinity and the component proportion of the salts affect not only the physiology and biochemistry of non-halophyte (glycophyte) plant and the water transport in the interface between soil and plant (Läuchli and Epstein, 1990; Hillel, 1998; Manchanda and Garg, 2008), but also soil structure and consequently water movement in soil (Hillel, 1998; Li, 2006).

Although a high soluble salt concentration in soil solution can result in some positive influences on plant growth, e.g., supplying necessary nutrients for plant growth and varying fruit taste or creating succulence (Botía et al., 2005), it usually imposes two adverse effects on plant growth both physiologically and chemically (Läuchli and Epstein, 1990; Manchanda and Garg, 2008).

High salinity in soil depresses soil water potential around the root system of plants. Therefore, the difference between the external and the internal water potentials of plants and the availability of water in the soil for plants is decreased, and in serious cases it can form "physiological drought" for the plant (Munns, 2002; Manchanda and Garg, 2008; Ben-Gal et al., 2009).

Moreover, some ions with high concentration such as Na+, Cl-, SO42-, and B in soil solution can directly impose deleterious effects on plant growth even though the total salt concentration is not very high (specific toxic effects) (U.S. Salinity Laboratory Staff, 1954; Keren, 2000; Li, 2006; Ben-Gal et al., 2009).

Soluble salt concentration and specific ions also can indirectly affect plant growth by the disturbance or deficiencies of some nutritional elements because of its or their existence with high proportion in soil solution (nutritional deficiency or imbalance) (U.S. Salinity Laboratory Staff, 1954; Rengasamy, 2006b).

For example, high salinity can result in Ca2+ deficiency (Munns, 2002; Manchanda and Garg, 2008), high Na+ concentration induces K+ and Ca2+ deficiencies (Breseler et al., 1982), and Cl- creates NO3- deficiency (Läuchli and Epstein, 1990; Li, 2006). Meanwhile, the interaction of osmotic potential and specific ion effect on plant growth also exists (Li, 2006; Ben-Gal et al., 2009). For example, the addition of Ca2+ and NO3- into the soil solution can relieve Na+ and Cl- toxicities, respectively, though it decreases the osmotic potential in soil.

However, the influence of soluble salt on water movement in soil is mainly by its effects on soil aggregate stability. High salinity imposes an active function on soil structure and subsequently improves soil hydraulic properties including conductivity and infiltration. On the other hand, univalent ions such as Na+ have deleterious effects on soil structure because of the dispersion and swelling of fine clay particles, especially when total soluble salt concentration is below the critical flocculation value of soil (Shainberg and Letey, 1984).

Clay dispersion will result in pore-conducting blockage, soil seal or crust formation, air and water movement inhibition, and finally the deterioration of growth conditions and the reduction of crop yield.

In some special cases, the effect of high Na^+ proportion in saline solution on plant growth by soil structure is more serious than its direct influence on plant biochemistry and physiology (Chassemi et al., 1995).

Classification and Characteristics of Saline/Sodic Soils

The soil with an electrical conductivity (ECe) of saturation paste extract more than 4 dS m-1 and exchangeable sodium percentage (ESP) less than 15 or sodium adsorption ratio (SAR) less than 13 at 25oc is referred to saline soil (U.S. Salinity Laboratory Staff, 1954; Chassemi et al., 1995; Qadir et al., 2000; Eynard et al., 2006). The pH value in the saline soil ordinarily is less than 8.5. The soil with ECe less than 4 dS m-1 and ESP greater than 15 or SAR greater than 13 is sodic soil, and its pH values usually range between 8.5 and 10. Saline-sodic soil refers to soil for which ECe is greater than 4 dS m-1 and ESP greater than 15 or SAR greater than 13, and its pH value is usually lower than 8.5.

Because of the presence of excess salts and the relative low exchangeable Na^+ amount in the exchangeable complex of the soil, saline soil generally is flocculated. The main task of saline soil reclamation is to remove the excessive soluble salts out of the plant root zone by leaching and to lower the salinity level of the root system zone below the endurable critical level for different growth stages of given plants.

In sodic soil, excessive exchangeable Na^+ in clay and relative low soluble salt concentration result in the easy dispersion and downward transportation of clay particles, and subsequently the blockage of pore-conducting. The removal of salt in such soil tends to increase the hydrolysis rate of exchangeable Na^+ and cause the rise of soil solution pH (Li and Keren, 2008), and hence causes more precipitation of Ca^{2+} and Mg^{2+} in soil solution at the presence of HCO_3^- (U.S. Salinity Laboratory Staff, 1954). This in turn deteriorates the soil structure stability further. The main task of sodic soil reclamation is to decrease the relative proportion of exchangeable Na^+ in exchangeable complex by ion exchange with divalent cations of Ca^{2+} and others. However, in many cases, maintaining soil structure stability and suitable soil permeability other than leaching excess salts usually is the key issue of successful reclamation of sodic soils. Too long reclamation time is difficult to accept by farmland managers in practical reclamation activity because of low reclamation efficiency and high cost.

Owing to relatively more salts in saline-sodic soil, this soil usually can maintain soil structure stability and suitable soil permeability, and hence is favorable to its reclamation. However, the physical and chemical characteristics of saline-sodic soil are easily turned into those of sodic soil once its reclamation management is not suitable. Therefore, maintaining a preferential reduction of the exchangeable Na^+ before the removal of excessive soluble salts out of the root zone of plants perhaps is critical for this soil reclamation.

Basic Reclamation Principles and Strategies

Based on the different characteristics of three categories for saline/sodic soils, the basic principle of saline soil reclamation is to decrease the salinity level of soil solution by leaching with high quality water. Therefore, a leaching method with a high efficient removal of soil salts should be chosen preferentially. Sufficient water sources with a high quality and good drainage conditions are a precondition for saline soil reclamation. With the leaching of soluble salt in soil, soil ESP level also can be decreased to a certain degree as a result of "valence dilution" effect and the supplement of Ca2+ or Mg2+ from leaching water or the dissolution of mineral salts in soil (Leffelaar and Sharma, 1977; Li, 2006). Deep tillage (plowing) with machines can enhance water infiltration and improve salt leaching to some degree (Prathapar et al., 2005a). The mechanical removal of salty spots also can alleviate the salinity in soils of surface layer (Ramussen et al., 1972). Sodic soil reclamation is to decrease soil ESP level by the aid of Ca2+. Therefore, sufficient Ca2+ sources, no matter from externals or from soils itself in which some Ca2+-containng minerals exist, is the prerequisite for the lowering of soil ESP level. Besides, because of poor physical conditions of sodic soils, reasonable soil permeability should be maintained during the reclamation process by providing a sufficiently high electrolyte concentration in soil solution to counter the disadvantageous influence of exchangeable Na+ on it. Generally, electrolyte concentration in soil solution and its SAR (or soil ESP) level impose an opposite role on soil permeability. The greater the electrolyte concentration in soil solution or the lower the SAR (or ESP) level, the greater the soil permeability (infiltration rate or hydraulic conductivity) (Shainberg and Letey, 1984; Li et al., 2004a; Zhang and Li, 2007; Li et al., 2010; Li and Zhang, 2010). Gypsum or other chemical amendments apply is a common method to get to the both aims at the same time. Sodic soil reclamation generally is more expensive than saline soil reclamation because of Ca2+ supply. It also is more difficult and takes longer time than the saline soil reclamation because of low soil permeability that usually occurs in sodic soil. Saline-sodic soil reclamation is to lower both soil ESP level and the salinity in soil solution. It is the basic principle for saline-sodic soil reclamation that the Na+ adsorbed in exchange complex is firstly turned into soluble Na+ by exchange with soluble Ca2+, and then the excessive salts in the soil solution are removed by leaching out of plant root zone. Therefore, the supplement or supply of sufficient soluble Ca2+ is important for successful reclamation of saline-sodic soil.

The guideline of soil reclamation has changed from the idea of "one-step accomplishment" on entire root zone to the present concept of "endurable level" in major root zone of plants. The traditional idea is that all excessive salts in the entire depth of root zone are removed by intensive leaching, which will result in a long ponding time (e.g., several months or more for saline virgin land), and consequently destroy the soil structure, inhibit soil aeration, and deplete soil nutrients (Keren, 1990). Meanwhile, this also results in the low efficiency of salt leaching, and consequently water waste and groundwater pollution. Generally, the efficiency of salt leaching decreases sharply when the depth of leaching water applied per unit depth of soil exceeds 0.5 – 0.75 (Leffelaar and Sharma, 1977; Keren, 1990; Sharma and Manchanda, 1996; Burt and Isbell, 2005) or when leaching water amount is greater than 2 – 3 pore volumes of soil (Li and Keren, 2008; Gharaibeh et al., 2009). Under a good monitoring and management condition on land use, leaching water amount and leaching time can be planned so as to decrease the salinity level in root zone to below the threshold value of crop according to crop's growth stage and species, and hence improves reclamation efficiency and avoid the disadvantageous impact of reclamation activity on environments.

Common Reclamation Methods and Reclamation Efficiency

Soluble Salt Leaching Methods and Leaching Efficiency

Leaching is a vital process for saline soil reclamation. All methods that can remove soluble salt such as continual flooding leaching, intermittent ponding leaching, alternate row (or border) leaching, sprinkling leaching, and surface flushing are available for saline soil reclamation. Leaching efficiency refers to the amount of salt in drainage water removed from root zone at a given friction of leaching water. Leaching method is one of the important factors affecting salt leaching efficiency.

Continual flooding leaching is a common leaching method in surface-irrigated area, and it is relative least expensive and usually also is low efficient for leaching, especially for structured soils. Groundwater water table and drainage condition affect leaching efficiency. The efficiency of salt leaching at deep groundwater table or nearby the laid drainage pipes (subsurface drainage) is high (Bahçeci and Nacar, 2009). On the contrary, lateral drainage, near the midpoint between drains, or in the stratified soils usually have poor leaching effects (Anapali et al., 2001).

Intermittent ponding leaching is suitable for the soils with low drainage rate and high groundwater table. It also available for soils that a surface seal is prone to be formed, in which the forming cracks following evaporation are helpful to water infiltration (Keren, 1990). Intermittent ponding leaching can minimize the adverse effects of bypass flow on salt removal and is more effective than continual ponding for salt leaching (James et al., 1982). Intermittent ponding leaching generally only needs half as much water as the continual flooding leaching to obtain the same leaching effects (Oster et al., 1972).

Furrow leaching or alternate row (or border) leaching belongs to partial leaching, and its purpose is to maintain a suitable salinity level in crop beds or planting area by transporting soil salts laterally into the unirrigated area. This method is suitable to high groundwater table. Similar to furrow leaching, drip irrigation leaching also can remove salt away from the infiltrating areas. However, under drip leaching, salt will accumulate seriously on the lateral edges of the wetted area, in crop bed ridges, or at the soil surface between emitters (Keren, 2000; Burt and Isbell, 2005). Drip irrigation leaching is helpful to the establishment of a new plant in highly salt-affected soil (Burt and Isbell, 2005). Partial leaching methods can maintain reasonable soil aeration and planting during soil reclamation. However, additional methods to remove or alleviate salinity level in the root zone of planting area such as rainfall leaching or local ponding leaching usually are required under this case.

Sprinkling leaching generally belongs to unsaturated leaching method because its water apply rate usually is lower than soil infiltration rate. Sprinkling leaching is more expensive but more efficient for salt leaching comparing with other methods (James et al., 1982). Similar to intermittent flooding leaching, sprinkling leaching usually uses less water to achieve the same degree of leaching in the same time period as compared with continual ponding leaching (James et al., 1982). Sprinkling leaching is suitable to unleveled field or regions where leaching water is scare or the drainage of saline water is constrained. However, wind speed is a limited factor for this method application.

At the presence of salt spot or crust with shallow groundwater table or low soil permeability, surface flushing method also can be used to remove the salts to nearby depressions by surface drainage (Nayak et al., 2008), but its effect is not prominent on salt leaching (Keren, 1990; Qadir et al., 1998).

In addition to leaching method and drainage conditions, soil texture, structure, salinity level, salt distribution, solute composition, and reclaimed soil management all affect the leaching efficiency of soil salts (Sharma and Manchanda, 1996; Keren, 2000). The Cl--dominated soil profile can be desalinized more efficiently than the SO42--dominated profile (Sharma and Manchanda, 1996). The leaching efficiency of clay loam is lower than that of sandy loam. The leaching of uniform soil generally is more efficient than that of stratified or spatially-variable soils (Keren, 1990; Keren, 2000).

Water Requirement for Salt Leaching

Even though many factors affect salt transportation in soil as discussed above, an empirical formula can be used to estimate the required water amount (D_L [L]) for salt leaching, based on initial soil salinity (C_0 [M L^{-3}]), the anticipated soil depth reclaimed (D_s [L]) and final soil salinity after leaching (C_f [M L^{-3}]), and the reclaimed soil property (k [-]):

$$\frac{C_f}{C_0} \times \frac{D_L}{D_s} = k \qquad (1)$$

in which k is an empirical coefficient that is related with soil property. The heavier the soil texture is, the greater the coefficient value is. k usually ranges from 0.1 ~ 0.3 (Keren, 2000; Bahçeci and Nacar, 2009). This formula is valid for $D_L/D_s>k$.

Calcium Sources of Sodic Soil Reclamation

Two processes occur during the conversion of a sodic soil into a normal agricultural soil: (1) the sodium ions on the exchange complex are replaced by soluble Ca2+, and (2) the excessive soluble salts in soil solution are leached from the system. The addition of chemical amendments into sodic soils has two roles: (1) supplying soluble Ca2+ for the displacement of adsorbed Na+ and subsequently the reduction of soil ESP level; (2) contributing a suitable electrolyte concentration to maintain feasible soil permeability and hence the continuation of reclamation activity. Therefore, the supplies of sufficient soluble Ca2+ and electrolyte concentration are requisite for the successful reclamation of sodic soils because of its low permeability at high soil ESP level.

Calcium ions usually are from the application of agricultural chemical amendments such as gypsum (CaSO4·2H2O) and calcium chloride (CaC12·2H2O) etc., this is called as external Ca2+ source. Gypsum includes mineral gypsum, desulfurized gypsum from power plants, and phosphogypsum from phosphate fertilizer industry (Sakai et al., 2004; Amezketa et al., 2005). Calcium chloride usually is industrial byproducts. Limestone (CaCO3) application also is feasible to improve the physical conditions of sodic soil (U. S. Salinity Laboratory Staff, 1954; Shainberg and Letey, 1984). However, its effect is little or doubtful on sodic soil reclamation because its slow dissolution rate usually cannot provide enough Ca2+ unless a acid or acident is applied concurrently (James et al., 1982; Shainberg and Letey, 1984; Keren, 1996). If there exists a high salt water containing Ca2+ near the reclamation location, successive dilutions with the salt water also can be used to reclaim sodic soils (Keren, 1996). The high-salt concentration in the water as a flocculant and the presence of divalent cations in the water as a Ca2+ source are the essential features of this reclamation method (Reeve and Bower, 1960; Li et al., 2004b).

Low-solubility CaCO3 and other Ca2+-containing minerals extensively exist in arid or semiarid regions. Therefore, calcareous sodic soil can be successfully ameliorated by Ca2+ release from the dissolution of Ca2+-containing minerals with the aid of applying acids (H2SO4 or HCl) or acid formers such as S (sulfur), FeS2 (pyrite), FeSO4·7H2O (ferrous sulfate), and CaS5 (lime-sulfur). This is called as native Ca2+ source. The application of acidic wastewater or acidic coal fly ash also is helpful to the dissolution of native calcareous minerals (Kumar and Singh, 2003). Under the presence of native CaCO3, only leaching without chemical amendment application also can reclaim sodic soil successfully if drainage condition is good and enough leaching water is available and inexpensive (Keren, 2000).

In calcareous sodic soils, the added acidents such as S, FeS2, and CaS5 need to be oxidized by soil microorganism firstly, and then their reaction products produce Ca2+ after a series of chemical reactions, which displaces the Na+ adsorbed in exchange complex. The relative effectiveness of various chemical amendments in supplying Ca2+ and their chemical reactions related sodic soil reclamation are listed by James et al. (1982) and Gupta and Abrol (1990).

The choice of amendments, the suitability of each amendment, and their combined application depend on soil characteristics and cost consideration. Calcium chloride and FeSO4 usually are expensive and are not suitable for a large scale application (Ghassemi et al., 1995).

Gypsum Requirement for Sodic Soil Reclamation

Gypsum is most commonly used and a traditional ameliorant for sodic soil reclamation because of its low cost, extensive availability, and easy handling. Gypsum requirement is related with many factors, and it can be estimated by initial soil ESP level (ESPi, %), the anticipated final soil ESP level (ESPf, %) after reclamation and reclaimed soil depth (Ds, m), and the cation exchange capacity (CEC, molc Mg-1) and the bulk density (γ, Mg m-3) of soil as follows (Li, 2006):

$$GR_p = k \times 0.8608 \times D_s \times \gamma \times CEC \times (ESP_i - ESP_f)/100 \qquad (2)$$

in which GPp is practical gypsum requirement (Mg ha-1); the value "0.8608" is a coefficient for unit conversion; k is a coefficient reflecting Ca2+-Na+ exchange efficiency or other uncertainties that is related with soil sodicity level and percolation rate. For example, at soil ESP level below 10, the exchange efficiency between applied Ca2+ and adsorbed Na+ is lower because a part of applied Ca2+ replaces exchangeable Mg2+ (Keren, 1996). The efficiency may also be low in fine-textured soil because of the difficult displacement of Na+ adsorbed inside soil structure. k value usually can be chosen as 1.25.

Sodic Soil Reclamation Efficiency

Besides the leaching efficiency of excessive soluble salts, sodic soil reclamation efficiency also closely depends on cation exchange efficiency. Soil properties, soil sodicity level, the dissolution rate of Ca2+-containing minerals, the chemical reaction process and reaction rate for Ca2+ supply, the water movement velocity in soil, and the applied place of chemical ameliorants all affect cation exchange efficiency, and consequently the amelioration efficiency of sodic soils.

The dissolution of chemical amendments in soil solution or irrigation water is the first step to produce soluble Ca2+, which subsequently replaces the Na+ adsorbed in exchange complex. Because of different solubilities, the Ca2+ supply rates from chemical amendments and their control factors for sodic soil reclamation are different. For example, CaC12 is soluble and its supply rate for Ca2+ is adequate. When CaC12 is used as an ameliorant, the main factor constraining reclamation efficiency is the exchange efficiency between Ca2+-Na+. Under this case, water movement velocity and soil texture and structure maybe are predominant factors for sodic soil reclamation. Calcium chloride usually is used to reclaim the seriously Na+-affected soils. On the other hand, for the amendments with a moderate or low solubility such as gypsum or limestone, sufficient Ca2+ supply maybe becomes a limited factor for sodic soil reclamation. However, these amendments are favorable to maintain a suitable soil permeability or soil reclamation for a longer period due to their continual supply for adequate electrolyte and Ca2+ concentrations, especially on the soil surface where sealing is prone to be formed.

Sodic soil reclamation efficiency generally increases with the increased solubility of amendments and soil sodicity level, but decreases with the increase of soil cation exchange capacity and water flow velocity (Oster and Frenkel, 1980; Keren and O'Connor, 1982; Nadler et al., 1996). Because of higher solubility and electrolyte level, CaC12 usually is more efficient than gypsum for the reclamation of soils with high ESP level (Prather et al., 1978; Keren, 1996; Gharaibeh et al., 2009). Moreover, CaC12 application on soil surface generally is more effective on soil permeability maintenance and soil reclamation than its mixing with soil (Gupta and Singh, 1988; Miyamoto and Enriquez, 1990). On the other hand, the effects of gypsum are more significant than that of CaC12 for the prolonged maintenances of soil permeability and reclamation activity, especially under rainfall conditions where the electrolyte release from gypsum dissolution can prevent crust formation and soil sealing (Shainberg and Letey, 1984; Miyamoto and Enriquez, 1990). For a given chemical amendment, high reclamation efficiency may be attained only when it can yield a high enough electrolyte concentration to overcome the adverse effects of Na+ on soil permeability at any depth of soil profile throughout the reclamation period (Miyamoto and Enriquez, 1990).

Sulfuric acid can dissolve CaCO3 in calcareous soil and other soil minerals to increase electrolyte concentration in soil, which enhances soil structure stability and subsequently water permeability increase and soil ESP reduction (Ghassemi et al., 1995). Lots of studies have been demonstrated that H2SO4 is more efficient than gypsum on calcareous sodic soil reclamation and its hydraulic property maintenance because of the supersaturated soluble Ca2+ relative to CaSO4 and the higher Mg2+ concentration under H2SO4 application (James et al., 1982; Miyamoto and Stroehlein, 1986; Mace et al., 1999; Amezketa et al., 2005). Sulfuric acid application also is superior to CaC12 application on reclaiming calcareous sodic soil (Miyamoto and Stroehlein, 1986) or maintaining its water permeability (Miyamoto and Stroehlein, 1986). Combining either CaC12 or H2SO4 with CaSO4 as an amendment usually can obtain a more effective reclamation than CaSO4 or CaC12 alone when soil ESP is very high (Prather et al., 1978; Gupta and Singh, 1988).

The application of gypsum mixing with sodic soil can increase gypsum dissolution and hence more Ca2+ supply and higher electrolyte concentration because the adsorbed Na+ is a sink of Ca2+ (Frenkel et al., 1989). Therefore, gypsum mixing with soil is more effective for sodic soil reclamation (Ghassemi et al., 1995). Gypsum is usually applied on soil surface and then incorporated with the soil by disking or ploughing (Ghassemi et al., 1995). To applied gypsum stone in irrigation water also is a traditional method, which can decrease the application expenditure of gypsum on its grinding and spreading to fields (Ahmad et al., 1979). The experimental result of Prathapar et al. (2005b) indicated that gypsum slotting application can obtain a greater reduction on soil salinity and sodicity levels compared with its broadcast application.

Gypsum dissolution rate depends on its particle size, electrolyte concentration, and water flow velocity. Gypsum dissolution rate increases with the increased electrolyte concentration (Gobran and Miyamoto, 1985), but decreases with the increased particle size of gypsum (Ahmad et al., 1979; Keren and Shainberg, 1981). Though gypsum dissolution rate coefficient increases with the increased water flow velocity, the net effect of water flow velocity on gypsum dissolution rate decreases because of the decreased contact time between an element volume of water and a unit of surface area of gypsum fragments (Ahmad et al., 1979; Keren and O'Connor, 1982). Industrial gypsum usually is more dissolvable than mineral gypsum, and hence it is more effective on sodic reclamation than the latter (Keren and Shainberg, 1981; Oster, 1982).

Management during and after Reclamation

Soil reclamation require a long ponding time, therefore avoiding water evaporation is an important measure to increase the utilization efficiency of leaching water during saline/sodic soil reclamation. Mulching with plastic sheet, plant straws, and other materials are helpful to decrease evaporation rate and hence enhance salt percolation downward. It also can lessen the raindrop impact on surface soil aggregate and maintain soil structure during rainfall.

With the progress of reclamation process, soil structure deteriorates gradually, and the reduction of soil infiltration capacity may become a critical factor for the continuation of soil reclamation, especially during rainfall. Gypsum application on soil surface is important to prevent soil sealing or crust formation under rainfall although this application method of gypsum is not favorable to Ca2+-Na+ exchange in soil profile. Gypsum application with a rate of 5 – 10 Mg ha-1 is suitable to the purpose of sealing prevention (Keren, 1990).

Continual flooding leaching for a prolonged time will result in the destruction of soil aggregate and the loss of soil nutrient. Therefore, organic fertilizer application or the planting of leguminous species after soil reclamation is important to enhance the formation or resumption of soil structure and the increase of soil fertility. On the contrary, partial leaching, intermittent flooding leaching, and unsaturated leaching reclamation usually do not need special measures to be adopted.

Possibility of Phytoremediation Combining with Management

Land with high salinity/sodicity levels usually is barren or abandoned wasteland with little production capacity. Although the land can be successfully reclaimed by applying chemical amendments, chemical reclamation method is being substituted by phytoremediation due to its high cost and potential hazard to environment (Qadir et al., 2006a). Lots of research results have demonstrated that phytoremediation for heavily saline and/or calcareous sodic soils is possible (Qadir and Oster, 2002; Mishra and Sharma, 2003; Mishra et al., 2004; Tripathi and Singh, 2005; Qadir et al., 2006a), but its reclamation effectiveness is low and the required reclamation time also is long, compared with chemical reclamation (Qadir et al., 2006a).

Afforestation of salt/Na+-endurable trees or shrub, or the planting of salt/Na+-tolerable vegetation species can yield fuelwood and alleviate energy burden for local farmers, or supply fodder for livestock and get to some economic return (Batra et al., 1997; Kumar et al., 2004), or serve as green manure to improve soil fertility (Akhter et al., 2003). Afforestation or plantation improves soil physical properties by root system activity and soil fertility by the addition of plant residuals into soil (Qadir et al., 1997; Mishra et al., 2004), and remove soil salts by plant biomass harvest (Kumar et al., 2004; Imada et al., 2009). It also decreases water and salt contents in soil profile by root system absorption, and consequently enhances water percolation, soil remediation, and crop yield (Singh et al., 1998; Mishra and Sharma, 2003; Mishra et al., 2004). Especially in the calcareous sodic soils, tree or vegetation growth increases soil CO2 concentration, in addition to root exudates, which enhances the dissolution of native CaCO3 and other minerals in soil and consequently the reduction of soil ESP level (Qadir et al., 1996b; Mishra et al., 2004; Qadir et al., 2005). However, for the phytoreclamation of sodic soils without the presence of native CaCO3, chemical amendment application is necessary although leaching process also can decrease soil soidicity level at a certain degree as discussed above. Even though the phytoreclamation of saline soils, sometimes the application of chemical amendments with some amount also is needed to maintain adequate hydraulic properties and to accelerate reclamation progress.

When phytoremediation method is adopted, land management and infrastructure construction on drainage system are important. Good drainage condition, low groundwater table, and level land all can improve reclamation efficiency and shorten reclamation time (Fang and Chen, 2007). Sometimes, a high-effective network system for surface drainage is necessary and cheap for the reclamation of heavily saline/sodic soils. The effective utilization of rainfall during rainfall season can enhance salt removal from soil profile though its effect possibly is not significant when soil salinity is high or rainfall amount is low (Tanton et al., 1995; Bahçeci and Nacar, 2009).

Existing Problems

Because of the increasing expenditure on amendments and transportation, chemical amendment application on a large scale is becoming unaffordable to farmers. Some industrial byproducts including gypsum or CaCl2 are preferable because they are relative cheaper, and their application on agriculture can be regarded as a disposal method for them too. Meanwhile, drainage water and soluble salt discharge at a great amount during reclamation period also are difficult to be accepted by local environmental protection agency. Especially when acids or acidents are used in calcareous sodic soil reclamation, the damage of acid corrosion on transport apparatus, irrigation equipments, operators, and environment should be carefully disposed. Besides, cultivation or planting activity during chemical reclamation process usually is impossible because of land flooding except for rice plantation. Although phytoremediation is an alternative, compared with chemical reclamation, its efficiency is low (Dubey and Mondal, 1993; Qadir et al., 1996a; Qadir et al., 1997). Phytoremediation of heavily saline/sodic soils will take too long reclamation time to be easily accepted. Phytoremediation with the aid of chemical amendments maybe is a promising option for primary saline/sodic soil reclamation.

Reclamation of Secondary Saline/Sodic Soils and Soil Management

Compared with primary saline/sodic soils, the seriousness degree of secondary salt/Na+-affected soils is less than the former. Secondary saline/sodic soil is usually induced by the improper manipulation during human production activities, and it can be lessened or avoided by the improvement of land management level. Therefore, the role of farmland management on the prevention of secondary saline/sodic soils and its reclamation are more important.

Formation Causes

Water seepage from earthen irrigation channel, irrigation with marginal water, the application of chemical fertilizers, the mobilization of salts in aquifer induced by excessive groundwater drawing, seawater invasion, and deforestation all can result in the formation of secondary saline/sodic soils. Of all the causes, groundwater table rising induced by poor drainage system and irrigation with poor quality water maybe are two key factors.

The construction of irrigation and drainage system in farmland is equivalent important for agricultural production in arid and semiarid region. The output from irrigated area with less than 20% of total farmland area accounts for about 30% – 40% of total agricultural yield, and the grain yield per unit acreage in the irrigated farmland is 2 times as much as or higher than that in rainfed agriculture worldwide (Letey, 1994; Clyma et al., 2008), which indicates the importance of irrigation. However, because of ignoring the importance of drainage water engineering or improper design, the drainage system in many irrigation regions is at poor condition especially in the developing countries, and hence results in the rising of groundwater table and salt accumulation in cultivated soil layer. Fresh water shortage in many regions in the world also compels agricultural irrigation to use more marginal water such as saline/sodic groundwater, municipal effluent, and farmland drainage water (Li, 2004). Irrigation with poor quality water is another important salt source for the increased salinity level in irrigation districts.

Responses of Crops to Saline/Sodic Soils

Soil salinity level affects crop growth and yield. Excessive soluble salt in soil will result in the reduction of crop yield and, at most seriously, crop failure. For most field crops, the response of crop yield to soil salinity can be described with a bilinear function. Crop yield generally is not affected by soil salinity when the salinity level is lower than some critical value for a given crop, and on the contrary, it will decrease linearly with the increase of soil salinity. Crop yield response function can be expressed as (Li, 2006):

$$\begin{aligned} Y_r &= 100 && EC_e \le A \\ Y_r &= 100 - B(EC_e - A) && EC_e > A \end{aligned} \qquad (3)$$

in which Yr is the relative crop yield, Yr = Yp / Ym, and Yp and Ym are crop yields in salt-affected soil and normal nonsaline soil, respectively; ECe is the electrical conductivity of saturated soil extract (dS m-1); A is the salinity threshold for a given crop (dS m-1); B is the percent yield decrease per unit salinity (dS m-1) increase. A and B values are listed by Breseler et al. (1982), Li (2006), and others.

The responses of crops to saline soils can be indicated by the sensitivity or endurable degree of crops to soil salinity level. Different crop species or growth stages have different endurances to salinity level. Field crops generally can be classified as endurable, moderately endurable, moderately sensitive, and sensitive crops for salinity (Breseler et al., 1982). Grain barley, cotton, and sugar beet belong to salt-tolerant crops; wheat, soybean, sorghum, oats, forage barley, and sudangrass are moderately salt-endurable crops; corns, rice, potato, tomato, peanut, clover, and alfalfa are moderately salt-sensitive crops; and some vegetables such as bean, onion, carrot, and some fruits are sensitive to salinity level. Based on this classification, suitable crop species can be chosen for phytoremediation or agricultural planting in salt-affected soils.

Similar to salinity, the responses of field crops to soil sodicity level also can be divided into tolerant, moderately tolerant, sensitive, and extremely sensitive. Sodicity-tolerant crops include barley, wheat, cotton, alfalfa, tomatoes, and beets, and they can grow in the soil ESP level of 40 – 60. Moderately sodicity-tolerant crops include rice, oats, and clover (ESP of 20 – 40). Sodicity-sensitive crops are beans etc. (ESP of 10 – 20), and extremely sodicity-sensitive crops mainly are some fruits. Based on this classification, suitable crop species also can be chosen for phytoremediation or agricultural planting in Na+-affected soils.

Reclamation Strategy, Phyto(Bio)-Remediation and Its Management

Basic principles and methods of soil reclamation discussed above are suitable for the reclamation of secondary saline/sodic soils, but they maybe are not most scientific and economical. Based on the characteristics of secondary saline/sodic soils which generally are affected moderately or slightly by salt/Na+, the reclamation strategy using phytoremediation (alive plant) or bioremediation (organic matter) in conjunction with management should be the first choice no matter from the viewpoints of environmental protection, farmer's affordability, economic benefit, or the effective utilization of scare fresh water resource.

On the basis of favorable physical and chemical conditions following crop plantation, suitably increasing irrigation quota or adjusting irrigation plan enhances the leaching of soluble salts in soil profile and decreases the salinity below the endurable level of given crops. Field leveling can maintain the even seepage of water downwards and hence is advantageous for salt leaching. Rainfall role on salt leaching cannot be ignored during the phytoremediation of secondary saline/sodic soils (Tanton et al., 1995; Li et al., 2005). To choose a deep-root crop, which can thrive in the given saline/sodic soil, is important for the improvement of soil physical properties and salt leaching. Usually, leguminous crops with N2-fixation function or cash crops with high value are better as reclamation plants due to their improvement role on soil fertility or economic return.

As discussed above, crop plantation plays an important role on calcareous sodic soil reclamation because it can supply suitable Ca2+ by the mobilization of native calcareous minerals in soil. The mechanisms of phytoremediation on calcareous sodic soils were discussed by Qadir and Oster (2002) and Qadir et al. (2002). During phytoremediation, soil CO2 concentration is critical to the dissolution of native calcareous minerals, soluble Ca2+ supply, and consequently reclamation success (Li and Keren, 2008). Therefore, any measure that can increase CO2 concentration or reduce pH value in soil is favorable to native CaCO3 dissolution and sodic reclamation (Robbins, 1986a; Li and Keren, 2008). The application of plant residues, farmyard or green manures, and acid cheese whey also is helpful to enhance calcareous sodic soil remediation (Robbins, 1986b; Jones et al., 1992; Chorom and Rengasamy, 1997; Li and Keren, 2009). However, phytoremediation with crop planting usually is better than bioremediation with organic matter application both on soil hydraulic properties and reclamation efficiency for calcareous sodic soils (Robbins, 1986b).

In the phytoreclamation of secondary sodic soils, special attention should be paid on the low soil permeability caused by low electrolyte concentration because, sometimes, it will control the success of sodic soil reclamation or not. Under rainfall condition, raindrop impact and electrolyte dilution on soil surface will result in dense sealing formation and low infiltration capacity. Soil infiltration rate depends on electrolyte concentration and sodicity level in the surface soil (McNeal and Coleman, 1966). The dissolution of soil minerals with moderate or slight solubility plays a significant role on the prevention of soil sealing formation, the stability of soil structure, and the improvement of soil hydraulic properties (Shainberg and Letey, 1984; Keren and Ben-Hur, 2003). If sealing is serious, a small quantity of gypsum can be spread on soil surface so as to overcome the adverse effects of low salt concentration on soil structure (Keren et al., 1983; Amezketa et al., 2005). Artificial polymers of polyacrylamide (PAM) also can improve soil structural stability (Levy and Agassi, 1995; Zhang and Miller, 1996; Peng et al., 2006; Li et al., 2010).

Soil Management to Prevent Salt Accumulation

Soil sodicity level usually is related with soil salinity level (Amezketa, 2007). In order to avoid the occurrence of secondary saline/sodic soils and the excessive accumulation of soluble salts in crop root zone, supplemental irrigation water is needed to maintain the salt balance in the root zone.

Soluble salt balance per unit area of soil in crop root zone can expressed as (Li, 2006):

$$(V_r c_r + V_i c_i + V_g c_g + M_S + M_a) - (M_p + M_c + V_d c_d) = \Delta M_{sw} \tag{4}$$

in which Vr, Vi, Vg, and Vd refer to rainfall volume, irrigation water volume, the water volume entering into crop root zone by capillarity from groundwater, and the water volume losing from the root zone by deep drainage in per unit area of soil [L3 L-2], respectively. cr, ci, cg, and cd refer to corresponding salt concentrations in rainfall, irrigation water, capillary water, and drainage water [M L-3], respectively. Ms and Ma refer to the increment of salt masses per unit area of soil [M L-2], respectively, by native mineral dissolution and agricultural production activity. Mp and Mc refer to the decrement of salt masses per unit area of soil [M L-2], respectively, by mineral precipitation and crop absorption. ΔMsw is the change of salt mass in liquid phase in a given period [M L-2].

When only considering the effects of irrigation and drainage and ignoring the influences of other factors, under salt balance, Eq. (4) can change into

$$V_i c_i = V_d c_d \tag{5}$$

or

$$D_i \times EC_i = D_d \times EC_d \tag{6}$$

in which Di and Dd refer to the depths of irrigation water and drainage water, respectively. ECi and ECd to the electrical conductivities in irrigation water and the drainage water through the bottom of given root zone, respectively.

Therefore, the drainage depth through the bottom of given root zone to maintain the salinity in the root zone below the critical salinity level of a given crop can be related with the depth of irrigation water, that is, leaching friction (LF) can be expressed as:

$$LF = (D_i - D_{ET}) / D_i = EC_i / EC_d \tag{7}$$

in which DET is the loss of water by crop transpiration and soil evaporation [L].

In order to avoid the salt hazard to a given crop species and stage, the required amount of irrigation water at a given water quality can be calculated out form Eq. 7.

The value of leaching friction is related with irrigation water quality, drainage condition, and irrigation method, and it usually is between 0.1 – 0.3 for surface drainage, 0.3 – 0.4 for sprinkler irrigation, and 0.5 for furrow irrigation (Breseler et al., 1982).

Soil Management to Improve Crop Growth

Overall soil reclamation or leaching sometimes is impossible because of the constraints of local conditions such as insufficient water source with good quality or poor drainage system. Farmland management is important to maintain a profitable agricultural cultivation. In addition to decrease the soil salinity level in crop root zone by the leaching of supplement water amount with normal irrigation practice, other measures also can be used to lessen the damage degree of soil salinity/sodicity level to crops or increase crop viability and endurance to given soil condition.

Engineering Measure

Good drainage system in farmland is necessary to maintain a healthy growth condition for crops no matter for air permeability or salt balance in soil profile, and its construction needs lots of investment. Based on the knowledge of water and salt movement, small works in farm field such as furrow bed etc. can be constructed to alleviate salt hazard on crops, especially during germination or seedling stages. The seedling stage of most crops usually is sensitive to salinity level. Therefore, frequent irrigation using sprinkler or dripper is favorable to prevent the temporal accumulation of salts in root zone and is helpful for seed germination or seedling growth. In addition, in order to good crop establishment, an extra irrigation before seeding (pre-sowing irrigation) is a feasible option in many seriously salt-affected soils. To increase the infiltration of rainfall during non-planting season, engineering measures such as land leveling and ridge or furrow-dike construction also is advantageous to the capture of rainfall and the leaching of soil salts (Assouline and Ben-Hur, 2003).

Agronomic Measure

Based on the crop's tolerance to soil salinity/sodicity level and the predicted crop yield as discussed above, to choose a suitable crop to given soil conditions is critical for the profitable return from agricultural plantation. Cotton, barley, wheat, and soybean are common agricultural crops in moderate saline/sodic soils, and alfalfa and clover are usually planted for forage growth or to improve soil fertility. In the regions with sufficient fresh water sources, saline/sodic soil phytoreclamation with rice plantation is popular (Chhabra and Abrol, 1977).

The rotation for crops with deep root system or the inter-plantation of grain crops with forage plants can prevent plant diseases and insect pests, improve soil physical and hydraulic properties and the utilizations of soil water and nutrients, and increase soil fertility (Ilyas et al., 1993). The application of farmyard manure or chopped crop residues also can improve soil physical properties and soil fertility. The increased soil fertility can lessen soil saline/sodic harms to crops.

Cultivation Measure

Deep plough can disrupt plow pan, and help water permeability and root system penetration deeper. If calcareous minerals are present in soil profile, this method is more beneficial to crop growth because the uplifted $Ca2+$-containing soil can enhance soil reclamation. Shallow tillage to break up the dense surface layer in some clay soils also is favorable to water infiltration (Sadiq et al., 2007).

ECONOMIC RETURN OF SOIL RECLAMATION

The purpose of soil reclamation is to increase grain and fiber outputs and obtain more economic return. However, the economic benefit of soil reclamation depends on not only local economic conditions but also social requirement. The expenditure on amendments, transport, labors, water source, and the price of products all affect the ratio of benefit to cost, and local social requirement on agricultural products may distort the evaluation of economic benefit on soil reclamation. Therefore, it is difficult to accurately estimate soil reclamation activity from economic aspect.

The results of cost-benefit analysis of Datta et al. (2000) indicated that financial and economic feasibility was favorable to the installation of drainage pipes when salt-affected land was reclaimed by subsurface drainage system, and an internal rate of return of 8% was obtained. According to the experimental results of Ghafoor et al. (2008), the reclamation of salt-affected soil obtained a net benefit of 3, 17.7, and 133 US$ ha-1 y-1, respectively, for gypsum (25% of requirement) combining with farmyard manure (25 Mg ha-1), farmyard manure (25 Mg ha-1), and gypsum (50% of requirement) under irrigation with saline-sodic water. However, labors were not included in it. Murtaza et al. (2009) thought the greatest net benefit was obtained from the combination treatment of gypsum (100% requirement), one irrigation with saline-sodic water, and one irrigation with fresh water, it was 484 and 216 US$ ha-1 y-1 higher than saline-sodic water irrigation treatment and freshwater irrigation treatment, respectively, under saline-sodic soil reclamation. The experimental results of Tyagi (1984) demonstrated that improving land surface uniformity was profitable in sodic soil reclamation. The economic benefit of phytoremediation or bioremediation generally is better than that of chemical reclamation because of the income of products in phytoremediation and the cost of chemical amendments in chemical reclamation.

Generally, soil reclamation is slightly profitable during the reclaiming period and its economic return will become greater along with cultivation. Soil reclamation cannot be considered only from economic aspect. As discussed above, turning barren land into cultivatable land or elevating soil with a low production capacity to higher level is related grain safety for every country. Soil reclamation is not farmers' affairs themselves, and it should be paid supreme attention and capital investment by all people and governments.

Reclamation Practice and Management Measures in China

China is a developing country with vast population, and the occupation amount per capita for farmland and water resources in China is far lower than the averaged level in the world. The farmland acreage per capita is 0.092 ha, and the occupation amount of water resource per capita is 2048 m3 in China in 2008[3], comparing with the corresponding values of 0.231 ha and 8135 m3 averagely in the world according to FAO statistic data in 2008[4]. Irrigated agriculture also plays an important role on food and fiber supply in China, and the irrigated farmland, which accounts for 44.9% of total farmland area[5], supplies about 75% of total grain output and 90% of total cotton yield.

Development of Irrigated Agriculture

Agricultural irrigation has a long history more than 4000 years in China. Dujiangyan Irrigation Project, established in 256 B.C., is the biggest irrigation area existing in China, and up to now still works well with an irrigation area bigger than 0.673 × 106 ha. The

[3] Statistical Bulletin on National Economy and Social Development in 2008. National Bureau of Statistics of P. R. China. 2009.

[4] http://faostat.fao.org/site/377/default.aspx#ancor.

[5] The Second General Investigation about Agriculture. National Bureau of Statistics of P. R. China. 2008

development of modern infrastructure engineering in farmland generally can divided into 3 phases in China, that is, fast development period from 1950s to 1980s, slow development period from 1980s to 1990s, and steady development period from 1990s.

After the large-scale construction of irrigation and drainage engineering, irrigated farmland area increases from 15.93 × 106 ha in 1950s to 48.89 × 106 ha in 1980s in China. After 1990s, the emphasis of farmland infrastructure construction has focused on the development of saving-water agriculture. Advanced irrigation techniques and reasonable irrigation scheme and irrigation norm according to crop species and growth stages are applied to increase water utilization. Open earthen irrigation channel is changed into pipe or anti-seepage disposed irrigation channel to decrease deep percolation of water. Drainage system is perfected, and considerable farmlands are leveled. All these measures improve water utilization efficiency on irrigated agriculture. Up to 2008, there are 447 irrigation regions with the irrigated area greater than 20 × 103 ha in China, and total area of effectively irrigated farmland is 58.47 × 106 ha. Of which, sprinkler and drip irrigation area accounts for 16.7%, the irrigation channel length with anti-seepage disposal is 35.1% of total length of main and branch irrigation channels. The averaged utilization coefficient of irrigation water is 0.475 for large-scale channel irrigation areas in 2008[6].

Development and Status of Secondary Saline/Sodic Soils

There is 9.2 × 106 ha of farmland affected by salt and/or sodicity in China, and it accounts for 6.62% of total cultivated land or 15% of irrigated farmland. The saline/sodic soils mainly distribute in Northwest, Northeast, and North China. Soil salinization/sodification is one of the main reasons for low soil production capacity.

With the development of irrigation agriculture in China, the area of secondary saline/sodic soils generally experiences increase, decrease, reexpansion, and steady decrease stages.

Because emphasis on irrigation system construction in order to supply sufficient food and ignoring the function of drainage, the area of secondary saline/sodic soils increased sharply after the construction of lots of irrigation water-conveying networks form 1950s to 1970s. About 6.6 × 106 ha of cultivated farmland degraded into salt/Na+-affected soils at different degrees during this period.

After then, the construction of drainage system was enhanced to control groundwater table in the irrigation area. By combined application of various reclamation measures, such as rice plantation, well irrigation and so on, the area of secondary salt/Na+-affected soils was reduced by 50% in Northeast Plain and Northern Plain of China up to 1980s. It also was controlled generally in Hetao irrigation area and Xinjiang Uygur Autonomous Region during 1970s – 1980s.

After 1980s, the management right of farmland was changed from collectivity to household in China, and the farmland was split into several small spots with the averaged area of about 0.1 ha or smaller in populous regions. The investment on farmland infrastructure construction and maintenance was reduced and management difficulty increased. Large numbers of infrastructure engineering in farmland were abandoned or destroyed, the ability to

[6] Statistical Bulletin on Water Resources. Ministry of Water Resources, P. R. China. 2009.

fight back drought and waterlogging disasters reduced, and correspondingly, the area of saline/sodic soils was enlarged during 1980s – 1990s.

From the middle of 1990s, a vast amount of capital from central and local governments has been invested on reconstruction and innovation of irrigation and drainage system for existing irrigation areas. It is expected that irrigation and drainage conditions will be improved, the area of saline/Na+-affected soils will be decreased, and soil growth capacity increased gradually.

Reclamation and Management Measures of Typical Irrigation Areas in China

The successful prevention and reclamation of secondary salt/Na+-affected soils needs to adopt comprehensive measures. To avoid the occurrence of secondary saline/sodic soils is better than its reclamation after occurrence, and this is the basic to develop sustainable irrigation agriculture. The experiences and lessons from the land management and soil reclamation in two typical irrigation areas in China maybe deserve to think about.

Renminshengliqu (RMSLQ) Irrigation Area

Renminshengliqu (RMSLQ) Irrigation Area is an example of successful reclamation of secondary salt/Na+-affected soil at a large scale in China. The RMSLQ Irrigation Area is the first established large-scale irrigation area in the lower reaches of Yellow River in 1952. It locates at the alluvium plain of Yellow River, north part of Henan Province in Northern China. The main crops in the irrigation area are winter wheat, corn, rice, and cotton, and the averaged annual precipitation is 620 mm. The topography of RMSLQ Irrigation Area is low-lying, and its drainage condition is poor. Groundwater table is 2 – 3 m below ground surface with an intensive lateral supplement from Yellow River. The dissolved salt content in groundwater is 1 – 2 g L-1. The RMSLQ Irrigation Area belongs to supplement irrigation according to the crop requirement for water, and its irrigation area covers 98 × 103 ha of farmland.

There was about 6.4×103 ha of salt/Na+-affected farmland in the RMSLQ Irrigation Area in 1957. After the establishment of irrigation area, secondary salt/Na+-affected farmland sharply increased to 18.8 × 103 ha in 1961 because of the rising of groundwater table from excessive irrigating quota, poor drainage system, and large-scale rice plantation. The depth of groundwater table generally was smaller than 1.5 m during this period.

The main reclamation measures that were adopted include: 1) Constructing drainage system and improving drainage condition. Drainage system was established and a networked drainage system is formed all over the irrigation area with the depths of main and branch drainage ditches of 3 – 4 m and 2 – 3 m, respectively. 2) Developing the irrigation with groundwater. Groundwater irrigation lowered groundwater level and decreased the amount of water-conveying. In the canal-irrigation area, a proportion that about 30% of irrigation water amount is withdrawn from groundwater can maintain a steady groundwater table. 3) Decreasing the water loss of deep percolation during the transport process by the anti-seepage disposal of water-conveying canal. 4) Decreasing irrigation norm from the maximum of about 1000 mm y-1 to 350 mm y-1. 5) Improving the soil physical properties of depression lands by the deposition of sediment carrying by Yellow River water combining rice plantation. About

5 × 103 ha of sodic soil farmland has been reclaimed successfully by rice plantation in the irrigation area. 6) The planting of green manure in the heavily salt/Na+-affected soils in the initial period of reclamation and crop rotation also contributed to the successful soil reclamation in this irrigation area.

Up to now, the saline/sodic soils in this irrigation area have been controlled and reclaimed effectively, the depth of groundwater table is controlled at 3.5 m, and all physical and chemical indexes of the soils in the irrigation area have changed into normal level. The effective utilization coefficient of irrigation water increases from 0.34 in 1954 to 0.45 in 2005. Grain and cotton yield per acreage in 2000 increase by 9.7 and 4 times than that in 1950s, respectively.

Hetao Irrigation Area

Hetao Irrigation Area locates at Inner Mongolia Autonomous Region with an averaged annual rainfall of 130 – 215 mm. It is the third largest irrigation area with an irrigation acreage of 0.574 × 106 ha in China, and its main channel engineering was established in 1961. The irrigation water of Hetao Irrigation Area comes from Yellow River. The drainage water discharges into Wuliangsuhai Lake with an area of 290 km2 inside the irrigation area, and then a little part (about 10%) of the drainage water enters into Yellow River again by drainage ditch. The topography in the irrigation area is a flat basin with surface slope gradient of 1/5000 – 1/8000. Underground runoff flows slowly, and groundwater mainly is up and down movement. Total dissolved solid salt usually is 1 – 3 g L-1, and it can be high to 5 – 8 g L-1 in some locations. The cultivated farmland acreage per capita in the irrigation area is 0.47 ha that is 5 times as much as the averaged value all over the country. The main crops in the irrigation area are spring wheat, corn, sunflower, and sugar beet.

Hetao Irrigation Area is famous in China because the area of salt/Na+-affected soils accounts for about 50% of total cultivated farmland in the irrigation area. In the end of 1950s, the area of saline/sodic soils accounted for about 14.1% of total cultivated farmland in the irrigation area, and the depth of groundwater was about 2 m. After the establishment of the irrigation area, the percentage of saline/sodic soils has increased sharply to the maximum of 74.3% and the groundwater depth decreased to about 1.5 m in 1980s. Hereafter, the area of salt/Na+-affected cultivated farmland did not increase significantly, but the area of discarded barren land increased obviously. The whole irrigation area showed a tendency of salt accumulation, and the area of secondary salt/Na+-affected farmland is increasing with a rate of 10 – 13×103 ha y-1.

With the financial aid of China governments at various levels and World Bank loan, the drainage system in the irrigation area was reconstructed and perfected from 1990s. The area of saline/sodic soils slightly shrinked and the percentage of the area of saline/sodic soils to total cultivated farmland decreased to 63.3% in 1990s and 45% in the end of the twentieth century. The grain yield per unit acreage increased by 5.6 times from 1960s to 1998.

The existing problems about Hetao Irrigation Area mainly are imperfect irrigation and drainage infrastructure engineering and poor management level. Because of constraint of natural conditions, poor drainage condition cannot be solved well and it yet is the main issue of soil reclamation in this irrigation area. A good option to solve drainage issue in this district maybe is to adopt well irrigation partially. Well irrigation can not only lower groundwater table but also decrease the loads of water-conveying and drainage, meanwhile it also increase available water amount for users in the lower reaches of Yellow River where economy generally is more developed.

The seepage of irrigation canals is another important issue for the Hetao Irrigation Area. The irrigation channel is an open earth canal almost without seepage prevention disposal, and more than half of the water amount (about 55%) taken from the Yellow River is lost during its transport. The effective utilization coefficient of irrigation water was 0.30 and 0.36, respectively, in 1998 and 2007, and this value is lowest in all irrigation areas at the same scale in China. Central and autonomous region governments in 2009 used special capital to plan and invest in the anti-seepage disposal of main and branch channels in this irrigation area. .

Possibly because of more arable farmland and less population compared with the average level in China, farmland management in the irrigation area is extensive with less expense and labor, small-size field engineering is less, farmland is rugged, irrigation homogeneity is not high, and soil fertility is low. In order to wash excessive salts in soils, an extra irrigation with a vast amount of water is being applied in late autumn in the Hetao Irrigation Area, and hence results in a high groundwater level of less than 1 m below the ground surface during the whole winter and early spring. This extra irrigation alone will consume about 1/3 of total irrigated water. Moreover, excessive irrigation quota also is popular in the irrigation area because of nonmodern irrigation techniques and poor management.

Existing Problems

Agricultural irrigation uses a lot of water; about 62.1% of the total water supply is consumed by irrigation in China. Low water utilization efficiency is a main obstacle for the development of sustainable irrigated agriculture. It not only wastes a lot of valuable fresh water resource but also is a basic reason for the development of secondary saline/sodic soils. Although government at various levels invests in main irrigation and drainage engineering, the investment and the maintenance for lots of field engineering and equipments is difficult to afford by farmers in China, which adversely affects the effective utilization of irrigation water and farmland management. Similar to the situation in many developing countries, crop plantation is a little profitable sector in China. Net benefit was about 412.9, 169.3, 203, and 716.6 US$/ha, respectively, for rice, wheat, corn, and cotton plantation in 2005[7], and average net income from crop plantation accounted for 17.96% of farmer's total income. Therefore, the affordability and enthusiasm of farmers to soil reclamation generally is not high. Moreover, the small management scale of farmland existing in China also increases the difficulty of soil reclamation activity. Cheap, little workload, and gradual phyto(bio)-reclamation in conjunction with management measures is a feasible method at present in China.

CONCLUSIONS

The development of irrigated agriculture is an important measure to supply sufficient food and fiber requirement for the increasing population and the improvement of living

[7] Annual Statistical Bulletin, Ministry of Agriculture, P. R. China, 2005.

quality. Agricultural production is forced to increase its yield per acreage although some barren saline/sodic lands also can be reclaimed and cultivated and other techniques related with agriculture are developed to contribute to total agricultural output. Reasonable design and high-level management for irrigation and drainage system is a key to avoid salt/Na+ accumulation in soil profile. With the intensifying of competition for fresh water, farmland proportion irrigated with saline water and disposed municipal effluent will increase and correspondingly the possibility of soil salinization and/or sodification increase. Because of low expenditure and being friendly to environments, the phytoremediation or bioremediation of moderate salt/Na+-affected soils with the aid of various management measures is suggested except for heavily salt/Na+-affected soils in which some chemical amendments may be needed. Generally, the application of combined reclamation measures of planting, organic material, chemical amendment, engineering, and management can get to more ideal effects. However, all the reclamation measures adopted must be coincided with local natural, management, and social conditions. To maintain a profitable, sustainable irrigated agriculture is our final aim.

REFERENCES

Ahmad, B., W. D. Kemper, G. Haider, and M. A. Niazi. 1979. Use of gypsum stones to lower the sodium adsorption ration of irrigation water. *Soil Sci. Soc. Am. J.* 43(4):698-702.

Akhter, J., K. Mahmood, K. A. Malik, S. Ahmed, and R. Murray. 2003. Amelioration of a saline sodic soil through cultivation of a salt-tolerant grass Leptochloa fusca. *Environmental Conservation* 30(2):168-174.

Amezketa, E. 2007. Use of an electromagnetic technique to determine sodicity in saline-sodic soils. *Soil Use and Management* 23(3):278-285.

Amezketa, E., R. Aragüés, and R. Gazol. 2005. Efficiency of sulfuric acid, mined gypsum, and two gypsum by-products in soil crusting prevention and sodic soil reclamation. *Agronomy Journal* 97(3):983-989.

Anapali, Ö., Ü. Sahin, T. Öztas, and A. Hanay. 2001. Defining effective salt leaching regions between drains. *Turkish Journal of Agriculture and Forestry* 25(1):51-56.

Armstrong, A. S. B., D. W. Rycroft, and T. W. Tanton. 1996. Seasonal movement of salts in naturally structured saline-sodic clay soils. *Agricultural Water Management* 32(1):15-27.

Assouline, S., and M. Ben-Hur. 2003. Effects of water applications and soil tillage on water and salt distribution in A Vertisol. *Soil Sci. Soc. Am. J.* 67(3):852-858.

Bahçeci, İ., and A. S. Nacar. 2009. Subsurface drainage and salt leaching in irrigated land in south-east Turkey. *Irrig. and Drain.* 58(3):346-356.

Batra, L., A. Kumar, M. C. Manna, and R. Chhabra. 1997. Microbiological and chemical amelioration of alkaline soil by growing Karnal grass and gypsum application. *Experimental Agriculture* 33(4):389-397.

Ben-Gal, A., H. Borochov-Neori, U. Yermiyahu, and U. Shani. 2009. Is osmotic potential a more appropriate property than electrical conductivity for evaluating whole-plant response to salinity? *Environmental and Experimental Botany* 65(2-3):232-237.

Botía, P., J. M. Navarro, A. Cerdá, and V. Martínez. 2005. Yield and fruit quality of two melon cultivars irrigated with saline water at different stages of development. *Europ. J. Agronomy* 23(3):243-253.

Breseler, E., B. L. McNeal, and D. L. Carter (eds.) 1982. *Saline and Sodic Soils - Principle-Dynamics-Modelling*. Springer-Verlag, Berlin.

Burt, C. M., and B. Isbell. 2005. Leaching of accumulated soil salinity under drip irrigation. *Transactions of the ASAE* 48(6):2115-2121.

Chassemi, F., A. J. Jakeman, and H. A. Nix (eds.) 1995. *Salinisation of Land and Water Resources - Human Causes, Extent, Management and Case Studies*. University of New South Wales Press, Sydney and CAB International, Oxford.

Chhabra, R., and I. P. Abrol. 1977. Reclaiming effect of rice grown in sodic soils. *Soil Science* 124(1):49-55.

Chorom, M., and P. Rengasamy. 1997. Carbonate chemistry, pH, and physical properties of an alkaline sodic soil as affected by various amendments. *Australian Journal of Soil Research* 35(1):149-161.

Clyma, W., M. S. Shafique, and J. V. Schilfgaarde. 2008. Irrigated agriculture: Managing toward sustainability. p. 542-547. In Trimble, S. W., et al. (eds.) *Encyclopedia of Water Science*, 2nd ed. Taylor and Francis Group, London.

Datta, K. K., C. de Jong, and O. P. Singh. 2000. Reclaiming salt-affected land through drainage in Haryana, India: a financial analysis. *Agricultural Water Management* 46(1):55-71.

Dubey, S. K., and R. C. Mondal. 1993. Sodic soil reclamation with saline water in conjunction with organic and inorganic amendments. *Arid Land Research and Rehabilitation* 7(3):219-231.

Eynard, A., R. Lal, and K. D. Wiebe. 2006. Salt-affected soils. p. 1538-1541. In Lal, R. (ed.) *Encyclopedia of Soil Science*. Taylor and Francis, New York.

Fang, S., and C. X. L. 1997. Using shallow saline groundwater for irrigation and regulating for soil salt-water regime. *Irrigation and Drainage Systems* 11(1):1-14.

Fang, S., and X. L. Chen. 2007. Developing drainage as the basis of comprehensive control of drought, waterlogging, salinity and saline groundwater. *Irrigation and Drainage* 56(Suppl. 1):S227-S244.

Frenkel, H., Z. Gerstl, and N. Alperovitch. 1989. Exchange-induced dissolution of gypsum and the reclamation of sodic soils. *Journal of Soil Science* 40(3):599-611.

Ghafoor, A., G. Murtaza, B. Ahmad, and T. H. M. Boers. 2008. Evaluation of amelioration treatments and economic aspects of using saline-sodic water for rice and wheat production on salt-affected soils under arid land conditions. *Irrigation and Drainage* 57(3):424-434.

Gharaibeh, M. A., N. I. Eltaif, and O. F. Shunnar. 2009. Leaching and reclamation of calcareous saline-sodic soil by moderately saline and moderate-SAR water using gypsum and calcium chloride. *Journal of Plant Nutrition and Soil Science* 172(5):713-719.

Ghassemi, F., A. J. Jakeman, and H. A. Nix (eds.) 1995. *Salinisation of Land and Water Resources: Human causes, Extent, Management and Case Studies*. University of New South Wales Press and CAB International, Sydney.

Gobran, G. R., and S. Miyamoto. 1985. Dissolution rate of gypsum in aqueous salt solutions. *Soil Science* 140(2):89-93.

Gupta, R. K., and R. R. Singh. 1988. A comparative evaluation of the reclamation efficiency of surface-concentrated versus internally incorporated calcium chloride and gypsum in an alkali soil. *Soil Science* 146(4):277-283.

Gupta, R. K., and I. P. Abrol. 1990. Salt-affected soils: their reclamation and management for crop production. p. 224-288. In Lal, R. and B. A. Stewart (eds.) *Advances in Soil Science.* Vol. 11. Soil Degradation. Springer-Verlag, New York.

Hillel, D. (ed.) 1998. *Environmental Soil Physics.* Academic Press, San Diego.

Ilyas, M., R. W. Miller, and R. H. Qureshi. 1993. Hydraulic conductivity of saline-sodic soil after gypsum application and cropping. *Soil Sci. Soc. Am. J.* 57(6):1580-1585.

Imada, S., N. Yamanaka, and S. Tamai. 2009. Effects of salinity on the growth, Na partitioning, and Na dynamics of a salt-tolerant tree, Populus alba L. *Journal of Arid Environments* 73(3):245-251.

James, D. W., R. J. Hanks, and J. J. Jurinak (eds.) 1982. *Modern Irrigated Soils.* A Wiley-Interscience Publication and John Wiley and Sons, New York.

Jones, S. B., C. W. Robbins, and C. L. Hansen. 1992. Sodic soil reclamation using cottage cheese (acid) whey. *Arid Land Research and Rehabilitation* 7(1):51-61.

Kendirli, B., B. Cakmak, and Y. Ucar. 2005. Salinity in the southeastern Anatolia project (GAP), Turkey: Issues and options. *Irrigation and Drainage* 54(1):115-122.

Keren, R. 1990. Reclamation of saline, sodic, and boron-affected soils. p. 411-431. In Tanji, K. T. (ed.) *Agricultural Salinity Assessment and Management.* American Society of Civil Engineers, New York.

Keren, R. 1996. Reclamation of sodic-affected soils. p. 353-373. In Agassi, M. (ed.) *Soil Erosion, Conservation, and Rehabilitation.* Marcel Dekker, Inc., New York.

Keren, R. 2000. Salinity. p. G-3-G-25. In Sumner, M. E. (ed.) *Handbook of Soil Science.* CRC Press LLC, Boca Raton, Florida, USA.

Keren, R., and I. Shainberg. 1981. Effect of dissolution rate on the efficiency of industrial and mined gypsum in improving infiltration of a sodic soil. *Soil Sci. Soc. Am. J.* 45(1):103-107.

Keren, R., and G. A. O'Connor. 1982. Gypsum dissolution and sodic soil reclamation as affected by water flow velocity. *Soil Sci. Soc. Am. J.* 46(4):726-732.

Keren, R., and M. Ben-Hur. 2003. Interaction effects of clay swelling and dispersion and CaCO3 content on saturated hydraulic conductivity. *Australian Journal of Soil Research* 41(5):979-989.

Keren, R., I. Shainberg, H. Frenkel, and Y. Kalo. 1983. The effect of exchangeable sodium and gypsum on surface runoff from loess soil. *Soil Sci. Soc. Am. J.* 47(5):1001-1004.

Kumar, D., and B. Singh. 2003. The use of coal fly ash in sodic soil reclamation. *Land Degradation and Development* 14(3):285-299.

Kumar, D., K. Singh, H. S. Chauhan, A. Prasad, S. U. Beg, and D. V. Singh. 2004. Ameliorative potential of Palmarosa for reclamation of sodic soils. *Communications in Soil Science and Plant Analysis* 35(9-10):1197-1206.

Läuchli, A., and E. Epstein. 1990. Plant responses to saline and sodic conditions. p. 113-135. In Tanji, K. T. (ed.) *Agricultural Salinity Assessment and Management.* American Society of Civil Engineers, New York.

Leffelaar, P. A., and R. P. Sharma. 1977. Leaching of a highly saline-sodic soil. *Journal of Hydrology* 32(3-4):203-218.

Letey, J. 1994. Is irrigated agriculture sustainable? p. 23-37. In Baker, R. S., et al. (eds.) *Soil and Water Sciences: Key to Understanding Our Global Environment*. SSSA Special Publication, No. 41. Soil Science Society of America, Inc., Madison, USA.

Levy, G. J. 2000. Sodicity. p. G-27-G-63. In Sumner, M. E. (ed.) *Handbook of Soil Science*. CRC Press LLC, Boca Raton, Florida, USA.

Levy, G. J., and M. Agassi. 1995. Polymer molecular weight and degree of drying effects on infiltration and erosion of three different soils. *Australian Journal of Soil Research* 33(6):1007-1018.

Li, F. H. 2004. Technique and policy issues on irrigation agriculture development and countermeasures in China. p. 710-715. In Huang G. H. and L. S. Pereira (eds.) *Land and Water Management: Decision Tools and Practices*. China Agriculture Press, Beijing.

Li, F. H. (ed.) 2006. *Physical Chemistry of Soil* (in Chinese). Chemical Industry Press, Beijing.

Li, F. H., and R. Keren. 2008. Native CaCO3 mineral dissolution and its contribution to sodic calcareous soil reclamation under laboratory conditions. *Arid Land Research and Management* 22(1):1-15.

Li, F. H., and R. Keren. 2009. Calcareous sodic soil reclamation as affected by corn stalk application and incubation: a laboratory study. *Pedosphere* 19(4):465-475.

Li, F. H., and L. J. Zhang. 2010. Combined effects of water quality and furrow gradient on runoff and soil erosion in north China. *Pedosphere* 20(1):35-42.

Li, F. H., R. Keren, and M. BenHur. 2004a. Effects of soil sodicity on soil water erosion. p. 710-715. In Huang, G. H. and L. S. Pereira (eds.) *Land and Water Management: Decision Tools and Practices*. China Agriculture Press Beijing.

Li, F. H., G. H. Huang, and J. Deng. 2005. Effects of effluent irrigation on the variation of chemical element concentrations of soil extractions under field conditions (in Chinese). *Transactions of the Chinese Society of Agricultural Engineering* 21(11):123-129.

Li, F. H., S. M. Yang, and C. Peng. 2010. Effects of domestic sewage water and ameliorant effectiveness on soil hydraulic conductivity. *Soil Sci. Soc. Am. J.* 74(2):461-468.

Li, Q. S., L. S. Willardson, L. X. Li, C. J. Liu, and L. Z. Wu. 2004b. Amelioration of saline-sodic soil with mildly saline water in the Songnen Plain, northeast China. *Soil Use and Management* 20(4):439-443.

Mace, J. E., C. Amrhein, and J. D. Oster. 1999. Comparison of gypsum and sulfuric acid for sodic soil reclamation. *Arid Soil Research and Rehabilitation* 13(2):171-188.

Manchanda, G., and N. Garg. 2008. Salinity and its effects on the functional biology of legumes. *Acta Physiol. Plant.* 30(5):595-618.

McNeal, B. L., and N. T. Coleman. 1966. Effect of solution composition on soil hydraulic conductivity. *Soil Sci. Soc. Am. Proc.* 30(3):308-312.

Minhas, P. S. 1996. Saline water management for irrigation in India. *Agricultural Water Management* 30(1):1-24.

Mishra, A., and S. D. Sharma. 2003. Leguminous trees for the restoration of degraded sodic wasteland in Eastern Uttar Pradesh, India. *Land Degradation and Development* 14(2):245-261.

Mishra, A., S. D. Sharma, R. Pandey, and L. Mishra. 2004. Amelioration of a highly alkaline soil by trees in northern India. *Soil Use and Management* 20(3):325-332.

Miyamoto, S., and J. L. Stroehlein. 1986. Sulfuric acid effects on water infiltration and chemical properties of alkaline soils and water. *Transactions of the ASAE* 29(5):1288-1296.

Miyamoto, S., and C. Enriquez. 1990. Comparative effects of chemical amendments on salt and NA leaching. *Irrigation Science* 11(2):83-92.

Munns, R. 2002. Comparative physiology of salt and water stress. *Plant, Cell and Environment* 25(2):239-250.

Murtaza, G., A. Ghafoor, G. Owens, M. Qadir, and U. Z. Kahlon. 2009. Environmental and economic benefits of saline-sodic soil reclamation using low-quality water and soil amendments in conjunction with a rice-wheat cropping system. *Journal of Agronomy and Crop Science* 195(2):124-136.

Nadler, A., G. J. Levy, R. Keren, and H. Eisenberg. 1996. Sodic calcareous soil reclamation as affected by water chemical composition and flow rate. *Soil Sci. Soc. Am. J.* 60(1):252-257.

Nayak, A. K., D. K. Sharma, V. K. Mishra, P. S. Minhas, and C. L. Verma. 2008. *Reclamation of saline-sodic soil under a rice–wheat system by horizontal surface flushing. Soil Use and Management* 24(4):337-343.

Oster, J. D. 1982. Gypsum usage in irrigated agriculture: a review. *Fertilizer Research* 3(1):73-89.

Oster, J. D., and H. Frenkel. 1980. The chemistry of the reclamation of sodic soils with gypsum and lime. *Soil Sci. Soc. Am. J.* 44(1):41-45.

Oster, J. D., L. S. Willardson, and G. J. Hoffman. 1972. Sprinkling and ponding techniques for reclaiming saline soils. *Transactions of the ASAE* 16(1):89-91.

Peng, C., F. H. Li, and X. Y. Pan. 2006. Effects of polyacrylamide application on hydraulic conductivity of sodic and nonsodic soils (in Chinese). *Acta Pedologica Sinica* 43(5):835-842.

Pimentel, D., O. Bailey, P. Kim, E. Mullaney, J. Calabrese, L. Walman, F. Nelson, and X. Yao. 1999. Will limits of the earth's resources control human numbers? *Environment, Development and Sustainability* 1(1):19-39.

Prathapar, S. A., M. Aslam, M. A. Kahlown, Z. Iqbal, and A. S. Qureshi. 2005a. Mechanically reclaiming abandoned saline soils in Pakistan. *Irrigation and Drainage* 54(5):519-526.

Prathapar, S. A., M. Aslam, M. A. Kahlown, Z. Iqbal, and A. S. Qureshi. 2005b. Gypsum slotting to ameliorate sodic soils of Pakistan. *Irrigation and Drainage* 54(5):509-517.

Prather, R. J., J. O. Goertzen, J. D. Rhoades, and H. Frenkel. 1978. Efficient amendment use in sodic soil reclamation. *Soil Sci. Soc. Am. J.* 42(5):782-786.

Qadir, M., and J. D. Oster. 2002. Vegetative bioremediation of calcareous sodic soils: history, mechanisms, and evaluation. *Irrigation Science* 21(3):91-101.

Qadir, M., R. H. Qureshi, and N. Ahmad. 1996a. Reclamation of a saline-sodic soil by gypsum and Leptochloa fusca. *Geoderma* 74(3-4):207-217.

Qadir, M., R. H. Qureshi, and N. Ahmad. 1997. Nutrient availability in a calcareous saline-sodic soil during vegetative bioremediation. *Arid Soil Research and Rehabilitation* 11(4):343-352.

Qadir, M., R. H. Qureshi, and N. Ahmad. 1998. Horizontal flushing: A promising ameliorative technology for hard saline-sodic and sodic soils. *Soil and Tillage Research* 45(1-2):119-131.

Qadir, M., A. Ghafoor, and G. Murtaza. 2000. Amelioration strategies for saline soils: A review. *Land Degradation and Development* 11(6):501-521.

Qadir, M., R. H. Qureshi, and N. Ahmad. 2002. Amelioration of calcareous saline sodic soils through phytoremediation and chemical strategies. *Soil Use and Management* 18(4):381-385.

Qadir, M., R. H. Qureshi, N. Ahmad, and M. Ilyas. 1996b. Salt-tolerant forage cultivation on a saline-sodic field for biomass production and soil reclamation. *Land Degradation and Development* 7(1):11-18.

Qadir, M., S. Schubert, A. Ghafoor, and G. Murtaza. 2001. Amelioration strategies for sodic soils: A review. *Land Degradation and Development* 12(4):357-386.

Qadir, M., J. D. Oster, S. Schubert, and G. Murtaza. 2006a. Vegetative bioremediation of sodic and saline-sodic soils for productivity enhancement and environment conservation. p. 137-146. In Öztürk, M., et al. (eds.) *Biosaline Agriculture and Salinity Tolerance in Plants*. Birkhäuser Verlag, Switzerland.

Qadir, M., Th. M. Boers, S. Schubert, A. Ghafoor, and G. Murtaza. 2003. Agricultural water management in water-starved countries: challenges and opportunities. *Agricultural Water Management* 62(3):165-185.

Qadir, M., A. D. Noble, J. D. Oster, S. Schubert, and A. Ghafoor. 2005. Driving forces for sodium removal during phytoremediation of calcareous sodic and saline-sodic soils: a review. *Soil Use and Management* 21(2):173-180.

Qadir, M., A. D. Noble, S. Schubert, R. J. Thomas, and A. Arslan. 2006b. Sodicity-induced land degradation and its sustainable management: Problems and prospects. *Land Degradation and Development* 17(6):661-676.

Qadir, M., A. D. Noble, A. S. Qureshi, R. K. Gupta, T. Yuldashev, and A. Karimov. 2009. Salt-induced land and water degradation in the Aral Sea basin: A challenge to sustainable agriculture in Central Asia. *Natural Resources Forum* 33(2):134-149.

Qureshi, A. S., P. G. McComick, M. Qadir, and Z. Aslam. 2008. Managing salinity and waterlogging in the Indus Basin of Pakistan. *Agricultural Water Management* 95(1):1-10.

Ramussen, W. W., D. P. Moore, and L. A. Alban. 1972. Improvement of a solonetzic (slick spot) soil by deep plowing, subsoiling, and amendments. *Soil Sci. Soc. Amer. Proc.* 36(137-142.

Reeve, R. C., and C. A. Bower. 1960. Use of high-salt waters as a flocculant and source of divalent cations for reclaiming sodic soils. *Soil Science* 90(2):139-144.

Rengasamy, P. 2006a. Sodic soils. p. 1586-1588. In Lal, R. (ed.) *Encyclopedia of Soil Science,* 2 ed. Taylor and Francis, New York.

Rengasamy, P. 2006b. World salinization with emphasis on Australia. *Journal of Experimental Botany* 57(5):1017-1023.

Robbins, C. W. 1986a. Carbon dioxide partial pressure in lysimeter soils. *Agronomy Journal* 78(1):151-158.

Robbins, C. W. 1986b. Sodic calcareous soil reclamation as affected by different amendments and crops. *Agronomy Journal* 78(5):916-920.

Sadiq, M., G. Hassan, S. M. Mehdi, N. Hussain, and M. Jamil. 2007. Amelioration of saline-sodic soils with tillage implements and sulfuric acid application. *Pedosphere* 17(2):182-190.

Sakai, Y., S. Matsumoto, and M. Sadakata. 2004. Alkali soil reclamation with flue gas desulfurization gypsum in China and assessment of metal content in corn grains. *Soil and Sediment Contamination* 13(1):65-80.

Shainberg, I., and J. Letey. 1984. Response of soils to sodic and saline condition. *Hilgardia* 52(2):1-57.

Sharma, B. R., and P. S. Minhas. 2005. Strategies for managing saline/alkali waters for sustainable agricultural production in South Asia. *Agricultural Water Management* 78(1-2):136-151.

Sharma, S. K., and H. R. Manchanda. 1996. Influence of leaching with different amounts of water on desalinization and permeability behaviour of chloride and sulphate-dominated saline soils. *Agricultural Water Management* 31(3):225-235.

Singh, G., H. Singh, and P. P. Bhojvaid. 1998. Amelioration of sodic soils by trees for wheat and oat production. *Land Degrad. Develop.* 9(5):453-462.

Tanton, T. W., D.W. Rycroft, and M. Hashimi. 1995. Leaching of salt from a heavy clay subsoil under simulated rainfall conditions. *Agricultural Water Management* 27(3-4):321-329.

Tripathi, K. P., and B. Singh. 2005. The role of revegetation for rehabilitation of sodic soils in semiarid subtropical forest, India. *Restoration Ecology* 13(1):29-38.

Tyagi, N. K. 1984. Effect of land surface uniformity on some economic parameters of irrigation in sodic soils under reclamation. *Irrigation Science* 5(3):151-160.

U. S. Salinity Laboratory Staff, (ed.) 1954. Diagnosis and improvement of saline and alkali soils. *USDA Agricultural Handbook* No. 60. U. S. Government Printing Office, Washington, D. C.

Willardson, L. S., D. Boels, and L. K. Smedema. 1997. Reuse of drainage water from irrigated areas. *Irrigation and Drainage Systems* 11(3):215-239.

Zhang, L. J., and F. H. Li. 2007. Combined effects of water quality and slope gradient on infiltration under simulated furrow irrigation for a soil in North China. p. 1197-1203. In Wang, Y. J., et al. (eds.) *Progress in Environmental Sciences and Technology*, Vol. I. Science Press and Science Press USA Inc., Beijing.

Zhang, X. C., and W. P. Miller. 1996. Polyacrylamide effect on infiltration and erosion in furrows. *Soil Sci. Soc. Am. J.* 60(3):866-872.

In: Handbook of Phytoremediation
Editor: Ivan A. Golubev
ISBN: 978-1-61728-753-4

Chapter 11

PHYTOREMEDIATION OF HEAVY METALS USING POPLARS (*POPULUS* SPP): A GLIMPSE OF THE PLANT RESPONSES TO COPPER, CADMIUM AND ZINC STRESS

***Fernando Guerra*[1], *Felipe Gainza*[2], *Ramón Pérez*[2] *and Francisco Zamudio*[3]**

[1]Departamento de Ciencias Forestales, Universidad Católica del Maule, Avenida San Miguel 3605, Talca, Chile

[2]Instituto de Biología Vegetal y Biotecnología, Universidad de Talca, 2 Norte 685, Talca, Chile

[3]Centro Tecnológico del Álamo, Universidad de Talca, 2 Norte 685, Talca, Chile

ABSTRACT

Heavy metals (HMs) as such as cadmium (Cd), copper (Cu) and zinc (Zn) have been widespread in soils by human activities (for example, mining, smelting and agriculture). These metals can affect the environmental quality and the health of people. The risk associated to their occurrence and the possibility to cleanup them using phytoremediation systems is increasing the interest for understanding the biological basis of metal tolerance and accumulation process in plants.

Species belonging to the *Populus* genus (poplars) are suitable candidates for phytoremediation. These trees have a high biomass production, extensive roots, high rates of transpiration and easy propagation. Also, the wide genetic diversity comprised within this genus and the development of multiple biotechnologies and information resources allow a genetic improvement based on traditional and biotechnological approaches. Studies carried out in different experimental conditions show that poplars exposed to Cu, Cd and Zn exhibit distinct tolerance levels and metal accumulation patterns. This response depends on specific genotypes. Some of them have been proposed as candidate for phytostabilization and phytoextraction.

Exposition of poplars to toxic concentrations of Cd, Cu and Zn triggers different effects on growth, biomass partitioning, metal allocation, photosynthesis, carbohydrate and nitrogen metabolism, reactive oxygen species (ROS) production, among others. Plants dispose different homeostatic mechanisms for coping with metal excess. These operate at different levels and their regulation determinates the ability of plant to restrict the metal uptake and (or) root to shoot transport, and compartmentalization. Biological mechanisms underlying metal homeostasis and tolerance in poplars and other tree species are only partially understood. Metal uptake in roots can be regulated by the exudation of organic acid anions, the binding effect of the cell wall and the flux of ions through plasmalem metal transporters. In cytoplasm, metals are chelated and/or transported toward organelles by peptidic chelators. Simultaneously, excesses of metallic ions can be directed to vacuole or apoplast by membrane transporters. Metals are mobilized through the xylem from roots to aerial structures in a process driven by transpiration. Inside leaf cells, a regulated network of membrane transporters and chelators directs metals to their final destination. A further defensive line against metal induced ROS involves enzymes and reducing metabolites. Response to metal stress also includes expression of general defense proteins and signaling elements as such as calcium and ethylene.

1. Introduction

Phytoremediation is a cleanup technology, which uses different plants and their associated microbes for treating environmental contaminants such as HMs, organic compounds or radioactive elements, in soil, groundwater or industrial wastes.

Heavy metals as for example, Cu, Cd, Pb, Hg, or Zn can affect the environmental quality and the health of people. Many metal ions are essential as trace elements, but at higher concentrations they become toxic.

Metals are not degradable and they can be accumulated and concentrated along the food chain. Metal pollution through human activities is a widespread problem around the world. The risk associated to their occurrence in the environment and the possibility to cleanup them using phytoremediation are increasing the understanding of the biological basis underlying plant behavior subjected to HM stress.

The genetic improvement of plants is an essential procedure to establish more efficient phytoremediation systems. In this sense, the development of plants able to tolerate HMs in toxic levels and accumulate them in different organs is a permanent objective in phytoremediation projects.

Species belonging to the genus *Populus*, including aspens, poplars, and cottonwoods (hereafter referred as poplars for simplicity), have been considered suitable candidates for phytoremediation of HM contaminated soils. Poplars have a high biomass production, extensive roots, high rates of transpiration and easy propagation, among other advantages. During last years, a growing amount of information has been generated from studies assessing the performance of poplars exposed to HMs under different conditions.

In this chapter, we analyze the response of poplars to three of the most widespread and studied HMs: Cd, Cu and Zn. Our main focus is on the effects of these metals on plant growth and physiology, metal distribution and the molecular mechanisms of metal homeostasis and tolerance.

2. Heavy Metals in the Environment and their Phytoremediation

2.1. Environmental Importance of Heavy Metals

Heavy metals are a group of elements with a density greater than 5 g/cm^3. Fifty three of the ninety naturally occurring elements are HMs (Schützendübel and Polle, 2002). Metals, such as Cd, Cu and Zn, are primarily of geogenic origin in soils, but anthropogenic activities such as, mining, smelting, metal-working industries, combustion of fossil fuels, phosphate fertilization, addition of sludge to soils, etc., lead to the emission of HMs and their accumulation in ecosystems. Contamination of soils by HMs is a critical environmental concern due to their potential adverse ecological effects. Heavy metals are potential threats for human health and the environment, through their accumulation in the soil, water and in the food-chain (Yadav, 2009). Heavy metals can enter in the human diet and accumulate gradually in the human body. It can result several adverse health effects (e.g. kidney damage or osteoporosis) (Wu et al., 2010). The regulatory limits of Cd, Cu and Zn in agricultural soils are 100, 600 and 1,500 mg kg^{-1}, respectively. Concentrations found in soils can exceed these limits, ranging from 100 - 345,000, 30 - 550,000 and 150 - 5,000,000 mg kg^{-1} for these three metals, respectively (Salt et al., 1998).

2.2. Phytoremediation of Heavy Metals

Phytoremediation is the use of plants and their associated microbes for environmental cleanup (Pilon-Smits, 2005). This technology has gained increasing attention in recent years due the possibility of remediating soils contaminated with HMs in a cost effective and environmentally-friendly way (Kotrba et al., 2009). Phytoremediation is based on the naturally occurring processes by which plants and their microbial rhizosphere flora sequester these pollutants.

Phytoremediation of HMs applies different strategies including phytoextraction, rhizofiltration and phytostabilization (Raskin and Ensley, 1999; Kotrba et al., 2009). In phytoextraction, metal-accumulating plants are used to concentrate pollutants in aboveground harvestable parts. Rhizofiltration (or phytofiltration) uses plant roots to absorb, concentrate and/or precipitate pollutants from contaminated effluents. Phytostabilization aims at using plants to prevent the migration of pollutants, rendering them harmless.

General advantages of phytoremediation include its relatively low cost (ten fold cheaper than methods such as soil excavation, soil washing or burning, or pump-and-treat systems) (Pilon-Smits, 2005) and the possibility of metal recycling. Phytoremediation is an *in situ* application, useful to a variety of contaminants, with public acceptance (Raskin and Ensley, 1999). In comparison to other biological alternatives, such as bioremediation using microorganisms, plants used in phytoremediation systems produce high biomass (with economical value sometimes) with low nutrient requirements and reduce the spread of pollutants through water and wind erosion (Kotrba et al., 2009). On the other hand, this technology exhibits some limitations. The plants that mediate the cleanup have to cope with toxicity levels and unfavorable climate conditions and soil properties. Phytoremediation can be also limited by root depth, time demand (phytoremediation may be slower than remediation methods like excavation or incineration) and bioavailability of the pollutants, especially when regulatory cleanup standards require that all the pollutant is removed (Pilon-Smits, 2005). Phytoremediation also involve some potential risks to the environment. For phytoextraction, risks include metal dispersal into adjacent environments, metal accumulation in topsoil and harmful effects of metals on herbivores (Langer et al., 2009).

2.3. Characteristics of Plants Used for Phytoremediation of Heavy Metals

Plants suitable for phytoremediation should possess a series of characteristics: (1) ability to accumulate metals preferably in the aboveground parts, (2) tolerance to metal concentration accumulated, (3) fast growth and high biomass, (4) widespread highly branched root system, (5) easy harvestability, and (6) non consumable by humans and animals (Arthur et al., 2005). However, plant species just can partially fulfill these conditions. For example, those few plants that can accumulate metals to exceptionally high concentrations in their shoots, with no adverse effects on their growth (hyperaccumulators), are both small and slow growing, and often they are rare species of limited population size and very restricted distributions (Pollard et al., 2002). On the other hand, high biomass producing species, such as trees and agricultural crops tend to take up relatively smaller amounts of heavy metals than hyperaccumulators. Comparing with agricultural species, trees have some advantages as for example their deep rooting favoring the metal extraction from deeper soil layers (Dos Santos and Wenzel, 2007). The phytoremediation of HM contaminated sites by trees has been reviewed in detail by Pulford and Watson (2003).

2.3.1. Heavy Metals in Plants. Functions and Negative Effects

Plants used for phytoremediation of HMs must be able to cope with negative effects of metal excess. Heavy metals occur naturally, but not all of them have a biological role (Schützendübel and Polle, 2002). Among HMs, only 17 may be bioavailable for cells and being of importance for organisms and ecosystems. For example, metals such as Cu, Zn, Ni, or Cr are toxic with high or low importance as trace elements. Cadmium, Pb or Ag has no known function as nutrients and seem to be more or less toxic to plants and micro-organisms. High concentrations of HMs in soils could be toxic for plants resulting in varied effects on plant physiology affecting its growth and survival. According to Hermle et al. (2006), Cd, Cu and Zn become toxic for sensitive plants if they reach values in the concentration range of 5 - 10 mg kg^{-1}, 15 - 20 mg $kg^{-1}$and 150 - 200 mg kg^{-1}, respectively.

The effects of Cd on plant physiology are only partially understood (Vollenweider et al., 2006). Visible effects of exposure to high Cd concentrations are growth reduction and leaf chlorosis (Clemens, 2006). Cadmium interferes with the uptake, transport, and use of different elements (e.g. Fe, Zn and Mg) (Pietrini et al., 2010a). This metal can disturb the plant water balance, inhibiting the stomatal opening and affecting the photosynthetic apparatus (Vollenweider et al., 2006). At the cell level, Cd can damage different organelles including chloroplasts, nucleus, vacuole and mitochondria. It inhibits or activates many enzymes, particularly those rich in sulfhydryl groups. Oxidative stress has been discussed as a primary effect of Cd exposure even though Cd is not a redox-active metal and it does not take part in Fenton and Haber-Weiss reactions (Clemens, 2006). Rather, symptoms of oxidative stress, such as lipid peroxidation are consequence of the activation or the inactivation of antioxidative enzymes (Vollenweider et al., 2006) or the depletion of glutathione (Clemens, 2006).

Copper is an essential trace element for plants as cofactor of various proteins. It plays an important role in process such as photosynthesis and respiration, carbon and nitrogen metabolism, oxidative stress protection, perception of ethylene, and cell wall synthesis. Copper functions as a redox agent in biochemical reactions. However, this property makes it also potentially toxic when plants grow under high concentrations. Cu excess induces stress and causes injury to plants leading to growth retardation and leaf chlorosis (Yadav, 2009). At cell level, Cu ions can catalyze the production of highly toxic hydroxyl radicals, in particular through Fenton chemistry, thus leading to the damage to macromolecules and disturbance of metabolic pathways (Hänsch and Mendel, 2009). In addition, Cu is highly reactive to thiols and can displace other essential metals in proteins (Burkhead et al., 2009). To balance needs and avoiding potential toxic excess, the cellular concentrations of Cu are tightly controlled (Pilon et al., 2009).

Zinc is an essential nutrient for plants. This element is a co-factor required for the structure and function of numerous proteins (Grotz and Guerinot, 2006), energy production and structural integrity of membranes (Hänsch and Mendel, 2009). High levels of Zn inhibit many plant metabolic functions resulting in retarded growth and senescence. Zinc toxicity in plants limits the growth of both roots and shoots and produces leaf chlorosis. Even though Zn is not redox active, too high levels of this metal are toxic because it can displace other metals (e.g. Fe, Mn and Cu) in the cell (Pilon et al., 2009; Yadav, 2009). Because this, Zn homeostasis is also strongly regulated in plant cells.

2.3.2. Mechanisms of Heavy Metal Homeostasis and Tolerance

Despite the negative effects that excess of HMs can produce, some plant species have developed ecotypes able to survive and grow on highly contaminated soil (Salt et al., 1998). Plants living in a contaminated environment can be roughly classified into three types (Hassinen et al., 2009): (1) excluders that tolerate metals by restricting uptake, (2) accumulators that have increased cellular detoxification mechanisms especially in the above-ground parts, and (3) indicators in which the elemental concentrations reflect the soil concentrations due to the lack of protective mechanisms. Within the second group, hyperaccumulating plants are an important case. As it was showed, they can accumulate metals to exceptionally high concentrations in their shoots, and without negative effects on their growth. The accepted shoot concentrations defining hyperaccumulation are (on a w/w basis) 0.01 % for Cd, 0.1 % for Cu and 1.0 % for Zn (Pollard et al., 2002).

Plants respond to negative effects of exposition to toxic levels of HMs developing different homeostatic mechanisms to maintain essential metallic ions in suitable concentrations within different cell compartments and minimize the damage caused by non-essential metallic ions. In this way, a regulated network of transport, chelation, traffic and compartmentation control the absorption, distribution and detoxification of the metallic ions (Clemens et al., 2002). The way in which is regulated determinate the ability of plants for restricting uptake and/or root to shoot transport, and sequestrating and compartmenting metals in organs and/or organelles.

2.3.3. Importance of the Plant Genetic Improvement

The operation of phytoremediation systems depends on both biological and environmental factors as well as on the interaction between them. The efficiency of phytoremediation can be improved through genetic selection (breeding of selected parental genotypes and progeny testing, hybridization between compatible species and direct clonal assessment of potentially useful pedigrees, etc.) of plants with the desired properties (tolerance level, metal accumulation patterns, biomass production, etc.) and by the application of the adequate agronomic practices (e.g. management of soil compaction, irrigation, fertilization, etc.). In terms of plant improvement, biotechnological approaches as such as the use of transgenic plants engineered for metal tolerance/accumulation or the marker assisted selection are key to complement traditional breeding techniques and developing plants with suitable phenotypes. In a similar way, plant growth, HM tolerance and accumulation can be also enhanced by the selection of the adequate plant interacting rhizospheric microorganisms.

3. Poplars and their Use in Phytoremediation

3.1. General Characteristics of Poplars

The genus *Populus* is a one of two members of the Salicaceae family. Poplars have a wide natural distribution in the Northern Hemisphere and a small representation in tropical Africa. Taxonomic classifications recognize 29 species that are grouped under six separate sections (Stettler et al.,1996). Poplars are dioecious, wind-pollinated and produce large amounts of small seeds that are dispersed by wind and water. They form a key component of riparian forests and are capable of rapidly invading disturbed sites. All poplars also have the capacity to reproduce asexually, mostly by sprouting from the root collar of cut trees or from abscised or broken branches that become embedded in the soil. Some poplars also propagate through sucker shoots that arise from horizontal roots (Bradshaw et al., 2000).

Poplars are part of fast growing tree species most cultivated worldwide. According to (Bradshaw et al., 2000), a biological system supports their growth and that begins with the elongation of a preformed shoot from its bud and continues to start and expand shoot segments and leaves throughout the growing season. Trees can reach 40 m in height in less than 20 years. The wood is diffuse-porous and light in weight. Poplars are cultivated in plantations for pulp and paper, veneer, engineered wood products, lumber, and biomass for energy. Growing at a commercial scale under intensive culture for 6 to 8-year rotations, production rates with hybrid poplar can be as high as 17 - 30 Mg/ha/y of dry woody biomass, comparable to the biomass produced by row crops such as corn. Historically, poplars have been widely used in windbreaks and for erosion control. Currently, they also are an important alternative for phytoremediation.

3.2. Poplars as a Model to Study the Biology of Trees

Poplars are regarded as a model tree in forest genetics and biotechnology studies. *Populus* provides opportunities to evaluate important plant processes absent or poorly developed in herbaceous plant genus (for example, *Arabidopsis*), such as wood formation, autumn senescence, and biotic interactions from a comparative point of view (Jansson and Douglas, 2007). Bradshaw et al. (2000), indicated the following strengths of *Populus* as a model system: (1) abundant genetic variation in natural populations, (2) ease of sexual propagation and inter-specific hybridization, (3) rapid and pronounced physiologic responses to environmental variables, (4) well-characterized molecular physiology, (5) cloning of individual tree genotype, and (6) closely related to other angiosperm model plants. Additionally, the poplar genome size is relative small (450 Mb, harboring around 40,000 genes) (Neale and Ingvarsson, 2008). A suite of critical genomic and molecular tools as such as EST collections, DNA microarrays, transformation protocols, etc. have been already developed (Jansson and Douglas, 2007). Its genome sequence is published (Tuskan et al., 2006) enabling the application of high-throughput genomics technology and easing comparative and evolutionary genomics studies, solidifying the role of poplars as a reference organism for the tree biology (Yang et al., 2009).

3.3. Phytoremediation with Poplars

Poplars are suitable for phytoremediation purposes due that they can remove contaminants in several ways, as for example, phytoextraction, phytostabilization and phytovolatilization. Advantageous characteristics of poplars include: quick establishment, fast growing, large biomass accumulation, extensive and deep root systems, high rates of transpiration, ease asexual propagation, exceptional growth on marginal lands, not part of food chain, long lived (25 - 30 years) and they can be harvested and then regrown (Sebastiani et al., 2004; Zalesny et al., 2008).

Additionally, poplars from phytoremediation systems are environmental acceptable sources of biomass for bioenergy (short rotation coppice cultures) as well as wood products (Laureysens et al., 2004; Licht and Isebrands, 2005).

A series of studies have been carried out for assessing the ability of poplars to clean up soil or water contaminated with petroleum hydrocarbons, landfill leachates, solvents, explosives, radionuclides and salts, among others (see table 3.1). Poplars have been also proposed as candidates for treating HM-polluted soils and producing economically biomass exploitable for energy production (Sebastiani et al., 2004).

Fast growing, moderate capacity to accumulate HMs as well as high biomass yields, extensive root systems and high transpiration rates are characteristics supporting that condition (Bissonnette et al., 2010). According to (Sebastiani et al., 2004), the phytoremediation of HMs with poplars could be possible using different approaches as for example, phytoextraction or phytostabilization.

Studies about the phytoextraction potential of poplars have demonstrated large variation in HM tolerance and in the partitioning of HMs within tree organs among species and clones (Pulford and Watson, 2003; Bissonnette et al., 2010). The high variability observed supports the strategy of selecting poplar genotypes for metal allocation in harvestable woody parts (Pietrini et al., 2010a). Besides of traditional genetic improvement approaches, biotechnologies such as genetic engineering are demonstrating to be effective improving the biomass production in HM-contaminated soils (Che et al., 2003).

Table 3.1. Selected examples of contaminants evaluated in studies involving poplars

Contaminant	Species or hybrids	Reference
2,4,6-trinitrotoluene (TNT)	*P. x canadensis* (DN34 clone); *P. tremula x P. tremuloides* (Etropole clone; transgenic plants expressing the bacterial nitroreductase gene *pnrA*)	(Thompson et al., 1998) (van Dillewijn et al., 2008)
Benzene, toluene, ethylbenzene and xylene (BTEX) compounds	*P. trichocarpa x P. deltoides* (Hoogvorst and Hazendans clones)	(Barac et al., 2009)
Boiler ash and biosolids	*P. nigra x P. maximowiczii* (NM6 clone)	(Cavaleri et al., 2004)
Boron contaminated wood-waste	*P. x canadensis* (Argyle and Selwin clones); *P. deltoides x P. yunnanensis* (Kawa clone); *P. euramericana x P. yunnanensis* (Toa clone); *P. alba x P. glandulosa* (Yeogi clone); *P. nigra x P. manimowic (*Shinsei clone)	(Robinson et al., 2007)
Methyl tert-butyl ether (MTBE) contaminated groundwater plume	*P. x canadensis*, (DN34 clone)	(Hong et al., 2001)
Ozone	*P. nigra*	(Omasa et al., 2000)
Perchlorate (Cl O_4^-)	*P. x canadensis*	(van Aken and Schnoor, 2002)
Petroleum hydrocarbons	(*P. trichocarpa x P. deltoides*) *x P. deltoides*; *P. deltoides x P. deltoides*; *P. deltoides x P. maximowiczii*; *P. nigra x P. maximowiczii*; *P. x canadensis*; *P. deltoides*	(Zalesny et al., 2005)
Polycyclic aromatic hydrocarbons	*P. x canadensis* (DN34 clone)	(Spriggs et al., 2005)
Polychlorinated biphenyls (PCBs)	*P. x canadensis* (DN34 clone)	(Liu and Schnoor, 2008) (Liu et al., 2009)
Solid waste landfill leachate	(*P. trichocarpa x P. deltoides*) x *P. deltoides*; *P. deltoides x P. maximowiczii*; *P. x canadensis*; *P. nigra x P. maximowiczii*	(Zalesny et al., 2007)
Trichloroethylene (TCE)	*P. trichocarpa x P. deltoides* (H11-11 and 50-189 clones); *P. trichocarpa x P. maximowiczii* (282-190 clones)	(Newman et al., 1997)

4. Responses of Poplars to Cd, Cu and Zn Excess

4.1. Plant Growth and Metal Distribution

The plant growth can be reduced by the application of toxic concentrations of HMs. Metals interfere with essential process as such as nutrient uptake and photosynthesis.

Exposition of poplars to Cd affects their biomass production. A reduction in root and leaf dry mass was observed by Pietrini et al. (2010a) in clones from different poplar hybrids/species (*P. x generosa*, *P. x canadensis*, *P. deltoides*, *P. nigra*, *P. alba* and *P. trichocarpa*) grown under Cd 50 μM in hydroponics. Relative to control plants, some clones decreased their root and leaf dry mass by near 80% and 65%, respectively. However, some of them increased their root to leaf ratio, suggesting a higher tolerance to Cd stress. Negative effect on total biomass would be dependant on Cd concentration. Experiments with *P. x canadensis* cultured in pots containing distinct sort of soils with increasing Cd concentrations (0 - 1.5 mg kg^{-1}) showed a negative tendency of total biomass production and Cd level (Wu et al., 2010).

A reduction of plant growth has been also observed by Cu excess in poplars. Borghi et al. (2007) reported a general reduction of plant biomass and growth variables when plants of *P. x canadensis* (Adda clone) were hydroponically cultured at Cu concentrations equal or higher than 100 μM. More Cu-sensitive poplars (e.g. *P. alba*, Villafranca clone) showed this sort of response (and other, as for example root thickening) at lower concentrations (< 25 μM) (Borghi et al., 2008).

Zinc toxicity also affects the biomass production of poplars. Plants of *P. x canadensis* (I-214 clone) treated with Zn (1 - 10 mM) in a hydroponic system decrease their shoot dry mass until five fold those observed in control plants (Di Baccio et al., 2005).

A series of experiments have assessed the effect of soils, containing a combination of metals, on the poplar growth. Plants of *P. tremula* grown in soils (in pots) with an added HM mix (Cu/Zn/Cd/Pb = 640/3,000/10/90 mg kg^{-1}) showed a negative effect on annual height increments and foliage mass and area (Hermle et al., 2006). Vamerali et al. (2009) evaluated the performance of three poplar species (*P. alba*, *P. nigra* and *P. tremula*) in soils contaminated with wastes from the sulphur extraction. Exposition of plants to contaminated soil (As/Co/Cu/Pb/Zn = 886/100/1,735/493/2,404 mg kg^{-1}) produced a general reduction in biomass (wood, roots and leaves). *P. alba* had highest growth, and *P. nigra* showed most growth inhibition compared with control plants. On the other hand, a positive effect on growth has been observed by some poplars exposed to HMs. *P. nigra* clones subjected to additions of Cd (4.45 μM), Zn (76.5 μM), or a metal cocktail (Cd/Zn/Cu/Pb = 4.45/76.5/7.87/24.1 μM) showed an increased root and foliar biomass in both single and cocktail treatments (Dos Santos et al., 2007). A positive effect in biomass has been registered for poplars cultured in pots with soil amended with industrial organic wastes (biosolids from tanneries) containing HMs (Fe/Zn/Cr/Cu/Cd = 54,000/10,300/14,800/102/4.4 mg kg^{-1}) (Sebastiani et al., 2004). Leaf, stem, root and woody cutting biomass of treated plants (*P. deltoides x P. maximowiczii*, Eridano clone, and *P. x canadensis* I-214 clone) were significantly greater than in controls. Metal concentrations did not exert any toxic effects on plants.

The distribution of HMs within poplar trees is a process depending on the specific element and the biological tolerance strategy of genotypes (Pulford and Watson, 2003). Exclusion and accumulation are two main tolerance mechanisms adopted by poplars to cope with HMs. The analysis of Cd distribution in plant organs was included by Pietrini et al. (2010a) in experiments in which poplar clones were subjected to Cd 50 μM. From the metal contents analysis, they identified three different accumulation patterns classifying the genotypes in low leaf accumulators, leaf accumulators and root accumulators (table 4.1).

The characterization of Cd phytoextraction efficiency of *P. x canadensis* in Cd contaminated soils was carried out by Wu et al. (2010) in a pot experiment assessing different Cd concentrations (0 - 1.5 mg kg^{-1}). Distinct Cd concentration in tissues was observed in plants growing in two soil types (purple and alluvial). Plants exhibited Cd transport from root to shoot in both soils regardless of Cd contamination levels, which were increased with increasing Cd in media. Cadmium concentration in poplar components showed the descendant order shoot > root > leaf (52.5 > 47.3 > 0.2, percentage of total Cd in plants cultured in alluvial soil), in a similar way than poplars categorized as "low leaf accumulators" by Pietrini et al. (2010a). Significant Cd accumulation in leaves and wood (stems and twigs) was registered by Hermle et al. (2006) in *P. tremula* plants, cultured in pots with a mix of soil and a HM combination (Cu/Zn/Cd/Pb = 640/3,000/10/90 mg kg^{-1}). Foliage and wood Cd concentrations (around 10 and 5 mg kg^{-1}, respectively) were higher those observed for other species (*Salix viminalis*, *Betula pendula* and *Picea abies*). On the other hand, the response of willows and poplars analyzed by Dos Santos and Wenzel (2007) showed that *P. nigra* clones included in the experiment had a limited Cd uptake and transfer to shoots (as non-accumulators species), with leaf concentrations ranging 37.5 - 48.7 mg kg^{-1}.

Table 4.1. Cadmium accumulation pattern observed in poplars (adapted from Pietrini et al., 2010a)

Type	Accumulation pattern	Genotypes
Low leaf accumulators	Low percentage of Cd in roots (< 67%) associated with high percentage of Cd in stem (> 31%) and low percentage in leaves (< 2%), showing limited transfer to leaves.	*P. nigra* (58-861 and Poli clones). *P. alba* (6K3 and 14P11 clones).
Leaf accumulators	Medium percentage of Cd in roots (67 - 76%) associated with medium percentage of Cd in stem (18 - 23%) and high percentage in leaves (> 4%), indicating high metal uptake and good transport to leaves.	*P. x generosa* (11-5 clone). *P. x canadensis* (I-214 and A4A clones). *P. deltoides* (Lux clone).
Root accumulators	High percentage of Cd in roots (> 83%) associated with low percentage of Cd in stem (< 13%) and medium percentage in leaves (1 - 3%), indicating high metal uptake but reduced transfer to leaves.	*P. x canadensis* (Luisa Avanzo clone) *P. trichocarpa* (Nisqually clone).

The distribution of Cu in poplars varies according the different tolerance strategies of species or hybrids. In the study of Borghi et al. (2007) with *P. x canadensis* (Adda clone), Cu was mainly accumulated in roots when plants were exposed to a range of Cu treatments.

Copper concentration in this organ (near 12,000 mg kg^{-1} at 1,000 μM) was significantly higher than in leaves and stem, and increased progressively along with the applied doses. A similar accumulation pattern, expressing a marked accumulation in roots, was observed by Guerra et al. (2009) in plants of a Cu-tolerant *P. deltoides* clone exposed to Cu 30 and 60 μM in hydroponics. Leaf and stem Cu contents did not differ with control plants, whereas roots accumulated around 6,000 mg kg^{-1} (at 60 μM), near 30 fold higher than in control plants. A prominent Cu accumulation was also registered by Sebastiani et al. (2004) in I-214 (*P. x canadensis*) and Eridano (*P. deltoides x P. maximowiczii*) clones grown in soils complemented with heavy metals-enriched organic wastes. Although Cu root content was not analyzed by Hermle et al. (2006) in *P. tremula,* plants cultured in pots with a mix of soil and a HM combination (Cu/Zn/Cd/Pb = 640/3,000/10/90 mg kg^{-1}), metal content in leaf and wood (stems and twigs) did not change significantly compared to control, suggesting also a low root-shoot Cu transport. However, Borghi et al. (2008) compared *P. x canadensis* (Adda clone) with *P. alba* (Villafranca clone) plants subjected to Cu 0.4, 25 and 75 μM observing different tendencies in metal allocation. In both hybrids, Cu concentration in the roots increased with higher doses. However, Cu allocation in leaves of *P. x canadensis* was relatively constant, whereas in *P. alba* it kept increased with Cu treatments. From results, clones of *P. x canadensis* and *P. alba* were suggested as species suitable for phytostabilization and biomonitoring, respectively.

The Zn accumulation pattern in poplars stressed by Zn excess has showed to be strongly influenced by the genetic background of plants, in a similar way to that described for Cd and Cu. Langer et al. (2009) analyzed the response of a metal accumulating *P. x canescens* (BOKU 01 AT-001 clone) grown under different Zn doses. Metal contents in tissues increased along with increasing Zn treatments. Leaf contents were higher than in stems and roots. They suggested the studied *P. x canescens* clone could be suitable for sites that contain up to 30 mg kg^{-1} of extractable Zn in soil, where leaf Zn concentration reached 1,000 mg kg^{-1}. Similar pattern was observed by Hermle et al. (2006) in *P. tremula* plants, whose Zn concentrations registered around 1,000 and 180 mg kg^{-1} for leaves and wood, respectively. On the other hand, experiments carried out by Di Baccio et al. (2003) and Di Baccio et al. (2009), in which *P. x canadensis* (I-214 clone) plants were treated with both low and high Zn concentrations, showed that Zn accumulation is depending on organ/tissue (and the age for leaves), time of exposure, and Zn treatment. A regulation involving complex structural, physiological and biochemical processes, attributed to both Zn excluders and accumulators allows to I-214 clone to cope with Zn toxicity, restricting the Zn transport towards young leaves and accumulation in old leaves or roots. This double strategy relative to Zn distribution was confirmed by Sebastiani et al. (2004) on *P. x canadensis* (I-214 clone) and *P. deltoides x P. maximowiczii* (Eridano clone) grown in soils complemented with HM-enriched organic wastes. Dos Santos and Wenzel (2007), in their experiment with willows and poplars subjected to both single and combined application of Cd, Zn (plus Cu and Pb), registered Zn leaf concentrations ranging 569 - 935 mg kg^{-1} for *P. nigra* clones, classifying them as non-accumulators.

Studies carried out with trees grown on contaminated areas have displayed a wide variability in metal distribution patters, which depends on species, metals and sites. Laureysens et al. (2004) assessed the HM accumulation in poplars established, under a short rotation coppice culture scheme, on a moderately polluted site (Al/Cd/Co/Cr/Cu/Fe/Mn/Ni/Pb/Zn = 32/1.60/16/71/43/22/210/38/171/486 mg kg^{-1}) in Belgium. The experiment included different poplar hybrids and species: *P. trichocarpa x P. balsamifera* (Balsam Spire clone); *P. trichocarpa x P. deltoides* (Beaupre, Hazendans, Hoogvorst, Raspalje and Unal clones), *P. trichocarpa* (Columbia River, Fritzi Pauley and Trichobel clones), *P. x canadensis* (Gaver, Gibecq and Primo clones) and *P. nigra* (Wolterson clone). Significant clonal differences in accumulation were found for most metals, although clones with the highest concentration of all metals were not found. Cadmium, Zn and Al were most taken up. The lowest concentration was found in wood and the highest concentrations were generally found in senescing leaves. Unterbrunner et al. (2007) analyzed the response of willow, poplar and birch species sampled in Cd and Zn contaminated sites in four sites in central Europe. They concluded that *P. tremula* (along with two willows) is one of species with the larger accumulation potential for both HMs. Cd and Zn concentrations were higher in leaves (the largest values were around to 40 mg kg^{-1} and 2,000 mg kg^{-1}, respectively) and fine roots than in wood, bark or other roots. On the other hand, Migeon et al. (2009) evaluated the HM accumulation in a 25 woody species planted on Cd, Zn and Pb-polluted sites in North of France. The highest Zn accumulators were poplar hybrids with an average concentration of 850 mg kg^{-1}. Zinc concentrations were 1.5 - 4 times lower in stems compared to leaves. Cadmium accumulation was the highest in poplars and willows with 13 - 44 mg kg^{-1} in leaves and 9 - 15 mg kg^{-1} in stems. Analysis of the bioconcentration factors (metal concentration in leaves/metal concentration in soils) in sampled poplars indicated the following order for Cd: *P. tremula x P. tremuloides* > *P. trichocarpa x P. deltoides* > *P. x canadensis* > *P. nigra* (values: 2.26 > 1.98 > 1.39 > 0.97). For Zn, observed bioconcentration factors were: *P. tremula x P. tremuloides* > *P. x canadensis* > *P. trichocarpa x P. deltoides* > *P. nigra* (values: 1.22 > 0.78 > 0.72 > 0.62).

4.2. Physiological Effects

Exposition of plants to excess of HMs alters important physiological process, such as photosynthesis, carbohydrate metabolism or nutrient uptake.

Photosynthesis is affected in poplars when they are subjected to toxic Cd, Cu and Zn concentrations. Cadmium can interfere with the whole photosynthetic process. Cd effects in photosynthetic parameters was investigated by Pietrini et al. (2010a) and Pietrini et al. (2010b) in 10 clones from poplar hybrids/species (*P. x generosa*, *P. x canadensis*, *P. deltoides*, *P. nigra*, *P. alba* and *P. trichocarpa*) subjected to Cd 50 μM. Plant response and Cd tolerance, as indicated by maintenance of photosynthesis with respect to control, varied among species, hybrids and clones. The concentration of photosynthetic pigments was affected by Cd treatment in all clones. Chlorophyll (total chlorophyll, chlorophyll a and b) and carotenoid contents were reduced in most of genotypes in comparison to control plants, suggesting an association of this effect with clonal variability for Cd tolerance. Photosynthetic parameters such as efficiency of photosystem II (PSII), fluorescence quantum yield of electron transport through PSII, photochemical and non-photochemical quenching of fluorescence, among others, were affected by Cd treatment in a differential way among clones. Additionally, most clones reduced their transpiration rate with respect to control, implying that Cd also affects plant water relations. In general terms, a high or low Cd uptake and translocation to leaves was associated with a strong reduction of photosynthesis.

The effects of Cu stress on the photosynthetic performance of poplars have analyzed recently by Borghi et al. (2008) in their studies with *P. x canadensis* (Adda clone) and *P. alba* (Villafranca clone). Measurement of parameters as such as chlorophyll content, light-saturated rate of electron transport and maximum rate of carboxylation, indicated that both clones had significantly different responses to Cu. Results suggested that the photosynthetic apparatus of *P. alba* is more sensitive to Cu than *P. x canadensis*, which would explain the reduction of growth reported in *P. alba*. Sensitivity to Cu also could be explained by the difference in the Cu concentration accumulated in leaves, which was increased only in *P. alba* with increasing Cu treatments. Symptoms of a decreased photosynthetic efficiency and a general foliar chlorosis in *Populus x canadensis* were observed only at Cu concentration of 1,000 μM (Borghi et al., 2007).

Di Baccio et al. (2005) and Di Baccio et al. (2009) investigated the effects of Zn on photosynthetic parameters in *P. x canadensis* (I-214 clone). Applied Zn (gradient 0.001 - 10 mM) negatively affected a series of variables including photosynthetic rate at saturation, maximum rate of carboxylation, light-saturated rate of electron transport, among others. These results allowed confirming the 1 mM concentration as the crucial dose for the clone-specific response to excess Zn. Zinc treatments also affected the chlorophyll a / chlorophyll b ratio in both young and old leaves (particularly those with 5 and 10 mM).

According to Stobrawa and Lorenc-Plucinska (2007), efficient carbohydrate metabolism is the basis of survival strategies of plants subjected to HM influence. In particular, the carbohydrate status of fine roots seems to be absolutely crucial. Their fast turnover rate requires systematic rebuilding of tissues, with an increased demand for energy and carbon atoms. Under stress conditions, the demand may also increase due the initiation of response mechanisms and secondary metabolism. Thus, the maintenance of primary metabolic pathways and the carbohydrate balance becomes fundamental in counteracting stress factors. Lorenc-Plucinska and Stobrawa (2004) investigated the effects of HMs on the carbohydrate metabolism in fine roots of *P. deltoides* growing at polluted site (Cd/Pb/Zn/Cu/Cr/Ni/Fe/Mn = 1.1/411.1/98.0/1,174.8/31.4/9.7/10,737/339.9 mg kg^{-1}) in Poland. Results showed that fine roots from polluted soils contained higher contents of total nonstructural carbohydrates, soluble sugars, starch and sucrose but lower hexoses level than roots from control sites. In a similar study, Stobrawa and Lorenc-Plucinska (2007) sampled 29 year-old plants of a *P. nigra* clone grown in contaminated soils near to a Cu smelter (Cu/Pb/Zn =1,174.8/411.1/98.0 mg kg^{-1}). They concluded that HMs in soils affected the carbohydrate metabolism in fine roots. Sucrose breakdown was enhanced and soluble total nonstructural carbohydrates level was decreased, but the lack of changes in glycolytic enzyme activities suggests that mobilized hexoses are not used in respiration. Thus, their possible uses might be sucrose re-synthesis, or synthesis of other carbohydrates, potentially including polysaccharides of the cell wall (callose and cellulose) or other secondary metabolites. No difference between control and polluted stands was observed in sucrose concentration. However, estimates of sucrolytic activity revealed markedly higher activities of sucrose synthase and invertases in the polluted stand than in the control. In contrast, the estimated glycolytic enzyme activities (hexokinase, fructokinase, glyceraldehyde 3-phosphate dehydrogenase) were not affected by the presence of HMs in soil.

The application of toxic concentrations of metals can induce growth reduction in plants because of the interference with nutrient uptake, and photosynthetic activity. In their studies about the responses of poplars clones to Cu stress, Borghi et al. (2007) analyzed variations of

N leaf contents. Reduced build up of nutrients to leaves was indicated by the strong decrease in total N contents, starting from the treatment with 100 μM of Cu in *P. x canadensis* clone Adda (Borghi et al., 2007). This tendency was confirmed in their comparative study of Adda and Villafranca (*P. alba*) clones, in which N content in leaves decreased through the treatments (0.4, 25, 75 μM) in both species. The interference of N uptake by Cu has been also suggested by Guerra et al. (2009) for roots of *P. deltoides* exposed to Cu 30 and 60 μM, in which a gene encoding a high affinity nitrate transporter was significantly down regulated by both doses.

4.3. Molecular Mechanisms of Metal Homeostasis and Tolerance

Heavy metals such as Cu and Zn (essentials) or Cd (non-essential) can be toxic to plants above a certain threshold. Plants have evolved a regulated network of uptake and distribution enabling an effective protection to the metabolic processes. In general, factors influencing the metal uptake and distribution in plants include: (1) mobilization from the soil, (2) uptake and sequestration by metal-complex formation and deposition in vacuoles for detoxification within roots, (3) metal translocation to shoots via xylem, and (4) distribution and sequestration in aboveground organs and tissues (Clemens et al., 2002). A further defensive line against HM effects is a series of antioxidant mechanisms against ROS produced by excess of metal ions. These include enzymes and reducing metabolites (Foyer and Noctor, 2005).

4.3.1. Metal Mobilization from the Soil

The mobilization of HMs from the soil involves in a first stage the ion absorption from the rhizosphere and its distribution along root cells. Different compounds have been described like metal ligands for transport and accumulation in tissues and sub-cellular compartments. Among these, organic acids (OAs) such as citrate, malate, and oxalate are predominant (Michael and Christopher, 2007). Additionally, OAs also have a protective role promoting the metal exclusion from roots.

An example is the aluminum (Al) tolerance mechanism in wheat, which avoids the Al uptake by the exudation of OAs and further formation of Al-OA complexes (Delhaize et al., 1993; Kochian et al., 2004). The exudation of OAs has been studied in roots of *P. tremula* exposed to HMs by (Qin et al., 2007). They showed that Cu induced root exudation of oxalate, malate and formate, while Zn induced root exudation of formate. These OAs could be associated to an exclusion mechanism decreasing the HM uptake by the ion chelation at the rhizosphere.

The relationship of plant roots and their mycorrhizal symbionts can influence the responses of plants to HMs significantly (Schützendübel and Polle, 2002). For example, ectomycorrhizal (ECM) fungi protect themselves and their hosts from heavy metal pollution by binding them into cell-wall components or by storing high amounts of HMs in their cytosol. The analysis of ECM fungal community on roots of *P. tremula* in HM contaminated soils in Europe (extractable metal fractions in mineral soils were 152 - 1,335 mg kg^{-1}, 10,686 - 58,773 mg kg^{-1}, 369 - 2,941 mg kg^{-1}, for Pb, Zn and Cd, respectively), showed an association of this poplar with a diverse ECM community (54 species), rich in *Basidiomycota*

(43 species), and dominated by *Cenococcum geophilum* and fungi with corticoid basidiomes (Krpata et al., 2008).

4.3.2. Metal Uptake, Traffic and Compartmentalization

At the cellular level, cell walls can bind metal ions regulating their influx toward cytoplasm by cationic exchange (Wang and Evangelou, 1995). Metals can be bound to pectine (Konno et al., 2005) or proteins as oxalate oxidase (Bringezu et al., 1999). Ions can diffuse into the apoplast of some root cells but its transport is blocked by the impermeable Casparian strip in the endodermal layer. At this point, plants have a series of metal transporters involved in metal uptake and homeostasis, which regulates its movement toward the symplast and subsequent loading into the vascular tissues (Palmer and Guerinot, 2009). Gene families encoding transporters are diverse and this diversity provide the high and low affinity systems needed to cope with varying metal availability in the soil, provide the specific requirements for transport at the different cellular membranes within the plant and to respond to stress conditions.

At plasmadesmata level, main metal transporters are heavy metal ATPases (HMAs or CPx-type) (Williams and Mills, 2005), Zrt- Irt-related protein (ZIP) (Grotz et al., 1998; Guerinot and Eidet, 1999), COPT-type transporters (Sancenón et al., 2003), and cation antiporters (Gaxiola et al., 2002). The knowledge about the structure and functioning of HM transporters comes mainly from species belonging to genus *Arabidopsis*. The characterization of HM transporters in poplars is very scarce. Uptake of Cd through the root cell plasma membrane occurs via a concentration-dependent process exhibiting saturable kinetics (Cutler and Rains, 1974; Cataldo et al., 1983; Mullins and Sommers, 1986a; Mullins and Sommers, 1986b; Blaudez et al., 2000). It is generally believed that Cd uptake by plants represents opportunistic transport by a carrier for other divalent cations such as Zn, Cu or Fe, or via cation channels for Ca and Mg. In fact, Cd and Zn are chemically very similar, suggesting that uptake and transport occurs by similar pathways (Obata and Umebayashi, 1993; Zhao et al., 2002). Copper uptake in *Arabidopsis* is dependent of the ability to be reduced by their respective plasma membrane transporter COPT1 (Sancenón et al., 2003). This metal is also transported by members of the ZIP family (ZIP2 and ZIP4) (Wintz et al., 2003). In the case of Zn, the regulation of uptake has been associated to the ZIP1-4 proteins in *Arabidopsis* (Grotz et al., 1998; Wintz et al., 2003).

Plants have evolved a suite of cytoplasmatic mechanisms that control and respond to the toxicity of both essential and nonessential HMs. In this way, there are two basic strategies for decreasing the toxicity of metals: chelation or efflux from the cytosol, either into the apoplast or by intracellular sequestration through specific ligands for HMs. Two of the best characterized HM binding ligands in plant cells are the phytochelatins (PCs) and metallothioneins (MTs).

Phytochelatins are a family of structures with increasing repetitions of the Glu-Cys dipeptide followed by a terminal Gly, (γ-Glu-Cys)-n-Gly, where n is generally between 2 to 11. Phytochelatins are present in a wide variety of plant species and in some microorganisms. These chelant molecules are structurally related to glutathione (GSH; γ-Glu-Cys-Gly). They are synthesized non-translationally from reduced GSH by the enzyme phytochelatin synthase. Synthesis of PCs in response to metals and formation of PC-metals complex is also well documented in literature (Cobbett and Goldsbrough, 2002). Information about PCs production in poplars is scarce. Phytochelatins has been proposed as bioindicator of Cu and

Ni pollution in adult poplars. According to Gawel et al. (2001), PCs concentrations in leaves of *P. alba* and *P. tremuloides* do not correlate with Cu and Ni levels in soils. Rather, PCs production in tree leaves correlated with the direct foliar uptake of metals. Pietrini et al. (2010b) analyzed the PC contents in *P. x canadensis* (A4A clone), *P. nigra* (Poli clone) and *Salix alba* (SS5 clone) plants exposed to Cd 50 μM. Total PC content in leaves of poplars was increased after Cd treatment. A similar induction level was observed in both poplars. Irrespective of Cd exposure, according to the percentage composition of the three main PCs in both poplar clones, the most abundant component was PC type 4.

Metallothioneins are characterized as low molecular weight, cysteine-rich, metal-binding proteins and may play a role in their intracellular sequestration (Cobbett and Goldsbrough, 2002). Although MTs have been proposed to play a role in HM detoxification or homeostasis, their precise role is not fully known. In an effort to understand processes that relate MTs to heavy metal sequestration, Kohler et al. (2004) characterized six metallothionein genes (PtdMTs) on *P. trichocarpa x P. deltoides*. Genes displayed differential expression patterns, which may be associated with the diverse roles and functions that PtdMTs have to cope with particular developmental and environmental signals. The heterologous expression in a Cd-hypersensitive yeast mutant showed the ability of PtdMT to confer Cd tolerance. The concentration of PtdMT mRNAs were increased by Zn, but not by Cu and Cd, suggesting a role more important of MTs in metabolism/detoxification of Zn rather than other metals. On the other hand, Hassinen et al. (2009) studied the metal uptake by *P. tremula x P. tremuloides* and its relationship with the foliar metallothionein 2b (MT2b) mRNA abundance. The levels of MT2b transcripts correlated with Cd and Zn concentrations in the leaves, demonstrating that increased *MT2b* expression is one of the responses of poplar to chronic metal exposure. The expression of MT genes was also analyzed by Guerra et al. (2009) in roots of a Cu tolerant *P. deltoides* clone exposed to four Cu stress treatments. Metallothionein genes (*Metallothionein 1a*, *Metallothionein 1b* and *Plant metallothionein, family 15*) were highly down regulated in all experimental conditions, suggesting limited participation of this type of metal binding molecules under the assessed treatments. The expression of MTs genes has been also studied in *Populus alba* (Villafranca clone) in vitro cultured shoots exposed to Zn stress (Castiglione et al., 2007). The MT gene expression was differentially affected by Zn in an organ-specific manner. In leaves, *MT1* and *MT3* mRNA levels were enhanced by Zn, while *MT2* transcripts were not affected.

Once transported to the proper tissue, metals are distributed toward the sub cellular compartments where they are requested or where they could safely be stored. The vacuole is emerging as an essential metal storage compartment in plant with a key role in the detoxification of HMs. In this sense, Zn is transported into the vacuole by members of the MTP (metal tolerance protein) family, belonging to the CDF (cation diffusion facilitator) proteins super family. Both MTP1 and MTP3 localize at vacuolar membrane (Desbrosses-Fonrouge et al., 2005; Gustin, 2009), and over expression of *MTP1* or *MTP3* confers resistance to high levels of Zn (Desbrosses-Fonrouge et al., 2005; Arrivault et al., 2006). One member of this family, *PtdMTP1*, has been characterized in *P. trichocarpa x P. deltoides* (Blaudez et al., 2003). *PtdMTP1* is expressed constitutively and ubiquitously. Heterologous expression in yeast showed that *PtdMTP1* was able to complement the hypersensitivity of mutant strains to Zn, but not to other metals, including Cd, Co, Mn, and Ni. PtdMTP1 localized to the vacuolar membrane, consistent with its function in the Zn sequestration. Over expression of *PtdMTP1* in *Arabidopsis* conferred it Zn tolerance. In the case of Cd, *AtHMA3*, a member of the Zn/Cd/Co/Pb P-type ATPases cluster would have a role in its accumulation in vacuole (Gravot et al., 2004; Puig and Peñarrubia, 2009). For Cu, transporters as such as PAA1 (HMA6), PAA2 (HMA8) and HMA1, members of the Cu-transporting P_{IB}-type ATPase family, are critical for Cu delivery to plastocyanin in the chloroplast (Shikanai et al., 2003; Abdel-Ghany et al., 2005). Cu is also transported to the mitochondria where is part of respiratory electron transport chain. Intracellular distribution of metals is performed by chaperones directing the metal to its final destination. Metal chaperones can act coordinately with ATPases in detoxification of HM in roots (Andres-Colas et al., 2006). Some metal chaperones characterized in *Arabidopsis* are AtCCH (Mira et al., 2001) and AtCOX17 (Balandin and Castresana, 2002) and PoCCH in the poplar hybrid *P. alba x P. tremula* var. glandulosa (Lee et al., 2005).

4.3.3. Metal Translocation to Shoots via Xylem

The root to shoot metal translocation involves at least two steps in roots, in which transmembrane transport is required. The first one involves the uptake from root surface to the epidermal tissue. Subsequently, metals are transported to pericycle or xylem parenchyma, and loaded into the xylem (Palmer and Guerinot, 2009).

Three transporter proteins, members of P_{IB} subfamily of the ATPases have been described as heavy metal ATPases (HMA) involved in Cd, Cu and Zn xylem loading in *Arabidopsis*. ATPases HMA2 and HMA4 are mainly expressed in vascular tissues. They are essential for root-to-shoot Zn transport, enhancing the xylem loading and the accumulation of Zn and Cd in shoots (Hussain et al., 2004; Hanikenne et al., 2008; Wong and Cobbett, 2009). In a similar way, the Cu transporter HMA5 also has been described in *Arabidopsis*, probably involved in Cu xylem loading (Puig et al., 2007; Kobayashi et al., 2008). None of these transporters have been neither isolated nor characterized in poplars, despite the recent release of the *P. trichocarpa* genome. According to metal accumulation function in other species, protein transporter regulation would have a key role on xylem loading for increasing translocation ratio from roots to shoots on poplar.

The root to shoot translocation of metals via the xylem sap involves a series of amino acids and organic acids. Ligands for Cd, Cu and Zn include citrate, malate, histidine and nicotianamine, among others (Pilon et al., 2009). The Cd-and Zn-citrate complexes are prevalent in leaves, even though malate is more abundant. In the xylem sap moving from roots to leaves, citrate, and histidine are the principal ligands for Cu and Zn (Yang et al., 2005; Curie et al., 2009). To our knowledge there is not information linking this sort of ligands and metals in the context of xylem transport in poplars.

4.3.4. Antioxidative System

The excess of HMs can cause oxidative stress and damages to exposed cells. The redox-active metals (e.g. Cu) as well as those non redox-actives (e.g. Cd and Zn) can cause direct or indirect oxidative damage. As a part of the defensive response of cells, an antioxidative system based on reducing metabolites (e.g. GSH, ascorbate [AA]) and enzymes (e.g. peroxidases, catalases, superoxide dismutases) is tightly regulated to keep their general redox balance.

Glutathione develops a series of roles in cell metabolism, including redox state regulation, oxidative stress control, and protection against HMs. Glutathione is synthesized from Glu, Cys, and Gly in two steps catalyzed by glutamylcysteine synthetase and GSH synthetase. Glutathione is the precursor of PCs. As a fundamental antioxidant molecule, GSH directly eliminates reactive oxygen radicals induced by HM ions in cells and provides reducing equivalents in the AA-GSH antioxidation cycle to maintain redox homeostasis for metabolism, signal transduction and gene expression (Foyer and Noctor, 2005). Glutathione can bind to several metals and metalloids (Verbruggen et al., 2009). On the other hand, AA has a similar role in the protection of cells against oxidative damage induced by ROS (Foyer and Noctor, 2005). Ascorbate is biosynthesized in high concentrations by plant cells from L-galactono-γ-lactone.

The effect of HMs on the biosynthesis and metabolism of GSH and AA has been assessed in poplars. Schützendübel et al. (2002) studied the effects of Cd and H_2O_2 exposure in the cellular redox control in roots of *P. x canescens*. Glutathione concentrations decreased, whereas AA remained unaffected by Cd. On the other hand, H_2O_2 caused GSH accumulation and loss of AA. Di Baccio et al. (2005), in *P. x canadensis* (I-214 clone), and Bittsánszky et al. (2005), in *P. nigra* and transgenic *P. x canescens* studied the role of GSH on the response of poplar to Zn. From the variations in GSH contents and the expression of genes coding enzymes participating in its biosynthesis and conjugation they conclude GSH would be important on the protective response of poplars to Zn excess. On the other hand, Guerra et al. (2009) established that genes coding enzymes of the GSH biosynthesis pathway were differentially regulated by Cu stress in a *P. deltoides* clone, suggesting a possible increase in the levels of two of GSH constituent amino acids (Glu and Gly), which could be related to an increasing demand of GSH driven by Cu excess.

A disturbance of antioxidative enzymes controlling the cellular redox control was observed by Schützendübel et al. (2002) in roots of *P. x canescens*. Cd exposure resulted in an inhibition of antioxidative enzymes superoxide dismutase, catalase, AA-peroxidase, monodehydroascorbate reductase, GSH-reductase, but had fewer effects on dehydroascorbate reductase.

The behavior of a set of antioxidative enzymes was also investigated by Stobrawa and Lorenc-Plucinska (2008) in the fine roots of *P. nigra* grown in Cu and Pb polluted soils. The stimulation or inhibition of important antioxidant enzymes such as catalase, superoxide dismutase, guaiacol, AA-peroxidases and GSH-reductase was detected in plants grown on polluted soils. At the same time, increasing malondialdehyde concentrations in roots also indicated the presence of lipid peroxidation product of the oxidative effects of metals. On the other hand, gene expression analysis of *P. deltoides* grown under Cu stress treatments (Guerra et al., 2009) also showed a differential regulation of genes associated to the antioxidant system. Peroxidases, Cu/Zn-superoxide dismutase and catalase genes were down regulated, suggesting a modulation of hydrogen peroxide contents by Cu applications. Monodehydroascorbate reductase gene was up regulated in almost all treatments, whereas cytosolic AA-peroxidase gene was repressed, suggesting the regulation of enzymes regenerating the active form of AA.

4.3.5. Other Mechanisms

New evidence supporting a positive role of other stress-protective molecules in the tolerance/adaptation to HMs in poplars has been reported during recent years. Particularly, polyamines (PAs), small organic polycations including putrescine, spermidine and spermine, occur both in free form and conjugated to phenolics compounds or proteins and cell wall constituents, would have a protective role under HM stress. An induction of the PA metabolism has been reported for micropropagated *P. alba* (Villafranca clone) plants exposed to Zn and Cu concentrations (Castiglione et al., 2007). Castiglione et al. (2009) from a study including a wide set of poplar clones grown on a field trial on heavy metal-polluted soil, established that leaf PA profiles correlated with tissue metal concentrations, depending on the clone, plant organ and metal. In particular, a high metal-accumulating *P. alba* (AL35 clone) exhibited a dramatically higher concentration of free and conjugated putrescine. The strong positive correlation between leaf conjugated putrescine and root Cu concentrations suggested that Cu, rather than Zn, would drive the long-term PA response.

The analysis of the root transcriptome of a Cu tolerant *P. deltoides* clone exposed to Cu stress carried out by Guerra et al. (2009) allowed to identify a series of genes that are part of cell response. Within them, an important part encoded defense and signaling proteins, as for example, genes of trypsin inhibitors and PR proteins, which were significantly up regulated in all stress treatments. The accumulation of this kind of transcripts has been reported in poplar subjected to biotic and abiotic stress agents (Gupta et al., 2005; Ralph et al., 2006; Rinaldi et al., 2007; Major and Constabel, 2008). In a similar way, a variety of genes encoding proteins participating to signal transduction pathways were significantly up or down regulated. Evidences about the participation of Ca^{2+} dependent signaling proteins (calmodulin and EF-proteins), MAP kinases and Rab small G protein (RAB GTP-binding protein) were detected in all treatments. Accumulation of transcripts coding enzymes such as catechol oxidase, allene oxide synthase, 1-aminocyclopropane-1-carboxylate oxidase and some ethylene responsive elements suggests participation of salicilic acid, jasmonic acid and ethylene in the response.

CONCLUSION

The efficiency of phytoremediation systems designed to clean-up HMs from contaminated soils is clearly determinate by the characteristics of plants and their interaction with biotic and abiotic environmental factors. The genetic diversity of poplars is evidenced by the wide variety of responses observed when they are exposed to different HM stress conditions.

The potential of poplars for phytoremediating HMs through distinct approaches is being confirmed under several experimental situations. Poplars are also an interesting biotechnological platform to complement and develop phytoremediation applications taking advantage of tolerance mechanisms identified in other biological systems.

Important advances have been done to characterize fundamental aspects of the response of poplars to HMs, as such as tolerance thresholds, metal distribution patterns, physiological adaptations, effects of genetic background and soil management, among others. However, important knowledge gaps remains to be covered, as for example at the biochemistry and molecular-genetic level. Recent advances in genomics and proteomics are very promising, in terms of the gain that we can achieve in next years to understand the biological basis underlying the HM tolerance and accumulation processes.

In this way, the genetic improvement of poplars by traditional and biotechnological approaches, besides of the optimization of agricultural practices, would allow to consolidate these trees as an important alternative for the phytoremediation of HM contaminated soils.

REFERENCES

Abdel-Ghany SE, Muller-Moule P, Niyogi KK, Pilon M, Shikanai T (2005) Two p-type atpases are required for copper delivery in *Arabidopsis thaliana* chloroplasts. *Plant Cell* 17: 1233-1251.

Andres-Colas N, Sancenón V, Rodríguez-Navarro S, Mayo S, Thiele D, Ecker J, Puig S, Peñarrubia L (2006) The *Arabidopsis* heavy metal P-type ATPase HMA5 interacts with metallochaperones and functions in copper detoxification of roots. *Plant Journal* 45: 225-236.

Arrivault S, Senger T, Krämer U (2006) The *Arabidopsis* metal tolerance protein AtMTP3 maintains metal homeostasis by mediating Zn exclusion from the shoot under Fe deficiency and Zn oversupply. *The Plant Journal* 46: 861-879.

Arthur EL, Rice PJ, Rice PJ, Anderson TA, Baladi SM, Henderson KLD, Coats JR (2005) Phytoremediation-an overview. *Critical Reviews in Plant Sciences* 24: 109 – 122.

Balandin T, Castresana C (2002) AtCOX17, an *Arabidopsis* homolog of the yeast copper chaperone COX17. *Plant Physiology* 129: 1852-1857.

Barac T, Weyens N, Oeyen L, Taghavi S, van der Lelie D, Dubin D, Spliet M, Vangronsveld J (2009) Field note: hydraulic containment of a BTEX plume using poplar trees. *International Journal of Phytoremediation* 11: 416 – 424.

Bissonnette L, St-Arnaud M, Labrecque M (2010) Phytoextraction of heavy metals by two Salicaceae clones in symbiosis with arbuscular mycorrhizal fungi during the second year of a field trial. *Plant and Soil.* DOI 10.1007/s11104-009-0273-x.

Bittsánszky A, Komives T, Gullner G, Gyulai G, Kiss J, Heszky L, Radimszky L, Rennenberg H (2005) Ability of transgenic poplars with elevated glutathione content to tolerate zinc($^{2+}$) stress. *Environment International* 31: 251-254.

Blaudez D, Botton B, Chalot M (2000) Cadmium uptake and subcellular compartmentation in the ectomycorrhizal fungus *Paxillus involutus*. *Microbiology* 146: 1109-1117.

Blaudez D, Kohler A, Martin F, Sanders D, Chalot M (2003) Poplar Metal Tolerance Protein 1 confers zinc tolerance and is an oligomeric vacuolar zinc transporter with an essential leucine zipper motif. *The Plant Cell* 15: 2911-2928.

Borghi M, Tognetti R, Monteforti G, Sebastiani L (2007) Responses of *Populus* × euramericana (*P. deltoides* × *P. nigra*) clone Adda to increasing copper concentrations. *Environmental and Experimental Botany* 61: 66-73.

Borghi M, Tognetti R, Monteforti G, Sebastiani L (2008) Responses of two poplar species (*Populus alba* and *Populus x canadensis*) to high copper concentrations. *Environmental and Experimental Botany* 62: 290-299.

Bradshaw HD, Ceulemans R, Davis J, Stettler R (2000) Emerging model systems in plant biology: poplar (*Populus*) as a model forest tree. *Journal of Plant Growth Regulation* 19: 306-313.

Bringezu K, Lichtenberger O, Leopold I, Neumann D (1999) Heavy metal tolerance of *Silene vulgaris*. *Journal of Plant Physiology* 154: 536-546.

Burkhead J, Gogolin Reynolds K, Abdel-Ghany S, Cohu C, Pilon M (2009) Copper homeostasis. *New Phytologist* 182: 799-816.

Castiglione S, Franchin C, Fossati T, Lingua G, Torrigiani P, Biondi S (2007) High zinc concentrations reduce rooting capacity and alter metallothionein gene expression in white poplar (*Populus alba* L. cv. Villafranca). *Chemosphere* 67: 1117-1126.

Castiglione S, Todeschini V, Franchin C, Torrigiani P, Gastaldi D, Cicatelli A, Rinaudo C, Berta G, Biondi S, Lingua G (2009) Clonal differences in survival capacity, copper and zinc accumulation, and correlation with leaf polyamine levels in poplar: A large-scale field trial on heavily polluted soil. *Environmental Pollution* **157**, 2108-2117.

Cataldo DA, Garland TR, Wildung RE (1983) Cadmium uptake kinetics in intact soybean plants. *Plant Physiology* 73: 844-848.

Cavaleri MA, Gilmore DW, Mozaffari M, Rosen CJ, Halbach TR (2004) Hybrid poplar and forest soil response to municipal and industrial by-products: a greenhouse study. *Journal of Environmental Quality* 33:1055-1061.

Clemens S (2006) Toxic metal accumulation, responses to exposure and mechanisms of tolerance in plants. *Biochimie* 88: 1707-1719.

Clemens S, Palmgren MG, Krämer U (2002) A long way ahead: understanding and engineering plant metal accumulation. *Trends in Plant Science* 7: 309-315.

Cobbett C, Goldsbrough P (2002). Phytochelatins and metallothioneins: roles in heavy metal detoxification and homeostasis. *Annual Review of Plant Biology* **53**, 159-182.Curie C, Cassin G, Couch D, Divol F, Higuchi K, Le Jean M, Misson J, Schikora A, Czernic P, Mari S (2009) Metal movement within the plant: contribution of nicotianamine and yellow stripe 1-like transporters. *Annals of Botany* 103: 1-11.

Cutler JM, Rains DW (1974) Characterization of cadmium uptake by plant tissue. *Plant Physiology*. 54: 67-71.

Che D, Meagher D, Heaton A, Lima A, Rugh C, Merkle S (2003) Expression of mercuric ion reductase in Eastern cottonwood (*Populus deltoides*) confers mercuric ion reduction and resistance. *Plant Biotechnology Journal* 1: 311-319.

Delhaize E, Ryan PR, Randall PJ (1993) Aluminum tolerance in wheat (*Triticum aestivum* L.) (II. Aluminum-stimulated excretion of malic acid from root apices). *Plant Physiol.* 103: 695-702.

Desbrosses-Fonrouge A-G, Voigt K, Schroder A, Arrivault S, Thomine S, Kramer U (2005) *Arabidopsis thaliana* MTP1 is a Zn transporter in the vacuolar membrane which mediates Zn detoxification and drives leaf Zn accumulation. *FEBS letters* 579: 4165-4174.

Di Baccio D, Kopriva S, Sebastiani L, Rennenberg H (2005) Does glutathione metabolism have a role in the defence of poplar against zinc excess? *New Phytologist* 167: 73-80.

Di Baccio D, Tognetti R, Minnocci A, Sebastiani L (2009) Responses of the *Populus × euramericana* clone I-214 to excess zinc: Carbon assimilation, structural modifications, metal distribution and cellular localization. *Environmental and Experimental Botany* 67: 153-163.

Di Baccio D, Tognetti R, Sebastiani L, Vitagliano C (2003) Responses of *Populus deltoides × Populus nigra* (*Populus × euramericana*) Clone I-214 to High Zinc Concentrations. *New Phytologist* 159: 443-452.

Dos Santos M, Wenzel W (2007) Cadmium and zinc accumulation in willow and poplar species grown on polluted soils. *Journal of Plant Nutrition and Soil Science* 170: 265-272.

Dos Santos M, Wieshammer G, Vega R, Wenzel WW (2007) Hydroponic screening for metal resistance and accumulation of cadmium and zinc in twenty clones of willows and poplars. *Environmental Pollution* 148: 155-165.

Foyer CH, Noctor G (2005) Redox homeostasis and antioxidant signaling: a metabolic interface between stress perception and physiological responses. *Plant Cell* 17: 1866-1875.

Gawel JE, Trick CG, Morel FoMM (2001) Phytochelatins are bioindicators of atmospheric metal exposure via direct foliar uptake in trees near Sudbury, Ontario, Canada. *Environmental Science and Technology* 35: 2108-2113.

Gaxiola RA, Fink GR, Hirschi KD (2002) Genetic manipulation of vacuolar proton pumps and transporters. *Plant Physiology* 129: 967-973.

Gravot A, Lieutaud A, Verret F, Auroy P, Vavasseur A, Richaud P (2004) AtHMA3, a plant P_{1B}-ATPase, functions as a Cd/Pb transporter in yeast. *FEBS letters* 561: 22-28.

Grotz N, Fox T, Connolly E, Park W, Guerinot ML, Eide D (1998) Identification of a family of zinc transporter genes from *Arabidopsis* that respond to zinc deficiency. *Proceedings of the National Academy of Sciences* 95: 7220-7224.

Grotz N, Guerinot ML (2006) Molecular aspects of Cu, Fe and Zn homeostasis in plants. *Biochimica et Biophysica Acta (BBA) - Molecular Cell Research* 1763: 595-608.

Guerinot ML, Eidet D (1999) Zeroing in on zinc uptake in yeast and plants. *Current Opinion in Plant Biology* 2: 244-249.

Guerra F, Duplessis S, Kohler A, Martin F, Tapia J, Lebed P, Zamudio F, González E (2009) Gene expression analysis of *Populus deltoides* roots subjected to copper stress. *Environmental and Experimental Botany* 67: 335-344.

Gupta P, Duplessis S, White H, Karnosky DF, Martin F, Podila GK (2005) Gene expression patterns of trembling aspen trees following long-term exposure to interacting elevated CO_2 and tropospheric O_3. *New Phytologist* 167: 129-142.

Gustin JL, Loureiro ME, Kim D, Na G, Tikhonova M, Salt DE (2009) MTP1-dependent Zn sequestration into shoot vacuoles suggests dual roles in Zn tolerance and accumulation in Zn-hyperaccumulating plants. *Plant Journal* 57: 1116-1127.

Hanikenne M, Talke IN, Haydon MJ, Lanz C, Nolte A, Motte P, Kroymann J, Weigel D, Kramer U (2008) Evolution of metal hyperaccumulation required cis-regulatory changes and triplication of *HMA4*. *Nature* 453: 391-395.

Hänsch R, Mendel RR (2009) Physiological functions of mineral micronutrients (Cu, Zn, Mn, Fe, Ni, Mo, B, Cl). *Current Opinion in Plant Biology* 12: 259-266.

Hassinen V, Vallinkoski V-M, Issakainen S, Tervahauta A, Kärenlampi S, Servomaa K (2009) Correlation of foliar *MT2b* expression with Cd and Zn concentrations in hybrid aspen (*Populus tremula* × *tremuloides*) grown in contaminated soil. *Environmental Pollution* 157: 922-930.

Hermle S, Günthardt-Goerg MS, Schulin R (2006) Effects of metal-contaminated soil on the performance of young trees growing in model ecosystems under field conditions. *Environmental Pollution* 144: 703-714.

Hong MS, Farmayan WF, Dortch IJ, Chiang CY, McMillan SK, Schnoor JL (2001) Phytoremediation of MTBE from a groundwater plume. *Environmental Science and Technology* 35: 1231-1239.

Hussain D, Haydon MJ, Wang Y, Wong E, Sherson SM, Young J, Camakaris J, Harper JF, Cobbett CS (2004) P-Type ATPase heavy metal transporters with roles in essential zinc homeostasis in *Arabidopsis*. *Plant Cell* 16: 1327-1339.

Jansson S, Douglas CJ (2007) *Populus*: a model system for plant biology. *Annual Review of Plant Biology* 58: 435-458.

Kobayashi T, Nakanishi H, Takahashi M, Mori S, Nishizawa N (2008) Generation and field trials of transgenic rice tolerant to iron deficiency. *Rice* 1: 144-153.

Kochian LV, Hoekenga OA, Piñeros MA (2004) How do crop plants tolerate acid soils? Mechanisms of aluminum tolerance and phosphorous efficiency. *Annual Review of Plant Biology* 55: 459-493.

Kohler A, Blaudez D, Chalot M, Martin F (2004) Cloning and expression of multiple metallothioneins from hybrid poplar. *New Phytologist* 164, 83-93.

Konno H, Nakato T, Nakashima S, Katoh K (2005) *Lygodium japonicum* fern accumulates copper in the cell wall pectin. *Journal of Experimental Botany* 56: 1923-1931.

Kotrba P, Najmanova J, Macek T, Ruml T, Mackova M (2009) Genetically modified plants in phytoremediation of heavy metal and metalloid soil and sediment pollution. *Biotechnology Advances* 27: 799-810.

Krpata D, Peintner U, Langer I, Fitz WJ, Schweiger P (2008) Ectomycorrhizal communities associated with *Populus tremula* growing on a heavy metal contaminated site. *Mycological Research* 112: 1069-1079.

Langer I, Krpata D, Fitz WJ, Wenzel WW, Schweiger PF (2009) Zinc accumulation potential and toxicity threshold determined for a metal-accumulating *Populus canescens* clone in a dose-response study. *Environmental Pollution* 157: 2871-2877.

Laureysens I, Blust R, De Temmerman L, Lemmens C, Ceulemans R (2004) Clonal variation in heavy metal accumulation and biomass production in a poplar coppice culture: I. Seasonal variation in leaf, wood and bark concentrations. *Environmental Pollution* 131: 485-494.

Lee H, Lee J, Bae E, Choi Y, Noh E (2005) Differential expression of a poplar copper chaperone gene in response to various abiotic stresses. *Tree Physiology* 25: 395-401.

Licht LA, Isebrands JG (2005) Linking phytoremediated pollutant removal to biomass economic opportunities. *Biomass and Bioenergy* 28: 203-218.

Liu J, Hu D, Jiang G, Schnoor JL (2009) In vivo biotransformation of 3,3',4,4'-tetrachlorobiphenyl by whole plants:poplars and switchgrass. *Environmental Science and Technology* 43: 7503-7509.

Liu J, Schnoor JL (2008) Uptake and translocation of lesser-chlorinated polychlorinated biphenyls (PCBs) in whole hybrid poplar plants after hydroponic exposure. *Chemosphere* 73: 1608-1616.

Lorenc-Plucinska G, Stobrawa K (2004) Acclimation of poplar trees to heavy metals in polluted habitats: Carbohidrate metabolism in fine roots of *Populus deltoides*. *Acta Societatis Botanicorum Poloniae* 74: 11-16.

Major IT, Constabel CP (2008) Functional analysis of the Kunitz trypsin inhibitor family in poplar reveals biochemical diversity and multiplicity in defense against herbivores. *Plant Physiology*. 146: 888-903.

Michael JH, Christopher SC (2007) Transporters of ligands for essential metal ions in plants. *New Phytologist* 174: 499-506.

Migeon A, Richaud P, Guinet F, Chalot M, Blaudez D (2009) Metal accumulation by woody species on contaminated sites in the north of France. *Water, Air, & Soil Pollution* **204**, 89-101.

Mira H, Martínez-Garcia F, Peñarrubia L (2001) Evidence for the plant-specific intercellular transport of the *Arabidopsis* copper chaperone CCH. *Plant Journal* 25: 521-528.

Mullins G, Sommers L (1986a) Cadmium and zinc influx characteristics by intact corn (*Zea mays* L.) seedlings. *Plant and Soil* 96: 153-164.

Mullins G, Sommers L (1986b) Characterization of cadmium and zinc in four soils treated with sewage sludge. *Journal of Environmental Quality* 15: 382-387.

Neale DB, Ingvarsson PK (2008) Population, quantitative and comparative genomics of adaptation in forest trees. *Current Opinion in Plant Biology* 11: 149-155.

Newman LA, Strand SE, Choe N, Duffy J, Ekuan G, Ruszaj M, Shurtleff BB, Wilmoth J, Heilman P, Gordon MP (1997) Uptake and biotransformation of trichloroethylene by hybrid poplars. *Environmental Science and Technology* 31: 1062-1067.

Obata H, Umebayashi M (1993) Production of SH compounds in higher plants of different tolerance to Cd. *Plant and Soil* 155-156: 533-536.

Omasa K, Tobe K, Hosomi M, Kobayashi M (2000) Absorption of ozone and seven organic pollutants by *Populus nigra* and *Camellia sasanqua*. *Environmental Science and Technology* 34: 2498-2500.

Palmer CM, Guerinot ML (2009) Facing the challenges of Cu, Fe and Zn homeostasis in plants. *Nature Chemical Biology* 5: 333-340.

Pietrini F, Zacchini M, Iori V, Pietrosanti L, Bianconi D, Massacci A (2010a) Screening of poplar clones for cadmium phytoremediation using photosynthesis, biomass and cadmium content analyses. *International Journal of Phytoremediation* 12: 105 – 120.

Pietrini F, Zacchini M, Iori V, Pietrosanti L, Ferretti M, Massacci A (2010b) Spatial distribution of cadmium in leaves and its impact on photosynthesis: examples of different strategies in willow and poplar clones. *Plant Biology* 12: 355-363.

Pilon-Smits E (2005) Phytoremediation. *Annual Review of Plant Biology* 56: 15-39.

Pilon M, Cohu CM, Ravet K, Abdel-Ghany SE, Gaymard F (2009) Essential transition metal homeostasis in plants. *Current Opinion in Plant Biology* 12: 347-357.

Pollard AJ, Powell KD, Harper FA, Andrew J, Smith C (2002) The genetic basis of metal hyperaccumulation in plants. *Critical Reviews in Plant Sciences* 21: 539 – 566.

Puig S, Mira H, Dorcey E, Sancenón V, Andrés-Colás N, Garcia-Molina A, Burkhead JL, Gogolin KA, Abdel-Ghany SE, Thiele DJ, Ecker JR, Pilon M, Peñarrubia L (2007) Higher plants possess two different types of ATX1-like copper chaperones. *Biochemical and Biophysical Research Communications* 354: 385-390.

Puig S, Peñarrubia L (2009) Placing metal micronutrients in context: transport and distribution in plants. *Current Opinion in Plant Biology* 12: 299-306.

Pulford ID, Watson C (2003) Phytoremediation of heavy metal-contaminated land by trees-a review. *Environment International* 29: 529-540.

Qin R, Hirano Y, Brunner I (2007) Exudation of organic acid anions from poplar roots after exposure to Al, Cu and Zn. *Tree Physiology* 27: 313-320.

Ralph S, Oddy C, Cooper D, Yueh H, Jancsik S, Kolosova N, Philippe Rn, Aeschliman D, White R, Huber D, Ritland Ce, Benoit F, Rigby T, Nantel A, Butterfield YsN, Kirkpatrick R, Chun E, Liu J, Palmquist D, Wynhoven B, Stott J, Yang G, Barber S, Holt Ra, Siddiqui A, Jones SjM, Marra Ma, Ellis Be, Douglas Cj, Ritland K, Bohlmann J

(2006) Genomics of hybrid poplar (*Populus trichocarpa x deltoides*) interacting with forest tent caterpillars (*Malacosoma disstria*): normalized and full-length cDNA libraries, expressed sequence tags, and a cDNA microarray for the study of insect-induced defences in poplar. *Molecular Ecology* 15: 1275-1297.

Raskin I, Ensley B (1999) Phytoremediation of toxic metals: using plants to clean up the environment. Wiley-Interscience New York. 304 p.

Rinaldi C, Kohler A, Frey P, Duchaussoy F, Ningre N, Couloux A, Wincker P, Le Thiec D, Fluch S, Martin F, Duplessis S (2007) Transcript profiling of poplar leaves upon infection with compatible and incompatible strains of the foliar rust *Melampsora larici-populina*. *Plant Physiology* 144: 347-366.

Robinson BH, Green SR, Chancerel B, Mills TM, Clothier BE (2007) Poplar for the phytomanagement of boron contaminated sites. *Environmental Pollution* 150: 225-233.

Salt DE, Smith RD, Raskin I (1998) Phytoremediation. *Annual Review of Plant Physiology and Plant Molecular Biology* 49: 643-668.

Sancenón V, Puig S, Mira H, Thiele DJ, Peñarrubia L (2003) Identification of a copper transporter family in *Arabidopsis thaliana*. *Plant Molecular Biology* 51: 577-587.

Schützendübel A, Polle A (2002) Plant responses to abiotic stresses: heavy metal-induced oxidative stress and protection by mycorrhization. *Journal of Experimental Botany* 53: 1351-1365.

Schützendübel A, Nikolovaa P, Rudolfa C, Polle A (2002) Cadmium and H_2O_2-induced oxidative stress in *Populus* × *canescens* roots. *Plant Physiology and Biochemistry* 40: 577-584.

Sebastiani L, Scebba F, Tognetti R (2004) Heavy metal accumulation and growth responses in poplar clones Eridano (*Populus deltoides* × *maximowiczii*) and I-214 (*P.* × *euramericana*) exposed to industrial waste. *Environmental and Experimental Botany* 52: 79-88.

Shikanai T, Muller-Moule P, Munekage Y, Niyogi KK, Pilon M (2003) PAA1, a P-Type ATPase of *Arabidopsis*, Functions in copper transport in chloroplasts. *Plant Cell* 15: 1333-1346.

Spriggs T, Banks MK, Schwab P (2005) Phytoremediation of polycyclic aromatic hydrocarbons in manufactured gas plant-impacted soil. *Journal of Environmental Quality* 34: 1755-1762.

Stettler R, Bradshaw T, Heilman P, Hinckley T (1996) Biology of *Populus* and its implications for management and conservation. NRC Research Press, Ottawa. 542 p.

Stobrawa K, Lorenc-Plucinska G (2007) Changes in carbohydrate metabolism in fine roots of the native European black poplar (*Populus nigra* L.) in a heavy-metal-polluted environment. *Science of The Total Environment* 373: 157-165.

Thompson PL, Ramer LA, Schnoor JL (1998) Uptake and transformation of tnt by hybrid poplar trees. *Environmental Science and Technology* 32: 975-980

Tuskan GA, DiFazio S, Jansson S, Bohlmann J, Grigoriev I, Hellsten U, Putnam N, Ralph S, Rombauts S, Salamov A, Schein J, Sterck L, Aerts A, Bhalerao RR, Bhalerao RP, Blaudez D, Boerjan W, Brun A, Brunner A, Busov V, Campbell M, Carlson J, Chalot M, Chapman J, Chen GL, Cooper D, Coutinho PM, Couturier J, Covert S, Cronk Q, Cunningham R, Davis J, Degroeve S, Dejardin A, dePamphilis C, Detter J, Dirks B, Dubchak I, Duplessis S, Ehlting J, Ellis B, Gendler K, Goodstein D, Gribskov M, Grimwood J, Groover A, Gunter L, Hamberger B, Heinze B, Helariutta Y, Henrissat B,

Holligan D, Holt R, Huang W, Islam-Faridi N, Jones S, Jones-Rhoades M, Jorgensen R, Joshi C, Kangasjarvi J, Karlsson J, Kelleher C, Kirkpatrick R, Kirst M, Kohler A, Kalluri U, Larimer F, Leebens-Mack J, Leple JC, Locascio P, Lou Y, Lucas S, Martin F, Montanini B, Napoli C, Nelson DR, Nelson C, Nieminen K, Nilsson O, Pereda V, Peter G, Philippe R, Pilate G, Poliakov A, Razumovskaya J, Richardson P, Rinaldi C, Ritland K, Rouze P, Ryaboy D, Schmutz J, Schrader J, Segerman B, Shin H, Siddiqui A, Sterky F, Terry A, Tsai CJ, Uberbacher E, Unneberg P, Vahala J, Wall K, Wessler S, Yang G, Yin T, Douglas C, Marra M, Sandberg G, Van de Peer Y, Rokhsar D (2006) The Genome of Black Cottonwood, *Populus trichocarpa* (Torr. and Gray). *Science* 313: 1596-1604.

Unterbrunner R, Puschenreiter M, Sommer P, Wieshammer G, Tlustos P, Zupan M, Wenzel WW (2007) Heavy metal accumulation in trees growing on contaminated sites in Central Europe. *Environmental Pollution* 148: 107-114.

Vamerali T, Bandiera M, Coletto L, Zanetti F, Dickinson NM, Mosca G. 2009. Phytoremediation trials on metal- and arsenic-contaminated pyrite wastes (Torviscosa, Italy). *Environmental Pollution* **157**, 887-894.

van Aken B, Schnoor JL (2002) Evidence of Perchlorate (ClO_4^-) Reduction in plant tissues (poplar tree) using radio-labeled $36ClO_4$. *Environmental Science and Technology* 36: 2783-2788.

van Dillewijn P, Couselo JL, Corredoira E, Delgado A, Wittich R-M, Ballester A, Ramos JL (2008) Bioremediation of 2,4,6-trinitrotoluene by bacterial nitroreductase expressing transgenic aspen. *Environmental Science and Technology* 42: 7405-7410.

Verbruggen N, Hermans C, Schat H (2009) Mechanisms to cope with arsenic or cadmium excess in plants. *Current Opinion in Plant Biology* 12: 364-372.

Vollenweider P, Cosio C, Günthardt-Goerg MS, Keller C (2006) Localization and effects of cadmium in leaves of a cadmium-tolerant willow (*Salix viminalis* L.): Part II Microlocalization and cellular effects of cadmium. *Environmental and Experimental Botany* 58: 25-40.

Wang J, Evangelou V (1995) Metal tolerance aspects of plant cell wall and vacuole. *In* M Pessarakli, ed, Handbook of plant and crop physiology. Marcel Dekker, Inc., New York, pp 695-717.

Williams L, Mills R (2005) P_{1B}-ATPases – an ancient family of transition metal pumps with diverse functions in plants. *Trends in Plant Science* 10: 491-502.

Wintz H, Fox T, Wu Y-Y, Feng V, Chen W, Chang H-S, Zhu T, Vulpe C (2003) expression profiles of *Arabidopsis thaliana* in mineral deficiencies reveal novel transporters involved in metal homeostasis. *J. Biol. Chem.* 278: 47644-47653.

Wong CKE, Cobbett CS (2009) HMA P-type ATPases are the major mechanism for root-to-shoot Cd translocation in *Arabidopsis thaliana. New Phytologist* 181: 71-78.

Wu F, Yang W, Zhang J, Zhou L (2010) Cadmium accumulation and growth responses of a poplar (*Populus deltoides* × *Populus nigra*) in cadmium contaminated purple soil and alluvial soil. *Journal of Hazardous Materials* 177: 268-273.

Yadav SK (2009) Heavy metals toxicity in plants: An overview on the role of glutathione and phytochelatins in heavy metal stress tolerance of plants. *South African Journal of Botany* 76: 167-179.

Yang X-E, Peng H-Y, Jiang L-Y, He Z-L (2005) Phytoextraction of copper from contaminated soil by *Elsholtzia splendens* as affected by EDTA, citric acid, and compost. *International Journal of Phytoremediation* 7: 69 – 83.

Yang X, Kalluri U, DiFazio S, Wullschleger S, Tschaplinski T, Cheng M, Tuskan G (2009) Poplar genomics: state of the science. *Critical Reviews in Plant Sciences* 28: 285-308.

Zalesny JA, Zalesny Jr RS, Coyle DR, Hall RB (2007) Growth and biomass of *Populus* irrigated with landfill leachate. *Forest Ecology and Management* 248: 143-152.

Zalesny JA, Zalesny Jr RS, Wiese AH, Sexton B, Hall RB (2008) Sodium and chloride accumulation in leaf, woody, and root tissue of *Populus* after irrigation with landfill leachate. *Environmental Pollution* 155: 72-80.

Zalesny R, Bauer E, Hall R, Zalesny J, Kunzman J, Rog C, Riemenschneider D (2005) Clonal variation in survival and growth of hybrid poplar and willow in an in situ trial on soils heavily contaminated with petroleum hydrocarbons. *International Journal of Phytoremediation* 7: 177-197.

Zhao F-J, Hamon RE, Lombi E, McLaughlin MJ, McGrath SP (2002) Characteristics of cadmium uptake in two contrasting ecotypes of the hyperaccumulator *Thlaspi caerulescens*. *Journal of Experimental Botany* 53: 535-543.

In: Handbook of Phytoremediation
Editor: Ivan A. Golubev

ISBN: 978-1-61728-753-4

Chapter 12

PHYTOREMEDIATION USING CONSTRUCTED MANGROVE WETLANDS: MECHANISMS AND APPLICATION POTENTIAL

***Lin Ke*[a,b] *and Nora F. Y. Tam*[b,*]**

[a]College of Environmental Science and Engineering,
South China University of Technology, Guangzhou, Guangdong 510006, China
[b]Department of Biology and Chemistry, City University of Hong Kong,
Tat Chee Avenue, Kowloon, Hong Kong SAR, China

1. INTRODUCTION

Phytoremediation is the "use of green plants and their associated microbiota, soil amendments, and agronomic techniques to remove, contain, or render harmless environmental contaminants" (Cunningham et al., 1996). It is an emerging technology which offers a potentially cost-effective and environmentally sound alternative to the environmentally destructive physical methods which are currently practiced for the cleanup of contaminated groundwater, terrestrial soils, sediments, and sludge (Shimp et al., 1993; Schnoor et al., 1995; Salt et al., 1998; Frick et al., 1999; Banks et al., 2000; Ke et al., 2003a, b; Bert et al., 2009).

In some definitions, phytoremediation is suggested to exclude constructed wetland treatment technology, as the former is the "use of living green plants for *in situ* risk reduction of contaminated soil, sludge, sediment, and ground water through contaminant removal, degradation, or containment" (USEPA, 1998), in which the scope of phytoremediation is strictly limited to *in situ* clean up areas that have been contaminated by past use; in contrast, the latter is the involvement of living plants for *ex situ* cleanup of a steady flow of wastewater (Cronk and Fennessy, 2001). In a broader sense however, wetland treatment technology also falls under phytoremediation, since both technologies take advantage of primary producers (i.e., photosynthetic plants or other autotrophic organisms in either terrestrial or aquatic

* Corresponding author. Address: Department of Biology and Chemistry, City University of Hong Kong, Tat Chee Avenue, Kowloon, Hong Kong SAR, China. Tel.: +852 2788 7793; fax: +852 2788 7406; email: bhntam@cityu.edu.hk

forms) to clean up and manage hazardous and non-hazardous contaminants, regardless of the fashion (i.e., *in situ* or *ex situ*) of application (Horne and Fleming-Singer, 2005). Actually, the first documented plant-based system installed in Germany over 300 years ago was designed to remove contaminants from municipal wastewater (reviewed by Cunningham et al., 1996). Since then, common designs such as reed-bed filters (Cooper et al., 1996), natural and constructed wetlands (Knight et al., 1992) and floating plant treatment systems (Buddhavarapu and Hancock, 1991) have been actively developed; these designs were primarily intended for purifying municipal sewage. In the past two decades, the initial concept of using plants in wastewater treatment has been expanded to remediate contaminated shallow groundwater (Shimp et al., 1993), air (Simonich and Hites, 1994), soil (Frick et al., 1999; Banks et al., 2000), and more recently, sediment (Ke et al., 2003a, b) and sludge (Bert et al., 2009). Within the current context, the broader scope of phytoremediation, which includes constructed wetlands treatment technology will be adopted.

Wetlands are "lands [that are] transitional between terrestrial and aquatic systems where the water table is usually at or near the surface or the land is covered by shallow water" (Cowardin et al., 1979). Wetlands are usually classified as natural and constructed wetlands, and have been considered as alternate wastewater treatment facilities in the recent decades (Sundaravadivel and Vigneswaran, 2001). Constructed wetlands, which are an application of the natural water purification functions of wetlands, are being developed all over the world for treating various types of wastewater (Sundaravadivel and Vigneswaran, 2001).

Mangrove wetlands are one of the three coastal wetland ecosystems dominant in the intertidal zone of tropical and subtropical regions (Mitsch and Gosselink, 2000). They fringe 70% of the coastline in these regions, and their ecological and socioeconomic significance has been recognized (Tam et al., 1997). Mangrove wetlands are an important buffer for adjacent marine ecosystems. They are vital for healthy coastal ecosystems, which not only offer a nursery ground for a number of commercially or ecologically important aquatic organisms, but also provide prime nesting and migratory sites for birds and wildlife (Day et al., 1987; Steinke and Ward, 1988; Woodroffe et al., 1988; Lee, 1990; Amarasinghe and Balasubramaniam, 1992; Tam and Wong, 2000a). Apart from the contributions to plants and animals, mangrove wetlands also are of importance in protecting coastal erosion and maintaining shore stability. They retain pollutants such as nitrogen, phosphorus and heavy metals from wastewater, and they can serve as a natural water and wastewater treatment plant (Kadlec, 1987; Tam et al., 1995; Tam and Wong, 1999a).

Mangrove plants are also of high standing biomass and productivity (Lugo, 1980). The plants are perennial, and have developed morphological, physiological and anatomical adaptations to cope with five major environmental problems of unstable substrata, anaerobic conditions, high salinity, establishment and desiccation (Chapman, 1976; Lugo and Snedaker, 1974; Mitsch and Gosselink, 2000). The adaptations, as summarized by Tam and Wong (2000a), include: (1) viviparous reproduction and production of numerous seeds to enhance reproductive success, (2) development of cable roots and pneumatophores (aerial roots) for gaseous exchange during periods of high tide, (3) formation of knee joints, buttress and prop root systems for aeration at high tide and anchorage in soft mud to stabilize the plant, (4) possession of salt glands, sclerophyllous tissues, sunken stomata, corky waterproof bark and thick waxy cuticle, hairy surface and succulent leaves to tolerate high salinity and to minimize excess water loss due to transpiration and evaporation during exposure at low tide and (5) being an evergreen and woody plant, with high primary production, fast decomposition rate and rapid nutrient turnover.

For phytoremediation initiatives, mangrove plants prevail over those vascular wetland plants, in that the latter requires frequent harvesting, which is not only labor-intensive and time-consuming but may also lead to poor plant growth and fluctuating treatment performance. In addition, disposal of plant biomass may generate a secondary pollution problem. Mangroves, as perennial plants not requiring frequent harvesting and annual planting, could be used as an alternative (Hammer and Bastian, 1989). As a novel pollution control technology, the feasibility of using constructed mangrove wetlands to remove wastewater-borne pollutants has been extensively explored in the past few decades (Tam et al., 2009), while their employment in contaminated sediments has received much less attention (Ke et al., 2003a, b). Despite the fact that sufficient, published evidence exists, the systematic summarization of information on roles and interactions among sediments, mangrove plants and microorganisms, and how they affect treatment and remediation efficiencies are lacking. The present review article, therefore, summarizes the recent progress in the research on the feasibility and potential of phytoremediation using constructed mangrove wetlands for wastewater and contaminated sediments. The focus is mainly on mechanistic aspects in phytoremediation, including the roles and involved processes for sediments, mangrove plants and microorganisms, and the physiological responses and tolerance of mangrove plants to pollutant toxicity. The linkage of functions to treatment efficiency, problems and prospective of this novel technology are also addressed.

2. Mangrove Wetlands as Pollutant Sinks

As a transit zone between terrestrial and marine environments, mangrove wetlands receive contaminants from tidal water, rivers and storm runoff (Tam and Wong, 1993, 1995a, 2000b). In addition, mangrove wetlands often suffered from negative anthropogenic impact due to the increased urban and industrial development in the surrounding areas. They have long been used as convenient sites for waste disposal and often inadvertently receive untreated sewage and livestock wastewater (Clough et al., 1983). Mangrove sediments have been reported as sinks or reservoirs of contaminants of various types, including nitrogen and phosphorus (e.g., Corredor and Morell, 1994; Tam and Wong, 1996b; Rivera-Monroy et al., 1999), heavy metals (e.g., Harbison, 1986; Silva et al., 1990; Tam and Wong, 1993, 1995a, 1996a, 1999a) and organic pollutants (e.g., Tam et al., 2001; Maskaoui et al., 2002). Mangrove plants are specially adapted to environmental extremes, which may explain, at least in part, their tolerance to low-moderate levels of pollutants. Their high productivity also indicates a great demand for nutrients in sediments.

The sediments, plant roots and associated large diversity of microbial communities suggest that chemical and biological transformation of pollutants, in addition to immobilization as insoluble precipitates or bound with organic matter, would occur in mangrove sediments (Mansell et al., 1985; Harbison, 1986; Lacerda et al., 1993; Tam and Wong, 1995a).

2.1. Nutrients

Different amounts of organic matter, total nitrogen and phosphorus have been found to accumulate in mangrove sediments worldwide, such as Southern China (Tam, 2006), India (Whigham et al., 2009), Japan (Meziane and Tsuchiya, 2002), Australia (Pitt et al., 2009) and the United States of America (Chen and Twilley, 1999). Nutrient and organic matter contents in sediments reflect the net results of interactions among many biogeochemical processes, including plant and microbial assimilation, litter decomposition and leaching (Davis et al., 2001; Meziane and Tsuchiya, 2002; Whigham et al., 2009). Domestic, industrial and agricultural inputs are also important contributors to the accumulation of nutrients in mangrove sediments (Richardson et al., 2000). For instance, the ammonium and nitrate concentrations in mangrove sediments at the Tamshui Estuary in Taiwan, which had been polluted by municipal sewage, were in the ranges from 0.15 to 17.10 and trace to 2.54 mg N kg^{-1} sediments, respectively, much higher than values reported elsewhere (Chiu and Chou, 1991; Chiu et al., 1996). Significant positive correlations found among organic matter, nitrogen and phosphorus in mangrove sediments give further evidence that the input sources of nutrients and organic matter were anthropogenic (Tam et al., 1993; Tam and Wong, 1998).

2.2. Heavy Metals

Elevated concentrations of heavy metals have also been found in mangrove sediments (e.g., Harbinson, 1986; Silva et al., 1990; Lacerda et al., 1993; Tam and Wong, 1993, 1995a, b, 1999a, 2000b), which is believed to be due to the anaerobic nature of sediments and the presence of high levels of sulfide, iron and organic matter (Ambus and Lowrance, 1991; Dunbabin and Bowmer, 1992). These properties favor precipitation and immobilization of heavy metals. Concentrations of heavy metals in mangrove sediments vary spatially and are more closely related to the degree of pollution than to other factors, including sediment properties and tidal flooding frequency and duration (reviewed by Tam, 2006). The degrees of contamination of six commonly detected heavy metals (i.e., Cu, Zn, Pb, Cd, Ni and Cr) in most mangrove sediments in China, Nigeria, Indonesia and Brazil vary from slight to moderate, with some "hot spots" (Neto et al., 2006; Tam, 2006; Amin et al., 2009; Essien et al., 2009). Concentrations of some heavy metals in mangrove sediments were higher than the ER-L (effects range-low) but lower than the ER-M (effects range median) values suggested by Long et al. (1995), implying that occasional adverse effects due to heavy metal contamination may exist; however, the occurrences should not be frequent and should not be a high risk.

2.3. Organic Pollutants

Compared to heavy metals, problems of toxic organic pollutants in mangrove sediments have received less attention. Anthropogenic activities such as discharge and dumping of wastes, oil spills, ship traffic, and dry and wet deposition of vehicle exhaust and industrial stack emission would lead to high levels of toxic organic pollutants, including polycyclic aromatic hydrocarbons (PAHs), polychlorinated biphenyls (PCBs) and organochlorine pesticides [e.g., dichlorodiphenyltrichloroethane (DDTs)] in coastal environments. The contamination of toxic organic pollutants in mangrove sediments worldwide varies greatly, with the total concentrations of PAHs, PCBs, DDTs and polybrominated diphenyl ethers (PBDEs) ranging from 8 to 15389, 0.1 to 184, 11 to 37 and 0.08 to 29 ng g^{-1} dw sediment, respectively (Table 1). Local anthropogenic activities were the main contributors to sediment contamination of these pollutants (Tam et al., 2001; Maskaoui et al., 2002; Ke et al., 2005). Similar to heavy metals, organic pollutants such as PAHs, PCBs and DDTs in mangrove sediments also show slight to moderate degree of contamination, with some "hot spots". Organic contamination problems were more severe in mangrove sediments in Brazil than those in other countries, especially for PCBs and DDTs (Table 1), and most of these values were located between the ER-L and the ER-M values, suggesting that detrimental biological effects were likely occur in mangrove sediments from this region. Although data on DDT concentrations in mangrove sediments are not available from countries other than Brazil, limited information on total DDT concentrations in coastal sediments in South China showed that they were above the ER-L value, implying that the DDT in mangrove sediments in China may also pose detrimental biological effects. Data on the concentrations of PBDEs in mangrove sediments are scarce, with only one report on the Sundarban mangrove wetland in India (Binelli et al., 2007), and their biological effects have never been evaluated. More attention is required on this new type of pollutant, as well as DDTs, in mangrove sediments.

3. Physiological Responses and Tolerance of Mangrove Plants to Pollutants

Some morphological, physiological and anatomical adaptations developed by mangrove plants to cope with environmental extremes such as fluctuated flooding, oligotrophic conditions and high salinity, may have additional merit in tolerating environmental toxicants. Progress in mechanistic research on the tolerance of mangrove species, as well as their physiological responses, to inorganic and organic toxicants has been made in the past few decades. It has become evident that exposure of plants to abiotic stresses would result in oxidative damage to plant cells due to the formation of reactive oxygen species (ROS) (Bartosz, 1997). Plants, including mangrove species, have developed robust mechanisms, along with enzymatic and non-enzymatic scavenging pathways to combat the deleterious effects of ROS production. Important antioxidative enzymes such as superoxide dismutase (SOD), peroxidase (POD) and catalase (CAT) in plants are responsible for scavenging ROS. SOD is involved in the first step of ROS elimination by catalyzing the conversion of superoxide radicals ($O_2^{\cdot-}$) to hydrogen peroxide (H_2O_2) and oxygen (O_2), and H_2O_2 can be further decomposed by CAT and POD (Parida et al., 2004). Research on antioxidative responses in mangrove plants to environmental stresses such as inorganic pollutants (e.g., nutrients and heavy metals), salinity, waterlogging, and oil contamination has been extensively conducted (Takemura et al., 2000; Ye and Tam, 2002; Ye et al., 2003, 2005; Parida et al., 2004; Ye and Tam, 2007; Zhang et al., 2007a, b; Caregnato et al., 2008).

Table 1. Concentrations of toxic organic pollutants (ng g^{-1} dry weight) in surface mangrove sediments worldwide (NR: not reported)

Locations	Total PAHs	Total PCBs	Total DDTs	Total PBDEs	References/Remarks
Hong Kong, China	685 - 11098	0.11-25.10	NR	NR	Tam et al. (2001); Tam and Yao (2002)
Shenzhen, China	238-726	NR	NR	NR	Zhang et al. (2004)
Jiulong River, Xiamen, China	59-1177	NR	NR	NR	Maskaoui et al. (2002)
Sundarban wetland, India	NR	0.18–2.33	NR	NR	Guzzella et al. (2005)
Sundarban wetland, India	241-1376	0.47–26.84	NR	0.08-29.03	Binelli et al. (2007, 2009)
Puerto Rico	500-6000	NR	NR	NR	Klekowski et al. (1994)
Caribbean island of Guadeloupe	103-1657	NR	NR	NR	Bernard et al. (1996)
Cocó River, Fortaleza, Brazil	720.7–2234.7	NR	NR	NR	Cavalcante et al. (2009)
Ceará River, Fortaleza, Brazil	96.4–1859.2	NR	NR	NR	Cavalcante et al. (2009)
Todos os Santos Bay, Salvador-Brazil	8-4163	NR	NR	NR	Venturini and Tommasi (2004)
Santos, São Paulo, Brazil	79.6–15389.1	NR	NR	NR	Medeiros and Bícego, (2004a)
São Sebastião channel, São Paulo, Brazil	20.4-200.3	NR	NR	NR	Medeiros and Bícego, (2004b)
Rio de Janeiro, Brazil		17.83-184.16	10.61-37.40	NR	de Souza et al. (2008)
ERL and ERM Guidelines[1]					
ERL	4022	22.7	1.58		Effect-Range Low
ERM	44792	180	46.1		Effect-Range Median

[1] Long et al. (1995)

The ultimate goal of these studies has been an attempt to establish the cause-effect relationships between stress and physiological responses and the subsequent linkage to biological endpoints, so as to develop valid biomarkers for "early-warning" of any potential threat to mangrove plants and the whole ecosystem. This information is also important in the selection of tolerant mangrove species for phytoremediation purposes, as well as in predicting the health status of plants and wetland systems.

3.1. Heavy Metals

Mangrove plants appear to be highly tolerant to heavy metals. Thomas and Eong (1984) found no adverse effects on the growth of hydroponic seedlings of *Rhizophora mucronata* Lam. and *Avicennia alba* Bl. treated with 10 - 500 μg ml^{-1} Zn and 50 - 250 μg ml^{-1} Pb. For *Kandelia obovata* Sheue, Liu & Yong [previously known as *Kandelia candel* (L.) Druce] seedlings, inhibition of leaf and root development was observed only at 400 mg Cu and Zn kg^{-1} sediment, the highest applied doses (Chiu et al., 1995). Similarly, MacFarlane and Burchett (2002) found Pb (0 - 800 μg g^{-1} sediment) had little negative effect on seedlings of *Avicennia marina* (Forsk.) Vierh. Cu and Zn also had relatively low toxicity in terms of emergence and biomass production and the LC_{50} for emergence and EC_{50} for biomass were 566 and 380 μg g^{-1} Cu, respectively, while the respective values for Zn were 580 and 392 μg g^{-1} Zn.

Like other wetland plants, roots of mangrove plants were found highly efficient in releasing excessive oxygen (radical oxygen loss, ROL) to oxidize the rhizosphere (Pi et al., 2009). The rhizosphere oxidizing capacity has been proven to play an important role in resisting flooding stress (Youssef and Saenger, 1996) and in excluding heavy metals through rhizosphere oxidation and fixation (Doyle and Otte, 1997; Ong Che, 1999; Machado et al., 2005). ROL has also been considered to be one of the most important factors affecting the formation of iron-plaque on root surface and in the rhizosphere (Mendellsohn et al., 1995; Pi et al., 2009, 2010), which may result in effectively detoxifying heavy metals (Machado et al., 2005). Another adaptive strategy to minimize the uptake of heavy metals may be attributed in part to the ability of mangrove plants to regulate the uptake of metals at the root level and to limit the translocation to the shoot (MacFarlane and Burchett, 2002). Although salt secretors were found to secrete metals from salt glands concomitantly with sodium, secretion did not significantly alter the overall distribution patterns of metals in leaf tissue, or when comparing secretors to non-secretors (MacFarlane et al., 2007). It has been reported that metals, such as Cu, Zn, Pb, Fe, Mn and Cd, were accumulated predominantly in root tissue, rather than in foliage, of numerous mangrove species in the field, such as *Avicennia*, *Rhizophora* and *Kandelia* (Peters et al., 1997). A comparative analysis of patterns of accumulation and partitioning of the heavy metals (Cu, Pb and Zn) in mangrove plants was recently examined by MacFarlane et al. (2007), who found that patterns of metal accumulation and partitioning for all metals examined were broadly similar across genera and families, with root bio-concentration factors (BCF; ratio of tissue metal to sediment metal concentration) ≤ 1 and translocation factors (TF; ratio of leaf metal to root metal concentration) of ~0.5 and ~0.3 for essential metals (Cu and Zn) and for Pb (a non-essential metal), respectively. These authors suggested that mangrove plants may act as excluders for non-essential metals and regulators for essential metals to prevent metal toxicity due to excessive uptake.

In addition to growth responses to heavy metal toxicity, other physiological responses such as root exudates, photosynthetic pigments and antioxidant enzymes, to heavy metal toxicity and their cause-effect relationships have been studied. Lu et al. (2007) reported that the production of root exudates (mono-, di- and tri-carboxylic acids) by *Kandelia obovata* under the stress of Cd lowered the bioavailability of Cd, and as a result, reduced its toxicity to the mangrove plant. MacFarlane (2002) suggested that POD activity may be an appropriate biomarker for Zn or total metal accumulation in leaf tissue of *Avicennia marina*, while the chlorophyll a/b ratio may be a suitable biomarker of Zn accumulation when the sediment was contaminated with Cu, Pb and Zn. Zhang et al. (2007b) found that when mangrove plants were subject to multiple heavy metal stress (five levels of Pb, Cd and Hg), the enzymatic and non-enzymatic responses to metal toxicity were species- and tissue-specific and proposed root and leaf POD may serve as a biomarker of heavy metal stress in *Kandelia obovata*, while malonialdehyde (MDA) content may be a biomarker in *Bruguiera gymnorrhiza*.

3.2. Organic Pollutants

Mangrove plants are susceptible to organic pollutants such as oil spills (Duke et al., 1997; Peters et al., 1997). The specialized pneumatophores, one of the physiological adaptations that mangrove plants have evolved to survive anaerobic sediments, are particularly vulnerable to smothering by oil (Teas et al., 1987; Böer, 1993). Penetration and long-term persistence of petroleum hydrocarbons in sediments, and plant and animal uptake would lead to lethal and sub-lethal toxic effects (Corredor et al., 1990; Garrity et al., 1994). Duke et al. (1998) summarized that around 5,000 tons of various types of oil had been spilled in the vicinity of mangrove habitats in Australia since 1970, resulting in the oiling of at least 220 hectares of mangrove wetlands and the mortality of 13 hectares of plants. In addition to the impact of oil spills, pesticides and herbicides also posed detrimental effects on mangrove plants. It is estimated that 41% (124,000 ha) of the total mangrove forest area of Vietnam experienced significant mortality due to wartime operations (reviewed by Peters et al., 1997).

Research on the physiological responses of mangrove plants to toxic, organic pollutants is relatively scarce and focused mostly on oil contamination. Ye and Tam (2007) found that *Avicennia marina* was more sensitive than *Aegiceras corniculatum* when exposed to spent (used) lubricating oil (5 L m^{-2}), and canopy-oiling resulted in more direct physical damage and stronger lethal effects than base-oiling. This study also reported various oil-induced physiological damages, including decreases in chlorophyll and carotenoid contents, nitrate reductase, POD and SOD activities, as well as an increase in MDA content. Zhang et al. (2007a) also showed that fresh and spent lubricating oil at a single initial dose of 5 L m^{-2} posed an oxidative stress to *Bruguiera gymnorrhiza* causing significant increases in $O_2^{\cdot-}$ release, MDA content and SOD, as well as inhibiting early growth, including height, leaf number and biomass of the seedlings. However, the oil pollution had no effects on the germination of mangrove seedlings. These authors further revealed the toxic effects of oil on mangrove plants in muddy sediments were more severe than in sandy sediments. Ke et al. (2010) compared the tolerance of four dominant mangrove species in South China to different doses of spent lubricating oil and found that *Bruguiera gymnorrhiza* was the most tolerant species and could survive at the highest oil dose (15 L m^{-2}), followed by *Acanthus ilicifolius*, and *Aegiceras corniculatum*, while *Kandelia obovata* was the most sensitive species, based on the results of the biomass production and various other physiological responses. Although it was suggested that salt excluders should have a higher tolerance than salt secretors to oil toxicity (Ye and Tam, 2007), the oil tolerance of mangrove plants appeared to have little connections to their salinity tolerance based on the observations by Ke et al. (2010), who found that *Bruguiera gymnorrhiza* was the most tolerant species while *Kandelia obovata* was the most sensitive one among the four mangrove species, and both *Kandelia obovata* and *Bruguiera gymnorrhiza* are known to be salt excluders. Similarly, *Avicennia marina* and *Aegiceras corniculatum* were salt-excreting species but the former one was more sensitive to spent lubricating oil (Ye et al., 2007). Some anatomical adaptations, such as the degree of development of xylem and phloem, and the degree of suberized structure of the root surface may be more important in the oil tolerance of mangrove plants than their mechanisms to tolerate salt.

4. Constructed Mangrove Wetlands in Phytoremediation and Remediation Capacity

4.1. Wastewater

Mangrove wetlands are highly efficient in adsorbing and absorbing wastewater-borne pollutants, including nitrogen, phosphorus, heavy metals and toxic organic pollutants (Clough et al., 1983; Gale et al., 1993; Corredor and Morell, 1994; Tam and Wong, 1995b, 1996; Yang et al., 2008). Clough et al. (1983) estimated that mangrove plants, through incorporation into the plant tissues, could annually immobilize around 150 to 250 kg N ha^{-1} and 10 to 20 kg P ha^{-1}. The removal efficiencies of nutrients and metals from the wastewater in a constructed mangrove wetland was found to range between 75-98% and 88-96%, respectively (Chu et al., 1998).

Mangrove wetlands have been used to filter shrimp pond effluents because of their close proximity to the ponds (Robertson and Phillips, 1995; Rivera-Monroy et al., 1999). Integrated pond-mangrove farming systems, with the wetland to shrimp pond ratios varying from 2 to 22, have been proposed (Robertson and Phillips, 1995). The ratio could be reduced to a range of 0.04-0.12 for the purpose of nitrogen removal if denitrification occurs in the wetland system (Rivera-Monroy et al., 1999). In addition to shrimp pond effluents, the feasibility of using mangrove wetlands to remove pollutants from municipal and livestock wastewater has been a focus of research since 1990. Sediments in a fringe mangrove forest were capable of removing nitrate in the effluent discharged from a sewage treatment plant via nitrification and denitrification processes (Corredor and Morell, 1994). Results from a three-year long field study in Futian National Nature Reserve, Shenzhen, China showed that primarily settled domestic sewage was purified if it was intermittently discharged to the landward region of the mangrove wetland during low tides, without contaminating the tidal water or posing any negative impacts on plant growth (Wong et al., 1995, 1997b). The diversity and abundance of macro-algae in this mangrove ecosystem and benthic invertebrates colonized in the surface mangrove sediments were also not affected by sewage discharge (Liu et al., 1995; Yu et al.,

1997). Greenhouse experiments demonstrated that the discharge of artificial municipal wastewater with high nutrient content and livestock wastewater collected from a local farm promoted the growth of four mangrove species, namely *Kandelia obovata*, *Aegiceras corniculatum*, *Avicennia marina* and *Bruguiera gymnorrhiza*, while nutrients in the wastewater were removed (Chen et al., 1995; Wong et al., 1997a; Ye and Tam, 2002). Mangrove wetlands in general have high assimilation/dissimilation capacities for nutrients. A field study in Futian, Shenzhen, China showed that at the end of a three-year study of the discharge of municipal sewage, only the surface sediments in the first two meters downstream the discharge points had a 20% increase in total Kjeldahl nitrogen and a 38% increase in phosphorus, while no significant change in nutrient concentrations was found in the sediments further away from the discharge points (Wong et al., 1997b). Around 27% of the nitrogen and 85% of the phosphorus from the wastewater were retained in mangrove sediments. Greenhouse experiments demonstrated that the percentage of nitrogen lost from the mangrove ecosystem was around 40% and the plant uptake varied from 12 to 68%, depending on the plant species and salinity (Ye et al., 2001). With continuous losses of nitrogen from the system, mangrove wetlands could be maintained as a sustainable wastewater treatment facility without saturation.

Environmental factors, such as salinity, affect the wastewater treatment efficiency of constructed mangrove wetlands. A greenhouse pot experiment showed that with the increase of salinity from 0 to 30‰, the removal percentages of dissolved organic carbon (DOC), ammonia-N and inorganic N, by a constructed mangrove wetland planted with *Aegiceras corniculatum*, decreased from 91% to 71%, 98% to 83% and 78% to 56%, respectively (Wu et al., 2008). The denitrification potential of sediments was also found to be retarded by high salinity. The treatment efficiency was affected by plant species. In a pilot-scale study using the constructed mangrove wetlands to treat municipal sewage in Futian, Shenzhen, China, the wetland planted with *Sonneratia caseolaris* had the highest COD removal (an average of six months of data was 75%), followed by *Aegiceras corniculatum* (64%) and *Kandelia obovata* (62%), and the removal of TN (46%), NH_3^--N (50%) and TP (60%) by *Sonneratia caseolaris* was also the highest (Yang et al., 2008).

4.2. Contaminated Sediments

In contrast to terrestrial soils, contaminated sediments using phytoremediation has been investigated mostly for *ex situ* remediation of dredged sediments disposed in landfills and for *in situ* remediation of shallow waters (Bert et al., 2009; Perelo, 2010). The effectiveness of remediation using plants was highly dependent on the types and concentrations of contaminants, as well as the environment and the plant species, all of which would lead to quite diverse outcomes. The outcomes ranged from high reduction (90%; Paquin et al., 2002) to no enhancement effects (Vervaeke et al., 2003; King et al., 2006), and in some cases, even negative effects (Smith et al., 2007). The case of negative effects was due to the presence of plant roots, which provided an oxidizing environment in the rhizosphere, which would interfere with the highly reducing conditions needed for the dechlorination process (Smith et al., 2007).

For mangrove wetlands, most of the studies have been focused on their feasibility of removing wastewater-borne pollutants (i.e., as treatment wetlands). Research on the potential of using mangrove plants for phytoremediation of contaminated sediments is very limited. Ke et al. (2003a) reported the high removal potential of constructed mangrove wetlands in phytoremediation of PAH-contaminated sediments and obtained greater than 90% removal of pyrene (a four-ring PAH) from contaminated sediments at the end of the six-month treatment, and the efficiency was slightly higher when planted with *Kandelia obovata* than with *Bruguiera gymnorrhiza*. The authors also investigated the effect of humic acid (HA) addition on the removal of pyrene and found that excessive HA in sediments (6.7% w/w) reduced both plant growth and pyrene removal from contaminated mangrove sediment, suggesting that pyrene binding to the HA limited its bioavailability (Ke et al., 2003b).

The presence of plants not only enhances aerobic degradation of organic pollutants in the rhizosphere, but also facilitates metal removal or immobilization in this area. Tables 2 and 3 show the performance of one-year old *Bruguiera gymnorrhiza* seedlings in phytoremediation of a sediment taken from Victoria Harbour, Hong Kong, which was heavily contaminated by heavy metals and PAHs. After five months of treatment under greenhouse conditions, concentrations of metals of barium, chromium, lead, iron, manganese and zinc were significantly lower in the rhizosphere as compared to those in the bulk and/or non-vegetated control sediments (Table 2).

Table 2. Concentrations of metals (μg g^{-1} dry weight for all metals except aluminum and iron which are in mg g^{-1} dry weight) in sediments after five months of a phytoremediation trial using mangrove microcosms. One-year old *Bruguiera gymnorrhiza* seedlings were used. Significant differences in the same row are marked with different superscripted lowercase letters according to one-way ANOVA at $p \leq 0.05$

	Vegetated		Non-vegetated control	
Metals	Bulk sediment	Rhizosphere	SK-KTACb [2]	KTACb [3]
Aluminum	14.0 ^{a}	27.2 ^{b}	12.8 ^{a}	14.2 ^{a}
Arsenic	ND [1]	ND	ND	ND
Barium	33.6 ^{a}	21.5 ^{b}	29.1 ^{a}	33.3 ^{a}
Cadmium	ND	0.8	ND	ND
Chromium	101.7 ^{ab}	45.3 ^{a}	83.3 ^{ab}	111.7 ^{b}
Copper	613.1 ^{a}	389.6 ^{a}	525.8 ^{a}	750.6 ^{a}
Iron	17 ^{a}	14.4 ^{b}	16.2 ^{ab}	17.1 ^{a}
Lead	122.1 ^{a}	69.7 ^{b}	113.5 ^{a}	126.3 ^{a}
Manganese	299.8 ^{a}	215.8 ^{b}	260 ^{ab}	286.5 ^{a}
Nickel	29.8 ^{a}	20.1 ^{a}	26.4 ^{a}	33.9 ^{a}
Silver	0.8 ^{a}	3.1 ^{b}	0.6 ^{a}	0.4 ^{a}
Zinc	386.8 ^{a}	263.8 ^{b}	335.7 ^{ab}	421.4 ^{a}

[1] Not detectable. Detection limits of arsenic and silver were 0.022 μg g^{-1} and 0.004 μg g^{-1} dry weight, respectively; [2] Sediments from Sai Keng, Hong Kong (SK, a mangrove swamp with low contamination in sediments, which were used for seedling raising) surrounded by contaminated sediments (from Kai Tak Approach Channel bridge, Hong Kong; KTACb). The reasons for including SK sediments were to

minimize damages to the delicate plant root system during transplanting and to resemble real phytoremediation practice; [3] Contaminated sediments (KTACb) only.

Table 3. Concentrations of PAHs (ng g^{-1} dry weight) in sediments after five months of a phytoremediation trial using mangrove microcosms. One-year old *Bruguiera gymnorrhiza* seedlings were used. Significant differences in the same row are marked with different superscripted lowercase letters according to one-way ANOVA at p ≤0.05

PAHs	Vegetated		Non-vegetated control	
	Bulk sediment	Rhizosphere	SK-KTACb [3]	KTACb [4]
LMW-PAHs [1]				
Naphthalene	46.3 ^{a}	38.2 ^{a}	149.5 ^{a}	63.6 ^{a}
Acenaphthylene	16.1 ^{a}	9.2 ^{b}	16.1 ^{a}	16.8 ^{a}
Acenaphthene	4.8 ^{a}	2.1 ^{a}	3.9 ^{a}	3.0 ^{a}
Fluorene	7.6 ^{a}	2.7 ^{b}	7.4 ^{a}	7.9 ^{a}
Phenanthrene	265.2 ^{a}	81.9 ^{b}	248.0 ^{a}	197.5 ^{ab}
Anthracene	93.3 ^{a}	32.6 ^{b}	93.8 ^{a}	119.0 ^{a}
Total	433.3 ^{a}	166.7 ^{b}	518.6 ^{a}	407.7 ^{a}
HMW-PAHs [2]				
Fluoranthene	194.5 ^{a}	100.6 ^{b}	143.1 ^{ab}	215.0 ^{a}
Pyrene	219.1 ^{a}	113.6 ^{b}	202.6 ^{a}	235.0 ^{a}
Benz[a]anthracene	55.4 ^{a}	21.4 ^{b}	59.3 ^{a}	51.9 ^{a}
Chrysene	109.6 ^{a}	46.1 ^{b}	100.0 ^{a}	112.8 ^{a}
Benzo[b]fluoranthene	88 ^{a}	40.2 ^{b}	106.4 ^{a}	88.1 ^{a}
Benzo[k]fluoranthene	94.3 ^{a}	27.2 ^{b}	97.8 ^{a}	86.9 ^{a}
Benzo[a]pyrene	49.5 ^{ab}	12.0 ^{a}	63.9 ^{b}	52.8 ^{b}
Indeno[1,2,3-c,d]anthracene	49.6 ^{a}	12.4 ^{b}	67.0 ^{a}	40.0 ^{a}
Dibenzo[a,h]anthracene	25.9 ^{a}	5.9 ^{b}	38.2 ^{a}	19.6 ^{b}
Benzo[g,h,i]perylene	121.4 ^{abc}	41.0 ^{a}	163.4 ^{b}	100.0 ^{c}
Total	812.7 ^{a}	319.8 ^{b}	898.5 ^{a}	787.2 ^{a}

[1] Low-molecular-weight PAHs; [2] High-molecular-weight PAHs; [3] Description of sediments same as in Table 2; [4] Description of sediments same as in Table 2.

Mangrove plants are not metal hyperaccumulators (MacFarlane, 2007). Therefore, the decrease in metal concentrations in the rhizosphere may be attributed to metal mobilization and precipitation on roots. On the contrary, metals of aluminum, silver and cadmium showed an opposite trend (i.e., with higher levels in the rhizosphere than in the bulk and non-vegetated control sediments). For the removal of PAHs, an enhancement effect was also significant in the rhizopshere (Table 3). It is suggested that mangrove plants may be used in phytoremediation of PAH-contaminated sediments, while their roles in metal remediation require further evaluation.

5. Mechanisms of Constructed Mangrove Wetlands in the Removal of Pollutants

Mangrove wetlands have inherent physical, chemical and biological properties for adsorption and/or utilization of nutrients and heavy metals. The sediments, plants and associated high diversity of microbial communities are important components in the retention and transformation of pollutants. The distributions of wastewater-borne pollutants in different mangrove wetland components are different and highly dependent on the types of pollutants. For P and heavy metals, the reduction is more likely to be attributed to sediment retention than to plant uptake; while for N, reduction of pollutants due to plant uptake is as important, or more important, than sediment retention (Tam, 2006). It is because the demand of N for plant growth is much greater than that for P and heavy metals. The mechanisms of phytoremediation of pollutants involved in a wetland system include phytostabilization, phytoextraction, biodegradation and biotransformation (Bert et al., 2009). The roles of different components of sediments, plants and microorganisms in phytoremediation are summarized as below.

5.1. Sediments

Wetland sediments are different from most terrestrial soils because they often undergo intermittent flooding and draining, and thus provide an alternating anaerobic and aerobic environment. Such conditions are particularly favorable for nitrification and denitrification processes, leading to the removal of N (Chiu and Chou, 1993; Tam et al., 2002). Ammonia volatilization, nitrification and denitrification have been generally accepted as the most important pathways to remove N in a wetland system. In mangrove wetlands, ammonia volatilization is negligible because of its acidic to neutral pH. Nitrification and denitrification processes become significant because mangrove sediments are also under alternating aerobic and anaerobic conditions due to periodical flooding by tidal water.

The presence of high levels of reducing sulfide, iron and manganese favor the precipitation and immobilization of heavy metals (Ambus and Lowrance, 1991; Dunbabin and Bowner, 1992). In addition, the clay-like nature of mangrove sediments could provide a physical trap for fine particulates and heavy metals. Organic pollutants, such as petroleum hydrocarbons, were found to persist for decades in mangrove sediments after an oil spill (Garrity et al., 1994). High organic matter, especially the humic substances, in mangrove sediment also provided strong adsorptive properties in binding heavy metals. Mansell et al. (1985) showed that phosphorus could be immobilized by complexation with humic substances and physiochemical adsorption on sites such as hydroxides and oxides of Al and Fe, carbonates of Ca and layer silicate minerals. These binding mechanisms would be further affected by the sediment properties such as texture, pH, reduction-oxidation reaction (redox) potential, proportion of organic matter, Fe and sulfide, etc., which have been extensively researched (Richardson, 1985; Harbison, 1986; DeBustamante, 1990; Lacerda et al., 1993). Patrick and Khalid (1974) found that sediments with lower redox potentials could bind more P from the aqueous phase than sediments with high redox potentials. In addition, variations in salinity are common in coastal wetlands which may also affect the binding of pollutants in sediments. Paalman et al. (1994) reported that when river water mixed with seawater, due to increments in chloride concentrations, heavy metals such as Cd mobilized from the sediment and became dissolved chloro-complexes (Comans and Van Dijk, 1988). Tam and Wong (1999) also reported that mangrove sediments receiving wastewater prepared in deionized water (i.e., freshwater) had slightly higher pollutant concentrations and larger enrichment factors than that treated with saline wastewater (salinity 1.5‰).

5.2. Plants

Vegetation is an indispensable component of a constructed wetland. The important roles of wetland plants in constructed wetlands have been well documented (Brix, 1994, 1997; Sundaravadivel and Vigneswaran, 2001). Similarly, mangrove plants affect the removal of wastewater-borne pollutants both physically and biochemically. The presence of plants provides various physical effects such as filtration, erosion control and provision of surface area for microorganisms to attach and colonize. The biochemical effects occur through plant metabolic processes, such as plant uptake. The role of mangrove plant uptake in nitrogen removal is significant. Chiu et al. (1996) showed that the ^{15}N-labeled ammonium added to the experimental pots disappeared rapidly, and around 20% of the N was taken up by *Kandelia obovata* after three months. The high demand of N for plant growth leads to the high N removal. Mangrove plants also play a significant role in immobilizing heavy metals, especially in roots. Higher amounts of heavy metals were found in the roots and leaf litter than in other aerial parts of *Kandelia obovata* receiving wastewater-borne heavy metals (Tam and Wong, 1997).

Water-logging is a typical phenomenon in wetland environments, which would result in oxygen and nutrient deficiency, low redox potential and accumulation of phytotoxins, such as Fe^{2+}, Mn^{2+}, H_2S and CH_4 in sediments (Gambrell et al., 1991). Wetland plants, including mangroves, have developed an adaption to water-logging by transporting oxygen from the atmosphere to shoots and to roots via extensive aerenchyma tissues, part of the oxygen is used for root respiration, while excessive oxygen is released from root and diffuses into the rhizosphere to create an oxygenated zone around the roots (Armstrong, 1979; Youssef and Saenger, 1996; McDonald et al., 2002; Jackson and Colmer, 2005). The oxygen release to the rhizosphere, called radial oxygen loss (ROL), can oxidize the reduced substances, such as phytotoxins, in the rhizosphere (Armstrong et al., 1992; Pedersen et al., 2004). ROL has the potential to significantly alter both microbial and chemical processes in rhizosphere, such as increasing aerobic respiration (Schussler and Longstreth, 1996), nitrification (Kirk and Kronzucker, 2005) and aerobic degradation and transformation of environmental pollutants, such as PAHs (St-Cyr and Campbell, 1996; Visser et al., 2000). ROL is very important in mangrove wetlands, as the sediments are often anaerobic or anoxic just few centimeters below the surface (Mitsch and Gosselink, 2000), which would inhibit PAH biodegradation, leading to elevated concentrations of PAHs in mangrove sediments (Tam et al., 2001). However, ROL has negative effects on the degradation of chlorinated compounds such as

PCBs because these compounds require a reducing environment for dechlorination (Smith et al., 2007).

ROL could affect redox potential in mangrove sediments (Koch, 1997), leading to changes in the bioavailability of trace elements through their combination with sulfides (Walsh et al., 1979; Lacerda et al., 1993) or by formation of iron plaque (Machado et al., 2005). The presence of iron plaque on the roots is also a striking feature of roots of some mangrove plants (Pi et al., 2009, 2010). Iron plague is composed mostly of iron hydroxides and other metals, such as manganese, that are mobilized and precipitated on the root surface. Their concentrations could reach 5-10 times the concentrations of the surrounding sediments (Sundby et al., 1998). For phytoremediation of heavy metals, ROL would lead to the mobilization of heavy metals, while the formation of iron plaque may provide some extra binding sites for heavy metals. Iron plaque also acted as a physical barrier to reduce plant uptake and metal toxicity to plants (Otte et al., 1989; Batty et al., 2000). It is also suggested that iron plaque is important in immobilizing P (Batty et al., (2000). Nevertheless, the role of iron plaque in the reduction of plant uptake is still debatable as the results were not always convincing and conclusive (Ye et al., 1998; Zhang et al., 1998). The net effect of the presence of plants, particularly roots, on the behavior of heavy metals, and even phosphorus (e.g., mobilization or immobilization), in the wetland system merits further study.

5.3. Microorganisms

Microorganisms may play a more significant role than sediments and plants in removing wastewater-borne nutrients and organic pollutants as they are involved in a variety of important processes, such as nitrification, denitrification, solubilization and biodegradation. Wetland sediments are different from most terrestrial soils because they often undergo intermittent flooding and draining, thus supporting both aerobic and anaerobic microbial communities to utilize a wide range of electron acceptors, including O_2, NO_3^-, Fe(III), SO_4^{2-}, CO_2 and organics during respiration (Mohanty and Dash, 1982). The anoxic layer in the mangrove sediment also favored the bacterial sulfate reduction to produce sulfide, which would then precipitate the soluble metal ions as metal sulfide (Harbison, 1986). There is increasing published evidence that mangrove sediments are high in microbial diversity (e.g., Kathiresan and Selvam, 2006; Gomes et al., 2008). Different functional species, such as nitrogen-fixing bacteria (Holguin and Bashan, 1996; Flores-Mireles et al., 2007; Zhang et al., 2008), phosphate solubilizing fungi and bacteria (Kothamasi et al., 2006), and hydrocarbon-degrading bacteria (Díaz et al., 2001; Yu et al., 2005) have been isolated from bulk mangrove sediments or the rhizosphere of mangrove plants. A total of 11 PAH-degrading strains belonging to four genera, *Mycobacterium*, *Sphingomonas*, *Terrabacter*, and *Rhodococcus*, were isolated from surface mangrove sediments in South China (Zhou et al., 2008). As discussed before, mangroves are subject to human interference and the microbial community in mangrove sediments is also susceptible to anthropogenic pollution. It has been reported that sediments contaminated by oil or PAHs reduced microbial diversity; however, the population of hydrocarbon-degrading bacteria increased under the pollution stress (El-Tarabily, 2002; Zhou et al., 2009).

It is sometimes difficult to distinguish the roles of plants and microorganisms in phytoremediation, as one benefits another. For instance, the highly productive and diverse microbial community in mangrove sediments rapidly degrades plant debris and dead materials to nutrients for plant growth. In turn, root exudates of mangrove plants, which consist of various organic acids, serve as a food source for microorganisms (Bashan et al., 2002). The plant roots also provide surface areas for microbial colonization. Kothanasi et al. (2006) found that up to 17% of arbuscular mycorrhizal fungus was colonized in the aerenchymatous cortex of mangrove roots as their survival was enhanced by the oxygen provided by mangrove plants, while the presence of phosphate solubilizing bacteria supported more plant growth by mobilizing insoluble phosphate.

6. Problems and Prospective of Constructed Mangrove Wetlands

Extensive greenhouse and pilot-scale studies have demonstrated that constructed mangrove wetlands have a high capacity for removing different types of wastewater-borne pollutants and the performance is comparable to, or even better than, other types of constructed wetlands such as cattail (*Typha latifolia*) (Tam et al., 2009). Although mangrove wetlands have a high potential to act as a sink for the non-degradable pollutants, such as heavy metals and phosphorus, these pollutants are mainly accumulated in sediments with some stored in roots because mangrove plants, like other wetland plants, are generally not hyperaccumulators (MacFarlane et al., 2007). The mechanical aspects of harvesting plants would be destructive to wetlands (Weis and Weis, 2004), thus immobilization of metals in sediments to be stored in below-ground plant tissues may be the preferable alternative. However, the continuous accumulation of these pollutants decreases their retention capacity and may pose long-term effects on the wetland plants. In addition, the dynamic nature of the mangrove wetlands, which is strongly influenced by factors like tidal flow, wave action, climate, salinity, redox potential and various biotic components (e.g., insect/fungal infestation), may cause the release of retained pollutants back to the aquatic environment, especially under extreme weather conditions. The possibility of the mangrove sediments becoming a secondary source of pollution has not been addressed and merits further research. Tam et al. (2009) pointed out that as a novel technology, more research must be conducted to understand the treatment mechanisms, the maximum capacity and saturation and the long-term adverse effects before the full-scale application of constructed mangrove wetlands as secondary wastewater treatment facilities is initiated.

In terms of phytoremediation of contaminated sediments, data on this aspect are very scarce (Ke et al., 2003a, b), and the successfulness may be highly dependent on types and degrees of contamination. Phytoremediation is only suitable for shallow and low- to mid-levels of contamination. Aerobic degradation efficiency may be low for persistent organic pollutants, such as PAHs, due to the anoxic conditions in mangrove sediments. Although ROL and root exudates may aid aerobic degradation by microorganisms in sediments, the relationships between these root features and biodegradation of persistent organic pollutants are still uncertain. On the other hand, the presence of plants may not be beneficial for certain chemical/biological processes. For instance, caution should be given to sediments contaminated by heavy metals and chlorinated compounds. The presence of mangrove plants may cause the mobilization of heavy metals by rhizosphere oxidation, Lacerda et al. (1993)

reported that *Avicennia* species of mangroves was able to oxidize the rhizosphere, thus reducing sulfides and enhancing metal concentrations in the exchangeable form. Although the formation of iron plaque may have positive effects on metal stabilization, the net effects of the presence of mangrove plants on the behaviors of heavy metals have never been evaluated. The presence of vegetation may also pose a negative impact on remediating chlorinated compounds; as dechlorination requires a highly reducing environment, in which plants would interfere (Smith et al., 2007). Compared to the role of mangrove wetlands in wastewater treatments, much less is known about their role in the phytoremediation of sediments. More fundamental studies on phytoremediation of metal- and chlorinated compound-contaminated sediments are needed.

SUMMARY

Mangrove wetlands have been demonstrated to be able to remove wastewater-borne pollutants, including nitrogen, phosphorus, heavy metals and toxic, organic pollutants. Mangrove sediments act as pollutant sinks and these pollutants are immobilized as insoluble precipitates or bound with clay and organic matter in sediments with some stored in root tissues. Mangrove plants have rapid growth rates and high primary productivity, and they exhibit an efficient conversion of nutrients to their biomass. They directly absorb and assimilate nutrients, particularly nitrogen in their aerial plant parts, while heavy metals are mainly accumulated in belowground tissues. Mangrove plants are capable of transferring oxygen from the aerial parts to the roots, which creates an aerobic rhizosphere and are favorable for aerobic degradation, nitrification and aerobic oxidation. These special features suggest that constructed mangrove wetlands can be developed as alternative wastewater treatment facilities with simple, cost-effective and low-maintenance properties. Although data are scarce, phytoremediation using constructed mangrove wetlands appears to be a potential management option for contaminated coastal sediments, as this system simulates natural wetlands, being not only green and environmental friendly, but also enhancing the aesthetic value and the biodiversity of the environment.

ACKNOWLEDGMENTS

The work described in this paper was supported by a research grant supported from Science, Industry, Trade and Information Technology Commission of Shenzhen Municipality (Project No. 2008-121).

REFERENCES

Amarasinghe, M.D., Balasubramaniam, S., 1992. Net primary productivity of two mangrove forest stands on the northwestern coast of Sri Lanka. *Hydrobiologia* 247, 37-47.

Ambus, R., Lowrance, R., 1991. Comparison of denitrification in two riparian soils. *Soil Science Society of America Journal* 55, 994-997.

Amin, B., Ismail, A., Arshad, A., Yap, C.K., Kamarudin, M.S., 2009. Gastropod assemblages as indicators of sediment metal contamination in mangroves of Dumai, Sumatra, Indonesia. *Water, Air, and Soil Pollution* 201, 9-18.

Armstrong, W., 1979. Aeration in higher plants. *Advance in Botany Research* 7, 225- 332.

Armstrong, J., Armstrong, W., Beckett, P.M., 1992. *Phragmites australis*: Venturiand humidity-induced convections enhance rhizome aeration and rhizosphere oxidation. *New Phytologist* 120, 197-207.

Banks, M.K., Govindaraju, R.S., Schwab, A.P., Kulakow, P.A., 2000. Part I: Field demonstration. In: Fiorenza, S., Oubre, C.L., Ward, C.H. (Eds.), *Phytoremediation of Hydrocarbon-Contaminated Soil*, CRC Press/Lewis Publishers, Boca Raton, FL, pp. 3-88.

Bashan, Y., Holguin, G., 2002. Plant growth-promoting bacteria: a potential tool for arid mangrove reforestation. *Trees* 16, 159-166.

Bartosz, G., 1997. Oxidative stress in plants. *Acta Physiologiae Plantarum* 19, 47-64.

Batty, L.C., Baker, A.J.M., Wheeler, B.D., Curtis, C.D., 2000. The effect of pH and plaque on the uptake of Cu and Mn in *Phragmites australis* (Cav.) *Trin ex. Steudel. Annals of Botany* 86, 647-653.

Bernard, D., Pascaline, H., Jeremie, J.J., 1996. Distribution and origin of hydrocarbons in sediments from lagoons with fringing mangrove communities. *Marine Pollution Bulletin* 34, 298-305.

Bert, V., Seuntjens, P., Dejonghe, W., Lacherez, S., Thuy, H.T.T., Vandecasteele, B., 2009. Phytoremediation as a management option for contaminated sediments in tidal marshes, flood control areas, and dredged sediment landfill sites. *Environmental Science and Pollution Research* 16, 745-764.

Binelli, A., Sarkar, S.K., Chatterjee, M., Riva, C., Parolini, M., deb Bhattacharya, B., Bhattacharya, A.K., Satpathy, K.K., 2007. Concentration of polybrominated diphenyl ethers (PBDEs) in sediment cores of Sundarban mangrove wetland, northeastern part of Bay of Bengal (India). *Marine Pollution Bulletin* 54, 1220-1229.

Binelli, A., Sarkar, S.K., Chatterjee, M., Riva, C., Parolini, M., deb Bhattacharya, B., Bhattacharya, A.K., Satpathy, K.K., 2009. Congener profiles of polychlorinated biphenyls in core sediments of Sunderban mangrove wetland (N.E. India) and their ecotoxicological significance. *Environmental Monitoring and Assessment* 153, 221-234.

Böer, B., 1993. Anomalous pneumatophores and adventitious roots of *Avicennia marina* (Forsk.) Vierh. Mangroves two years after the 1991 Gulf War oil spill in Saudi Arabia. *Marine Pollution Bulletin* 27, 207-211.

Brix, H., 1994. Functions of macrophytes in constructed wetlands. *Water Science and Technology* 29, 71-78.

Brix, H., 1997. Do macrophytes play a role in constructed treatment wetlands? *Water Science Technology* 35, 11-17.

Buddhavarapu, L.R., Hancock, S.J., 1991. Advanced treatment for lagoons using duckweed. *Water Environmental and Technology* 3, 41-44.

Caregnato, F.F., Koller, C.E., MacFarlane, G.R., Moreira, J.C.F., 2008. The glutathione antioxidant system as a biomarker suite for the assessment of heavy metal exposure and effect in the grey mangrove, *Avicennia marina* (Forsk.) Vierh. *Marine Pollution Bulletin* 56, 1119-1127.

Cavalcante, R.M., Sousa, F.W., Nascimento, R.F., Silveira, E.R., Freire, G.S.S., 2009. The impact of urbanization on tropical mangroves (Fortaleza, Brazil): Evidence from PAH distribution in sediments. *Journal of Environmental Management* 91, 328-335.

Chapman, V.J., 1976. *Mangrove Vegetation*. ISBS, Inc., Forest Grove, Oregon, 447pp.

Chen, G.Z., Miao, S.Y., Tam, N.F.Y., Wong, Y.S., Li, S.H., Lan, C.Y., 1995. Effect of synthetic wastewater on young *Kandelia candel* plants growing under greenhouse conditions. *Hydrobiologia* 295, 263-273.

Chen, R., Twilley, R.R., A simulation model of organic matter and nutrient accumulation in mangrove wetland soils. *Biogeochemistry* 44, 93-118.

Chiu, C.Y., Chou, C.H., 1991. The distribution and influence of heavy metals in mangrove forests of the Tamshui Estuary in Taiwan. *Soil Science and Plant Nutrition* 37, 659-669.

Chiu, C.Y., Chou, C.H., 1993. Oxidation in the rhizosphere of mangrove *Kandelia candel* seedling. *Soil Science and Plant Nutrition* 39, 725-731.

Chiu, C.Y., Hsiu, F.S., Chen, S.S., Chou, C.H., 1995. Reduced toxicity of Cu and Zn to mangrove seedlings (*Kandelia candel* (L.) Druce.) in saline environments. *Botany Bulletin of Academy Sinica* 36, 19-24.

Chiu, C.Y., Lee, S.C., Juang, H.T., Hur, M.T., Hwang, Y.H., 1996. Nitrogen nutritional status and fate of applied N in mangrove soils. *Botanical Bulletin of Academia Sinica* 37, 191-196.

Chu, H.Y., Tam, N.F.Y., Lam, S.K.S., Wong, Y.S., 2000. Retention of pollutants by mangrove soil and the effects of pollutants on *Kandelia candel*. *Environmental Technology* 21, 755-764.

Clough, B.F., Boto, K.G., Attiwill, P.M., 1983. Mangroves and sewage: a re-evaluation. In: Teas, H. (Ed.), *Tasks for Vegetation Science*, vol. 8. Dr. W. Junk Publishers, The Hague, pp. 151-161.

Comans, R.N.J., Van Dijk, C.P.J., 1988. Role of complexation processes in cadmium mobilization during estuarine mixing. *Nature* 336, 151-154.

Corredor, J.E., Morell, J.M., 1994. Nitrate depuration of secondary sewage effluents in mangrove sediments. *Estuaries* 17, 295-300.

Corredor, J.E., Morell, J.M., Del Castillo, C.E., 1990. Persistence of spilled crude oil in a tropical intertidal environment. *Marine Pollution Bulletin* 21, 385-388.

Cowardin, L.M., Carter, V., Golet, E.C., LaRoe, E.T., 1979. Classification of wetlands and deepwater habitats of the United States. US Fish and Wildlife Service FWS/OBS 79/31. 103 pp.

Cooper, P.F., Job, G.D., Green, M.B., Shutes, R.B.E., 1996. Reed Beds and Constructed Wetlands for Wastewater Treatment. WRC Publications, Wiltshire, 184 pp.

Cronk, J.K., Fennessy, M.S., 2001. *Wetland Plants: Biology and Ecology*. CRC Press/Lewis Publishers, Boca Raton, FL. 440 pp.

Cunningham, S.D., Anderson, T.A., Schwab, A.P., Hsu, F.C., 1996. Phytoremediation of soils contaminated with organic pollutants. *Advances in Agronomy* 56, 55-114.

Davis, S.E., Childers, D. L., Day Jr., J.V., Rudnick, D.T., Sklar, F.H., 2001. Wetland-water column exchanges of carbon, nitrogen, and phosphorus in a Southern Everglades dwarf mangrove. *Estuaries* 24, 610-622.

Day, J.W., Conner, W.H., Leylou F., Day, R.H., Navarro, A.M., 1987. The productivity and composition of mangrove forests, Laguna De Terminos, Mexico. *Aquatic Botany* 27, 267-284.

DeBustamante, I., 1990. Land application: its effectiveness in purification of urban and industrial wastewaters in La Mancha, Spain. *Environmental Geology and Water Science* 16, 179-185 .

de Souza, A.S., Torres, J.P.M., Meire, R.O., Neves, R.C., Couri, M.S., Serejo, C.S., 2008. Organochlorine pesticides (OCs) and polychlorinated biphenyls (PCBs) in sediments and crabs (Chasmagnathus granulata, Dana, 1851) from mangroves of Guanabara Bay, Rio de Janeiro State, Brazil. *Chemosphere* 73, S186-S192.

Díaz, M.P., Grigson, S.J.W., Peppiatt, C.J., Burgess, J.G., 2000. Isolation and characterization of novel hydrocarbon-degrading Euryhaline consortia from crude oil and mangrove sediments. *Marine Biotechnology* 2, 522-532.

Doyle, M.O., Otte, M.L., 1997. Organism-induced accumulation of iron, zinc and arsenic in wetland soils. *Environmental Pollution* 96, 1-11.

Duke, N.C., Pinzón, Z.S., Prada, M.C., 1997. Large scale damage to mangrove forests following two large oil spills in Panama. *Biotropica* 29, 2-14.

Duke, N.C., Ellison, J.C., Burns, K.A., 1998. Surveys of oil spill incidents affecting mangrove habitat in Australia: a preliminary assessment of incidents, impacts on mangroves, and recovery of deforested areas. In: Beck, C. (Ed.), *APPEA Journal, Australian Petroleum Production and Exploration Association Canberra*, pp. 646-654.

Dunbabin, S., Bowmer, K.H., 1992. Potential use of constructed wetlands for treatment of Industrial wastewaters containing metals. *The Science of the Total Environment 111*, 151-168.

El-Tarabily, K.A., 2002. Total microbial activity and microbial composition of a mangrove sediment are reduced by oil pollution at a site in the Arabian Gulf. *Canadian Journal of Microbiology* 48, 176-172.

Essien, J.P., Antai, S.P., Olajire, A.A., 2009. Distribution, seasonal variations and ecotoxicological significance of heavy metals in sediments of cross river estuary mangrove swamp. *Water, Air, and Soil Pollution* 197, 91-105.

Flores-Mireles, A.L., Winans, S.C., Holguin G., 2007. Molecular characterization of diazotrophic and denitrifying bacteria associated with mangrove roots. *Applied and Environmental Microbiology* 73, 7308-7321.

Frick, C.M., Farrell, R.E., Germida, J.J., 1999. Assessment of Phytoremediation as an In-Situ Technique for Cleaning Oil-Contaminated Sites, Petroleum Technology Alliance Canada, Calgary.

Gale, P.M., Reddy, K.R., Graetz, D.A., 1993. Nitrogen removal from reclaimed water applied to constructed and natural wetland microcosms. *Water Environment Research* 65, 162-168.

Gambrell R.P., Delaune, R.D., Patrick, W.H., 1991. Redox processes in soils following oxygen depletion. In: Jackson, M.B., Davies, D.D., Lambers, H. (Eds.), *Plant Life under Oxygen Deprivation: Ecology, Physiology and Biochemistry*, SPB Academic Publishing, The Hague, pp.101-117.

Garrity, S.D., Levings, S.C., Burns, K.A., 1994. The Galeta oil spill. I. Long-term effects on the physical structure of the mangrove fringe. *Estuarine, Coastal and Shelf Science* 38, 327-348.

Gomes, N.C.M., Borges, L.R., Paranhos, R., Pinto, F.N., Mendonca-Hagler, L.C.S., Smalla, K., 2008. Exploring the diversity of bacterial communities in sediments of urban mangrove forests. *FEMS Microbiology Ecology* 66, 96-109.

Guzzella, L., Roscioli, C., Viganò, L., Saha, M., Sarkar, S. K., Bhattacharya, A., 2005. Evaluation of the concentration of HCH, DDT, HCB, PCB and PAH in the sediments along the lower stretch of Hugli estuary, West Bengal, northeast India. *Environment International,* 31, 523-534.

Hammer, D.A., Bastian, R.K., 1989. Wetland ecosystems: Natural water purifiers? In: *Constructed wetlands for wastewater treatment.* D.A. Hammer, (Ed.), Lewis publishers, Chelsea, Michigan, pp. 5-20.

Harbison, P., 1986. Mangrove muds – a sink and a source for trace metals. *Marine Pollution Bulletin* 17, 246-250.

Holguin, G., Bashan, Y., 1996. Nitrogen-fixation by *Azospirillum brasilense* Cd is promoted when co-cultured with a mangrove rhizosphere bacterium *Staphylococcus* sp. *Soil Biology and Biochemistry* 28, 1651-1660.

Horne, A.J., Fleming-Singer, M., 2005. Phytoremediation using constructed treatment wetlands: an overview. In: Fingerman, M., Nagabhushanam, R. (Eds.), *Bioremediation of Aquatic and Terrestrial Ecosystems*, Science Publishers, Enfield, USA, pp. 329-377.

Jackson, M.B., Colmer, T.D., 2005. Response and adaptation by plants to flooding stress. *Annals of Botany* 96, 501-505.

Kadlec, J. A., 1987. Nutrient dynamics in wetlands. In: Reddy, K.R. and Smith, W.H. (Eds.), *Aquatic Plants for Water Treatment and Resource Recovery.* Magnolia Publishing Inc., Florida, pp. 393-413.

Kathiresan, K., Selyam, M.M., 2006. Evaluation of beneficial bacteria from mangrove soil. *Botanica Marina* 49, 86-88.

Ke, L., Wang, W.Q., Wong, T.W.Y., Wong, Y.S., Tam, N.F.Y., 2003a. Potential of mangrove microcosms to remove pyrene from contaminated sediments. *Chemosphere* 51, 25-34.

Ke, L., Wong T.W.Y., Wong, A.H.Y., Wong, Y.S., Tam, N.F.Y., 2003b. Negative effects of humic acid addition on phytoremediation of pyrene-contaminated mangrove wetland. *Chemosphere* 52, 1581-1591.

Ke, L., Wong, Y.S., Tam, N.F.Y., 2004. Determination of polycyclic aromatic hydrocarbons in mangrove sediments: comparison of two internal standard surrogate methods and quality-control procedures. *International Journal of Environmental Analytical Chemistry* 84, 661-675.

Ke, L., Yu, K.S.H., Wong, Y.S., Tam, N.F.Y., 2005. Spatial and vertical distribution of polycyclic aromatic hydrocarbons (PAHs) in mangrove sediments. *Science of the Total Environment* 340, 177-187.

Ke, L., Zhang, C., Wong, Y.S., Tam, N.F.Y., 2010. Dose and accumulative effects of spent lubricating oil on mangrove seedlings. In preparation.

King, R.F., Royle, A., Putwain, P.D., Dickinson, N.M., 2006. Changing contaminant mobility in a dredged canal sediment during a three-year phytoremediation trial, *Environmental Pollution* 143, 318-326.

Kirk, G.J.D., Kronzucker, H.J., 2005. The potential for nitrification and nitrate uptake in the rhizosphere of wetland plants: A modelling study. *Annals of Botany* 96, 639-646.

Klekowski, E.J., Corredor, J.E., Lowenfeld, R., Klekowski, E.H., Morell, J.M., 1994. Using mangroves to screen for mutagens in tropical marine environments. *Marine Pollution Bulletin* 28, 346-350.

Knight, R.L., Kadlec, R.H., Reed, S., 1992. Wetland treatment database. In: Proceedings of Water Environment Federation: 65th Annual Conference and Exposition, No. C2009. Water Environment Federation, Alexandria, VA, pp. 25-35.

Koch, M.S., 1997. *Rhizophora mangle* L. seedling development into the sapling stage across resource and stress gradients in subtropical Florida. *Biotropica* 29, 427-439.

Kothamasi, D., Kothamasi, S., Bhattacharyya, A., Kuhad, R.C., Babu, C.R., 2006. Arbuscular mycorrhizae and phosphate solubilising bacteria of the rhizosphere of the mangrove ecosystem of Great Nicobar island, India. *Biology and Fertility of Soils* 42, 358-361.

Lacerda, L.D., Carvalho, C.E.V., Tanizaki, K.F., Ovalle, A.R.C., Rezende, C.E., 1993. The biogeochemistry and trace metals distribution of mangrove rhizospheres. *Biotropica* 25, 252-257.

Lee, S.Y., 1990. Primary productivity and particulate organic matter flow in an estuarine mangrove-wetland in Hong Kong. *Marine Biology* 106, 453-463.

Liu, Y., Chen, G.Z., Wong, Y.S., Tam, F.Y., 1995. Study on the population structure and pollution ecology of algae in mangrove area in Futian, Shenzhen. *China Environmental Science* 15, 171-176.

Long, E.R., MacDonald, D.D., Smith, S.L., Calder, F.D., 1995. Incidence of adverse biological effects within ranges of chemical concentrations in marine and estuarine sediments. *Environmental Management* 19, 81-97.

Lu, H., Yan, C., Liu, J., 2007. Low-molecular-weight organic acids exuded by mangrove (*Kandelia candel* (L.) Druce) roots and their effect on cadmium species change in the rhizosphere. *Environmental and Experimental Botany* 61, 159-166.

Lugo, A.E., 1980. Mangrove ecosystems-successional or steady-state. *Biotropica* 12, 65-72.

Lugo, A.E., Snedaker, S.C., 1974. The ecology of mangroves. *Annual Review of Ecology and Systematics* 5, 39-64.

McDonald, M.P., Galwey, N.W., Colmer, T.D., 2002. Similarity and diversity in adventitious root anatomy as related to root aeration among a range of wetland and dryland grass species. *Plant, Cell and Environment* 25, 441-451.

MacFarlane, G.R., 2002. Leaf biochemical parameters in *Avicennia marina* (Forsk.) Vierh as potential biomarkers of heavy metal stress in estuarine ecosystems. *Marine Pollution Bulletin* 44, 244-256.

MacFarlane, G.R., Burchett, M.D., 2002. Toxicity, growth and accumulation relationships of copper, lead and zinc in the grey mangrove *Avicennia marina* (Forsk.) *Vierh. Mar. Environ. Res*. 54, 65–84.

Machado, W., Gueiros, B., Lisboa-Filho, S.D., Lacerd, L.D., 2005. Trace metals in mangrove seedlings: role of iron plaque formation. *Wetlands Ecology and Management* 13, 199-206.

Mansell, R.S., McKenna, P.J., Flag, E., Hall, M., 1985. Phosphate movement in columns of sandy soil from a wastewater irrigated site. *Soil Science* 140, 59-68.

Maskaoui, K., Zhou, J.L., Hong, H.S., Zhang, Z.L., 2002. Contamination of polycyclic aromatic hydrocarbons in the Jiulong River Estuary and Western Xiamen Sea, China. *Environmental Pollution* 118, 109-122.

Medeiros, P.M., Bícego, M.C., 2004a. Investigation of natural and anthropogenic hydrocarbon inputs in sediments using geochemical markers. I. Santos, SP-Brazil. *Marine Pollution Bulletin* 49, 761-769.

Medeiros, P.M., Bícego, M.C., 2004b. Investigation of natural and anthropogenic hydrocarbon inputs in sediments using geochemical markers II. São Sebastião, SP-Brazil. *Marine Pollution Bulletin* 49, 892-899.

Mendelssohn I.A., Kleiss B.A., Wakeley, J.S., 1995. Factors controlling the formation of oxidized root channels: a review. *Wetlands* 15, 37-46.

Meziane, T., Tsuchiya, M., 2002. Organic matter in a subtropical mangrove-estuary subjected to wastewater discharge: Origin and utilisation by two macrozoobenthic species. *Journal of Sea Research* 47, 1-11.

Mitsch, W.J., Gosselink, J.G., 2000. *Wetlands*, John Wiley and Sons, Inc., New York, pp. 920.

Mohanty, S.K., Dash, R.N., 1982. The chemistry of waterlogged soils. In: Wetlands – *Ecology and Management*. Natural Institute of Ecology and International Scientific Publications Jaipur India, pp. 389-396.

Neto, J.A.B., Gingele, F.X., Leipe, T., Brehme, I., 2006. Spatial distribution of heavy metals in surficial sediments from Guanabara Bay: Rio de Janeiro, Brazil. *Environmental Geology* 49, 1051-1063.

Ong Che, R.G., 1999. Concentration of 7 heavy metals in sediments and mangrove root samples from Mai Po, Hong Kong. *Marine Pollution Bulletin* 39, 269-279.

Otte, M.L., Rozema, J., Koster, L., Haarsma, M.S., Broekman, R.A., 1989. Iron plaque on roots of *Aster tripolium* L.: interaction with zinc uptake. *New Phytologist* 111, 309-317.

Paalman, M.A.A., Van der Weijden, C.H., Loch, J.P.G., 1994. Sorption of cadmium on suspended matter under estuarine conditions: competition and complexation with major sea-water ions. *Water, Air, and Soil Pollution* 73, 49-60.

Paquin, D., Ogoshi, R., Campbell, S., Li, Q.X., 2002. Bench-scale phytoremediation of polycyclic aromatic hydrocarbon-contaminated marine sediment with tropical plants, *International Journal of Pytoremediation* 4, 297-313.

Parida, A.K., Das, A.B., Mohanty, P., 2004. Defense potentials to NaCl in a mangrove, *Bruguiera parviflora*: Differential changes of isoforms of some antioxidative enzymes. *Journal of Plant Physiology* 161, 531-542.

Patrick Jr., W.H., Khalid, R.A., 1974. Phosphate release and sorption by soils and sediments: Effects of aerobic and anaerobic conditions. *Science* 186, 53-55.

Pedersen, O., Binzer, T., Borum, J., 2004. Sulphide intrusion in eelgrass (*Zostera marina* L.). *Plant, Cell and Environment* 27, 595-602.

Perelo, L.W., 2010. Review: In situ and bioremediation of organic pollutants in aquatic sediments. *Journal of Hazardous Materials* 177, 81-89.

Peters, E.C., Gassman, N.J., Firman, J.C., Richmond, R.H., Power, E.A., 1997. Ecotoxicology of tropical marine ecosystems. *Environmental Toxicology and Chemistry* 16, 12-40.

Pi, N., Tam, N.F.Y., Wu, Y., Wong, M.H., 2009. Root anatomy and spatial pattern of radial oxygen loss of eight true mangrove species. *Aquatic Botany* 90, 222-230.

Pi, N., Tam, N.F.Y., Wong, M.H., 2010. Effects of wastewater discharge on formation of Fe plaque on root surface and radial oxygen loss of mangrove roots. *Environmental Pollution* 158, 381-387.

Pitt, K.A., Connolly, R.M., Maxwell, P., 2009. Redistribution of sewage-nitrogen in estuarine food webs following sewage treatment upgrades. *Marine Pollution Bulletin* 58, 573-580.

Richardson, C.J., 1985. Mechanisms controlling phosphorus retention capacity in freshwater constructed wetlands. *Science* 228, 1424-1427.

Richardson, B.J., Lam, P.K.S., Wu, R.S.S., 2000. The coast of Hong Kong. In: Sheppard, C. (Ed.), Seas at the Millennium: An Environmental Evaluation. Elsevier Science, pp. 171-183.

Rivera-Monroy, V.H., Torres, L.A., Bahamon, N., Newmark, F., Twilley, R.R., 1999. The potential use of mangrove forests as nitrogen sinks of shrimp aquaculture pond effluents: the role of denitrification. *Journal of the World Aquaculture Society* 30, 12-25.

Robertson, A.I., Phillips, M.J., 1995. Mangroves as filters of shrimp pond effluent: predictions and biogeochemical research needs. *Hydrobiologia* 295, 311-321.

Salt, D.E., Smith, R.D., Raskin, I., 1998. Phytoremediation. *Annual Reviews on Plant Physiology and Plant Molecular Biology* 49, 643-668.

Schnoor, J.L., Licht, L.A., McCutcheon, S.C., Wolfe, N.L., Carreira, L.H., 1995. Phytoremediation of organic and nutrient contaminants. *Environmental Science and Technology* 29, 318A-323A.

Shimp, J.F., Tracy, J.C., Davis, L.C., Lee, E., Huang, W., Erickson, L.E., Schnoor, J.L., 1993. Beneficial effects of plants in the remediation of soil and groundwater contaminated with organic materials. *Critical Reviews in Environmental Science and Technology* 23, 41-77.

Silva, C.A.R., Lacerda, L.D., Rezende, C.E., 1990. Heavy metal reservoirs in a red mangrove forest. *Biotropica* 22, 339-345.

Simonich, S.L., Hites, R.A., 1994. Importance of vegetation in removing polycyclic aromatic hydrocarbons from the atmosphere. *Nature* 370, 49-51.

Smith, K.E., Schwab, A.P., Banks, M.K., 2007. Phytoremediation of polychlorinated biphenyl (PCB)-contaminated sediment: a greenhouse feasibility study. *Journal of Environmental Quality* 36, 239-244.

St-Cyr, L., Campbell, P.G.C., 1996. Metals (Fe, Mn, Zn) in the root plaque of submerged aquatic plants collected in situ: relations with metal concentrations in the adjacent sediments and in the root tissue. *Biogeochemistry* 33, 45-76.

Sundaravadivel, M., Vigneswaran, S., 2001. Constructed wetlands for wastewater treatment. *Critical Reviews in Environmental Science and Technology* 31, 351-409.

Sundby, B., Vale, C., Cacador, I., Catarino, F., Madureira, M.J., Caetano, M., 1998. Metal-rich concretions on the roots of salt marsh plants: mechanism and rate of formation. *Limnology and Oceanography* 43, 245-252.

Takemura, T., Hanagata, N., Sugihara, K., Baba, S., Karube, I., Dubinsky, Z., 2000. Physiological and biochemical responses to salt stress in the mangrove, *Bruguiera gymnorrhiza*. *Aquatic botany* 68, 15-28.

Tam, N.F.Y., 1998. Effects of wastewater discharge on microbial populations and enzyme activities in mangrove soils. *Environmental Pollution* 102, 233-242.

Tam, N.F.Y., 2006. Pollution studies on mangroves in Hong Kong and Mainland China. In: Wolanski, E. (Ed.) *The Environment in Asian Pacific Harbours*. Springer, Netherlands, pp. 147-163.

Tam, N.F.Y., Wong, Y.S., 1993. Retention of nutrients and heavy metals in mangrove sediment receiving wastewater of different strengths. *Environmental Technology* 14, 719-729.

Tam, N.F.Y., Wong, Y.S., 1995a. Mangrove soils as sinks for wastewater-borne pollutants. *Hydrobiologia* 295, 231-242.

Tam, N.F.Y., Wong, Y.S., 1995b. Spatial and temporal variations of heavy metal contamination in sediments of a mangrove swamp in Hong Kong. *Marine Pollution Bulletin* 31, 254-261.

Tam, N.F.Y., Wong, Y.S., 1996a. Retention and distribution of heavy metals in mangrove soils receiving wastewater. *Environmental Pollution* 94, 283-291.

Tam, N.F.Y., Wong, Y.S., 1996b. Retention of wastewater-borne nitrogen and phosphorus in mangrove soils. *Environmental Technology* 17, 851-859.

Tam, N.F.Y., Wong, Y.S., 1998. Variations of soil nutrient and organic matter content in a subtropical mangrove ecosystem. *Water, Air and Soil Pollution* 103, 245-261.

Tam, N.F.Y., Wong, Y.S., 1999a. Mangrove swamps as pollutant sink and wastewater treatment facility. In: Chua, T.E., Bermas, N., (Eds.), Challenges *and opportunities in managing pollution in the East Asian Seas*. Quezon City, Philippines, pp. 471-483.

Tam, N.F.Y., Wong, Y.S., 1999b. Mangrove soils in removing pollutants from municipal wastewater of different salinities. *Journal of Environmental Quality* 28, 556-564.

Tam, N.F.Y., Wong, Y.S., 2000a. *Hong Kong Mangroves*. City University of Hong Kong Press, Hong Kong, 148 pp.

Tam, N.F.Y., Wong, Y.S., 2000b. Spatial variation of heavy metals in surface sediments of Hong Kong mangrove swamps. *Environmental Pollution* 110, 195-205.

Tam, N.F.Y., Yao, M.W.Y., 2002. Concentrations of PCBs in coastal mangrove sediments of Hong Kong. *Marine Pollution Bulletin* 44, 642-651.

Tam, N.F.Y., Vrijmoed, L.L.P., Wong, Y.S., 1993. The chemical properties of soil and its association with litter and plant production in a sub-tropical mangrove community in Hong Kong. In Morton, B. (Ed.), The Marine Biology of Hong Kong and the South China Sea, *Proceedings of the 1st International Conference on the Marine Biology of Hong Kong and the South China Sea*. Hong Kong University Press, Hong Kong, pp. 521-541.

Tam, N.F.Y., Li, S.H., Lan, C.Y., Chen, G.Z., Li, M.S., Wong, Y.S., 1995. Nutrients and heavy metal contamination of plants and sediments in Futian mangrove forest. *Hydrobiologia* 295, 149-158.

Tam, N.F.Y., Ke, L., Wang, X.H., Wong, Y.S., 2001. Contamination of polycyclic aromatic hydrocarbons in surface sediments of mangrove swamps. *Environmental Pollution* 114, 255-263.

Tam, N.F.Y., Chu, I.H.Y., Wong, Y.S., 2002. Constructed mangrove tide tank systems as wastewater treatment facility. In: Kocasoy, G., Atabarut, T., Nuhoglu, I. (Eds.), *Appropriate Environmental and Solid Waste Management and Technologies for Developing Countries*, Vol. 3. Bogazici Universitesi, Istanbul, Turkey, pp. 1577-1584.

Tam, N.F.Y., Wong, Y.S., Wong, M.H., 2009. Novel technology in pollutant removal at source and bioremediation. Ocean and Coastal Management 52, 368-373.

Teas, H.J., Duerr, E.O., Wilcox, J.R., 1987, Effects of south Louisiana crude oil and dispersants on Rhizophora mangroves. *Marine Pollution Bulletin* 18, 122-124.

Thomas, C., Eong, O.J., 1984. Effect of heavy metals zinc and lead on *Rhizophora mucronata* LAM. and *Avicennia alba* Bl. seedlings. In: Soepadmo, E., Rao, A.N., MacIntosh, D.J. (Eds.), *Proceedings of the Asian Symposium on Mangrove Environment, Research and Management,* 25-29 August 1980. University of Malaya Press, Kuala Lumpur, pp. 568-574.

U.S. Environmental Protection Agency (USEPA), 1998. *A Citizen's Guide to Phytoremediation.* 6 pp. Report No. EPA 542-F-98-011. Washington, D.C. Office of Solid Waste and Emergency Response.

Venturini, N., Tommasi, L.R., 2004. Polycyclic aromatic hydrocarbons and changes in the trophic structure of polychaete assemblages in sediments of Todos os Santos Bay, Northeastern, Brazil. *Marine Pollution Bulletin* 48, 97-107.

Vervaeke, P., Luyssaert, S., Mertens, J., Meers, E., Tack, F.M.G., Lust, N., 2003. Phytoremediation prospects of willow stands on contaminated sediment: a field trial, *Environmental Pollution* 126, 275-282.

Visser, E.J.W., Colmer, T.D., Blom, C.W.P.M., Voesenek, L.A.C.J., 2000. Changes in growth, porosity, and radial oxygen loss from adventitious roots of selected mono- and dicotyledonous wetland species with contrasting types of aerenchyma. *Plant Cell and Environment* 23, 1237-1245.

Walsh, G.E., Ainsworth, K.A., Rigby, R., 1979. Resistance of red mangrove (*Rhizophora mangle* L.) seedlings to lead, cadmium and mercury. *Biotropica* 11, 22-27.

Wang, B.S., Zn, Q.J., Zhang, W.Y., Wang Y.J., 2003. Accumulation and cycle of heavy metals in *Sonneratia apetala* and *S. caseolaris* mangrove community at Futian of Shenzhen, China. *Marine Science Bulletin* 5, 59-68.

Weis, J.S., Weis, P., 2004. Metal uptake, transport and release by wetland plants: implications for phytoremediation and restoration. *Environment International* 30, 685-700.

Whigham, D.F., Verhoeven, J.T.A., Samarkin, V., Megonigal, P.J., 2009. Responses of *Avicennia germinans* (Black Mangrove) and the soil microbial community to nitrogen addition in a hypersaline wetland. *Estuaries and Coasts* 32, 926-936.

Wong, Y.S., Lan, C.Y., Chen, G.Z., Li, S.H., Chen, X.R., Liu, Z.P., Tam N.F.Y., 1995. Effect of wastewater discharge on nutrient contamination of mangrove soils and plants. *Hydrobiologia* 295, 243-254.

Wong, Y.S., Tam, N.F.Y., Chen, G.Z., Ma H., 1997a. Response of Aegiceras corniculatum to synthetic sewage under simulated tidal conditions. *Hydrobiologia* 352, 89-96.

Wong, Y.S., Tam, N.F.Y., Lan, C.Y., 1997b. Mangrove wetlands as wastewater treatment facility: a field trial. *Hydrobiologia* 352, 49-59.

Woodroffe, C.D., Bardsley, K.N., Ward, P.J., Hanley, J.R., 1988. *Production of mangrove litter in a macrotidal embayment*, Darwin Harbour, N.T., Australia. Estuarine, Coastal and Shelf Science 26, 581-598.

Wu, Y., Tam, N.F.Y., Wong, M.H., 2008. Effects of salinity on treatment of municipal wastewater by constructed mangrove wetland microcosms. *Marine Pollution Bulletin* 57, 727-734.

Yang, Q., Tam, N.F.Y., Wong, Y.S., Luan, T.G., Su, W.S., Lan, C.Y., Shin, P.K.S., Cheung, S.G., 2008. Potential use of mangroves as constructed wetland for municipal sewage treatment in Futian, Shenzhen, China. *Marine Pollution Bulletin* 57, 735-743.

Ye, Z., Bakera, A.J.M., Wong, M.H., Willis, A.J., 1998. Zinc, lead and cadmium accumulation and tolerance in *Typha latifolia* as affected by iron plaque on the root surface. *Aquatic Botany* 61, 55-67.

Ye, Y., Tam, N.F.Y., 2002. Growth and physiological responses of *Kandelia candel* and Bruguiera gymnorrhiza to livestock wastewater. *Hydrobiologia* 479, 75-81.

Ye, Y., Tam, N.F.Y., 2007. Effects of used lubricating oil on two mangroves *Aegiceras corniculatum* and *Avicennia marina*. *Journal of Environmental Sciences* 19, 1355-1360.

Ye, Y., Tam, N.F.Y., Wong, Y.S., 2001. Livestock wastewater treatment by a mangrove pot-cultivation system and the effect of salinity on the nutrient removal efficiency. *Marine Pollution Bulletin* 42, 512-520.

Ye, Y., Tam, N.F.Y., Wong, Y.S., Lu, C.Y., 2003. Growth and physiological responses of two mangrove species (*Bruguiera gymnorrhiza* and *Kandelia candel*) to waterlogging. *Environmental and Experimental Botany* 49, 209-221.

Ye, Y., Tam, N.F.Y., Lu, C.Y., Wong, Y.S., 2005. Effects of salinity on germination, seedling growth and physiology of three salt-secreting mangrove species. *Aquatic Botany* 83, 193-205.

Youssef T., Saenger, P., 1996. Anatomical adaptive strategies to flooding and rhizosphere oxidation in mangrove seedlings. *Australian Journal of Botany* 44, 297-313.

Yu, R.Q., Chen, G.Z., Wong, Y.S., Tam, N.F.Y., Lan, C.Y., 1997. Benthic macrofauna of the mangrove swamp treated with municipal wastewater. *Hydrobiologia* 347, 127-137.

Yu, S.H., Ke, L., Wong, Y.S., Tam, N.F.Y., 2005. Degradation of polycyclic aromatic hydrocarbons (PAHs) by a bacterial consortium enriched from mangrove sediments. *Environment International* 31, 149-154.

Zhang, X.K., Zhang, F.S., Mao, D.R., 1998. Effect of iron plaque outside roots on nutrient uptake by rice (*Oryza sativa* L.). Zinc uptake by Fe-deficient rice. *Plant and Soil* 202, 33-39.

Zhang, J., Cai, L., Yuan, D., Chen, M., 2004. Distribution and sources of polynuclear aromatic hydrocarbons in mangrove surficial sediments of Deep Bay, China. *Marine Pollution Bulletin* 49, 479-486.

Zhang, C.G., Leung, K.K., Wong, Y.S., Tam, N.F.Y., 2007a. Germination, growth and physiological responses of mangrove plant (*Bruguiera gymnorrhiza*) to lubricating oil pollution. *Environmental and Experimental Botany* 60, 127-136.

Zhang, F.Q., Wang, Y.S., Lou, Z.P., Dong J.D., 2007b. Effect of heavy metal stress on antioxidative enzymes and lipid peroxidation in leaves and roots of two mangrove plant seedlings (*Kandelia candel* and *Bruguiera gymnorrhiza). Chemosphere* 67, 44-50.

Zhang, Y., Dong, J. Yang, Z., Zhang, S., Wang, Y., 2008. Phylogenetic diversity of nitrogen-fixing bacteria in mangrove sediments assessed by PCR–denaturing gradient gel electrophoresis. *Archives of Microbiology* 190, 19-28.

Zheng, W.J., Zhang, F.Z., Lian, Y.W., Lin, P., 1996. Accumulation and dynamics of Cu, Pb, Zn and Mn elements in *Kandelia candel* (L.) Druce mangrove community of Jiulong River Estuary of Fujian. *Acta Botanica Sinica* 38, 227-233.

Zhou, H.W., Luan, T.G., Zou, F., Tam, N.F.Y., 2008. Different bacterial groups for biodegradation of three- and four-ring PAHs isolated from a Hong Kong mangrove sediment. *Journal of Hazardous Materials* 152, 1179-1185.

Zhou, H.W., Wong, A.H.Y., Yu, R.M.K., Park, Y.D., Wong, Y.S., Tam, N.F.Y., 2009. Polycyclic aromatic hydrocarbon-induced structural shift of bacterial communities in mangrove sediment. *Microbial Ecology* 58, 153-160.

In: Handbook of Phytoremediation
Editor: Ivan A. Golubev

ISBN: 978-1-61728-753-4

Chapter 13

USE OF LEGUME-MICROBE SYMBIOSES FOR PHYTOREMEDIATION OF HEAVY METAL POLLUTED SOILS: ADVANTAGES AND POTENTIAL PROBLEMS

***V. I. Safronova*[1]*, G. Piluzza*[2]*, S. Bullitta*[2] *and A. A. Belimov*[1]**

[1]All-Russia Research Institute for Agricultural Microbiology, Podbelskogo Sh., 3, Pushkin-8, 196608, St.-Petersburg, Russian Federation.

[2]ISPAAM-CNR u.o.s. Sassari, Traversa La Crucca 3, Località Baldinca, 07100 Li Punti-Sassari, Italy

ABSTRACT

There is evidence that many legume species of the flowering plant family *Fabaceae* may be efficiently used in phytoremediation of heavy metal polluted soils, particularly for revegetation and phytostabilization of mine soils. For such purposes, a number of legume species were used and this chapter gives an updated glimpse on scientific experiences dealing with microbial effects on several legume species growing in heavy metal polluted soils. Legume species are able to form symbiosis with various beneficial microorganisms, such as nitrogen-fixing nodule bacteria, arbuscular mycorrhizal fungi and plant growth-promoting bacteria. Such plant microbe associations have implications in plant growth, nutrition and disease control. The symbioses between legumes and microorganisms provide nutrients for the plant, stimulate plant growth, exert antistress effects on plants, improve soil fertility, and restore ecosystem biodiversity and functions. This makes legumes very tempting subjects for phytoremediation purposes, particularly for the development of ecologically friendly phytostabilization technologies, since many of HM polluted soils are characterized by low nutrients and degenerated biocenosis. Moreover, symbiotrophic microorganisms possess a number of mechanisms which may be involved in improving tolerance of plants to environmental stresses, including those caused by heavy metals. As a consequence, the use of legume species for phytoremediation purposes should be considered in the context of their interactions with symbiothrophic microorganisms. Several plant species from the family *Fabaceae* and their performances in combination with microorganisms on heavy metal polluted soils or hydroponics are reported in this chapter. Particular attention is drawn on the effects of symbiotrophic microorganisms on legumes in the presence of heavy metals in conditions of monoinoculation and in combined inoculations. Intraspecific variability of plant species in their interactions with microorganisms is also discussed as well as the perspectives for phytoremediation with genetically modified legumes and symbiotrophic microorganisms. Successful attempts to increase tolerance to and accumulation of HMs by legume plants via genetic modifications and selection are mentioned. Although the presence of literature reports on the use of legume plants for phytoremediation purposes, it is undoubtedly wise to state that their potential for phytoremediation has not yet been adequately explored. Aim of this chapter is the discussion of advantages and problems in the application of legume-microbe systems for restoration and phytoremediation of polluted soils.

INTRODUCTION

It is a commonly accepted opinion that for efficient phytoremediation of heavy metal (HM) polluted sites it is essential to use plants having high biomass and fast growth rate, increased metal tolerance and metal accumulating capabilities and easily cultivable and harvestable. Most of the commonly known plants recommended for phytoremediation belong to the family Brassicaceae, because a number of cruciferous species are HM tolerant metallophytes and hyperaccumulators. However the growth and metal uptake may be significantly inhibited in extremely polluted sites even for tolerant species. Metal hyperaccumulating species have small biomass and growth rate, and their introduction in view of endemicity, and harvest is complicated. On the other hand, the species having high biomass production and easily cultivated, such as agricultural crops, as a rule are less tolerant to HMs compared to hyperaccumulators or metallophytes.

The family Fabaceae is one of the largest families of flowering plants and combines about 20000 species of 674 genera (Allen and Allen, 1980). Although many legume species are less tolerant to HMs as compared to cruciferous, cereals and grasses, they can produce high biomass, have fast growth rate and hence they possess rather high metal accumulating capability. Legumes are widely used as agricultural crops in a large scale of climate and soil conditions. A feature of the family Fabaceae is the ability to form nitrogen-fixing symbiosis with nodule bacteria of the order Rhizobiales, resulting in symbiotrophic nitrogen nutrition. The legumes also form obligate symbiosis with arbuscular mycorrhizal fungi (AMF), which mainly supply the plant with phosphorus, and associative symbiosis with plant growth-promoting rhizobacteria (PGPR) and endophytic microorganisms exerting multiple effects on plant growth, nutrition and disease control. An advanced symbiotrophic potential of legumes is of the utmost significance for improvement of soil fertility, biodiversity and activity of soil biota, soil genesis and hence for maintenance and restoration of healthy ecosystems. There is evidence that legumes may be efficiently used in phytoremediation of HM pollutes sites, particularly for revegetation and phytostabilization of mine soils. For these purposes a number of legume species such as *Anthyllis vulneraria* (Frerot *et al.*, 2006), *Coronilla varia* (Evanylo *et al.*, 2005), Lotus corniculatus (McGrath, 1998), Lupinus albus (Vazquez *et al.*, 2006), *Trifolium repens* (Bidar *et al.*, 2009) and *Vicia faba* (Pichtel and Bradway, 2008) were used. Although the results of these attempts are encouraging, it is proposed that up to date the phytoremediation potential of legume plants has not been adequately explored (Vamerali *et al.*, 2009; Sinha *et al.*, 2007).

On the one hand, the ability of legumes to form plant-microbe symbioses suggests that their growth and nutrition significantly depends on interactions with beneficial microorganisms. On the other hand, symbiotrophic microorganisms possess a number of mechanisms, which may be involved in improving tolerance of plants to environmental stresses, including those caused by HMs, and may play an important role for improving phytoremediation technologies (Kamaludeen and Ramasamy, 2008; Rajkumar *et al.*, 2009; Gamalero *et al.*, 2009; Giasson *et al.*, 2008; Göhre and Paszkowski, 2006; Gadd, 2004; Wenzel, 2009). Therefore the use of legume plants for agriculture and for remediation technologies should be considered in the context of their interactions with symbiotrophic microorganisms. The aim of this chapter is to discuss possibilities for application of legume-microbe systems as a biological tool for phytoremediation of HM polluted soils and restoration of healthy ecosystems.

Effects of Symbiotrophic Microorganisms on Legumes in the Presence of HMs

Mono-Inoculations with Different Types of Microorganisms

It is well documented that arbuscular mycorrhizal fungi (AMF) play an important role in tolerance to and accumulation of HMs by plants grown in polluted soils (Giasson *et al.*, 2008; Göhre and Paszkowski, 2006; Khan, 2006). This makes symbiosis between plants and AMF an ecologically safe and efficient biological instrument for the improvement of different phytoremediation processes, particularly phytostabilization, revegetation and restoration of healthy ecosystems. The described effects of inoculations solely with AMF on legume plants are outlined in Table 1. In most cases the mycorrhized plants showed better growth and uptake of phosphorus and other nutrients, suggesting increased HM tolerance and buffered HM-induced stress in different plant genera and species. More specific effects such as maintenance of high photosynthetic activity (Rivera-Becerril, *et al.*, 2002) or reduced free proline accumulation (Andrade *et al.*, 2009) were also observed. The important observation was that inoculation with AMF resulted in stimulation of nodule formation by native symbiotic nitrogen fixers (Andrade *et al.*, 2004; Andrade *et al.*, 2009; Lin *et al.*, 2007).

Case studies of the effects of nodule bacteria on legumes grown in polluted soils clearly showed significant positive effects of inoculations on plant growth (Table 2). Stimulation of nodulation frequency and increased biomass of nodules were described for several plants such as *Cicer arietinum* (Wani *et al.*, 2008c; Gupta *et al.*, 2004), *Lens culinaris* (Wani *et al.*, 2008b) and *Pisum sativum* (Wani *et al.*, 2008a). The higher seed and shoot N content (Wani *et al.*, 2008a; Wani *et al.*, 2008b; Jian *et al.*, 2009) and leghemoglobin content in nodules of the inoculated plants (Wani *et al.*, 2008b) supported that formation and function of nitrogen-fixing symbiosis in the presence of HMs was improved by the introduced rhizobia. Alleviation of oxidative stress by the inoculated *Prosopis juliflora* plants grown in multi metal polluted fly ash was also described (Sinha *et al.*, 2005).

Table 1. Effects of mono-inoculations with AMF on legume plants grown in HM polluted soils

Plant	Mycorrhizal fungi	Conditions	HMs	Microbial effects on plants	Reference
Anthyllis cytisoides	*Glomus macrocarpum, Glomus mosseae*	G, MAS	Pb, Zn	Increased shoot growth. *G. macrocarpum* increased, but *G. mosseae* decreased shoot Pb and Zn content.	Diaz *et al.*, 1996
Astragalus sinicus	*Glomus mosseae, Glomus intraradices*	G, MS	Cd	Increased shoot growth, root and shoot Cd content.	Li *et al.*, 2009
Canavalia ensiformis	*Glomus etunicatum*	G, MAS	Zn	Increased shoot growth, nodulation frequency. Decreased shoot Zn content.	Andrade *et al.*, 2009
Glycine max	*Glomus macrocarpum*	G, MAS	Pb	Increased P accumulation, nodulation frequency. Decreased shoot Pb content.	Andrade *et al.*, 2004
Leucaena leucocephala	*Glomus spp.*	G, MT	Pb, Zn	Increased shoot growth, accumulation of N, P and K. Reduced mobility of Pb and Zn in soil.	Ma *et al.*, 2006
Medicago truncatula	*Glomus intraradices*	G, MPS	Cd, Zn	Increased shoot growth, shoot Cd and Zn uptake. Decreased shoot Cd content.	Redon *et al.*, 2009
Pisum sativum	*Glomus intraradices*	G, MAS	Cd	Increased shoot growth, shoot Cd content. Stimulation of photosystesis. Decreased root Cd content.	Rivera-Becerril, *et al.*, 2002
Pisum sativum	*Glomus intraradices*	G, MPS	Cd	Increased shoot growth, seed yield, seed Cd content.	Engqvist *et al.*, 2006
Sesbania rostrata, S. cannabina, Medicago sativa	*Glomus mosseae*	G, MPS	Cu, Zn	Increased shoot growth, accumulation of N and P, nodulation frequency. Reduced translocation of Cu and Zn from root to shoot.	Lin *et al.*, 2007
Trifolium pratense	*Glomus mosseae*	G, MAS	Zn	Decreased shoot Zn content. Increased pH and reduced Zn mobility in soil.	Li and Christie, 2001
Trifolium pratense	*Glomus mosseae*	G, MAS	Pb	Increased shoot and root growth, N and P uptake, nodule number and AMF infection, shoot Pb content.	Vivas *et al.*, 2003a
Trifolium repens	*Glomus mosseae*	G, MPS	Fe, Cd, Pb, Zn	Increased shoot and root growth. Increased shoot P, K, Fe, B, Mo, Al, Cd, Zn, Cu,Cr, Mn and Ni content.	Azcon *et al.*, 2006
Trifolium repens	*Glomus mosseae*	G, MAS	Cd	Increased shoot and root growth, nodulation frequency, shoot N, P and Cd content.	Vivas *et al.*, 2003b

Table 1. (Continued)

Plant	Mycorrhizal fungi	Conditions	HMs	Microbial effects on plants	Reference
Trifolium repens	*Glomus mosseae* or indigenous strains	G, MAS	Cd	Increased root and shoot growth, nodulation frequency, shoot P content. Decreased shoot Fe, Zn, Mn, Cu, Ni and Mo content.	Vivas *et al.*, 2005
Trifolium repens	*Brevibacillus brevis*	G, MAS	Zn	Increased shoot growth, N and P accumulation, nodule number and AMF infection. Decreased shoot Zn content.	Vivas *et al.*, 2006
Trifolium subterraneum	*Glomus mosseae*	G, MAS	Cd	Decreased shoot Cd content, increased root Cd content. Immobilization of Cd in fungal hyphaes.	Joner and Leyval, 1997

Abbreviations: G, greenhouse; F, field; FA, fly-ash; MT, Mine tailings; MAS, metal amended soil; MPS, metal polluted soil.

Table 2. Effects of mono-inoculations with nodule bacteria on legume plants grown in HM polluted soils

Plant	Nodule bacteria	Conditions	HMs	Microbial effects on plants	Reference
Cicer arietinum	*Mesorhizobium sp.*	G, MAS	Cr	Increased shoot growth, seed yield, grain protein, nodule number and biomass, root and shoot N content. Decreased shoot Cr content.	Wani *et al.*, 2008c
Cicer arietinum	*Rizobium sp.*	G, FA	Zn, Cu, Cr, Cd, Fe	Increased root and shoot growth, seed biomass, nodulation frequency, shoot HM contents.	Gupta *et al.*, 2004
Lens culinaris	*Rhizobium leguminosarum*	G, MAS	Zn	Increased shoot growth, seed yield, grain protein, nodule number and biomass, leghemoglobin content in nodules. Decreased shoot Zn content.	Wani *et al.*, 2008b
Lotus edulis	*Mesorhizobium loti*	G, MPS	Cd, Pb, Zn	Increased nodule number, shoot Ca and Mg content.	Safronova *et al.*, 2010
Lotus ornithopodioides	*Mesorhizobium loti*	G, MPS	Cd, Pb, Zn	Increased nodule number. Decreased shoot K, Ca and Cu content, Zn translocation factor.	Safronova *et al.*, 2010

Table 2. (Continued)

Plant	Nodule bacteria	Conditions	HMs	Microbial effects on plants	Reference
Lupinus albus, L. luteus	*Bradyrhizobium* sp., *Ochrobactrum* sp.	G, MPS	Cd, Cu, Pb, Zn	Increased shoot growth. Decreased shoot Cd, Cu and Pb content.	Pajuelo *et al.*, 2008
Medicago ciliaris	*Sinorhizobium* sp.	G, MPS	Cd, Pb, Zn	Decreased shoot Cd and Cu content.	Safronova *et al.*, 2010
Mimosa pudica	*Cupriavidus taiwanensis*	G, MPS	Pb, Cu, Cd	Increased shoot growth and shoot HM content and uptake.	Chen *et al.*, 2008
Pisum sativum	*Rhizobium sp.*	G, MAS	Ni, Zn	Increased shoot growth, nodule numbers, root and shoot N, seed yield, leghemoglobin content in nodules, grain protein. Decreased shoot Ni and Zn content.	Wani *et al.*, 2008a
Pisum sativum	*Rhizobium leguminosarum*	G, MPS	Cd	Increased seed P content.	Engqvist *et al.*, 2006
Prosopis juliflora	*Rhizobium sp.*	F, FA	Fe, Mn, Cu, Zn, Cr	Increased plant biomass and shoot HM contents.	Rai *et al.*, 2004
Prosopis juliflora	*Rhizobium sp.*	F, FA	Fe, Mn, Cu, Zn, Cr	Increased plant biomass, content of photosynthetic pigments, protein content, accumulation of HMs. Alleviation of oxidative stress.	Sinha *et al.*, 2005
Sesbania cannabina, S. grandiflora, S. rostrata, and S. sesban	*Azorhizobium caulinodans*	G, MT	Pb, Zn	Increased plant growth.	Chan *et al.*, 2003
Sesbania rostrata	*Azorhizobium caulinodans*	G, MAS	Pb, Zn	Increased plant height, stem basal diameter, biomass, leaf chlorophyll content, shoot N content and accumulation.	Jian *et al.*, 2009
Vigna radiata	*Ochromobactrum intermedium*	H, MAS	Cr	Increased shoot growth (in hydroponics only). Lowered Cr toxicity by reduction of Cr(VI) to Cr(III). Decreased shoot Cr content.	Faisal and Hasnain, 2006

Abbreviations: G, greenhouse; F, field; FA, fly-ash; H, hydroponics; MT, Mine tailings; MAS, metal amended soil; MPS, metal polluted soil.

The role of PGPR in tolerance of plants to HMs and in microbial assisted phytoremediation of polluted soils has been recently reviewed by several authors (Jing *et al.*,

2007; Kamaludeen and Ramasamy, 2008; Khan *et al.*, 2009; Rajkumar *et al.*, 2009; Saleem *et al.*, 2007). It was concluded that a number of PGPR activities may counteract negative effects of HMs on plant growth and nutrition through various mechanisms. Some of these growth-promoting mechanisms are more universal and may be involved in plant growth promotion under various environmental conditions while others are more specific in relation to plant-metal interactions. Many PGPR stimulate plant growth due to production of phytohormones (auxins, cytokinins, gibberellins), and may mitigate disturbances in the hormonal status of plants caused by HMs. Inhibition of plant nutrient uptake induced by HM toxicity may be alleviated through microbially-mediated biogeochemical processes such as biological nitrogen fixation, bacterial phosphate solubilization or siderophore production, and through specific effects on nutrient uptake and transport systems in plants. Being inhabitants of the rhizosphere, PGPR may reduce HM solubility and modify speciation in the root zone via production of metal binding substances, sorption to microbial cell walls and exopolymers, intercellular sequestration and precipitation, and for some HM reductive precipitation (Gadd, 2004, Wenzel, 2009). However several microbial processes may enhance mobilization of HM and hence increase their phytoavailability and toxicity. Mobilization of HMs may be mediated by bacterial siderophores and other chelating substances, degradation of plant and soil metal binding compounds, and acidification of the rhizosphere as a result of bacterial metabolism (Gadd, 2004, Wenzel, 2009). The reports describing response of legume plants to PGPR in the presence of elevated HM concentrations are outlined in Table 3. In all cases, along with plant growth promotion, the bacteria decreased HM contents in the inoculated plants. This suggested that metal mobilization processes governing by PGPR in the rhizosphere of these plant species were of little importance.

Some PGPR contain enzyme 1-aminocyclopropane-1-carboxylate (ACC) deaminase and may possess a peculiar anti-stress activity through lowering the HM induced evolution of phytohormone ethylene that inhibits plant growth (Gerhardt *et al.*, 2006; Arshad *et al.*, 2007). However little is known about the role of bacterial ACC deaminase in the response of legumes to elevated HM concentrations. Occurrence of this enzyme in PGPR strains listed in Table 3 was not studied, except three ACC-utilizing strains *Pseudomonas brassicacearum*, *P. marginalis* and *Rhodococcus* sp., which were used for inoculation of *Pisum sativum* (Table 3). The important observation was that *Rhodococcus* sp. only had no ACC deaminase activity in vitro in the presence of Cd and lost its ability to stimulate plant growth in Cd-supplemented soil (Safronova *et al.*, 2006). In another study the same strain *P. brassicacearum* had no effect on pea growth, but increased seed Cd content (Engqvist *et al.*, 2006).

We have found no growth promoting effects of ACC utilizing *Variovorax paradoxus* on *Lotus edulis*, *L. ornithopodioides* and *Medicago ciliaris* grown in HM polluted mine waste (Safronova *et al.*, 2010), however the bacteria changed element composition of the inoculated plants (see Table 3).

In parallel with PGPR there were repeatedly described various positive effects of nodule bacteria on non-legume plants, suggesting that rhizobia may act as PGPR. Recently the related reports were reviewed by Mehboob *et al.* (2009) and demonstrated clearly that nodule bacteria, in the same manner as PGPR, are capable of producing numerous biologically active substances (phytohormones, antibiotics, siderophores, Nod factors, lumichrome, rhiboflavin), solubilising phosphates, improving nutrient uptake, containing ACC deaminase and possessing biocontrol activity. Interestingly, an experience by Belimov *et al.* (1999), showed that the PGPR strain DR65, which dominated in the rhizosphere of barley and was applied successfully as biofertilizer for increasing barley yield, was initially misidentified by numeric taxonomy as *Pseudomonas denitrificans*, but then it was reclassified to *Sinorhizobium* sp. by 16S rRNA gene sequence (accession number HM002636).

Table 3. Effects of mono-inoculations with PGPR on legume plants grown in HM polluted soils

Plant	PGPR	Conditions	HMs	Microbial effects on plants	Reference
Cajanus cajan	*Proteus vulgaris*	G, MAS	Cu	Increased root and shoot growth, root length, leaf chlorophyll content. Decreased root and shoot Cu content.	Rani *et al.*, 2008
Cicer arietinum	Unidentified PGPR	G, MAS	Ni	Increased shoot growth. Decreased shoot Ni content.	Tank and Saraf, 2009
Lotus edulis	*Variovorax paradoxus*	G, MPS	Cd, Pb, Zn	Increased shoot Ca and Mg content.	Safronova *et al.*, 2010
Lotus ornithopodioides	*Variovorax paradoxus*	G, MPS	Cd, Pb, Zn	Decreased shoot K, Ca and Cu content, Zn translocation factor.	Safronova *et al.*, 2010
Medicago ciliaris	*Variovorax paradoxus*	G, MPS	Cd, Pb, Zn	Decreased shoot Cd and Cu content.	Safronova *et al.*, 2010
Phaseolus vulgaris	*Pseudomonas putida*	G, MAS	Cd, Pb	Increased root and shoot growth, chlorophyll content. Decreased shoot Cd and Pb content.	Tripathi *et al.*, 2005.
Pisum sativum	*Pseudomonas brassicacearum*, *P. marginalis*, *Rhodococcus* sp.	G, MAS	Cd	Increased root and shoot growth, uptake of N, P, K, Ca, S and Fe. Decreased shoot Cd content. The growth-promoting effect varied depending on plant genotype and bacterial strain.	Safronova *et al.*, 2006
Pisum sativum	*Pseudomonas brassicacearum*	G, MPS	Cd	Increased seed Cd content.	Engqvist *et al.*, 2006
Trifolium repens	*Bacillus cereus*	G, MPS	Fe, Cd, Pb, Zn	Increased root growth, shoot Al, Cd, Zn, Cu,Cr, Mn and Ni content.	Azcon *et al.*, 2006
Trifolium repens	*Brevibacillus brevis*	G, MAS	Zn	Increased shoot growth, N and P accumulation, nodule number and AMF infection. Decreased shoot Zn content.	Vivas *et al.*, 2006

Table 3. (Continued)

Plant	PGPR	Conditions	HMs	Microbial effects on plants	Reference
Trifolium repens	*Brevibacillus brevis*	G, MAS	Cd	Increased root and shoot growth, nodulation frequency, shoot Cd, Cr, Mo, Ni and Cu content. Decreased shoot K content.	Vivas *et al.*, 2005
Trifolium pratense	*Brevibacillus* sp.	G, MAS	Pb	Increased shoot and root growth, N and P accumulation, nodule number and AMF infection, shoot Pb content.	Vivas *et al.*, 2003a
Trifolium repens	*Brevibacillus* sp.	G, MAS	Cd	Increased shoot and root growth, nodulation frequency, shoot N, P and Cd content.	Vivas *et al.*, 2003b
Vigna radiata	*Bacillus cereus*	H, MAS	Cr	Increased shoot growth, pod length and number, seed number. Lowered Cr toxicity by reduction of Cr(VI) to Cr(III). Decreased shoot Cr content.	Faisal and Hasnain, 2006

Abbreviations: G, greenhouse; F, field; FA, fly-ash; H, hydroponics; MT, Mine tailings; MAS, metal amended soil; MPS, metal polluted soil.

It should be taken into account that along with symbiotic nitrogen fixation the introduced nodule bacteria may exert growth promoting effects and act as PGPR in the rhizosphere of legume plants. Moreover, nodule bacteria may have very high HM tolerance (Smith and Giller, 1992; Chaintreuil *et al.*, 2007; Ahmad *et al.*, 2001; El-Aziz *et al.*, 1991), accumulate and detoxify HMs (Pereira *et al.*, 2006). Although it is difficult to differentiate symbiotic and rhizosphere effects of nodule bacteria on legume plants, it would be important to understand the potential of these bacteria for mitigation of HM stress in plants by mechanisms typical for PGPR. Application of nod-minus mutants of legume plants may be a promising approach for elucidations of the mechanisms involved.

Inoculation with AMF might exert opposite effects on the HM content in legume plant tissues (Table 1), and this is in agreement with variable effects of mycorrhiza on HM uptake by other plant species (Leyval *et al.*, 1997). Increase in shoot Zn (Andrade *et al.*, 2009) and shoot Cd content (Rivera-Becerril, *et al.*, 2002), as well as root Cd content (Joner and Leyval, 1997; Li *et al.*, 2009) of mycorrhized plants was described. However, in two latter reports the increased root Cd content was accompanied by decreased Cd content in shoots. Although the content of HMs in mycorrhized plants was generally decreased (Table 1), the total HM accumulation might be increased due to plant growth promotion (Diaz *et al.*, 1996; Li *et al.*, 2009; Redon *et al.*, 2009). The decreased content of HM in over-ground mycorrhized plant tissues might be mediated by dilution of metal concentration in the increased plant biomass and by immobilization of metals in fungal hyphas (Giasson *et al.*, 2008; Gohre and Paszkowski, 2006). There was evidence that effects of AMF on translocation of HMs from root to shoot was opposite depending on mycorhizal species (Diaz *et al.*, 1996). Symbiotic nodule bacteria also had negative effects on the uptake of toxic HMs by legume plants (Table 2). The decrease in HM contents was observed after inoculations with different rhizobial species of *Cicer arietinum* (Gupta *et al.*, 2004; Wani *et al.*, 2008c), *Lens culinaris* (Wani *et al.*, 2008b), *Medicago ciliaris* (Safronova *et al.*, 2010), *Mimosa pudica* (Chen *et al.*, 2008), *Lupinus albus* and *L. luteus* (Pajuelo *et al.*, 2008), *Pisum sativum* (Wani *et al.*, 2008a) and *Vigna radiata* (Faisal and Hasnain, 2006). In parallel with effects of AMF and rhizobia, various PGPR reduced the content of HMs in legume plants grown in polluted soils (Table 3). Such effects were found for *Cajanus cajan* (Rani *et al.*, 2008), *Cicer arietinum* (Tank and Saraf, 2009), *Lupinus albus* (Pajuelo *et al.*, 2008), *Medicago ciliaris* (Safronova *et al.*, 2010), *Phaseolus vulgaris* (Tripathi *et al.*, 2005), *Pisum sativum* (Safronova *et al.*, 2006), *Trifolium repens* (Vivas *et al.*, 2006) and *Vigna radiata* (Faisal and Hasnain, 2006). Immobilization in the rhizosphere, biosorption by bacterial cells, production of siderophores and modulation of metal uptake systems in plant roots may be potential mechanisms involved in decreased HM uptake by the inoculated plants (Gadd, 1990; Safronova *et al.*, 2006; Khan *et al.*, 2009).

Combined Inoculations with Different Types of Microorganisms

A combined application of microorganisms possessing different beneficial traits is considered as a promising approach for enhancement of inoculation efficiency. Combinations of beneficial microbial traits may exert multiple effects on plants. Positive interactions between the introduced or/and aboriginal microorganisms may increase their activity and persistence, and facilitate symbiotic relations with plants. As a result, additive and synergistic effects on plant growth and nutrition may be expected. Although numerous studies confirming this view were performed with combinations of different microorganisms, such as AMF, nodule bacteria and/or PGPR (Belimov and Kozhemyakov, 1992; Dobbelaere *et al.*, 2003; Vessey, 2003; Artursson *et al.*, 2006; Frey-Klett *et al.*, 2007), application of this approach for phytoremediation of HM polluted soils received little attention. Case reports with legume plants showed, that when *Trifolium repens* was cultivated in Cd-supplemented soil, co-inoculation with AMF *Glomus mosseae* and PGPR *Brevibacillus brevis* had additive effects on plant growth, accumulation of nutrient elements and toxic Cd (Vivas *et al.*, 2003a). Moreover, this PGPR strain stimulated nodulation on roots by native rhizobia present in soil. The observed effects of *B. brevis* were suggested to be due to the indole acetic acid produced by PGPR bacteria (Vivas *et al.*, 2005). Similar results were obtained with *Glomus mosseae* and PGPR strain *Bacillus cereus* (Azcon *et al.*, 2009). Positive interactions between *Pisum sativum* and single or combined cultures of *G. intraradices*, *R. leguminosarum* bv. *viciae* and ACC-utilizing PGPR *Pseudomonas brassicacearum* were more pronounced in Cd amended soil as compared to non polluted one (Engqvist *et al.*, 2006). In such pot experiment only *G. intraradices* increased shoot biomass and seed yield, and increased seed Cd content was found in plants inoculated with *G. intraradices* or *P. brassicacearum*. Significant growth promotion, increased P uptake and nodulation, but decreased shoot Pb content in *Glycine max* plants inoculated with AMF *G. macrocarpum* and rhizobia *Bradyrhizobium* sp. were evident

(Andrade *et al.*, 2004), however it was not possible to estimate synergism of these microbes, because uninoculated controls were not included into the experiment. Our recent results demonstrated that a combined inoculation with PGPR *Variovorax paradoxus* containing enzyme ACC deaminase and the respective strains of nodule bacteria *Mesorhizobium loti* had synergistic and additive effects on nodulation frequency, plant growth, mineral nutrition, and accumulation of Cd, Pb and Zn in shoots of *Lotus edulis* and *L. ornithopodioides* (Safronova *et al.*, 2010). Synergistic effects on growth and the content of P and N in *Anthyllis cytisoides* were found after inoculation with mixtures containing several strains of AMF, rhizobia and PGPR (Requena *et al.*, 1997). The result showed that this plant-microbe model increased tolerance of plants to stress caused by aridity and nutrient deficiency and might be useful for revegetation of semi-arid ecosystems, however no information was given about HM pollution of that soil. Taking into account that polluted sites often contain a mixture of toxic metals and are subjected to other stress factors (aridity, low nutrients, erosion and extreme pH values), application of microbial compositions having a set of complemented beneficial traits, which counteract different stress factors, offer promise for improvement of phytoremediation processes. However, more efforts should be given to substantiate this hypothesis.

The literature data suggest that additive or synergistic effects of co-inoculation with different types of described microorganisms on lowering HM contents in plants should be expected. Up to date the only result that confirmed this hypothesis and showed synergistically decreased Cd, Cu and Zn contents was observed in *Lupinus albus* plants inoculated with *Bradyrhizobium lupini*, *Ochrobactrum* sp. and *Pseudomonas* sp. (Pajuelo *et al.*, 2008). Contrary to this, our recent results showed that no further decrease in HM contents in plants occurred after combined inoculations of *Lotus edulis* or *L. ornithopodioides* with nodule bacteria *M. loti* and PGPR *V. paradoxus* (Safronova *et al.*, 2010). When seed Cd content in *Pisum sativum* plants was increased by *G. intraradices* or *P. brassicacearum*, no further changes in this parameter was evident after combined inoculation. These case results suggest that more experimental data are needed for estimation of interactions between the introduced microbes in polluted soils and the resulting effects on HM uptake by plants.

Chemically Assisted HM Accumulation by Legume-Microbe Symbioses

There is evidence that legume plants are capable of actively accumulate HMs from polluted soils and hydroponics. For example, *Cassia fistula* accumulated Cr, Cu, Zn and Mn (Gupta and Sinha, 2007), *Medicago sativa* actively accumulated Cd, Cu, Ni and Zn (Peralta-Videa *et al.*, 2002) and Cd, Cr and Ni (Bonfanceschi *et al.*, 2009), and *Prosopis juliflora* accumulated Cd and Cu (Senthilkumar *et al.*, 2005). However, comparison studies showed that legume species are characterised by relatively low translocation of HMs from roots to shoots and can be assigned to the excluder type (Kuboi *et al.*, 1986; Pettersson, 1977; Zwarich and Mills, 1982). Metal hyperaccumulation trait was not found for plants of the family Fabaceae, and the exception is that *Sesbania drummondii* was described as Pb-hyperaccumulator having 40 mg Pb per g of dried shoot biomass (Sahi *et al.*, 2002). Recently we have found that the root-shoot translocation factor of Pb for *Lotus ornithopodioides* was above 1, suggesting that this plant showed hyperaccumulating trait for such element

(Safronova *et al.*, 2010). In addition, a relatively high HM sensitivity of legumes may restrict metal accumulation in aboveground parts via both induction of mechanisms counteracting translocation of toxicants and growth inhibition.

A frequently observed negative effect of symbiotrophic microorganisms on the content of HMs in legumes should be taken into account for application of these plants in phytoremediation. On the one hand this phenomenon may play beneficial role to grazing animals when legumes are utilised in phytostabilization and revegetation technologies. On the other hand, this restricts accumulation of HMs in harvested plant parts and output of pollutants from soil resulting in decreased phytoextraction efficiency. It should be mentioned that microorganisms possess several mechanisms of metal mobilization and may increase availability of HMs in the rhizosphere resulting in enhanced metal uptake by plants (Gadd, 1990; Wenzel, 2009). Therefore, selection of microorganisms associated with legumes and harbouring traits for increasing HM availability and/or stimulating metal uptake and transport systems in plants may be a promising approach for improved phytoextraction. Low metal availability in the rhizosphere was shown to be a limitation factor for HM accumulation by legume plants (Rodriguez *et al.*, 2007). Although accelerated HM uptake may cause toxic effects and inhibit plant growth, particularly of relatively sensitive plants like legumes, their HM extraction potential might be significantly enhanced through increasing the metal availability in the rhizosphere. Chemically-assisted phytoextraction is known as an efficient approach for enhancement of HM uptake by plants (Lasat, 2000; Wenzel *et al.*, 2003; Singh, 2007). It was demonstrated that addition of chelating substances, such as EDTA, raised the content of Pb in shoots of *Medicago sativa* (Lopez *et al.*, 2005), *Pisum sativum* (Piechalak *et al.*, 2003), *Sesbania drummondii* (Ruley *et al.*, 2006) and Vigna radiata (Shen *et al.*, 2002). Similar results were obtained in chelate-assisted extraction of Cd, Cu, Pb and Zn by *Lupinus albus* (Penalosa *et al.*, 2007) and *Phaseolus vulgaris* (Luo *et al.*, 2005). It is worth to estimate experimentally the phytoextraction potential of legume plants treated with chemical chelating agents and beneficial microorganisms in combinations. In this respect it should be taken into account that many microorganisms are capable of degrading and/or producing their own metal chelating and metal binding organic compounds.

Intraspecific Variability of Plants in their Interactions with Microorganisms

It is well known that plants significantly differ in their tolerance to and accumulation of HMs and intraspecific genetic variation of these traits exists. There are several reports that describe variability for such traits in legume plants. For example, cultivars of *Glycine max* differed in Zn (White *et al.*, 1979) and Cd (Sugiyama *et al.*, 2007) tolerance, inbred lines of *Lotus purshianus* differed in Cu tolerance (Lin and Wu, 1994), cultivars of *Phaseolus vulgaris* differed in Zn and Cu tolerance (Polson and Adams, 1970), and cultivars of *Vigna unguiculata* differed in Mn tolerance (Horst, 1983). Cultivars of *V. unguiculata* (Horst, 1983) and *G. max* (White *et al.*, 1979) varied in the capacity to take up Mn and Zn, respectively. Differences in Cd content were found among cultivars of *Arachis hypogaea* and *P. vulgaris* (Bell *et al.*, 1997), *G. max* (Bell *et al.*, 1997; Keck and Redlich, 1975; Sugiyama *et al.*, 2007) and *Trifolium fragiferum* (Jauert *et al.*, 2002).

A decreased root-shoot transport of Cd, Mn and Zn was observed in a population of *Bituminaria bituminosa* collected in a polluted site, as compared to that originated from a non polluted site (Walker *et al.*, 2007). Experiments by Belimov *et al.* (2003) showed significant genotypic variability in Cd tolerance and accumulation of different HMs (Cd, Cr, Cu, Ni, Pb, Sr and Zn) among 99 *Pisum sativum* varieties. A negative correlation between Cd tolerance and shoot Cd content was found, suggesting that Cd exclusion and limited translocation from roots to shoots are important mechanisms of tolerance. In the same experiments, Cd-sensitive varieties with low and Cd-tolerant varieties with high shoot Cd content were identified, demonstrating the existence of genotypic differences in mechanisms of tolerance and accumulation of toxic Cd in this plant species. No correlations were found between plant biomass and Cd tolerance or shoot HM contents. These results suggested that the relationships between tolerance and accumulation traits are complex and depend on the plant genotype. The lack of such correlations indicates the existence of independent genetic control of these traits. This provides a possibility for breeding varieties combining increased tolerance to and modified accumulation of HMs in one genotype efficient in biomass production.

However the question on how variability of these traits may be involved and affects interactions of plants with microbes in the presence of HMs received little attention. The pea varieties described above (Belimov *et al.*, 2003) were also studied for their interactions with AMF *Glomus sp.*, and significant intraspecific variability in the response of plants to inoculation with AMF was described (Jacobi *et al.*, 2000). This made possible to find relationships between polymorphism in the response to Cd toxicity and the efficiency of mycorrhizal symbiosis in the absence of toxic Cd. It was found that Cd tolerance was negatively correlated with the positive effects of *Glomus* sp. on biomass of roots, straw and individual seeds, suggesting higher ability of Cd-sensitive varieties to form efficient symbiosis (Belimov and Wenzel, 2009). A negative correlation between Cd content in Cd-treated plants and the effect of *Glomus* sp. on seed P content suggested that the Cd-excluding varieties are more efficient in using P from the symbiosis with AMF. Mycorrhiza was shown to alleviate phytotoxic effects of HMs, associated with intracellular chelating of metal ions by polyphosphates present in fungal hyphaes as one of protective mechanisms (Leyval *et al.*, 1997; Gohre and Paszkowski, 2006). Therefore it may be proposed that Cd tolerant pea varieties are less efficient in exploring the protective potential of symbiosis with AMF, but Cd sensitive varieties are capable of compensating their deficient metal tolerance through mycorrhizal symbiosis. Interestingly, a similar situation was evident when Cd tolerance of *Brassica juncea* varieties (Belimov *et al.*, 2007) was plotted against the effect of PGPR *V. paradoxus* 5C-2 on shoot biomass (Belimov and Wenzel, 2009). A negative correlation was found and suggested lower ability of Cd-tolerant varieties to benefit from this bacterium.

There is evidence that, the modern cultivars of legume crops have lower potential for biological nitrogen nutrition in symbiosis with nodule bacteria compared to wild-growing varieties as a result of auto-selection of genotypes that efficiently assimilate combined nitrogen from fertilizers (Provorov and Tikhonovich, 2003). Our experiments with 64 genotypes of *Brassica juncea* revealed a negative correlation between growth parameters and Cd tolerance (Belimov *et al.*, 2007) and supported the hypothesis about increased energy expenditure for operation of the mechanisms of metal tolerance resulting in slower growth and lower biomass production of metal tolerant plants as compared with their non tolerant counterparts (Wu 1990).

These observations should be taken into account in the studies aimed at selection of legume genotypes combining the traits for high metal tolerance, excessive metal accumulation and efficient symbiotrophic interactions with beneficial microorganisms.

Relationships between HM Tolerance of Symbiotic Partners

It is known that many of the AMF are well adapted to environments characterized by high concentrations of HMs and survive for long periods in polluted soils, but negative effects of HMs on root colonization and mycorrhizal structures in roots were also described (Leyval *et al.*, 1997; Leyval and Joner 2001, Ouziad *et al.*, 2005; Giasson *et al.*, 2008; Andrade *et al.*, 2004). These micro-symbionts developed a number of mechanisms of HM tolerance such as: (1) extracellular or intracellular metal sequestration and precipitation with organic acids and other ligands, polyphoshates and metallothioneins; (2) metal biosorption by protein glomalin; (3) metal binding to cell walls and intracellular metal chelation; (4) reduced uptake or increased efflux of HMs by fungal cells. However there are also observations showing significant inhibition of mycorrhizal root colonization by the presence of HMs in soil. Since AMF are obligate symbionts, their HM tolerance depends on the host plant (plant metabolism and nutrient status, decreased contact with soil as a spatial arrangement of hyphas in roots) and mediated by their effects on the plant metal tolerance (decreased plant responses to HM and oxidative stress, changes in plant gene expression). This means that mycorrhizal colonization of roots and proper function of AMF-plant symbiosis strongly depends on the capability of the plant to maintain metabolic homeostasis and to counteract disturbances of the processes related to symbiosis formation and function.

It was reported that development of legume-rhizobia symbiosis may be tolerant to the presence of elevated concentrations of HMs in soils. The *T. repens* plants cultivated in soil, originated from mining site and extremely polluted with 220 μg g-1 Cd, 30000 μg g^{-1} Pb and 20000 μg g^{-1} Zn, had healthy nodules and their potential for nitrogen fixation (80 g N ha^{-1} h^{-1}) was high (Rother *et al.*, 1983). However as a rule, nodulation and symbiotic nitrogen fixation was sensitive to HMs and inhibited in polluted soils, resulting in nitrogen deficiency and plant growth limitation. For example, significant reduction of nodule formation and nitrogen fixation caused by elevated HM concentrations was described for *Glycine max* (Chen *et al.*, 2003), *Leucaena leucocephala* (Cheung *et al.*, 2000), *Lotus purshianus* (Wu and Lin, 1990) and *Lupinus albus* (Pastor *et al.*, 2003). Moreover, it was proposed to use nodulation process as a bioindicator to test the toxicity of HM polluted soils (Neuman *et al.*, 1998; Manier *et al.*, 2009). Symbiotic interaction between *Vigna unguiculata* and rhizobia were more sensitive to Cu toxicity than both partners separately (Kopitte *et al.*, 2007). In line with these reports we propose that the plant genotype is of prime consideration, because plants, being on a higher evolution level, are less tolerant to metal toxicity as compared to microorganisms. Our results showed that a minimum growth inhibiting concentration of Cd for *R. leguminosarum* bv. *viciae* varied between 15 and 120 μM (Belimov and Wenzel, 2009). Strong toxicity symptoms and growth inhibition for hydroponically grown pea genotypes, considered as the most Cd-tolerant (Belimov *et al.*, 2003), were evident in the presence of 5 μM Cd (Metwally *et al.*, 2005). At 0.5 μM Cd plant growth was not affected, but nodulation frequency with *R.*

leguminosarum bv. *viciae* CIAM1066 having threshold growth inhibiting concentration of 60 μM Cd, decreased by a factor of four (Belimov and Wenzel, 2009). It is likely that the plant was affected by Cd to a greater extent compared to bacteria, resulting in loss of symbiotic capability. These results demonstrate that processes of plant-microbe interactions may be very sensitive to HM toxicity and can be disturbed at metal concentrations below threshold toxicity levels determined for each partner separately. However there are situations where even metal tolerant microorganisms loose their growth promoting activity in the presence of HMs. For example, a Cd tolerant PGPR *Rhodococcus* sp. Fp2 containing ACC deaminase stimulated growth of *Pisum sativum* cultivated in uncontaminated soil, but not in Cd-spiked soil, likely due to its inability to degrade ACC in the presence of Cd (Safronova *et al.*, 2006).

It may be assumed that for creation of legume-microbe systems having high phytoremediation potential, the basic challenge is the high sensitivity of symbiotic interactions to HMs. Main attention should be given to understanding mechanisms of HM toxicity on development and function of symbioses and to elaborate approaches for efficient integration of legume plants with beneficial microorganisms in the presence of HMs. For this purpose, combined selection of complementary metal tolerant pairs of micro- and most notably macro-partners holds promise.

Phytoremediation with Genetically Modified Legumes and Symbiotrophic Microorganisms

It is assumed that for successful phytoremediation technologies, plants having high metal tolerance, metal uptake potential, biomass production and growth rate are required. However, no natural metalliferous and hyperaccumulating species neither agricultural crops possess sufficient level of all these characteristics. One promising method of attack and overcome these shortcomings is the creation of genetically modified plants via transgenic techniques and mutagenesis. A number of genetically modified plants were generated in order to modify their tolerance to and accumulation of HMs, and the related reports were repeatedly reviewed (Kramer and Chardonnens, 2001; Pilon-Smith and Pilon, 2002; Vassilev *et al.*, 2004; Zhang *et al.*, 2006; Goel *et al.*, 2009). Different approaches in genetic manipulations with plants such as transferring or mutagenising the genes responsible for HM tolerance, uptake, cellular and long-distance transport, binding and chelation, as well as transformation and volatilization were applied. In most cases the target plants were *Arabidopsis thaliana*, *Nicotiana tabacum* and *Brassica juncea*, however only few studies were devoted to transformation of legumes, using their genes for transformation of other plants or to mutagenesis of legume plants.

Introduction of Arabidopsis metallothionein genes *AtMT1* and *AtMT2* to guard cells of *Vicia faba* resulted in reduction of the level of reactive oxygen species and thereby increased tolerance to supplemented Cd (Lee *et al.*, 2004). The gene encoding selenecystein (Se-Cys) methyltransferase was isolated from *Astragalus bisulcatus* and overexpressed in *A. thaliana* and *Brassica juncea* resulting in significant increase in Se tolerance and volatilization in transgenic plants (LeDuc *et al.*, 2004). Overexpression of pea (*Pisum sativum*) metallothionein gene *PsMTA* in *A. thaliana* enhanced its capability of Cu uptake (Murphy and Taiz, 1995). Increased Cd tolerance and decreased root Cd content was observed in *Nicotiana tabacum* plants after transformation of stress related gene *PvSR2* cloned from *Phaseolus vulgaris* (Chai *et al.*, 2003).

Welch and LaRue (1990) isolated *Pisum sativum* mutant named E107 (*brz*), with an abnormally high uptake of Fe and characteristic necrotic spots on leaves due to Fe toxicity. The roots of the E107 released Fe(III)-reducing substances to the surrounding medium at higher rates than the wild type Sparkle, suggesting that the mutant acts functionally as a Fe-deficient plant. This mutant excessively accumulated Al and manifested symptoms typical of Al toxicity (Guinel and LaRue, 1993). More recently it was shown that in soil culture the mutant E107 actively accumulated other metal ions including Ca, Cu, Mg, Mn, Zn, and particularly Pb, which is usually present in soil as insoluble component (Chen and Huang, 2007). When the soil was supplemented with EDTA, the genotypic differences between the E107 and wild type plants were not manifested, suggesting that metal availability in the root zone was a crucial factor mediating excessive metal accumulation. Another mutant was obtained on *Medicago truncatula* and characterized by a recessive mutation *raz*, defined as "requires additional zinc" (Ellis *et al.*, 2003). The *raz* mutant showed Zn deficiency symptoms (characteristic necrotic spots on leaves) in the presence of this micronutrient in soil and accumulated Zn, Mn and Cu more actively compared to wild type plants. Recently the first plant mutant SGECdt characterized by both increased Cd-tolerance and Cd-accumulation was isolated using chemical mutagenesis of *Pisum sativum* (Tsyganov *et al.* 2007). Comparative analysis of physiological, nutritional and biochemical characteristics of SGECd^t showed lower levels of Cd-stress and demonstrated capability to cope well with increased Cd levels in roots, shoots, leaves and mesophyll protoplasts. Inoculation of SGECdt with *R. leguminosarum* bv. *viciae* in hydroponics demonstrated its ability to form symbiotic nodules in the presence of 2 μM Cd, whereas nodulation of wild type plants was completely terminated at 1.5 μM Cd (Tsyganov *et al.* 2005). Significant disturbances of nodule histological organization and bacteroid differentiation were observed even at 0.5 μM Cd, but in wild type only. This mutant provides promising new genetic material for the study of the mechanisms underlying plant-microbe interactions under stressed conditions caused by HMs and for phytoremediation technologies based on plant-microbe systems.

Several attempts were made to generate genetically modified microorganisms associated with legume plants. The *AtPCS* gene encoding phytochelatin synthase was introduced to *Mesorhizobium* sp. and *M. huakuii* subsp. *rengei* and increased by several times the accumulation of Cd in bacterial cells (Sriprang *et al.*, 2003). Inoculation of *Astragalus sinicus* with transformed mesorhizobia increased accumulation of Cd in nodules. Similar results were obtained with *Astragalus sinicus* grown in pollutedsoil and inoculated with *M. huakuii* subsp. *rengei* expressing a human metallothionein gene *MTL4* (Sriprang *et al.*, 2002). This symbiotic system was applied for phytoremediation of paddy soil polluted with Cd (Ike *et al.*, 2007). Increased accumulation of Cd in nodules and roots was observed resulting in removing about 10% Cd from the soil after two months of plant cultivation. After expression of the Ni resistance genetic system *ncc-nre* from *Ralstonia metallidurans* in endophytic bacteria *Burkholderia cepacia*, the transformed strain was capable of accumulating and precipitating Ni from the growth medium in vitro and increased root Ni content of the inoculated *Lupinus luteus* plants grown in Ni-supplemented perlite (Lodewyckx *et al.*, 2001). However no bacterial effects on plant growth or shoot Ni content were detected. Gupta *et al.* (2002) generated mutants of PGPR *Pseudomonas* sp. having increased resistance to high concentrations of Cd, Cr and Ni, and the growth promoting effect of these mutants, but not of the wild type strain, was detected on *Glycine max* plants cultivated in metal amended soil.

The overview of these few reports clearly points out that genetic modifications of legume plants and symbiotrophic microorganisms aimed at increased metal tolerance, modified metal uptake and efficient functioning under stressed conditions, holds great promise for the improvement of phytoremediation technologies using legume-microbe symbioses. Taking into account that HM sensitive symbiotic interactions may be a limiting factor for performance of legume plants cultivated in polluted soils, it is worth to develop genetic engineering approaches for targeting particularly plant-microbe symbiosis.

CONCLUSION

In the bibliographic survey of the literature regarding experimental research on phytoremediation of HMs done by Vamerali *et al.* (2010) over the period 1995-2009, it was found that cruciferous (Brassicaceae) and cereals (Poaceae) were the most cited plants, while fewer citations were made for the legumes (Fabaceae). Among the 27 legume species of 18 genera cited in this chapter, many are field crops, while others are wild species. Considering that only few of the 20000 species of Fabaceae are field crops and that various plant types such as herbs, shrubs and trees are representatives, it is evident the underexploited potential of such plant family. Comparison of legumes with other plants for efficiency of phytoremediation processes was outside of this chapter. However in some of the cited reports such evaluation was undertaken, and generally legumes showed relatively high phytoremediation potential comparable with the other species tested. Future research work is needed to ascertain the value of many legume species in terms of phytoremediation efficiency in polluted environments.

The important challenge for successful application of legumes in phytoremediation technologies is the enhancement of their metal tolerance. For improvement of plant adaptation to stressful environments such as HM polluted soils it is undoubtedly advisable to exploit beneficial plant-associated microorganisms. This approach is of particular importance for legume plants, since they possess very high symbiotrophic potential. The gained experience clearly demonstrated that inoculations of legumes with AMF, nodule bacteria or PGPR significantly promote plant growth in the presence of toxic HM concentrations in soils. Moreover, positive synergistic and additive effects of different microorganisms on plant growth and nutrition after combined inoculations support perspectives of using microbial associations expressing multiply effects on plants and rhizosphere processes related to function of microbial community and HM transformation. Although the beneficial effects of microorganisms on the growth of plants subjected to HM stress is well documented, the mechanisms underlying these growth-promoting effects are scarcely understood. More attention should be given to biodiversity of beneficial microorganisms inhabiting polluted environments, interactions between microorganisms in the rhizosphere, and selection of metal tolerant strains having high potential for development of efficient symbioses with plants under stressful conditions. However, the improvement of HM tolerance of macro-symbiont is of crucial importance, since as a rule the plant is more sensitive to HM stress and most probably the plant genotype controls development of symbiosis in the presence of HM toxicity to a greater extent, compared to microorganisms. Screening of natural plant genotypes and genetic manipulations aimed at enhancement of HM tolerance and uptake should be performed together with estimation and improvement of their symbiotic potential. Serious efforts should be aimed toward the understanding of limiting steps in development and efficient functioning of symbiotic plant-microbe interactions in the presence of HM stress.

The literature analysis revealed that very often inoculation of plants with symbiotrophic microorganisms had negative effect on the HM contents in over-ground plant parts, although total accumulation increased due to plant growth promotion. Understanding the mechanisms of this phenomenon and monitoring of the HM transformation and translocation in the rhizosphere are important challenges for the process to be controlled. Taking into account that legumes are not hyperaccumulators of metals, enhancement of HM uptake in such systems would be desirable for both phytoextraction and phytostabilisation processes. One way for increasing HM accumulation is application of chelating substances, which is however of limited application. An alternative approach may be the intensification of microbiological processes providing increased HM availability in the rhizosphere and stimulation of metal uptake systems in plants by specific microorganisms. We totally agree with Wenzel (2009) that for the enhancement of phytoremediation technologies it is required a deep understanding of the complex interactions in the rhizosphere involving a number of biological, biochemical and physico-chemical processes.

Basically the available results with legumes were originated from pot experiments in greenhouses using soil artificially amended with one or two metals, and the transfer of results to open field conditions was not available. There is no doubt that mechanistic studies under controlled environmental conditions are absolutely necessary, particularly in those experiments, where the plants and soils are inoculated with different types of microorganisms. The reasons for this are: (1) preliminary testing and caution should be taken for a large scale introduction of microorganisms in to open environment; (2) investigation of the mechanisms underlying plant-microbe interactions and screening for efficient plant-microbe associations needs application of a complex and multifactor experimental design. Nevertheless, it is essential that site specific field results should be produced for proper evaluation of the laboratory findings. According to the literature review made in this chapter, it is also fundamental to consider the experimental scale, as microbial treatments successfully performed at the bench and pot experiment level might fail when applied to contaminated soils in field experiments. Emphasis should be put on evaluating results obtained in simplified bench and pot experiments compared to heterogeneous, multiple polluted field sites and the functioning of phyto/rhizoremediation systems under various ecological conditions. A deeper knowledge of plants and microorganisms control of metal bioavailability in the contaminated soil is recommended in order to develop integrated approaches particularly suitable in multiple contaminated soils. Also, taking into account that polluted sites often contain a mixture of toxic metals and are subjected to other stress factors (aridity, low nutrients, erosion and extreme pH values), application of microbial compositions having a set of complemented beneficial traits, which counteract different stress factors, offer promise for improvement of phytoremediation processes. However, more efforts should be given to substantiate this hypothesis. The case results reported in this chapter suggest that more experimental data are needed for estimation of interactions between the introduced microbes in polluted soils and the resulting effects on HMs uptake by plants.

We propose that the mere selection of metal tolerant legume plants or metal accumulator legume plants is not sufficient for the development of efficient phytoremediation strategies, and this is particular pertinent to phytoextraction technologies. According to Van Nevel *et al.* (2007), in spite of an "explosion" of literature addressing phytoextraction of metals and metalloids during the past decade, there is still limited evidence for satisfactory extraction rates even for the most active accumulators and hyperaccumulators. However the use of legume plants for phytostabilization and revegetation technologies is particularly intriguing, basically due to their high potential to form symbioses with various beneficial microorganisms. Surely, more experimental data are needed for the estimation of interactions between partners of plant-microbe symbioses in polluted soils and the resulting effects of microorganisms on the plant HM tolerance and uptake. In this respect, the selection and genetic engineering of HM tolerant legume-microbe symbioses and the rhizosphere engineering based on such symbioses provide unique possibilities and offer promise for successful phytoremediation of polluted sites and for ecologically safe restoration of healthy ecosystems.

REFERENCES

Ahmad, I; Hayat, S; Ahmad, A; Inam, A; Samiullah. Metal and antibiotic resistance traits in *Bradyrhizobium* sp. (cajanus) isolated from soil receiving oil refinery wastewater. *World J. Microbiol*, 2001, 17, 379-384.

Allen, ON; Allen, EK. *The Leguminosae: a source book of characteristics, uses, and nodulation*. Madison: The University of Wisconsin Press; 1981, 812 pp.

Andrade, SAL; Abreu, CA; de Abreu, MF ; Silveira, APD. Influence of lead additions on arbuscular mycorrhiza and Rhizobium symbioses under soybean plants. *Appl. Soil Ecol*, 2004, 26, 123-131.

Andrade, SA; Gratão, PL; Schiavinato, MA; Silveira, AP; Azevedo, RA; Mazzafera, P. Zn uptake, physiological response and stress attenuation in mycorrhizal jack bean growing in soil with increasing Zn concentrations. *Chemosphere*, 2009, 75, 1363-1370.

Arshad, M; Saleem, M; Hussain, S. Perspectives of bacterial ACC deaminase in phytoremediation. *Trends Biotechnol*, 2007, 25, 356-362.

Artursson, V; Finlay; RD, Jansson, J. Interactions between arbuscular mycorrhizal fungi and bacteria and their potential for stimulating plant growth. *Environ. Microbiol*, 2006, 8, 1–10.

Azcón, R; Medina, A; Roldán, A; Biró, B; Vivas, A. Significance of treated agrowaste residue and autochthonous inoculates (Arbuscular mycorrhizal fungi and *Bacillus cereus*) on bacterial community structure and phytoextraction to remediate soils contaminated with heavy metals. *Chemosphere*, 2009, 75, 327-334.

Belimov, AA; Kozhemiakov, AP. Use of mixed cultures of nitrogen fixing bacteria in agriculture. *Agricultural Biology*, 1992, 5, 77-87 (in Russian).

Belimov, AA; Ivanchikov, AY; Yudkin, LV; Khamova, OF; Postavskaya, SM; Popolzukhina, PV; Shmakova, AA, Kozlova, GY. New strains of associative growth-stimulating bacteria dominating the rhizoplane of barley seedlings: characterization and introduction. *Microbiologiya*, 1999, 68, 337-342 (in Russian).

Belimov, AA; Safronova, V; Demchinskayaa, SV; Dzyuba, OO. Intraspecific variability of cadmium tolerance in hydroponically grown Indian mustard (*Brassica juncea* (L.) Czern.) seedlings. *Acta Physio.l Plant*, 2007, 29, 473-478.

Belimov, AA; Safronova, VI; Tsyganov, VE; Borisov, AY; Kozhemyakov, AP; Stepanok, VV; Martenson, AM; Gianinazzi-Pearson, V; Tikhonovich IA. Genetic variability in tolerance to cadmium and accumulation of heavy metals in pea (*Pisum sativum* L.). *Euphytica*, 2003, 131, 25-35.

Belimov, AA; Wenzel, WW. The role of rhizosphere microorganisms in heavy metal tolerance of higher plants. *Aspects Appl. Biol*, 2009, 98, 81-90.

Bell, MJ; McLaughlin, MJ; Wrigh, GC; Cruickshank, J. Inter- and intra-specific variation in accumulation of cadmium by peanut, soybean, and navy bean. *Australian J. Agric. Res*, 1997, 48, 1151-1160.

Bidar, G; Pruvot, C; Garçon, G; Verdin, A; Shirali, P; Douay, F. Seasonal and annual variations of metal uptake, bioaccumulation, and toxicity in *Trifolium repens* and *Lolium perenne* growing in a heavy metal-contaminated field. *Environ. Sci. Pollut Res. Int*, 2009, 16, 42-53.

Bonfanceschi, BA; Flocco, CG; Donati, ER. Study of the heavy metal phytoextraction capacity of two forage species growing in an hydroponic environment. *J. Hazard Mater*, 2009, 165, 366-371.

Chaintreuil, C; Rigault, F; Moulin, L; Jaffré, T; Fardoux, J; Giraud, E; Dreyfus, B; Bailly, X. Nickel resistance determinants in bradyrhizobium strains from nodules of the endemic new caledonia legume *Serianthes calycina*. *Appl. Environ. Microbiol*, 2007, 73, 8018-8022.

Chan, GY; Ye, ZH; Wong, MH. Comparison of four *Sesbania species* to remediate Pb/Zn and Cu mine tailings. *Environ. Management*, 2003, 32, 246-251.

Chai, T; Chen, Q; Zhang, Y; Dong, J; An C. Cadmium resistance in transgenic tobacco plants enhanced by expressing bean heavy metal-responsive gene *PvSR2*. *Sci. China*, 2003, 46, 623-630.

Chen, J; Huang, JW. Increased lead accumulation in a single gene mutant of pea (*Pisum sativum* L.). *Bull. Environ. Contam. Toxicol*, 2007, 79, 25-28.

Chen, WM; Wu, CH; James, EK; Chang, JS. Metal biosorption capability of *Cupriavidus taiwanensis* and its effects on heavy metal removal by nodulated *Mimosa pudica*. *J. Hazard. Mater*, 2008, 151, 364-371.

Chen, YX; He, YF; Yang, Y; Yu, YL; Zheng, SJ; Tian, GM; Luo, YM; Wong, MH. Effect of cadmium on nodulation and N2-fixation of soybean in contaminated soils. *Chemosphere*, 2003, 50, 781-787.

Cheung, KC; Wong, JP; Zhang, ZQ; Wong, JW; Wong, MH. Revegetation of lagoon ash using the legume species *Acacia auriculiformis* and *Leucaena leucocephala*. *Environ. Pollut*, 2000, 109, 75-82.

Diaz, G; Azcon-Aguilar, C; Honrubia, M. Influence of arbuscular mycorrhizae on heavy metal (Zn and Pb) uptake and growth of *Lygeum spartum* and *Anthyllis cytisoides*. *Plant Soil*, 1996, 180, 241-249.

Dobbelaere, S; Vanderleyden, J; Okon, Y. Plant growth-promoting effects of diazotrophs in the rhizosphere. *Crit. Rev. Plant Sci*, 2003, 22, 107-149.

El-Aziz, R; Angle, JS; Chaney, RL. Metal tolerance of *Rhizobium meliloti* isolated from heavy-metal contaminated soils. *Soil Biol. Biochem*, 1991, 8, 795-798.

Ellis, DR; Lopez-Millan, AF; Grusak, MA. Metal physiology and accumulation in a *Medicago truncatula* mutant exhibiting an elevated requirement for zinc. *New Phytol*, 2003, 158, 207–218.

Engqvist, LG; Mårtensson, A; Orlowska, E; Turnau, K; Belimov, AA; Borisov, AY; Gianinazzi-Pearson, V. For a successful pea production on polluted soils, inoculation with beneficial microbes requires active interaction between the microbial components and the plant. *Acta Agric. Scand. B,* 2006, 56, 9-16.

Evanylo, GK; Abaye, AO; Dundas, C; Zipper, CE; Lemus, R; Sukkariyah, B; Rockett, J. Herbaceous vegetation productivity, persistence, and metals uptake on a biosolids-amended mine soil. *J. Environ. Qual*, 2005, 34, 1811-1819.

Faisal, M; Hasnain, S. Growth stimulatory effect of *Ochrobactrum intermedium* and *Bacillus cereus* on *Vigna radiata* plants. *Lett. Appl. Microbiol*, 2006, 43, 461-466.

Frey-Klett, P; Garbaye, J; Tarkka, M. The mycorrhiza helper bacteria revisited. *New Phytol*, 2007, 176, 22–36.

Frérot, H; Lefèbvre, C; Gruber, W; Collin, C; Dos Santos, A; Escarré J. Specific interactions between local metallicolous plants improve the phytostabilization of mine soils. *Plant Soil*, 2006, 282, 53-65.

Gadd, GM. Heavy metal accumulation by bacteria and other microorganisms. *Experientia*, 1990, 46, 834-840.

Gadd, GM. Microbial influence on metal mobility and application to bioremediation. *Geoderma*, 2004, 122, 109-119.

Gamalero, E; Linqua, G; Berta, G; Glick, BR. Beneficial role of plant growth promoting bacteria and arbuscular mycorrhizal fungi on plant responses to heavy metal stress. *Can. J. Microbiol*, 2009, 55, 501-514.

Gerhard, KE; Gerrnberg, BM; Glick, BR. The role of ACC deaminase in facilitating the phytoremediation of organics, metals and salt. *Cur. Trends Microbiol*, 2006, 2, 61-73.

Goel, S; Malik, JA; Nayyar, H. Molecular approach for phytoremediation of metal-contaminated sites. *Arch. Agron. Soil Sci*, 2009, 55, 451-475.

Gohre, V; Paszkowski, U. Contribution of the arbuscular mycorrhizal symbiosis to heavy metal phytoremediation. *Planta*, 2006, 223, 1115–1122.

Guinel, FC; LaRue, TA. Excessive aluminium accumulation in pea mutant E107 (*brz*). *Plant Soil*, 1993, 157, 75-82.

Gupta, A; Meyer, JM; Goel, R. Development of heavy metal-resistant mutants of phosphate solubilizing *Pseudomonas* sp. NBRI 4014 and their characterization. *Curr. Microbiol*, 2002, 45, 323-327.

Gupta, DK; Rai, UN; Sinha, S; Tripathi, RD; Nautiyal, BD; Rai, P; Inouhe, M. Role of *Rhizobium* (CA-1) inoculation in increasing growth and metal accumulation in *Cicer arietinum* L. growing under fly-ash stress condition. *Bull. Environ. Contam. Toxicol*, 2004, 73, 424-431.

Horst, WJ. Factors responsible for genotypic manganese tolerance in cowpea (*Vigna unguiculata*). *Plant Soil*, 1983, 72, 213-218.

Ike, A; Sriprang, R; Ono, H; Murooka, Y; Yamashita, M. Bioremediation of cadmium contaminated soil using symbiosis between leguminous plant and recombinant rhizobia with the MTL4 and the PCS genes. *Chemosphere*, 2007, 66, 1670-1676.

Jacobi, LM; Kukalev, AS; Ushakov, KV; Tsyganov, VE; Naumkina, TS; Provorov, NA; Borisov, AY; Tikhonovich, IA. Polymorphism of garden pea (*Pisum sativum* L.) for the efficiency of symbiosis with endomycorrhizal fungus *Glomus* sp. under the conditions of rhizobia inoculation. *Agricultural Biology*, 2000, 3, 94-102 (In Russian).

Jauert, P; Schumacher, TE; Boe, A; Reese, RN. Rhizosphere acidification and cadmium uptake by strawberry clover. *J. Environ. Qual*, 2002, 31, 627-633.

Jian, S; Shen, W; Yang, Z. Enhanced adaptability of *Sesbania rostrata* to Pb/Zn tailings via stem nodulation. *J. Environ. Sci* (China), 2009, 21, 1135-1141.

Jing, Y; He, Z; Yang, X. Role of rhizobacteria in phytoremediation of heavy metal contaminated soils. *J. Zhejiang Univ. Sci. B*, 2007, 8, 192-207.

Joner, EJ; Leyval, C. Uptake of 109Cd by roots and hyphae of a *Glomus mosseae/Trifolium subterraneum* mycorrhiza from soil amended with high and low concentrations of cadmium. *New Phytol*, 1997, 135, 353-360.

Khan, MS; Zaidi, A; Wani, PA; Oves, M. Role of plant growth promoting rhizobacteria in the remediation of metal contaminated soils. *Environ.* Chem Lett, 2009, 7, 1-19.

Kamaludeen, SPB; Ramasamy, K. Rhizoremediation of metals: harnessing microbial communities. *Ind. J. Microbiol*, 2008, 48, 80–88.

Giasson, P; Karam, A; Jaouich, A. *Arbuscular mycorrhizae and alleviation of soil stresses on plant growth.* In Siddiqui ZA; *et al.*, eds. Mycorrhizae: Sustainable Agriculture and Forestry, Springer Science + Business Media B.V.; 2008; 99–134.

Keck, RW; Redlich, L. Cadmium tolerance in soybeans. *Plant Physiol*, 1975, 56, Suppl. 13, (Abstract 68).

Khan, AG. Mycorrhizoremediation - an enhanced form of phytoremediation. *J. Zhejiang Univ. SCI B*, 2006, 7, 503-514.

Kopittke, PM; Dart, PJ; Menzies, NW. Toxic effects of low concentrations of Cu on nodulation of cowpea (*Vigna unguiculata*). *Environ. Pollut*, 2007, 145, 309-315.

Kramer, U; Chardonnens, AN. The use of transgenic plants in the bioremediation of soils contaminated with trace elements. *Appl. Microbiol. Biotechnol*, 2001, 55, 661-672.

Kuboi, T; Noguchi, A; Yazaki, J. Family-dependent cadmium accumulation characteristics in higher plants. *Plant Soil,* 1986, 92, 405-415.

Lasat, MM. Phytoextraction of metals from contaminated soil: a review of plant/soil/metal interaction and assessment of pertinent agronomic issues. *J. Hazard. Subst. Res*, 2000, 2, 1-25.

LeDuc, DL; Tarun, AS; Montes-Bayon, M; Meija, J; Malit, MF; Wu, CP. Overexpression of selenocysteine methyltransferase in *Arabidopsis* and *Indian mustard* increases selenium tolerance and accumulation. *Plant Physiol*, 2004, 135, 377-383.

Lee, J; Shim, D; Song, W-Y; Hwang, I; Lee, Y. Arabidopsis metallothioneins 2a and 3 enhance resistance to cadmium when expressed in *Vicia faba* guard cells. *Plant Mol. Biol*, 2004, 54, 805-815.

Leyval, C; Turnau, K; Haselwandter, K. Effect of heavy metal pollution on mycorrhizal colonization and function: physiological, ecological and applied aspects. *Mycorrhiza*, 1997, 7, 139-153.

Leyval, C; Joner, EJ. *Bioavailability of heavy metals in the mycorrhizosphereIn*: In Gobran GR; Wenzel WW; Lombi E. eds. Trace elements in the rhizosphere. Boca Raton, FL; CRC; 2001; 165–185.

Lin, S-L; Wu, L. Effects of copper concentration on mineral nutrient uptake and copper accumulation in protein of copper-tolerant and non tolerant *Lotus purshianus* L. *Ecotoxicol Environ. Safety*, 1994, 29, 214-228.

Li, X; Christie, P. Changes in soil solution Zn and pH and uptake of Zn by arbuscular mycorrhizal red clover in Zn-contaminated soil. *Chemosphere*, 2001, 42, 201–207.

Li, Y; Peng, J; Shi, P; Zhao, B. The effect of Cd on mycorrhizal development and enzyme activity of *Glomus mosseae* and *Glomus intraradices* in *Astragalus sinicus* L. *Chemosphere*, 2009, 75, 894-899.

Lin, A-J; Zhang, X-H; Wong, M-H; Ye, Z-H; Lou, L-Q; Wang, Y-S; Zhu, Y-G. Increase of multi-metal tolerance of three leguminous plants by arbuscular mycorrhizal fungi colonization. *Environ. Geochem. Health*, 2007, 29, 473-481.

Lodewyckx, C; Taghavi, S; Mergeay, M; Vangronsveld, J; Clijsters, H; Van Der Lelie, D. The effect of recombinant heavy metal-resistante bacteria on heavy metal uptake by their host plant. *Int. J. Phytoremediation*, 2001, 3, 173–187.

Lopez, ML; Peralta-Videa, JR; Benitez, T; Gardea-Torresdey, JL. Enhancement of lead uptake by alfalfa (*Medicago sativa*) using EDTA and a plant growth promoter. *Chemosphere*, 2005, 61, 595-598.

Luo, C; Shen, Z; Li, X. Enhanced phytoextraction of Cu, Pb, Zn and Cd with EDTA and EDDS. *Chemosphere*, 2005, 59, 1-11.

Ma, Y, Dickinson, NM, Wong, MH. Beneficial effects of earthworms and arbuscular mycorrhizal fungi on establishment of leguminous trees on Pb/Zn mine tailings. *Soil Biol. Biochem*, 2006, 38, 1403-1412.

Manier, N; Deram, A; Broos, K; Denayer, FO; Van Haluwyn, C. White clover nodulation index in heavy metal contaminated soils- a potential bioindicator. *J. Environ. Qual*, 2009, 38, 685-692.

McGrath, SP. *Phytoextraction for soil remediation*. In: Brooks RR, eds. Plants that hyperaccumulate heavy metals: their role in phytoremediation, microbiology, archaeology, mineral exploration and phytomining. Wallingford, Oxon, UK; NY: CAB International; 1998; 261-287.

Mehboob, I; Naveed, M; Zahir, ZA. Rhizobial association with non-legumes; mechanisms and applications. *Crit. Rev. Plant Sci*, 2009, 28, 432-456.

Metwally, A; Safronova, VI; Belimov, AA; Dietz, K-J. Genotypic variation of the response to cadmium toxicity in *Pisum sativum* L. *J. Exp. Bot*, 2005; 56: 167-178.

Murphy, A; Taiz, L. Comparison of metallothionein gene expression and non protein thiols in ten *Arabidopsis* ecotypes (correlation with copper tolerance). *Plant Physiol*, 1995, 109, 945-954.

Neumann, H; Bode-Kirchhoff, A; Madeheim, A; Wetzel, A. Toxicity testing of heavy metals with the Rhizobium-legume symbiosis: High sensitivity to cadmium and arsenic compounds. *Environ. Sci. Pollut. Res. Int*, 1998, 5, 28-36.

Ouziad, F; Hildebrandt, U; Schmelzer, E; Bothe, H. Differential gene expressions in arbuscular mycorrhizal-colonized tomato grown under heavy metal stress. *J. Plant Physiol*, 2005, 162, 634–649.

Pajuelo, E; Dary, M; Palomares, AJ; Rodriguez-Llorente, ID; Carrasco, JA; Chamber, MA. Biorhizoremediation of heavy metals toxicity using Rhizobium-legume symbioses. In: Dakora DF, *et al.*, eds. Biological Nitrogen Fixation: Towards Poverty Alleviation through sustainable Agriculture. *Springer Science and Business Media BV*; 2008; 101-104.

Pastor, J; Hernández, AJ; Prieto, N; Fernández-Pascual, M. Accumulating behaviour of *Lupinus albus* L. growing in a normal and a decalcified calcic luvisol polluted with Zn. *J. Plant Physiol*, 2003, 160, 1457-1465.

Peñalosa, JM; Carpena, RO; Vázquez, S; Agha, R; Granado, A; Sarro, MJ; Esteban, E. Chelate-assisted phytoextraction of heavy metals in a soil contaminated with a pyritic sludge. *Sci. Total Environ*, 2007, 378, 199-204.

Peralta-Videa, JR; Gardea-Torresdey, JL; Gomez, E; Tiemann, KJ; Parsons, JG; Cerillo, G. Effect of mixad cadmium, copper, nickel and zinc at different pHs upon alfalfa growth and heavy metal uptake. *Environ. Pollut*, 2002, 119, 291-301.

Pereira, SIA; Lima, AIG; Figueira, AMAP. Screening possible mechanisms mediating cadmium resistance in *Rhizobium leguminosarum* bv. *viciae* isolated from contaminated Portuguese soils. *Microbial Ecol*, 2006, 52, 176-186.

Pettersson, O. Differences in cadmium uptake between plant species and cultivars. *Swedish J. Agric. Res,* 1977, 7, 21-24.

Pichtel, J; Bradway, DJ. Conventional crops and organic amendments for Pb, Cd and Zn treatment at a severely contaminated site. *Bioresour. Technol*, 2008, 99, 1242-1251.

Piechalak, A; Tomaszewska, B; Barałkiewicz, D. Enhancing phytoremediative ability of *Pisum sativum* by EDTA application. *Phytochem*, 2003, 64, 1239-1251.

Pilon-Smith, E; Pilon, M. Phytoremediation of metal using transgenic plants. *Crit. Rev. Plant Sci*, 2002, 21, 439-456.

Polson, DE; Adams, MV. Differential response of navy beans (*Phaseolus vulgaris* L.) to Zn. I. Differential growth and element composition at excessive Zn levels. *Agron. J*, 1970, 62, 557-560.

Provorov, NA; Tikhonovich, IA. Genetic resources for improving nitrogen fixation in legume-rhizobia symbiosis. *Genet. Resour. Crop Evol*, 2003, 50, 89-99.

Rai, UN; Pandey, K; Sinha, S; Singh, A; Saxena, R; Gupta, DK. Revegetating fly ash landfills with *Prosopis juliflora* L.: impact of different amendments and *Rhizobium* inoculation. *Environ. Int*, 2004, 30, 293-300.

Rajkumar, M; Narasimha, M; Prasad, V; Freitas, H; Ae, N. Biotechnological applications of serpentine soil bacteria for phytoremediation of trace metals. *Crit. Rev. Biotechnol*, 2009, 29, 120-130.

Rani, A; Shouche, YS; Goel, R. Declination of copper toxicity in pigeon pea and soil system by growth-promoting *Proteus vulgaris* KNP3 strain. *Curr. Microbiol*, 2008, 57, 78-82.

Redon, PO; Béguiristain, T; Leyval, C. Differential effects of AM fungal isolates on *Medicago truncatula* growth and metal uptake in a multimetallic (Cd, Zn, Pb) contaminated agricultural soil. *Mycorrhiza*, 2009, 19, 187-195.

Requena, N; Jimenez, I; Toro, M; Barea, JM. Interactions between plant growth-promoting rhizobacteria (PGPR), arbuscular mycorrhizal fungi and *Rhizobium* spp. In the rhizosphere of *Anthyllis cytisoides*, a model legume for revegetation in mediterranean semi-arid ecosystems. *New Phytol*, 1997, 136, 667-677.

Rivera-Becerril, F; Calantzis, C; Turnau, K; Caussanel, JP; Belimov, AA; Gianinazzi, S, Strasser, RJ; Gianinazzi-Pearson, V. Cadmium accumulation and buffering of cadmium-induced stress by arbuscular mycorrhiza in three *Pisum sativum* genotypes. *J. Exp. Bot*, 2002, 53, 1177-1185.

Rodriguez, L; Rincón, J; Asencio, I; Rodríguez-Castellanos, L. Capability of selected crop plants for shoot mercury accumulation from polluted soils: phytoremediation perspectives. *Int. J. Phytoremediation*, 2007, 9, 1-13.

Rother, JA; Millbank, JW; Tornton, I. Nitrogen fixation by white clover (*Trifolium repens*) in grasslands on soils contaminated with cadmium, lead and zinc. *J. Soil Sci*, 1983, 34, 127-136.

Ruley, AT; Sharma, NC; Sahi, SV; Singh, SR; Sajwan, KS. Effects of lead and chelators on growth, photosynthetic activity and Pb uptake in *Sesbania drummondii* grown in soil. *Environ. Pollut*, 2006, 144, 11-18.

Safronova, VI; Stepanok, VV; Engqvist, GL; Alekseyev, YV; Belimov AA. Root-associated bacteria containing 1-aminocyclopropane-1-carboxylate deaminase improve growth and nutrient uptake by pea genotypes cultivated in cadmium supplemented soil. *Biol. Fertil. Soil*, 2006, 42, 267-272.

Safronova, VI; Piluzza, G; Belimov, AA; Bullitta S. Combined inoculation with rhizosphere and nodule bacteria improves growth, nutrition and heavy metal uptake of pasture legumes grown in polluted mine waste. *Biol. Fertil. Soil* 2010, (BFSO-D-10-00240, under review).

Sahi, SV; Bryant, NL; Sharma, NC; Singh, SR. Characterization of a lead hyperaccumulator shrub, *Sesbania drummondii*. *Environ. Sci. Technol*, 2002, 36, 4676-4680.

Saleem, M;·Arshad, M;·Hussain, S; Bhatti, AS. Perspective of plant growth promoting rhizobacteria (PGPR) containing ACC deaminase in stress agriculture. *J. Ind. Microbiol, Biotechnol*, 2007, 34, 635-648.

Senthilkumar, P; Prince, WS; Sivakumar, S; Subbhuraam, CV. *Prosopis juliflora* - a green solution to decontaminate heavy metal (Cu and Cd) contaminated soils. *Chemosphere*, 2005, 60, 1493-1496.

Shen, ZG; Li, XD; Wang, CC; Chen, HM; Chua, H. Lead phytoextraction from contaminated soil with high-biomass plant species. *J. Environ. Qual*, 2002, 31, 1893-1900.

Sinha, S; Rai, UN ; Bhatt, K ; Pandey, K ; Gupta, AK. Fly-ash-induced oxidative stress and tolerance in *Prosopis juliflora* L. grown on different amended substrates. *Environ. Monit. Assess,* 2005, 102, 447-457.

Sing, RP; Dhania, G; Sharma, A; Jaiwal, PK. *Biotechnological approaches to improve phytoremediation efficiency for environment contaminants,* In: Singh SN, Tripathi RD, eds. Biotechnological Approaches in Phytoremediation. Berlin Heidelberg: Springer; 2007; 223-258.

Sriprang, R; Hayashi, M; Yamashita, M; Ono, H; Saeki, K; Murooka, Y. A novel bioremediation system for heavy metals using the symbiosis between leguminous plant and genetically engineered rhizobia. *J. Biotechnol*, 2002, 99, 279-293.

Sriprang, R; Hayashi, M; Ono, H; Takagi, M; Hirata, K; Murooka, Y. Enhanced accumulation of Cd2+ by a *Mesorhizobium* sp. transformed with a gene from *Arabidopsis thaliana* coding for phytochelatin synthase. *Appl. Environ. Microbiol*, 2003, 69, 1791-1796.

Sugiyama, M; Ae, N; Arao, T. Role of roots in differences in seed cadmium concentration among soybean cultivars - proof by grafting experiment. *Plant Soil*, 2007, 295, 1-11.

Tank, N; Saraf, M. Enhancement of plant growth and decontamination of nickel-spiked soil using PGPR. *J. Basic Microbiol*, 2009, 49, 195-204.

Tripathi, M; Munot, HP; Shouche, Y; Meyer, JM; Goel, R. Isolation and functional characterization of siderophore-producing lead- and cadmium-resistant *Pseudomonas putida* KNP9. *Curr. Microbiol*, 2005, 50, 233-237.

Tsyganov, VE; Belimov, AA; Borisov, AY; Safronova, VI; Georgi, M; Dietz, K-J; Tikhonovich, IA. A chemically induced new pea (*Pisum sativum* L.) mutant SGECd^t with increased tolerance to and accumulation of cadmium. *Ann. Bot*, 2007, 99, 227-237.

Tsyganov, VE; Zhernakov, AI; Khodorenko, AV; Kisutin, PY; Belimov, AA; Safronova, VI; Naumkina, TS; Borisov, AY; Lindblad, P; Dietz, K-J; Tikhonovich, IA. Mutational analysis to study the role of genetic factors in pea adaptation to stresses during development its symbioses with *Rhizobium* and mycorrhizal fungi. In: Wang YP, *et al.*, eds. *Biological Nitrogen Fixation, Sustainable Agriculture and the Environment. the Netherlands: Springer;* 2005; 279-281.

Vamerali, T; Bandiera, M; Mosca, G. Field crops for phytoremediation of metal-contaminated land: a review. *Environ. Chem. Lett*, 2010; 8: 1–17.

Van Nevel, L; Mertens, J; Oorts, K; Verheyen, K. Phytoextraction of metals from soils: how far from practice. *Environ. Pollut*, 2007, 150, 31-40.

Van Loon, LC; Bakker, PAH; Pieterse, CMJ. Systemic resistance induced by rhizosphere bacteria. *Annu. Rev. Phytopathol*, 1998, 36, 453–483.

Vassilev, A; Schwitzguebel, JP; Thewys, T; Van Der Lelie, D; Vangronsveld, J. The use of plants for remediation of metal-contaminated soils. *Sci. World J*, 2004, 16, 9-34.

Vázquez, S; Agha, R; Granado, A; Sarro, MJ; Esteban, E; Peñalosa, JM; Carpena, RO. Use of white lupin plant for phytostabilization of Cd and As polluted acid soil. *Water Air Soil Pollut*, 2006, 177, 349-365.

Vessey, JK. Plant growth promoting rhizobacteria as biofertilizers. *Plant Soil*, 2003, 255, 571-586.

Vivas, A; Azcón, R; Biró, B; Barea, JM; Ruiz-Lozano, JM. Influence of bacterial strains isolated from lead-polluted soil and their interactions with arbuscular mycorrhizae on the growth of *Trifolium pratense* L. under lead toxicity. *Can. J. Microbiol*, 2003a, 49, 577-588.

Vivas, A; Barea, JM; Azcón, R. Interactive effect of *Brevibacillus brevis* and *Glomus mosseae*, both isolated from Cd contaminated soil, on plant growth, physiological mycorrhizal fungal characteristics and soil enzymatic activities in Cd polluted soil. *Environ. Pollut*, 2005, 134, 257-266.

Vivas, A; Vörös, I; Biró, B; Campos, E; Barea, JM; Azcón, R. Symbiotic efficiency of autochthonous arbuscular mycorrhizal fungus (*G. mosseae*) and *Brevibacillus* sp. isolated from cadmium polluted soil under increasing cadmium levels. *Environ. Pollut*, 2003b, 126, 179-189.

Vivas, A; Biró, B; Ruíz-Lozano, JM; Barea, JM; Azcón, R. Two bacterial strains isolated from a Zn-polluted soil enhance plant growth and mycorrhizal efficiency under Zn-toxicity. *Chemosphere*, 2006, 62, 1523-1533.

Walker, DJ; Bernal, MP; Correal, E. The influence of heavy metals and mineral nutrient supply on *Bituminaria bituminosa*. *Water Air Soil Pollut*, 2007, 184, 335-345.

Wani, PA; Khan, MS; Zaidi A. Effect of metal-tolerant plant growth-promoting *Rhizobium* on the performance of pea grown in metal-amended soil. *Arch. Environ. Contam. Toxicol*, 2008a, 55, 33-42.

Wani, PA; Khan, MS; Zaidi, A. Impact of zinc-tolerant plant growth-promoting rhizobacteria on lentil grown in zinc-amended soil. *Agron. Sustain. Dev*, 2008b, 28, 449-455.

Wani, PA; Khan, MS; Zaidi, A. Chromium-reducing and plant growth-promoting *Mesorhizobium* improves chickpea growth in chromium-amended soil. *Biotechnol. Lett*, 2008c, 30, 159-163.

Welch, RM; LaRue, TA. Physiological characteristics of Fe accumulation in the 'Bronze' mutant of *Pisum sativum* L., cv 'Sparkle' E107 (*brz brz*). *Plant Physiol*, 1990, 93, 723-729.

Wenzel, WW; Unterbrunner, R; Sommer, P; Sacco, P. Chelate assisted phytoextraction using canola (*Brassica napus* L.) in outdoors pot and lysimeter experiments. *Plant Soil*, 2003, 249, 83-96.

Wenzel, WW. Rhizosphere processes and management in plant-assisted bioremediation (phytoremediation) of soils. *Plant Soil*, 2009, 321, 385-408.

White, MC; Decker, AM; Chancy, RL. Differential cultivar tolerance in soybean to phytotoxic levels of soil Zn. I. Range of cultivar response. *Agron. J*, 1979, 71, 121-126.

Wu, L; Lin, S-L. Copper tolerance and copper uptake of *Lotus purshianus* (Benth.) Clem. and Clem and its symbiotic *Rhizobium loti* derived from a copper mine waste population. *New Phytol*, 1990, 116, 531-539.

Wu, L. *Colonization and establishment of plants in contaminated environments*. In: Shaw AJ, eds. Heavy metal tolerance in plants: evolutionary aspects. Boca Raton, FL: CRC; 2000; 269-284.

Zhang, RQ; Tang, CF; Wen, SZ; Liu, YG; Li, KL. Advances in research on genetically engineered plants for metal resistance. *J. Integr. Plant Biol*, 2006, 48, 1257−1265.

Zwarich, MA; Mills, JG. Heavy metal accumulation by some vegetable crops grown on sewage-sludge-amended soils. *Can. J. Soil Sci*, 1982, 62, 243-247.

In: Handbook of Phytoremediation
Editor: Ivan A. Golubev

ISBN: 978-1-61728-753-4

Chapter 14

PHYTOREMEDIATION TECHNOLOGIES FOR THE REMOVAL OF TEXTILE DYES - AN OVERVIEW AND FUTURE PROSPECTS

Sanjay P. Govindwar* ***and Anuradha N. Kagalkar***
Department of Biochemistry, Shivaji University,
Kolhapur-416 004, India

ABSTRACT

Phytoremediation which involves the use of plants and rhizospheric organisms for the removal of pollutants is an emerging technology for the clean up of contaminated sites. The removal of textile dyes mediated by plants has been one of the most neglected areas of phytoremediation research. Dyes, which are primary constituents of the wastes from textile industry effluents, constitute a group of recalcitrant compounds, many of which are known to have toxic and carcinogenic effects. Hence, the review focuses on the studies of the mechanisms adopted by plants in the removal of textile dyes and the future scope for research in this area which will help in broadening the horizons of phytoremediation technologies. Plant species many a times referred to as 'green livers', are known to possess a wide range of detoxifying and biotransforming enzymes some of which may also be secreted extracellularly in the rhizosphere and can bring about the transformation of organic pollutants such as textile dyes.

The use of *in vitro* plants for phytoremediation studies can help to explore the enzymatic status and the products of metabolism of the dye, thus providing a new dimension to phytoremediation studies. The use of transgenic plants with microbial genes can combine the advantages of both plant and microbial systems for enhanced dye degradation. Biotechnological approaches involving the development of hairy roots and suspension cultures may find good utility in phytoremediation studies. The ultimate aim of phytoremediation involves applying these well studied plant systems at the contaminated sites which may constitute the development of constructed wetlands for on-site treatment of industrial effluents.

* Department of Biochemistry, Shivaji University, Kolhapur- 416 004, India. Email: spgovindwar@rediffmail.com; spg_biochem@unishivaji.ac.in ; Tel: +91-231-2609152. Fax: +91-231-2691533.

INTRODUCTION

Since pre historic times innumerable uses of plants as sources of food, shelter, fuel etc., have been known to mankind. But, the newer approach of phytoremediation which involves the use of plant systems and/or rhizospheric organisms to remove content, inactivate or degrade harmful environmental contaminants and to revitalize contaminated sites (Vangronsveld et al, 2009), is an upcoming research area in the field of environmental biotechnology. Conventional techniques for bioremediation that involve the digging up of contaminated soils and disposal of the wastes to a landfill, lead to contamination elsewhere and can create significant risks in excavation, handling and transport of hazardous materials (Vidali, 2001). The chemical treatment methods used have multiple disadvantages such as their high cost, coupled with the formation of a large amount of sludge and the emission of toxic substances (Senan and Abraham, 2004), because of which bioremediation methodologies can be used as alternative technologies for the removal of industrial wastes. Microbial bioremediation processes for the removal of hazardous compounds, have received quite a lot of focus from researchers all over the world because of the high potentiality of prokaryotic systems to perform a variety of functions. But, the use of phytoremediation processes for the removal toxicants (especially textile dyes) is comparatively an unexplored methodology since the fact that plants also possess some inherent metabolic pathways that can breakdown a wide range of toxicants (Chaudhry et al, 2005) was much less realized. Since researchers have now begun to realize the potential of plant systems as effective remediating agents, this new area of phytoremediation has started gaining importance from academicia and industry (Cluis, 2004). Since plants are autotrophic systems of large biomass and require little nutrient input, phytoremediation technologies are easier to manage than microbial bioremediation systems and offer cost effective and aesthetically appealing options for environmental clean up (Cluis, 2004). Afforestation is one of the prescribed ways for minimizing the green house gases in the environment and reducing the effects of global warming since, plants have been known for their consumption of CO_2 and more recently of other gaseous industrial by products. Therefore, the value of plants to counterbalance the hazards of industrialization processes is being appreciated (Cumnningham and Ow, 1996). Phytoremediation can thus serve dual purposes.

The release of large amount of toxic wastes into water bodies is one of the consequences of increasing urbanization and industrialization in the modern world. A variety of organic (pesticides, explosives such as TNT, petrochemicals, chlorinated solvents, etc.) and inorganic (radionuclides, heavy metals such as mercury, lead, etc.) wastes which have toxic effects on the ecosystem have been contaminating our natural resources (Cluis, 2004). Out of the different types of pollutants released, dyes which are released by textile, dyestuff and dyeing industries constitute one recalcitrant group and are known to have carcinogenic and mutagenic effects with a potential toxicity to all life forms (Bafana et al, 2009). Most of the research involving phytoremediation technologies has been focused on the removal of heavy metals and a few organic compounds such as pesticides, polycyclic aromatic hydrocarbons etc. from the environment. The removal of textile dyes mediated by plant systems is still a much unexplored area of phytoremediation research. Hence, the article aims at reviewing the basic research and mechanisms involving the removal of dyes by plants and the application of these technologies at the dye contaminated sites with an insight into the future perspectives of research in this area.

Dyes-Toxicity and Need for Phytoremediation

Dyes are known to have complex structures that are difficult to degrade (Nilratnisakorn et al, 2007). With the advancement of technologies, enhancement has been made in dye properties so that they provide resistance to fading, provide improved delivery to fabrics and have increased variety of shades. These additional properties make them highly resistant to environmental degradation, thus increasing pollution (Togo et al, 2008). Sulfonated anthraquinones are generally the parent compounds for a vast array of dyes and thus the waste waters of textile industries are likely to contain these compounds which are recalcitrant and toxic (Page and Schwitzguébel, 2009). The difference in the chromophoric groups of dyes facilitates their classification into different types such as azo, triphenylmethane, anthraquinone, indophenol, diazonium, quinone dyes etc. Moreover, the nature of substituents attached to the basic aromatic ring structure also differs because of which they are not uniformly susceptible to bioremediation (Aubert and Schwitzguébel, 2004). In case of sulfonated dyes, the organosulfonate group plays an important role in altering the solubility and dispersion properties of the xenobiotic molecule and increases its recalcitrance to environmental breakdown, because of the thermodynamically stable carbon-sulfur bond (Duc et al, 1999). The mechanisms for the carcinogenicity of azo dyes that have been identified include metabolic activation to reactive electrophilic intermediates that covalently bind to DNA. Triphenylmethane dyes are known to cause reproductive abnormalities in rabbits and fish (Chen et al, 2010). Sometimes the products formed after the processing of these dyes themselves are toxic. In the environment, azo bonds of these dyes are reduced to liberate benzidine and other aromatic amines, which may cause adverse systemic health effects or cancer. Urinary bladder cancer is the most common form of cancer caused by exposure to benzidine. Stomach, kidneys, brain, mouth, esophagus, liver, and gall bladder might also be targets (Bafana et al, 2009). Toxicity of reactive dyes has been reported at concentrations as low as 5.2 mg/l (Nilratnisakorn et al, 2007). Hence, it is extremely important to implement technologies that completely remove such recalcitrant compounds from the environment or biotransform them into products that have reduced toxicity. Hence, for the targeted removal of such hazardous wastes from industrial effluents, plants can be used as efficient systems.

Selection of Plant Systems for Remediation of Textile Dyes

For the removal of textile dyes from the environment, the selection of an appropriate plant with certain desirable characteristics is one of the most important preliminary steps in phytoremediation research. Though several plants have shown the ability to remediate contaminated soils; non edible plants are generally selected to be applied onto dye contaminated sites. Most of the studies on phytoremediation of textile dyes demonstrate their removal through either degradation of the dye or the adsorption and/or accumulation of the dye. Accumulation of organic compounds such as sulfonoaromatics has been shown in Rhubarb species (Duc et al, 1999). Compounds accumulated in the plant roots could be further translocated to shoots and leaves. To prevent the accumulated dye compounds or their metabolites from entering the food chain, the use of non edible plants is always preferred. Different types of grasses, ferns, weeds or agricultural wastes have been suggested and tested for the removal of dyes. *Phragmites* species have shown immense potential to remediate textile waste waters. *Phragmites australis*, a reed which is a component of the wetland community has been extensively studied for remediation of textile effluents and mainly with respect to the removal of the dye, Acid Orange 7 (Carias et al, 2007). Many native populations of *Phragmites autralis* are benign in that they pose little or no threat to other species. Among 10 different species of macrophytes screened, *Phragmites karka* was found to have broad amplitude of pH tolerance and was found to be growing well in alkaline, neutral and acidic textile wastewaters resulting in considerable shoot density and biomass to achieve maximum translocation of water and assimilation of nutrients. This makes the plants highly suitable for the treatment of textile waste waters that may be contaminated with different types of acidic as well as basic dyes. The good growth of underground organs in these species thereby provides maximum surface area to assimilate pollutants (Sharma et al, 2005). In addition, the plant should be fast growing and should have a deep rooted system that enables it to reach the pollutants easily. Larger biomass and surface area of the plant system can facilitate more efficient removal of the dye. Kagalkar et al have demonstrated that increase in plant biomass in terms of increasing number of plants used, gave higher % decolorization values. Reports have shown the efficient degradation of the dye Direct Red 5B (DR5B) with *Blumea malcolmii*, a deep rooted and fast growing plant system that forms sufficiently large biomass and can grow in soils with little nutrient availabilities (Kagalkar et al, 2009). Thus, a plant that has the ability to remove dye molecules from the environment and also have a good biomass can prove to be potent for phytoremediation. The plant *Typhonium flagelliforme* which has recently been reported for degradation of the dye Brilliant Blue R, exhibits dye degradation capacity when used even in distilled water, devoid of any other nutrients. The use of such plants can help to reduce the overall cost of the experiments (Kagalkar et al, 2010). Not all plants will be able to demonstrate similar responses or have similar removal rate for all dyes. In addition, for the phytotreatment of textile dyes, additional perspectives that should be kept in mind while selecting the plant species include high uptake rate of the pollutant, high translocation factor (TF) in case of phytoextraction, presenting the ability to translocate contaminants to the shoot and high tolerance level towards the dye (Zabłudowska et al, 2009). Hence, extensive screening of different plant species can help us to understand the selective abilities of a particular plant to remove a dye or a group of dyes. Moreover, textile effluents are generally mixtures of different dyes because of which plants that can be potent phytoremediators of textile industry wastes will be the ones that will be able to demonstrate the capacity to remove a large number and a diverse group of textile dyes eg., the species *B. malcolmii* showed the capacity to decolorize five different dyes, namely Direct Red 5B, Reactive Red 2, Methyl Orange, Malachite Green and Golden Yellow HER to varying extents (Kagalkar et al, 2009). *Typhonium* plantlets that have been studied for their phytoremediation potentialities, showed the additional advantage of remediating textile effluents and synthetic mixture of dyes along with 8 individual dyes, out of which maximum decolorization was obtained for the dye Brilliant Blue R (BBR) which was about 80%. To quantitate the % removal of color from dye effluents or synthetic dye mixtures, American Dye Manufacturers'

Institute (ADMI 3WL) tristimulus filter method is used. The % removal of ADMI in case of mixture of dyes was 47% while in case of textile effluents was found to be 28% (Kagalkar et al, 2010). Many edible plants have been known to possess dye decolorizing abilities because of the rich enzymatic status of these plants. Though such plants are unsuitable for field applications, their enzymes can be extracted and used for degradation of various textile dyes. The species of *Sorghum vulgare*, *Phaseolous mungo* and *Brassica juncea* have shown the potential to decolorize the dye Reactive Red 2 and have also demonstrated to possess abilities to decolorize and detoxify textile effluents (Ghodake et al, 2009). Even though different plants are capable of degrading the same dye molecule, the products formed after degradation are likely to be different indicating that the pattern of transformation of the xenobiotic molecule is dependent upon the plant species (Page and Schwitzguébel, 2009). Plants that degrade the dye molecules into non toxic products are preferable for phytoremediation. Hence, the selection of the plant will depend upon the genetic make up of the plant that manifests in terms of varied enzyme activities in the plant and differential absorptive capacities resulting into variable patterns for the removal of dyes. Moreover, all the plants selected should be in the same stage of growth and should have almost equivalent dry weights and should have almost similar root and shoot lengths (Kagalkar et al, 2009) that can help to achieve reproducibility of results. Further, the plants selected should preferably be from the same area since factors such as age of the plant, soil conditions, nutrient status, light availability etc. are factors that can affect the removal of the dye. In addition, the use of flowering plants for the removal of textile dyes would offer aesthetically appealing systems and will serve dual purposes of bioremediation and will also allow the flowers to be used for decorative purposes, thus serving economic benefits.

Plant Mechanisms for the Removal of Dyes

A) Mechanisms Involving Adsorption and/or Accumulation of Textile Dyes

Plant mechanisms behind the removal of textile dyes, which are a group of organic pollutants, may be diverse. Though phytodegradation or phytotransformation are the most predominantly observed mechanisms adopted by plants for the degradation of organic compounds, the removal of textile dyes by plants also utilizes the mechanisms of adsorption and accumulation on plant surfaces. It has been established that the binding of xenobiotics to roots occurs by adsorption followed by its absorption into the plants (Davies at al, 2005). *Posidonia oceanica* leaf sheaths have shown effective adsorptive removal of the textile dye, Reactive Red 228. Moreover, the authors also demonstrated changes in the adsorptive capacities with changes in factors such as temperature and pH. Increase in temperature favored the better adsorptive removal of the dye which was probably because of the greater movement of the adsorbent material. Besides, the highest dye removal efficiency was found at pH 5 which might correspond to the rate of dissociation of the studied dye with maximum ionization of the molecule. The use of orange peels, banana peels, neem leafs, peanut hulls and agricultural wastes have also been suggested for the removal of various dyes (Ncibi et al, 2007). Untreated pulverized plant leaves of *Salsola vermiculata* revealed interesting adsorptive properties for Methylene Blue and iodine from aqueous solutions. Further, the activation of these plant leaves by treatment with zinc chloride gave better adsorptive properties. In this plant too, the effect of pH on the adsorption capacity of the chemically activated plant was investigated which showed a decrease in the adsorption capacity at lower pH values which may be because of the competition of protons with the dye molecules for available adsorption sites. The iodine number value which determines the capacity of the adsorbent to remove color from the solution, evaluated in terms of adsorption of iodine from the adsorbent pointed out that significant additional surface area can be achieved through zinc chloride activation and that microporosity contributes considerably to the total surface area of the prepared material making it a very good adsorbent for small compounds (Bestani et al, 2008). Thus, the inherent adsorptive capacity of the plant material can be enhanced through such activation techniques. The significant adsorption of Methylene Blue by this plant shows its potentiality in the remediation of textile dyes and effluents. Adsorptive removal of the dye Malachite Green by the roots of *Blumea malcolmii* was also reported where the % adsorption of the dye by the plant was approximately 45% (Kagalkar et al, 2009). The adsorption of a number of dyes was reported on the roots of *Typhonium flagelliforme* plantlets that were used for dye degradation experiments (Kagalkar et al, 2010). Phytoremediation processes not only involve the adsorption of the dye on the root or shoot systems of the plant but also involve the accumulation of dyes into plant tissues. Narrow-leaved Cattails (*Typha angustifolia* Linn.) that has the capacity to absorb a large amount of nutrients, has been demonstrated for the removal of the commercial diazo reactive dye, Reactive Red 141. The plant demonstrates the ability of 60% removal of the dye. After 28 days of exposure of the plant to synthetic reactive waste water containing the dye, the intercellular space of the plant showed the presence of the dye which was confirmed with transmission electron microscopy connected with electron dispersive X ray spectroscopy (TEM-EDX). Moreover, the plant also shows the precipitation of metal complexes with the dye which according to the authors are probably mechanisms to avoid damage to the plant (Nilratnisakorn et al, 2007). Rhubarb (*Rheum rabarbarum*) species have shown to accumulate synthetic anthraquinones which are starting materials for the production of a large number of synthetic dyes. The transpiration stream concentration factor (TSCF) was determined to detect the concentration of the accumulated compound in the xylem sap of these plants (Aubert and Schwitzguébel, 2004). When this value exceeds unity, the movement of the compound is faster than water. Among the different sulfonated anthraquinones the TSCF value obtained for anthraquinone-1-sulfonic acid was 2.5 which helped the authors to conclude that the movement of the pollutant was faster than water. Anthraquinone-2-sulfonic acid and anthraquinone-2,6-disulfonic acid also showed values higher than unity. Plant screening showed that *Rheum rabarbarum* and *Rumex hydrolapatum* were most efficient for the accumulation of the five selected sulphonated anthraquinones (determined by the transpiration stream coefficient factor) which are precursors for many different dyes (Aubert and Schwitzguébel, 2004).

B) Plant Stress Response and Mechanisms for the Degradation of Dyes

Though adsorption and accumulation of dyes are important ways of phytoremediation, these processes lead to the mere concentration of pollutants from textile effluents onto and/or into plant surfaces and do not lead to the complete eradication of the pollutant. Thus, the phytotransforming abilities of a plant are of a greater significance since they can be employed to either completely degrade the dye or to transform it into products which are non toxic and can be safely released into the ecosystem. Because plants are static and live in a competitive and sometimes hostile environment, they have evolved mechanisms that protect them from environmental abiotic stress, including the detoxification of xenobiotic compounds (Page and Schwitzguébel, 2009). Plant mechanism is diverse and can be used to treat compounds not degradable by bacteria. Different aromatic compounds such as derivatives of sulfonated anthraquinones occur naturally in several plant genera and thus these plants are likely to possess enzymes that can accept these aromatic compounds as substrates and process them (Aubert and Schwitzguébel, 2004). An important step in the removal of sulfonated anthraquinones appears to be involving the action of dioxygenases adding oxygen across the double bond bearing the sulfonate group leading to its elimination (Schwitzguébel et al., 2002). Plant degradation of textile dyes may either be intracellular involving enzyme systems inside plant tissues or it may involve degradation with the help of extracellular enzymes secreted by the plant in rhizosphere regions. The hydrophobicity of a compound can affect its uptake or translocation. Moderately hydrophobic dyes can be most readily taken up by the plant or translocated within the plant. Hydrophobic compounds can also be bound to root surfaces or partition into roots but cannot be further translocated into the plant (EPA, 2000). Thus, phytodegradation of the dye outside the plant will not depend upon plant uptake.

Plant detoxification pathways comprise of three phases with specific enzymes. Phase I enzymes like cytochrome P450 and peroxidases transform xenobiotics (mainly by oxidation reactions) in order to allow the conjugation of the oxidized xenobiotic with glutathione catalyzed by glutathione S-transferase, in phase II. Phase III involves the translocation of these conjugates into vacuoles (Carias et al, 2008). The degradation process of dyes has shown to involve a significant role of peroxidases that are enzymes which are typically activated as an enzymatic stress response and are comparatively more extensively studied in plants than the other enzymes that can have a role in textile dye degradation. This strategy where peroxidases are activated as stress response appears to be very interesting as plants not only allow the pretreatment of specific recalcitrant compounds by changing their physicochemical properties and making them more amenable for treatment but also in their transformation into innocuous products (Carias, 2008). Lignin peroxidases (LiP) are the primary enzymes that are involved in the lignoloysis of wood and they oxidize lignin structures by one electron yielding cation radical intermediates that undergo spontaneous fission reactions. Studies with lignin model compounds have shown that LiP cleaves the predominant aryl glycerol β-aryl ether substructure of lignin which accounts for about half the total polymer, between Cα and Cβ of its propyl side chain (Sarkanen et al, 1991). They are known to act upon a variety of xenobiotic compounds and mediate the symmetric or asymmetric cleavage of many dyes at their C-C linkages. Thus, this enzyme is found to be one of the most predominant enzymes involved in the transformation and/or degradation of dyes. Peroxidases are heme containing enzymes able to oxidize a wide range of organic and inorganic compounds, using hydrogen peroxide as a co-substrate. They are non specific and can use a broad range of electron donor substrates. The first step of action of this enzyme leads to the cleavage of hydrogen peroxide molecule with the concomitant production of water and incorporation of one of the oxygen atoms of hydrogen peroxide into the initial compound. The next two steps involve the reduction of enzyme in order to regenerate it. Pollutants like azo dyes act as electron donor substrates (Davies et al, 2005). Many plant species have shown the presence of peroxidases in their tissues. The involvement of these enzymes was shown in curly dock plant that was used for the decolorization of the dye Remazol Brilliant Blue R where it was found that addition of hydrogen peroxide into the cultivation medium of the plant greatly stimulated the decolorization of this dye (Takashi et al, 2005). Similar observations were noted in case of degradation of the dye Acid Orange 7 by *Phragmites australis*. On the addition of H_2O_2 at 192 h of exposure of the plant to the dye, a significant reduction in the absorbance of the dye was found after 48 h and 120 h of H_2O_2 addition (Davies et al, 2005). As a response to the polydye R-478, the MPH-4 clonal lines of *Mentha pulegium* showed an increase in the guaicol peroxidase activity while decrease in the phenolic content which probably indicates that the phenolics had been used up for the lignin biosynthesis process (Strycharz and Shetty, 2002a), which could be stimulated because of the increased peroxidase activities. Similar results with phenolic content and guaicol peroxidase activity were demonstrated in case of oregano (*Origanum vulgare* L.) cell lines (Strycharz and Shetty, 2002b). In these studies, the authors speculate that the dye might be used as a substrate for peroxidase cross linking and might participate in lignification process though there is no evidence provided. An induction in intracellular peroxidase activities in the roots was also found during the degradation of the dye DR5B, mediated by *Blumea malcolmii* after 3 days of exposure to the dye when the plant showed 60% decolorization of the dye. The role of peroxidase in the degradation of DR5B, mediated by *Blumea*, can be predicted in catalyzing the asymmetric C-C cleavage of the intermediates formed during processing of the dye by the plant enzyme systems (Kagalkar et al, 2009). The root tissues of the plant *Typhonium flagelliforme* also showed induction of intracellular and extracellular peroxidase values upon exposure to the dye Brillaint Blue R (Kagalkar et al, 2010). The exposure of the dye Reactive Red 198, to *Tagetes patula* L. hairy roots also demonstrated a significant induction in peroxidase activities. Both *Typhonium* and marigold peroxidases can also be predicted to be having a similar role in the degradation of Brilliant Blue R and Reactive Red 198 respectively, where the enzyme probably catalyzes the asymmetric cleavage of the original dye molecule itself (Patil et al, 2009; Kagalkar et al, 2010). The presence of oxidoreductive enzymes was assessed in three types of plants, Alfalfa, Mustard and Cresswhich which showed that peroxidases were the most dominating enzyme species found in the root, shoot and exudates of these plants (Gramss and Rudeschko, 1998).

Cytochrome P450 monooxygenases represent a multigenic family of enzymes, involved in the detoxification process of many xenobiotic compounds. In addition to activating xenobiotics, cytochrome P450 plays an important role in the normal secondary metabolism of plants, which produce compounds involved in cell signaling and defense mechanisms. A significant activity of the enzymes was detected in the leaves of Rhubarb with different anthraquinones as substrates. Lower activities of these enzymes were also detected in roots and petioles. In contrast, when the authors used common sorrel no significant difference was found in the activities of the plants which were exposed and those which were unexposed to the anthraquinones. The higher activities of the enzyme in the leaf tissues of Rhubarb indicated that the sulfonated anthraquinones were taken up by the plant and were translocated in the leaf tissues (Page and Schwitzguébel, 2009). The ability of these plants to accumulate sulfonated anthraquinones has been confirmed by the studies performed by Aubert and Schwitzguébel (Aubert and Schwitzguébel, 2004). This assumption was supported by the data obtained with capillary electrophoresis which confirms the presence of anthraquinones in leaves of Rhubarb plants (Aubert and Schwitzguébel, 2002). Further, new metabolites were also found to be present in leaf tissues of plants exposed to the anthraquinones which indicate the biotransformation of these compounds, but the metabolites have not yet been identified (Aubert and Schwitzguébel, 2004). These results indicate that Rhubarb probably has the ability to not only accumulate sulfonated anthraquinones but also biotransform them.

Another detoxifying enzyme that has been studied in plants is glutathione S-transferase. The best known role of this enzyme is in the detoxification of endobiotic and nucleophilic xenobiotic compounds by covalently linking GSH to a broad variety of reactive electrophilic and hydrophobic substrates, which results in the formation of more polar and less reactive conjugates. An increase in the activities of this enzyme has been studied in *Phragmites australis* in response to Acid Orange 7 (Carias et al, 2008). The plant has shown the potential to degrade the dye with a removal efficiency of 68+8% when used in constructed wetlands. The presence of a dye in the vicinity of a plant can offer stress conditions which the plant will try to overcome with inherent stress response mechanisms. Plants react to stress conditions by increasing the concentration of reactive oxygen species using enzymes like NADPH oxidase in order to signal plant defences. This in turn leads to the activation of antioxidant scavenging enzymes to remove the reactive oxygen species. Superoxide dismutase (SOD) converts the reactive oxygen species into hydrogen peroxide which will then be converted into water and oxygen, by the action of peroxidase, catalase and ascorbate peroxidase. An induction in the activities of these antioxidative and detoxifying enzymes was observed following exposure to the dye and these enzymes may play an active role in the degradation mechanism of this dye. The authors speculated that this induction could be attributed either to the activation of the antioxidant scavenging enzyme battery or could also be due to the *de novo* synthesis of enzymatic proteins to face the stress conditions. Molecular studies performed shows that Acid Orange 7 acts as a chemical stressor agent for *Phragmites australis*, activating the gene expression for Cu/Zn SOD, Mn SOD, glutathione peroxidase and catalase isoforms, which leads to the conclusion that this gene over expression is related with the sudden production of reactive oxygen species after exposure to high concentrations of the dye Acid Orange 7 (Davies et al, 2009). Increase in the content of dehydroascorbate reductase which is involved in the regulation of ascorbate glutathione pathway was also observed in *Phragmites* plants exposed to the dye. The dye seems to be incorporated in the cytosol of the plants, where via the detoxification pathway; it is modified (Phase I), conjugated (Phase II) and translocated (Phase III) into the vacuoles. Increase in the activity of glutathione S-transferase indicates that the dye is being compartmentalized in the cells (Carias et al, 2008).

Laccases constitute another major class of degradative enzymes that have been studied. They constitute one class of polyphenol oxidases that catalyze the oxidation of various substituted phenolic compounds by using molecular oxygen as an electron acceptor. They catalyze the removal of hydrogen atom from the hydroxyl group of ortho- and para-substituted mono and polyphenolic substrates and from aromatic amines by one-electron abstraction to form free radicals capable of undergoing further depolymerization, repolymerization, demethylation or quinone formation (Abadulla et al., 2000). The inherent ability of plant systems to produce these enzymes probably stems from the role of these enzymes in plant development and lignification. Though laccases have been found in various plants such as peach, sycamore, tobacco and poplar, they have not been characterized or used extensively because their detection and purification is often difficult as crude plant extracts often contain a large number of oxidative enzymes with broad substrate specificities (Sharma et al, 2007). Laccases can have immense potential in the detoxification and degradation of wastes form textile industries. Fungal and bacterial laccases have been well characterized and their role in bacterial and fungal dye decolorization has been well illustrated. Though studies on the involvement of plant laccases in phytoremediation of dyes are few, recently Ghodake et al reported the significant induction of intracellular laccase in roots of *Brassica juncea* upon exposure to the dye Reactive Red 2 and textile effluent (Ghodake et al, 2009). Laccase was also found to be induced in marigold hairy roots during decolorization of the dye Reactive Red 198 (Patil et al, 2009). An interesting feature of these studies indicated the absence of this enzyme in roots unexposed to the dye.

Tyrosinases are also a group of copper containing polyphenol oxidases that catalyze two type of reactions, the o-hydroxylation of some monophenols (monophenolase, cresolase) and the oxidation of o-diphenols to o-quinones (diphenolase, catecholase) using molecular oxygen (Chen and Flurkey, 2002). These enzymes have been studied for their role in microbial dye degradation but dye degradation mechanisms associated with plants have been poorly studied in context with these enzymes. The presence of the dye Direct Red 5B, in the medium has shown an induction in intracellular tyrosinase activities while the presence of Reactive Red 198 in the medium has resulted into an induction in both the intracellular and extracellular activities of the enzyme which indicates the vital role of these enzymes in dye degradation processes (Kagalkar et al, 2009; Patil et al, 2009).

The other enzymes that were found to be induced in *Blumea malcolmii* during the degradation of Direct Red 5B include riboflavin reductase, azoreductase and NADH-DCIP reductase which indicates their involvement in dye degradation processes. Azoreductase is known to catalyze the break down of azo linkages in azo dye structures. Direct Red 5B being an azo dye can be predicted to be acted upon by azoreductase from *Blumea* species leading to the degradation of the dye into simpler molecules (Kagalkar et al, 2009). Enzymes such as laccase, catechol 2,3-dioxygenase, ascorbate oxidase have also been detected in many other plants such as Alfalfa, Mustard and Cresswhich but, their application for dye removal has not been studied in these plants (Gramss and Rudeschko, 1998). Thus, the enzymatic status and the mechanisms of dye removal are found to be highly variable with different plant species. An evidence for this is given by the studies on two different plant systems, *Blumea malcolmii* and *Typhonium flagelliforme* which even upon exposure of the same dye molecule exhibited different enzyme activity patterns. Azoreductase activity was found to be absent in *Typhonium* species whereas the presence and induction in its activity was reported in *Blumea*, when both the plants were exposed to the same dye, DR5B. Similarly intracellular laccase was found to be present in *Typhonium* root tissues but absent in *Blumea* roots. The activities of riboflavin reductase which were found in *Blumea* species after exposure to Direct Red 5B was not detected in *Typhonium* species exposed to the same dye molecule. Moreover, the enzyme activities in the same species can be found to vary upon exposure to two different dyes. Enzyme activity in *Typhonium flagelliforme* plantlets exposed to two different dyes, DR5B and BBR show higher induction in the values of peroxidase and laccase in case of plantlets exposed to BBR than DR5B, which according to the authors could be the reason behind the better removal of BBR by *Typhonium* than DR5B. *Blumea* plantlets exhibited induction in azoreductase activities on exposure to DR5B, while no activity of the enzyme was detected for BBR (Kagalkar et al, 2009; Kagalkar et al, 2010).

Since reports suggest that plant species are capable of adsorbing as well as accumulating dyes, it is of great importance to confirm that the decolorization of the dye is not because of the adsorption or accumulation of the dye onto and/or into plant species but it is because of the transformation or degradation of the dye into different products. HPLC analyses showing the different retention times observed for peaks of the individual dye and the extracted products formed after the treatment of the dye, with the selected plant species, help to confirm the degradation of the dye. Simpler techniques such as thin layer chromatography can also be used for such analyses. The studies carried out on the plants *B. juncea*, *B. malcolmii* (Figure. 3) and marigold hairy roots have confirmed the degradation of the dye, detected through techniques such as high performance liquid chromatography (Ghodake et al, 2009; Kagalkar et al, 2009; Patil et al, 2009).

Analysis of Products Formed After the Phytodegradation of Dyes

A) Analysis of Products With Respect to their Chemical Structures

It is indeed very important to predict the chemical nature of the degradation products which can be done by correlating the analysis of gas chromatography mass spectroscopy (GCMS) techniques with fourrier transfrom infrared spectroscopy (FTIR). The FTIR spectra obtained helps us to predict the changes occurring in the functional groups of the original dye molecules, after degradation by the plant system. FTIR spectral data can further help to give confirmatory evidences in favor of the formation of products that have been predicted as a result of the degradation of the dye, by GCMS techniques. The probable products formed after the degradation of Direct Red 5B by *Blumea* species have been predicted to be 4-(4-amino-phenylazo)-benzene sulfonic acid, 3-amino-7-carboxyamino-4-hydroxy-naphthalene-2-sulfonic acid and 7-carboxyamino-naphthalene-2-sulfonic acid (Figure. 1) (Kagalkar et al, 2009). Sometimes, the exudates secreted by the plant can be misinterpreted as metabolites formed after the degradation of the dye, thus giving false positive results with high performance liquid chromatography (HPLC), FTIR and GCMS techniques. Hence it is indeed very important to run controls that will help us to clearly differentiate between the products formed after the degradation of the dye molecule and those compounds which are normally secreted as plant exudates. The FTIR and HPLC analyses of three different samples including the dye Direct Red 5B, the exudates of the plants along with the dye and the degraded sample clearly helped to select peaks that were neither of the plant exudates nor of the original dye molecules and thus could be easily concluded to be the peaks representing the formation of new products owing to the metabolism of the dye (Figure. 2 and 3) (Kagalkar et al, 2009).

Similarly, the products formed after the degradation of the dye Reactive Red 2 by B. juncea species were found to be naphthalene sufamide and 2-amino-4, 6-dichlorotriazine while the products formed after the degradation of Reactive Red 198 by marigold hairy roots were predicted to be 2-aminonaphthol, p-aminovinylsulfone ethyl disulfate and 1-aminotriazine, 3-pyridine sulfonic acid (Ghodake et al, 2009; Patil et al, 2009). The identification of these chemical structures also helps to portray the role played by different degradative enzymes in the sequential metabolism of dye structures that can help us to design the probable pathway of metabolism of the dye. Shaffiq et al also analyzed the products formed after the degradation of various dyes by *Ipomoea* and *Saccharum* peroxidases by HPLC techniques and concluded that the products were not aromatic amines which are known to be toxic (Shaffiq et al, 2002).

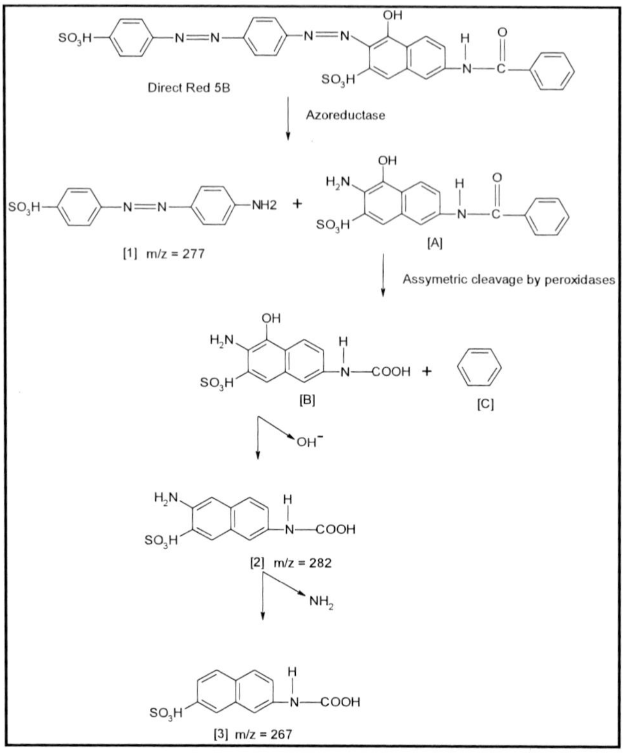

Figure 1. Proposed pathway for the phytotransformation of the dye Direct Red 5B by *B. malcolmii*. The compounds represented by alphabets have not been found, but their existence is rationalized as necessary intermediates for the final products found. The compounds in Arabic numbers have been found in reaction mixture (Kagalkar et al, 2009).

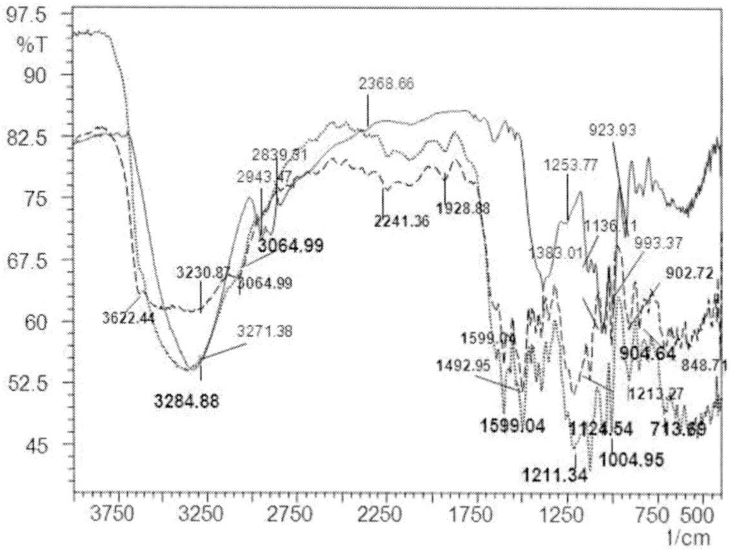

Figure 2. FTIR spectral analysis of Direct Red 5B (---), extracts of *Blumea malcolmii* exudates along with the dye (···) and products formed after the degradation of the dye by *Blumea malcolmii* (–) (Kagalkar et al, 2009).

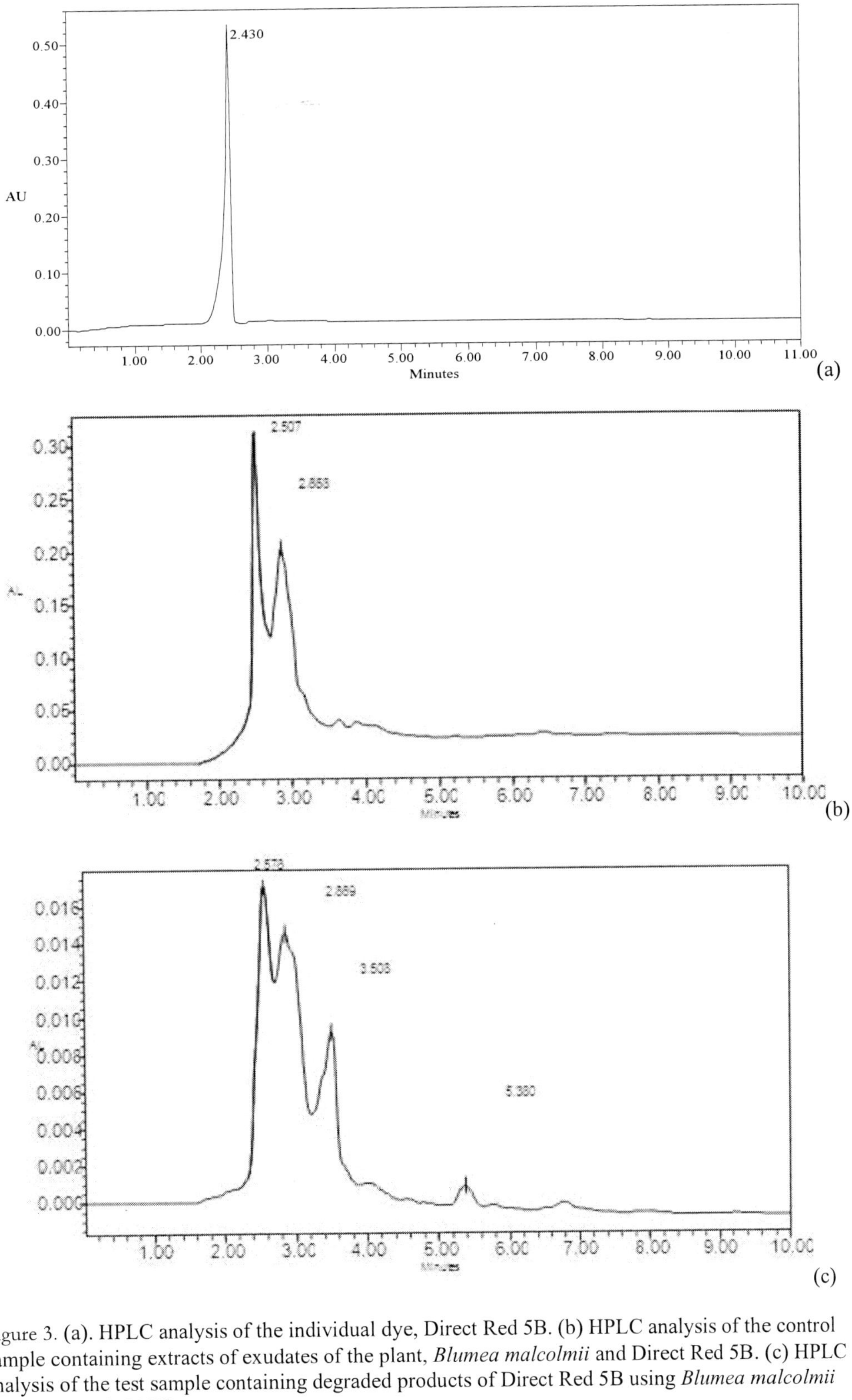

Figure 3. (a). HPLC analysis of the individual dye, Direct Red 5B. (b) HPLC analysis of the control sample containing extracts of exudates of the plant, *Blumea malcolmii* and Direct Red 5B. (c) HPLC analysis of the test sample containing degraded products of Direct Red 5B using *Blumea malcolmii* (Kagalkar et al, 2009).

Analysis of Products Formed With Respect to Their Toxicity

Before applying any plant system at the contamination site, it is very important to have a detailed knowledge about the basic mechanisms underlying the removal of the dye by the plant species. Moreover, it is also very essential to determine the nature of the products formed with the degradation of the dye in terms of the chemical nature of the product and also its toxicity to different life forms including microorganisms, aquatic animals like fishes which are very frequently exposed to contaminated water supplies, plant systems etc. The phytotoxicity studies of the products formed after the degradation of Reactive Red 198 and BBR, after treatment with marigold hairy roots and *Typhonium* plantlets, towards the plants *Phaseolus mungo* L. and *Triticum aestivum* L. revealed the non toxic nature of the metabolites formed (Patil et al, 2009; Kagalkar et al, 2010). Such studies then indicate that the plant systems can be well applied in soil or water ecosystems and their application will only be beneficial to the ecosystem and will not lead to the formation of products that will be toxic for any life form. In addition, it is also important to analyze the toxicity of the dyes to be used towards the plants that have been selected for their application in dye removal. When narrow leaved cattails was subjected to high concentration of dyes from 100 to 300 mg/1, the plants responded by showing symptoms such as green wilting, then a yellow spot was observed which is a symptom of necrosis after 48 h of exposure to the dye. Thus, higher dye concentration was found to be toxic to the plant and at concentrations of 300 mg/1; there was no survival of the plant. The toxicity of the dye to the plant was found to be at and above the concentration of 25.33 mg/l. Such type of studies helped the authors to ascertain a dye concentration of 20 mg/l for further experiments (Nilratnisakorn et al, 2007). Kagalkar et al also reported the effect of increasing concentration of the dye Direct Red 5B on the percentage decolorization values. Increasing concentrations of the dye led to a decrease in the percentage decolorization values, indicating that higher dye concentrations could be toxic to the plant (Kagalkar et al, 2009). When the degradation of textile effluents or synthetic dye mixtures is studied the reduction BOD (Biological Oxygen Demand), COD (Chemical Oxygen Demand), TS (Total Solids) and TDS (Total Dissolved Solids) etc. also constitute important parameters in the assessment of toxicity of the waste waters. *Typhonium flagelliforme* plantlets reduced the BOD of the textile effluent as well as of the synthetic mixture of dyes. Similarly, the COD value of the industrial effluent was reduced (Kagalkar et al, 2010). Reduction in these values indicates the reduced toxicities of these effluents.

The Use of Hydroponics and Plant Tissue Culture Technologies for Dye Degradation

The cultivation of plants and their further experiments with dye degradation can be carried out using hydroponic solutions. These solutions provide a nutrient status which is close to that of the soil in which the plant usually grows. Thus, such solutions are enriched with various macro and micro nutrients and can be used for the cultivation and/or maintenance of plants for phytoremediation. The use of hydroponics provides a cost effective method for phytoremediation of dyes. Aubert and Schwitzguébel carried out the screening of plant species (*Rheum rabarbarum*, *Rumex acetosa*, *Rumex hydrolapatum* and *Apium graveolens*), in hydroponic solutions for the removal of sulfonated anthraquinones. Many plant species have the capacity to absorb large quantities of water form hydroponic solutions. The water absorption capacity of a plant is a factor that should be taken into consideration while performing studies in hydroponic solutions because it reflects the overall health of a plant. Lower water absorption capacity for the plant *Rumex acetosa* in hydroponic solution indicated that the plant was not in optimum health under hydroponic conditions and thus the metabolism and transpiration was probably reduced as compared to soil grown plants. Though difficult, it is quite possible to grow adult terrestrial plants such as Rhubarb and common sorrel under hydroponic conditions (Aubert and Schwitzguébel, 2004). But, research that has been involving the cultivation and experimentation with plants in such systems also portrays some major disadvantages of these systems. Pege and Schwitzguébel found it impossible to collect leaves of the same age and same stage of growth and development in case of plants grown in hydroponics. It has been found that the level of enzymes like cytochrome P450 changes with the growth of the plants since they play a role in several physiological functions of the plant. The same is true with peroxidases. This methodology of work makes it difficult to exactly confirm the role of these enzymes in the detoxification of dyes (Pege and Schwitzguébel, 2009). Moreover, the enzyme activities of a plant may also be affected by conditions such as nutritional status, dark and shade requirements, effect of microbial contaminants etc. These factors make it very difficult to get reproducible results. To overcome these problems with wild plants grown in hydroponics or in wetlands, the importance of tissue culture based technologies has been stressed by a few researchers. Though tissue culture involves processes that require a high cost, the use of these technologies has been suggested for basic research which lays the foundation for the application of plant systems in wetland conditions. Our current ability to exploit phytoremediation technologies for the treatment of dyes is restricted by the fact that the knowledge regarding the basic mechanisms and pathways involving the decolorization of dyes is limited. The advantage of using tissue culture based technologies is that the plants can be grown in controlled conditions and once established, they can be propagated indefinitely and are available on demand as contrast to whole plants that have a limited life span. Moreover, in vitro culture techniques offer an environment that is totally free of microbial contamination and can be used to distinguish the responses and capabilities of plant cells from microbes present in the rhizhospheric regions in dye contaminated sites. They also offer conditions that are controlled in terms of nutrient levels, phytohormone level, light requirements etc. (Doran, 2009). Thus, Kagalkar et al have reported studies on the decolorization of various dyes using *in vitro* plants. These techniques have given reproducible results and have led the authors to analyze the role of different enzymes involved in the degradation of the dye DR5B and BBR (using the plants, *Blumea malcolmii* and *Typhonium flagelliforme* respectively), also predict the probable pathway behind the metabolism of these dyes (Kagalkar et al, 2009; Kagalkar et al, 2010). Tissue culture technologies also help to manipulate the plant in order to obtain callus cultures, suspension cultures and hairy roots. Hairy roots grow relatively quickly and do not require exogenous hormones in the medium (Doran, 2009). These advantages have led to the use of marigold hairy roots in the degradative analysis of the dye Reactive Red 198 (Patil et al, 2009). Experiments with separately cultured organs of a plant can evaluate the accumulation and/or biotransformation abilities thus minimizing the interference of translocation effect of dyes (Doran, 2009).

Though plant tissue culture technologies offer multiple advantages, their use is feasible only for studying the basic processes and not for the actual applications of dye degradation in the field. Moreover, the characteristic features of the plants that are demonstrated in hydroponics or in vitro culture conditions could be different than those observed in the soil because of the complexity of the soil environment which exposes the plant to different biotic (microorganisms) and abiotic (other contaminants) soil elements (Zabłudowska et al, 2009). Hence such experiments should always be accompanied with field trials of the plant.

SYNERGISTIC APPROACHES FOR DYE DEGRADATION

Rhizhosphere remediation constitutes an interesting branch of phytoremediation technologies involving the use of plants along with rhizospheric microorganisms to remediate contaminated soils. In addition to the root zone (rhizosphere), where the microbial biomass can be one order of magnitude or more higher than that in bulk soil, bacteria can colonize the interior of their host plant without causing symptoms of disease (Weyens et al, 2009a). Researchers have shown that plants can be able to degrade a wide variety of organic pollutants in association with microbes (Peng et al, 2009). The limitations to remediation processes can be overcome by utilizing the dynamic synergy between plants and rhizhospheric organisms. Soil microorganisms are also known to produce certain biosurfactant compounds that may further facilitate the removal/degradation of organic pollutants by increasing their availability to plants. The genera commonly found in rhizospheres include *Rhizobium*, *Azotobacter* and *Pseudomonas* as well as a number of root-associated fungi, including those involved in the formation of symbiotic *mycorrhizas* (Chaudhry et al, 2005). Root exudates are compounds produced by the plant and released by the plant roots. These exudates contain water soluble, insoluble, and volatile compounds including sugars, amino acids, organic acids, nucleotides, flavonones, phenolic compounds and certain enzymes. The microbial populations and activity in the rhizosphere can be increased due to the presence of these exudates, and can result in increased organic contaminant biodegradation in the soil (EPA, 2000). In return many microbes have shown the potential to boost plant growth directly by secretion of different phytohormones or by fixing and solubilizing nutrients that are unavailable to plants and indirectly by competing with plant pathogens for availabilities of nutrients and space (Weyens et al, 2009b). Microbes have been known for their potential to degrade a variety of dyes with diverse structures and properties. There have been reports of the presence of different enzymes such as lignin peroxidase, laccase, tyrosinase, DCIP reductase, azoreductase etc in microbial species (Kalme et al, 2006). It is quite likely that the degradation of a dye mediated by individual microbial or plant systems could lead to partial degradation. Use of microbial and plant systems together with their diverse inherent enzyme producing abilities can thus be used together in an attempt to obtain the mineralization of the dye molecule and return to the nature its elements in their native forms. Strycharz et al showed that when oregano cell lines were inoculated with *Pseudomonas* Z strain, hyperhydricity was prevented, thus improving the quality of the plant. An increase in the activity of peroxidase was found when *Pseudomonas* was used in combination with the dye without a substantial decrease in the phenolics. This suggests that this *Pseudomonas* species may have the potential to degrade the dye. Moreover, the authors speculate that as many *Pseudomonas* species have azoreductase activities, they could break the azo linkages in the dye and convert it into a more biologically available form that can be degraded by the plant. If plant-bacterial combinations are to be used to the fullest advantage of remediating the soil, it will be good to establish tolerant clonal lines before planting on contaminated soil (Strycharz, 2002b). Research involving synergistic approaches for phytoremediation is still at the very initial stage which is probably because of the lack of knowledge behind the basic mechanistic processes of plant dye degradation.

APPLICATIONS OF PHYTOREMEDIATION TECHNOLOGIES FOR DYE DEGRADATION

A) Constructed Wetlands

Most of the research involving the degradation of different pollutants is limited to laboratory conditions and very few of it is actually applied in the field. The use of constructed wetlands can take us a step closer to the application of potent plant species on the actual sites of contamination. Experiments performed in the laboratory are performed under controlled conditions and the behavior and efficiency of the system when applied at the actual site of contamination remains a question. Constructed wetlands are engineered systems to treat waste water (Barbera et al, 2009) and can be designed to mimic conditions close to those prevalent at the dye contaminated sites. Moreover, unlike natural wetlands, depuration processes in constructed wetlands are performed under environments which are more controlled and thus can assure greater efficiency and regular depuration activity across the entire bed (Barbera et al, 2009). Biodegradation of less-degradable pollutants generally requires combination of anaerobic and aerobic processes. For example, azo dyes, sixty to seventy percent of dyes used in the textile industry, are mineralized aerobically only after the azo-linkage is broken anaerobically (Ong et al., 2005). To treat such pollutants with constructed wetlands, therefore, anaerobic processes should properly incorporated to wetland systems. A vertical flow constructed wetland was designed so as to work in intermittent feeding mode (8 feeding cycles per day) which enhanced the characteristics like constant hydraulic permeability and maximized the oxygen transfer rate and was tested for the removal efficiency of the dye Acid Orange 7. The flow rate of the effluent must be critically maintained in such experiments. The use of sprinklers for feeding the dye solutions on the beds allows good distribution of the dye over the entire surface area (Davies et al, 2005). The main role of wetland vegetation is attributed to the modification of soil texture, hydraulic conductivity and soil chemistry. For phytoremediation technologies to be applied in constructed wetlands, parameters such as BOD, COD, TOC content, hardness, alkalinity etc of the effluent should be evaluated before and after treatment of the industrial effluent. The nature of the products formed should also be determined. The effluent treated by *Phragmites australis* plants in constructed wetlands showed an efficient reduction in the COD and TOC levels (Davies et al, 2005). Thus, constructed wetlands can be used as economic technologies that can treat enormous amounts of wastes through batch or continuous processes. The efficiency of a wetland system depends largely on the basic biological, physicochemical processes induced by the interaction of plants, microorganisms, substrates and pollutants (Barbera et al, 2009).

B) Purified Enzymes for Phytoremediation

Plants remain indispensable sources for the production of a number of biologically active chemical substances that are difficult to synthesize because of their complicated structures (Rudrappa et al, 2005). The uses of degradative enzymes that have been studied for dye colorization are seen as attractive options in the development of effective strategies for the biological treatment of waste water. The catalytic action of enzymes is highly specific and efficient as compared to chemical catalysis because of the higher reaction rates, milder reaction conditions and greater stereospecificity (Maddhinni et al, 2006). One of the major drawbacks of using these enzymes for remediation purposes is the low yield and high cost of production as compared to bacterial and fungal enzymes. These processes can be made economic either by reducing the production cost or by extracting these enzymes form cheaply available plant sources and increasing the purification fold and percentage of recovery after purification (Shaffiq et al, 2002). Enzymes can be immobilized on suitable carriers after purification and can be used for remediating dye effluents. Immobilization techniques offer several advantages such as easy separation from the soluble reaction products and the untreated substrate. Moreover, immobilization techniques allow repeated usage and can help to reduce the overall cost of the process. They allow the continuous removal of toxic metabolites thus simplifying the work (Matto et al, 2009). The stability of carrier enzyme binding is an important factor in the application of immobilized enzymes. Moreover, matrices such as silica, polyvinyl alcohol, polyacrylamides etc., allow the adsorption of dye molecules on the matrix which may lead to the inactivation of the enzyme (Shaffiq et al, 2002). Further, it is very important that the enzyme retains its activity after immobilization. Once the enzyme is successfully immobilized it can be used in bioreactors. The plant systems, *Ipomoea palmata* and *Saccharum spontaneum* served as good sources of peroxidase. The enzyme purified from these sources was found to decolorize various dyes. But, the pH optima for degradation of acidic dyes by *Ipomoea* peroxidase was found to be between 4.0 and 6.0 while the pH optima for the degradation of basic dyes was found to be between 6.0 and 8.0. The dye Supranol Green (25 mg/l) was most efficiently degraded to about 84% within 4 h of treatment with the purified enzyme followed by Brilliant Green (25 mg/l) which showed 54% degradation for the same dye concentration. Similarly, the optimum pH for the decolorization of various dyes by *Saccharum* peroxidase was different. Supranol Green (25 mg/l) was again the most efficiently removed to get 99% degradation within just 20 min of reaction time with the enzyme. *Saccharum* peroxidase being more efficient in the degradation of dyes was immobilized using modified polyethylene which is a hydrophobic matrix and thus prevents the adsorption of dyes onto the matrix. The immobilized enzyme retained 20% of its activity and could efficiently decolorize Procion Green HE-4BD, Supranol Green, Procion Brilliant Blue H-7G and Procion Navy Blue HER at 50 mg/l concentrations of the dyes when used in a batch reactor (Shaffiq et al, 2002). Horseradish peroxidase has also been immobilized to be used for treating the effluents of paper and textile. Peralta-Zamora et al have recommended the photoenzymatic decolorization of textile effluents where the previous photochemical treatment of the effluent before exposure to the enzyme led to sufficient enhancement of the biological decolorization process (Peralta-Zamora et al, 1998). Gel entrapment methods have been tried out for immobilization of horseradish peroxidase and gel immobilized enzyme was found to be more efficient in dye removal than the free enzyme (Maddhinni et al, 2006). Reactions with enzymes have to be optimized with respect to pH, temperature, enzyme concentration and dye concentration values. The results with the removal of the dye Direct Yellow-12 in the presence of horseradish peroxidase indicated that the decolorization values obtained with the enzyme were lower above the dye concentration of 25 mg/l which can be acknowledged as the cut-off value. The optimum pH for dye removal by this enzyme was found to be 4.0 (Maddhinni et al, 2006). Matto and Hussain purified bitter gourd peroxidase by ammonium sulfate fractionation and further immobilized it on the surface of concavalin A layered calcium alginate-starch beads where it was found that the immobilized enzyme retained 69% of its original activity. Jack bean extract was used as a source of concavalin A which acts as a cheap source and helps in reducing the overall cost of the system. The authors have demonstrated very interesting studies on the effects of redox mediators on the effluent decolorization. The enzyme oxidizes the mediator and the oxidized mediator further oxidizes the substrate. Among the various redox mediators used, 1-hydroxybenzotriazole was found to be the most efficient, in the presence of which the free enzyme showed 28% decolorization and the immobilized enzyme showed 70% decolorization within 1 h. The authors further have also used a two reactor system for the decolorization of textile effluent, one reactor containing the immobilized enzyme and the other containing activated silica. This system was found to decolorize more than 90% of the textile effluent within 3 h of incubation in a batch process whereas the free form decolorized only 48% which might be because of the stability provided by immobilization techniques.

Though the dye removal efficiency of the system was found to decrease with time even after 60 days the removal efficiency was 40% (Matto et al, 2009). Salt fractionated bitter gourd peroxidase was also used for the decolorization of two water insoluble disperse dyes which were Disperse Red 171 and Disperse Brown 1. Here too, 1-hydroxybenzotriazole was found to be the most efficient redox mediator which facilitated 90% removal of Disperse Red 171 and 65% removal of Disperse Brown 1. Similarly, partially purified potato and brinjal polyphenol oxidases (PPO) have also been reported for the removal of textile and non textile dyes. Most efficient degradation with potato PPO was obtained at pH 3.0, while as the pH increased, the percentage decolorization values were found to decrease. Brinjal PPO showed no decolorization at pH 5.0. Thus, potato PPO mediated decolorization at broader pH ranges. The decolorization of dyes decreased after the time span of 1 h which may be because of the inhibition caused by the products. Potato PPO gave the highest decolorization (93%) for Reactive Blue 160 within 30 min of treatment with the enzyme. The enzymes were capable of decolorizing different dye mixture combinations containing 4 different dyes. A comparison of the two enzymes reveals that potato PPO's are more efficient than brinjal PPO's for the decolorization of textile dyes (Khan and Husain, 2007). Since purified enzymes can prove to be efficient systems for bioremediation, more work can be done in terms of the use of various inducers and redox mediators for making the decolorization processes more efficient. Moreover, if redox mediators are needed for better efficiency of the enzyme, an attempt could be made to use natural sources such as agricultural wastes as redox mediators. Similarly, cheaper processes and matrices should be tried out for immobilization which will help to reduce the overall cost of the dye removal process. Though purified and immobilized enzymes have shown the potential to degrade a variety of dyes, their use for phytoremediation is obviously costlier than the application of whole plant systems at the dye contaminated sites.

Table 1. The use of purified enzymes for the decolorization of dyes

Plant source	Enzyme	Purification techniques	Dyes decolorized	Reference
Ipomoea palmata	Peroxidase	Ion exchange chromatography, gel filtration chromatography	Methyl Orange, Chrysoidine, Supranol Green, Brilliant Green, Direct Blue, Crystal Violet	Shaffiq et al, 2002
Saccharum spontaneum	Peroxidase	Ion exchange chromatography, gel filtration chromatography	Chrysoidine, Blue MR, Porcion, Brilliant Blue HER, Supranol Green, Porcion Green HE-4BD, Direct Blue.	Shaffiq et al, 2002
Horseradish	Peroxidase	Dialysis	Direct Yellow-12, Direct Yellow 11, Basazol 46L, Azure B, Poly R-478, Remazol Brilliant Blue R, Crystal violet, Textile effluent.	Maddhinni et al, 2006; Knutson et al, 2005; Gramms and Rudeschko, 1998.
Soybean	Peroxidase		Direct Yellow 11, Basazol 46L, Azure B, Poly R-478, Remazol Brilliant Blue R, Crystal Violet,	Knutson et al, 2005; Gramms and Rudeschko, 1998.
Brinjal	Polyphenol oxidase	Ammonium sulfate precipitation, dialysis	Reactive Blue 160, Reactive Blue 171, Reactive Red 11, Reactive Orange 4, Reactive Yellow 84, Reactive Orange 86, Reactive Blue 4, Reactive Red 120, PAGE Blue 83, Commasie Brilliant Blue G 250, Comassie Brilliant Blue R 250, Methylene Blue, Naphthaquinone-4-sulphonic acid, Tropaeolin, Evans Blue, Dye mixtures	Khan and Husain, 2007
Potato	Polyphenol oxidase	Ammonium sulfate precipitation, dialysis	Reactive Blue 160, Reactive Blue 171, Reactive Red 11, Reactive Orange 4, Reactive Yellow 84, Reactive Orange 86, Reactive Blue 4, Reactive Red 120, PAGE Blue 83, Commasie Brilliant Blue G 250, Comassie Brilliant Blue R 250, Methylene Blue, Naphthaquinone-4-sulphonic acid, Tropaeolin, Evans Blue, Dye mixtures.	Khan and Husain, 2007
Bitter gourd	Peroxidase	Ammonium sulfate fractionation	Disperse Red 17, Disperse Brown 1.	Matto and Husain, 2009
Turnip (*Brassica rapa*)	Peroxidase	Ammonium sulfate fractionation	Acid Blue 92, Acid Red 97, Acid Yellow 42, Acid Black 1, Acid Black 210.	Kulshrestha and Husain, 2007.

FUTURE PROSPECTS

The entire literature survey of dye decolorizing processes using plants makes one realize the immense potential of plants as systems for the efficient removal of dyes. Though plant systems have many advantages over bacteria that have been studied for dye degradation, there are many dimensions and processes involving the removal of dyes mediated by plants that are still unexplored. Hence, this area demands a lot of attention from researchers over the world for broadening the horizons of this technology.

Similar to metal phytoremediation, there is a need for extensive screening of different plant species that have the capacity to remove dyes that will help the application of more and more efficient systems for remediation of dyes. But, plant mediated removal of textile dyes has some limitations too. Unlike microbial degradation processes plant systems are known to be slower. Moreover, their capacity to tolerate high dye concentrations is also limited in comparison to microbes. Phytoremediation is restricted to the sites of contamination as deep as plant roots. Some of these limitations can be overcome by using genetic engineering techniques. Genetic modification techniques can be used to over express the enzymes involved in the existing metabolic pathways of the plant. Moreover, newer pathways can also be incorporated into the plants. If the plant does not have the ability to produce one or more enzymes that can degrade dyes or if the activities of these enzymes are low, related microbial genes can be inserted into the plant so as to make it a more potent system for phytoremediation. Transgenic poplar tress expressing the mammalian cytochrome P450 enzyme has been shown to be 640 fold faster than wild poplar trees for the removal of trichloroethylene. These enzymes also have a role in dye degradation. Hence, such transgenic plants should also be tested for the removal of textile wastes (Aken 2008). Such genetic modifications can aim at achieving the complete mineralization of the dye molecules and probably help to make the plants tolerant to very high dye concentrations. In addition, similar to the studies involving plant-microbe associations for the degradation of textile dyes, we also need to consider approaches involving the synergistic and/or cumulative effects caused because of the use of different plant species (Cheema et al, 2010). Out of the many plants used for the degradation of textile dyes, no two plants can be found to be exactly similar with respect to characteristics such as enzymatic response, capacity for accumulation and/or adsorption of dyes etc. Combinatorial approaches using different plant species will therefore involve the merits of all the plant species used and can lead to enhanced degradation of the dye contaminated sites. Molecular studies need to be done to find out whether the induction in the enzyme activities upon exposure to dyes could be attributed to the over expression of the genes responsible for the production of these enzymes.

Thus, a more detailed understanding of plant processes for dye removal can help us to manipulate these systems in order to achieve more efficient remediation of the dye contaminated sites. Till today, many biological alternatives have been suggested for the removal of dyes from industrial effluents but due to various hindrances very few have actually been used on a large scale for the detoxification of effluents. Phytoremediation is a more applicable technology as compared to other bioremediation processes for dye removal and thus adequate research in this area will definitely lead to the actual application of these systems to remediate dye contaminated sites.

References

Abadulla, E., Tzanov, T., Costa, S., Robra, K.H., Cavaco-Paulo, A., Gubitz, G.M., 2000. Decolorization and detoxification of textile dyes with a laccase from *Trametes hirsuta*. *Appl. Environ. Microbiol.* 66, 3357-3362.

Aken, B.V., 2008. Transgenic plants for phytoremediation: helping nature to clean up environmental pollution. *Trends Biotechnol.* 26, 25-27.

Aubert, S., Schwitzguébel, J.P., 2002. Capillary electrophoretic separation of sulphonated anthraquinones in a variety of matrices. *Chromatographia* 56, 693-697.

Aubert, S., Schwitzguebel, J.P., 2004. Screening of plant species for the phytotreatment of wastewater containing sulfonated anthraquinones. *Water Res.* 38, 3569-3575.

Bafana, A., Jain, M., Agrawal, G., Chakrabarti, T., 2009. Bacterial reduction in genotoxicity of Direct Red 28 dye. *Chemosphere.* 74, 1404-1406.

Barbera, A.C., Cirelli, G.L., Cavallaro, V., Silvestro, I.D., Pacifici, P., Castiglione, V., Toscano, A., Milani, M., 2009. Growth and biomass production of different plant species in two different constructed wetland systems in Sicily. *Desalination.* 246, 129-136.

Bestani, B., Benderdouche, N., Benstaali, B., Belhakem, M., Addou, A., 2008. Methylene Blue and iodine adsorption onto an activated desert plant. *Bioresour. Technol.* 99, 8441-8444.

Carias, C.C., Novais, J.M., Martins-Dias, S., 2007. *Phragmites australis* peroxidases role in the degradation of an azo dye. *Water Sci. Technol.* 56, 263-269.

Carias, C.C., Novais, J.M., Martins-Dias, S., 2008. Are *Phragmites australis* enzymes involved in the degradation of the textile azo dye Acid Orange 7? *Bioresour. Technol.* 99, 243-251.

Chaudhry, Q., Zandstra, M.B., Gupta, S., Joner, E.J., 2005. Utilizing the synergy between plants and rhizosphere organisms to enhance breakdown of organic pollutants in the environment. *Environ. Sci. Pollut. Res.* 12, 34-48.

Cheema, S.A., Khan, M.I., Shen, C., Tang, X., Farooq, M., Chen, L., Zhang, C., Chen,Y., 2010. Degradation of phenanthrene and pyrene in spiked soils by single and combined plants cultivation. *J. Hazard. Mater.* 177, 384-389.

Chen, C.H., Chang, C.F., Liu, S.M., 2010. Partial degradation mechanisms of Malachite Green and Methyl Violet B by *Shewanella decolorationis* NTOU1 under anaerobic conditions. *J. Hazard. Mater.* 177, 281-289.

Chen, L., Flurkey, W.H., 2002. Effect of protease inhibitors on the extraction of crimini mushroom tyrosinase isoforms. *Curr. Top. Phytochem.* 5, 109-120.

Cluis, C., 2004. Junk-greedy Greens: phytoremediation as a new option for soil decontamination. *Bio. Teach. J.* 2, 61-67.

Cunningham, S.D., Ow, D.W., 1996. Promises and prospects of phytoremediation. *Plant Physiol.* 110, 715-719.

Davies, L.C., Cabrita, G.J.M., Ferreira, R.A., Carias, C.C., Novais, J.M., Martins-Dias, S., 2009. Integrated study of the role of *Phragmites australis* in azo-dye treatment in a constructed wetland: From pilot to molecular scale. *Ecol. Eng.* 35, 961-970.

Davies, L.C., Carias, C.C., Novais, J.M., Martins-Dias, S., 2005. Phytoremediation of textile effluents containing azo dye by using *Phragmites australis* in a vertical flow intermittent feeding constructed wetland. *Ecol. Eng.* 25, 594-605.

Doran, P.M., 2009. Application of plant tissue cultures in phytoremediation research: Incentives and limitations. *Biotechnol. Bioeng.* 103, 60-76.

Duc, R., Vanek, T., Soudek, P., Schwitzguebel, J.P., 1999. Accumulation and transformation of sulfonated aromatic compounds by Rhubarb cells (*Rheum palmatum*). *Int. J. Phytorem.* 1, 255-271.

EPA, 2000. *Introduction to phytoremediation.* EPA/600/R-99/107. National Risk Management Research Laboratory, Cincinnati, OH.

Ghodake, G.S., Telke, A.A., Jadhav, J.P., Govindwar, S.P., 2009. Potential of *Brassica juncea* in order to treat textile effluent contaminated sites. *Int. J. Phytorem.* 11, 297-312.

Gramss, G., Rudeschko, O., 1998. Activities of oxidoreductase enzymes in tissue extracts and sterile root exudates of three crop plants, and some properties of the peroxidase component. *New Phytol.* 138, 401-409.

Kagalkar, A.N., Jagatap, U.B., Jadhav, J.P, Bapat, V.A., Govindwar, S.P., 2009. Biotechnological strategies for phytoremediation of the sulphonated azo dye Direct Red 5B using *Blumea malcolmii* Hook. *Bioresour. Technol.* 100, 4104-4110.

Kagalkar, A.N., Jagatap, U.B., Jadhav, J.P., Govindwar, S.P., Bapat, V.A., 2010. Studies on phytoremediation potentiality of *Typhonium flagelliforme* for the degradation of Brilliant Blue R. *Planta.* 232, 271-285.

Kalme, S.D., Parshetti, G.K., Jadhav, S.U., Govindwar, S.P., 2006. Biodegradation of benzidine-based dyes Direct Blue 6 by *Pseudomonas desmolyticum* NCIM 2112. *Bioresour. Technol.* 98, 1405-1410.

Khan, A.A., Husain, Q., 2007. Potential of plant polyphenol oxidases in the decolorization and removal of textile and non-textile dyes. *J. Environ. Sci.* 19, 396-402.

Knutson, K., Kirzan, S., Ragauskas, A., 2005. Enzymatic bioleaching of two recalcitrant paper dye with horseradish and soybean peroxidase. *Biotechnol. Lett.* 27, 753-758.

Maddhinni, V.L., Vurimindi, B.H., Yerramilli, A., 2006. Degradation of azo dye with horseradish peroxidase (HRP). *J. Indian Inst. Sci.* 86, 507-514.

Matto, M., Husain, Q., 2009. Decolorization of textile effluent by bitter gourd peroxidase immobilized on concanavalin A layered calcium alginate–starch beads. *J. Hazard. Mater.* 164, 1540-1546.

Ncibi, M. C., Mahjoub, B., Seffen, M., 2007. Adsorptive removal of textile reactive dye using *Posidonia oceanica* (L.) fibrous biomass. *Int. J. Environ. Sci. Tech.* 4, 433-440.

Nilratnisakorn, S., Thiravetyan, P., Nakbanpote, W., 2007. Synthetic reactive dye wastewater treatment by narrow-leaved cattails (*Typha angustifolia* Linn.): Effects of dye, salinity and metals. *Sci. Total Environ.* 384, 67-76.

Ong, S.A., Toorisaka, E., Hirata, M., Hano, T., 2005. Treatment of azo dye Orange II in aerobic and anaerobic-SBR systems. *Process Biochem.* 40, 2907-2914.

Page, V., Schwitzguébel, J.P., 2009. The role of cytochromes P450 and peroxidases in the detoxification of sulphonated anthraquinones by Rhubarb and common sorrel plants cultivated under hydroponic conditions. Environ. *Sci. Pollut. Res.* 16, 805-816.

Patil, P., Desai, N.S, Govindwar, S., Jadhav, J.P., Bapat, V., 2009. Degradation analysis of Reactive Red 198 by hairy roots of *Tagetes patula* L. (Marigold). *Planta* 230, 725-735.

Peng, S., Penga, S., Zhoua,Q., Caia, Z., Zhanga, Z., 2009. Phytoremediation of petroleum contaminated soils by *Mirabilis jalapa* L. in a greenhouse plot experiment. *J. Hazard. Mater.* 168, 1490-1496.

Peralta-Zamora, P., Esposito, E., Pelegrini, R., Groto, R., Reyes, J., Duran, N., 1998. Effluent treatment of pulp and paper and textile industries using immobilized horse radish peroxidase. *Environ. Technol.* 19, 55-63.

Rudrappa, T., Neelwarne, B., Kumar, V., Lakshmanan, V., Venkataramareddy, S.R., Aswathanarayaana, R.G., 2005. Peroxidase production from hairy root cultures of red beet (*Beta vulgaris*). *Electron. J. Biotechnol.* 8, 185-196.

Sarkanen, S., Razals, R.A., Piccariellos, T., Yamamotoa, E., Lewiss, N.G., 1991. Lignin peroxidase: toward a clarification of its ole *in vivo*. *J. Biol. Chem.* 266, 3636-3643.

Schwitzguébel, J.P., Aubert, S., Grosse, W., Laturnus, F., 2002. Sulphonated aromatic pollutants: limits of microbial degradability and potential of phytoremediation. *Environ. Sci. Pollut. Res.* 9, 62-72.

Senan, R.C., Abraham, E.T., 2004. Bioremediation of textile azo dyes by aerobic bacterial consortium. *Biodegradation.* 15, 275-280.

Shaffiqu, T.S., Roy, J.J., Nair, R.A., Abraham, T.E., 2002. Degradation of textile dyes mediated by plant peroxidases. *Appl. Biochem. Biotechnol.* 102-103, 315-326.

Sharma, K.P., Sharma, K., Kumar, S., Grover, R., Soni, P., Bharadwaj, S.M., Chaturvedi, R.K., Sharma, S., 2005. Response of selected aquatic macrophytes towards textile dye wastewaters. *Indian J. Biotechnol.* 4, 538-545.

Sharma, P., Goel, E.R., Capalash, N., 2007. Bacterial laccases. *World J. Microbiol. Biotechnol.* 23, 823-832.

Strycharz, S., Shetty, K., 2002a. Peroxidase activity and phenolic content in elite clonal lines of *Mentha pulegium* in response to polymeric dye R-478 and *Agrobacterium rhizogenes*. *Process Biochem.* 37, 805-812.

Strycharz, S., Shetty, K., 2002b. Response of oregano (*Origanum vulgare* L.) clonal lines to *Pseudomonas* sp. Z strain and polydye R-478 and implications for hyperhydricity prevention in tissue culture. *Process Biochem.* 38, 343-350.

Takahashi, M., Tsukamoto, S., Kawaguchi, A., Sakamoto, A., Morikawa, H., 2005. Phytoremediators from abandoned rice field. *Plant Biotechnol.* 22, 167-170.

Togo, C.A., Mutambanengwe, C.C.Z., Whiteley, C.G., 2008. Decolorisation and degradation of textile dyes using a sulphate reducing bacteria (SRB)–biodigester microflora co-culture. *Afr. J. Biotechnol.* 7, 114-121.

Vangronsveld, J., Herzig, R., Weyens, N., Boulet, J., Adriaensen, K., Ruttens, A., Thewys, T., Vassilev, A., Meers, E., Nehnevajova, E., Lelie, D., Mench, M., 2009. Phytoremediation of contaminated soils and groundwater: lessons from the field. *Environ. Sci. Pollut. Res.* 16, 769-794.

Vidali M., 2001. Bioremediation: An overview. *Pure Appl. Chem.* 73, 1163–1172.

Weyens, N., Taghavi, S., Barac, T., Lelie, D., Boulet, J., Artois, T., Carleer, R., Vangronsveld, J., 2009a. Bacteria associated with oak and ash on a TCE-contaminated site: characterization of isolates with potential to avoid evapotranspiration of TCE. *Environ. Sci. Pollut. Res.* 16, 830-843.

Weyens, N., Lelie, D.,Taghavi, S., Newman, L., Vangronsveld, J., 2009b. Exploiting plant–microbe partnerships to improve biomass production and remediation. *Cell.* 27, 591-598.

Zabłudowska, E., Kowalska, J., Jedynak, L., Wojas, S., Skłodowska, A., Antosiewicz, D.M., 2009. Search for a plant for phytoremediation–What can we learn from field and hydroponic studies? *Chemosphere.* 77, 301-307.

In: Handbook of Phytoremediation
Editor: Ivan A. Golubev

ISBN: 978-1-61728-753-4

Chapter 15

ANALYTICAL STRATEGIES TOWARDS THE STUDY OF METALLOPHYTES PLANTS GROWING IN CU-NI MINING AREAS IN BOTSWANA

***Dikabo Mogopodi*[1], *Kabo Mosetlha*[1], *Bonang Nkoane*[1], *Edward Mmatli*[1], *Nelson Torto*[2] *and Berhanu Abegaz*[1]**

[1] Department of Chemistry, University of Botswana, P/ Bag 00704, Gaborone, Botswana
[2] Department of Chemistry, Rhodes University ,Grahamstown, 6140, South Africa

ABSTRACT

Metallophytes have the ability to tolerate extreme metal concentrations. This unique property commends them to be exploited in technologies such as biogeochemical and biogeobotanical prospecting as well as phytoremediation. Although there are many publications on metallophytes and their potential use in phytoremediation, in Botswana such studies are in their infancy, albeit the country having numerous mining activities. This paper discusses the chemical studies of metallophytes from mineralized zones and other vulnerable areas in Botswana as well as their potential use in phytoremediation. The metallophytes dealt with include Helichrysum candolleanum, Blespharis aspera, Tephrosia longipes and Indigofera melanadenia some of whose capacity for multiple metal accumulation is investigated. A number of analytical methods have been applied in these studies. These include attractive sample preparation techniques such as microdialysis and solid phase extraction as well as chromatographic methods such as size exclusion chromatography and online liquid chromatography-solid phase extraction-nuclear magnetic resonance which are particularly employed for speciation studies. These techniques have demonstrated a lot of potential for metallophytes research.

1. INTRODUCTION

In recent years a lot of attention has been drawn to the study of metal tolerant plants. The number of metal tolerant species identified has risen from ten of species to over 450 species [1]. It is therefore likely that many thousand species remain to be added to existing inventory on metal tolerant plants. These plant species, described as metallophytes, have evolved particularly efficient mechanisms which enable them to grow in metalliferrous environments [1,2]. Metallophytes are of immense use for biogeochemical prospecting and geochemical exploration [3-6]. Biogeochemical prospecting and geochemical exploration involve chemical analysis of plant in order to identify mineral deposits buried under a thick cover of post-mineralisation material such as glacial till and thick soil. The presence of these plants can thus indicate the presence of a specific mineralization. Pujari demonstrated that leaves of *Terminalia alata* have strong biogeochemical signatures which reflect Cu values in the soil associated with mineralization [4]. The Zambian copper flower *Becium centraliafricanum* has also been used to indicate the presence of Cu in Shaba Province of Zaire and the Zambian Cu belt [7]. Metallophytes can also be used for phytomining where plants are used to harvest metals of low grade ore that cannot be processed economically by other means [1].

Metallophytes are also excellent candidates for phytoremediation [8]. Phytoremediation is an emerging technology that involves the use of higher plants to remove, destroy or sequester pollutants from contaminated sites [8-11]. This technique is emerging as a viable alternative to conventional methods, such as soil excavation and incineration, due to its many advantages. It is solar driven and can be carried out in situ hence it is a cost effective and environmentally friendly clean-up technique [8-11]. The biological properties and physical structure of the soil is maintained and thus the technique avoids dramatic landscape disruption and preserves the ecosystem [11]. In order for phytoremediation to gain more recognition and to become a technically viable at a commercial level a better understanding of the molecular, biochemical and physiological processes that characterize accumulation is required.

2. Plants Response towards Metal Toxicity

Mechanisms of tolerance vary considerably according to the metal accumulated and of course the plant species. There is much contention in the literature over the possible mechanisms of metal tolerance in plants. This reflects the complex nature of higher plant responses to metal toxicity [12]. There are a number of strategies that plants could possibly employ to combat high external metal concentrations, these can be classified into two basic main categories: exclusion and accumulation [12,14]. Plants that restrict uptake or transport of metals through the roots by either precipitating metal by increasing the pH of the rhizosphere or by excreting anions such as phosphate which complex metals in the root environment, are known as excluders [1,14].

Accumulators, on the other hand can concentrate metals in their above the ground tissues to levels far exceeding those present in the soil. Generally the concentrations in the above ground tissues also exceed those in underground tissues [1,12,14]. Resistance mechanisms in the accumulator plants include a high turnover of organic acids such as phytate, citrate and malate and the induction and activation of antioxidant enzymes such as glutathione peroxidase [15].

Accumulators are further divided into indicators and hyperaccumulators. The division is dependent on the extent of metal accumulation. Indicators show a linear relationship between elemental content in the plant and the concentration of the same element in the soil [13]. In an ideal situation, the concentration of the metal in the aboveground plant tissues of indicators reflects the concentration of the metal in the soil. Plants with extreme level of metal tolerance, which can accumulate metals in concentrations 100 fold in their shoots, compared to normal plants without any toxicity symptoms are referred to as hyperaccumulators [13,16]. Typically the metal concentration in hyperaccumulator plants are > 100 mg kg -1 for Cd, >1000 mg kg -1 for Ni, Pb and Cu and 10 000 mg kg -1 for Zn and Mn. Hyper-accumulator status has also been suggested for plants that concentrate metalloids rather than metals i.e., selenium and arsenic.

The best-known metal hyper-accumulator is Thlaspi caerulescens (alpine pennycress). While most plants show toxicity symptoms at Zn accumulation of about 100 ppm, Thlaspi caerulescens is known to accumulate up to 26000 ppm without showing any signs of toxicity [17]. Thlaspi caerulescens has also been recorded to hyper-accumulate Zn, Ni, Cd, and possibly Pb, and appears to have the capacity to hyper-accumulate Mn and Co under laboratory conditions [17]. Hyperaccumulators in particular have a tremendous potential for application in remediation of metal ions [14,17].

3. Screening of Metallophytes in Metalliferous Areas in Botswana

Botswana's economy is largely dependant on mining industry which generates 70 % of the country's revenue. Botswana is among the Africa's top three mineral producers by value [18]. The identification of accumulator plants is thus particularly important to Botswana. Mining activities leave behind a vast amount of mine spoils and tailings which become the source of metal pollution.

The effects are air, soil and water pollution and the degradation of the cultivated forest or the grazing land with concominant reduction in production. These harm the biodiversity and economic wealth. Metallophytes can fulfil the objective of pollution control, visual improvement and removal of threats to mankind [17].

The commercial potential of metallophytes in phytoremediation and biogeochemical exploration is an incentive to carry out research regarding plant communities growing in mineralized areas in Botswana. Moreover very little research has been done with regard to the identification of metal tolerant plants in Botswana [19-27].

As such, there is a need to carry out comprehensive studies in order to identify indigenous accumulator plants. These studies are also critical for conservation of indigenous species particularly that site restoration of mining area through revegetation can be best achieved using native and local species.

Studies that have been carried out in Botswana date back to the late 1970's and these include the use of Helicrysum leptolepis, to indicate Cu mineralization in areas of shallow overburden in the Ghanzi area [19], and Ecbolium lugardae to identify Cu mineralization hidden under wind-blown sand in Maun area [20,21].

The latter species were also consistently found in the other mines of high Cu concentration in Botswana [21].

Over a period of years a research project has been going on at University of Botswana that involves chemical studies of plants that have accumulating capabilities from mineralised zones and other vulnerable areas in Botswana. The plants studied were collected from the Cu and Ni mineralized areas in the North Eastern part Botswana. Helichrysum candolleanum and

Blepharis diversispinia were initially shown by Takuwa et al. [22] to accumulate high concentrations of Cu and Ni.

Further studies by Nkoane et al. [24] revealed Helichrysum candolleanum as a Cu and Ni indicator which could be used in biogeochemical or biogeobotanical prospecting. Tephrosia longipes and Indigofera melanadenia plant species also revealed accumulation of high concentrations of Cu and Ni [23]. These studies require multi disciplinary and coordinated research efforts that combine analytical aspects and organic aspects of chemistry.

Tephrosia longipes is an annual non-climbing shrub, low growing plant approximately 10 cm in height with a woody base, an extensive root system and reddish purple flowers; Indigofera melanadenia is an annual non-climbing herb with a long soft stem of approximately 50 cm, compound alternate leaves, fibrous root system and pinkish red flowers.

Both plants belong to the Fabaceae family. Blepharis diversispina (Nees) C. B. Clarke and Blepharis aspera Obermeyer both of the Acanthaceae family, and Helichrysum candolleanum H. Buek of the Asteraceae family were also studies.

Three critical steps to success in this research work include sample collection, sample preparation as well as sample detection. The development of meticulous procedures regarding these latter aspects of analysis is required thus methods were developed aimed at perfecting these latter aspects.

3.1. Sample Collection

All plant and soil samples were collected from four mineralized areas in the North-Eastern part of Botswana: Selkirk, Nakalakwana, Malaka and Thakadu (figure 1).

Selkirk is an active copper mine; Nakalakwana is an abandoned copper excavation site; Thakadu and Malaka are abandoned copper mines in the Matsitama area. Selkirk is underlain by meta-gabbro, containing actinolite, hornblende, plagioclase, biotite, chlorite and epidote (the geology of the other areas were not available).

For only Selkirk mine, the sampling site was divided into 25 equal quadrants of dimensions 20m by 20m (figure 2) for sampling purpose.

The plants species were collected from quadrants A1, A2, B2, D2, D4 and E1 as indicated in figure 2, where they were predominantly distributed. From the same quadrants and the same location points of plants, soil samples were collected from the surface to a depth of about 15 cm.

For other sampling site there was no division of the sites, plants species were collected randomly where they were predominantly distributed. From the same location points of plants, soil samples were collected from the surface to a depth of about 15 cm. Table 1 shows the concentrations of Cu and Ni of these sampling sites [24].

The results indicate higher than normal concentration i.e. higher than 5-100 μg/g Cu and 20 μg/g. The soils from Selkirk and Thakadu showed severe Cu mineralization and more Cu than Ni.

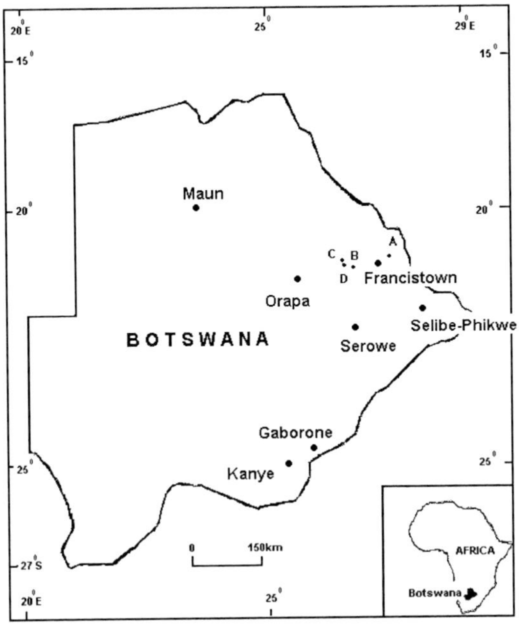

Legend:

A Selkirk, about 35 km north east of Francistown; GPS location: S 21° 18′, E 27° 44′, Altitude: 1018 m.

B Nakalakwana, about 60 km from Francistown, along Francistown-Orapa road; GPS location: S 21° 05′, E 26° 59′, Altitude: 1117 m.

C D:Thakadu and Malaka, respectively, are about 82 km from Francistown, 5 km off Francistown Orapa road, and 2 km apart.

GPS locations: Thakadu S 21° 03′, E 26° 46′, Altitude: 1058 m, Malaka: S 21° 02′, E 26° 45′, Altitude: 1044 m.

Figure 1. Map of Botswana showing Cu-Ni mineralized area in North Eastern part of Botswana.

Table 1. Sampling area information and metal concentration range in soils (% w/w), (n=5) collected from Selkirk, Thakadu, Malaka and Nakalakwana , where plants were sampled

Place	Status	Soil concentration range		Where collected	Comment
		Ni (% w/w)	Cu (% w/w)		
Selkirk	Active Cu-Ni mine	0.08-0.3	0.3-4.3	Collected at the foot of the ore hill	Severe mineralization
Thakadu	Abandoned Cu-Ni mining site	0.009-0.015	1.5-4.3	Collected along the ore (20-30m)	Severe mineralization
Malaka	Abandoned Cu-Ni mining site	0.007-0.014	0.004-0.05	Collected 15-20 m from the ore body	Slightly mineralised
Nakalakwana	Abandoned Cu-Ni excavation site	0.006-0.009	0.004-0.008	Collected 15-20 m from the ore body	Within normal conc. For Cu. Higher than normal conc.for Ni

3.2. Sample Preparation for Analysis of Metals

By estimates 60 % of analysis time is spent on sample preparation [28] and it is the most probable source of inaccuracy and imprecision that can advertantly be introduced into the entire analytical procedure and often the cause of the largest variability in the analytical results. Hence sample preparation has always been a challenge to analytical chemists. Sample preparation eliminates possible substance interference, concentrates and stabilizes the analytes that are in solution leaving them in optimal conditions for instrumental analysis [29]. There are various forms of sample preparation methods, with their effectiveness dependant on the specific nature of the analytes of interest, and the matrix involved. Hence special emphasis should be placed on fast, efficient and well designed sample preparation procedures.

Any method proposed for sample preparation must meet several criteria if it is to be an attractive choice for the analytical chemist. The ideal sample preparation method must be able to selectively separate the molecules of interest from unwanted and interfering matrix molecules. This clean-up effect is closely related to the signal-to-noise ratio (S/N) in the subsequent analytical steps because it reduces the chemical background noise. In addition, elimination of macromolecules allows quantification to very low levels and prevents deterioration of the analytical instruments, thereby enhancing stability. In order to get a strong enough detector signal for the analytes of interest present in low amounts (below the detection limit), it is necessary to enrich their concentrations. Enrichment along with sample clean-up are often required to achieve sensitive and accurate qualitative and quantitative analyses [30]. The ideal sample preparation method should also have the option of being used in an automated fashion to minimize contamination and to reduce costly and tiresome labor. Automation is the foundation of high sample throughput, which in turn is a requirement for routine analysis [30].One way of designing fully automated analysis systems, where the samples are automatically fed at one end and the reports are produced at the other end is to have a sample preparation method that can be coupled to other analytical instruments. Hyphenation prevents cross contamination of the analyte after sample preparation and also cuts on labour [30].

3.2.1. Sample Preparation for Determination of Metals in Environmental Samples

For total metal determination, the sample preparation method used prior to instrument detection is digestion. The purpose of this pretreatment is to attain a complete digestion of organic matrix, decrease viscosity, increase homogeneity and release of analytes from various compounds and phases which they might be bound to and can otherwise have adverse effects on the analytical signal [31-33]. In our research studies oxidizing agents which include nitric acid, hydrochloric acid, perchloric acid, hydrofluoric acid and hydrogen peroxide, as well as mixtures of these reagents, with simultaneous heating, were used to decompose plant and soil samples [23-27]. A classical open digestion method was employed in the preliminary screening of plants [23]. This method requires hours to ensure complete digestion and consumes a lot of acids, despite its disadvantages this method is a good inexpensive alternative in the absence of modern instruments of digestion. For further studies microwave digestion was employed [27]. Microwave is regarded as fast and efficient method of digestion due to combined speed of microwave heating with elevated temperatures and pressure achieved in sealed Teflon PFA vessels. This results in decomposition of the sample matrix, high penetration by the extractant, and rapid transfer rates of the analyte into solution [33,34].

In order to achieve effective decomposition and hence precise and consistent results critical parameters which were optimized were; digestion acid combinations, microwave power, extraction times, digestion temperature and sample weight. The optimized conditions are shown in table 2.

Table 2. Optimised parameters for microwave digestion

Parameter	Optimized condition
Microwave power	1000 W
Extraction times	30 Minutes
Digestion temperature	200°C
Sample weight	1 g
Acid combinations	8 ml HNO_3, 3 ml HF acid and 1 ml H_2O_2

3.2.2. Slurry Sampling Electrothermal Atomic Absorption Spectrometry

The introduction of a slurry (solid suspended in a liquid diluent) into the electrothemal atomic absorption spectrophotometer and subsequent analysis constitutes what is commonly known as slurry sampling ETAAS. Slurry sampling ETAAS provides a unique combination of minimal sample preparation, high sample throughput, accuracy, low cost and operational simplicity of the instrumentation, rapid and unattended sample throughput and minimal sample contamination as there is the possibility of performing the sample decomposition inside the graphite furnace [35,36]. This technique has been extensively used for the analysis of biological tissues, solid environmental samples such as soils and plant material [36].

Preparation of a slurry sample is an in important aspect of slurry sampling and if it is not adequately addressed then, errors due to sedimentation of dense particles or floatation of lighter particles are observed. Nkoane et al. [24] used slurry sampling ETAAS methodology in the analysis of plants parts of *Blepharis diversispinia* and *Helichrysum condelleanum* along with the host soil [24]. In these studies a method for slurry developed by Takuwa et al. [22] was employed. The slurry was prepared by weighing 1.0-4.0 mg of finely ground sample material (less than 63μm) into autosampler vials with subsequent addition of 1ml of the liquid diluents; 5% v/v HNO_3 [24]. In order to ensure that a representative aliquot of slurry is injected into the furnace using an autosampler, the slurry must be either stabilized or homogenized. In these studies stabilization was achieved by addition of 0.05 % Triton X-100 while homogenization has been achieved by using a hand held ultra sonic probe [24]. Other dispersing agents that can be used for the purpose of stabilization include glycerol, Viscalex HV-30, sodium hexametaphosphate, and nitric acid, magnetic stirring, gas bubbling, and vortex mixing [35,36].

3.2.3. Microdialysis Sampling of Metals

3.2.3.1. Practical Aspects of Microdialysis Sampling

Microdialysis is based upon size-selective diffusion of molecules across a semi-permeable membrane driven by the concentration gradient established between the liquid in the membrane i.e. the perfusion liquid and the sample medium. Microdialysis sampling is accomplished by means of a microdialysis probe. Several probe designs are commercially available and can be fabricated according to requirement. The most common ones include the hollow fibre probe, the planar probe, the concentric probe and several others reviewed in the literature [37,38]. Amongst several geometries of microdialysis probes, the most commercially available and preferred probe design for several applications for environmental applications are of the concentric design [40-44]. The tunable concentric probe consists of the inner cannula placed inside the outer cannula with the dialysis membrane glued to one end of the outer cannula (see Figure 2). This probe is made of stainless steel allowing for repeated use and sterilisation. It consists of an inner and an outer cannula (See Figure 2). The outer cannula is fitted with the membrane and this is sealed with glue at both ends. The perfusion liquid is continuously pumped into the inner cannula. The inner cannula directs the perfusion liquid from the inlet to its distal end. Here the perfusion liquid comes in contact with the membrane where diffusion takes place driven by the concentration gradient generated between the solution outside the probe and the perfusion liquid. The direction of the flow is then reversed and the dialysate moves to the proximal end and flows through the outlet. The dialysate can be analysed on-line or off-line after fraction collection. The protruding length of the inner cannula into the membrane where diffusion takes place is referred to as effective dialysis length and it is one of the important parameters that influence analyte recovery [38]. In order to maximise the flux and thus obtain a high analyte recovery, effective dialysis length should be optimised [38]. The in situ tunable concentric probe has an inner cannula which is movable lengthwise making it possible to adjust the membrane effective dialysis length.

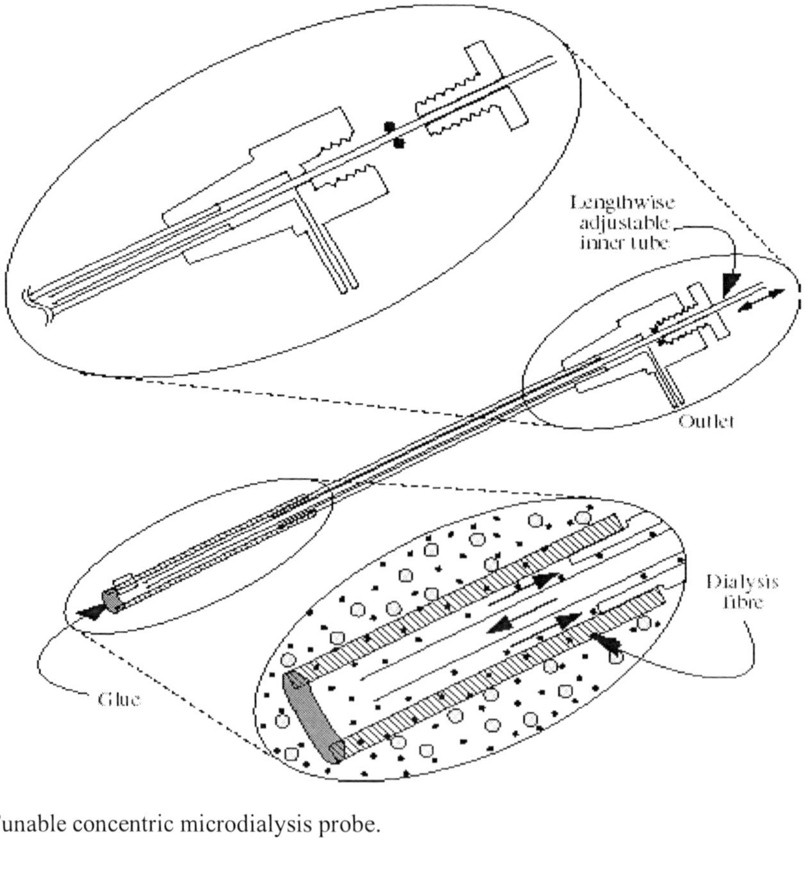

Figure 2. Tunable concentric microdialysis probe.

3.2.3.2. Applications of Microdialysis

Microdialysis, as a sampling technique, has been routinely used in neuroscience [45], pharmacology [46] and drug metabolism studies [47]. Recently it has found use in the area of biotechnology for studies of enzymatic bioprocesses [39] and fermentation broths [48]. Microdialysis offers a plethora of advantages which have greatly contributed to its increasing use evidenced by the rapid increase in publications.

One of the most distinctive features of microdialysis is the ability to perform sampling and sample clean up in one step [43-46]. The immediate consequence of this is the elimination of time-consuming and tedious steps associated with traditional sample clean-up procedures which utilise large amounts of toxic organic solvents.

Microdialysis yields a clean enough dialysate free of particles and macromolecules and of well defined volume feasible for chromatographic or capillary electrophoretic or optical flow-through analysis without need for further sample pretreatment. It can be efficiently coupled on-line with many analytical detection techniques [49,50]. Microdialysis has the inherent ability to cope with very complex matrix such as those found in enzymatic hydrolysates [49] fermentation broths [50] and plant slurry [51].

Therefore microdialysis is a potential alternative to conventional extraction techniques for isolating complicated matrix samples. With the presented advantages the use of microdialysis continues to grow.

Very recently the idea of using microdialysis in environmental samples has been realized and now microdialysis stands as a new tool in environmental sampling. One recent interesting application of microdialysis is in the area of environmental monitoring [40-44]. This includes the characterization of carbohydrates in storage septic tanks [40], the monitoring of metal uptake by plants [52],the monitoring of metals in waste water [42,44] predicting heavy-metal pollution in soils and current metal bioavailability [53-55], identifying solid phase associations and monitoring both soil-plant fluxes and release rates of metal ions under natural events or anthropogenic occurrences [53-55].

These should be highlighted as promising approaches which denote versatility and potentialities of microdialysis for in situ sampling of environmental samples.

3.2.3.3. Microdialysis Sampling of Cu and Ni in Plant Suspension

The enormous and still partially unexploited potential of microdialysis technique in the sampling of metals has been recently summarized [44].

Recently Mosetlha et al. evaluated the applicability of microdialysis sampling for use as a sampling and sample clean-up technique for metals in complex matrices of plant samples [25]. In these studies microdialysis sampling was paired with acid digestion method which is routinely employed as a sample preparation technique.

Microdialysis was performed at optimized condition of 3 μL/min with the incorporation of 0.05% w/v humic acid in the perfusion liquid. Table 3 and 4 show summaries of the concentrations of Cu and Ni respectively determined after microdialysis sampling and acid digestion for flower samples of the Blepharis aspera species sampled from Selkirk mineralised area.

The versatility of microdialysis sampling as an in-situ sampling and sample clean up technique was demonstrated by the detection of Cu and Ni in all the six plant flower samples with high reproducibility.

Table 2. Comparison of Cu concentrations obtained by microdialysis sampling and acid digestion for six plant flower samples

Plant, n=5 for each soil sample	pH of solution before sampling	Cu determination by microdialysis sampling Concentration in (μg/g)	Total Cu determination by acid digestion Concentration in (μg/g)	Concentration ratio (microdialysis sampling/ acid digestion)
A	5.47	6.18 (5.44)	451.10	0.0137
B	5.40	3.50 (14.73)	230.26	0.0152
C	5.62	2.53 (6.87)	199.21	0.0127
D	5.56	6.59 (5.33)	477.56	0.0138
E	5.49	1.73 (14.11)	124.50	0.0139
F	5.55	1.73 (3.10)	120.14	0.0144

Numbers in parenthesis are % RSD.

Table 3. Comparison of Ni concentrations obtained by microdialysis sampling and acid digestion for six plant flower samples

Plant, n=5 for each soil sample	pH of solution before sampling	Ni determination by microdialysis sampling Concentration in (μg/g)	Total Ni determination by acid digestion Concentration in (μg/g)	Concentration ratio (microdialysis sampling/ acid digestion)
A	5.47	13.67 (17.12)	309.20	0.0442
B	5.40	7.91 (13.34)	163.40	0.0484
C	5.62	6.09 (15.56)	145.73	0.0418
D	5.56	11.36 (10.03)	269.15	0.0422
E	5.49	10.90 (11.02)	230.44	0.0473
F	5.55	7.13 (9.11)	169.82	0.0420

Numbers in parenthesis are %RSD.

Although all the Cu and Ni concentrations obtained by microdialysis sampling were lower than the concentrations determined after acid digestion a linear relationship was observed linking the metal concentrations determined after microdialysis sampling and acid digestion as shown in Figure 3. A constant ratio of 0.0138 and 0.0440 was obtained for Cu and Ni respectively from the slopes of the plots. It is from these metal concentration ratios that microdialysis sampling with its advantages can be used to predict the total concentrations of metals in plant samples, as was for Blepharis aspera flowers using Eqs (1) and (2).

$$\text{Total Cu concentration} = \frac{M_{Cu}}{0.0140} \quad (1)$$

$$\text{Total Ni concentration} = \frac{M_{Ni}}{0.0440} \quad (2)$$

where M_{Cu} and M_{Ni} are the Cu and Ni concentrations obtained by microdialysis sampling in µg/g.

After the constant metal concentration ratios are known, there is no need for plant samples to be digested by acid in order to know their total metal concentrations. Instead only microdialysis sampling will be carried out on the plant samples and from the Cu and Ni concentration obtained, the total concentrations of Cu and Ni in the plant samples can be calculated by using the latter equations. Thus microdialysis sampling has the potential to be used in the indirect determination of metals in plant samples.

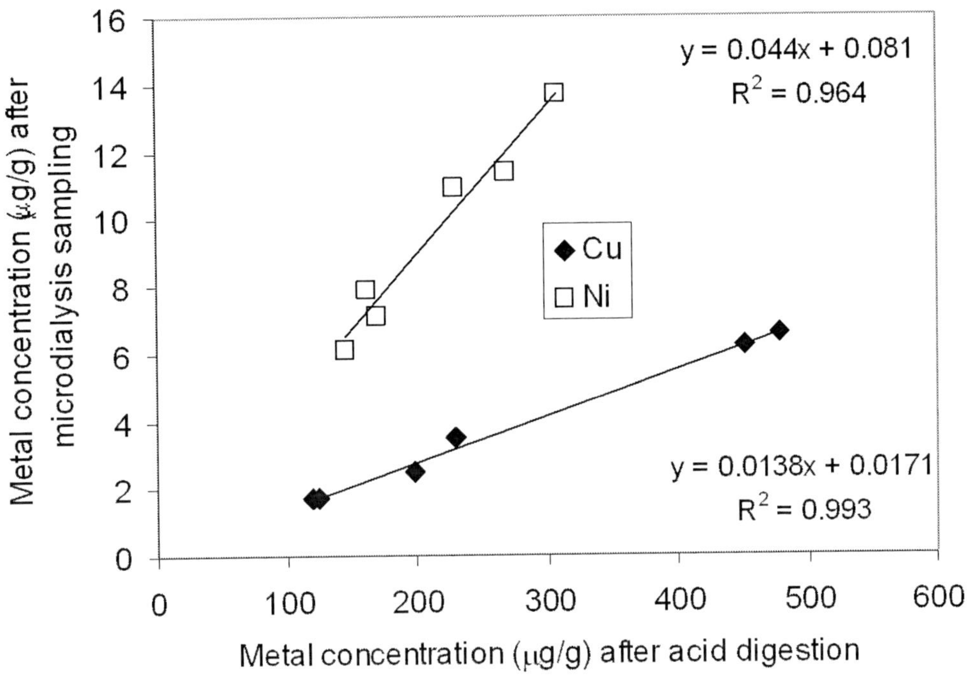

Figure 3. Relation between the concentrations of Cu and Ni determined after microdialysis sampling and acid digestion of plant flower samples.

3.3. Instrumental Approaches to Detection of Metal

The techniques employed for detection of metals in these studies are the flame atomic absorption spectrometer (FAAS), Electrothermal atomic absorption spectrometer (ETAAS), inductively coupled plasma mass spectrometer (ICP-MS) and ICP-OES. ICP-MS and ICP-AES are detection techniques which are being routinely used in metal studies. These techniques have several qualities making them preferred methods such as high sensitivity and a great capacity for simultaneous rapid and precise determination with a large dynamic range. They can be applied to the determination of elements across the periodic table; lithium to actinides. These techniques can be combined with numerous sample introduction accessories or separation techniques on-line or off-line for sensitive element specific detection. The virtual independence of elemental signal intensity of the coeluting matrix makes ICP-MS a valuable detection technique to screen biological extracts for the presence of metal-containing fractions. ICP-MS is therefore is also an ideal technique to spot heteroelement-containing species.

While it will be preferable to use powerful instrumental techniques with multi elemental detection capabilities, such as ICP MS, (FAAS) still has a place in some analytical laboratories, especially in developing countries and it can be used as a screening test prior to more sophisticated techniques. FAAS analytical technique is remarkable for its sensitivity, its speed and relatively low operational cost. It is one the most extensively used analytical techniques for determining various element with significant precision and accuracy. It is widely used for analysis of elements in a wide variety of complex sample matrices including biota, soils and water.

3.4. Accumulation Patterns of Cu and Ni for Indigofera Melanadenia and Tephrosia Longipes Plant Species

Mogopodi et al. employed FAAS in the preliminary studies of accumulation patterns of Tephrosia longipes and Indigofera melanedinia collected from Selkirk mines [23]. The site was divided into 25 equal quadrants of dimensions 20m by 20m (Fig. 3). The plants species were collected from quadrants A1, A2, B2, D2, D4 and E1 as indicated in figure 3, where they were predominantly distributed.

N

Up Hill (Ore source)

	1	2	3	4	5
E	Tephrosia longipes				
D		Tephrosia longipes		Tephrosi a longipes	
C					
B		Indigofera melanadenia			
A	Indigofera melanadenia	Indigofera melanadenia			

Down Hill

Figure 3. Schematic diagram showing the orientation of the sampling area situated in Selkirk mine, Botswana. Indigofera melanadenia plants species were collected from quadrants A1, A2 and B2. Tephrosia longipes plants species were collected from quadrants D2, D4 and E1.

Figures 4 and 5 show the concentration values of Cu and Ni in different plant parts i.e. roots, stem and leaves of Indigofera melanadenia. The highest concentrations of both Cu and Ni were found in the leaves showing preferential accumulation in the leaves. The observed high concentrations of Cu and Ni in the leaves could suggest that Indigofera melanadenia has an efficient translocation of metals from roots to shoots which is a recognized characteristic of accumulator plants.

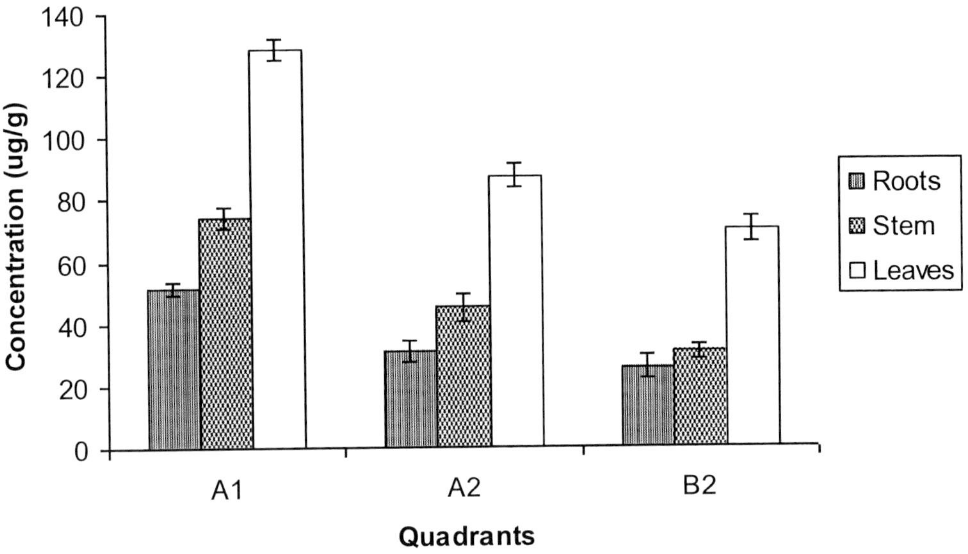

Figure 4. Concentration (μg/g) of Cu in different parts of *Indigofera melanadenia* collected from quadrants A1, A2 and B2 (n=6).

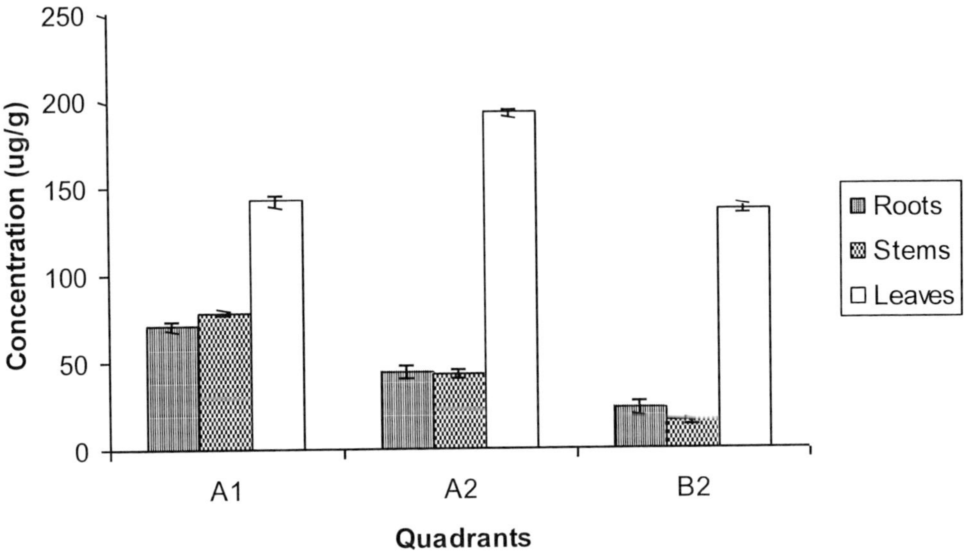

Figure 5. Concentration (μg/g) of Ni in different parts of *Indigofera melanadenia* collected from A2, A2 and B2 (*n = 6)*.

In the preliminary studies plant species collected from different quadrant did not show a similar trend. This prompted further studies using the ICP-MS. Figure 6 shows Tephrosia longipes sampled from quadrants D2, D4 and E1.

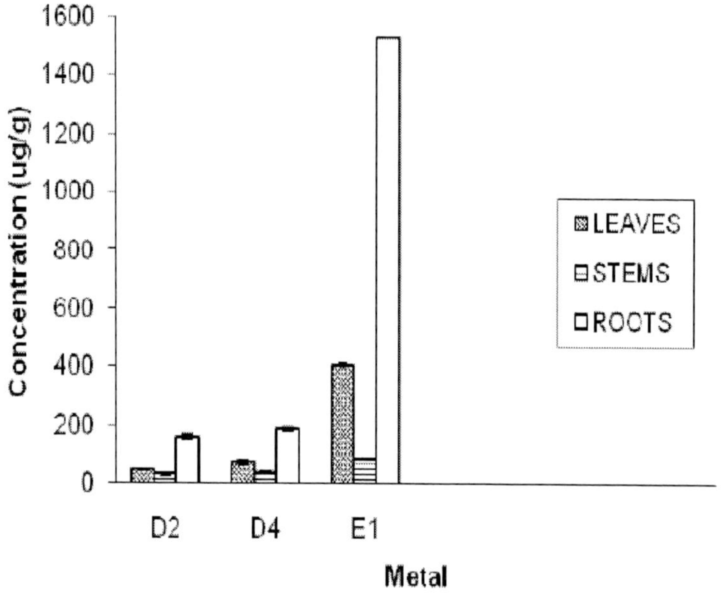

Figure 6. Concentration (μg/g) of Cu in different parts of *Tephrosia longipes* collected from Selkirk mine (n=6).

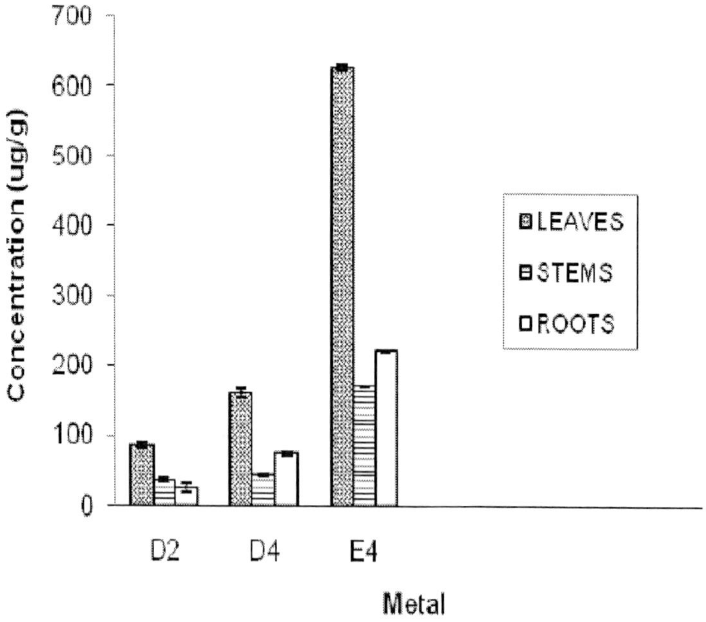

Figure 7. Concentration (μg/g) of Ni in different parts of *Tephrosia longipes* collected from Selkirk mine (n=6).

The concentration of Cu was highest in the roots of Tephrosia longipes sampled from all the quadrants with highest concentration value of ~1500 μg/g observed in samples collected closest to the ore body i.e. quadrant E1. The high concentration values of Cu in the roots suggest that Cu is not efficiently translocated to the above ground tissues and also suggest that Trephosia longipes species are Cu excluders. Figure 7 shows that for all the quadrants; D2, D4, E1, Tephrosia longipes accumulated highest concentration of Ni in the leaves suggesting its tendency to translocate these metal ions readily to the leaves hence a possible accumulator for Ni.

3.5. Accumulation Patterns of H. Candolleanum, B. Diversispina and B. Aspera: Multi-Element Study

In spite of the multi-element capabilities of modern instruments few studies have been carried out for multi-elemental accumulation. Multi elemental determination is of importance since some metallophytes hyper-accumulate more than one metal, e.g., Thlapsi species such as Thlapsi caerulescens (Cd, Ni, Pb and Zn), Thlapsi ochroleucum (Ni, Zn) and Thlapsi rotundifolium (Ni, Pb and Zn) [57,58]. Nkoane et al. investigated the multi-element accumulation capabilities of the three plants, H. candolleanum, B. diversispina and B. aspera for a total of 61 elements using two complementary analytical techniques, ICP-MS and ICP-AES and compared the accumulation patterns of the said species [27]. The levels and distribution of the metals in the roots, stem and leaves of these plants were determined. The host soils were also analysed.

Tables 4 and 5 give the metal concentrations in the plant parts and the corresponding concentrations of the host soils. Seven elements (Al, Co, Cr, Cu, Fe, Ni and Ti) were found in high concentrations in the plant parts of B. aspera, (Table 4) with the highest concentrations found in the stem, except for Cd. For B. diversispina which was collected from Malaka and Nakalakwana, large amounts of Al, Fe, and Ti were found (Table 4). H. candolleanum plants collected from both Selkirk and Thakadu had high concentrations of Al, Cd, Cr, Cu, Fe and Ti. All the concentrations were 10-100 times higher than the normal values. In addition, the H. candolleanum from Thakadu accumulated higher concentrations of Ba (≈ 30 times), while the Selkirk plant had about 10 times higher than normal concentrations of Co and Ni– their host soils also had higher than normal concentrations of these elements (Table 5). All the elements (except Cd) that were found in the plants in higher than normal concentrations had elevated concentrations in the soils as well. The highest Al concentrations were found in the stem of B. aspera (0.22-0.33%), the roots of B. diversispina (0.12-0.14%) and the leaves of H. candolleanum (0.18-0.33%). Plants in this study took up and translocated Al to the leaves, and therefore are Al-accumulators but the concentrations are not in the hyperaccumulation range. By definition, hyperaccumulators have 100-fold more metal than normal plants [59,60].

This study suggests that the three plant species have different ways of dealing with excessive levels of metals in the soil and also suggests that the accumulation pattern is independent of the sampling location.

Table 4: Metal concentrations[a] with the standard deviations (n=6, in µg/g) of plant parts of *B. diversispina* and *B. aspera*, and their host soils

Plant	Elemental concentration (mean ± SD) µg/g									
	Ba	Cd	Co	Cr	Cu	Ni	V	Ti	Al[b]	Fe[b]
B. diversispina										
a) Malaka										
Soil	930 ± 90	0.1±0.004	9.5±0.6	41±3	72.6 ± 7	13±0.2	24 ± 2	1026 ± 38	59±0.7	46±4
Root	32±2	0.6±0.05	0.9±0.05	**62±5**	19.9±0.3	13±0.2	3.1 ± 0.1	47 ± 4	**1.2±0.005**	1.5±0.07
Stem	17±0.7	0.3±0.02	0.6±0.05	7±0.6	9.±0.5	1.9 ± 0.01	1.1 ± 0.01	28 ± 0.5	0.6±0.02	0.6±0.01
Leaves	64±0.6	2.9±0.3	0.7±0.07	0.4±0.02	20.9±0.2	3.1±0.2	1.6 ± 0.06	45 ± 3	**0.8±0.02**	0.8±0.001
b) Nakalakwana										
Soil	650 ± 20	0.2±0.001	4.1±0.06	31±1	75 ±4.1	11± 1	27± 3	950 ± 16	46±0.3	45±1
Root	32±0.09	0.3±0.02	1.1±0.1	21±2	16.8±0.2	17±2	2.2 ± 0.1	**54 ± 1**	**1.4±0.08**	1.4±0.05
Stem	17±1	0.2± 0.04	0.6±0.005	11±1	6.6±0.4	20±0.2	1.1 ± 0.1	37 ± 4	**0.9±0.01**	0.7±0.02
Leaves	24±0.6	3.6±0.2	1.1±0.09	2.7±0.05	15±0.001	4.70±0.5	0.9 ± 0.06	39 ± 4	0.7±0.02	0.6±0.009
B. aspera **Selkirk**										
Soil: Plant 1	56 ± 3	0.3±0.001	53±2	270±3	**5.4±0.3**[b]	**2300±50**	33 ± 0.4	540 ± 22	21±2	99±0.6
Plant 2	70 ± 2	0.2±0.01	63±1	140±4	**18±0.4**[b]	**320±10**	13± 0.5	360± 9	123±9	86±4
Root: Plant 1	14± 0.6	0.3 ±0.04	**41 ±1**	19 ±0.4	**500 ±3**	**310 ± 6**	2.3 ± 0.2	43 ± 2	**1.9 ±0.03**	4.4 ±0.1
Plant 2	4.9±0.2	0.3±0.01	**30 ±2**	14 ±1	160±1	**360±8**	**6.4±0.2**	**78±2**	**1.9±0.02**	2.9±0.02
Stem: Plant 1	21± 0.6	0.2±0.003	**52 ±0.02**	24 ±2	**490 ±40**	**610±20**	2.4 ± 0.1	**83 ± 2**	**2.2 ±0.1**	4.7 ±0.3
Plant 2	14±0.05	0.8±0.03	**38±0.9**	24± 1	**240±1**	**570±20**	**10±0.05**	**140±3**	**3.4±0.05**	4.3±0.04
Leaves: Plant 1	16±1	5.5 ±0.09	10 ±1	3.5±0.2	140±9	**210±6**	0.7±0.0001	**52±3**	0.5±0.04	1.2±0.06
Plant 2	10±0.1	4.6 ±0.1	11 ±0.4	3.6 ±0.03	44±4	**420±10**	1.5±0.02	21±0.3	0.5±0.03	0.5±0.05

[a] For plants, concentrations are for elements that were present in concentrations higher than normally found in plants. (**Bold** numbers represent concentrations more than 10 times higher than normal), n=6 (pooled mean used for 2 plants) except for *B. aspera* plant where n=3 for each plant

[b] Concentration in mg/g

Table 5. Metal concentrations[a] with the standard deviations (n=3, in μg/g) of plant parts of H. candolleanum and the host soils

Plant	Elemental concentration (mean ± SD) μg/g									
H. candolleanum	Ba	Cd	Co	Cr	Cu [b]	Ni	V	Ti	Al[b]	Fe [b]
a) Selkirk										
Soil	170±6	0.1±0.01	71±2	160±8	**16±0.5**	**1400±11**	26±0.8	410±5	82±7	70±4
Root	23 ± 1	4.7±0.1	4.8±0.1	4.8±0.002	**0.9±0.04**	41 ± 0.4	1.1±0.004	**58 ± 0.3**	0.7±0.03	1.1±0.06
Stem	8 ± 2	12±1	5.6±0.2	2.5±0.08	**0.2±0.005**	21 ± 0.5	0.8 ± 0.09	**69± 5**	0.3±0.02	0.6±0.03
Leaves	19±0.04	21±0.5	18±0.4	21±0.4	**1.5±0.06**	**110 ± 1**	4.4 ± 0.2	**230±10**	**3.3±0.1**	**5.5±0.2**
b) Thakadu										
Soil	3.4±0.2[b]	0.9±0.03	7.8±0.4	67±3	**39±1**	18±0.1	54±4	1300±50	16±0.5	46±0.7
Root	**2.9±0.1b**[b]	4.8±0.1	0.3±0.09	2.7±0.4	**1.2±0.002**	2.4±0.3	**6.3±0.07**	35±1	0.3±0.02	0.7 ± 0.03
Stem	**2.4±0.1**[b]	23±0.9	0.9±0.07	10±0.9	**0.7 ± 0.002**	5.7±0.09	**7.9 ± 0.9**	**150±1**	0.4±0.01	1.0±0.07
Leaves	**2.3±0.4**[b]	22±1	4.4±2	29±0.6	**2.2 ±0.04**	7.2±0.6	**31 ± 0.3**	**390±6**	**1.8±0.03**	3.0±0.02

[a] For plants, concentrations are for elements that were present in concentrations higher than normally found in plants. (**Bold** numbers represent concentrations more than 10 times higher than normal)

[b] Concentration in mg/g.

From Figure 8 which illustrates metal concentrations (logarithmic scale) in the plant parts of: A) B. diversispina collected from Malaka (BD-M) and Nakalakwana (BD-N); B) H. candolleanum collected from Selkirk (HC-S) and from Thakadu (HC-T); and C) H. candolleanum (HC-S) and B. aspera (BA-S) collected from Selkirk, it can be seen that the accumulation levels for most metals were within the same order of magnitude for the B. diversispina plants collected from two different places. This trend was also observed for H. candolleanum (Figure 8), except for Ba which was two orders of magnitude higher for the Thakadu plant, and Ni which was one order of magnitude higher for the Selkirk plant.

In the same studies Nkoane et al also studied accumulation of rare earth elements [27]. The results indicated that the concentrations of REEs in the roots, stems and leaves were above normal for all plants and most of the REEs were found in the leaves for B. aspera indicating that for almost all the plants REEs were taken up and translocated to the upper parts and mostly in the roots for H. candolleanum.

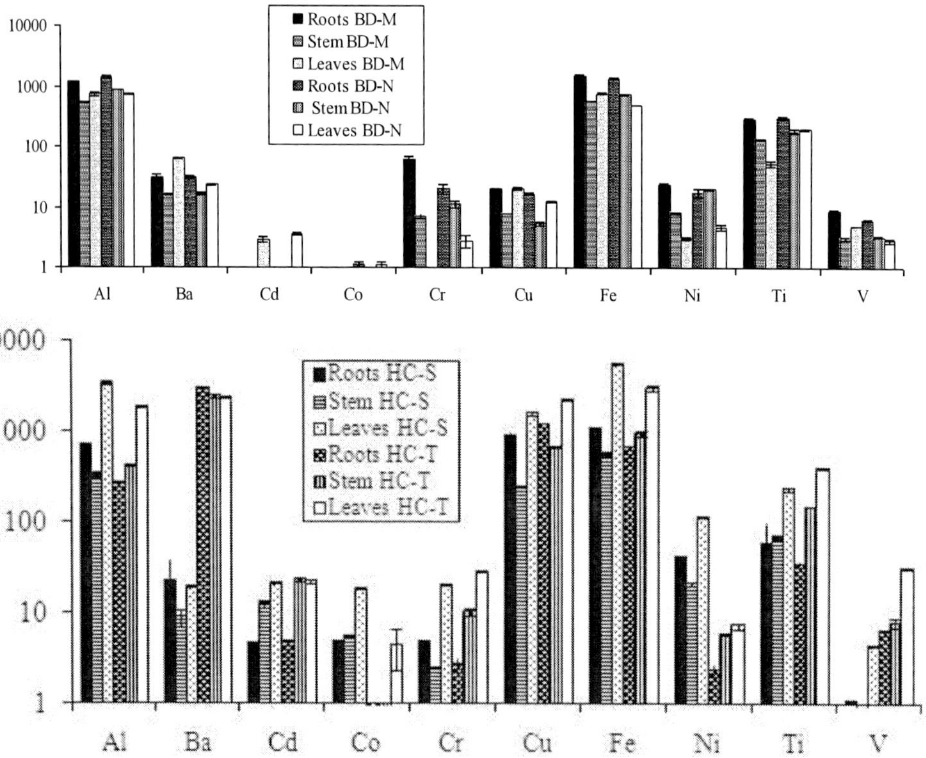

Figure 8. Concentration (μg/g) of elements in different plant parts.

4. ROLE OF PHYTOCHELATINS (PCS) IN METAL TOLERANCE

The key to understanding accumulation is identification and characterization of corresponding ligands. One recurrent mechanism for heavy metal detoxification is chelation by ligand. A number of chelation ligands such as PCs, glutathione (GSH) and metallothionines (MTs) ligands have now been recognized in plants.

A well known mechanism for enhancing metal tolerance is the expression of metal-binding PCs [61-63]. These are enzymatically synthesized from GSH in response to excessive uptake of metal ions during a reaction catalysed by enzyme γ-Glu-Cys dipeptididyl transpeptidase (PC synthase) [61,62]. The polypeptide consists of repeating sequence of two amino acids; glutamic acid and cysteine and it is terminated by glycine. Thus the basic structure is (γ-Glu-Cys)n-Gly where n is generally in the range of two-five but can be as high as eleven [61-63]. The activity of PC synthase has been identified in cultured cells of Silene vulgaris [63]. The metal inducibility of PCs has been demonstrated in a number of plants [63-65]. The Indian mustard Brassica juncea which accumulates high concentrations of Cd has been shown to have increased concentrations of GSH and PCs [65].

4.1. Analytical Methods Used to Study PCs

Very recent studies have demonstrated speciation as very important aspect. Different analytical approaches were used to study speciation in plants. A large number of determination methods have been frequently employed for identification, quantification and structural elucidation of thiol containing compounds such as PCs, MTs and GSH [63-79]. These include spectroscopic [65-69], separation and hyphenated methods [67,68] as well as electrochemical methods [70-72].

4.1.1. High Performance Liquid Chromatography (HPLC)

HPLC has found extensive application in the field of PCs analysis and with its high resolution capabilities represents the most common and arguably effective method of quantifying thiol containing compounds in complex media [68,69]. PCs are commonly purified by liquid chromatography based on gel filtration and strong anion exchange [68]. With subsequent application of reverse phase HPLC linked to a specific detector of thiol containing compounds, it becomes possible to detect different isoforms coexisting in a sample [68]. Chromatographic detection has been mostly achieved by ultraviolet and fluroscence detectors.

In order to obtain compounds suitable for detection by UV or fluorimetry post column derivatisation of the SH groups of metal free PCs is carried out with a suitable reagent such as Ellman's reagent (5,5'-dithiobis-2-nitorobenzoic acid) [68]. Raab [79] studied the stability and chromatographic behaviour of GSH complexes with trivalent arsenic at different pH and temperatures values. HPLC coupled to metal detector such as ICP-MS has also been employed [65,68,73,74]. Leopold et al demonstrated the use of HPLC coupled on-line to ICP-MS for determination of heavy metal binding properties of PCs in Silene vulgaris cell culture [74]. Their studies showed that Cu binds most stably to these PCs under in vitro and in vivo condition. In their work they characterised heavy metal-PC complexes with n=2.

4.1.2. Electrospray Ionisation Mass Spectrometer (ES-MS)

ES-MS is a soft ionisation technique that offers several advantages which have conspired to favour its role at the forefront of biochemical and environmental research [73-77]. The formation of multiply charged complexes with little or no fragmentation enables the determination of the mass of an intact complex with high degree of accuracy [75]. Complexes of biological molecules such as amino acids, peptides and proteins with transition metals are readily transferred into the gas phase by electrospray ionization and their gas-phase and solution properties can be correlated [76]. ES-MS is compatible with liquid chromatography [75]. Whereas HPLC techniques frequently require derivatization for sensitive detection, ES-MS provides detection method independent of the formation of chemical derivatives or of the UV absorption and fluorescence properties of the molecule [75].

The ability to analyze for heavy metals associated with different ligands in biological samples and to estimate the binding stability of these complexes is very important for a better understanding of the physiological role of metal binding peptides hence the interest to study complexes of PCs, MTs and GSH with metal ions has increased significantly. ES-MS is well suited to the study of metal peptide interactions and has been successfully employed in the study of PCs [75,76]. These complexes were shown to be successfully separated by reverse phase chromatography. Schmoger and cowokers have successfully characterised complexes of As III with PCs using mass spectrometer coupled with HPLC [80]. In a similar work Chassaigne directly detected metal-PC clusters with by electrospray mass spectrometer [81]. Yen et al used nano-ES-MS/MS and capillary liquid chromatography/electrospray ion tandem MS methods to analyse, identify and elucidate nature of PC-Cd complexes isolated from plant extracts of Datura innoxia [73].

4.1.2.1. Probing Metal-Glutathione Interaction Using Electrospray Ionisation Mass Spectrometry

The role of PCs in detoxification of metals in plants is an area of interest for this research group and is the basis of these studies. In view of the importance of GSH as building blocks for PCs preliminary experiments were carried out in order to investigate GSH interaction with metal ions [56]. The observations of these studies could provide a foundation in the understanding of the role of PCs in metal detoxification.

Using ES-MS a study of the complexation of GSH with Cu2+ was carried out at different pHs and adjusted stoichiometry. Because of the known affinity of Cu2+ the choice of these metals is very idoneous to study the binding property of GSH. From the formed complexes, the sites of binding and the formation of mononuclear versus netted complexes should be observed. Metal-peptide complexes were prepared using Cu and GSH at complexing ratios of 1:1 and 1: 2 of the metal to peptide. Both full scan and single ion monitoring were carried out.

The mass spectra confirmed the metal ligand ratio, 1:1 and 2:1 for Cu2+ GSH complexes. Complexes were detected in their protonated forms. The protonated GSH [GSH + H]+ molecule has a molecular weight of 308 and a corresponding peak was observed in the spectra as the base peak (100%) at m/z = 307.9 in the positive ion spectrum. A peak at m/z = 613 was observed and this peak corresponds to a proton bound cluster of glutathione. Figure 9 shows the mass spectrum of GSH under acidic conditions. Peaks at m/z = 288 and 316 are a result of the glacial acetic acid that was used for pH adjustments.

Although the spectrum observed for the GSH-Cu complexes did not show any additional peaks, the use of selected ion monitoring to investigate the presence of peaks corresponding to the molecular weights of the complexes represented quantitatively, complexation of GSH with Cu. The formula and the corresponding molecular weights of the complexes investigated are reported in Table 6. There are numerous factors that influence the affinity of ligands to metal ions and hence the stability of a complex, such as the coordination geometry of the complex, ligand field stabilization effects, concentration of other competing metals, the nature of the reacting species and pH of the media [82-84].

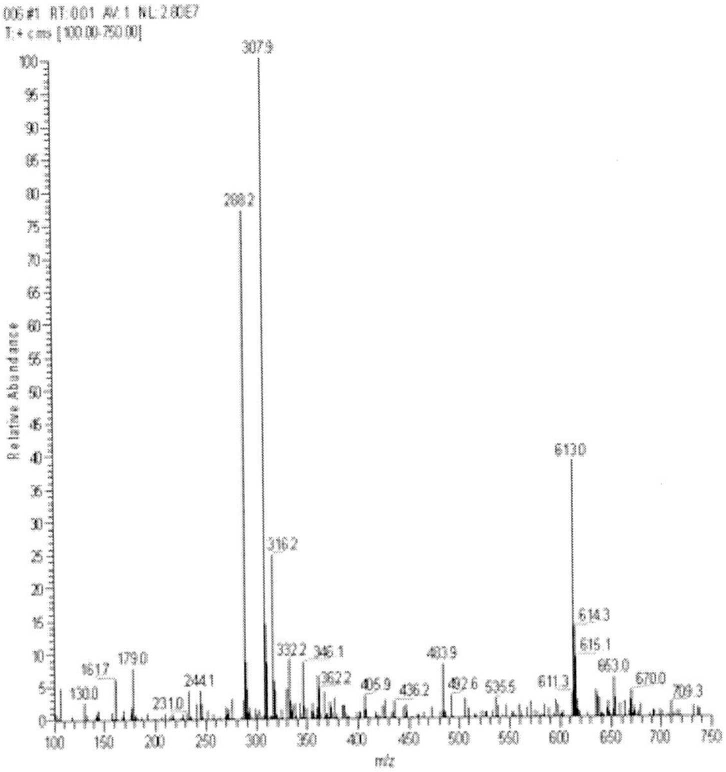

Figure 9. Typical mass spectrum of glutathione.

Table 6. Summary of the molecular formulae and molecular weights for the complexes detected for Cu^{2+}

Molecular formula	Molecular weight of Copper
$[M + GSH + H]^+$	369
$[2M + GSH + H]^+$	432
$[M + (GSH)_2 + H]^+$	674
$[2M + (GSH)_2 + H]^+$	736

Table 7. pKa values of the dissociating groups on glutathione

Functional group	pKa values
Carboxylate	2.1
Carboxylate	3.5
Thiol	8.7
Amino	9.6

Since pH is an important parameter in metal complexation, particularly where removal of protons from ligand is required, the study of interactions at different pH can yield substantial information about features of the complexes formed. In order to cover several complexing conditions which can yield different complexes and to ensure specificity to the binding functional group, metal complexation was carried out at the respective pH's corresponding to pKa values shown in Table 7. The binding of metal with GSH was shown to be pH dependent with complexation being high at pH 2.1. The high ion count could be attributed to the fact that at low acidic pH the peptides are less folded and allow more complexation [84]. At this pH complexation is largely through the deprotonated the carboxylate group. It is known that in certain metal ion peptide complexes, the metal ion promotes the ionisation of peptide protons with subsequent binding to the protonated site [83,84]. At pH 9.6, high complexation is also observed indicated by high ionisation count. This is expected as all sites are deprotonated and are thus available for binding to metal ions. This suggests GSH oxidises Cu2+ to Cu+. If both species are present in solution the soft Cu+ will bind to the deprotonated thiol while the harder Cu2+ will bind to the harder carboxyl groups. Cu has also been shown to have a strong affinity for the nitrogen group [82,83].

Though this work has been carried out and some interesting points raised, there is still some work that can be carried to support the findings reported herein [56]. Fragmentation by tandem mass spectrometry (MS/MS) should be carried out in order to evaluate the specific functional groups of GSH involved in coordination. These studies demonstrate the potential of ES-MS as a technique for studying metal protein interaction.

4.1.2.2. Probing metal-glutathione interaction using ^{1}H-NMR

^{1}H-NMR was used complimentary to the ES-MS in order to ascertain that complexation has taken place and also to identify the possible binding site based on the proton shifts. The samples were prepared in deuterated solvents. The spectrum of GSH was obtained and compared to that of the complexes. The values for the proton shifts from 1H-NMR spectrum of GSH at a molar ratio of 1:2 are shown in Table 8. These assignments were confirmed by the 2D COSY correlation (data not shown).

Table 8 ^{1}H NMR value for protons of GSH (δ in ppm)

α CH_2 of gly	α CH of cys	β CH_2 of cys	α CH of glu	β CH_2 of glu	δ CH_2 of glu
3.85 singlet	3.75	2.90 dd	2.52	2.16	4.53
	3.73	2.89	2.49	2.14	4.57
	3.71	2.88	2.48	2.11	4.48
		2.87		2.09	
				2.06	

^{1}H NMR chemical shift data indicate that Cu is well able to bind to coordination sites of GSH. These studies demonstrate the potential of ^{1}H NMR and ESI-MS as techniques for elucidating the binding of metal ions by biological molecules.

In the presence of metal a shift to higher frequencies was observed for all peaks. The major shift in the spectrum was in the signal of cysteinyl βCH2 indicating that complexation is largely through the sulfhydryl group of the cysteine residue and possibly to the peptide linkage between the cysteinyl and glycyl residues. Other resonances were not significantly affected by presence of metals. The small downfield shift observed in the δ CH2 of glutamic acid residue indicate that binding to the carboxylic acid group is possible. The binding of carboxylic group to the hard acid metals and borderline metals is known [83,84]. The success in determination of stoichiometry of metal GSH complex and elucidation of the possible structures formed is dependent on carrying out ^{1}H NMR experiments at several pHs. This prompts further investigation using deteriorated acids.

5. Identification of Major Complexing Compounds in Blepharis Aspera

Understanding chelation mechanism is an important aspect in developing plants as agents of phytoremediation for contaminated sites. Recent advances have been made by Mmatli et al to study aspects of chelation in Blepharis Aspera as plausible mechanism for detoxification using LC-SPE-NMR [27]. In these studies reverse phase HPLC with UV detection was used to isolate compounds of the plant extracts. SPE column packed with porous graphite carbon was employed to enable multiple trappings of the target compound and hence improve NMR sensitivity. The major compounds were identified by NMR as phenyl propanoids verbascoside and isoverbascoside and these were present in 0.7% w/w and 0.2 % w/w dry weight respectively [27]. The potential of these compounds; verbascoside and isoverbascoside to complex with metals (Cu2+, Ni2+ and Fe2+) was further investigated using ES-MS and UV-Vis spectrometer. Through the significant shift in absorbances the UV-Vis spectroscopy results confirmed complexation of compounds with Cu2+, Ni2+ and Fe2+ cations. ES-MS studies, showed m/z values that could correspond to [(verbascoside2-Me]2+ complexes with Me= Ni2+ and Fe2+ but no m/z values corresponding to verbascoside-Cu complex were observed.

6. Metal Speciation in Metallophytes

It is now generally accepted by environmental chemists, nutritionists and toxicologists that the knowledge of the total elemental concentrations of a sample does not suffice to assess the environmental hazard, essentiality and bioavailability of elements as these could be present in a variety of forms [85-87]. Information regarding both the total content and the chemical forms of species present in a complex matrix is thus required in order to have a true reflection of potential toxicity, bioavailability, bioaccumulation and transport of a particular element. It is thus important in our research studies not only to know the total concentration of metals accumulated in metallophytes but also to know the nature of various metals species present in different parts of the plant. A comprehensive knowledge of chemical forms of metals that plants accumulate is essential as this could shed some light on the processes involved and in detoxification mechanisms metallophytes employ.

This distribution of an element among defined chemical species in a system or the elucidation of the various physicochemical forms of a given element is referred to as speciation [88]. Speciation studies can provide an insight into the chemical behaviour of elements and their interaction with different biological systems [89]. The increasing awareness of the importance of elemental speciation is resulting in a growing demand from research and routine laboratories for analytical methods that can be directly used to measure or deduce the different species of the total dissolved metal concentration. Instrumental techniques that have been employed and are now well established methods in the study of metal speciation include liquid chromatography with size exclusion chromatography (SEC) and ion exchange [30,90,91].

6.1. SPE towards Metal Speciation in Metallophytes

Solid phase extraction using small prepacked cartridges containing up to 500 mg sorbent have been applied in speciation and sample clean up for metal analysis [85-87]. In SPE the sample solution is passed through a preconditioned SPE column driven by a positive pressure, centrifugal force, or most commonly vaccum. The general mechanism of this technique is the physical adsorption of analytes between mobile and stationary phase, appropriate washing of the cartridge for the further removal of impurities/interfering substances without loss of analyte and then finally the complete elution of the desired analytes selectively by a suitable elution solvent [92-95]. Alternatively an extraction column may be chosen which retains interferences in the sample but allows analytes to pass through unretained. The transfer is stimulated by the selection of appropriate optimal conditions in the system of three major components; liquid phase, analyte and sorbent. Packing materials are mostly based on silica particles and they cover a wide selection of sorbent chemistries such as, the polymeric materials based on styrene-divinylbenzene, C8 or C18 organic group among others and the carbon or ion exchange materials which can comprise strong or weak anion or cation exchangers bound to silica support material [92]. The chromatographic sorbents can be packed into mini-columns, which are well suited for on-line applications. Mixed mode sorbents containing both non-polar and strong ion exchange functional groups [92] and restricted access matrix sorbents which combine size exclusion and reverse phase mechanisms have also been introduced [92].

In some of our work SPE was used to probe chemical speciation of plants [56]. In these studies SCX, chelate iminodiacetate, SAX and C18 SPE cartridges were employed for speciation off-line with ICP-MS. After testing the retention performance of these cartridges against control samples the catridges were employed in plant extracts at an optimum flow rate of 2 ml/min. Heavy metals from ground plant samples were extracted ultrasonically for 1 hour, using ultra pure water followed by filtration through 0.45 μm pore size filter. Water was used for extraction in order to minimize the possibility of affecting the species. The parameters that were optimized for extraction include the extraction time, the volume of water needed for extraction, and the mass of plant sample with respect to volume. The optimum mass was 0.5 g using 100 ml and the optimal extraction time was 1 hour and this was used for the rest of the experiment. The plant extracts were analysed by ICP-MS.

Table 9 shows concentration of metals in the leaf and root extracts of Tephrosia longipes. The concentration of Cu was not significantly different in the leaf and root extracts with values ~ 30 for Cu. The concentration of Ni was higher in the leaf extracts (~80 μg/g) than in root extract where it was found to be ~12 μg/g.

Table 9. concentration of Cu and Ni in the leaf and root extracts of *Tephrosia longipes*

Metal	Concentration (μg/g) in leaves	Concentration (μg/g) in roots
Cu	29.72	27.2
Ni	79.8	12.3

6.1.3. Distribution of Cu and Ni Species in Different Parts of Tephrosia Longipes

Results obtained for speciation of metals in leaf and roots extracts of Tephrosia longipes using various cartridges are shown in Table 10. Cationic Cu species present in the leaves on average was 57 % whereas that in roots was 16 %. 10 % Ni cationic species were present in the leaves and 48 % Ni cations were present in the roots. The results show that a high percentage of cations are present in the roots compared to leaves. 56 % of anionic Cu was found in the leaves and 45 % anionic Cu was found in the roots.

The results showed that the leaves had a higher percentage of anionic species relative to roots. Less than 3 % of Ni in the leaf extracts of Tephrosia longipes were retained by the reverse phase C18 indicating the virtual absence of metal-organic species of Ni in the leaves. A similar observation was made in the roots with Ni barely exceeding 5 % an indication of virtual absence of metal-organic forms. 16 % Cu in the leaf extract is organically bound whereas 60 % in the roots are organically bound.

It was found that the organically bound metal species were higher in the roots compared to leaves. The results were reproducible with RSD less than 3%. Further work has to be done for identification and characterization of these species ES-MS.

Table 10. Retention of Cu and Ni species on different SPE cartridges

	% Cationic Species		% Anionic Species		% Organic Species	
Plant part	Roots	Leaves	Roots	Leaves	Roots	Leaves
Cu	57	16	45	56	60	60
Ni	48	10	69	85	-	-

6.2. Size Exclusion Chromatography (SEC) towards Metal Speciation in Metallophytes

SEC allows fractionation of samples when metal-containing fractions are discriminated from others by on line ICP-MS detection. Although chromatographic purity of fractions is usually low and metal-binding species are not usually identified, SEC, being robust, is well suited to direct injection of complexes without any extensive pre treatment and thus remains the first chromatographic step for multi dimensional chromatographic approach. Fractions isolated by SEC can be further fractionated by an independent separation mechanism with and objective to produce a more detailed map and to achieve a degree of purity of metal species sufficient for their characterisation with molecular mass spectrometer. In case of a relatively pure sample, the resolution of SEC may be sufficient to accomplish separation of metallo proteins from potential low molecular weight impurities prior to ICP MS detection. Nkoane et al. analysed water extracts of Helichrysum candolleanum SEC-UV-ICP-MS to reveal molecular size of the metal complex moiety [29]. The organic material was detected at 254 nm and the metals detected from and on line SEC-ICP-MS.

Conclusions

Metalliferous areas in Botswana support plant species that can be classified as metallophytes. These include; Tephrosia longipes, Indigofera melanadenia, Helichrysum candolleanum, Blepharis diversispinia and Blepharis aspera. These plant species represent a resource that is valuable to scientific research and have tremendous potential in various applications such as phytoremediation and mineral exploration. The success of these technologies depends on the development of analytical methods sensitive enough for the detection and quantification of metals in plant matrices thus studies in this research were aimed at development of such methods.

The capabilities of different analytical techniques are demonstrated for the study complex plant matrices. The suitability of sample handling protocols based on microdialysis and slurry ETAAS sampling was demonstrated for sampling of complex matrices of plant samples. Microdialysis showed the potential to be used in predicting the total concentrations of Cu and Ni in plant flower suspensions through its direct and linear relation with the acid digestion method. SPE was shown to be a convenient, simple and rapid method for reliable speciation of metals in plant extracts.

The approach was shown to offer the possibility of fast screening of metal species. The results provide information on plant species which could shed some light at fundamental mechanisms of metal accumulation.

Additionally the coupling of SEC to ICP-MS and LC-SPE-NMR were employed to screen organic compounds that could be associated with metals in the plants. These studies also demonstrate the potential of 1H NMR and ES-MS as techniques for elucidating the binding of metal ions by biological molecules. The insights gained could benefit the study of PCs. However there is still need to identify more plants in other mining areas. Moreover there is a challenge to develop more analytical protocols for the study of metallophytes.

Acknowledgments

The authors are grateful to the Norwegian Universities Committee for Development Research and Education (NUFU) and the University of Botswana for financial assistance.

REFERENCES

[1] Whiting S.N.; Reeves R.D.; Baker A.J.M.; 2002. *Mining environmental management.* 10, 11-16.

[2] Baker, A.J.M.; 1981. Accumulators and excluders - strategies in the response of plants to heavy metals. *Journal of Plant Nutrition*, 3, 643-654.

[3] Nkoane, B.B.M.; Sawula, G.M.; Wibetoe, G.; Lund, W.; 2005. Identification of Cu and Ni indicator plants in Botswana. *Journal of Geochemical Exploration*, 86,130-142.

[4] Pujari, G.N.; Shrivastava, J.P.; 2001. High bioassay values in *Terminalia alata* leaves: indication of Cu mineralisation in Malanjkhand Granitoid, Central India. *Chemical Speciation and Bioavailability,* 13, 97-111.

[5] Bennett, L.E; Burkhead J.L; Hale K.L; Terry, N.; Pilon, M.; Pilon-Smits, E.A.H.; 2003. Analysis of transgenic Indian mustard plants for phytoremediation of metal-contaminated mine tailings. *Journal of Environmental Quality*, 32, 432-440.

[6] Pratas, J.; Prasad, M.N.V.; Freitas, H.; Conde, L.; 2005. Plants growing in abandoned mines of Portugal are useful for biogeochemical exploration of arsenic, antimony, tungsten and mine reclamation. *Journal of Geochemical Exploration*, 85, 99–107.

[7] Brummer, J. J.; Woodward, G. D.; 1999. A history of the 'Zambian copper flower,' Becium centraliafricanum (B-homblei). *Journal of Geochemical Exploration*, *65*, 133-140.

[8] Cherian, S.; Oliveira, M.M.; 2005. Transgenic Plants in Phytoremediation: Recent Advances and New Possibilities. *Environmental Science Technology*, 39, 9377-9390.

[9] Zhuang, P.; Ye, Z.H.; Lan, C.Y.; Xie, Z.W.; Shu, W.S.; 2005. Chemically Assisted Phytoextraction of Heavy Metal Contaminated Soils using Three Plant Species. *Plant and Soil*, 276, 153-162.

[10] LeDuc, D.L.; Terry, N., 2005. Phytoremediation of toxic trace elements in soil and water. *Journal of Industrial Microbiology Biotechnology*, 32, 514-520.

[11] Gosh M.; Singh S.P.; A review on phytoremediation of heavy metals and utilization of its products. *Applied ecology and environmental research* 3 (11):1-18.

[12] Baker, A. J. M., 1981. Accumulators and excluders - strategies in the response of plants to heavy metals. *Journal of Plant Nutrition*, *3*, 643-654.

[13] Prasad, M.N.V.; Freitas, H.M.O.; 2003. Metal hyperaccumulation in plants - Biodiversity prospecting for phytoremediation technology. *Electronic Journal of Biotechnology* 6, 284-321

[14] Dahmani-Muller, H.; van Oort, F.; Gelie, B.; Balabane, M. 2000. Strategies of heavy metal uptake by three plant species growing near a metal smelter. *Environmental Pollution*, *109*, 231-238.

[15] Memon, A.R.; Aktoprakligül, D.; Zdemür, A.; Vertii A., 2001. Heavy Metal Accumulation and Detoxification Mechanisms in Plants. *Turkish Journal of Botany*, 25, 111-121.

[16] Baker, A.J.M.; Brooks, R.R., 1989. Terrestrial higher plants which hyper-accumulate metallic elements - a review of their distribution, ecology and phytochemistry. *Biorecovery*, 1, 81-126.

[17] Lasat, H.H. 2000. Phytoextraction of metals from contaminated soil: A review of plant/soil/metal interaction and assessment of pertinent agronomic issues. *Journal of Hazardous Substance Research* 2(5), 1-25.

[18] Ekosse G., Van Den Heever D., De Jager L., Totolo O., 2003. Environmental mineralogy of soils around Selibe Phikwe Ni-Cu plant, *Botswana International Journal Environ. Stat* 60, 251-262.

[19] Cole, M. M.; Le Roex, H. D. 1978. The role of geobotany, biogeochemistry and geochemistry in mineral exploration in South West Africa and Botswana — a case history. *Transactions of the Geological Survey of South Africa*, *81*, 277–317.

[20] Cole, M. M. 1971. The importance of environment in biogeographical/geobotanical and biogeochemical investigations. *Geochemical Exploration*, *11*, 1-12.

[21] Kausel, G. 1991. Study of heavy metal tolerant flora in Botswana. *Botswana Notes and Records*, *23*, 158-174.

[22] Takuwa, D. T.; Sawula, G.; Wibetoe, G.; Lund, W. 1997. Determination of cobalt, nickel and copper in flowers, leaves, stem and roots of plants using ultrasonic slurry sampling electrothermal atomic absorption spectrometry. *Journal of Analytical Atomic Spectrometry*, *12*, 849-854.

[23] Mogopodi, D., Mosetlha, K., Torto, N., Wibetoe, G. Assessing accumulation patterns of Cu and Ni for *Indigofera melanadenia* and *Tephrosia longipes* plant species growing in a Cu-Ni mining area in Botswana.

[24] Nkoane, B.B.M.; Sawula, G.M.; Wibetoe, G.; Lund, W.; 2005. Identification of Cu and Ni indicator plants in Botswana. *Journal of Geochemical Exploration*, 86,130-142.

[25] Mosetlha, K., Torto, N., Wibetoe, G. 2007. Determination of Cu and Ni in plant slurry by microdialysis sampling: Comparison of dialyzable fractions with total metal content. Talanta, 71, 766-770.

[26] Nkoane, B.B.M, Wibetoe, G., Lund, W.; 2007. *A multi-element study on metallophytes from mineralized areas in Botswana using ICP-OES and ICP-MS.* Geochemistry: Exploration, Environment, Analysis.

[27] Mmatli E.E., Mallerod, H., Wilson, S.R.; Abegaz, B., Greibbok, T., Lundanes, E.; Malterud, K.E.; Petersen, D.; Rise, F 2007. Identification of major complexing compounds in Blepharis Aspera. *Analytical Chimica Acta* 597, 24-31.

[28] Buldini, P.L.; Ricci, L.; Sharma J. L.; 2002. Recent applications of sample preparation techniques in food Analysis. *Journal of Chromatography* A, 975, 47–70.

[29] Torto N., Mmualefe, Mwatseteza J.F., Nkoane B., Chimuka L., Nindi M.M., Ogunfowokan, 2007. Sample preparation for chromatography: An African perspective. *Journal of Chromatography* A, 1153, 1-3.

[30] Łobi´nski, R.; Szpunar, J.; 1999. Biochemical speciation analysis by hyphenated techniques. *Analytica Chimica Acta*, 400, 321–332.

[31] Melaku, S.; Dams, R.; Moens, L. 2005. Determination of trace elements in agricultural soil samples by inductively coupled plasma-mass spectrometry: Microwave acid digestion versus aqua regia extraction. *Analytica Chimica Acta*, *543*, 117-123.

[32] Medved, J.; Kubova, J.; Chmielewska, E.; Stresko, V. 2001. Evaluation of sample pre-treatment procedures for the determination of Cr, Ni and V in biological matrices by ETAAS. *Turkish Journal of Chemistry*, *25*, 323-331.

[33] Chen, M.; Ma, L. Q. 2001. Comparison of three aqua regia digestion methods for twenty Florida soils. *Soil Science Society of America Journal*, *65*, 491-499.

[34] Tighe, M.; Lockwood, P.; Wilson, S.; Lisle, L. 2004. Comparison of digestion methods for ICP-OES analysis of a wide range of analytes in heavy metal contaminated soil samples with specific reference to arsenic and antimony. *Communications in Soil Science and Plant Analysis*, *35*, 1369-1385.

[35] Oliveira, E. 2003. Sample preparation for atomic spectroscopy: evolution of future trends. *Journal of the Brazilian Chemical society* 14(2), 174-182.

[36] Morita, Y., Tanaka, A., Isozaki, A., Okutani, T. 1999. Direct determination of iron in silicon carbide and silicon nitride by electrothermal atomic absorption spectrometry with slurry-introduction. *Analytical sciences* 15, 1145-1148.

[37] Wisniewski, N.; Torto, N.; 2002. Optimisation of microdialysis sampling recovery by varying inner cannula geometry. *Analyst*, 127, 1129-1134.

[38] Torto, N.; Mikeladze, E.; Gorton, L.; Csoregi, E.; Laurell, T.; 1999. Maximising Microdialysis Sampling by Optimising the Internal Probe Geometry. *Analytical Communication*, 36, 171-174.

[39] Torto, N.; Laurell, T.; Gorton, L.; Marko-Varga, G.; 1998. Recent trends in the application of microdialysis in bioprocesses. *Analytica Chimica Acta*, 374, 111-135.

[40] Torto, N.; Lobelo, B.; Gorton, L.; 2000. The determination of saccharides in wastewater from beverage industry by microdialysis sampling, microbore HPAEC and integrated pulsed electrochemical detection. *Analyst*, 125, 1379-1381.

[41] Mogopodi, D.; Torto, N.; 2003. Enhancing microdialysis recovery of metal ions by incorporating poly-l-aspartic acid and poly-l-histidine in the perfusion liquid. *Analytica Chimica Acta,* 482, 91-97.

[42] Mogopodi, D.; Torto, N.; 2005. Maximising metal ions flux across a microdialysis membrane by incorporating poly-l-aspartic acid, poly-l-histidine, 8-hydroxyquinoline and ethylenediaminetetraacetic acid in the perfusion liquid. *Analytica Chimica Acta*, 534, 239-246.

[43] Mosetlha, K.; Torto, N.; 2006. Enhancing the microdialysis recovery for sampling of Cu and Ni by incorporating humic acid in the perfusion liquid. *Analytica Chimica Acta*, 562, 158-163.

[44] Mogopodi, D.; Torto, N.; 2004. Opportunities in microdialysis sampling of metal ions. *Trends in Analytical Chemistry*, 23, 109-115.

[45] Kissinger P.T.; 1991. *Microdialysis and liquid chromatography*. In: Microdialysis in the Neurosciences (Eds. T.E.Robinson and J.B. Justice Jr.). Elsevier, p. 103-115.

[46] Ungerstedt, U.; Pycock, C.; 1974. *Functional correlates of dopamine transmission*. Bulletin der Schweizerischen Akademie der Medizinischen Wissenschaften, 1278, 1-13.

[47] Davies, M.I.; 1999. A review of microdialysis sampling for pharmacokinetic applications. *Analytica Chimica Acta,* 379, 227-249.

[48] Marko-Varga, G.; Buttler, T.; Gorton, L.; Olsson, L.; Durand, G.; Barceló, D.; 1994. Qualitative and quantitative carbohydrate analysis of fermentation substrates and broths by liquid chromatographic techniques. *Journal of Chromatography A*, 665, 317-332.

[49] Okatch H.; Torto, N.; Amarteifio, J.; 2003. Characterisation of carbohydrate hydrolysates from legumes using enzymatic hydrolysis, microdialysis sampling and high performance anion exchange chromatography with electrospray ionization mass spectrometry. *Journal of Chromatography* A, 992, 62-74.

[50] Marko-Varga, G.; Buttler, T.; Gorton, L.; Olsson, L.; Durand, G.; Barceló, D.; 1994. Qualitative and quantitative carbohydrate analysis of fermentation substrates and broths by liquid chromatographic techniques. Journal of Chromatography A, 665, 317-332.

[51] Mosetlha, K., Torto, N., Wibetoe, G. 2007. Determination of Cu and Ni in plant slurry by microdialysis sampling: *Comparison of dialyzable fractions with total metal content.* Talanta, 71, 766-770.

[52] Torto, N.; Mwatseteza, J.; Sawula, G.; 2002. A study of microdialysis sampling of metal ions. *Analytica Chimica Acta*, 456, 253-261.

[53] Miró, M.; Jimoh M.; Frenzel, W.; 2005. A novel dynamic approach for automatic microsampling and continuous monitoring of metal ion release from soils exploiting a dedicated flow-through microdialyser. *Analytical Bioanalytical Chemistry*, 382, 396–404.

[54] Miró, M.; Frenzel, W.; 2003. A novel flow-through microdialysis separation unit with integrated differential potentiometric detection for the determination of chloride in soil samples. *The Analyst*, 128, 1291–1297.

[55] Miró, M.; Frenzel, W.; 2005. The potential of microdialysis as an automatic sample-processing technique for environmental research. *Trends in Analytical Chemistry*, 24, 324-333.

[56] Mogopodi D., 2006. PhD Thesis. Microdialysis sampling and SPE strategies with spectrometric detection towards multielement speciation in Indigofera melanadenia and Tephrosia longipes plants growing in Cu-Ni mining areas in Botswana.

[57] Baker, A. J. M, McGrath, S. P., Reeves, R. D. and Smith, J. A. C. 2000. Metal Hyperaccumulator Plants: A Review of the ecology and physiology of a biological resource for phytoremediation of metal-polluted soils. *In*: Terry N. and Banuelos G. S. (eds) *Phytoremediation of Contaminated Soil and Water*. CRC Press Inc, Boca Raton, FL, USA, 85-107.

[58] Prasad, M. N. V., Freitas, H., 2003. Metal hyperaccumulation in plants – Biodiversity prospecting for phytoremediation technology. *Electronic Journal of Biotechnology*, 6, 285-319.

[59] Brooks, R. R. 1995. Specific indicator plants. *In*: Brooks, R. R., Dunn, C. E. and Hall, G. E. M. (eds) *Biological Systems in Mineral Exploration and Processing*. Ellis Horwood, New York, 39-70.

[60] Brooks, R.R. 1998. *Biogeochemistry and hyperaccumulators*. *In*: Brooks, R.R (ed.) Plants that Hyperaccumulate Heavy Metals. CAB International, Oxon UK, 95-118.

[61] Memon, A.R.; Aktoprakligül, D.; Zdemür, A.; Vertii A., 2001. Heavy Metal Accumulation and Detoxification Mechanisms in Plants. *Turkish Journal of Botany*, 25, 111-121.

[62] Vacek, J.; Kizek, R.; Klejdus, B.; Havel, L. 2003. Phytochelatins and their role in heavy metal detoxification: biosynthesis, regulation, and transport. *Biologicke Listy*, 68, 133-153.

[63] Sneller, F.E.C.; Heerwaarden, L.M.; Schat, H.; Verkleij, J.A.C.; 2000. Toxicity, metal uptake, and accumulation of phytochelatins in *Silence vulgaris* exposed to mixtures of cadmium and arsenate. *Environmental Toxicology and Chemistry*,19, 2982-2986.

[64] Kertulis, G.M.; Ma, L.Q.; MacDonald, G.E.; Chen, R.; Winefordnerc, J.D.; Cai, Y.; 2005. Arsenic speciation and transport in *Pteris vittata* L. and the effects on phosphorus in the xylem sap. *Environmental and Experimental Botany*, 54, 239–247.

[65] Montes-Bayon, M.; Meija, J.; LeDuc, D. L.; Terry, N.; Caruso, J. A.; Sanz-Medel, A., 2004. HPLC - ICP - MS and ESI - Q - TOF analysis of biomolecules induced in Brassica juncea during arsenic accumulation. *Journal Analytical Atomic Spectrometry*, 19, 153-158.

[66] Cheng, Y., Yan, Y., Liu, J., 2005.Spectroscopic characterization of metal bound phytochelatin analogue (Glu–Cys)4–Gly. *Journal of Inorganic Biochemistry* 99, 1952–1962.

[67] Murasugi, A., Wada, C., Hayashi, Y., 1981. Purification and unique properties in UV and CD spectra of Cd-binding peptide 1 from Schizosaccharomyces pombe. *Biochemical and Biophysical Research Communications*, 103, 1021-1028.

[68] Braga, P., Montes-Bayon, M., Alvarez, J., Lopez, J.M., Sanz-Medel, A., 2004. Characterization, biological interactions and in-vivo detection of selenotrisulfide derivatives of glutathione, cysteine and homocysteine by HPLC -ICP-MS. *Journal of Analytical Atomic Spectrometry*, 19, 1128-1133.

[69] Aldini, G., Granata, P., Carini, M., 2002. Detoxification of cytotoxic, unsaturated aldehydes by carnosine: characterization of conjugated adducts by electrospray ionization tandem mass spectrometry and detection by liquid chromatography /mass spectrometry in rat skeletal muscle. *Journal of Mass Spectrometry*, 37, 1219-1228.

[70] Vacek, J., Petrek, J., Kizek, R., Havel, L., Klejdus, B., Trnkova, L., Jelen, F., 2004. *Electrochemical determination of lead and glutathione in a plant cell culture. Bioelectrochemistry*, 63, 347-351.

[71] Fang, C.; Zhou, X., 2002. Electrochemical properties of glutathione monolayer on mercury surface. *Electroanalysis*, 14, 711-714.

[72] Vasquez, B.H.C., Diaz-Cruz, J.M., Arino, C., Esteban, M., 2003. Differential pulse polarography of the Zn^{2+} complexation by glutathione fragments Cys-Gly and gamma-Glu-Cys. *Electroanalysis*, 15, 1177-1184.

[73] Yen, T.Y.; Villa, J.A.; DeWitt, J.G.; 1999. Analysis of phytochelatin-cadmium complexes from plant tissue culture using nano-electrospray ionization tandem mass spectrometry and capillary liquid chromatography/electrospray ionization tandem mass spectrometry. *Journal of Mass Spectrometry*, 34, 930-41.

[74] Leopold, I., Gunther, D., 1997. Investigation of the binding properties of heavy-metal-peptide complexes in plant cell cultures using HPLC-ICP-MS. *Fresenius' Journal of Analytical Chemistry*, 359, 364-370.

[75] Yen, T.Y.; Villa, J.A.; DeWitt, J.G.; 1999. Analysis of phytochelatin-cadmium complexes from plant tissue culture using nano-electrospray ionization tandem mass spectrometry and capillary liquid chromatography/electrospray ionization tandem mass spectrometry. *Journal of Mass Spectrometry*, 34, 930-41.

[76] Chassaigne, H.; Mounicou, S.; Casiot, C.; Łobi´nski, R.; Potin-Gautier, M.; 2000. Detection and identification of rabbit liver metallothionein-2 subisoforms by capillary zone electrophoresis – ICP-MS and microbore HPLC - electrospray mass spectrometry. Analusis, 28, 357-360.

[77] Mounicou, S.; Vacchina, V.; Szpunar, J.; Potin-Gautier, M.; Łobi´nski, R.; 2001. Determination of phytochelatins by capillary zone electrophoresis with electrospray tandem mass spectrometry detection (CZE-ES MS/MS). *Analyst*, 126, 624-632.

[78] Vacchina, V.; Chassaigne, H.; Łobi´nski, R.; Oven, M.; Zenk, M.H.; 1999. Characterisation and determination of phytochelatins in plant extracts by electrospray tandem mass spectrometry. *Analyst*, 124, 1425-1430.

[79] Raab, A.; Meharg, A. A.; Jaspars, Marcel; Genney, David R.; Feldmann, Joerg. Arsenic - glutathione complexes - their stability in solution and during separation by different HPLC modes. *J. Anal. Atomic Spectrom.* (2004), 19(1), 183-190

[80] Schmoger, M.E.V., Oven, M., Grill, E., 2000. Detoxification of arsenic by phytochelatins in plants. *Plant Physiology*, 122, 793–801.

[81] Chassaigne, H.; Mounicou, S.; Casiot, C.; Łobi´nski, R.; Potin-Gautier, M.; 2000. Detection and identification of rabbit liver metallothionein-2 subisoforms by capillary zone electrophoresis – ICP-MS and microbore HPLC - electrospray mass spectrometry. Analusis, 28, 357-360.

[82] Yang, W.; Gooding, J.J.; Hibbert, D.B.; 2001. Redox voltammetry of sub-parts per billion levels of Cu^{2+} at polyaspartate-modified gold electrodes. *The Analyst*, 126, 1573-1577.

[83] Gooding, J.J.; Hibbert, D.B.; Yang, W.; 2001. Electrochemical metal ion sensors: exploiting amino acids and peptides as recognition elements. *Sensor* 1, 75-79.

[84] Sigel, H.; Martin, R.B.; 1982. Coordinating properties of the amide Bond. Stability and structure of metal ion complexes of peptides and related ligands. *Chemical Reviews*, 82, 385-426.

[85] Yu, C.; Cai, Q.; Guo, Z.; Yang, Z.; Khoo, S.B.; 2002. Antimony speciation by inductively coupled mass spectrometry using solid phase extraction cartridges. *Analyst*, 127, 1380-1385.

[86] Yu, C.; Cai, Q.; Guo, Z.; Yang, Z.; Khoo, S.B.; 2003. Speciation analysis of tellurium by solid-phase extraction in the presence of ammonium pyrrolidine dithiocarbamate and inductively coupled plasma mass spectrometry. *Analytical and Bioanalytical Chemistry*, 376, 234-242.

[87] Chwastowska, J.; Skwara, W.; Sterli´nska, E.; Pszonicki, L.; 2005. Speciation of chromium in mineral waters and salinas by solid-phase extraction and graphite furnace atomic absorption spectrometry. *Talanta*, 66, 1345–1349.

[88] Templeton, D.; Ariese, F.; Cornelis, R., Danielsson, L.G., Muntau, H., Leeuven, H.P.; Łobi´nski, R., 2000. Guidelines for terms related to chemical speciation and fractionation of elements. Definitions, structural aspects, and methodological approaches. *Pure and Applied Chemistry*, 72, 1453-1470.

[89] Sanz-Medel A., 1998. Toxic trace metal speciation: importance and tools for environmental and biological analysis. *Pure and Applied Chemistry,* 70, 2281-2285.

[90] Yathavakilla, K.S.V.; Shah, M.; Mounicou, S.; Caruso, J.A.; 2005. Speciation of cationic selenium compounds in Brassica juncea leaves by strong cation-exchange chromatography with ICP-MS. *Journal of Chromatography* A, 1100, 153-159.

[91] Łobi´nski, R.; Szpunar, J.; 2003. *Hyphenated techniques in speciation analysis. Chromatography Monographs*. The Royal Society of Chemistry, UK, pp.53-54.

[92] Poole, C.F.; 2003. New trends in solid-phase extraction. *Trends in Analytical Chemistry*, 22,362-373.

[93] Fritz, J.S.; *Analytical Solid-Phase Extraction*, 1999, pp140-165.

[94] Hyötyläinen, T.; Riekkola, M.; 2005. Solid-phase extraction or liquid chromatography coupled on-line with gas chromatography in the analysis of biological samples, *Journal of Chromatography* B, 817, 13–21.

[95] Hennion M.C.; 1999. Solid phase extraction: method development, sorbents, and coupling with liquid chromatography. *Journal of Chromatography A,* 856, 3-54.

In: Handbook of Phytoremediation
Editor: Ivan A. Golubev

ISBN: 978-1-61728-753-4

Chapter 16

PHYTOREMEDIATION OF CD, PB AND CR BY WOODY PLANTS

Alex-Alan F. de Almeida, Marcelo S. Mielke, Fábio P. Gomes, Luana Mahé C. Gomes, Pedro A. O. Mangabeira and Raúl R. Valle
Departamento de Ciencias Biologicas, Universidade Estadual de Santa Cruz, Illheus, B. A. Brazil

ABSTRACT

High concentrations of metallic elements as Cd, Pb and Cr can cause harmful effects to the environment. These highly toxic pollutants constitute a risk for the aquatic and terrestrial life, especially plants, animals and humans. They are associated to diverse bioavailable geochemical fractions, such as the water-soluble fraction and the exchangeable fraction, and to non-available fractions such as those associated with the crystalline net of clays and silica minerals.

Depending upon its chemical and physical properties different mechanisms of metals toxicity in plants can be distinguish, such as production of reactive oxygen species from the auto-oxidation, blocking and/or displacement of essential functional groups or metallic ions of biomolecules, changes in the permeability of cellular membranes, reactions of sulphydryl groups with cations, affinity for reactions with phosphate groups and active groups of ADP or ATP, substitution of essential ions, induction of chromosomal anomalies and decrease of the cellular division rate.

To deal with heavy metal pollution, remediation using plants, including woody species, is becoming a widespread practice. Phytoremediation is an environmentally friendly technology and the use of woody species presents advantageous characteristics as an economic and ecologically viable system that becomes an appropriate, practical and successful technology.

Phytoremediator woody species, with (i) high biomass production, (ii) deep root system, (iii) high growth rate, (iv) high capacity to grow in soils with low nutrient availability and (v) high capacity to allocate metals in the trunk, can be an alternative for the recovery of degraded soils due to excess of metallic elements.

INTRODUCTION

Heavy metals are natural constituent of the lithosphere, but the human action has promoted an increase of these elements in ecosystems (Sebastiani et al., 2004). High concentrations of Cd, Pb and Cr originated from mining (Prasad and Freitas, 2003) or by anthropogenic actions such as discharges of toxic residues in rivers, lakes, maritime coast and in the air, industrial activities, farm use of fertilizers and pesticides, incineration of urban and industrial residues, among others sources, have been causing harmful effects to the environment over decades (Ahluwalia and Goyal, 2007; Jadia and Fulekar, 2009). That situation has become more severe since there are neither controls nor adequate environmental norms (Pilon-Smits, 2005).

The increase of heavy metals in soil is dangerous because they remain in the environment for long periods, changing soil fertility. More disturbing, however, they could be absorbed by plants, affecting agricultural production (Gratão et al., 2005). Plants can function as a transference mechanism of contaminants from the soil to higher levels of the food chain and eventually affect human health ((Khan et al., 2000; Schützendübel and Polle, 2002; Benavides et al., 2005).

Metallic elements, isolated or in group, are commonly used in industrial processes of diverse sectors like paper and cellulose, petrochemical, chemical products, fertilizers, oil refining, steel production, non-iron metals, spare parts of vehicles, plain glass and cement, textile and leather products and manufacture of other devices (Sanitá di Toppi and Gabrielli, 1999).

In the soil solution, the pH is one of the most important factors in the control of the concentration of metals (King, 1988; Henning et al., 2001; Yap et al., 2009). Metals have different soil behaviors, but as a general rule, the formation of complexes is favored at pH values next to neutrality, because, under acid conditions the ligands are protonated, whereas, under alkaline conditions the metals can precipitate in the form of hydroxides (McCarthy and Perdue, 1991). Thus, the assimilation of trace elements by plants varies a great deal as a function of soil conditions, also on the concentration and speciation of the metal in the soil solution, on its successive movement from the soil to the root surface and from the root to the aerial part (Clemens et al., 2002; Patra et al., 2004).

Toxics metallic ions penetrate cells using the same absorption processes of essential micronutrient ions. There is lack of uptake specificity and distribution systems for these metallic elements, leading to their accumulation, i.e. cadmium, a non-essential element (Clemens et al., 2002). The translocation and accumulation of heavy metals to the aerial part depends on the plant species, the particular element and the environmental conditions (Liu et al., 2007). The accumulation of heavy metals in vascular plants provokes significant biochemical and physiological responses, modifying several metabolic processes (MacFarlane et al., 2003; Zhang et al., 2010). The genotoxic effects of the metals depend on their oxidation state, concentration and duration of exposition. The effects are more pronounced at high concentrations and after long exposition time (Cosio et al., 2005).

Plant ecotypes tolerant to metals are a classic example of local adaptation and microevolution restricted to species with appropriate genetic variability (Lindegaard and Barker, 1997). These rapid evolutionary changes in plants can occur by the rapid rate of metal pollution in the environment, which can induce an increase in the strength of selection

(Bondada and Ma, 2003). Several plant species had developed tolerance to metals in a relatively short period of about thirty years (Hall, 2002).

Several technologies have been developed to reduce and/or remove the presence of heavy metals from polluted areas, such as industrial water treatment, soil excavation, solidification/stabilization (S/S) technology, soil washing using physical separation techniques or chemical agents (Ahluwalia and Goyal, 2007; Dermont et al, 2007; Jadia and Fulekar, 2009). However, these techniques are expensive, difficult to use on a large scale and, sometimes, cause great environmental impact because affect biological activity and soil structure and fertility (Pulford and Watson, 2003). In recent years there has been a great interest for phytoremediation of metallic pollutants in the soil, in view of its low cost, sustainability and ecological viability, without soil removal, deposition or destruction of the biological and functional integrity of the soil (Pulford and Watson, 2003; Dickison and Pulford, 2005; Pilon-Smits, 2005).

This review has as main objectives to describe the main effects of Cd, Pb and Cr on the whole plant physiology, the resistance/tolerance mechanisms to metals in plants and the importance of the use of woody species in the process of phytoremediation of soils with high indices of contamination.

Metals and Plant Metabolism

Cadmium

Among the heavy metals, cadmium (Cd) is a major environmental pollutant due to its high water solubility, and high toxicity to animals and plants (Zacchini et al., 2009). Cause significant disorders in the organisms even at low concentrations, because it is a non-essential element (Pinto et al., 2004). In some species, it can promote decreases up to 50% in dry matter production, with cases of decreases in root dry mass by around 80% (Pietrini et al, 2010). Furthermore, it is easily absorbed and translocated to different plant parts (Oliveira et al, 2001; Souza et al., 2009). Different plant species show highly variable capacity to accumulate Cd in relation to the substrate concentration in which they grow (Vassilev et al., 2002). Even among cultivars of the same species a wide variation in the absorption and translocation of this element can occur (Sanitá di Toppi and Gabbrielli, 1999, Guimarães et al., 2008). When absorbed, it binds to the cell wall constituents and to other macromolecules in the cell interior (Vassilev et al., 2002).

Cadmium concentration in plant tissues increases with the increment of its concentration in nutrient solutions and with time of exposure (Oliveira et al., 2001; Souza, 2007), the concentration in the roots being higher than in the aerial part. According to Souza (2007) the increase in Cd concentration in the roots is not due to the increase in the absorption of this element, but to the concomitant decrease in dry matter accumulation. Although there is a high Cd concentration in roots it is also found in leaves and stems, demonstrating that this metallic element is not totally immobilized in the root portion, but translocated to the aerial part (Unterbrunner et al., 2007).

The presence of Cd in the growth substrate of plants can influence the concentration of other mineral elements in the plant tissues, disturbing its mineral nutrition. Cadmium can accumulate in tissue and cell compartments, hampering the general metabolism of the plant. The presence of Cd in cells can affect the content of polyvalent cations through competition for binding sites of proteins or transporters. Decreases of calcium (Ca) content in different plant species occur in the presence of Cd (Gussarson et al., 1996; Sandalio et al., 2001). Cadmium produced a decrease in the contents of Ca, Cu, Fe, Mn, and Zn in species such birch (*Betula pendula*), sunflower (*Helianthus annuus*) and pea (*Pisum sativum*) (Gussarson et al., 1996; Azevedo, 2005; Rodríguez-Serrano et al., 2009). According to Gussarsson (1994), the mineral composition of *Betula pendula* roots is different of the aerial part when exposed to Cd, increasing Cu and Mo concentrations, despite the reduction of K, Ca, Mg and Mn contents. Cadmium can affect long-distance transport of Fe rather than in acquisition of Fe by roots in poplar (*Populus alba*) plants (Fodor et al., 2005).

Leaf chlorosis is the most common effect of Cd phytotoxicity, followed by the decrease in photosynthetic rate, may inhibit respiration, mitochondrial electron transport, and enzymatic activity (Sanità di Toppi and Gabbiella, 1999; Pietrini et al., 2003, Soares et al., 2005). Chlorosis is one of the symptoms of Cd toxicity caused by competition of both Fe and Cd elements for the same absorption site in the plasma membrane (Sanitá di Toppi and Gabbrielli, 1999). In high Cd concentrations, chlorosis probably is associated to the decrease in Fe translocation to leaves (Wong et al., 1984). On the other hand, Root et al. (1975) suggest that chlorosis induced by Cd may be due to alterations in the Fe/Zn relation, and not properly to Fe deficiency, since plants treated with Cd showed greater concentration of this micronutrient. The Cd effects on Fe and Zn absorption have presented conflicting results.

Table 1. Main symptoms of Cd toxicity in plants

Symptoms	References
Inhibition or growth reduction of the aerial part and the root system.	Schützendübel and Polle (2002), Mendelssohn et al. (2001).
Induction of phytochelatins production.	Cobbett and Goldsbrough (2002)
Interference in the activity of specific enzymes, such as peroxidase, ascorbate peroxidase, catalase, glutathione synthetase, glutathione reductase, dehydroascorbate reductase, superoxide dismutase, guaiacol peroxidase, mono-dehydroascorbate reductase.	Vassilev et al. (2002)
Induction of oxidative stress.	Souza (2007)
Induction of apoptotic bodies and oligonucleosomal DNA fragments.	Souza (2007)
Damages in chloroplasts and interference in chlorophyll biosynthesis.	Vollenweider et al. (2006), Vassilev et al. (2002)
Reduction of transpiration and photosynthetic rates.	Sanitá di Toppi and Gabbrielli (1999)
Induction of leaf premature senescence and chlorosis.	Souza (2007)
Stimulation of the secondary metabolism, lignification and cellular death.	Schützendübel and Polle (2002)

Thus, depending on the species, the presence of Cd in the growth media can increase (Wong et al., 1984), decrease (Gussarsson, 1994) or not affect (Souza, 2007) the absorption of Fe in plants.

Lead

Plants absorb and accumulate lead (Pb) in all parts, including roots, stems, leaves, root nodules and seeds. Uptake of Pb in plants is regulated by pH, particle size and soil cation exchange capacity, as well as by exudation and other physico-chemical parameters (Sharma and Dubey, 2005). Lead accumulation in plant tissues depends on the increment of Pb levels in the substrate (Patra et al., 2004).

Great part of the absorbed Pb accumulates in roots, and only a small fraction is translocated to the aerial part (Patra et al., 2004). The retention of Pb in the roots is due to binding sites of exchange ions and the extracellular precipitation, mainly in the form of Pb carbonates, both mechanisms occurring in the cell wall (Jarvis and Leung, 2002).

However, not always Pb penetrates the root endoderm and enters the stele. Therefore, the endoderm acts as a barrier to Pb absorption and penetration to the interior of the stele and its transport to the aerial plant part (Weis and Weis, 2004). Lead is found at higher levels in the cell wall of root cells (MacFarlane and Burchett, 2000), of cells in tissue culture, in the intercellular spaces, in vacuoles and in dictyosomes (Jarvis and Leung, 2002).

Once absorbed by plants, Pb causes multiple indirect and direct effects on growth and metabolism (Sharma and Dubey, 2005). Its effects depend on the concentration, salt type, pH and the plant species involved. Lead effects are more pronounced at high concentrations and duration of exposition. However, in some cases, Pb is able to stimulate metabolic processes when at low concentrations (Patra et al., 2004).

The visible symptoms of Pb toxicity include chlorotic spots and necrotic lesions at the leaf surface, growth retardation and leaf senescence - promoted by reduction of chlorophyll, DNA, RNA, protein and dry biomass, decreases in the activity ratio of acid:alkaline pyrophosphatases and drop in the activities of protease and RNase (Patra et al., 2004).

Chlorosis and necrosis could be due to disruption of thylacoid and stromal membranes, resulting in photosynthesis decrease and, therefore, reduction in the availability of photosynthates for biomass accumulation (Sharma and Dubey, 2005). Oxidative stress induced by Pb can generate great amounts of reactive oxygen species, such as superoxide, hydroxide, peroxide and oxygen singlet (Sharma and Dubey, 2005), that involve all areas of the aerobic metabolism and usually are also associated to damages in membranes and the rebuilding of lipid peroxidation (Smirnoff, 1995). The effectiveness of Pb in displacing some cationic metals from roots is known, which suggests that Pb could play a role in desestabilization of physiological barriers for the movement of solutes in the roots and, in this form, limits the availability of nutrients to plants (Sharma and Dubey, 2005). Studies have shown that Pb in the substrate can decrease the absorption and transport of macronutrients in plants (Godbold and Kettner, 1991). Macronutrients plant deficiencies are, many times, a manifestation of toxic effects due to heavy metals (Siedleska, 1995). Some macroelements, including Ca, Mg and P play a protective role against the toxic effects of heavy metals (Rashid and Popovic, 1990). Lead competes with Ca for the same coupling site in the cell

(Godbold and Kettner, 1991). Moreover, Pb can be transported through Ca channels to the symplast (Tomsig and Suszkiw, 1991).

Table 2. Main symptoms of Pb toxicity in plants

Symptoms	References
Affects germination of seeds.	Fargasova (1994)
Stunted growth, chlorosis and blackening of root system.	Sharma and Dubey (2005)
Promotes reductions in stomatal conductance and stomata size (however, it increases its number and the resistance to water vapor).	Xiong (1997)
Reduces the activity of some enzymes.	Patra et al. (2004)
Inhibits photosynthesis due to disturbs in electron transfer reactions.	Sharma and Dubey (2005)
Reduces the respiration rate.	Romanowska et al. (2002)
Upsets mineral nutrition and water balance, changes hormonal status and affects membrane structure and permeability.	Sharma and Dubey (2005)

Chromium

Chromium (Cr) has not been recognized as an essential element for plant growth, however, some stimulant effects has been reported (Samantaray et al., 1998), with no specific mechanism for its absorption (Shanker et al., 2005). In some cases, plant growth is stimulated at low Cr concentrations; however, at high concentrations it shows a definitive retarding effect (Samantaray et al., 1998, Barbosa et al., 2007). Chromium toxicity affects the length of the primary roots and promotes changes in the architecture of the entire root system (Samantaray et al., 1996). The inhibitory effect of Cr in root growth (Barbosa et al., 2007) and its toxic effects in cell division result from the fixation of Cr^{3+} by plant tissue and disturbs of the osmotic relations that promote restrictions to Ca^{+2} transport through the plasma membrane to the cytoplasm (Liu et al., 1992).

It is difficult to separately analyze the effects of Cr^{3+} and Cr^{6+} in the plant since both can be interconverted (Shanker et al., 2005) and immobilized in the soil (Cervantes et al., 2001). Both chromate and dichromate are negatively charged with limited chance of adsortion by organic materials (Panda and Choudhury, 2005). According to Panda and Choudhury (2005), Cr^{6+}, in contrast to Cr^{3+}, is absorbed by the plant due to its natural soil mobility. The Cr^{6+} is a biologically toxic oxidation state and thus far, there is no evidence indicating any potential biological role in plants (Von Burg and Liu, 1993). The information concerning the form in which this element is extracted and translocated in the plant is contradictory. Generally, this phenomenon is attributed to the different cultural techniques, bioavailability of Cr^{3+} and CrO_4^{2-} as a function of pH and to the concentration of others ions in the root substrata (McGrath, 1982).

The solubility of Cr^{3+} can be increased or diminished in the presence of other elements in the soil-plant system. This fact can cause interactions between Cr^{3+} and other essential elements that can have a significant effect in the concentration of nutrients and their plant distribution, as well as modifications in some physiological and morphological plant processes (Panda and Choudhury, 2005). Chromium in the soil solution is absorbed by roots through transporters used for absorption of essential metals. Its toxic effects depend primarily on its speciation, which in turn determines its absorption, translocation and accumulation (Shanker et al., 2005). The mechanism of Cr^{6+} transport is active, involving transporters of essential anions like sulphate (Cervantes et al., 2001). Elements like Fe, S and P compete with Cr when they bind to the transporter (Samantaray et al., 1998).

Chromium stress can induce metabolic modifications in plants, such as alterations in photosynthesis (Barbosa et al., 2007), degradation of photosynthetic pigments and induction of oxidative stress (Panda and Choudhury, 2005). Furthermore, Cr promotes reduction of leaf area and biochemical changes responsible for the inhibition of chlorophyll synthesis (Vajpayee et al., 1999) and disorganization of the chloroplast ultrastructure (Panda and Choudhury, 2005). Chromium stress also causes leaf chlorosis and necrosis (Barbosa et al., 2007), oxidative damages in biomolecules such as lipids and proteins (Vajpayee et al., 2002), disturbances in the mineral nutrition (Barbosa et al., 2007), increase in glutathione and ascorbic acid production (Shanker, 2003), alterations in the metabolic pool that intermediates the production of phytochelatins and hystidine, interference in the activity of nitrate reductase (Panda and Patra, 2000), root Fe^{3+} reductase (Shanker et al., 2004), plasma membrane H^+-ATPase (Dietz et al., 2001), Na^{2+}/K^+ dependent ATPase (Pauls et al., 1980), Ca^{2+} dependent ATPase (Serpersu et al., 1982), alkaline phosphatases (Viola et al., 1980), superoxide dismutase, catalase (Shanker et al., 2003) and peroxidase (Samantaray et al., 2001) and, eventually, plant growth reduction, hindering its development and, finally, being able to cause its death (Barbosa et al., 2007).

The effect of Cr ions in photosynthesis and in the transference of excitation energy can also be due to abnormalities in the ultrastructure of chloroplasts linked to the development of the lamellar system, with an ample thylocoidal space and little grana (Van Assche and Clijsters, 1983). The disorganization of the chloroplast ultrastructure, the inhibition of the electron transport process due to Cr and the electron deviation from the electrons donor site of photosystem 1 (PS-1) to Cr^{6+} are possible explanations for the decrease in photosynthetic rates induced by Cr (Shanker et al., 2005). It is possible that the electrons produced by the photochemical process are not necessarily used for carbon fixation (Shanker et al., 2005), as indicated by the low photosynthetic rate shown by plants stressed by Cr.

Due to its structural similarity with some essential elements, Cr can affect the mineral nutrition of plants in a complex way (Shanker et al., 2005). Once accumulated and distributed in the interior of the plant, it can interact with other essential elements and significantly affect the concentration and distribution of nutrients in the plant, as well as modify its morphology and some physiological processes (Barbosa et al., 2007). Formation of complexes of Cr with organic acids can play an important role in the inhibitory and stimulatory effects of Cr in the translocation of different mineral nutrients (Panda and Choudhury, 2005). The excess of Cr interferes in the absorption of Na, Fe, Mn, Cu, N, P, K and Mg (Barbosa et al., 2007).

One of the reasons for the decrease in the absorption of some nutrients in Cr stressed plants is the inhibition of the plasma membrane H^+-ATPase activity (Shanker, 2003). Chromium strongly inhibits the incorporation of P, K, Ca, Mg, Fe, Mn, Zn and Cu in different cellular constituent in *Cocos nucifera* (Biddappa and Bopaiah, 1989). The inhibitory effects of Cr in plants growth are the result of specific interaction between Cr and P, Cr and Fe or Cr and Cu (Barbosa et al., 2007). This could be associated to the chemical properties of these metals, for example the charge (Cr^{3+} and Fe^{3+}) and the effective ionic radius (Cr and Cu).

Leaf chlorosis promoted by Cr^{3+} could be caused either by the inhibition of Fe absorption or the reductions of N transport (Barbosa et al., 2007). High Cr concentration can disturb the chloroplast ultrastructure thereby disturbing the photosynthetic process (Panda and Choudhury, 2005). The decrease in the ratio chlorophyll a:b (Shanker, 2003) induced by Cr, indicates that the toxicity of Cr probably reduces the size of the peripheral parts of the antenna complex. The decrease in chlorophyll a can be due to desestabilization and degradation of proteins of the peripheral part. The inactivation of enzymes involved in the chlorophyll biosynthetic pathway can also contribute to the general reduction in chlorophyll content in the majority of Cr stressed plants (Shanker et al., 2005).

Resistance or Tolerance to Metals in Plants

Physiological and genetic factors determine which species can or cannot evolve tolerance (Baker and Proctor, 1990). Genetic evidences exist for multiple independent evolutionary origins of tolerant populations to heavy metals (Vekemans and Lefèbvre, 1997). The populations only develop tolerance for different metals present at high concentrations in its soil of origin. This suggests that the genes for different types of tolerances are different and that selection acts to increase the frequency of genes that give tolerance to a particular metal, present in a determined local (Macnair, 1993).

There is also information about co-tolerance, where tolerance to a metal confers, somehow, tolerance to other metals that are not present in toxic concentrations in the soils in which the plants are growing (Schat and Vooijs, 1997).

It has been observed that tolerant species possess defense mechanisms linked to cellular antioxidants and to antioxidant enzymes that protect several vital physiological processes against damages promoted by oxygen reactive forms produced by metallic stresses (Panda and Choudhury, 2005). Information has been reported about the hyperactivity of oxidant enzymes and the accumulation of cellular antioxidants in several plants species under Cu and Pb stress (Ali et al., 2003). Several species resistant to Cu had been found in contaminated and uncontaminated areas (Liu et al., 2004). According to De Vos et al. (1992), tolerance to Cu is related to the function of glutathione as an antioxidant substance against free radicals and hydrogen peroxide formed by Cu excess.

Tissue culture studies have demonstrated that multiple resistance to metals appeared in mature trees exposed to heavy metals in different contaminated areas (Watmough and Dickinson, 1996). Characteristics of resistance to metals can be induced in suspension cell cultures through successive exposures and gradual increases of the metal concentration in the growth media (Dickinson et al., 1992). Rooted cuttings of *Salix* sp. can be acclimated to metallic stresses in hydroponic conditions (Punshon and Dickinson, 1997).

These studies have contributed to explain how the plants survive and grow in potentially toxic environments (Dickinson et al., 1992; Turner and Dickinson, 1993b). Plants tolerance and/or resistance to metallic stress can be associated to one or more mechanisms (Table 3). As a result of these tolerance and/or resistance mechanisms (alone or in combination), some plants can grow in environments contaminated with metals, in which other species cannot survive (Hall, 2002).

Table 3. A summary of the main plant mechanisms of resistance and/or tolerance to metals

Mechanisms of resistance and/or tolerance	References
Excretion of complexed compounds that reduce the availability of the metal in the soil or water; exclusion of the metal through selective absorption of elements; retention of the metal in roots, preventing its translocation to the aerial part; chelation or sequestration of heavy metals by ligands, compartmentalization, biotransformation and mechanisms of cellular repair; development of enzymes tolerant to the metal.	Hall (2002), Cobbett and Goldsbrough (2002), Patra et al. (2004)
Increase of production of intracellular compounds linked to the metal.	Sharma and Dietz (2006)
Immobilization of the metal in the cellular wall.	Cosio et al. (2005)
Homeostatic cellular mechanisms to regulate the concentration of metal ions inside the cell.	Clemens et al. (2005), Benavides et al. (2005)
Induction of heat-shock proteins.	Heckathorn et al. (2004)
Release of phenols from roots.	Jung et al. (2003)
Increase of tolerance to mineral deficiency or the decrease of nutritional requirements; increase in absorption of certain macronutrients; development of the capacity to absorb and to use minerals in the presence of heavy metal.	Meda et al. (2007)

In the case of biotransformation, the metal toxicity in plants can be decreased by chemical reduction of the element and/or by its incorporation into organic compounds (Salt et al., 1998). Intraspecific and intravarietal differences exist regarding to tolerance to Cr excess that can be controlled by different genes, through diverse biochemical pathways (Samantaray et al., 1998). In the root system of certain plant species Cr also can be reduced chemically from Cr^{6+} to Cr^{3+}, as part of a detoxification mechanism (Shanker et al., 2005).

Inside plant cells, the metals in excess, together with those not used in the metabolism, need to be stored to prevent its toxicity (Briat and Lebrun, 1999). Several potential storage mechanisms, at the cellular level, can be involved in the detoxification and tolerance to metal stress (Cobbett and Goldsbrough, 2002). Moreover, some plant species are capable to accumulate great amounts of metals in the aerial part, while others accumulate them in the roots (Barbosa et al., 2007). It has been verified that Cr, for example, accumulates mainly in the roots and little is carried to the aerial part (Shanker et al., 2005). Possibly, that is due to its immobilization in the root cell vacuoles, becoming less toxic (Arduini et al., 1996). This could be a natural plant response to cope with its toxicity (Shanker et al., 2004).

Chelation of metallic ions by specific ligands of high affinity diminishes the concentration of free ions in the solution. The main ligands associated to metals found in plant tissue include amino acids, oligo and polypeptides (glutathione, phytochelatins, metallothioneins) (Patra et al., 2004), macrocyclical agents (porphyrins, cobalamines, chlorophylls), polysaccharides and glycosides (ramnogalacturonana), nucleobases, oligopolynucleosides and nucleotides (DNA fragments) (Lobinski and Potin-Gautier, 1998). Several of these bioligands associated to metals has been localized in the plant vascular system (Briat and Lebrun, 1999). When these systems are overloaded, defense mechanisms to oxidative stress are activated (Patra et al., 2004).

Metallothioneins and phytochelatins are two representative classes of chelant peptides of heavy metals existing in plants (Zenk, 1996; Cobbett, 2000). Genes directly codify metallothioneins, which have low molecular weights, are rich in cysteine polypeptides and are induced by Cu (Cobbett and Goldsbrough, 2002).

Phytochelatins possess low molecular weights, are enzimatically synthesized, have peptides rich in cysteine and bind to various metals including Cd, Cu (Inouhe, 2005) and Pb (Kahle, 1993) via the sulphydryl and carboxyl residues, but their biosyntheses are controlled preferentially by Cd (Inouhe, 2005). Moreover, they are associated to the intra- and extra-cellular precipitation of Pb as carbonates, sulphates and phosphates, playing an important role in the detoxification of this metal in plant tissues (Salt et al., 1998). Metallothioneins and phytochelatins have been indicated as possible Cu chelants in the cytosol (Van Hoof et al., 2001). The manipulation of phytochelatins gene expression is one of the potential mechanisms to increase the capacity of plants for phytoremediation (Cobbett and Goldsbrough, 2002).

Tolerance mechanisms to Cd include exclusion and accumulation in high amounts in their tissues (Sun et al., 2009). There are a variety of complex mechanisms for tolerance to metals, suggesting that these strategies serve to control the uptake and accumulation of heavy metals (Hasan et al., 2009). According to Steffens (1990), phytochelatins, whose synthesis is induced by the heavy metal, can sequester and detoxify the excess of Cd ions. The accumulation of phytochelatins in plant cells exposed to Cd has been reported in diverse species as effective protection or tolerance mechanisms for the stress effect (Grill et al., 1985; Schützendübel and Polle, 2002; Pietrini et al., 2010).

The compartmentalization of Cd and phytochelatins occurs at the vacuole level (Sanità di Toppi and Gabbrielli, 1999; Vassilev et al., 2002; Cobbett and Goldsbrough, 2002), contributing to the protection from heavy metals toxicity in several plant species (Schützendübel and Polle, 2002). The increase in content of thiolic groups (sulphydrylic and SH groups of cysteines) in phytochelatins, responsible for the complexation of heavy metals in these peptides, are proportional to the increment of Cd absorption by plant roots (Grill et al., 1985). Lead and Cd induced chromosomal aberrations and disturbed mitotic divisions in a *Pinus sylvestris* population (Prus-Glowacki et al., 2006). However, plants show considerable constitutional tolerance to Pb and, in some cases, reach levels of induced tolerance (Sharma and Dubey, 2005). High constitutional tolerance for Pb is associated to high levels of Ca in the tissue during the administration of Pb and with high tolerance to Ca deficiency (Patra et al., 2004). Besides, oxalate compounds secreted by roots can reduce the Pb bioavailability (Sharma and Dubey, 2005).

PHYTOREMEDIATION

Phytoremediation can be applied to organic and inorganic pollutants present in solid, liquid and air substrata. It is applied mainly in soils contaminated with heavy metals, oil hydrocarbons, pesticides, explosives, chlorinated solvents and industrial toxic byproducts (Prasad and Freitas, 2003; Pilon-Smits, 2005). Metal tolerance, and consequently the protection of the integrity and functionality of the primary physiological and metabolic processes (Pietrini et al. 2003), is an essential pre-requisite for a plant to be utilized in phytoremediation (Zacchini et al., 2009). Phytoremediation can be done through several processes, for example: (i) phytoextraction, which consists in the use of plants that accumulate pollutants, like metals or organic compounds that concentrate in the plant part that would be harvested; (ii) phytodegradation that is associated to microorganisms degrading organic pollutants; (iii) rhizofiltration, in which vegetables that employ its roots to absorb, concentrate and adsorb pollutant are used; (iv) phytostabilization, mainly of metals in waters and sewers, to diminish the bioavailability of pollutants to the environment; and (v) phytovolatilization to volatilize pollutant (Pulford and Watson, 2003; Weis and Weis, 2004; Pilon-Smits, 2005). Phytoextraction is adopted for long-term remediation. The time required for extraction depends on the contamination levels, but usually is between one to 20 years (Kumar et al., 1995). It is estimated that plants can remove from 180 to 530 kg Pb ha^{-1} $year^{-1}$ (Huang and Cunningham, 1996).

After the harvest, the volume of the contaminated plant material can be later reduced by incineration, composting or stored as dangerous material or, if economically viable, used for the metal recovery (Gardea-Torresdey et al., 2005; Yang et al., 2005). In the case of woody species, such as Princess-tree (*Paulownia tomentosa*), the wood can be industrialized (Pilon-Smits, 2005, Doumett et al., 2008). Two basic strategies of phytoextraction have been developed: (i) phytoextraction assisted by synthetic chelants, called induced phytoextraction (Huang et al., 1997; Salt et al., 1998; Wu et al., 1999); and (ii) continuous, long run phytoextraction (Salt et al., 1998).

Induced phytoextraction consists of two basic processes that involve metal release to the soil solution combined with metal transport, via xylem, to the aerial part of the plant that will be harvested (Salt et al., 1998). This type of phytoextraction is more advanced and currently commercially implemented (Nascimento and Xing, 2006). A good example of this sort of phytoremediation is that reported for Pb in soil in which EDTA was applied (Salt et al., 1998). However, the main limitation for the use of synthetic chelants in the field, especially EDTA, is its low biodegradation. This results in maintenance of high contents of soluble metals in the soil for long periods, which increases the lixiviation risks (Meers et al., 2004).

In continuous phytoextraction, the metal absorption is carried out by hyperaccumulator plants that grow in soils rich in heavy metals (Salt et al., 1998). These plants are naturally capable to accumulate metals in more than 1% of its aerial part dry biomass (Huang et al, 1997; Sun et al. 2009). This process is based on the genetic and physiological capacities of some plants to accumulate, translocate and resist high metal concentrations. However, have as disadvantages the low biomass production and slow growth, as well as, lack of hyperaccumulator plants for the more important metallic pollutants in the environment like Pb and Cd (Jarvis and Leung, 2002). Even so, some plant species are actually used for phytoextraction of Cd, Cu, Pb and Cr (Baker et al., 1991).

High production of biomass, deep root system, high growth rate, capacity to grow in soils poor in nutrients and to concentrate metals, associated to the characteristics of resistance to metals, are lacking factors of plant species to the method of soil decontamination (Pilon-Smits, 2005; Yang et al., 2005).

Currently, about 450 hyperaccumulator species of metals pertaining to the Asteraceae, Brassicaceae, Caryophyllaceae, Cyperaceae, Cunouniaceae, Fabaceae, Flacourtiaceae, Lamiaceae, Poaceae, Violaceae and Euphorbiaceae families exist (Prasad and Freitas, 2003; Maestri et al, 2010). However, most of these species shows low biomass production (Prasad and Freitas, 2003). Surprisingly, it is scarce the knowledge regarding the responses of woody plants to toxic metal levels (Kukkola et al., 2000). However, in some woody plants such as *Pinus radiata* D. Don. (West, 1979), *Salix* and *Populus* species (Dickinson and Pulford, 2005; Giachetti and Sebastiani, 2006) and *Paulownia tomentosa* (Doumett et al., 2008) responses have been reported.

Woody species are important primary producers in local food chains and long-lived organisms, which can take up trace elements from the environment and store them for a long time. (Domínguez et al., 2008). In recent years, the interest in the potential use of trees to cover the soil and for phytoremediation of soil contaminated by heavy metals has increased due to the high biomass production and wide genetic variability of some species (Dickinson and Pulford, 2005; Unterbrunner et al., 2007). There are much evidences of the natural establishment of trees in contaminated soil, and that some types of trees can survive under severe adverse conditions (Turner and Dickinson, 1993a, 1993b). Several species of *Salix* are explored in programs of despolution of soil Cd (Dickinson and Pulford, 2005; Unterbrunner et al., 2007). In these species, Cd concentration in the aerial part tends to increase as soil Cd concentration increases (Vandecasteele et al., 2002; Unterbrunner et al., 2007).

The interest also is extended to fast growing woody species that would be utilized in high density cropping systems for extraction of soil metals through absorption and harvesting of the aerial part, using successive prunings (Pulford and Watson, 2003). The accelerated growth and the regular pruning are associated to the fast translocation of nutrients and, consequently, of soil heavy metals (EPA, 1999). Clones of *Populus*, resulting from crosses of *P. deltoides* x *P. maximowiczii* (Erídano) and *P. deltoides* x *P. euramericana* (I-214), if cultivated in a population density of 10,000 plants ha^{-1} (forest of short rotation), in soils with high heavy metal concentrations, could produce about 119 tons of stem dry biomass ha^{-1} in a cycle of 11 years (Bonari, 2001). This would correspond to 902 and 962 g ha^{-1} of Cu and 2,700 and 2,058 g ha^{-1} of Cr in the stems of I-214 and Erídano clones, respectively. These figures are preliminary results of absorbed heavy metals by these species obtained under greenhouse conditions (Sebastiani et al., 2004).

A hindrance to rapid selection of genotypes tolerant to heavy metals is the long growth period of the trees. Wide tree genomes and facultative tolerance, such as roots redistribution in less contaminated soil zones, make possible the survival of determined woody species in soils polluted by heavy metals, even with reduced growth indices (Dickinson et al., 1992). The true tolerance requires the development of one or more genetically based physiological mechanisms (Dickinson et al., 1991). However, genetic stability of tolerance is questionable, since it can either be induced or inhibited in woody species. Therefore, the capacity of acclimation to fluctuating stresses, due to pollution, comes to be more important for the species survival (Dickinson et al., 1991). Furthermore, other factors, as soil fertility, can increase the resistance to the metal (Pulford et al., 2002).

The physical phytostabilization of soils contaminated by heavy metals is one of the main benefits of the use of trees in phytoremediation processes (Dickinson and Lepp, 1997). Therefore, besides the direct stabilization of the soil by roots, the cover vegetation decreases the risk of soil loss due to erosion (Jadia and Fulekar, 2009). On the other hand, the trees senescence also produces an increase in the metal levels by the loss of fluids (Pulford and Watson, 2003). Although there are seasonal variations in metal concentration in woody plants, mainly at the foliar level (Ehlin, 1982), leaf fall adds significant amount of organic matter to the surface soil layers, promoting nutritional cycling, soil aggregation and water retention capacity.

CONCLUSIONS

Plant species show different allocation patterns for Cd, Pb and Cr, which translocation from roots to the aerial part and its release from foliar tissue can be an important step for the metal flow in ecosystems. Contamination by these metals affects growth, distribution and the biological cycle of plant species, promoted by several different toxicity mechanisms, like alterations (i) in carbohydrate and N metabolism; (ii) in the activity of certain metalloenzymes; (iii) in protein synthesis; (iv) in the reduction of photosynthetic activity; (v) in the production of oxygen reactive species by auto-oxidation; (vi) in the obstruction of functional groups and (vii) in the displacement of metallic ions essential to biomolecules. Furthermore, it promotes changes (i) in the permeability of cellular membranes; (ii) in the reactions of sulphydrylic groups with cations; (iii) in the affinity for reactions with phosphate groups and active groups of ADP or ATP; (iv) in the substitution of essential ions; (v) in the induction of chromosomal anomalies; and (vi) decrease in the rate of cellular division. Plant tolerance to these metallic elements can be associated to one or more mechanisms, such as (i) the excretion of complexed compounds that reduce the availability of the metal in the soil or water; (ii) the metal exclusion through selective absorption of elements; (iii) the retention of the metal in roots preventing its translocation to the aerial part; (iv) the immobilization of the metal in the cellular wall; (v) the chelation or sequestration of heavy metals by ligands; (vi) the compartmentalization; (vii) the biotransformation and cellular repair mechanisms; (viii) the production increase of intracellular compounds that bind to the metal; (ix) the development of tolerant enzymes to the metal; (x) the increase of tolerance to mineral deficiency; (xi) the decrease of nutritional requirements; (xii) the increase in the absorption of certain macronutrients; and (xiii) the capacity to absorb and use minerals in the presence of heavy metal. Phytoremediator woody species, with (i) high biomass production, (ii) deep root system, (iii) high growth rate, (iv) high capacity to grow in soils with low nutrient availability and (v) high capacity to allocate metals in the trunk, can be an alternative for the recovery of degraded soils due to excess of metallic elements. Phytoremediation using woody species is ecological and economically viable due to the low cost of implantation, promoting soil stabilization that limits the propagation of metallic contaminants.

This technology is emergent and in development. Regarding the phytoremediation strategy, it would be necessary to better understand the absorption, transport and tolerance of metals in woody plants, since they are of great importance for the planning of large-scale application of this technique, under field conditions.

Acknowledgments

Alex-Alan F. de Almeida and Marcelo S. Mielke gratefully acknowledges the Conselho Nacional de Desenvolvimento Científico e Tecnológico (CNPq), Brazil, for the concession of a fellowship of scientific productivity.

References

Ahluwalia SS, Goyal D (2007) Microbial and plant derived biomass for removal of heavy metals from wastewater. *Bioresource Technol.* 98: 2243-2257.

Albasel N, Cottenie A (1985) Heavy metal contamination near major highways industrial and urban areas in Belgian grassland. *Wat. Air, Soil Poll.* 24: 103-109.

Ali MB, Vajpayee P, Tripathi RD, Rai UN, Singh SN, Singh SP (2003) Phytoremediation of lead, nickel and copper by *Salix acmophylla* Boiss.: role of antioxidant substances. *Bull. Environ. Contam. Toxicol.* 70: 462-469.

Arduini I, Godbold DL, Onnis A (1995) Influence of copper on root growth and morphology of *Pinus pinea* L. and *Pinus pinaster* Ait. seedlings. *Tree Physiol.* 15: 411-415.

Arduini I, Godbold DL, Onnis A (1996) Cadmium and copper uptake and distribution in Mediterranean tree seedlings. *Physiol. Plantarum* 97: 111-117.

Azevedo H, Pinto CGG, Fernandes F, Loureiro S, Santos C (2005) Cadmium effects on sunflower growth and photosynthesis. *J. Plant Nutr.*, 28: 2211–2220.

Baker AJM, Proctor J (1990) The influence of cadmium, copper, lead, and zinc on the distribution and evolution of metallophytes in the British Isles. *Plant Syst. Evol.* 173: 91-108.

Baker AJM, McGrath SP, Reeves RD (1991) *In situ decontamination of heavy metal polluted soils using crops of metal-accumulating plants - a feasibility study.* In: Abstracts of the International Symposium on In Situ and On-Site Bioreclamation, San Diego, California, USA, pp.19-21.

Barbosa RMT, Almeida A-AF, Mielke MS, Loguercio LL, Mangabeira PAO, Gomes FP (2007) A physiological analysis of *Genipa americana* L.: a potential phytoremediator tree for chromium polluted watersheds. *Environ. Exp. Bot.* (In press).

Benavides MP, Gallego SM, Tomaro, ML (2005) Cadmium toxicity in plants. *Braz. J. Plant Physiol.* 17: 21-34.

Biddappa CC, Bopaiah MG (1989) Effect of heavy metals on the distribution of P, K, Ca, Mg and micronutrients in the cellular constituents of coconut leaf. *J. Plantation Crops.* 17: 1-9.

Bonari E (2001) Potenzialità e problematiche agronomiche della silvicoltura a breve rotazione come coltura da energia negli ambienti mediterranei. *Riv. Agron.* 35: 188-199.

Bondada BR, Ma LQ (2003) *Tolerance of heavy metals in vascular plants: Arsenic hyperaccumulation by Chinese brake fern (Pteris vittata L.).* In: Pteridology in New Millennium. S. Chandra, M. Srivastava, eds. Kluwer Academic Publishers, The Netherlands. pp. 397-420.

Burke DJ, Weis JS, Weis P (2000) Release of metals by the leaves of the salt marsh grasses *Spartina alterniflora* and *Phragmites australis*. *Estuar Coast Shelf. Science.* 51: 153-159.

Cervantes C, Campos-Garcia J, Devars S, Gutierrez-Corona F, Loza-Tavera H, Torres-Guzman JC, Moreno-Sanchez R (2001) Interactions of chromium with microorganisms and plants. FEMS *Microbiol. Rev.* 25: 335-347.

Clemens S (2001) Molecular mechanisms of plant metal tolerance and homeostasis. *Planta.* 212: 475-486.

Clemens S, Palmgren MG, Krämer U (2002) A long way ahead: understanding and engineering plant metal accumulation. *TRENDS in Plant Science*, 7(7):309-315.

Cobbett CS (2000) Phytochelatins and Their Roles in Heavy Metal Detoxification. *Plant Physiol.* 123:825–832.

Cobbett C, Goldsbrough P (2002) Phytochelatins and metallothioneins: roles in heavy metal detoxification and homeostasis. *Ann. Rev. Plant Biol.* 53: 159-182.

Cosio C, De Santis L, Frey B, Diallo S, Keller C (2005) Distribution of cadmium in leaves of *Thlaspi caerulescens*. *J. Exp. Bot.* 56: 765-775.

De Vos CHR, Vonk M.J, Vooijs H, Schat H (1992) Glutathione depletion due to copper-induced phytochelatin synthesis causes oxidative stress in *Silene cucubalus*. *Plant Physiol.* 98: 853-858.

Demirevska-Kepova K, Simova-Stoilova L, Stoyanova Z, Hölzer R, Feller U (2004) Biochemical changes in barley plants after excessive supply of copper and manganese. *Environ. Exp. Bot.* 52: 253-266.

Dermont G, Bergeron M, Mercier G, Richer-Lafl`eche M (2007) Soil washing for metal removal: A review of physical/chemical technologies and field applications. *J. Hazard. Mat.* 152:1-31.

Dickinson NM, Turner AP, Lepp NW (1991) How do trees and others long-lived plants survive in polluted environments? *Funct. Ecol.* 5: 5-11.

Dickinson NM, Turner AP, Watmough SA, Lepp NW (1992) Acclimation of trees to pollution stress: cellular metal tolerance traits. *Ann. Bot.* 70: 569-572.

Dickinson NM, Lepp NW (1997) *Metals and trees: impacts, responses to exposure and exploitation of resistance traits.* In: Prost, R. (ed). Contaminated soils: The 3rd International Conference on the Biogeochemistry of Trace Elements. Paris, France, pp. 247-54.

Dickinson NM, Pulford ID (2005) Cadmium phytoextraction using short-rotation coppice Salix: the evidence trail. *Environ. Int.* 31: 609-613.

Dietz KJ, Tavakoli N, Kluge C, Mimura T, Sharma SS, Harris GC, Chardonnens AN, Golldack D (2001) Significance of the V-type ATPase for the adaptation to stressful growth conditions and its regulation on the molecular and biochemical level. *J. Exp. Bot.* 52: 1969-1980.

Doumett S, Lamperi L, Chcchini L, Azzarello E, Mugnai S, Mancuso S, Petruzzelli G, Del Bubba M (2008) Heavy metal distribution between contaminated soil and *Paulownia tomentosa*, in a pilot-scalc assisted phytoremediation study: Influence of different complexing agents. *Chemosphere* 72:1481–1490.

Domínguez MT, Marañón T, Murillo JM, Schulin R, Robinson BH (2008) Trace element accumulation in woody plants of the Guadiamar Valley, SW Spain: A large-scale phytomanagement case study. *Environ. Pollut.*, 152:50-59.

Dučić T, Polle A (2005) Transport and detoxification of manganese and copper in plants. Braz. *J. Plant Physiol.* 17: 103-112.

Elliott HA, Liberati MR, Huang CP (1986) Competitive adsorption of heavy metals by soils. *J. Environ. Qual.* 15: 214-217.

EPA-(US Environmental protection Agency) (1999*) Phytoremediation resource guide.* Washington: EPA. 542-B-99-003.

Fargasova A (1994) Effect of Pb, Cd, Hg, As, and Cr on germination and root growth of *Sinapis alba* seeds. Bull. Environ. *Contam. Toxicol.* 52: 452-456.

Fernandes JC, Henriques FS (1991) Biochemical, physiological, and structural effects of excess copper in plants. *Bot. Rev.* 57: 247-273.

Fodor F, Gáspár L, Morales F, Gogorcena Y, Lucena JJ, Cseh E, Kröpfl K, Abadía J, Sárvári É (2005) Effects of two iron sources on iron and cadmium allocation in poplar (Populus alba) plants exposed to cadmium. *Tree Physiol.* 25:1173–1180.

Gardea-Torresdey JL, Peralta-Videa JR, Rosa G, Parsons JG (2005) Phytoremediation of heavy metals and study of the metal coordination by X-ray absorption spectroscopy. *Coord. Chem. Rev.* 249: 1797-1810.

Godbold DL, Kettner C (1991) Lead influences root growth and mineral nutrition of *Picea abies* seedlings. *J. Plant Physiol.* 139: 95-99.

Grill E, Winnacker EL, Zenk MH (1985) Phytochelatins: the principal heavy-metals complexing peptides of higher plants. *Science.* 230: 674-676.

Guimarães MA, Santana TA, Silva EV, Zenzen IL, Loureiro ME (2008) Toxicidade e tolerância ao cádmio em plantas. Rev. Tróp.– Ciênc. Agr. e Biol. 3(1):56.

Gratão PL, Prasad MNV, Cardoso PF, Lea PJ, Azevedo RA (2005) Phytoremediation: green technology for the cleanup of toxic metals in the environment. *Braz. J. Plant Physiol.* 17(1):53-64.

Gussarsson M (1994) Cadmium-induced alterations in nutrient composition and growth of Betula pendula seedlings: the significance of the fine roots as a primary target for cadmium toxicity. *J. Plant Nutr.* 17: 2151-2163.

Gussarson M, Asp H, Adalsteinsson S, Jensén P (1996) Enhancement of Cd effects on growth and nutrient composition of birch (*Betula pendula*) by buthionine sulphoxinine (BSO). *J. Exp. Bot.* 47: 211–215.

Hall JL (2002) Cellular mechanisms for heavy metal detoxification and tolerance. *J. Exp. Bot.* 53: 1-11.

Hasan SA, Fariduddin Q, Ali B, Hayat S, Ahmad A (2009) Cadmium: Toxicity and tolerance in plants. *J. Environ. Biol.*, 30(2):165-174.

Heckathorn SA, Mueller JK, LaGuidice S, Zhu B, Barrett T, Blair B, Dong Y (2004) Chloroplast small heat-shock proteins protect photosynthesis during heavy metal stress. *Am. J. Bot.* 91: 1312-1318.

Henning BJ, Snyman HG, Aveling TAS (2001) Plant-soil interactions of sludge-borne heavy metals and the effect on maize (*Zea mays* L.) seedling growth. Water SA 27(1):71-78.

Hsu PH (1989) Aluminum oxides and oxyhydroxides. In: Dixon JB, Weed SB (eds), Minerals in soil environments, pp.331-378. *Soil Sci. Soc. Am.*, Madison.

Huang JW, Cunningham SD (1996) Lead phytoextraction: species variation in lead uptake and translocation. *New Phytol.* 134: 75-84.

Huang JWW, Chen JJ, Berti WR, Cunningham SD (1997) Phytoremediation of lead-contaminated soils: role of synthetic chelates in lead phytoextraction. *Environ. Sci. Technol.* 31:800-805.

Inouhe M (2005) Phytochelatins. Braz. J. *Plant Physiol.* 17: 65-78.

Jadia CD, Fulekar MH (2009) Phytoremediation of heavy metals: Recent techniques. *Afr. J. Biotechnol.* Vol. 8 (6): 921-928.

Jarvis MD, Leung DWM (2002) Chelated lead transport in Pinus radiata: an ultrastructural study. Environ. *Exp. Bot.* 48: 21-32.

Jung C, Maeder V, Funk F, Frey B, Sticher H, Frossard E (2003) Release of phenols from *Lupinus albus* L. roots exposed to Cu and their possible role in Cu detoxification. *Plant Soil* 252: 301-312.

Kahle H (1993) Response of roots of trees to heavy metals. Environ. *Exp. Bot.* 33: 99-119.

Kalbitz K, Wenrich R (1998) Mobilization of heavy metals and arsenic in polluted wetland soils and its dependence on dissolved organic matter. *Sci. Total Environ.* 209: 27-39.

Khan AG, Kuek C, Chaudhry TM, Khoo CS, Hayes WJ (2000) Role of plants, mycorrhizae and phytochelators in heavy metal contaminated land remediation. *Chemosphere* 41:197-207.

King LD (1988) Retention of metals by several soils of the southeastern United States. *J. Environ. Qual.* 17: 239-246.

Kraus ML (1988) Accumulation and excretion of five heavy metals by the salt marsh grass Spartina alterniflora. *Bull. N. J. Acad. Sci.* 33: 39-43.

Kukkola E, Rautio P, Huttunen S (2000) Stress indicators in copper- and nickel-exposed Scots pine seedlings. *Environ. Exp. Bot.* 43: 197-210.

Kumar PBAN, Dushenkov V, Motto H, Raskin I (1995) Phytoextraction: the use of plants to remove heavy metals from soils. *Environ. Sci. Technol.* 29: 1232-1238.

Leyval C, Turnau K, Haselwandter K (1997) Effect of heavy metal pollution on mycorrhizal colonization and function: physiological, ecological and applied aspects. *Mycorrhiza* 7: 139-153.

Lidon FC, Henriques FS (1991) Limiting step in photosynthesis of rice plants treated with varying copper levels. *J. Plant Physiol.* 138: 115-118.

Liu D, Jiang W, Li M (1992) Effect of trivalent and hexavalent chromium on root growth and cell division of *Allium cepa. Hereditas* 117: 23-29.

Liu J, Xiong Z, Li T, Huang H (2004) Bioaccumulation and ecophysiological responses to copper stress in two populations of *Rumex dentatus* L. from Cu contaminated and non-contaminated sites. *Environ. Exp. Bot.* 52: 43-51.

Liu Y-J, Zhu Y-G, Ding H (2007) Lead and cadmium in leaves of deciduous trees in Beijing, China: development of a metal accumulation index (MAI). Environ. Pollut. 145: 387-390.

Lindegaard KN, Barker JHA (1997) Breeding willows for biomass. *Asp. Appl. Biol.* 49: 155-162.

Lobinski R, Potin-Gautier M (1998) Metals and biomolecules- bioinorganic analytical chemistry. *Analusis Magazine*. 26: 21- 24.

Lombardi L, Sebastıani L (2005) Copper toxicity in *Prunus cerasifera*: growth and antioxidant enzymes responses of in vitro grown plants. *Plant Sci.* 168: 797-802.

MacFarlane GR, Burchett MD (2000) Cellular distribution of copper, lead and zinc in the grey mangrove, *Avicennia marina* (Forsk.) Vierh. *Aquat. Bot.* 68: 45-59.

MacFarlane GR, Pulkownik A, Burchett MD (2003) Accumulation and distribution of heavy metals in the grey mangrove, Avicennia marina (Forsk.) Vierh.: biological indication potential. *Environ. Pollut.* 123: 139-151.

Macnair MR (1993) The genetics of metal tolerance in vascular plants. *New Phytol.* 124: 541-559.

Maksymiec W (1997) Effect of copper on cellular processes in higher plants. *Photosynthetica.* 34: 321-342.

McCarthy P, Perdue EM (1991) *Complexation of metal ions by humic substances: fundamental considerations.* In: Bolt FH, DeBoot MF, Hayes MHB, McBride MB (eds), Interactions at the soil colloid-soil solution interface, pp.469-489. Kluwer Academic Publishers, Dordrecht, The Netherlands.

McGrath SP (1982) The uptake and translocation of tri- and hexavalent chromium and effects on the growth of oat in flowing nutrient solution and in soil. *New Phytol.* 92: 381-390.

Meda AR., Scheuermann E B., Prechsl U E., Erenoglu B, Schaaf G, Hayen H, Weber G, von Wirén N (2007) Iron acquisition by phytosiderophores contributes to cadmium tolerance. *Plant Physiol.* 143:1761-1773.

Mendelssohn IA, McKee KL, Kong T (2001) A comparison of physiological indicators of sublethal cadmium stress in wetland plants. *Environ. Exp. Bot.* 46: 263-275.

Meers E, Hopgood M, Lesage E, Vervaeke P, Tack FMG, Verloo MG (2004) Enhanced phytoextraction: in search of EDTA alternatives. *Inter. J. Phytorem.* 6:95-109.

Nascimento CWA, Xing B (2006) Phytoextraction: A review on enhanced metal availability and plant accumulation. *Sci. Agric.* 63:299-311.

Neumann D, zur Nieden U, Lichtenberger O, Leopold I (1995) How does *Armeria maritima* tolerate high heavy metal concentrations? *J. Plant Physiol.* 146:704–717.

Oliveira JA, Cambraia J, Cano MAO, Jordão CP (2001) Absorção e acúmulo de cádmio e seus efeitos sobre o crescimento relativo de plantas de aguapé e de salvínia. *R. Bras. Fisiol.* Veg., 13(3): 329-341.

Panda SK, Patra HK (2000) Nitrate and ammonium ions effect on the chromium toxicity in developing wheat seedlings. P. *Natl. Acad. Sci. India* B. 70: 75-80.

Panda SK, Choudhury S (2005) Chromium stress in plants. Braz. J. Plant Physiol. 17: 95-102.

Patra M, Bhowmik N, Bandopadhyay B, Sharma A (2004) Comparison of mercury, lead and arsenic with respect to genotoxic effects on plant systems and the development of genetic tolerance. *Environ. Exp. Bot.* 52: 199-223.

Pauls H, Bredenbröcker B, Schoner W (1980) Inactivation of (Na+ + K+)- ATPase by chromium (III) complexes of nucleotide triphosphates. *Eur. J. Biochem.* 109: 523-533.

Pietrini F, Iannelli MA, Pasqualini S, Massacci A (2003) Interaction of Cadmium with glutathione and photosynthesis in developing leaves and chloroplasts of *Phragmites australis* (Cav.) Trin. ex Steudel. *Plant Physiol.*, 133:829–837.

Pietrini F, Zacchini M, Iori V, Pietrosanti L, Bianconi D, Massacci A (2010) Screening of poplar clones for cadmium phytoremediation using photosynthesis, biomass and cadmium content analyses. *Intern J. Phytorem.*, 12:105–120.

Pilon-Smits E (2005) Phytoremediation. *Annual Review Plant Biology* 56: 15-39.

Pinto AP, Mota AM, Varennes A. de, Pinto FC (2004) Influence of organic matter on the uptake of cadmium, zinc, copper and iron by sorghum plants. *Sci. Total Environ.* 326:239–247.

Prasad MNV, Freitas HMO (2003) Metal hyperaccumulation in plants – Biodiversity prospecting for phytoremediation technology. *Electronic J. Biotechnology* 6: 285-321.

Prasad SM, Singh JB, Rai LC, Kumar HD (1991) Metal-induced inhibition of photosynthetic electron transport chain of the cyanobacterium *Nostoc muscorum*. *FEMS Microbiol. Lett.* 82: 95-100.

Prus-Glowacki W, Chudzińska E, Wojnicka-Półtorak A, Kozacki L, Fagiewicz K (2006) Effects of heavy metal pollution on genetic variation and cytological disturbances in the Pinus sylvestris L. population. *J. Appl. Genet.* 47: 99-108.

Pulford ID, Riddell-Black D, Stewart C (2002) Heavy metal uptake by willow clones from sewage sludge-treated soil: the potential for phytoremediation. *Int. J. Phytoremediat.* 4: 59-72.

Pulford ID, Watson C (2003) Phytoremediation of heavy metal-contaminated land by trees – a review. *Environ. Int.* 29: 529-540.

Punshon T, Dickinson NM (1997) Acclimation of *Salix* to metal stress. *New Phytol.* 137: 303-314.

Quartacci MF, Pinzino C, Sgherri CLM, Dalla Vecchia F, Navari-Izzo F (2000) Growth in excess copper induces changes in the lipid composition and fluidity of PSII-enriched membranes in wheat. *Physiol. Plant.* 108: 87-93.

Ramachandran V, D'Souza TJ, Mistry KB (1980) Uptake and transport of chromium in plants. *J. Nucl. Agric. Biol.* 9: 126-129.

Rashid A, Popovic R (1990) Protective role of $CaCl_2$ against Pb^{+2} inhibition in Photosystem II. *FEBS Lett.* 271: 181-184.

Rodríguez-Serrano M, Romero-Puertas MC, Pazmiño DM, Testillano PS, Risueño MC, del Río LA, Sandalio LM (2009) Cellular response of pea plants to cadmium toxicity: cross talk between reactive oxygen species, nitric oxide, and calcium. *Plant Physiol.* 150(1):229-243.

Romanowska E, Igamberdiev AU, Parys E, Gardestrom P (2002) Stimulation of respiration by Pb^{2+} in detached leaves and mitochondria of C3 and C4 plants. *Physiol. Plant.* 116: 148-154.

Root AR, Miller RJ, Koeppe DE (1975) Uptake of cadmium - its toxicity and effect on the iron ratio in hydroponically grown corn. *J. Environ. Qual.* 4: 473-476.

Salt DE, Smith RD, Raskin I (1998) Phytoremediation. *Ann. Rev. Plant Physiol. Plant Mol. Biol.* 49: 643-668.

Samantaray S, Rout GR, Das P (1996) Root growth of *Echinochloa colona*: Effects of heavy metals in solution culture. *Fresnius. Environ. Bull.* 5: 469-473.

Samantaray S, Rout GR, Das P (1998) Role of chromium on plant growth and metabolism. Acta Physiol. *Plant.* 20: 201-212.

Samantaray S, Rout GR, Das P (2001) Induction, selection and characterization of Cr and Ni-tolerant cell lines of *Echinochloa colona* (L.) in vitro. *J. Plant Physiol.* 158: 1281-1290.

Sandalio LM, Dalurzo HC, Gómez M, Romero-Puertas MC, del Río LA (2001) Cadmium-induced changes in the growth and oxidative metabolism of pea plants. *J. Exp. Bot.* 52:2115–2126.

Sanità di Toppi L, Gabbrielli R (1999) Response to cadmium in higher plants. *Environ. Exp. Bot.* 41: 105 - 130.

Schat H, Vooijs R (1997) Multiple tolerance and co-tolerance to heavy metals in *Silene vulgaris*, a co-segregation analysis. *New Phytol.* 136: 489-496.

Schützendübel A, Polle A (2002) Plant responses to abiotic stresses: heavy metal-induced oxidative stress and protection by mycorrhization. *J. Exp. Bot.* 53:1351-1365.

Sebastiani L, Scebba F, Tognetti R (2004) Heavy metal accumulation and growth responses in poplar clones Eridano (*Populus deltoides* x *maximowiczii*) and I-214 (*P.* x *euramericana*) exposed to industrial waste. *Environ. Exp. Bot.* 52: 79-88.

Serpersu EH, Kirch U, Schoner W (1982) Demonstration of a stable occluded form of Ca2+ by the use of the chromium complex of ATP in the Ca2+-ATPase of sarcoplasmic reticulum. *European J. Biochem.* 122: 347–354.

Shainberg O, Rubin B, Rabinowitch HD, Tel-Or E (2001) Loading beans with sublethal levels of copper enhances conditioning to oxidative stress. *J. Plant Physiol.* 158: 1415-1421.

Shanker AK (2003) *Physiological, biochemical and molecular aspects of chromium toxicity and tolerance in selected crops and tree species*. PhD Thesis, Tamil Nadu Agricultural University, Coimbatore, India.

Shanker AK, Sudhagar R, Pathmanabhan G (2003) *Growth, phytochelatin SH and antioxidative response of sunflower as affected by chromium speciation*. In: Abstracts of the 2nd International Congress of Plant Physiology on sustainable plant productivity under changing environment, New Delhi, India.

Shanker AK, Djanaguiraman M, Sudhagar R, Chandrashekar CN, Pathmanabhan G (2004) Differential antioxidative response of ascorbate glutathione pathway enzymes and metabolites to chromium speciation stress in green gram (*Vigna radiate* (L) R. Wilczek, cv CO 4) roots. *Plant Sci.* 166: 1035-1043.

Shanker AK, Cervantes C, Loza-Tavera H, Avudainayagam S (2005) Chromium toxicity in plants. *Environ. Int.* 31: 739-753.

Sharma P, Dubey RS (2005) Lead toxicity in plants. Braz. *J. Plant Physiol.* 17: 35-52.

Sharma SS, Dietz K-J (2006) The significance of amino acids and amino acid-derived molecules in plant responses and adaptation to heavy metal stress. *J. Exp. Bot.* 57: 711-726.

Siedleska A (1995) Some aspects of interactions between heavy metals and plant mineral nutrients. *Acta Soc. Bot. Pol.* 64: 265–272.

Skeffington RA, Shewry PR, Peterson PJ (1976) Chromium uptake and transport in barley seedling (*Hordeum vulgare* L.). *Planta* 132: 209-214.

Smirnoff N (1995) *Antioxidant systems and plant response to the environment.* In: Smirnoff N (ed), Environment and Plant Metabolism: Flexibility and Accumulation, pp.217-243. BIOS Scientific Publishers, Oxford.

Soares CRFS, Siqueira JO, De Carvalho JG, Moreira FMS (2005) Fitotoxidez de cádmio para *Eucalyptus maculata* e *E. urophylla* em solução nutritiva. Rev. Árvore 29(2): 175-183.

Souza VL (2007) Expressão gênica, respostas morfo-fisiológicas e morte celular induzidas por cádmio em *Genipa americana* L. (Rubiaceae). Ilhéus, Universidade Estadual de Santa Cruz, Dissertação de Mestrado.

Souza, VL, Silva DC, Santana KB, Mielke, MS, Almeida, A-A F, Mangabeira, P A O, Rocha, E A R (2009) Efeitos do cádmio na anatomia e na fotossíntese de duas macrófitas aquáticas. *Acta bot. bras.* 23(2): 343-354.

Steffens JC (1990) The heavy metal-binding peptides of plants. Annu. Rev. Plant Physiol. *Mol. Biol.* 41: 553-575.

Strange J, Macnair MR (1991) Evidence for a role for the cell membrane in copper tolerance of *Mimulus guttatus* Fisher ex D.C. *New Phytol.* 119: 383-388.

Sun Y, Zhou Q, Wang L, Liu W (2009) Cadmium tolerance and accumulation characteristics of *Bidens pilosa* L. as a potential Cd-hyperaccumulator. *J Hazard Mater.*161(2-3): 808-814.

Tavares TM, Carvalho FM (1992) Avaliação da exposição de populações humanas a metais pesados no ambiente: exemplos do Recôncavo Baiano. *Quim. Nova*, 15: 147-153.

Tomsig JL, Suszkiw JB (1991) Permeation of Pb+2 through calcium channels: fura-2 measurements of voltage- and dihydropiridine-sensitive Pb^{+2} entry in isolated bovine chromaffin cells. Biochim. *Biophys. Acta.* 1069: 197-200.

Turner AP, Dickinson NM (1993a) Copper tolerance of *Acer pseudoplatanus* L. (sycamore) in tissue culture. *New Phytol.* 123: 523-530.

Turner AP, Dickinson NM (1993b) Survival of *Acer pseudoplatanus* L. (sycamore) seedlings on metalliferous soils. *New Phytol.* 123: 509-521.

Unterbrunner R, Puschenreiter M, Sommer P, Wieshammer G, Tlustoš P, Zupan M, Wenzel WW (2007) Heavy metal accumulation in trees growing on contaminated sites in Central Europe. *Environ. Pollution* 148: 107-114.

Vajpayee P, Sharma SC, Tripathi RD, Rai UN, Yunus M (1999) Bioaccumulation of chromium and toxicity to photosynthetic pigments, nitrate reductase activity and protein content of *Nelumbo nucifera* Gaertn. *Chemosphere* 39: 2159-2169.

Vajpayee P, Rai UN, Ali MB, Tripathi RD, Yadav V, Sinha S, Singh SN (2002) Chromium-induced physiologic changes in *Vallisneria spiralis* L. and its role in phytoremediation of tannery effluent. *Bull. Environ. Contam. Toxicol.* 67: 246-256.

Van Assche F, Clijsters H (1983) *Multiple effects of heavy metals on photosynthesis.* In: Marcelle R (ed), Effects of Stress on Photosynthesis, pp. 371-382. The Hague: Nijhoff/Junk.

Van Assche F, Clijsters H (1990) Effects of the metals on enzyme activity in plants. *Plant Cell Environ.* 13: 195-206.

Van Hoof NALM, Hassinen VH, Hakvoort HWJ, Ballintijn KF, Schat H, Verkleij JAC, Ernst WHO, Karenlampi SO, Tervahauta AI (2001) Enhanced copper tolerance in *Silene vulgaris* (Moench) Garcke populations from copper mines is associated with increased transcript levels of a 2b-type metallothionein gene. *Plant Physiol.* 126: 1519-1526.

Vandecasteele B, De Vos B, Tack FMG (2002) Cadmium and zinc uptake by volunteer willow species and elder rooting in polluted dredged sediment disposal sites. *Sci. Total Environ.* 299: 191-205.

Vassilev A, Vangronsveld J, Yordanov I (2002) Cadmium phytoextraction: present state, biological backgrounds and research needs. Bulg. *J. Plant Physiol.* 28: 68–95.

Vekemans X, Lefèbvre C (1997) On the evolution of heavy metal tolerant populations in Armeria maritima, evidence from allozyme variation and reproductive barriers. *J. Evol. Biol.* 10: 175-191.

Viola RE, Morrison JF, Cleland WW (1980). Interaction of metal (III)-adenosine 5'-triphosphate complexes with yeast hexokinase. *Biochemistry.*19:3131–3137.

Vollenweider P, Cosio C, Gunthardt-Goerg MS, Keller C (2006) Localization and effects of cadmium in leaves of a tolerant *Salix viminalis* L. Part II. Microlocalization and cellular effect of cadmium. *Environ. Exp. Bot.* 58: 25-40.

Von Burg R, Liu D (1993) Chromium and hexavalent chromium. J. Appl. Toxicol. 13: 225-230.

Wallace A, Cha JW (1989) Interactions involving copper toxicity and phosphorus deficiency in bush bean plants grown in solutions of low and high pH. *Soil Sci.* 147: 430-431.

Watmough SA, Dickinson NM (1996) Variability of metal resistance in *Acer pseudoplatanus* L. (sycamore) callus tissue of different origins. *Environ. Exp. Bot.* 36: 293-302.

Weis JS, Weis P (2004) Metal uptake, transport and release by wetland plants: implications for phytoremediation and restoration. *Environ. Int.* 30: 685-700.

Wisniewski L, Dickinson NM (2003) Toxicity of copper to *Quercus robur* (English Oak) seedlings from a copper-rich soil. *Environ. Exp. Bot.* 50: 99-107.

Wong MK, Chuan GK, Koh LL, Ang KP, Hew CS (1984) The uptake of cadmium by *Brassica chinensis* and its effect on plant zinc and iron distribution. *Environ. Exp. Bot.* 24: 189-195.

Wu J, Hsu FC, Cunningham SD (1999) Chelate-assisted Pb phytoremediation: Pb availability, uptake, and translocation constraints. *Environ. Sci. Technol.* 33:1898-1904.

Yap DW, Adezrian J, Khairiah J, Ismail BS, Ahmad-Mahir R (2009) The uptake of heavy metals by paddy plants (*Oryza sativa*) in Kota Marudu, Sabah, Malaysia. American-Eurasian *J. Agric. and Environ. Sci.*, 6 (1): 16-19.

Yang X, Feng Y, He Z, Stoffella PJ (2005) Molecular mechanisms of heavy metal hyperaccumulation and phytoremediation. *Journal of Trace Elements in Medicine and Biology* 18: 339-353.

Yruela I (2005) Copper in plants. Braz. *J. Plant Physiol.* 17: 145-156.

Zenk MH (1996) Heavy metal detoxification in higher plants- a review. *Gene*, 179: 21-30.

Zhang X, Liu C, Nan Z (2009) Effects of cadmium stress on growth and anti-oxidative systems in *Achnatherum inebrians* symbiotic with *Neotyphodium gansuense. J. Hazard. Mat.* 175:703–709.

Zacchini M, Pietrini F, Mugnozza GS, Iori V, Pietrosanti L, Massacci A (2009) Metal tolerance, accumulation and translocation in poplar and willow clones treated with cadmium in hydroponics. *Water Air Soil Pollut* 197:23–34.

In: Handbook of Phytoremediation
Editor: Ivan A. Golubev

ISBN: 978-1-61728-753-4

Chapter 17

IMPACT OF PLANT GROWTH PROMOTING RHIZOBACTERIA *PSEUDOMONAS* IN PHYTOREMEDIATION PROCESS

T. V. Siunova, T. O. Anokhina, O. I. Sizova, V. V.Kochetkov and A. M.Boronin

Institute of Biochemistry and Physiology of Microorganisms,
Russian Academy of Sciences (RAS), Russia

ABSTRACT

Plant growth promoting rhizobacteria (PGPR) *Pseudomonas P. aureofaciens*, *P. chlororaphis*, *P. fluorescens* and their plasmid-bearing variants: destructors of polycyclic aromatic hydrocarbons (PAH) (naphthalene, phenanthrene), strains resistant to heavy metals (cobalt, nickel) and metalloids (arsenic), and multifunctional ones combined both characteristics, were used to estimate their impact in the phytoremediation process. All used bacterial strains that possessed ability to produce phytohormone indole acetic acid, various antifungal compounds, and suppressed phytopathogens. The PGPR strain's ability to degrade naphthalene and phenanthrene was shown to be stable in the rhizosphere at different conditions. The introducing of PGPR destructors in the rape rhizosphere increased the naphthalene biodegradation efficiency up to 90% in comparison with control without bacteria at gnotobiotic system in 7 days cultivation. The arsenite resistant PAH-destructors *P. aureofaciens* BS1393(pBS216,pKS1) and *P. chlororaphis* PCL1391(pBS216,pKS1) also promoted mostly complete naphthalene degradation at the same experiments supplemented arsenite (15 mg/kg). It was shown, that the most active strains *P. fluorescens* 38a(pBS216) and *P. aureofaciens* OV17(pOV17) in the barley rhizosphere decreased the phenanthrene concentration 2 and 3 times respectively in 28 days in pot experiments. The impact of rhizosphere strains in plant accumulation of heavy metals/metalloids was tested in pot experiments. The cobalt-nickel resistant strain *P. aureofaciens* BS1393(pBS501) promoted growth of barley plants and protected from chlorosis contrary to the sensitive strain *P. aureofaciens* BS1393 in soil containing 235–940 mg Ni/kg. In one month growing the nickel accumulation in plant biomass increased by 5.6 and 2.5 times in the case of sensitive and resistant strain, respectively, compared to non-treated plants. The sorghum plants, inoculated by the resistant *P. aureofaciens*

BS1393(pKS1) and phosphate-dissolving *P. aureofaciens* BS1393(pUCP22:*gltA*) strains accumulated arsenic in plant biomass on an average of 25% more than non-treated plants in one month growing on arsenic contaminated soil (100 mg/kg). Nevertheless, the amount of bacteria in the plant rhizosphere varied, depending on bacterial species, plasmids occurrence and experiment conditions, but PGPR inoculation of plants protected them against PAH and metal/metalloid phytotoxicity, promoted seed germination and plant biomass.

INTRODUCTION

Phytoremediation, the use of plants and their associated microbiota to remediate environmental contamination, is a cost-effective technique that includes several techniques, such as rhizoremediation, phytostabilization, phytoextraction, and phytovolatilization [1, 2]. The plant rhizosphere (the immediate region around plant roots) frequently contains highly enriched bacterial populations in comparison with unvegetated soils [3, 4]. Fluorescent pseudomonads are typical inhabitants of rhizosphere and rhizoplane. Some rhizosphere Pseudomonas may stimulate plant growth by the synthesis of phytohormones, improve mineral nutrition of plants and protect plants from phytopathogenic fungi due to the synthesis of antibiotics and siderophores [5–7]. These strains are referred to as plant growth promoting rhizobacteria (PGPR) Pseudomonas. In addition to the ability to promote plant growth and suppress growth of soil borne pathogens, PGPR Pseudomonas are also promising candidates for the bioremediation of soils contaminated by oil, polycyclic aromatic hydrocarbons (PAHs), heavy metals and other pollutants [8]. To obtain PGPR Pseudomonas beneficial for phytoremediation, the introduction of plasmids responsible for organic pollutants degradation and for resistance to heavy metals and other toxicants can be used. Of peculiar interest are microorganisms combining plant growth promotion properties and the ability to degrade PAHs and accumulate/detoxificate inorganic compounds in soil.

APPLICATION OF PGPR FOR PAHs DEGRADATION

Due to its adverse environmental and health effects, oil pollution possesses a significant hazard to natural ecosystems. Quantitatively, the most important constituents of petroleum pollution are polycyclic aromatic hydrocarbons (PAHs). Because of their toxic, mutagenic, and carcinogenic properties, PAHs represent serious and chronic environmental contaminants [10–12]. The biodegradation of PAHs in soils is a complex process that depends on their physical and chemical properties, as well as on the physical characteristics of soil. Although PAHs are hydrophobic, a variety of microorganisms are able to degrade and mineralize low molecular weight PAHs (such as naphthalene, phenanthrene, and anthracene) into carbon dioxide and water [13–15]. The rhizosphere has been shown to have an important role in the biodegradation of different organic compounds [16, 17]. For example, the mineralization of PAHs occurs faster in planted soil than in soil without plants [18–20]. The most likely explanation of this phenomenon is that roots exude a variety of soluble organic compounds, which facilitate the growth of microbes and cometabolic breakdown of contaminants [3, 16]. It is known that some *Pseudomonas* are able to degrade various xenobiotics, including PAHs

[21, 22]. The PAHs biodegradation genes are usually located in plasmids. A number of conjugative degradation plasmids has been described [22–24]. As a rule, the naphthalene catabolic genes are organized in three operons. *Nah1*-operon encoding for the upper-pathway enzymes is involved in the conversion of naphthalene to salicylate, *nah2*-operon encoding for the lower-pathway enzymes is involved in the oxidation of salicylate via the plasmid-encoded catechol *meta*-cleavage pathway to acetaldehyde and pyruvate. The third operon encodes for regulatory protein. Certain interest represents studying of plant–PGPR *Pseudomonas* associations in the remediation of PAHs contamination. We used wild type PGPR strains (laboratory collection) and constructed plasmid-bearing variants capable of biodegrading naphthalene and phenanthrene for this purpose [25, 26] (Table 1).

Table 1. Bacterial strains

Strain	Phenotype description	Source
P. fluorescens		
38a	Producer of pyoluteorin, Phn$^-$Nah$^-$Sal$^-$	IBPM RAS[a]
38a(pBS216)	Producer of pyoluteorin, Phn$^+$Nah$^+$Sal$^+$	[26]
P. chlororaphis		
PCL1391	Producer of phenazine-1-carboxamide, Phn$^-$Nah$^-$Sal$^-$	[28]
PCL1391(pBS216)	Producer of phenazine-1-carboxamide Phn$^+$Nah$^+$Sal$^+$	[26]
PCL1391(pOV17)	Producer of phenazine-1-carboxamide Phn$^+$Nah$^+$Sal$^+$	[26]
P. putida		
53a	Producer of indole-3-acetic acid, Phn$^-$Nah$^-$Sal$^-$	IBPM RAS
53a(pBS216)	Producer of indole-3-acetic acid, Phn$^+$Nah$^+$Sal$^+$; loss the C1,2O and C2,3O activities	[26]
53a(pOV17)	Producer of indole-3-acetic acid, Phn$^+$Nah$^+$Sal$^+$	[26]
P. aureofaciens		
BS1393	Producer of phenazine antibiotics, Phn$^-$Nah$^-$Sal$^-$	IBPM RAS
BS1393(pBS216)	Producer of phenazine antibiotics, Phn$^+$Nah$^+$Sal$^+$	[25]
BS1393(pOV17)	Producer of phenazine antibiotics, Phn$^+$Nah$^+$Sal$^+$	[26]
Strain	Phenotype description	Source
OV17(pOV17), wild type	Producer of phenazine antibiotics, Phn$^+$Nah$^+$Sal$^+$	IBPM RAS
OV17(pBS216)	Producer of phenazine antibiotics, Phn$^+$Nah$^+$Sal$^+$	[26]

Phn+, Nah+, Sal+ – ability to grow on phenanthrene, naphthalene, salicylate, respectively. C1,2O and C2,3O–catechol-1,2- and catechol-2,3-dioxygenases activities, respectively.

[a]Strains from the collection of Laboratory of Plasmid Biology, G.K. Skryabin Institute of Biochemistry and Physiology of Microorganisms, Russian Academy of Sciences.

The application of rhizosphere microorganisms in phytoremediation technologies is based on their ability to degrade, transform, and accumulate pollutants (direct mechanism); and promote growth and stress resistance of plants (mediated mechanism). Earlier, the biopreparate "Pseudobacterin-2" was developed on the basis of *Pseudomonas aureofaciens* BS1393 which possessed high biological efficiency (65–96%) against bacterial and fungal phytopathogens. Besides, "Pseudobacterin-2" had a high growth-stimulating activity and allowed us to obtain reliable rises in the yield of cereals (2–10 centners per ha) and open-ground vegetables (18-120 centners per ha) [27]. We supposed that the application of *P. aureofaciens* BS1393 and other PGPR *Pseudomonas* will be beneficial for clean-up of contaminated soils.

Naphthalene Degradation

Sterile model systems were used to study the effect of plasmid-bearing PGPR strains on plant growth and naphthalene degradation [29]. The model systems represented the closed plastic vessels containing of sterile sand, naphthalene (200 μg g-1 sand) and rape seedlings (Brassica napus) inoculated by bacterial cultures. As the negative/positive controls used bacteria-free plant grown with/without naphthalene, respectively.

The measurement of shoot length and the total dry plant biomass after week cultivation of rape seedlings in the presence of naphthalene demonstrated that naphthalene had a strong phytotoxic effect (Figure 1a).

The shoot length in negative control was in average by 80% shorter than in positive control. Free-plasmid strains did not protect the plants from PAH. In this case biometric data were similar to negative control (Figure 1b). Treatment of seedlings with plasmid-bearing rhizobacteria led to a pronounced protective effect from naphthalene (Figure 1b).

The exception were the seedlings treated with P. putida 53a(pBS216) that has not catechol dioxygenases activities [30]. In this case dark brown pigment was accumulated in the sand and no seedlings development was observed (Figure 2). We assume that the strong phytotoxic effect could be connected with accumulation of catechol oxidation products in substrate. The similar results describing toxic action of catechol-related metabolites on the P. putida NCIB 9186-4 strain have been described earlier [31].

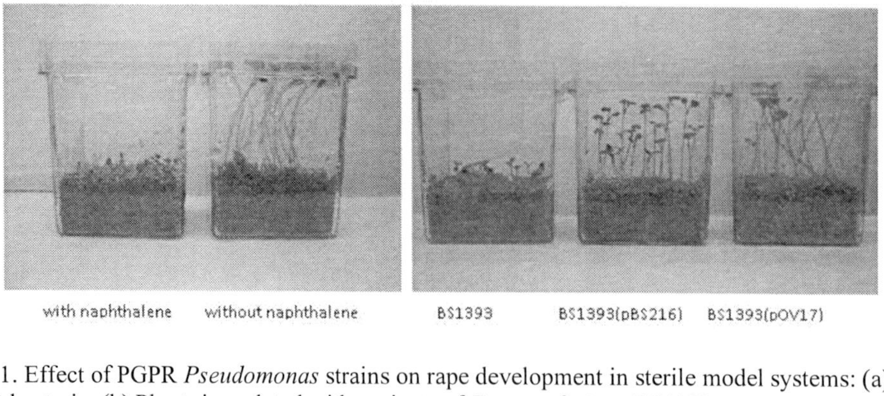

Figure 1. Effect of PGPR *Pseudomonas* strains on rape development in sterile model systems: (a) Plants without bacteria; (b) Plants inoculated with variants of *P. aureofaciens* BS1393.

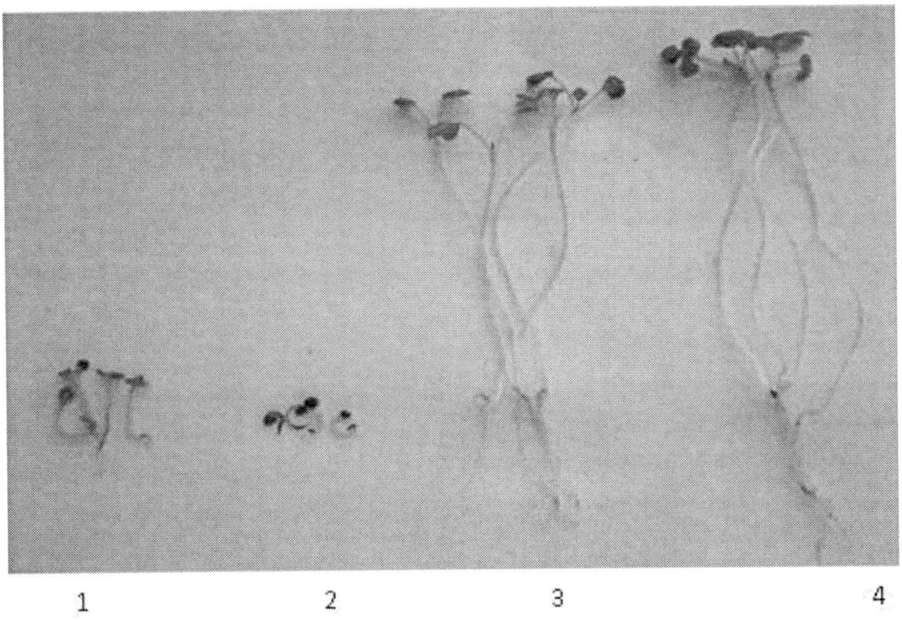

Figure 2. Effect of variants *P. putida* 53a on rape growth in sterile model system.

Table 2. Effect of rape inoculation by PGPR *Pseudomonas* on naphthalene biodegradation

Experiment [a]	Naphthalene, μg/g sand[b]
Without bacteria	
Zero point (on 1 day)	196.4
Final point (on 7 day)	91.15
With plants (on 7 day)	95.66
Plasmid-free bacteria	
P. aureofaciens BS1393	97.15
P. chlororaphis PCL1391	103.3
Bacteria-destructors	
P. chlororaphis PCL1391(pBS216)	7.13
P. chlororaphis PCL1391(pOV17)	7.95
P. aureofaciens BS1393(pBS216)	9.85
P. aureofaciens BS1393(pOV17)	8.25
P. aureofaciens OV17(pBS216)	2.83
P. aureofaciens OV17(pOV17)	3.44
P. putida 53a(pBS216)	4.85
P. putida 53a(pOV17)	4.53
P. fluorescens 38a(pBS216)	2.40

[a]Seedlings, sterile or inoculated with plasmid-free or plasmid-bearing naphthalene-degrading PGPR *Pseudomonas* strains, were grown for 7 days in model systems with naphthalene (200 μg g-1 sand). The sand was extracted with methanol, and HPLC was used to analyze samples of the methanol fractions.

[b]Standard deviation was not more than 20% in all variants.

Plants inoculated with (1) free-plasmid *P. putida* 53, (2) *P. putida* 53a(pBS216), (3) *P. putida* 53a(pOV17) cultivated in the naphthalene presence; (4) control plants inoculated with *P. putida* 53. cultivated in cleane sand. It is worthy of note that abundance of introduced strains in rhizoplane increased at an average by one order in the presence of naphthalene. For examples the titer of *P. chlororaphis* PCL1391 and PCL1391(pBS216) increased from 2×10^8 (0 day) to 5×10^9 CFU/g root (7 day of cultivation). It is possible to explain the date either naphthalene use as an energy source (for plasmid-bearing strains) or changes in root exudates composition, as reaction of plants to naphthalene (for plasmid-free strains).

Independently of naphthalene presence the stability of pBS216 and pOV17 plasmids in all strains was considerably above (75–100%) in rhizosphere, than it was in lab cultivation [26]. It can be explained more long time of bacteria generation in rhizosphere. All plasmid-bearing strains were able to degrade naphthalene in model systems. It was shown that naphthalene concentration in sand decreased in 10–30 times in comparison with non-inoculated plants (Table 2).

Earlier it has been shown that bacteria utilizing naphthalene via the *meta*-pathway of catechol oxidation grew faster than those which utilize naphthalene via *ortho*-pathway, i.e. the *meta*-pathway proves to be more efficient than *ortho*-pathway in batch culture in excess of naphthalene [32]. The PGPR *Pseudomonas* strains, containing the plasmid pOV17, possessed higher activity catechol-2,3-dioxygenase (*meta*-pathway) in comparison with one pBS216 (*ortho*-pathway), but these strains did not differ by efficiency of naphthalene degradation in rape rhizosphere.

Phenanthrene Degradation

Phenanthrene represents relatively immobile organic compound consisting of three fused benzene rings that is common industrial pollutant. The effect of rhizosphere strains on phenanthrene degradation was estimated in the pot experiments with peat mixture under natural conditions [33]. We used wild type Nah^+Phe^+ strains IC7, OV29, OV25 and OV17(pOV17) isolated from the rhizosphere of randomly selected cereals growing on oil-contaminated soil in West Siberia, Russia and mentioned above *P. aureofaciens* BS1393(pBS216) and *P. fluorescens* 38a(pBS216). The barley (*Hordeum sativum)* seeds were used for bacteria inoculation.

It is known that in the polluted soils the content of individual PAHs can exceed the maximum allowable concentration in hundreds, and even thousands times. We used 5 mg/g of phenanthrene because the barley growth was inhibited by this concentration. It is known that the content of PAHs in soil can be reduced by abiotic processes. In our experiment, the phenanthrene concentration in bacteria-free variant after 28 days of incubation was 1.2 mg/g).

All phenanthrene-degrading bacteria improved the plants growth. Throughout the experiment, the noninoculated plants lagged behind the control plants (clean environment) and plants inoculated with the Phe^+Nah^+ strains (Figure 3). Our studies did not reveal any difference between obtained and wild type strains. At batch cultivating mentioned above strains were characterized by slow growth at presence phenanthrene as sole carbon and energy source (unpublished data). Despite the abundance of inoculated strains in the presence of phenanthrene was lower (about 10^5 CFU/g rhizosphere) than in the presence of naphthalene, nevertheless, the phenanthrene degradation in peat mixture reached 50% and more for *P. aureofaciens* (OV17(pOV17) and *P. fluorescens* 38a(pBS216) strains in comparison with peat mixture without barley (1), and without bacteria inoculation (2) (Figure 4).

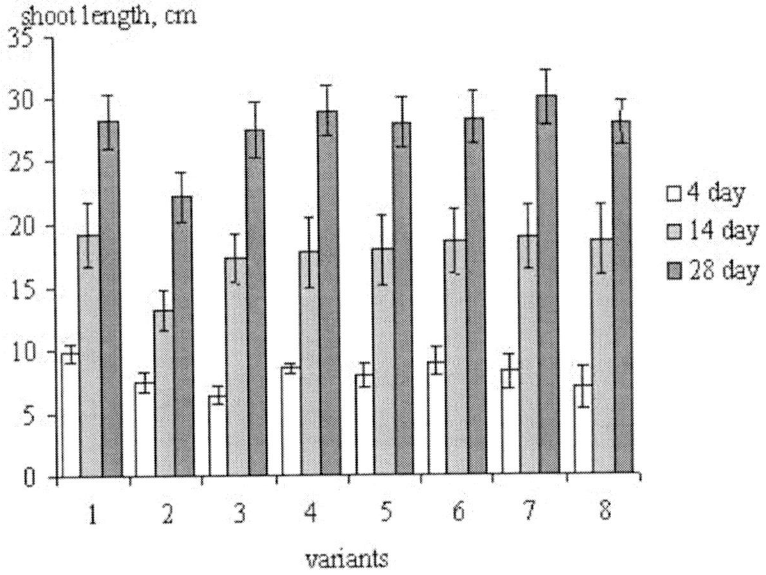

Figure 3. Effect of inoculation of barley seeds with phenanthrene-degrading rhizosphere bacteria on plant growth (cm) in a peat mixture containing 5 mg/g phenanthrene: (1) bacteria-free plants without phenanthrene; (2) bacteria-free plants with phenanthrene; plants inoculated with (3) BS1393(pBS216); (4) 38a(pBS216); (5) IC7; (6) OV29; (7) OV17(pOV17); (8) OV25 in the phenanthrene presence.

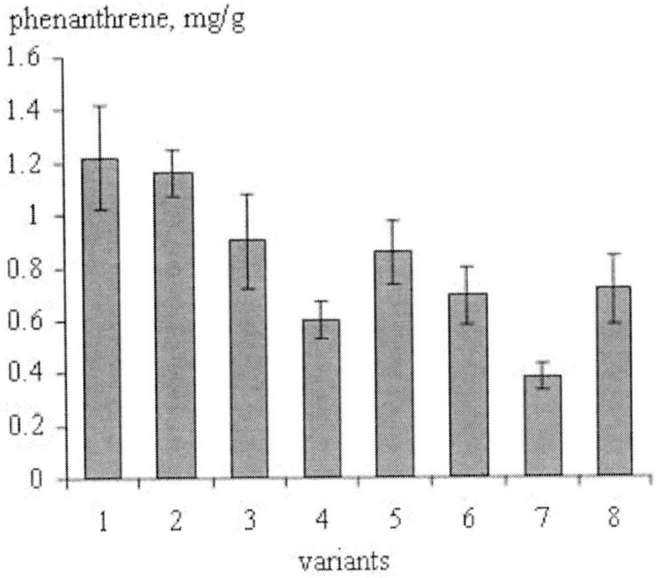

Figure 4. Phenanthrene concentration (mg/g) in peat mixture after 28 days according to HPLC data:(1) peat mixture without barley plants; (2) bacteria-free plants; plants inoculated with (3) BS1393(pBS216); (4) 38a(pBS216); (5) IC7; (6) OV29; (7) OV17(pOV17); (8) OV25.

Nowadays there is a number of data concerning positive influence of plants [20, 34, 35] or bacteria-degraders [36, 37] on PAHs utilization in sterile system and pot experiments. Nevertheless, estimation of degradation efficiency of these plant-microbe associations represents certain difficulty owing to various conditions of experiments (duration, type of soils, PAHs concentration and structure, inoculation and sampling methods etc.). In a number of works the contribution of bacteria-degraders is not considered at all [20, 38]. According to researchers' data efficiency of phenanthrene degradation varies from 30 [39] to 87 [37] and even to 98% [40]. We used PGPR *Pseudomonas* which not only degraded naphthalene and phenanthrene, but also promoted plants growth and development via various mechanisms (Table 1). The phytotoxicity of contaminated soil is determined by both the direct action of organic pollutants (PAHs, oil hydrocarbons, pesticides, herbicides) and the influence of various microbial toxins. The main group of microorganisms producing toxic metabolites is presented by non-symbiotic micromycetes belonging to the genera *Mucor*, *Aspergillus*, *Penicilum*, *Fusarium* [41]. Thus we assume that application of pollutant-degrading PGPR *Pseudomonas* will be perspective for remediation contaminated soils.

APPLICATION OF PGPR *PSEUDOMONAS* FOR NICKEL REMOVAL

Restoration of the heavy metal polluted soils is a priority direction in many countries. Soil phytoremediation and protection of agricultural crops from heavy metals are actual problems of modern biotechnology. Bacteria are actively involved in global cycles of metals circulation in biosphere. Heavy metal contamination in soil leads to dynamic instability of certain groups of microorganisms in soil [42], and predominance of phytopathogens decreases the efficiency of metabolism in plants and plant-associated microorganisms and as a result, impairs phytoremediation. Hence, PGPR resistant to heavy metals can be used to optimize phytoremediation [43]. Microorganisms possess different mechanisms of resistance to heavy metal divalent cations. The efflux of heavy metals from cells realizes with help of trans-membrane protein complexes [44, 45], cation diffusion facilitators [46], specific ATPase [47]. The mechanism of resistance via efflux provides detoxification only cytoplasm of bacterial cell, therefore usage of such bacteria in remediation technologies is problematic. However bacteria can carry out biochemical reactions, such as precipitation of metals by carbon dioxide, which evolutes during bacteria growth or xenobiotics degradation [48], bacteria can precipitate heavy metals in form phosphates and sulfides [49], by siderophores [50] and cystein-rich proteins [51, 52], etc. It was shown that Co/Ni resistant strain *P. aureofaciens* BS1393(pBS501) possesses *cnr*-like efflux system [53] and additionally forms granules of bound cobalt on the cell surface [54]. The investigations of bacteria, expressed various efflux systems of heavy metals, as soil remediators, is limited. It is known that *Lupinus luteus* L, when grown on a nickel-enriched substrate and inoculated with endophytic bacterium *Burkholderia cepacia* L.S.2.4:*ncc-nre*, showed a significant increase (30%) of the nickel concentration in the roots, whereas the nickel concentration in the shoots remained comparable with that of the control plants [55].

We have concentrated our attention on nickel which is carried to a class of highly-dangerous substances for live organisms along with mercury, selenium, zinc, fluorine and benz[a]pyrene. The plasmid pBS501 from *Comamonas* sp. BS501 determining the cobalt/nickel resistance was transferred to PGPR *Pseudomonas* resulting their cobalt/nickel resistant variants possessing lower level of resistance in comparison with parental strain [54]. The effect of PGPR—sensitive *P. aureofaciens* BS1393 and resistant *P. aureofaciens* BS1393(pBS501) variants—on nickel accumulation by barley plants was demonstrated in pot experiments with nickel-supplemented soil when the metal concentrations exceeded the maximum permissible concentration by 50–200 times (235, 470 and 940 mg Ni/kg soil). Non-inoculated (PGPR-free) plants grown in soil without metal and with corresponding Ni concentrations were used as control 1 and 2, respectively [56].

The study of dynamics of introduced bacteria has shown that the resistant strain *P. aureofaciens* BS1393(pBS501) revealed higher survival in barley rhizosphere (6.2×10^8 CFU/cm root) in comparison with the sensitive strain *P. aureofaciens* BS1393 (5.4×10^6 CFU/cm root) in two weeks of growing in the presence of 235 mg Ni/kg. However, both strains abundance in rhizosphere fell drastic in increasing of nickel concentration in two or four times. The increase in PGPR abundance in rhizosphere led to enhancement of the plant tolerance index (TI) (Table 3). The inoculated plants were more tolerant to nickel (235 mg Ni/kg soil) and possessed of IT more 100% despite from PGPR-free plants (58 and 76%). The increase in nickel concentration in soil resulted vitality decrease of both PGPR strains in rhizosphere and therefore TI became similar in all variants [56].

The cultivating of noninoculated plants in the presence of nickel (235 mg/kg) led to decrease of plant biomass on 28%C, whereas both PGPR strains provided a biomass increase on 10 and 17%C in comparison with control 1 (PGPR-free plants in clean soil). Moreover, in comparison with control 2 (PGPR-free plants in Ni-supplemented soil), the sensitive and resistant strains promoted the plant biomass on 53 and 64%M, respectively, and at increasing of nickel concentration in two or four times—up to 20%M (Table 4) [56].

The excessive accumulation of nickel in plant shoots lead to leaves discoloration (chlorosis) and growth suppression. At the nickel concentration of 470 mg/kg soil we observed too fast development of chlorosis in variants without inoculation and with sensitive strain (from 20 to 80% of discolored plants). The resistant BS1393(pBS501) strain protected barley leaves from colorless (7–50%) in first–fourth weeks, respectively, and at the increasing in metal concentration two times–only in the first week (30% of discolored plants). Furthermore, shoots of non-inoculated plants was dried in the presence of 470 and 940 mg Ni/kg soil by the end of the fourth and first week, respectively.

Table 3. The tolerance index (TI)* of barley growing on Ni-contaminated soil (%)

Treatment	Plant part	TI, %		
		235*	470**	940*
BS1393	shoot	100	76	61
	root	135	51	42
BS1393(pBS501)	shoot	113	81	63
	root	129	61	55
PGPR-free	shoot	76	73	53
	root	58	58	45

* TI=WM/WC×100%, where WM–plant weight on soil with Ni, WC–plant weight on Ni- and PGPR-free soil. ** mg Ni/kg soil.

Table 4. The plant weight variance (±ΔW, %)

Inoculation	±ΔW, %		
	235*	470	940
$\%_C = W_M/W_{C1}\times 100\%$			
BS1393	+10	−30	−44
BS1393(pBS501)	+17	−24	−39
PGPR-free	−28	−30	−49
$\%_M = W_M/W_{C2}\times 100\%$			
BS1393	+53	+1	+10
BS1393(pBS501)	+64	+10	+20

Notes. WM, WC1, and WC2–Weight of plants grown on Ni-supplemented soil, in control 1 (PGPR-free, clean soil), and control 2 (PGPR-free, Ni-supplemented soil), respectively.
*– mg Ni/kg.

As it was shown in pot experiments (greenhouse or growth chamber), free-living PGPR promoted or did no effect on the metal accumulation by plants. For example, *A. radiobacter* 10 and *A. mysorens* 7 enhanced the lead accumulation by barley plants [57] and *A. lipoferum* 137 enhanced the cadmium uptake in barley roots [58]. *Agrobacterium* sp., *Pseudomonas* sp., *Stenotrophomas* sp. increased cadmium, copper, lead, nickel, zinc uptake in maize, bacteria addition did no effect on metal uptake by lupin, pea and rye [59]. The inoculation of *T. caerulescens* with *Enterobacter cancerogenes*, *Microbacterium saperdae*, *Pseudomonas monteilii* two- and four-fold increased of zinc concentration in roots and shoots, respectively, while the inoculation of *Thlaspi arvense* did no effect on metal accumulation [60]. *Kluyvera ascorbata* SUD165 no increased the nickel, lead and zinc uptake in canola, tomato and Indian mustard [61].

We have shown that rhizosphere bacteria stimulate of nickel accumulation in barley plant; however degree of such influence in variants with resistant and sensitive strains differed essentially. The shoots of plants inoculated both strains accumulated equal amount of nickel (80 and 100 mg Ni/kg), but the resistant BS1393(pBS501) strain possess decreasing of nickel accumulation in roots two times (188 mg Ni/kg) to compare to sensitive BS1393 strain (330 mg Ni/kg) at the nickel concentration of 235 mg Ni/kg soil (Table 5).

As it was mentioned above the resistant strain BS1393(pBS501) formed the granules of bound cobalt on the cell surface, and their abundance in rhizoplane was two orders higher in comparison with the sensitive strain, therefore, we proposed that binding of nickel (similarly to cobalt) on the cell surface increased and it became less available for root system. The differences in amount of accumulated Ni between plants inoculated both tested strains were negligible when the metal concentration in soil increased two--four times.

The percentage of nickel removing by plants in variant with sensitive strain (0.47%) was more 2 times than in variant with resistant strain (0.25%) at the concentration 235 mg Ni/kg soil. Besides, shoots removed similar amount of nickel (0.18 and 0.16%) and roots–0.29 and 0.09% for BS1393- and BS1393(pBS501)- inoculated plants, respectively. At this nickel concentration the absorb ability of roots decreased, and therefore minimal percentage of nickel removing was characteristic for non-inoculated plants (0.06%) (Table 6).

Table 5. The amount of nickel in plants

Treatment of plants	Ni, mg/ kg of soil	Ni, mg/kg plant dry weight[a]		
		Shoots	Roots	Plant
BS1393	235	100	330	430
BS1393(pBS501)		80	108	188
Non-inoculated		30	46	76
BS1393	470	160	1490	1650
BS1393(pBS501)		160	1110	1270
Non-inoculated		130	940	1070
BS1393	940	220	3310	3530
BS1393(pBS501)		210	2080	2290
Non-inoculated		320	1900	2220

[a]Standard deviation was not more than 5% in all variants.

Table 6. Nickel removal by plants from soil (%)

Plant inoculation	Ni*	Shoots		*Roots*		Plant
		Weight, mg	Ni removing, %*	Weight mg	Ni removing, %	Ni removing %
BS1393	235	1680	0.18	840	0.29	0.47
BS1393(pBS501)		1920	0.16	800	0.09	0.25
Non-inoculated		1300	0.04	360	0.02	0.06
BS1393	470	1300	0.11	320	0.25	0.36
BS1393(pBS501)		1380	0.11	380	0.22	0.33
Non-inoculated		1240	0.08	360	0.18	0.26
BS1393	940	1040	0.06	320	0.22	0.28
BS1393(pBS501)		1080	0.06	340	0.18	0.24
Non-inoculated		900	0.07	280	0.17	0.24

Percentage of nickel removing was calculated as (Ni amount in plants / Ni amount in soil) × 100%; where Ni amount in plants = Ni concentration in plant x weight of 20 plants in one vessel, Ni amount in soil = Ni concentration in soil × weight of soil in one vessel.

*Ni concentration in soil, mg/kg.

It is known the rhizosphere bacteria promoted of Zn accumulation in shoots to 1500 mg/kg dry weight in hyperaccumulator plant *Thlaspi caerulensces* by means Zn-chelating metalophores [60].

The hyperaccumulator plants possess of low productivity of biomass and therefore they are not effective for phytoremediation. The barley is not hyperaccumulator plant, but it is a wide-spread cereal in Russia and possess of high productivity of biomass. Despite the fact that our experiment was carried out during a short vegetation period (4 weeks), at the concentration 940 mg Ni/kg soil the barley inoculated of the sensitive strain accumulated up to 3530 mg Ni/kg dry weight. The plants inoculated of the resistant strain and plants without bacteria accumulated equal amount of nickel (2290 and 2220 mg Ni/kg dry weight, accordingly). The nickel accumulation by BS1393(pBS501)- and BS1393- inoculated plants was 2.5 and 5.6 times more, respectively, in comparison with non-inoculated plants. Thus, nickel resistant *P. aureofaciens* BS1393(pBS501) strain showed higher survival in barley rhizosphere, had greater plant growth-promoting effect and protected plants from chlorosis and excessive nickel accumulation by plant shoots in comparison with the sensitive strain *P. aureofaciens* BS1393 in contaminated soil.

Application of PGPR *Pseudomonas* for Arsenic Accumulation

Soils often contain high concentrations of various natural and man-made compounds of arsenic (As). As is considered moderately phytotoxic, because like Se, Cd, Zn, Mn, and Cr ions [62]. Arsenic is a ubiquitous trace metalloid and is found in virtually all environmental media. Arsenites [As(III)] and arsenates [As(V)] play the most important role in interaction with soil biota. Arsenites are powerful inhibitors of sulfhydryl groups. They inactivate microbial enzymes and attack plant-cell membranes, thus suppressing root function on contact with roots or causing a rapid necrosis on contact with leaves. Arsenates do not damage membranes, because they do not react with sulfhydryl groups; however, arsenates affect phosphorylation in mitochondria [63]. Microorganisms resistant to As(III)/As(V) occur among members of various taxonomic groups. The mechanisms of As(III)/As(V) resistance, determined by plasmid as well as chromosomal genes, have been described [64–67]. Two mechanisms of bacterial resistance to arsenic are known. The first mechanism is associated with expression of the *ArsRBC* operon: arsenate reductase ArsC, reduces As(V) to As(III) in the cytoplasm, reduced arsenic is excreted from the cell via a special membrane protein (porin) ArsB, and ArsR is the transcription regulator. The second mechanism is associated with arsenite oxidase, which oxidizees As(III) to As(V), and then As(V) excreted from the cell via membrane protein [68, 69]. Some pseudomonades can solubilize phosphates [70]. The enzyme citrate synthase encoding *gltA* gene exists in nearly all living cells and stands as a pace-making enzyme in the first step of the Citric Acid Cycle. The citric acid dissolves the phosphates, increases the bioavailability of soil arsenates and stimulates the supply of arsenic to plants [71]. Some plants are able to absorb (consume) arsenic in biomass. Earlier we have indicated that sorghum (*Sorhum sacharatum*) and sunflower are regarded as the most perspective crops for cleanup of soil from As-containing compounds. [72].

To receive of PGPR *Pseudomonas* resistant to high concentrations of arsenite/arsenate and enable to transform bounded arsenic into a form available to plants, the genes *arsRBC* and *gltA* from *P. aeruginosa* PA01 were cloned in the vector pUCP22 yielding the plasmids pUCP22:*arsRBC* (later named as pKS1) and pUCP22:*gltA*, which were used to transfer in *P. aureofaciens* BS1393 [73].

We assumed that inoculation of plant seeds with the recombinant strains would enhance the arsenic accumulation by plant via increasing the solubility of soil's arsenates. For this purpose sorghum plants inoculated by recombinant strains were grown in pot trials with arsenite-supplemented (100 mg/kg) gray forest soil. It was shown that inoculation by recombinant strains supplied higher seed germination and plant growth in comparison with control without inoculation. Evidently, the resistant strain *P. aureofaciens* BS1393 (pUCP22:*arsRBC*) had a selective advantage in rizosphere in 35 days unlike the sensitive strain BS1393.

The sorghum seeds inoculated with the sensitive BS1393 strain did not germinate. The amount of arsenic in plants inoculated with the resistant BS1393(pUCP22::*ars*RBC) strain was 20% higher than in control. The strain increased the content of the available arsenic in soil (due to the activity of arsenate reductase).

Under the conditions of phosphorus deficiency, however, the plants grew more slowly and, most likely, their accumulation of arsenic was weak. The strain *P. aureofaciens* BS1393(pUCP22:*glt*A) supplied the decrease of arsenic content in soil about 30% and increase it in plant biomass about 40% in comparison with control. These data may be assigned the production of citric acid, which favors dissolving of bound soil phosphorus, thereby increasing biological availability of arsenic and stimulating its absorption by plants [73]. Control plants without bacteria accumulated arsenic too, but most of them became dead for 10 days (Table 7).

Table 7. Arsenic content in soil and sorghum plants

Variants	Arsenic, μg/g dry mass	
	soil	plant
Control (noninoculated plants)	35.78	81.27
P. aureofaciens BS1393 (pUCP22:gltA)	26.17	114.16
P. aureofaciens BS1393 (pUCP22:arsRBC)	33.06	104.57

APPLICATION OF PGPR *PSEUDOMONAS* FOR AT COMPLEX CONTAMINATION

Biosphere pollution of organic xenobiotics and heavy metals/metalloids in the last years becomes one of our actual environmental problems. The areas adjoining the enterprises oil-extracting and a petroleum-refining and chemical industry, sewage, storage site, agricultural lands treated with arsenic-containing pesticides, chemical weapon destruction polygons, etc., are exposed to the greatest danger of complex contamination. [2]. In cocontaminated sites, metal toxicity inhibits the activity of organic-degrading microorganisms [74–78]. Besides, the structure of microbe population changes in these soils that can lead to domination of the phytopathogenic fungi considerably reducing efficiency of phytoremediation. Approaches for cleaning of cocontaminated sewage are developed with use of associations of the microorganisms including bacteria, able to precipitate heavy metals from the environment and microorganisms-degraders. The introduction of two strains: the Cd-resistant *Pseudomonas* sp. H1, capable precipitate cadmium, and the Cd-sensitive *Ralstonia eutropha* JMP134 strain-degrader of 2,4-dichlorphenoxyacetic acid led to increasing of xenobiotic degradation in the reactor in the presence of cadmium (60 μg/l) [79].

The combination of genetic systems of biodegradation and resistance in bacterial cell may be one of the directions for the clean-up of cocontaminated soils. Use of natural plasmids of degradation and resistance for these purposes is environmentally friendly approach, unlike the use of genetically-modified construct. Moreover, the expression of such plasmids in PGPR can essentially accelerate the phytoremediation of cocontaminated soils. It is known that the plasmids pMOL30 and pMOL28 containing resistance operons *czc* (Co^r Zn^r Cd^r) and *cnr* (Co^r Ni^r) stably co-existed with the degradation plasmids of polychlorbiphenyl pSS50 (Bph+/Cbp+) and 2,4-dichlorphenoxyacetic acid pJP4 (Tfd+) in *Alcaligenes eutrophus*. Two-plasmid strains effectively degraded toxic compounds in the presence of high concentration of nickel and cadmium [80, 81]. Nevertheless the data concerning interaction of different genetic systems and their effect on physiology of bacteria are insufficient. Earlier we reported, for example, that the transfer of naphthalene biodegradation plasmid pBS216 in PGPR *P. putida* BS1380 strain led to increase in synthesis of the phytohormone indolil-3-acetic acid [82]; and the presence of the resistance plasmid pBS501 in *P. aureofaciens* BS1393 strain effect indirectly on production of phenazine antibiotics in culture medium with nickel or cobalt [54]. The possibility of using microbial–plant associations is considered as one of phytoremediation strategies aimed at fighting against complex pollution of soil. We propose to use associations of plants with PGPR *Pseudomonas*, combining several properties in these strains, such as the resistance to metals/metalloids and the ability to degrade PAHs.

Nickel Resistant PGPR Pseudomonas Degrading Naphthalene

In the lab experiments modeling complex pollution (naphthalene and heavy metals) is shown that resistance level of two-plasmid strains for cobalt and nickel was four-eight and two-four times higher in comparison with sensitive degraders. Nickel in concentration of 100 μM has no effect on key enzymes activity of naphthalene biodegradation pathway in the P. chlororaphis PCL1391(pBS216,pBS501) strain. Toxic effect of nickel on sensitive PCL1391(pBS216) strain was accompanied by decrease in respiratory activity and endogenous NADH consumption, and, as consequence lack of enzymes activity in the end of exponential phase, accumulation of more toxic intermediates in the medium, falling of bacteria viability and efficiency of naphthalene biodegradation. The cnr-like operon in the resistant strain PCL1391(pBS216,pBS501), apparently, provided an active efflux of nickel from cells, therefore metal has no inhibiting effect. It was shown that efficiency of naphthalene degradation of resistant strain was up to 100% for 21 hours, whereas one of sensitive strain only 12% for 36 hours [83]. Furthermore P. chlororaphis PCL1391(pBS216,pBS501) dominated in comparison with the sensitive PCL1391(pBS216) strain in rhizosphere of sorghum in 3 weeks of cultivation in pot trials on soil polluted by naphthalene (1 g/kg) and nickel (400 mg/kg). The abundances of the two-plasmid and sensitive strains were 2.4×104 and 4.8×103 CFU/cm root, respectively. The resistant strain promoted the weight of shoots and roots 1.5 and 2 times; while the sensitive strain, contrary, almost did not stimulate plant growth.

Arsenic Resistant PGPR Pseudomonas Degrading Naphthalene

To obtain two-plasmid strains able to degrade naphthalene in the presence of arsenic, the previously obtained strains *P. chlororaphis* PCL1391(pBS216) and *P. aureofaciens* BS1393(pBS216) were transformed with the arsenic resistance plasmid pKS1 [84]. The transformants BS1393(pBS216,pKS1) and PCL1391(pBS216,pKS1) were able to grow in the presence of arsenite (300 mg/l) in a mineral medium containing naphthalene as the only source of carbon. It has been shown that strain *P. chlororaphis* PCL1391(pBS216,pKS1) was more resistant than strain *P. aureofaciens* BS1393(pBS216,pKS1). This difference is determined by the higher resistance of the initial strain PCL1391 to arsenic compared to strain BS1393.

The effect of two-plasmid strains on the naphthalene mineralization was studied in the sterile model system with rape (*Brassica napus* ssp. *Oleifera* L.), grown in the presence of naphthalene (200 μg g^{-1} sand) and sodium arsenite (15 μg g^{-1} sand) [84]. The residual naphthalene content was determined on 7 day. The analysis of the naphthalene content showed that up to 50% of added naphthalene was volatilized naturally during the experiment (Table 8).

Plants are capable to consume and metabolize a number of organic compounds including pesticides and herbicides as well as aliphatic, monocyclic, and polycyclic hydrocarbons. The ability of plants to degrade compounds containing aromatic rings was demonstrated in experiments with plants grown under field and sterile conditions [2].

However, according to our research the contribution of rape plants to naphthalene mineralization was insignificant. When rape plants were inoculated with the plasmid-free strains, the residual naphthalene content after the end of the experiment was also comparable to that in the control.

The inoculation of plants with the strains carrying the pBS216 plasmid significantly reduced the residual naphthalene content in sand. When the experiment was finished, the content of naphthalene in these samples accounted for 8–10% of the control.

In the experiments with sand containing both naphthalene and arsenic, the residual content of naphthalene in the variants with the arsenic-sensitive strains was higher than in the experiments with sand containing only naphthalene, which was due to the toxic effect of arsenic on these bacteria.

Table 8. Residual naphthalene content after cultivating of PGPR-inoculated rape

Variants[a]	Residual naphthalene content (μg g^{-1} sand[b])	
	Naph contamination	(Naph + As) contamination
Zero point (on 1 day)	196.4	
Final point (on 7 day)	91.0	91.15
PGPR-free (on 7 day)	95.66	97.0
BS1393	97.15	119.0
BS1393(pBS216)	9.85	27.9
BS1393(pBS216,pKS1)	6.78	2.77
PCL1391	103.3	106.2
PCL1391(pBS216)	8.56	14.1
PCL1391(pBS216,pKS1)	7.13	3.21

[a]Seedlings, sterile or inoculated with plasmid-free or plasmid-bearing PGPR *Pseudomonas* strains, were grown for 7 days in model systems with naphthalene (200 μg g^{-1} sand) or naphthalene (200 μg g^{-1} sand) and sodium arsenite (15 μg g^{-1} sand). The sand was extracted with methanol, and HPLC was used to analyze samples of the methanol fractions.

[b]Standard deviation was not more than 15% in all variants.

However, when plants were inoculated with the arsenic-resistant multifunctional strains, the residual content of naphthalene in sand accounted for about 3% of the naphthalene content in control, which testifies to the efficiency of the function of these strains under the complex pollution conditions. The effect of multifunctional strains on sorghum development was estimated in the pot experiments with the grey forest soil contaminated with naphthalene (1 g/kg), phenanthrene (0.2 g/kg) and sodium arsenite (50 mg/kg). It was shown that the arsenic resistant PAH-degrading strains prevailed in sorghum rhizosphere and improved the plants growth. Throughout the experiment, the noninoculated plants and plants inoculated with arsenic sensitive strains lagged behind the control and plants treated with the multifunctional strains (Figure 5).

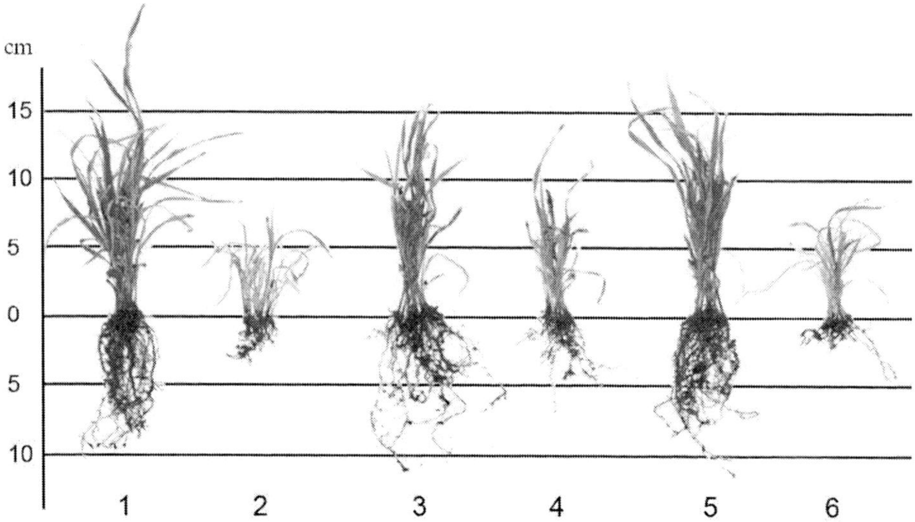

Figure 5. Effect of various *Pseudomonas* strains on sorghum development in a soil containing naphthalene (1 g/kg), phenanthrene (0.2 g/kg) and sodium arsenite (50 mg/kg): (1) bacteria-free plants in clean soil; (2) bacteria-free plants in contaminated soil. Plants inoculated with (3) *P. chlororaphis* PCL1391(pBS216,pKS1); (4) *P. chlororaphis* PCL1391(pBS216); (5) *P. aureofaciens* BS1393(pBS216,pKS1); (6) *P. aureofaciens* BS1393(pBS216) in contaminated soil.

The Plasmid Stability in Multifunctional PGPR Strains

For constructing of variants multifunctional strains of PGPR: *P. chlororaphis* PCL1391, *P. aureofaciens* BS1393, *P. aureofaciens* OV17, *P. fluorescens* 38a have been used mentioned above catabolic plasmids: pBS216, pOV17 (wild-type) and resistance plasmids: pKS1 (construct pUCP22::arsRBC), pBS501 (wild-type).

The stable coexistence of catabolic and resistance plasmids is very important when the multifunctional strains are introduced in rhizosphere. Determination of plasmid stability in a lab non-selective environment revealed that natural plasmids are more stable than genetically-constructed pKS1, and beside that the plasmid stability depended from host strain. For example, the stability both plasmids pBS216 and pBS501 in P. chlororaphis PCL1391 was about 100% whereas in other recipients less than 50% after 10 passages [83].

Approximately 50% of the cell population retained the plasmid in strain BS1393(pBS216) after seven passages. The presence of the second plasmid pKS1 in strain BS1393(pBS216,pKS1) decreased the stability of plasmid pBS216 in this strain. The stability of plasmid pKS1 in strains BS1393(pBS216,pKS1) and PCL1391(pBS216,pKS1) accounted for 5 and 25%, respectively [83].

It is interesting to note that in the rhizosphere of plants the stability of plasmids pBS216 and pKS1 (90 and 70%, respectively) was considerably higher in 7 days of cultivation, than in batch culture. It is likely that these differences of stability can be explained with selective pressure (arsenic and naphthalene).

Conclusion

Due to wide spreading anthropogenous contamination of the environment, the necessity of development of effective biologically safe techniques for cleanup and restoration of soils increases. The use of plants and their associated microbiota to remediate environmental contamination is a cost-effective technique. PGPR *Pseudomonas* can facilitate the development of plants via various mechanisms; however, rhizobacteria are often sensitive to various pollutants and their application in phytoremediation may be restricted. The obtained data shows an example of new strategy of development of beneficial PGPR *Pseudomonas* with required properties with use of natural plasmids, bacteria and ways of carrying over of the genetic information. In addition to the ability to actively colonize rhizosphere, suppress soil-borne pathogens, promote plant growth in contaminated environment, these bacteria supplied more effective recovery of metal/metalloid or PAHs biodegradation by plant–microbe associations. Moreover, multifunctional strains were able to degrade PAH at complex contamination. We expect that high colonizing ability of PGPR *Pseudomonas* and stability of introduced plasmids will be the important factors in the use of plant–microbe associations in field experiments. Despite the bacteria being tested in pot trials for a short vegetation period, the proposed approach may be used in the long term in field experiments for cleanup and restoration of contaminated soils.

Acknowledgments

Our investigations were partially supported by the Development of Scientific Potential of Higher School departmental scientific program (project 2.1.1/4341) and by Federal Targeted Program, State contract no. 02.512.11.2337 and 02.740.11.0682.

References

[1] Salt DE, Smith RD, Raskin I. Phytoremediation. *Annu Rev. Plant. Physiol. Plant Mol. Biol.* 1998 Jun;49:643–668.

[2] Pilon-Smits E. Phytoremediation. *Annu Rev. Plant Biol.* 2005 Jun 11;56:15–39.

[3] Lynch JM, Whipps JM. Substrate flow in the rhizosphere. *Plant and Soil.* 1990 Dec;129(1):1–10.

[4] Anderson TA, Coats, JR. An overview of microbial degradation in the rhizosphere and its implications for bioremediation. Skipper HD, Turco RF, editors. Bioremediation: Science and Application. *Madison: Soil Sci. Soc. Am. Spec. Pub.* 43. 1995. p. 135–143.

[5] Bloemberg GV, Lugtenberg BJ. Molecular basis of plant growth promotion and biocontrol by rhizobacteria. *Curr. Opin. Plant Biol.* 2001;4(4):343–350.

[6] Compant S, Duffy B, Nowak J, Clement C, Barka EA. Use of plant growth-promoting bacteria for biocontrol of plant diseases: principles, mechanisms of action, and future prospects. *Appl. and Environ Microbiol.* 2005 Apr;71(4):4951–4959.

[7] Haas D, Defago G. Biological control of soil-born pathogens by fluorescent Pseudomonads. *Nature Reviews Microbiology*. 2005;3(4):307–319.

[8] Glick BR. Phytoremediation: synergistic use of plants and bacteria to clean up the environment. *Biotechnol. Adv.* 2003 Aug;21(5):383–393.

[9] Lucy M, Reed E, Glick BR. *Applications of free living plant growth-promoting rhizobacteria*. Antonie van Leeuwenhoek. 2004 Jul;86(1):1–25.

[10] Edwards NT. Polycyclic aromatic hydrocarbons (PAH's) in the terrestrial environment–a review. *J. Environ Qual.* 1983;12:427–441.

[11] Harvey RG. *Polycyclic aromatic hydrocarbons: chemistry and carcinogenicity*. New York, Port Chester, Melbourne, Sydney: Cambridge University Press; 1991.

[12] Flowers-Geary L, Bleczinki W, Harvey RG, Penning TM. Cytotoxity and mutagenicity of policiclic aromatic hydrocarbon *orto*-quinones produced by dihydrodiol dehydrogenase. *Chem. Biol. Interact.* 1996;99(1–3):55–72.

[13] Cerniglia CE. Biodegradation of polycyclic aromatic hydrocarbons. *Curr. Opin Biotechnol.* 1993;4:331-338.

[14] Sutherland JB, Rafii F, Khan AA, Cerniglia CE. Mechanisms of polycyclic aromatic hydrocarbon degradation. In: Young LY, Cerniglia CE, editors. *Microbial transformation and degradation of toxic organic chemicals*. New York: Wiley; 1995. p. 269–305.

[15] Mrozik A, Piotrowska-Seget Z, Labużek S. Bacterial degradation and bioremediation of polycyclic aromatic hydrocarbons. *Polish Journal of Environmental Studies*. 2003;12(1):15–25.

[16] Flether JS, Hedge RS. Release of phenols by perrennial plant roots and they potential importance in bioremediation. *Chemosphere.* 1995;31:3009–3016.

[17] Siciliano SD, Germida JJ, Banks K, Greer CW. Changes in microbial community composition and function during a polyaromatic hydrocarbon phytoremediation field trial. *Appl. and Environ. Microbiol.* 2003 Jun;69(1):483–489.

[18] Reilley KA, Banks MK Schwab AP. Dissipation of polycyclic hydrocarbons in the rhizosphere. *J. Environ. Qual.* 1996;25:212-219.

[19] Wetzel SC, Banks MK, Schwab AP. Rhizosphere effects on the degradation of pyrene and anthracene in soil. In: Kruger EL, Anderson TA, Coast JR, editors. *Current Protocols in molecular Biology*. Washington: American Chemical Society; 1996. p. 241–245.

[20] Liste HH, Alexander M. Plant-promoted pyrene degradation in soil. *Chemosphere.* 2000;40(1):7–10.

[21] Kiyohara H, Nagao K. The catabolism of phenanthrene and naphthalene by bacteria. *J. Gen. Microbiol.* 1978;105:69–75.

[22] Yen KM, Serdar CM. Genetics of naphthalene catabolism in pseudomonads. *CRC Crit. Rev. Microbiol.* 1988;15:247–267.

[23] Kochetkov VV, Boronin AM. Comparative study of plasmids controlling naphthalene degradation by *Pseudomonas. Mikrobiologiya.* 1984 Jul-Aug;53(4):639–644. [in Russian]

[24] Habe H, Omori T. Genetics of polycyclic aromatic hydrocarbon metabolism in diverse aerobic bacteria. *Biosci. Biotechnol. Biochem.* 2003;67(2):225–243.

[25] Kochetkov VV, Balakshina VV, Mordukhova EA, Boronin AM. Plasmids of naphthalene degradation in rhizosphere *Pseudomonas. Mikrobiologiia.* 1997;66:211–216. [in Russian]

[26] Anokhina TO, Volkova OV, Puntus IF, Filonov AE, Kochetkov VV, Boronin AM. Plant growth-promoting *Pseudomonas* bearing catabolic plasmids: naphthalene degradation and effect on plants. *Process Biochemistry.* 2006 Sept;41(5): 2417–2423.

[27] Boronin AM, Kochetkov VV. (). *Effective biofungicide for plants protection. AGRO* XXI. 2003;1–6:64–66. [in Russian]

[28] Chin-A-Woeng TFC, Bloenberg GV, van der Bij AJ, van der Drift KMGM, Schripsema J, Kroon B, Scheffer RJ, et al. Biocontrol by phenazine-1-carboxamide-producing *Pseudomonas chlororaphis* PCL1391 of tomato root rot caused by *Fusarium oxysporum* f. sp. *radicis-lycopersici. Mol. Plant-Microbe Interact.* 1998;11:1069–1077.

[29] Simons M, van der Biji AJ, Brand I, de Wager LE, Wijffelman CA, Lutenberg BJJ. Gnotobiotic system for studying rhizosphere colonization by plant growth promoting *Pseudomonas* bacteria. *Mol. Plant-Microbe Interact.* 1996;9(7): 600-607.

[30] Volkova OV, Anokhina TO, Puntus IF, Kochetkov VV, Filonov AE, Boronin, AM. Effects of naphthalene degradative plasmids on the physiological characteristics of rhizosphere bacteria of the genus *Pseudomonas. Appl. Biochem. and Microbiol.* 2005;41(5):460–464.

[31] Park W, Jeon CO, Cadillo H, DeRito C, Madsen EL. Survival of naphthalene-degrading *Pseudomonas putida* NCIB 9816-4 in naphthalene-amended soils: toxicity of naphthalene and its metabolites. *Appl. Microb. and Biotech.* 2004 Apr;64(3):429–435.

[32] Boronin AM, Filonov AE, Gayazov RR, Kulakova AN, Mshensky YN. Growth and plasmid-encoded naphthalene catabolism of *Pseudomonas putida* in batch culture. *FEMS Microbiology Letters.* 1993;113:303–308.

[33] Anokhina TO, Kochetkov VV, Zelenkova NF, Balakshina VV, Boronin AM. Biodegradation of phenanthrene by *Pseudomonas* bacteria bearing rhizospheric plasmids in model plant-microbial associations. *Appl. Biochem. and Microbiol.* 2004;40(6): 568–572.

[34] Olson PE, Castro A, Joern M, DuTeau NM, Pilon-Smits EA, Reardon KF. Comparison of plant families in a greenhouse phytoremediation study on an aged polycyclic aromatic hydrocarbon-contaminated soil. *J. Environ Qual.* 2007;36(5):1461–1469.

[35] Muratova A, Hūbner T, Tischer S, Turkovskaya O, Möder M, Kuschk P. Plant-rhizosphere-microflora association during phytoremediation of PAH-contaminated soil. *Int. J. Phytoremediation.* 2003;5(2):137–151.

[36] Kuiper I, Bloemberg GV, Lugtenberg BJ. Selection of a plant-bacterium pair as a novel tool for rhizostimulation of polycyclic aromatic hydrocarbon-degrading bacteria. *Mol. Plant Microbe Interact.* 2001;14(10):1197–1205.

[37] Gao Y, Yu XZ, Wu SC, Cheung KC, Tam NF, Qian PY, et al. Interaction of rice (*Oryza sativa* L.) and PAH-degrading bacteria (*Acinetobacter* sp.) on enhanced dissipation of spiked phenanthrene and pyrene in waterlogged soil. *Sci. Total Environ.* 2006;372(1):1–11.

[38] Aprill W, Sims RC. (). Evaluation of the use of prairie grasses for stimulating polycyclic aromatic hydrocarbon treatment in soil. *Chemosphere.* 1990;20:253–265.

[39] Fang C, Radosevich M, Fuhrmann JJ. Atrazine and phenanthrene degradation in grass rhizosphere soil. *Soil Biology and Biochemistry.* 2001;33:671–678.

[40] Xu SY, Chen YX, Wu WX, Wang KX, Lin Q, Liang XQ. Enhanced dissipation of phenanthrene and pyrene in spiked soils by combined plants cultivation. *Sci. Total. Environ.* 2006;363(1–3):206–215.

[41] Nazarov AV, Illarionov SA. Potential of use plant-microbe interaction for bioremediation. *Biotechnology.* 2005;5:54–62. [in Russian]

[42] Smit E, Leeflang P, Wernars K. Detection of shifts in microbial community structure and diversity in soil caused by copper contamination using amplified ribosomal DNA restriction analysis. *FEMS Microbiol. Ecol.* 1997;23:249–261.

[43] Bruins MR, Kapil S, Oehme FW. Microbial resistance to metals in invironment. *Ecotoxicology and Environmental safety.* 2000;45:198–207.

[44] Saier MHJ, Tam R, Reizer A, Reizer J. Two novel families of bacterial membrane proteins concerned with nodulation, cell division and transport. *Mol. Microbiol.* 1994;11:841–847.

[45] Johnson JM, Church GM. Alignment and structure prediction of divergent protein families: periplasmic and outer membrane proteins of bacterial efflux pumps. *J. Mol. Biol.* 1999; 287:695–715.

[46] Saier MHJ. A functional-phylogenetic classification system for transmembrane solute transporters. *Microbiol. Mol. Biol. Rev.* 2000;46:354–411.

[47] Herrman L, Schwan D, Garner R, Mobley HLT, Haas R, Schafer KP, et al. Helicobacter pylori cadA encodes an essential Cd(II)-Zn(II)-Co(II) resistance factor influencing urease activity. *Mol. Microbiol* 1999;33:524–536.

[48] Nies DH. Efflux-mediated heavy metal resistance in prokaryotes. *FEMS Microbiol. Rev.* 2003;27(3):314–341.

[49] Kamashwaran SR, Crawford DL. Mechanisms of cadmium resistance in anaerobic bacterial enrichments degrading pentachlorophenol. *Can. J. Microbiol.* 2003;49(7):418–424.

[50] Gilis A, Corbisier P, Baeyens W, Taghavi S, Mergeay M, van der Lelie DJ. Effect of the siderophore alcaligin E on the bioavailability of Cd to *Alcaligenes eutrophus* CH34. *Ind. Microbiol. Biotechnol.* 1998;20(1):61–68.

[51] Silver S. Plasmid-determined metal resistance mechanisms to heavy metals. *Plasmid.* 1992; 27:1–3.

[52] Rouch DA, Lee BTD, Morby AP. Understanding cellular responses to toxic agents: A model for mechanism choice in bacterial metal resistance. *J. Ind. Microbiol.* 1995;14:132–141.

[53] Siunova TV, Siunov AV, Kochetkov VV, Boronin AM. The *cnr*-like operon in strain *Comamonas* sp. encoding resistance to cobalt and nickel. *Genetika*, 2009;45(3):336–341. [in Russian].
[54] Siunova TV, Kochetkov VV, Validov ShZ, Suzina NE, Boronin AM. The production of phenazine antibiotics by the *Pseudomonas aureofaciens* strain with plasmid-controlled resistance to cobalt and nickel. *Mikrobiologiia*. 2002;71(6):778–785.
[55] Lodewyckx C, Tadhavy S, Mergeay M, Vangronsveld J, Clijsters H, van der Lelie D. The effect of recombinant heavy metal resistant endophytic bacteria in heavy metal uptake by their host plant. *Int. J. Phytoremediat*. 2001;3:173–187.
[56] Siunova TV, Kochetkov VV, Boronin AM. Effect of rhizosphere bacteria on nickel accumulation by barley plants. *Agrokhimiya*. 2006;(10):80-84. [in Russian]
[57] Belimov AA, Kunakova AM, Kozhemiakov AP, Stepanok VV, Yudkin LY. Effect of associative bacteria on barley grown in heavy metal contaminated soil. In: *Proceedings of the International Symposium on Agro-environmental issues and future strategies: Towards the 21st Century*; 1998. Faisalakad, Pakistan, May 25–30.
[58] Belimov AA, Dietz K. Effect of associative bacteria on element composition of barley seedlings grown in solution culture at toxic cadmium concentrations. *Microbiol. Res.* 2000;155:113–121.
[59] Hoflich G, Metz R. Interactions of plant-microorganism associations in heavy metal containing soils from sewage farms. *Bodenkultur*. 1997;48:239–247.
[60] Whiting SN, De Souza MP, Terry N. Rhizosphere bacteria mobilize Zn for hyperaccumulation by *Thlaspi caerulescens*. *Environ. Sci. Technol.* 2001;35: 3144–3150.
[61] Burd GI, Dixon DG, Glick BR. Plant growth-promoting bacteria that decrease heavy metal toxicity in plants. *Can. J. Microbiol.* 2000;46:237–245.
[62] Alexeyev YuV. *Heavy Metals in Soil and in Plants*. Agropromizdat Publ., Leningrad, 1987. [in Russian]
[63] Gamayurova VS. *Arsenic in Ecology and in Biology*. Nauka Publ., Moscow, 1993. [in Russian]
[64] Kaur P, Rosen BP. Plasmid-encoded resistance to arsenic and antimony. *Plasmid.* 1992;27:29–40.
[65] Carlin A, Shi W, Dey S, Rosen BP. The *ars* operon of *E. coli* confers arsenical and antimonial resistance. *J. Bacteriol.* 1995;177(4):981–986.
[66] Rosenstein P, Peschel A, Wieland B, Gotz F. Expression and regulation of the antimone, arsenite and arsenate resistance operon of *Staphylococcus xylosus* plasmid pSX267. *J. Bacteriol.* 1992;174(11):3676–3683.
[67] Cai J, Salmon K, DuBow MS. A chromosomal *ars* operon homologue of *Pseudomonas aeruginosa* confers increased resistance to arsenic and antimony in *Escherichia coli. Microbiology*. 1998;144(10).2705–2713.
[68] Anderson GR, Williams J, Hille R. Purification and Characterization of Arsenite Oxidase; a Molybdenum Hydroxylase. *J. Biol. Chem.* 1992;267(33):23674–3682.
[69] Santini J, Sly L, Schnagl R, and Macy J. A new chemolithautotrophic arsenite-oxidizing bacterium isolated from a gold mine: phylogenetic, physiological and preliminary biochemical studies. *Appl. Environ. Microbiol.* 2000;66(1):92–97.
[70] Kumar V, Singh KP. Enriching vermicompost by nitrogen fixing phosphate solubilizing bacteria. *Bioresour Technol.* 2001;76(2):173–175.

[71] Peryea FJ. Phosphate starter fertilizer temporarily enhances soil arsenic uptake by apple trees grown under field conditions. *Hortscience*. 1989;33(3):826–829.

[72] Sizova OI, Kochetkov VV, Validov ShZ, Boronin AM, Kosterin PV, Lyubun EV. Arsenic-contaminated soils: genetically modified *Pseudomonas* spp. and their arsenic-phytoremediation potential. *J. Soil. Sediments*. 2002 Mart;2(1):19–23.

[73] Sizova OI, Kochetkov VV, Boronin AM. The arsenic-phytoremediation potential of genetically modified *Pseudomonas* spp. In: Morel JL et al. editors. Phytoremediation of metal-contaminated soils.Netherlands: Springer; 2006. p. 327–34.

[74] Pickering IJ, Prince RC, George MJ, Smith RD, George GN, Salt DE. Reduction and coordination of arsenic in Indian Mustard. *Plant. Physiol.* 2000;122(4):1171–1177.

[75] Mukhopadhyay R, Rosen BP, Phung Le T, Silver S. Microbial arsenic: from geocycles to genes and enzymes. *FEMS Microbiol. Rev.* 2002;26:311–325.

[76] Said WA, DA. Lewis. Quantitative assessment of the effects of metals on microbial degradation of organic chemicals. *Appl. Environ. Microbiol.* 1991 May;57:1498–1503.

[77] Sokhn J, De Leij FA, Hart TD, Lynch JM. Effect of copper on the degradation of phenanthrene by soil microorganisms. *Lett. Appl. Microbiol.* 2001 Aug;33(2):164–168.

[78] Sandrin TR, Maier M. Impact of metals on biodegradation of organic pollutants. *Environmental Health Perspectives*. 2003;111(8):1093–1101.

[79] Roane TM, Josephson KL, Pepper IL. Dual-bioaugmentation strategy to enhance remediation of co-contaminated soil. *Appl. Environ. Microbiol.* 2001 Jul;67(7):3208–3215.

[80] Springael D, Diels L, Hooyberghs L, Kreps S, Mergeay M. Construction and characterization of heavy metal-resistant haloaromatic-degrading Alcaligenes eutrophus strains. *Appl. and Environ. Microbiol.* 1993 Jan; 59(1): 334–339.

[81] Collard JM, Corbisier P, Diels L, Dong Q, Jeanthon C, Mergeay M, et al. Plasmids for heavy metal resistance in Alcaligenes eutrophus CH34: mechanisms and applications. *FEMS Microbiol Rev.* 1994 Aug;14(4):405-14. Review.

[82] Mordukhova EA, Sokolov SL, Kochetkov VV, Kosheleva IA, Zelenkova ZF, Boronin AM. Involvement of naphthalene dioxygenase in indole-3-acetic acid biosynthesis by *Pseudomonas putida*. *FEMS Microbiol. Lett.* 2000 Sep 15;190(2):279-85.

[83] Siunova TV, Anokhina TO, Mashukova AV, Kochetkov VV, Boronin AM. Rhizosphere strain of *Pseudomonas chlororaphis* capable of degrading naphthalene in the presence of cobalt/nickel. *Mikrobiologiia*. 2007 Apr;76(2).212–218. [in Russian].

[84] Sizova OI, Kochetkov VV, Boronin AM. Rhizosphere bacteria *Pseudomonas aureofaciens* and *Pseudomonas chlororaphis* oxidizing naphthalene in the presence of arsenic. *Applied Biochemistry and Microbiology*. 2010 Jan;46(1):38–43. [in Russian].

In: Handbook of Phytoremediation
Editor: Ivan A. Golubev

ISBN: 978-1-61728-753-4

Chapter 18

ARSENIC IN THE ENVIRONMENT: PHYTOREMEDIATION USING AQUATIC MACROPHYTES

M. Azizur Rahman[1*], M. Mahfuzur Rahman[2], Ismail M. M. Rahman[1,3] and Hiroshi Hasegawa[1]

[1]Graduate School of Natural Science and Technology, Kanazawa University
Kakuma, Kanazawa 920-1192, Japan
[2]Department of Botany, Jahangirnagar University
Savar, Dhaka 1342, Bangladesh
[3]Department of Chemistry, Faculty of Science, University of Chittagong
Chittagong 4331, Bangladesh

ABSTRACT

A large number of sites worldwide are contaminated by arsenic (As) as a result of human activities as well as from natural sources. Arsenic is a vital environmental and health concern due to its known chronic and epidemic toxicity. The main arsenic exposures to humans are through water pathway and food contamination originates from natural processes. Many of the available remediation technologies lost economic favor and public acceptance because of some unavoidable limitations of those technologies. Therefore, phytoremediation, a plant-based green technology, becomes an emerging and alternative technology that aims to extract or inactivate As in the environment. However, two approaches have been proposed in literature for the phytoremediation of arsenic: continuous or natural phytoremediation, and chemically enhanced phytoremediation. The first one is based on the use of natural hyperaccumulator plants having the ability to accumulate very high concentration of As in their shoots with exceptionally higher tolerance to As toxicity. On the other hand, As uptake in high biomass crop plants is increased using some chelating ligands in chemically enhanced phytoremediation technology.

Freshwater and seawater around the world have been contaminated by As from various anthropogenic activities and natural sources over time. Therefore, remediation of

* E-mail: rahmanmazizur@gmail.com; Tel/Fax +81-76-234-4792

As-contaminated aquatic systems is important as it is for terrestrial system. Aquatic macrophytes could be used to remediate the aquatic system. The use of aquatic macrophytes or other floating plants in phytoremediation technology is commonly known as phytoextraction. This cleanup process involves biosorption and accumulation of As. Recently, aquatic macrophytes and some other small floating plants such as *Spirodela polyrhiza* L., *Lemna spp.*, *Azolla pinnata*, *Salvinia natans*, *Eichhornia crassipes* have been investigated for the remediation of As-contaminated aquatic systems. Compared to the As-phytoremediation in terrestrial system, less work has been done in aquatic systems. In this chapter, process and prospect of As phytoremediation by aquatic macrophytes is discussed.

INTRODUCTION

Arsenic, the name came from Latin *arsenicum* and Greek *arsenikon* meaning *yellow orpiment* (pigment), occurs between the metals and nonmetals in the periodic table. Arsenic is a member of the nitrogen family with both metallic and nonmetallic properties, and is ubiquitous in the environment (soil, water, air and all living matters) [1]. Because of the poisonous character of arsenic, it has been used as herbicide, insecticides, wood preservatives, cattle and sheep dips. The biological toxicity and redistribution of arsenic in the environment make it evoking public concern.

Natural activities such as volcanic action, erosion of rocks, and forest fires introduce arsenic into the environment. Anthropogenic sources include arsenic sources added to the soil plant system as insecticides, herbicides, pesticides, livestock dips and wood preservatives. Indiscriminate use of arsenical pesticides during the early to mid-1900s led to an extensive contamination of soils worldwide [2]. Mining and smelting processes contribute to arsenic contamination because arsenic is a natural component of lead, zinc, copper and gold ores.

The average concentration of arsenic in the earth's crust is 2–5 mg kg^{-1} [1] though in regions with abundant volcanic rocks or sulfidic ores, its concentration is elevated. Weathering of arsenic contaminated igneous and sedimentary rocks liberates arsenic in the form of inorganic compounds including arsenic trioxide, arsenate, and arsenite [1]. Weathering of rocks is considered to be the major natural source of arsenic, estimated to release about 45,000 metric-tons $year^{-1}$ [3]. Moreover, microorganisms have been shown to increase the rate of arsenic release from sulfidic ores by catalyzing the oxidation of sulfide to sulfate and ferrous to ferric iron. Precipitation from the atmosphere (about 63,600 metric-tons $year^{-1}$) and the application of agricultural products such as herbicides and desiccants (about 4,560 and 12,000 metric-tons $year^{-1}$) are also two major sources of arsenic influx to soil [1]. Arsenicals can, thus, cause surface soil contamination of 600 mg kg^{-1} or more [4]. Worldwide, the median soil concentration is 6.0 mg kg^{-1} with a typical range of 0.1 to 40 mg kg^{-1} [5]. In the United States, surface soils contain an average of 7.2 mg kg^{-1} (range <0.1 to 97 mg kg^{-1}) [6].

Arsenic concentrations in seawater are relatively homogeneous, while it varies widely in freshwater (rivers and lakes) [7, 8]. Arsenic contamination in ground water has turned into the gravest natural disaster with spatial extent encompassing Bangladesh and West Bengal, India, where arsenic concentration in both drinking and irrigation waters has exceeded the safe level (50 and 100 μg L^{-1} for drinking and irrigation water, respectively) [9, 10]. Use of arsenic-contaminated water in irrigation contaminates not only the agricultural soils but also increases the possibility of its entry into the food chains of natural ecosystem [11–15].

CHEMISTRY AND MINERALOGY OF ARSENIC

Arsenic is an elusive element, with a mysterious ability to change color, behavior, reactivity, and toxicity. The chemical variability of arsenic stems from its electronic structure and bonding properties, which give rise to a variety of forms in the solid, aqueous, and gas states. [16]. It is an element of group VA in the periodic classification with properties that allow it to form alloys with various metals and covalent bonds with carbon, hydrogen, oxygen, and sulfur [3]. As arsenic is seated beneath nitrogen and phosphorus in the periodic table, it has an excess of electrons and unfilled orbitals that stabilize formal oxidation states from +5 to –3 [16]. The oxidation states and electron orbitals are similar between arsenic and phosphate [1]. Although, the electronegativity of arsenic is greater than that of nitrogen and similar to that of phosphorus by most measures [17], arsenic has a greater oxidation potential than nitrogen and phosphorus, which increases its cationic character. Arsenic also bonds readily to a variety of ligands, which strongly influences its chemical behavior [18]. While arsenic can combine with many other elements to form covalent compounds, it most commonly bonds to oxygen and sulfur in nature.

Arsenic is very similar chemically to its predecessor, phosphorus. Like phosphorus, it forms colorless, odorless, crystalline oxides As_2O_3 and As_2O_5 which are hygroscopic and readily soluble in water to form acidic solutions. Arsenic (V) acid, like phosphorous acid, is a weak acid. Like phosphorus, arsenic forms an unstable, gaseous hydride: arsine (AsH_3). The similarity is so great that arsenic will partly substitute for phosphorus in biochemical reactions and is thus poisonous. In aqueous solutions, arsenic forms the oxo-anions arsenite ($H_3As^{3+}O_3$) and arsenate ($H_3As^{5+}O_4$). The aqueous arsenic species in most natural waters (pH 4–10) are the neutral species H_3AsO_3 for As^{3+}, and H_2AsO^{3-} and $HAsO_4^{2-}$ for As^{5+} [19].

The redox potential of arsenic oxo-anions is such that arsenite is expected to be the stable aqueous form under moderately reducing conditions, while arsenate is stable in oxic aqueous systems [20, 21]. Arsenic is rarely found as its native form because of its affinity to bond with other elements and species. The simple As^{3+} oxides, arsenolite and claudetite are polymorphs with similar thermodynamic stability, with claudetite thought to be slightly more stable at standard conditions [21]. These minerals form naturally as secondary weathering products of arsenic sulfides but are more commonly found as the oxidation products of the roasting of arsenic-bearing ore minerals or coal [18]. Arsenic trioxides are moderately soluble in water and are still used as an ingredient in insecticides.

Arsenate minerals comprise a large class with extensive substitution and solid solution. Mineralogically, arsenates are usually considered a subclass of the phosphate mineral group because of the similarity in size and charge of the phosphate and arsenate anionic unit [18]. Similar to phosphate minerals found in soils and surface environments, arsenate minerals occur in a variety of soil and oxidized environments with a range of waters and hydration, depending on the degree of desiccation. They are also commonly found as weathering products of arsenic-bearing ore deposits, where sulfidic ore minerals are often coated with surface layers of oxidized and hydrated arsenate minerals [22]. The widespread and extensive use of arsenate and arsenite compounds in agriculture and industry has added a significant load of arsenic that may participate in biogeochemical activity to surface environments.

Arsenopyrite, orpiment, and realgar are the most common arsenic sulfide minerals, occurring primarily in hydrothermal and magmatic ore deposits. Other transition metals, such as Co, Ni, and Cu, also combine with arsenic and sulfur to form a variety of minor sulfides and sulfosalts, often with extensive solid solution [16]. Compared to the structure and bonding properties found in oxide minerals, arsenic sulfides are quite different from the oxides and even differ substantially within the group of sulfides [18].

Similar to nitrogen and phosphorus, arsenic has rich organic chemistry. Its ability to bind to a variety of organic ligands with different coordination geometries and its greater redox potential than phosphorus help to explain why mechanisms of arsenic toxicity in living organisms are still incompletely understood despite much investigation [23]. A large number of organoarsenic compounds are found in the environment as a consequence of a variety of biomethylation and other biosynthetic pathways [19]. One of the most common classes of natural organoarsenic compounds are methylated forms of As^{3+} and As^{5+}, such as the mono– and dimethyl oxoacids in their simplest forms, which can be generated by replacing a hydroxyl (–OH) ligand by a methyl ($-CH_3$) group in inorganic arsenate and arsenite structures [18]. Many organoarsenic compounds have arsenic substitution for nitrogen in an organic moiety, for example arsenobetaine and arsenocholine, which are major forms of arsenic found in marine animals. With improvements in analytical detection methods, a large number of previously unidentified organoarsenic compounds have been found throughout the biosphere in microorganisms, plants, invertebrates, and vertebrates.

Arsenic Compounds in the Environment

Rocks and Minerals

Arsenic is the 20th most abundant elements in earth's crust, with an average concentration of 2–3 mg kg^{-1} [19, 24] and is primarily associated with igneous sedimentary rocks in the form of inorganic arsenic compounds. It has been estimated that of the total arsenic contained in the various natural reservoirs (rocks, oceans, soils, biota, atmosphere, etc.), more than 99% is associated with rocks and minerals [25]. Arsenopyrite (FeAsS) in the most abundant arsenic-containing mineral and the other important minerals are arsenolite (As_2O_3), olivenite ($CuOHAsO_4$), cobaltite (CoAsS), and proustite (Ag_3AsS_3) [26]. Weathering of arsenic-containing rocks is considered as the major natural source of arsenic (45,000 metric-tons $year^{-1}$), which liberates arsenic in the form of inorganic compounds including arsenic trioxide, arsenite, and arsenate [3].

Seawaters and Freshwaters

Although open-ocean systems are relatively constant, the possibility of anthropogenic additions of arsenic can result in higher arsenic concentrations in coastal areas and fresh water [7]. The concentration of arsenic is between 1.4 and 1.8 μg L^{-1} in deep Pacific and Atlantic oceans [27]. In contrast, arsenic concentrations in freshwaters (rivers and lakes) vary widely with the geological composition of the drainage area and the extent of anthropogenic input. The geometric mean of arsenic concentration is about 1.4 μg L^{-1} in European and some North and South American rivers with a range of 0.5–75 μg L^{-1} [19]. It is extremely difficult to suggest typical arsenic levels in freshwater systems due to very high variations, but most values are in the μg L^{-1} range [19].

Arsenate is the major form of arsenic in most seawater samples, and arsenite can occur at significant levels as a consequence of reduction by marine phytoplankton and bacteria [28]. Methylarsonate and dimethylarsinate are not particularly stable in seawater though they can exist as significant species in surface waters where primary productivity is high [27, 29]. Seasonal changes and factors associated with biological activity might also influence the amounts of methylarsonate and dimethylarsinate in seawaters [30–32]. There have been reports of unknown arsenic compounds, often referred to as *hidden arsenic*, in seawater [33, 34] and freshwater [34, 35].

In marine algae, arsenosugars are the major arsenic species. The concentrations of arsenosugars in brown algae are much higher than those in red and green algae [36]. Arsenate occurs generally as a minor compound, and arsenite and methylarsonate appear to be trace constituents only. Marine algae also contained a considerable amount of lipid soluble arsenic [37]. On the other hand, freshwater algae have been little studied, but their pattern of arsenic compounds appears to be similar [38].

Terrestrial Environments

Terrestrial plants contain mainly inorganic arsenic. With few exceptions, either arsenite or arsenate has been reported as the major species of arsenic in terrestrial plant species. Methylarsonate (MMAA) and dimethylarsinate (DMAA) are also reported in those plants [36]. Arsenobetaine, arsenocholine and arsenosugars have also been reported in several plant species but the levels are low [39, 40]. Arsenic contents in terrestrial animals (including those used for human food) are low and the dominant species are arsenite and arsenate [26].

Soil and Sediments

Arsenic extraction from soil is very difficult because most analytical techniques for determining arsenic species are performed on water-based solutions of the analytes (extractable arsenic). Thus, the results obtained for soil and sediments represent only a small proportion of the total arsenic present. However, in the extractable fractions of arsenic in soil, arsenate and arsenite dominate [19]. Methylarsonate and dimethylarsinate are also reported in soil [41]. These four arsenicals are also commonly found in sediments and interstitial water (pore water) of the sediments [27, 42].

Air

The major sources of arsenic in air are metal smelters, coal burning, volcanoes, etc. Particulate arsenic trioxide (As_2O_3) is the main form of arsenic in air [26]. Moreover, recent works has revealed the presence of arsine (AsO_3) and methylated arsenic ($MeAsH_2$, Me_2AsH, and Me_3As) as trace constituents of the air sample, particularly over sites of higher biological activity [43–45].

ARSENIC CONTAMINATION

Soil

A natural background level of 6.3 mg kg^{-1} was reported for agricultural soils in Ontario, Canada [46]. Arsenic occurs in 20 minerals, approximately 60% are arsenate. Arsenic compounds are not common in hypergenic environments. This element is highly associated with deposits of many metals and therefore is known as a good indicator in geochemical prospecting surveys [47].

The background arsenic levels in top soils are generally low, although they exceed those in rocks several times. The range of arsenic in uncontaminated soils worldwide ranges from <1 to 95 mg kg^{-1}. The grand mean of soil arsenic is calculated to be 8.7 mg kg^{-1}. The podzol and sandy soils contain 5.8 mg kg^{-1} arsenic in Canada [46], 4.0 mg kg^{-1} in Japan and 4.6 mg kg^{-1} in Korea [48]. The loamy and clay soils contain 4.8 mg kg^{-1} arsenic in Canada [46] and 12.8 mg kg^{-1} in Thailand [48]. The fluvisols contain 25 mg kg^{-1} arsenic in Grate Britain [49] and histosols contain 13.6 mg kg^{-1} in Canada [46]. The chernozems of Bulgaria contain 8.2 mg kg^{-1} arsenic [50]. The forest soils of Norway contain 2.2 mg kg^{-1} arsenic [51].

Water

Arsenic concentrations in groundwater are of increasing environmental concern because of the risk arsenic poses to plants, animals, and human health. The EPA is in the process of setting the new arsenic standard for drinking water at 10 μg L^{-1} to protect humans against the effects of long-term, chronic exposure to arsenic in drinking water. Roughly, 5% of community water systems serving 11 million people will have to take corrective action to lower the current levels of arsenic in their drinking water [52].

Higher levels of arsenic are found in groundwater sources than in surface-water sources. In parts of the southern San Joaquin Valley, California; and parts of Arizona and the middle Rio Grande Basin, New Mexico; oxic, alkaline groundwater contains high arsenic concentrations that may result from desorption from iron oxide. In the southwest, high arsenic concentrations are associated with iron-rich groundwater (which is consistent with dissolution of iron oxide as a source of arsenic). High pH groundwater in felsic volcanics contains high concentrations of arsenic in parts of the Willamette Basin, Oregon. In the upper Midwest, glaciated quaternary sediments appear to be associated with high arsenic concentrations in groundwater. Arsenic in groundwater is mainly inorganic with arsenate comprising about 50% of the total [53].

Many countries (including Taiwan, Argentina, India, Bangladesh, Mexico, Hungary, and Chile) have reported extensive arsenic contamination of their groundwater supplies [54, 55]. In the United States, arsenic-contaminated groundwater has been reported in New England [56], the Mid-west [57], Oklahoma [58], Nevada [59] and California [60]. Groundwater in Bangladesh is currently contaminated by up to 2 mg L^{-1} As with reports of widespread arsenic-related health effects on millions of people [53]. Use of this contaminated water for irrigation of crops has led to elevated concentrations of arsenic in agricultural soils.

Out of the groundwater investigated in the quaternary loess aquifers in northern La Pampa Province of central Argentina, 95% exceeded the World Health Organization (WHO) guideline value of 10 μg L^{-1} As. High concentration of As in the aquifer resulted from desorption of arsenic from Fe and Mn oxides; weathering of primary silicate minerals, and apatite; high pH and alkalinity from silicate and carbonate reactions [55].

Arsenic Toxicity

A large number of sites worldwide have been contaminated by arsenic from natural and anthropogenic sources. Elevated levels of arsenic in agricultural soil could pose a serious threat to plants and human health and the environment through the food chain pathways [12, 13]. Epidemiological studies show a direct relationship between environmental exposure of human to inorganic arsenic and cancer of the skin and lungs. Millions of people have been exposed to arsenic poisoning in Bangladesh and West Bengal, India [10]. The similarity of arsenic to phosphorus and its ability to form covalent bonds with sulfur are the two reasons for arsenic toxicity. Arsenate is an analogue of phosphate and is taken up via the phosphate transport system by the most organisms, which replaces phosphate in the energy transfer phosphorylation reactions [1].

Therefore, remediation of arsenic contaminated soil and water is an important concern worldwide. Various technologies are in place to clean up arsenic or to reduce exposure to arsenic from contact with (or ingestion of) arsenic-contaminated soil and water. Technologies for remediation of arsenic contaminated soils include excavation, immobilization, verification, soil washing/flushing and phytoremediation. Treatment technologies applicable for arsenic-contaminated water include precipitation, membrane filtration, adsorption, ion exchange, permeable reactive barriers and biological treatment.

Arsenate inhibits ATP synthesis by uncoupling oxidative phosphorylation leading to the breakdown of energy metabolism. Arsenate may also replace phosphate in substituted monosaccharides [1]. The toxicity of arsenic in aquatic plants is a function of pH and usually decreases with increasing pH. In addition, phosphate addition produces an opposite effect on the toxicity of arsenic to the aquatic plants. This is thought to be a result of competition of phosphate with arsenate for uptake.

A large increase in arsenic concentration in the crops at 50 to 250 mg kg^{-1} rates of arsenic application resulted in a marked yield reduction, suggesting arsenic in crops at this concentration may have caused physiological damaged to plants. A level of arsenic concentration in crops can be considered to be critical over which it will cause significant yield reduction. The arsenic concentration at which, rice yield decreased by 10% is judged to be the maximum allowable limit or critical content of arsenic in soil. Rice production reduced by 10% at 25 mg As kg^{-1} in soil [61]. Thus, 25 mg As kg^{-1} soils is critical level for rice plant. Phytotoxicity studies have shown that 1, 7 and 2 mg kg^{-1} soluble arsenic cause injury to cowpeas, rice and barley, respectively.

PHYTOREMEDIATION

A large number of sites worldwide are contaminated by heavy metals as a result of human activities. Due to some unavoidable limitations, the traditional remediation technologies lost economic and public acceptance worldwide. During the 1980s, the US government initiated a large scale program for the development of environmental clean-up technologies, which has accelerated the growth of a new productive research field worldwide. *Phytoremediation*, a plant based green technology, received huge attention from scientific community for its low cost of implementation and environmental benefits that aims to extract or inactivate metals in soils [62–64].

Phytoremediation is an organic, low input, and solar-energy powered remediation technique that is applicable to sites with surficial and low to medium levels of contamination. Plants that can grow under these conditions and contain high concentrations of metals are hyperaccumulators. It is very useful for treating a wide variety of environmental contaminants at one time. It has the same mass transfer limitations as other bioremediation systems, and determining fate or transformation of contaminants is complicated. Phytoremediation process might be seasonal if plants slow down growth or become dormant at certain times of the year. Phytoremediation is appropriate for metal, pesticide, solvent, explosives, and crude oil contamination [65].

There are various aspects of phytoremediation *e.g.*, phytoextraction, phytodegradation, rhizofiltration, phytostabilization and phytovolatilization. *Phytoextraction* involves using hyperaccumulating plants to remove the contaminant from the contaminated media and concentrate it in their aboveground plant tissues, which is periodically harvested. The metal-enriched plant residue can be disposed of as hazardous material and if economically feasible, used for metal recovery [62–65].

It has been estimated that the market for phytoremediation of metals from soils in the USA alone was approximately 1–2 million dollars in 1997, with a potential to increase to 15–25 million dollars by 2000 and 70–100 million dollars in by 2005 [66].

Two approaches have been proposed in literature [64] for the phytoremediation of heavy metals: continuous or natural phytoremediation, and chemically enhanced phytoremediation.

Continuous or Natural Phytoremediation

Continuous or Natural Phytoremediation is based on the use of natural hyperaccumulator plants with an exceptional metal-accumulation capacity. These plants have the ability to accumulate very high concentration of metals in their shoots, and are exceptionally higher tolerance to metal toxicity [68, 69]. Mechanisms of metal accumulation involving extracellular and intracellular metal chelation, precipitation, compartmentalization and translocation in the vascular system are poorly understood. Metal contaminants in the soil are usually bound to organics or clay, or are present as insoluble precipitates. Plants can secrete metal-chelating molecules into the rhizosphere. *Graminaceous* species secrete mugeneic acid in response to metal deficiencies. Roots can reduce these metals by root plasma membrane reductase and can lower the pH of the soil by releasing protons. These actions are aided by soil microorganisms [69, 70]. Remediation of heavy metals and radionuclides requires efficient removal from contaminated sites. Plants can extract inorganics, but effective phytoextraction requires plants that produce high biomass and possess high uptake capacity [71].

Generally, phytoremediation systems use hyperaccumulators that tend to be slow growing. This biomass is usually considered a hazardous waste that requires specialized disposal. Commercially important fast growing hardwoods are generally not hyperaccumulators, but there is a potential that they can match the extraction of metals by hyperaccumulators in the same time span due to fast growth and high biomass production. For example, trees of the genus *Populus* have been recognized as being important for phytoremediation. *Populus* trees have deep roots and high transpiration rates, and can transpire between 150 and 900 L per day due to fast growth. This water movement towards the surface can help prevent both metal contamination and downward movement of metals from the soil [68]. The concentration of the contaminant in the wood has to be diluted enough for it to be used as a fuel or mulch source [72].

Chemically Enhanced Phytoremediation

Natural phytoremediation has some common drawbacks such as low biomass and slow growth rate of the hyperaccumulators, lower translocation of metals from roots to shoots, and also the immobility or insolubility of the target metals in soil. Chemically enhanced phytoremediation has been developed to overcome these limitations [73–76].

In this technology, chemicals are used to increase metal mobility is soil and accumulation in high-biomass crop plants. Several chelating ligands such as EDTA, CDTA, DTPA, EGTA, EDDHA, NTA, etc. have been studied for their ability to mobilize metals and increase metal uptake in different plant species [77, 78]. Chelating ligands not only increase metal uptake but also increase uptake of many essential nutrients in crop plants [79].

In the past decade, chelant-enhanced phytoremediation has received much attention from the scientific community. The most promising, effective, and therefore, commonly used chelating ligand is EDTA for the increase solubility of heavy metals for plant uptake during phytoremediation [62].

Despite the success of this technology, the enhanced mobility of metals in soil by EDTA and their potential risk of leaching to groundwater is an important concern [78]. The persistence of metal-EDTA complexes in contaminated soils is also disquiet of the technology.

Phytoremediation of Arsenic

The development of phytoremediation technology needs a thorough understanding of the underlying processes at the genetic, molecular, biochemical, physiological and agronomic levels. Plants initially accumulate arsenic into their roots through phosphate uptake pathway *i.e.*, active apoplastic or symplastic mechanisms and translocate to the above ground parts (shoots and leaves). The amount of arsenic translocated from roots to shoots indicates the phytoremediation efficiency of that plant. However, more then 90% of total arsenic accumulated into the plant is stored in roots. A very few number of plants have the ability to translocate considerable amount of arsenic from roots to shoots. Although *Agrostis castellana*, *Agrostis delicatula*, *Bidens cynapiifolia*, and *Pityrogramma calomelanos* L. have been reported as arsenic hyperaccumulators, Chinese brake fern (*Pteris vittata* L.) has shown the highest ability to accumulate and translocate arsenic from roots to the shoots [79]. The discovery of *Pteris vittata* L. is a milestone in the development of arsenic phytoremediation technology.

Tossell et al. [80] reported that *Eucalyptus camaldulensis* and *Tamarix parviflora* DC can tolerate elevated concentrations of dissolved As and sodium in groundwater found at Palo Alto, CA. Hydraulic control of the contaminant was assessed for use in conjunction to a slurry wall, or as a substitute.

Phytoremediation Using Aquatic Plants

Freshwater as well as seawater systems are being contaminated by heavy metals from various anthropogenic activities and natural sources over time. Therefore, remediation of metal-contaminated aquatic systems is as important as it is for terrestrial systems. As compared to the metal-phytoremediation of terrestrial systems, less work has done of aquatic systems. The use of aquatic macrophytes or other floating aquatic plants in phytoremediation technology is commonly known as phytoextraction. This cleanup process involves biosorption and accumulation of pollutants.

Recently, aquatic macrophytes and some other small floating plants have been investigated for the remediation of wastewater contaminated with Cu (II), Cd(II) and Hg(II) [82–84]. Water fern (*Salvinia natans* L.) is a free floating freshwater macrophyte, which grows rapidly in ponds, lakes, ditches, and wastewater bodies mostly in southern Asian countries affected by arsenic especially in Bangladesh and West Bengal, India. Previously, the *Salvinia natans* L. was tested for Hg (II) [82], and Cu (II) [85] removal. The encouraging results of metal uptake capacity by aquatic plants [83–88] gained the attention of researchers and scientists to conduct more studies on metal-phytoremediation in aquatic systems.

Arsenate is the predominant species in the oxic water and arsenate and arsenite are phytoavailable forms to the plants in the aquatic systems [89]. The dynamics of arsenate exchange between water and adsorbing colloids are analogous to those of phosphate, though the competition for exchange sites favors phosphate over arsenate [90]. Thus, aquatic macrophytes can be a good tool for the remediation of arsenic contaminated aquatic systems, because a few species have already been reported to accumulate arsenic from water [91, 92].

The *Lemna gibba* L. and the *Lemna minor* L. are the most studied species of *Lemnaceae* family in phytoremediation and ecotoxicology [90–91].

Rahman et al. [93–95] studied on arsenic uptake by duckweed (*Spirodela polyrhiza* L.) and found that duckweed accumulated a formidable amount of arsenic. Results showed that from culture solutions containing 4.0 μM of arsenate or DMAA and 0.02 μM of phosphate, the plant accumulated 0.353±0.003 μmol and 7.65±0.27 nmol g^{-1} dry weight of arsenic, respectively, after 6 days of incubation. It indicates that the macrophyte uptake 79% higher arsenic from arsenate solution compared to DMAA solution. *Spirodela polyrhiza* L. grows in rice field, and prevents the growth of harmful aquatic weeds. Under natural conditions with sufficient nutrients, this plant forms a mat-like covering on the water surface of ponds, lakes and ditches. In addition, *Spirodela polyrhiza* L. is characterized by fast growth, wide distribution and stability to the environmental changes. Under ideal conditions, its biomass doubles in 24 h.

Arsenate is the predominant species in the oxic water and arsenate and arsenite are bioavailable forms to the plants in the aquatic systems [89]. The studies of Rahman et al. [93–95] reveal that *Spirodela polyrhiza* L. accumulates a good amount of arsenic, and the macrophyte could be a good option for phytoremediation.

ARSENIC UPTAKE MECHANISMS IN AQUATIC PLATS

The accumulation of arsenic in aquatic macrophyte (*Spirodela polyrhiza* L.) decreased with the increase of phosphate concentration [95]. AsO_4^{3-} is a sorption analog of PO_4^{3-}, and competes with PO_4^{3-} for uptake carriers in the plasmalemma [90]. Therefore, more arsenate is expected to be desorbed in the solution with the increase phosphate concentration [96]. The arsenate uptake in aquatic macrophyte (*Lemna gibba* L.) occurs through the phosphate uptake pathway [91] because of similar chemical behavior between AsO_4^{3-} and PO_4^{3-}, and the same result was observed for *Spirodela polyrhiza* L. [95].

Physico-chemical adsorption, an alternative mechanism for arsenic accumulation into aquatic plants, has been proposed in the literature [97]. In this mechanism, suspended oxides of iron (Fe-plaque) on the aquatic plant surfaces adsorb and accumulate arsenic. Robinson et al. [97] also reported a positive correlation between arsenic and iron concentrations in aquatic plants, since the arsenic is adsorbed on iron oxides on plant surfaces. But, which species of arsenic is mostly adsorbed by iron oxides was not clear from their studies. Rahman et al. [95] suggested arsenate as the predominant species in such incorporation, because the correlation between arsenic and iron concentrations in tissues of *Spirodela polyrhiza* L. was significantly higher when the macrophyte was exposed to arsenate solution than that of DMAA. They also observed that arsenic and iron concentrations in *Spirodela polyrhiza* L. of arsenate treated phosphate deficient solution were highly correlated, while they were not significantly correlated in phosphate-sufficient solution. This can be attributed to the adsorption of arsenic on iron plaque of plant surfaces in phosphate-deficient solution, which was blocked by phosphate in phosphate-sufficient solution. Thus, aquatic macrophytes might accumulate arsenic onto the roots by physico-chemical adsorption, and into the roots via the phosphate uptake pathway.

REFERENCES

[1] Tamaki, S.; Frankenberger, W. T. J., *Environmental Chemistry of Arsenic*. Springer-Verlag: New York, 1992; Vol. 124.

[2] Smith, E. R. N.; Alston, A. M., Arsenic in the soil environment: A Review. *Adv. Agron.* 1998, *64*, 149–195.

[3] Ferguson, J. F.; Gavis, J., A review of the arsenic cycle in natural waters. *Water Res.* 1972, *6*, 1259–1274.

[4] Adriano, D. C., *Trace Elements in the Terrestrial Environment*. Springer-Verlag: New York, 1986.

[5] Bowen, H. J. M., *Environmental Chemistry of the Elements*. Academic Press: New York, 1979.

[6] Shacklette, H. T.; Boerngen, J. G. *Element Concentrations in Soils and Other Surficial Materials of the Conterminous United States*; United States government printing office: Washington DC, 1984.

[7] Sanders, J. G., Arsenic cycling in marine systems. *Mar. Environ. Res.* 1980, *3*, 257–266.

[8] Chilvers, D. C.; Peterson, P. J., Global cycling of arsenic. In *Lead, Mercury, Cadmium, and Arsenic in the environment*, Hutchinson, T. C.; Meema, K. M., Eds. Jhon Wiley: New York, 1987; pp 279–301.

[9] Karim, M. M., Arsenic in groundwater and health problems in Bangladesh. *Water Res.* 2000, *34*, 304–310.

[10] Paul, P. C.; Chattopadhyay, A.; Dutta, S. K.; Mazumder, D. N.; Santra, A., Histopathology of skin lesions in chronic arsenic toxicity-grading of changes and study of proliferative markers. *Indian J. Pathol. Microbiol.* 2000, *43*, 257–264.

[11] Rahman, M. A.; Hasegawa, H.; Rahman, M. M.; Miah, M. A. M.; Tasmin, A., Arsenic accumulation in rice (Oryza sativa L.): human exposure through food chain. *Ecotoxicol. Environ. Saf.* 2008, *69*, 317–324.

[12] Duxbury, J. M.; Mayer, A. B.; Lauren, J. G.; Hassan, N., Food chain aspects of arsenic contamination in Bangladesh: effects on quality and productivity of rice. *J. Environ. Sci. Health Part A Toxic-Hazard. Subst. Environ. Eng.* 2003, *38*, 61–69.

[13] Bruce, S. L.; Noller, B. N.; Grigg, A. H.; Mullen, B. F.; Mulligan, D. R.; Ritchie, P. J.; Currey, N.; Ng, J. C., A field study conducted at Kidston Gold Mine, to evaluate the impact of arsenic and zinc from mine tailing to grazing cattle. *Toxicol. Lett.* 2003, *137*, 23–34.

[14] Shariatpanahi, M.; Anderson, A. C., Distribution and toxicity of monosodium methanearsonate following oral administration of the herbicide to dairy sheep and goats. *J. Environ. Sci. Health Part B Pestic. Food Contam. Agric. Wastes* 1984, *19*, 427–39.

[15] Thornton, I.; Webb, J. S., Geochemistry and health in the United Kingdom. *Philos. Trans. R. Soc. Lond. B. Biol. Sci.* 1979, *288*, 151–168.

[16] O'day, P. A.; Vlassopoulos, D.; Root, R.; Rivera, N., The influence of sulfur and iron on dissolved arsenic concentrations in the shallow subsurface under changing redox conditions. *Proc. Natl. Acad. Sci. U. S. A.* 2004, *101*, 13703–13708.

[17] Allred, A. L.; Hensley Jr, A. L., Electronegativities of nitrogen, phosphorus, arsenic, antimony and bismuth. *J. Inorg. Nucl. Chem.* 1961, *17*, 43–54.
[18] O'day, P., Chemistry and mineralogy of arsenic. *Elements* 2006, *2*, 77–83.
[19] Cullen, W. R.; Reimer, K. J., Arsenic speciation in the environment. *Chem. Rev.* 1989, *89*, 713–764.
[20] Inskeep, W. P.; McDermott, T. R.; Fendorf, S., Arsenic (V)/(III) cycling in soils and natural waters: chemical and microbiological processes. In *Environmental Chemistry of Arsenic*, Frankenberger, W. T. J., Ed. Marcel Dekker: New York, 2002; pp 183–215.
[21] Nordstrom, D. K.; Archer, D. G., Arsenic thermodynamic data and environmental geochemistry. In *Arsenic in Groundwater: Geochemistry and Occurrence*, Welch, A. H.; Stollenwerk, K. G., Eds. Kluwer Academic Publishers: Boston, 2003; pp 1–25.
[22] Anthony, J. W.; Bideaux, R. A.; Bladh, K. W.; Nichols, M. C., *Handbook of Mineralogy Vol. III: Arsenates, Phosphates, Vanadates*. Mineral Data Publishing: Tucson, 2000.
[23] Carter, D. E.; Aposhian, H. V.; Gandolfi, A. J., The metabolism of inorganic arsenic oxides, gallium arsenide, and arsine: a toxicochemical review. *Toxicol. Appl. Pharmacol.* 2003, *193*, 309–334.
[24] Tanaka, T., Distribution of arsenic in natural environment with an emphasis on rocks and soils. *Appl. Organomet. Chem.* 1988, *2*, 283–295.
[25] Bhumbla, D. K.; Keefer, R. F., Arsenic mobilization and bioavailability in soils. In *Arsenic in the Environment, Part I: Cycling and Characterization*, Nraigu, J. O., Ed. Wiley: New York, 1994; pp 51–82.
[26] Leonard, A., Arsenic. In *Metals and Their Compounds in the Environment*, Merian, E., Ed. Weinheim: VCH, 1991; pp 751–774.
[27] Andreae, M. O., Arsenic speciation in seawater and interstitial waters: The influence of biological-chemical interactions on the chemistry of a trace element. *Limnol. Oceanogr.* 1979, *24*, 440–452.
[28] Francesconi, K. A.; S., E. J., Arsenic and marine organisms. *Adv. Inorg. Chem.* 1997, *44*, 147–189.
[29] Andreae, M. O., Distribution and speciation of arsenic in natural waters and some marine algae. *Deep-Sea Res.* 1978, *25*, 391–402.
[30] Santosa, S. J.; Wada, S.; Tanaka, S., Distribution and cycle of arsenic compounds in the ocean. *Appl. Organomet. Chem.* 1994, *8*, 273–283.
[31] Millward, G. E.; Ebdon, L.; Walton, A. P., Seasonality in estuarine sources of methylated arsenic. *Appl. Organomet. Chem.* 1993, *7*, 499–511.
[32] Hasegawa, H., Seasonal changes in methylarsenic distribution in Tosa Bay and Uranouchi inlet. *Appl. Organomet. Chem.* 1996, *10*, 733–740.
[33] Howard, A. G.; Comber, S. D. W., The discovery of hidden arsenic species in coastal waters. *Appl. Organomet. Chem.* 1989, *3*, 509 514.
[34] Hasegawa, H.; Matsui, M.; Okamura, S.; Hojo, M.; Iwasaki, N.; Sohrin, Y., Arsenic speciation including 'hidden' arsenic in natural waters. *Appl. Organomet. Chem.* 1999, *13*, 113–119.
[35] Bright, D. A.; Dodd, M.; Reimer, K. J., Arsenic in subArctic lakes influenced by gold mine effluent: the occurrence of organoarsenicals and 'hidden' arsenic. *Sci. Total Environ.* 1996, *180*, 165–182.

[36] Francesconi, K. A.; Kuehnelt, D., Arsenic compounds in the environment. In *Environmental Chemistry of Arsenic*, Frankenberger, W. T. J., Ed. Marcel Dekker: New York, 2002; pp 51–94.

[37] Morita, M.; Shibata, Y., Chemical form of arsenic in marine macroalgae. *Appl. Organomet. Chem.* 1990, *4*, 181–190.

[38] Lai, V. W. M.; Cullen, W. R.; Harrington, C. F.; Reimer, K. J., The characterization of arsenosugars in commercially available algal products including a nostoc species of terrestrial origin. *Appl. Organomet. Chem.* 1997, *11*, 797–803.

[39] Kuehnelt, D.; Lintschinger, J.; Goessler, W., Arsenic compounds in terrestrial organisms. IV. Green plants and lichens from an old arsenic smelter site in Austria. *Appl. Organomet. Chem.* 2000, *14*, 411–420.

[40] Koch, I.; Feldmann, J.; Wang, L.; Andrewes, P.; Reimer, K. J.; Cullen, W. R., Arsenic in the Meager Creek hot springs environment, British Columbia, Canada. *Sci. Total Environ.* 1999, *236*, 101–117.

[41] Braman, R. S.; Foreback, C. C., Methylated forms of arsenic in the environment. *Science* 1973, *182*, 1247–1249.

[42] Reimer, K. J.; Thompson, J. A. J., Arsenic speciation in marine interstitial water: the occurrence of organoarsenicals. *Biogeochemistry* 1988, *6*, 211–237.

[43] Feldmann, J.; Grumping, R.; Hirner, A. V., Determination of volatile metal and metalloid compounds in gases from domestic waste deposits with GC–ICP-MS. *Fresenius. J. Anal. Chem.* 1994, *350*, 228–234.

[44] Feldmann, J.; Hirner, A. V., Occurrence of volatile metal and metalloid Species in landfill and sewage gases. *Int. J. Environ. Anal. Chem.* 1995, *60*, 339–359.

[45] Hirner, A. V.; Feldmann, J.; Krupp, E.; Grumping, R.; Goguel, R.; Cullen, W. R., Metal(loid)organic compounds in geothermal gases and waters. *Org. Geochem.* 1998, *29*, 1765–1778.

[46] Frank, R.; Braun, H. E.; Ishida, K.; Suda, P., Persistent organic and inorganic pesticide residues in orchard soils and vineyards of Southern Ontario. *Can. J. Soil Sci.* 1976, *56*, 463 – 484.

[47] Kabata-Pendias, A.; Pendias, H., *Trace Element in Soil and Plants*. CRC Press: Florida, 1985.

[48] Kitagishi, K.; Yamane, I., *Heavy Metals Pollution in Soils of Japan*. Japan Science Society Press: Tokyo, 1981.

[49] Thornton, P.; Webb, J. S., Trace elements in soils and surface waters contaminated by past metalliferous mining in parts of England. In *Trace Substances in Environmental Health, IX : Proceedings of University of Missouri's 9th Annual Conference on Trace Substances in Environmental Health*, Hemphill, D. D., Ed. University of Missouri: Columbia, 1975.

[50] Petrov, I. I.; Tsalev, D. L.; Lyotchev, I. S., Investigation on arsenic and cadmium content in soils. *Higiena Zdravyeopazv* 1979, *22*, 574.

[51] Låg, J.; Steinnes, E., Soil selenium in relation to precipitation. *Ambio* 1974, *3*, 237–238.

[52] EPA *Drinking Water Standards for Arsenic*; United States Environmental Protection Agency: 2001.

[53] Abedin, M. J.; Cresser, M. S.; Meharg, A. A.; Feldmann, J.; Cotter-Howells, J., Arsenic accumulation and metabolism in rice. *Environ. Sci. Technol.* 2002, *36*, 962–968.

[54] Nikolaidis, N. P.; Dobbs, G. M.; J., C.; Lackovic, J. A., Arsenic mobility in contaminated lake sediments. *Environ. Pollut.* 2004, *129*, 479–487.
[55] Smedley, P. L.; Nicolli, H. B.; Macdonald, A. J.; Barros, A. J.; Tullio, J. O., Hydrogeochemistry of arsenic and other inorganic constituents in groundwaters from La Pampa, Argentina. *Appl. Geochem.* 2002, *17*, 259–284.
[56] Peters, S. C.; Blum, J. D.; Klaue, B.; Karagas, M. R., Arsenic occurrence in New Hampshire drinking water. *Environ. Sci. Technol.* 1999, *33*, 1328–1333.
[57] Welch, A. H.; Westjohn, D. B.; Helsel, D. R.; Wanty, R. B., Arsenic in ground water of the United States: occurrence and geochemistry. *Ground Water* 2000, *38*, 589–604.
[58] Schlottmann, J. L.; Breit, G. N., Mobilization of As and U in the central Oklahoma aquifer, USA. In *7th International Symposium on Water-Rock Interaction*, Kharaka, Y. K.; Maest, A. S., Eds. Balkema Publishers: Rotterdam, 1992.
[59] Welch, A. H.; Lico, M. S., Factor controlling As and U in shallow groundwater, southern Carson desert, Nevada. *Appl. Geochem.* 1998, *13*, 521–539.
[60] Wilkie, J. A.; Hering, J. G., Rapid oxidation of geothermal arsenic (III) in streams of the eastern Sierra Nevada. *Environ. Sci. Technol.* 1998, *32*, 657–662.
[61] Nriagu, J. O., *Arsenic in the Environment: Part I.* John Willey: London, 1994.
[62] Salt, D. E.; Smith, R. D.; Raskin, I., Phytoremediation. *Annu. Rev. Plant Physiol. Plant Mol. Biol.* 1998, *49*, 643–668.
[63] McGrath, S. P., Phytoextraction for soil remediation. In *Plants that Hyperaccumulate Heavy Metals*, R., B. R., Ed. CAB Int.: Wallingford, UK, 1998; pp 261–287.
[64] Salt, D. E.; Smith, R. D.; Raskin, I., Phytoremediaiton. *Annu. Rev. Plant Physiol. Plant Mol. Biol.* 1998, *49*, 643–668.
[65] USEPA. *A Citizen's Guide to Phytoremediation Technology*; Washington, DC., 1998.
[66] Glass, D. J., Economic potential of phytoremediation. In *Phytoremediation of Toxic Metals*, Raskin, I.; Ensley, B. D., Eds. John Wiley: New York, 2000; pp 15–31.
[67] Ma, L. Q.; Komar, K. M.; Tu, C.; Zhang, W.; Cai, Y., A fern that hyperaccumulates arsenic. *Nature* 2001, *409*, 579.
[68] Baker, A. J. M.; McGrath, S. P.; Reeves, R. D.; Smith, J. A. C., Metal hyperaccumulator plants: a review of the ecology and physiology of a biochemical resource for phytoremediation of metal-polluted soils. In *Phytoremediation of Contaminated Soil and Water*, Terry, N.; Banuelos, G., Eds. Lewis Publication: Boca Raton, FL, 2000; pp 85–107.
[69] Chaney, R. L.; Malik, M.; Li, Y. M.; Brown, S. L.; Angle, J. S.; Baker, A. J. M., Phytoremediation of soil metals. *Curr. Opin. Biotechnol.* 1997, *8*, 279–284.
[70] Raskin, I.; Kumar, P.; Dushenkov, S.; Salt, D., Bioconcentration of heavy metals by plants. *Op. Biotechnology* 1994, *5*, 285–290.
[71] Ow, D.; Shewry, P.; Napier, J.; Davis, P., Prospects of engineering heavy metal detoxification genes in plants. In *Symposium of the Industrial Biochemistry and Biotechnology Group of the Biochemical Society*, Bristol, UK, 1998.
[72] Alker, G. R.; Rockwood, D. L.; Ma, L. Q.; Cardellino, R. W., Heavy metal contaminated land for woody biomass production: Moving technology into the market place. In *The 9th Biennial Bioenergy Conference*, Buffalo, New York, 2000.
[73] Huang, J. W.; Cunningham, S. D., Lead phytoextraction: species variation in lead uptake and translocation. *New Phytol.* 1996, *134*, 75–84.

[74] Blaylock, M. J., Field demonstration of phytoremediation of lead contaminated soils. In *Phytoremediation of Contaminated Soil and Water*, Terry, N.; Banuelos, G., Eds. Lewis Publication: Boca Raton, FL, 2000; pp 1–12.

[75] Blaylock, M. J.; Salt, D. E.; Dushenkov, S.; Zakharova, O.; Gussman, C.; Kapulnik, Y.; Ensley, B. D.; Raskin, I., Enhanced accumulation of Pb in Indian mustard by soil-applied chelating agents. *Environ. Sci. Technol.* 1997, *31*, 860–865.

[76] Huang, J. W.; Chen, J.; Berti, W. B.; Cunningham, S. D., Phytoremediation of lead-contaminated soils: role of synthetic chelates in lead phytoextraction. *Environ. Sci. Technol.* 1997, *31*, 800–805.

[77] Huang, J. W.; Blaylock, M. J.; Kapulnik, Y.; Ensley, B. D., Phytoremediation of uranium contaminated soils: role of organic acids in triggering uranium hyperaccumulation in plants. *Environ. Sci. Technol.* 1998, *32*, 2004–2008.

[78] Cooper, E. M.; Sims, J. T.; Cunningham, S. D.; Huang, J. W.; Berti, W. R., Chelate-assisted phytoextraction of lead from contaminated soils. *J. Environ. Qual.* 1999, *28*, 1709–1719.

[79] Madrid, F.; Liphadzi, M. S.; Kirkham, M. B., Heavy metal displacement in chelate-irrigated soil during phytoremediation. *J. Hydrol.* 2003, *272*, 107–119.

[80] Tossell, R. W.; Binard, K.; Sangines-Uriarte, L.; Rafferty, M. T.; Morris, N. P., Evaluation of Tamarisk and Eucalyptus transpiration for the application of phytoremediation. In *First International Conference on Remediation of Chlorinated and Recalcitrant Compounds*, Wickramanayake, G. B.; Hinchee, R. E., Eds. Battelle Press: Columbus, 1998.

[81] Ma, L. Q.; Komar, K. M.; Tu, C.; Zhang, W.; Cai, Y., A fern that hyperaccumulates arsenic. *Nature (Lond.)* 2001, *409*, 579.

[82] Sen, A. K.; Mondal, N. G., *Salvinia natans* as the scavenger of Hg (II). *Water, Air, Soil Pollut.* 1987, *34*, 439–446.

[83] Selvapathy, P.; Sreedhar, P., Heavy metals removal by water hyacinth. *J. Indian Pub. Health Eng.* 1991, *3*, 11–17.

[84] Alam, B.; Chatterjee, A. K.; Duttagupta, S., Bioaccumulation of Cd(II) by water lettuce. *Pollut. Res.* 1995, *14*, 59–64.

[85] Sen, A. K.; Mondal, N. G., Removal and uptake of copper (II) by *Salvinia natans* from wastewater. *Water, Air, Soil Pollut.* 1990, *49*, 1–6.

[86] Sen, A. K.; Bhattacharyya, M., Studies on uptake and toxic effects of lead on *Salvinia natans*. *Indian J. Environ. Health* 1993, *35*, 308–320.

[87] Low, K. S.; Lee, C. K.; Tai, C. H., Biosorption of copper by water hyacinth roots. *J. Environ. Sci. Health, Part A* 1994, *29*, 171–188.

[88] Reddy, K. R.; Busk, W. F. D., Nutrient removal potential of selected aquatic macrophytes. *J. Environ. Qual.* 1985, *14*, 459–462.

[89] Sizova, O. I.; Kochetkov, V. V.; Validov, S. Z.; Boronin, A. M.; Kosterin, P. V.; Lyubun, Y. V., Arsenic-contaminated soils: genetically modified *Pseudomonas* spp. and their arsenic-phytoremediation potential. *J. Soils Sediments* 2002, *2*, 19–23.

[90] Mkandawire, M.; Lyubun, Y. V.; Kosterin, P. V.; Dudel, E. G., Toxicity of arsenic species to *Lemna gibba* L. and the influence of phosphate on arsenic bioavailability. *Environ. Toxicol.* 2004, *19*, 26–35.

[91] Mkandawire, M.; Dudel, E. G., Accumulation of arsenic in *Lemna gibba* L. (ducweed) in tailing waters of two abandoned uranium mining sites in Saxony, Germany. *Sci. Total Environ.* 2005, *336*, 81–89.

[92] Robinson, B.; Duwing, C.; Bolan, N.; Kannathasan, M.; Saravanan, A., Uptake of arsenic by New Zealand watercress (*Lepidium sativum* L.). *Sci. Total Environ.* 2003, *301*, 67–73.

[93] Rahman, M. A.; Hasegawa, H.; Ueda, K.; Maki, T.; Okumura, C.; Rahman, M. M., Arsenic adsorption on iron plaque and uptake in duckweed (*Spirodela polyrhiza* L.) affected by chemical species. In *International Conference on Environmental Science and Technology (ICEST)*, Starrett, S. K.; Hong, J.; Wilcock, R. J.; Li, Q.; Carson, J. H.; Arnold, S., Eds. American Science Press: Texas, Houston, USA, 2007; Vol. 1, pp 319–325.

[94] Rahman, M. A.; Hasegawa, H.; Ueda, K.; Maki, T.; Okumura, C.; Rahman, M. M., Arsenic accumulation in duckweed (*Spirodela polyrhiza* L.): A good option for phytoremediation. *Chemosphere* 2007, *69*, 493–499.

[95] Rahman, M. A.; Hasegawa, H.; Ueda, K.; Maki, T.; Rahman, M. M., Arsenic uptake by aquatic macrophyte *Spirodela polyrhiza* L.: Interactions with phosphate and iron. *J. Hazard. Mater.* 2008, *160*, 356–361.

[96] Smith, S. E.; Read, D. J., *Mycorrhizal symbiosis*. Academic press: San Diego, 1997.

[97] Robinson, B.; Kim, N.; Marchetti, M.; Moni, C.; Schroeter, L.; Dijssel, C.; Van Den Milne, G.; Clothier, B., Arsenic hyperaccumulation by aquatic macrophytes in the Taupo Volcanic Zone, New Zealand. *Environ. Exp. Bot.* 2006, *58*, 206–215.

In: Handbook of Phytoremediation
Editor: Ivan A. Golubev
ISBN: 978-1-61728-753-4

Chapter 19

HAIRY ROOT STUDIES IN PHYTOREMEDIATION AND PHYTOMINING

Pauline M. Doran*
School of Biological Sciences and Department of Chemical Engineering
Monash University, VIC 3800, Australia

ABSTRACT

Because plant roots are in direct contact with pollutants in contaminated soil or water, their responses to toxic substances are of particular importance in phytoremediation and phytomining research. Genetically transformed hairy roots offer many practical advantages in experimental studies, such as ease of initiation, culture, and maintenance, indefinite propagation of material derived from the same parent plant, and genotypic and phenotypic stability. Hairy roots have been applied mainly in metabolic studies of xenobiotic biotransformations and degradation in plants, and for determining the responses of plant tissues to toxic heavy metals. The aim of this chapter is to review the applications of hairy roots in phytoremediation and phytomining research. Experimental results are also presented to demonstrate the capacity of hairy root cultures to hyperaccumulate heavy metals such as cadmium and nickel, allowing practical examination of the biological mechanisms responsible for elevated heavy metal tolerance in hyperaccumulator plant species.

INTRODUCTION

Understanding the biochemical and physiological mechanisms used by plants to tolerate and detoxify organic and heavy metal pollutants is a major challenge in phytoremediation research. Plant tissue cultures have been adopted in many studies as a convenient laboratory tool for investigating the biological phenomena responsible for phytoremediation (Doran, 2009). Because plant roots are in direct contact with pollutants in contaminated soil or water, their responses to toxic substances are of particular importance. As organized tissues, roots

* Phone: +61-3-9905-1373, Fax: +61-3-9905-5613, Email: pauline.doran@sci.monash.edu.au

are closer in structure and function to the organs of whole plants than de-differentiated callus and suspended cells. They thereby offer a greater degree of authenticity with regard to their biological behavior and properties after exposure to pollutants, and are relatively free of somaclonal variations that can alter the properties of de-differentiated plant cultures with time. The study of roots isolated from the aerial parts of the plant can also reveal relevant root-based responses without interference from translocation effects.

Genetically transformed hairy roots offer many practical advantages over whole plants in experimental studies of phytoremediation and are being applied with increasing frequency in phytoremediation research. As a laboratory tool, hairy roots offer a range of benefits such as ease of initiation, culture and maintenance, indefinite propagation, and genotypic and phenotypic stability. The experimental conditions applied to hairy root cultures are more easily controlled using a broader range of variables than can be imposed on soil-grown plants. As hairy roots are routinely cultivated as axenic cultures, the effects of microbial activity can be distinguished readily from observations of plant behavior. The principal advantage of hairy roots over untransformed root cultures is their autotrophy in plant hormones, which greatly simplifies their culture medium requirements. However, not all plant species are easily transformed to produce hairy roots and this technology is restricted in practical terms to a range of dicotyledonous plants.

The principal applications of hairy roots have been in metabolic studies of xenobiotic biotransformation and degradation in plants, and for determining the responses of plant tissues to toxic heavy metals. Being themselves products of genetic transformation, hairy roots are also readily amenable to genetic modification to test whether foreign gene expression can improve plant tolerance and metabolism of pollutants, and to allow rapid screening of transformants for improved phytoremediation traits.

Initiation, Culture, and Properties of Hairy Roots

Infection of plants with *Agrobacterium rhizogenes* results in hairy root disease, which is characterized by the production of rapidly growing roots at the sites of bacterial infection. These genetically transformed hairy roots contain part of the bacterial Ri (root-inducing) plasmid. After removal from the infected plant, excised hairy roots will continue to grow *in vitro* on hormone-free plant tissue culture medium.

The size of the Ri plasmid T-DNA (transferred DNA) inserted into plant cell chromosomes during bacterial infection varies widely from about 15 kb to more than 40 kb. Transformation results in the modification of amino acid metabolism in the root cells so that special metabolites, opines, are produced. However, opine production can be unstable in hairy roots and may disappear after a few serial passages (Tepfer, 1984; Kamada et al., 1986).

Detailed procedures for initiating hairy root cultures from leaf or stem tissue have been described by Hamill and Lidgett (1997). To illustrate the consequences of *A. rhizogenes* infection, an *Atropa belladonna* (deadly nightshade) seedling with infected stem showing prolific hairy root formation at the wounding site is shown in Figure 1. Of the many *A. rhizogenes* strains available, strains A4, 15834 and LBA9402 are generally the most successful in inducing hairy root formation.

Figure 1. Prolific hairy root development on an *Atropa belladonna* seedling after wounding and infection of the stem with *A. rhizogenes*.

After infection of plant tissue with bacteria and emergence of hairy roots, the excised roots can be cultured indefinitely *in vitro* in liquid medium. Initially, hairy root cultures are treated with cefotaxime antibiotic to inhibit bacterial growth; however, after several passages in the presence of antibiotic, the roots can be grown in antibiotic- and hormone-free medium as axenic cultures. Some hairy roots are sensitive to mechanical forces, which can disrupt the differentiated state of the tissues during shake-flask or bioreactor culture, leading to callus formation and culture disintegration.

Hairy roots typically exhibit profuse growth of root hairs and a high level of lateral branching. Figure 2a shows a single *A. belladonna* hairy root observed using scanning electron microscopy and revealing the dense layer of root hairs and mucilage covering the root tissue. Figure 2b shows the morphology of *Solanum aviculare* (kangaroo apple) hairy roots grown in liquid medium in a Petri dish, indicating extensive lateral root development. The presence of many lateral root tips generally translates into rapid culture growth, as meristemic cells in the tip divide to produce cells behind the meristemic zone that expand and differentiate to form mature, vacuolated root cells. Hairy roots are essentially young roots without secondary thickening or cell wall lignification. An important characteristic of hairy roots is that they are generally more genotypically and phenotypically stable than de-differentiated plant tissue cultures such as callus and cell suspensions. Integration of the Ri T-DNA into the plant chromosome is also stable. However, because integration of T-DNA is random with respect to both copy number and position on the chromosomes (Byrne et al., 1983; Ambros et al., 1986), variations are often observed between hairy root clones generated at different infection sites on the same parent plant material.

Hairy roots readily regenerate into whole plants which transmit the Ri T-DNA to their progeny. The presence of Ri T-DNA in plants is associated with a variety of morphological and physiological characteristics such as leaf wrinkling, shortened internodes, reduced apical dominance in roots and stems, reduced fertility, altered flowering, and plagiotropic root development (Tepfer, 1989).

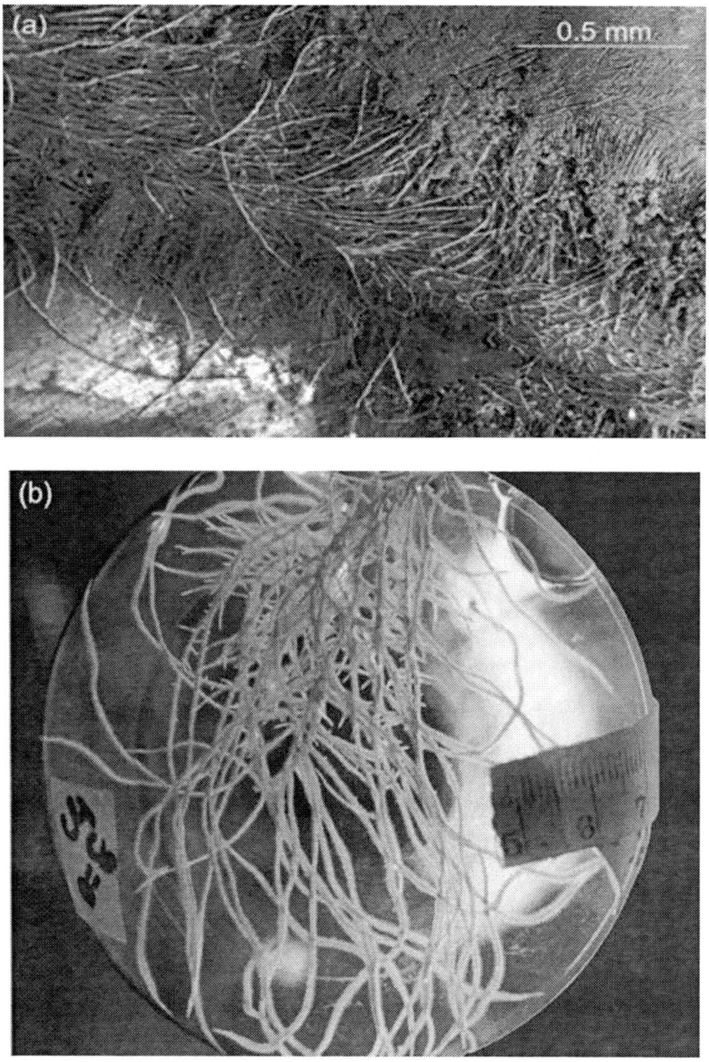

Figure 2. (a) Cryo-scanning electron micrograph of a single *A. belladonna* hairy root cultured in liquid medium. The root lies diagonally across the entire photograph, oriented downwards from left to right, with the root tip near the bottom right-hand corner. The sample has been partially dehydrated to disrupt the thick layer of root hairs and mucilage coating the root tissue. (Photo: G. Williams) (b) Hairy roots of *Solanum aviculare* cultured in liquid medium in a Petri dish showing profuse lateral branching. The culture was initiated using a single root tip and is shown after 26 days of root growth. (Photo: M.A. Subroto)

Hairy Root Studies of Phytoremediation of Organic Compounds

Several organic compounds present in industrial effluents are major environmental hazards because of their carcinogenicity and recalcitrance to degradation. Contamination of soil with toxic xenobiotics is also of concern because of the possibility that these compounds will enter the food chain. Plants are capable of removing organic chemicals from the environment and can enzymatically transform many toxic organic substrates either within the plant tissues or externally in conjunction with microorganisms in the rhizosphere. Xenobiotic compounds are not usually completely degraded or mineralized by plants; rather, organic residues are formed as a result of plant-based biocatalysis.

At the present time, our understanding of the metabolism of most xenobiotics in plants is limited. Hairy root cultures have been used in many studies aimed at elucidating the responses of plants and their interactions with toxic organic substrates. In general terms, detoxification of organic chemicals is considered to proceed via three steps (Coleman et al., 1997; Sandermann, 1994; van Eerd et al., 2003). In Step I, the compound is chemically transformed by oxidation, reduction, or hydrolysis, using enzymes such as P450 monooxygenases, peroxidases, reductases, dehydrogenases, and esterases, to give products that typically retain some level of toxicity. Overexpression of Step I enzymes in hairy root cultures has been used to enhance this element of xenobiotic metabolism (Banerjee et al., 2002; Wevar Oller et al., 2005). Detoxification occurs largely during Step II as the metabolites are conjugated by transferases to sugar residues, glutathione, or amino acids, resulting in the formation of stable, water-soluble compounds. Step III involves the export of conjugated derivatives from the cytosol by ATP- or proton-dependent membrane transporters or exocytosis for compartmentation and immobilization in the vacuole or apoplast; further enzymatic conversion or binding to cell wall components such as lignin, hemicellulose, and pectin may also occur. According to this general scheme, plant metabolism of toxic organics results in the formation of stable, conjugated products that are stored within plant tissues.

Phenols, Chlorinated Phenols, and Polychlorinated Biphenyls

Phenol and its chlorinated derivatives enter the environment in bio-wastes produced during the manufacture of industrial and agricultural products such as resins, herbicides, and pesticides. Enzymatic treatment of phenols using peroxidase enzymes is a potentially useful technique for bioremediation of these substances. Oxidation of phenolic compounds by peroxidase generates phenoxy radicals that conjugate together to form water-insoluble oligomers that can then be removed from the enzyme reaction mixture by filtration or sedimentation. A critical limitation in enzymatic phenol removal is deactivation of the enzyme; continuous synthesis of active peroxidases in living tissues such as plants offers a possible remedy for this problem. As plant peroxidases are located in the cell walls, they are available for phenol detoxification in plant roots contacting the environment.

Hairy roots of *Daucus carota* (carrot), *Brassica napus* (rape), *B. juncea* (Indian mustard), *Lycopersicon esculentum* (tomato) and *Nicotiana tabacum* (tobacco) have been employed in studies of phenol phytoremediation (de Araujo et al., 2002; Agostini et al., 2003; Wevar Oller et al., 2005; González et al., 2006; Singh et al., 2006; Coniglio et al., 2008; Sosa Alderete et al., 2009; Talano et al., 2010). In terms of growth, *D. carota* hairy roots were more tolerant of phenol than its chlorinated derivatives, with the toxicity of the derivatives increasing with the number of chlorine substitutions (de Araujo et al., 2002). Uptake of phenols from the medium was accompanied by induction of peroxidase activity. Browning of *B. napus* hairy roots in the presence of dichlorophenol and hydrogen peroxide (H_2O_2) was taken to indicate the formation of phenolic polymer detoxification products bound to the root surface. Peroxidase isozyme profiles were modified in *B. napus* and *L. esculentum* hairy roots when the roots were re-used in repeated phenol or dichlorophenol oxidation cycles (Agostini et al., 2003; González et al., 2006; Coniglio et al., 2008). Exposure to phenol increased endogenous levels of peroxidase activity and H_2O_2 in *B. juncea* hairy roots; phenol removal also occurred without addition of H_2O_2 to the cultures (Singh et al., 2006). Exposure of *N. tabacum* hairy roots to dichlorophenol changed the pattern of lignin deposition in the roots, suggesting the formation of lignin-type transformation products and their localization in the cell walls (Talano et al., 2010). In other work, a native peroxidase was overexpressed in tomato hairy roots to enhance phenol detoxification. Application of a transgenic root clone expressing high peroxidase activity increased the amount of phenol removed from aqueous solution by about 20% compared with wild-type tomato hairy roots (Wevar Oller et al., 2005). Hairy roots of transgenic tobacco expressing basic peroxidases from tomato achieved 12–15% higher phenol removal than wild-type tobacco hairy roots (Sosa Alderete et al., 2009).

Polychlorinated biphenyls (PCBs) are present in soils as a result of anthropogenic activities, including the production of electrical transformers and capacitors, turbines, pumps, and paints. Hairy roots of *Solanum nigrum* (black nightshade) have been applied in studies of PCB degradation in plant cultures (Macková et al., 1997a, 1997b; Kucerová et al., 2000; Rezek et al., 2007). The presence of PCBs in the culture medium increased the total peroxidase activity in the roots and induced new peroxidase isozymes (Macková et al., 1997a), suggesting the involvement of peroxidases in PCB transformation. Root-produced conversion products from the degradation of six different dichlorophenyls were identified as monohydroxydichlorobiphenyls (Kucerová et al., 2000). It was found using dichlorinated, trichlorinated, tetrachlorinated, and pentachlorinated PCBs that the number of metabolites formed decreased as the number of chlorine atoms per molecule of PCB increased, with tetrachlorinated and pentachlorinated PCBs typically giving no transformation products (Rezek et al., 2007). In most cases, the degree of chlorination of the root metabolites was the same as that of the parent compound (Rezek et al., 2007).

Trichloroethylene

Trichloroethylene (TCE) is a widespread pollutant in the environment and originates from a range of industrial activities. Hairy roots of *A. belladonna* were found to be capable of oxidizing TCE to trichloroethanol (TCOH) (Banerjee et al., 2002). *A. belladonna* hairy roots genetically engineered to express the mammalian liver enzyme, P450 2E1, produced 5–10-fold greater TCOH levels than control cultures.

Ketones

Enantio-selective reduction of prochiral ketones has been observed in *D. carota* hairy root cultures (Caron et al., 2005). Acetophenone added to the medium was reduced to (*S*)-1-phenylethanol over 7 days with a yield of 96%, and the roots were found to retain their biocatalytic activity after repetitive application cycles. Several other aromatic ketones and keto esters and a simple aliphatic ketone were also transformed by the hairy roots with yields of 25–90%.

Two xenobiotic diketones have been reported to undergo stereo- and regio-selective reduction in *B. napus* hairy root cultures with relatively high biotransformation levels of > 78% (Orden et al., 2006). The observed stereo- and regio-selection for the 5-acetyl group was considered a particular advantage associated with hairy root transformation, as the production of secondary alcohols with regio-control is not feasible using typical chemical reductive agents. *B. napus* hairy roots were also capable of glycosylating phenolic hydroxyl groups. In other work, a series of prochiral alkylaryl ketones was applied to *Raphanus sativus* (radish) hairy roots for stereo-selective reduction with high yield and enantio-selectivity (Orden et al., 2009). *R. sativus* roots demonstrated the capacity for stereo-selective transfer of hydrogen without the need to add exogenous NAD(P)H coenzyme to the medium.

Explosives

Contamination of soils with explosives occurs mainly as the result of munitions production and handling. The ability of plants to take up and transform 2,4,6-trinitrotoluene (TNT) was investigated using *Catharanthus roseus* (Madagascar periwinkle) hairy roots (Hughes et al., 1997). The roots converted TNT into aminated nitrotoluenes, which accumulated in the culture medium and root biomass. *C. roseus* hairy roots were also used to study the metabolism of two polynitramine explosives, hexahydro-1,3,5-trinitro-1,3,5-triazine (RDX) and octahydro-1,3,5,7-tetranitro-1,3,5,7-tetrazocine (HMX) (Bhadra et al., 2001). Both these substances were more recalcitrant to accumulation in and transformation by plant tissues than TNT. However, whereas HMX was largely unaffected by the roots, RDX was taken up from aqueous solution and RDX-derived ^{14}C was confirmed to accumulate in the hairy root biomass.

Hairy roots of *Armoracia rusticana* (horseradish) were treated with 2,4-dinitrotoluene (DNT), TNT, and the TNT-degradation products, aminodinitrotoluenes (ADNTs) and diaminonitrotoluenes (DANTs), to determine the effect of these compounds on the activities of glutathione *S*-transferase and peroxidase enzymes in the cells (Nepovím et al., 2004). TNT and ADNT markedly induced two glutathione *S*-transferase isoenzymes in the hairy root cultures. In contrast, peroxidase activity was inhibited by TNT, while treatment with DNT, ADNT and DANT induced peroxidase activity in the order ADNT < DANT < DNT. In general, exposure to nitroaromatic compounds resulted in severe stress responses in the hairy roots and the initiation of effective defensive reactions.

Dyes

Azo dyes are used widely in the manufacture of textiles. Discharge of strongly colored dye effluents into waterways creates a significant environmental impact, including reduced light penetration, altered pH, and increased chemical and biological oxygen demand. Hairy roots of *Tagetes patula* (marigold) have been investigated for phytoremediation and decolorization of the triazinic-ring-containing azo dye, Reactive Red 198 (Patil et al., 2009). Treatment with the dye led to an increase in the activities of several intracellular and extracellular biotransformation enzymes, such as peroxidases and reductases, in the hairy root cultures. Reactive Red 198 was metabolized by the roots into several non-toxic compounds that were identified using GC–MS. *T. patula* hairy roots were also effective in decolorizing five other structurally diverse textile dyes with efficiencies of greater than 62%.

Pesticides

Pesticide contamination of the environment occurs mainly as a result of agricultural application of pesticides to combat insect and other small-animal infestations. Although now banned in most countries, 1,1,1-trichloro-2,2-bis-(4′-chlorophenyl)ethane (DDT) remains in soil and continues to have an impact on wildlife and human health. Hairy root cultures of *B. juncea* and *Cichorium intybus* (chicory) have been used to investigate DDT uptake and degradation in plants (Suresh et al., 2005). The results suggested that root enzymes could play a role in DDT breakdown to yield several possible degradation products identified using thin layer chromatography.

Pharmaceuticals

Antibiotics used in the livestock industry enter ecosystems in waste water from animal feeding operations and farm run-off. To minimize the development of antibiotic resistance in microorganisms, there is a need to prevent or control the widespread availability of antibiotics in the environment. *Helianthus annus* (sunflower) hairy roots have used to study phytoremediation of tetracycline and oxytetracycline (Gujarathi et al., 2005; Gujarathi and Linden, 2005). Transformation of these compounds occurred in cell-free exudates from the hairy root cultures, suggesting the involvement of root-secreted agents. Oxytetracycline was oxidized through the activity of reactive oxygen species released into the medium by hairy roots under oxidative stress conditions. The products formed in this oxidative transformation process no longer exhibited antibiotic activity (Gujarathi and Linden, 2005).

Acetaminophen (paracetamol) is a commonly used analgesic, anti-pyretic, and anti-inflammatory agent in human medicine. Concern about its excretion through sewage into the aquatic food chain and accumulation in water supplies has prompted investigation of its removal using phytoremediation. Transformation of acetaminophen by hairy roots of *A. rusticana* was found to follow metabolic pathways similar to those in mammalian cells, with the formation of acetaminophen–glutathione, acetaminophen–glucoside and acetaminophen–cysteine conjugates (Huber et al., 2009). Acetaminophen–glucoside was the dominant metabolite and is of particular interest, as glucosides of xenobiotic compounds are known precursors of insoluble residues responsible for immobilizing toxic organic substances in stable, indigestible forms bound to lignins in plant cell walls.

HAIRY ROOT STUDIES OF HEAVY METAL PHYTOREMEDIATION AND PHYTOMINING

Heavy metals from various sources such as sewage sludge, fertilizers, fossil fuel combustion, mining tailings, and manufacturing waste are significant contributors to environmental pollution. Unlike organic compounds that can be metabolically degraded, remediation of heavy metals requires their removal from the environment or conversion into biologically inert forms. Plants have a remarkable ability to accumulate and concentrate metals. However, the mechanisms involved are complex and still only partially understood. Of particular interest for phytoremediation applications are 'hyperaccumulators' of heavy metals: these are plant species capable of tolerating and storing very high concentrations of metals without deleterious physiological effects. The mechanisms of hyperaccumulation are not fully known and appear to vary significantly for different plant and metal species.

As well as phytoremediation, plants capable of removing heavy metals from the environment are important for the development of phytomining. In phytomining operations, plants growing in high-mineral environments extract and concentrate metals from the soil. The crop is then harvested and dried, and the plant biomass is treated for metal enrichment and recovery. Phytomining is a relatively cheap technology for mineral extraction and has the potential to allow economic exploitation of mineralized soils or low-grade surface ores that are too metal-poor for conventional mining operations.

Hairy root cultures have been used in several studies of heavy metal uptake and tolerance by non-hyperaccumulator and hyperaccumulator plant species. They have also been investigated for the development of appropriate processing strategies to recover metals from plants used in phytomining.

Non-Hyperaccumulators of Heavy Metals

Hairy roots of *N. tabacum*, *Beta vulgaris* (sugar beet) and *Calystegia sepium* (morning glory) were applied as models to evaluate the availability of Cd in anaerobically digested sewage sludge (Metzger et al., 1992). Growth of *B. vulgaris* roots was found to be highly sensitive to Cd. The level of Cd accumulation in the roots varied with plant species and source of the sludge. A 5-day, non-sterile bioassay procedure was developed to measure Cd accumulation in the root biomass. Cd uptake by tobacco hairy roots was lower than for whole tobacco plantlets; this was attributed to the absence of transpiration in the hairy root cultures.

The effect of experimental conditions on Cd accumulation was investigated using hairy roots of *S. nigrum* (Macek et al., 1994, 1997). Cd uptake by the roots was biphasic, with an initial phase of rapid uptake due to physical adsorption followed by slower, metabolically-sponsored accumulation. Adsorption equilibria between the root surfaces and media containing 2–50 mg L^{-1} Cd were reached after 1–3 h of exposure; however, uptake of Cd into the biomass continued by other mechanisms for a further 5–20 h. Cd accumulation was dependent on the solution temperature, composition, pH, and initial Cd concentration. Although root growth and morphology were little affected at the low Cd concentration of 2 mg L^{-1}, growth was almost completely inhibited at 50 mg L^{-1} Cd and the roots became dark in color and began to callus (Macek et al., 1997).

Hairy roots of *Rubia tinctorum* (madder) were used to study the response of this plant species to Cd (Maitani et al., 1996). Cd at a concentration of 100 μM inhibited growth of the roots. The biomass was found to be saturated with Cd after 1 day of exposure; further Cd accumulation occurred as the roots grew. The percentage of biomass-associated Cd recovered in the supernatant fraction of homogenized roots increased from 33% to 77% during the 14-day culture period, indicating a change in the subcellular distribution of Cd. Phytochelatins, which are peptides of general structure $(\gamma\text{-Glu-Cys})_n\text{-Gly}$, $n = 2\text{–}11$, involved in the detoxification of heavy metals in plants, were induced in the hairy roots by Cd treatment. Cu as well as Cd was incorporated into the phytochelatins as a metal constituent.

The response of an endangered plant species, *Adenophora lobophylla*, to Cd was investigated using hairy root cultures (Wu et al., 2001). The properties of *A. lobophylla* hairy roots were compared with those of a related but non-endangered species, *A. potaninii*, to determine if the two species differed in their ability to tolerate Cd soil pollution. Growth of the roots was inhibited by 50–400 μM Cd applied in liquid medium. Cd treatment at concentrations of 10–200 μM increased root protein contents; *A. lobophylla* was found to accumulate higher levels of Cd per mg of protein than *A. potaninii*. Endogenous levels of reduced glutathione (GSH) and cysteine were also higher in *A. lobophylla* hairy roots than in *A. potaninii*. Overall, the results suggested that these two species employ different metabolic strategies for Cd detoxification.

Rhizofiltration of U has been studied using hairy roots of *B. juncea* and *Chenopodium amaranticolor* (Eapen et al., 2003). Concentrations of uranyl nitrate from 25 to 5000 μM were tested. Root growth was reduced at U concentrations of 1000–5000 μM; growth of *B. juncea* was retarded to a much greater extent than *C. amaranticolor*. U uptake was 2–4-fold greater using *B. juncea* hairy roots than with *C. amaranticolor*.

Hairy roots of *A. rusticana* were treated with Pb, Ni and Cd to determine the effect on glutathione *S*-transferase and peroxidase enzyme activities (Nepovím et al., 2004). The presence of heavy metals decreased peroxidase activity in the roots compared with untreated controls. The expression of constitutive glutathione *S*-transferases may have been affected by the toxicity of the heavy metals and the resulting cellular stress response.

Hairy roots of *Hyptis capitata* (knobweed), *Polycarpaea longiflora* and *Euphorbia hirta* (asthma weed or hairy spurge) were studied for their tolerance and accumulation of Cu (Nedelkoska and Doran, 2000a). Whereas the short-term (9-h) Cu uptake capacities of *H. capitata* and *P. longiflora* roots were similar, the Cu content of *E. hirta* roots was lower than for the other species. In longer-term culture experiments (28 days), growth of *H. capitata* roots was not significantly affected by 20 ppm Cu in the presence of an equimolar concentration of disodium ethylene diaminetetraacetate dihydrate (EDTA). Cu uptake by *H. capitata* roots was biphasic with initial rapid accumulation followed by a slower increase in Cu content.

Hyperaccumulators of Heavy Metals

Approximately 400 plant species are known to hyperaccumulate heavy metals such as Ni, Cd, Zn, Co, Pb, and Cu (Brooks et al., 1998). Hyperaccumulators are of particular importance for the development of phytoremediation and phytomining technologies, which depend on the ability of plants to accumulate large quantities of metals in their tissues. In order to fully exploit the capacities of hyperaccumulators, or to endow other species with hyperaccumulator traits through genetic engineering, an understanding of the biological mechanisms responsible for hyperaccumulation is required. Efficient transport of heavy metals across the plasma membrane or into the vacuoles of plant cells has been demonstrated to occur in hyperaccumulator species (Lombi et al., 2001; Pence et al., 2000; Persans et al., 2001). In other studies, metal-chelating agents such as organic and amino acids have been implicated in heavy metal transport and tolerance in hyperaccumulator plants (Krämer et al., 1996, 2000; Sagner et al., 1998). To maintain cellular integrity and function in the presence of high external and internal concentrations of heavy metals, hyperaccumulators must possess a means of neutralizing the toxic effects of metal ions as soon as they contact the cells.

Cadmium

Hairy roots of the Cd hyperaccumulator, *Thlaspi caerulescens* (alpine pennycress), were used to study metal tolerance and hyperaccumulation in this species (Nedelkoska and Doran, 2000b). *T. caerulescens* hairy roots exhibited superior tolerance to Cd than hairy roots of the non-hyperaccumulator, *N. tabacum*. Whereas growth of the *T. caerulescens* roots was essentially unaffected by 20–50 ppm Cd, in contrast, *N. tabacum* roots turned dark brown after exposure to 20 ppm Cd and growth was severely retarded.

As high concentrations of heavy metals generate stress responses in plants, including oxidative stress, hairy roots of *T. caerulescens* were applied to test the hypothesis that antioxidative defences play a key role in Cd tolerance in this species (Boominathan and Doran, 2003a). The activities of three antioxidative enzymes, superoxide dismutase (SOD), catalase (CAT), and ascorbate peroxidase (APX), were measured in *T. caerulescens* and *N. tabacum* hairy roots as a function of culture time. Figure 3 shows results for maximum enzyme activities measured in the biomass in the absence and presence of 20 ppm (178 μM) Cd. There were significant differences between *T. caerulescens* and *N. tabacum* roots in terms of their antioxidative enzyme activities and responses to Cd. Without Cd, endogenous SOD activities were higher in *T. caerulescens* than *N. tabacum*; in addition, Cd treatment resulted in a decrease in SOD activity in *N. tabacum* (Figure 3a). The greatest distinction between *T. caerulescens* and *N. tabacum* was evident in the results for CAT (Figure 3b). Maximum endogenous CAT activities were more than two orders of magnitude higher in *T. caerulescens* hairy roots than in *N. tabacum*. Although CAT activity in *N. tabacum* increased when Cd was added to the cultures, the induced CAT levels remained substantially lower than in *T. caerulescens*. APX activities in *T. caerulescens* were somewhat lower than in *N. tabacum* and were not significantly altered by Cd treatment in either species (Figure 3c). Overall, the results shown in Figure 3, particularly those for CAT, suggest that *T. caerulescens* is equipped with superior antioxidative defences compared with the non-hyperaccumulator, *N. tabacum*.

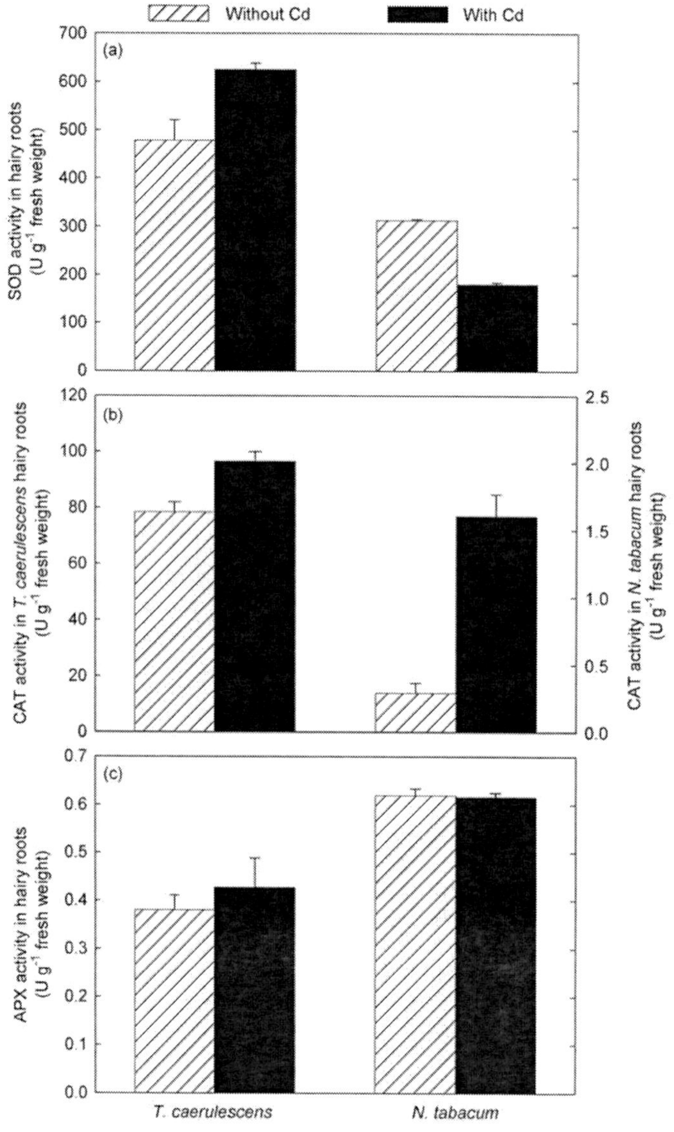

Figure 3. Maximum antioxidative enzyme activities in hairy roots of *T. caerulescens* (Cd hyperaccumulator) and *N. tabacum* (non-hyperaccumulator) cultured for 21–28 days with and without 20 ppm Cd. (a) Superoxide dismutase (SOD) activity; (b) catalase (CAT) activity; (c) ascorbate peroxidase (APX) activity. The error bars represent standard errors from triplicate cultures. Data from Boominathan and Doran (2003a).

Sustained build-up of H_2O_2 is a typical plant cell response to toxic concentrations of Cd (Stroinski and Zielezinska, 1997; Schützendübel et al., 2001). To assess the ability of *T. caerulescens* and *N. tabacum* to control the oxidative burst induced by Cd treatment, H_2O_2 levels were measured in hairy roots of these species (Boominathan and Doran, 2003a). As shown in Figure 4, maximum H_2O_2 concentrations in the *T. caerulescens* roots were not significantly different with and without Cd, suggesting that *T. caerulescens* exerted tight control over H_2O_2 accumulation. In contrast, Cd elicited a substantial increase in H_2O_2 levels in *N. tabacum* roots, indicating that *N. tabacum* was not as capable as *T. caerulescens* of preventing the build-up of reactive oxygen species. The ability of *T. caerulescens* to keep H_2O_2 concentrations in check may be an important survival mechanism for this hyperaccumulator species.

Nickel

Hairy roots of three Ni hyperaccumulators, *Alyssum bertolonii*, *A. tenium* and *A. troodii*, were applied in an investigation of Ni tolerance and uptake in plant tissues (Nedelkoska and Doran, 2001). Whereas *A. bertolonii* hairy roots grew and remained healthy in appearance in the presence of 20–100 ppm Ni, hairy roots of the non-hyperaccumulator, *N. tabacum*, turned dark brown at 20 ppm Ni and growth was negligible. Enhanced Ni tolerance in *A. bertolonii* was thus demonstrated independent of the presence of shoots. However, although hairy roots were shown in these studies to be a useful experimental tool, comparisons between *A. tenium* hairy roots and regenerated plants showed that there were significant differences in Ni uptake capacity and tolerance between these forms of plant culture. Whole *A. tenium* plants in hydroponic culture were much more tolerant of Ni and capable of accumulating higher Ni concentrations than hairy roots of the same species, suggesting that functional root–shoot translocation plays a significant role in the detoxification of Ni in whole plants.

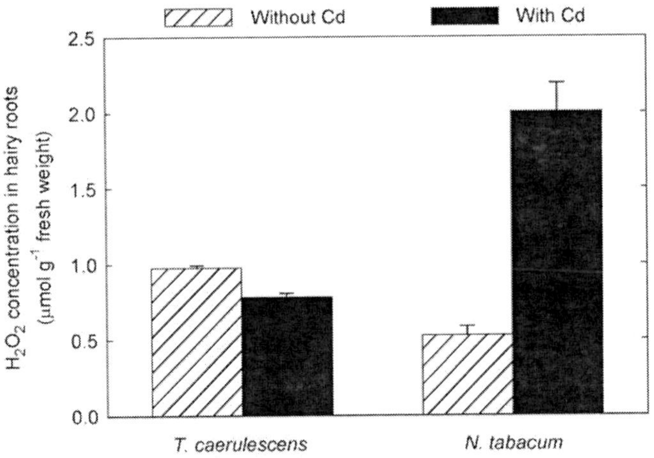

Figure 4. Maximum H_2O_2 concentrations in hairy roots of *T. caerulescens* (Cd hyperaccumulator) and *N. tabacum* (non-hyperaccumulator) cultured for 21–28 days with and without 20 ppm Cd. The error bars represent standard errors from triplicate cultures. Data from Boominathan and Doran (2003a).

Analysis of Ni uptake, complexation and distribution in *A. bertolonii* hairy roots showed that 85–95% of the Ni present in the biomass was in the cell symplasm and not associated with the cell walls (Boominathan and Doran, 2003b). About 28% of Ni in the roots was complexed with three organic acids: citric acid, malic acid and malonic acid. Hairy roots of *A. bertolonii* were used to study the antioxidative responses of this hyperaccumulator species to Ni (Boominathan and Doran, 2002). As shown in Figure 5a, without Ni, the maximum endogenous SOD activity in *A. bertolonii* hairy roots was significantly greater than in *N. tabacum*. After treatment with 25 ppm (426 μM) Ni, SOD activity levels in *A. bertolonii* roots were about twice those in *N. tabacum*. Without Ni, maximum endogenous CAT activities in *A. bertolonii* roots were about 280 times greater than in *N. tabacum*; CAT activity was not induced by Ni in either species (Figure 5b).

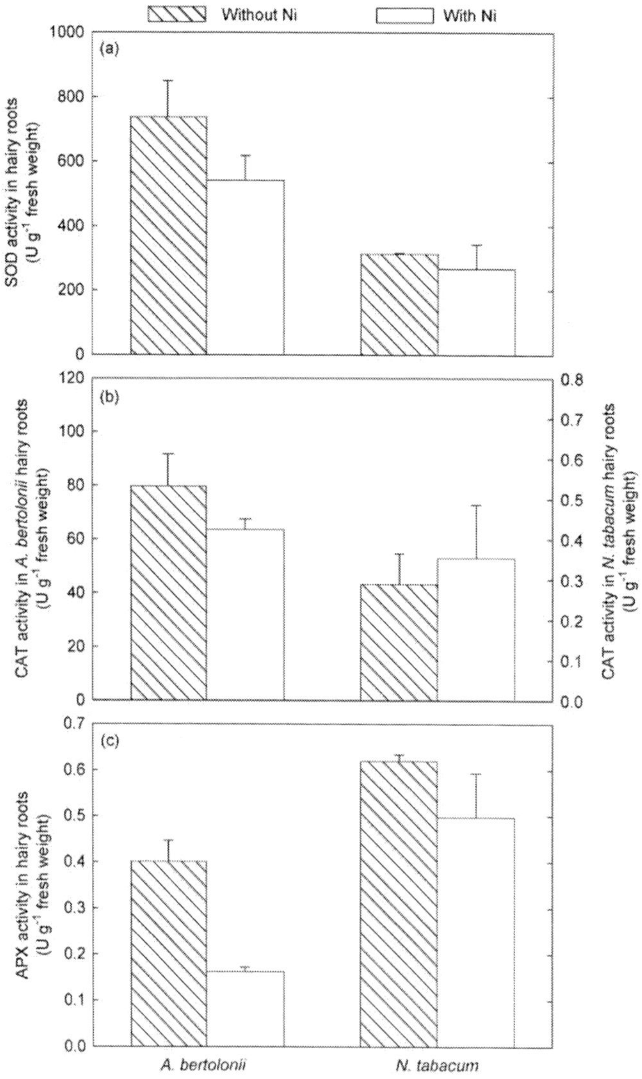

Figure 5. Maximum antioxidative enzyme activities in hairy roots of *A. bertolonii* (Ni hyperaccumulator) and *N. tabacum* (non-hyperaccumulator) cultured for 21–28 days with and without 25 ppm Ni. (a) Superoxide dismutase (SOD) activity; (b) catalase (CAT) activity; (c) ascorbate

peroxidase (APX) activity. The error bars represent standard errors from triplicate cultures. Data from Boominathan and Doran (2002).

Without Ni, maximum endogenous APX activities in *A. bertolonii* roots were lower than in *N. tabacum* and declined significantly after exposure to Ni (Figure 5c). The maximum APX activity in Ni-treated *N. tabacum* roots was about three times that in Ni-treated *A. bertolonii*. Endogenous H_2O_2 concentrations were similar in the two species (Figure 6); however, *A. bertolonii* hairy roots were better able to control H_2O_2 accumulation after Ni treatment than *N. tabacum*, even though H_2O_2 levels rose in both cultures. These results indicate that the much higher CAT activities in the hyperaccumulator roots are likely to give *A. bertolonii* an advantage over *N. tabacum* for combating Ni-induced oxidative stress (Boominathan and Doran, 2002).

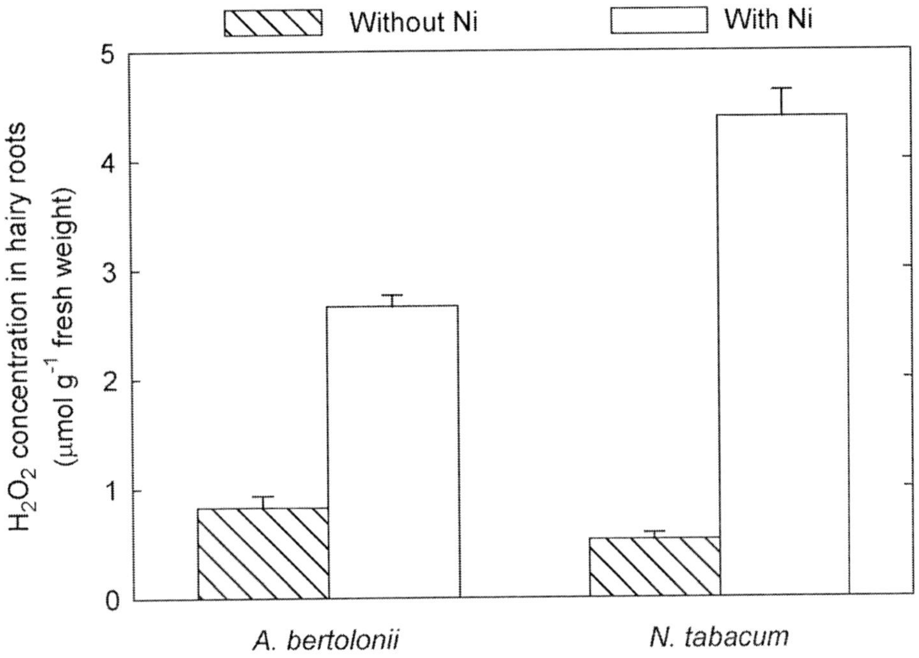

Figure 6. Maximum H_2O_2 concentrations in hairy roots of *A. bertolonii* (Ni hyperaccumulator) and *N. tabacum* (non-hyperaccumulator) cultured for 21–28 days with and without 25 ppm Ni. The error bars represent standard errors from triplicate cultures. Data from Boominathan and Doran (2002).

Phytomining

In phytomining, metal-laden plant biomass harvested from mineral-rich soil must be converted into a 'bio-ore' suitable for further processing and metal recovery. Little work has been carried out to investigate methods for generating bio-ores or to examine whether plant-derived feedstocks are suitable for conventional mineral processing operations such as smelting, flotation or acid leaching.

In initial work in this area, hairy roots of the Ni hyperaccumulator, *A. bertolonii*, were exposed to $NiCl_2{\cdot}6H_2O$ in medium solutions to generate Ni-rich root biomass (Boominathan et al., 2004). The roots were then dried, ground to a powder and placed in a laboratory-scale horizontal-tube furnace operated at 1200°C open to the atmosphere. The weight loss of the samples was about 94% for furnace treatment times of 3 to 17 h and the Ni enrichment factor was 15–18. The resulting furnace residue contained 78–82% Ni by weight, which is a much higher Ni content than the 1–2% Ni typically found in mined ores. The surface morphology of the furnace residue showed the formation of crystalline structures with a dark gray, metallic appearance. Surface energy-dispersive spectroscopy (EDS) indicated that the major elements remaining in the residue other than Ni were Ca, P and Mg. *A. bertolonii* hairy roots provided a convenient model system in this work for investigating the production of plant-derived Ni-bearing bio-ore.

TRANSGENIC HAIRY ROOT STUDIES IN PHYTOREMEDIATION

Genetic modification is a potential solution to the inherent kinetic and metabolic limitations of plants applied for environmental clean-up. The creation of new plants with improved degradative or accumulative capabilities is of great commercial interest. Hairy roots are a quick and easy means for testing the effects of genetic transformation on plants. Introduction of foreign genes into *A. rhizogenes* bacteria allows genetic modifications to be effected at the same time as hairy roots are generated. Alternatively, transgenic hairy roots can be initiated from already-transformed plant material. Transgenic hairy root cultures are a useful tool in metabolic studies or for screening genetic transformants prior to regeneration of whole plants with enhanced phytoremediation potential.

A. belladonna seedlings were used to develop hairy roots expressing rabbit P450 2E1 (Banerjee et al., 2002). P450 2E1 is a mammalian liver enzyme involved in the metabolism of a range of endogenous and xenobiotic substances, including trichloroethylene (TCE). The gene for P450 2E1 was incorporated into *A. belladonna* after introducing an appropriate binary vector into *A. rhizogenes* for generation of hairy roots. P450 2E1 expressed in the roots remained with the biomass and was not found in the medium. Wild-type *A. belladonna* hairy roots were capable of oxidizing TCE to trichloroethanol (TCOH); however, TCOH levels were 5–10-fold higher in transgenic hairy roots expressing P450 2E1. As activity of the enzyme requires complexation with two additional membrane proteins, cytochrome P450 oxidoreductase and cytochrome B5, and since transformation with the P450 2E1 gene alone enhanced TCE biotransformation in roots, the results suggest that mammalian P450 2E1 expressed in plants was able to form complexes with plant homologs of oxidoreductase and cytochrome B5 proteins.

Transgenic hairy roots have been used in studies to enhance the capacity of plants for enzymatic detoxification of phenol. The practical feasibility of using enzymes to remove phenols from the environment is limited by the rapid deactivation of peroxidase after contact with the substrate. A potential strategy for overcoming this problem is to develop new plants with enhanced expression of large quantities of peroxidase enzyme. *L. esculentum* hairy roots were found to contain relatively high levels of peroxidase and were therefore capable of removing phenols from aqueous solution. This capacity was increased further by introducing the gene for an endogenous basic (pI = 9.6) peroxidase into tomato plants using *A. tumefaciens* transformation and then initiating hairy roots from leaf and stalk segments of the transgenic plants (Wevar Oller et al., 2005). Transgenic hairy root clones were selected for high specific peroxidase activity, high growth rate, and stable morphology. Phenol removal by the transgenic hairy roots was higher than that achieved using wild-type hairy roots; however, although peroxidase activities were up to 2.6-fold higher in crude extracts of the transgenic roots, phenol removal was improved by only about 20%. H_2O_2-independent enzymatic reactions for phenol removal and other non-specific physical and reactive processes such as phenol adsorption to the biomass accounted for 40–45% of the phenol removal observed in all types of tomato hairy root culture. Peroxidase activity was reduced by about 50% in extracts of both the wild-type and transgenic hairy roots after exposure to phenol. In other work, transgenic tobacco hairy roots expressing basic peroxidases from tomato achieved 12–15% higher phenol removal relative to wild-type tobacco hairy roots (Sosa Alderete et al., 2009).

Conclusion

Hairy roots are being applied with increasing frequency in phytoremediation research. As a laboratory tool, hairy roots offer a range of advantages such as ease of initiation, culture and maintenance, indefinite propagation, and genotypic and phenotypic stability. The principal applications of hairy roots have been in metabolic studies of xenobiotic biotransformation and degradation in plants, and for determining the responses of plant tissues to toxic heavy metals. Being themselves products of genetic transformation, hairy roots are readily amenable to genetic modification to allow rapid screening of transformants for improved phytoremediation traits and to test whether foreign gene expression can improve plant tolerance to and/or metabolism of pollutants.

Acknowledgments

Financial support for the author's research was provided by the Australian Research Council (ARC).

References

Agostini, E., Coniglio, M. S., Milrad, S. R., Tigier, H. A., Giulietti, A. M. (2003) Phytoremediation of 2,4-dichlorophenol by *Brassica napus* hairy root cultures. *Biotechnol. Appl. Biochem*. 37, 139–144.

Ambros, P. F., Matzke, A. J. M., Matzke, M. A. (1986) Localization of *Agrobacterium rhizogenes* T-DNA in plant chromosomes by *in situ* hybridization. *EMBO J*. 5, 2073–2077.

Banerjee, S., Shang, T. Q., Wilson, A. M., Moore, A. L., Strand, S. E., Gordon, M. P., Doty, S. L. (2002) Expression of functional mammalian P450 2E1 in hairy root cultures. *Biotechnol. Bioeng.* 77, 462–466.

Bhadra, R., Wayment, D. G., Williams, R. K., Barman, S. N., Stone, M. B., Hughes, J. B., Shanks, J. V. (2001) Studies on plant-mediated fate of the explosives RDX and HMX. *Chemosphere* 44, 1259–1264.

Boominathan, R., Doran, P. M. (2002) Ni-induced oxidative stress in roots of the Ni hyperaccumulator, *Alyssum bertolonii*. *New Phytol.* 156, 205–215.

Boominathan, R., Doran, P. M. (2003a) Cadmium tolerance and antioxidative defenses in hairy roots of the cadmium hyperaccumulator, *Thlaspi caerulescens*. *Biotechnol. Bioeng.* 83, 158–167.

Boominathan, R., Doran, P. M. (2003b) Organic acid complexation, heavy metal distribution and the effect of ATPase inhibition in hairy roots of hyperaccumulator plant species. *J. Biotechnol.* 101, 131–146.

Boominathan, R., Saha-Chaudhury, N. M., Sahajwalla, V., Doran, P. M. (2004) Production of nickel bio-ore from hyperaccumulator plant biomass: applications in phytomining. *Biotechnol. Bioeng.* 86, 243–250.

Brooks, R. R., Chambers, M. F., Nicks, L. J., Robinson, B. H. (1998) Phytomining. *Trends Plant Sci.* 3, 359–362.

Byrne, M. C., Koplow, J., David, C., Tempé, J., Chilton, M. -D. (1983) Structure of T-DNA in roots transformed by *Agrobacterium rhizogenes*. *J. Mol. Appl. Genet.* 2, 201–209.

Caron, D., Coughlan, A. P., Simard, M., Bernier, J., Piché, Y., Chênevert, R. (2005) Stereoselective reduction of ketones by *Daucus carota* hairy root cultures. *Biotechnol. Lett.* 27, 713–716.

Coniglio, M. S., Busto, V. D., González, P. S., Medina, M. I., Milrad, S., Agostini, E. (2008) Application of *Brassica napus* hairy root cultures for phenol removal from aqueous solutions. *Chemosphere* 72, 1035–1042.

Coleman, J. O. D., Blake-Kalff, M. M. A., Davies, T. G. E. (1997) Detoxification of xenobiotics by plants: chemical modification and vacuolar compartmentation. *Trends Plant Sci.* 2, 144–151.

de Araujo, B. S., Charlwood, B. V., Pletsch, M. (2002) Tolerance and metabolism of phenol and chloroderivatives by hairy root cultures of *Daucus carota* L. *Env. Pollut.* 117, 329–335.

Doran, P. M. (2009) Application of plant tissue cultures in phytoremediation research: incentives and limitations. *Biotechnol. Bioeng.* 103, 60–76.

Eapen, S., Suseelan, K. N., Tivarekar, S., Kotwal, S. A., Mitra, R. (2003) Potential for rhizofiltration of uranium using hairy root cultures of *Brassica juncea* and *Chenopodium amaranticolor*. *Env. Res.* 91, 127–133.

González, P. S., Capozucca, C. E., Tigier, H. A., Milrad, S. R., Agostini, E. (2006) Phytoremediation of phenol from wastewater, by peroxidases of tomato hairy root cultures. *Enzyme Microb. Technol.* 39, 647–653.

Gujarathi, N. P., Haney, B. J., Park, H. J., Wickramasinghe, S. R., Linden, J. C. (2005) Hairy roots of *Helianthus annuus*: a model system to study phytoremediation of tetracycline and oxytetracycline. *Biotechnol. Prog.* 21, 775–780.

Gujarathi, N. P., Linden, J. C. (2005) Oxytetracycline inactivation by putative reactive oxygen species released to nutrient medium of *Helianthus annuus* hairy root cultures. *Biotechnol. Bioeng.* 92, 393–402.

Hamill, J. D., Lidgett, A. J. (1997) *Hairy root cultures – opportunities and key protocols for studies in metabolic engineering*. In: P. M. Doran (Ed.), Hairy Roots: Culture and Applications, pp. 1–29. Amsterdam: Harwood Academic.

Huber, C., Bartha, B., Harpaintner, R., Schröder, P. (2009) Metabolism of acetaminophen (paracetamol) in plants – two independent pathways result in the formation of a glutathione and a glucose conjugate. *Environ. Sci. Pollut. Res.* 16, 206–213.

Hughes, J. B., Shanks, J., Vanderford, M., Lauritzen, J., Bhadra, R. (1997) Transformation of TNT by aquatic plants and plant tissue cultures. *Environ. Sci. Technol.* 31, 266–271.

Kamada, H., Okamura, N., Satake, M., Harada, H., Shimomura, K. (1986) Alkaloid production by hairy root cultures in *Atropa belladonna*. *Plant Cell Rep.* 5, 239–242.

Krämer, U., Cotter-Howells, J. D., Charnock, J. M., Baker, A. J. M., Smith, J. A. C. (1996) Free histidine as a metal chelator in plants that accumulate nickel. *Nature* 379, 635–638.

Krämer, U., Pickering, I. J., Prince, R. C., Raskin, I., Salt, D. E. (2000) Subcellular localization and speciation of nickel in hyperaccumulator and non-accumulator *Thlaspi* species. *Plant Physiol.* 122, 1343–1353.

Kucerová, P., Macková, M., Chromá, L., Burkhard, J., Tríska, J., Demnerová, K., Macek, T. (2000) Metabolism of polychlorinated biphenyls by *Solanum nigrum* hairy root clone SNC-9O and analysis of transformation products. *Plant Soil* 225, 109–115.

Lombi, E., Zhao, F. J., McGrath, S. P., Young, S. D., Sacchi, G. A. (2001) Physiological evidence for a high-affinity cadmium transporter highly expressed in a *Thlaspi caerulescens* ecotype. *New Phytol.* 149, 53–60.

Macek, T., Kotrba, P., Suchova, M., Skacel, F., Demnerova, K., Ruml, T. (1994) Accumulation of cadmium by hairy-root cultures of *Solanum nigrum*. *Biotechnol. Lett.* 16, 621–624.

Macek, T., Kotrba, P., Ruml, T., Skácel, F., Macková, M. (1997) *Accumulation of cadmium ions by hairy root cultures.* In: P. M. Doran (Ed.), Hairy Roots: Culture and Applications, pp. 133–138. Amsterdam: Harwood Academic.

Mackova, M., Macek, T., Ocenaskova, J., Burkhard, J., Demnerova, K., Pazlarova, J. (1997a) Biodegradation of polychlorinated biphenyls by plant cells. *Int. Biodeter. Biodegrad.* 39, 317–325.

Macková, M., Macek, T., Kucerová, P., Burkhard, J., Pazlarová, J., Demnerová, K. (1997b) Degradation of polychlorinated biphenyls by hairy root culture of *Solanum nigrum*. *Biotechnol. Lett.* 19, 787–790.

Maitani, T., Kubota, H., Sato, K., Takeda, M., Yoshihira, K. (1996) Induction of phytochelatin (class III metallothionein) and incorporation of copper in transformed hairy roots of *Rubia tinctorum* exposed to cadmium. *J. Plant Physiol.* 147, 743–748

Metzger, L., Fouchault, I., Glad, C., Prost, R., Tepfer, D. (1992) Estimation of cadmium availability using transformed roots. *Plant Soil* 143, 249–257.

Nedelkoska, T. V., Doran, P. M. (2000a) Characteristics of heavy metal uptake by plant species with potential for phytoremediation and phytomining. *Miner. Eng.* 13, 549–561.

Nedelkoska, T. V., Doran, P. M. (2000b) Hyperaccumulation of cadmium by hairy roots of *Thlaspi caerulescens*. *Biotechnol. Bioeng.* 67, 607–615.

Nedelkoska, T. V., Doran, P. M. (2001) Hyperaccumulation of nickel by hairy roots of *Alyssum* species: comparison with whole regenerated plants. *Biotechnol. Prog.* 17, 752–759.

Nepovím, A., Podlipná, R., Soudek, P., Schröder, P., Vaněk, T. (2004) Effects of heavy metals and nitroaromatic compounds on horseradish glutathione *S*-transferase and peroxidase. *Chemosphere* 57, 1007–1015.

Orden, A. A., Bisogno, F. R., Cifuente, D. A., Giordano, O. S., Kurina-Sanz, M. (2006) Asymmetric bioreduction of natural xenobiotic diketones by *Brassica napus* hairy roots. *J. Mol. Catal. B Enzym.* 42, 71–77.

Orden, A. A., Magallanes-Noguera, C., Agostini, E., Kurina-Sanz, M. (2009) Anti-Prelog reduction of ketones by hairy root cultures. *J. Mol. Catal. B Enzym.* 61, 216–220.

Patil, P., Desai, N., Govindwar, S., Jadhav, J. P., Bapat, V. (2009) Degradation analysis of Reactive Red 198 by hairy roots of *Tagetes patula* L. (marigold). *Planta* 230, 725–735.

Pence, N. S., Larsen, P. B., Ebbs, S. D., Letham, D. L. D., Lasat, M. M., Garvin, D. F., Eide, D., Kochian, L. V. (2000) The molecular physiology of heavy metal transport in the Zn/Cd hyperaccumulator *Thlaspi caerulescens*. *Proc. Natl. Acad. Sci. USA* 97, 4956–4960.

Persans, M. W., Nieman, K., Salt, D. E. (2001) Functional activity and role of cation-efflux family members in Ni hyperaccumulation in *Thlaspi goesingense*. *Proc. Natl. Acad. Sci. USA* 98, 9995–10000.

Rezek, J., Macek, T., Mackova, M., Triska, J. (2007) Plant metabolites of polychlorinated biphenyls in hairy root culture of black nightshade *Solanum nigrum* SNC-9O. *Chemosphere* 69, 1221–1227.

Sagner, S., Kneer, R., Wanner, G., Cosson, J. -P., Deus-Neumann, B., Zenk, M. H. (1998) Hyperaccumulation, complexation and distribution of nickel in *Sebertia acuminata*. *Phytochemistry* 47, 339–347.

Sandermann, H. (1994) Higher plant metabolism of xenobiotics: the 'green liver' concept. *Pharmacogenetics* 4, 225–241.

Schützendübel, A., Schwanz, P., Teichmann, T., Gross, K., Langenfeld-Heyser, R., Godbold, D. L., Polle, A. (2001) Cadmium-induced changes in antioxidative systems, hydrogen peroxide content, and differentiation in Scots pine roots. *Plant Physiol.* 127, 887–898.

Singh, S., Melo, J. S., Eapen, S., D'Souza, S. F. (2006) Phenol removal using *Brassica juncea* hairy roots: role of inherent peroxidase and H_2O_2. *J. Biotechnol.* 123, 43–49.

Sosa Alderete, L. G., Talano, M. A., Ibáñez, S. G., Purro, S., Agostini, E., Milrad, S. R., Medina, M. I. (2009) Establishment of transgenic tobacco hairy roots expressing basic peroxidases and its application for phenol removal. *J. Biotechnol.* 139, 273–279.

Stroinski, A., Zielezinska, M. (1997) Cadmium effect on hydrogen peroxide, glutathione and phytochelatin levels in potato tuber. *Acta Physiol. Plant.* 19, 127–135.

Suresh, B., Sherkhane, P. D., Kale, S., Eapen, S., Ravishankar, G. A. (2005) Uptake and degradation of DDT by hairy root cultures of *Cichorium intybus* and *Brassica juncea*. *Chemosphere* 61, 1288–1292.

Talano, M. A., Frontera, S., González, P., Medina, M. I., Agostini, E. (2010) Removal of 2,4-dichlorophenol from aqueous solutions using tobacco hairy root cultures. *J. Hazard. Mater.* 176, 784–791.

Tepfer, D. (1984) Transformation of several species of higher plants by *Agrobacterium rhizogenes*: sexual transmission of the transformed genotype and phenotype. *Cell* 37, 959–967.

Tepfer, D. (1989) *Ri T-DNA from Agrobacterium rhizogenes: a source of genes having applications in rhizosphere biology and plant development, ecology, and evolution*. In T. Kosuge, E. W. Nester (Eds.), Plant–Microbe Interactions: Molecular and Genetic Perspectives, vol. 3, pp. 294–342. New York: McGraw-Hill.

van Eerd, L. L., Hoagland, R. E., Zablotowicz, R. M., Hall, J. C. (2003) Pesticide metabolism in plants and microorganisms. *Weed Sci*. 51, 472–495.

Wevar Oller, A. L., Agostini, E., Talano, M. A., Capozucca, C., Milrad, S. R., Tigier, H. A., Medina, M. I. (2005) Overexpression of a basic peroxidase in transgenic tomato (*Lycopersicon esculentum* Mill. cv. Pera) hairy roots increases phytoremediation of phenol. *Plant Sci*. 169, 1102–1111.

Wu, S., Zu, Y., Wu, M. (2001) Cadmium response of the hairy root culture of the endangered species *Adenophora lobophylla*. *Plant Sci*. 160, 551–562.

In: Handbook of Phytoremediation
Editor: Ivan A. Golubev
ISBN: 978-1-61728-753-4

Chapter 20

APPLICATION OF PHYTOREMEDIATION FROM EXPERIMENTAL STAGE TO PRACTICAL STAGE: COMPARATIVE STUDY IN THE SOUTHERN PART AND THE NORTHERN PART OF THE EUROPEAN REGION

Ryunosuke Kikuchi[1*], Tamara T. Gorbacheva[2] and Romeu Gerardo[3]
[1,3]ESAC – Polytechnic Institute of Coimbra, Bencanta, Coimbra, Portugal
[2]Institute of the North Industrial Ecological Problem – Kola Science Center, Apatity, Russia

ABSTRACT

Phytoremediation is the use of plants to remove contaminants from the environment or render them harmless. Current engineering-based technologies to clean up soils are costly, and most considerations usually state that soil phytoremediation will be cheaper than alternatives such as soil washing. However, phytoremediation is a comparatively new field and not all of its applications are well understood. Most metal-contaminated soils contain more than one metal. For example, combinations of Pb and Zn are common in urban soils, while Pb, Zn, Cd and Cu are all often present in the vicinity of a metallurgic smelter. There will be minimal economic value in a technology that can efficiently remove one metal from a soil but leave most of another behind. However, most of the experiments on phytoremediation only address a single metal contaminant. Two field surveys were carried out in order to understand the multiple-metal effect on phytoremediation.

A 2-ha survey was performed over 2 years to study how plants such as eucalypts would remove lead and zinc from the abandoned mine at Sanguinheiro (40°30'N and 8°18'W in Portugal). The average comparison of metal content in leaves is summarized as follows (cf. remediation zone vs. background zone): Pb – 2.9 vs. 3.6 mg kg^{-1} and Zn – 29.7 vs. 14.1 mg kg^{-1}. Another 8-ha survey using willow was also performed over 2 years under conditions of continuous metal deposition near the Monchegorsk smelter (68°02'N and 34°48'E in the most northern part of the European fringe of the Russian Federation). The average comparison of metal content in leaves is summarized as follows (cf. remediation zone with fertilization vs. background zone): K – 6781 vs. 7635 mg kg^{-1}, Mn

[*] Tel.: +351 239 802940; fax: +351 239 802979. E-mail: kikuchi@mail.esac.pt

– 43 vs. 845 mg kg^{-1}, P – 2303 vs. 2856 mg kg^{-1}, and Zn – 109 vs. 161 mg kg^{-1}. The results obtained from the metal analysis (Cu, Ni, Fe, Pb, etc.) indicate high efficiencies of phytoremediation (i.e. preferable effect of phytoextraction), but a clear relation of leaf chemistry with soil chemistry could not be obtained.

Both field tests in Portugal and Russia suggest that the root system is more important than the leaf system in the evaluation of remediation efficiency. The data presented in this chapter may help the planning of a commercial application of phytoremediation in cases of multiple-metal stress. However, long-term observation is also necessary to confirm reliable feasibility for underpinning the design of a large-scale phytoremediation project.

Introduction

BATNEEC means "the best available technology not entailing excessive costs", and NEEC means "not entailing excessive costs." The BATNEEC concept sets out the balance between environmental benefit and financial cost (review in [Macken, 1995]). It is now gaining attention in Europe in the area of environment management (e.g. EU air framework directive, EU dangerous substance directive, etc.). This concept will require the adaptation of an on-going program of environmental management and control that will focus on continuing improvements aimed at prevention, elimination and/or progressive reduction (review in [Macken, 1995]).

From the NEEC viewpoint, cost is a thorny subject in a remediation project, and the balance between the promised effect and the financial cost is often highlighted. In the past, chemical pollution in soil has been treated using a number of methods [Semple et al., 2001; Houghton, 1996]: physical and chemical processes are expensive; in addition, thermal treatment and removal to landfill are also expensive; by contrast, biological methods are comparatively cheap. It is only a few years since the strategies of bioremediation were adopted; as a result, there is a lack of general information as well as a limited number of pollutants (or pollutant matrixes) treated [Semple et al., 2001]. It seem to be ecumenically advantageous to use a biological method, but there is doubt as to whether biological methods are promising in all cases.

Phytoremediation is a biologic treatment technique which uses plants; this technique has the advantage of not only purifying or removing pollutants but also producing plants [Mueller et al., 1999a]. Current engineering-based technologies to clean up soils are costly, and most considerations usually state that soil phytoremediation will be cheaper than alternatives such as soil remediation, but phytoremediation is a comparatively new field and not all of its applications are well understood [Mueller et al., 1999a]. Therefore, the present chapter mainly discusses the treatment performance of phytoremediation on metal-contaminated land. This leads to the suggestion of possibly linking NEEC with phytoremediation.

Background

It seems impossible to restore metal-contaminated areas without a great amount of financial aid: e.g. the Sudbury project in Canada spent C$15 million (ca. 11 million Euro) to restore a polluted area of 3,700 ha during 1978-1993 [Winterhalder, 2000]. More than 50,000 metal-contaminated sites await remediation in the U.S. alone [Ensley, 2000]. Approximately 80% of the U.S. superfund sites (designated by the U.S. Environmental Protection Agency as priority sites for cleanup) contain heavy metals, often mixed with organic pollutants [Ensley, 2000]. Conventional remediation methods for metal-contaminated substrates include soil washing, excavation, and reburial of soil, and pump and treat systems for water [Glass, 1999]. The present costs of U.S. remediation are US$7 to 8 billion per year, of which about 35% involves metals remediation [Glass, 1999, 2000]. The use of plants for remediation of metals offers an attractive alternative, because it is solar driven and can be carried out in situ, minimizing cost and human exposure [Salt et al., 1998].

Metal Contamination on Land

Environmental problems associated with metals are well reviewed as follows [Pilon-Smits and Pilon, 2002]. Metals are present naturally in the Earth's crust at various levels (cf. [Angelone and Bini, 1992]). Mining, industry, and agriculture lead to accelerated release of metals into ecosystems, causing serious environmental problems and posing a threat to human health (e.g. [Ross, 1994]). Although many metals (Cu, Fe, Mn, Ni, Zn, etc.) are essential for cells, all metals are toxic at higher concentrations [Marschner, 1995]. One reason metals may become toxic is that they may cause oxidative stress. Especially redox active transition metals, which can take up or give off an electron (e.g. $Fe^{2+/3+}$, $Cu^{+/2+}$), can give rise to free radicals that cause damage [Li and Trush, 1993], but other metals can cause oxidative stress as well [Baccouch et al., 1998; Cho and Park, 2000].

Approach of Phytoremediation to Metal Treatment

Phytoremediation applications are classified based on contaminant fate, degradation, extraction, contaminant type or a combination of these. In the soil-plant-atmosphere continuum, a specific contaminant can be remediated at specific points along this continuum by the different phytoremediation mechanisms. This is illustrated in figure 1.

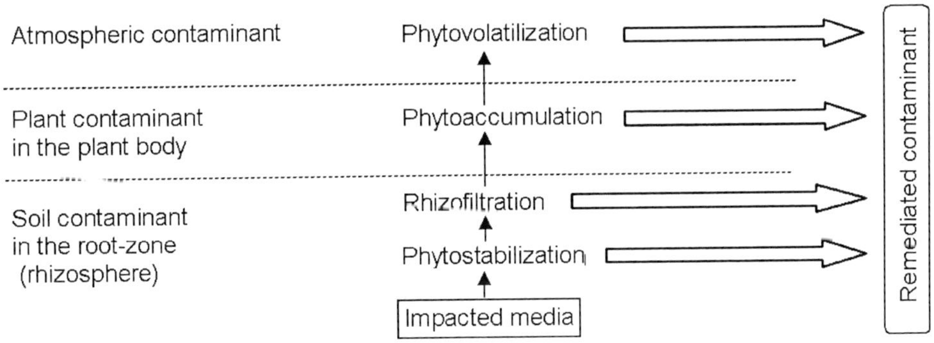

Figure 1. Fates of inorganic contaminants in the soil-plant-atmosphere continuum (redrawn from [Mueller et al., 1999b]).

Unlike organic contaminants, metals cannot be degraded. Instead, phytoremediation strategies for metals are based on stabilization, accumulation, and in some cases volatilization [EPA, 1998]. Phytostabilization of metals may simply involve the prevention of leaching through the upward water flow created by plant transpiration, reduced runoff due to above-ground vegetation, and reduced soil erosion via stabilization of soil by plant roots [Berti and Cunningham, 2000]. In some cases of phytostabilization, metals may be transformed to less bioavailable and therefore less toxic forms. For instance, many wetland plants reduce metals to insoluble precipitates on their root surface [Horne, 2000]. Thus, in phytostabilization, mobilization of metals is prevented; although metal concentrations are not reduced, the metal becomes less of a risk to the environment.

Accumulation of metals by roots in a hydroponic setup, followed by harvesting of the plant biomass, is termed rhizofiltration [Dushenkov and Kapulnik, 2000]. The accumulation of metals in shoot tissue, followed by harvesting of shoot biomass, is called phytoextraction [Blaylock and Huang, 2000]. After harvesting of the root and/or shoot biomass, the plant material may be ashed, followed by recycling of the metals if economically feasible [Chaney et al., 2000], or the disposal of the ash in landfill. Alternatively, the plant material may be used for non-food purposes, for example, cardboard or wood products.

Certain metals/metaloids can be converted by plants into a gaseous form and emitted into the atmosphere [Hansen et al., 1998]. The use of plants for volatilization of contaminants is called phytovolatilization.

Each of these metal phytoremediation technologies has already been shown to be effective. To give a few examples: a rhizofiltration system uses sunflowers to remove radioactive U from contaminated wastewater to levels below regulatory limits (95% removal in 24 hours [Dushenkov and Kapulnik, 2000]); chelator-assisted phytoextraction using Brassica juncea lowered soil Pb levels from 2055 mg kg^{-1} to 960 mg kg^{-1} in three crops [Blaylock and Huang, 2000]; constructed wetlands routinely remove over 90% of metals from various wastewater streams [Horne, 2000].

Phytoextraction, phytovolatilization, phytostabilization, and rhizofiltration are not exclusive technologies; e.g. in a constructed wetland, phytoextraction, phytostabilization, and phyto-volatilization may be used simultaneously [Hansen et al., 1998].

POTENTIAL PROBLEMS IN PHYTOREMEDIATION

Considerable effort is devoted towards metal phytoremediation. This effort is largely academic in nature, as the technology is still under development [Slater, 2000]. It is valuable to consider some of the problems that may limit the future commercial application of phytoremediation, as understanding of the problems may be of value during this R&D phase; the problems are summarized as follows [Slater, 2000]: (i) metal bioavailability – the members of the European Union currently regulate acceptable soil metal thresholds with regard to total metal levels, but the plant-available fraction of soil metal is almost always less than the total; (ii) biomass – metal phytoextraction relies on the accumulation of metal into biomass, but biomass per unit area is fixed within fairly narrow limits; (iii) biomass disposal – metal-rich biomass must still be regarded as a contaminated material that requires safe disposal; (iv) economic cost – a commercial problem concerns the guarantees that will need to be granted concerning performance, as in the event of a failed project the technology provider could well be responsible for the ultimate disposal of the contaminated soil; (v) time scale – most commercial remediation projects require a rapid solution to a soil contamination problem. Few developers are willing to wait for a long-term project to be completed; and (vi) multiple metal contamination – most metal-contaminated soils contain more than one metal. For example, combinations of lead and zinc are common in urban soils, while lead, zinc, cadmium and copper are all often present in the vicinity of a zinc smelter. There will be minimal economic value in a technology that can efficiently remove one metal (e.g. zinc) from a soil but leave most of another (e.g. lead) behind. However, most of the experiments on phytoextraction only address a single metal contaminant.

FIELD SURVEYS OF PHYTOREMEDIATION APPLICABILITY

The above-mentioned problems (iv), (v) and (v) are connected with each other: that is, technical performance influences cost performance (i.e. the issue of guarantee) and time scale. In practice, this technical performance should be evaluated from the viewpoint of multiple-metal pollution. Several types of phytoremediation are being commercially used or are in an advanced stage of development especially in USA, Canada and Australia [Marmiroli, 2000]. However, unless phytoremediation can overcome multiple metal pollution, its applicability will not be generally recognized in the commercial world. Therefore, field surveys were carried out in the southern part (Portugal) and the northern part (Russia) of the European region (see figure 2). These parts are climatologically quite different, so the comparative study promises to contribute to evaluating the general applicability of phytoremediation in the European region. Basic information about these surveys is briefly presented in table 1.

Table 1. Phytoremediation survey in different parts of the European region

Description	Southern part (Portugal)	Northern part (Russia)
Location	Sanguinheiro near Coimbra in central Portugal, 40°18'N and 8°21'W.	Metallurgy area near Monchegorsk on the Kola Peninsula, 67°51'N and 32°48'E. cf. Arctic Circle is defined as the latitude line of 66°51'N.
Duration	October 2006 to April 2008.	August 2003 to August 2005
Metal source	Zn-Pb Mine abandoned by Shist-Metagrayawacke complex 40 years ago.	The surrounding region is currently suffering from smelter pollution.
Soil type	Podzolic soil with 10-60 cm depth [Cardoso et al., 1971].	Podzolic soil with 30-50 cm depth [Koptski and Koptski, 2001].
Local vegetation	Mainly Eucalyptus globulus and Pinus pinaster (see figure 2A).	Willow and various deciduous trees (see figure 2B).
Field size (area)	2 ha	8 ha
Climate	Csa climate (i.e. maritime temperate climate), 17.4°C of annual mean temperature, and 48.4 mm of annual mean precipitation.	Subarctic climate, annual mean temperature of -1.0°C, and snowmelt takes place during April to June.
Phytoremediator	*Eucalyptus globulus*	Willow

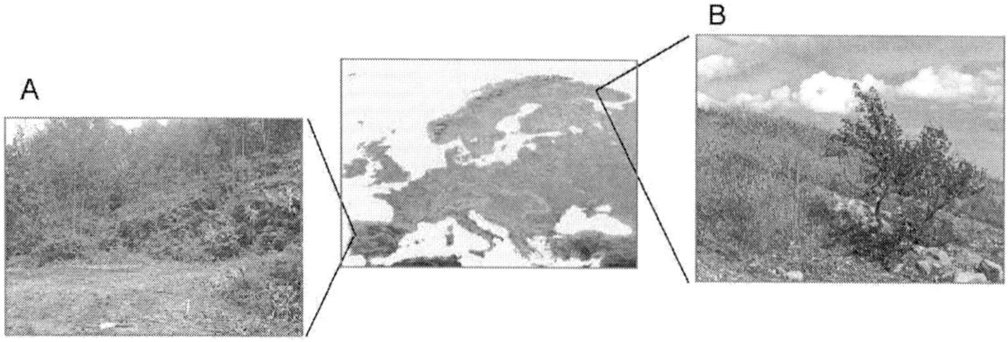

Figure 2. Locations of the test fields: (A) Sanguinheiro in Portugal; and (B) Monchegorsk on the Kola Peninsula.

Phytoremediation Test in the Southern Part (Portugal) of the European Region

The study area is situated in Sanguinheiro (figure. 2A) where galena and sparelite mainly occur as a mineralized vein. The test field chosen for the phytoremediation test was divided into three sites (cf. figure 3): (i) site #1 (S1) — main mine tailing with presence of young *Eucalyptus globulus* (abbreviated as *E. globulus*) and with a high slope (~75%); (ii) site #2 (S2) — secondary mine tailing with presence of vegetation including *E. globulus*; and (iii) Site #3 (S3) — control site situated out of the mining area but separated only by a stream and with presence of vegetation. The field test consisted of soil sampling (n = 22), plant sampling (n = 30), analysis of soil properties and analysis of heavy metals in *E. globulus* leaves.

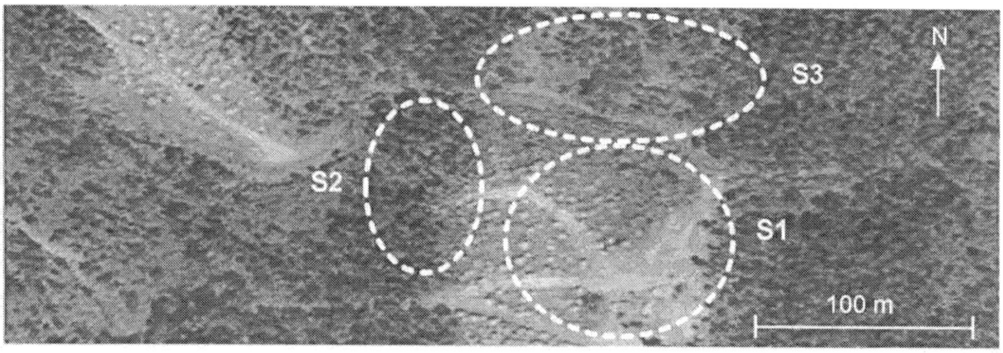

Figure 3. Division of the test field. See the above-mentioned text for detail of site description.

Sampling and Analytical Method

Using a stony auger, soil samples were taken in each site at a depth of 0–20 cm. After the sampled soil was homogenized and stored in plastic bags, each sample was transported to a chemical laboratory. The samples were dried at room temperature and oven-dried at 30ºC. All soil samples were put through a sieve with a 2-mm grid.

Soil was mixed with distillated water at a rate of 1.0 (soil) to 2.5 (water). After 1 hour of shaking, the pH value of the soil supernatant was determined by the glass electrode method.

Soil organic carbon was determined by the Tinsley method using dichromate mixtures [Tinsley, 1950]. Metal elements were extracted from the sieved sample by acqua regia (HCl + HNO_3): i.e. after 3.0 g soil was mixed with 7.0 ml concentrated HNO_3 and 21.0 ml concentrated HCl, this mixture was left overnight and boiled for 2 hours. Then, the extracted metals from the soil sample were filtrated and determined by atomic absorption spectrometry.

Leaf samples were washed with tap water and deionized water, then oven-dried at 105°C for 12 hours. These dried samples were grinded and then calcinated at 550°C for 12 hours. Each calcinated sample was mixed with 20% HCl (v/v) at a rate of 1.0 (sample) to 5.0 (HCl) in a beaker, and this beaker was covered with a watch glass and gently heated in a hot-water bath for 1 hour. The digested sample was filtrated, and metal elements were determined by atomic absorption spectrometry.

Test Results

The soil properties in the field survey are summarized in table 2. The pH of the soils ranged from 4.5 to 4.8, indicating the presence of acid soil in each site. Organic carbon showed low values, and its value ranged from 2.2% in site #3 to 4.1% in site #2 in the test field. Although mining activity ended 40 years ago, the contents of Pb and Zn in the sampled soil were comparatively high in site #1 with ~75% slope.

Table 2. Soil characteristics in the test field (mean values during the field survey)

Parameter	Site #1 (S1)		Site #2 (S2)		Site #3 (S3)	
pH	4.7	(±0.1)	4.5	(±0.1)	4.8	(±0.2)
Organic carbon (%)	3.5	(±0.8)	4.1	(±0.8)	2.2	(±0.2)
Pb (mg kg^{-1})	165.3	(±84.6)	46.9	(±6.2)	67.7	(±30.7)
Zn (mg kg^{-1})	108.6	(±30.0)	97.1	(±29.2)	128.6	(±30.8)
Numbers in parentheses represent the value of standard deviation.						

E. globulus is the most representative arborous species in the test field. Figure 4 shows the amounts of Pb and Zn in the measured leaves. There is no clear difference in the Pb amounts, but the leaves in site #1 (S1) contained a rich amount of Zn.

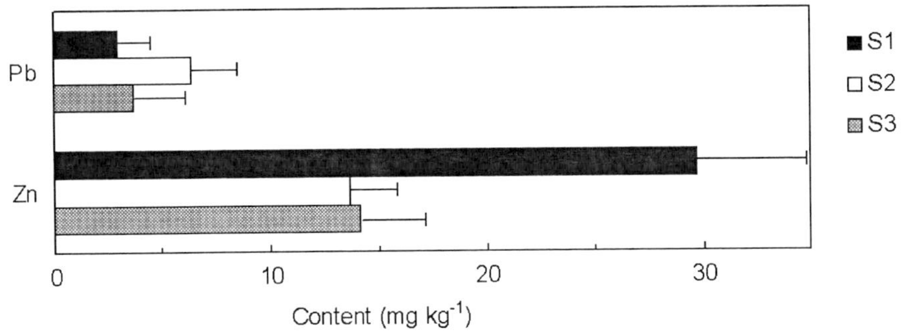

Figure 4. Metals in *E. globulus* leaves in different sites of the test field (mean values measured in 2008).

Consideration

The pH value affects the solubility of several mineral elements [Costa, 1995]; in particular, Zn is greatly affected by the value of soil pH [McGrath et al., 1988; Turner, 1994]. It is considered that heavy metals in the study are mobile and are available for plant uptake. Though *E. globulus* prefers a soil pH greater than 5.0 [Oliver, 1995], the population continues to grow in the study area. This implies that *E. globulus* is tolerant to mobile metals and acidic soil.

The Pb content of leaves in site #1 (S1) is lower than that in site #2 (S2) and in site #3 (S3). This trend may be interpreted as follows: there is a younger population of *E. globulus* in S1; Pb is not an essential element to plant development, so several mechanisms prevent entry of Pb to the plant body – e.g. the epidermis acts as a strong barrier [Weis and Weis, 2004]; even if the Pb levels in the soil are high, those in leaves remain low because the leaves demonstrate the low translocation of this metal from the soil to plants.

The influence of Zn on leaf growth is important for young plants in particular. Zn deficiency may lead to a decrease in the growth of leaves, resulting in a decrease of photosynthetic activity. A younger population of E. globulus is mainly situated in S1. It is therefore considered that the Zn content of leaves in S1 is higher than the contents in S2 and S3.

Certain heavy metals such as Cu, Zn, Pb, Fe and Cd tend to accumulate in the tissues of roots rather than in leaf tissues [Peters et al., 1997]. There is a possibility that great amounts of Zn and Pb accumulate in tissues of eucalyptus roots. Metal analysis of root tissues is a subject for future study.

Remarks

Though several studies have demonstrated the potential of eucalyptus in the phytoremediation of contaminated soils (review in [Pyatt, 2001; Rockwood et al., 2006]), feasibility studies of eucalyptus-based phytoremediation are rare in Portugal. The present results show that eucalyptus can survive in soil contaminated with Pb and Zn, suggesting that plant roots can effectively immobilize heavy metals. This immobilization seems to reduce the environmental risk associated with metal contamination.

Phytoremediation Test in the Northern Part (Kola) of the European Region

Most smelter emissions come from the Nickel, Zapolyarnyy, and Monchegorsk complexes on the Kola Peninsula located in the most northern region of the European fringe of Russia [Kashulina and Reimann, 2002]. The Monchegorsk smelter complex produces refined Ni, Cu and Co metal, sulfuric acid and noble-metal sludge (see figure 5).

Industrial barrens are classified to be the bleak open landscapes that evolve around the point sources of industrial pollution due to deposition of airborne pollutants, with small patches of vegetation (cover usually reduced to 10% or less relative to control) surrounded by bare land with illuvial horizon or even the rock exposed due to intensive soil erosion (relict soil cover usually less than 20%) [Kozlov and Zvereva, 2007]. First records of an industrial barren near Monchegorsk smelter were made in the early 1960s. The barren or semibarren area is estimated to be 21,000 to 44,000 ha (estimation in the 1990s).

A field test for rehabilitation of forest land was performed while manufacture was continuing nearby; that is, the survey fields are now suffering from pollution emitted by the metallurgical industry and moreover this suffering may continue in the future.

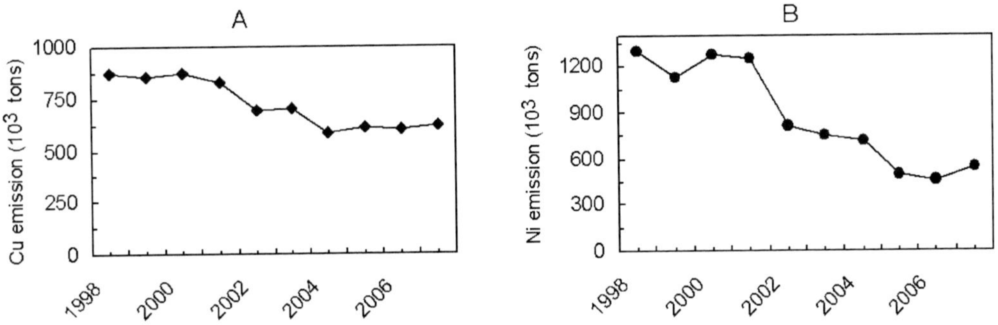

Figure 5. Emission trends in the Monchegorsk smelter complex [ОАО Кольская ГМК, 2010]: (A) Cu annual emission and (B) Ni annual emission.

Nutrient deficiency adversely affects plant life in an industrial barren. In industrial barrens near Monchegorsk, the number of vascular plant species ranged 0–25% of the number recorded in unpolluted (control) sites. In addition, ground layer vegetation is highly fragmented [Kozlov and Zvereva, 2007]. In industrial barrens nearby, vegetation cover and species richness are higher in the bottom of the valley, near the small stream, compared with slopes that are only some 50–80 m above [Kozlov and Zvereva, 2007]. Although abundance of most species in industrial barrens is extremely low, some plants and animals flourish in these habitats. In particular, this applies to some willow species that are much more abundant in barren sites than in unpolluted forests. The trees which manage to survive in industrial barrens generally demonstrate bush-like or even creeping growth forms. Two willow species (*Salix borealis* (Fries.) Nasar. and *S. caprea* L.) in industrial barrens have more epiciormic shoots than in unpolluted forests [Zvereva and Kozlov, 2001]. An increased light availability (due to forest decline) may also have contributed to higher branching and formation of bush-like crowns in woody plants surviving in barren sites [Zvereva and Kozlov, 2001].

Preparation and Test Design — Different Test Sites

In the late autumn of 2003 (before snow cover), an artificial substratum from sewage sludge was formed in 4 ha of the test field. In the late autumn of 2004 (before snow cover), a turf layer of 15 cm was formed in 1 ha of the test field, and this turf was withdrawn from territories adjacent to the smelter. In the late autumn of 2004 (before snow cover), fertilization of the industrial barren ground was carried out in 3 ha of the test field.

Test site #1: sewage sludge as artificial substratum — sewage sludge composting was carried out to prepare the artificial substratum. Freestanding piles (lower than 3 m height) of sewage sludge were built on level well-aerated spots, and they were occasionally turned for the purpose of material homogenization and re-oxygenation for a period of about 1 month (July to August 2003). The resulting compost was mixed with additives: 1,200 tons ha^{-1} sewage sludge deposit, 1,200 tons ha^{-1} sawdust, and 600 tons ha^{-1} sand. Furthermore, 2,000 tons ha^{-1} compost made from sewage sludge was added to the graded field, and 2 tons ha^{-1} dolomite was scattered.

Test site #2: turf and fertilizer as artificial substratum — each survey field was bulldozed and big stones were removed, then turf (10-15 cm layer) was introduced to the prepared fields. After this, dolomite (2 tons ha^{-1} and 0.75 tons ha^{-1} NPK fertilizer) was scattered on the fields prior to afforestation in September 2004.

Test site #3: simple fertilization of contaminated ground — each survey field was bulldozed and big stones were removed, then dolomite (2 tons ha^{-1} and 0.75 tons ha^{-1} NPK fertilizer) was scattered on the fields prior to afforestation in September 2004.

Afforestation (common to all test sites) — willows were used for afforestation in the autumn of 2003 and 2004 after the above-mentioned artificial substratums had been prepared according to each test site design. Willows had been left untrimmed since 1995 in an abandoned farm situated within the tolerance zone, so seedlings of these trees were collected from the farm. Each seedling (~1 m height) was put in a polyethylene sack where the local soil encircled its root zone in order to protect against root dry during transplantation. The seedlings prepared in this way were carefully transported to the test field. As soon as the seedlings arrived in the test field, they were taken out from the sacks and were planted 4 abreast at regular intervals of 2 m in a 10 m-wide strip; in addition, grasses were laid in the gaps between the seedlings (free space) to facilitate soil aggregation.

Sampling, Analytical Methods and Data Processing

Artificial substratum samples were taken in each plot of the test field in autumn 2005: n = 34 in test site #1, n = 5 in test site #2, and n = 20 in test site #3. The above-mentioned sampling was carried out as follows (Russian standard N 3–15/582): the soil or ground was taken from the top 10 cm to avoid additional mixing with the contaminated soil layer. Sampling points were chosen on each single plot according to the "envelope" method (i.e. approximately 5 points per plot). Each sample from each single point of a single plot was incorporated into a single polyethylene bag. The mass of the sample was about 1.0 kg.

All soil samples were dried at room temperature (20°C) and put through a sieve with a 1 mm grid. After this pretreatment, chemical analysis was carried out as follows: (i) metal elements were extracted from the pre-treated sample using 1M ammonium acetate (i.e. extraction of bio-available forms) [Halonen et al., 1983], and the following elements were determined by atomic absorption spectrometry — K, Ca, Mg, Al, Zn, Fe, Cu, Ni and Mn; (ii) the sieved sample was treated with sulfuric acid and nitric acid, and phosphorous composition was then determined by molybdate colorimetry; (iii) 10.0 g of the sieved sample was mixed with 25 ml of distilled water, and then the pH value of the sample was measured by the glass electrode method. Atomic absorption spectrometry was used for the quantification of metals.

Leaves of willow were taken in an abandoned farm situated within the tolerance zone in September 2003 (n = 12), in the background field in September 2003 (n = 12), and in the rehabilitation test field in August 2005 (n = 34 in test site #1, n = 5 in test site #2, and n = 20 in test site #3). This leaf sampling was carried out at the end of the growing season. Leaf samples were dried at room temperature, and they were digested with concentrated nitric acid to destroy the matrix and dissolve metals. Metals in a sample solution were determined by atomic absorption spectrometry, and phosphorous composition was determined by molybdate colorimetry.

Significant levels (P values) were calculated to statistically evaluate the significant difference, and standard deviation values were also calculated to show the dispersion of a set of measured values. The Pearson correlation coefficient (r) was applied to evaluate the statistical relations. The obtained coefficient is interpreted in this paper as follows: $0 \leq |r| < 0.2$ indicates little or no association; $0.2 \leq |r| < 0.4$ indicates weak association; $0.4 \leq |r| < 0.7$ indicates good association; and $0.7 \leq |r| \leq 1.0$ indicates strong association.

Results

The substratum in test site #3 was more acidic: the mean pH values in test sites #1, #2 and #3 were 5.3, 5.4 and 4.6, respectively. Artificial grounds were limed by dolomite in each site, but more comprehensive results were received in test sites #1 and #2. In order of test sites #1, #2 and #3, the contents of available Ca forms were ~1,010 mg kg^{-1}, ~1,060 mg kg^{-1} and ~405 mg kg^{-1}; those of available Mg forms were ~120 mg kg^{-1}, ~140 mg kg^{-1} and 30 mg kg^{-1}; those of available K forms were ~37 mg kg^{-1}, ~65 mg kg^{-1} and ~41 mg kg^{-1}; those of available P forms were ~16 mg kg^{-1}, ~8 mg kg^{-1} and ~8 mg kg^{-1}. Less difference was observed in Mn and Zn availabilities: i.e. 18 mg kg^{-1} Mn, 8 mg kg^{-1} Mn, 8 mg kg^{-1} Mn; and 14 mg kg^{-1} Zn, 17 mg kg^{-1} Zn and 7 mg kg^{-1} Zn in test sites #1, #2 and #3, respectively. Erosion and dusting of neighboring territories resulted in high levels of Al and Fe forms: 426 mg kg^{-1} Al, 287 mg kg^{-1} Al, 889 mg kg^{-1} Al; and 117 mg kg^{-1} Fe, 62 mg kg^{-1} Fe and 293 mg kg^{-1} Fe in test sites #1, #2 and #3, respectively.

The contents of available Cu forms were ~290 mg kg^{-1}, ~280 mg kg^{-1} and ~420 mg kg^{-1}; and those of available Ni forms were ~46 mg kg^{-1}, ~56 mg kg^{-1} and ~86 mg kg^{-1} in test sites #1, #2 and #3, respectively. According to Russian regulations (SanPin 2.1.7.1287-03), there are limit values in the soil (as content of available forms): lithological level (clarke) — e.g. Ni < 4.0 mg kg^{-1} and Cu < 3.0 mg kg^{-1}. Considering the lithological level, Cu and Ni in the artificial substratum exceed the limit values.

Table 3 shows the content of each element in willow leaves as results of survival under conditions of pollution on planting material. Based on the data presented in table 3, variations in P values were calculated to statistically evaluate each significant difference, and these values are summarized in table 4. Since a P value of < 0.05 is taken as statistically significant, it is considered that most elements had variations in most elements regardless of the site type.

Table 3. Leaf elements (mean values) of willow in the test field and the background field

Element ($mg\ kg^{-1}$)	2003		2005		
	Background	Original	Test site #1	Test site #2	Test site #3
P	2856	1680	1909	2576	2303
K	7635	10371	9030	9559	6781
Ca	10727	11966	17236	18306	18072
Mg	2034	4154	3629	3728	3387
Cu	2.3	19	338	461	379
Zn	161	238	180	122	109
Mn	845	1264	173	43	43
Fe	54	137	230	191	248
Ni	3.9	52	457	569	575
Al	44	69	170	128	146

See the text for detail of test sites.

Table 4. P-value analysis of leaf elements in different types of test field

Element	Test site #1	Test site #2	Test site #3
P	†	†	***
K	*	*	***
Ca	***	†	**
Mg	†	*	***
Cu	**	†	***
Zn	*	*	***
Mn	†	*	***
Fe	***	***	***
Ni	***	***	***
Al	†	***	**

† represents no significance ($P \geq 0.05$); * represents significance ($P < 0.05$); ** represents high significance (P value < 0.01); *** represents extreme significance ($P < 0.001$). See the text for detail of test sites.

As seen in table 3, the leaf contents of P, Ca and Fe increased; by contrast, those of K and Mg decreased. Compared with the contents in the background, nutrient deficiency could not be clearly observed.

Willows survived on the metal-contaminated land under the current metal pollution, and their contents of Cu and Ni considerably increased during the observation period: 18–25 times greater in Cu content and 9–11 times greater in Ni content.

Though the test field is currently suffering from metal pollution (see figure 4), the results obtained from leaf diagnosis indicate the possibility of willow-based phytoremediation in the Arctic (or Subarctic) region.

Consideration

Ni and Cu statistically increased in willow leaves in all test sites (table 4); however, attention should be paid to excess amounts of elements rather than deficiencies because microelements are required in very small amounts (a few mg kg^{-1}) in plant tissue, being one or more orders of magnitude lower than for essential elements [Food and Fertilizer Technology Center, 2001].

Excluding essential elements, willows in the test field generally contained much greater amounts of metals than those in the background (table 3). It is hence considered that willow can survive even if the test field is suffering from metal pollution.

The data obtained during the field test suggest that Ni, Cu and the other metals can be treated by willow-based phytoremediation.

However, more detailed research is necessary to assess whether metal superaccumulation has really taken place in the test field because there are basically two types of metal loading: (i) metals merely adhered to leaf surfaces (deposition of metal particles) and (ii) metals contained within plant tissues through air-foliar pathways and soil-root pathways (metal absorption). To put it differently, it is important to correctly determine the correlation between plant chemistry and ground chemistry.

Tables 5–7 show such coefficients (r values) obtained in test sites. It follows from these tables that there are weak relations between leaf chemistry and soil chemistry. Based on field observation during 2003-2005, the obtained data showed similarity of leaf chemistry despite differences in the nutritional regime of the ground.

The chemistry of local soil surrounding the root zone of willow seedlings is proposed to be a more important factor than artificial grounds (see also the section presenting the Sanguinheiro field test in Portugal).

Metals adhering to leaf surfaces will be the preferential means of testing pollutant accumulation in planting material rather than root system proliferation in the vicinity of the planting area (a restricted area).

Table 5. Correlation coefficients of leaf elements with soil elements in test site #1

Soil\leaves	Ca	Mg	K	Al	Fe	Mn	Zn	Ni	Cu	P
Ca	-0.08	-0.03	-0.19	-0.09	-0.08	-0.03	-0.32	-0.10	-0.02	-0.55
Mg	-0.19	0.12	-0.31	-0.07	-0.04	0.08	-0.34	-0.05	0.02	-0.55
K	0.26	0.18	0.19	-0.20	-0.14	0.04	-0.03	0.13	0.10	-0.13
Na	0.26	-0.30	0.23	-0.08	-0.04	-0.24	0.28	-0.10	-0.07	-0.17
Al	0.13	-0.16	0.23	-0.12	-0.08	-0.32	0.11	-0.03	-0.03	-0.31
Fe	-0.11	-0.13	0.33	0.28	0.15	0.28	0.03	-0.13	-0.17	0.28
Mn	-0.03	-0.03	-0.14	0.01	0.03	-0.05	-0.27	-0.07	0.00	-0.50
Zn	0.47	-0.39	0.17	0.24	0.18	-0.21	0.37	-0.38	-0.22	0.15
Ni	-0.10	0.11	-0.09	-0.10	-0.01	-0.17	-0.25	0.04	0.12	-0.53
Cu	-0.02	-0.01	-0.02	-0.09	-0.03	-0.28	-0.21	-0.08	0.03	-0.34
P	0.51	-0.44	0.13	0.12	0.03	0.06	0.61	-0.46	-0.30	0.52

$n = 34$, and see the text for detail of test sites.

Table 6. Correlation coefficients of leaf elements with soil elements in test site #2

Soil\leaves	Ca	Mg	K	Al	Fe	Mn	Zn	Ni	Cu	P
Ca	-0.53	0.58	0.64	0.23	0.01	0.73	0.70	-0.17	0.05	0.79
Mg	-0.62	0.39	0.87	-0.33	-0.61	0.84	0.51	-0.72	-0.58	0.63
K	-0.27	0.14	0.66	-0.78	-0.94	0.48	0.01	-0.88	-0.93	0.09
Na	-0.41	-0.48	0.50	-0.69	-0.91	0.37	-0.18	-0.91	-0.95	-0.06
Al	0.13	-0.16	0.30	-0.78	-0.81	-0.06	-0.52	-0.67	-0.86	-0.47
Fe	-0.32	-0.43	0.24	-0.79	-0.88	0.30	-0.07	-0.80	-0.87	-0.03
Mn	-0.71	0.24	0.71	-0.53	-0.75	0.91	0.65	-0.81	-0.68	0.72
Zn	0.12	-0.18	0.20	-0.87	-0.86	-0.03	-0.43	-0.68	-0.87	-0.42
Ni	-0.36	0.38	0.71	-0.68	-0.84	0.64	0.28	-0.80	-0.79	0.34
Cu	0.02	0.80	0.39	0.20	0.16	0.25	0.31	0.11	0.19	0.32
P	-0.42	0.60	0.30	-0.48	-0.46	0.73	0.86	-0.40	-0.29	0.78

$n = 5$, and see the text for detail of test sites.

Table 7. Correlation coefficients of leaf elements with soil elements in test site #3

Soil\leaves	Ca	Mg	K	Al	Fe	Mn	Zn	Ni	Cu	P
Ca	0.20	0.24	-0.04	0.29	0.26	-0.20	-0.12	-0.14	-0.05	-0.04
Mg	0.38	0.24	-0.27	0.21	0.33	-0.20	0.08	-0.03	0.08	-0.07
K	0.58	-0.39	0.26	0.62	0.82	-0.32	0.36	0.39	0.71	0.35
Na	-0.37	0.15	-0.20	-0.09	-0.12	0.26	-0.47	-0.48	-0.12	-0.45
Al	-0.27	0.05	-0.19	-0.44	-0.20	0.36	-0.38	-0.39	-0.03	-0.32
Fe	-0.34	0.33	-0.35	-0.60	-0.67	0.23	-0.52	-0.38	-0.65	-0.29
Mn	0.06	0.23	-0.25	-0.33	-0.27	-0.01	0.01	0.23	-0.38	0.22
Zn	0.25	-0.37	0.21	0.33	0.45	-0.48	0.04	0.18	0.44	0.20
Ni	0.27	-0.06	0.09	0.55	0.61	-0.14	0.00	0.15	0.38	0.02
Cu	-0.05	-0.45	0.18	0.02	0.41	-0.25	-0.01	0.30	0.53	0.19
P	0.27	-0.25	0.44	0.67	0.34	-0.02	0.30	0.00	0.32	0.23

$n = 20$, and see the text for detail of test sites.

Remarks

Willows have adapted to a broad range of climates and site conditions, and their properties are useful as a phytoremediator – rapid growth rate, high biomass production, reliable coppicing ability, tolerance of high planting density, a low nutrient requirement, tolerance to low pH, resistance to air pollutants, tolerance to drought and heat, resistance to chemical contaminants and so on [Kuzovkina and Volk, 2009]. Therefore, willows were used for phytostabilization in the 8-ha industrial barren. Though the results obtained from the metal analysis (Cu, Ni, Fe, and Pb) indicate efficient phytoremediation (i.e. preferable effect of phytoextraction), some questions remain: e.g. week correlation of leaf chemistry with soil chemistry.

CONCLUSION

Most metal-contaminated soils contain more than one metal. For example, combinations of Pb and Zn are common in urban soils, while Pb, Zn, Cd and Cu are all often present in the vicinity of a metallurgic smelter [Slater, 2000]. Since most of the experiments on phytoextraction only address a single metal contaminant, attention was focused on multiple-metal stress. The performed field surveys suggest that eucalyptus and willows act as pyhtoremediators in multiple-metal conditions.

The data presented in this chapter may help the planning of a commercial application of phytoremediation in cases of multiple-metal stress. However, some questions remain — e.g. assessment of root systems and week correlation between leaf chemistry and soil chemistry; it is therefore considered that a long-term and detailed observation is necessary to confirm reliable feasibility on a large scale to enable the proper design of a phytoremediation project. This future work will establish a close link between phytoremediation technology and the concept of NEEC (i.e. the balance between environmental benefit and financial cost).

ACKNOWLEDGMENTS

Parts of the Monchegorsk field survey are supported by Monchegorsk Forestry Enterprise and Severonikel Co., and organized by the Institute of the North Industrial Ecology Problems (Russian Academy of Science). Parts of the Sanguinheiro field survey are supported by the soil laboratory of ESAC/IPC. The authors are grateful to Ms C. Lentfer for English review.

REFERENCES

Angelone, M. and Bini, C. (1992). *Trace element concentrations in soils and plants of Western Europe*. In: Adriano, D.C. (ed.), Biogeochemistry of Trace Metals (pp. 19–60). Boca Raton: Lewis.

Baccouch, S., Chaoui, A. and El Ferjani, E. (1998). Nickelinduced oxidative damage and antioxidant responses in Zea mays shoots. *Plant Physiology and Biochemistry* 36, 689–694.

Blaylock, M.J. and Huang, J.W. (2000). *Phytoextraction of metals*. In: Raskin, I. and Ensley, B.D. (eds.), Phytoremediation of toxic metals — using plants to clean up the environment (pp. 53–70). New York: Wiley.

Berti, W.R. and Cunningham, S.D. (2000). *Phytostabilization of metals*. In: Raskin, I. and Ensley, B.D., (eds.), Phytoremediation of toxic metals — using plants to clean up the environment (pp. 71–88). New York: Wiley.

Cardoso J.C, Teixeira, B. and Marado M.B. (1971). Carta dos solos de Portugal (Soil chart of Portugal). Agronomia Lusitana 33, 481–602 [in Portuguese].

Chaney, R.L., Li, Y.M., Brown, S.L., Homer, F.A., Malik, M., Angle, J.S., Baker, A.J.M., Reeves, R.D. and Chin, M. (2000). *Improving metal hyperaccumulator wild plants to develop commercial phytoextraction systems: approaches and progress*. In: Terry, N. and Banuelos, G. (eds.), Phytoremediation of contaminated soil and water (pp. 129–158). Boca Raton: Lewis.

Cho, U.H. and Park, J.O. (2000). Mercury-induced oxidative stress in tomato seedlings. *Plant Science* 156, 1–9.

Costa, J.B. (1995). Caracterização e constituição do solo (Characterization and constitution of soil). Lisbon: Fundação Calouste Gulbenkian [in Portuguese].

Dushenkov, S. and Kapulnik, Y. (2000). *Phytofiltration of metals*. In: Raskin, I. and Ensley, B.D. (eds.), Phytoremediation of Toxic Metals — Using Plants to Clean Up the Environment (pp. 89–106). New York: Wiley.

Ensley, B.D. (2000). *Rationale for use of phytoremediation*. In: Raskin, I. and Ensley, B.D. (eds.), Phytoremediation of toxic metals — using plants to clean up the environment. (pp. 1–12). New York: Wiley.

EPA – Environmental Protection Agency (1998). *A citizen's guide to phytoremediation*. (#542–F-98–011). Washington DC: EPA Publications.

Food and Fertilizer Technology Center (2001). *Micronutrients and crop production* (report #bc 51009). Taipei, Taiwan: Food and Fertilizer Technology Center.

Glass, D.J. 2000. *Economic potential of phytoremediation*. In: Raskin, I. and Ensley, B.D. (eds.), Phytoremediation of toxic metals — using plants to clean up the environment. (pp. 15–33). New York: Wiley.

Halonen O.,Tulkki H. and Derome J. (1983). *Nutrient analysis methods*. Metsantutkimuslaitoksen Tiedonantoja 121, 1–28.

Hansen, D., Duda, P.J., Zayed, A. and Terry, N. (1998). Selenium removal by constructed wetlands: role of biological volatilization. *Environmental Science and Technology* 32, 591–97.

Horne, A.J. 2000. Phytoremediation by constructed wetlands. In: Terry, N. and Banuelos, G. (eds.), Phytoremediation of contaminated soil and water (pp. 13–40). Boca Raton: Lewis.

Houghton, J. (1996). *Sustainable Use of Soil*. London: Royal Commission on Environmental Pollution.

Kashulina, G. and Reimann, C. (2002). Sulfur in the Arctic environment - results of multimedium regional mapping, *Environmental Pollution* 116, 337–350.

Koptski, S.V. and Koptski, G.N. (2001). *Soil pollution pattern in terrestrial ecosystem of the Kola Peninsula*. In: Stott, D., Mohtar, R. and Steinhardt, G. (eds.), Sustainable the global farm (pp. 212–216). Ashland, OH: Purdue University Press.

Kozlov M.V. and Zvereva E.L. (2007). Industrial barrens: Extreme habitats created by non-ferrous metallurgy. *Reviews in Environmental Science and Biotechnology* 6, 231–259.

Kuzovkina Y.A. and Volk T.A. (2009). The characterization of willow (Salix L.) varieties for use in ecological engineering applications: Co-ordination of structure, function and autecology. *Ecological Engineering* 35, 1178–1189.

Li, Y. and Trush, M.A. (1993). DNA damage resulting from the oxidation of hydroquinone by copper: role for a Cu(II)/Cu(I) redox cycle and reactive oxygen generation. *Carcinogenes* 7, 1303–1311.

Macken, K. (1995). *BATNEEC guidance note* (EPA #LC7). Wexford: Irish Environmental Protection Agency.

McGrath, S.P., Sanders, J.R. and Shalaby, M.H. (1988). The effects of soil organic matter levels on soil solution concentrations and extractabilities of manganese, zinc and copper. *Geoderma* 42, 177–188.

Marmiroli, N. (2000). *Bringing phytoremediation to a successful technology in Europe Where we go from now?* In: abstract book of Phytoremediation – state of the art in Europe (pp. 6-8). Hersonissos, 6-8 April: 1st Scientific workshop of COST action #837.

Marschner, H. (1995). *Mineral Nutrition of Higher Plants*. London: Academic Press.

Mueller, B., Goswami, D., Rock, S. and Arguello, R. (1999a). *Phytoremediation decision tree* (#PHYTO-1). Washington DC: Interstate Technology and Regulatory Council.

Mueller, B. Strauss, P., Gatchett, A., Berti, B., Thuraisingham, R., Compton, H., Hirsh, S., Arguello, R., Ellis, B., Douglas, T., Olson, K., Greene, K.A. and David Tsao, D. (1999b). *Phytoremediation decision tree*. Washington DC: Interstate Technology and Regulatory Council.

ОАО Кольская ГМК (2010). *Экология* (Ecology). http://www.kolagmk.ru/rus/ecology [in Russian].

Oliver, J.M.M. (1995). El eucalipto (the eucalypt). Madrid: Ediciones Mundi-Prensa [in Spanish].

Peters, E.C., Gassman, N.J., Firman, J.C., Richmond, R.H. and Power, E.A. (1997). Ecotoxicology of tropical marine ecosystems. *Environmental Toxicology Chemistry* 16, 2–40.

Pilon-Smits, E. and Marinus Pilon, M. (2002). Phytoremediation of Metals Using Transgenic Plants. *Critical Reviews in Plant Sciences* 21(5), 439–456.

Pyatt, F.B. (2001). Copper and Lead Bioaccumulation by Acacia retinoides and Eucalyptus torquata in Sites Contaminated as a Consequence of Extensive Ancient Mining Activities in Cyprus. *Ecotoxicology and Environmental Safety* 50, 60–64.

Rockwood, D.L., Cárter, D.R., Langholtz, M.H. and Striker, J.Á. (2006). Eucalyptus and Populus short rotation woody crops for phosphate mined lands in Florida. *Biomass and Bioenergy* 30, 728–734.

Ross, S.M. (1994). Toxic Metals in Soil-Plant Systems. *Chichester* (UK): Wiley.

Salt, D.E., Smith, R.D. and Raskin, I. (1998). Phytoremediation. *Annual Review of Plant Physiology and Plant Molecular Biology* 49, 643–668.

Semple, K. T., Reid, B. J. and Fermor, T. R. (2001). Impact of composting strategies on the treatment of soil contaminated with organic pollutants. *Environmental Pollution* 112, 269-283.

Slater, A. (2000). *Some possible future limitations on the commercial application of metal phytoremediation*. In: abstract book of Phytoremediation – state of the art in Europe (pp. 2-3). Hersonissos, 6-8 April: 1st Scientific workshop of COST action #837.

Tinsley, J. (1950). *The determination of organic carbon in soils by dichromate mixtures*. In: Transactions of the 4th International Congress of Soil Science (pp. 161–164). Amsterdam: Hoitsema Brothers.

Turner, A.P. (1994). *The responses of plants to heavy metals*. In: Ross, S.M. (ed.), Toxic metals in soil-plant systems (pp. 153-18). Chichester: John Wiley and Sons.

Weis, J.S. and Weis, P. (2004). Metal uptake, transport and release by wetland plants: implications for phytoremediation and restoration. *Environmental Pollution* 30, 685–700.

Winterhalder K. (2000). *Landscape degradation by smelter emissions near Sudbury, Canada and subsequent amelioration and restoration*. In Innes J.L. and Oleksyn J. (eds), Forest Dynamics in Heavily Polluted Regions (pp. 87–119). London: CAB International.

Zvereva E.L. and Kozlov M.V. (2001). Effects of pollution induced habitat disturbance on the response of willows to simulated herbivory. *Journal of Ecology* 89, 21–30.

In: Handbook of Phytoremediation
Editor: Ivan A. Golubev

ISBN: 978-1-61728-753-4

Chapter 21

GENETIC BIODIVERSITY OF MAIZE AND SUNFLOWER COMMERCIAL CULTIVARS, AND THEIR PHYTOEXTRACTION CAPABILITY OF A MULTI-METAL ARTIFICIALLY POLLUTED SOIL

Daniela Baldantoni, Angela Cicatelli and Stefano Castiglione
Dipartimento di Chimica,
Università degli Studi di Salerno,
Via Ponte Don Melillo - 84084 Fisciano, SA (Italy)

ABSTRACT

Phytoextraction of heavy metals (HMs) is a promising technology that uses plants to remove pollutants from soil. Two high biomass yield crops, maize and sunflower, with their ability to accumulate HMs, have been widely used to remediate contaminated soils. Nine commercial cultivars of maize and three of sunflower were characterized for their Genetic Bio-Diversity (GBD) using two different molecular approaches: Random Amplified Polymorphic DNA (RAPD) and Amplified Fragment Length Polymorphism (AFLP).

A pot experiment was subsequently carried out to estimate the phytoextraction capacity of three cultivars for each plant species grown on multi-metal (copper and zinc) artificially contaminated soil. The HM accumulation was estimated in all three plant organs: root, stem and leaf.

The results of the molecular analysis showed a considerable GBD among all tested cultivars. Moreover, a highly significant difference was observed among cultivars for their HM extraction capability. In both species, the highest metal concentration was detected in roots, followed by stems and leaves; sunflower cultivars exhibited the highest potential for the removal of HMs from a multi-metal polluted soil.

Keywords: Genetic Bio-Diversity (GBD), phytoextraction, multi-metal, sunflower, maize.

BIODIVERSITY AND PHYTOREMEDIATION

Biodiversity is the core of Charles Darwin's theory of evolution (Darwin, 1859). The term "biodiversity" comprises the wealth of forms of life present on Earth, with their billions of plants, animals and micro-organisms, the genes that they contain, and the complex ecosystems that they form in the biosphere. The importance of biodiversity for mankind was ratified in the "Earth Summit of Rio de Janeiro" in 1992. Biodiversity can be observed and estimated at several levels: biomes, ecology, taxonomy, phylogeny, species, and genetics (Grassi *et al.*, 2005). In the last two decades, Genetic Bio-Diversity (GBD) has received particular attention in the scientific community, due to the fact that it can be estimated at all levels, population, variety, race and strain, by means of different molecular tools acting either at locus-specific (*e.g.*, SSRs, SNPs, gene sequences) or at random (RAPDs, AFLPs, ISSRs) levels (Agarwal *et al.*, 2008; Jones *et al.*, 2009). Indeed, molecular methods have proven to be valuable tools to estimate the level and patterns of GBD within, or among, populations or cultivars in an ample variety of plant species (Agarwal *et al.*, 2008; Jones *et al.*, 2009). The possibility to estimate GBD is of primary importance both in the case of fundamental and applied research. In fact, if no genetic difference is present in populations, or in varieties of a certain species, it might be possible that the response of each single individual, to any kind of stress, is not effective, and the populations, or the varieties, and, consequently, the species will suffer for the alteration of the environment where they live (Booy *et al.*, 2000). Biodiversity of natural populations, or of the parental lines employed in breeding programmes, is essential for artificial selection of ameliorated individuals, as recently demonstrated in the case of poplar clones tolerant to copper and zinc (Castiglione *et al.*, 2009).

Phytoremediation is an attractive environment friendly and cost-effective method, well accepted by the public opinion, to reclaim soils, waters and air contaminated by different kinds of organic or inorganic pollutants (Pilon-Smits, 2005). Phytoremediation shows several advantages, and a few limited disadvantages (*e.g.*, the relatively long time required for clean-up) in comparison with traditional methods that require soil excavation and dumping, containment methods (*e.g.*, vitrification, stabilization), or soil washing/flushing, which are generally costly and destructive of soil properties (Mulligan *et al.*, 2001). Therefore, phytoremediation and phytoextraction, the latter especially of heavy metals (HMs), from contaminated soils appears to be a useful alternative approach (Komarek *et al.*, 2007). Several plant species showing HM tolerance, a well developed root apparatus, high biomass production, and good transpiration rates, such as sunflower (*Helianthus annuus*), maize (*Zea mays*), Indian mustard (*Brassica juncea*), *Salix* spp and *Populus* spp, as regarded as good candidates for phytoextraction (Pilons-Smits, 2005). Maize and sunflower (Figure 1) have received particular attention for phytoextraction purposes from biologists and agriculturalists due to a huge available germoplasm (http://www.sementi.it/registri_varietali/Agrarie_09.pdf), well defined agricultural practices (Lasat, 2002) and, in some cases, a high capacity to extract and accumulate large amounts of HMs from contaminated soils (Lin *et al.*, 2003; Madejón *et al.*, 2003; Nehnevajova *et al.*, 2005, Lin *et al.*, 2008; Murakami and Ae, 2009; Nehnevajova *et al.*, 2009). Shoot metal accumulation depends on three key factors: metal solubility, metal absorption by roots, and metal translocation from roots to shoots (Vassilev *et al.*, 2002).

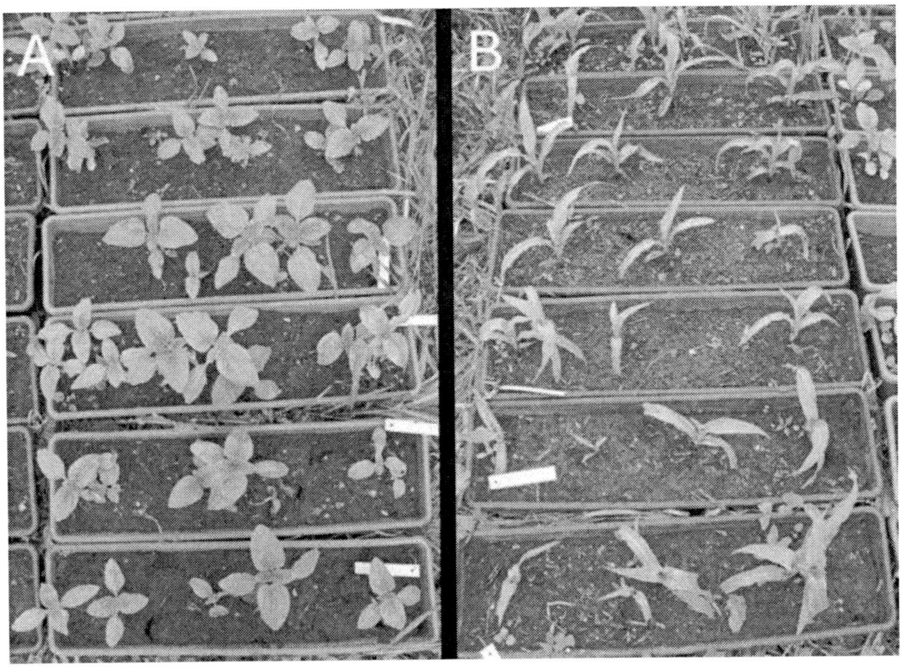

Figure 1. Sunflower (A) and maize (B) commercial varieties used for pot experiments.

Soil factors, such as pH, cation exchange capacity (CEC), and organic matter content, strongly affect the bio-available metal fractions (Chaney *et al.*, 1993; Tiller *et al.*, 1995). Since a negligible amount of HMs has been detected in seeds of sunflower and maize plants grown on contaminated soil, the use of extracted oil for technical purposes (Madejón *et al.*, 2003) could also be considered.

The aims of our pot-based sunflower (*Helianthus annuus* L.) and maize (*Zea mays* L.) screening study were as follows:

- to estimate the GBD, through molecular tools (RAPD or AFLP), in 3 and 9 commercial cultivars of sunflower and maize, respectively;
- to investigate Cu and Zn accumulation in the different plant organs of three cultivars of both species, and to assess the effect of nitrate supplied with the metals;
- to identify the sunflower and maize cultivar with the best total metal accumulation/phytoextraction capacity in relation to biomass production, and to check the beneficial effect of fertilization.

Maize and Sunflower Gentic Bio-Diversity

Several studies have been conducted with wild and cultivated sunflower, and with maize accessions using different molecular markers, such as RAPDs (Berry *et al.*, 1994; Iqbal *et al.*, 2008; Leal *et al.*, 2010), AFLPs (Rönicke *et al.*, 2005; Dong *et al.*, 2007), SSRs (Yu *et al.*, 2002; Chen *et al.*, 2006; Leal *et al.*, 2010; Eschholz *et al.*, 2010), SNPs (Liu and Burke, 2006), and TRAPs (Yue *et al.*, 2009). The present study, RAPD or AFLP markers were used to investigate GBD in several sunflower and maize cultivars (Table 1) supplied by different seed companies, and grown in pots. Both molecular tools require a purified genomic DNA of good quality (i.e., lacking in contaminating proteins, polysaccharides, and secondary metabolites); to this purpose, the DNeasy Plant Mini Kit, commercialized by Qiagen Italia, (Milano-Italy) was used.

Table 1. List of commercial maize and sunflower cultivars used in the present study. Asterisks indicate the cultivars employed for the phytoremediation experiments

Maize cultivars	Seed company
DKC 5783	Monsanto
DKC 6818	Monsanto
DK 440 *	Monsanto
NX 7234 *	Syngenta
NX 7464	Syngenta
NX 8441 *	Syngenta
AGRISTER	KWS
AZUAGA	KWS
BELGRANO	KWS
Sunflower cultivars	Seed company
Isar *	KWS
Pretor *	KWS
Trisun *	KWS

Table 2. RAPD results of maize cultivars. Number of loci, Polymorphic DNA fragments and % of polymorphic loci

Primer (annealing T: 40 °C)	Number of loci	Polymorphic	% Polymorphic
Wpms 20 REV	8	8	100
Ha1287 FOR	8	5	63
Ha1287 REV	11	6	55
Ha1608 FOR	9	7	78
Ha1608 REV	9	8	89
Wpms18 FOR	8	4	50
Wpms18 REV	7	3	43
Udo 36 REV	7	6	86

To perform the molecular characterization, fresh leaves were collected from control plants (no HMs added), and the DNA immediately extracted and purified for PCR reactions. The RAPD method utilizes selected decamers, or twentymers, to amplify gene loci randomly distributed on genomic DNA (Williams *et al.*, 1990, Castiglione *et al.*, 1993). The RAPD technique is the cheapest reliable molecular tool used for efficient DNA fingerprinting. Once stable and reproducible PCR conditions are established, the method is quick and easy, and requires, as a template, only a very limited amount (5-10 ng) of total genomic DNA. The thermal profile and the PCR mixtures used for RAPD reactions useful to the characterization of the maize cultivars analyzed in our study were essentially similar to those described by Castiglione and co-workers (1993) for the twentymer primers.

To screen the primers suitable for genotyping, a set of primers was tested using a subset of purified genomic DNA samples. The primers (Table 2) producing reproducible and stable profiles in time and in different experiments, as well as distinctive and sharp DNA bands, were chosen to characterize all the maize cultivars used in our pot experiment.

The AFLP technique, essentially generated as described by Vos *et al.* (1995), is based on the restriction digestion of a DNA template, followed by the ligation of adapters to DNA digested fragments, and by two subsequent PCR amplifications of a subset of the restricted fragments, using specific primers (Table 3).

On the basis of the molecular data obtained, a cluster analysis was performed. The presence or absence of AFLP and RAPD bands were coded as 1 or 0, respectively, and used to create a binary data matrix. To discriminate the level of polymorphisms between maize cultivars, the genetic similarity matrix was elaborated using Jaccard's similarity coefficient (Jaccard, 1908). On the similarity matrix, a cluster analysis was constructed by means of the Unweighted Pair Group Mean with Arithmetical Averages (UPGMA) method using NTSYSpc (Numerical Taxonomy System, version 2.1 software - http://www.exeter software.com/cat/ntsyspc/ntsyspc.html). In accordance with other studies on maize, such as those of Pejic *et al.* (1998), de Souza *et al.* (2008) and Leal *et al.* 2010, the RAPD marker turned out to be an appropriate method to study the germplasm as compared to the use of other molecular markers, such as SSRs. Both SSRs and RAPDs were used to study genetic diversity in inbred popcorn lines, providing consistent data, and a high correlation between genetic distances (Leal *et al.*, 2010). In particular, some RAPD primers yielded a larger number of polymorphic bands as compared to the SSR loci.

In the present study, the results of the RAPD analysis show that primer Ha 1258 REV yields the highest number (11) of amplified DNA bands, and WPMS 18 REV and Udo 36 REV the lowest (7 bands). Primer WPMS 20 REV has the greatest number (8) of polymorphic bands, and the highest percentage of polymorphism (100%), while WPMS 18 REV primer has the lowest (43%). The selected primers are able to discriminate among maize accessions, producing a total of 67 RAPD multiple bands. A total of 47 polymorphic markers were obtained through the RAPD analysis using these 8 different primers (70% of polymorphic bands).

Table 3. AFLPs primer combinations, Number of loci, Polymorphic DNA fragments and % of polymorphic loci of sunflower cultivars

Primer combinations		Number of loci	Polymorphic	% Polymorphic
EcoRI-AGG	MseI-AAC	48	16	33
EcoRI-ATA	MseI-CCT	49	28	57
EcoRI-AGC	MseI-CCA	43	23	53
EcoRI- ACC	MseI-AGC	73	26	36
EcoRI- AGA	MseI-CGG	70	34	49
EcoRI-ACC	MseI-ACA	58	32	55
EcoRI-ATA	MseI-ACT	44	17	39

Fragment ranged in size from 500 to 3000 bp (Figure 2). The level of polymorphism observed in this study (70%) was higher than in some other maize studies (Melo *et al.*, 2001), but lower than that obtained on maize inbred lines (80% - Bruel *et al.*, 2006; Vilela *et al.*, 2008; Leal *et al.*, 2010), although different cultivars were tested. The level of polymorphism depends on the degree of divergence among the genotypes studied, but also on the primers employed since these could contribute to increase the level of DNA polymorphism.

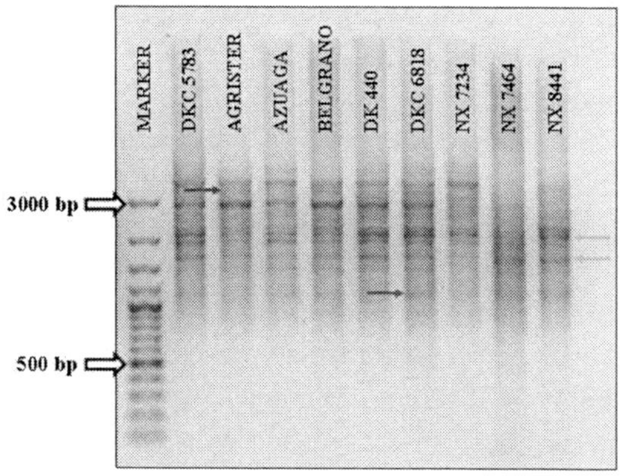

Figure 2. RAPD pattern obtained with Ha1608REV primer in maize cultivars. White arrows show molecular weights of GeneRuler™ 100bp DNA Ladder [Fermentas Italia (Milano, Italy)]. Red and yellow arrows show polymorphic and not polymorphic DNA bands, respectively.

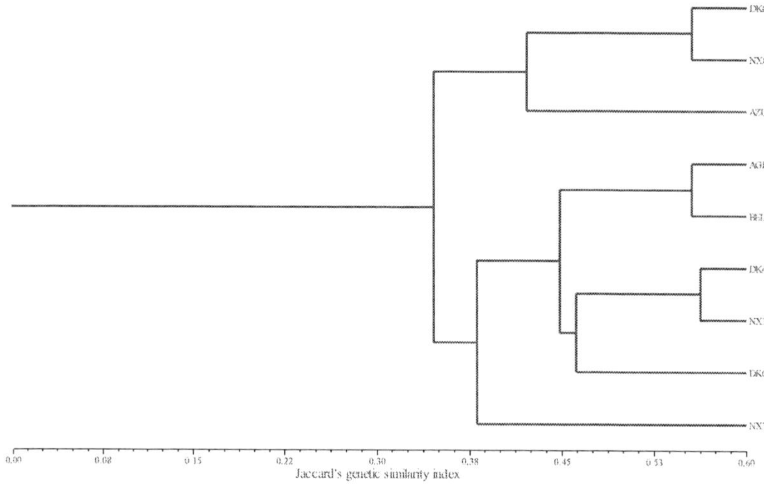

Figure 3. Dendrogram of maize cultivars generated on the basis of Jaccard's similarity coefficient and UPGMA cluster analysis on RAPD marker data elaborated with NTSys PC software package.

A cluster analysis was performed on the RAPD data, and the output dendrogram (Figure 3) illustrates a high genetic diversity between the 9 maize cultivars. The genetic similarity ranges from 0.34 to 0.56. Similar values were obtained using RAPDs with other cultivated and hybrid maize lines (Sobrinho *et al.*, 2001; Leal *et al.*, 2010). A careful observation of Figure 3 suggests the presence of two major clusters, and shows that the associations among cultivars were not influenced by the supplier. The first cluster includes the accessions DKC 5383, NX 8441 and Azuga; the second one includes other two subgroups, while the NX 7464 accession has a basal position as outlier. One cluster of the second subgroup comprises Agrister and Belgrano, while the other one includes DK 440 and NX 7234, showing a genetic similarity value of 0.56.

The AFLP technique has been used to estimate GBD in several crops including poplar (Castiglione *et al.*, 1993), barley (Becker *et al.*, 1995), lentil (Sharma *et al.*, 1996), soybean (Maughan *et al.*, 1996), cotton (Pillay and Myers, 1999), and pepper (Joy *et al.*, 2007), etc. In the present study, AFLPs generated a large number of reproducible and unambiguous molecular markers useful for fingerprinting the commercial sunflower cultivars used in this comparative study. Seven primer pair combinations (Table 3) were used to assay the GBD of the three sunflower accessions (Table 1).

Between 40 and 70 scorable DNA bands (varying from 50 to 400 bps) were detected after selective PCR amplification with each primer combination. The complete data set includes a total of 333 bands; 177 fragments, corresponding to 53% of the total DNA bands, were polymorphic. AFLP similarity, based on Jaccard's index, separates the three sunflower accessions into two clusters (Figure 4).

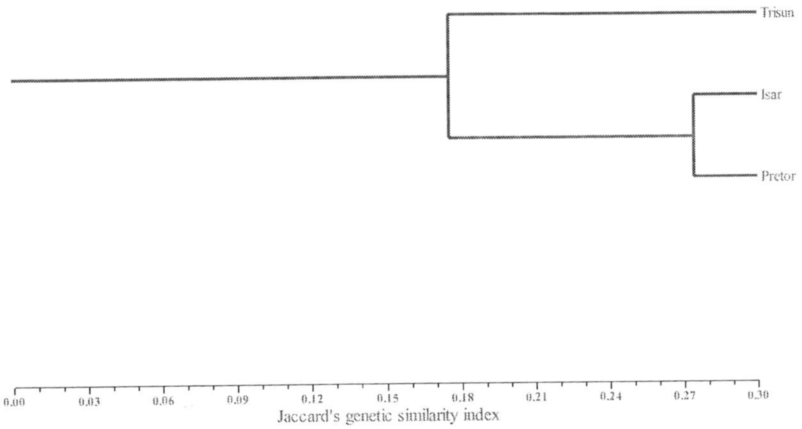

Figure 4. Dendrogram of sunflower cultivars generated on the basis of Jaccard's similarity index and UPGMA cluster analysis on sunflower cultivars AFLP marker data elaborated with NTSys PC software package.

Isar and Pretor varieties are clustered together, showing a more similar banding pattern. Trisun is clearly separated from the other two accessions, with a similarity index value of 0.18. The genetic similarity among the three analyzed sunflower accessions is not high

(<20%), and was lower than that observed in previous studies on sunflower. Although different molecular markers and data analysis methods were employed, the genetic similarities ranged from 0.58 to 0.98, with an average of 0.70, among 25 oilseed sunflower lines (Rönicke *et al.*, 2005), and from 0.70 to 0.91 among other sunflower lines investigated using AFLP marker loci (Hongtrakul *et al.*, 1997).

The AFLP and RAPD analyses performed in this work were able to easily discriminate among accessions obtained from the germplasm collection of three different seed companies, and allowed us to identify, from a genetic point of view, the most divergent cultivars. Molecular screenings provide valuable information on the genetic diversity in commercial cultivars, which could have been dramatically narrowed during domestication, and in subsequent cycles of selection by plant breeders.

Moreover, they provide a framework to select adequate genotypes with good agronomic performance (Mokrani *et al.*, 2002; Bert *et al.*, 2003; Al-Chaarani *et al.*, 2005; Flavel, 2010; Ortiz *et al.*, 2010), also under stress (Kiani *et al.*, 2007), or with interesting traits, such as tolerance, accumulation and extraction efficiency towards HMs (Nehnevajova *et al.*, 2009). In fact, in a field trial conducted on a highly Cu and Zn contaminated soil using 168 different clones of *Populus alba*, *P. nigra* and *P.* x *canadensis* with a high level of GBD, 5 clones were found to be particularly suitable for phytoremediation purposes (Castiglione *et al.*, 2009).

Maize and Sunflower Phytoextraction

Pot experiments were carried out in order to evaluate the phytoextraction capability of 3 out of the 9 maize cultivars assayed for their GBD (cultivars for the pot experiments are identified by * in Table 1), and of the 3 sunflower cultivars (Table 1). Three seeds per cultivar were sown in pots filled with soil collected, at a depth of about 10-20 cm, from an area cultivated with chestnut and hazel in the surroundings of the campus of the University of Salerno (Fisciano, SA-Italy). Pot filling soil (control soil) was characterized for total and available Cu and Zn concentrations.

For Cu and Zn determination in soil and plant organs, the soil granulometric fraction (2 mm) and plant roots were pulverized in an agate mortar (Eatchs, Retsch, Germany); dried stems were reduced to ashes at 550 °C for 2 h in a muffle furnace (Nabertherm, Controller B170, Germany); leaves were pulverized in liquid nitrogen and redried, at 75 °C to eliminate all traces of water. Soil and plant matrices were digested with an acid mixture (HNO_3 65%:HF 50% = 2:1 = v:v) in a microwave oven (Ethos, Milestone, Shelton, CT-USA), as described by Baldantoni *et al.* (2009). The Lindsay and Norvell (1978) method was used to estimate the soil Cu and Zn bioavailable fractions. Metal concentrations were determined by means of an atomic absorption spectrophotometer (AAnalyst 100, Perkin Elmer, Wellesley, MA-USA), via a graphite atomizer (Cu), or air-acetylene flame (Zn). Two replicates of each analysis were carried out to evaluate sample variability. Standard reference materials (calcareous loam soil BCR CRM 141R - European Commission, 1996, and pine needles 1575a - NIST, 2004) were also analyzed to obtain accurate data, and Cu and Zn concentrations were calculated considering the recoveries of the certified metals.

The data were processed by statistical tests using the Stata10.1 software package (http://www.stata.com/company). The significance of differences in biomass production, and in Cu and Zn accumulation between species and among cultivars was tested by Nested ANOVA; the overall significance of differences in metal accumulation (both Cu and Zn) between species and among cultivars was also assayed by Nested MANOVA.

Soil used for pot filling was moderately contaminated by Cu (Table 4), since a 3-fold higher concentration relative to the maximum values of the ranges reported for natural soils by Allen (1989) and Markert (1992), i.e., 5-80 μg g^{-1} d.w. and 1-80 μg g^{-1} d.w., respectively, were measured. Total Zn concentration measured in pot soil was, on the contrary, within the background ranges: 20-300 μg g^{-1} d.w. (Allen, 1989) and 3-300 μg g^{-1} d.w. (Markert, 1992). Available Cu and Zn concentrations were also within the ranges measured for these elements in natural soils in Southern Italy (Marchetti *et al.*, 2010), equal to 11.7-68.7 μg g^{-1} d.w. for Cu and 1.8-17.8 μg g^{-1} d.w. for Zn. Owing to these low metal availabilities (Table 4), and to better understand the phytoextraction potential of the studied sunflower and maize cultivars, half of the pots used in this experiment, were supplied with a surplus of available Cu and Zn at the moment of plant sowing and during the early stages of growth.

Table 4. Mean total and available Cu and Zn concentrations ± standard deviations (μg g^{-1} d.w.) in soil used for pot filling

	Total	Available
Cu	234.59 ± 2.98	48.91 ± 1.17
Zn	92.53 ± 3.36	9.26 ± 0.11

In order to estimate the Cu and Zn phytoextraction capability of the 6 cultivars, the Translocation Factor (TF) and BioConcentration Factor (BCF) were calculated after half of the pots were supplied with Cu and Zn, in the available form of nitrate (NO_3^-). To reach the total concentration of 150 μg g^{-1} soil (d.w.) for Cu and of 450 μg g^{-1} soil (d.w.) for Zn, pots were irrigated with the metal solutions at the time of sowing, and later during early plant growth (10 and 20 days after sowing).

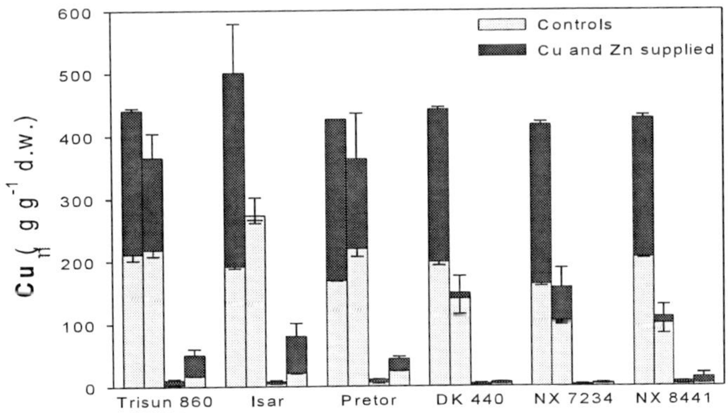

Figure 5. Mean Cu concentrations measured in soil and in plant organs of three sunflower (Trisun 860, Isar, Pretor) and three maize (DK 440, NX 7234, NX 8441) cultivars, grown on control or metal-supplemented soils. For each cultivar, the vertical bars correspond, from left to right to: soil, root, stem and leaf Cu concentrations. The error bars represent standard errors.

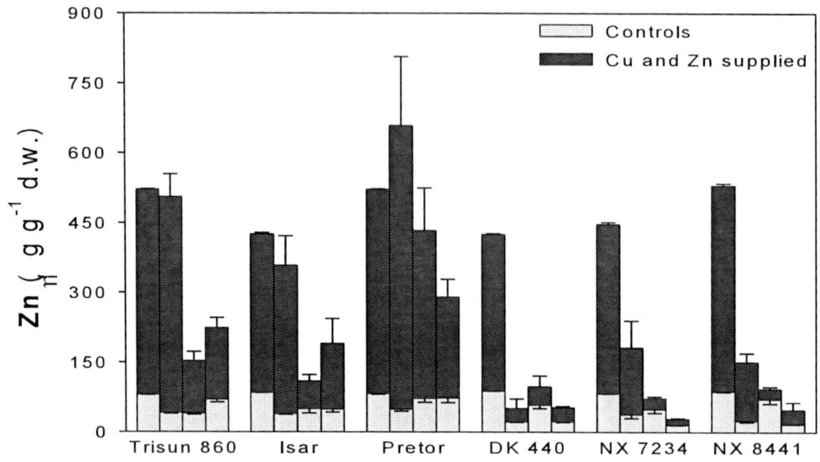

Figure 6. Mean Zn concentrations measured in soil and in plant organs of three sunflower (Trisun 860, Isar, Pretor) and three maize (DK 440, NX 7234, NX 8441) cultivars, grown on control or metal-supplied soils. For each cultivar, the vertical bars correspond, from left to right to: soil, root, stem and leaf Zn concentrations. The error bars represent standard errors.

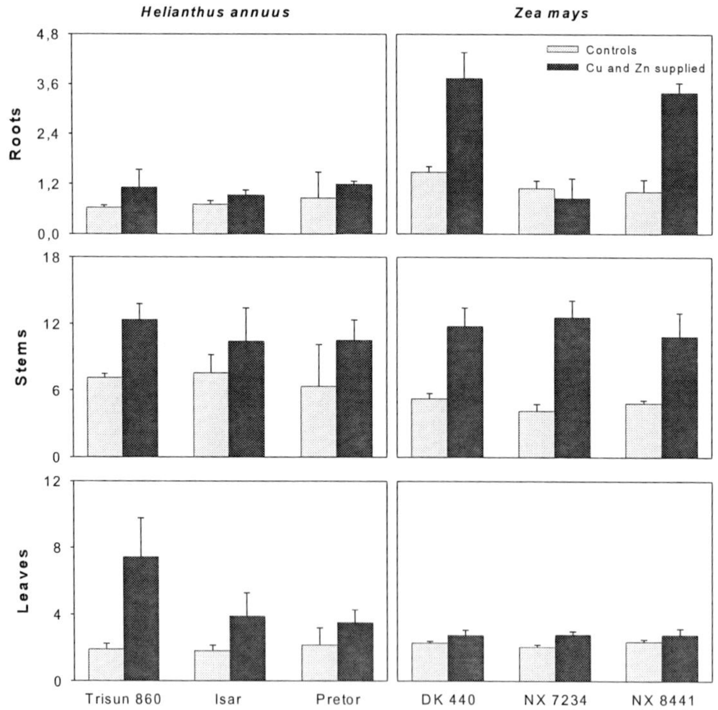

Figure 7. Biomass (g d.w. plant^{-1}) of roots, stems and leaves of sunflower and maize plants grown either on unsupplemented (control) or on Cu- and Zn-supplemented soil. The error bars represent standard errors.

When the plants reached maturity (ca. 130 days after sowing), they were harvested, and tap water washed roots, stems and leaves were collected separately. At the time of biological material harvesting, soil samples were also collected. The dry weights of roots, stems and leaves were measured after drying at 75 °C for at least 24 h. Roots, stems and leaves of the mature plants were analyzed for their Cu and Zn concentrations as described above, and metal concentrations in plants were compared to those measured in pot soils at harvest time (Figures 5 and 6). Both in maize and sunflower plants the biomass (Figure 7) varied in the following order: stems > leaves > roots; confirming the suitability of these species for phytoremediation purposes.

Maize and sunflower plants did not show significant differences in biomass production, contrary to results by Usman and Mohamed (2009), who found a higher biomass production (both roots and shoots) in maize than in sunflower plants.

On control soil, there were no significant differences in biomass production between the three cultivars of each studied species. Maize and sunflower plants grown on soil supplied with Cu and Zn showed, on average, significant differences in the biomass of roots, stems and leaves in comparison with the plants grown on control soil. This may be due to the positive effects on plant growth of the nitrate supplied with the metals, and would also indicate that under well fertilized conditions, the high bioavailable concentrations of Cu and Zn in this soil did not inhibit plant growth and development, as also observed by Madejón *et al.*, (2003) in mature sunflower plants grown on a spill-affected soil. This constitutes an important point, because the potential of high biomass plants for the phytoremediation of polluted soils depends not only on their ability to accumulate HMs, but also on their capacity to tolerate high soil metal concentrations, while maintaining a fast growth rate (Hernández-Allica *et al.*, 2008).

Copper and Zinc Accumulation

In maize and sunflower plants, the root was the organ with the highest Cu concentrations, followed by leaf and stem (Figure 5) on both control and supplied soils. Between the two species, sunflower plants showed, on average, the higher Cu concentrations in all organs (Fig. 5, $P<0.01$), but no statistically significant differences among cultivars were found.

The three organs of maize and sunflower, with the exception of Isar roots, exhibited higher Cu concentrations when grown on soil supplied with bio-available metals compared with control soil, highlighting the capability of these plants to take up available Cu from soil, and to accumulate it. Even the plants grown on metal-supplemented soil, as in the case of the controls, showed significant differences ($P<0.001$) in Cu accumulation between species, but not among cultivars. However, after plants were harvested, soils supplied with metals still had Cu concentrations that were on average 2-fold higher than in control soils for both species. While sunflower roots, excluding those of Isar, accumulated Cu concentrations comparable to those measured in the respective soils, maize roots had Cu concentrations that were lower than in the soil. Thus, on both metal-supplemented and non supplemented soil, maize plants exhibited a reduced phytoextraction capability in comparison with sunflower plants, also in the presence of high bioavailable soil Cu concentrations.

Among the three analyzed organs, roots accumulated the highest Zn concentrations both in maize (with the exception of DK 440) and sunflower plants (Figure 6), followed by stems

(NX 7234 and NX 8441 maize cultivars and Pretor sunflower cultivar) and leaves (Trisun 860 and Isar sunflower cultivars). As in the case of Cu, Zn concentrations were also higher in sunflower ($P<0.05$) in comparison to maize plants, but no significant differences among cultivars were found. All the organs of maize and sunflower exhibit the highest Zn concentrations in the plants grown on soil supplied with HMs, as reported by Cui *et al.* (2004) for maize.

This observation confirms the high capacity of both maize and sunflower plants to absorb available Zn from soil, and to accumulate it in all plant organs. Plants grown on metal supplemented soil, as in the case of control soil, showed significant differences ($P<0.05$) in Zn accumulation between species, but not among cultivars. After plant harvesting, the soils supplied with HMs had Zn concentrations that were on average 3-fold higher than in control soils. The three sunflower cultivars had Zn root concentrations comparable to those measured in their respective soils (Figure 6); on the contrary, Zn concentrations in maize roots were always lower than in soil.

In both maize and sunflower plants, higher Cu and Zn concentrations in roots than in shoots are often reported, with values comparable (Usman and Mohamed, 2009) to those found in the present study. On the contrary, Nehnevajova *et al.* (2009) found Zn concentrations in sunflower plants distributed in the order: leaves > stems > roots, and on a soil heavily contaminated by Zn (total concentration equal to 2739 $\mu g\ g^{-1}$ d.w. and available concentration equal to 182 $\mu g\ g^{-1}$ d.w.), Solhi *et al.* (2005) found comparable Zn concentrations in roots and in shoots of sunflower (cv. Hybride, hysun 25) plants, amounting to 433 and 463 $\mu g\ g^{-1}$ d.w., respectively.

In a pot experiment, Hernández-Allica *et al.* (2008) found in the shoots of maize cv. Ranchero, grown on soil artificially contaminated with Zn at the dose of 100 $\mu g\ g^{-1}$ d.w. of soil, a concentration equal to 951 ± 47 $\mu g\ g^{-1}$ d.w., possibly indicating a higher Zn phytoextraction capability in this cultivar compared with our maize cultivars DK 440, NX 7234 and NX 8441 even if experimental conditions were completely different (soil characteristics, metal availability, climate and so on).

Copper and Zinc Phytoextraction Capability in the Whole Plant

On the whole, sunflower plants showed higher metal (Cu+Zn) concentrations in the three organs in comparison with maize plants ($P<0.01$); no significant differences among the tested cultivars were found.

Considering the biomass of the different organs and the Cu concentrations found in each organ, roots of both maize and sunflower plants accumulated the highest amount of Cu, followed by stems and leaves (Figure 8). In the whole plant, each of the three sunflower cultivars accumulated more Cu than maize cultivars: on metal-supplied soils, Trisun 860 accumulated 911 μg Cu $plant^{-1}$, Isar 662 μg Cu $plant^{-1}$ and Pretor 702 μg Cu $plant^{-1}$.

Different from Cu, stems of both maize and sunflower plants accumulated the highest amount of Zn, followed by leaves and roots (Figure 9). Sunflower plants showed a higher uptake of Zn than maize plants and, in the HM-supplied soils, the three sunflower cultivars accumulated, on the whole dry plant, a huge amount of this HM: 4227 μg Zn $plant^{-1}$ in Trisun 860, 2231 $\mu g\ plant^{-1}$ in Isar and 5981 $\mu g\ plant^{-1}$ in Pretor.

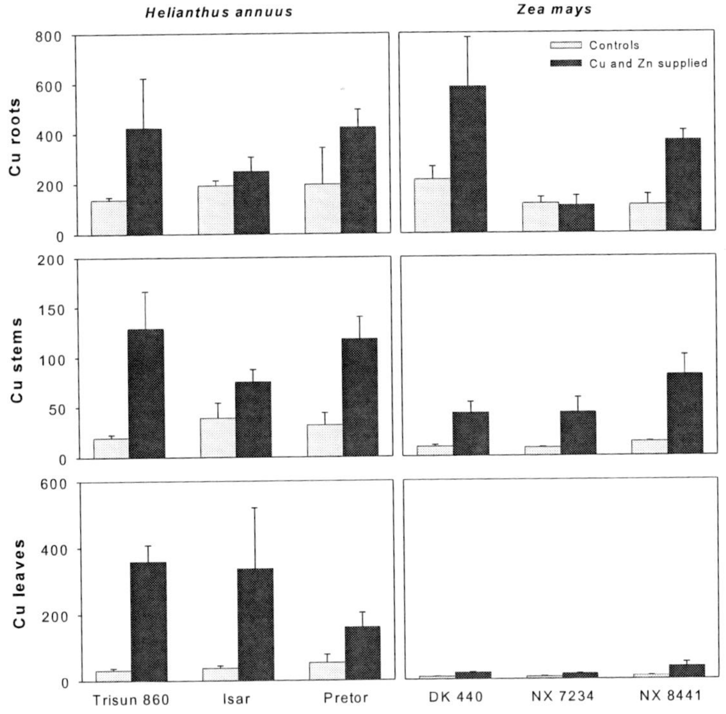

Figure 8. Cu accumulation (μg Cu d.w. plant^{-1}) in roots, stems and leaves of sunflower and maize plants grown either on unsupplemented (control) or on Cu- and Zn-supplemented soil. The error bars represent standard errors.

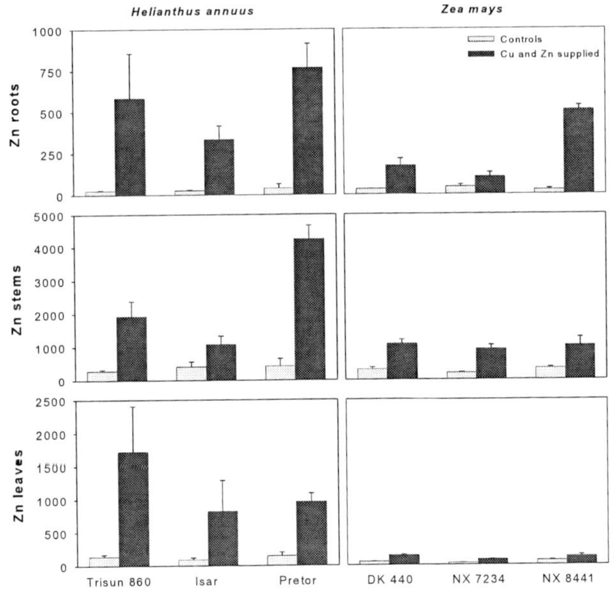

Figure 9. Zn accumulation (μg Zn d.w. plant^{-1}) in roots, stems and leaves of sunflower and maize plants grown either on unsupplemented (control) or on Cu- and Zn-supplemented soil. The error bars represent standard errors.

Comparing the metal phytoextraction capability of four agronomic plant species with high annual biomass yield (*Zea mays* cv. Impact, *Helianthus annuus* cv. Giganteus, *Brassica rapa* Durmelander, *Cannabis sativa* cv. Chameleon), grown on a moderately contaminated soil, Meers *et al.* (2005) found that maize plants exhibited the lowest shoot concentrations for all the HMs analyzed (Cu, Zn, Cd, Ni, Pb).

The Zn TF and BCF (measured in the aerial part of the plant, i.e., stem+leaves) was higher in the plants grown on control soil than in those grown on HM supplied soils (0.55<TF<1.1 – 0.23<BCF<1.39 – Table 5). On the other hand, an opposite trend was observed in the case of Cu (0.06<TF<0.34 – 0.02<BCF<0.18; Table 5).

We might assume that, in the case of Cu, a large quantity of the same metal may exert a stimulatory effect on the transporters responsible for root-to-shoot translocation of this HM, while an excess of available Zn in the soil might lead, on the contrary, to an inhibition of the Zn transporters. (Jeong and Connelly, 2009)

Table 5. Ratios between Cu and Zn concentrations measured in both stems and leaves (shoots) respect to those measured in roots (TF) and between Cu and Zn concentrations measured in shoots respect to those measured in soil (BCF) of both control (white) and metal supplied (grey) maize and sunflower plants

	Cu TF shoots/roots	Zn TF shoots/roots	Cu BCF shoots/soil	Zn BCF shoots/soil
Trisun 860	0.09	2.66	0.09	1.36
Trisun 860	0.17	0.75	0.14	0.72
Isar	0.09	2.52	0.13	1.14
Isar	0.34	0.84	0.18	0.70
Pretor	0.14	2.99	0.18	1.76
Pretor	0.15	1.10	0.13	1.39
DK 440	0.04	3.65	0.03	0.91
DK 440	0.08	2.95	0.03	0.35
NX 7234	0.06	1.69	0.04	0.77
NX 7234	0.06	0.55	0.02	0.23
NX 8441	0.07	3.68	0.03	1.00
NX 8441	0.20	0.93	0.05	0.27

CONCLUSIONS

This pilot study on phytoremediation of maize and sunflower cultivars, supplied by various seed companies, shows that the considerable intra-specific GBD in both plant species, does not reflect some remarkable and evident differences in the aptitude to reclaim HM polluted soils of diverse varieties. As a matter of fact, no statistically significant difference in phytoextraction or phytostabilization (the ability of any plant to accumulate metal ions in roots) was observed among the studied sunflower and maize cultivars. However differences in metal uptake and accumulation capacity suggest that great gains can be obtained from the use of different plant species with a wide genetic background. In our experimental conditions, given the low TFs and BCFs, for all cultivars, with the exception of Pretor (TF and BCF 1.10 and 1.39, respectively), sunflower and maize can be regarded as plant species with a marked phytostabilization rather than phytoextraction capacity. The possibility of using intercrops to ameliorate phytoremediation has been recently considered (see Li *et al.*, 2009). Since the use of maize and sunflower in combination with perennial trees is a well consolidated agronomic practice, the phytoremediation of multi-metal contaminated soils could be achieved through the intercrops of these annual high-biomass plants with long-life trees, such as poplar or salix, equally promising for use in soil clean-up from pollutants.

ACKNOWLEDGMENTS

We thank the didactic area of Environmental Sciences, of the Science Faculty of the University of Salerno (84084 Fisciano SA-IT) for the fund support to the conducted research, the Osservatorio Appennino Meridionale (Via Ponte Don Melillo – 84084 Fisciano SA-IT) for space availability necessary to realize all the phytoremediation experiences and Prof. Stefania Biondi for a careful reading of the manuscript and for referee's revision (Dipartimento di Biologia e.s., Università di Bologna, via Irnerio, 42, 40126 Bologna, Italy).

REFERENCES

Agarwal M, Shrivastava N, Padh H. 2008. Advances in molecular marker techniques and their applications in plant sciences. *Plant Cell Reports* 27: 617-631.

Al-Chaarani GR, Gentzbittel L, Wedzony M, Sarrafi A. 2005. Identification of QTLs for germination and seedling development in sunflower (*Helianthus annuus* L.). *Plant Science* 169: 221-227.

Allen SE. 1989. *Chemical analysis of ecological materials*. Second edition. S.E. Allen (ed.). Blackwell Scientific Publications.

Baldantoni D, Ligrone R, Alfani A. 2009. Macro- and trace-element concentrations in leaves and roots of *Phragmites australis* in a volcanic lake in Southern Italy. *Journal of Geochemical Exploration* 101: 166-174.

Barakat MN, Al Doss AA, Moustafa KA, Ahmed EI, Elshafei AA. 2010. Morphological and molecular characterization of Saudi wheat genotypes under drought stress. *Journal of Food Agriculture & Environment* 8: 220-228.

Becker J, Vos P, Kuiper M, Salamini F, Heun M. 1995. Combined mapping of AFLP and RFLP markers in barley. *Molecular & General Genetics* 249: 65-73.

Berry ST, Allen RJ, Barnes SR, Caligari PDS. 1994. Molecular marker analysis of *Helianthus annuus* L. Restriction-Fragment-Length-Polymorphism between inbred lines of cultivated sunflower. *Theoretical and Applied Genetics* 89: 435-441.

Bert PF, Jouan I, de Labrouhe DT, Serre F, Philippon J, Nicolas P, Vear F. 2003. Comparative genetic analysis of quantitative traits in sunflower (*Helianthus annuus* L.). 2. Characterisation of QTL involved in developmental and agronomic traits. *Theoretical and Applied Genetics* 107: 181-189.

Booy G, Hendriks RJJ, Smulders MJM, Van Groenendael JM, Vosman B. 2000. Genetic diversity and the survival of populations. *Plant Biology* 2: 379-395.

Bruel DC, Carpentieri-Pipolo V, Gerage AC, Fonseca ND, Prete CEC, Ruas CDF, Ruas PM, de Souza SGH, Garbuglio DD . 2006. Genetic distance estimated by RAPD markers and its relationship with hybrid performance in maize. *Pesquisa Agropecuaria Brasileira* 41: 1491-1498.

Castiglione S, Wang G, Damiani G, Bandi C, Bisoffi S, Sala F. 1993. RAPD fingerprints for identification and for taxonomic studies of elite poplar (*Populus* Spp) clones. *Theoretical and Applied Genetics* 87: 54-59.

Castiglione S, Todeschini V, Franchin C, Torrigiani P, Gastaldi D, Cicatelli A, Rinaudo C, Berta G, Biondi S, Lingua G. 2009. Clonal differences in survival capacity, copper and zinc accumulation, and correlation with leaf polyamine levels in poplar: A large-scale field trial on heavily polluted soil. *Environmental Pollution* 157: 2108-2117.

Chaney RL, Li YM, Schneiter CE, Green CE, Miller JF, Hopkins DG. 1993. *Progress in developing technologies to produce low Cd concentration sunflower kernels*. In: Proceedings of the 15th Sunflower Research Workshop, pp. 80–92, Fargo, ND, January 14–15, National Sunflower Assoc.

Chen JF, Hu JG, Vick BA, Jan CC. 2006. Molecular mapping of a nuclear male-sterility gene in sunflower (*Helianthus annuus* L.) using TRAP and SSR markers. *Theoretical and Applied Genetics* 113: 122-127.

Cui YS, Dong YT, Li HF, Wang QR. 2004. Effect of elemental sulphur on solubility of soil heavy metals and their uptake by maize. *Environment International* 30: 323-328.

Darwin C. 1859. Sulla origine delle specie per elezione naturale, ovvero conservazione delle razze perfezionate nella lotta per l'esistenza di Charles Darwin. Traduzione italiana di Giovanni Canestrini (1933). Edizioni A. Barion. Casa per Edizioni popolari. Sesto San Giovanni (MI-Italia).

de Souza SGH, Carpentieri-Pipolo V, Ruas CD, Carvalho VDP, Ruas PM, Gerage AC. 2008. Comparative analysis of genetic diversity among the maize inbred lines (*Zea mays* L.) obtained by RAPD and SSR markers. *Brazilian Archives of Biology and Technology* 51: 183-192.

Dong GJ, Liu GS, Li KF. 2007. Studying genetic diversity in the core germplasm of confectionary sunflower (*Helianthus annuus* L.) in China based on AFLP and morphological analysis. *Russian Journal of Genetics* 43: 627-635.

Eschholz TW, Stamp P, Peter R, Leipner J, Hund A. 2010. Genetic structure and history of Swiss maize (*Zea mays* L. ssp *mays*) landraces. Genetic Resources and Crop Evolution 57: 71-84.

European Commission 1996. The certification of the total contents (mass fractions) of Cd, Co, Cr, Cu, Hg, Mn, Ni, Pb and Zn and the aqua regia soluble contents (mass fractions) of Cd, Co, Cr, Cu, Hg, Mn, Ni, Pb and Zn in a calcareous loam soil; CRM 141R. Bureau Community of Reference, Bruxelles, EUR 16890 EN.

Flavell R. 2010. From genomics to crop breeding. *Nature Biotechnology* 28: 144-145.

Grassi F, Labra M, Sala F. 2006. Introduzione alla biodiversità del mondo vegetale, Piccin, Padova.

Hernández-Allica J, Becerril JM, Garbisu C. 2008. Assessment of the phytoextraction potential of high biomass crop plants. *Environmental Pollution* 152: 32-40.

Hongtrakul V, Huestis GM, Knapp SJ. 1997. Amplified fragment length polymorphisms as a tool for DNA fingerprinting sunflower germplasm: genetic diversity among oilseed inbred lines. *Theoretical and Applied Genetics* 95: 400-407.

Iqbal MA, Sadaqat HA, Khan IA. 2008. Estimation of genetic diversity among sunflower genotypes through random amplified polymorphic DNA analysis. *Genetics and Molecular Research* 7: 1408-1413.

Jaccard P. 1908. Nouvelle recherches sur la distribution florale. *Bulletin Société Vaudoise des Sciences Naturelles* 44: 223-270.

Jeong J, Connolly EL. 2009. Iron uptake mechanisms in plants: Functions of the FRO family of ferric reductases. *Plant Science* 176: 709-714.

Jones N, Ougham H, Thomas H, Pasakinskiene I. 2009. Markers and mapping revisited: finding your gene. *New Phytologist* 183: 935-966.

Joy N, Abraham Z, Soniya EV. 2007. A preliminary assessment of genetic relationships among agronomically important cultivars of black pepper. *Bmc Genetics* 8: Art. 42.

Kiani SP, Talia P, Maury P, Grieu P, Heinz R, Perrault A, Nishinakamasu V, Hopp E, Gentzbittel L, Paniego N, Sarrafi A. 2007. Genetic analysis of plant water status and osmotic adjustment in recombinant inbred lines of sunflower under two water treatments. *Plant Science* 172: 773-787.

Komarek M, Tlustos P, Szakova J, Chrastny V, Ettler V. 2007. The use of maize and poplar in chelant-enhanced phytoextraction of lead from contaminated agricultural soils. *Chemosphere* 67: 640-651.

Lasat MM. 2002. Phytoextraction of toxic metals: A review of biological mechanisms. *Journal of Environmental Quality* 31: 109-120.

Leal AA, Mangolin CA, do Amaral ATD, Goncalves LSA, Scapim CA, Mott AS, Eloi IBO, Cordoves V, da Silva MFP. 2010. Efficiency of RAPD versus SSR markers for determining genetic diversity among popcorn lines. *Genetics and Molecular Research* 9: 9-18.

Li NY, Li ZA, Zhuang P, Zou B, McBride M. 2009. Cadmium uptake from soil by maize with intercrops. *Water Air and Soil Pollution* 199: 45-56.

Lin JX, Jiang WS, Liu DH. 2003. Accumulation of copper by roots, hypocotyls, cotyledons and leaves of sunflower (*Helianthus annuus* L.). *Bioresource Technology* 86: 151-155.

Lin Q, Shen KL, Zhao HM, Li WH . 2008. Growth response of *Zea mays* L. in pyrene-copper co-contaminated soil and the fate of pollutants. *Journal of Hazardous Materials* 150: 515-521.

Lindsay WL, Norvell WA. 1978. Development of a DTPA soil test for zinc, iron, manganese, and copper. *Soil Science Society of America Journal* 42: 421-428.

Liu AZ, Burke JM. 2006. Patterns of nucleotide diversity in wild and cultivated sunflower. *Genetics* 173: 321-330.

Madejón P, Murillo JM, Maranon T, Cabrera F, Soriano MA. 2003. Trace element and nutrient accumulation in sunflower plants two years after the Aznalcollar mine spill. *Science of the Total Environment* 307: 239-257.

Marchetti M, Tognetti R, Lombardi F, Chiavetta U, Palumbo G, Sellitto M, Colombo C, Iovieno P, Alfani A, Baldantoni D, Barbati A, Ferrari B, Bonacquisti S, Capotorti G, Copiz R. Blasi C.. Ecological portrayal of old-growth forests and persistent woodlands in the Cilento and Vallo di Diano National Park (southern Italy). *Plant Biosystems* 144 (1): 130-147.

Markert B. 1992. Presence and significance of naturally-occurring chemical-elements of the periodic system in the plant organism and consequences for future investigations on inorganic environmental chemistry in ecosystems. *Vegetatio* 103: 1-30.

Maughan PJ, Maroof MAS, Buss GR, Huestis GM. 1996. Amplified fragment length polymorphism (AFLP) in soybean: Species diversity, inheritance, and near-isogenic line analysis. *Theoretical and Applied Genetics* 93: 392-401.

Meers E, Ruttens A, Hopgood M, Lesage E, Tack FMG. 2005. Potential of *Brassica rapa*, *Cannabis sativa*, *Helianthus annuus* and *Zea mays* for phytoextraction of heavy metals from calcareous dredged sediment derived soils. *Chemosphere* 61: 561-572.

Melo W, Pinho R, Santos J, Ferreira D. 2001. Use of morphoagronomical characters and molecular markers to evaluate genetic divergence among maize hybrids. *Revista Ceres* 48: 195-207.

Mokrani L, Gentzbittel L, Azanza F, Fitamant L, Al Chaarani G, Sarrafi A. 2002. Mapping and analysis of quantitative trait loci for grain oil content and agronomic traits using AFLP and SSR in sunflower (*Helianthus annuus* L.). *Theoretical and Applied Genetics* 106: 149-156.

Mulligan CN, Yong RN, Gibbs BF . 2001. Surfactant-enhanced remediation of contaminated soil: a review. *Engineering Geology* 60: 371-380.

Murakami M, Ae N. 2009. Potential for phytoextraction of copper, lead, and zinc by rice (*Oryza sativa* L.), soybean (*Glycine max* [L.] Merr.), and maize (*Zea mays* L.). *Journal of Hazardous Materials* 162: 1185-1192.

Nehnevajova E, Herzig R, Federer G, Erismann KH, Schwitzguebel JP. 2005. Screening of sunflower cultivars for metal phytoextraction in a contaminated field prior to mutagenesis. *International Journal of Phytoremediation* 7: 337-349.

Nehnevajova E, Herzig R, Bourigault C, Bangerter S, Schwitzguebel JP. 2009. Stability of enhanced yield and metal uptake by sunflower mutants for improved phytoremediation. *International Journal of Phytoremediation* 11: 329-346.

NIST 2004. Certification of NIST Standard Reference Material 1575a Pine Needles and results of an international laboratory comparison. *Special Publication*: 260-156.

Ortiz R, Taba S, Chavez VH, Mezzalama M, Xu YB, Yan JB, Crouch JH. 2010. Conserving and enhancing maize genetic resources as global public goods-a perspective from CIMMYT. *Crop Science* 50: 13-28.

Pejic I, Ajmone-Marsan P, Morgante M, Kozumplick V, Castiglioni P, Taramino G, Motto M. 1998. Comparative analysis of genetic similarity among maize inbred lines detected by RFLPs, RAPDs, SSRs, and AFLPs. *Theoretical and Applied Genetics* 97: 1248-1255.

Pillay M, Myers GO. 1999. Genetic diversity in cotton assessed by variation in ribosomal RNA genes and AFLP markers. *Crop Science* 39: 1881-1886.

Pilon-Smits E. 2005. Phytoremediation. *Annual Review of Plant Biology* 56: 15-39.

Rönicke S, Hahn V, Friedt W. 2005. Resistance to *Sclerotinia sclerotiorum* of 'high oleic' sunflower inbred lines. *Plant Breeding* 124: 376-381.

Sharma SK, Knox MR, Ellis THN. 1996. AFLP analysis of the diversity and phylogeny of Lens and its comparison with RAPD analysis. *Theoretical and Applied Genetics* 93: 751-758.

Sobrinho FS, Ramalho MAP, de Souza JC. 2001. Genetic diversity and inbreeding potential of maize commercial hybrids. *Maydica* 46: 171-175.

Solhi M, Shareatmadari H, Hajabbasi M. 2005. Lead and zinc extraction potential of two common crop plants, *Helianthus annuus* and *Brassica napus*. *Water Air and Soil Pollution* 167: 59-71.

Tiller KG, Oliver DP, McLaughlin MJ, Merry RH, Naidu R. 1995. *Managing cadmium contamination of agricultural land.* In: Biochemistry of Trace Elements 2. Advances in Environmental Science (Adriano, D.C., Ed.). Northwood, UK, Science Reviews.

Usman ARA, Mohamed HM. 2009. Effect of microbial inoculation and EDTA on the uptake and translocation of heavy metal by corn and sunflower. *Chemosphere* 76: 893-899.

Vassilev A, Vangronsveld J, Yordanov I. 2002. Cadmium phytoextraction: Present state, biological backgrounds and research needs. *Bulgarian Journal of Plant Physiology* 28: 68–95.

Vilela FO, do Amaral AT, Pereira MG, Scapim CA, Viana AP, Freitas SDP. 2008. Effect of recurrent selection on the genetic variability of the UNB-2U popcorn population using RAPD markers. *Acta Scientiarum-Agronomy* 30: 25-30.

Vos P, Hogers R, Bleeker M, Reijans M, Vandelee T, Hornes M, Frijters A, Pot J, Peleman J, Kuiper M, Zabeau M. 1995. AFLP - A new technique for DNA-fingerprinting. *Nucleic Acids Research* 23: 4407-4414.

Williams JGK, Kubelik AR, Livak KJ, Rafalski JA, Tingey SV. 1990. DNA polymorphisms amplified by arbitrary primers are useful as genetic-markers. *Nucleic Acids Research* 18: 6531-6535.

Yu JK, Mangor J, Thompson L, Edwards KJ, Slabaugh MB, Knapp SJ. 2002. Allelic diversity of simple sequence repeats among elite inbred lines of cultivated sunflower. *Genome* 45: 652-660.

Yue B, Cai XW, Vick BA, Hu JG. 2009. Genetic diversity and relationships among 177 public sunflower inbred lines assessed by TRAP markers. *Crop Science* 49: 1242-1249.

In: Handbook of Phytoremediation
Editor: Ivan A. Golubev

ISBN: 978-1-61728-753-4

Chapter 22

ECOLOGICAL ASPECTS OF SELENIUM PHYTOREMEDIATION

Colin F. Quinn, Stormy D. Lindblom and Elizabeth A. H. Pilon-Smits
Biology Department, Colorado State University,
Fort Collins, CO 80523, USA

ABSTRACT

Selenium is essential for many organisms but is toxic at elevated concentrations. The window between nutritious and toxic levels of Se is narrow, and both Se deficiency and toxicity are problems worldwide. For plants Se serves no known essential function, and uptake of Se by plants can lead to toxicity due to the similarity of Se to sulfur (S) and the incorporation of Se into S proteins. However, many plants readily take up Se and can benefit from increased Se due to increased growth and/or as an elemental defense.

In relation to Se, plants can be classified into three categories: 1) non-Se accumulators 2) Se accumulators, and 3) Se hyperaccumulators. Non-Se accumulators do not accumulate Se, or only accumulate trace concentrations of Se, even when growing on seleniferous soils, Se accumulators can accumulate up to 1,000 mg Se kg-1 and Se hyperaccumulators accumulate upwards of 1,000 mg Se kg-1 and as much as 15,000 mg Se kg-1.

Elevated tissue Se levels can protect plants from a variety of herbivores and pathogens, including fungi, arthropods and mammals. This elemental plant defense may act as a convenient pesticide when using plants for Se phytoremediation, and may also help prevent toxic Se concentrations from entering the ecosystem. Selenium as a defense has been disarmed in at least one instance, by a population of diamondback moth (Plutella xylostella), and probably has been disarmed on other occasions. Understanding the mechanisms that have led to the disarmament of Se as a defense is important to better understand how plant Se may enter higher trophic levels. In addition, many decomposers in seleniferous environments appear to have evolved Se tolerance, resulting in increased decomposition rates of Se-rich plant material and possibly faster release of Se into soil. Selenium may also influence pollination. There is evidence that Se accumulation changes flower phenotype characteristics and that important reproductive tissues, such as pistils, stamens, nectar and pollen, accumulate Se. Another interesting ecological aspect of plant Se accumulation is the role of rhizosphere and endophytic-microbes in Se (hyper) accumulation; there is evidence that rhizosphere microbes can increase plant Se accumulation and volatilization.

Investigating the ecological implications of Se accumulation in plants is crucial to managing phytoremediation of Se-polluted sites. Moreover, studies on the effects of Se on plant ecology may serve as a model for ecological implications of plant accumulation of other elements during phytoremediation or production of fortified foods.

INTRODUCTION

Selenium is an essential micronutrient for many organisms including humans, but is toxic at elevated concentrations. The gap between Se deficiency and toxicity is narrow and both are problems worldwide. Selenium is essential in the active site of redox-active selenoproteins such as the enzyme glutathione peroxidase, which protects cells from free radicals [Steinbrenner and Sies 2009]. Sufficient Se helps prevent a variety of cancers such as lung and prostate cancer, assists with the detoxification of heavy metals such as lead and mercury, and is essential for thyroid function [Clark et al. 1996; Birringer et al. 2000; Shin et al. 2007; Kato et al. 2010]. The recommended dietary intake of Se for humans is 55 -100 μg Se a day [Lyon et al. 1989; Sriram and Lonchyna 2009]. Selenium deficiency can lead to Keshan disease, a lethal heart disease named after the county in northeast China of the same name where the disease was first observed [Chen et al. 1980]. Long before the essential function of Se was discovered, Se was famous for its toxicity.

Selenium is toxic due to its similarity to sulfur (S). Selenium readily replaces S in essential S proteins, interfering with their function [Stadtman 1990]. In humans, chronic Se intake of more than 400 μg Se a day can lead to toxic symptoms, which include loss of hair and nails, gastrointestinal complications and eventually death [Oliveira 2007; Steinbrenner and Sies 2009]. In the Western United States, where soils have elevated Se concentrations, chronic ingestion of high-Se plants by livestock has been reported to result in $330 million in losses annually [Rosenfield and Beath 1964; Wilbur 1980]. A one-time ingestion of upwards of 1000 μg Se for a healthy human adult can lead to acute Se poisoning, and even death [Rosenfield and Beath 1964]. Famously, in 2009, 21 Argentinean polo horses mysteriously died shortly before their match in the U.S. Open Polo Championship. The death of the horses, whose value was estimated at $2 million collectively, was a result of acute Se poisoning due to accidental elevated concentrations of Se in their vitamins.

Selenium serves no known essential function in plants, although Se is beneficial to many plants because it increases growth and antioxidant activity and protects against a wide variety of herbivores and pathogens [Hanson et al. 2003; Freeman et al. 2006a]. Plants are classified as either non Se accumulators, plants that do not take up Se when grown on seleniferous sites, Se accumulators, plants that take up to 1000 mg Se kg-1 when grown on seleniferous sites, and Se hyperaccumulators, plants that accumulate upwards of 1000 mg Se kg-1 at seleniferous sites and have been show to accumulate up to 15,000 mg Se kg-1 [Terry et al. 2000]. Selenium accumulators have traditionally been used for phytoremediation more often than Se hyperaccumulators because they yield more biomass, grow faster and some are crop species (e.g. Brassica spp.). However, Se hyperaccumulators, found in the families Asteraceae, Brassicaceae and Fabaceae, are gaining popularity in phytoremediation as a result of increased understanding of their physiology and taxonomy (see Figure 1 for an example).

Figure 1. A field of the Se hyperaccumulator Stanleya pinnata that was seeded in seleniferous soil in the Fort Collins, CO, USA as part of a restoration project after the construction of an irrigation pipe (Pine Ridge Natural Area).

Transgenic crops expressing genes from hyperaccumulators that are responsible for Se tolerance, uptake and volatilization are also promising for Se phytoremediation [for a review see Pilon-Smits and Leduc, 2009].

Most soils contain low Se concentrations: less than 1 mg Se kg-1. However, natural Se deposits and human activity both contribute to Se pollution and can cause widespread health problems and economic devastation. During the warm Cretaceous period (approximately 100 million years ago) oceans covered many of the lower elevations of the earth's continents. When these oceans retreated they left shale high in Se concentration. Use of these seleniferous soils for agriculture leads to accelerated release of this naturally occurring Se into the environment. Mining, burning of seleniferous coal, and refining and burning of seleniferous oil also contribute to Se pollution of water, soil and air [Diaz et al. 1996; Blagojevic et al. 1998; Senesi et al. 1999; Lemly et al 2004; Xu et al. 2005]. In the early 1980's agricultural drainage water with high Se concentrations was responsible for the death of fish and migrant bird species in the Kesterson reservoir in California's central agriculture valley [Ohlendorf et al.1986; Saiki and Lowe 1987].

Plants are an effective tool to clean up Se-polluted soil or water.In a constructed wetland system established in the San Joaquin Valley of California, selenate-contaminated agricultural drainage water was treated effectively by stands of cattail (Typha latifolia) or rabbitfoot grass, reducing Se levels by about 90% [Lin et al., 2000]. A similar Se reduction level was found for a constructed wetland used to treat selenite-containing industrial wastewater from an oil refinery in the San Francisco Bay Area [Hansen et al., 1998]. In both cases the Se removal was due to accumulation in plant tissues, immobilization in sediment, and volatilization. In another study Se-polluted water passing through a constructed wetland of common reed (Phragmites australis) lost 100% of Se in 25 days, and constructed wetlands with broadleaf cattail removed over 50% of Se in the same time period [Shardendu et al. 2002]. Growing Se-accumulating terrestrial plants such as members of the Brassica genus (Indian mustard, canola) has been shown to effectively clean up Se-polluted agricultural fields in the San Joaquin Valley in California [Banuelos et al., 2002a].

Since Se is not only a toxin at high levels but also an essential nutrient at low levels, phytoremediation of high-Se areas provides a unique opportunity to use plants to remove a toxic element from one area and use the concentrated Se as a mineral in Se deficient areas. Prior to using this technology it is important to consider biotic and abiotic ecological consequences of the large-scale growth of Se accumulating plants. This is the main focus of this chapter.

MOVEMENT OF SELENIUM

Into and within Plants

In soils, Se is most commonly found as selenate (SeO42-), which plants readily take up and assimilate utilizing sulfate transporters and the S assimilation pathway [for a review see Pilon-Smits and Quinn, 2009]. In short, selenate is taken up and reduced to selenite and selenide, respectively, and then combined with O-acetylserine into selenocysteine (SeCys). SeCys can be further converted to selenomethionine (SeMet). Both SeCys and SeMet can be non-specifically incorporated into proteins, which is toxic. Hyperaccumulator plants can methylate SeCys, and accumulate most of their Se as methyl-SeCys [Neuhierl et al., 1999]. Most non-hyperaccumulator plants accumulate primarily selenate when supplied with selenate; the reduction of selenate to selenite appears to be a rate-limiting step, as plants can quickly reduce selenite to organic SeCys [de Souza et al. 1998].

Both selenate and SeCys are toxic, the latter more so due to its inadvertent incorporation into S amino acids, which leads to a loss of function [Stadtman 1990]. Plants also can convert SeMet to volatile Se as dimethyselenide (DMSe), a large component of atmospheric Se; they can also absorb atmospheric Se [Lewis et al. 1966; Haygarth et al 1995]. Selenium hyperaccumulating plants differ from non-hyperaccumulators in that they preferentially take up Se over S, have increased biomass when grown with elevated Se, show positive chemotropism of their roots to Se in soil and can accumulate up to 15,000 mg Se kg-1 from soils with Se concentrations as low as 2-5 mg Se kg-1 [Freeman et al. 2006a; Lyons et al. 2009; Pilon-Smits et al. 2009].

The ability to accumulate such high levels of Se and avoid toxicity is due to the unique Se metabolism of Se hyperaccumulating plants: as mentioned above, they store Se as non-toxic methylselenocysteine (MeSeCys), which is not incorporated into proteins [Neuhierl et al. 1999; Pilon-Smits and Quinn 2010] (see Figure 2 for a comparison of Se accumulators and hyperaccumulators).

Se Accumulating Plants

- Se concentrations up to 1,000 mg Se kg^{-1}
- Usually fast growing, high biomass producing species
- Often grows in multiple habitats
- Se is primarily stored as selenate (SeO_4)
- Volatilizes Se as dimethylselenide (DMSe)

Se Hyperaccumulating Plants

- Se concentrations as high as 15,000 mg Se kg^{-1}
- Usually slow growing, low biomass producing species
- Se indicating plants that are often restricted to seleniferous habitats
- Se is primarily stored as methylselenocysteine (MeSeCys)
- volatilizes Se as dimethyldiselenide (DMDSe)

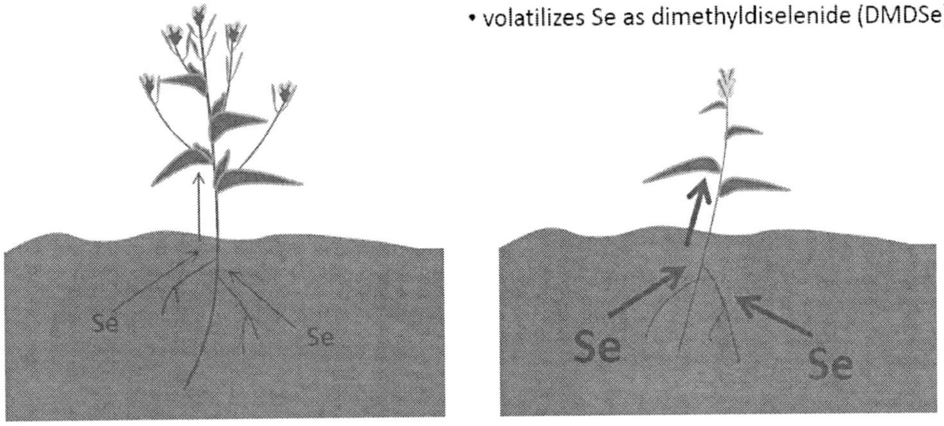

Figure 2. Comparison of traits associated with Se accumulating plants (left column) and Se hyperaccumulating plants. Although Se hyperaccumulating plants take up more Se, Se accumulating plants have traditionally been used for phytoremediation due to their fast growth and high biomass production.

Hyperaccumulating plants can also convert MeSeCys to volatile Se as dimethyldiselenide (DMDSe), which is not often found in the atmosphere because it is unstable and returns to the soil as organic Se soon after volatilization [Martens and Suarez 1999; Kubachka et al. 2007]. Adding microbes to terrestrial and aquatic ecosystems play an important role in Se accumulation and volatilization, which is discussed in more detail below.

Within plants, Se levels are usually similar in shoot and root, particularly if the plant uses selenate as a source of Se (the predominant bioavailable form of Se in terrestrial habitats) and younger leaves have higher Se levels than older leaves [de Souza et al. 1998]. In most plants the Se in leaves is highest in vascular tissues [Freeman et al. 2006b].

Interestingly, the distribution of Se in organs and tissues of Se hyperaccumulating plants is different from non-hyperaccumulators, and may point to the functional significance of Se hyperaccumulation as a defensive mechanism. Hyperaccumulator Stanleya pinnata stores Se in globular structures along the margin of leaves and primarily in epidermal cells, and hyperaccumulator A. bisulcatus stores Se primarily in trichomes [Freeman et al. 2006b]. Thus, in leaves of hyperaccumulators Se is concentrated in areas and cells that are often the first part of the plant to be consumed by generalist herbivores, and are generally associated with defensive functions. In as seasonal field study, younger leaves of these same two hyperaccumulator species had higher Se concentrations than older leaves and Se concentration in leaves peaked during early spring, when plants invest most in growth and development. Selenium levels steadily decreased in leaves until senescence in the fall months, when the Se appeared to be redistributed to the reproductive tissues and back to the roots [Galeas et al. 2007]. Related non-hyperaccumulators growing on the same site showed a peak in leaf Se levels in mid-summer, and S levels also peaked in summer for both non-hyperaccumulators and hyperaccumulators [Galeas et al., 2007].

Flowers of Se hyperaccumulating plants, particularly male and female sex organs, the pistil and stamen, respectively, have the highest concentrations of Se, together with the seeds [Quinn and Pilon-Smits, unpublished results]. Flowers of the hyperaccumulator Stanleya pinnata have more than twice as much Se as leaves, and flowers of A. bisulcatus have 1.5 times as much Se in flowers compared to leaves; S levels were comparable in both organs. Flowers of non-Se hyperaccumulator species, like B. juncea, have less Se than leaves. Thus, it appears that hyperaccumulating plants are able to distinguish between Se and S, and specifically partition Se to plant parts that are most valuable, particularly plant parts essential to plant fitness, and to tissues that are most effective locations for storage of defense compounds.

From Plants to the Abiotic Environment

When utilizing phytoremediation technologies to clean up Se-polluted areas it is important to consider the role the plants being used play in the ultimate fate of Se. As a result of their Se accumulation and volatilization, plants play a vital role in Se ecosystem cycling [Wen and Carignan 2007].

High-Se plants, particularly Se hyperaccumulators, create small pockets of elevated Se within seleniferous habitats due to their uptake and redistribution of Se in the soil, decomposition of high-Se plant material and Se volatilization. Soil collected from under the canopy of hyperaccumulating species A. bisulcatus and S. pinnata has higher Se concentrations than bulk soil or soil collected under the canopy of comparable non hyperaccumulator species from the same site [Quinn and Pilon-Smits, unpublished results].

Moreover, soil under decomposing A. bisulcatus leaf material had higher Se concentrations than soil under decomposing leaf material from non hyperaccumulator species [Quinn and Pilon-Smits, unpublished results]. In addition, the Se concentration in rhizospheric soil from A. bisulcatus was higher than in surface soil under the canopy of A. bisulcatus and higher than in rhizospheric soil of comparable non-hyperaccumulator species at the same site. This type of redistribution of Se in soil creates high-Se micro-habitats that may be toxic to many organisms, but also may create a niche for Se tolerant organisms. The interactions between organisms and high-Se micro habitats are discussed in more detail later in this chapter.

The ability of plants to volatilize Se contributes to atmospheric Se, which is becoming an increasingly important pollution problem due to the continued burning of seleniferous coal. Atmospheric Se leads to Se deposition in aquatic and terrestrial ecosystems [Wen and Carignan 2007].

Plants may help remove Se from the atmosphere through absorption and metabolism of atmospheric Se, creating a valuable Se sink [Zieve and Peterson 1986]. During phytoremediation, it is important to consider the amount of volatile Se being released by the plant compared to how much Se the plant removes from the atmosphere. While plants may produce both DMSe and DMDSe, most organic atmospheric Se from plants is probably DMSe since DMDSe is unstable and its ultimate fate is likely to either be metabolized by nearby plants or re-deposited in nearby soil [Martens and Suarez 1999].

To minimize negative effects of translocation and redistribution of Se caused by plants used for phytoremediation it is best to remove high-Se biomass from the site, to optimize Se removal and prevent further Se pollution.

ECOLOGICAL PARTNERS

Microbes

While Se is toxic to most microbes, some bacteria and fungi live in the rhizosphere of Se hyperaccumulating plants where Se can be upwards of 100 mg Se kg-1. These microbes appear to have evolved mechanisms to overcome the toxic effects of Se. Some of these rhizosphere microbes may also play a role in plant Se accumulation. It has been shown that the presence of rhizosphere bacteria enhances Se accumulation and volatilization in Indian mustard, a Se accumulating plant [de Souza et al. 1999a] as well as certain wetland species [de Souza et al. 1999b]. The activity of microbes in the rhizosphere may make Se more bioavailable to plants, stimulate plant Se uptake and assimilation and stimulate plant root growth leading to a larger Se uptake capacity. Some plant-associated microbes are microbial decomposers that break down dead plant material that contains Se and release it back into the soil for reuptake by the plant.

In addition to free living microbes in the soil surrounding Se accumulators, there are microbes that live inside of the plant tissue: these are called endophytes. Most plants tested so far contain multiple bacterial and fungal endophyte species , which can colonize all plant tissues and be transmitted horizontally (to neighboring plants) or vertically (via the seeds) [for reviews see Saikkonen 1998; Sturz 2000]. Endophytes have been found in Se hyperaccumulators [Lindblom and Pilon-Smits, unpublished results]. It is feasible that endophytes with high Se tolerance, accumulation or volatilization facilitate plant Se accumulation, volatilization or tolerance. Increased understanding of the role microbes play in plant Se accumulation and hyperaccumulation will prove to be a valuable tool when designing phytoremediation projects and when working towards biofortification of crops with Se.

Some microorganisms living on or inside plants are plant pathogens, or can become pathogenic under conditions of plant stress. Similar to plant-herbivore interactions, plants and microbial pathogens participate in an arms race. Plants often produce chemical defenses that microbes evolve to disarm. Selenium may function as a plant defense compound against microbial pathogens. Plants that were treated with Se had reduced disease when infected with the fugal pathogens Alternaria alternata and Fusarium sp. [Hanson et al. 2003]. This may have important implications for Se phytoremediation: when growing Se-accumulating plants there may be less need for chemical antimicrobial pesticides and less biomass loss due to microbial pathogens. The total losses in the US of barley and wheat crops due to Fusarium head blight and seedling rot between 1991 and 1996 have been estimated at $3 billion [Brewing Microbiology 2003]. In addition, it is estimated that at least 20% of agricultural loss can be attributed to Alternaria plant pathogens. Pseudomonas syringae, a prevalent bacterial pathogen on plants has also been shown to be Se-sensitive [Lindblom and Pilon-Smits, unpublished results]. It is encouraging for Se phytoremediation and biofortification projects that plants with elevated Se are protected from a wide range of devastating fungal and bacterial pathogens. However, in view of the chemical arms race between plants and microbial pathogens there are likely also many microbial pathogens that have evolved to overcome the toxic effects of Se. Such microbes may have co-evolved with Se hyperaccumulating plants. There is indeed a report of a Fusarium sp. isolated from a hyperaccumulator plant that is extremely Se tolerant and may even grow better in the presence of Se [Wangeline 2007]. The further investigation of the effects of Se on positive and negative plant-microbial interactions and, conversely, the effects of microbes on plant Se accumulation and volatilization will be an interesting area of further study.

Herbivores and Higher Trophic Levels

Since Se is toxic to many herbivores at concentrations found in hyperaccumulator plants, it has been hypothesized that the functional significance of Se hyperaccumulation is elemental defense – termed the elemental defense hypothesis by Boyd and Martens [1992] (see Figure 3 for an overview of Se uptake and interaction with ecological partners).

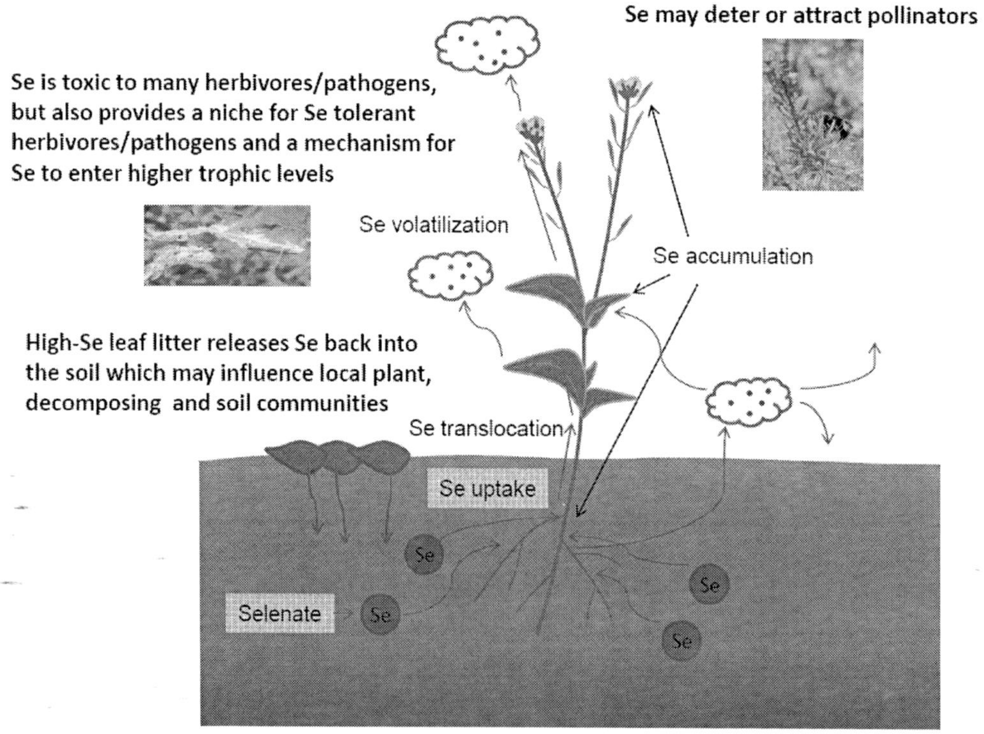

Figure 3. Overview of Se uptake, translocation and release of Se by plants and the effect high-Se plants have on ecological partners.

There is mounting evidence that Se serves as an elemental defense against many herbivores. Brassica juncea, a Se accumulator and an important crop species often used for phytoremediation, was protected by elevated Se from important arthropod pests as well as prairie dog herbivory [Bañuelos et al. 2002b; Hanson et al. 2003, 2004; Quinn et al. 2008]. The Se hyperaccumulators A. bisulcatus and S. pinnata have also been shown to be protected from arthropod and mammalian herbivory [Freeman et al. 2006a, 2009]. Selenium added to an artificial diet proved to be toxic to the herbivore Spodoptera exigua (beet armyworm) and the fly detrivore Megaselia scalaris [Vickerman and Trumble 1999; Jensen et al. 2005]. Moreover, a field study comparing hyperaccumulating plants with non-hyperaccumulating plants found more arthropod individuals and arthropod species living on non-hyperaccumulating plants than hyperaccumulating plants [Galeas et al. 2008]. Herbivores may use the distinct smell of volatile Se as an indicator of plants with elevated Se, and avoid them; they may also dislike the taste of Se-rich plants [Hanson et al. 2004; Freeman et al. 2007].

Like most arms races between plants and their herbivores/pathogens this elemental defense has been disarmed: Se-tolerant organisms have evolved that are able to occupy the niches provided by high-Se plants. Micro-arthropods and microbes responsible for decomposition were found living in Se hyperaccumulator leaf litter with toxic Se concentrations [Quinn and Pilon-Smits, unpublished results].

In addition, a possible specialist diamondback moth (Plutella xylostella) has been discovered that feeds on S. pinnata in the seleniferous Western United States [Freeman et al. 2006a].

Selenium tolerance studies comparing this population of diamondback moth with a population of diamondback moth collected from a non-seleniferous habitat in the Eastern United States confirmed that the population of diamondback moth found thriving on S. pinnata was Se tolerant and the diamondback moth from the non-seleniferous habitat was Se sensitive. The mechanism of Se tolerance was revealed through Se speciation studies. The Se-tolerant diamondback was shown to accumulate Se in the form of MeSeCys, the same form found in the plants, which is not toxic. The Se-sensitive diamondback moth accumulates SeCys, which is toxic.

This Se-tolerant diamondback moth has the ability to feed on high-Se plants that are toxic to many generalist herbivores, which may decrease browsing competition and provide a browsing niche. The elevated concentration of Se in the Se-tolerant diamondback moth may even protect it from Se-sensitive predators – this has not been studied yet. Interestingly, a parasitic wasp (Diadegma insulare) was found living on the Se-tolerant diamondback moth. This parasitic wasp also accumulates Se concentrations that are toxic to many organisms, and appears to have developed the same Se tolerance mechanism as the Se tolerant diamondback moth because it also accumulates MeSeCys [Freeman et al. 2006a].

Understanding the interactions between high-Se plants and their herbivores, in addition to how Se affects higher trophic levels, aids in management of Se phytoremediation sites. Selenium has the ability to act as an elemental pesticide in Se-polluted areas. However, if a Se-tolerant herbivore/pathogen evolves this may reduce the productivity of Se phytoremediation plants. Furthermore, if Se specialists prefer high-Se plant material then it is possible that monocultures of Se phytoremediation plants risk eradication.

Pollinators

Since Se is toxic to many herbivores and pathogens, it may also affect pollination of plants with elevated Se. For plants, pollination is a key process in passing down genes and evolution through natural selection [Parra-Tabla and Vargas 2004]. Elevated floral Se concentrations may act as a deterrent to Se sensitive pollinators, like it does too many herbivores. Alternatively, if pollinators benefit from Se (e.g. as a nutrient or antioxidant, or as a defense against pathogens or predators) or another characteristic unique to high-Se plants, Se may act as a cue to certain pollinators.

The role Se plays in pollination has large economic and ecological implications for Se phytoremediation. With the recent increase of sudden hive death syndrome in honey bee hives the possible effect Se has on pollinators has gained increased interest, especially in the seleniferous central valley of California that economically relies on honeybee pollination [Reilly 2009].

Pollinators of high-Se plants are likely exposed to Se because both hyperaccumulating and non-hyperaccumulating plants accumulate Se in flowers and flower parts that are particularly important to pollinators, specifically in pollen grains and nectar. Initial speciation studies have shown that S. pinnata flowers accumulate primarily the less toxic MeSeCys [Quinn and Pilon-Smits, unpublished results]. There are currently no data on Se speciation in flowers of non-Se hyperaccumulators, but it is likely that they accumulate primarily selenate, as found in their other tissues [Freeman et al. 2006b]. Pollinator studies investigating the role Se plays in pollination found that honeybee and other pollinator visits were the same on B. juncea flowers with less than 10 mg Se kg-1 and B. juncea flowers with 230 mg Se kg-1 [Quinn and Pilon-Smits, unpublished results]. Future studies exploring how Se affects flower characteristics and pollination are important for phytoremediation.

Plant-Plant Interactions

To date, little is known about the role Se plays in plant-plant interactions. Selenium pollution may prevent many plant species from living on a once habitable ecosystem, particularly anthropogenic Se pollution because plants are unlikely to have previously evolved Se tolerance on these sites. Introducing plants that remediate Se pollution may also play an important role in redistributing Se in the soil and/or utilizing Se in allelopathic chemicals.

Since it is known that plants with elevated Se increase local soil Se concentrations around the plant and around the rhizosphere of the roots, it is reasonable to assume that this increased Se will affect Se-sensitive plant species growing in the same area. Indeed, Arabidopsis thaliana, a Se-sensitive plant, had lower germination rates and grew less biomass on high-Se soil that was collected from under the canopy of Se hyperaccumulating plants than on low-Se soil collected under the canopy of non-Se hyperaccumulators from the same site [El-Mehdawi, Quinn and Pilon-Smits, unpublished results]. Managing Se phytoremediation sites requires consideration of how high-Se plants will interact with other plant species and alter soil characteristics that may affect the local vegetation.

FUTURE DIRECTIONS

There are many potential ecological impacts of Se phytoremediation that deserve further study. The interactions between Se accumulating or hyperaccumulating plants and neighboring plants are only beginning to be investigated. Hyperaccumulators contain extremely high Se levels in seeds, and Se released from germinating seeds may inhibit the germination of other, Se-sensitive species nearby, offering an additional selective advantage to Se-tolerant species. Another ecological question related to Se accumulation is the role volatile Se plays in plant-plant interactions. If Se accumulation is an induced defense then plants may volatilize Se as a signaling compound alerting nearby plants of eminent attack by pathogens or herbivores. This is an area that has not yet been studied, but will be a very interesting and ecologically significant research topic.

It is becoming more evident that many plant species can tolerate elevated Se in their tissues without suffering negative effects. Future studies may focus on the movement of this plant Se through the ecosystem, particularly to higher trophic levels. It is essential to know how these Se (hyper)accumulating plants may alter the ecosystems that they inhabit prior to utilizing them for phytoremediation.

Interactions between Se accumulating plants and microbes also deserves receiving further attention. We have just begun to grasp an understanding of the relationships between microbes and Se accumulating plants. Very little is known about the mechanism of Se detoxification in microbes that have evolved Se tolerance. Only very few microbes have been identified that associate intimately with Se accumulating or hyperaccumulating plants. Studies in this area are hampered by the fact that very few microbes can be cultured in artificial media: perhaps less than 1% of soil microbes are culturable [Torsvik and Ovreas 2002].

Molecular tools such as DNA and RNA amplification and sequencing will be valuable in identifying many unique microbes never before identified or characterized. The possible importance of such microbes for Se tolerance, accumulation and volatilization in the plants they associate with has been documented by de Souza et al [1999a, 1999b] where rhizosphere bacteria enhanced plant Se accumulation and volatilization. More studies are needed to identify microbes that can enhance Se accumulation and volatilization, and the mechanisms responsible. Gaining this understanding has the potential to greatly increase the success of phytoremediation projects by simple inoculation with beneficial microbes. A potential additional benefit of Se phytoremediation besides cleaning up excess Se from the environment is the creation of Se-enriched food products. Since Se is an essential nutrient for humans and other mammals, Se-enriched plants can be used to combat Se deficiency worldwide.

The crops themselves have the potential for use as food for humans, or for the production of supplements, or for being processed into feed for animals. There are already many crops that are being fortified with Se, such as wheat, garlic and broccoli [Lintschinger et al. 2000; McSheehy et al. 2000; Roberge et al. 2003]. Future research investigating Se concentration and speciation in crops used for phytoremediation will aid in understanding how best to use Se-enriched food products.

CONCLUSION

Elevated Se concentrations in soil, watersheds and the atmosphere occur naturally but are increasingly due to human activities such as agricultural practices and burning of seleniferous fossil fuels. Selenium toxicity can have devastating effects on ecosystems as was seen at the Kesterson Reservoir in central California, USA in the early 1980's when several fish and migratory bird species died due to Se poisoning. Phytoremediation can be an effective and inexpensive tool to clean up Se- polluted terrestrial and aquatic habitats. In terrestrial habitats plants can remove Se from soil through accumulation in plant shoots and volatilization. In aquatic systems constructed wetlands can remove Se from drainage or surface waters. Phytoremediation of Se is especially attractive because Se-enriched plants can help combat Se deficiency in low-Se areas. Like any remediation strategy, it is important to consider ecological interactions and ecosystem consequences when utilizing Se phytoremediation.

Traditionally, Se accumulating plants have been preferred over Se hyperaccumulating plants for phytoremediation because they are typically fast growing and are sometimes crop species. However, Se hyperaccumulating plants also have potential uses for phytoremediation because of their ability to accumulate extremely high concentrations of Se, in a form that is highly anti-carcinogenic (MeSeCys). The genes of hyperaccumulator plants are also useful to help develop transgenic crop plants. Both Se accumulating and hyperaccumulating plants may change the distribution of Se in seleniferous habitats, and the Se accumulated in these plants has been shown to influence their interactions with ecological partners. Elevated Se concentrations in plants may increase biomass yield and Se removal from a site because plant-accumulated Se acts as a pesticide through both deterrence of and toxicity to a variety of generalist herbivores and pathogens; moreover Se can be a beneficial element for plants, promoting plant growth and stress resistance. However, high-Se plants provide a niche for Se tolerant herbivores/pathogens which may even prefer to feed on Se rich plant material and have the potential to cause devastating biomass losses to Se phytoremediation plants.

Se tolerant herbivores also provide a mechanism for Se to enter higher trophic levels in the ecosystem. In addition to herbivores and pathogens, high-Se plants may influence soil microbial communities and local plant communities. By creating micro-habitats of Se rich areas plants force microbial communities to adapt Se tolerance or move to another location lower in Se. Similarly, plants that are Se sensitive may not be able to survive around Se rich plants that have concentrated Se in a small area. Selenium phytoremediation has a promising future and we are beginning to understand the interactions between Se-accumulating plants and their ecological partners. Selenium may provide a useful model element to aid in understanding how phytoremediation of other inorganics affect local ecosystems.

REFERENCES

Bañuelos, G.S., Lin, Z.-Q., Wu, L., and Terry, N. (2002a) Phytoremediation of selenium-contaminated soils and waters: fundamentals and future prospects. *Reviews on Environmental Health* 4: 291-306.

Bañuelos, G.S., Vickerman, D.B., Trumble, J.T., Shannon, M.C., Davis, C.D., Finley, J.W., and Mayland, H.F. (2002b) Biotransfer possibilities of selenium from plants used in phytoremediation. *International Journal of Phytoremediation*, 4, 315-331.

Birringer, C.I.P.M., Block, E., Kotrebai, M., Tyson, J.F., Uden, P.C., and Lisk, D.J. (2002) Chemical speciation influences comparative activity of selenium-enriched garlic and yeast in mammary cancer prevention. *Journal of Agriculture and Food Chemistry*, 48, 2062.

Blagojevic, S., Jakovljevic, M., and Zarkovic, B, (1998) Influence of long-term fertilization on the selenium content of calcareous chernozem soil. *Journal of Environmental Pathology and Toxicology*, 17, 183–187.

Boyd, R.S., and Martens, S.N. (1992) *The raison d'être for metal for metal hyperaccumulation by plants*. In: Baker AJM, Proctor J, Reeves RD (eds) The vegetation of ultramafic (Serpentine) soils. Intercept, Andover, UK, pp 279-289.

Chen, X.S., Yang, G.Q., Chen, X.C., Wen, Z.M., and Ge, K.Y. (1980) Studies on the relations of selenium and Keshan disease. *Biological Trace Element Research*, 2, 91-107.

Clark, L.C., Combs, Jr. G.F., Turnbull, B.W., Slate, E.H., Chalker, D.K., Chow, J., Davis, L.S., Glover, R.A., Graham, G.F., and Gross E.G. et al. (1996) Effects of selenium supplementation for cancer prevention in patients with carcinoma of the skin: a randomized controlled trial. *Journal of the American Medical Association*, 276, 1957–1963.

de Souza, M.P., Pilon-Smits, E.A.H., Lytle, C.M., Hwang, S., Tai, J.C., Honma. T.S.U., Yeh, L., and Terry, N. (1998) Rate-limiting steps in selenium volatilization by Brassica juncea. *Plant Physiology*, 117, 1487-1494.

de Souza, M.P., Chu, D., Zhao, M., Zayed, A.M., Ruzin, S.E., Schichnes, D., and Terry, N. (1999a) Rhizosphere bacteria enhance selenium accumulation and volatilization by Indian mustard. *Plant Physiology*, 119, 565-573.

de Souza, M.P., Huang, C.P.A., Chee, N., and Terry, N. (1999b) Rhizosphere bacteria enhance the accumulation of selenium and mercury in wetland plants. *Planta*, 209, 259-263.

Diaz, J. P., Navarro, M., L´opez, H., and L´opez, M. C. (1996) Selenium (IV) and (VI) levels in potable, irrigation and waste waters from an industrial zone in southeastern Spain. *Science of the Total Environment*, 186, 231–236.

Freeman, J.L., Quinn, C.F., Marcus, M.A., Fakra, S., and Pilon-Smits, E.A.H. (2006a) Selenium-tolerant diamondback moth disarms hyperaccumulator plant defense. *Current Biology,* 16, 2181-2192.

Freeman, J.L., Zhang, L.H., Marcus, M.A., Fakra, S., and Pilon-Smits, E.A.H. (2006b) Spatial imaging, speciation and quantification of selenium in the hyperaccumulator plants Astragalus bisulcatus and Stanleya pinnata. *Plant Physiology*, 142, 124-134.

Freeman, J.L., Lindblom, S.D., Quinn, C.F., Fakra, S., Marcus, M.A., and Pilon-Smits, E.A.H. (2007) Selenium accumulation protects plants from herbivory by orthoptera due to toxicity and deterrence. *New Phytologist*, 175, 490-500.

Freeman, J.L., Quinn, C.F., Lindblom, S.D., Klamper, E.M., and Pilon-Smits, E.A.H. (2009) Selenium protects the hyperaccumulator Stanleya pinnata against black-tailed prairie dog herbivory in native seleniferous habitats. *American Journal of Botany*, 96, 1075-1085.

Galeas, M.L., Zhang, L.H., Freeman, J.L., Wegner, M., and Pilon-Smits, E.A.H. (2007) Seasonal fluctuations of selenium and sulfur accumulation in selenium hyperaccumulators and related non-accumulators. *New Phytologist,* 173, 517-525.

Galeas, M.L., Klamper, E.M., Bennett, L.E., Freeman, J.L., Kondratieff, B.C., and Pilon-Smits, E.A.H. (2008) Selenium hyperaccumulation affects plant arthropod load in the field. *New Phytologist*, 177, 715-724.

Hansen, D., Duda, P.J., Zayed, A.,and Terry, N. (1998) Selenium removal by constructed wetlands: role of biological volatilization. *Environmental Science and Technology* 32: 591–597.

Hanson, B., Garifullina, G.F., Lindblom, S.D., Wangeline, A., Ackley, A., Kramer, K., Norton, A.P., Lawrence, C.B., and Pilon-Smits, E.A.H. (2003) Selenium accumulation protects Brassica juncea from invertebrate herbivory and fungal infection. *New Phytologist*, 159, 461-469.

Hanson, B., Lindblom, S.D., Loeffler, M.L., and Pilon-Smits, E.A.H. (2004) Selenium protects plants from phloem feeding aphids due to both deterrence and toxicity. *New Phytologist*, 162, 655-662.

Haygarth, P.M., Harrison, A.F., and Jones, K.C. (1995) Plant Selenium from Soil and the Atmosphere. *Journal of Environmental Quality*, 24, 768-771.

Jensen, P.D., Rivas, M.D., and Trumble, J.T. (2005) Developmental responses of a terrestrial insect detritivore, *Megaselia scalaris* (*Loew*) *to four selenium species Ecotoxicology*. 14, 313-322.

Kato, M.A., Finley, D.J., Lubitz, C.C., Zhu, B.X., Moo, T.A., Loeven, M.R., Ricci, J.A., Zarnegar, R., Katdare, M., and Fahey, T.J. (2010) Selenium Decreases thyroid cancer cell growth by increasing expression of GADD153 and GADD34. *Nutrition and Cancer-An International Journal*. 62, 66-73.

Kubachka, K.M., Meija, J., LeDuc, D.L., Terry, N., and Caruso, J.A. (2007) Selenium volatiles as proxy to the metabolic pathways of selenium in genetically modified Brassica juncea. *Environmental Science and Technology*, 41, 1863–1869.

Lemly, D.A. (2004) Aquatic selenium pollution is a global environmental safety issue. *Ecotoxicology and Environmental Safety*, 59, 44–56.

Lewis, B.G., Johnson, C.M., and Delwiche, C.C. (1966) Release of volatile selenium compounds by plants: collection procedures and preliminary observations. *Journal of Agriculture and Food Chemistry*, 14, 638–640.

Lin Z-Q, Schemenauer RS, Cervinka V, Zayed A, Lee A, Terry N (2000) Selenium volatilization from a soil-plant system for the remediation of contaminated water and soil in the San Joaquin Valley. *J Environ Qual* 29: 1048–1056

Lintschinger, J., Fuchs, N., Moser, J., Huehnelt, D., and Moosham, K.F. (2000) Selenium-enriched sprouts – A raw material for fortified cereal based diets. *Journal of Agriculture and Food Chemistry*. 48, 5362-5368.

Lyon, T.D.B., Fell, G.S., Hall, S.D.J., Clark, J., and McKenna, F. (1989) Determination of nine inorganic elements in human autopsy tissue. *Journal of Trace Element Electrolytes and Health Disorders*, 31, 109-118.

Lyons, G.H., Genc, Y., Soole, K., Stangoulis, J.C.R., Liu, F., and Graham, R.D. (2009) Selenium increases seed production in Brassica. *Plant Soil*, 318, 73–80.

Martens, D.A., and Suarez, D.L. (1999) Transformations of volatile methylated selenium in soil. *Soil Biology and Biochemistry*, 31, 1355–1361.

McSheehy, S., Yang, W.J., Pannier, F., Szpunar, J., Lobinski, R., Auger, J., and Potin-Gautier, M. (2000). Speciation analysis of selenium in garlic by two-dimensional high-performance liquid chromatography with parallel inductively coupled plasma mass spectromic and electrospray tandem mass spectrometric detection. *Analytica Chimica Acta,* 421, 147-153.

Neuhierl, B., Thanbichler, M., Lottspeich, F., and Bo¨ck, A. (1999) A family of S-methylmethionine dependent thiol/selenol methyltransferases. Role in selenium tolerance and evolutionary relation. *Journal of Biological Chemistry*, 274, 5407–5414.

Ohlendorf, H.M., Hoffman, D.J., Salki, M.K., and Aldrich, T.W. (1986) Embryonic mortality and abnormalities of aquatic birds: apparent impacts of selenium from irrigation drain water. *Science of the Total Environment*, 52, 49–63.

Oliveira, K.D., Franca, T.N., Nogueira, V.A., and Peixoto P.V. (2007) *Diseases associated with selenium poisoning in animals*. Pesquisa Veterinaria Brasileira, 27, 125-136.

Parra-Table, V. and Vargas, C.F. (2004) Phenology and Phenotypic Natural Selection on The Flowering Time of a Deceit-pollinated Tropical Orchid, Myrmecophila christen. *Annals of Botany*, 94, 243-250.

Pilon-Smits, E.A.H., and Leduc, D.L. (2009) Phytoremediation of selenium using transgenic plants. *Current Opinion in Biotechnology*. 20, 207-212.

Pilon-Smits, E.A.H., Quinn, C.F., Tapken, W., Malagoli, M., and Schiavon, M. (2009) Physiological functions of beneficial elements. *Current Opinion in Plant Biology*, 12, 267-274.

Pilon-Smits, E.A.H., and Quinn, C.F. (2010) *Selenium Metabolism in Plants*. In: "Cell Biology of Metal and Nutrients", Hell R, Mendel R, eds. In press.

Quinn, C.F., Freeman, J.L., Galeas, M.L., Klamper, E.M., Pilon-Smits, E.A.H. (2008) The role of selenium in protecting plants against prairie dog herbivory: implications for the evolution of selenium hyperaccumulation. *Oecologia*, 155, 267-275.

Reilly, M. (2009) Toxic Pollen, Nectar Could Sting Bees. Discovery News, *http://dsc.discovery.com/news/2009/07/29/bees-selenium.html*

Roberge, M.T., Borgerdingand, A.J., Finley, J.W. (2003) Speciation of selenium compounds from high selenium broccoli is affected by the extracting solution. *Journal of Agriculture and Food Chemistry*, 51, 4191-4197.

Rosenfield, I. and Beath, O.A. (1964) *Selenium, Geobotany, Biochemistry, Toxicity, and Nutrition*, Academic, New York.

Saiki, M.K., and Lowe, T.P. (1987) Selenium in aquatic organisms from subsurface agricultural drainage water, San Joaquin Valley, California. *Archives of Environmental Contamination Toxicology*, 19, 496-499.

Saikkonen, K., Faeth, S.H., Helander, K., and Sullivan, T.J. (1998) Fungal endophytes: A Continuum of Interactions with Host Plants. *Annual Review of Ecology and Systematics*, 29, 319-343.

Senesi, G.S., Baldassarre, G., Senesi, N., and Radina, B. (1999) Trace element inputs into soils by anthropogenic activities and implications for human health. *Chemosphere*, 39, 343–377.

Shardendu., Salhani, N., Boulyga, S. F., Stengel, E. (2002). Phytoremediation of selenium by two helophyte species in substrate flow constructed wetland. *Chemosphere*, 50, 967-973.

Shin, S.H., Yoon, M.J., Kim, M., Kim, J.I., Lee, S.J., Lee, Y.S., and Bae, S. (2007) Enhanced lung cancer cell killing by the combination of selenium and ionizing radiation. *Oncology reports.* 17, 209-216.

Sriram, K. and Lonchyna, V.A. (2009) Micronutrient Supplementation in Adult Nutrition Therapy: Practical Considerations. Journal of Parental and Eternal Nutrition, 33, 548-562.

Stadtman, T.C. (1990) Selenium biochemistry. *Annual Reviews of Biochemistry*, 59, 111-127.

Steinbrenner, H., and Sies, H. (2009) Protection against reactive oxygen species by selenoproteins. *Biochimica et Biophysica Acta – General Subjects.* SI, 1478-1485.

Stones, M. (2010) Nutra ingrediants-usa.com: *http://www.nutraingredients-usa.com /Industry/New-selenium-enriched-garlic-product-reaches-US-market*

Sturz, A. V., Christie, B.R., Nowak, J. (2000). Bacterial Endophytes: Potential Role in Developing Sustainable Systems of Crop Production. *Critical Reviews in Plant Science.* 19, 1-30.

Terry, N., Zayed, A.M., de Souza, M.P., and Tarun, A.S. (2000) Selenium in higher plants. *Annual Review of Plant Physiology and Plant Molecular Biology*, 51, 401-432.

Torsvik, V., and Ovreas, L. (2002) Microbial diversity and function in soil: from genes to ecosystems. *Current Opinion in Microbiology*, 5, 240–245.

Wen, H. and Carignan, J. (2007) *Reviews on atmospheric selenium: Emissions, speciation and fate Atmospheric Environment*, 41, 7151–7165.

Vickerman, D.B., and Trumble, J.T. (1999) Feeding preferences of Spodoptera exigua in response to form and concentration of selenium. *Archives of Insect Biochemistry*, 42, 64-73.

Wangeline, A. (2003) *Fungi from seleniferous habitats and the relation of selenium to fungal oxidative stress.* PhD dissertation.

Wilbur, G.C. (1980) Toxicology of selenium: A review. Clinical Toxicology, 17, 171–230.

Zieve, R, and Peterson, P.J. (1986) An assessment of the atmosphere as a source of plant selenium: Application of stable and radioactive selenium isotopes. *Toxicological and Environmental Chemistry*, 11, 313 – 318.

Xu, W.D., Ceng, Y.S., Ye, D.N., Querol, X. (2005) Distributions and environmental impacts of selenium in waste of coal from a power plant. *Environmental Science*, 26, 64–68.

In: Handbook of Phytoremediation
Editor: Ivan A. Golubev
ISBN: 978-1-61728-753-4

Chapter 23

EXPERIMENTAL SYSTEMS IN AGROCHEMICALS-CONTAMINATED SOILS PHYTOREMEDIATION RESEARCH

L. J. Merini*[1], *V. Cuadrado and A. M. Giulietti*

Cátedra de Microbiología Industrial y Biotecnología – Facultad de Farmacia y Bioquímica – Universidad de Buenos Aires - Junín 956, 6 piso. Buenos Aires – Argentina

ABSTRACT

Contamination of soil and water has grown as an environmental problem along with the increase of human activities. Among the main pollutants involved, agrochemicals are of major concern, since millions of tons are applied every year for crops and forestry, expecting that nature would take care of them. Although there are many effective physical-chemical methods for soil and water decontamination, the application cost of such techniques and the wide expanse of moderately polluted areas make them inappropriate. In this context, phytoremediation has arisen as an environmentally friendly, low cost and effective alternative for this kind of pollution. Nevertheless, the effectiveness of the process depends on the particular characteristics of the soil, the contaminant and the environmental conditions and their interactions, which makes phytoremediation a site-specific technology. The field-scale applicability of the results obtained at lab research mainly depends on the accuracy of the selected experimental system. In this way, there are two divergent positions: on one hand, a simplified system (cell cultures, organ cultures, hydroponics) where the variables are reduced at minimum and fully controlled, gives precise information about the mechanisms involved in the remediation process. On the other hand, a complex experimental system (microcosms) gives information closely related to real scale, but having less control over the experimental variables involved. We have designed and optimized experimental systems of different complexity for studying phytoremediation of soils contaminated with agrochemicals. Azinphos–methyl, 2,4-dichlorophenoxyacetic acid, 2,4-dichlor-

---

[1] These authors contributed equally to this work.

* Corresponding author: Cátedra de Microbiología Industrial y Biotecnología – Facultad de Farmacia y Bioquímica – Universidad de Buenos Aires. Junin 956, 6° piso (1113) – Buenos Aires, Argentina. Tel: ++54-11-4964-8270. E-mail: agiule@ffyb.uba.ar.

ophenoxybutyric acid and atrazine were selected since they are among the most controversial agrochemicals, because of their toxicity and potential as environmental pollutant. In the designed experimental systems, the biodegradation potential of model and novel tolerant plant species and their influence on soil microflora was observed. At the same time, the systems were used to investigate the mechanisms involved in plant tolerance to herbicides. The soil, contaminant, microflora and plant interactions observed in lab scale experiments and the degradation profiles of the different agrochemicals will be discussed. Conclusions about the influence of experimental system complexity on mechanisms elucidation and reliability of the scaling-up will be presented.

1. Introduction

The rising food demand associated with the constantly growing world population has led to an increase in agronomical exploded areas. In addition, in the last decades, new technologies in crops and farming practices have been implemented. Although a great improvement in crops productivity was achieved, the environmental cost associated with the intensive use of agrochemicals is an evident drawback. In contrast to notorious pollution events such as oil spills, agrochemical contamination is a silent threat, especially considering that their long-term effects are still poorly known. In short-term, water and soil contamination with this kind of compound has been proved to negatively affect human health and wildlife (Fan et al. 2007; HSDB 2005).

Although there are many effective physical-chemical strategies for soil decontamination, including excavation and incineration, off-site storage and soil washing, the application of such techniques usually requires soil transportation which imply a risk of contaminant spreading and human exposure. In addition, the application cost of conventional techniques can be as much as \$1500/ton (Table 1) which makes them inappropriate for wide extensions of moderately polluted soils (Gerhardt et al. 2009).

Table 1. Costs comparison between biological strategies and conventional physical-chemical techniques for soil remediation (from: Gerhardt et al. 2009, Glass 1999)

Technology	Cost (\$/ton)
Phytoremediation	10 - 35
In situ Bioremediation	50 - 150
Soil Venting	20 - 220
Indirect Thermal	120 - 300
Soil washing	80 - 200
Solidification / Stabilisation	240 - 340
Solvent extraction	360 - 440
Incineration	200 - 1500

In this context, phytoremediation has arisen as an environmentally friendly alternative method for this kind of pollution. The in-situ use of plants and their associated microflora for soil decontamination is a relatively low-complexity technique, having low cost of implementation and minimal maintenance (Table 1). In agricultural soils, the improvement of soil structure, fertility and physical-chemical properties generated by planting is of special interest. Moreover, the sole presence of plants, independently of their remediation capacity per se, is an advantage since they generate a microenvironment favorable for microbial activity and usually reduce horizontal (run-off) and vertical (leaching) migration of the contaminant. Finally, the esthetic value and public acceptance of plants presence is an additional unscientific advantage.

Research on phytoremediation has gained interest during the last two decades, in correlation with the success of experimental studies carried out in the lab and greenhouse. Many authors have shown the specific degrading capabilities of different plant species and the mechanisms involved (Burken 2003, Pilon-Smits 2005, Shang et al. 2003). Although mineralization of organics can be achieved in some plant tissues, it has been recognized that a combination of vegetal and microbial metabolic activities is the most efficient strategy for a successful remediation (Schwab and Banks, 1994). In this way, most scientific research is currently focused on microbe-assisted phytoremediation of organic environmental contaminants (Kuiper et al. 2004, Shaw and Burns 2004, Yateem et al. 2008).

Despite the promising results obtained at lab scale, they are not in correlation with the number of successful projects carried out at field scale. When the lab data is intended to be scaled up, researchers face a whole new universe to cope with. On one hand, the impossibility of controlling the complex network of variables involved in field occurring processes, sets up the main difference with lab experimentation. On the other hand, dealing with large amounts of soil implies engineering knowledge and heavy duty machinery use. Although phytoremediation ought to be an essentially interdisciplinary technology, the successful combination of the different involved areas is probably the most challenging objective.

At the moment of going through a phytoremediation process, it is necessary to deeply explore the mechanisms involving the plant, soil and microflora and their interactions during the contaminant dissipation. A key step at lab research on these subjects is the selection of the experimental system. In this way, there are two divergent positions (Figure 1): on one hand, a simplified system where the variables are reduced at minimum and fully controlled, gives precise information about the mechanisms involved in the remediation process. Controlled experimental systems like cell cultures, organ cultures and hydroponics can be used to determine contaminant tolerance levels and mechanisms, to elucidate metabolic pathways and fate of the contaminants, to measure degradation rates and optimize greenhousing conditions.

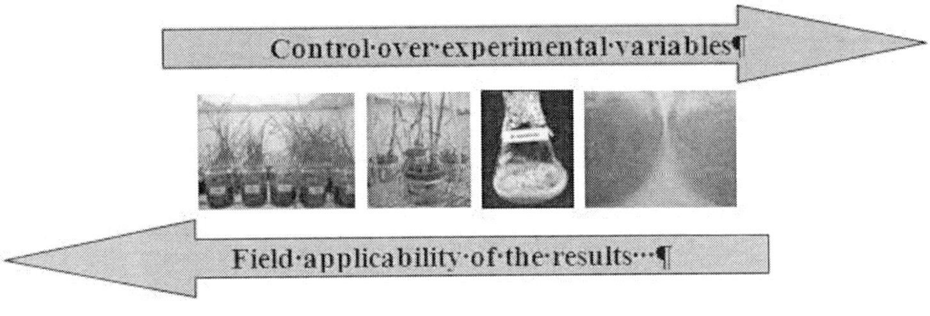

Figure 1. Scheme of main differences among low and high-complexity experimental systems.

On the other hand, a complex experimental system gives information closely related to real scale, but having less control over the experimental variables involved. Microcosms can help to understand the behavior of the remediation system determined by the complex interactions between soil, plant, microorganisms and the contaminant working as a whole entity. The field-scale applicability of the results obtained at lab research requires a thoughtful selection of the experimental system, designed to closely represent real conditions. In addition, it is also necessary to accomplish with some treatability criteria for the field application of a phytoremediation technology, considering the specific contamination event.

2. Applicability Criteria for a Phytoremediation Process

The complexity of the events occurring during a phytoremediation process makes it a site-specific technology. Before starting a phytoremediation research, there are some major criteria to accomplish, which should be considered case by case. They include the physical-chemical properties of the contaminant, such as: water solubility, vapor pressure, octanol-water partition coefficient (Kow), pKa, molecular weight, commercial product formulation; soil properties, such as: pH, salinity, texture, humidity; characteristics of the site, such as: weather, topography, remediation urgency, contamination degree and deepness reached into the soil profile. Besides, in agrochemicals-contaminated soils phytoremediation, it is important to take into account that some of the compounds are specifically designed to kill plants, which represents an additional difficulty to be considered.

Public organisms of international reference like the US Environmental Protection Agency (EPA) and the Interstate Technology and Regulatory Council (ITRC) have summarized the mentioned major criteria into guidelines or decision trees (Figure 2) (ITRC 1999, ITRC 2009, USEPA 2000). These publications are helpful in the first approach for new projects. Some of the decision steps included necessarily require previous lab experimentation.

As a first stage, a screening for plant species selection should be carried out based on their agronomical properties and potential use. This is a necessary consideration since agrochemicals-contaminated soils phytoremediation is usually applied in agricultural exploded areas.

Once the plant specie was selected, its applicability and effectiveness for phytoremediation purposes can be experimentally assessed throw:

- Tolerance assays.
- Dose-response tests.
- Degradation tests.
- Assessment of plant and microbes roles in contaminant removal.
- Assessment of the degradation degree (mineralization or partial degradation).
- Elucidation of the mechanism involved in contaminant dissipation.
- Analysis of the degradation products or metabolites fate.

As it was already mentioned, the thoughtful selection of the experimental system will ensure the results suitability.

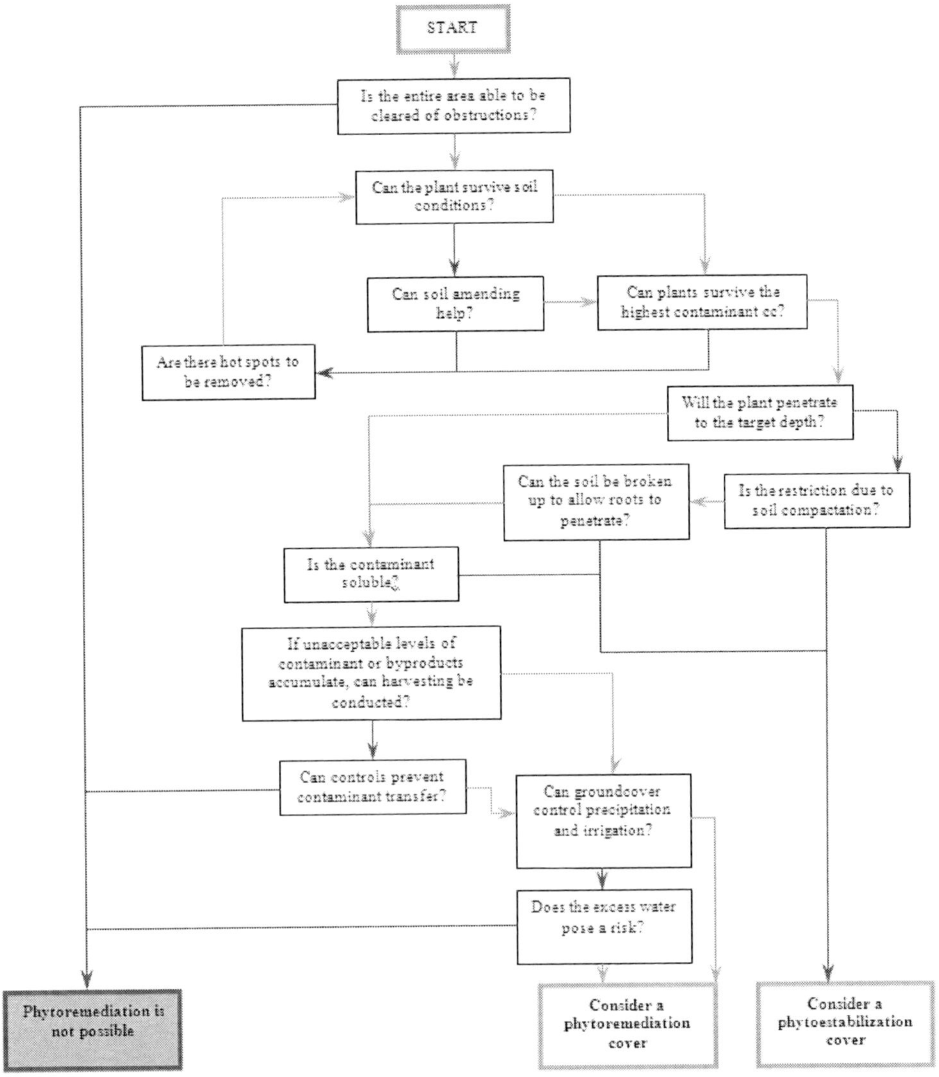

Figure 2. Soil/Sediment decision tree (from: ITRC 2009).

3. EXPERIMENTAL SYSTEMS

When designing lab experiments for studying phytoremediation related events and mechanisms, it is necessary to clearly understand the applicability context of the results obtained.

An experimental system for phytoremediation research in a controlled environment could be dissected into four elements: contaminant, plant, microflora and soil. The number of these elements present at the experimental system will determine its complexity. When optimizing the experimental systems, some important considerations about these four components should be taken into account:

- The soil: its represents, analytical and microbiologically, one of most complex matrices to work with. Thus, it is necessary to assess the texture, the mineral fraction, the organic matter content and composition as well as the history of contamination. Besides, microbial population and contaminant content should be evaluated (including secondary contaminants that could be involved). These considerations ensure a proper experimental design, suitable selection of the analytical methods and lab handling.
- The contaminant: among physical-chemical properties of the contaminant, Kow represents a major criterion to be considered since it allows estimation of the contaminant mobility across the soil profile and the capacity to be absorbed and transported along plant tissues. Additional properties such as chemical structure and reactive sites should be taken into account. In the special case of agrochemicals, the particular characteristics of the commercial product (formulation, associations or mixtures with other chemicals) must be considered as they usually modify key properties of the compound such as water solubility and half life.
- The microflora: soil degradation capacity mainly depends on microorganisms metabolic activity. Among soil microflora, fungi and bacteria are the most important actors on xenobiotics remediation. Considering the complexity of soil microbial communities and the fact that less than 1% of the total population can be cultured, the combined application of both molecular and conventional techniques will allow a comprehensive evaluation (Morgante et al. 2010).
- In phytoremediation processes, the microflora close or associated to plant roots in the rhizospheric environment is the main responsible for degradation. Its structure, diversity and metabolic activity is deeply influenced by the vegetal specie, plant age, the nature and concentration of the contaminant, the soil properties and history of agronomical practices. In this respect, there is evidence about significant differences in soil microbial communities between herbicides chronically-exposed and pristine soils (Cuadrado et al. 2008, Cuadrado 2009).
- The plant: once performed the first screening for plant specie selection, the plant phytoremediation potential must be initially determined by tolerance assessment. For no-native plants, compatibility with the ecosystem where it will be implanted should be an additional consideration.

When the contaminant removal process involves absorption and transformation into plant tissues, it is important to take into account both the toxicity and fate of metabolites or degradation products. The final disposition of the vegetal material must be also considered, since contamination could be extended and even magnified if toxic metabolites get into the trophic chain. The outcome of assembling these four elements is the design and implementation of different experimental systems.

3.1. Cell Cultures

In vitro plant cell cultures have been used in plant physiology research for decades. Although phytoremediation research was initially focused on experimentation with entire plants, cell cultures have gained interest in the last years for performing basic studies on plant physiology and genetics. The reasons include some advantages related to the use of undifferentiated cells and the need of understanding the molecular mechanisms involved in plants degradation activity. The knowledge obtained at cellular scale experimentation provides new perspectives for improving phytoremediation performance. Methodologically, in vitro plant cell cultures present differential advantages in comparison to more complex experimental systems:

- Independence of weather and environmental conditions.
- Low facility requirements.
- Use of simple and well defined culture media, which allows the use of standard commercial formulations like Murashige and Skoog (Murashige and Skoog 1962), Gamborg B5 (Gamborg et al. 1968) and others.
- Higher growth rate in comparison with entire plants.
- Undifferentiated cell lines confer high results reproducibility.
- Reduction in experimental variability.
- Simplicity in detection, identification and potential isolation of metabolites or degradation products.
- Low complexity matrix for chemical analyses.
- Possibility of performing toxicity tests on metabolites or degradation products released to culture media or previously isolated.

These characteristics make the cell cultures a suitable experimental system to perform studies on xenobiotics metabolism and explore the mechanisms that underlie a phytoremediation process (Mackova et al. 2007). The understanding of metabolic pathways and mechanisms helps to improve the performance of plants in remediation technologies (Macek et al. 2000). Moreover, when a plant cell culture collection is available, in vitro liquid or agar cell culture is a fast and useful tool for tolerance, resistance or degradation screening studies (Chroma et al. 2002). In addition, cell culture is usually a necessary step when performance improvements are intended to be achieved by both genetic modifications and metabolic engineering (Perassolo et al. 2007, Quevedo et al. 2010). The genetic strategy has been used for enhancing herbicides tolerance in poplar trees (Gullner et al. 2001).

3.2. Organ Cultures

Among the cultivable plant organs, hairy roots culture is the most broadly employed experimental system for phytoremediation research (Figure 3). Hairy roots are produced when Agrobacterium rhizogenes infects a wounded higher plants tissue, transferring a DNA segment (T-DNA) that induces a change in hormonal balance and the formation of adventitious roots from the wounded site. The resulting hairy roots have differential characteristics such as: high proliferation rate, hormone independent growth, lack of geotropism, lateral branching and an important genetic and phenotypic stability (Guillon et al. 2006). Besides, the simplicity of obtaining and culturing hairy roots, makes this experimental system an attractive option.

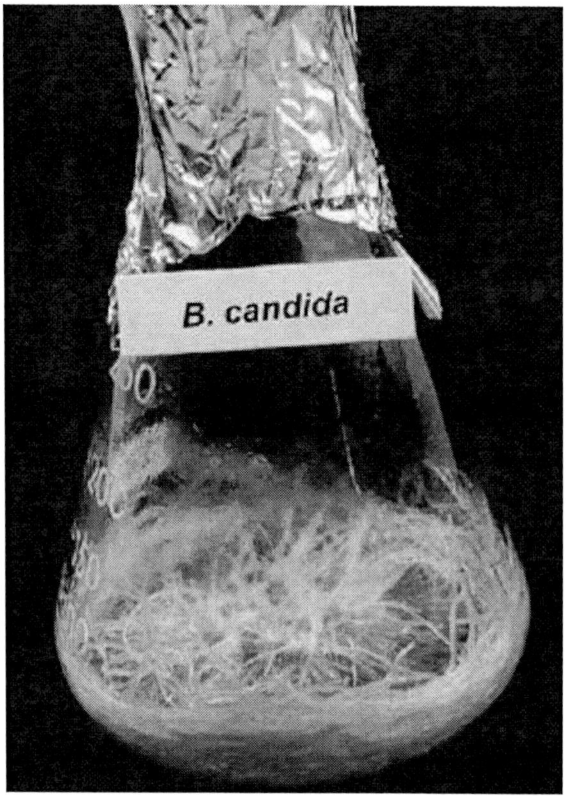

Figure 3. Hairy root culture of *Brugmansia candida* obtained by infection with *Agrobacterium rhizogenes*.

In lab phytoremediation research, hairy roots share the advantages of cell cultures for studying metabolic pathways and the resulting degradation products. In addition, the tissue arranged cells provide certain structural complexity, without the interference of other plant organs.

Their particular characteristics and the possibility of scaling up make hairy roots cultures suitable biotechnological tools for industrial effluents phytoremediation processes. In this way, recent reports have shown the hairy roots detoxification capacity of different plant species (Agostini et al. 2003, Gujarathi et al. 2005, Suresh et al. 2005). The removal of the contaminant 2,4-dichlorophenol (2,4-DCP), usually generated by wood preservers, pesticides and biocides industries (Quan et al. 2003), by Brassica napus hairy roots was studied by Agostini et al. (2003). The mechanisms of 2,4-DCP degradation, involving peroxidase activity, under different H2O2 concentrations and pHs were delineated (Figures 4A and 4B) (Agostini et al. 2003). Moreover scaling-up of this hairy root system was achieved and successfully applied at bench top reactor scale (Figure 5) (Busto et al. 2005).

Hairy roots metabolic engineering has been extensively applied for secondary metabolites production (Zhang et al. 2004). This approach can be also useful for improving phytoremediation processes.

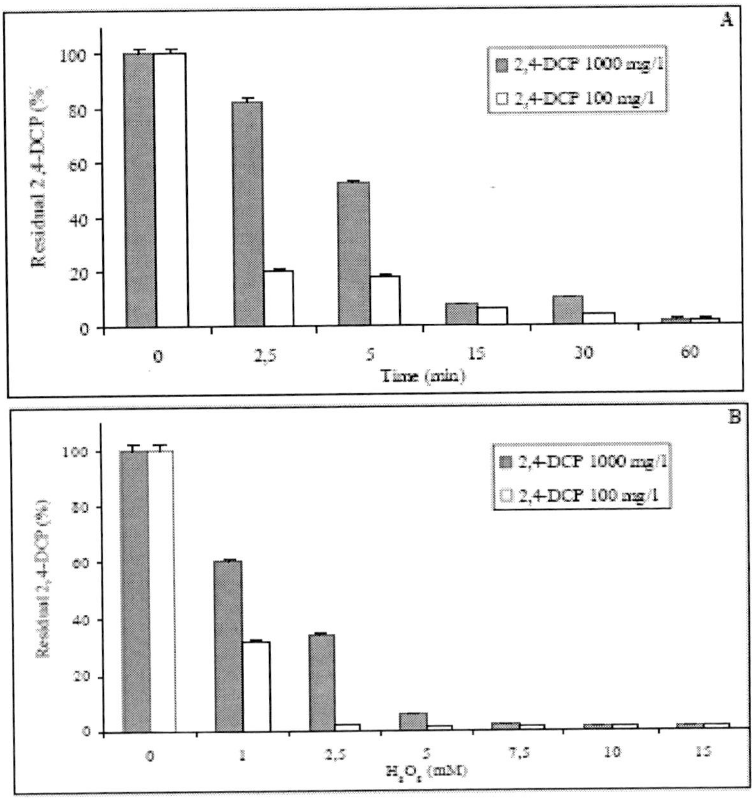

Figure 4. Effect of incubation time (A) and H_2O_2 concentration (B) on the removal of 2,4-DCP from *Brassica napus* hairy roots cultures, at two initial 2,4-DCP concentrations (100 and 1000 mg/l) (from: Agostini et al. 2003).

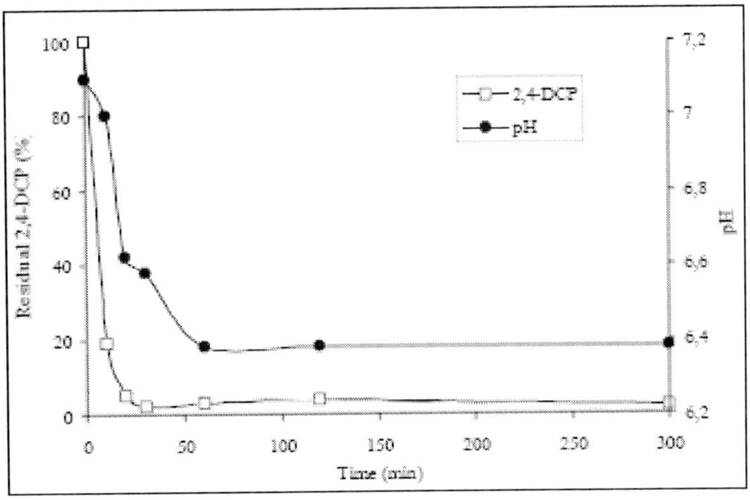

Figure 5. 2,4-DCP removal and pH variation during *Brassica napus* hairy roots culture in a bench top reactor, with 2,4-DCP 100 mg/l and H_2O_2 5 mM as initial conditions (from: Busto et al. 2005).

3.3. Plant Cultures

Including the entire plant in phytoremediation experimental systems means a significant change in complexity and outcome terms. Methodologically, the relatively simple nutrient solution where plants are grown allows performing chemical analysis without any serious matrix interference and ensures complete bioavailability of the contaminant. These characteristics remain from the simplest experimental systems. However, the use of the entire plant has differential aspects which contribute to a better understanding of a real scale process. Among them, the most important for phytoremediation purposes are:

- Cell differentiation: cells arrangement into tissues implies differential phenotypic characteristics, which leads to plant functions compartmentalization. Occurrence of different metabolic activities in different plant organs, can affect the final degradation state and/or fate of the contaminant.
- Phloem and xylem transport: the contaminant mobility across the different plant organs depends on this internal transport system. This is particularly important when the contaminant transformation includes multi-organ metabolic steps.
- Photosynthesis: although this physiological process can occur in cell cultures, it becomes significant when the experimental system include the entire plant. In this way, it is important to consider in the experimental design, the light intensity and quality provided to the plants. Occurrence of photosynthesis is essential for phytoremediation studies where the contaminant directly affects photosynthetic processes. As an example, studies on photosystem II inhibitors such as triazinic herbicides can be mentioned (Alvi et al., 2003, Cherifi et al., 2001).
- Presence of complex structural tissues: a common detoxification mechanism of organics phytoremediation is the final disposition of the transformed contaminant into the cell wall lignin matrix.
- Physiological processes: there are some physiological processes that require the presence of the whole plant and deeply influence the evolution of a phytoremediation process. In this way, germination or plant age affect the detoxification metabolic activities of many vegetal species, which should be carefully considered in the experimental design and results analysis.

Experimental systems for plant cultures include plant growing on different artificial supports:

3.3.1. Semisolid Supports

Plants culture in nutrient medium with low agar concentrations, allows germination and development of seedlings and adult plants. This kind of culture is easy to implement either in plates or flasks with simple sterilization procedures and ensures a good contaminant bioavailability, even with insoluble compounds, and easiness in harvesting.

Semisolid plant cultures have been specially used for germination tests and tolerances assays. Germination tests can be used to asses the germination ratio of different vegetal species, hybrids or even new seed batches. When studies are conducted with a given vegetal species for the first time, it is also important to test plant development and growth in this semi solid agar media. Through these tests, helpful information such as germination lag, development stages, growth rate and biomass production can be obtained. In this way, during plant selection for phytoremediation purposes, germination tests were performed on different species. A photograph of Lotus glaber germination test is presented (Figure 6).

Figure 6. Germination test of *Lotus glaber* during plant selection screening for phytoremediation purposes.

Since semi solid agar is a transparent medium, additional caution must be taken regarding the effects that light could have over germination, the contaminant dissipation and roots physiology.

For implementing a phytoremediation process, it is essential that the selected plant species is able to tolerate the presence of the contaminant. With this purpose, tolerance assays should be carried out. Tolerance is usually evaluated employing plant cultures in semi solid media containing a contaminant concentration range selected according to the levels expected to be found at the contaminated site. This experimental system presents low matrix interference and variability, and provides better contaminant bioavailability than soil. Plant tolerance can be evaluated through simple physiological parameters such as: germination ratio, shoot length, root length and branching, fresh weight, dry weight and leaf wetability. Considering plant survival and the physiological indicators results, the tolerated concentration range and inhibitory concentration 50% (CI50) can be calculated. These tolerance parameters are not intended to be directly extrapolated to field condition but to establish the concentrations to be tested in soils assays.

During plant species selection for phytoremediation of phenoxy herbicides and atrazine, tolerance assays were carried out on Zea mais, Medicago sativa, Lotus spp. Quenopodium quinoa and Lolium multiflorum. Shoot length and fresh weight resulted to be the most sensitive indicators of the herbicide effect on plant physiology. Results of Z. mais tolerance assays with atrazine (Figure 7) and 2,4-dichlorophenoxyacetic acid (2,4-D) (Figure 8) are presented (Cuadrado 2009, Merini 2009). These assays were crucial for selecting the plant species able to stand the herbicide concentrations found in both, agronomical practices residues and "hot spots". The phytoremediation applicability of selected species and tolerance mechanisms were further tested in more complex experimental systems (Section 3.3.2 and Section 3.4).

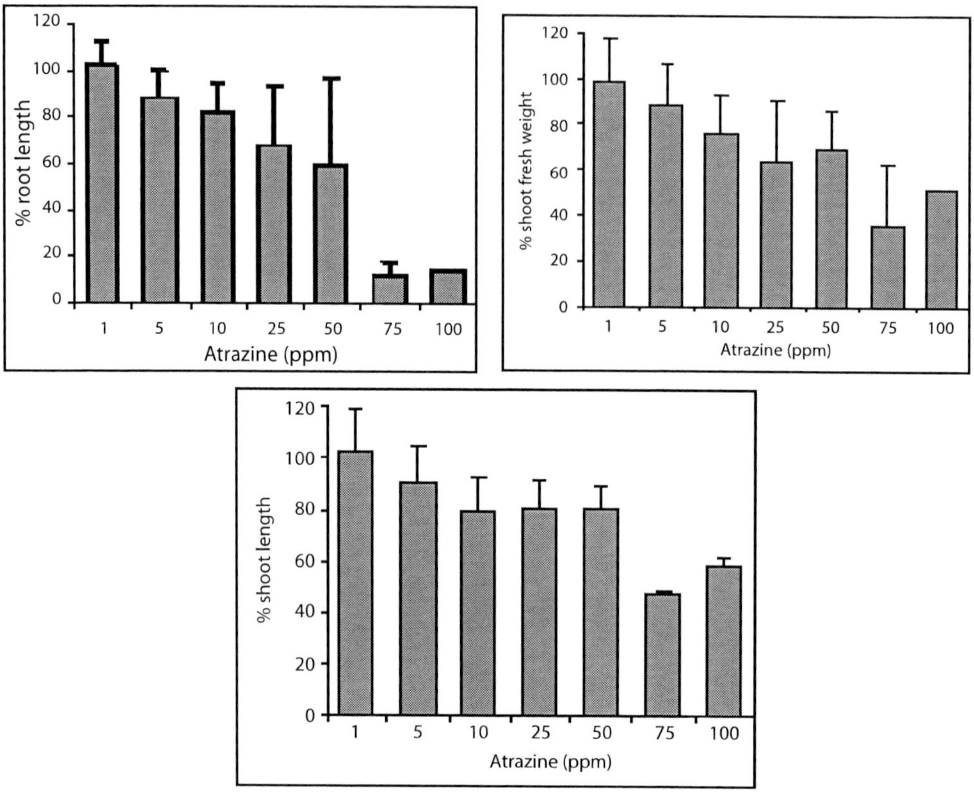

Figure 7. In vitro semisolid agar assay to evaluate Z. mais tolerance to the herbicide atrazine. Results are shown as a percentage of the control without herbicide. Mean ± SD; n=10 (Cuadrado 2009, Merini 2009).

3.3.2. Hydroponics

Hydroponic cultures (from the Greek Hydro = water, ponos = labor, i.e., working water) are based on plants capacity of growing in a minimum mineral aqueous medium without mechanical support. Hydroponics has been used as experimental systems for phytoremediation research because of the comparative advantages over other plant culture systems, including:

- Medium recycling: hydroponics design usually includes pump systems which constantly provide fresh medium, ensuring an optimum environment for plant growth. This dynamic system also allows variations on the medium composition along the time course of the experiment, including plant exposure to different contaminant concentrations and/or other chemical agents.
- Since pump media recycling system is not common in lab equipment, alternative media recycling procedures can be applied by manual replacing the growing medium at proper time intervals. The medium replacing frequency will be established according to plant species requirements and the experimental set up.
- Aeration: since root systems is submerged, it is necessary to provide a constant air supply. This can be achieved by air pumping or shaking. Either by the bubbles

generation or agitation, concentration gradients in the liquid medium are avoided. This represents a remarkable advantage over semi solid media systems.

- Minimum matrix interference: besides the maximum contaminant bioavailability, the simple medium composition facilitates the detection and potential isolation of compounds released by plant roots. Studies on such root exudates constitute an essential step on understanding the mechanisms involved in plant-assisted microbial degradation during a phytoremediation process.

These characteristics make the hydroponic cultures versatile tools for studying the removal of different organic compounds and the physiological events underlying the phytoremediation process (Narayanan et al. 1995, Narayanan et al. 1999). In this way, hydroponics has been successfully applied for studying the removal of the insecticide azinphos methyl by M. sativa plants (Flocco et al. 2004). It was estimated the disappearance kinetics of azinphos methyl (half-life reduced from 10,8 to 3,4 days in the presence of plants) and the plants physiological status was analyzed during the removal assay.

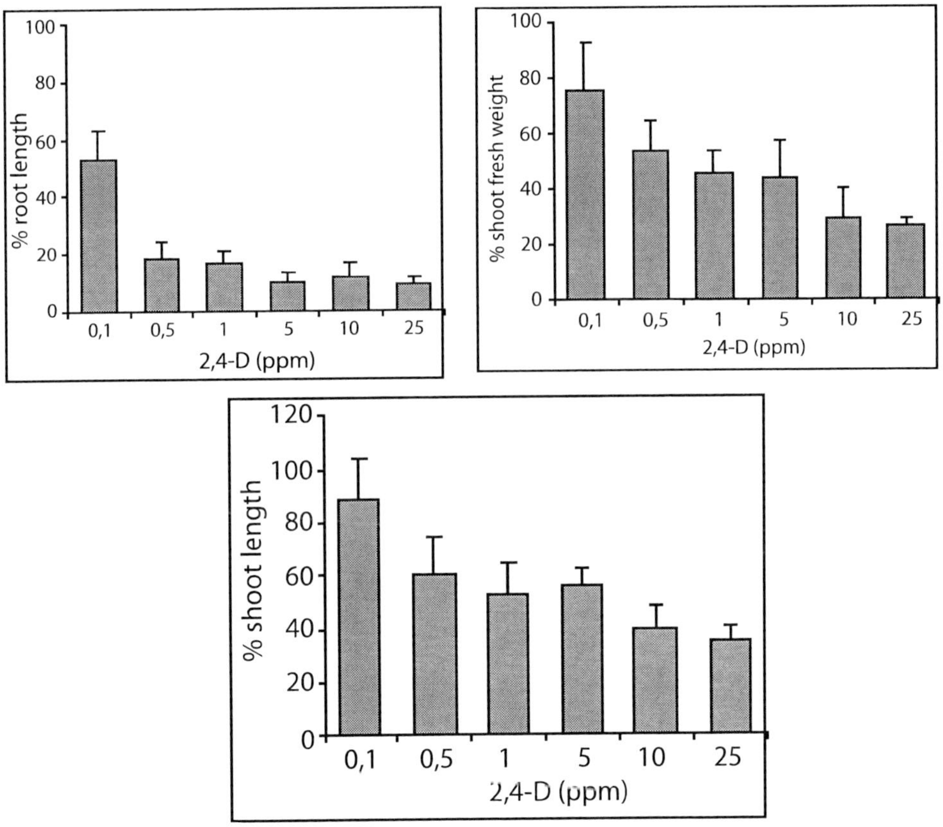

Figure 8. In vitro semisolid agar assay to evaluate *Z. mais* tolerance to the herbicide 2,4-D. Results are shown as a percentage of the control without herbicide. Mean ± SD; n=10 (Cuadrado 2009, Merini 2009).

Besides removal assays, hydroponics is a helpful system for elucidating the mechanisms involved in plant tolerance to agrochemicals, including enzymatic reactions, genetic modifications and mechanisms implying chemically active compounds. During studies on the atrazine tolerance mechanisms in L. multiflorum, hydroponic experimental systems were set up. It was possible to determine the role of the oxidases system P450 in atrazine tolerance by exposing 21 days old plants to the presence of 1-amonobenzotriazole (P450 inhibitor), atrazine and their combination (Figure 10) (Merini et al. 2009).

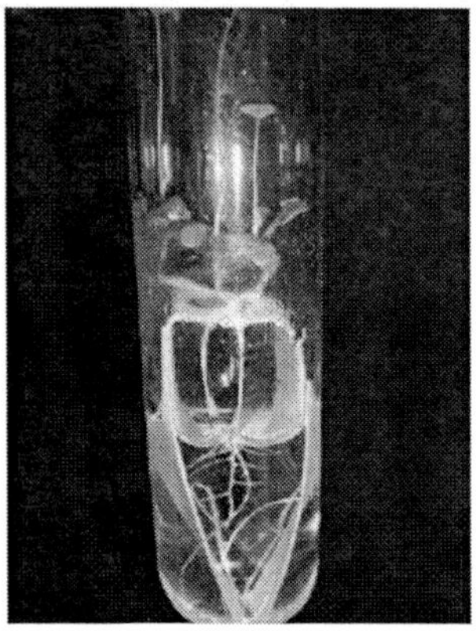

Figure 9. Hydroponic experimental systems employed for obtaining root exudates from *M. sativa* plants (Cuadrado 2009).

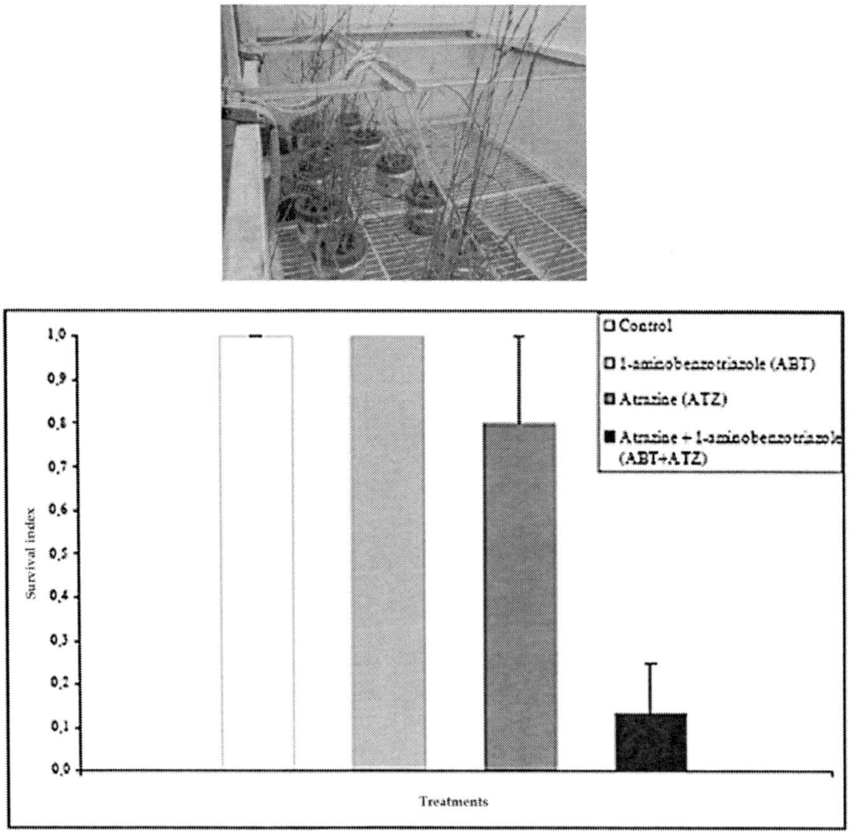

Figure 10. Response of *L. multiflorum* to atrazine in presence of 1-aminobenzotriazole (P_{450} inhibitor) in hydroponics cultures at 21 d after treatment. Mean ± SD; n=3 (from Merini et al. 2009).

In addition, batch hydroponic experimental systems were employed for obtaining root exudates from M. sativa plants (Figure 9) in order to asses their chemical composition and influence over an atrazine degrading consortium activity (Cuadrado 2009).

Despite most of plants can grow without any root mechanical support, some specific experimental designs require its presence. In this case, different inert solid matrices like quartz sand, vermiculite, perlite, glass beads or glass wool are useful supports. Nevertheless, special considerations on sterilization procedures, sterile conditions maintenance, matrix adsorption properties and the influence of root mechanical contact on plant physiology, should be taken into account.

3.4. Microcosms

Microcosms have been defined as small ecosystems in containers, acting as a bridge between theory and nature (Fraser 1999). The presence of the four elements: soil, plant, microflora and contaminant, makes the microcosm a useful experimental system for determining the role of biotic and abiotic factors on a phytoremediation process, as well as their interactions (Cuadrado 2009, Merini 2009). The soil inclusion sets the main difference between microcosms and the simpler experimental systems previously described (Sections 3.1 to 3.3) and represents an important approach to understanding the complexity of real scale processes. In this way, there are particular methodological considerations that must be taken into account during the experimental design and set up:

- Regarding the experimental model: the experimental model design starts with the container selection, which should be based on the soil amount to be contained, the particular characteristics of the selected vegetal species, the number of plants per microcosm and their final size. If sterility conditions are required, the container material and the sterilization procedure should be carefully selected. Finally, the simplicity of handling and sampling without disturbing initial experimental conditions is desirable.
- Regarding the soil: the microstructure of the soil matrix causes certain difficulties in achieving long lasting sterility conditions without drastically modifications in soil properties. The soil sterilization procedures usually employed are: γ-radiation, steam and chemical treatment with $HgCl_2$, chloroform or methyl bromide. Determination of the soil water holding capacity (WHC) is necessary to adjust watering procedures and soil humidity during the experiment.
- Regarding the plant: depending on the specific characteristics of the selected plant specie, previous seeds treatments could be necessary. In this way, surface seeds sterilization is a needed practice when evaluating the sole plant effects (Cuadrado 2009, Merini 2009). Microbes attached to seed surface and endophytes can play an active role in contaminant removal and/or affect plant physiology. A recommended additional procedure on seeds is the overnight incubation in distilled water at 4°C before seeding, in order to broke dormancy and synchronize germination. Additionally, some kinds of seeds may require a scarification treatment (legumes such as *Lotus*, *Trifolium* and *Medicago spp.*).

- Regarding the analytical methods: the extraction, detection and quantification of the contaminants and their degradation products from the soil matrix can be the bottle neck of the experimental set up. Soil properties like clay and humic matter contents drastically affect the contaminant extractability. Besides these well known interfering variables, some other aspects such as soil humidity and contaminant vehicle used at the moment of the soil spiking should be also considered. In this way, an optimized protocol for phenoxy herbicides extraction and quantification from agricultural soils with high humic mater content was developed (Merini et al. 2008). Effects of the kind and volume of the solvent used for 2,4-D spiking in a microcosms experiment over the herbicide extractability are shown (Table 2, Figure 11).

Table 2. Recovery ratios for 2,4-D in soils samples, using different spiking solvents (from: Merini et al. 2008)

Spiking solvent	Recovery (%)
Methanol*	81 ± 1
Water	49 ± 2

Mean ± SD; n=6.

*Recovery calculated for 2 ml of vehicle.

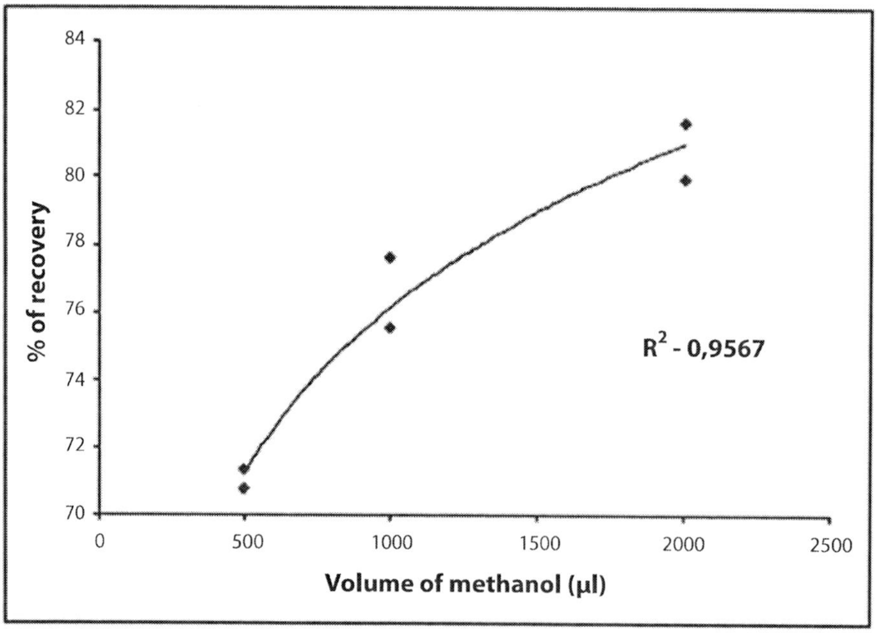

Figure 11. 2,4-D recovery ratio from soil samples using different volumes of methanol as spiking solvent (from: Merini et al. 2008).

Aiming to assess the dissipation of organochlorine herbicides from agricultural soils and evaluate the contribution of biotic and abiotic factors to the process, different microcosms studies were carried out. A general experimental design established for this purpose is presented in Figure 12.

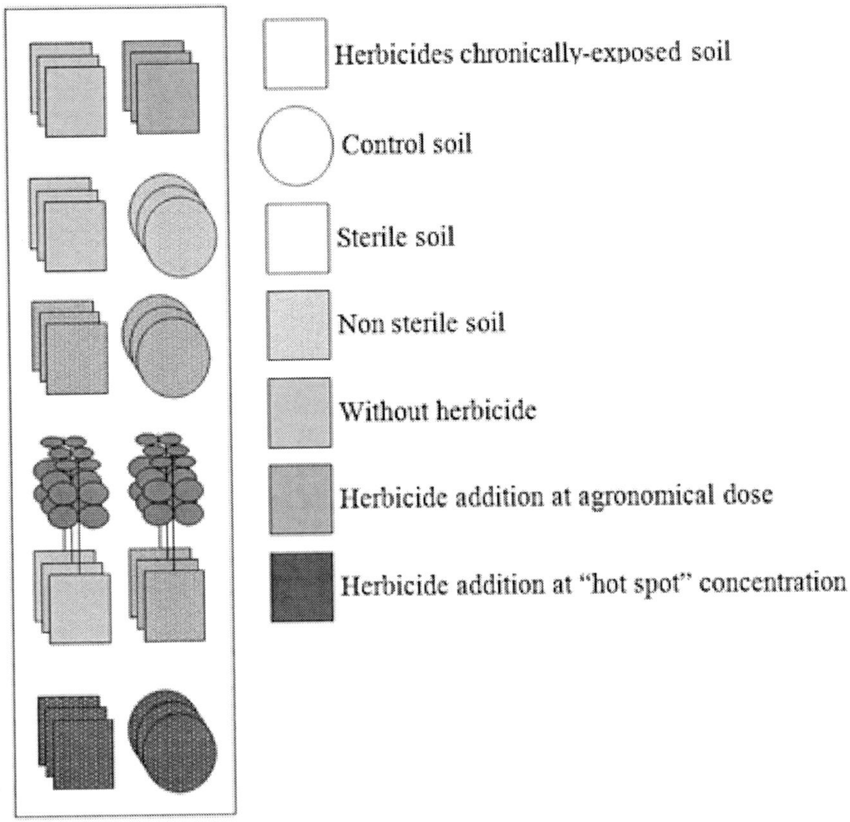

Figure 12. Scheme of the experimental design of microcosms assays employed for studying organochlorine herbicides dissipation from agricultural soils (Cuadrado 2009).

Microcosms experiments employing M. sativa plants were carried out for assessing the 2,4-D and 2,4-dichlorophenoxybutiric acid (2,4-DB) dissipation.

Results for 2,4-D dissipation assay showed that the herbicide was rapidly degraded, and the permanence of the first metabolite (2,4-DCP) in soil depended on the presence of plants and soil microorganisms. Synergism between plants and microorganisms occurred, since it was observed an enhanced dissipation of the herbicide in the ''non-sterile with plants'' microcosms (Figure 13A). The analysis of 2,4-DCP soil content suggested a quantitative transformation of the originally added 2,4-D into its first metabolite (Figure 13B). The removal of 2,4-D and detection of 2,4-DCP in ''sterile without plants'' microcosms evidenced that abiotic dissipation of the herbicide was contributing to the general process (Figure 13A and 13B) (Merini et al. 2007).

Results from the 2,4-DB microcosms assay showed that no abiotic processes were involved in the herbicide dissipation from agricultural soils of the Argentinean Humid Pampa region. The added 2,4-DB was completely degraded by soils with and without history of herbicide use, both at agronomic (5 ppm) and "hot spots" doses. At 5 ppm 2,4-DB addition, the chronically exposed soil showed a faster degradation and increased herbicide-degrading bacteria population in comparison with the pristine soil (Figure 14). However, at the "hot spot" dose (500 ppm), both soils showed similar degradation rates and degrading bacteria numbers. The most important difference was related to quantitative changes in Total Heterotrophic Bacteria (THB) numbers (Figure 15).

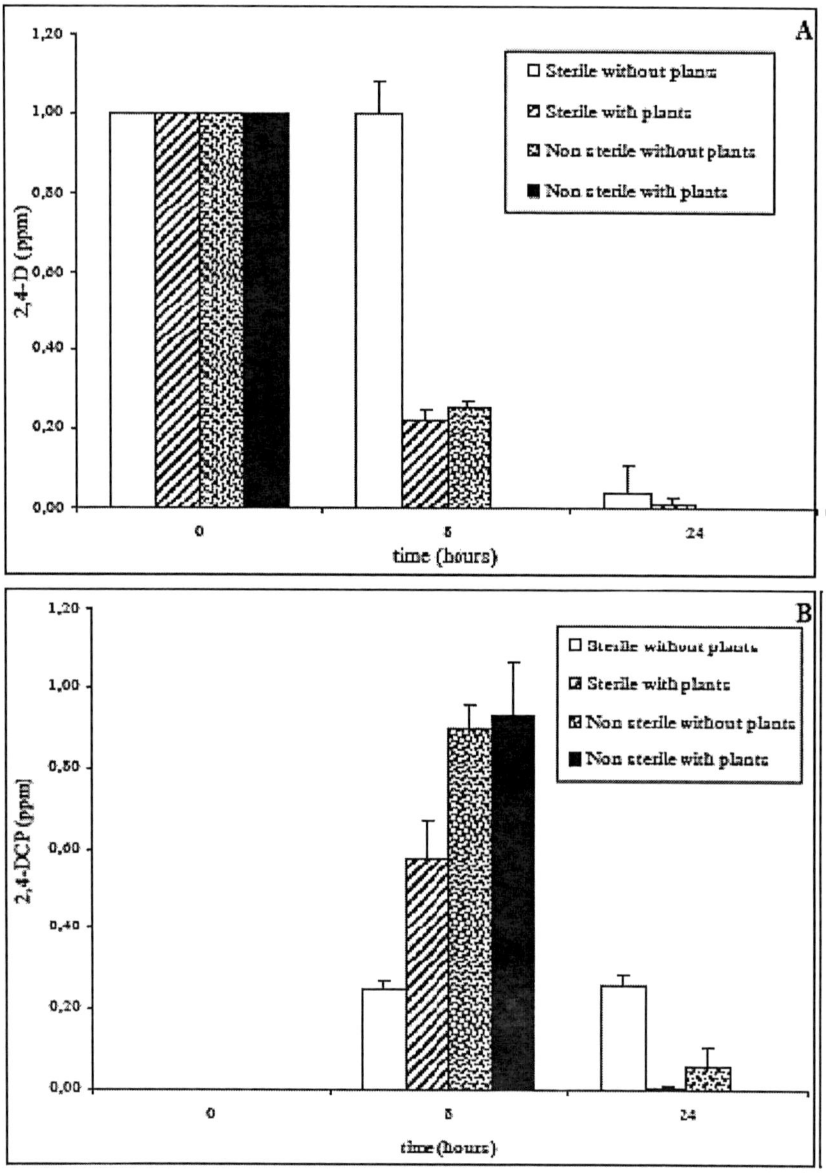

Figure 13. Residual content of 2,4-D (A) and 2,4-DCP (B) in soils under different experimental conditions. Mean ± SD; n=3.

The presence of M. sativa plants did not affect 2,4-DB degradation rate but influenced the metabolite (2,4-DCP) accumulation pattern (Figure 16) (Cuadrado et al. 2008).

The main factor affecting phenoxy herbicides degradation rate would be the history of application, as a consequence of the adaptation of the indigenous microflora to the presence of herbicides in the field. Otherwise, microcosms assays were carried out with atrazine as model herbicide, in order to evaluate the contribution of the novel tolerant specie L. multiflorum, to the herbicide dissipation in agricultural soils. It was demonstrated that L. multiflorum is able to germinate and grow in soils containing a high agronomical dose of atrazine (1 ppm). Implanted soil presented faster initial degradation rate and a 20% lower atrazine concentration by the end of the assay, in comparison with the unplanted soil (Figure

17). The rapid initial atrazine removal in L. multiflorum presence suggests its potential application in run-off avoiding strategies (Merini et al. 2009).

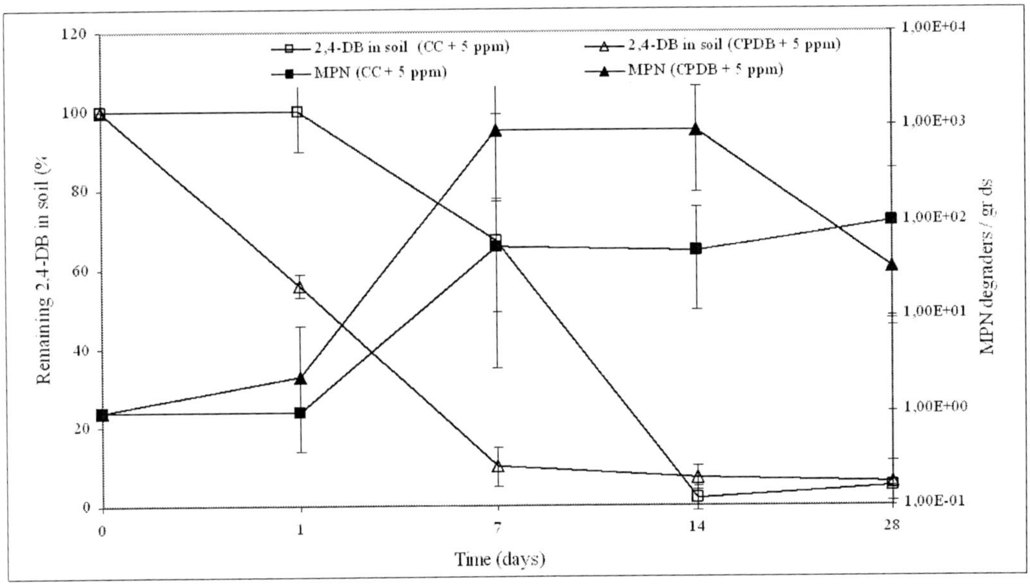

Figure 14. 2,4-DB removal and most probable number (MPN) of herbicide degrading bacteria in a herbicides chronically-exposed (CPDB) and a pristine soil (CC), during a microcosms experiment with agronomical herbicide dose (5 ppm). Bars represent 1 SD for 2,4-DB analysis and 95% confidence intervals in MPN determination. (from: Cuadrado et al. 2008).

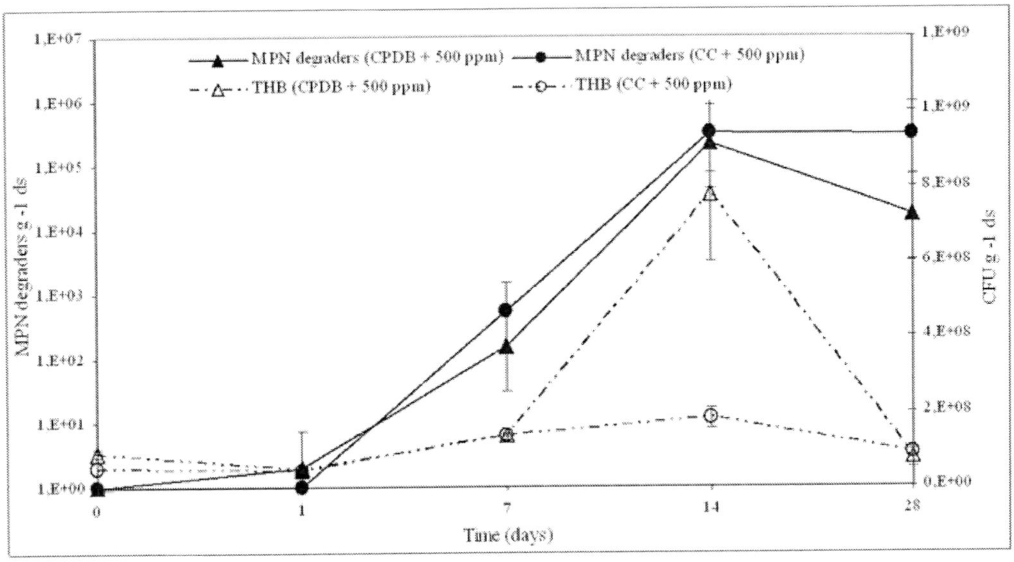

Figure 15. Total heterotrophic bacteria (THB) number and MPN of herbicide degrading bacteria in a herbicides chronically-exposed (CPDB) and a pristine (CC) soil during exposure to 500 ppm of 2,4-DB in a microcosms experiment. Bars represent 1 SD for THB counts and 95% confidence intervals in MPN determination. (from: Cuadrado et al. 2008).

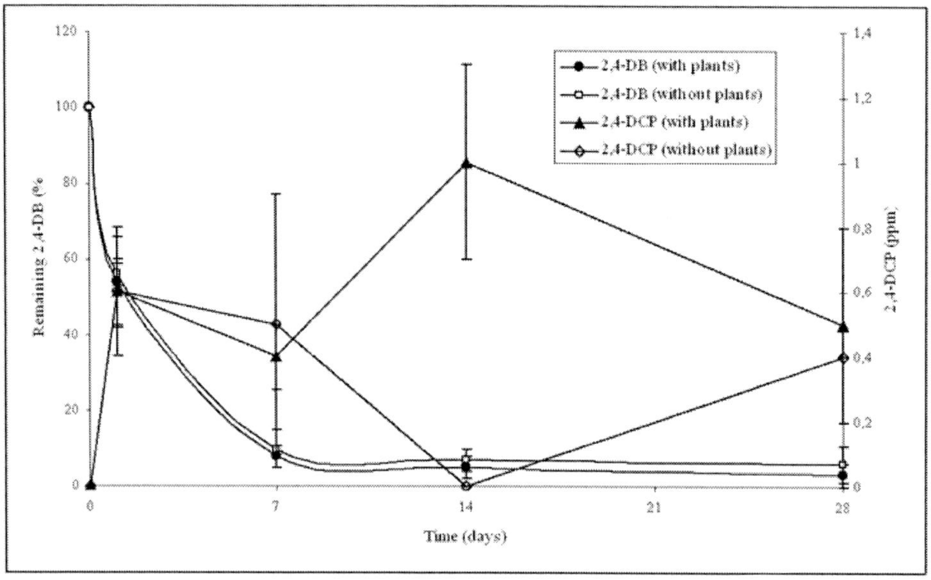

Figure 16. Dissipation of 2,4-DB and appearance of the metabolite 2,4-DCP in herbicides chronically-exposed soil (CPDB) after a 5 ppm dose of herbicide. Bars represent one SD (n=3) (from: Cuadrado et al. 2008).

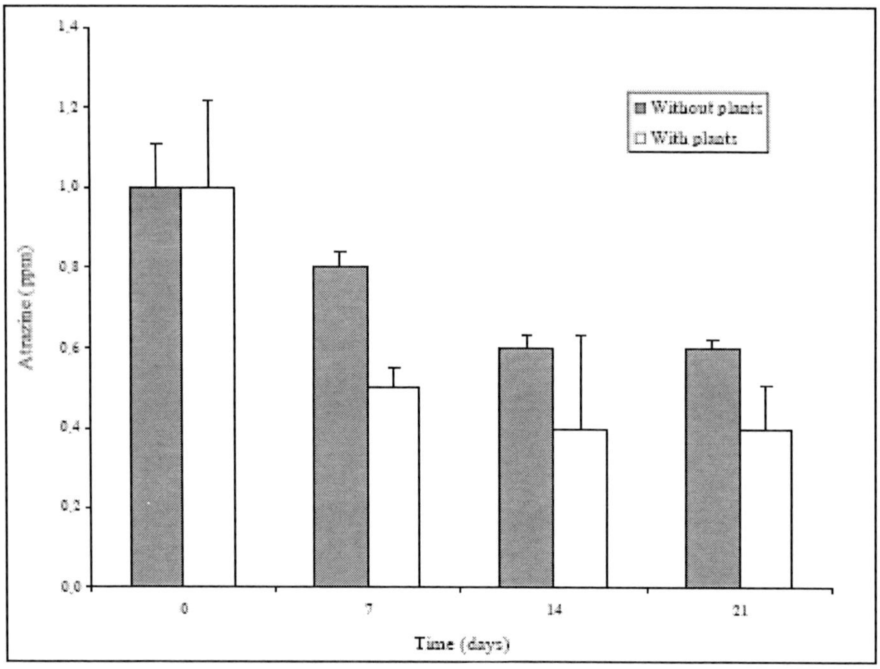

Figure 17. Effect of L. multiflorum plants on atrazine dissipation in a soil microcosms assay. Mean ± SD; n=3.

Outputs obtained by using different experimental systems and especially microcosms are helpful to understand the complex interactions between plant, soil, microorganisms and contaminant in a phytoremediation process. Moreover, from the analysis of the results obtained at microcosms experimentation, specifically designed field application strategies can be generated.

CONCLUSION

Although phytoremediation is a promising technology, it is highly dependant on the ecosystem where it will be implemented, which makes it site specific. In this way, the previous experimentation necessarily implies a case by case approach.

The appropriate selection of the experimental system will allow obtaining essential information for understanding the different aspects involved in the complex phytoremediation process. On one hand, a simplified system, gives precise information about the mechanisms involved in the remediation process. On the other hand, the use of a complex experimental system helps us to understand the interactions between soil, plant, microorganisms and the contaminant working as a whole entity. Since both kinds of systems bring about useful information from different perspectives, the ideal approach would be their integrative application, defining the reach of the inferences made from the results obtained with every experimental system.

Considering the gap in scale and complexity, field scale phytoremediation still requires additional considerations, including a different application, analytical and statistical perspective than the one used at lab scale, as well as ecological and engineering aspects exclusive of this scale.

ACKNOWLEDGMENTS

This work was supported by Grant ACCESS (ICA4-CT-2002-10011 - European Community) and PICT (2005) - 33166. LJ Merini and V Cuadrado are fellows of CONICET. AM Giulietti is researcher of CONICET.

REFERENCES

Agostini, E; Coniglio, MS; Milrad, S; Tigier, HA; Giulietti, AM. Phytoremediation of 2,4-dichlorophenol by *Brassica napus* hairy root cultures. *Biotechnol. Appl. Biochem.*, 2003 37,139-144.

Alvi, S; Perveen, R; Naqvi, II; Shaukat, SS. Effect of atrazine on absortion and translocation of P32, chlorophyll, carbohydrates, protein and potassium contents in bean *Vigna radiate (L.)* Wilczek. *Pakistan J. Biol. Sci.*, 2003 6,249-251.

Burken, JG. Uptake and metabolism of organic compounds: green-liver model In: S. C. McCutcheon, Schnoor, J.L., ed. *Phytoremediation: Transformation and Control of Contaminants*. Wiley, New York. 2003; 59-84.

Busto, VD; Angelini, V; Orejas, J; Tigier, H; Milrad, S; Agostini, E. Fitorremediación de 2,4-diclorofenol por raíces transformadas de nabo: escalado del proceso y caracterización de enzimas involucradas. Proceedings of International Congress REDBIO BAIRESBIOTEC 2005; 392-393.

Cherifi, M; Raveton, M; Picciocchi, A; Ravanel, P; Tissut, M. Atrazine metabolism in corn seedlings. *Plant Physiol. Biochem.*, 2001 39,665−672.

Chroma, L; Macek, T; Demnerova, K; Mackova, M. Decolorization of RBBR by plant cells and correlation with the transformation of PCBs. *Chemosphere*, 2002 49,739–748.

Cuadrado, V; Merini, L; Flocco, C; Giulietti, AM. Degradation of 2,4-DB in Argentinean agricultural soils with high humic matter content. *Appl. Microbiol. Biotechnol.*, 2008 77,1371–1378.

Cuadrado, V. Estrategias de rizorremediación de suelos contaminados con herbicidas organoclorados. Doctoral thesis – University of Buenos Aires, 2009.

Fan, W; Yanase, T; Morinaga, H; Gondo, S; Okabe, T; Nomura, M; Komatsu, T; Morohashi, K; Hayes, TB; Takayanagi, R; Nawata, H. Atrazine-induced aromatase expression is SF-1 dependent: implications for endocrine disruption in wildlife and reproductive cancers in humans. *Environ. Health Perspect.*, 2007 115,720-727.

Flocco, CG; Carranza, MP; Carvajal, LG; Loewy, RM; Pechén de D'Angelo, AM; Giulietti, AM. Removal of azinphos methyl by alfalfa. *Sci. Tot. Environ.*, 2004 327,31-39.

Fraser, UH. The use of microcosms as an experimental approach to understanding terrestrial ecosystem functioning. *Adv. Space Res.*, 1999 24,297-302.

Gamborg, O; Miller, R; Ojima, K. Nutrient requirement suspensions cultures of soybean root cells. *Exp. Cell Res.*, 1968 50,151-158.

Gerhardt, KE; Huang, X-D; Glick, BR; Greenberg, BM. Phytoremediation and rhizoremediation of organic soil contaminants: Potential and challenges. *Plant Sci.*, 2009 176, 20-30.

Glass, D. U.S. And International Markets for Phytoremedation, 1999-2000. David Glass Assoc., Inc., Needham, Massachusetts, USA; 1999.

Gonzalez, P; Agostini, E; Milrad, S. Application of *Brassica napus* hairy root cultures for phenol removal. *Chemosphere*, 2008 72, 1035-1042.

Guillon, S; Trèmouillaux-Guiller, J; Pati, PK; Rideau, M; Gantet, P. Hairy root research: recent scenario and exciting prospects. *Curr. Opin. Plant Biol.*, 2006 9, 341-346.

Gujarathi, NP; Haney, BJ; Park, HJ; Wickramasinghe, SR; Linden, JC. Hairy roots of *Helianthus annuus*: a model system to study phytoremediation of tetracycline and oxytetracycline. *Biotechnol. Prog.*, 2005 21, 775-780.

Gullner, G; Kömives, T; Rennenberg, H. Enhanced tolerance of transgenic poplar plants overexpressing γ-glutamylcysteine synthetase towards chloroacetanilide herbicides. *J. Exp. Bot.*, 2001 52,971-979.

HSDB. 2005. Hazardous Substances Data Bank. U.S. National Library of Medicine [Online] http://sis.nlm.nih.gov

ITRC. Phytoremediation decision tree. The Interstate Technology and Regulatory Cooperation Work Group - Phytoremediation Work Team; 1999.

ITRC. Phytotechnology Technical and Regulatory Guidance and Decision Trees; Revised. The Interstate Technology and Regulatory Council - Phytotechnologies Team; 2009.

Kuiper, I; Lagendijk, EL; Bloemberg, GV; Lugtenberg, BJ. Rhizoremediation: a beneficial plant-microbe interaction. *Mol. Plant Microbe Interact.*, 2004 17, 6-15.
Macek, T; Macková, M; Kás, J. Exploitation of plants for the removal of organics in environmental remediation. *Biotechnol. Adv.*, 2000 18, 23–34.
Mackova, M; Vrchotova, B; Francova, K; Sylvestre, M; Tomaniova, M; Lovecka, P; Demnerova, K; Macek, T. Biotransformation of PCBs by plants and Bacteria - consequences of plant-microbe interactions. *Eur. J. Soil Biol.*, 2007 43, 233-241.
Merini, LJ; Bobillo, C; Cuadrado, V; Corach, D; Giulietti, AM. Phytoremediation potential of the novel atrazine tolerant *Lolium multiflorum* and studies on the mechanisms involved. *Environ. Pollut.*, 2009 157, 3059-3063.
Merini, LJ; Cuadrado, V; Giulietti, AM. Spiking solvent, humidity and their impact on 2,4-D and 2,4-DCP extractability from high humic matter content soils. *Chemosphere*, 2008 71,2168–2172.
Merini, LJ; Cuadrado, V; Flocco, CG; Giulietti, AM. Dissipation of 2,4-D in soils of the Humid Pampa region, Argentina: A microcosm study. *Chemosphere*, 2007 68, 259-265.
Merini, LJ. Fitorremediacion de herbicidas organoclorados: diseño de estrategias biotecnológicas para su aplicación en la producción agrícola. Doctoral thesis – University of Buenos Aires; 2009.
Morgante, V; López-López, A; Flores, C; González, M; González, B; Vásquez, M; Rosselló-Mora, R; Seeger, M. Bioaugmentation with *Pseudomonas* sp. strain MHP41 promotes simazine attenuation and bacterial community changes in agricultural soils. *FEMS Microbiol. Ecol.*, 2010 71, 114–126.
Murashige, T; Skoog, F. A revised medium for rapid growth and bioassay with tobacco tissue cultures. *Plant Physiol.*, 1962 15, 473-497.
Narayanan, M; Davis, LC; Tracy, JC; Erickson, LE; Green, RM. Experimental and modeling. *J. Hazard. Mater.*, 1995 41, 229-249.
Narayanan, M; Erickson, LE; Davis, LC. Simple plant-based design strategies for volatile organic pollutants. *Environ. Prog.*, 1999 18, 231-242.
Perassolo, M; Quevedo, C; Busto, V; Ianone, F; Giulietti, AM; Rodríguez Talou, J. Enhance of anthraquinone production by effect of proline and aminoindan-2-phosphonic acid in *Rubia tinctorum* suspension cultures. *Enz. Microb. Technol.*, 2007 41, 181-185.
Pilon-Smits, E. Phytoremediation. *Annu. Rev. Plant Biol.*, 2005 56, 15-39.
Quan, X; Shi, H; Wang, J; Qian, Y. Biodegradation of 2,4-dichlorophenol in sequencing batch reactors augmented with immobilized mixed culture. *Chemosphere*, 2003 50, 1069-1074.
Quevedo, C; Perassolo, M; Alechine, E; Corach, D; Giulietti, AM; Rodriguez Talou, J. Increasing anthraquinone production by overexpression of 1-deoxy-D-xylulose-5-phosphate synthase in transgenic cell suspension cultures of *Morinda citrifolia. Biotechnol. Lett.*, 2010, In press.
Schwab, AP; Banks, MK. Biologically mediated dissipation of polyaromatic hydrocarbons in the root zone. In: T. A. Anderson, and Coats, J.R., ed. Bioremediation Through Rhizosphere Technology. American Chemical Society, Washington, DC; 1994; 132-141.
Shang, TQ; Newman, LA; Gordon, MP. Fate of tricholorethylene in terrestrial plants. In: S. C. McCutcheon, Schnoor, J.L., ed. *Phytoremediation: Transformation and Control of Contaminants*. Wiley, New York; 2003; 529–560.

Shaw, LJ; Burns, RG. Enhanced mineralization of [U-(14)C]2,4-dichlorophenoxyacetic acid in soil from the rhizosphere of *Trifolium pratense*. *Appl. Environ. Microbiol.*, 2004 70, 4766-74.

Suresh, B; Sherkhane, PD; Kale, S; Eapen, S; Ravishankar, GA. Uptake and degradation of DDT by hairy root cultures of *Cichorium intybus* and *Brassica juncea*. *Chemosphere*, 2005 61, 1288-1292.

USEPA. Introduction to Phytoremediation; 2000.

Yateem, A; Al-Sharrah, T; Bin-Haji, A. Investigation of microbes in the rhizosphere of selected trees for the rhizoremediation of hydrocarbon-contaminated soils. *Int. J Phytorem.*, 2008 10, 311-24.

Zhang, L; Ding, R; Chai, Y; Bonfill, M; Moyano, E; Oksman-Caldentey, K-M; Xu, T; Pi, Y; Wang, Z; Zhang, H; Kai, G; Liao, Z; Sun, X; Tang, K. Engineering tropane biosynthetic pathway in *Hyoscyamus niger* hairy root cultures. *PNAS.*, 2004 101, 6786-6791.

In: Handbook of Phytoremediation
Editor: Ivan A. Golubev

ISBN: 978-1-61728-753-4

Chapter 24

SILVER NANOPARTICLES PRODUCED BY LIVING PLANTS AND BY USING PLANT EXTRACTS

Richard G. Haverkamp
Massey University, New Zealand

INTRODUCTION

The practice of phytoremediation to remove unwanted elements from soils can be turned to a different application – to generate nanoparticles of a wanted element. The same processes are at work but the goal is different. In phytoremediation the task is to remove a contaminant from soil, whereas in phytomining it is to concentrate a valuable element and for phytosynthesis it is to synthesise a particular form, for example nanoparticles.

The understanding of the formation of nanoparticles (generally noble metals) by plants also contributes to the understanding of the uptake and accumulation of specific elements by plants, which may then be applied to phytoremediation and to phytomining. This chapter describes the use of plants to produce silver nanoparticles and the understanding that has been developed around the mechanism underlying the nanoparticle formation.

1. METAL NANOPARTICLE FORMATION IN PLANTS

1.1. Accumulation of Metals

The analysis of metals in plants has been used as a technique for prospecting for minerals since at least the 1930's on the basis that the local plants will reflect the chemical environment and may provide a more sensitive or more representative reflection of the local chemistry (Garrett et al., 2008). Gold is perhaps one of the more widely prospected element using this method (Brooks, 1982) but silver (Warren et al., 1984), platinum, cobalt and nickel are also prospected for by this means. Bacteria have been shown to be able to be used in a similar way for mineral prospecting (Klaus et al., 1999).

For high value metals it has been proposed to use plants to concentrate the metals from soil or mine tailings by using hyperaccumulating plants. The attractive feature of this idea is that deposits which are at such low levels that traditional extraction methods may not be economic may instead be amenable to this technique. Whereas traditional extraction might require the whole volume of the soil to be dug up and chemically processed, extraction with hyperaccumulating plants would require a crop or successive crops to be sown on the area and later harvested. The harvested crop, which is effectively a higher concentration ore, is then processed to extract the metal.

Phytomining has been proposed for gold extraction and has been subjected to field trials in Brazil, New Zealand and the US (Anderson et al., 2005). Unfortunately gold is not accumulated at sufficiently high levels in plants when grown naturally on soils containing gold at levels which are below those commercially viable for traditional mining techniques (this varies with the gold price but is generally less than around 0.1 ppm). Therefore it is necessary to add a complexing agent to the soil to assist in the dissolution of the metal so that it may be more readily taken up by the plants growing on the soil. For gold the use of cyanide or thiocyanate are the obvious candidates to assist in dissolution with thiocyanate having been used in large scale field trials.

In addition to the rather more passive nature of this form of "mining", where the soil does not need to be shifted, phytomining has the advantage of being aesthetically more pleasing: a field full of crops is more appealing to most people than mounds of mine feedstock and mine tailings. The environmental impact of phytomining may also perhaps be less than traditional mining, although it must be recognised that chemicals that may not be desirable in runoff are added and ploughing, fertilising and harvesting, and industrial cropping procedures, are necessary operations.

So far field trials for phytomining have only been applied to gold, although there is mention in the literature of the possibility of also applying this to silver (Sheoran et al., 2009). However, this is unlikely to happen in the near future because of the significantly lower value of silver than gold.

Accumulation of metals can also occur externally around living organisms, and may be the origin for some native gold ore deposits (Southam et al., 2009). It has recently has been shown that metallic copper nanoparticles may also form in this way (Weber et al., 2009).

If the emphasis is changed from using plants to collect and concentrate a valuable element from soil, to the plant as a processing agent to convert an element from one chemical or physical form to another (for example from a metallic salt to a metal nanoparticle) then it is sensible to provide the elements for processing. It is also preferable to therefore grow the plants hydroponically, which enables very good control of the solution concentrations of metallic ions available to the plant. Hydroponic growth also enables very high concentrations of metals in plants to be achieved.

The accumulation of metallic elements in plants does not necessitate either the formation of the metal element in a reduced (elemental) form or the formation of the metal into elemental nanoparticles. Most studies of metal accumulation in plants analyse the plants for the total amount of the element present, not differentiating between the metal salt and the elemental form. Since this chapter is about metal nanoparticle formation, we need to distinguish between accumulation of the element and formation of metallic nanoparticles of the element, in this case silver.

1.2. Silver Nanoparticle Formation

The first deliberate formation of reduced silver in plants resulted from studies of the activity of the chloroplast (Weber, 1938). These studies generally used fresh pieces of plant but not whole growing plants. Nanoparticles of silver, roughly 50 nm in diameter, were formed in a number of plant species and identified by the scanning electron microscope (Brown et al., 1962).

Most of the work on silver nanoparticle formation in plants has been in vascular land plants. Generally fast growing, soft tissue plants have been used, primarily since these are most amenable to laboratory study, but also because they would be easy to cultivate as a crop on a large scale. The list of species that have been used to make silver nanoparticles are contained in Table 1.

Silver concentrations in plants grown hydroponically increase rapidly in the first few hours (Haverkamp and Marshall, 2009) to reach a maximum after less than 24 hours for *Brassica juncea* and perhaps 48 hours for *Medicago sativa* depending on the hydroponic concentration of silver (Harris and Bali, 2008). Plants have been shown to be able to accumulate silver to concentrations in the whole plant of up to 12.4% for *Brassica juncea* and 13.6% for *Medicago sativa* (Harris and Bali, 2008). With a growth medium containing not only silver, but also gold and copper, alloy nanoparticles can be formed (Haverkamp et al., 2007). It is perhaps surprising that plants are able to take up significant quantities of silver considering the toxicity of silver to many organisms.

Table 1. Silver nanoparticles formed in plants

Plant species	Growth substrate	Conc. in plant	Particle size	Confirmed as metal?	Reference
Trifolium repens, Hymenocallis occidentalis, Anthephora pubescens, Aristida ternipes	Not growing: cut pieces of plant in solution		~50nm	No	(Brown *et al.*, 1962)
Brassica juncea	soil	730 ppm	5 – 50 nm	Partially: STEM (alloy)	(Haverkamp *et al.*, 2007)
Brassica juncea, Medicago sativa	hydroponic	12.4% 13.6%	~ 50 nm	No	(Harris and Bali, 2008)
Brassica juncea	hydroponic	1.2%	2 – 35 nm	Yes: XAS	(Haverkamp and Marshall, 2009)

1.3. Silver Nanoparticle Identification

There is a variety of methods for observing silver nanoparticles in plants. The formation of silver nanoparticles can be seen by TEM where dense particles show up as dark features because of their electron density. Better still, using electron backscattering, regions that contain high atomic weight elements show up as bright features. Both of these techniques give a good indication that the particles formed contain silver. However, an alternative explanation for these particles, which is not eliminated by this method, is that they consist not of Ag metal but of a precipitate of Ag such as AgCl, $AgNO_3$ or some other compound.

Therefore more specific techniques are required to positively identify the composition of the observed nanoparticles. Energy dispersive X-ray spectroscopy (EDS or EDAX) can be used to positively identify the presence of silver in nanoparticles in plant tissue. However, the technique is usually limited to heavier elements so that it is not always possible to tell with certainty that the silver is not precipitated with a light element as a salt. However, it gives a more reliable indication than electron density alone.

X-ray diffraction of the whole tissue could be used to identify the presence of nanoparticles of silver. This has been used for gold nanoparticles in plants, and enables not only the confirmation of the identity of metallic gold based on the diffraction pattern, but also provides the average crystallite size (which for small particles is likely to be the same as the particle size). It does not appear yet to have been successfully applied to silver nanoparticles in plants.

Electron diffraction and high resolution imaging of silver nanoparticles under the TEM can be used to determine the lattice spacing of the crystals observed and therefore distinguish between compounds of silver and metallic silver (Haverkamp et al., 2007).

UV-visible absorption can be used to indicate the presence of silver nanoparticles and to give an indication of particle size and is widely used for this purpose.

A robust technique for identifying the chemical state of the silver present, and quantifying the proportions in different chemical states is X-ray absorption spectroscopy (XAS). Both XANES (X-ray Absorption Near Edge Spectroscopy) and EXAFS (Extended X-ray Absorption Fine Structure) fall under this technique. XAS typically requires the use of synchrotron radiation, although laboratory scale XAS instruments are now available. XAS is able to positively indentify silver metal in plants and distinguish the metal from salts of silver by the shape of the XANES spectrum for the different chemical states of silver (Figure 1).

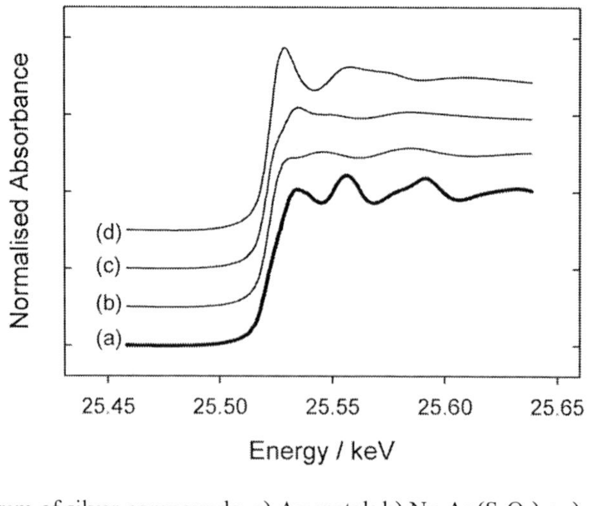

Figure 1. XAS spectrum of silver compounds. a) Ag metal; b) $Na_3Ag(S_2O_3)_2$; c) $Ag(NH_3)_2NO_3$; d) $AgNO_3$.

One of the first applications of XAS for the identification of metals in the reduced state in plants was for gold (Gardea-Torresdey et al., 2002; Marshall et al., 2007) but more recently it has been applied to silver (Haverkamp and Marshall, 2009).

1.4. Identification of Metallic Silver

In the TEM of sections of *Brassica juncea* leaves, after uptake of $AgNO_3$ to levels of 1.1% Ag in the leaves (Haverkamp and Marshall, 2009), particles which are relatively opaque to electron transmission are visible (Figure 2). The composition of these particles can be partially determined by energy dispersive X-ray analysis which can identify the presence of silver (e.g. (Haverkamp et al., 2007)) however this method alone does not distinguish between silver metal and small particles of a silver salt (e.g. AgCl or $AgNO_3$ crystals).

To positively identify the presence of the reduced form of the metal it is necessary to use a technique that gives information on the chemical state of the silver, such as XAS, or the crystalline form such as XRD. Using XAS, reduced silver has been positively identified as the form of 50% of the silver in *Brassica juncea* when a level of 1.1% Ag is present (Figure 3a). The balance consisted of silver nitrate which had not been reduced and probably some other unidentified component since the fit was not very good using only these two components.

From the XANES analysis (Figure 3) it is found that when *Brassica juncea* plants are grown hydroponically with different metal salts in the growth medium different proportions of the silver are reduced to the metal. With $AgNO_3$ taken up to a level of 1.1% Ag in the dried plant only 50% is reduced to metal, with $Ag(NH_3)_2NO_3$ taken up to a level of 0.9% Ag 30% is reduced and with $Na_3Ag(S_2O_3)_2$ taken up to a level of 0.3% Ag 10% is reduced. This difference in the amount of silver reduced is in keeping with the thermodynamic driver for the reactions as will be explained below.

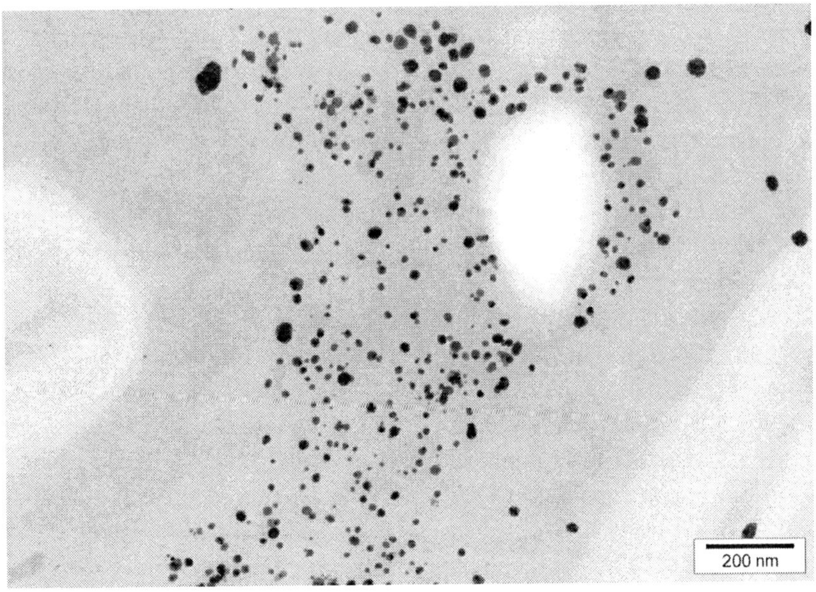

Figure 2. Silver nanoparticles in *Brassica juncea* after $AgNO_3$ uptake.

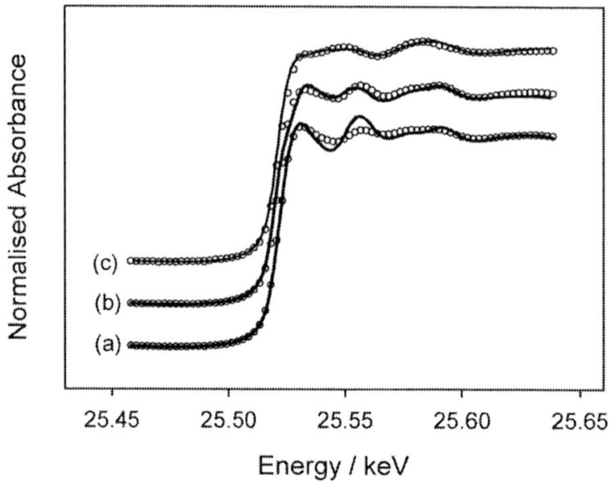

Figure 3. XANES spectrum of dried *Brassica juncea* leaf after uptake of a) $AgNO_3$; b) $Ag(NH_3)_2NO_3$; c) $Na_3Ag(S_2O_3)_2$. Circles, data; Solid lines, fitted curves formed from linear combinations of standard spectra.

2. Mechanisms of Silver Nanoparticle Formation

2.1. Chelating Agents and Silver Uptake

Silver ions may exist as simple hydrated ions in solution or may form complexes with a variety of complexing agents. EDTA is one of the more widely used chelating agents for a variety of metal ions and has been used in an attempt to increase the accumulation of silver in plants (Harris and Bali, 2008). However, although the quantitative results were not disclosed, it was noted that EDTA resulted in a lower uptake of silver than with uncomplexed silver.

Two other complexed forms of silver have been used to investigate the mechanism of silver uptake. These are $Ag(NH_3)^{2+}$ and $Ag(S_2O_3)_2^{3-}$ (Haverkamp and Marshall, 2009). The first is a cation, so might be subject to similar absorption pathways to Ag^+. The second is an anion, which may be absorbed through different biochemical pathways, and it is generally understood that large anions are not as readily absorbed by many plants as cations (Wright and Diamond, 1977).

It has been found that the rate of total silver uptake by *Brassica juncea* is similar for both the cations (Ag^+ and $Ag(NH_3)^{2+}$), just slightly lower for the larger complex. The rate of uptake of the anionic complex $Ag(S_2O_3)_2^{3-}$, as might have been predicted, is much slower (Figure 4). The other difference between these complexes and Ag^+ is that the Gibbs energy to reduce the silver greater for the complexes. We can think of it as the energy to dissociate the complex in addition to the energy to reduce the silver, although the mechanism may not follow this path, nevertheless the energy can be estimated this way.

One of the driving forces for the uptake of silver metal ions is the concentration gradient between the silver ions in the soil solution and the silver ions in the plant fluid. A high concentration in the soil water (or hydroponic solution) will increase this driving force. A low concentration in the plant will also increase this driving force. The concentration in the soil water may sometimes be increased by using an agent to enhance the solubility the metal, and for this reason complexing agents are used in phytoremediation for various partially soluble metals. Generally this is not an issue with hydroponic plant growth.

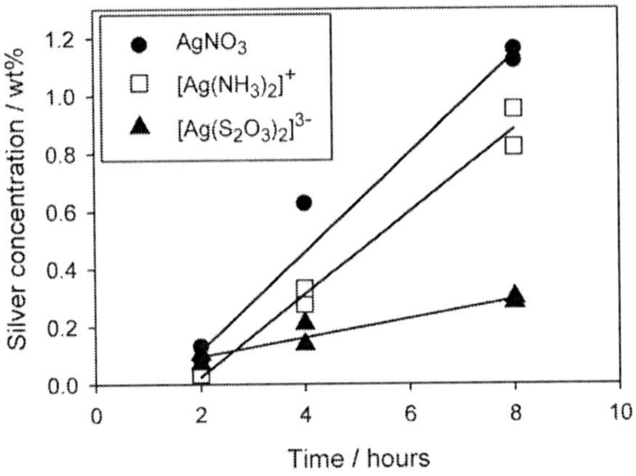

Figure 4. Silver concentration in *Brassica juncea* determined by AAS after uptake of 10g/L Ag in the form of three salts (adapted from (Haverkamp and Marshall, 2009)).

The concentration in the plant fluid can be decreased by removing the dissolved silver by reducing it to the metal. This reduction to the metal serves to lower the concentration of the initially taken up species, and therefore to maintain a concentration driving force for further uptake. These may be represented by the following rate equations and equilibria. An equilibrium between the concentration of silver ions in the soil water and the plant fluid (Eqn. 1) is being approached which leads to the rate equation (Eqn. 2) for the rate of uptake by the plant. Here K_1 is the equilibrium constant, and k_1 is the rate constant.

$$Ag^{+}_{(aq)soil} \xleftrightarrow{K_1} Ag^{+}_{(aq)plant} \quad (1)$$

$$Rate = k_1([Ag^{+}_{(aq)soil}] - [Ag^{+}_{(aq)plant}]) \quad (2)$$

Within the plant the silver is being reduced by a reductant, R, to the metal driving ultimately towards and equilibrium being established (Eqn. 3) where K_2 is the equilibrium constant:

$$Ag^{+}_{(aq)plant} + R \xleftrightarrow{K_2} Ag^{0} \quad (3)$$

This leads to the rate equation for the reduction of Ag^{+} to Ag^{0} (Eqn. 4) where k_2 is the rate constant:

$$Rate = -k_2 \frac{[Ag^{+}_{(aq)plant}][R]}{[Ag^{0}_{(s)plant}]} \quad (4)$$

Since the metal nanoparticles are solid they are considered to have a concentration of one and are removed from the equation so that the rate of removal of silver ions is proportional to the concentration of the silver ion and the reductant (Eqn. 5).

$$Rate = -k_2[Ag^+_{(aq)\,plant}][R] \tag{5}$$

This is a very simplified view, since transport within the plant is also very important of course, and transport across the root membrane is unlikely to be quite as simple as this. However, to a first approximation, the principle that a diminishing of the concentration of Ag^+ in the plant by reduction to Ag^0 metal nanoparticles may assist the rate of uptake of silver ions by the plant is illustrated.

2.2. Thermodynamic Drivers and Limits

Silver metal nanoparticles are formed in plants from silver salts. Therefore, the process necessarily requires a reduction of the salt to metal. A reducing agent must be present or formed in the plant in order to effect this reduction. Plants contain a complex mix of organic compounds capable of reducing metal ions, with the specific combination depending on the particular species and the state of the plant.

These include a variety of sugars. We will take a broad view of the process of reduction within plants, rather than considering the specific organic compounds involved. The fluid within a plant will have, at any specific region chosen, a reduction potential, using the word potential in the electrochemical sense. This reduction potential, or Eh, dictates the electrochemical reactions that are possible, or rather what state an electrochemical equilibrium will occupy. For those not familiar with Eh, this is a similar concept to pH, where the pH of a system dictates the acid base equilibrium of any components within that system.

While the Eh within a plant may vary between the site within a plant, plant species, and at different times or growing conditions, there is likely to be a broad similarity between plants. With this assumption, we can look at published experimental data of metal ion reactions within plants to get an approximation of the Eh conditions that may be found inside living plants. A range of reactions with metallic elements that have been observed to occur, as well some that were not observed but would have been observed had they occurred, are included in Table 2. A further simplifying assumption is made that the relevant electrochemical potentials are the potentials at standard conditions. It can be seen from the reactions that have been observed that the Eh within plants is greater than approximately 0 V relative to the standard hydrogen electrode (SHE). Silver is able to be reduced from the simple ion Ag^+ (at 0.80 V), the ammonia complex $Ag(NH_3)^{2+}$ at 0.37 V and the thiosulfate complex $Ag(S_2O_3)_2^{3-}$ at 0.04 V. However, these E^o values are for 1 mol L^{-1} concentrations so that at lower concentrations the E^o for the reaction is higher. Another way of expressing this is to say that close to the limit of Eh of the plant a smaller portion of the ion will be reduced.

Table 2. Reactions observed to occur in living plants with the reduction potentials. Adapted from (Haverkamp and Marshall, 2009), standard electrochemical potentials from (Aylward and Findlay, 2002; Bard *et al.*, 1985; Hubin *et al.*, 2004). A higher electrochemical potential indicates a reduction reaction more easily achieved

Ion	E° (V) SHE	References
Reactions observed		
$Cr^{IV} \rightarrow Cr^{III}$	2.10	(Bluskov *et al.*, 2005)
$Cr^{V} \rightarrow Cr^{IV}$	1.34	(Aldrich *et al.*, 2003)
$Cr^{V} \rightarrow Cr^{III}$	1.33	(Aldrich *et al.*, 2003)
$AuCl_4^- \rightarrow Au^0$	1.0	(Gardea-Torresdey *et al.*, 2002)
$Ag^+ \rightarrow Ag^0$	0.80	(Haverkamp and Marshall, 2009; Gardea-Torresdey *et al.*, 2003; Harris and Bali, 2008)
$Na_2SeO_3 \rightarrow$ Se-S (approx)	0.74	(Montes-Bayon *et al.*, 2002)
$Au(CN)_2^- \rightarrow Au^0$	0.66	(Marshall *et al.*, 2007)
$As^{IV} \rightarrow As^{III}$	0.56	(Bard *et al.*, 1985; Brooks, 1992)
$Ag(NH_3)^{2+} \rightarrow Ag^0$	0.37	(Haverkamp and Marshall, 2009)
$Cu^{2+} \rightarrow Cu^0$	0.35	(Haverkamp *et al.*, 2007; Manceau *et al.*, 2008)
$Cu^{2+} \rightarrow Cu^+$	0.17	(Polette *et al.*, 2000)
$Ag(S_2O_3)_2^{3-} \rightarrow Ag^0$	0.04	(Haverkamp and Marshall, 2009)
Reactions not observed		
$Pb^{2+} \rightarrow Pb^0$	–0.13	(Sharma *et al.*, 2004)
$Ni^{2+} \rightarrow Ni^0$	–0.24	(Kramer *et al.*, 2000)
$Zn^{2+} \rightarrow Zn^0$	–0.24	(Salt *et al.*, 1999)
$Tl^{I} \rightarrow Tl^0$	–0.336	(Scheckel *et al.*, 2004)
$Cd^{2+} \rightarrow Cd^0$	–0.40	(Pickering *et al.*, 1999; De La Rosa *et al.*, 2004)

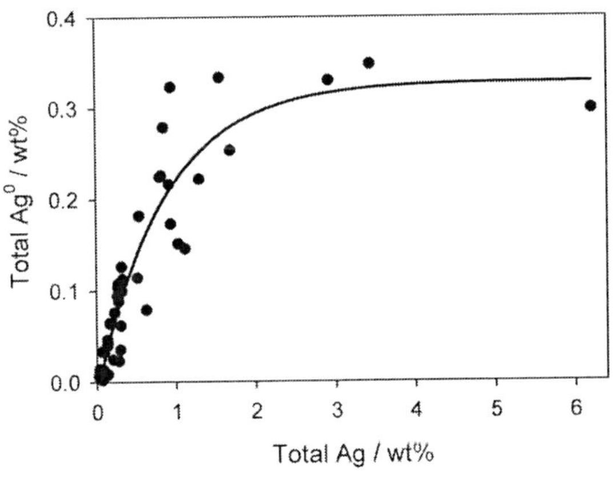

Figure 5. Ag and total silver (by dry weight) in *Brassica juncea* after $AgNO_3$ uptake (Adapted from (Haverkamp and Marshall, 2009)).

In addition to the Eh at one point in time or one concentration of silver ions, in the production of silver nanoparticles it is necessary to consider the quantity of reaction. Just as a pH measurement at one point in time tells us only about the present state of equilibrium and not the quantity of acid present, so the Eh tells only the present reduction potential and not the capacity for reduction. Therefore we must also consider the capacity of the plant to reduce a quantity of material. This may be due to the amount of reducing agent or agents present and the amount that may be generated by biological activity.

In a study of *Brassica juncea* a limit on the total amount of silver nanoparticles produced was observed with uptake of $AgNO_3$ (Figure 5), regardless of the total amount of silver ions taken up by the plant after a limit was reached, in this case of 0.3 % Ag^0 by dry weight. This perhaps gives a measure of the quantity of reducing agents that the plant had available which were able to reduce to a potential of 0.80 V or more. It could be that there are a range of reducing agents present, each with their own reduction potentials, so that the amount of any particular ion that can be reduced depends on its reduction potential.

2.3. Biological Mechanisms

The thermodynamic explanation proposed for the reduction of silver within plants tells us whether metal nanoparticles will form, how many will form and gives an indication of the rate of formation. It does not, however, give a detailed description of the biochemical pathways for transport and reduction. There is a range of enzymes that have been identified in ion transport and reduction processes and this is a topic that will not be covered in detail here. Readers are referred to the review by Schutzendubel and Polle (2002) for an insight into this area.

2.4. Particle Size Control

For technological uses of silver nanoparticles the average particle size and size distribution may be very important. Therefore any method of producing silver nanoparticles should preferably have the ability to exercise some control over particle size.

The use of different complexes of silver provides some control over particle size. Simple silver nitrate has been shown to produce metal particles in *Brassica juncea* of 10 – 35 nm. The silver complexes $Ag(NH_3)^{2+}$ and $Ag(S_2O_3)_2^{3-}$, however, produce much smaller particles of around 2 nm. It has also been observed in these experiments that the simple Ag^+ ion is reduced the most to Ag^0 (50%), while only 30% of the $Ag(NH_3)^{2+}$ is reduced and 10% of the $Ag(S_2O_3)_2^{3-}$. These differences in the proportion or the absolute amount of the salts produced could be due to the different reduction potentials for each of these salts as described above.

One speculative reason for this difference in particles size is that the silver complexes are more slowly reduced and therefore have more time to become distributed around the tissues of the plant rather than forming higher concentrations near vascular channels. In contrast the silver nitrate may begin to be reduced rapidly and therefore form regions with strong particle growth at sites of early transport. This is a topic which has not yet been extensively studied.

3. Silver Nanoparticle Formation Using Plant Based Extracts

An alternative to getting living plants to produce silver metal nanoparticles is to use plant material as a reducing reagent with a silver salt. This has been widely studied with a range of plant mush and extracts and under a range of conditions. In general, the detailed mechanism of formation of the nanoparticles has not been determined and these methods are empirical. The first work of this kind was published in 1921 ((Weber, 1938) and references cited therein). Interest in silver nanoparticles has been revived in recent years and many new studies have been conducted. Aqueous leaf extracts are the most common method for plant based reduction of silver salts. Leaf extracts from a variety of plants have been used, with silver nitrate as the source of silver (Table 2).

Table 2. Plant extracts used for silver nanoparticles synthesis

Plant species	Common name	Part of plant	Particle Size (nm)	Reference
Pelargonium graveolens	geranium	Aqueous leaf extract	16 – 40	(Shankar *et al.*, 2003)
Azadirachta indica	neem	Aqueous leaf extract		(Shankar *et al.*, 2004; Tripathi, 2009)
Emblica officinalis	Indian gooseberry, amla	Aqueous fruit extract	10 – 20	(Ankamwar *et al.*, 2005)
Aloe vera		Aqueous leaf extract	10 – 20	(Chandran *et al.*, 2006)
Cinnamomum camphora		Aqueous extract of dried leaf	55 – 80	(Huang *et al.*, 2007)
Phyllanthus amarus		Ether extraction of leaf (phyllanthin)	30	(Kasthuri *et al.*, 2009)
Meicago sativa	alfalfa	Aqueous extract of dried plant	4 – 50	(Tavera-Davila *et al.*, 2007)
Carica papaya	papaya	Callus aqueous extract	60 – 80	(Mude *et al.*, 2009)
Cinnamon zeylanicum	cinnamon	Powdered bark	31 – 100	(Sathishkumar *et al.*, 2009)
Pinus desiflora *Diopyros kaki* *Ginko biloba* *Magnolia kobus* *Platanus orientalis*	pine persimmon ginkgo magnolia platanus	Aqueous leaf extract	10 – 80	(Song and Kim, 2009)
Camellia senensis	tea	Aqueous and ethyl acetate extracts of black tea	20	(Begum *et al.*, 2009)
Chlorella vulgaris	(an alga)	Aqueous extract	28	(Xie, 2007)

These include an aqueous extract of geranium leaves (boiled) (Shankar et al., 2003); neem leaf broth (Shankar et al., 2004; Tripathi, 2009); an aqueous extract from dried leaves of *Cinnamomum camphora* (Huang et al., 2007); pine, persimmon, ginkgo, magnolia, and platanus (Song and Kim, 2009); and black tea (Begum et al., 2009). Silver particle sizes made from these extracts range from 10 to 80 nm. *Aloe vera* leaf extract has been used to reduce silver ions in a silver nitrate ammonia mixture (Chandran et al., 2006).

An aqueous extract from one month old callus of *Carica papaya* (Mude et al., 2009) has also successfully been used to produce silver nanoparticles, although one may wonder why.

Extracts of other parts than the leaf have been used. An extract from the fruit, Indian Gooseberry (*Emblica officinalis*) reacted with silver sulfate produces silver nanoparticles (Ankamwar et al., 2005). The powdered bark of *Cinnamon zeylanicum* known as the spice cinnamon can be reacted directly with silver nitrate solution to for nanoparticles (Sathishkumar et al., 2009). An aqueous extract of the dried whole plant of alfalfa produces nanoparticles, with pH a controlling factor in determining the particle size of the silver (Tavera-Davila et al., 2007). Organic solvents can also be used, for example an ether extraction of the leaf of *Phyllanthus amarus* with some other processing produces the compound phyllanthin with reduces silver ions (Kasthuri et al., 2009).

Lastly, it addition to vascular land plants, an extract of the alga *Chlorella vulgaris* has been used to reduce silver nitrate (Xie, 2007).

4. Extraction of Silver Nanoparticles from Plants

Plants containing silver nanoparticles are unlikely to be directly useful, rather it is the nanoparticles that are formed that are of most interest. Therefore a method to extract the silver nanoparticles, or to concentrate them, is required.

To extract metal nanoparticles from plants, degradation of the plant material is required. Ashing of plants will obviously remove the organic component and leave a residue of metals and other refractory elements or oxides. However, with the high temperatures involved metallic nanoparticles may be fused or sintered to form larger particles or agglomerates no longer having the desirable structure or properties of the original nanoparticles.

It is conceivable that the other "ash" components in the plant will serve to keep the metal nanoparticles separate despite high temperatures, but this has not been directly investigated yet. A low temperature method to concentrate silver nanoparticles is therefore desirable. Enzymatic degradation has been used in an attempt to concentrate metallic gold from plants (*Brassica juncea*) (Marshall et al., 2007). This resulted in an increase in concentration of the metal of only about 50% although it should have the potential to do better than this.

Concentration of silver nanoparticles produced from plant extracts is a simpler exercise since the solutions are, in many cases, free from other suspended solids. Phase transfer into chloroform using the cationic surfactant octadecylamine has been demonstrated as a way of separating the nanoparticles from the preparative solution (Ankamwar et al., 2005). Centrifugation can be used to remove the nanoparticles from solution, followed by freeze drying (Mude et al., 2009; Song et al., 2009; Tripathi, 2009).

5. Reactivity and Bioactivity of Plant Derived Silver Nanoparticles

While a lot has been written about producing silver nanoparticles based on plants much less has been done on practical uses for the particles once produced. Silver nanoparticles (produced by any means) have some practical applications. Silver as Ag^+ is a good antibacterial agent, and has long been used in small scale drinking water treatment. Silver metal nanoparticles may be used as a slow release depository of silver ions to provide antibacterial action. The silver nanoparticles have been incorporated into clothing to mitigate unpleasant odours, particularly with sports clothing; silver nanoparticle containing bandages are marketed; silver containing food containers providing better hygiene are available; and many other applications exist or are proposed as evidenced in other chapters of this volume. Silver nanoparticles may also exhibit some catalytic properties.

Plant derived silver nanoparticles have the potential to be used in many of the applications that have been developed for inorganically produced silver nanoparticles. However, it does not appear that any commercial applications have yet been established. It is hard to envisage how the plant materials will show any greater reactivity or bioactivity than other silver nanoparticles, and may show lower specific activity since the particles are largely contained within plant cells in the case of living plant derived nanoparticles. However, it is conceivable that the plant-derived silver nanoparticles will have the advantage of greater long term stability as they are already encapsulated in an organic matrix. The plant may provide additional chemicals which have a synergistic effect for a particular application.

Bactericidal activity has been shown in two studies. With silver nanoparticles derived from neem leaf extract antibacterial activity was observed (Tripathi, 2009), however, in the study it was not compared with controls of just the neem extract with no silver (neem is known as an antibacterial plant) nor with silver nanoparticles in the absence of neem. Therefore the antimicrobial activity that can be ascribed to those nanoparticles and their relative effectiveness has not been determined.

Bactericidal activity was also shown in silver nanoparticles produced from powdered cinnamon bark (Sathishkumar et al., 2009). Cinnamon is also known to have antimicrobial properties. Again, appropriate controls were not used, which limits what we can infer about the contribution from the silver.

Conclusion

Phytoremediation principles can be applied to the production of metal nanoparticles. Silver nanoparticles can be successfully produced by living plants and with a wide range of plant extracts. Concentrations of up to several percent total silver in dried plant material have been achieved and silver metal nanoparticle concentrations of 0.3% in leaves have been recorded. In the future, better size control of the nanoparticles may be developed. A higher conversion of silver salt to silver metal would also be desirable. Separation processes, in order to be able to make commercial use of the silver nanoparticles, have largely been avoided and also need to be developed further. The field still contains more that is unknown or barely known than is well understood.

References

Aldrich, M. V., Gardea-Torresdey, J. L., Peralta-Videa, J. R. and Parsons, J. G. (2003). Uptake and Reduction of Cr(VI) to Cr(III) by Mesquite (Prosopis spp.): Chromate-Plant Interaction in Hydroponics and Solid Media Studied Using XAS. *Environmental Science and Technology* 37(9): 1859-1864, ISSN 0013-936X.

Anderson, C., Moreno, F. and Meech, J. (2005). A Field Demonstration of Gold Phytoextraction Technology. *Minerals Engineering* 18(4): 385-392, ISSN 0892-6875.

Ankamwar, B., Damle, C., Ahmad, A. and Sastry, M. (2005). Biosynthesis of gold and silver nanoparticles using Emblica officinalis fruit extract, their phase transfer and transmetallation in an organic solution. *Journal of Nanoscience and Nanotechnology* 5(10): 1665-1671, ISSN 1533-4880.

Aylward, G. and Findlay, T. (2002). SI *Chemical Data.* John Wiley and Sons, ISBN 0470800445, Milton, Australia.

Bard, A. J., Parsons, R. and Jordan, J. (1985). *Standard Potentials in Aqueous Solution.* Marcel Dekker, ISBN 0824772911, New York.

Begum, N. A., Mondal, S., Basu, S., Laskar, R. A. and Mandal, D. (2009). Biogenic synthesis of Au and Ag nanoparticles using aqueous solutions of Black Tea leaf extracts. *Colloids and Surfaces B-Biointerfaces* 71: 113-118, ISSN 0927-7765.

Bluskov, S., Arocena, J. M., Omotoso, O. O. and Young, J. P. (2005). Uptake, Distribution, and Speciation of Chromium in Brassica Juncea. *International Journal of Phytoremediation* 7(2): 153-165, ISSN 1522-6514.

Brooks, R. R. (1982). Biological Methods of Prospecting for Gold. *Journal of Geochemical Exploration* 17(2): 109-122, ISSN 0375-6742.

Brooks, R. R. (1992). *Noble metals and biological systems; their role in medicine, mineral exploration, and the environment.* CRC, ISBN 0849361648, Boca Raton.

Brown, W. V., Johnson, C. and Mollenhauer, H. (1962). An Electron Microscope Study of Silver Nitrate Reduction in Leaf Cells. *American Journal of Botany* 49(1): 57-63, ISSN 0002-9122.

Chandran, S. P., Chaudhary, M., Pasricha, R., Ahmad, A. and Sastry, M. (2006). Synthesis of gold nanotriangles and silver nanoparticles using Aloe vera plant extract. *Biotechnology Progress* 22: 577-583, ISSN 8756-7938.

De La Rosa, G., Peralta-Videa, J. R., Montes, M., Parsons, J. G., Cano-Aguilera, I. and Gardea-Torresdey, J. L. (2004). Cadmium Uptake and Translocation in Tumbleweed (Salsola kali), a Potential Cd-Hyperaccumulator Desert Plant Species: ICP/OES and XAS Studies. *Chemosphere* 55(9): 1159-1168, ISSN 0045-6535.

Gardea-Torresdey, J., Parsons, J., Gomez, E., Peralta-Videa, J., Troiani, H., Santiago, P. and Yacaman, M. (2002). Formation and growth of Au nanoparticles inside live alfalfa plants. *Nano Letters* 2(4): 397-401, ISSN 1530-6984.

Gardea-Torresdey, J. L., Gomez, E., Peralta-Videa, J. R., Parsons, J. G., Troiani, H. and Jose-Yacaman, M. (2003). Alfalfa Sprouts: A Natural Source for the Synthesis of Silver Nanoparticles. *Langmuir* 19: 1357-1361, ISSN 1520-5827.

Garrett, R. G., Reimann, C., Smith, D. B. and Xie, X. (2008). From geochemical prospecting to international geochemical mapping: a historical overview. *Geochemistry-Exploration Environment Analysis* 8: 205-217, ISSN 1467-7873.

Harris, A. T. and Bali, R. (2008). On the formation and extent of uptake of silver nanoparticles by live plants. *Journal of Nanoparticle Research* 10(4): 691-695, ISSN 1388-0764.

Haverkamp, R. G. and Marshall, A. T. (2009). The mechanism of metal nanoparticle formation in plants: limits on accumulation. *Journal of Nanoparticle Research* 11(6): 1453-1463, ISSN 1388-0764.

Haverkamp, R. G., Marshall, A. T. and van Agterveld, D. (2007). Pick Your Carats: Nanoparticles of Gold-Silver-Copper Alloy Produced in Vivo. *Journal of Nanoparticle Research* 9(4): 697-700, ISSN 1388-0764.

Huang, J., Li, Q., Sun, D., Lu, Y., Su, Y., Yang, X., Wang, H., Wang, Y., Shao, W., He, N., Hong, J. and Chen, C. (2007). Biosynthesis of silver and gold nanoparticles by novel sundried Cinnamomum camphora leaf. *Nanotechnology* 18: 105104, ISSN 0957-4484.

Hubin, A., Simons, W., Pauwels, L. and Vereecken, J. (2004). Reduction of Silver Complexes: Towards an Ever More Elaborated Insight in the Influence of the Type of Ligand on the Reaction Rate. *Journal of Electroanalytical Chemistry* 572(2): 399-408, ISSN 0022-0728.

Kasthuri, J., Kathiravan, K. and Rajendiran, N. (2009). Phyllanthin-assisted biosynthesis of silver and gold nanoparticles: a novel biological approach. *Journal of Nanoparticle Research* 11(5): 1075-1085, ISSN 1388-0764.

Klaus, T., Joerger, R., Olsson, E. and Granqvist, C.-G. (1999). *Silver-based crystalline nanoparticles, microbially fabricated*. Proceedings of the National Academy of Science, U.S.A. 96(24): 13611-13614, ISSN 0027-8424.

Kramer, U., Pickering, I. J., Prince, R. C., Raskin, I. and Salt, D. E. (2000). Subcellular localization and speciation of nickel in hyperaccumulator and non-accumulator Thlaspi species. *Plant Physiology* 122(4): 1343-1353, ISSN 0032-0889.

Manceau, A., Nagy, K. L., Marcus, M. A., Lanson, M., Geoffroy, N., Jacquet, T. and Kirpichtchikova, T. (2008). Formation of metallic coppernanoparticles at the soil-root interface. *Environmental Science and Technology* 42(5): 1766-1772, ISSN 0013-936X.

Marshall, A. T., Haverkamp, R. G., Davies, C. E., Parsons, J. G., Gardea-Torresdey, J. L. and van Agterveld, D. (2007). Accumulation of gold nanoparticles in Brassic juncea. *International Journal of Phytoremediation* 9(1-3): 197-206, ISSN 1522-6514.

Montes-Bayon, M., Yanes, E. G., De Leon, C. P., Jayasimhulu, K., Stalcup, A., Shann, J. and Caruso, J. A. (2002). Initial Studies of Selenium Speciation in Brassica juncea by LC With ICPMS and ES-MS Detection: an Approach for Phytoremediation Studies. *Analytical Chemistry* 74(1): 107-113, ISSN 0003-2700.

Mude, N., Ingle, A., Gade, A. and Rai, M. (2009). Synthesis of silver nanoparticles using callus extract of Carica papaya - a first report. *Journal of Plant Biochemistry and Biotechnology* 18(1): 83-86, ISSN 0971-7811.

Pickering, I. J., Prince, R. C., George, G. N., Rauser, W. E., Wickramasinghe, W. A., Watson, A. A., Dameron, C. T., Dance, I. G., Fairlie, D. P. and Salt, D. E. (1999). X-Ray Absorption Spectroscopy of Cadmium Phytochelatin and Model Systems. *Biochimica Et Biophysica Acta-Protein Structure and Molecular Enzymology* 1429(2): 351-364, ISSN 0167-4838.

Polette, L. A., Gardea-Torresdey, J. L., Chianelli, R. R., George, G. N., Pickering, I. J. and Arenas, J. (2000). XAS and Microscopy Studies of the Uptake and Bio-Transformation of Copper in Larrea tridentata (Creosote Bush). *Microchemical Journal* 65(3): 227-236, ISSN 0026-265X.

Salt, D. E., Prince, R. C., Baker, A. J. M., Raskin, I. and Pickering, I. J. (1999). Zinc Ligands in the Metal Hyperaccumulator Thlaspi Caerulescens as Determined Using X-Ray Absorption Spectroscopy. *Environmental Science and Technology* 33(5): 713-717, ISSN 0013-936X.

Sathishkumar, M., Sneha, K., Won, S. W., Cho, C.-W., Kim, S. and Yun, Y.-S. (2009). Cinnamon zeylanicum bark extract and powder mediated green synthesis of nano-crystalline silve particles and its bactericidal activity. *Colloids and Surfaces B-Biointerfaces* 73: 332-338, ISSN 0927-7765.

Scheckel, K. G., Lombi, E., Rock, S. A. and McLaughlin, N. J. (2004). In vivo synchrotron study of thallium speciation and compartmentation in lberis intermedia. *Environmental Science and Technology* 38(19): 5095-5100, ISSN 0013-936X.

Schutzendubel, A. and Polle, A. (2002). Plant Responses to Abiotic Stresses: Heavy Metal-Induced Oxidative Stress and Protection by Mycorrhization. *Journal of Experimental Botany* 53(372): 1351-1365, ISSN 0022-0957.

Shankar, S. S., Ahmad, A. and Sastry, M. (2003). Geranium leaf assisted biosynthesis of silver nanoparticles. *Biotechnology Progress* 19(6): 1627-1631, ISSN 8756-7938.

Shankar, S. S., Rai, A., Ahmad, A. and Sastry, M. (2004). Rapid synthesis of Au, Ag, and bimetallic Au core-Ag shell nanoparticles using Neem (Azadirachta indica) leaf broth. *Journal of Colloid and Interface Science* 275(2): 496-502, ISSN 0021-9797.

Sharma, N. C., Gardea-Torresdey, J. L., Parsons, J. and Sahi, S. V. (2004). Chemical Speciation and Cellular Deposition of Lead in Sesbania Drummondii. *Environmental Toxicology and Chemistry* 23(9): 2068-2073, ISSN 0730-7268.

Sheoran, V., Sheoran, A. S. and Poonia, P. (2009). *Phytomining*: A review. Minerals Engineering 22(12): 1007-1019, ISSN 0892-6875.

Song, J. Y., Jang, H. K. and Kim, B. S. (2009). Biological synthesis of gold nanoparticles using Magnolia kobus and Diopyros kaki leaf extracts. *Process Biochemistry* 44(10): 1133-1138, ISSN 1359-5113.

Song, J. Y. and Kim, B. S. (2009). Rapid biological synthesis of silver nanoparticles using plant leaf extracts. *Bioprocess and Biosystems Engineering* 32: 79-84, ISSN 1615-7605.

Southam, G., Lengke, M. F., Fairbrother, L. and Reith, F. (2009). *The Biogeochemistry of Gold. Elements* 5(5): 303-307, ISSN 1811-5209.

Tavera-Davila, L., Liu, H. B., Herrera-Becerra, R., Canizal, G., Balcazar, M. and Ascencio, J. A. (2007). *Analysis of Ag Nanoparticies Synthesized by Bioreduction*. In International Conference on Bionano Science (ICONBS 2007), 1785-1791, ISBN 1533-4880, Taipei, Taiwan.

Tripathi, A., Chandrasekaran, N., Raichur, A.M., Mukherjee, A. (2009). Antibacterial Applications of Silver Nanoparticles Synthesized by Aqueous Extract of Azadirachta Indica (Neem) Leaves. *Journal of Biomedical Nanotechnology* 5(1): 93-98, ISSN 1550-7033.

Warren, H. V., Horsky, S. J. and Barakso, J. J. (1984). Prelimiary Studies of the Biogeochemistry of Silver in British-Columbia. *Cim Bulletin* 77(863): 95-98, ISSN 0317-0926.

Weber, F. (1938). Silber-Reduktion der Chloroplasten. *Protoplasma* 29(1): 427-434.

Weber, F. A., Voegelin, A., Kaegi, R. and Kretzschmar, R. (2009). Contaminant mobilization by metallic copper and metal sulphide colloids in flooded soil. *Nature Geoscience* 2(4): 267-271, ISSN 1752-0894.

Wright, E. M. and Diamond, J. M. (1977). Anion Selectivity in Biological Systems. *Physiological Reviews* 57: 109-156, ISSN 0031-9333.

Xie, J. (2007). *Silver nanoplates: from biological to biomimetic synthesis*. ACS Nano 1(5): 429-439, ISSN 1936-086X.

In: Handbook of Phytoremediation
Editor: Ivan A. Golubev

ISBN: 978-1-61728-753-4

Chapter 25

ECO-ENVIRONMENTAL CONSEQUENCES ASSOCIATED WITH CHELANT-ASSISTED PHYTOREMEDIATION OF METAL-CONTAMINATED SOIL

Ismail M. M. Rahman[1,2*], M. Mosharraf Hossain[3,4], Zinnat A. Begum[1], M. Azizur Rahman[1], Hiroshi Hasegawa[1]

[1]Graduate School of Natural Science and Technology, Kanazawa University
Kakuma, Kanazawa 920-1192, Japan
[2]Department of Chemistry, Faculty of Science, University of Chittagong
Chittagong 4331, Bangladesh
[3]Graduate School of Engineering, Osaka University
2-1 Yamadaoka, Suita, Osaka 565-0871, Japan
[4]Institute of Forestry and Environmental Sciences, University of Chittagong
Chittagong 4331, Bangladesh

ABSTRACT

Phytoremediation utilizes different plant species as a media of containment, destruction, or extraction of contaminants from different matrices including soil and water. Plants require essential metals i.e. Cu, Mn, Fe, Zn, Mo, *etc.* for growth and as such they are capable of accumulating these metals. Plants can also accumulate Cd, Cr, Pb, Co, Ag, Se, Hg, *etc.*, which are apparently non-essential for their growth and survival. This metal-accumulating property of plants has made them very popular in recent days in the remediation of metal-contaminated soil. This approach of remediation has the benefit of cost savings compared to the conventional treatment options. Plants capable of concentrating metal pollutants at enhanced rate - the hyperaccumulators - are commonly used for metal-polluted soil remediation. But, the bioavailability of the metals limits the performance of hyperaccumulators since a large proportion of metals in contaminated soils exist in 'non-labile' state. There came the application of synthetic chelants to enhance the mobility and phytoavailability of metals to remediating plants. Various

* E-mail: I.M.M.Rahman@gmail.com; Tel/Fax +81-76-234-4792

chelants are available which forms bioavailable and water-soluble stable metal complexes facilitating phytoextraction of these metals at enhanced rates by plants. While chelants are used because of their powerful metal solubilizing properties, it is the same characteristic which gives them the potential of becoming an eco-environmental threat. Environmental concerns are evoked due to the high persistency and poor photo-, chemo- and biodegradability of metal-chelant complexes. Different approaches have been proposed to combat the eco-environmental concerns raised by the use chelants in phytoremediation. Within the scope of this chapter, we will focus on the chelant assisted phytoremediation approaches for the removal of heavy metal contaminants from soil and eco-environmental consequences associated with it.

INTRODUCTION

Soil heavy metal burden is becoming a major environmental concern throughout the world mainly due to acid mine drainage, tailings embankments, mining rock dumps and metallurgical waste piles [1–3]. Non-biodegradability gives heavy metals an indefinite life-time [4–7] though their bioavailability depends on their interactions with the various soil constituents [8].

When bio-available at high concentrations, these non-essential metals displace essential metal ions or block the essential functional groups or modify the active conformation of biomolecules [9]. Exposure to high concentration of heavy metals causes significant toxic effects to humans, animals, microorganisms and plants [10–13].

Heavy metals assimilation into the food chain becomes a risk when agricultural lands get low to medium range heavy metal contamination. In some areas of the world stricter environmental laws have been imposed to prevent any such occurrence by limiting the food production on contaminated lands. For example, a European Union Council Directive limited the concentrations of heavy metals in cultivable soils to be 3 mg kg^{-1} for Cd, 140 mg kg^{-1} for Cu, 75 mg kg^{-1} for Ni, 300 mg kg^{-1} for Pb, 300 mg kg^{-1} for Zn, and 1.5 mg kg^{-1} for Hg [14, 15]. The directive identified several million ha of agricultural lands as polluted by heavy metals in Europe [16]. As estimated, total clean-up cost to restore the cultivability of the contaminated sites only in Europe may cost approximately between EUR 59 and 109 billion [17], which is prohibitingly high even for developed nations as evident from the case of Germany where only one-third of the total soils from the contaminated sites is treated in soil remediation facilities [2].

Development of low-cost and environmentally friendly heavy metal removal techniques to reduce the risk from metal-contaminated soil is thus gaining enhanced research focus. Some of the soil clean-up techniques, for example, soil washing using particle size separation, chemical extraction with aqueous solutions of surfactants and mineral acids are in full-scale use. Soil washing using chelant solution or chelant-enhanced phytoextraction of heavy metals from contaminated soils received considerable interest by the researchers in recent days [18]. When added to the contaminated soil matrices, chelant materials increase the bioavailability of metals to metal accumulating plants for being removed through phytoextraction. The approach is economic but enhanced bioavailability of metals due to metal-chelant interactions may pretense impending risks to the eco-environs as discussed in detail in the following sections.

METAL SEQUESTRATION WITH PLANTS

Plants naturally uptake metals, such as Cu, Fe, Mo, Mn, Ni, and Zn, as essential mineral nutrients for their growth and survival. Plants also accumulate some other metals *e.g.*, Cd, Cr, Pb, Co, Ag, Se, and Hg which have no known biological function [3, 19, 20]. This innate ability to absorb wide range of metals makes plants amenable to be used for metal sequestration from contaminated medium. The technique is popularly known as 'phytoremediation' where plants are used for the removal of pollutants to restore the pristine environment. Phytoremediation covers several technological subsets i.e. phytohydraulics, phytodegradation, phytoextraction, phytostabilization, phytovolatilization, rhizodegradation and rhizofiltration, which aim at utilizing the plants' ability to remediate contaminated soil and water [3, 21–26].

In 'phytoextraction', pollutant-accumulating plants remove metals or organics from soil by concentrating them in their harvestable parts [26–29] either *continuously* or in an *induced* manner [26]. Continuous phytoextraction is associated with the natural ability of some plants, known as 'hyperaccumulators', which accumulate and translocate metal contaminants over the growth cycles [21, 26]. Metal hyperaccumulation is an ecophysiological adaption to metalliferous soils which is hypothetically described as a defense mechanism of plants against plant pathogens and predation [2, 30–32]. In induced phytoextraction, metal uptake ability of high biomass common crop plants are exploited by the application of chelants in the soil contaminated with the metal species of low bioavailability such as, Pb, Cr, U, Hg [18, 33]. Chelant substances exclusively participate as the mobilizing agent in such chemically induced phytoextraction process [18]. Phytoextraction is usually limited to fast growing, deep-rooted plants having high shoot-root ratio and capable of yielding high biomass. Such plants exhibit high metal uptake capacity, possess tolerance to high levels of metal contamination and propagate easily to accumulate target metals [2, 27, 34–36]. Phytoremediation is a sequential process involving – desorption of metal(s) from soil matrix, diffusion or mass flow of soluble metals to roots, uptake of metals via roots and translocation to shoots [25].

Although about 400 species of hyperaccumulating plants have been identified, their metal removal competences are somehow limited by the slow growth and inadequate biomass production. Hyperaccumulators, in spite of low biomass, possesses much greater metal-accumulation capacity than the non-hyperaccumulators provided that the metal is available in uptake-ready form [2, 37–39].

The determining factor in the successful phytoremediation is the ability to cultivate a high biomass yielding hyperaccumulators in soil with high metal content [2]. Genetic modification of some plants to enhance their phytoextraction capabilities has been tried. Such modifications aimed at converting slow-growing, low-biomass hyperaccumulators into fast-growing, high-biomass species [40]. Enhanced tolerance to metals was observed in transgenic plants containing gene related to phytochelatins (metal inducible cysteine-rich peptides, a natural plant internal chelant for heavy metal detoxification) [37, 41, 42]. However, public concern about the introduction of transgenic plants turned out to be the main obstacle to allow such approach. Instead of gene-modification approaches, the conventional plant breeding approaches with the emphasis on the search of natural hyperaccumulators are opt to be accepted [43, 44]. The most feasible strategy to develop a low-cost, environment-friendly remediation process lies in the enhancement of metal accumulation in the existing high yielding crop plants without diminishing their yield. In the purview of this strategy, chelant-induced phytoextraction has been proposed in order to overcome the low phytoavailability of heavy metals in the contaminated source [25].

Chelant-Assisted Phytoextraction of Metals

Major barrier in the efficiency of phytoextraction process is the accessibility of heavy metals to plant roots [45] owing mainly to the innate bioavailability of heavy metal species. Soil pH, soil cation exchange capacity, soil organic matter content, the ionic strength of soil solution, and the speciation of the metals control the availability and transportation of metals from soil to plant roots [46–52]. High percentage of 'non-labile' constituents in the metal contaminants of soil reduces their bioavailability to plant [25]. Based on the bioavailability of metals for plant uptake, Alkorta et al. [53] proposed three categories of – *readily bioavailable* (*e.g.*, Cd, Ni, Zn, As, Se, Cu), *moderately bioavailable* (*e.g.*, Co, Mn, Fe), and *hardly bioavailable* (*e.g.*, Pb, Cr, U) metals.

Chelants, when added, form soluble complexes with both 'labile' and 'non-labile' metals in the soil solution via desorption of sorbed species and dissolution of precipitated compounds [54]. Formation of chelant-metal complexes hinders re-precipitation and re-sorption of metals and the metals become bioavailable [28]. Addition of certain chelants increase the mobility of metals from soil to plants to the maximum and helps plant to overcome the bioavailability barrier against phytoextraction rendering phytoremediation technologically feasible [55].

However, the properties of soil in question and the nature of the chelant applied determine the amount of bioavailable metals in soil matrix [56–58]. Stability constants, K_s, of chelant-metal complexes are the decisive feature in selecting a chelant or rank different chelants for chelant assisted phytoremediation. The chemical characteristics of the chelant itself and the metal speciation in the soil matrix also influence the effectiveness of different chelant in the separation process [58–60].

A wide range of synthetic chelants have been tested for chelant-induced phytoextraction. Aminopolycarboxylate chelants (APCs) are the most commonly used chelants. Ethylenediaminetetraacetic acid (EDTA), diethylenetriaminepentaacetic acid (DTPA), N–(hydroxyethyl)ethylenediaminetriacetic acid (HEDTA), nitrilotriacetic acid (NTA), Ethylenediamine–N,N'–disuccinic acid (EDDS), methylglycinediacetic acid (MGDA), glutamic acid diacetic acid (GLDA) are examples of typically used APCs for metal phytoextraction. In particular, EDTA is the most widely used APC owing to its ability to form strong water-soluble chelant complexes with most toxic metals [18, 61, 62]. The efficiency of the phytoextraction process depends in large part on how the chelant is applied to the contaminated soil. The chelant is applied to the soil matrix either in a single dose after the optimum growth of the accumulator crop, or in small multiple doses gradually during the growth cycle [63]. Phytoextraction efficiency is reportedly improved by the combined application of different chelants to the metal-contaminated soil. A summary of the different combination recipe is available from Leštan et al. [18]. In combined application approach, two different chelants are assorted together to increase the metal solubility either by lowering the soil pH, or by differential interaction between metals and chelants where one chelant minimizes the effect from other metals in soil to increase the solubility of the target metals by the other chelant. The utilization of one chemical to destroy the plant root formation to assist the direct uptake of metal-chelants followed by the translocation into the shoots is another approach tested [18]. A hypothetical protocol for the chelant-induced phytoextraction based on the remarks of Salt et al. [26], Alkorta et al. [53] and McGrath et al. [64] is shown schematically in Figure 1. The proper selection of plant species, the precise choice of the chelants as well as the determination of the most-suitable application combination is necessary to achieve the optimum separation effectiveness from phyto-treatment of a particular metal-contaminated site.

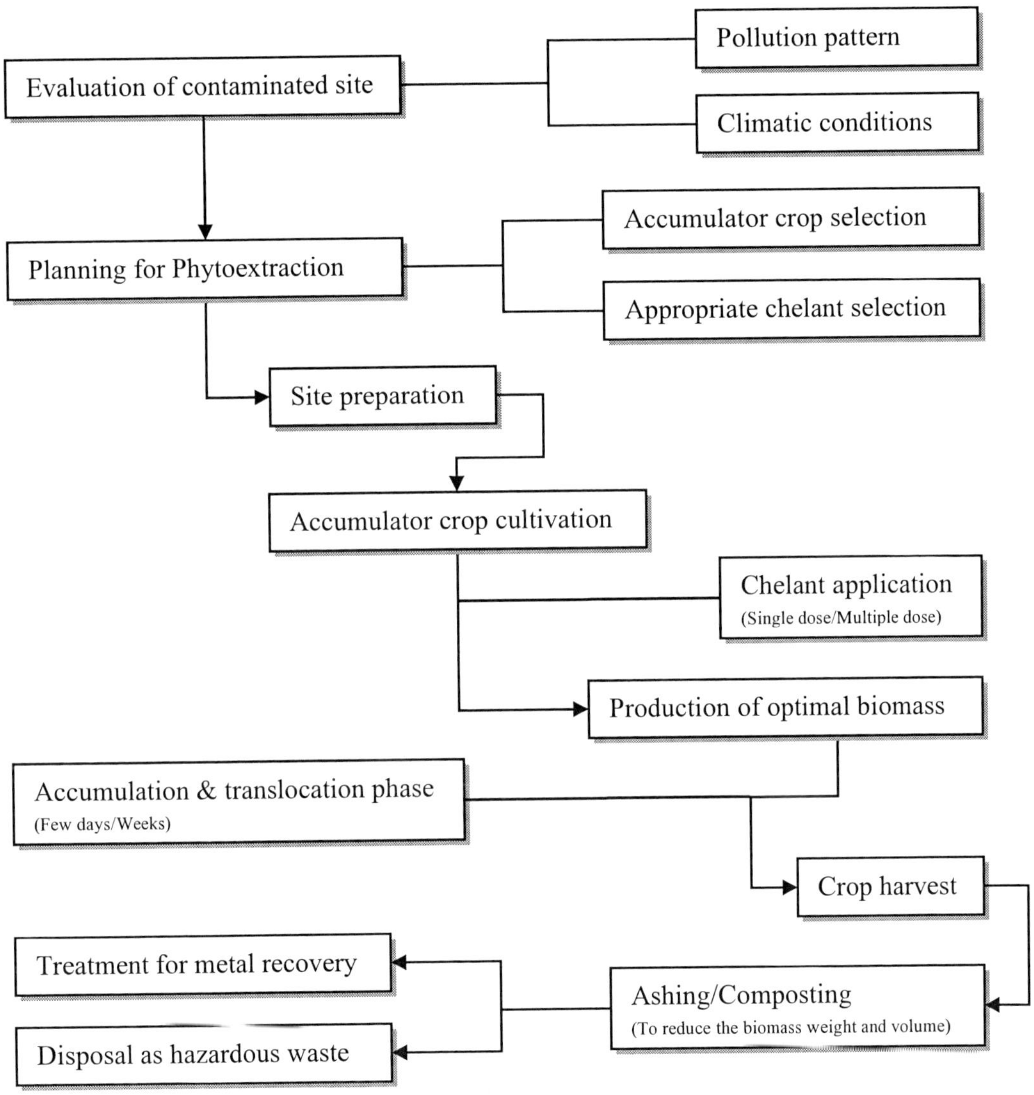

Figure 1. Schematic protocol for chelant-assisted phytoextraction.

Eco-Environmental Consequences

Chelants can have multifarious effects from ecological and environmental points of view. They bring about changes in the soil chemical properties, affect the soil micro-flora and micro-fauna and subsequently affect plants and its ecosphere. Ultimately, other components of the ecosphere including animals depending on plants get affected. Plants, when exposed to chelants, suffer from foliar necrosis, leaf wilt and abscission, shoot desiccation, reduced biomass resulting in stunted growth and in some cases plant death [25, 34, 65]. Foliar necrosis in plants is augmented by the presence of free protonated chelant in leaves [66]. Leaf wilt followed by abscission, shoots desiccation and reduced biomass reduce the net metal removal by plants [66, 67]. Phyto-toxicity leading to immature plant death is caused by the excess concentrations of soluble metal-chelant complexes and/or free chelants. Excess remainings of bioavailable metal in the soil matrix as metal-chelant complexes also reduce the plant growth during the follow-up cultivations in successive crop rotations [25, 65].

The soil ecosystem, which includes soil microbes and soil biota, is influenced when chelants are added to persuade the phytoextraction process. The corollary is exerted from the metal-ligand complexes, free chelates and existing plants collectively [25, 34, 68, 69].

A competition arises between the essential plant nutrient elements (*e.g.*, Ca^{2+}, Fe^{3+}, Mg^{2+}, Al^{3+}) and the toxic heavy metals for the binding sites of applied chelants which necessitates the application of an excess amount of chelant to ensure the adequate desorption of toxic metal-contaminants from soil [18]. This may lead to the leaching of essential plant nutrients (*e.g.*, Fe, Ca and Mg) together with the leaching of heavy metals [25].

In the chelant-induced phyto-removal process, the concentrations of bioavailable metals in the soil profile usually become too high to be accumulated and sequestered through a single-season cropping. This leads to the leaching of soluble metal-chelant complexes to the surrounding ecosystems with time [25].

Wherever chelants (*e.g.*, APCs) have been introduced into the natural environment, the aqueous transport of metals in the form of stable complexes with chelants has occurred [62, 70]. Such movement of metals extends the residence time of the metals and as they are remobilized from soils to get translocated into the surface and ground water systems - secondary pollution becomes inevitable [48, 54, 71, 72]. Excess chelant also increases the total nitrogen content and the phosphate solubility in the receiving water bodies [54, 71, 73–75]. Such escalated bioavailability of nitrogen and phosphate supposed to enhance the eutrophication in interstitial water by stimulating algal growth [72].

Free-form of chelants, particularly APCs, exhibit poor photo–, chemo– and biodegradability in the environment [61, 70, 76–80] and, in most cases, metal complexation raises the threshold values for toxic effects of metals [72, 81, 82]. Metal-complexes with varying reactivity may form depending on the metals and speciation of chelating agents which is an important factor to determine their eco-environmental fate [76, 83].

Chelant materials are non-toxic to many forms of life in terms of acute toxicity while chronic toxicity effects are unknown [84, 85]. EDTA, which is widely used among the chelants, have low toxicity profile for human health as concluded from the extensive evaluation of different legislative bodies [86]. However, excretion of metals and cell membrane permeability in mammals is affected at extreme ingestion of EDTA [87, 88].

Addressing Eco-Environmental Issues Associated with Chelants

In order to minimize the related environmental threats and considering the after-treatment providence of the metal-loaded plants, the technology is suggested to combine with a profit-making operation such as forestry and bio-energy production [36]. Alternative chelant formulations which combines high biodegradability with good chelating strength are also suggested [52, 68, 69]. More innovative agronomic practices may also minimize the associated risk factors [20]. However, apart from all the proposed alternatives and suggestions, more field trial is required to understand the real potential and related risks of this technology.

Mode of application of chelant is also crucial to the chelant induced phytotoxicity. In a single (0.5 g kg^{-1}) and split (0.25 + 0.25 g kg^{-1}) EDTA application experiment, Gabos et al. [89] showed enhanced vertical movement of lead through the soil mostly for the single EDTA application whereas the split EDTA application decreased the metal leaching and minimized environmental contamination from excess Pb.

Cr and Ni bioremediation using citric acid and histidine as biodegradable chelants was effective over EDTA based phytoremediation. At a concentration of 0.05 mol L^{-1}, citric acid allowed better Cr mobilization keeping the soil mineral integrity intact [90, 91]. NTA is another biodegradable chelant [92, 93] that augments Pb bioremediation with reduced environmental consequences compared to non biodegradable EDTA, though there is concern about the potential carcinogenicity of NTA [94]. EDDS is also a candidate biodegradable chelant which enhances Pb mobilization by complexation [95] and increase solubilization of Cu [96]. Organic additives to EDTA, such as citric acid, acetic acid, oxalic acid resulted in better phytoremediation of Zn and Cd and minimized the risk of leaching of liberated metals to underground water [97]. If non-biodegradable EDTA is used for chelation, the EDTA-metal complexes can be broken down by using bacterial digestion of such complexes [98], or by electrochemical advanced oxidation process as reported in case of Cu bioremediation [99].

One approach of combating the problems associated with the release of remediated metals back to the environment is to use phytoextraction as *phytomining* i.e. to use plants to accumulate metals into their tissue followed by purification of those metals after harvesting the trees [100–102]. Chelating agents enhances the metal bioavailability, and excess liberated metals create eco-environmental problems if unabsorbed by plants. But this excess metals can be absorbed by hyperaccumulators and then extracted as bio-ore or pure metal from their tissue. Phytomining has so far been reported for Ni, Ti, Au, Ag, Co by different plant species [100, 103–107]. Such approach is better in addressing chelant mediated phytoremediation as it helps in metal recycling and converts waste into wealth.

Better understanding of the molecular mechanism associated with the phytoextraction and plant-chelant interactions will help us in making the process more efficient in eco-environmental terms. Bohuslavek **et** al. [108] reported the genes responsible for the initial steps of EDTA and NTA degradation. Better understanding of the genes and proteins involved in phytoextraction along with the signaling pathways that control metal and chelant processing in plant cells will make us more capable of taking advantage of inherent genetic machinery of the plant hyperaccumulators in phytoremediation of heavy metals from soil with minimal effect on the environment from chelating agents.

Conclusion

Despite the eco-environmental consequences, phyto-treatment of metal contaminated soil is, so far, the most commercially viable option for reclaiming the heavy metal contaminated soil as it offers approximately one order of magnitude lower operational cost compared to the conventional, predominantly chemical, approaches. However, there are eco-environmental concerns regarding the release of high amount of soluble metal species into the soil by chelants and their subsequent leaching into other systems including the underground water. Persistency of chelants in the remediated soil is another concern. These concerns are fairly addressable by a number of approaches i.e. microbial, chemical or electrochemical digestions, phytomining based extraction or use of biodegradable chelating agents *etc.* During the phytoremediation of a metal-contaminated site, the eventual behaviour of a chelant or a chelant-metal complex depends on the extent of metal concentration and possible interaction between metals and chelants. Extent of metal contamination, metal bioavailability, plant uptake pattern, toxicity profile, dosing strategy of chelant, transport and translocation of metals and metal chelant complexes are some factors which require considerable attention to achieve maximum net metal removal from the plant-based separation process in an eco-environment friendly manner. In any scheme of using phytoremediation for cleaning up contaminated soils, eco-environmental consequences should be given the highest priority to make sure that solving of one problem does not end up in creating another.

References

[1] Xia, W.-b.; Li, X.; Gao, H.; Huang, B.-r.; Zhang, H.-z.; Liu, Y.-g.; Zeng, G.-m.; Fan, T., Influence factors analysis of removing heavy metals from multiple metal-contaminated soils with different extractants. *J. Cent. South Univ. T.* 2009, *16*, 108–111.

[2] Evangelou, M. W. H.; Ebel, M.; Schaeffer, A., Chelate assisted phytoextraction of heavy metals from soil. Effect, mechanism, toxicity, and fate of chelating agents. *Chemosphere* 2007, *68*, 989–1003.

[3] Raskin, I.; Kumar, P.; Dushenkov, S.; Salt, D., Bioconcentration of heavy metals by plants. *Curr. Opin. Biotech.* 1994, *5*, 285–290.

[4] Romo-Kröger, C. M.; Morales, J. R.; Dinator, M. I.; Llona, F.; Eaton, L. C., Heavy metals in the atmosphere coming from a copper smelter in Chile. *Atmos. Environ.* 1994, *28*, 705–711.

[5] Steinnes, E.; Allen, R. O.; Petersen, H. M.; Rambæk, J. P.; Varskog, P., Evidence of large scale heavy-metal contamination of natural surface soils in Norway from long-range atmospheric transport. *Sci. Total Environ.* 1997, *205*, 255–266.

[6] Tong, S. T. Y.; Lam, K. C., Home sweet home? A case study of household dust contamination in Hong Kong. *Sci. Total Environ.* 2000, *256*, 115–123.

[7] Martley, E.; Gulson, B. L.; Pfeifer, H. R., Metal concentrations in soils around the copper smelter and surrounding industrial complex of Port Kembla, NSW, Australia. *Sci. Total Environ.* 2004, *325*, 113–127.

[8] Doumett, S.; Lamperi, L.; Checchini, L.; Azzarello, E.; Mugnai, S.; Mancuso, S.; Petruzzelli, G.; Del Bubba, M., Heavy metal distribution between contaminated soil and *Paulownia tomentosa*, in a pilot-scale assisted phytoremediation study: influence of different complexing agents. *Chemosphere* 2008, *72*, 1481–1490.

[9] Schutzendubel, A.; Polle, A., Plant responses to abiotic stresses: heavy metal-induced oxidative stress and protection by mycorrhization. *J. Exp. Bot.* 2002, *53*, 1351–1365.

[10] Wagner, G. J., Accumulation of cadmium in crop plants and its consequences to human health. *Adv. Agron.* 1993, *51*, 173–212.

[11] Gaetke, L. M.; Chow, C. K., Copper toxicity, oxidative stress, and antioxidant nutrients. *Toxicology* 2003, *189*, 147–163.

[12] Hernández-Ochoa, I.; García-Vargas, G.; López-Carrillo, L.; Rubio-Andrade, M.; Morán-Martínez, J.; Cebrián, M. E.; Quintanilla-Vega, B., Low lead environmental exposure alters semen quality and sperm chromatin condensation in northern Mexico. *Reprod. Toxicol.* 2005, *20*, 221–228.

[13] Bodar, C. W. M.; Pronk, M. E. J.; Sijm, D. T. H. M., The European Union risk assessment on zinc and zinc compounds: the process and the facts. *Integr. Environ. Assess. Manag.* 2005, *1*, 301–319.

[14] EC, On the protection of the environment, and in particular of the soil, when sewage sludge is used in agriculture. In *Council Directive 86/278/EEC, In: EC Official Journal L181*, EC, Ed. European Community: Brussels, Belgium, 1986.

[15] Grčman, H.; Velikonja-Bolta, Š.; Vodnik, D.; Kos, B.; Leštan, D., EDTA enhanced heavy metal phytoextraction: metal accumulation, leaching and toxicity. *Plant Soil* 2001, *235*, 105–114.

[16] Flathman, P. E.; Lanza, G. R., Phytoremediation: current views on an emerging green technology. *J. Soil Contam.* 1998, *7*, 415–432.

[17] EC, Communication from the Commission to the Council, the European Parliament, the European Economic and Social Committee and the Committee of the Regions: Towards a Thematic Strategy on the Urban Environment. In *COM(2004)60 final*, EC, Ed. Commission of the European Communities: Brussels, Belgium, 2004.

[18] Leštan, D.; Luo, C. L.; Li, X. D., The use of chelating agents in the remediation of metal-contaminated soils: a review. *Environ. Pollut.* 2008, *153*, 3–13.

[19] Baker, A. J. M.; Brooks, R. R., Terrestrial higher plants which hyperaccumulate metallic elements – a review of their distribution, ecology and phytochemistry. *Biorecovery* 1989, *1*, 81–126.

[20] Lasat, M. M., Phytoextraction of toxic metals: a review of biological mechanisms. *J. Environ. Qual.* 2002, *31*, 109–120.

[21] Garbisu, C.; Alkorta, I., Phytoextraction: a cost-effective plant-based technology for the removal of metals from the environment. *Bioresource Technol.* 2001, *77*, 229–236.

[22] Meagher, R. B., Phytoremediation of toxic elemental and organic pollutants. *Curr. Opin. Plant Biol.* 2000, *3*, 153–162.

[23] Susarla, S.; Medina, V. F.; McCutcheon, S. C., Phytoremediation: an ecological solution to organic chemical contamination. *Ecol. Eng.* 2002, *18*, 647–658.

[24] EPA *Introduction to Phytoremediation (EPA/600/R-99/107)*; National Risk Management Research Laboratory, Office of Research and Development, U.S. Environmental Protection Agency: Cincinnati, Ohio, 2000.

[25] Song, J.; Luo, Y. M.; Wu, L. H., Chelate-enhanced phytoremediation of heavy metal contaminated soil. In *Biogeochemistry of Chelating Agents*, Nowack, B.; VanBriesen, J. M., Eds. American Chemical Society: Washington, DC, 2005; pp 366–382.

[26] Salt, D. E.; Smith, R. D.; Raskin, I., Phytoremediation. *Annu. Rev. Plant Phys.* 1998, *49*, 643–668.

[27] Garbisu, C.; Allica, J. H.; Barrutia, O.; Alkorta, I.; Becerril, J. M., Phytoremediation: a technology using green plants to remove contaminants from polluted areas. *Rev. Environ. Health* 2002, *17*, 173–188.

[28] Salt, D. E.; Blaylock, M.; Kumar, N. P. B. A.; Dushenkov, V.; Ensley, B. D.; Chet, I.; Raskin, I., Phytoremediation: a novel strategy for the removal of toxic metals from the environment using plants. *Nat. Biotechnol.* 1995, *13*, 468–474.

[29] Memon, A. R.; Schröder, P., Implications of metal accumulation mechanisms to phytoremediation. *Environ. Sci. Pollut. R.* 2009, *16*, 162–175.

[30] Maywald, F.; Weigel, H. J., Biochemistry and molecular biology of heavy metal accumulation in higher plants (Zur biochemie und molekularbiologie der schwermetallaufnahme und -speicherung bei höheren pflanzen). *Landbauforsch. Volk.* 1997, *47*, 103–126

[31] Boyd, R. S.; Shaw, J. J.; Martens, S. N., Nickel hyperaccumulation defends *Streptanthus polygaloides* (Brassicaceae) against pathogens. *Am. J. Bot.* 1994, *81*, 294–300.

[32] Sagner, S.; Kneer, R.; Wanner, G.; Cosson, J.-P.; Deus-Neumann, B.; Zenk, M. H., Hyperaccumulation, complexation and distribution of nickel in *Sebertia acuminata*. *Phytochemistry* 1998, *47*, 339–347.

[33] Shen, Z.-G.; Li, X.-D.; Wang, C.-C.; Chen, H.-M.; Chua, H., Lead phytoextraction from contaminated soil with high-biomass plant species. *J. Environ. Qual.* 2002, *31*, 1893–1900.

[34] Römkens, P.; Bouwman, L.; Japenga, J.; Draaisma, C., Potentials and drawbacks of chelate-enhanced phytoremediation of soils. *Environ. Pollut.* 2002, *116*, 109–121.

[35] Robinson, B.; Mills, T.; Petit, D.; Fung, L.; Green, S.; Clothier, B., Natural and induced cadmium-accumulation in poplar and willow: implications for phytoremediation. *Plant Soil* 2000, *227*, 301–306.

[36] Robinson, B.; Fernández, J.-E.; Madejón, P.; Marañón, T.; Murillo, J. M.; Green, S.; Clothier, B., Phytoextraction: an assessment of biogeochemical and economic viability. *Plant Soil* 2003, *249*, 117–125.

[37] McGrath, S. P.; Zhao, F.-J., Phytoextraction of metals and metalloids from contaminated soils. *Curr. Opin. Biotech.* 2003, *14*, 277–282.

[38] Cunningham, S. D.; Berti, W. R.; Huang, J. W., Phytoremediation of contaminated soils. *Trends Biotechnol.* 1995, *13*, 393–397.

[39] Comis, D., Green remediation: using plants to clean the soil. *J. Soil Water Conserv.* 1996, *51*, 184–187.

[40] Cunningham, S. D.; Ow, D. W., Promises and prospects of phytoremediation *Plant Physiol.* 1996, *110*, 715–719.

[41] Schäfer, H. J.; Haag-Kerwer, A.; Rausch, T., cDNA cloning and expression analysis of genes encoding GSH synthesis in roots of the heavy-metal accumulator *Brassica juncea* L.: evidence for Cd-induction of a putative mitochondrial γ-glutamylcysteine synthetase isoform. *Plant Mol. Biol.* 1998, *37*, 87–97.

[42] Liang Zhu, Y.; Pilon-Smits, E. A. H.; Jouanin, L.; Terry, N., Overexpression of glutathione synthetase in indian mustard enhances cadmium accumulation and tolerance. *Plant Physiol.* 1999, *119*, 73–80.
[43] Baker, A. J. M.; Whiting, S. N., In search of the holy grail: a further step in understanding metal hyperaccumulation? *New Phytol.* 2002, *155*, 1–4.
[44] Chaney, R. L.; Angle, J. S.; Broadhurst, C. L.; Peters, C. A.; Tappero, R. V.; Sparks, D. L., Improved understanding of hyperaccumulation yields commercial phytoextraction and phytomining technologies. *J. Environ. Qual.* 2007, *36*, 1429–1443.
[45] Felix, H., Field trials for *in situ* decontamination of heavy metal polluted soils using crops of metal-accumulating plants. *Z. Pflanz. Bodenkunde* 1997, *160*, 525–529.
[46] Reddy, K. J.; Wang, L.; Gloss, S. P., Solubility and mobility of copper, zinc and lead in acidic environments. *Plant Soil* 1995, *171*, 53–58.
[47] Schmidt, U., Enhancing phytoextraction: the effect of chemical soil manipulation on mobility, plant accumulation, and leaching of heavy metals. *J. Environ. Qual.* 2003, *32*, 1939–1954.
[48] Li, Z.; Shuman, L. M., Heavy metal movement in metal-contaminated soil profiles. *Soil Sci.* 1996, *161*, 656–666.
[49] Atanassova, I., Competitive effect of copper, zinc, cadmium and nickel on ion adsorption and desorption by soil clays. *Water Air Soil Poll.* 1999, *113*, 115–125.
[50] Dijkshoorn, W.; Lampe, J.; Van Broekhoven, L., The effect of soil pH and chemical form of nitrogen fertilizer on heavy-metal contents in ryegrass. *Nutr. Cycl. Agroecosys.* 1983, *4*, 63–74.
[51] Albasel, N.; Cottenie, A., Heavy metals uptake from contaminated soils as affected by peat, lime, and chelates. *Soil Sci. Soc. Am. J.* 1985, *49*, 386–390.
[52] Kulli, B.; Balmer, M.; Krebs, R.; Lothenbach, B.; Geiger, G.; Schulin, R., The influence of nitrilotriacetate on heavy metal uptake of lettuce and ryegrass. *J. Environ. Qual.* 1999, *28*, 1699–1705.
[53] Alkorta, I.; Hernández-Allica, J.; Becerril, J. M.; Amezaga, I.; Albizu, I.; Onaindia, M.; Garbisu, C., Chelate-enhanced phytoremediation of soils polluted with heavy metals. *Rev. Environ. Sci. Biotechnol.* 2004, *3*, 55–70.
[54] Norvell, W. A., Comparison of chelating agents as extractants for metals in diverse soil materials. *Soil Sci. Soc. Am. J.* 1984, *48*, 1285–1292.
[55] Blaylock, M. J.; Haung, J. W., Phytoextraction of metals. In *Phytoremediation of Toxic Metals: Using Plants to Clean Up the Environment*, Raskin, I.; Ensley, B. D., Eds. John Wiley and Sons, Inc.: New York, 1999; pp 53–70.
[56] Kos, B.; Lestan, D., Chelator induced phytoextraction and in situ soil washing of Cu. *Environ. Pollut.* 2004, *132*, 333–339.
[57] Tandy, S.; Bossart, K.; Mueller, R.; Ritschel, J.; Hauser, L.; Schulin, R.; Nowack, B., Extraction of heavy metals from soils using biodegradable chelating agents. *Environ. Sci. Technol.* 2004, *38*, 937–944.
[58] Luo, C.; Shen, Z.; Li, X., Enhanced phytoextraction of Cu, Pb, Zn and Cd with EDTA and EDDS. *Chemosphere* 2005, *59*, 1–11.
[59] Elliott, H. A.; Brown, G. A., Comparative evaluation of NTA and EDTA for extractive decontamination of Pb-polluted soils. *Water Air Soil Poll.* 1989, *45*, 361–369.

[60] Huang, J. W.; Chen, J.; Berti, W. R.; Cunningham, S. D., Phytoremediation of lead-contaminated soils: role of synthetic chelates in lead phytoextraction. *Environ. Sci. Technol.* 1997, *31*, 800–805.

[61] Egli, T., Biodegradation of metal-complexing aminopolycarboxylic acids. *J. Biosci. Bioeng.* 2001, *92*, 89–97.

[62] Nowack, B.; VanBriesen, J. M., Chelating agents in the environment. In *Biogeochemistry of Chelating Agents*, Nowack, B.; VanBriesen, J. M., Eds. American Chemical Society: Washington, DC, 2005; pp 1–18.

[63] Wenzel, W. W.; Unterbrunner, R.; Sommer, P.; Sacco, P., Chelate-assisted phytoextraction using canola (*Brassica napus* L.) in outdoors pot and lysimeter experiments. *Plant Soil* 2003, *249*, 83–96.

[64] McGrath, S. P.; Zhao, F. J.; Lombi, E., Phytoremediation of metals, metalloids, and radionuclides. *Adv. Agron.* 2002, *75*, 1–56.

[65] Lombi, E.; Zhao, F. J.; Dunham, S. J.; McGrath, S. P., Phytoremediation of heavy metal-contaminated soils: natural hyperaccumulation versus chemically enhanced phytoextraction. *J. Environ. Qual.* 2001, *30*, 1919–1926.

[66] Vassil, A. D.; Kapulnik, Y.; Raskin, I.; Salt, D. E., The role of EDTA in lead transport and accumulation by Indian mustard. *Plant Physiol.* 1998, *117*, 447–453.

[67] Chen, H.; Cutright, T., EDTA and HEDTA effects on Cd, Cr, and Ni uptake by *Helianthus annuus*. *Chemosphere* 2001, *45*, 21–28.

[68] Kos, B.; Leštan, D., Influence of a biodegradable ([S,S]-EDDS) and nondegradable (EDTA) chelate and hydrogel modified soil water sorption capacity on Pb phytoextraction and leaching. *Plant Soil* 2003, *253*, 403–411.

[69] Grčman, H.; Vodnik, D.; Velikonja-Bolta, Š.; Leštan, D., Ethylenediaminedissuccinate as a new chelate for environmentally safe enhanced lead phytoextraction. *J. Environ. Qual.* 2003, *32*, 500–506.

[70] Means, J. L.; Kucak, T.; Crerar, D. A., Relative degradation rates of NTA, EDTA and DTPA and environmental implications. *Environ. Pollut. B* 1980, *1*, 45–60.

[71] Erel, Y.; Morgan, J. J., The relationships between rock-derived lead and iron in natural waters. *Geochim. Cosmochim. Ac.* 1992, *56*, 4157–4167.

[72] Sillanpää, M. E. T., Distribution and fate of chelating agents in the environment. In *Biogeochemistry of Chelating Agents*, Nowack, B.; VanBriesen, J. M., Eds. American Chemical Society: Washington, DC, 2005; pp 226–233.

[73] Hering, J. G.; Morel, F. M. M., Kinetics of trace metal complexation: role of alkaline-earth metals. *Environ. Sci. Technol.* 2002, *22*, 1469–1478.

[74] Li, Z.; Shuman, L. M., Extractability of zinc, cadmium, and nickel in soils amended with EDTA. *Soil Sci.* 1996, *161*, 226–232.

[75] Horstmann, U.; Gelpke, N., Algal growth stimulation by chelatisation risks associated with complexants in P-free washing agents. *Rev. Intl. Oceanogr. Med.* 1991, *260*, 101–104.

[76] Nowack, B., Environmental chemistry of aminopolycarboxylate chelating agents. *Environ. Sci. Technol.* 2002, *36*, 4009–4016.

[77] Bolton Jr., H.; Li, S. W.; Workman, D. J.; Girvin, D. C., Biodegradation of synthetic chelates in subsurface sediments from the southeast coastal plain. *J. Environ. Qual.* 1993, *22*, 125–132.

[78] Kari, F. G.; Giger, W., Modeling the photochemical degradation of ethylenediaminetetraacetate in the river Glatt. *Environ. Sci. Technol.* 1995, *29*, 2814–2827.

[79] Kari, F. G.; Hilger, S.; Canonica, S., Determination of the reaction quantum yield for the photochemical degradation of Fe(III)-EDTA: implications for the environmental fate of EDTA in surface waters. *Environ. Sci. Technol.* 1995, *29*, 1008–1017.

[80] Nörtemann, B., Biodegradation of chelating agents: EDTA, DTPA, PDTA, NTA, and EDDS. In *Biogeochemistry of Chelating Agents*, Nowack, B.; VanBriesen, J. M., Eds. American Chemical Society: Washington, DC, 2005; pp 150–170.

[81] Sillanpää, M.; Oikari, A., Assessing the impact of complexation by EDTA and DTPA on heavy metal toxicity using microtox bioassay. *Chemosphere* 1996, *32*, 1485–1497.

[82] Sorvari, J.; Sillanpää, M., Influence of metal complex formation on heavy metal and free EDTA and DTPA acute toxicity determined by *Daphnia magna*. *Chemosphere* 1996, *33*, 1119–1127.

[83] Nowack, B., Chelating agents and the environment. *Environ. Pollut.* 2008, *153*, 1–2.

[84] Christian, M. S.; York, R. G.; Hoberman, A. M.; Fisher, L. C.; Brown, W. R., Oral (drinking water) two-generation reproductive toxicity study of bromodichloromethane (BDCM) in rats. *Int. J. Toxicol.* 2002, *21*, 115–146.

[85] Sillanpää, M., Environmental fate of EDTA and DTPA. *Rev. Environ. Contam. Toxicol.* 1997, *152*, 85–111.

[86] Grundler, O. J.; van der Steen, A. T. M.; Wilmot, J., Overview of the European risk assessment on EDTA. In *Biogeochemistry of Chelating Agents*, Nowack, B.; VanBriesen, J. M., Eds. American Chemical Society: Washington, DC, 2005; pp 336–347.

[87] Braide, V. B., Calcium EDTA toxicity: renal excretion of endogenous trace metals and the effect of repletion on collagen degradation in the rat. *Gen. Pharmacol.-Vasc. S.* 1984, *15*, 37–41.

[88] Hancock, R. E. W., Alterations in outer membrane permeability. *Annu. Rev. Microbiol.* 1984, *38*, 237–264.

[89] Gabos, M. B.; Abreu, C. A.; Coscione, A. R., EDTA assisted phytorremediation of a Pb contamined soil: metal leaching and uptake by jack beans. *Sci. Agri.* 2009, *66*, 506–514.

[90] Jean, L.; Bordas, F.; Gautier-Moussard, C.; Vernay, P.; Hitmi, A.; Bollinger, J.-C., Effect of citric acid and EDTA on chromium and nickel uptake and translocation by *Datura innoxia*. *Environ. Pollut.* 2008, *153*, 555–563.

[91] Jean, L.; Bordas, F.; Bollinger, J. C., Chromium and nickel mobilization from a contaminated soil using chelants. *Environ. Pollut.* 2007, *147*, 729–736.

[92] Quartacci, M. F.; Irtelli, B.; Baker, A. J. M.; Navari-Izzo, F., The use of NTA and EDDS for enhanced phytoextraction of metals from a multiply contaminated soil by *Brassica carinata*. *Chemosphere* 2007, *68*, 1920 1928.

[93] Freitas, E. V. S.; do Nascimento, C. W. A., The use of NTA for lead phytoextraction from soil from a battery recycling site. *J. Hazard. Mater.* 2009, *171*, 833–837.

[94] Peters, R. W., Chelant extraction of heavy metals from contaminated soils. *J. Hazard. Mater.* 1999, *66*, 151–210.

[95] Sarkar, D.; Andra, S. S.; Saminathan, S. K. M.; Datta, R., Chelant-aided enhancement of lead mobilization in residential soils. *Environ. Pollut.* 2008, *156*, 1139–1148.

[96] Coscione, A.; Abreu, C.; Santos, G., Chelating agents to solubilize heavy metals from oxisols contaminated by the addition of organic and inorganic residues. *Sci. Agri.* 2009, *66*, 64–70.

[97] Wu, Q.; Deng, J.; Long, X.; Morel, J.; Schwartz, C., Selection of appropriate organic additives for enhancing Zn and Cd phytoextraction by hyperaccumulators. *J. Environ. Sci.* 2006, *18*, 1113–1118.

[98] Satroutdinov, A. D.; Dedyukhina, E. G.; Chistyakova, T. I.; Witschel, M.; Minkevich, I. G.; Eroshin, V. K.; Egli, T., Degradation of metal-EDTA complexes by resting cells of the bacterial strain DSM 9103. *Environ. Sci. Technol.* 2000, *34*, 1715–1720.

[99] Pociecha, M.; Lestan, D., EDTA leaching of Cu contaminated soil using electrochemical treatment of the washing solution. *J. Hazard. Mater.* 2009, *165*, 533–539.

[100] Sheoran, V.; Sheoran, A. S.; Poonia, P., Phytomining: a review. *Miner. Eng.* 2009, *22*, 1007–1019.

[101] Gleba, D.; Borisjuk, N.; Borisjuk, L.; Kneer, R.; Poulev, A.; Skarzhinskaya, M.; Dushenkov, S.; Logendra, S.; Gleba, Y.; Raskin, I., Use of plant roots for phytoremediation and molecular farming. *P. Natl. Acad. Sci. USA* 1999, *96*, 5973–5977.

[102] Brooks, R.; Chambers, M.; Nicks, L.; Robinson, B., Phytomining. *Trends Plant Sci.* 1998, *3*, 359–362.

[103] Leblanc, M.; Petit, D.; Deram, A.; Robinson, B.; Brooks, R., The phytomining and environmental significance of hyperaccumulation of thallium by *Iberis intermedia* from southern France. *Econ. Geol.* 1999, *94*, 109–114.

[104] Msuya, F.; Brooks, R.; Anderson, C.; Substrate, A.; Experiments, P., Chemically-induced uptake of gold by root crops: its significance for phytomining. *Gold Bull.* 2000, *33*, 134–137.

[105] Robinson, B.; Chiarucci, A.; Brooks, R.; Petit, D.; Kirkman, J.; Gregg, P.; De Dominicis, V., The nickel hyperaccumulator plant *Alyssum bertolonii* as a potential agent for phytoremediation and phytomining of nickel. *J. Geochem. Explor.* 1997, *59*, 75–86.

[106] Anderson, C.; Brooks, R.; Stewart, R.; Simcock, R., Harvesting a crop of gold in plants. *Nature* 1998, *395*, 553–554.

[107] Boominathan, R.; Saha-Chaudhury, N.; Sahajwalla, V.; Doran, P., Production of nickel bio-ore from hyperaccumulator plant biomass: applications in phytomining. *Biotechnol. Bioeng.* 2004, *86*, 243–250.

[108] Bohuslavek, J.; Payne, J. W.; Liu, Y.; Bolton Jr, H.; Xun, L., Cloning, sequencing, and characterization of a gene cluster involved in EDTA degradation from the bacterium BNC1. *Appl. Environ. Microb.* 2001, *67*, 688–688.

In: Handbook of Phytoremediation
Editor: Ivan A. Golubev

ISBN: 978-1-61728-753-4

Chapter 26

Salt Marshes: An Interesting Ecosystem to Study Phytoremediation

I. Caçador and B. Duarte
Institute of Oceanography, Faculty of Sciences, University of Lisbon,
Campo Grande, 1749-016 Lisbon, Portugal

1. Salt Marshes and Metal Retention

Salt marshes located in estuaries frequently receive large inputs of nutrients (Caçador, et al., 1993; Tobias et al., 2001), and also of particulate and dissolved organic matter.

Salt marsh plants retain suspended particles and associated anthropogenic metals transported by the tides.

This high nutrient input makes salt marsh one of the most productive ecosystems of the planet. In highly industrialized estuaries, along with this nutrient input there is also a large input of heavy metals (Figure 1) which will be accumulated in salt marsh sediments (Caçador at al., 1996; Doyle and Otte, 1997).

These high inputs make salt marshes key zones for the biogeochemistry of the estuary, but also for metal cycling (Caçador et al., 2000).

When accumulated in salt marsh sediments, metals can become adsorbed to the sediment constituents, uptaken by plant roots and translocated to their aboveground organs. With this a new cycle is initiated: plants uptake metals, translocate them, eventually decay and die, remobilizing the metals in the sediments in a new speciation process with microbial intervention.

This natural phytoremediation processes are very important to understand and these ecosystems are ideal for these studies, allowing a better comprehension of the mechanisms in a natural situation for further application in manipulated phytoremediation trials.

This will be the major focus of this chapter, targeting all the steps of the phytoremediative processes occurring in salt marshes.

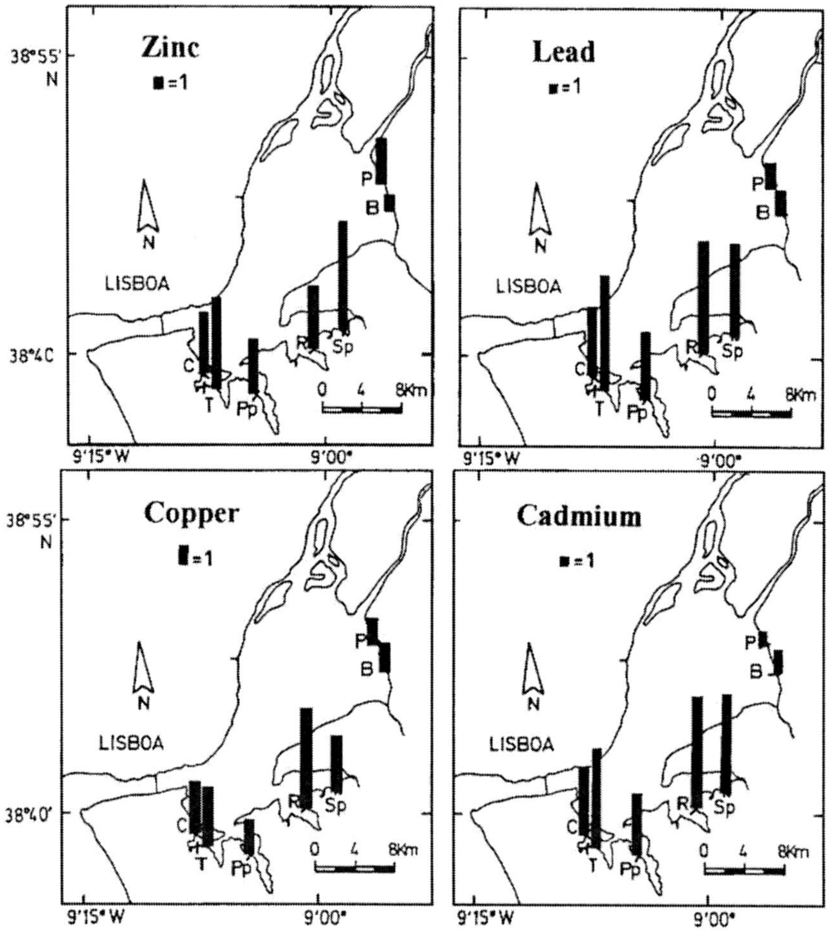

Figure 1. Heavy metal enrichment factors (0-5 cm): heavy metal concentration)(45-55 cm) in μg μg^{-1}, for Zn, Pb, Cu, and Cd in several Tagus estuary salt marshes: Pancas (P), Barco (B), Sarilhos Pequenos (S), Rosário (R), Paio Pires (Pp), Talaminho (T), and Corroios (C).

2. Different Halophytes, Different Influences

Two of the most important and abundant species present in Tagus salt marshes are Halimione portulacoides (Chenopodiaceae family) and Spartina maritima (Poaceae family). Previous studies (Caçador et al., 2000; Padinha et al., 2000) showed that the major pool of metals in the salt marsh is the sediment and the major living pool is the root tissue of the halophytes. Although this general compartimentation, there are important differences among the halophytes. Given the effect of biomass, pools of Cu, Cd and Pb in the stems and leaves of H. portulacoides often are significantly higher when compared to the same parts of S. maritima (Figure 2). Other studies also found lower values in Spartina alterniflora when compared with Phragmites australis (Windham et al., 2003). As a result of this distribution, the overall amount of metals in above ground parts of plants from those areas colonised by H. portulacoides was significantly higher than in areas colonized by S. maritima (Reboreda and Caçador, 2007a).

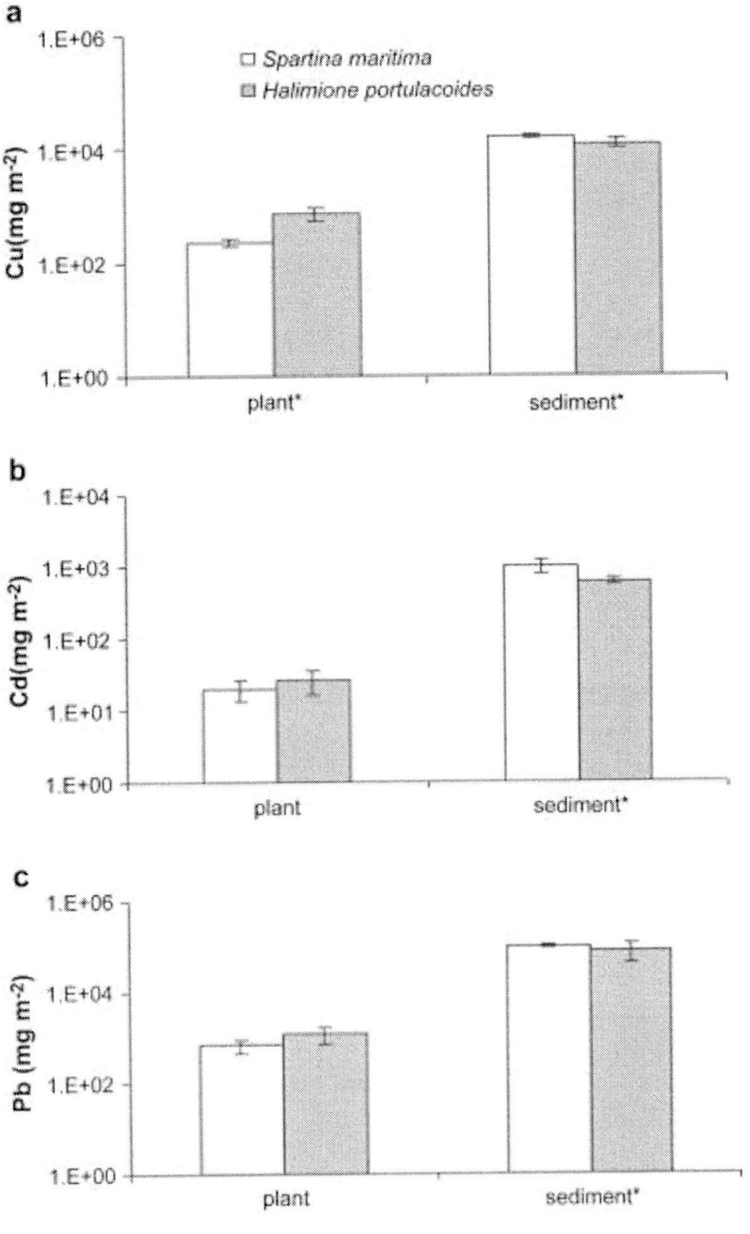

Figure 2. Cu, Cd and Pb content (mean ± SD) in plants (belowground + aboveground) and sediment (sediment + pore water) (mg m^{-2}) in a 0–40 cm depth layer in areas colonized by Halimione portulacoides and Spartina maritima. (in Reboreda and Caçador, 2007a).

Althought for phytoremediation proposes the more important sink to be considered and studied is the aboveground organs of the plants, the accumulation in belowground tissues contributes more effectively to the phytostabilization of the metals in the salt marsh and its function as sink, because they become less bioavailable (Caçador et al., 1996a). Doyle and Otte (1997) demonstrated that the soil pools of Fe, Zn and As associated to the sum of rhizosphere (sediments between the roots), iron plaque (metal iron oxides precipitated on the root surface) and roots of Spartina townsendii and Halimione portulacoides could be substantial. Thus, not only the accumulation of metals in roots, but also the induced accumulation in the surrounding sediment plays an important role in the retention of metals, reducing their availability to the ecosystem. Therefore, those areas dominated by S. maritima seem to be more effective sinks, at least for Cu and Cd, than areas dominated by H. portulacoides (Reboreda and Caçador, 2007a). To understand the role of both species on the stabilisation of metals in the salt marsh we need to consider not only the amount of metal in the sediment, but also the chemical form of the metal, i.e., availability for the plant uptake (Otte et al., 1993). It is well known that under reductive conditions sulphates are reduced to sulphides by soil microorganisms inducing the formation of insoluble metal sulphides that accumulate in the sediment (Williams et al., 1994). The capacity of halophyte plants to influence the concentration and speciation of metals in the sediment within the roots is well documented (Almeida et al., 2004; Caçador et al., 1996b; Doyle and Otte, 1997; Lacerda et al., 1997; Otero and Macıas, 2002b; Sundby et al., 2003). It is also important to account for the different position of the two species in the salt marsh. Halimione portulacoides occurs in the middle marsh and S. maritima in the low marsh. This fact is important because it enhances the differences between the chemical characteristics of the sediment, related to different submersion times (Reboredo, 1992; Williams et al., 1994).

3. The Halophyte Influence on Metal Speciation

For this chemical assurance of the bounds established between heavy metals and sediment matrix, studies on metal speciation become very important. Several protocols for determining metal bounds in other sediment constituents have been developed (Tessier et al., 1979, BCR). When using these protocols several metal fractions are defined according to the availability and strength of the established bounds, from the weaker and more bioavailable fractions to the tightly bound metals and almost unavailable fraction. As these bounds are highly dependent on the sediment matrix and its physic-chemical characteristics, it leads to major differences between the sediments of different colonizing marsh species. Chemical associations of Cu, Zn and Pb to the operationally defined fractions 'exchangeable', 'carbonates', 'Mn oxides', 'organic complexes', 'amorphous Fe oxides', 'crystalline Fe oxides' and 'residual' in sediments between the roots of S. maritima and H. portulacoides are shown in Figure 2. The higher redox potential observed in areas colonised by H. portulacoides may, in part, explain the observed differences in the speciation of Zn and Pb. Mortimer and Rae (2000) concluded that redox conditions in the salt marsh controlled the iron cycling (iron and manganese oxides) and this, in turn, controlled trace metal association. Our results also show a higher association of metals to Fe and Mn oxides in sediments with a higher redox potential, i.e., those colonized by H. portulacoides. Otero et al. (2000) also found different chemical associations of metals in salt marsh sediments under different redox conditions. Differences in redox characteristics of sediments between the roots of both species might be explained by two different reasons. On one hand, the different position of these species on the salt marsh contributes to variations in chemical and physical properties between their roots, such as changes in the position of the redoxcline (limit between oxic and anoxic zone) due to different tidal immersion times. S. maritima is situated in the lower salt marsh and therefore immersion time of the sediment is higher than in H. portulacoides, limiting oxygen diffusion from the atmosphere. H. portulacoides establishes in this well aerated sediments because as a dicotyledon it has a poorly developed arenchyma, limiting the transport of oxygen from aerial parts to roots (Ingold and Havill, 1984; Rozema et al., 1985; Sanchez et al., 1998). S. maritima has a well developed arenchyma and therefore higher capacity to oxidize the rhizosphere, being able to colonize anoxic environments. On the other hand, sediment redox potential may be influenced by differences on the amount and chemical composition of root exudates of the two species. The considerable differences observed in Zn and Pb fractionation between sediments colonized by S. maritima and H. portulacoides could indicate that both species may influence the fractionation of these metals in different ways, which in addition seemed to be metal specific.

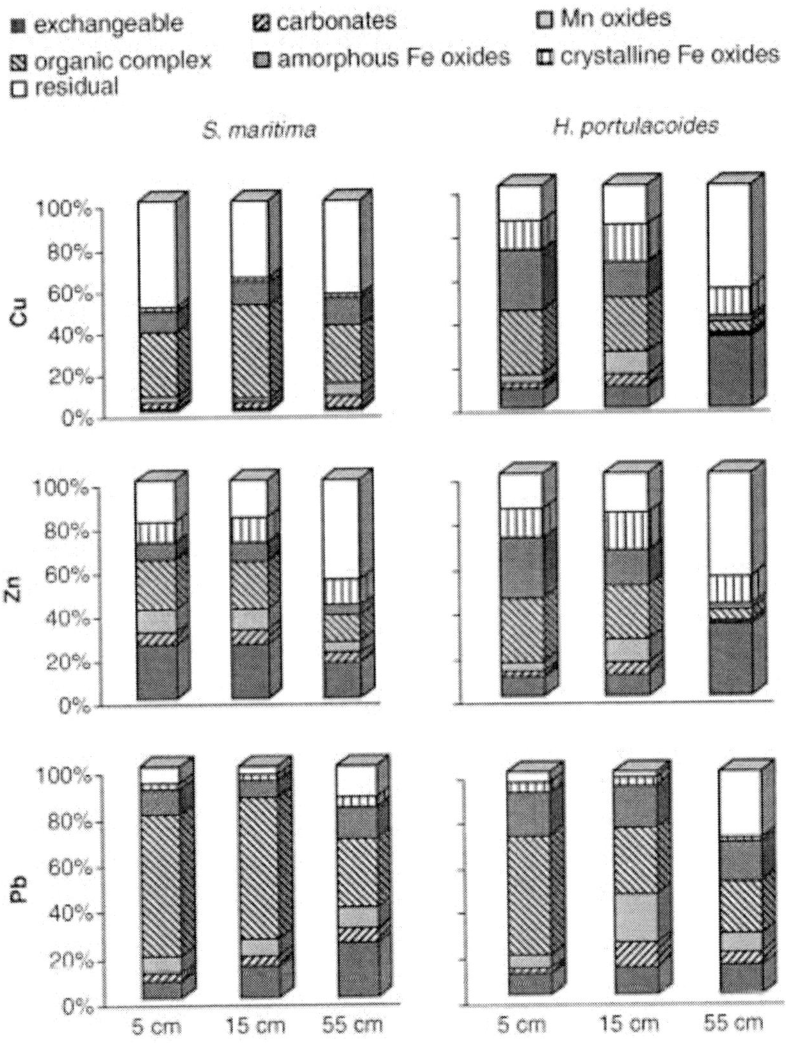

Figure 2. Chemical associations of Cu, Zn and Pb at surface, root layer and deep layer in sediments colonised by Spartina maritima and Halimione portulacoides. (in Reboreda and Caçador, 2007)

Plant roots can synthesize, accumulate and secrete many compounds (Walker et al., 2003), such as organic acids (Jócsák et al., 2005) that will modify the soil and metals physical and chemical characteristics (Nigam et al., 2000), participating in nutrient uptake and in heavy metal detoxification (Ma et al., 2001). It is known that some of these organic compounds exudated by roots are acids such as citric, malic, oxalic and succinic acids (Dakora and Phillips, 2002). A previous study (Duarte et al., 2007) focused on the role of a specific organic acid (citric acid) detected on root exudates from H. portulacoides individuals when exposed to cadmium and nickel. There are also bibliographic references that indicate the increase of the citric acid among some salt marsh plants, such as Juncus maritimus (Mucha et al., 2005). While for nickel the application of this chelator showed no advantage to phytoremediative processes, it was very useful for Cd treatment. In the presence of citric acid, Cd uptake was enhanced and, most importantly, the translocation to the higher parts of the plant also increased, when citric acid was applied at a concentration similar to the one detected on root exudates (Figure 3).

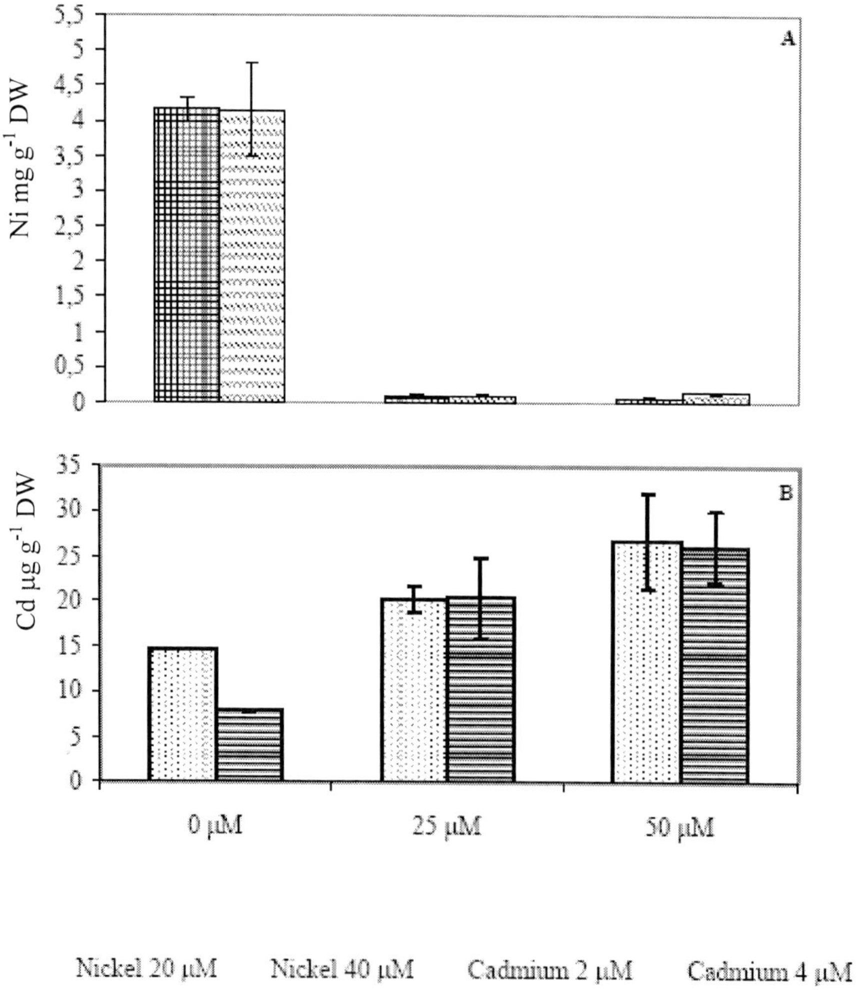

Figure 3. Nickel and cadmium root concentrations conjugated with different citric acid treatments (0 mM, 25 mM and 50 mM) and metal additions (from Duarte et al., 2007).

4. SALT MARSH METAL CYCLING

All these aspects are important to consider when analyzing a plant community with such a high degree of complexity for understanding their role in natural phytoremediation processes. For a global vision of the salt marsh role in phytoremediation proposes all these matrixes (aerial organs, roots, sediments) have to be considered along a complex temporal and environmental gradient. The pools of metals bioaccumulated by the four more abundant species (H. portulacoides, Sarcocornia fruticosa, Sarcocornia perennis and S. maritima) were found to be much lower in the aboveground organs, as seen by the lower translocation rates and by the metal concentrations, especially for Pb, probably due to its low solubility in less oxic environments (Sundby et al., 2005). The higher metal budgets in roots corroborate their increased ability to heavy metal accumulation. The calculated root decomposition rates suggest that these metal pools are quite mobile in particular in S. fruticosa and in S. maritima. This mobility is very important since creates a cycle of metals between the sediment and the root system. Although it was found a higher fraction of biomass losses in the aboveground organs, their low metal concentration makes the detritus generated by the aboveground less contributing for the metal budgets. Conversely, the comparatively low losses of biomass in the root system generate less necromass, but with very high concentrations of metals. This necromass becomes important to de metal budget of the sediment, not only due to its input of heavy metals, but also due to the increase of organic matter content of this matrix. Therefore, the metal cycling between the two major sinks (roots and sediments) was also evaluated throughout the root turnovers and cycling coefficients. The turnover values were above 0.5, being the highest detected in S. maritima, indicating that the cycling of metals between roots and sediments occurs at a rather slow rhythm. Sediments colonized by S. maritima, were different from those in the upper marsh where the other species are dominant. Low marsh sediments are subjected to larger periods of submersion that could have potentially prevailing anoxic conditions. These conditions are only counteracted by the active pumping of oxygen from the atmosphere to the rhizosphere, by S. maritima (Duarte et al., 2009). In these sediments the metals are highly bound to sulfides and/or organic matter, being easily stabilized in the solids (Reboreda and Caçador, 2007a). Moreover, root system of S. maritima has lower specific area than the other plant species, which leads to minor root–sediment interaction (Lee et al., 1999). Consequently, less metal sulphides are oxidised and transported towards their roots (Caetano et al., 2008). Although low metal turnover rates from root to sediment were found in the other salt marsh plants roots are the most important sink of metals in salt marshes. The bioaccumulation of these elements in the belowground organs is rather mobile, being able to return to the sediment matrix due to necromass generation and mineralization processes subdue. The return of metals due to these decaying processes and consequent input of metals into the sediments, although in rather lower concentrations comparatively to the existent in this matrix, is very important to be considered not because of the amount of metal released through this process but by the metal forms it introduces into the sediment. These organic bound metals were already reported as being one of the most important fractions of metals present in these sediments, being subjected to microbial degradation processes (Duarte et al., 2008), that can lead to more bioavailable metal forms and contamination of the pore waters. The metal budgets can be very important when assessing remediation projects. Both the high root turnovers and cycling coefficients verified in S. maritima for the majority of the analyzed of metals can be useful for understanding metal cycling. The low cycling of the elements from sediment to plant clearly points out that this specie act as phytostabilizer. For phytoremediation proposes it should be considered the use of species with low root turnovers and low cycling coefficients. Coupled with both these characteristics, the low root necromass generation makes S. perennis the more suitable of the studied species, for remediation projects. From the ecosystem point of view, these findings help understanding the role of plant detritus in the trophic transport of metals and also the processes involved in the contamination of a salt marsh and the adjoining estuarine areas. This should be regarded as a starting point to better understand the metals cycling within the salt marsh, not only the flux between plant and sediment, but also through the trophic chain. Furthermore, the results presented here support the observations from previous studies that point out salt marshes as a sink of heavy metals, clarifying the major cycles of metals between the two more important metal retaining matrixes (Figure 4).

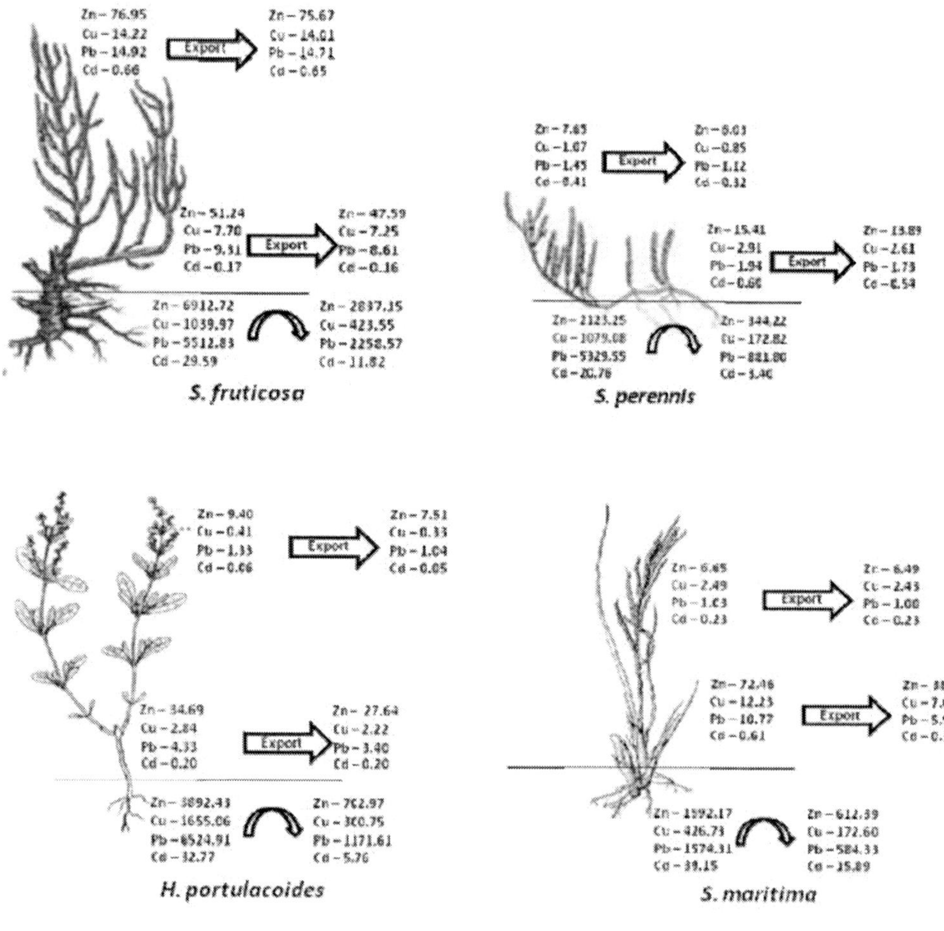

Figure 4. Metal Primary Accumulation (MPA, mg) and losses due to litter generation for the studied species and plant organs (Sf – S. fruticosa, Sp – S. perennis, Hp – H. portulacoides, Sm – S. maritima) (in Duarte et al., 2010)

5. The Decomposers Impact on Phytoremediation

Although this completes the role of the plants in phytoremediation processes in salt marsh metal cycling, these processes don't stop here and another matrix has to be considered, the decomposers. When the above ground organs enter in senescence and fall, the generated litter can go two ways: accumulate in the topsoil layer or be exported to the water column due to tidal inundation. Either way their contribution for metal contamination of the topsoil or of the water column is very low, when compared with the highly contaminated root necromass generation. In this case the necromass stays buried in the sediment and starts to be decomposed by several extracellular enzymes. Going back to the metal speciation process already referred, one of the major chemical sinks of metals is the organic fraction. This organic fraction comprises organic debris, proteins, cellular garbage as well as coatings (Tessier et al., 1979). It will be here that is found the metals associated to the decomposing necromass generated by root senescence. Considering the organic-bound fraction of all metals and the organic matter content and humic acids content in the sediment, it was possible to observe a similar seasonal pattern for these three parameters. Based on this, it is possible to infer that a depletion of the organic-bound fraction of metals may be caused by a decrease in the organic matter content in the sediment, and not by desorption of these metals from the organic matter. As previously demonstrated, there are several enzymes involved in the decomposition and breakdown of organic matter. These enzymes exhibited different patterns of activity throughout the seasons, having peaks of activity in different periods of the year and consequently leading to a differential degradation of the organic components of the sediment in different periods of the year. Every activity of enzymes is greatly affected by medium conditions, in this case the pH, Eh and salinity of the sediment. Two different periods of organic matter cycling seem to occur during the year. High protein degradation activities were also evident. During the warm seasons (spring and summer), high β-N-acetylglucosaminidase and phenol oxidase activities were also observed. β-N-Acetylglucosaminidase degrades chitin and the release of this enzyme is associated with the ecdysis process. As for phenol oxidase, it catalyzes the degradation of recalcitrant phenolic materials, such as lignin (Freeman et al., 2004). Both these enzymes degrade large polymers that are structural components of animals and plants. As previously mentioned, chitin exoskeletons are known to be able to accumulate toxic metals (Bergey and Weis, 2007). There are also reports (Sousa et al., 2008) that lignin from H. portulacoides can accumulate small amounts of heavy metals. As chitin is a protein, its degradation is due not only to β-N-acetylglucosaminidase but also to protease, explaining the simultaneous high activity of both these enzymes. Comparing the peaks of activity of these two enzymes with the organic-bound fraction of metals found, it is noticeable that, along with this degradation process, there is also a high depletion of the organic-bound fraction of metals. This indicates that these elements were probably bound to chitin like proteins. Phenol oxidase presented its maximum activity during summer and a total inhibition in the seasons hereafter. This inhibition is probably due to the obligatory need of molecular oxygen (Freeman et al., 2004) for the activity of this enzyme, scarce in the rhizosediments during the cold seasons, as indicated by the negative redox values verified during these periods. Together with a decrease in the activity of protease and β-N-acetylglucosaminidase, there was a peak of phenolic degradation in Summer, keeping the organic-bound metal fractions in low values, compared with Autumn, when all enzymes were found to be inactive.

This indicates a degradation of plant residues, releasing the percentage of metals associated to phenolics, as it can be observed by the increase in the labile metals fraction. This is also supported by the results of phenolic content, which only show a considerable increase in the concentration of phenolic substances in Winter, the season without phenol oxidase activity. Whilst in Spring and Winter an increase of the percentage of metals in the residual fraction together with a depletion of metals in the organic-bound fraction could be observed, the same did not happen in Summer. This may be due to the peaks of sulphatase activity detected in spring and winter. Some authors (Hullebusch et al., 2005) point out that high sulphatase activity can lead to the conversion of the sulphate produced by this enzyme into sulphides, by sulphate reducing bacteria. Sulfides can chemically reduce metals into a stable form for extended periods of time (Tabak et al., 2005), increasing therefore the metal concentrations in the residual fraction of this extraction scheme. This could be observed when comparing the amount of metals in the more available form with sulphatase activity. There seemed to be a second major period of organic matter depletion, most significantly observed in Winter. In this season all enzymes except sulphatase, peroxidase and protease are inactive or inexistent. Extracellular peroxidase is known to be produced mostly by ligninolytic fungi in order to degrade plant litter (Johnsen and Jacobsen, 2008). This enzyme catalyzes the degradation of ligninocellulosic litter in the presence of hydrogen peroxide. Due to this mechanism, peroxidase can operate even when the Eh values are low (as it was verified in Winter) without the need of molecular oxygen contrarily to phenol oxidase. It is also known that, in salt marshes, saprophytic fungi are majorly present in leaves and stems as epiphytes (Castro and Freitas, 2000). As previously stated, low Eh values prevailing in Winter most probably do not allow the activity of phenol oxidase, being the peroxidase activity the principal source of organic matter recycling during Winter. In this season, there are great inputs of plant litter due to leaf and stem senescence. Metals retained in these decaying parts become available for peroxidase activity, the principal source of organic matter recycling during this period. The degradation of these compounds produces phenolic substances of lower molecular weight, which are accumulated in the sediment due to the absence of phenol oxidase activity, as verified by the phenol concentration results. Together with this intense activity there is also a high protease activity, due to the degradation of lignin and associated components. This renders protein content more accessible for protease degradation. The breakdown of these bounds releases metals in the surrounding environment, and as verified in Spring, the high sulphatase activity detected could be responsible for the depletion of the labile metal fraction contrarily to which would be expected. These modifications in metal speciation due to EEA are important to be considered within the entire ecosystem. Organic matter is known to be an important sink of heavy metals, by the strong ligations that metals establish with organic compounds. The hydrolysis and breakdown of this organic compounds lead to the release of these metals to the surrounding medium, and consequently change their speciation. These changes are also important to be considered. Metals that were previously bound to organic matter suffer after its breakdown changes in their availability and also in their mobility, affecting therefore the community. This availability can be increased if these metals do not establish stable connections with any other sediment components, meaning that can then be uptaken by plants and consequently enter the food chain. Comparing these decomposition insights with the mechanism verified under a more phytostabilizer species, like S. maritma, it was found exactly the same mechanism (Duarte et al., 2009) only with a different temporal pattern of the enzyme actuation, clearly associated to an also different seasonality of the halophyte life cycle.

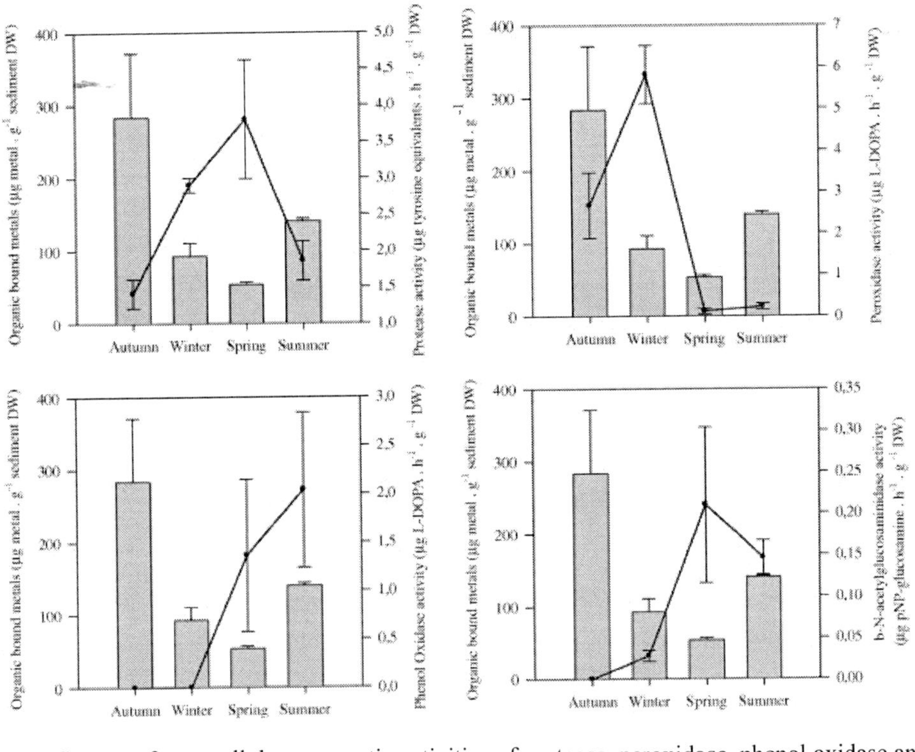

Figure 5. Influence of extracellular enzymatic activities of protease, peroxidase, phenol oxidase and β-N-acetylglucosaminidase (line) on total organic-bound metals (bars).

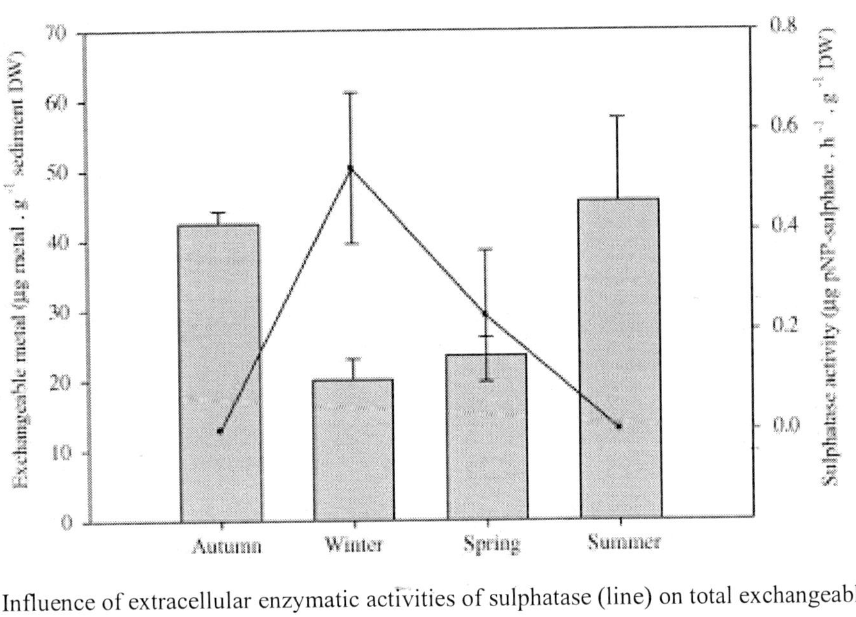

Figure 6. Influence of extracellular enzymatic activities of sulphatase (line) on total exchangeable metals (bars).

Conclusion

Salt marsh vegetation influences the dynamics of the estuarine ecosystem and efficiently retains anthropogenic metals discharged to the system. The salt marshes vascular plants are determinant to the dynamics of the estuarine ecosystem and strongly influence the processes of accumulation and retention of heavy metals there. Typically, these influences include in summary the taking up metals from contaminated sediments during the growth season and accumulating them in the plant tissues, mostly within the root system. These findings lead to inevitable conclusion that vascular plants may act as temporary sinks for heavy metals. Although this temporal sinking characteristic, they continue to influence the biogeochemistry of the surrounding environment. By releasing oxygen and organic compounds to the rhizosphere salt marsh plants critically impact the biogeochemistry of the sediments, modifying dramatically the soil characteristics and with this metal speciation and bioavailability. This will lead us to the starting point with possible increases or decreases of uptake. The cycle becomes complete with the senescence and necromass generation and consequent re-input of metals to the surrounding environment in a different chemical form from the uptake. Another key factor at this point will be the biogeochemical cycling by the microbial community inhabiting the rhizosphere, decomposing organic matter and this way acting on the bonds established with heavy metals and again the chemical speciation is affected. With this a new uptake is enhanced or delayed and the process restarted. In such a complex environment with all these processes acting, it becomes evident that all the variables are gathered to make salt marshes ideal ecosystems to study phytoremediation and its possibilities of application in other contaminated areas.

References

Almeida, C., Mucha, A. and Vasconcelos, M., 2004. Influence of the sea rush Juncus maritimus on metal concentration and speciation in estuarine sediment colonized by the plant, *Environmental Science and Technology* 38, 3112–3118.

Caçador, I., Vale, C. and Catarino, F., 1993. Effects of plants on the accumulation of Zn, Pb, Cu and Cd in sediments of the Tagus estuary salt marshes, Portugal J-P. Vernet (ed). *Studies In Environment Sciences Environment Contamination*, Elsevier Science Publisher B.V., Amsterdam. 55:355-364.

Caçador, I., Vale, C. and Catarino, F., 1996. Accumulation of Zn, Pb, Cu, Cr and Ni in Sediments Between Roots of the Tagus Estuary Salt Marshes, Portugal. Estuarine, *Coastal and Shelf Science* 42:393-403.

Caçador, I., Vale, C. and Catarino, F., 1996a. The influence of plants on concentration and fractionation of Zn, Pb, and Cu in salt marsh sediments (Tagus Estuary, Portugal). *Journal of Aquatic Ecosystem Health* 5:193-8.

Caçador, I., Vale, C., Catarino, F., 2000. Seasonal variation of Zn, Pb, Cu and Cd concentrations in the root–sediment system of Spartina maritima and Halimione portulacoides from Tagus estuary salt marshes. *Marine Environmental Research* 49, 279–290.

Caetano, M., Vale, C., Cesário, R. and Fonseca, N., 2008. Evidence for preferential depths of metal retention in roots of salt marsh plants. *Sci. Total Environ*, 390, 466-474.

Castro, P. and Freitas, H., 2000. Fungal biomass and decomposition of Spartina maritima leaves in the Mondego salt marsh (Portugal). *Hydrobiologia* 428, 171-177.

Dakora, F.D., Phillips, D., 2002. Root exudates as mediators of mineral acquisition in low-nutrient environments. Plant Soil 245, 35-47.

Doyle, M. and Otte, M., 1997. Organism-induced accumulation of iron, zinc and arsenic in wetlands soils. *Environmental Pollution*, 96-97, 1-11.

Duarte, B., Caetano, M., Almeida, P., Vale, C., Caçador, I., 2010. Accumulation and biological cycling of heavy metal in the root-sediment system of four salt marsh species, from Tagus estuary (Portugal). Environmental Pollution 158, 1661-1668.

Duarte, B., Almeida, P., Caçador, I., 2009. Spartina maritima (cordgrass) rhizosediment extracellular enzymatic activity and its role in organic matter decomposition processes and metal speciation. *Mar. Ecol. Evol. Persp.* 30, 65-73.

Duarte, B., Delgado, M., Caçador, I., 2007. The role of citric acid in cadmium and nickel uptake and translocation, in Halimione portulacoides. *Chemosphere* 69, 836–840.

Duarte, B., Reboreda, R. and Caçador, I., 2008. Seasonal Variation of extracellular enzymatic activity (EEA) and its influence on metal speciation in a polluted salt marsh. *Chemosphere* 73, 1056-1063.

Freeman, C., Ostle, N., Fenner, N. and Kang, H., 2004. A regulatory role for phenol oxidase during decomposition in peatlands. *Soil Biol. and Biochem.* 36, 1663-1667.

Hullebusch, E., Utomo, S., Zandvoort, M. and Lens, P., 2005. Comparison of three sequential extraction procedures to describe metal fractioning in anaerobic granular sludges. *Talanta* 65, 549-558.

Ingold, A. and Havill, D., 1984. The influence of sulphide on the distribution of higher plants in salt marshes. *Jornal of Ecology* 72, 1043–1054.

Jócsák, I., Végvéri, G., Droppa, M., 2005. *Heavy metal detoxification by organic acids exudation in barley seedlings.* Proceedings of the 8th Hungarian Congress of Plant Physiology and the 6th Hungarian Conference on Photosynthesis.

Johnsen, A. and Jacobsen, O., 2008. A quick and sensitive method for the quantification of peroxidase activity of organic surface soil from forests. *Soil Biol. Biochem.* 40, 814-821.

Lacerda, L., Freixo, J. and Coelho, S., 1997. The effect of Spartina alterniflora Loisel on trace metals accumulation in inter-tidal sediments. *Mangroves Salt Marshes* 1, 201–209.

Lee, R., Kraus, D., Doeller, J., 1999 Oxidation of sulfide by Spartina alterniflora roots. *Limnology Oceanography* 44, 1155-1159.

Ma, J.F., Ryan, P.R., Delhaize, E., 2001. Aluminium tolerance in plants and the complexing role of organic acids. *Trends Plant Sci.* 6, 273-278.

Mortimer, R. and Rae, J., 2000. Metal speciation (Cu, Zn, Pb, Cd) and organic matter in oxic to suboxic salt marsh sediments, Severn Estuary, Southwest Britain. *Marine Pollution Bulletin* 40, 377–386.

Mucha, A.P., Almeida, C.M., Bordalo, A.A. Vasconcelos, M.T., 2005. Exudation of organic acids by a marsh plant and implications on trace metal availability in the rizosphere of estuarine sediments. Estuar. *Coasta. Shelf S.* 65, 191-198.

Nigam, R., Srivastava, S., Prakash, S., Srivastava, M.M., 2000. Effect of organic acids on the availability of cadmium in wheat. *Chem. Spec. Bioavailab.* 12, 125-132.

Otero, X. and Macías, F., 2002. Variation with depth and season in metal sulfides in salt marsh soils. *Biogeochemistry* 61, 247–268.

Otero, X., Huerta-Diaz, M. and Macías, F., 2000. Heavy metal geochemistry of salt marsh soils from Ría of Ortigueira (mafic and ultramafic areas, NW Iberian Peninsula), *Environmental Pollution* 110, 285–296.

Otte, M., Haarsma, M., Broekman, R. and Rozema, J., 1993. Relation between heavy metal concentrations in salt-marsh plants and soil. *Environmental Pollution* 82, 13–22.

Padinha, C., Santos, R. and Brown, M., 2000. Evaluating environmental contamination in Ria Formosa (Portugal) using stress indexes of Spartina maritima. *Marine Environmental Research* 49, 67–78.

Reboreda, R. and Caçador, I., 2007a. Halophyte vegetation influences in salt marsh retention capacity for heavy metals. *Environmental Pollution* 146, 147-154.

Reboreda, R., Caçador, I., 2007b. Copper, zinc and lead speciation in salt marsh sediments colonized by Halimione portulacoides and Spartina maritima. *Chemosphere* 69, 1655–1661.

Reboredo, F., 1992. How differences in the field influence Cu, Fe and Zn uptake by Halimione portulacoides and Spartina maritima. *Science of Total Environment* 133, 11–132.

Rozema, J., Lupper, E. and Broekman, R., 1985. Differential response of salt marsh species to variation of iron and manganese. *Vegetatio* 62, 293–301.

Sánchez, J., Otero., X and Izco, J., 1998. Relationships between vegetation and environmental characteristics in a salt marsh system on the coast of Northwest Spain. *Plant Ecology* 136, 1–8.

Sundby B, Caetano M, Vale C, Gobeil C, Luther G, Nuzzio D. 2005. Root-induced cycling of lead in salt marsh sediments. *Environ. Sci. Technol* 39, 2080–2086.

Sundby, B., Caetano, M., Vale, C., Gobeil, C., Luther, G. and Nuzzio, D., 2005. Root-induced cycling of lead in salt marsh sediments. *Environmental Science and Technology* 39, 2080–2086.

Tabak, H., Lens, P., Hullebush, E. and Dejonghe, W., 2005. Developments in bioremediation of soils and sediments polluted with metals and radionuclides. – 1. Microbial processes and mechanisms affecting bioremediation of metal contamination and influencing metal toxicity and transport. *Reviews in Environmental Science and Bio/Technology* 4, 115-156.

Tessier, A., Campbell, P. and Bisson, M., 1979. Sequential extraction procedure for the speciation of particulate trace metals. *Analytical Chemistry* 51, 844–851.

Tobias C., Macko S., Anderson I. and Canuel E., 2001. Tracking the fate of a high concentration groundwater nitrate plume through a fringing marsh: a combined groundwater tracer and in situ isotope enrichment study. *Limnology and Oceanography* 46, 1977–1989.

Walker, T.S., Bais, H.P., Grotewold, E., Vivanco, J.M., 2003. Root exudation and rhizosphere biology. *Plant Physiol.* 132, 44-51.

Williams, T., Bubb, J. and Lester, J., 1994. The occurrence and distribution of trace metals in halophytes. *Chemosphere* 28, 1189–1199.

Windham, L., Weis, J. and Weis, P., 2003. Uptake and distribution of metals in two dominant salt marsh macrophytes, Spartina alterniflora (cordgrass) and Phragmites australis (common reed), Estuarine, *Coastal and Shelf Science* 56, 63–72.

In: Handbook of Phytoremediation
Editor: Ivan A. Golubev

ISBN: 978-1-61728-753-4

Chapter 27

PRIOR TO A SUCCESSFUL PHYTOEXTRACTION: POT EXPERIMENTS AND FIELD SCALE STUDIES ON THE TOTAL REMOVAL CAPACITY BY GARDEN FLOWERS GROWN IN CADMIUM-CONTAMINATED SOILS IN TAIWAN

Hung-Yu Lai[1] ***and Zueng-Sang Chen***[2][1]
[1]Department of Post-Modern Agriculture, MingDao University
[2]Department of Agricultural Chemistry, National Taiwan University

ABSTRACT

In Taiwan, many heavy metals (HMs)-contaminated arable soils have been founded since 1980. Agricultural irrigation system was mixed with river waste water contaminated with HM is the primary reason for the contamination of cropping lands. Soil turnover/attenuation technique, which mixes the surface 30 cm layer of contaminated soils with deeper clean soil layer, was the most popular technique to be used to dilute the HM-contaminated soils to meet the soil regulation of HM in the soil contamination site. Phytoextraction technique was also regarded as another candidate technique to remove the HMs from the HMs-contaminated sites. Seedlings of various native garden flowers of Taiwan were planted either in-situ in HM-contaminated sites or in pot experiments artificially spiked soils to investigate their tolerance and removal capacity from the sites. These sites were primary contaminated with cadmium (Cd), lead (Pb), zinc (Zn), or mixed-combined with them. The total removal of HMs plays an important role prior to conduct a successful phytoextraction and decontamination. Although some of the selected plant species can accumulate higher concentration of HM in their shoots, they are small biomass and thus just can remove little amounts of HM from the contaminated soils.

The accumulation and growth of a specific plant in-situ grown in contaminated site is quite different compared with that of pot experiments. This paper summarizes the total removal capacity of high potential super accumulator garden flowers and estimated the period needed to cleanup the Cd from the contamination sites.

[1] No.1, Sec.4, Roosevelt Rd., Taipei 10617, Taiwan, E-mail: soilchen@ntu.edu.tw

Keywords: heavy metals (HMs)-contaminated soils, turnover/attenuation technique, Phytoextraction technique, garden flowers, removal capacity, high potential super accumulator.

1. INTRODUCTION

Heavy metals accumulated in the soil is primary as a totality of the weathering of parent materials and human activities, which include mining, smelting, application of sludge and fertilizers, and discharge of waste water from the industrial park, etc. (Kabata-Pendias and Pendias, 2001).

There are many sources for Cd to be added into the soil, including anthropogenic emission, phosphate fertilizers, and sewage sludge, etc. (Alloway, 1995; Alloway and Steinnes, 1999).

Anthropogenic emission is the most dominate among them and was about 56% of the total input per year (Singh and McLaughlin, 1999).

The total concentration of Cd in the crust levels from 0.3 to 11 mg kg-1 but less than 1 mg kg-1 mostly with an average of 0.2 mg kg-1 (Alloway, 1995; Lalor, 2008). Soil pH and oxidation potential are the most important factors control its mobility in the soil system (Kabata-Pendias and Pendias, 2001).

Cadmium often presents as Cd2+ ion (Alloway, 1995) and has high mobility make it easier to transfer from soils to plants compared with other heavy metals (Kloke et al., 1984; Podlesáková et al., 2001).

The uptake of Cd was affected by plant species, growth stage, soil pH, redox potential, organic matter content, clay content, and oxide minerals (Singh and McLaughlin, 1999). The increase of soil pH value decreased the Cd concentration in both the leaves of soybean and seedling of corn (Kitagishi and Yamane, 1981).

The rice grown in the flooded soil of reduced condition accumulated lower concentration of Cd compared with that in the condition of oxidation (Bingham et al., 1976). The mean concentration of Cd ranged from 0.013 to 0.22 mg kg-1 for cereal grains, 0.07 to 0.27 mg kg-1 for grasses, and 0.08 to 0.28 mg kg-1 for legumes (Kabata-Pendias and Pendias, 2001).

Excess concentration of Cd in the tissues of plants will inhibit their photosynthesis and respiration (Chugh and Sawhney, 1999; Garbisu and Alkorta, 2001; Schmidt, 2003; Schwartz et al., 2003; Mishra and Tripathi, 2008), reduce their uptake of nutrients (Hernandez et al., 1996; Obata and Umebayashi, 1997), and reduce the fixation of nitrogen (Hernandez et al., 1995).

The growth of root and shoot was inhibited when Cd was in the toxic level (Jiang et al., 2001; Liu et al., 2003). Roots accumulated higher concentration of Cd compared with those in other tissues (Uraguchi et al., 2006; Baghour et al., 2001).

The maximum intake of Cd for one person recommended by World Health Organization (WHO) is 400-500 μg/week/p, which is equal to 60-70 μg/day/p (WHO, 1989). There were links between Cd exposure and human cancer of the bladder, kidney, lung, and prostate (Lalor, 2008).

2. HEAVY METALS CONTAMINATION IN TAIWAN

In Taiwan, HMs soil pollution was produced by the illegal discharges of many electroplating plants. According to the survey of Environmental Protection Administration (EPA) of Taiwan, there are approximately 470 ha of cropped lands been contaminated with HMs till the end of 2008 (Taiwan EPA, 2010).

Most of them were located in central Taiwan and contaminated mainly with copper (Cu) and nickel (Ni), followed by Zinc (Zn) and cadmium (Cd).

These contaminated lands resulted from the use of contaminated irrigating waste water should be conducted further soil remediation to decrease the total concentrations, digested by aqua regia, of HMs to below the soil regulation (SR) announced by the Soil and Groundwater Pollution Remediation Act (SGWPR Act).

The soil remediation techniques was used in Taiwan to treat the HMs-contaminated soils include acid washing and soil turnover or attenuation. During 2003 to 2007, Taiwan's EPA spent approximately US$ 9.4 million to treat these HMs-contaminated lands (Table 1).

Table 1. Area and cost of HMs-contaminated soils which had been decontaminated in Taiwan

	Area		Cost		
	T/A	AW	Fallow compensation	Soil Remediation practices	Soil Fertility recovery
	ha		US$		
Northern Taiwan	49.7	1.4	255,722	1,709,300	6,815
Central Taiwan	177	21.5	279,964	5,495,455	829,333
Southern Taiwan	15.4	3.9	21,641	712,041	40,697
Total	242	26.8	557,327	7,916,796	876,845

T/A: Turnover/Attenuation; AW: Acid washing.
The cost was produced based on the data of Taiwan EPA (2010).

The soil turnover/attenuation method was the most popular technique to be used in Taiwan, which just mixed the contaminated soils in surface layer (about 20 cm depth) with appropriate amounts of subsurface layer (20 to 60 cm or deeper) to reduce the concentration of HM to meet the soil regulation of HMs.

The soil depth of subsurface layer needed for dilution can be calculated by using the equations [1] and [2].

While the total amount of HMs in soil remains unchanged and soil characteristics manifest changed. Further application of fertilizers or organic materials, however, is necessary to recover the soil fertility for framer's sustainable agriculture (Lai et al., 2007).

$$X = \frac{A \times D \times B_d \times C \times 1000}{1 + W} \quad [1]$$

where

X: Total amounts of HM (mg)

A: Total area of HM-contaminated soils (m^2)

D: Soil depth (m)

B_d: Bulk density (g cm^{-3})
C: Concentration of HM (mg kg^{-1})
W: Water content

$$S = \frac{X_1 + X_2}{[X_1 / C_1 + X_2 / C_2]} \times E. \qquad [2]$$

where
S: Target concentration (mg kg^{-1})
E: safety coefficient = 1.2~1.3

Acid washing technique was conducted to remove HM in contaminated soils with the advantages of high efficiency and short period needed. Coarse particles can be separated before the treating process and without acid washing because most of the HMs was distributed in the fine particles, such as the clay particle.

However, the electrical conductivity was increased and the soil pH was decreased to the levels of 2-3 after acid washing treated, which is not suitable for the growth of all the crops.

The application of lime materials is always necessary to raise the soil pH to the levels of 5.5-7.0 (Lai et al., 2007). Additional cost in also needed in treating the wastewaters.

Pilot study is necessary to obtain the suitable conditions before using soil acid washing technique to remediate the HM-contaminated soils which including HM-concentration distribution of different soil particles and suitable agent for soil acid washing, etc.

In previous study in Taiwan, contaminated soils were whole soaked and mixed in the treating system using 0.2M HCl and the maximum treating amounts are about 1.8 m3 day-1. Used waste liquids were drained into the waste water treating system in the end of the day, the recycling use as the diluting water after treated and the produced sludge was packed and treated legally.

One treatment of acid washing technique can significantly and rapidly decrease the soil Cd concentration from 9.58±1.39 to 5.02±0.35 mg kg-1 and the treating efficiency is about 52%.

Experimental result showed that the addition of lime materials into the acidified washing-treated soils can increase the germination percentage of Chinese kitam.

Further experiments are necessary to assess the effects of applying different soil amendments on the soil fertility of acid washing-treated soils.

3. Phytoextraction of Cadmium-Contaminated Soils

Phytoextraction is the use of plants to remove HMs from contaminated soils, with the advantages of related lower costs compared with traditional techniques, avoiding the excavation of soil, no advised effect on the soil quality, and can be used in the contaminated soil of large area (Chaney et al., 1997; Wenzel et al., 1999; Garbisu and Alkorta, 2001).

Hyperaccumulator is defined as the plants that can accumulated more than 100 mg Cd kg-1 in their tissues (Baker et al., 2000) and their TF ratio (translocation factor; ratio of shoot concentration to root concentration) and BCF ratio (bioconcentration factor; ratio of shoot concentration to soil concentration) are all more than 1 (Sun et al., 2009).

Thlaspi caerulescens is a famous hyperaccumulator for Cd and Zn and the accumulated Cd concentration in the tissues was in the levels of 164～2,800 mg kg-1 (Baker et al., 1994; Brown et al., 1994; Robinson et al., 1998; Lombi et al., 2000).

Over 420 plant species of hyperaccumulator from all over the world were discovered according to Baker et al. (2000), but many of them have a low growth rate and very low biomass. The application of chelating agents and genetic engineering were the two strategies to be developed to increase the uptake of HM by plants.

The synthesized chelating agents, such as EDTA (ethylenedinitrilo tetraacetic acid), DTPA (diethylenetrinitrilo pentaacetic acid), HEDTA (hydroxyl- ethylenediamine triacetic acid), CDTA (trans-1,2-cyclohexylenedinitrilotetraacetic acid), EGTA (ethylenebis (oxyethylenetrinitrilo) tetraacetic acid), and EDDS (S,S-ethylenediamine disuccinic acid) were applied to HM-contaminated soil to increase their mobility and bioavailability and thus the amount of HMs were accumulated in the upper parts of plants (Huang and Cunningham, 1996; Huang et al., 1997; Blaylock et al., 1997; Ebbs and Kochian, 1998; Wu et al., 1999; Lai and Chen, 2004; 2005; Luo et al., 2006; Tandy et al., 2006).

However, synthetic chelating agents at high concentrations can also be toxic to plants and also have the risk of leaching to groundwater (Cooper et al., 1999; Lai and Chen, 2006).

4. Bioconcentration of Cadmium by Garden Flowers Grown in the Pot Experiments of Cd-Contaminated Soils

Leafy vegetables can accumulate high concentration of Cd after it was grown in artificially Cd-contaminated soils for 35 days (Chen et al., 2010). There is a linear relationship between Cd concentration in soil and that in the edible parts of Pak-chio, which is a most often consuming vegetable in Taiwan (Figure 1).

When Pak-chio was grown in artificially spiked soils with different Cd concentrations, the edible parts of Pak-chio accumulated 1.82-fold more Cd concentration than that of soil which the Cd concentration in soils were less that the soil regulation for cropped soil (5 mg kg-1).

The European Union's soil regulation for leafy vegetables is 0.2 mg Cd kg-1 (fresh weight) (Howe et al., 2005). If the water content is 80-90% of the leafy vegetable, the EU regulation should be less than 2 mg kg-1 (dry weight). Although the spiked Cd has higher bioavailability compared with the contaminated sites, planting foliar vegetables in soils with lower concentration of Cd is not appropriate when the consumption risk was considered.

In such situation, planting garden flowers seems to be a suitable alternatively method because the flower farmers can sell them to the markets and also provides the contaminated soils covered the flowers to decrease the soil erosion of high rainfall and wind and thus decrease the exposing risk to the farmers and community people. The contaminants in soils can be removed by phytoextraction technique if the total metals removal by garden flowers is large.

To increase the total removal of Cd, plants should have high accumulated Cd concentration accompany with large biomass. There were many studies using pot experiments to assess the capacity on the removal of heavy metals by different garden flowers in Taiwan.

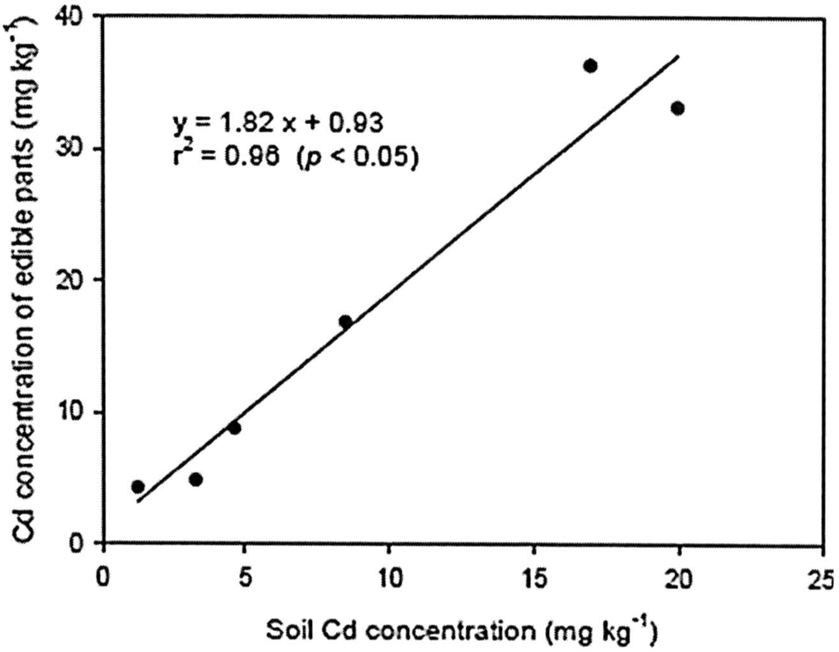

Figure 1. Linear relationship between the Cd total concentration in soil and that in the edible of Pak-chio.

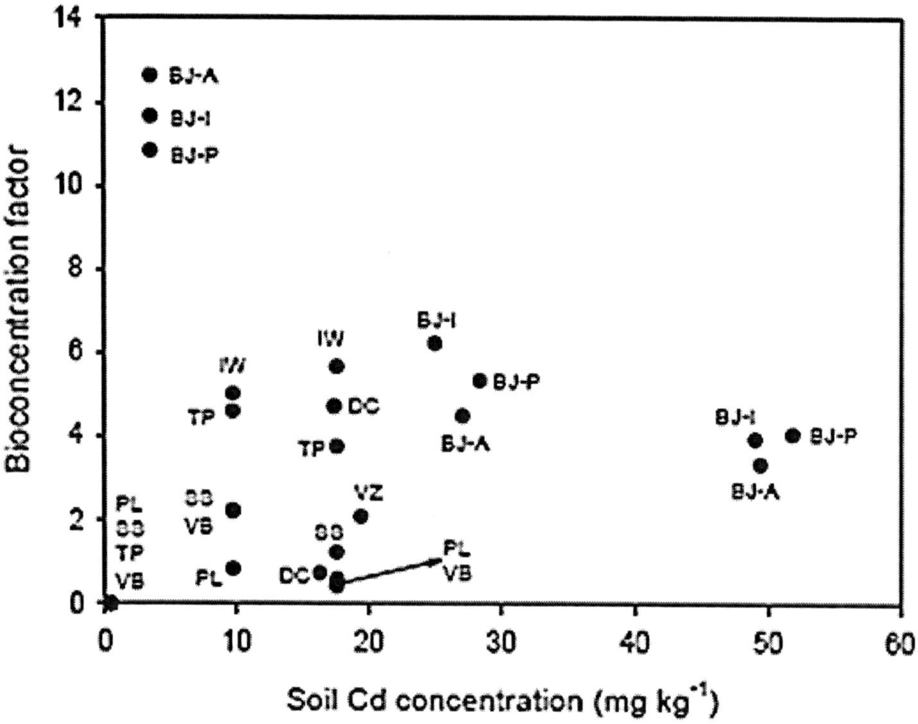

Figure 2. Bioconcentration factor (BCF) of different plant species grown in artificially Cd-contaminated soils for 35 to 50 days. (TP—French marigold; VB—Garden verbena; IW—Impatiens; PL—Star cluster; SS—Scarlet sage; BJ-A—India mustard Afghanistan; BJ-I—India mustard India; BJ-P—India mustard Paskistan, and DC—Rainbow pink)

These plants included French marigold (Tagetes patula L., coded as TP), Garden verbena (Verbena bipinnatifida Nutt., coded as VB), Impatiens (Impatiens walleriana Hook. f., coded as IW), Star cluster (Pantas lanceolata Deflers., coded as PL), Scarlet sage (Salvia splendens Ker-Gawl., coded as SS) (Lin et al., 2010), India mustard (Brassica juncea Afghanistan, coded as BJ-A; Brassica juncea India, coded as BJ-I; and Brassica juncea Paskistan, coded as BJ-P) (Lai et al., 2008), and rainbow pink (Dianthus chinensis, coded as DC) (Lai and Chen, 2004). BCF value was used as one of the thresholds to identify the potential of hyperaccumulator and the plants suitable for phytoextraction should have more than unity in BCF (Sun et al., 2009).

The BCF of the plants species which has been tested in Taiwan was shown in Figure 2. The BCF values of DC, PL, SS, TP, and VB plant species were less than unity because of the lower Cd concentration in soil (0.43 mg kg-1) and their lower translocation capacity.

The plant species which can accumulate four times of BCF as more Cd concentration were in soil including the species of BJ-A (BCF = 3.36-12.6), BJ-I (BCF = 3.96-11.7), BJ-P (BCF = 4.07-10.8), DC (BCF = 4.72), IW (BCF = 5.03-5.68), and TP (BCF = 3.77-4.61) species.

5. Total Removal of Garden Flowers Grown in the Pot Experiments of Cd-Contaminated Soils

Although some of the tested plant species can accumulated high concentration of Cd in their tissues and their BCF values were all more than unity, however, they always has lower biomass and thus increase the period needed for removal of pollutants to meet the soil regulation. The total removal of Cd is more an important factor for the assessment of the period for decontamination and experimental results of the pervious experiments conducted in Taiwan (Figure 3). If four blocks were divided in Figure 3, plant species locating at A3 area and especially at A4 area that have lower total removal capacity (<200 μg plant-1), which were evaluated as "not to be as candidate" for phytoextraction species of Cd-contaminated soils. The main limiting factor is resulted from the lower biomass. For example, three species of India mustard can accumulate high concentrations of Cd (37.8-211 mg kg-1) and high BCF values (3.36-12.6), which are higher values in relative to other tested plants (Lai et al., 2008).

Some of them are regarded as a Cd-hyperaccumulator according to the accumulated concentration and BCF value, however, the total removal of Cd of India mustard was still located at A3 and A4 region (Figure 3) and are not regarded as an appropriate candidate for phytoextraction. Longer period is needed when they are recommended to be planted to phytoextraction of Cd-contaminated soils. Quartacci et al. (2006) used a sandy soil with total Cd concentration of 10 mg kg-1 to test the uptake capacity of India mustard. Similar result was obtained and the total removal was less than 20 μg plant-1, which were only 20-55% of those at A4 region even chemical enhanced with citrate or NTA (nitriltriacetate acid).

The total removal capacity of plant species located at A1 and A2 region of Figure 3 were the most suitable species used for phytoextraction in the field scale because they can remove up to 200-1,000 μg Cd plant-1 from Cd-contaminated soils. No plant species can not found at A2 region because the soil regulation for normal lands of SGWPR Act in Taiwan is 20 mg Cd kg-1 and most of the study sites were conducted in this concentration.

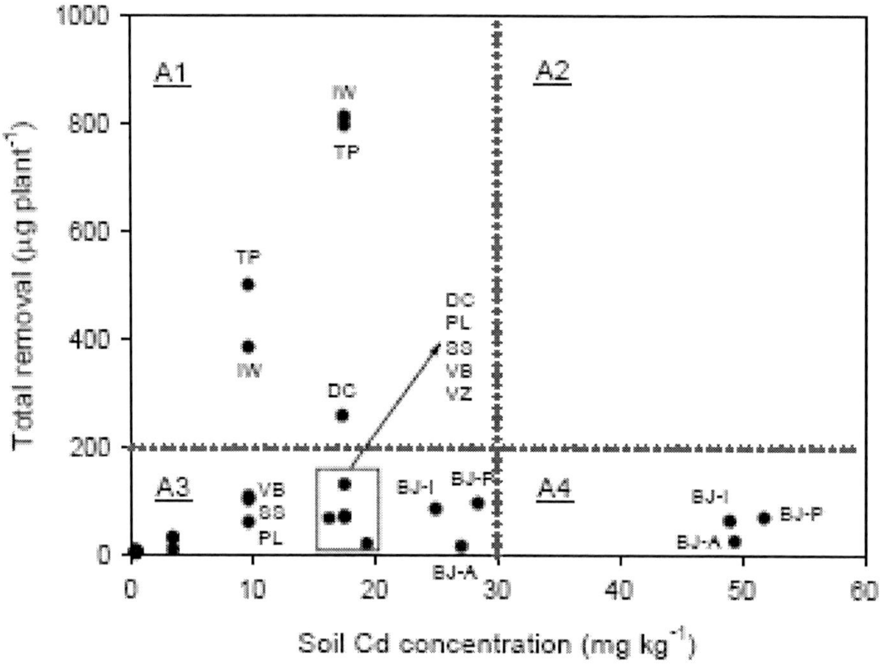

Figure 3. Total removal of Cd by different plant species grown in artificially Cd-contaminated soils for 35 to 50 days. (Codes are the same as in Figure 2).

More pot experiments were conducted by using the plant species within A1 region to test their accumulation and removal capacity when they are growing in the Cd-contaminated soils with more than 40 mg Cd/kg, but the studies are not finished yet.

When we only consider the plant species at A1 region (Figure 3), the total removal of Cd of pervious studies was shown in Figure 4. Seedlings of rainbow pink (Dianthus chinensis, coded as DC) bought from a flowering market were planting in the pots of Cd-contaminated soils with total concentration of 19.9 mg kg-1 (Lai and Chen, 2004). After growing for 50 days, shoots of rainbow pink (DC) species can accumulate 82.2 mg Cd kg-1 and its total removal can reach 260 μg Cd plant-1. In an ideal situation without considering the changes of phytoavailability of Cd to calculate the period needed to decrease the current Cd concentration to lower the soil regulation for cropped lands, it will take at least 24 yrs for clean up the Cd from the contaminated sites.

Lin et al. (2010) planted seedlings of five garden flowers in pots and artificially contaminated soils with various Cd concentrations to test their accumulation capacity. Experimental results showed that the total removals of TP (French marigold, Tagetes patula L.P) and IW (Impatiens, Impatiens walleriana Hook. f.) species have higher removal capacity compared with other three plants. After growing in the soils with total Cd concentrations of 9.73 and 17.6 mg kg-1 for 35 days, IW species can accumulate 48.9 and 100 mg kg-1 in their tissues and the total removals can reach to 387 and 815 μg Cd plant-1, respectively (Figure 4). For TP species grown in similar soil Cd concentrations for 35 days, the accumulated Cd concentrations reached 44.9 and 66.3 mg kg-1 and the total removals was 503 and 799 μg Cd plant-1, respectively. IW species can accumulate higher Cd concentration and higher BCF value (5.03-5.68) compared with TP species (3.77-4.61) (Figure 2).

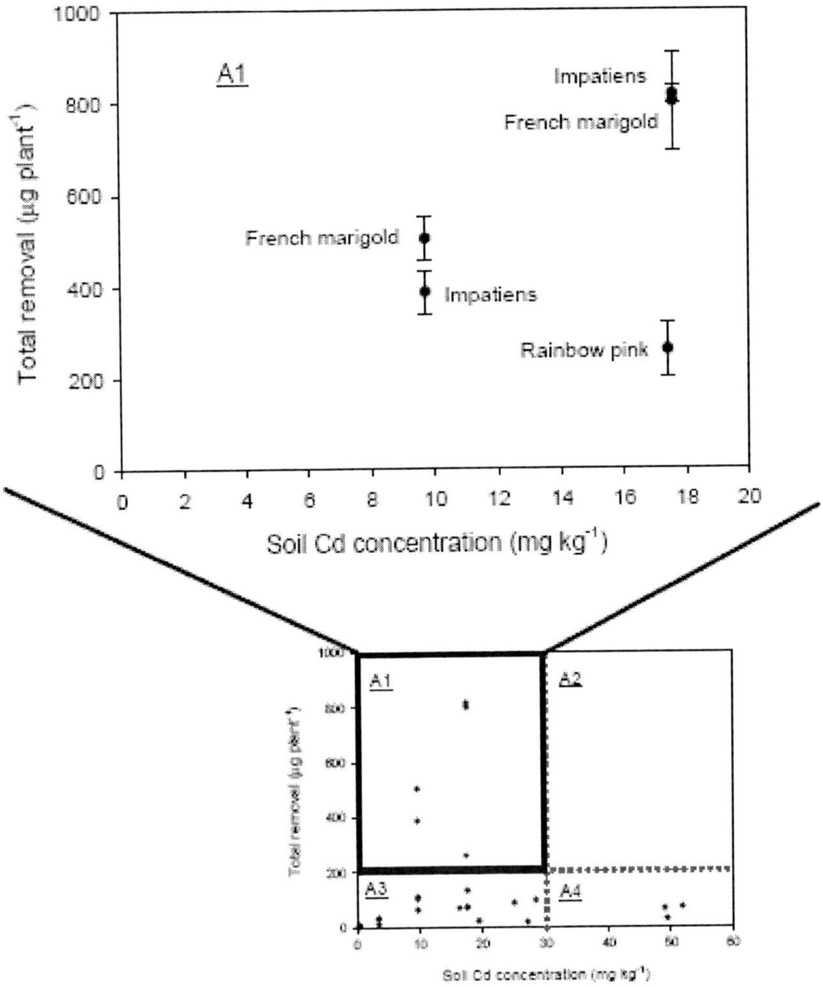

Figure 4. Total removal of Cd by different plant species were grown in artificially Cd-contaminated soils for 35 to 50 days for it only was considered at the A1 region of Figure 3.

In an ideal condition to estimate the period needed to decrease the current Cd concentration (9.73 and 17.6 mg kg-1) to the soil regulation for cropped lands, it will take approximately 5.0-6.4 yrs and 3.9-6.5 yrs for contaminated soils before planting IW and TP species, respectively. These five plant species can remove Cd from contaminated soils within a reasonable period (4-7 yrs) and the tested species also have higher BCF values, which was in the levels of 3.7-5.7 (Figure 2).

6. Total Removal of Garden Flowers In-Situ Grown in the Field of Cd-Contaminated Sites

Chen and Lee (1997) in-situ planted 22 species of garden flowers in the Cd-contaminated sites in the northern Taiwan to test their accumulation capacity of Cd. Experimental results showed that the plant species can remove more than 200 μg Cd plant-1 including the species of Cock-comb (Celosia cristata, CC), DC, IW, PL, and TP species (Figure 5).

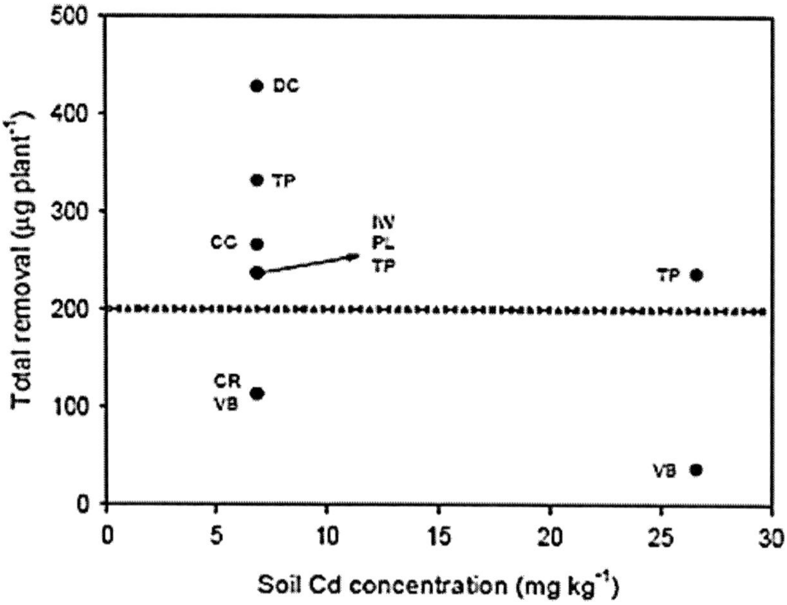

Figure 5. Total removal of Cd by different plant species grown in Cd-contaminated sites for 35 to 42 days. (CC—Cock-comb; CR—Rose periwinkle; TP—French marigold; VB—Garden verbena; IW—Impatiens; and DC—Rainbow pink).

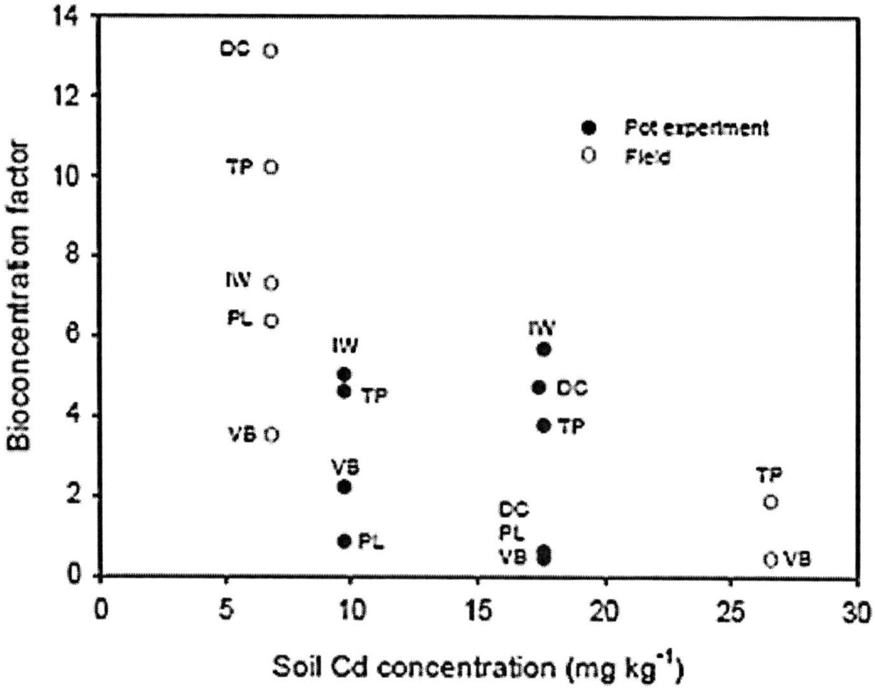

Figure 6. Bioconcentration factor (BCF) of different pant species grown in artificially Cd-contaminated soils or in-situ grown in Cd-contaminated sites for 35 to 50 days. (Codes are the same as in Figure 2).

It will need 2-5 yrs to remove the Cd from contaminated site by the plant species of CC, DC, IW, PL, and TP in an ideal situation to decrease the current Cd concentration (6.85 mg kg-1) to the soil regulation for cropped lands. For another site that with total Cd concentration of 26.6 mg kg-1 and it will take 32 yrs by the plant species of TP to reduce the Cd concentration to meet the soil regulation for cropped lands.

The differences of the total Cd removal capacity per plant between pot experiments and field study was difficult to evaluate removal capacity because the soil Cd concentration and soil characteristics were quite different. If the BCF value was used as an indicator, the same plant species in-situ grown in Cd-contaminated sites have higher BCF compared with that in the pot experiments (Figure 6). Plants grown in Cd-contaminated field sites can remove more amounts of Cd from soils because they have more root space and good growth condition if they have similar biomass.

CONCLUSION

The selection of appropriate plant species is the fist step prior to a successful phytoextraction of HM-contaminated soils in Taiwan. Many of the literatures focused on the accumulated concentration of Cd in the shoots of plants, however, total removal capacity (μg Cd plant-1) is more important because the plant can accumulate high concentration of HM but they are always have very low biomass. Total removal capacity of high potential super accumulator garden flowers is a total performance of uptake which summarizes the effect of soil Cd concentration on the growth of biomass and accumulation of Cd in plant. The period needed for these plant species to remove Cd from contaminated soils was estimated by using total removal without considering the change of bioavailable concentration of Cd in the soils. However, the bioavailability of Cd in the soil is not a constant concentration and decreases with increasing time in the site. Further study is needed to investigate the change of bioavailability of metals on the total removal capacity of Cd to accurately estimate the period needed for decontamination.

REFEREBCES

Alloway, B.J. 1995. *Soil processes and the behavior of metals*. In: Alloway B.J., editor. Heavy Metals in Soils. Glasgow, UK: Blackie Academic and Professional, p. 11-37.

Alloway, B.J. and Steinnes, E. 1999. *Anthropogenic additions of cadmium to soils*. In: AcLaughlin M.J. and Singh B.R., editors. Cadmium in Soils and Plants. Dordrecht, Netherlands: Kluwer Academin Publishers, p. 97-123.

Baghour, M., Moreno, D.A., Villora, G., Hernández, J., Castilla, N. and Romero, L. 2001. Phytoextraction of Cd and Pb and physiological effects in potato plants (Solanum Tuberosum Var. Spunta) importance of root temperature. *J. Agric. Food Chem*. 49, 5356-5363.

Baker, A.J.M., Reeves, R.D. and Hajar, A.S.M. 1994. Heavy metal accumulation and tolerance in British populations of the metallophyte Thlaspi caerulescens J. and C. Presl (Brassicaceae). *New Phytol*. 127, 61-68.

Baker, A.J.M., McGrath, S.P., Reeves, R.D. and Smith, J.A.C. 2000. Metal hyperaccumulator plants: A review of the ecology and physiology of a biological resource for phytoremediation of metal-polluted soils. In: Terry N. and Bañuelos G., editors. *Phytoremediation of Contaminated Soil and Water*. USA: CRC Press LLC; p. 85-107.

Bingham, F.T., Page, A.L., Mahler, R.J. and Ganje, T.J. 1976. Yield and cadmium accumulation of forage species in relation to cadmium content of sludge amended soils. *J. Environ. Qual.* 5, 57-60.

Blaylock, M.J., Salt, D.E., Dushenkov, S., Zakharova, O., Gussman, C., Kapulnik, Y., Ensley, B.D. and Raskin, I. 1997. Enhanced accumulation of Pb in Indian mustard by soil-applied chelating agents. *Environ. Sci. Technol.* 31, 860-865.

Brown, S.L., Chaney, R.L., Angel, J.S. and Baker, A.J.M. 1994. Phytoremediation potential of Thlaspi caerulescens and bladder campion for zinc- and cadmium-contaminated soil. *J. Environ. Qual.* 23,1151-1157.

Chaney, R.L., Malik, M., Li, Y.M., Brown, S.L., Brewer, E.P., Angle, J.S. and Baker, A.J.M. (1997). Phytoremediation of soil metals. *Environ. Biotechnol.* 8, 279-284.

Chen, Z.S. and Lee, D.Y. 1997. *Evaluation of remediation techniques on two cadmium-polluted soils in Taiwan.* In: Iskandar I.K. and Adriano D.C., editors. Remediation of soils contaminated with metals. Northwood, UK: Science Reviews, p. 209-223.

Chen, H.L., Lai, H.Y., Wang, S.M., Kuo, Y.C. and Lu, C.J. 2010. Effect of biosolid and the Cd/Pb interaction on the growth and Cd accumulation of Brassica rapa grown in the Cd-contaminated soils. *Water Air and Soil Pollut.* 206, 385-394.

Chugh, L.K. and Sawhney, S.K. 1999. Photosynthetic activities of Pisum sativum seedlings grown in presence of cadmium. *Plant Physiol. Biochem.* 37, 297-303.

Cooper, E.M., Sims, J.T., Cunningham, S.D., Huang, J.W. and Berti, W.R. 1999. Chelate-assisted phytoextraction of lead from contaminated soils. *J. Environ. Qual.* 28, 1709-1719.

Ebbs, S.D. and Kochian, L.V. 1998. Phytoextraction of zinc by oat (Avena sativa), barley (Hordeum vulgare), and Indian mustard (Brassica juncea). *Environ. Sci. Technol.* 32, 802-806.

Garbisu, C. and Alkorta, I. 2001. Phytoextraction: a cost-effective plant-based technology for the removal of metals from the environment. *Biores. Technol.* 77, 229-236.

Hernandez, L.E., Garate, A. and Carpena-Ruiz, R. 1995. Effect of cadmium on nitrogen fixing pea plants grown in perlite and vermiculite. *Plant Nutr.* 18, 287-303.

Hernandez, L.E., Carpena-Ruiz, R. and Garate, A. 1996. Alterations in the mineral nutrition of pea seedlings exposed to cadmium. *Plant Nutr.* 19, 1581-1598.

Howe, A., Hoofung, L., Lalor, G.C., Rattray, R. and Vutchkov, M. 2005. Elemental composition of Jamaican foods: A survey of five food crop categories. *Environ. Geochem. Health* 17, 19-30.

Huang, J.W. and Cunningham, S.D. 1996. Lead phytoextraction: species variation in lead uptake and translocation. *New Phytol.* 134, 75-84.

Huang, J.W., Chen, J., Berti, W.R. and Cunningham, S.D. 1997. Phytoremediation of lead-contaminated soils: Role of synthetic chelates in lead phytoextraction. *Environ. Sci. Technol.* 31, 800-805.

Jiang, W.S., Liu, D.H. and Hou, W.Q. 2001. Hyperaccumulation of cadmium by roots, bulbs and shoots of garlic (Allium sativum L.). *Biores. Technol.* 76, 9-13.

Kabata-Pendias, A. and Pendias, H. 2001. *Trace Element in Soils and Plants*. 3rd edition. Boca Raton, FL, USA: CRC Press.

Kitagishi, K. and Yamane, I. 1981. *Heavy Metal Pollution in Soils of Japan*. Tokyo, Japan: Japan Sci. Soc. Press.

Kloke, A., Saurebeck, D.R. and Vetter, H. 1984. *The contamination of plants and soils with heavy metals and the transport of metals in terrestrial food chains*. In: Nriagu J.O., editor. Changing Metal Cycles and Human Health. Berlin, Germany: Springer-Veriag; p. 113-141.

Lai, H.Y. and Chen, Z.S. 2004. Effects of EDTA on solubility of cadmium, zinc, and lead and their uptake by rainbow pink and vetiver grass. *Chemosphere* 55, 421-430.

Lai, H.Y. and Chen, Z.S. 2005. The EDTA effect on phytoextraction of single and combined metals-contaminated soils using rainbow pink (Dianthus chinensis). *Chemosphere* 60, 1062-1071.

Lai, H.Y. and Chen, Z.S. 2006. The influence of EDTA application on the interactions of cadmium, zinc, and lead and their uptake of rainbow pink (Dianthus chinensis). *J. Hazard. Mater.* 137, 1710-1718.

Lai, H.Y., Su, S.W., Lin, C.C. and Chen, Z.S. 2007. A study on the recovery of soil fertility of two metal-contaminated soils after treating with soil acid washing. *J. Sci. Technol.* 16, 39-46. (in Chinese with English abstract).

Lai, H.Y., Chen, S.W. and Chen, Z.S. 2008. Pot experiment to study the uptake of Cd and Pb by three Indian mustards (Brassica juncea) grown in artificially contaminated soils. *Int. J. Phytoremediat.* 10, 91-105.

Lalor, G.C. 2008. Review of cadmium transfers from soil to humans and its health effects in the Jamaican environment. *Sci. Total Environ.* 400, 162-172.

Lin, C.C., Lai, H.Y. and Chen, Z.S. 2010. Bioavailability assessment and accumulation by five garden flower species grown in artificially cadmium-contaminated soils. *Int. J. Phytoremediat.* (In press)

Liu, D.H., Jiang, W.S. and Gao, X.Z. 2003. Effects of cadmium on root growth, cell division and nucleoli in root tip cells of garlic. *Biol. Plant.* 47, 79-83.

Lombi, E., Zhao, F.J., Dunham, S.J. and McGrath, S.P. 2000. Cadmium accumulation in populations of Thlaspi caerulescens and Thlaspi goesingense. *New Phytol.* 145, 11-20.

Luo, C.L., Shen, Z.G., Lou, L.Q. and Li, X.D. 2006. EDDS and EDTA-enhanced phytoextraction of metals from artificially contaminated soil a residual effects of chelant compounds. *Environ. Pollut.* 144, 862-871.

Mishra, V.K. and Tripathi, B.D. 2008. Concurrent removal and accumulation of heavy metals by the three aquatic macrophytes. *Biores. Technol.* 99, 7091-7097.

Obata, H. and Umebayashi, M. 1997. Effects of cadmium on mineral nutrient concentrations in plants differing in tolerance for cadmium. *J. Plant Nutr.* 20, 97-105.

Podlesáková, E., Nemecek, J. and Vácha, R. 2001. *Mobility and bioavailability of trace elements in soils*. In: Iskander I.K. and Kirkham M.B., editors. Trace Elements in Soils. FL, USA: Lewis Publisher, p. 21-41.

Quartacci, M.F., Argilla, A., Baker, A.J.M. and Navari-Izzo, F. 2006. Phytoextraction of metals from a multiply contaminated soil by India mustard. *Chemosphere* 63, 918-925.

Robinson, B.H., Leblanc, M., Petit, D., Brooks, R.R., Kirkman, J.H. and Gregg, P.E.H. 1998. The potential of Thlaspi caerulescens for phytoremediation of contaminated soils. *Plant Soil* 23, 47-56.

Schmidt, U. 2003. Enhancing phytoremediation: The effect of chemical soil manipulation on mobility, plant accumulation, and leaching of heavy metals. *J. Environ. Qual.* 32, 1939-1954.

Schwartz, C., Echevarria, G. and Morel, J.L. 2003. Phytoextraction of cadmium with Thlaspi caerulescens. *Plant Soil* 249, 27-35.

Singh, B.R. and McLaugnlin, M.J. 1999. *Cadmium in soils and plants.* In: McLaugnlin M.J. and Singh B.R., editors. Cadmium in Soils and Plants. Dordrecht, Netherlands: Kluwer Academic Publishers. p. 257-267.

Sun, Y.B., Zhou, Q.X., Wang, L. and. Liu, W.T. 2009. Cadmium tolerance and accumulation characteristics of Bidens pilosa L. as a potential Cd-hyperaccumulator. *J. Hazard. Mater.* 161, 808-814.

Taiwan EPA (Environmental Protection Administration of Taiwan). 2010. Status of soil pollution and groundwater pollution in Taiwan. Taiwan EPA website: *http://sgw.epa.gov.tw/public/En/index.htm.*

Tandy, S., Schilin, R. and Nowack, B. 2006. The influence on the uptake of heavy metals in hydroponically grown sunflowers. *Chemosphere* 62, 1454-1463.

Uraguchi, S., Watanabe, I., Yoshitomi, A., Kiyono, M. and Kuno, K. 2006. Characteristics of cadmium accumulation and tolerance in novel Cd-accumulating crops Avena strigosa and Crotalaria juncea. *J. Exper. Bot.* 57, 2955-2965.

Wenzel, W.W., Adriano, D.C., Salt, D. and Smith, R. 1999. *Phytoremediation: A plant-microbe-based remediation system.* In: Adriano D.C., Bollag, J.-M., Frankenberger Jr. W.T. and Sims, R.C., editors. Bioremediation of Contaminated Soils. Madison, Wisconsin, USA: American Society of Agronomy, Inc.; p. 457-508.

World Health Organization. 1989. *Evaluation of certain food additives and contaminants.* In: WHO Technical Series, no. 837, Geneva 124. Thirty-Third Report of the Joint FAO/WHO Expert Committee on Food Additives.

Wu, J., Hsu, F.C., and Cunningham, S.D. 1999. Chelate-assisted Pb phytoremediation: Pb availability, uptake, and translocation constraints. *Environ. Sci. Technol.* 33, 1898-1904.

In: Handbook of Phytoremediation
Editor: Ivan A. Golubev

ISBN: 978-1-61728-753-4

Chapter 28

THE ROLE OF ARBUSCULAR MYCORRHIZAL FUNGI IN PHYTOSTABILIZATION AND PHYTOEXTRACTION OF HEAVY METAL CONTAMINATED SOILS

Mehdi Zarei and Jamal Sheikhi
Department of Soil Science, College of Agriculture,
University of Shiraz, Shiraz, Iran

ABSTRACT

Of the various physico-chemical and biological technologies that have been used for remediation of heavy metals (HMs) contaminated soils, all methods are expensive and totally destroy physical, chemical and biological properties of treated soils, reduce yield of plant growth and disrupt ecosystems. Therefore, it is best to develop suitable, natural, cheaper and in situ technologies to recover degraded land. Phytoremediation is an alternative to physico-chemical methods and is emerging as a promising environmentally friendly method for detoxification and /or deactivation and removal of elements from polluted soils. It is possible to improve the capabilities of plants in different types of phytoremediation processes by inoculating with appropriate soil microorganisms especially arbuscular mycorrhizal fungi (AMF). Some AMF species occur naturally and form a symbiosis with plant roots in the HMs polluted soils. In some cases, AMF have generally such a strong influence on plant biomass and can increase HMs uptake and root-to shoot transport (phytoextraction), while in other cases AMF contribute to HMs stabilization within the soil/root and reduce their uptake (phytostabilization). In this chapter, some knowledge concerning the role of AMF in phytoextraction and phytostabilization of HMs contaminated soil was summarized and discussed.

Keywords: Arbuscular mycorrhizal fungi, phytoextraction, phytostabilization, Heavy metals, polluted soils.

INTRODUCTION

Soil as a part of the biosphere has an important role in environment sustainability and food production. Geological origins and increasing populations and their various agricultural, industrial and military activities are continuously raising the concentration of heavy metals (HMs) that pose detrimental effects on the environment, soil ecosystem and human health. Therefore, awareness of soil pollutants and attempting to remove them from soil is important and needs much more attention and investigation (Khan, 2005). The polluted soil has included very large areas that can be suitable for farming and other advantageous applications. HMs have the ability to translocate from soil to plant and so reduce yield and the safety of food chains in the environment and for humans. Therefore, removal and deletion and/or alleviation of their effects in the environment based on scientific and applicable methods are necessary and an important goal. Heavy metals are not degradable chemically and usually they should be removed physically or immobilized (Gaur and Adholeya, 2004). Different physical, chemical and biological technologies have been developed for remediation of HMs polluted soil; each one is site specific and has advantages and disadvantages. The physico-chemical techniques for soil remediation damage the soil ecosystem for plant growth as they remove all biological activities including useful microorganisms such as mycorrhiza (Khan et al., 2000). Among biological technologies, the use of metal accumulating plants to clean up polluted soils, as a natural, alternative, in situ, cheaper and environmentally friendly method is an innovative technique known as phytoremediation.

There are different types of phytoremediation such as phytoextraction, phytostabilization, phytovolatilization, rhizofiltration and phytodegradation. Phytoremediation has several approaches such as using high hyperaccumulator plants, high biomass and fast growing trees (Wang et al., 2005). It takes a long time and is more effective where pollutants are present at low to medium levels. Phytoremediation of HM contaminants is based on phytostabilization, phytoextraction, phytovolatilization and rhizofiltration processes (Table 1).

The scientists have attempted phytostabilization more for reducing the entrance of HMs in food chains and the environment. This type of phytoremediation is suitable for areas where phytoextraction is not applicable and is effective at shallow depth and where rapid immobilization is needed. In this method, high levels of HMs concentration, easy translocation and its toxicity for plants are the main limitations. Phytoextraction means to reduce soil pollution by hyperaccumulator plants that can take up over 100 mg Cd, 1000 mg Co, Cr, Cu and Pb and/or 1000 mg Ni and Zn per kg dry weight of their shoots.

Table 1. Different types of phytoremediation of HMs polluted soils

Type	Definition
Phytostabilization	The use of plants to transform HMs to less toxic forms
Phytoextraction	The use of plants to extract and remove HMs from soils
Phytovolatilization	The use of plant to convert HMs to volatile species
Rhizofiltration	The use of plants to remove HMs from flowing water

However, there are some restrictions as follows: rareness of metal hyperaccumulation phenomena, small biomass and slow growth of terrestrial plants, pollution possibility of food chains and also the need for proper disposal or burying of harvested biomass.

Several methods to enhance phytoremediation efficiency have been proposed. Soil microorganisms can influence phytoremediation directly and/or indirectly. Therefore, they should be considered in remediation programs. Among them, use of arbuscular mycorrhizal symbiosis as healthy biological source has multidirectional effects such as excretion of chelating agents, producing of plant growth promoting factors and increasing of plant biomass, extending of soil rhizosphere (mycorrhizosphere) and increasing of uptake per unit surface area.

There are conflicting reports about the influences of arbuscular mycorrhizal fungi (AMF) in HM uptake and translocation. Some reports showed higher levels of HMs in plants (Killham and Firestone, 1983; Weissenhorn and Leyval, 1995; Weissenhorn et al., 1995), while others indicated reduced concentrations in plants due to mycorrhizal colonization (Gildon and Tinker, 1983; Heggo et al., 1990; Leyval et al., 1991; Hetrick et al., 1994; Hildebrandt et al., 1999). Different results about AMF in HMs uptake may be due to inherent capacity of plants for elements uptake, root density, plant growth condition, type species and characteristics of fungi and sorption and desorption properties of soils (Leyval et al., 1997). Diaz et al. (1996) evaluated effects of two AMF species *Glomus mosseae* and *Glomus macrocarpum* on Zn and Pb uptake and growth of *Lygeum spartum* and *Anthyllis cytisoides*. Their results showed that mycorrhizal plants had identical and/or high amounts of Pb and Zn in comparison with non-inoculated plants at low soil metal concentration, whereas under high concentrations, plants inoculated with *Glomus mosseae* had lower and those inoculated with *Glomus macrocarpum* had identical and/or high Pb and Zn concentrations than control treatments. In a pot experiment, it was observed that the concentrations of Zn and Ni in the mycorrhizal plant tissues of lentil and soybean were significantly higher than non-inoculated ones. Among two plants, mycorrhizal soybean plant tissues had higher Zn and Ni than those in the mycorrhizal lentil plant tissues (Jamal et al. 2002). It has drawn attention to studies about utility of AMF in stabilization and extraction processes of HMs.

The Role of AMF in Phytostabilization

Phytostabilization is immobilization of HMs in soil through absorption, accumulation by roots, and adsorption onto roots or precipitation within the root zone. In phytostabilization, plant root controls HMs uptake, inactives them in the rhizospheric soil and reduces the toxicity of elements in the environment and for humans. The presence of plants also reduces erosion, leaching and soil dispersion. Effective phytostabilization requires the establishment of condensed plant cover in a minimum time. The goal of phytostabilization is establishment of a sustainable ecosystem with the essential condition of plant diversity and functional sustainability. In this case the AMF may be very helpful and enhance stabilization of HMs in the roots and soil. It was proved that the AMF are effective in immobilization of metals in the plant rhizosphere, and help in HMs stabilization by their accumulation in a non-toxic form in plant roots and extracellular mycelia. There are similar strategies in decreasing of the toxic effects of HMs for fungi and host plants that include immobilization of these elements by the fungal exudations, their deposit in polyphosphate granular, adsorption of elements on the cell wall and chelation in the fungal organs. Glomalin is a glycoprotein produced abundantly on hyphae and spores of AMF in soil and in roots and is able to link with HMs and extract them from the soil, so a positive correlation has been proven between the amount of soil glomalin and the amount of binding HMs. Therefore, it can be said the fungal strains that secrete more glomalin are more suitable for biological stabilization (Göhre and Paszkowski, 2006). Binding heavy metals with chitin in the cell wall of fungal organs reduces their concentration in the soil solution and broad absorption surface of extraradical mycelia are considered an important source of discharged HMs from the soil solution. The vesicles of fungi also have a role in accumulation of toxic compounds and in this way can help in the detoxification of metals. It was shown that with increasing of HMs concentration, the number of vesicles in the colonized roots increased (Zarei et al., 2008). In addition, HMs tolerant fungi have more performance for stabilizing metals within the roots and are more appropriate in phytostabilization (Chen et al., 2003). In a highly polluted soil, mycorrhizal colonization increased the uptake of cadmium, zinc and copper in oat roots but their transport to the shoot reduced (Loth and Hofner, 1995).

A *Glomus* strain isolated from violet plant was effective in accumulation of HMs to non-toxic form in the colonized roots of corn, barley and alfalfa plants and helped to plants complete their life cycle in highly polluted soils (Hildebrandt et al., 1999). The AMF originated from rhizosphere of violet plant increased the cadmium and zinc concentrations in the roots up to 8 and 3 times respectively without it made significant effect on concentration of elements and shoot biomass (Wang et al., 2007c). In the evaluation of the role of *Glomus mosseae* in uptake of zinc by the red clover in a sterilized calcareous soil at levels of 0, 600 and 1200 mg Zn per kg, it was observed that at the high levels of zinc, mycorrhizal colonization was not reduced and similar to low levels of zinc was up to 45-46 percent. With increasing of zinc levels, dry weight of root and shoot decreased, but was not significant difference between treatments. Mycorrhizal fungi inoculum increased phosphorus uptake in clover but not effected from zinc levels. Overall, phosphorus concentration of mycorrhizal clover was higher up to 2 to 3 times than non-mycorrhizal treatment. With increasing zinc levels, zinc uptake in plant roots and shoots at presence or absence of mycorrhizal fungi treatment was increased. Increasing zinc concentration in shoots of mycorrhizal plants was less than non-mycorrhizal ones but in roots was higher (Chen et al., 2003).

High concentration of zinc in mycorrhizal roots than non-mycorrhizal ones showed that the fungus could maintain zinc in surface and/or within mycelia. Zinc concentration in fungal mycelia in comparison with plant tissues was reported more than 10 times (Chen et al., 2001). Different distribution pattern of elements in mycorrhizal and non-mycorrhizal maize was reported by the researchers. The results showed that most of zinc accumulated in the fungal tissue, such as vesicles inside the cells of root cortex (Kaldorf et al., 1999). So it can be concluded that the immobilization of zinc in fungal tissue is one of the mechanisms of reducing zinc toxicity in mycorrhizal plants. In an experiment, the direct role of mycorrhizal fungus hyphae in zinc uptake of red clover with three levels of 0, 50 and 200 mg zinc per kg, was evaluated. Mycorrhizal colonization decreased with increasing of zinc levels to some extent, but there was no significant difference between shoot dry weight in mycorrhizal and non-mycorrhizal plant. Root dry weight mycorrhizal plants decreased at all zinc levels in mycorrhizal than non-mycorrhizal treatment except at level of 50 mg per kg. Zinc concentration in roots and shoots of non-mycorrhizal plants with increasing levels of zinc was not changed, while mycorrhizal infection significantly intensified zinc concentration in shoots and roots of plant, especially at higher concentrations. They eliminated the dilution effect with addition of 300 mg phosphorus per kg in non-mycorrhizal fungus treatments, and through they indicated the role of mycorrhizal hyphae in zinc uptake. Based on this experiment it was observed that at the level of 50 mg per kg, the amount of zinc uptake by fungus, was 62 μg (equivalent to 22 percent of total zinc absorption) and at the level of 200 mg per kg, was 21 μg (equivalent to 8.1 percent of total zinc absorption). Increased maintenance of zinc in the plant roots colonized with AMF has been reported in various studies (Chen et al., 2001; Chen et al., 2003). These result showed the beneficial role of AM symbiosis in alleviation and reduction of zinc toxicity.

The effect of *Glomus intraradices* in the uptake of uranium has been investigated in a greenhouse experiment (Rufyikiri et al., 2004).The results showed that with increasing the amount of soil uranium, the uranium concentration increased in roots and shoots of mycorrhizal plants and control treatment. AMF increased the phosphorus concentration and dry weight roots and shoots in comparison with control treatment. At the highest level of uranium added to the soil (87 mg per kg), the uranium concentration in the shoots of control plants were 1.7 times more than mycorrhizal plants. The results of this experiment showed the accumulation of uranium in the plant root of mycorrhizal plant that were exposed to the high levels of uranium and supportive effect of this fungus for the host plant (Rufyikiri et al., 2004).

The results of a greenhouse experiment showed that *Acaulospora mellea* increased the uptake of copper by maize in the soils with copper deficiency, while at the high levels of copper reduced the concentration of this element in the shoots. It was caused that the toxic effects of this metal decreased and plant growth increased. These authors, attempting to the high uptake of copper by mycorrhizal roots, showed the possible role of this fungus in phytostabilization (Wang et al., 2007b; Wang et al., 2007c).

In a pot experiment the effects of three isolates of *Glomus intraradices* in accumulation and transportation of lead in maize was studied. The isolates, included 1- isolate originated from the non-polluted soils (reference isolate), 2- isolate originated from the lead polluted soils that was propagated exclusively in the polluted substrate, 3- isolate originated from the non-polluted soils (reference isolate) that for 45 months was cultured in the polluted substrate. The results showed that the tolerance and colonization of three isolates were different in the lead polluted substrate. The root colonization of maize plant with isolates numbers 1, 2 and 3 was 38, 75 and 85 percent, respectively. The abundance changes of arbuscule and vesicle in the roots was similar to root colonization. Two indigenous isolates produced highest extracellular mycelium and NADPH activity in their mycelium was higher than reference isolate.These researchers showed that the accumulation of phosphorus and lead among the three isolates was different. Increasing lead concentration in the roots and decreasing of its transport to shoots of plants inoculated with isolate numbers 2 and 3 showed lead immobilization within the roots and the protective role of the fungi in the soils polluted with lead (Sudova and Vosatka, 2007).

The results of studies in this field indicated the role of AMF in immobilization of HMs in the soil and roots, and showed this fact that AMF play the important and useful role in phytostabilization. The best option for this purpose is indigenous fungi originated from the polluted soil (Göhre and Paszkowski, 2006).

The Role of AMF in Phytoextraction

Phytoextraction is a technology with high performance for clean up HMs from soils with low or moderate contamination. In this method, hyperaccumulator plants with high root-to-shoot transfer accumulating high amounts of HM in above-ground parts and plants producing high biomass with normal concentrations of HM are great promise. Normally hyperaccumulator plants can accumulate HM without displaying toxicity symptoms.

Finally, cultured plants in the polluted areas can be harvested, dried and burned, and/or HMs can be again recaptured or phytominig (Kramer, 2005). In this approach, the soil clean up is slow process and its implementation require to cultivation of these plants in the several years. Furthermore, the availability of metals in the soil is an important factor in phytoextraction. Based on using mathematical calculation that it was required for reducing 1 mg cadmium per kg soil by *Thlaspi caerulescens* about 2 years time and concluded that phytoextraction of the highly polluted soil is not suitable. Also, the researchers reported that the amounts of metals transported from soil-to-plant highly correlated to available forms of HM, for example, long-term cultivation of *Salix*, reduced available Cd up to 30-40 percent while had no effect on the total concentration of this element (Nevel et al., 2007). Most hyperaccumulator plants produce modest biomass and/or their biomass reduce over time, due to nutrient discharge, incidence of disease or occurrence of stress factors such as decreasing pH, low aeration and other environmental stress. In addition, most of soils are polluted to several metals that intensively influence on plant productivity. Plant species can rarely extract more than one metal because of this, the decontamination process need a longer time. There are possibilities of high pollution of topsoil layers nearby non-polluted soils due to distribution of polluted remains of plants. The results of these studies showed that phytoextraction was applicable for arsenic and nickel because of most hyperaccumulator plants belonged to these elements. Plant species that could accumulate high amounts of zinc, lead, cadmium, cobalt, arsenic and copper were report less (Nevel et al., 2007).

AMF increased the uptake and accumulation of arsenic in hyperaccumulator plant of *Pteris vittata*. At the concentration of 100 mg arsenic per kg soil, control plants accumulated 60.4 mg arsenic per kg and plants inoculated with native AMF form the arsenic mine accumulated 88.1 mg arsenic per kg plant dry matter. The results accompanied by improving nutrition and increasing growth of colonized plants as a result phosphorus absorption up to 257 mg per pot in comparison with 36.3 mg phosphate per pot in control plants. In these situations accumulation and recovery of arsenic increased (Leung et al., 2006). *Berkheya coddii* belonging to Asteraceae family like to *Pteris vittata* was known as an effective plant in phytoextraction and this plant was used for phytominig. The biomass of this nickel hyperaccumulator plant in mycorrhizal treatments was two times more than non-mycorrhizal plants and nickel accumulation was higher in mycorrhizal plants up to 30 percent (Göhre and Paszkowski, 2006).Non-hyperaccumulator plants will be suitable for the phytoextraction if it can tolerate HMs and produce high biomass. It was reported that root and shoot biomass of mycorrhizal tomato plant were up to 30 percent in arsenic concentration of 75 mg per kg soil higher than non-mycorrhizal plant and accompanied by this ,phosphorus uptake also increased (Göhre and Paszkowski, 2006). It was shown dynamic and mobilization of zinc and transferring to shoots of corn and clover colonized by AMF at zinc deficiency conditions (Chen et al., 2003). The results showed beneficial role of mycorrhizal symbiosis in improving zinc nutrition as a micronutrient. The uptake of HMs by *Elsholtzia splendens* inoculated with two inocula of AMF were studied in the HMs polluted soils with copper, zinc, lead and cadmium in pot and field experiment (Wang et al., 2005). First inoculum containing only one AM fungus *Glomus caledonium* significantly increased root colonization of *Elsholtzia splendens* than the second inoculum consisting of *Gigaspora*, *Acaulospora*, *Scutellospora* and *Glomus*. The both inocula increased root and shoot dry weight. The concentration of copper, lead, zinc and phosphorus increased in shoots of plants inoculated with mixture of AMF while the first mycorrhizal treatment decreased the concentration of lead and zinc, increased phosphorus but did not alter Cu. The results showed that the AMF consortium was more effective in HM phytoextraction and therefore play role in phytoremediation. At another experiment the effect of two inocula noted above were studied in the uptake of heavy metals by corn in the non-sterilized polluted soil (Wang et al., 2007a). The results showed that concentrations of copper, zinc, lead and cadmium in plant shoots inoculated with second inoculum increased while other decreased. None of inocula influenced significantly stems and roots dry weight but increased phosphorus concentration in the roots. The uptake of heavy metals increased in shoot of plant inoculated with second inoculum. The second inoculum was more effective and appropriate in HMs phytoextraction (Wang et al., 2007a). In a field experiment it was studied the effect of microbial inocula, including AMF and two species of *Penicillium*, with and without chitosan (Chitosan is a linear polysaccharide composed of randomly distributed β-(1-4)-linked D-glucosamine (deacetylated unit) and N-acetyl-D-glucosamine (acetylated unit)) in phytoextraction of copper, zinc, lead and cadmium by *Elsholtzia splendens*. Microbial inocula increased plant biomass especially shoots dry weight and also concentrations of lead, zinc and copper in the shoots. In comparison with microbial inocula, application of chitosan alone has not effect on plant growth, but increased concentrations of copper, zinc, lead and cadmium in the shoots. The results showed synergetic effects of chitosan and microbial inocula in HMs phytoextraction (Wang et al., 2007b). *Solanum nigrum* can accumulate large amounts of zinc up to 1622 mg per kg in its tissue in the zinc polluted soil. Inoculation with *Glomus claroideum* and *Glomus intraradices* originated from the HMs polluted soil increased zinc uptake and accumulation of zinc in this plant up to 83 and 49 percent respectively (Marques et al., 2007). These results indicated that using tolerant fungi inocula strengthened the extraction and accumulation capacity of zinc in the plants. Dominant AMF species including *Glomus mosseae*, *G. intraradices* and *G. versiforme* from soils with high and moderate levels of HMs were isolated and propagated by monospecies and trap culture, and their effectiveness in phytoremediation of zinc polluted soils (5 levels of zinc 0, 10, 50, 100 and 500 mg per kg) using maize and tall fescue as host plants were studied in a greenhouse experiment (Zarei, 2008). The difference between the effects of three studied fungal species on both host plants and their reactions against the Zn content of soil, were more obvious at higher level of soil Zn. At the highest level of this heavy metal (500 mg per kg), the concentration of Zn were higher in aerial parts and lower in the root of plants inoculated with *G. mosseae*, in comparison with two other AMF species (Figures 1 and 2).

The inverse partitioning of Zn was observed in case of plants inoculated with *G. intraradices*, out of which the amount of Zn in the roots was higher than that of shoots in comparison with plants inoculated with *G. mosseae*. In general, for both plants and under the high soil pollution (500 mg per kg), *G. mosseae* was the most effective fungal species in Zn extraction and translocation while, *G. intraradices* had the highest effectiveness for accumulation of Zn in the roots. The overall situation of *G. versiforme* was mostly between the two other fungal species (Zarei, 2008).

Generally, the results showed that plant colonized by AMF can increase the uptake and accumulate of heavy metals in plant shoots or phytoextraction and in this context more attention to the composition of plant-fungus and suitable conditions of edaphic and climatic is required.

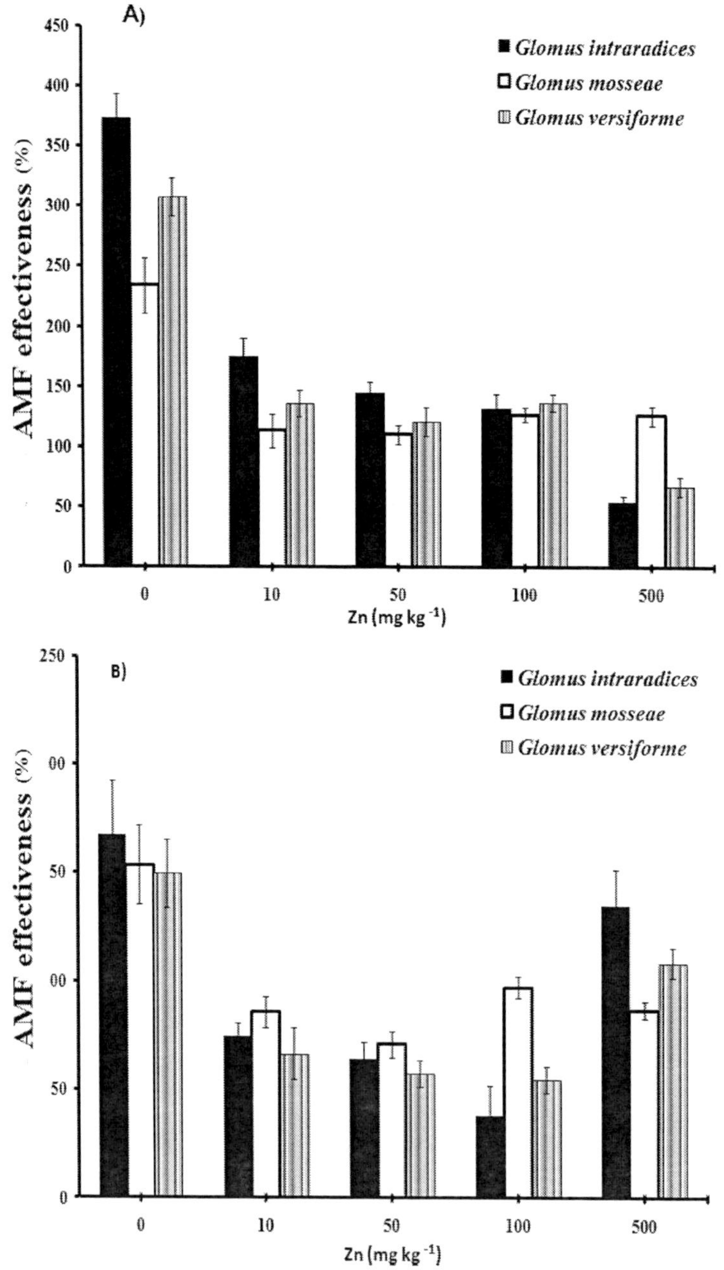

Figure 1. AMF effectiveness percent on Zn uptake (mean ± SE) in shoot (A) and root (B) of maize plants under different treatments. Bars represent the standard error (Zarei, 2008).

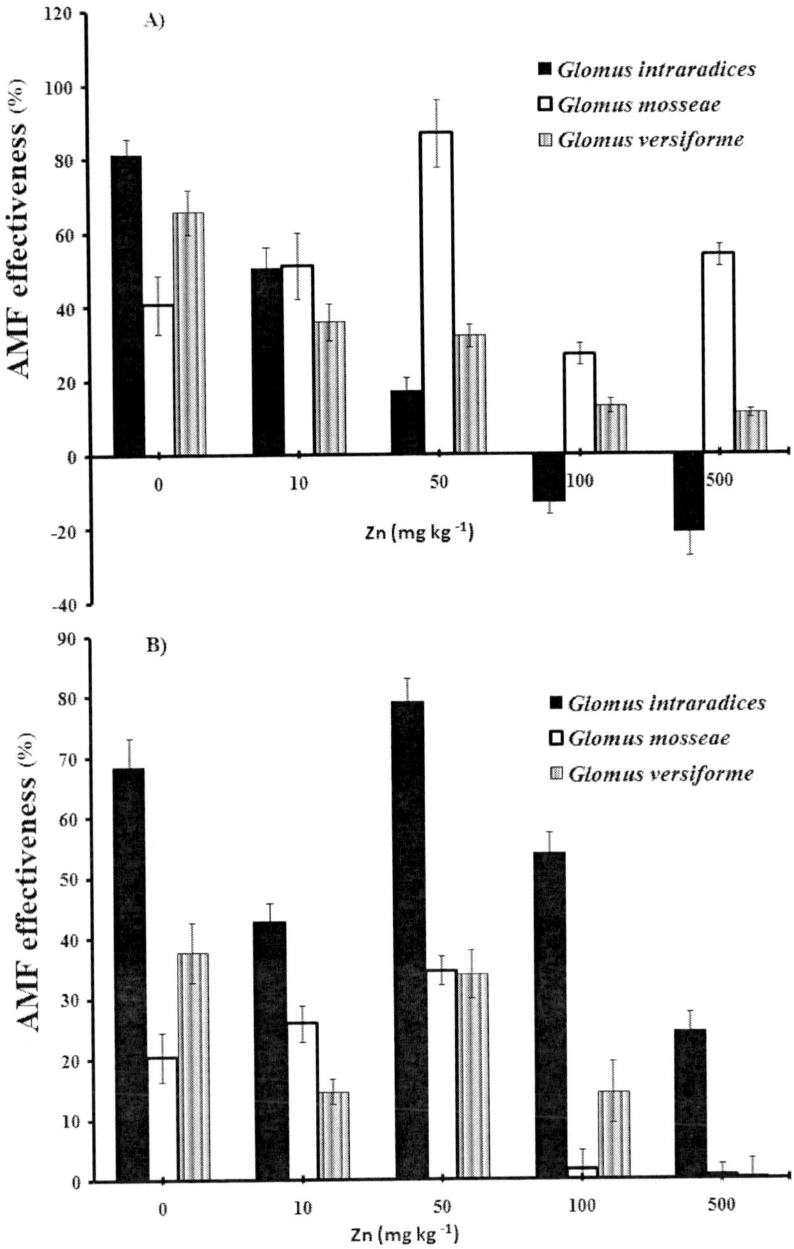

Figure 2. AMF effectiveness percent on Zn uptake (mean ± SE) in shoot (A) and root (B) of Festuca plants under different treatments. Bars represent the standard error (Zarei, 2008).

CONCLUSION

Polluted soils are complex ecosystems; therefore, the importance of selection of an appropriate combination of plant-fungus and optimizing the productivity and interaction among soil components, environmental conditions and microbial communities that positively or negatively influence on effectiveness of phytoremediation should be recognized and success of most phytoremediation applications will be dependent on this dynamic relationship (Olsson et al. 2003). Among soil microorganisms, AMF provide a direct link between soil and roots and have proved to be especially useful in phytoextraction and phytostabilization of HMs polluted soils. Considering that the AM fungus is a non-host specific symbiont, and that the host plant always plays the important role in the symbiosis, more attention to host plants and more systematic researches on the mechanisms for how AM fungus are involved in HMs absorption and transportation in plants and the tolerance of AMF to HMs is necessary for the specific restoration and re-vegetation programs (Chen et al. 2003, Göhre and Paszkowski, 2006).

ACKNOWLEDGMENT

We would like to thank Dr. Srefan Hempel for reviewing the chapter and for valuable comments.

REFERENCES

Chen, B. D., Li, X. L., Tao H. Q., Christie, P. and Wong, M. H. (2003). The role of arbuscular mycorrhiza in zinc uptake by red clover growing in a calcareous soil spiked with various quantities of zinc. *Chemosphere*, 50, 839-846.

Chen, B., Christie, P. and Li, X. (2001). A modified glass bead compartment cultivation system for studies on nutrient and trace metal uptake by arbuscular mycorrhiza. *Chemosphere*, 42,185-192.

Diaz, G., Azcon-Aguilar, C. and Hornubia, M. (1996). Influence of arbuscular mycorrhizae on heavy metal (Zn and Pb) uptake and growth of *Lygeum spartum* and *Anthyllis cystoides*. *Plant and Soil*, 180, 241-249.

Gaur, A. and Adholeya, A. (2004). Prospects of arbuscular mycorrhizal fungi in phytoremediation of heavy metal contaminated soils. *Current Science*, 86, 528-534.

Gildon, A. and Tinker, P. B. (1983). Interactions of vesicular-arbuscular mycorrhizal infection and heavy metals in plants. The effects of heavy metals on the development of vesicular-arbuscular mycorrhizas. *New Phytologist*, 95, 247-261.

Gohre, V. and Paszkowski, U. (2006). Contribution of the arbuscular mycorrhizal symbiosis to heavy metal phytoremediation. *Planta*, 223, 1115-1122.

Heggo, A., Angle, A. and Chaney, R. L. (1990). Efects of vesicular arbuscular mycorrhizal fungi on heavy metal uptake by soybeans. *Soil Biology and Biochemistry*, 22, 865-869.

Hetrick, B. A. D., Wilson, G. W. T. and Figge, D. A. H. (1994). The influence of mycorrhizal symbiosis and fertilizer amendments on establishment of vegetation in heavy metal mine spoil. *Environmental Pollution*, 86, 171-179.

Hildebrandt, U., Kaldorf, M. and Bothe, H. (1999). The zinc violet and its colonization by arbuscular mycorrhizal fungi. *Journal of Plant physiology*, 154, 709–717.

Jamal, A., Ayub, N., Usman, M. and Khan, A. G. (2002). Arbuscular mycorrhizal fungi enhance zinc and nickel uptake from contaminated soil by soybean and lentil. *International Journal of Phytoremediation*, 4(3), 205-221.

Kaldorf, M., Kuhn, A., Schroder, W.H., Hildebrandt, U. and Bothe, H. (1999). Selective element deposits in maize colonized by a heavy metal tolerance conferring arbuscular mycorrhizal fungus. *Journal of Plant physiology*, 154, 718-728.

Khan, A. G., Kuek, C., Chaudhry, T. M., Khoo, C. S. and Hayes,W. J. (2000). Role of plants, mycorrhizae and phytochelators in heavy metal contamination land remediation. *Chemosphere*, 41, 197-207.

Khan, A.G. (2005). Role of soil microbes in the rhizosphere of plants growing on trace metal contaminated soils in phytoremediation, *Journal of Trace Elements in Medicine and Biology*, 18, 355-364.

Killham, K. and Firestone, M. K. (1983). Vesicular arbuscular mycorrhizal mediation of grass response to acidic and heavy metal depositions. *Plant and Soil*, 72, 39-48.

Kramer, U. (2005). Phytoremediation: novel approaches to cleaning up polluted soils. *Current Opinion in Biotechnology,* 16, 133-141.

Leung, H. M., Ye, Z. H. and Wong, M. H. (2006). Interactions of mycorrhizal fungi with *Pteris vittat*a (as hyperaccumulator) in as-contaminated soils. *Environmental Pollution*, 139, 1-8.

Leyval, C., Berthelin, J., Schontz, D., Weissenhorn, I. and Morel, L. (1991). Influence of endomycorrhizas on maiz uptake of Pb, Cu and Cd applied as mineral salts or sewage sludge. In *J. G. Farmer* (Eds.), Heavy metals in the environment (pp 204-207). Edinburgh, CEP Consultants.

Leyval, C., Turnau, K. and Haselwandter, K. (1997). Effect of heavy metal pollution on mycorrhizal colonization and function: physiological, ecological and applied aspects. *Mycorrhiza*, 7, 139-153.

Loth, F. G. and Hofner, W. (1995). Einfluss der VA-mycorrhiza auf die schwermetallaafnahme von hafer (*Avena sativa* L.) in Abhangigheit vom contaminations gradder Boden. *Pflanzenernaehr Bodenk*, 158, 339-345.

Marques, A. P. G. C., Oliveira, R. S., Samardjieva, K. A., Pissarra, J., Rangel, A. O. S. S. and Castro, P. M. L. (2007). *Solanum nigrum* grown in contaminated soils. Effects of AMF on zinc accumulation and histolocalisation. *Environmental Pollution*, 145, 691-699.

Nevel, L. V., Mertens, J., Oorts, K. and Verheyen, K. (2007). Phytoextraction of metals from soils: how far from practice?. *Enviromental Pollution*, 150, 34-40.

Raskin, I. (1995). Phytoremediation: a novel strategy for the removal of toxic metals from the environment using plants. *Biotechnology*, 13, 468.

Rufyikiri, G., Huysmans, L., Wannijin, J., Hees, M. V., Leyval, C. and Jakobsen, I. (2004). Arbuscular mycorrhizal fungi can decrease the uptake of uranium by subterranean clover grown at high levels of uranium in soil. *Environmental Pollution*, 130, 427-436.

Sudova, R. and Vosatka, M. (2007). Differences in the effects of three arbuscular mycorrhizal fungal strains on P and Pb accumulation by maize plants. *Plant and Soil*, 296, 77-83.

Tonin, C., Vandenkoornhuyse, P., Joner, E. J., Straczek, J. and Leyval, C. (2001). Assessment of arbuscular mycorrhizal fungi diversity in the rhizosphere of *Viola calaminaria* and effect of these fungi on heavy metal uptake by clover. *Mycorrhiza*, 10, 161-168.

Wang, F. Y., Lin, X. G. and Yin, R. (2004). Heavy metal uptake by arbuscular mycorrhizas of *Elsholtzia splendens* and the potential for phytoremediation of contaminated soil. *Plant and Soil*, 269, 225-232.

Wang, F. Y., Lin, X. G. and Yin, R. (2007a). Effect of arbuscular mycorrhizal fungal inoculation on heavy metal accumulation of maize grown in a naturally contaminated soil. *International Journal of Phytoremediation*, 9(4), 345-353.

Wang, F. Y., Lin, X. G. and Yin, R. (2007b). Inoculation with arbuscular mycorrhizal fungus *Acaulospora mellea* decrease Cu phytoextraction by maize from Cu-contaminated soil. *Pedobiologia,* 51, 99-109.

Wang, F. Y., Lin, X. G. and Yin, R. (2007c). Role of microbial inoculation and chitosan in phytoextraction of Cu, Zn, Pb and Cd by *Elsholtzia splendens* - a field case. *Environmental Pollution*, 147, 248-255.

Weissenhorn, I. and Leyval, C. (1995). Root colonization of maize by a Cd-sensitive and a Cd-tolerant *Glomus mosseae* and cadmium uptake in sand culture. *Plant and Soil*, 175, 233-238.

Weissenhorn, I., Leyval, C., Belgy, G. and Berthelin, J. (1995). Arbuscular mycorrhizal contribution to heavy metal uptake by maize (*Zea mays* L.) in pot culture with contaminated soil. *Mycorrhiza*, 5, 245-252.

Zarei M., S. König, Hempel S., Khayyam Nekouei M., Savaghebi Gh. and Buscot F. (2008) Community structure of arbuscular mycorrhizal fungi associated to *Veronica rechingeri* at Anguran zinc and lead mining region. *Environmental Pollution*, 156, 1277-1283.

Zarei, M. (2008). *Diversity of arbuscular mycorrhizal fungi in heavy metal pollution soils and their roles in phytoremediation*. Ph.D. dissertation in Soil Science (Soil Biology and Biotechnology), Agricultural Faculty, University of Tehran, Tehran, Iran, 219 pp (In Farsi).

In: Handbook of Phytoremediation
Editor: Ivan A. Golubev

ISBN: 978-1-61728-753-4

Chapter 29

BACTERIAL ACC DEAMINASE AND IAA: INTERACTIONS AND CONSEQUENCES FOR PLANT GROWTH IN POLLUTED ENVIRONMENTS

Elisa Gamalero[1] and Bernard R. Glick[2]
[1]Università del Piemonte Orientale ''Amedeo Avogadro'',
Dipartimento di Scienze dell'Ambiente e della Vita,
Viale Teresa Michel 11, 15121 Alessandria, Italy.
[2]University of Waterloo, Department of Biology, Waterloo,
ON, Canada N2L 3G1

ABSTRACT

Plant growth-promoting bacteria are soil bacteria that are involved in a beneficial association with plants; these bacteria use a variety of mechanisms to facilitate plant growth. The major mechanisms used by plant growth-promoting bacteria include the functioning of the enzyme 1-aminocyclopropane-1-carboxylate (ACC) deaminase which cleaves the compound ACC, the immediate precursor of the phytohormone ethylene in all higher plants, and synthesis of the plant hormone indoleacetic acid (IAA). Plant growth-promoting bacterial strains that contain ACC deaminase and produce IAA provide a wide range of different plant species with a significant level of protection from the damage caused by various environmental stresses including heavy metals and the presence of organic environmental contaminants. Here we discuss how bacterial ACC deaminase and IAA work synergistically to facilitate plant growth during the phytoremediation of metals and/or organics.

INTRODUCTION

The soil surrounding plant roots (i.e., the rhizosphere) is the main source of bacteria with plant-beneficial activities. These bacteria, generally defined as plant growth-promoting bacteria (PGPB) (Bashan and Holguin, 1998), promote plant growth by both direct and indirect methods, the latter consisting of the inhibition of the deleterious effects of phytopathogenic microorganisms (Glick, 1995).

The direct stimulation of plant growth by PGPB is mediated by a range of mechanisms including improvement of mineral nutrition (e.g., through nitrogen fixation, phosphate solubilization and iron chelation); phytohormone production (including auxins such as indoleacetic acid (IAA), cytokinins and gibberellins), with IAA generally considered to be the most important of the phytohormones provided by bacteria; and the enhancement of plant tolerance to biotic and abiotic stress and modification of root architecture by the modulation of plant ethylene levels as a consequence of the activity of the enzyme 1-aminocyclopropane-1-carboxylate (ACC) deaminase (Glick, 1995; Glick et al., 1999; Gamalero et al., 2010).

This brief chapter aims to describe the role of bacterial ACC deaminase and IAA, in concert, in protecting plants against the damage caused by environmental stresses such as the presence of heavy metals and/or organic contaminants. Herein, the complex synergistic interaction between ACC deaminase and IAA is emphasized, along with the consequences of that interaction in terms of its effect on plant growth in polluted soils. The first section provides a brief description of how environmental contaminants (metals and organics) cause plant stress, with particular reference to ethylene synthesis. In the second section, the mechanism of plant growth promotion induced by ACC deaminase and IAA, both separately and together, is detailed. Phytoremediation experiments where bacteria with ACC deaminase and IAA were used to promote plant growth are summarized. Finally, the effect of ACC deaminase on plant gene expression is discussed briefly.

Impact of Environmental Contaminants on Plants: Emergence of Plant Stress

Plants growing in polluted environments may respond in a variety of ways according to the nature of the contaminant and the ability of the plant to tolerate the xenobiotic. These responses may lead to the degradation or breakdown of organic contaminants by internal and external metabolic processes of the plant by itself (phytodegradation) or in conjunction with root associated microflora (rhizodegradation); to the uptake of a pollutant and its release into the atmosphere (phytovolatilization); to the uptake of an inorganic pollutant and, either its accumulation in the roots (phytostabilization) or its translocation into plant tissues (phytoextraction). Irrespective of the contaminant type, the exposure of plants to stressful conditions raises the ethylene level, leading to inhibition of root elongation and to a stress senescence response (Deikman 1997). Ethylene is a gaseous hormone, involved in several phases of plant growth (i.e., fruit ripening, flower senescence, leaf and petal abscission). In addition, this phytohormone is also important for its role in plant responses to biotic and abiotic stresses (Abeles et al., 1992). The term "stress ethylene" (Abeles, 1973), describes the increase in ethylene synthesis associated with environmental stresses such as extremes of temperature, high light, flooding, drought, the presence of toxic metals and organic pollutants, radiation, wounding, insect predation, high salt, and various pathogens including viruses, bacteria and fungi (Morgan and Drew, 1997).

In higher plants, ethylene is synthesized from L-methionine via the intermediates, S-adenosyl-L-methionine (SAM) and ACC (Yang and Hoffman 1984) through the enzymes SAM synthetase (Giovanelli et al., 1980) and ACC synthase, respectively. Finally, the enzyme ACC oxidase converts ACC to ethylene, carbon dioxide, and cyanide (John, 1991).

According to a model originally developed to describe the response of plants to pathogen stress (Abeles et al., 1992; Robinson et al., 2001; Glick et al., 2007), plants exposed to a stress show an initial small peak of ethylene a few hours after the onset of a stress. Then, ACC synthase genes are transcribed and, usually one to three days later, as a consequence of the synthesis of additional ACC, a second much larger ethylene peak is observed. The first ethylene peak is believed to be involved in the initiation of a protective response by the plant, activating processes such as transcription of pathogenesis-related (PR) genes and those involved in acquired resistance (Ciardi et al., 2000; Van Loon and Glick, 2004), while the second (deleterious) ethylene peak induces processes such as senescence, chlorosis and abscission leading to a general inhibition of plant growth and survival.

Promotion of Plant Growth in Polluted Environments, Mediated by Bacterial IAA and ACC Deaminase

PGPB colonize the seed or root of a developing plant and, in response to the exudation of tryptophan and other small molecules (Bayliss et al., 1997; Penrose and Glick, 2001), the bacteria synthesize and secrete IAA (Patten and Glick, 1996, 2002) that can influence plant cell and tissue division, extension and differentiation; stimulate seed and tuber germination; increase the rate of xylem development; control processes of vegetative growth; initiate the formation of lateral and adventitious roots; mediate plant responses to light and gravity; modulate florescence and fructification; increase photosynthesis; and facilitate resistance to stressful conditions (Tsakelova et al., 2006). Production of IAA is widespread among soil bacteria, and it has been estimated ~80% of all soil bacteria are able to synthesize IAA (Patten and Glick, 1996).

Different IAA concentrations have diverse effects on the physiology of plants, with plant responses being a function of the type of plant, the particular tissue involved and the developmental stage of the plant (Audus, 1959). The actual range of effective IAA concentrations varies according to the plant species and the sensitivity of the plant tissue to IAA; levels of IAA below a critical range have no effect whereas higher concentrations inhibit growth (Audus, 1959, Peck and Kende, 1995). Moreover, the endogenous pool of IAA in the plant is affected by soil microorganisms that can synthesize this phytohormone, and the impact of bacterial IAA on plant development ranges from positive to negative effects according to the amount of IAA available to the plant and to the sensitivity of the host plant to the phytohormone. In addition, the level of IAA synthesized by the plant itself may be important in determining whether bacterial IAA will stimulate or suppress plant growth. In plant roots, endogenous IAA may be suboptimal or optimal for growth (Pilet and Saugy, 1987) and additional IAA from bacteria could alter the IAA level to become either optimal or supraoptimal, resulting in plant growth promotion or inhibition, respectively. One of the main effects of bacterial IAA is the enhancement of lateral and adventitious rooting leading to improved mineral and nutrient uptake and root exudation that, in turn, stimulates bacterial proliferation on the roots (Dobbelaere et al., 1999; Lambrecht et al., 2000; Steenhoudt and Vanderleyden, 2000). In addition, a low concentration (up to 10^{-8} M) of exogenous IAA can enhance nodulation by Rhizobia on *Medicago truncatula* and *Phaseolus vulgaris*, while higher concentrations inhibit nodulation (Plazinski and Rolfe, 1985; van Noorden et al., 2006).

It has been suggested that one mechanism that PGPB use to prevent some of the deleterious effects of environmental stresses is the synthesis of IAA (Lindberg et al., 1985; Frankenberger and Arshad, 1995). For example, IAA has been reported to stimulate lengthening of the roots and shoots of wheat seedlings exposed to high levels of salt (Egamberdieva, 2009).

In addition to stimulating plant growth, bacterial IAA can also induce the synthesis of plant ACC synthase, thereby increasing the conversion of SAM to ACC, hence favoring the synthesis of ethylene and the formation of a plant stress response (Figure 1). A portion of the newly produced ACC is exuded from plant seeds or roots (Bayliss et al., 1997; Penrose and Glick 2001), taken up by the bacteria, and converted by ACC deaminase to ammonia and α-ketobutyrate (Figure 1). As a result of this activity, the amount of ethylene produced by the plant is reduced. Therefore, root colonization by bacteria that synthesize ACC deaminase prevents limits ethylene levels that might otherwise be growth inhibitory (Glick, 1995). The main visible effect of seed inoculation with ACC deaminase-producing bacteria, under gnotobiotic conditions, is the enhancement of root elongation (Glick et al., 1995; Hall et al., 1996; Shah et al., 1997).

In the model (Glick, 1995; Glick et al., 2007) shown in Figure 1, the amino acid tryptophan is exuded by plant cells (usually roots) along with other small molecules, and is subsequently taken up by PGPB that are bound to the roots, where it is converted into IAA. The bacterially-produced IAA is excreted and then taken up by the plant cells and, together with the plant's endogenous pool of IAA, stimulates an auxin signal transduction pathway (which includes auxin response factors) which results, on the one hand, in cell growth and proliferation, and on the other, in increased transcription of the gene for ACC synthase, eventually yielding an increased concentration of ACC. Various biotic and abiotic stresses can also either increase the synthesis of IAA (e.g. Ribaut and Pilet, 1994; Dell'Amico et al., 2008) or stimulate the transcription of the gene for ACC synthase (e.g. Nakajima et al., 1990; Van der Straeten et al., 1990; Mathooko et al. 1999). The ACC may either be converted to ethylene by the plant enzyme ACC oxidase or taken up the bacterium that is bound to the plant, where the ACC is degraded by the enzyme ACC deaminase to ammonia and α-ketobutyrate (both readily metabolizable products). In the latter instance, the ACC deaminase-containing bacterium acts as a sink for ACC with the consequence that less ethylene is formed by the plant and the stress response of the plant is decreased. As the level of ethylene in a plant increases, the transcription of auxin response factors is inhibited (Dharmasiri and Estelle 2004; Glick et al., 2007; Czarny et al., 2007). In the absence of bacterial ACC deaminase, by limiting transcription of auxin response factors, ethylene limits both cell growth and proliferation, and (important for plant survival) IAA stimulation of the synthesis of additional ethylene. In the presence of bacterial ACC deaminase, less ethylene is formed because some of the ACC is degraded by ACC deaminase. Thus, when ACC deaminase is present, transcription of auxin response factors is not inhibited, and IAA can stimulate cell growth and proliferation without simultaneously causing a buildup of ethylene. In this model, ACC deaminase both decreases ethylene inhibition of plant growth and allows IAA to maximally promote plant growth, both in the presence and absence of plant stress.

Bacteria that synthesize ACC deaminase and IAA facilitate plant growth under a variety of ethylene-producing environmental stresses including pollution by organic contaminants such as polycyclicaromatic hydrocarbons (PAHs), polycyclic biphenyls (PCBs) and total petroleum hydrocarbons (TPHs) (Saleh et al., 2004; Huang et al., 2004a, b; Reed and Glick, 2005) and by heavy metals and metalloids such as nickel, lead, zinc, copper, cadmium, cobalt and arsenic (Burd et al., 1998, 2000; Belimov et al., 2001, 2005; Nie et al., 2002; Glick, 2003; Reed and Glick, 2005; Farwell et al., 2006; Rodriguez et al., 2008).

In addition to the studies mentioned above, where the direct involvement of ACC deaminase in facilitating phytoremediation has been demonstrated (by comparing the behavior of wild-type strains with ACC deaminase to mutant strains that lack this activity), in several other studies it is purported that ACC deaminase (either by itself or together with IAA) is responsible for the positive effects of added bacteria on phytoremediation. In these studies, the involvement of ACC deaminase is never demonstrated per se, rather, it is inferred that the presence of ACC deaminase is de facto sufficient proof of its involvement (e.g., Idris et al., 2004; Dell'Amico et al., 2005; Di Gregorio et al., 2006; Jiang et al. 2008; Kumar et al., 2008; Ganesan, 2008; Rajkumar and Freitas, 2008; Rajkumar and Freitas, 2008b; Sheng et al., 2008a; Sheng et al., 2008b; He et al., 2009; Ma et al., 2009; Kumar et al., 2009). This caveat notwithstanding, it is likely that in most instances, ACC deaminase-containing bacteria provide plants with a significant growth advantage under stressful conditions.

Effect of ACC Deaminase and Nickel on Plant Gene Expression

Transcriptional profiling, using *Arabidopsis thaliana* 60-mer long oligonucleotide microarrays that represent more than 80% of the genome, of non-transformed canola and transgenic canola that expresses a bacterial ACC deaminase gene, in the presence and absence of nickel, was performed (Czarny, Shah and Glick, submitted for publication). The results indicate that transcription of plant stress response genes is consistently lower in transgenic compared to non-transformed plants, both in the absence and presence of nickel, while at the same time, the transcription of genes involved in secondary metabolism is higher in the transgenic plants. Moreover, the transcription of genes for auxin (IAA) signaling is up-regulated in transgenic plants, consistent with the above mentioned model of ethylene dependent feedback regulation of auxin signaling. The presence of nickel causes down-regulation of the transcription of canola genes involved in photosynthesis, and up-regulation of the transcription of genes involved in the plant's stress response and amino acid, lipid and sugar breakdown, effects which are reduced in magnitude when the ACC deaminase transgene is present. Overall, the data indicate that in the presence of nickel, ethylene contributes to plant stress by reducing photosynthesis, auxin signaling and secondary metabolism. Moreover, the expression of a bacterial ACC deaminase transgene helps to mitigate these effects.

While the experiments discussed above include transgenic canola with an ACC deaminase transgene expressed under the control of a root specific promoter, the behavior of these transgenic plants is physiologically very similar to that of plants whose roots or seeds have been treated with ACC deaminase-expressing plant growth-promoting bacteria. However, the bacterially treated plants typically benefit to a greater extent from the treatment. This likely reflects the fact that the bacteria provide the plant with much more than just ACC deaminase, i.e. plant growth-promoting bacteria introduce IAA and other compounds such as siderophores that may also stimulate plant growth.

CONCLUSION AND FUTURE PROSPECTS

The complexities of facilitating effective phytoremediation protocols notwithstanding, certain conditions may be employed to optimally facilitate the phytoremediation of organic contaminants. In addition to choosing plants that are most effective for specific soils and contaminants these include selecting or engineering bacteria with the following traits: (1) the ability to degrade soil contaminants; (2) plant growth-promoting activity that is based on the activity of ACC deaminase and the synthesis of IAA; and (3) endophytic bacteria that are not subject to the vagaries of the soil environment (Weyens et al., 2009). For the phytoremediation of metals to be successful, besides using endophytic bacteria that produce ACC deaminase and IAA, it is necessary to find practical solutions to the problem of the limited bioavailability of many metals in the soil. Finally, although most current efforts are focused on selecting and utilizing naturally-occurring bacteria to assist phytoremediation, the use of genetically engineered bacteria may greatly simplify the search for the most effective bacterial strain(s). In this regard, the major hurdles to the successful use of genetically engineered bacteria in the environment are political and regulatory, rather than scientific.

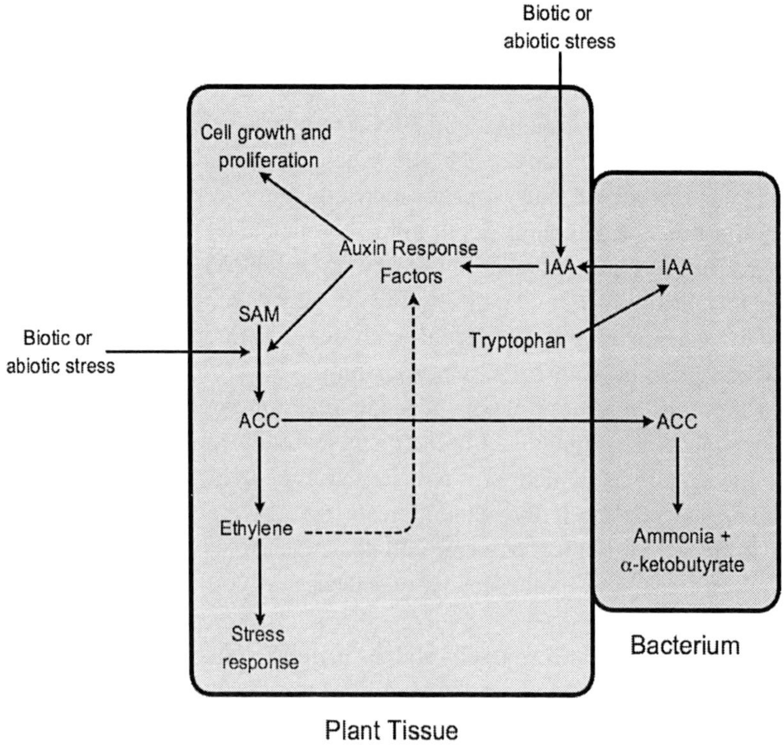

Figure 1. Schematic representation of the lowering of plant stress levels by a PGPB bound to a plant tissue.

ACKNOWLEDGMENTS

The work from our laboratories that is reviewed here was supported by funds from the Italian Ministry of University and Research and the Natural Sciences and Engineering Research Council of Canada.

REFERENCES

Abeles F.B. (1973). *Ethylene in Plant Biology*. Academic Press, New York. pp.302.

Abeles F.B., Morgan P.W., Saltveit M.E. Jr. (1992). *Ethylene in Plant Biology*, 2nd ed., Academic Press, New York.

Bashan Y., Holguin G. (1998). Proposal for the division of plant growth-promoting rhizobacteria into two classifications: biocontrol-PGPB (Plant Growth-Promoting Bacteria) and PGPB. *Soil Biol. Biochem.* 30: 1225-1228.

Bayliss C., Bent E., Culham D.E., MacLellan S., Clarke A.J., Brown G.L., Wood J.M. (1997). Bacterial genetic loci implicated in the *Pseudomonas putida* GR12-2R3-canola mutualism: identification of an exudate-inducible sugar transporter. *Can. J. Microbiol.* 43: 809-818.

Belimov A.A., Safronova V.I., Sergeyeva T.A., Egorova T.N., Matveyeva V.A., Tsyganov V.E., Borisov A.Y., Tikhonovich I.A., Kluge C., Preisfeld A., Dietz K.J., Stepanok V.V. (2001). Characterization of plant growth promoting rhizobacteria isolated from polluted soils and containing 1-aminocyclopropane-1-carboxylate deaminase. *Can. J. Microbiol.* 47: 642-652.

Belimov A.A., Hontzeas N., Safronova V.I., Demchinskaya S.V., Piluzza G., Bullitta S., Glick B.R. (2005). Cadmium-tolerant plant growth-promoting bacteria associated with the roots of Indian mustard (*Brassica juncea* L. Czern.). *Soil Biol. Biochem.* 37: 241-250.

Burd G.I., Dixon D.G., Glick B.R. (1998). A Plant Growth-Promoting Bacterium that decreases nickel toxicity in seedlings. *Appl. Environ. Microbiol.* 64: 3663-3668.

Burd G.I., Dixon D.G., Glick B.R. (2000). Plant growth promoting bacteria that decrease heavy metal toxicity in plants. *Can. J. Microbiol.* 46: 237-245.

Ciardi J.A., Tieman D.M., Lund S.T., Jones J.B., Stall R.E., Klee H.J. (2000). Response to *Xanthomonas campestris* pv. *vesicatoria* in tomato involves regulation of ethylene receptor gene expression. *Plant Physiol.* 123: 81-92.

Czarny J.C., Shah S., Glick B.R. (2007). *Response of canola plants at the transcriptional level to expression of a bacterial ACC deaminase in the roots*. In: Advances in Plant Ethylene Research, (Eds.) A. Ramina, C. Chang, J. Giovannoni, H. Klee, P. Perata and E. Woltering, Springer, Dordrecht, Netherlands, pp. 377-382.

Deikman J. (1997). Molecular mechanisms of ethylene regulation of gene transcription. *Physiol. Plant.* 100: 561-566.

Dell'Amico E., Cavalca L., Andreoni V. (2005). Analysis of rhizobacterial communities in perennial Graminaceae from polluted water meadow soil, and screening of metal-resistant, potentially plant growth-promoting bacteria. *FEMS Microbiol. Ecol.* 52: 153-162.

Dell'Amico E., Cavalca L., Andreoni V. (2008). Improvement of *Brassica napus* growth under cadmium stress by cadmium-resistant rhizobacteria. *Soil Biol. Biochem.* 40: 74-84.

Dharmasiri N., Estell M. (2004). Auxin signaling and regulated protein degradation. *Trends Plant Sci.* 9: 302-308.

Di Gregorio S., Barbafieri M., Lampis S., Sanangelantoni A.M., Tassi E., Vallini G. (2006). Combined application of Triton X-100 and *Sinorhizobium* sp Pb002 inoculum for the improvement of lead phytoextraction by *Brassica juncea* in EDTA amended soil. *Chemosphere* 63: 293-299.

Dobbelaere S., Croonenborghs A., Thys A., Vande Broek A., Vanderleyden J. (1999). Phytostimulatory effect of *Azospirillum brasilense* wild type and mutant strains altered in IAA production on wheat. *Plant Soil* 212: 155-164.

Egamberdieva D. (2009) Alleviation of salt stress by plant growth regulators and IAA producing bacteria in wheat. *Acta Physiol. Plant.* 31: 861-864.

Farwell A.J., Vesely S., Nero V., Rodriguez H., Shah S., Dixon D.G., Glick B.R. (2006). The use of transgenic canola (*Brassica napus*) and plant growth-promoting bacteria to enhance plant biomass at a nickel-contaminated field site. *Plant Soil* 288: 309-318.

Frankenberger J.W.T., Arshad M. (1995). *Microbial synthesis of auxins.* In: Frankenberger WT, Arshad M (eds) Phytohormones in soils. Marcel Dekker, New York, pp 35-71.

Gamalero E., Glick B.R. *Mechanisms Used by Plant Growth-Promoting Bacteria*". In: PGPR: Emergence in Agriculture and Biotechnology (Vol. 2) I. K. International Ltd. Publisher, New Delhi. In press.

Ganesan V. (2008) Rhizoremediation of cadmium soil using a cadmium-resistant plant growth-promoting rhizopseudomonad. *Curr. Microbiol.* 56: 403-407.

Giovanelli J., Mudd S.H., Datko A.H. (1980). *Sulfur amino acids in plants.* In Amino Acids and Derivatives, The Biochemistry of Plants: A Comprehensive Treatise, Vol. 5, (ed) Miflin BJ, Academic Press, New York. pp. 453-505.

Glick B.R. (1995). The enhancement of plant growth by free-living bacteria. *Can. J. Microbiol.* 41: 109-117.

Glick B.R., Karaturovíc D., Newell P. (1995). A novel procedure for rapid isolation of plant growth-promoting rhizobacteria. *Can. J. Microbiol.* 41: 533-536.

Glick B.R. (2003). Phytoremediation: Synergistic use of plants and bacteria to clean up the environment. *Biotechnol. Adv.* 21: 383-393.

Glick B.R., Cheng Z., Czarny J., Duan J. (2007). Promotion of plant growth by ACC deaminase-containing soil bacteria. *Eur. J. Plant Pathol.* 119: 329-339.

Glick B.R., Patten C.L., Holguin G., Penrose D.M. (1999). *Biochemical and genetic mechanisms used by plant growth promoting bacteria*, Imperial College Press, London

Hall J.A., Peirson D., Ghosh S., Glick B.R. (1996). Root elongation in various agronomic crops by the plant growth promoting rhizobacterium *Pseudomonas putida* GR 12-2. Isr. *J. Plant Sci.* 44: 37-42.

He L-Y., Chen Z-J., Ren G-D., Zhang Y-F., Qian M., Sheng X-F. (2009). Increased cadmium and lead uptake of a cadmium hyperaccumulator tomato by cadmium-resistant bacteria. *Exotoxicol. Environ. Safety* 72: 1343-1348.

Huang X.D., El-Alawi Y., Penrose D.M., Glick B.R., Greenberg B.M. (2004a). Responses of three grass species to creosote during phytoremediation. *Environ. Poll.* 130: 453-463.

Huang X.D., El-Alawi Y., Penrose D.M., Glick B.R., Greenberg B.M. (2004b). A multi-process phytoremediation system for removal of polycyclic aromatic hydrocarbons from contaminated soils. *Environ. Poll.* 130: 465-476.

Idris R., Trifanova R., Puschenreiter M., Wenzel W.W., Sessitsch A. (2004). Bacterial communities associated with flowering plants of the nickel hyperaccumulator *Thlapsi goesingense. Appl. Environ. Microbiol.* 70: 2667-2677.

Jiang C-Y., Sheng X-F., Qian M., Wang Q-Y. (2008). Isolation and characterization of a heavy metal-resistant Burkholderia sp. from heavy metal-contaminated paddy field soil and its potential in promoting plant growth and heavy metal accumulation in metal-polluted soil. *Chemosphere* 72: 157-164.

John P. (1991). How plant molecular biologists revealed a surprising relationship between two enzymes, which took an enzyme out of a membrane where it was not located, and put it into the soluble phase where it could be studied. *Plant Mol. Biol. Rep.* 9: 192-194.

Kumar K.V., Singh N., Behl H.M., Srivastava S. (2008). Influence of plant growth promoting bacteria and its mutant on heavy metal toxicity in *Brassica juncea* grown in fly ash amended soil. *Chemosphere* 72: 678-683.

Kumar K.V., Srivastava S., Singh N., Behl H.M. (2009). Role of metal resistant plant growth promoting bacteria in ameliorating fly ash to the growth of *Brassica juncea. J. Hazard. Mater.* 170: 51-57.

Lambrecht M., Okon Y., Vande Broek A., Vanderleyden J. (2000). Indole-3-acetic acid: a reciprocal signalling molecule in bacteria-plant interactions. *Trends Microbiol.* 8: 298-300.

Lindberg T., Granhall U., Tomenius H. (1985). Infectivity and acetylene reduction of diazotrophic rhizosphere bacteria in wheat (*Triticum aestivum*) seedlings under gnotobiotic conditions. *Biol. Fertil. Soils* 1: 123-129.

Ma Y., Rajkumar M., Freitas H. (2009). Inoculation of plant growth promoting bacterium *Achromobacter xylosoxidans* strain Ax10 for the improvement of copper phytoextraction by Brassica juncea. *J. Environ. Managem.* 90: 831-837.

Mathooko F.M., Mwaniki M.W., Nakatsuka A., Shiomi S., Kubo Y., Inaba A., Nakamura R. (1999). Expression characteristics of CS-ACS1, CS-ACS2 and CS-ACS3, three members of the 1-aminocyclopropane-1-carboxylate synthase gene family in cucumber (Cucumis sativus L.) fruit under carbon dioxide stress. *Plant Cell Physiol.* 40: 164-172.

Morgan P.W., Drew C.D. (1997). Ethylene and plant responses to stress. *Physiol. Plant.* 100: 620-630.

Nakajima N., Mori H.,Yamazaki K., Imaseki H. (1990). Molecular cloning and sequence of a complementary DNA encoding 1-aminocyclopropane-1-carboxylate synthase induced by tissue wounding. *Plant Cell Physiol.* 31:1021-1029.

Nie L., Shah S., Rashid A., Burd G.I., Dixon D.G., Glick B.R. (2002). Phytoremediation of arsenate contaminated soil by transgenic canola and the plant growth-promoting bacterium *Enterobacter cloacae* CAL2. *Plant Physiol. Biochem.* 40: 355-361.

Patten C., Glick B.R. (1996). Bacterial biosynthesis of indole-3-acetic acid. *Can. J. Microbiol.* 42: 207-220.

Patten C.L., Glick B.R. (2002). The role of bacterial indoleacetic acid in the development of the host plant root system. *Appl. Environ. Microbiol.* 68: 3795-3801.

Peck S.C., Kende H. (1995). Sequential induction of the ethylene biosynthetic enzymes by indole-3-acetic acid in etiolated peas. *Plant Mol. Biol.* 28: 293-301.

Penrose D.M., Glick B.R. (2001). Levels of ACC and related compounds in exudate and extracts of canola seeds treated with ACC deaminase containing plant growth-promoting bacteria. *Can. J. Microbiol.* 47: 368-372.

Pilet P.E., Saugy M. (1987). Effect on root growth of endogenous and applied IAA and ABA. *Plant Physiol.* 83: 33-38.

Plazinski J., Rolfe B.G. (1985). Influence of *Azospirillum* strains on the nodulation of clovers by *Rhizobium* strains. *Appl. Environ. Microbiol.* 49: 984-989.

Rajkumar M., Freitas H. (2008a). Effects of inoculation of plant-growth promoting bacteria on Ni uptake by Indian mustard. *Biores. Technol.* 99: 3491-3498.

Rajkumar M., Freitas H. (2008b). Influence of metal resistant-plant growth-promoting bacteria on the growth of Ricinus communis in soil contaminated with heavy metals. *Chemosphere* 71: 834-842.

Reed M.L.E., Glick B.R. (2005). Growth of canola (*Brassica napus*) in the presence of plant growth-promoting bacteria and either copper or polycyclic aromatic hydrocarbons. Can. *J. Microbiol.* 51: 1061-1069.

Ribaut J.M., Pilet P.E. (1994). Water stress and indolyl -3-acetic acid content of maize roots. *Planta* 193: 502-507.

Robinson M.M., Griffith M., Pauls K.P., Glick B.R. (2001). Dual role for ethylene in susceptibility of tomato to verticillium wilt. *J. Phytopathol.* 149: 385-388.

Rodriguez H., Vessely S., Shah S., Glick B.R. (2008). Effect of a nickel-tolerant ACC deaminase-producing *Pseudomonas* strain on growth of non-transformed and transgenic canola plants. *Curr. Microbiol.* 57: 170-174.

Saleh S., Huang X.-D., Greenberg B.M., Glick B.R. (2004). *Phytoremediation of Persistent Organic Contaminants in the Environment.* In: Singh A., Ward O. (Eds.), Applied Bioremediation and Phytoremediation, (Series: Soil Biology,Vol. 1), pp. 115-134. Springer-Verlag, Berlin.

Shah S., Li J., Moffatt B.A., Glick B.R. (1997). *ACC deaminase genes from plant growth promoting rhizobacteria.* In: Plant Growth-Promoting Rhizobacteria. Present Status and Future Prospects. Ogoshi A., Kobayashi K., Hemma Y., Kodema F., Kondo N., Akino S. (eds) Organization for Economic Cooperation and Development, Paris pp 320-324.

Sheng X-F, Xia J-J, Jiang C-Y, He L-Y, Qian M. (2008a). Characterization of heavy metal-resistant endophytic bacteria from rape (*Brassica napus*) roots and their potential in promoting the growth and lead accumulation of rape. *Environ. Poll.* 156: 1164-1170.

Sheng X-F., Jiang C-Y., He L-Y. (2008b). Characterization of plant growth-promoting *Bacillus edaphicus* NBT and its effect on lead uptake by Indian mustard in a lead-amended soil. *Can. J. Microbiol.* 54: 417-422.

Steenhoudt O., Vanderleyden J. (2000). *Azospirillum*, a free-living nitrogen-fixing bacterium closely associated with grasses: genetic, biochemical and ecological aspects. *FEMS Microbiol. Rev.* 24: 487-506.

Tsavkelova E.A., Klimova S.Y., Cherdyntseva T.A., Netrusov A.I. (2006). Microbial producers of plant growth stimulators and their practical use: a review. *Appl. Biochem. Microbiol.* 42: 117-126.

Van der Straeten D., Van Wiemeersch L., Goodman H.M., Van Montagu M. (1990). Cloning and sequence of two different cDNAs encoding 1-aminocyclopropane-1-carboxylate synthase in tomato. *Proc. Natl. Acad. Sci.* U.S.A. 87: 4859-4863.

Van Loon L.C., Glick B.R. (2004). *Increased plant fitness by rhizobacteria.* In: Sandermann H (Ed.). Molecular ecotoxicology of plants, Ecological Studies Berlin/Heidelberg Springer-Verlag Vol. 170 pp 177-205.

van Noorden G.E., Ross J.J., Reid J.B., Rolfe B.G., Mathesius U. (2006). Defective long distance auxin transport regulation in the *Medicago truncatula* super numerary nodules mutant. *Plant Physiol.* 140: 1494-1506.

Weyens N., van der Lelie D., Taghavi S., Vangronsveld J. (2009). Phytoremediation: plant–endophyte partnerships take the challenge. *Curr. Opin. Biotechnol.* 20: 248-254.

Yang S.F., Hoffman N.E. (1984). Ethylene biosynthesis and its regulation in higher plants. *Annu. Rev. Plant Physiol.* 35: 155-189.

In: Handbook of Phytoremediation
Editor: Ivan A. Golubev

ISBN: 978-1-61728-753-4

Chapter 30

CONSIDERATIONS ON CHEMICALLY-ENHANCED PHYTOEXTRACTION OF PB USING EDTA UNDER FIELD CONDITIONS

Reinhard W. Neugschwandtner[1 2 *], Pavel Tlustoš[1], Michael Komárek[1], Jiřina Száková[1] and Lucie Jakoubková[1]

[1]Department of Agro-Environmental Chemistry and Plant Nutrition, Faculty of Agrobiology, Food and Natural Resources, Czech University of Life Sciences Prague, Kamýcká 129, 165 21 Prague 6 – Suchdol, Czech Republic

[2]Institute of Agronomy and Plant Breeding, Department of Applied Plant Sciences and Plant Biotechnology, University of Natural Resources and Life Sciences Vienna, Gregor Mendel Straße 33, A-1190 Vienna, Austria

INTRODUCTION

Large areas of agricultural land have been contaminated with potentially toxic metals like Pb by smelting activities in the last centuries (Loska et al., 2004). The possible negative impacts on the environment and human health demand the need for remediation of contaminated sites. Conventional remediation techniques for heavily contaminated soils like excavation or soil-flushing are very cost intensive and not appropriate for large areas of low or medium contaminated agricultural land. Furthermore, these technologies result in a removal of topsoil and in many cases also subsoil needed for agricultural production or in the decrease of its fertility. Discussion and research has therefore focused on in-situ remediation technologies which seem to be cost-effective and environmentally acceptable. The use of plants for the remediation of potentially toxic metals, so-called phytoextraction, could be an alternative.

Phytoextraction uses the ability of plants to accumulate potentially toxic metals from the soil in their above-ground harvestable biomass. The idea behind that system is that the harvested biomass containing the potentially toxic metals can be removed from the site. The

* Corresponding author. Tel.: +43-1-47654-3311; fax: +43-1-47654-3342; E-mail address: reinhard.neugschwandtner@boku.ac.at (R. Neugschwandtner)

ash can be stored at a landfill or the potentially toxic metals can be recycled after burning the biomass which generates at best some extra green energy. There are two possible ways of phytoextraction: hyperaccumulation and chemically-enhanced phytoextraction. Most hyperaccumulators are adapted to metalliferous soils and accumulate a higher amount of potentially toxic metals than other plants. Metal hyperaccumulators (like *Arabidopsis halleri* and *Thlaspi caerulescens*) are generally slow-growing and low-yielding plants and there is a lack of large-scale cultivation techniques for them. Most hyperaccumulators are therefore not suitable for large-scale remediation activities (Raskin et al., 1997). Therefore, research has focused especially in the 1990s and in the first years of the new millennium on chemically-enhanced phytoextraction. Synthetic chelating agents have the potential to remobilize metals and to form strong soluble complexes with them. They thereby overcome the low mobility of major potentially toxic metals (like Pb) which is caused by their primarily binding to organic, oxide and residual fractions in soils. Ethylenediaminetetraacetic acid (EDTA) proved to be the most effective chelating agent among several tested in increasing Pb desorption from soils (Komárek et al., 2007). Increased concentrations of metal–chelant complexes in the soil solution promote the metal uptake by plants and the translocation of potentially toxic metals from roots to shoots and their accumulation in the harvestable parts of the plants.

A large number of papers dealing with the chemically-enhanced phytoextraction technique have been published in high-ranking scientific journals over the years partly presenting promising results. Anyhow, almost all of the presented experiments were performed under model conditions like in pot experiments. Therefore, several authors stated up to a few years ago that the evaluation of the chemically-enhanced phytoextraction technique under field conditions is missing (e.g., Kos et al., 2003; Hajiboland, 2005; Li et al., 2005; Grispen et al., 2006; Komárek et al., 2007). Some recently published papers and our own three-year experiment (results of the first experimental year are published in Neugschwandtner et al., 2008, 2009) close this gap and give a more realistic view on the chemically-enhanced phytoextraction technique. Our experiment was performed in the close vicinity of Příbram, a historical smelting and mining town located approximately 60 km SW of Prague, one of the most polluted regions in the Czech Republic due to atmospheric deposition of potentially toxic metals from a Pb smelter (Pb content in the arable layer of the experimental field: 544 mg kg^{-1}).

Considerations derived from recently gained experiences involve (i) differences of the phytoextraction efficiency under model and under field conditions, (ii) discharge of potentially toxic metals which may result in groundwater pollution, (iii) co-mobilization of macro- and micronutrients next to the mobilization of target potentially toxic metals, (iv) limitations of EDTA application under field conditions and (v) economic costs of EDTA application under field conditions.

I. Differences of Phytextraction Efficiency under Model and under Field Conditions

Several authors reported promising results for the phytoextraction efficiency from pot experiments. The phytoextraction efficiency represents the percentage of an element removed by the plant dry aboveground biomass from the total element content in the soil during one cropping season. Anyhow, these results may not be transferable to field conditions mainly due to the higher plant/soil ratio in the field than that commonly occurring in pot experiments. In our own experiment, we had in the first experimental year one plant of *Zea mays* per 1.25 kg of soil in the pot experiment but just one plant for more than 40 kg of topsoil in the field one. Consequently, the maximal phytoextraction efficiency was 40-times higher for Pb in the pot experiment compared to the field one due to the broader plant/soil ratio in the field experiment. We found in our experiment that phytoremediation efficiency for Pb was too low in the field for a successful remediation in a reasonable time frame. For example, calculating with the maximal obtained remediation factor it would take approximately 300 cropping seasons to obtain Czech threshold values for Pb in loamy or clay soils (220 mg kg^{-1} soil). Other authors even reported longer durations for a successful remediation of some potentially toxic metals (e.g., Meers et al., 2004).

A further limitation is the possible biomass yield decrease after EDTA application and the generally too low target metal uptake rates of the plants. Huang et al. (1997) calculated that phytoextraction of Pb can only be feasible if systems can be developed in which more than 1% of Pb in the soil is accumulated in the shoots and more than 20 tonnes of biomass is produced per hectare and year. In our experiment the addition of 9 mmol EDTA kg^{-1} resulted in a maximum Pb uptake of 0.30% (pot experiment) and 0.0049% (field experiment), respectively, of the total element content in the soil in the first experimental year. Dry harvestable biomass production of *Zea mays* was decreased from 9.4±0.4 to 2.1±0.2 t ha^{-1}. All these values are far lower than those stated by Huang et al. (1997).

II. Discharge of Potentially Toxic Metals

Addition of EDTA to the soil considerably increases the water-soluble concentrations of potentially toxic metals. Several authors have already stated that the low biodegradability of EDTA does not make it suitable for large-scale field applications (e.g., Kos and Leštan, 2004; Tandy et al., 2004). Neugschwandtner et al. (2008) reported in a field experiment a 326-fold increase of the water-soluble Pb concentration in the upper soil layer (0–5 cm) 58 d following the application of 9 mmol EDTA kg^{-1} soil. Water-soluble Pb was still 4.3-fold higher in the depth of 45–50 cm compared to the variant un-treated with EDTA. Only a small percentage of that mobilized water-soluble fraction can be taken up by the plants. Meers et al. (2005) reported in a comparison between four different crops that a maximum of 1.1% of Pb mobilized by EDTA could be recovered in the plant shoots. The rest of the metal–chelant complexes remains in the soil and is subjected to degradation processes and leaching.

Monitoring of the water-soluble Pb concentration in our experiment showed a higher concentration even two years after the application of EDTA (Neugschwandtner et al., unpublished data). This clearly indicates the danger of groundwater contamination especially in humid areas with a positive water balance. Saifullah et al. (2009) stated that for these reasons the scientific community is now distancing itself from the concept of using persistent chelating agents like EDTA in the field. Anyhow, further research for areas where the danger of groundwater pollution is low (e.g., due to a negative water balance) certainly is reasonable. Additionally, adapted application techniques could limit the danger of groundwater pollution. Sun et al. (2001) reported that the mode of EDTA addition is one of the main factors controlling the behavior of metal leaching. Split applications can reduce the risk of groundwater contamination (Lai and Chen, 2007; Neugschwandtner et al., 2008). Anyhow, phytoextraction efficiency is lower when split applications are performed (Neugschwandtner et al., 2008).

III. Co-Mobilization of Macro- and Micronutrients

The knowledge of the speciation of the applied chelating agent in soil solution is critical in enhanced phytoextraction. This includes the co-solubilization of non-target metals (e.g., Ca, Fe, Mg) by the chelant as clearly not all chelant added to the soil binds the target metals (Nowack et al., 2006). Anyhow, the presence of major cations is neglected in most phytoremediation studies when choosing the reagent concentration for removing target potentially toxic metals from soils (Manouchehri et al., 2006).

A high co-mobilization of macronutrients (Ca, K, Mg) and micronutrients (Fe, Mn) by EDTA that caused competition to target potentially toxic metals during the phytoextraction process was reported by Neugschwandtner et al. (2009) in a field experiment (pH of Na_2EDTA-solution: 4.6). An increase of water-soluble nutrient concentrations of 12-times (Fe), 101-times (Mn), 3.7-times (Ca), 1.6-times (Mg), and 1.2-times (K), respectively, in the upper soil layer (0–5 cm) following 40 d after the application of 9 mmol EDTA kg^{-1} was observed. In a depth of 25–30 cm, total water-soluble Fe and Ca were still increased 1.7-fold; whereas total water-soluble fractions of all other nutrients were in the same range in the control and the EDTA treatments. Especially Fe is an important competitor to target metals for chelating agents due to its high concentration in soils and its relatively high complex formation stability constant.

Mobilization of macro- and micronutrients is caused by formation of metal-EDTA complexes. Additionally, EDTA is also recognized to complex with soil components such as oxides and carbonates resulting in their dissolution (Tsang et al., 2007). This way of mobilization is especially likely for metals with a lower complex formation stability constant with EDTA (e.g. Ca). Mobilization of the negatively charged ion phosphorus has been reported (Hovsepyan and Greipsson, 2005; Neugschwandtner et al., 2009) supporting the observation that dissolution processes occur. Mobilisation of P may result from the dissolution of metal phosphates by EDTA as P cannot be complexed by EDTA which is present as an anion itself (Neugschwandtner et al., 2009).

Mobilization of nutrients may result in leaching and loss of soil fertility. Anyhow, adding EDTA to less fertile soils could have a positive effect on plant nutrition due to the increased uptake of deficient ions (e.g. Fe) (Komárek et al., 2007).

IV. Technical Considerations on EDTA Application under Field Conditions

Remaining questions involve the technical realization of the chelant application under field conditions. The chelant solution has to be applied to well established plant stands to guarantee a high uptake of target metals and to minimize stress effects on germination and early stage growth. Anyhow, established plant stands will hinder or even inhibit the passing over the field with agricultural field machinery.

Therefore, applying the solution by irrigation has to be discussed. Drip irrigation would highly increase the costs of application. A possible way could be sprinkler irrigation which is cheaper. Anyhow, here the disadvantage remains that the solution cannot be applied directly to the soil and therefore salt could remain unproductive on the leaves. To overcome this limitation the solution would have to be applied either before rain or irrigation of water after the solution application would have to be performed.

Another possible way could be to apply EDTA with fertilizer equipment. Dispersing the salt is not recommendable due to the danger of wind dispersion. Therefore, the use of EDTA granules should be tested. These can be applied relatively cheap like fertilizer granules with common fertilizer equipment into growing plant stands. Li et al. (2005) reported the use of slow-release coated EDTA granules which release the chelating agent over a longer time period. A disadvantage of the use of slow-release coated granules would be that they cannot exploit the positive effect of physiological stress on the increased uptake of metal–chelant complexes (Schaider et al., 2006). Neugschwandtner et al. (2008) have shown that physiological stress (evident as necrosis) is higher when applying a certain dose of EDTA at once instead of applying the same amount in several smaller doses over time.

V. Economic Costs of EDTA Application under Field Conditions

The limitations for the use of EDTA under field conditions are furthermore economic ones. An example: The application of 1 mmol Na_2EDTA (molar mass: 372.24 g mol^{-1}) to one square meter of agricultural soil (300 kg of soil m^{-2}; arable layer depth: 25 cm; bulk density: 1.2 g cm^{-3}) would consume about 112 g of EDTA. This would result in a need of more than 1.1 tonnes of EDTA per ha for applying the low doses of 1 mmol EDTA kg^{-1}. The German trading company Omikron offers 25 kg bags of Disodium EDTA Dihydrate with a purity of over 98.0% for 298 Euro including taxes (March 2010). Therefore, applying 1 mmol EDTA kg^{-1} would result in costs for the chemical of more than 13.300 Euro per ha. Considering the fact that a higher dose of EDTA would be needed and that for several years the costs of enhanced phytoextraction would increase dramatically making it not feasible for economic reasons. However, if a large amount of chelating agents is ordered for large-scale remediation activities a reduced price may be expected.

Conclusion

Based on the recently published papers it can be stated that the initial euphoria for enhanced phytoextraction using EDTA for remediation of Pb contaminated soils has gone. Phytoextraction seems not to be practically feasible at the present state of knowledge for most trace elements (Van Nevel et al., 2007). Our own three-year experiment supports the fact that major limitations for EDTA-enhanced phytoextraction of Pb exist under field conditions. These limitations include the fact that the phytoextraction efficiency observed under model conditions cannot be transferred to field conditions, the discharge of potentially toxic metals which may result in groundwater pollution, co-mobilization of macro- and micronutrients next to the mobilization of target potentially toxic metals, technical limitations of EDTA application under field conditions and the high economic costs of EDTA application under field conditions.

Few commercial phytoextraction operations have been carried out so far despite the intensive research on the subject. These commercial activities are presently limited to phytomining, a subset of phytoextraction (Evangelou et al., 2007). Phytomining aims to produce a commercial 'bio-ore' by planting a hyperaccumulator crop over a low grade-ore body or mineralized soil (McGrath and Zhao, 2003). Commercially viable phytomining technologies using high biomass Ni hyperaccumulators have already been developed (Broadhurst et al., 2004). Economic aspects of phytomining for nickel, thallium and gold using hyperaccumulators have been discussed by Anderson et al. (1999).

Evangelou et al. (2007) concluded that it is questionable if further research on enhanced phytoextraction using chelating agents like EDTA will lead to promising solutions. The suggestion is supported for EDTA-enhanced phytoextraction of Pb by our experiment (Neugschwandtner et al., 2008). If further research is conducted it has to deal with and to solve the presented limitations.

References

Anderson, C.W.N., Brooks, R.R., Chiarucci, A., LaCoste, C.J., Leblanc, M., Robinson, B.H., Simcocke, R., Stewart, R.B., 1999. Phytomining for nickel, thallium and gold. *Journal of Geochemical Exploration* 67, 407–415.

Broadhurst, C.L., Chaney, R.L., Angle, J.S., Maugel, T.K., Erbe, E.F., Murphy, C.A., 2004. Simultaneous hyperaccumulation of nickel, manganese, and calcium in *Alyssum* leaf trichomes. *Environmental Science and Technology* 38, 5797–5802.

Evangelou, M.W.H., Ebel, M., Schaeffer, A., 2007. Chelate assisted phytoextraction of heavy metals from soil. Effect, mechanism, toxicity, and fate of chelating agents. *Chemosphere* 68, 989–1003.

Grispen, V.M.J., Nelissen, H.J.M., Verkleij, J.A.C., 2006. Phytoextraction with *Brassica napus* L.: a tool for sustainable management of heavy metal contaminated soils. *Environmental Pollution* 144, 77–83.

Hajiboland, R., 2005. An evaluation of the efficiency of cultural plants to remove heavy metals from growing medium. *Plant Soil Environment* 51, 156–164.

Hovsepyan, A., Greipsson, S., 2005. EDTA-enhanced phytoremediation of lead-contaminated soil by corn. *Journal of Plant Nutrition* 28, 2037–2048.

Huang, J.W., Chen, J., Berti, W.R., Cunningham, S.D., 1997. Phytoremediadon of lead-contaminated soils: role of synthetic chelates in lead phytoextraction. *Environmental Science and Technology* 31, 800–805.

Komárek, M., Tlustoš, P., Száková, J., Chrastný, J., Ettler, V., 2007. The use of maize and poplar in chelant enhanced phytoextraction of lead from contaminated agricultural soil. *Chemosphere* 67, 640–651.

Kos, B., Leštan, D., 2003. Influence of a biodegradable ([S,S]-EDDS) and nondegradable (EDTA) chelate and hydrogel modified soil water sorption capacity on Pb phytoextraction and leaching. *Plant and Soil* 253, 403–411.

Kos, B., Grčman, H., Leštan, D., 2003. Phytoextraction of lead, zinc and cadmium from soil by selected plants. *Plant Soil Environment* 49, 548–553.

Lai, H.-Y., Chen, Z.-S., 2007. Multi-dose applying EDTA to decrease the potential groundwater contamination using rainbow pink (*Dianthus chinensis*) for enhanced phytoextraction. *Desalination* 210, 236–247.

Li, H., Wang, Q., Cui, Y., Dong, Y., Christie, P., 2005. Slow release chelate enhancement of lead phytoextraction by corn (*Zea mays* L.) from contaminated soil—a preliminary study. *Science of the Total Environment* 339, 179–187.

Loska, K., Wiechulła, D., Korus, I., 2004. Metal contamination of farming soils affected by industry. *Environment International* 30, 159–165.

Manouchehri, N., Besancon, S., Bermond, A., 2006. Major and trace metal extraction from soil by EDTA: Equilibrium and kinetic studies. *Analytica Chimica Acta* 559, 105–112.

McGrath, S.P., Zhao, F.J., 2003. Phytoextraction of metals and metalloids from contaminated soil. *Current Opinion in Biotechnology*. 14, 277–282.

Meers, E., Hopgood, M., Lesage, E., Vervaeke, P., Tack, F.M.G., Verloo, M.G., 2004. Enhanced Phytoextraction: In Search of EDTA Alternatives. *International Journal of Phytoremediation* 6, 95–109.

Meers, E., Ruttens, A., Hopgood, M., Lesage, E., Tack, F.M.G., 2005. Potential of *Brassica rapa*, *Cannabis sativa*, *Helianthus annuus* and *Zea mays* for phytoextraction of heavy metals from calcareous dredged sediment derived soils. *Chemosphere* 61, 561–572.

Neugschwandtner, R.W., Tlustoš, P., Komárek, M., Száková, J., 2009. Nutrient mobilization and nutrient contents of *Zea mays* in response to EDTA additions to heavy-metal-contaminated agricultural soil. *Journal of Plant Nutrition and Soil Science* 172, 520–527.

Neugschwandtner, R.W., Tlustoš, P., Komárek, M., Száková, J., 2008. Phytoextraction of Pb and Cd from a contaminated agricultural soil using different EDTA application regimes: Laboratory versus field scale measures of efficiency. *Geoderma* 144, 446–454.

Nowack, B., Schulin, R., Robinson, B., 2006. Critical assessment of chelant enhanced metal phytoextraction. *Environmental Science and Technology* 40, 5225–5236.

Raskin, I., Smith, R.D., Salt, D.E., 1997. Phytoremediation of metals: using plants to remove pollutants from the environment. *Current Opinion in Biotechnology* 8, 221–226.

Saifullah, Meers, E., Qadir, M., de Caritat, P., Tack, F.M.G., Du Laing, G., Zia, M.H., 2009. EDTA-assisted Pb phytoextraction. *Chemosphere* 74, 1279–1291.

Schaider, L.A., Parker, D.R., Sedlak, D.L., 2006. Uptake of EDTA-complexed Pb, Cd and Fe by solution and sand-cultured *Brassica juncea*. *Plant and Soil* 286, 377–391.

Sun, B., Zhao, F.J., Lombi, E., McGrath, S.P., 2001. Leaching of heavy metals from contaminated soils using EDTA. *Environmental Pollution* 113, 111–120.

Tandy, S., Schulin, R., Nowack, B., 2006. Uptake of metals during chelant assisted phytoextraction with EDDS related to the solubilized metal concentration. *Environmental Science and Technology* 40, 2753–2758.

Tsang, D.C.W., Irene, M., Lo., C., Chan, K.L., 2007. Modeling the Transport of Metals with Rate-Limited EDTA-Promoted Extraction and Dissolution during EDTA-Flushing of Copper-Contaminated Soils, *Environmental Science and Technology* 41, 3660–3666.

Van Nevel, L., Mertens, J., Oorts, K., Verheyen, K., 2007. Phytoextraction of metals from soils: How far from practice? *Environmental Pollution* 150, 34–40.

In: Handbook of Phytoremediation
Editors: Ivan A. Golubev

ISBN: 978-1-61728-753-4

Chapter 31

FIELD-SCALE RHYZOREMEDIATION OF A CONTAMINATED SOIL WITH HEXACHLOROCYCLOHEXANE (HCH) ISOMERS: THE POTENTIAL OF POPLARS FOR ENVIRONMENTAL RESTORATION AND ECONOMICAL SUSTAINABILITY

D. Bianconi[a], M.R. De Paolis[a], A.C. Agnello[a], D. Lippi[a], F. Pietrini[a], M. Zacchini[a], C. Polcaro[b], E. Donati[b], P. Paris[a], S. Spina[c] and A. Massacci[*] [a]

[a]Institute of Agro-environment and Forest Biology,
Rome Research Area, Monterotondo Scalo (Rome)
[b]Institute of Chemical Methodologies ,
Rome Research Area, Monterotondo Scalo (Rome)
[c]Commissariato Starordinario per l'emergenza nel bacino del Fiume Sacco,
Regione Lazio (Rome)

ABSTRACT

Three pre-selected poplar clones and two soil HCH degrader micro-organisms have been experimentally applied in a contaminated agricultural soil in the basin of Fiume Sacco near to Rome for its reclamation. The aim was to successfully associate soil cleaning by rhizoremediation with an economically sustainable biomass for energy production of large poplar plantations. Plants and micro-organisms were selected for the best association with bacteria to obtain 1) the maximum HCH concentration reduction in soil, 2) the minimum plant contamination with HCH, and 3) the maximum biomass production. Results showed that an association between all these traits is possible in a specific poplar clone inoculated with a selected HCH degrader bacterium. The need for a pre-remediation phase in situ to select best candidate plants and bacteria with lowest HCH accumulation in its organs is emphasized. Rhyzoremediation associated with the

* Corresponding author: angelo.massacci@ibaf.cnr.it ,Institute of Agro-environment and Forest Biology, Rome Research Area, Via Salaria km 29, 00015 Monterotondo Scalo (Rome)

safe thermo-convertible biomass production is confirmed as a sustainable recovery of soils interdicted to food-agricultural activities.

Introduction

Rhizoremediation is a type of phytoremediation that takes place in the rhizosphere by a complex interaction of roots, root exudates and microorganisms with contaminants (Kuiper et al., 2001; Walker et al., 2003). The rhizosphere is by definition the volume of soil influenced by roots that extends approximately 1–3 mm from their surface (Shimp, et al., 1993; Schnook, 1998). Besides being capable to metabolically degrade some organics and extract some heavy metals, roots provide a niche and key nutrients or other substances that stimulate microorganism growth and remediation activities (Cardon and Gage, 2006). These substances include relatively small molecules such as amino acids, organic acids, sugars, phenolics, and large molecules, such as mucilage (high molecular weight polysaccharides) and enzymatic proteins (Cardon and Gage, 2006). Some of these molecules affect the rhizosphere by modifying its pH or its redox status and by significantly contributing to an easy to use and essential carbon pool for microorganism growth (El Shatnawi and Makhadmeh, 2001). Nevertheless, other molecules can specifically induce or enhance the enzymatic breakdown of complex organic contaminants finally providing further components to the carbon pool (Walker et al., 2003). The proliferation of soil organisms in this soil volume can be 3 or 4 orders of magnitude greater than the population of soil microorganisms in non-vegetated soils (ITRC, 2001). The rhizosphere is thus a crowded and active chemical laboratory where breakdown of organic contaminants and the complexation of heavy metals are facilitated.

Besides breakdown activities, the microbiological consortium perform several other important functions in the rhizosphere such as promotion of plant growth, protection against plant pathogens, and the production of chelators for absorbing and translocating essential nutrients to the plant (Walker et al., 2003; El Shatnawi and Makhadmeh, 2001). To this latter process are known to strongly concur mycorrhizal fungi associated with roots of some plants (e.g. poplars). The mycorrhizal infection of roots is usually a mutualistic relationship, with the fungi receiving sugars from the host plant in exchange for tremendously increasing the plant's absorption surface of essential mineral nutrients and eventually bioavailable contaminants (Quoreshi and Khasa, 2008). All of these rhizosphere activities involving the indigenous microbial population can be indicated as natural attenuation. Recent research has shown that by the use of specific amendments both growth and activities of these microorganisms can be stimulated to accelerate or enhance the natural attenuation. In alternative, microorganisms isolated from the soil and proved to degrade an organic contaminat can be re-injected into the rhizosphere as a concentrated culture (bioaugmentation). The application of either of these processes to cope with environmental contamination is still limited. In the few cases of "in situ" preliminary assessments the remediation results were promising despite some post-remediation problems that have still to be solved. It is evident that rhizoremediation bears most positive aspects out of all phytotechnologies. It is, for example, a relatively low cost, mostly energy sun-driven and eco-sustainable technology. It has no size restrictions for sites, it uses organic biodegradable materials, it has a discrete versatility to treat a diverse range of contaminants. Furthermore, it can potentially be employed in any geographical area that supports plant growth. On the other hand, the so far evidenced drawbacks of rhizoremediation are mainly related to factors expected from plant characteristics. It takes time that tree plants, as being the ones with most useful root apparatus, grow and develop a large rhizosphere to obtain acceptable effects. Many contaminants, especially the organic ones, diffuse little and, if outside of the rhizosphere, are scarcely bioavailable to degrader microorganisms. These microorganisms are, besides, aerobic and their growth is limited by the scarcity of oxygen deep below 0.3-0.4 m in the soil. In addition, stressors affecting rhizoremediation in the field are numerous and unforeseen (Wenzel, 2009). Nevertheless, substantial research efforts have improved and can continue to improve efficiency of the rhizosphere processes by eliminating or at least minimizing these drawbacks. For example, bioavailability of contaminants, one of the factors that most limit their degradation, can be improved by investigating the use of suitable surfactants, amphipathic molecules with both a hydrophobic and a hydrophilic part. These molecules are able to lower the surface tension, and can form micelles where substances that are generally insoluble in water, such as the chlorinated organics, for example, may be accumulated at interfaces and then solubilized. In turn, the solubilization renders these molecules more available for degradation by microorganisms (Lafrance and Lapointe, 2007). However, chemical surfactants are themselves a source of further and persistent pollution. By contrast, many microorganisms (Pseudomonas aeruginosa, Bacillus subtilis, Pichia pastoris, etc.) can produce diversified structurally surface-active agents, referred to as biosurfactants. Biosurfactants are much less harmful for natural ecosystems due to their low toxicity and biodegradable nature. A further help to improve bioavailability of some contaminants is to exploit the anastomosis ability of some plants. Poplars that have been selected for high density plantation (i.e. short rotation coppicing, SRC) with 7-8,000 plants ha^{-1} can root to a depth between 0.80 and 2.43 m (Jordahl et al., 1997; Trapp and Karlson 2001; Zacchini et al., 2008) and already at the end of the first growth season they form a complex root mat within rows (Coleman et al., 2004). This mat is useful as a diffusion pathway for exudates, microorganisms, nutrients, oxygen and contaminants that become more biovailable to rhizodegradation activities..

RHIZOREMEDIATION WITH POPLARS

Poplar trees have specific characteristics that, beside their important rooting ability, make them suitable for rhizoremediation. They are perennial, easily propagated, fast growing especially in the juvenile phase, large biomass producers and when they are planted at high density for short rotation coppicing (SRC) poplar trees can be used for producing heat and/or electric power. Furthermore, poplars show a high leaking capacity of organic carbon into the rhizosphere. According to Lynch (1987) plants release about 30% of photosynthates to the soil. A similar or even slightly higher percentage (35%) has been estimated for poplars. This amount of released carbon is one of the most likely reasons that make poplars a very good host of microorganisms in the rhizosphere and thus capable of higher rhizoremediation of organic xenobiotics as compared to other trees (Jordahl et al., 1997). Indeed, field and greenhouse studies showed that poplars have already themselves an important role in the uptake and degradation of particular chlorinated organics. They can degrade trichloroethylene (TCE) to trichloroethanol, di and trichloroacetic acid, or complete mineralization to CO_2

(Gordon et al., 1998). In addition, they can uptake, hydrolyze, and dealkylate atrazine to less toxic metabolites (Burken and Schnoor 1996, 1997). It has also been shown that they can take up and transpire dioxane in both hydroponic and soil experiments (Aitchison et al., 2000). Further, they can take up and translocate from root to shoots lesser-chlorinated polychlorinated biphenyls (PCBs), being the translocation of PCBs to stems inversely related to PCB hydrophobicity (Liu and Schnoor, 2008). Moreover, *in vivo* metabolism of 3,3′,4,4′-tetrachlorobiphenyl through hidroxilation was observed (Liu et al., 2009). It is noteworthy to report that poplars are among the few tree genera that can associate in a mutualistic symbiosis with both types of mycorrhizal fungi (ectomycorrhizae and vesicular arbuscular mycorrhizae), which have a fundamental role in increasing the root surface up to 800 fold. However, it is not clear if the fungi can directly degrade the contaminant or if they stimulate bacterial degradation by enhancing plant growth (McCutcheon and Schnoor, 2003; Karlinski et al., 2009; Ma et al., 2009). Lastly, poplars have a high intraspecific genetic polymorphism and genetic diversity. As a consequence, a number of different genotypes could be available with a high degree of adaptability to a given climate and able to tolerate and uptake/degrade a given contaminant. Exploring the poplar diversity can provide specific material that can be applied in most environmental conditions and for rhizodegrading a variety of organic contaminants.

The Case Study of Poplar Rhizoremediation in the Valle Del Sacco Area Near to Rome

A land of about 8 Kmq in the Rome and Frosinone provinces of South Lazio Region of Italy is contaminated by hexachlorocyclohexane (α, β, γ and δ HCH) isomers with concentrations that are spotty above law soil limits fixed for industrial or green areas (figure 1). The area is 100 m wide along the both sides of the Sacco river and was part of agricultural lands used for growing forage. Sediments of the river were polluted by nearby industrial landfill percolations, probably in the eighties, containing co-products of the lindane (the γ-HCH) synthesis. The forage crop irrigation with the river water and river floodings that occurred few times have widespread the contaminated sediments in that stretch of land. In the year 2005 a high concentration of the β-HCH was detected in the milk produced in local farms and further analysis in the surrounding soils confirmed the presence of this and other isomers. These results immediately led to the interdiction of the above-indicated large area. Such a decision aiming to reduce health risk to the population caused an important economic loss to farmers. Health risks and socio-economic problems thereby induced environmental authorities to include the area of Valle del Sacco in the list of national interest contaminated sites. The HCH isomer contamination has become the target of many activities, including the removal of active sources and the cleaning of agricultural soils.

It is important to consider that HCH isomers show marked differences in some physicochemical properties and their relative diffusion pathway in the environment, their relative persistence, and toxicity are different. They are produced by the photochemical chlorination of benzene during the manufacturing of the technical lindane (t-HCH). The production of one ton of lindane is accompanied with the production of 6-10 tons of other isomers (IHPA, 2006) characterized by different chlorine atom orientations in the cyclohexane molecule (axial or equatorial position). This structural difference seems to significantly affect the individual chemical-physical and biological properties of some isomers. For example, it has been shown that isomers with the chlorine atoms in axial position are less persistent than β-HCH having this latter all its six chlorine atoms in the equatorial orientation (Bachmann et al. 1988b; Beurskens et al. 1991). It can be argued that this can be the reason because the β-HCH comprises above 80% of HCH residues of technical lindane in a ten year contaminated field surrounding an industrial landfill in Germany (ATSDR, 2005). It has to be considered also that biotic and abiotic factors may lead to the isomerization process that enables the interconversion between HCH chemical isomers (Malaiyandi et al. 1982; Huhnerfuss et al. 1992).

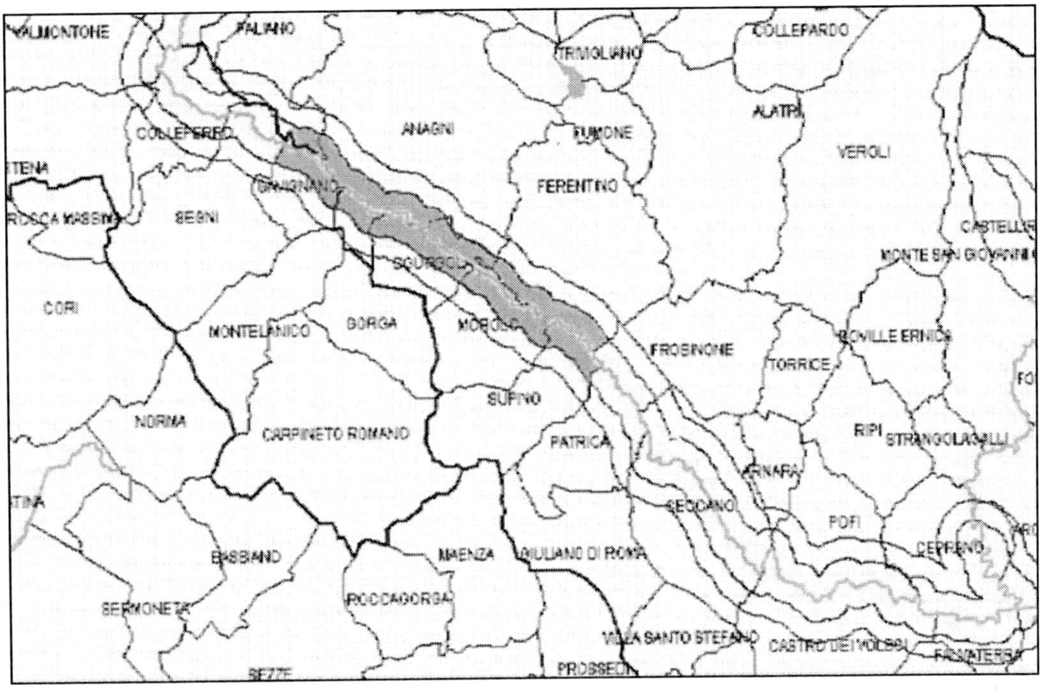

Figure 1. The contaminated surface (red color) in the Valle del Sacco area.

Further, it is to underline that the β-HCH transportation through the environment doesn't much follow the diffusion water path. Its octanol-water partition coefficient (log K_{OW}) is, in fact, close to 4. Remarkably, its vapour pressure at 25°C is 6.3 10^{-5} mmHg. This means that in a moist soil surface at this temperature a significant amount of β-HCH can be released to the atmosphere and can probably move for long distances carried by wind bound to small dust particles (Wania et al., 1999). Luckily, this isomer can be degraded in the atmosphere by reacting with photochemically produced hydroxyl radicals. An important target of HCH isomer phytoremediation is then the degradation of the recalcitrant β-isomer. Degradation of HCHs has been observed under both oxic and anoxic conditions (Phillips et al., 2005; Nagata et al., 2007). However, the mineralization (complete degradation of the pollutant to CO_2) generally occurs in oxic systems. Over time, several species of microorganisms (bacteria and fungi) have demonstrated the ability to degrade specific HCH isomers in the two mentioned conditions (summarized by Phillips et al., 2005). It has also been reported that the aerobic bacteria Arthrobacter citreus can degrade γ-HCH and use it as a sole source of carbon (Datta et al., 2000). In certain cases, one single strain might not have all the enzymes involved in the complete catabolic pathway of a pollutant. A consortium of bacteria, each with different parts of the degradation route has been reported to be very efficient (Rahman et al., 2002).

THE EXPERIMENTAL APPLICATION

The HCH remediation in the Valle del Sacco area has been designed considering present knowledge in the use of poplars in association with microorganisms to reduce chlorinated organic concentration from the soil and to produce biomass for energy. Preliminary studies have been performed to select candidate bacteria and hybrid poplars that combine the most efficient contaminant reduction in the soil, the highest biomass production and the uncontamination of harvestable biomass. Bacteria isolated from contaminated soil samples of Valle del Sacco and those isolated and characterized for previous studies on organic xenobiotic metabolism have been tested and selected to obtain an efficient β-HCH degradation. Figure 2 reports the growth curves of two Arthrobacter strains in culture with the β-HCH isomer as the sole carbon source. This persistent isomer was degraded by about 50% (N:P:K) in 48 h (De Paolis et al., 2009) and given its highest persistence in the environment the two strains were chosen for association studies with six hybrid poplar clones. These clones were among those more successful in terms of biomass production in the Valle del Sacco area outside the contaminated soils. These studies were performed in the glasshouse and three candidate clones were finally selected for an experimental "in situ" application of 0.4 Ha in the contaminated area. I-214, AF2 and Monviso were those showing growth characteristics and response to contaminants that have been considered useful for a rhizodegradation application. I-214 showed itself good β-HCH degradation ability without bacterial inoculation, while AF-2 showed the highest uncontaminated biomass production, and Monviso both the highest HCH degradation with inoculated bacteria and a high biomass production. It is remarkable that bacteria inoculated in the potted contaminated soil without poplar cutting degraded 50% less HCH isomers than in association with Monviso. This indicates that this hybrid can provide rhizosphere conditions stimulating the HCH degradation of the two agrobacters.

Cuttings of these three hybrid poplar clones were mechanical planted in 2008 in a site of about 0.5 ha. Plant density in 5 replicated parcels of 10x9 m for each treatment was about 8000 ha^{-1}. Treatments were: a) poplar clone alone, b) poplar clone inoculated with bacteria, c) poplar clone on the soil amended with commercial compost and with oxygen release compound (ORC, IXPER® 75C Calcium Peroxide, Solvay, S.A). The addition of the compost (produced on plant wastes) was due to assess whether its high microbic count could be alternative to the inoculation, which is technologically complicated and a bit more expensive than composting. The simultaneous presence of ORC served to enhance the degrading microbiological activity by the continuum oxygen supply of this compound in presence of water and by its own degradation ability of chlorinated organics (Arienzo, 2000; Cassidy and Irvine, 1999). A drip irrigation system was placed to provide evapotranspired water restitution twice a week. Before plantation the soil physico-chemical and biological characteristics were analysed on samples collected according to regulated sampling procedure for contaminated soils. Data are reported in the table 1. The soil appears equilibrated regarding the C/N ratio required by optimal plant and bacteria growth. A high carbon content would in fact disadvantage the bacterial use of contaminant carbon skeleton by competition with more easily metabolized organic compounds and consequently this would reduce degradation efficiency. A typical fertilization adopted for poplar plantation in SRC in the Valle del Sacco soils with ammonium nitrate at a rate of 150 Kg ha^{-1} was applied during soil mechanical preparation. Bacteria concentrated solution (10^6 CFU per plant) was inoculated at root level in early June after 60 days from plantation. At the end of the growing season (figure 3) in October soil and plants were sampled for HCH isomers analysis. Table 2 shows results of HCH degradation under the various treatments and evidences that the field application of rhizoremediation significantly reduced HCH content in the soil planted with Monviso and inoculated with the two agrobacter strains. With respect to the glasshouse studies the field degradation was lower in percentage (25:40) probably reflecting the different glasshouse and field conditions. Furthermore, in pots the interaction between degrading bacteria and contaminants is certainly facilitated. Interestingly, the compost and ORC treatment also reduced β-HCH content by a significant extent. However, the HCH chromatogramme showed that the disappearance of HCH isomers was compensated by the detection of a new, not yet identified, peak at high molecular weight. Further studies are in course to identify the chemical characteristic and toxicity of this compound.

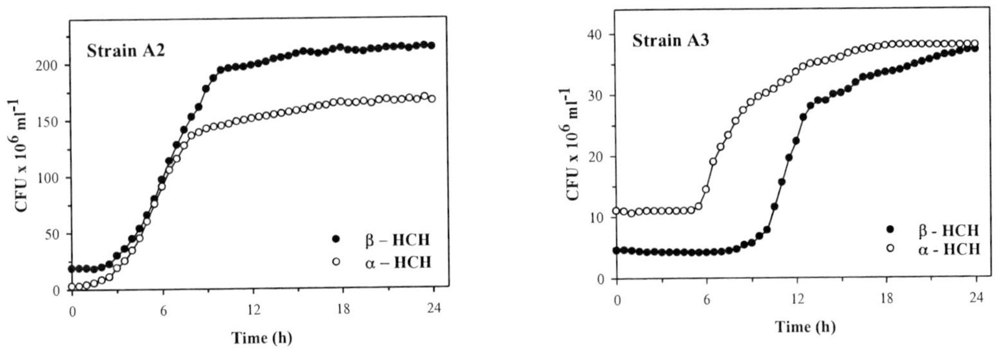

Growth on α-HCH	Growth on β-HCH
$\mu_{max} = 0.84 \pm 0.02$	$\mu_{max} = 0.73 \pm 0.02$
Lag = 0.72 ± 0.12	Lag = 3.03 ± 0.11
$t_d = 0.82 \pm 0.02$	$t_d = 0.95 \pm 0.02$

Growth on α-HCH	Growth on β-HCH
$\mu_{max} = 0.86 \pm 0.07$	$\mu_{max} = 1.13 \pm 0.05$
Lag = 6.38 ± 0.15	Lag = 9.73 ± 0.07
$t_d = 0.80 \pm 0.06$	$t_d = 0.61 \pm 0.03$

Figure 2. Growth curves and parameters of *Arthrobacter* strains on α- and β-HCH. Two *Arthrobacter* strains, A2 e A3, were grown in mineral medium B_7^+ containing α-HCH (10 mg/l) or β-HCH (5 mg/l) as sole carbon source. 3, 7 and 14 days old-cultures were analyzed for capacity to utilize HCH. Moreover bacterial growth curves on α- e β-HCH were recorded and their growth parameters were calculated by means of software MicroFit (ver. 1.0), following the mathematic model of Baranyi.

It can be expected that in the second growth season in the field, when the root system will be more extended and deep enough to completely explore the soil volume down to 0.8 m, the interaction of bacteria with contaminants will increase using the more complex and extended root as diffusion pathway.

Table 1. Some physico-chemical and biological characteristics of soil from experimental site.

Sand (%)	75.25
Silt + Clay (%)	24.75
pH (H_2O)	7.72
Total N (g / Kg)	0.16
Organic C (g / Kg)	12.24
Organic matter (g / Kg)	21.11
Electrical conductivity (mS / cm)	462.0
Bacteria (CFU x 10^7 g^{-1})	58.80
Actinomycetes (CFU x 10^7 g^{-1})	13.10
Fungi (CFU x 10^5 g^{-1})	18.00

CFU: colony-forming unit.

Figure 3. The experimental plantations with inter-row distances of 200 cm and a spacing of 50 cm between cuttings within the row.

Another important result obtained in this experimental application was that the biomass production of the three clones (figure 4) was as good as in the uncontaminated areas of the Valle del Sacco, with a minimum of 5 and a maximum of 10 tons d.m. ha^{-1} $year^{-1}$ with I-214 and AF-2 showing the minimum and maximum yield, respectively. Besides it is remarkable that the harvestable plant parts were only slightly contaminated. Only 10% of samples were contaminated by a 0.03 mg Kg^{-1} of the β-HCH isomer (data not shown).

Table 2. Mean of individual HCH isomers and total HCH ($\alpha + \beta + \gamma + \delta$) concentrations (± S.E.) in soil and rhizosphere samples (ppm) under three treatments. Means (n = 3) with different letters indicate significant differences at $p < 0.05$ level (Anova). S.E.: standard error.

SAMPLE	α-HCH	β-HCH	γ-HCH	δ-HCH	Total HCH
Soil before poplar cuttings planting	0.044 ±0.031 ab	0.058 ±0.052 a	0.017 ±0.002 abc	0.006 ±0.001 a	0.125 ±0.008 a
I-214	0.044 ±0.074 ab	0.046 ±0.015 abcd	0.011 ±0.001 cd	0.002 ±0.001 c	0.103 ±0.009 abc
I-214 + bacteria	0.042 ±0.015 ab	0.047 ±0.058 abcd	0.016 ±0.001 abc	0.004 ±0.003 abc	0.108 ±0.007 ab
I-214 + ORC + compost	0.047 ±0.035 a	0.054 ±0.055 abc	0.020 ±0.003 a	0.006 ±0.027 ab	0.126 ±0.006 a
AF-2	0.044 ±0.061 ab	0.039 ±0.019 bcd	0.014 ±0.001 abcd	0.004 ±0.006 abc	0.102 ±0.008 abc
AF2 + bacteria	0.045 ±0.035 ab	0.053 ±0.044 abc	0.018 ±0.001 ab	0.005 ±0.001 abc	0.120 ±0.008 ab
AF2 + ORC + compost	0.039 ±0.054 ab	0.038 ±0.063 cd	0.012 ±0.002 bcd	0.003 ±0.001 abc	0.093 ±0.010 bcd
Monviso	0.038 ±0.062 abc	0.057 ±0.015 ab	0.013 ±0.001 bcd	0.004 ±0.001 abc	0.112 ±0.008 ab
Monviso + bacteria	0.025 ±0.007 c	0.032 ±0.099 d	0.010 ±0.001 cd	0.002 ±0.001 bc	0.070 ±0.011 d
Monviso + ORC +compost	0.032 ±0.012 bc	0.030 ±0.020 d	0.013 ±0.001 bcd	0.003 ±0.001 abc	0.079 ±0.002 cd

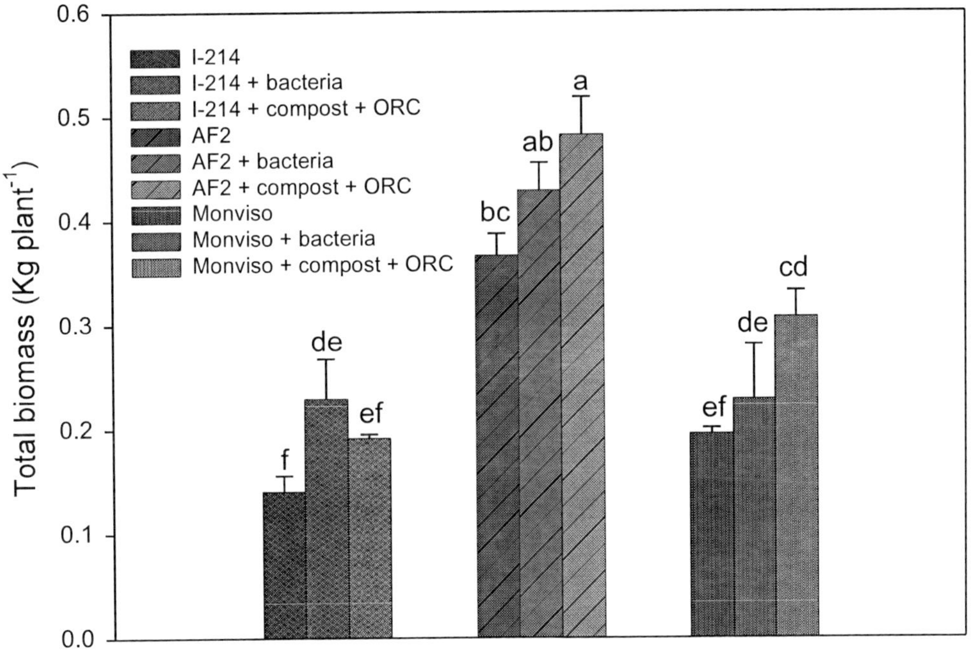

Figure 4. Dry biomass (Kg plant^{-1}) measured in plants of three poplar clones at the end of the first year from plantation. Means values with a same letter were not significantly different (P < 0.05, ANOVA; LSD mean comparisons test). Values are the mean of five replicates. Error bars indicate standard error.

CONCLUSION

Some important aspects of the poplar rhizoremediation and biomass production in the Valle del Sacco contaminated area have still to be dealt with. It is, for example, important to test a large scale bacteria injection into the rhizosphere. Adaptation of commercial agricultural machines is under evaluation as well as the distribution of inoculums via the buried drip irrigation system. The choice will be affected by the ongoing studies on the agrobacters persistence in the soil after the inoculation. It is evident that the improvement of the inoculation techniques providing a higher distribution of degrading bacteria in the contaminated soil will be an important goal of experimental activities that will increase the HCH bioavailability and bacterial degradation capacity. Results achieved are positive indications that the adopted strategy can be effective in reducing soil HCH contamination, offering local farmers to produce bioenergy as an important payback of incomes lost by the interdiction of the agricultural activities. It is evident that the thermoconvertible biomass production adds an important economic value. So far, the problem of remedying HCH isomer contamination in the Valle del Sacco seems to be more of answering the question: does the biomass produced in a contaminated site have to be considered a waste or is it thermoconvertible as good biomass?

References

Aitchison E.W., Kelley S.L., Alvarez P.J.J., Schnoor J.L. (2000). Phytoremediation of 1,4-dioxane by hybrid poplar trees. *Water Environ Fed* 72(3): 313-321.

Arienzo M. (2000). Degradation of 2,4,6-trinitrotoluene in water and soil slurry utilizing a calcium peroxide compound. *Chemosphere* 40(4):331-337.

Bachmann A., Wijnen P., DeBruin W., Huntjens J.L.M., Roelofsen W., Zehnder A.J.B. (1988b). Biodegradation of alpha- and beta-hexachlorocyclohexane in a soil slurry under different redox conditions. *Appl Environ Microbiol* 54: 143–149

Beurskens J.E.M., Stams A.J.M., Zehnder A.J.B., Bachmann A. (1991). Relative biochemical reactivity of three hexachlorocyclohexane isomers. *Ecotoxicol Environ Safety* 21: 128–136.

Burken J.G., Schnoor, J.L. (1996). Phytoremediation: Plant uptake of atrazine and role of root exudates. *J Envir Engrg* 122(11): 958-963.

Burken J.G., Schnoor, J.L.(1997). Uptake and metabolism of atrazine by poplar trees. *Environ Sci Technol* 31(5): 1399-1406.

Cardon Z.G., Gage D.J. (2006). Resource Exchange in the Rhizosphere: Molecular Tools and the Microbial Perspective. *Annu Rev Ecol Evol Syst* 37: 459-488.

Cassidy D.P., Irvine R.L. (1999). Use of calcium peroxide to provide oxygen for contaminant biodegradation in a saturated soil. *Journal of Hazardous Materials* 69(1): 25-39.

Coleman M.D., Friend A.L., Kern C.C. (2004). Carbon allocation and nitrogen acquisition in a developing *Populus deltoides* plantation. *Tree Physiology* 24: 1347–1357.

Datta J., Maiti A.K., Modak D.P., Chakrabartty P.K., Bhattacharyya P., Ray P.K. (2000). Metabolism of γ-hexachlorocyclohexane by Arthrobacter citreus strain BI-100: Identification of metabolites. *J Gen Appl Microbiol* 46: 59-67.

El-Shatnawi M.K.J., Makhadmeh I.M. (2001). Ecophysiology of the Plant–Rhizosphere System. *Journal of Agronomy and Crop Science* 187 (1): 1 – 9.

Gordon M., Choe N., Duffy J., Ekuan G., Heilman P., Muiznieks I., Ruszaj M., Shurtleff B.B., Strand S., Wilmoth J., Newman L.A. (1998). Phytoremediation of trichloroethylene with hybrid poplars. *Environ Health Perspect* 106(4):1001-4

Huhnerfuss H., Faller J., Konig W.A., Ludwig P. (1992). Gas chromatographic separation of the enantiomers of marine pollutants. 4. Fate of Hexachlorocyclohexane isomers in the Baltic and North Sea. *Environ Sci Technol* 26: 2127–2133.

Jordahl J.L., Foster L., Schnoor J.L., Alvarez P.J.J. (1997). Effect of hybrid poplar trees on microbial populations important to hazardous waste bioremediation. *Environ Toxicol Chem* 16: 1318-1321.

Karlinski L., Rudawska M., Kieliszewska-Rokicka B., Leski T. (2009). Relationship between genotype and soil environment during colonization of poplar roots by mycorrhizal and endophytic fungi. *Mycorrhiza* (Epub ahead of print).

Kuiper I., Lagendijk E.L., Bloemberg G.V., and Lugtenberg B.J.J. (2004). Rhizoremediation: A Beneficial Plant-Microbe Interaction. *Molecular Plant-Microbe Interactions* 17(1): 6-15.

Lafrance P., Lapointe M. (2007). Mobilization and Co-Transport of Pyrene in the Presence of Pseudomonas aeruginosa UG2 biosurfactants in Sandy Soil Columns. *Ground Water Monitoring & Remediation* 18(4): 139 – 147.

Liu J., Schnoor J.L. (2008). Uptake and translocation of lesser-chlorinated polychlorinated biphenyls (PCBs) in whole hybrid poplar plants after hydroponic exposure. *Chemosphere* 73(10):1608-16.

Liu J., Hu D., Jiang G., Schnoor J.L. (2009). In vivo biotransformation of 3,3′,4,4′-tetrachlorobiphenyl by whole plants−poplars and switchgrass. *Environ Sci Technol* 43(19): 7503–7509.

Lynch J.M. (ed.) (1987).*The rhizosphere*. Chichester: Wiley Interscience.

Ma L., Wu X.Q., Zheng L. (2009). *Relationship between plant hormone level excreted by ectomycorrhizal fungi and growth of poplar* NL-895. Front For China 4(2): 236-241.

Malaiyandi M., Shah S.M., Lee P. (1982). Fate of alpha- and gamma-hexachlorocyclohexane isomers under simulated environmental conditions. *J Environ Sci Health, Part A* A17: 283–297.

McCutcheon S.C., Schnoor J.L. (eds) (2003). *Phytoremediation. Transformation and control of contaminants*. Wiley Interscience.

Nagata Y., Endo R., Ito M., Ohtsubo Y., Tsuda M. (2007). Aerobic degradation of lindane (γ-hexachlorocyclohexane) in bacteria and its biochemical and molecular basis. *Appl Microbiol Biotechnol* 76: 741-752.

Phillips T.M., Seech A.G., Lee H., Trevors J.T. (2005). Biodegradation of hexachlorocyclohexane (HCH) by microorganisms. *Biodegradation* 16: 363-392.

Quoreshi A.M., Khasa D.P. (2008). Effectiveness of mycorrhizal inoculation in the nursery on root colonization, growth, and nutrient uptake of aspen and balsam poplar. *Biomass and Bioenergy* 32: 381-391.

Rahman K.S., Rahman T., Lakshmanaperumalsamy P., Banat I.M. (2002). Occurrence of crude oil degrading bacteria in gasoline and diesel station soils. *J Basic Microbiol* 42(4): 284-91.

Shimp J.F., Tracy J.C., Davis L.C., Lee E., Huang W., Erickson L.E., Schnoor J.L. (1993). Beneficial effects of plants in the remediation of soil and groundwater contaminated with organic materials. *Critical Reviews in Environmental Science and Technology* 23(1): 41 – 77.

Trapp S., Karlson U. (2001). Aspects of phytoremediation of organic pollutants. *J Soils Sediments* 1(1): 37-43.

Walker T.S., Bais H.P., Grotewold E., Vivanco J.M. (2003). Root Exudation and Rhizosphere Biology. *Plant Physiol* 132: 44-51.

Wania F., Mackay D., LI Y.-F., Bidleman T.F., Strand A. (1999). Global chemical fate of a-hexachlorocyclohexane. 1. Evaluation of a global distribution model. *Environmental Toxicology and Chemistry* 18(7): 1390–1399.

Wenzel W.W. (2009). Rhizosphere processes and management in plant-assisted bioremediation (phytoremediation) of soils. *Plant Soil* 321: 385-408.

Zacchini M., Pietrini F., Scarascia Mugnozza G., Iori V., Pietrosanti L., Massacci A.(2008). Metal tolerance, accumulation and translocation in poplar and willow clones treated with cadmium in hydroponics. *Water Air Soil Pollut* 197: 23-34.

INDEX

A

B

C

D

E

F

G

H

I

J

K

L

N

O

P

Q

R

T

X

Y